Study and Solutions Guide for

CALCULUS
EARLY TRANSCENDENTAL FUNCTIONS

THIRD EDITION

Larson/Hostetler/Edwards

Volume I
Chapters P-9
and
Appendix A

Bruce H. Edwards
University of Florida

HOUGHTON MIFFLIN COMPANY Boston New York

Editor-in-Chief: Jack Shira
Managing Editor: Cathy Cantin
Development Manager: Maureen Ross
Development Editor: Laura Wheel
Assistant Editor: Rosalind Martin
Supervising Editor: Karen Carter
Project Editor: Patty Bergin
Editorial Assistant: Meghan Lydon
Production Technology Supervisor: Gary Crespo
Marketing Manager: Michael Busnach
Marketing Assistant: Nicole Mollica
Senior Manufacturing Coordinator: Jane Spelman

Printed in the United States of America

ISBN 0-618-22309-6

5 6 7 8 9 – VG – 06 05

Preface

This *Study and Solutions Guide* is designed as a supplement to *Calculus: Early Transcendental Functions*, Third Edition, by Ron Larson, Robert P. Hostetler, and Bruce H. Edwards. All references to chapters, theorems, and exercises relate to the main text. Solutions to every odd-numbered exercise in the text are given with all essential algebraic steps included. Although this supplement is not a substitute for good study habits, it can be valuable when incorporated into a well-planned course of study. For suggestions that may assist you in the use of this text, your lecture notes, and this *Guide*, please refer to the student web site for your text at *college.hmco.com*.

I have made every effort to see that the solutions are correct. However, I would appreciate hearing about any errors or other suggestions for improvement. Good luck with your study of calculus.

Bruce H. Edwards
University of Florida
Gainesville, Florida 32611
(bc@math.ufl.edu)

CONTENTS

CHAPTER P
Preparation for Calculus

CHAPTER P
Preparation for Calculus

Section P.1 Graphs and Models

Solutions to Odd-Numbered Exercises

1. $y = -\frac{1}{2}x + 2$

x-intercept: $(4, 0)$

y-intercept: $(0, 2)$

Matches graph (b)

3. $y = 4 - x^2$

x-intercepts: $(2, 0), (-2, 0)$

y-intercept: $(0, 4)$

Matches graph (a)

5. $y = \frac{3}{2}x + 1$

x	-4	-2	0	2	4
y	-5	-2	1	4	7

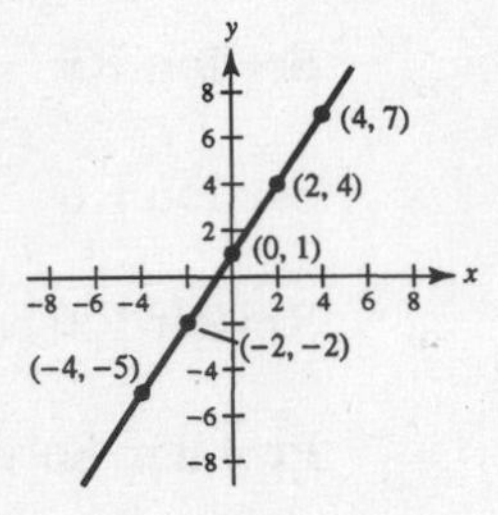

7. $y = 4 - x^2$

x	-3	-2	0	2	3
y	-5	0	4	0	-5

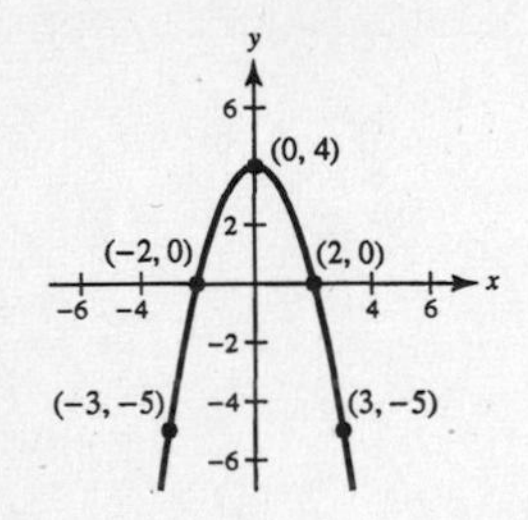

9. $y = |x + 2|$

x	-5	-4	-3	-2	-1	0	1
y	3	2	1	0	1	2	3

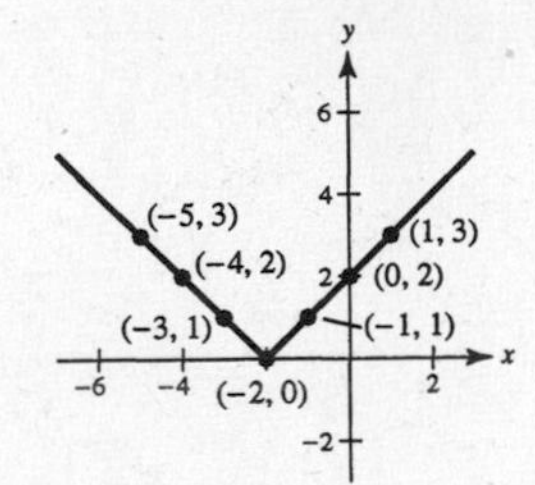

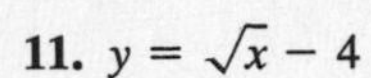

11. $y = \sqrt{x} - 4$

x	0	1	4	9	16
y	-4	-3	-2	-1	0

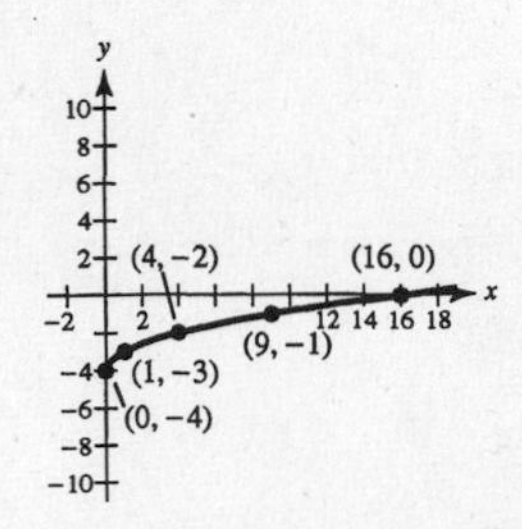

13.

Xmin = -3
Xmax = 5
Xscl = 1
Ymin = -3
Ymax = 5
Yscl = 1

Note that $y = 4$ when $x = 0$.

15.

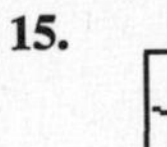

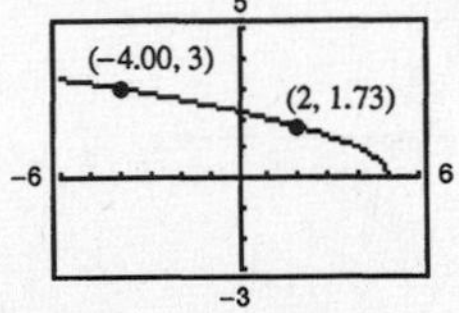

(a) $(2, y) = (2, 1.73)$

$\left(y = \sqrt{5 - 2} = \sqrt{3} \approx 1.73\right)$

(b) $(x, 3) = (-4, 3)$

$\left(3 = \sqrt{5 - (-4)}\right)$

17. $y = x^2 + x - 2$

y-intercept: $y = 0^2 + 0 - 2$

$y = -2;\ (0, -2)$

x-intercepts: $0 = x^2 + x - 2$

$0 = (x + 2)(x - 1)$

$x = -2, 1;\ (-2, 0), (1, 0)$

19. $y = x^2\sqrt{25 - x^2}$

y-intercept: $y = 0^2\sqrt{25 - 0^2}$

$y = 0;\ (0, 0)$

x-intercepts: $0 = x^2\sqrt{25 - x^2}$

$0 = x^2\sqrt{(5 - x)(5 + x)}$

$x = 0, \pm 5;\ (0, 0);\ (\pm 5, 0)$

21. $y = \dfrac{3(2 - \sqrt{x})}{x}$

y-intercept: None. x cannot equal 0.

x-intercepts: $0 = \dfrac{3(2 - \sqrt{x})}{x}$

$0 = 2 - \sqrt{x}$

$x = 4;\ (4, 0)$

23. $x^2y - x^2 + 4y = 0$

y-intercept: $0^2(y) - 0^2 + 4y = 0$

$y = 0;\ (0, 0)$

x-intercept: $x^2(0) - x^2 + 4(0) = 0$

$x = 0;\ (0, 0)$

25. Symmetric with respect to the y-axis since

$$y = (-x)^2 - 2 = x^2 - 2.$$

27. Symmetric with respect to the x-axis since

$$(-y)^2 = y^2 = x^3 - 4x.$$

29. Symmetric with respect to the origin since

$$(-x)(-y) = xy = 4.$$

31. $y = 4 - \sqrt{x + 3}$

No symmetry with respect to either axis or the origin.

33. Symmetric with respect to the origin since

$$-y = \frac{-x}{(-x)^2 + 1}$$

$$y = \frac{x}{x^2 + 1}.$$

35. $y = |x^3 + x|$ is symmetric with respect to the y-axis since

$$y = |(-x)^3 + (-x)|$$

$$= |-(x^3 + x)|$$

$$= |x^3 + x|.$$

37. $y = -3x + 2$

Intercepts: $\left(\frac{2}{3}, 0\right), (0, 2)$

Symmetry: none

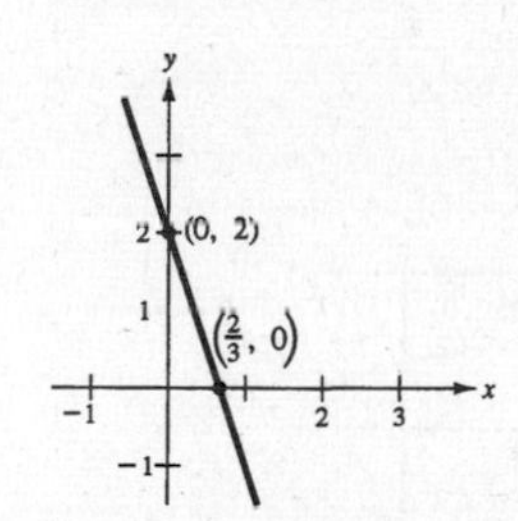

39. $y = \dfrac{x}{2} - 4$

Intercepts: $(8, 0), (0, -4)$

Symmetry: none

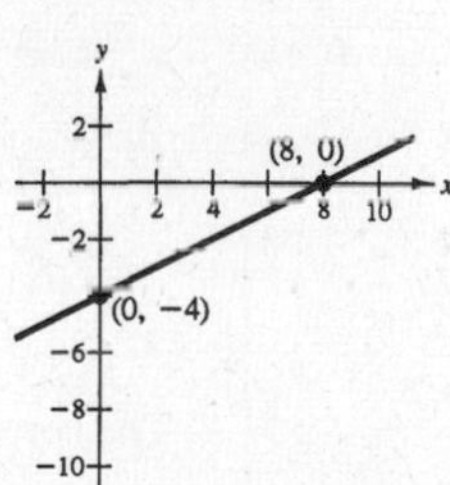

41. $y = 1 - x^2$

Intercepts: $(1, 0), (-1, 0), (0, 1)$

Symmetry: y-axis

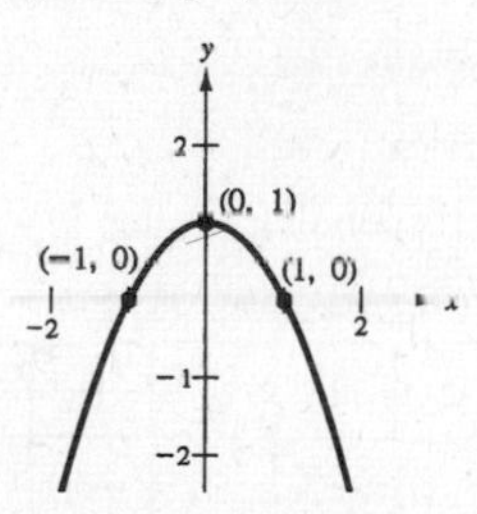

43. $y = (x + 3)^2$

Intercepts: $(-3, 0), (0, 9)$

Symmetry: none

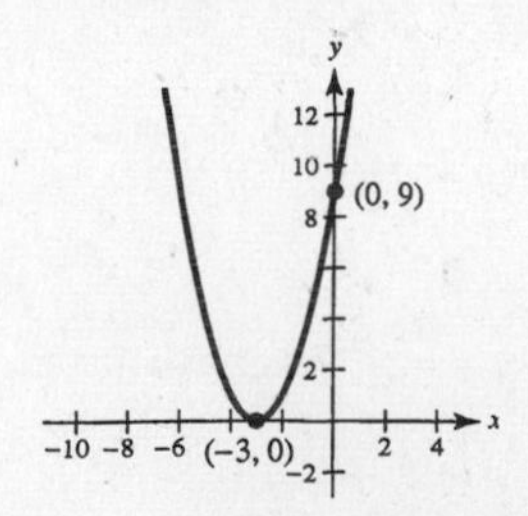

45. $y = x^3 + 2$

Intercepts: $\left(-\sqrt[3]{2}, 0\right), (0, 2)$

Symmetry: none

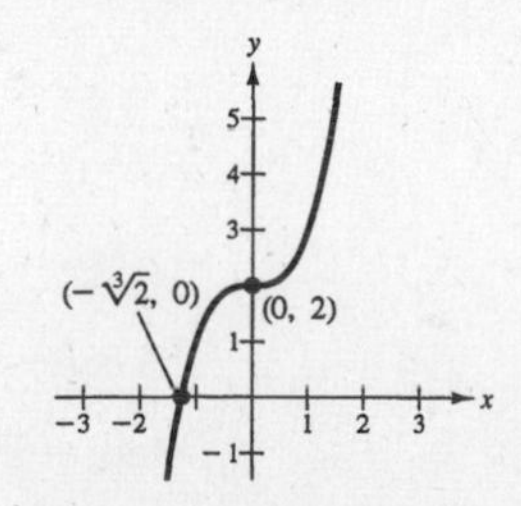

47. $y = x\sqrt{x + 2}$

Intercepts: $(0, 0), (-2, 0)$

Symmetry: none

Domain: $x \geq -2$

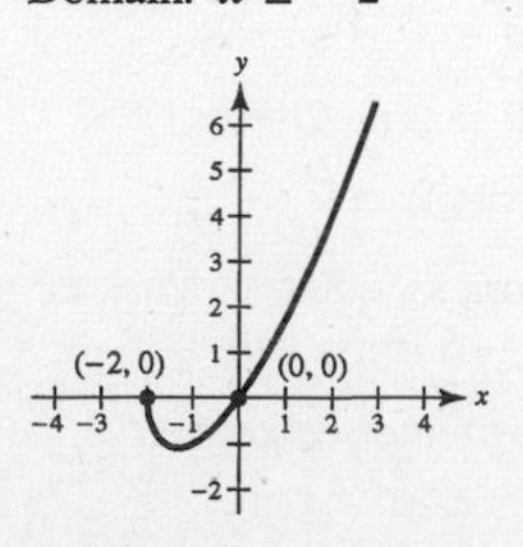

49. $x = y^3$

Intercepts: $(0, 0)$

Symmetry: origin

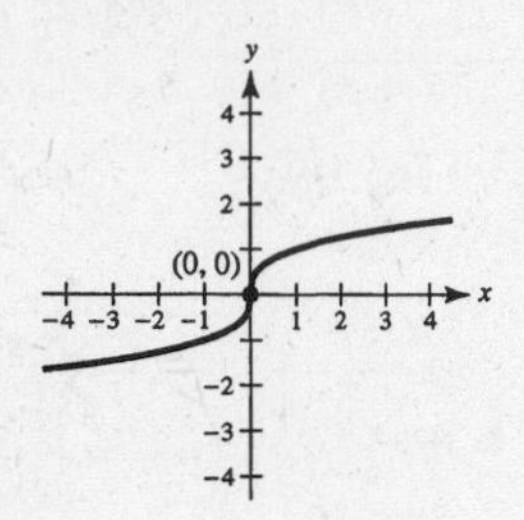

51. $y = \frac{1}{x}$

Intercepts: none

Symmetry: origin

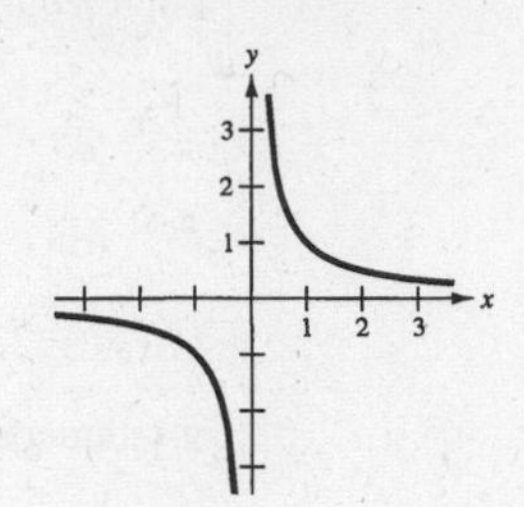

53. $y = 6 - |x|$

Intercepts: $(0, 6), (-6, 0), (6, 0)$

Symmetry: y-axis

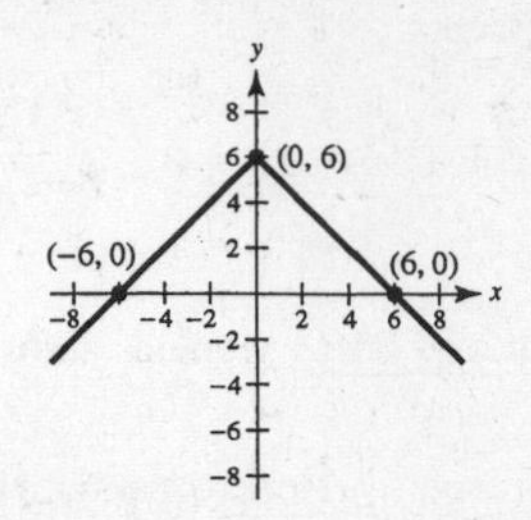

55. $y^2 - x = 9$

$y^2 = x + 9$

$y = \pm\sqrt{x + 9}$

Intercepts:

$(0, 3), (0, -3), (-9, 0)$

Symmetry: x-axis

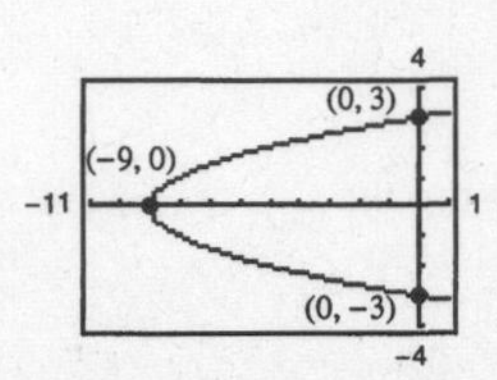

57. $x + 3y^2 = 6$

$3y^2 = 6 - x$

$y = \pm\sqrt{2 - \frac{x}{3}}$

Intercepts:

$(6, 0), (0, \sqrt{2}), (0, -\sqrt{2})$

Symmetry: x-axis

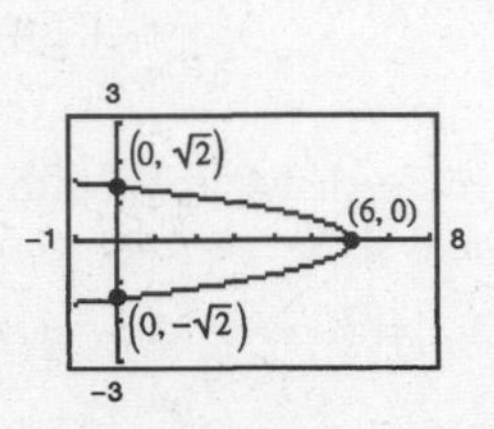

59. $y = (x + 2)(x - 4)(x - 6)$ (other answers possible)

61. Some possible equations: $y = x$

$y = x^3$

$y = 3x^3 - x$

$y = \sqrt[3]{x}$

63. $x + y = 2 \Rightarrow y = 2 - x$

$2x - y = 1 \Rightarrow y = 2x - 1$

$2 - x = 2x - 1$

$3 = 3x$

$1 = x$

The corresponding y-value is $y = 1$.

Point of intersection: $(1, 1)$

65. $x + y = 7 \Rightarrow y = 7 - x$

$3x - 2y = 11 \Rightarrow y = \frac{3x - 11}{2}$

$7 - x = \frac{3x - 11}{2}$

$14 - 2x = 3x - 11$

$-5x = -25$

$x = 5$

The corresponding y-value is $y = 2$.

Point of intersection: $(5, 2)$

67. $x^2 + y = 6 \Rightarrow y = 6 - x^2$

$x + y = 4 \Rightarrow y = 4 - x$

$6 - x^2 = 4 - x$

$0 = x^2 - x - 2$

$0 = (x - 2)(x + 1)$

$x = 2, -1$

The corresponding y-values are $y = 2$ (for $x = 2$) and $y = 5$ (for $x = -1$).

Points of intersection: $(2, 2), (-1, 5)$

69. $x^2 + y^2 = 5 \Rightarrow y^2 = 5 - x^2$

$x - y = 1 \Rightarrow y = x - 1$

$5 - x^2 = (x - 1)^2$

$5 - x^2 = x^2 - 2x + 1$

$0 = 2x^2 - 2x - 4 = 2(x + 1)(x - 2)$

$x = -1$ or $x = 2$

The corresponding y-values are $y = -2$ and $y = 1$.

Points of intersection: $(-1, -2), (2, 1)$

71.
$$y = x^3$$
$$y = x$$
$$x^3 = x$$
$$x^3 - x = 0$$
$$x(x + 1)(x - 1) = 0$$
$$x = 0, x = -1, \text{ or } x = 1$$

The corresponding y-values are $y = 0, y = -1$, and $y = 1$.

Points of intersection: $(0, 0), (-1, -1), (1, 1)$

73.

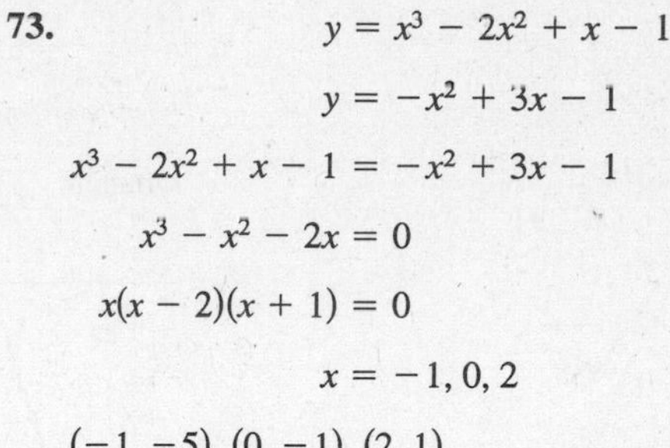

$$y = x^3 - 2x^2 + x - 1$$
$$y = -x^2 + 3x - 1$$
$$x^3 - 2x^2 + x - 1 = -x^2 + 3x - 1$$
$$x^3 - x^2 - 2x = 0$$
$$x(x - 2)(x + 1) = 0$$
$$x = -1, 0, 2$$
$$(-1, -5), (0, -1), (2, 1)$$

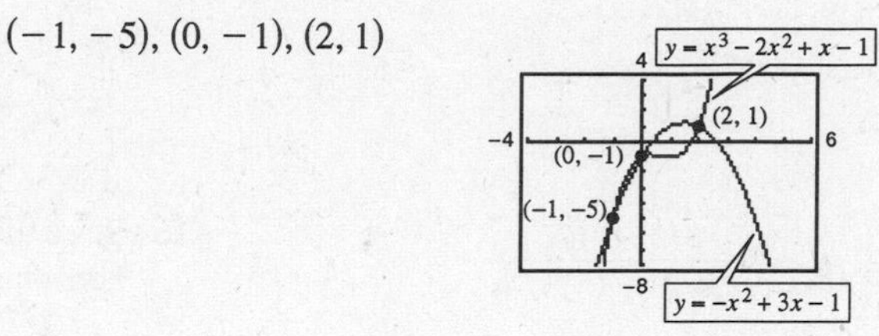

75.
$$C = R$$
$$5.5\sqrt{x} + 10{,}000 = 3.29x$$
$$\left(5.5\sqrt{x}\right)^2 = (3.29x - 10{,}000)^2$$
$$30.25x = 10.8241x^2 - 65{,}800x + 100{,}000{,}000$$
$$0 = 10.8241x^2 - 65{,}830.25x + 100{,}000{,}000 \quad \text{Use the Quadratic Formula.}$$
$$x \approx 3133 \text{ units}$$

The other root, $x \approx 2949$, does not satisfy the equation $R = C$. This problem can also be solved by using a graphing utility and finding the intersection of the graphs of C and R.

77. (a) Using a graphing utility, you obtain

$$y = -0.0153t^2 + 4.9971t + 34.9405$$

(b)

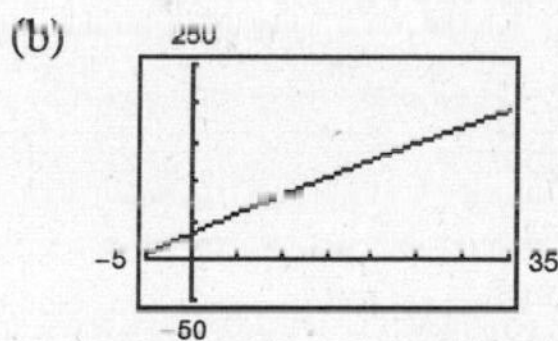

(c) For the year 2004, $t = 34$ and $y \approx 187.2$ CPI.

79. $y = \dfrac{10{,}770}{x^2} - 0.37$

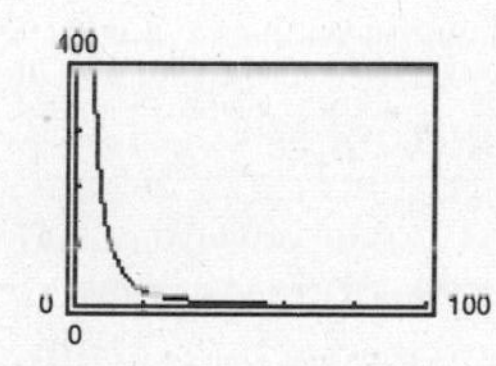

If the diameter is doubled, the resistance is changed by approximately a factor of $\left(\frac{1}{4}\right)$. For instance, $y(20) \approx 26.555$ and $y(40) \approx 6.36125$.

81. False; x-axis symmetry means that if $(1, -2)$ is on the graph, then $(1, 2)$ is also on the graph.

83. True; the x-intercepts are

$$\left(\frac{-b \pm \sqrt{b^2 - 4ac}}{2a}, 0\right).$$

85.
$$\text{Distance from the origin} = K \times \text{Distance from } (2, 0)$$
$$\sqrt{x^2 + y^2} = K\sqrt{(x - 2)^2 + y^2}, K \neq 1$$
$$x^2 + y^2 = K^2(x^2 - 4x + 4 + y^2)$$
$$(1 - K^2)x^2 + (1 - K^2)y^2 + 4K^2x - 4K^2 = 0$$

Note: This is the equation of a circle!

Section P.2 Linear Models and Rates of Change

1. $m = 1$

3. $m = 0$

5. $m = -12$

7.

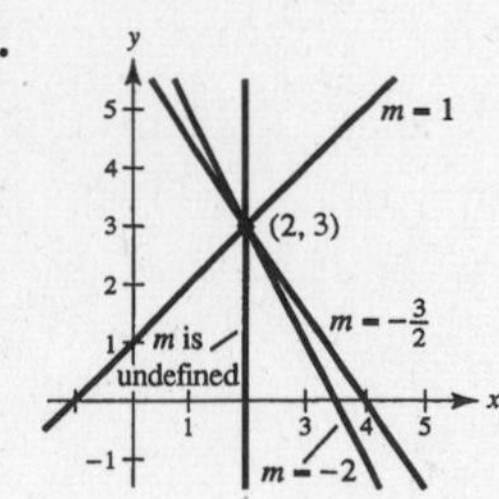

9. $m = \dfrac{2 - (-4)}{5 - 3} = \dfrac{6}{2} = 3$

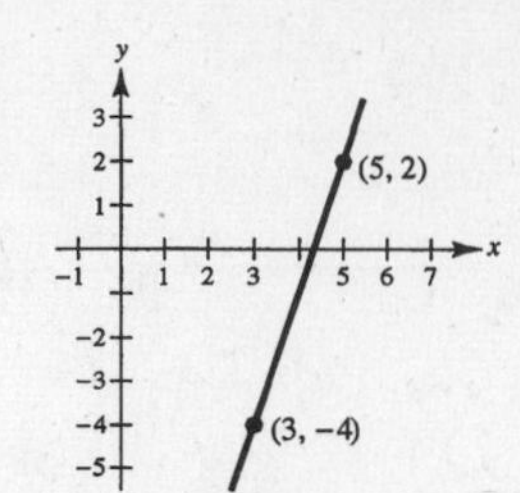

11. $m = \dfrac{5 - 1}{2 - 2} = \dfrac{4}{0}$

undefined

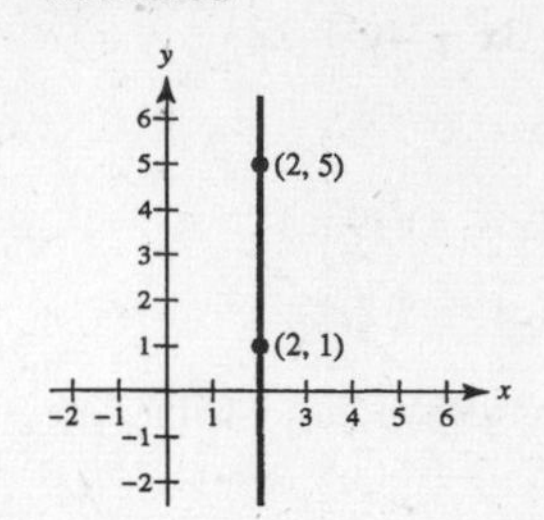

13. $m = \dfrac{2/3 - 1/6}{-1/2 - (-3/4)}$

$= \dfrac{1/2}{1/4} = 2$

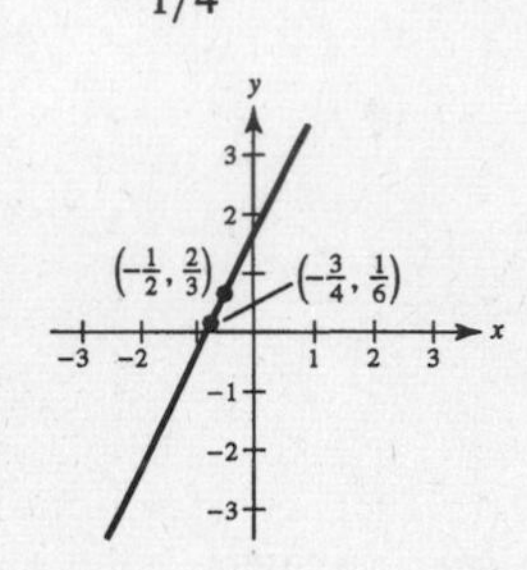

15. Since the slope is 0, the line is horizontal and its equation is $y = 1$. Therefore, three additional points are (0, 1), (1, 1), and (3, 1).

17. The equation of this line is

$$y - 7 = -3(x - 1)$$

$$y = -3x + 10.$$

Therefore, three additional points are (0, 10), (2, 4), and (3, 1).

19. Given a line L, you can use any two distinct points to calculate its slope. Since a line is straight, the ratio of the change in y-values to the change in x-values will always be the same. See Section P.2 Exercise 93 for a proof.

21. (a)

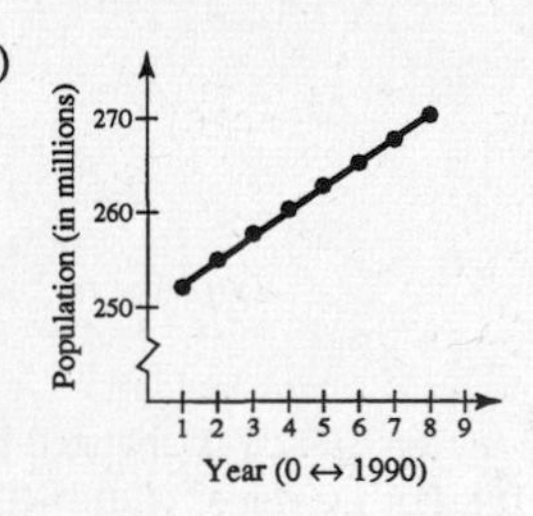

(b) The slopes of the line segments are

$$\frac{255.0 - 252.1}{2 - 1} = 2.9$$

$$\frac{257.7 - 255.0}{3 - 2} = 2.7$$

$$\frac{260.3 - 257.7}{4 - 3} = 2.6$$

$$\frac{262.8 - 260.3}{5 - 4} = 2.5$$

$$\frac{265.2 - 262.8}{6 - 5} = 2.4$$

$$\frac{267.7 - 265.2}{7 - 6} = 2.5$$

$$\frac{270.3 - 267.7}{8 - 7} = 2.6$$

The population increased most rapidly from 1991 to 1992.

$(m = 2.9)$

23. $x + 5y = 20$

$$y = -\tfrac{1}{5}x + 4$$

Therefore, the slope is $m = -\frac{1}{5}$ and the y-intercept is $(0, 4)$.

25. $x = 4$

The line is vertical. Therefore, the slope is undefined and there is no y-intercept.

27. $y = \frac{3}{4}x + 3$

$4y = 3x + 12$

$0 = 3x - 4y + 12$

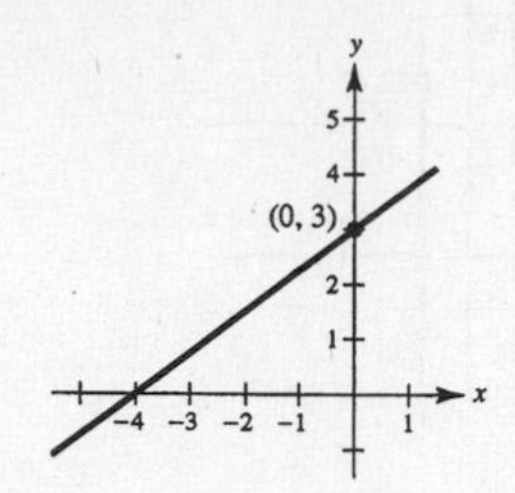

29. $y = \frac{2}{3}x$

$3y = 2x$

$2x - 3y = 0$

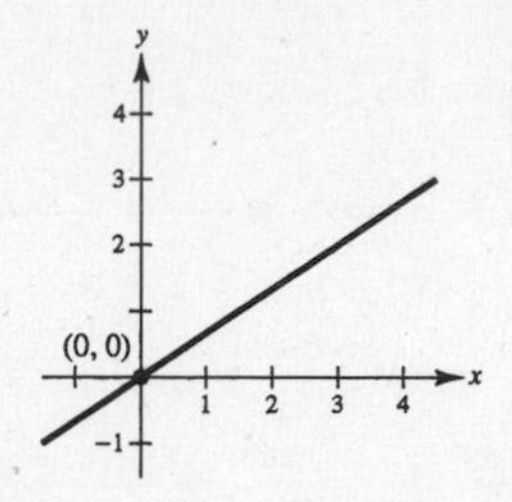

31. $y + 2 = 3(x - 3)$

$y + 2 = 3x - 9$

$y = 3x - 11$

$y - 3x + 11 = 0$

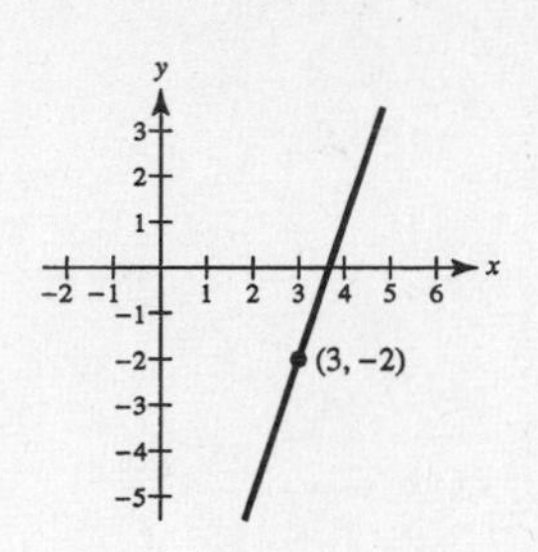

33. $m = \dfrac{6 - 0}{2 - 0} = 3$

$y - 0 = 3(x - 0)$

$y = 3x$

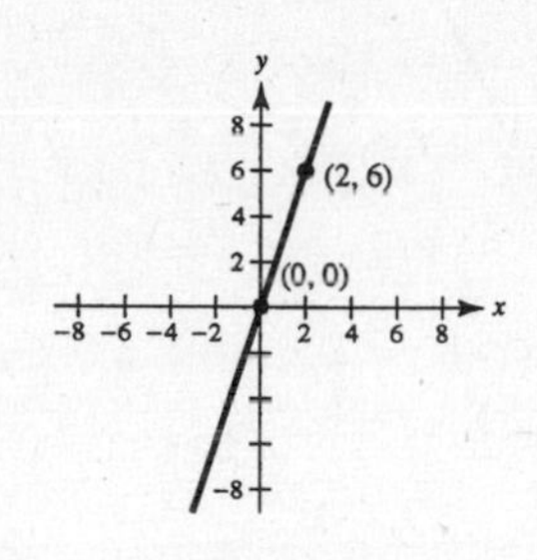

35. $m = \dfrac{1 - (-3)}{2 - 0} = 2$

$y - 1 = 2(x - 2)$

$y - 1 = 2x - 4$

$0 = 2x - y - 3$

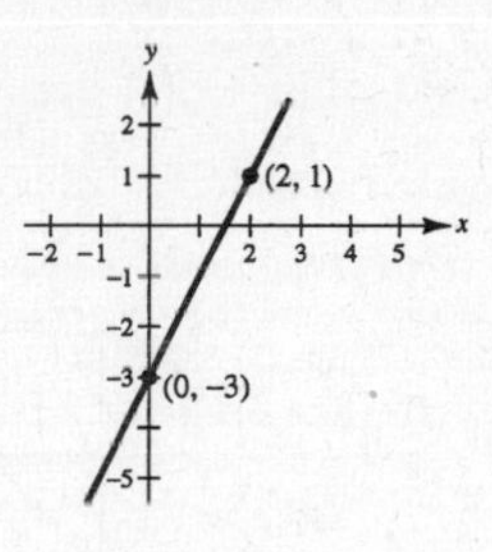

37. $m = \dfrac{8 - 0}{2 - 5} = -\dfrac{8}{3}$

$y - 0 = -\dfrac{8}{3}(x - 5)$

$y = -\dfrac{8}{3}x + \dfrac{40}{3}$

$3y + 8x - 40 = 0$

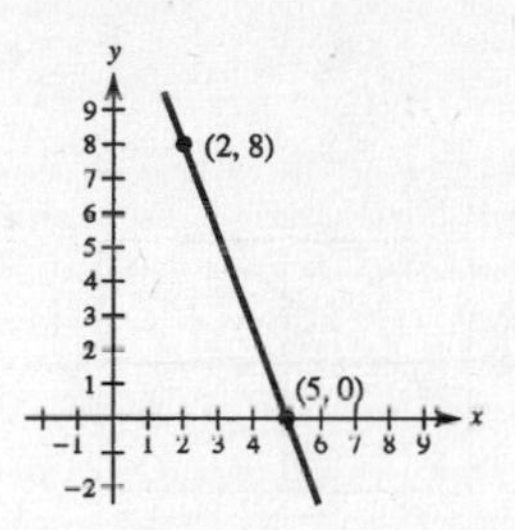

39. $m = \dfrac{8 - 1}{5 - 5}$ Undefined.

Vertical line $x = 5$

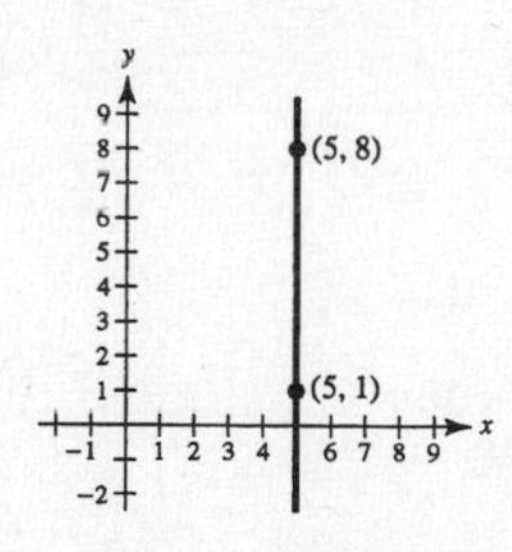

41. $m = \dfrac{7/2 - 3/4}{1/2 - 0} = \dfrac{11/4}{1/2} = \dfrac{11}{2}$

$y - \dfrac{3}{4} = \dfrac{11}{2}(x - 0)$

$y = \dfrac{11}{2}x + \dfrac{3}{4}$

$22x - 4y + 3 = 0$

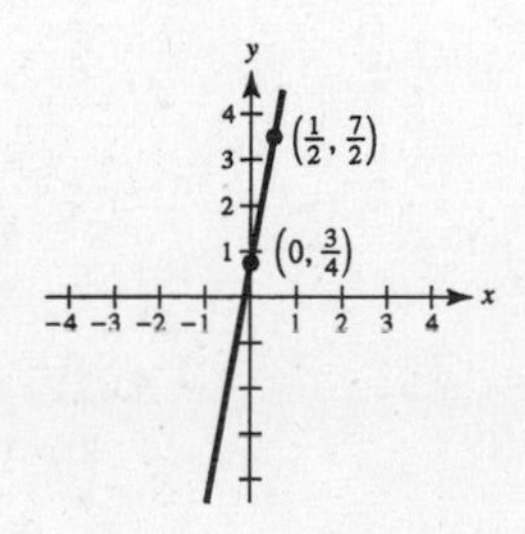

43. $x = 3$

$x - 3 = 0$

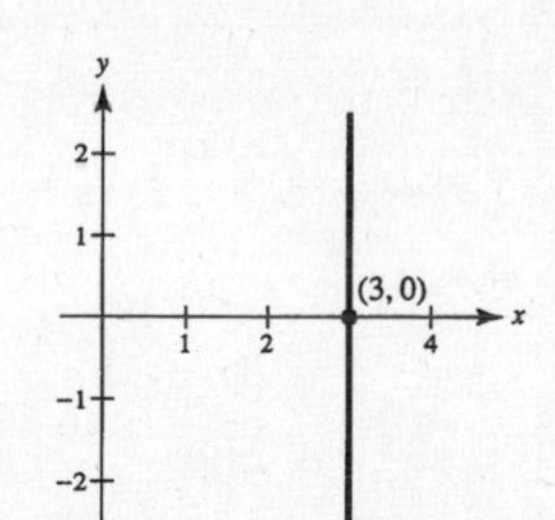

45. $\dfrac{x}{2} + \dfrac{y}{3} = 1$

$3x + 2y - 6 = 0$

47. $\dfrac{x}{a} + \dfrac{y}{a} = 1$

$\dfrac{1}{a} + \dfrac{2}{a} = 1$

$\dfrac{3}{a} = 1$

$a = 3 \Rightarrow x + y = 3$

$x + y - 3 = 0$

49. $y = -3$

$y + 3 = 0$

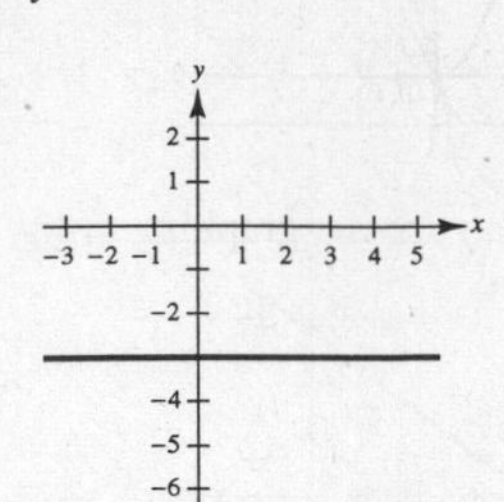

51. $y = -2x + 1$

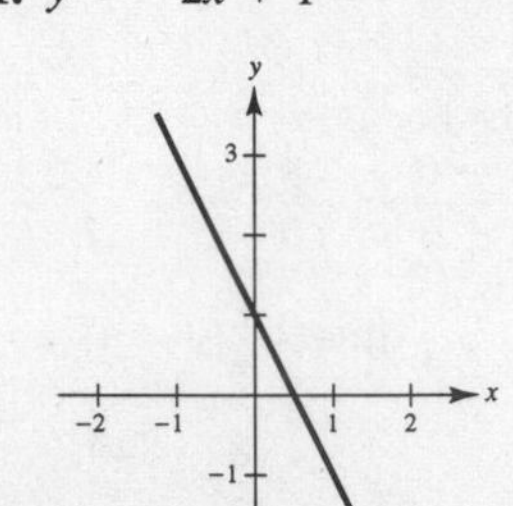

53. $y - 2 = \frac{3}{2}(x - 1)$

$y = \frac{3}{2}x + \frac{1}{2}$

$2y - 3x - 1 = 0$

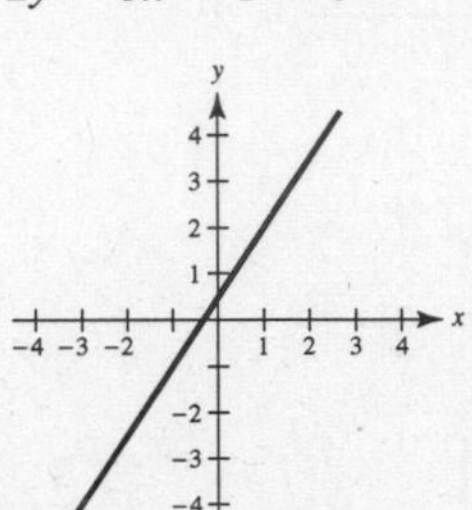

55. $2x - y - 3 = 0$

$y = 2x - 3$

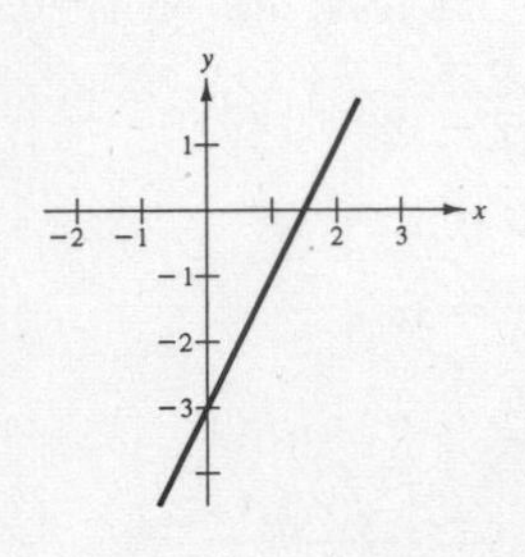

57. (a)

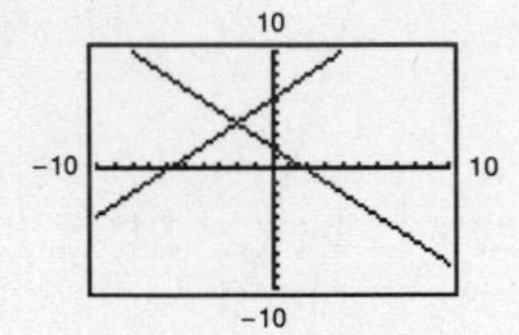

The lines do not appear perpendicular.

(b)

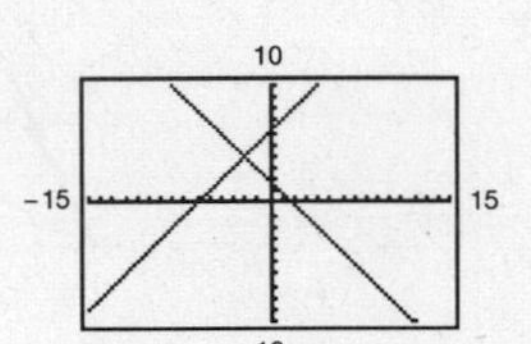

The lines appear perpendicular.

The lines are perpendicular because their slopes 1 and -1 are negative reciprocals of each other. You must use a square setting in order for perpendicular lines to appear perpendicular.

59. $4x - 2y = 3$

$y = 2x - \frac{3}{2}$

$m = 2$

(a) $y - 1 = 2(x - 2)$

$y - 1 = 2x - 4$

$2x - y - 3 = 0$

(b) $y - 1 = -\frac{1}{2}(x - 2)$

$2y - 2 = -x + 2$

$x + 2y - 4 = 0$

61. $5x - 3y = 0$

$y = \frac{5}{3}x$

$m = \frac{5}{3}$

(a) $y - \frac{7}{8} = \frac{5}{3}\left(x - \frac{3}{4}\right)$

$24y - 21 = 40x - 30$

$24y - 40x + 9 = 0$

(b) $y - \frac{7}{8} = -\frac{3}{5}\left(x - \frac{3}{4}\right)$

$40y - 35 = -24x + 18$

$40y + 24x - 53 = 0$

63. The given line is vertical.

(a) $x = 2 \Rightarrow x - 2 = 0$

(b) $y = 5 \Rightarrow y - 5 = 0$

65. The slope is 125. Hence,

$V = 125(t - 1) + 2540$

$= 125t + 2415$

67. The slope is -2000. Hence,

$V = -2000(t - 1) + 20{,}400$

$= -2000t + 22{,}400$

69.

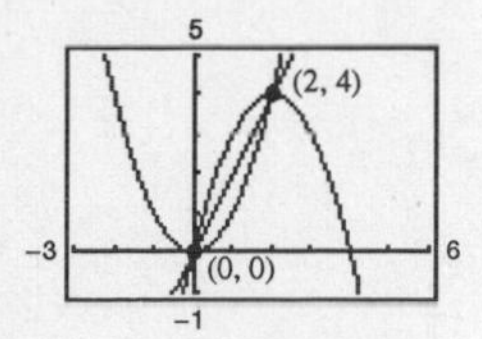

You can use the graphing utility to determine that the points of intersection are (0, 0) and (2, 4). Analytically,

$$x^2 = 4x - x^2$$
$$2x^2 - 4x = 0$$
$$2x(x - 2) = 0$$
$$x = 0 \Rightarrow y = 0 \Rightarrow (0, 0)$$
$$x = 2 \Rightarrow y = 4 \Rightarrow (2, 4).$$

The slope of the line joining (0, 0) and (2, 4) is $m = (4 - 0)/(2 - 0) = 2$. Hence, an equation of the line is

$$y - 0 = 2(x - 0)$$
$$y = 2x.$$

71. $m_1 = \dfrac{1 - 0}{-2 - (-1)} = -1$

$$m_2 = \frac{-2 - 0}{2 - (-1)} = -\frac{2}{3}$$

$$m_1 \neq m_2$$

The points are not collinear.

73. Equations of perpendicular bisectors:

$$y - \frac{c}{2} = \frac{a - b}{c}\left(x - \frac{a + b}{2}\right)$$

$$y - \frac{c}{2} = \frac{a + b}{-c}\left(x - \frac{b - a}{2}\right)$$

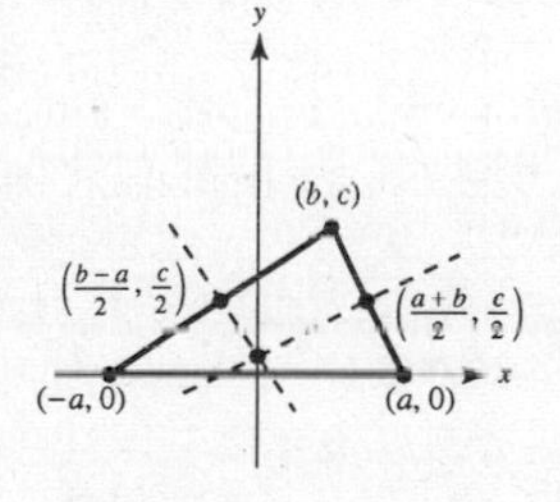

Setting the right-hand sides of the two equations equal and solving for x yields $x = 0$.

Letting $x = 0$ in either equation gives the point of intersection:

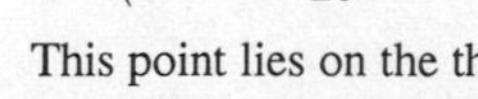

$$\left(0, \frac{-a^2 + b^2 + c^2}{2c}\right).$$

This point lies on the third perpendicular bisector, $x = 0$.

75. Equations of altitudes:

$$y = \frac{a - b}{c}(x + a)$$

$$x = b$$

$$y = -\frac{a + b}{c}(x - a)$$

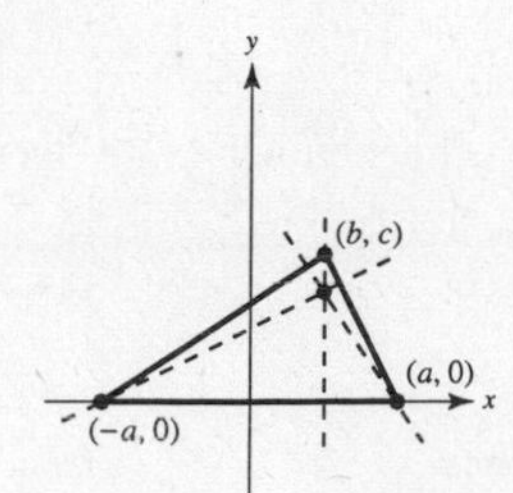

Solving simultaneously, the point of intersection is

$$\left(b, \frac{a^2 - b^2}{c}\right).$$

77. Find the equation of the line through the points (0, 32) and (100, 212).

$$m = \tfrac{180}{100} = \tfrac{9}{5}$$

$$F - 32 = \tfrac{9}{5}(C - 0)$$

$$F = \tfrac{9}{5}C + 32 \quad \text{or} \quad C = \tfrac{1}{9}(5F - 160)$$

$$5F - 9C - 160 = 0$$

For $F = 72°$, $C \approx 22.2°$.

79. (a) $W_1 = 0.75x + 12.50$

$W_2 = 1.30x + 9.20$

(b) 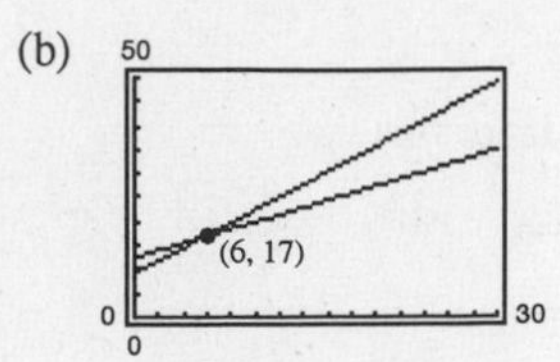

Using a graphing utility, the point of intersection is (6, 17). Analytically,

$$0.75x + 12.50 = 1.30x + 9.20$$

$$3.3 = 0.55x \Rightarrow x = 6$$

$$y = 0.75(6) + 12.50 = 17.$$

(c) Both jobs pay \$17 per hour if 6 units are produced. For someone who can produce more than 6 units per hour, the second offer would pay more. For a worker who produces less than 6 units per hour, the first offer pays more.

81. (a) Two points are (50, 580) and (47, 625). The slope is

$$m = \frac{625 - 580}{47 - 50} = -15.$$

$$p - 580 = -15(x - 50)$$

$$p = -15x + 750 + 580 = -15x + 1330$$

$$\text{or } x = \tfrac{1}{15}(1330 - p)$$

(b) 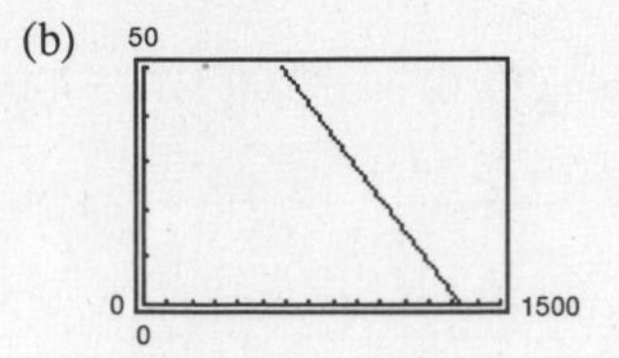

If $p = 655, x = \frac{1}{15}(1330 - 655) = 45$ units.

(c) If $p = 595, x = \frac{1}{15}(1330 - 595) = 49$ units.

83. $4x + 3y - 10 = 0 \Rightarrow d = \dfrac{|4(0) + 3(0) - 10|}{\sqrt{4^2 + 3^2}} = \dfrac{10}{5} = 2$

85. $x - y - 2 = 0 \Rightarrow d = \dfrac{|1(-2) + (-1)(1) - 2|}{\sqrt{1^2 + 1^2}} = \dfrac{5}{\sqrt{2}} = \dfrac{5\sqrt{2}}{2}$

87. A point on the line $x + y = 1$ is (0, 1). The distance from the point (0, 1) to $x + y - 5 = 0$ is

$$d = \frac{|1(0) + 1(1) - 5|}{\sqrt{1^2 + 1^2}} = \frac{|1 - 5|}{\sqrt{2}} = \frac{4}{\sqrt{2}} = 2\sqrt{2}.$$

89. If $A = 0$, then $By + C = 0$ is the horizontal line $y = -C/B$. The distance to (x_1, y_1) is

$$d = \left|y_1 - \left(\frac{-C}{B}\right)\right| = \frac{|By_1 + C|}{|B|} = \frac{|Ax_1 + By_1 + C|}{\sqrt{A^2 + B^2}}.$$

If $B = 0$, then $Ax + C = 0$ is the vertical line $x = -C/A$. The distance to (x_1, y_1) is

$$d = \left|x_1 - \left(\frac{-C}{A}\right)\right| = \frac{|Ax_1 + C|}{|A|} = \frac{|Ax_1 + By_1 + C|}{\sqrt{A^2 + B^2}}.$$

(Note that A and B cannot both be zero.)

The slope of the line $Ax + By + C = 0$ is $-A/B$. The equation of the line through (x_1, y_1) perpendicular to $Ax + By + C = 0$ is:

$$y - y_1 = \frac{B}{A}(x - x_1)$$

$$Ay - Ay_1 = Bx - Bx_1$$

$$Bx_1 - Ay_1 = Bx - Ay$$

—CONTINUED—

89. —CONTINUED—

The point of intersection of these two lines is:

$$Ax + By = -C \implies A^2x + ABy = -AC \quad (1)$$

$$Bx - Ay = Bx_1 - Ay_1 \implies B^2x - ABy = B^2x_1 - ABy_1 \quad (2)$$

$$(A^2 + B^2)x = -AC + B^2x_1 - ABy_1 \quad \text{(By adding equations (1) and (2))}$$

$$x = \frac{-AC + B^2x_1 - ABy_1}{A^2 + B^2}$$

$$Ax + By = -C \implies ABx + B^2y = -BC \quad (3)$$

$$Bx - Ay = Bx_1 - Ay_1 \implies -AB_x + A^2y = -ABx_1 + A^2y_1 \quad (4)$$

$$(A^2 + B^2)y = -BC - ABx_1 + A^2y_1 \quad \text{(By adding equations (3) and (4))}$$

$$y = \frac{-BC - ABx_1 + A^2y_1}{A^2 + B^2}$$

$$\left(\frac{-AC + B^2x_1 - ABy_1}{A^2 + B^2}, \frac{-BC - ABx_1 + A^2y_1}{A^2 + B^2}\right) \text{ point of intersection}$$

The distance between (x_1, y_1) and this point gives us the distance between (x_1, y_1) and the line $Ax + By + C = 0$.

$$d = \sqrt{\left[\frac{-AC + B^2x_1 - ABy_1}{A^2 + B^2} - x_1\right]^2 + \left[\frac{-BC - ABx_1 + A^2y_1}{A^2 + B^2} - y_1\right]^2}$$

$$= \sqrt{\left[\frac{-AC - ABy_1 - A^2x_1}{A^2 + B^2}\right]^2 + \left[\frac{-BC - ABx_1 - B^2y_1}{A^2 + B^2}\right]^2}$$

$$= \sqrt{\left[\frac{-A(C + By_1 + Ax_1)}{A^2 + B^2}\right]^2 + \left[\frac{-B(C + Ax_1 + By_1)}{A^2 + B^2}\right]^2}$$

$$= \sqrt{\frac{(A^2 + B^2)(C + Ax_1 + By_1)^2}{(A^2 + B^2)^2}}$$

$$= \frac{|Ax_1 + By_1 + C|}{\sqrt{A^2 + B^2}}$$

91. For simplicity, let the vertices of the rhombus be $(0, 0)$, $(a, 0)$, (b, c), and $(a + b, c)$, as shown in the figure. The slopes of the diagonals are then

$$m_1 = \frac{c}{a + b} \text{ and } m_2 = \frac{c}{b - a}.$$

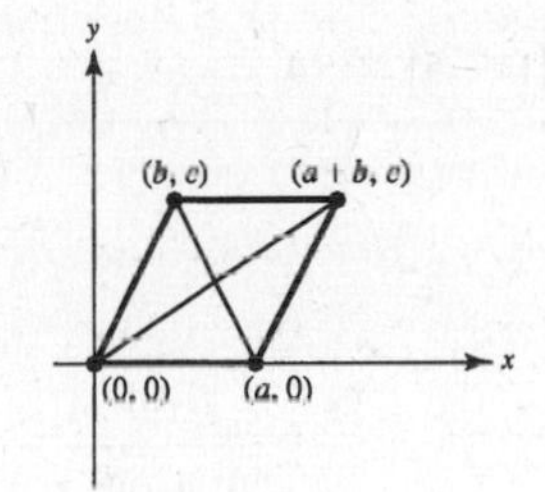

Since the sides of the Rhombus are equal, $a^2 = b^2 + c^2$, and we have

$$m_1m_2 = \frac{c}{a + b} \cdot \frac{c}{b - a} = \frac{c^2}{b^2 - a^2} = \frac{c^2}{-c^2} = -1.$$

Therefore, the diagonals are perpendicular.

93. Consider the figure below in which the four points are collinear. Since the triangles are similar, the result immediately follows.

$$\frac{y_2^* - y_1^*}{x_2^* - x_1^*} = \frac{y_2 - y_1}{x_2 - x_1}$$

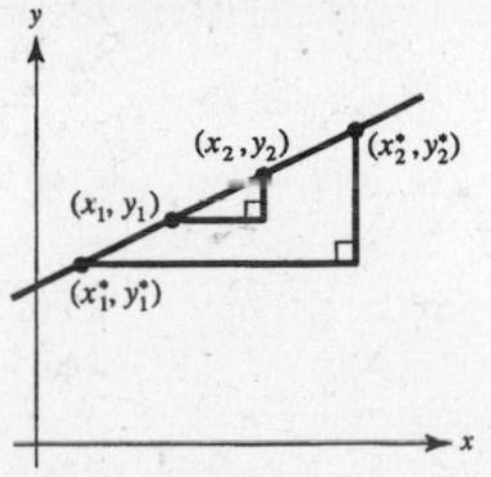

95. True

$$ax + by = c_1 \implies y = -\frac{a}{b}x + \frac{c_1}{b} \implies m_1 = -\frac{a}{b}$$

$$bx - ay = c_2 \implies y = \frac{b}{a}x - \frac{c_2}{a} \implies m_2 = \frac{b}{a}$$

$$m_2 = -\frac{1}{m_1}$$

Section P.3 Functions and Their Graphs

1. (a) $g(0) = 3 - 0^2 = 3$

(b) $g(\sqrt{3}) = 3 - (\sqrt{3})^2 = 3 - 3 = 0$

(c) $g(-2) = 3 - (-2)^2 = 3 - 4 = -1$

(d) $g(t - 1) = 3 - (t - 1)^2 = -t^2 + 2t + 2$

3. $$\frac{f(x + \Delta x) - f(x)}{\Delta x} = \frac{(x + \Delta x)^3 - x^3}{\Delta x}$$
$$= \frac{x^3 + 3x^2\Delta x + 3x(\Delta x)^2 + (\Delta x)^3 - x^3}{\Delta x}$$
$$= 3x^2 + 3x\Delta x + (\Delta x)^2, \quad \Delta x \neq 0$$

5. $$\frac{f(x) - f(2)}{x - 2} = \frac{(1/\sqrt{x - 1} - 1)}{x - 2}$$
$$= \frac{1 - \sqrt{x - 1}}{(x - 2)\sqrt{x - 1}} \cdot \frac{1 + \sqrt{x - 1}}{1 + \sqrt{x - 1}} = \frac{2 - x}{(x - 2)\sqrt{x - 1}(1 + \sqrt{x - 1})} = \frac{-1}{\sqrt{x - 1}(1 + \sqrt{x - 1})}, x \neq 2$$

7. $h(x) = -\sqrt{x + 3}$

Domain: $x + 3 \geq 0 \Rightarrow [-3, \infty)$

Range: $(-\infty, 0]$

9. $f(t) = \sec \dfrac{\pi t}{4}$

$\dfrac{\pi t}{4} \neq \dfrac{(2k + 1)\pi}{2} \Rightarrow t \neq 4k + 2$

Domain: all $t \neq 4k + 2$,
k an integer

Range: $(-\infty, -1], [1, \infty)$

11. $f(x) = \dfrac{1}{x}$

Domain: $(-\infty, 0), (0, \infty)$

Range: $(-\infty, 0), (0, \infty)$

13. $f(x) = \begin{cases} 2x + 1, x < 0 \\ 2x + 2, x \geq 0 \end{cases}$

(a) $f(-1) = 2(-1) + 1 = -1$

(b) $f(0) = 2(0) + 2 = 2$

(c) $f(2) = 2(2) + 2 = 6$

(d) $f(t^2 + 1) = 2(t^2 + 1) + 2 = 2t^2 + 4$

(Note: $t^2 + 1 \geq 0$ for all t)

Domain: $(-\infty, \infty)$

Range: $(-\infty, 1), [2, \infty)$

15. $f(x) = \begin{cases} |x| + 1, x < 1 \\ -x + 1, x \geq 1 \end{cases}$

(a) $f(-3) = |-3| + 1 = 4$

(b) $f(1) = -1 + 1 = 0$

(c) $f(3) = -3 + 1 = -2$

(d) $f(b^2 + 1) = -(b^2 + 1) + 1 = -b^2$

Domain: $(-\infty, \infty)$

Range: $(-\infty, 0] \cup [1, \infty)$

17. The slope between $(-2, 0)$ and $(-1, 1)$ is $\dfrac{1 - 0}{-1 - (-2)} = 1.$

The segement has equation $y - 0 = 1(x + 2) \Rightarrow y = x + 2, -2 \leq x \leq -1.$

The slope between $(-1, 1)$ and $(3, -1)$ is $\dfrac{-1 - 1}{3 - (-1)} = -\dfrac{1}{2}.$

The segment has equation $y - 1 = -\dfrac{1}{2}(x + 1) \Rightarrow y = -\dfrac{1}{2}x + \dfrac{1}{2}, -1 \leq x \leq 3.$

Finally,

$$f(x) = \begin{cases} x + 2, & -2 \leq x \leq -1 \\ -\dfrac{1}{2}x + \dfrac{1}{2}, & -1 < x \leq 3 \end{cases}$$

19. $f(x) = 4 - x$

Domain: $(-\infty, \infty)$

Range: $(-\infty, \infty)$

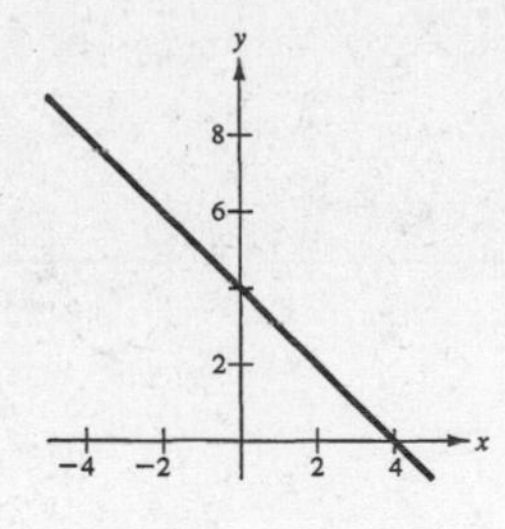

21. $h(x) = \sqrt{x - 1}$

Domain: $[1, \infty)$

Range: $[0, \infty)$

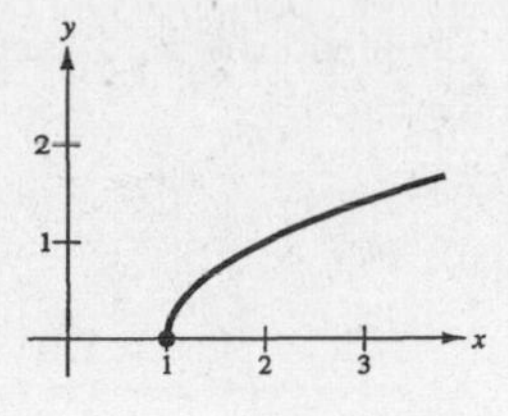

23. $f(x) = \sqrt{9 - x^2}$

Domain: $[-3, 3]$

Range: $[0, 3]$

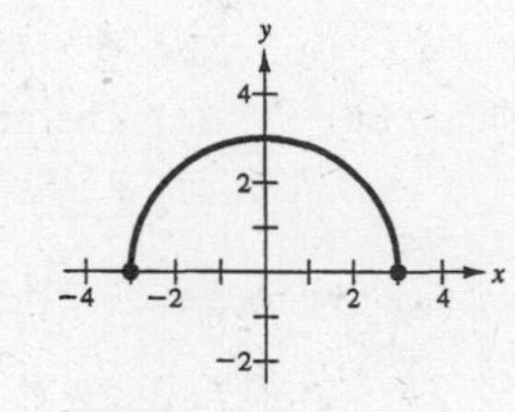

25. $g(t) = 2 \sin \pi t$

Domain: $(-\infty, \infty)$

Range: $[-2, 2]$

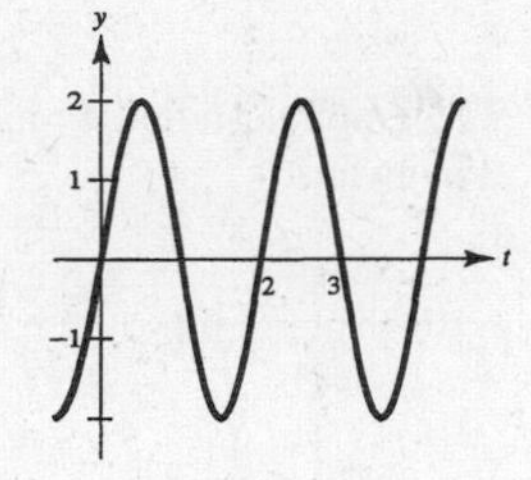

27. $x - y^2 = 0 \Rightarrow y = \pm\sqrt{x}$

y is not a function of x. Some vertical lines intersect the graph twice.

29. y is a function of x. Vertical lines intersect the graph at most once.

31. $x^2 + y^2 = 4 \Rightarrow y = \pm\sqrt{4 - x^2}$

y is not a function of x since there are two values of y for some x.

33. $y^2 = x^2 - 1 \Rightarrow y = \pm\sqrt{x^2 - 1}$

y is not a function of x since there are two values of y for some x.

35. $f(x) = |x| + |x - 2|$

If $x < 0$, then $f(x) = -x - (x - 2) = -2x + 2 = 2(1 - x)$.

If $0 \le x < 2$, then $f(x) = x - (x - 2) = 2$.

If $x \ge 2$, then $f(x) = x + (x - 2) = 2x - 2 = 2(x - 1)$.

Thus,

$$f(x) = \begin{cases} 2(1 - x), & x < 0 \\ 2, & 0 \le x < 2. \\ 2(x - 1), & x \ge 2 \end{cases}$$

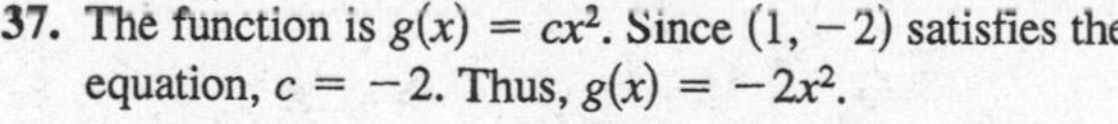
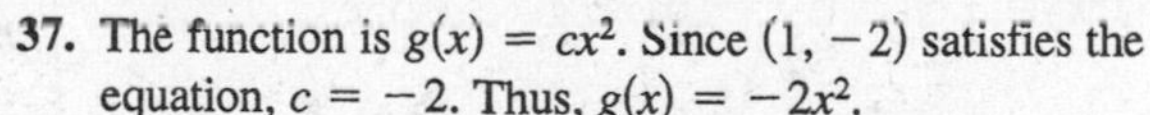

37. The function is $g(x) = cx^2$. Since $(1, -2)$ satisfies the equation, $c = -2$. Thus, $g(x) = -2x^2$.

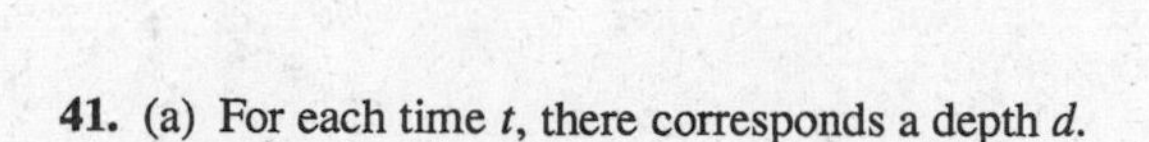

39. The function is $r(x) = c/x$, since it must be undefined at $x = 0$. Since $(1, 32)$ satisfies the equation, $c = 32$. Thus, $r(x) = 32/x$.

41. (a) For each time t, there corresponds a depth d.

(b) Domain: $0 \le t \le 5$

Range: $0 \le d \le 30$

(c)

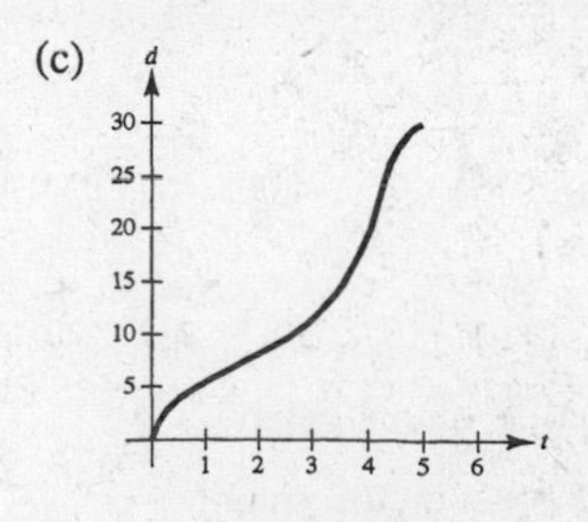

43.

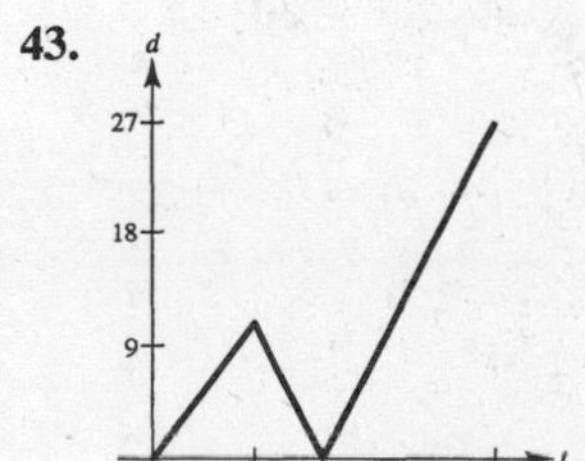

45. (a) The graph is shifted 3 units to the left.

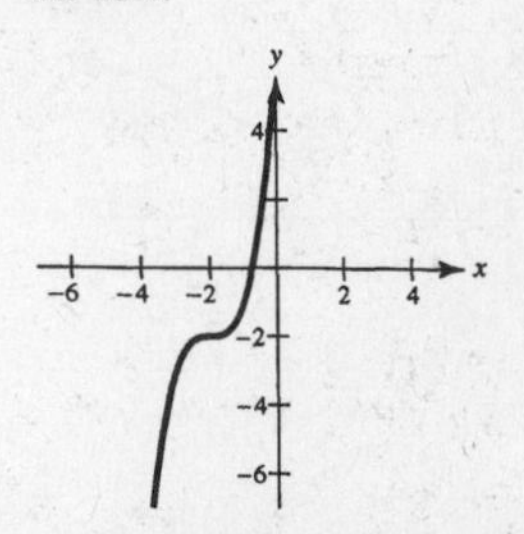

(b) The graph is shifted 1 unit to the right.

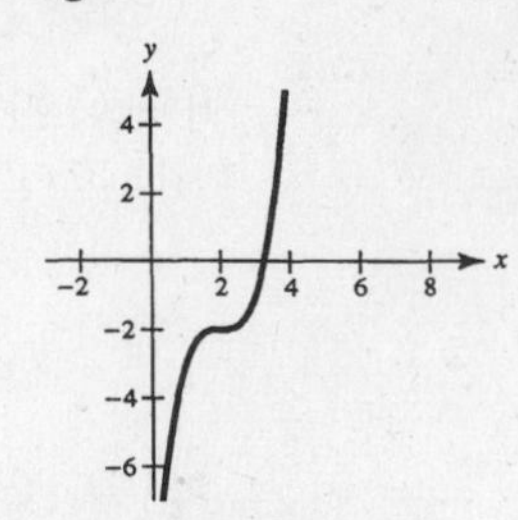

(c) The graph is shifted 2 units upward.

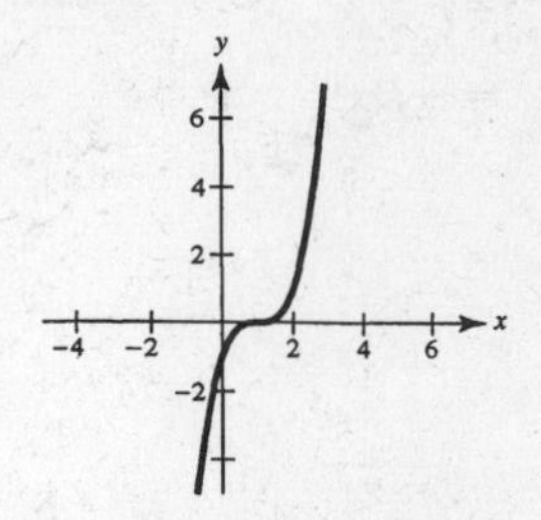

(d) The graph is shifted 4 units downward.

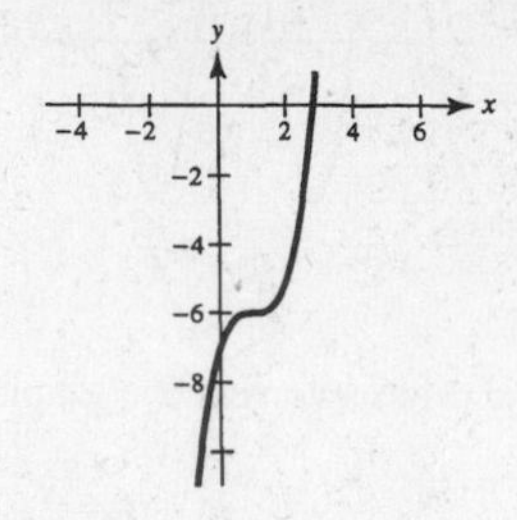

(e) The graph is stretched vertically by a factor of 3.

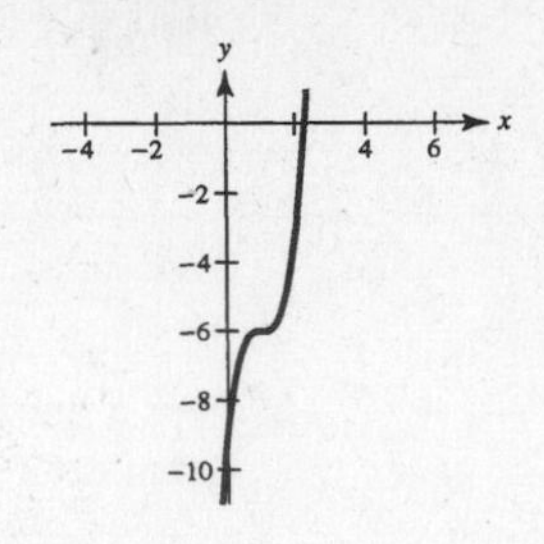

(f) The graph is stretched vertically by a factor of $\frac{1}{4}$.

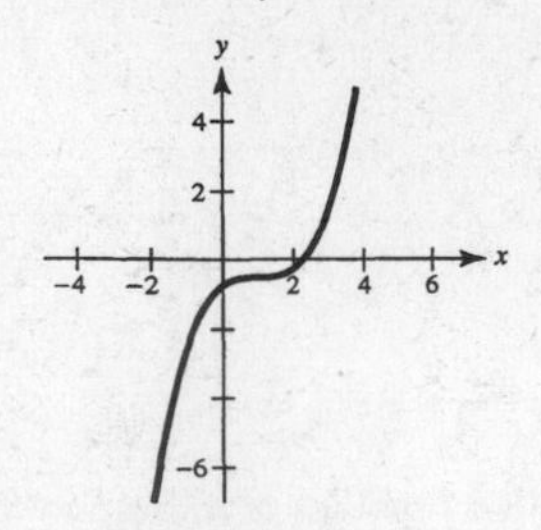

47. (a) $y = \sqrt{x} + 2$

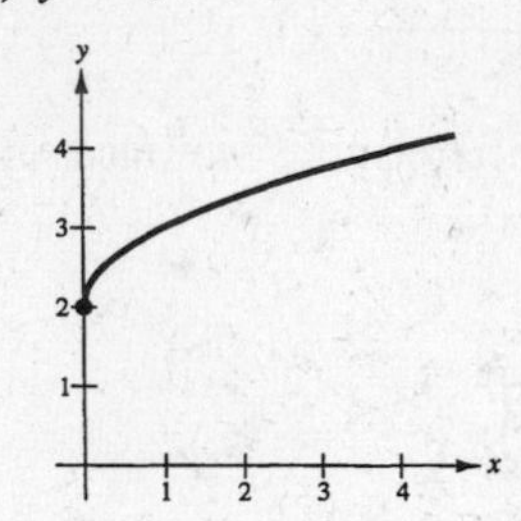

Vertical shift 2 units upward

(b) $y = -\sqrt{x}$

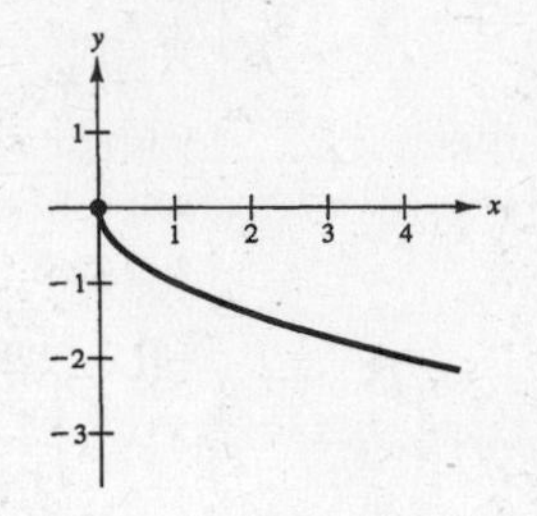

Reflection about the x-axis

(c) $y = \sqrt{x - 2}$

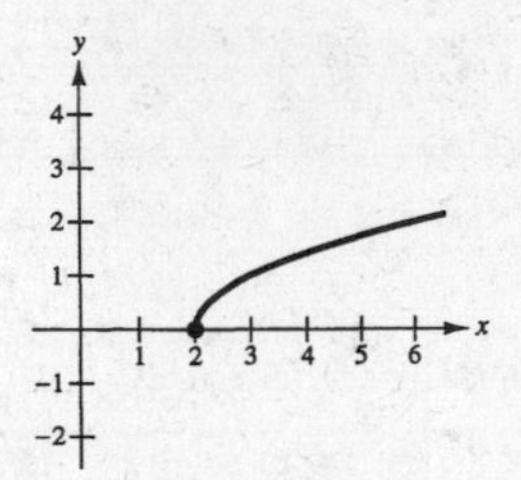

Horizontal shift 2 units to the right

49. (a) $T(4) = 16°$, $T(15) \approx 23°$

(b) If $H(t) = T(t - 1)$, then the program would turn on (and off) one hour later.

(c) If $H(t) = T(t) - 1$, then the overall temperature would be reduced 1 degree.

51. $f(x) = x^2$, $g(x) = \sqrt{x}$

$(f \circ g)(x) = f(g(x)) = f(\sqrt{x}) = (\sqrt{x})^2 = x, \quad x \geq 0$

Domain: $[0, \infty)$

$(g \circ f)(x) = g(f(x)) = g(x^2) = \sqrt{x^2} = |x|$

Domain: $(-\infty, \infty)$

No. Their domains are different. $(f \circ g) = (g \circ f)$ for $x \geq 0$.

53. $f(x) = \dfrac{3}{x}$, $g(x) = x^2 - 1$

$(f \circ g)(x) = f(g(x)) = f(x^2 - 1) = \dfrac{3}{x^2 - 1}$

Domain: all $x \neq \pm 1$

$(g \circ f)(x) = g(f(x)) = g\left(\dfrac{3}{x}\right) = \left(\dfrac{3}{x}\right)^2 - 1 = \dfrac{9}{x^2} - 1 = \dfrac{9 - x^2}{x^2}$

Domain: all $x \neq 0$

No, $f \circ g \neq g \circ f$.

55. $(A \circ r)(t) = A(r(t)) = A(0.6t) = \pi(0.6t)^2 = 0.36\pi t^2$

$(A \circ r)(t)$ represents the area of the circle at time t.

57. $f(-x) = (-x)^2(4 - (-x)^2) = x^2(4 - x^2) = f(x)$

Even

59. $f(-x) = (-x)\cos(-x) = -x\cos x = -f(x)$

Odd

61. (a) If f is even, then $\left(\frac{3}{2}, 4\right)$ is on the graph.

(b) If f is odd, then $\left(\frac{3}{2}, -4\right)$ is on the graph.

63. $f(-x) = a_{2n+1}(-x)^{2n+1} + \cdots + a_3(-x)^3 + a_1(-x)$

$= -[a_{2n+1}x^{2n+1} + \cdots + a_3x^3 + a_1x]$

$= -f(x)$

Odd

65. Let $F(x) = f(x)g(x)$ where f and g are even. Then

$$F(-x) = f(-x)g(-x) = f(x)g(x) = F(x).$$

Thus, $F(x)$ is even. Let $F(x) = f(x)g(x)$ where f and g are odd. Then

$$F(-x) = f(-x)g(-x)$$
$$= [-f(x)][-g(x)] = f(x)g(x) = F(x).$$

Thus, $F(x)$ is even.

67. $f(x) = x^2$ and $g(x) = 4 - x^2$ are even.

$f(x)g(x) = x^2(4 - x^2) = 4x^2 - x^4$ is even.

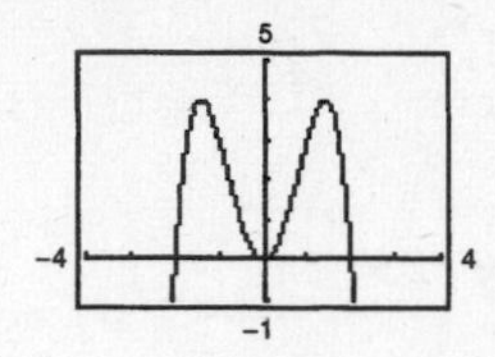

$f(x) = x$ is odd and $g(x) = 4 - x^2$ is even.

$f(x)g(x) = x(4 - x^2) = 4x - x^3$ is odd.

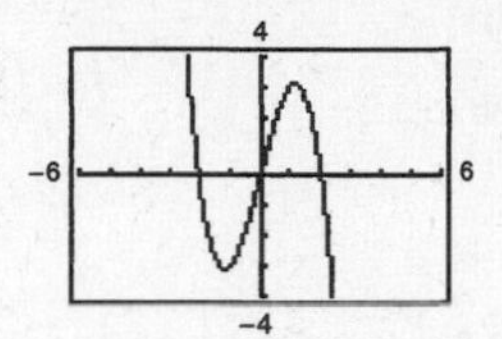

69. (a)

x	*length and width*	*volume V*
1	$24 - 2(1)$	484
2	$24 - 2(2)$	800
3	$24 - 2(3)$	972
4	$24 - 2(4)$	1024
5	$24 - 2(5)$	980
6	$24 - 2(6)$	864

The maximum volume appears to be 1024 cm^3.

(c) $V = x(24 - 2x)^2 = 4x(12 - x)^2$

Domain: $0 < x < 12$

(b)

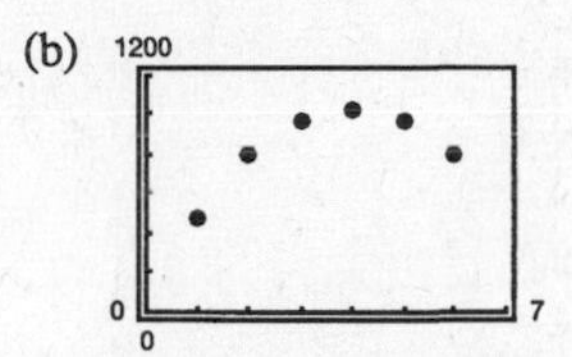

Yes, V is a function of x.

(d)

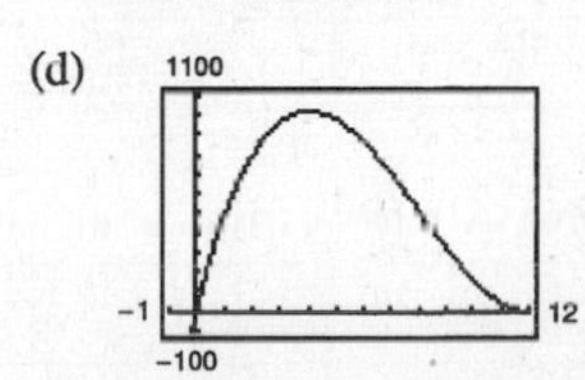

Maximum volume is $V = 1024$ cm^3 for box having dimensions $4 \times 16 \times 16$ cm.

71. False; let $f(x) = x^2$. Then $f(-3) = f(3) = 9$, but $-3 \neq 3$.

73. True, the function is even.

Section P.4 Fitting Models to Data

1. Quadratic function

3. Linear function

5. (a), (b)

Yes. The cancer mortality increases linearly with increased exposure to the carcinogenic substance.

(c) If $x = 3$, then $y \approx 136$.

7. (a) $d = 0.066F$ or $F = 15.1d + 0.1$

(b)

The model fits well.

(c) If $F = 55$, then $d \approx 0.066(55) = 3.63$ cm.

9. (a) Let x = per capita energy usage (in millions of Btu)

y = per capita gross national product (in thousands)

$y = 0.0764x + 4.9985 \approx 0.08x + 5.0$

$r = 0.7052$

(b)

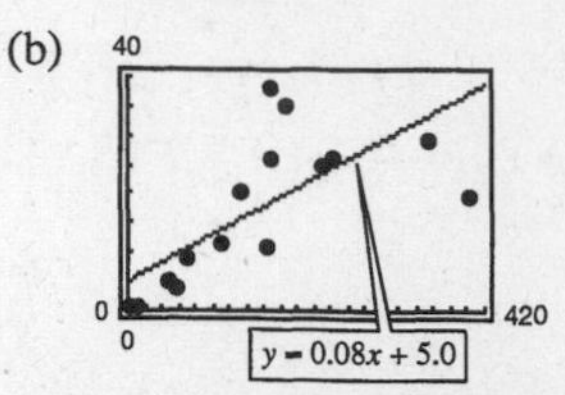

(c) Denmark, Japan, and Canada

(d) Deleting the data for the three countries above,

$y = 0.0959x + 1.0539$

($r = 0.9202$ is much closer to 1.)

11. (a) $y_1 = 0.0343t^3 - 0.3451t^2 + 0.8837t + 5.6061$

$y_2 = 0.1095t + 2.0667$

$y_3 = 0.0917t + 0.7917$

(b)

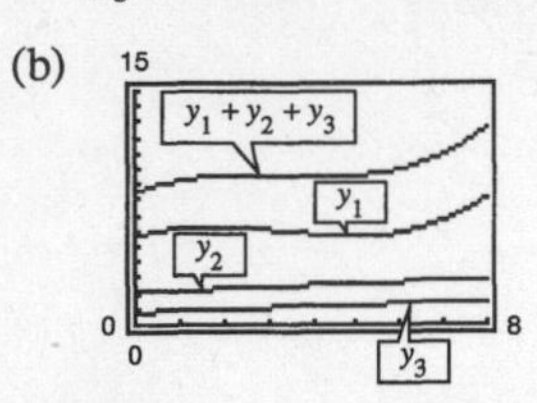

For 2002, $t = 12$ and $y_1 + y_2 + y_3 \approx 31.06$ cents/mile.

13. (a) $y_1 = 4.0367t + 28.9644$

$y_2 = -0.0099t^3 + 0.5488t^2 + 0.2399t + 33.1414$

(b)

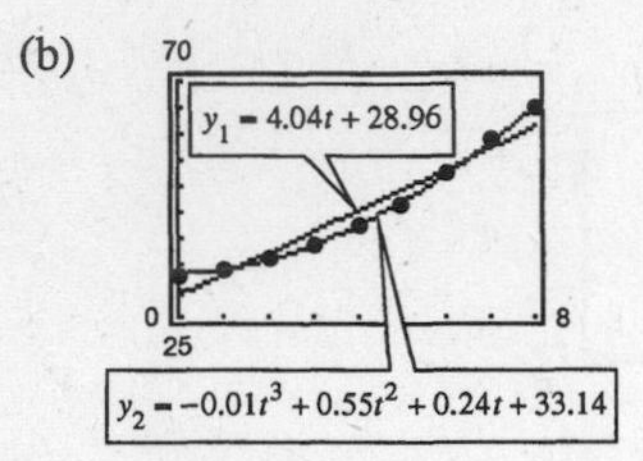

(c) The cubic model is better.

(d) $y_3 = 0.4297t^2 + 0.5994t + 32.9745$

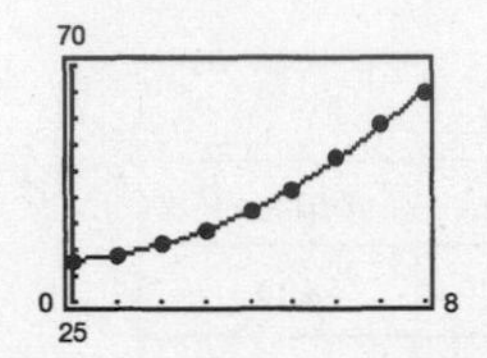

(e) The slope represents the average increase per year in the number of people (in millions) in HMOs.

(f) For 2000, $t = 10$, and $y_1 \approx 69.3$ million. (linear)

$y_2 \approx 80.5$ million (cubic)

15. (a) $y = -1.81x^3 + 14.58x^2 + 16.39x + 10$

(b)

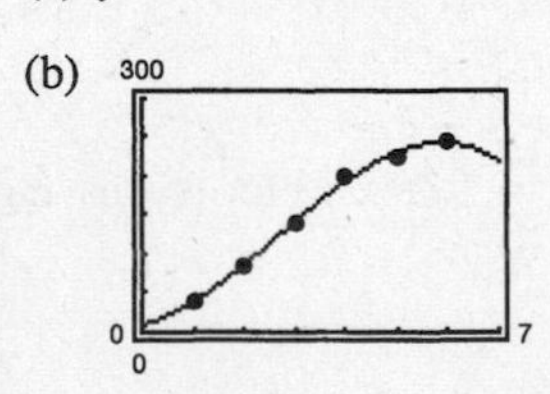

(c) If $x = 4.5$, $y \approx 214$ horsepower.

17. (a) Yes, y is a function of t. At each time t, there is one and only one displacement y.

(b) The amplitude is approximately $(2.35 - 1.65)/2 = 0.35$.

The period is approximately $2(0.375 - 0.125) = 0.5$.

(c) One model is $y = 0.35 \sin(4\pi t) + 2$.

(d)

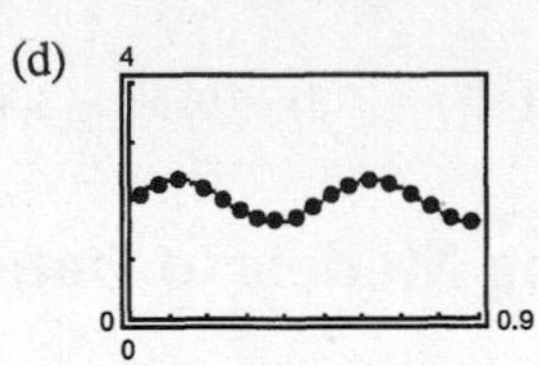

19. Answers will vary.

Section P.5 Inverse Functions

1. (a) $f(x) = 5x + 1$

$$g(x) = \frac{x - 1}{5}$$

$$f(g(x)) = f\left(\frac{x - 1}{5}\right) = 5\left(\frac{x - 1}{5}\right) + 1 = x$$

$$g(f(x)) = g(5x + 1) = \frac{(5x + 1) - 1}{5} = x$$

(b)

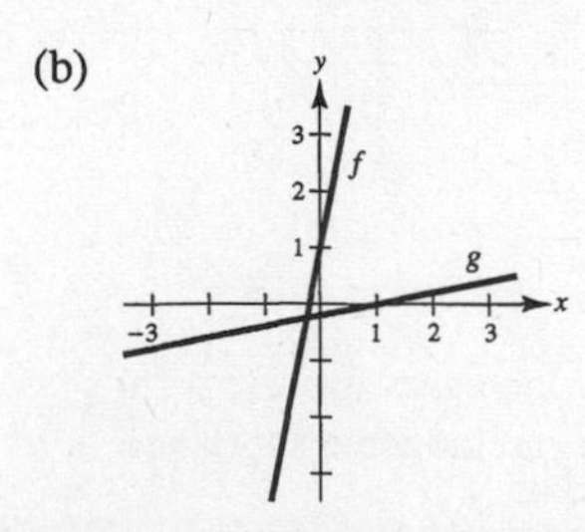

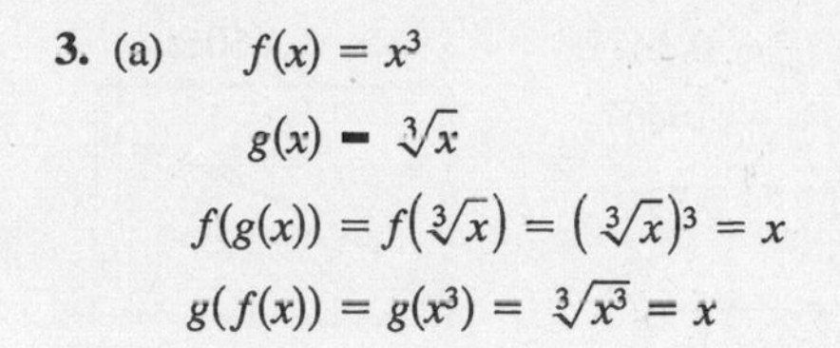

3. (a) $f(x) = x^3$

$g(x) = \sqrt[3]{x}$

$f(g(x)) = f(\sqrt[3]{x}) = (\sqrt[3]{x})^3 = x$

$g(f(x)) = g(x^3) = \sqrt[3]{x^3} = x$

(b)

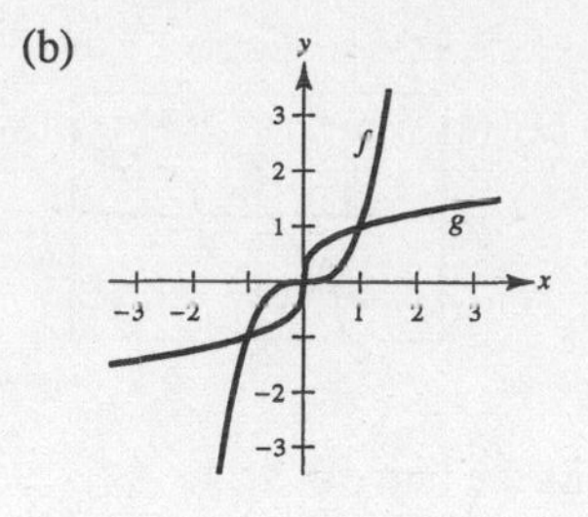

5. (a) $f(x) = \sqrt{x-4}$

$g(x) = x^2 + 4,\ x \geq 0$

$f(g(x)) = f(x^2 + 4)$

$= \sqrt{(x^2+4) - 4} = \sqrt{x^2} = x$

$g(f(x)) = g(\sqrt{x-4})$

$= (\sqrt{x-4})^2 + 4 = x - 4 + 4 = x$

(b)

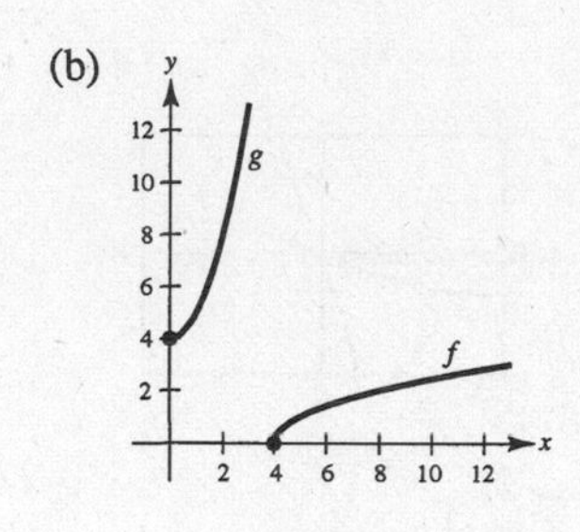

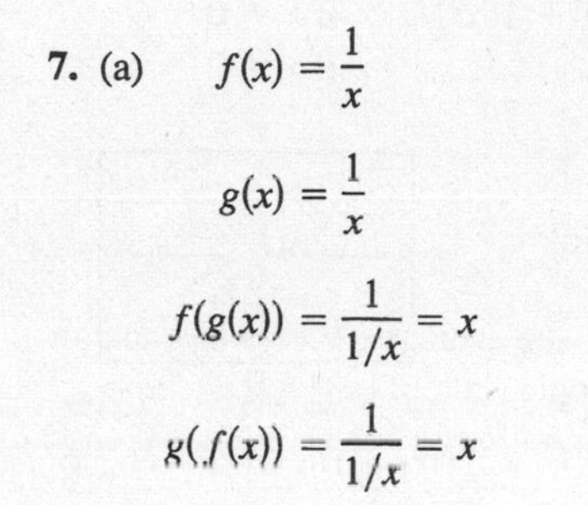

7. (a) $f(x) = \dfrac{1}{x}$

$g(x) = \dfrac{1}{x}$

$f(g(x)) = \dfrac{1}{1/x} = x$

$g(f(x)) = \dfrac{1}{1/x} = x$

(b)

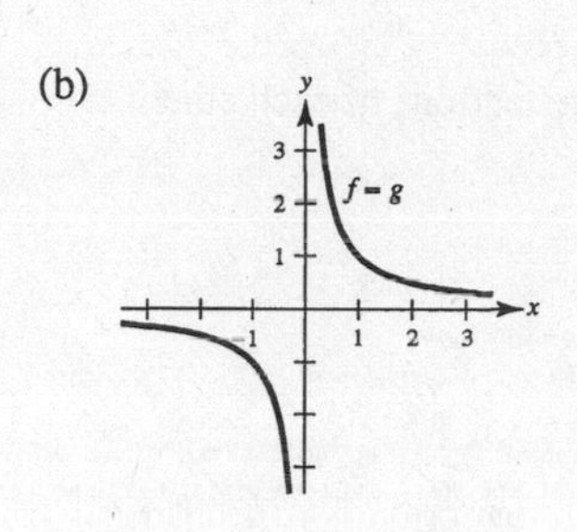

9. Matches (c)

11. Matches (a)

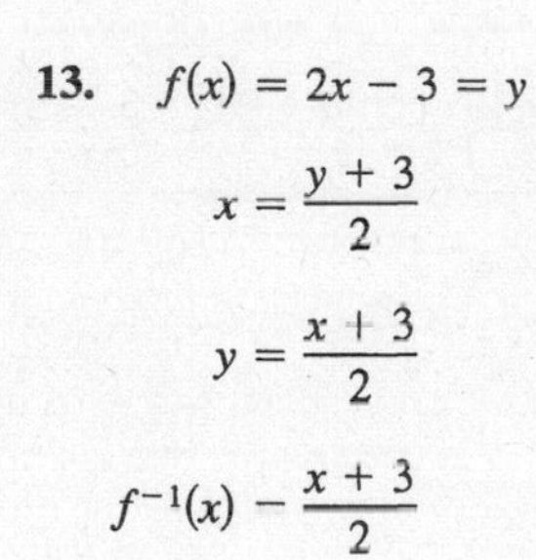

13. $f(x) = 2x - 3 = y$

$x = \dfrac{y+3}{2}$

$y = \dfrac{x+3}{2}$

$f^{-1}(x) = \dfrac{x+3}{2}$

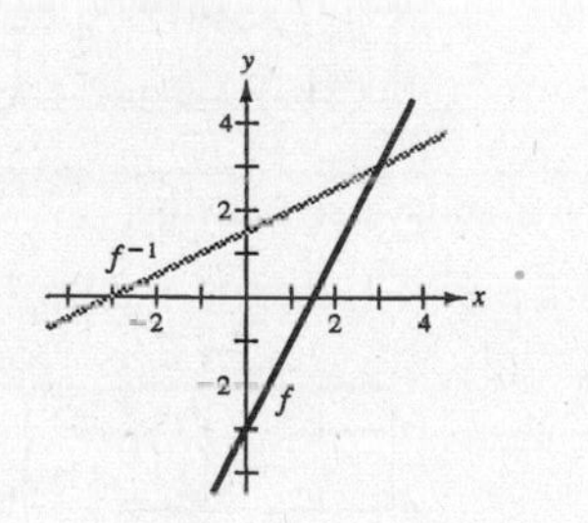

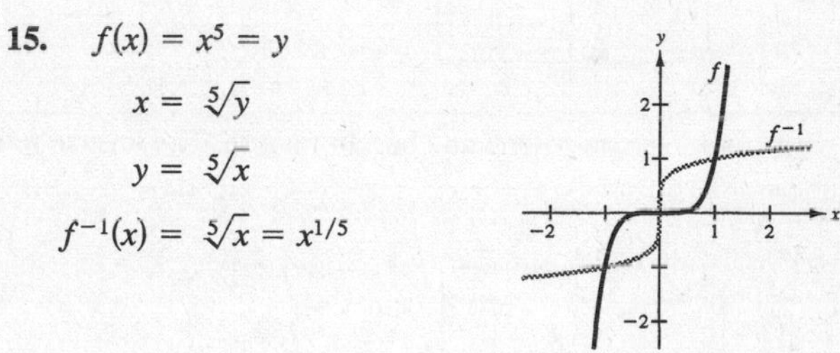

15. $f(x) = x^5 = y$

$x = \sqrt[5]{y}$

$y = \sqrt[5]{x}$

$f^{-1}(x) = \sqrt[5]{x} = x^{1/5}$

17. $f(x) = \sqrt{x} = y$

$x = y^2$

$y = x^2$

$f^{-1}(x) = x^2,\ x \geq 0$

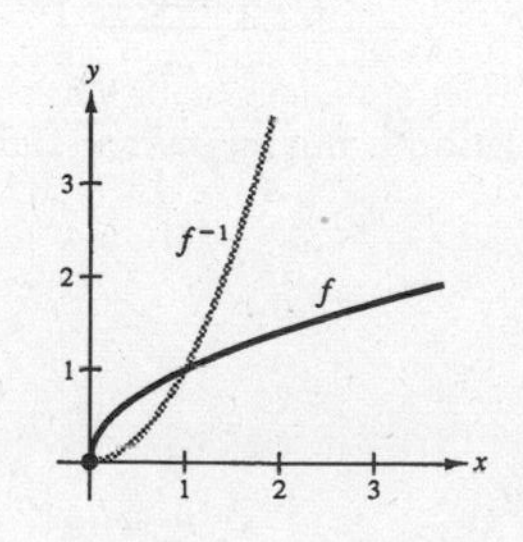

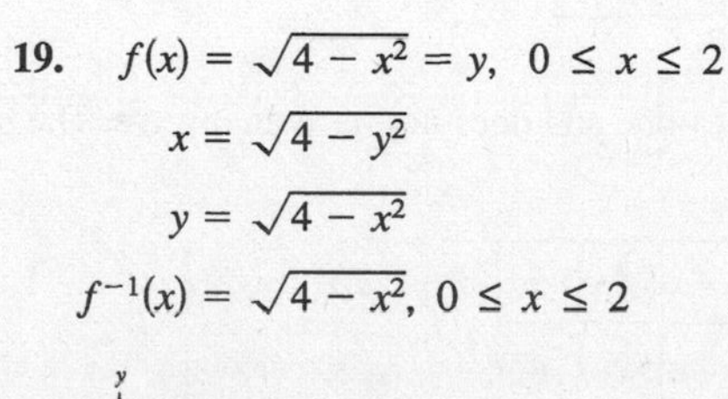

19. $f(x) = \sqrt{4 - x^2} = y,\ 0 \leq x \leq 2$

$x = \sqrt{4 - y^2}$

$y = \sqrt{4 - x^2}$

$f^{-1}(x) = \sqrt{4 - x^2},\ 0 \leq x \leq 2$

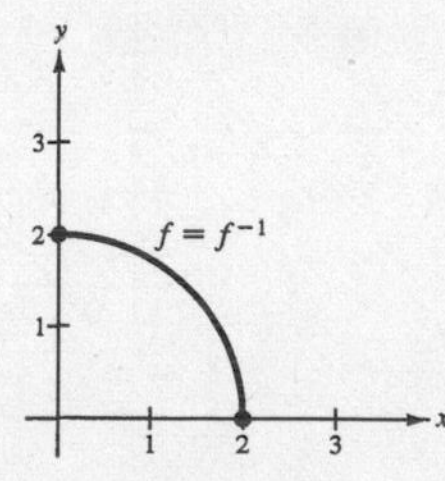

21. $f(x) = \sqrt[3]{x-1} = y$

$x = y^3 + 1$

$y = x^3 + 1$

$f^{-1}(x) = x^3 + 1$

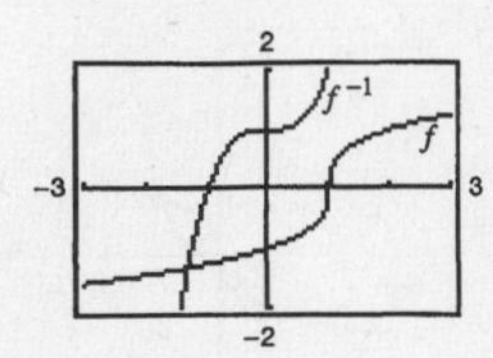

The graphs of f and f^{-1} are reflections of each other across the line $y = x$.

23. $f(x) = x^{2/3} = y, \quad x \geq 0$

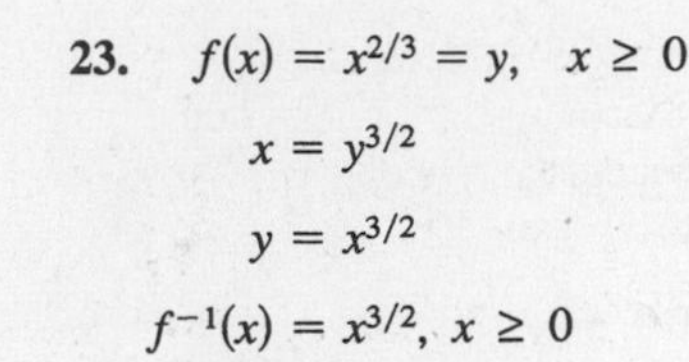

$x = y^{3/2}$

$y = x^{3/2}$

$f^{-1}(x) = x^{3/2}, \; x \geq 0$

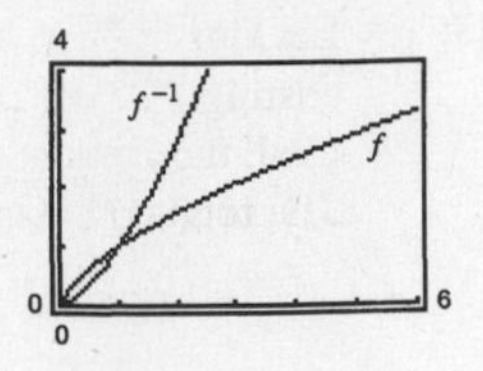

The graphs of f and f^{-1} are reflections of each other across the line $y = x$.

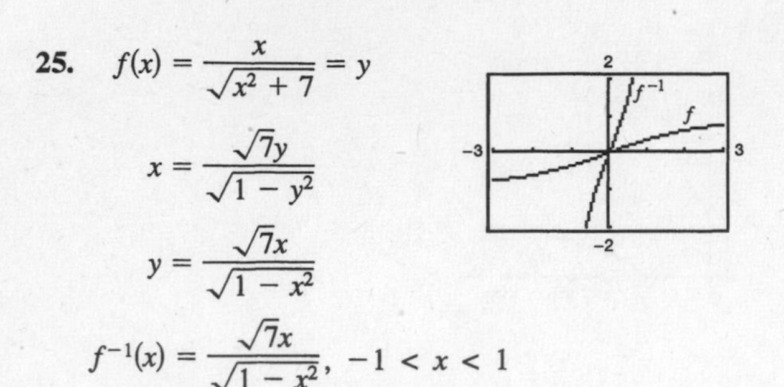

25. $f(x) = \dfrac{x}{\sqrt{x^2+7}} = y$

$x = \dfrac{\sqrt{7}y}{\sqrt{1-y^2}}$

$y = \dfrac{\sqrt{7}x}{\sqrt{1-x^2}}$

$f^{-1}(x) = \dfrac{\sqrt{7}x}{\sqrt{1-x^2}}, \quad -1 < x < 1$

The graphs of f and f^{-1} are reflections of each other across the line $y = x$.

27. $f(x) = \dfrac{x}{x^2-4} = y$ on $(-2, 2)$

$x^2y - 4y = x$

$x^2y - x - 4y = 0$

$a = y, b = -1, c = -4y$

$x = \dfrac{1 \pm \sqrt{1-4(y)(-4y)}}{2y} = \dfrac{1 \pm \sqrt{1+16y^2}}{2y}$

$y = f^{-1}(x) = \begin{cases} \left(1 - \sqrt{1+16x^2}\right)/2x, & \text{if } x \neq 0 \\ 0, & \text{if } x = 0 \end{cases}$

Domain: all x

Range: $-2 < y < 2$

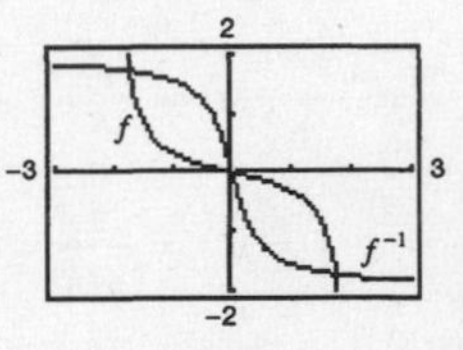

The graphs of f and f^{-1} are reflections of each other across the line $y = x$.

29. (a)

(b)

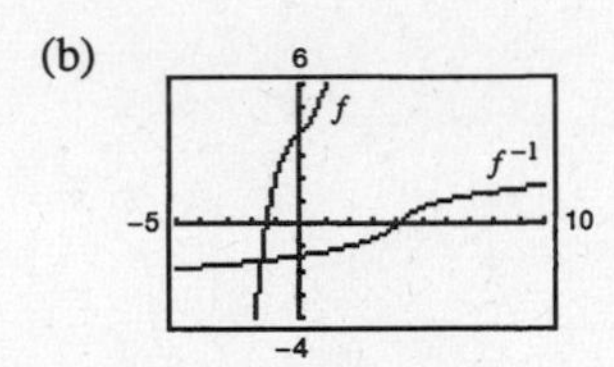

(c) Yes, f is one-to-one and has an inverse. The inverse relation is an inverse function.

31. (a)

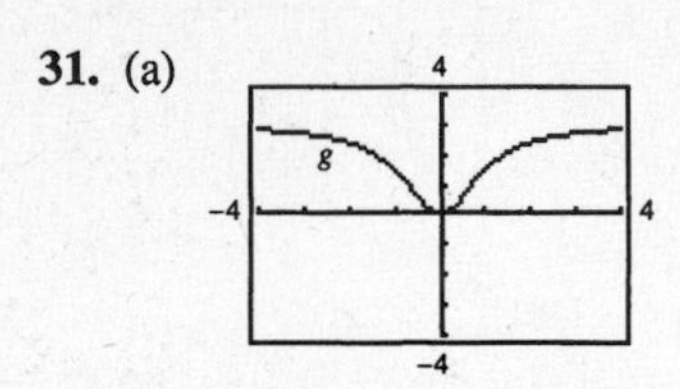

(b)

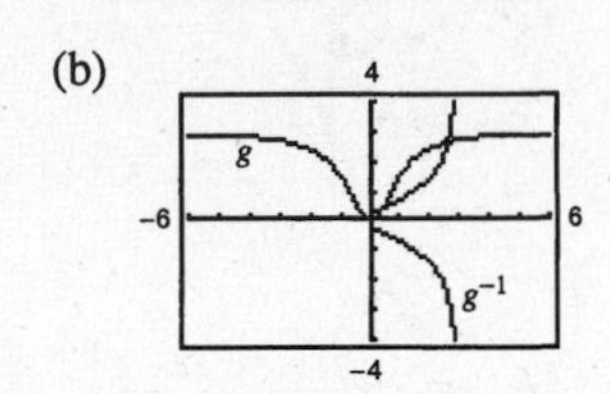

(c) g is not one-to-one and does not have an inverse. The inverse relation is not an inverse function.

33.

x	1	2	3	4
$f^{-1}(x)$	0	1	2	4

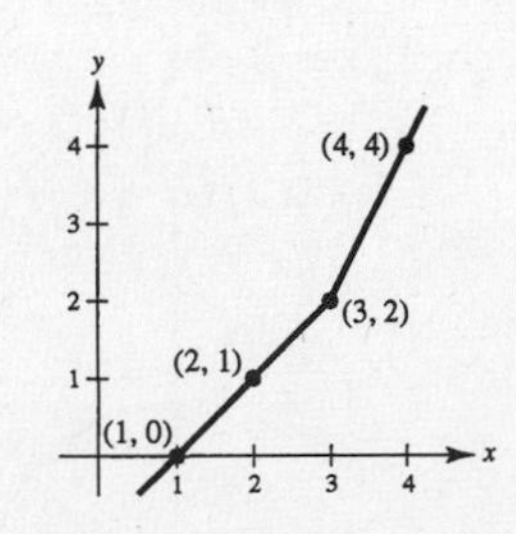

35. (a) Let x be the number of pounds of the commodity costing 1.25 per pound. Since there are 50 pounds total, the amount of the second commodity is $50 - x$. The total cost is

$$\begin{aligned} y &= 1.25x + 1.60(50 - x) \\ &= -0.35x + 80 \qquad 0 \le x \le 50. \end{aligned}$$

(b) We find the inverse of the original function:

$$\begin{aligned} y &= -0.35x + 80 \\ 0.35x &= 80 - y \\ x &= \tfrac{100}{35}(80 - y) \end{aligned}$$

Inverse: $y = \frac{100}{35}(80 - x) = \frac{20}{7}(80 - x)$.

x represents cost and y represents pounds.

(c) Domain of inverse is $62.5 \le x \le 80$.

(d) If $x = 73$ in the inverse function,

$$y = \tfrac{100}{35}(80 - 73) = \tfrac{100}{5} = 20 \text{ pounds.}$$

37. $f(x) = \dfrac{3}{4}x + 6$

One-to-one; has an inverse

39. $f(\theta) = \sin\theta$

Not one-to-one; does not have an inverse

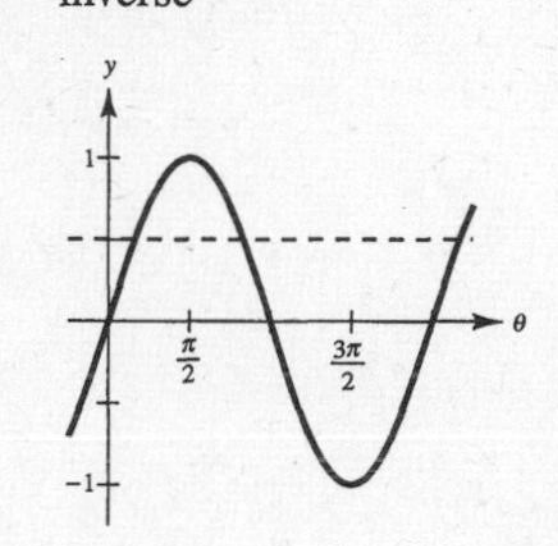

41. $h(s) = \dfrac{1}{s - 2} - 3$

One-to-one; has an inverse

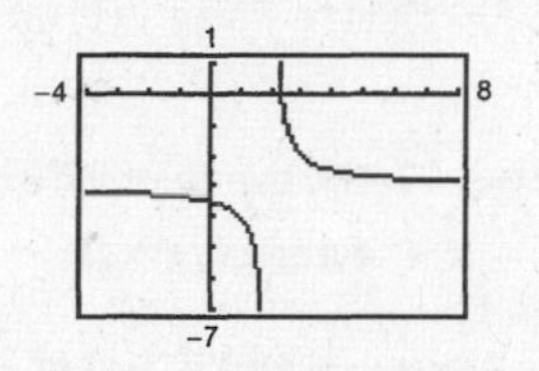

43. $f(x) = \dfrac{1}{1 + x^2}$

Not one-to-one; does not have an inverse

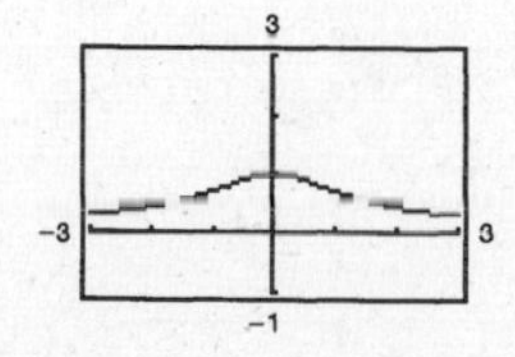

45. $g(x) = (x + 5)^3$

One-to-one; has an inverse

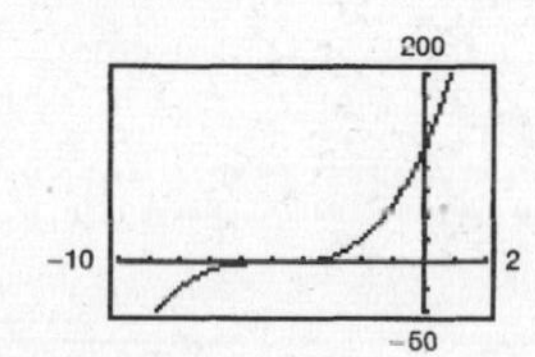

47. $f(x) = (x + a)^3 + b$

f is one-to-one; has an inverse.

49. $f(x) = \dfrac{x^4}{4} - 2x^2$

Not one-to-one; f does not have an inverse.

51. $f(x) = 2 - x - x^3$

One-to-one; has an inverse.

53. $f(x) = (x - 4)^2$ on $[4, \infty)$

f passes the horizontal line test on $[4, \infty)$; hence, one-to-one.

55. $f(x) = \dfrac{4}{x^2}$ on $(0, \infty)$

f passes horizontal line test on $(0, \infty)$; hence, one-to-one.

57. $f(x) = \cos x$ on $[0, \pi]$

f passes horizontal line test on $[0, \pi]$; hence, one-to-one.

59. $f(x) = \sqrt{x - 2}$, Domain: $x \ge 2$

f is one-to-one; has an inverse

$$\begin{aligned} \sqrt{x - 2} &= y \\ x - 2 &= y^2 \\ x &= y^2 + 2 \\ y &= x^2 + 2 \\ f^{-1}(x) &= x^2 + 2, x \ge 0 \end{aligned}$$

61. $f(x) = |x - 2|, \quad x \le 2$

$$\begin{aligned} &= -(x - 2) \\ &= 2 - x \end{aligned}$$

f is one-to-one; has an inverse

$$\begin{aligned} 2 - x &= y \\ 2 - y &= x \\ f^{-1}(x) &= 2 - x, \ x \ge 0 \end{aligned}$$

63. $f(x) = (x-3)^2$ is one-to-one for $x \geq 3$.

$$(x-3)^2 = y$$
$$x - 3 = \sqrt{y}$$
$$x = \sqrt{y} + 3$$
$$y = \sqrt{x} + 3$$
$$f^{-1}(x) = \sqrt{x} + 3,\ x \geq 0$$

65. $f(x) = |x+3|$ is one-to-one for $x \geq -3$.

$$x + 3 = y$$
$$x = y - 3$$
$$y = x - 3$$
$$f^{-1}(x) = x - 3,\ x \geq 0$$

67. $f(x) = x^3 + 2x - 1$

$f(1) = 2 = a \Rightarrow f^{-1}(2) = 1$

69. $f(x) = \sin x$

$f\left(\dfrac{\pi}{6}\right) = \dfrac{1}{2} = a \Rightarrow f^{-1}\left(\dfrac{1}{2}\right) = \dfrac{\pi}{6}$

71. $f(x) = x^3 - \dfrac{4}{x}$

$f(2) = 6 = a \Rightarrow f^{-1}(6) = 2$

73. $(f^{-1} \circ g^{-1})(1) = f^{-1}(g^{-1}(1)) = f^{-1}(1) = 32$

75. $(f^{-1} \circ f^{-1})(6) = f^{-1}(f^{-1}(6)) = f^{-1}(72) = 600$

In Exercises 77–80, use the following.

$f(x) = x + 4$ and $g(x) = 2x - 5$

$f^{-1}(x) = x - 4$ and $g^{-1}(x) = \dfrac{x+5}{2}$

77.
$$(g^{-1} \circ f^{-1})(x) = g^{-1}(f^{-1}(x))$$
$$= g^{-1}(x-4)$$
$$= \frac{(x-4)+5}{2}$$
$$= \frac{x+1}{2}$$

79.
$$(f \circ g)^{-1} = [f(g(x))]^{-1}$$
$$= [f(2x-5)]^{-1}$$
$$= [2x-1]^{-1}$$
$$= \frac{x+1}{2}$$

81. $y = \arcsin x$

(a)

x	-1	-0.8	-0.6		-0.2	0	0.2	0.4	0.6	0.8	1
y	-1.571	-0.927	-0.644	-0.412	-0.201	0	0.201	0.412	0.644	0.927	1.571

(b)

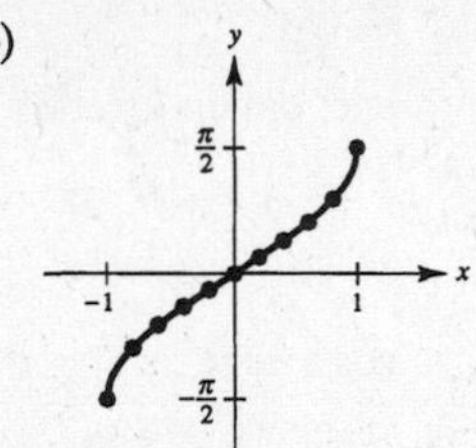

(c)

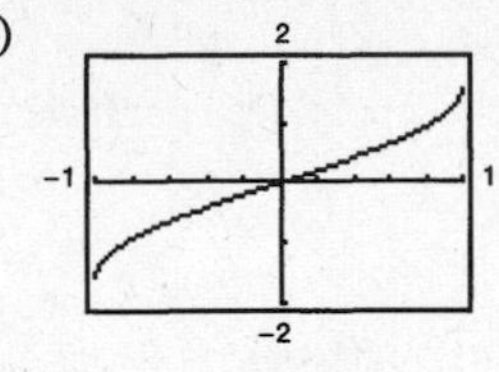

(d) Symmetric about origin: $\arcsin(-x) = -\arcsin x$

Intercept: $(0, 0)$

83. Let $y = f(x)$ be one-to-one. Solve for x as a function of y. Interchange x and y to get $y = f^{-1}(x)$. Let the domain of f^{-1} be the range of f. Verify that $f(f^{-1}(x)) = x$ and $f(f^{-1}(x)) = x$.

Example: $f(x) = x^3$

$$y = x^3$$
$$x = \sqrt[3]{y}$$
$$y = \sqrt[3]{x}$$
$$f^{-1}(x) = \sqrt[3]{x}$$

85. Answers will vary.

Example: $y = x^4 - 2x^3$ on $(-\infty, \infty)$ does not have an inverse.

87. If the domains were not restricted, then the trigonometric functions would not be one-to-one and hence would not have inverses.

89. $y = \arcsin x$ represents the inverse of the restricted sine function where $-1 \le x \le 1$ and $-\pi/2 \le y \le \pi/2$.

91. $\arcsin \frac{1}{2} = \frac{\pi}{6}$

93. $\arccos \frac{1}{2} = \frac{\pi}{3}$

95. $\arctan \frac{\sqrt{3}}{3} = \frac{\pi}{6}$

97. $\text{arccsc}(-\sqrt{2}) = -\frac{\pi}{4}$

99. $\arccos(-0.8) \approx 2.50$

101. $\text{arcsec}(1.269) = \arccos\left(\frac{1}{1.269}\right)$

≈ 0.66

103. $f(x) = \tan x$

$g(x) = \arctan x$

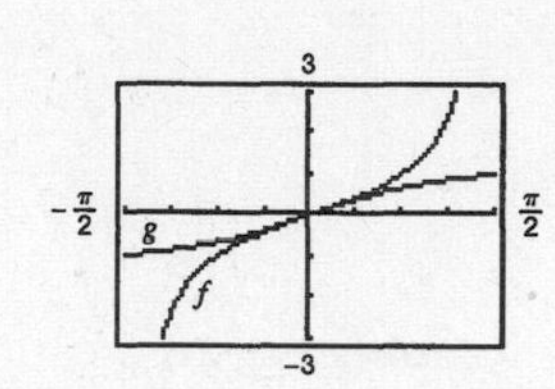

105. $\cos[\arccos(-0.1)] = -0.1$

107. (a) $\sin\left(\arcsin \frac{1}{2}\right) = \sin\left(\frac{\pi}{6}\right)$

$= \frac{1}{2}$

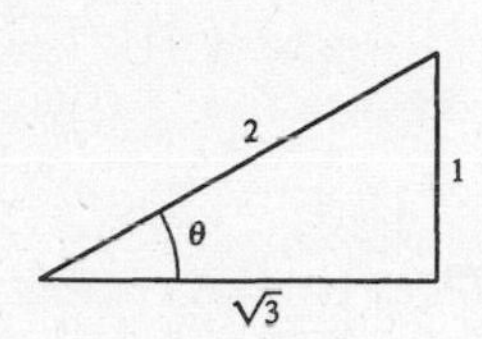

(b) $\cos\left(\arcsin \frac{1}{2}\right) = \cos\left(\frac{\pi}{6}\right)$

$= \frac{\sqrt{3}}{2}$

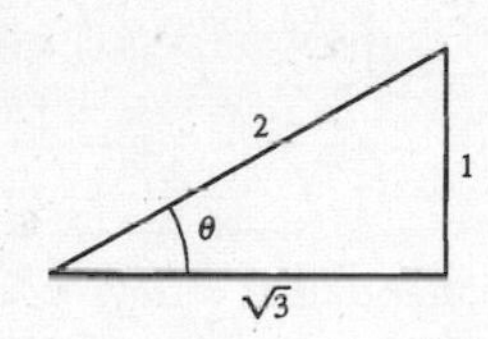

109. (a) $\sin\left(\arctan \frac{3}{4}\right) = \frac{3}{5}$

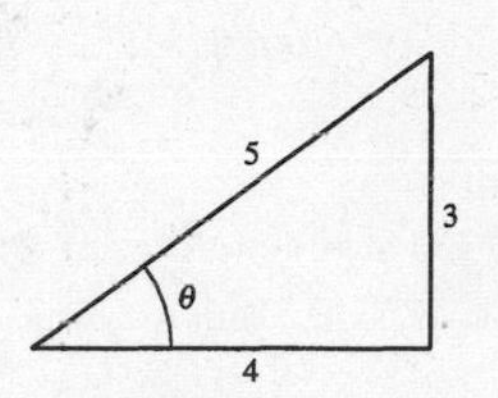

(b) $\sec\left(\arcsin \frac{4}{5}\right) = \frac{5}{3}$

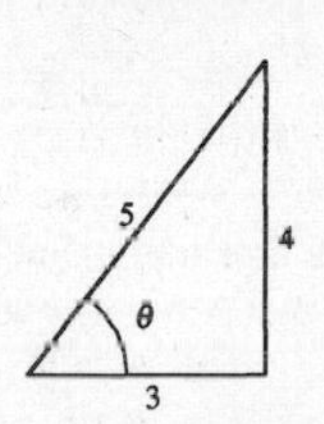

111. (a) $\cot\left[\arcsin\left(-\frac{1}{2}\right)\right] = \cot\left(-\frac{\pi}{6}\right) = -\sqrt{3}$

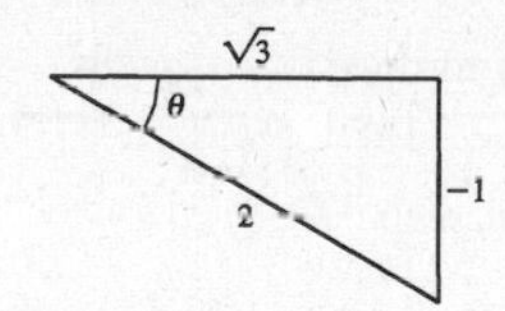

(b) $\csc\left[\arctan\left(-\frac{5}{12}\right)\right] = -\frac{13}{5}$

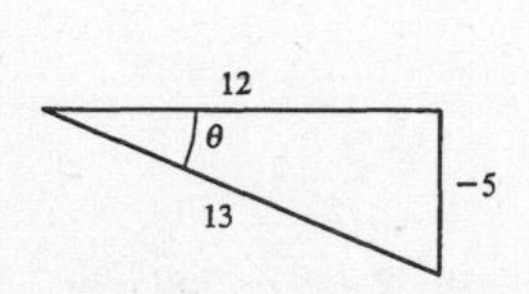

113. $\arcsin(3x - \pi) = \frac{1}{2}$

$3x - \pi = \sin\left(\frac{1}{2}\right)$

$x = \frac{1}{3}\left[\sin\left(\frac{1}{2}\right) + \pi\right] \approx 1.207$

115. $\arcsin \sqrt{2x} = \arccos \sqrt{x}$

$\sqrt{2x} = \sin\left(\arccos \sqrt{x}\right)$

$\sqrt{2x} = \sqrt{1 - x},\ 0 \le x \le 1$

$2x = 1 - x$

$3x = 1$

$x = \frac{1}{3}$

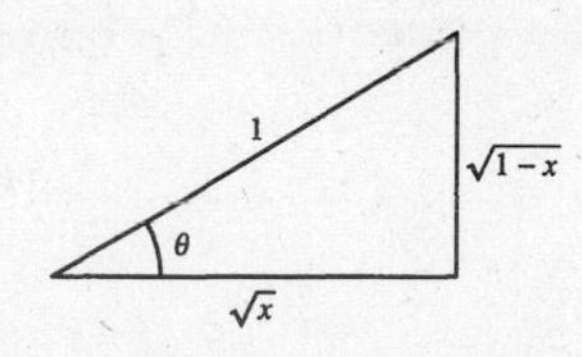

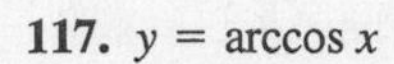

117. $y = \arccos x$

$y = \arctan x$

The point of intersection is given by

$f(x) = \arccos x - \arctan x = 0,\ \cos(\arccos x) = \cos(\arctan x).$

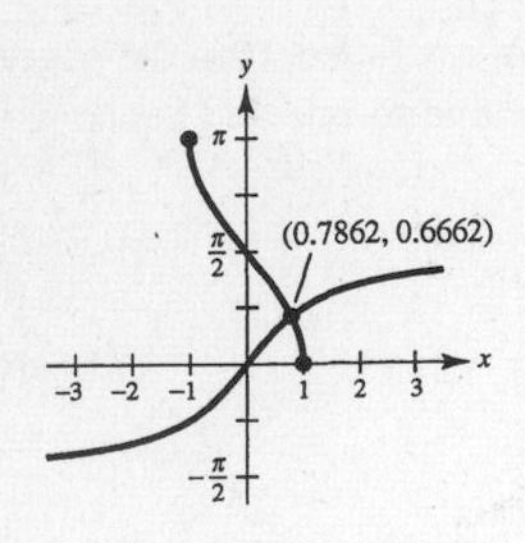

$$x = \frac{1}{\sqrt{1 + x^2}}$$

$$x^2(1 + x^2) = 1$$

$$x^4 + x^2 - 1 = 0 \text{ when } x^2 = \frac{-1 + \sqrt{5}}{2}.$$

Therefore, $x = \pm\sqrt{\dfrac{-1 + \sqrt{5}}{2}} \approx \pm 0.7862.$

Point of intersection: $(0.7862, 0.6662)$ [Since $f(-0.7862) = \pi \neq 0.$]

119. $y = \tan(\arctan x)$

$\theta = \arctan x$

$y = \tan\theta = x$

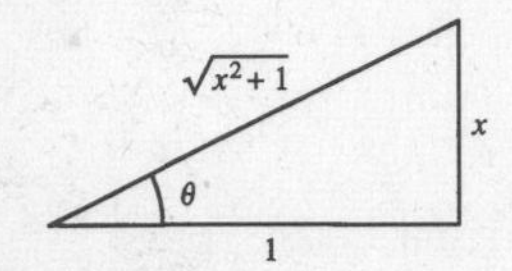

121. $y = \cos(\arcsin 2x)$

$\theta = \arcsin 2x$

$y = \cos\theta = \sqrt{1 - 4x^2}$

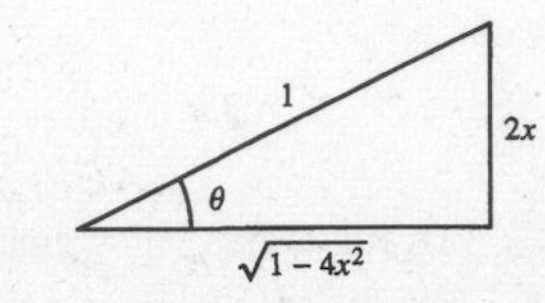

123. $y = \sin(\operatorname{arcsec} x)$

$\theta = \operatorname{arcsec} x,\ 0 \le \theta \le \pi,\ \theta \neq \dfrac{\pi}{2}$

$y = \sin\theta = \dfrac{\sqrt{x^2 - 1}}{|x|}$

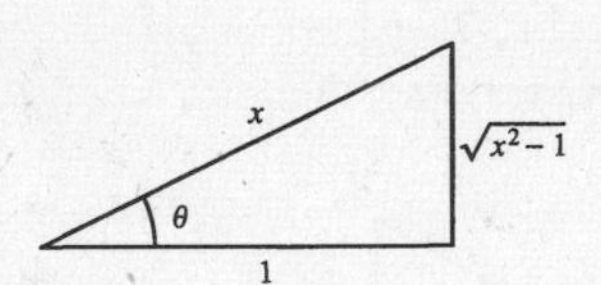

The absolute value bars on x are necessary because of the restriction $0 \le \theta \le \pi$, $\theta \neq \pi/2$, and $\sin\theta$ for this domain must always be nonnegative.

125. $y = \tan\left(\operatorname{arcsec}\dfrac{x}{3}\right)$

$\theta = \operatorname{arcsec}\dfrac{x}{3}$

$y = \tan\theta = \dfrac{\sqrt{x^2 - 9}}{3}$

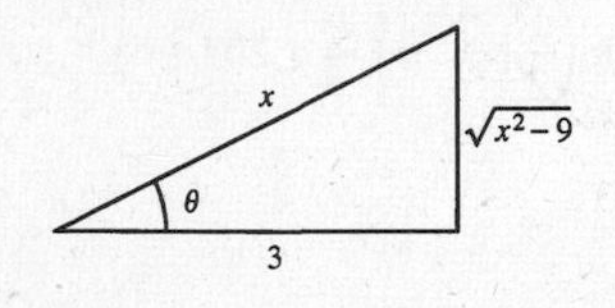

127. $y = \csc\left(\arctan\dfrac{x}{\sqrt{2}}\right)$

$\theta = \arctan\dfrac{x}{\sqrt{2}}$

$y = \csc\theta = \dfrac{\sqrt{x^2 + 2}}{x}$

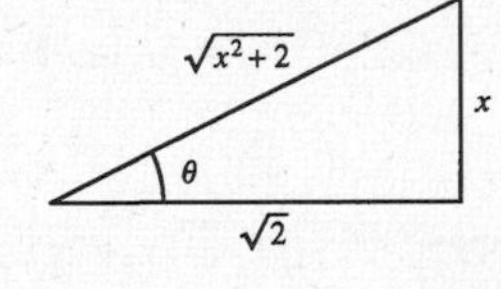

129. $\arctan\dfrac{9}{x} = \arcsin\dfrac{9}{\sqrt{x^2 + 81}}$

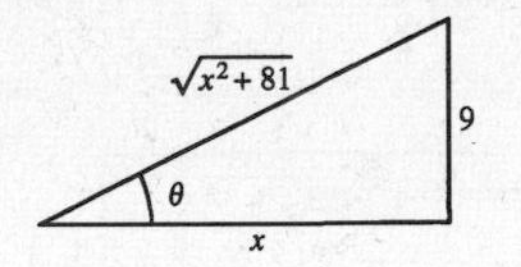

131. (a) $\operatorname{arccsc} x = \arcsin\dfrac{1}{x},\ |x| \ge 1$

Let $y = \operatorname{arccsc} x$. Then for

$$-\frac{\pi}{2} \le y < 0 \text{ and } 0 < y \le \frac{\pi}{2},$$

$$\csc y = x \Rightarrow \sin y = \frac{1}{x}.$$

Thus, $y = \arcsin\left(\dfrac{1}{x}\right)$. Therefore, $\operatorname{arccsc} x = \arcsin\left(\dfrac{1}{x}\right)$.

(b) $\arctan x + \arctan\dfrac{1}{x} = \dfrac{\pi}{2},\ x > 0$

Let $y = \arctan x + \arctan\dfrac{1}{x}$. Then

$$\tan y = \frac{\tan(\arctan x) + \tan[\arctan(1/x)]}{1 - \tan(\arctan x)\tan[\arctan(1/x)]}$$

$$= \frac{x + (1/x)}{1 - x(1/x)} = \frac{x + (1/x)}{0}$$

which is undefined.

Thus, $y = \dfrac{\pi}{2}$. Therefore, $\arctan x + \arctan\left(\dfrac{1}{x}\right) = \dfrac{\pi}{2}$.

133. $f(x) = \arcsin(x - 1)$

$x - 1 = \sin y$

$x = 1 + \sin y$

Domain: $[0, 2]$

Range: $[-\pi/2, \pi/2]$

$f(x)$ is the graph of $\arcsin x$ shifted right 1 unit.

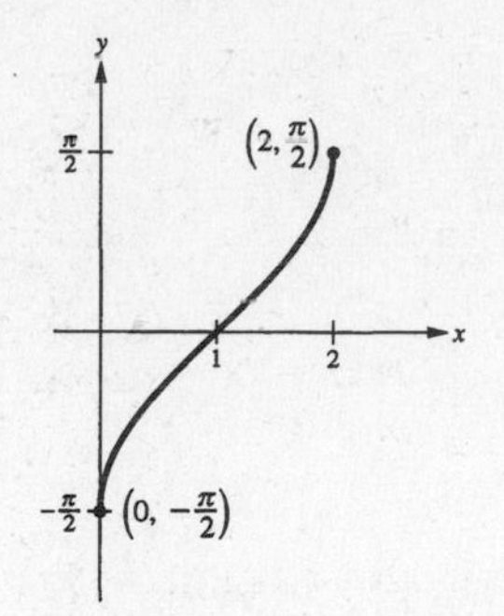

135. $f(x) = \operatorname{arcsec} 2x$

$2x = \sec y$

$x = \frac{1}{2}\sec y$

Domain: $\left(-\infty, -\frac{1}{2}\right], \left[\frac{1}{2}, \infty\right)$

Range: $\left[0, \frac{\pi}{2}\right), \left(\frac{\pi}{2}, \pi\right]$

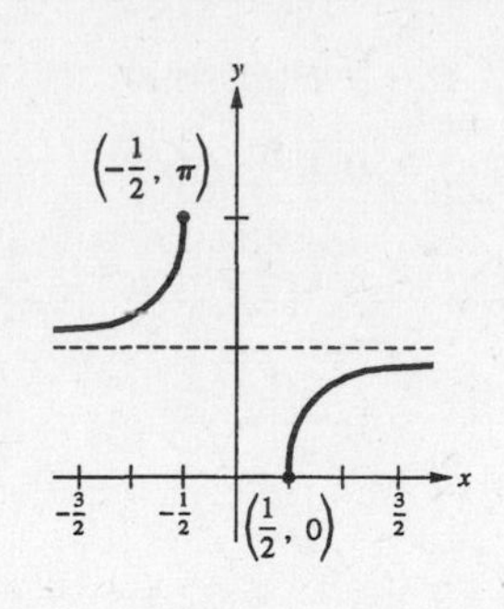

137. Let $(f \circ g)(x) = y$ then $x = (f \circ g)^{-1}(y)$. Also,

$$(f \circ g)(x) = y$$
$$f(g(x)) = y$$
$$g(x) = f^{-1}(y)$$
$$x = g^{-1}(f^{-1}(y))$$
$$= (g^{-1} \circ f^{-1})(y)$$

Since f and g are one-to-one functions, $(f \circ g)^{-1} = g^{-1} \circ f^{-1}$.

139. Suppose $g(x)$ and $h(x)$ are both inverses of $f(x)$. Then the graph of $f(x)$ contains the point (a, b) if and only if the graphs of $g(x)$ and $h(x)$ contain the point (b, a). Since the graphs of $g(x)$ and $h(x)$ are the same, $g(x) = h(x)$. Therefore, the inverse of $f(x)$ is unique.

141. False

Let $f(x) = x^2$.

143. False

$$\arcsin^2 0 + \arccos^2 0 = 0 + \left(\frac{\pi}{2}\right)^2 \neq 1$$

145. True

147. $\tan(\arctan x + \arctan y) = \dfrac{\tan(\arctan x) + \tan(\arctan y)}{1 - \tan(\arctan x)\tan(\arctan y)} = \dfrac{x + y}{1 - xy}, \quad xy \neq 1$

Therefore,

$$\arctan x + \arctan y = \arctan\left(\frac{x + y}{1 - xy}\right), \quad xy \neq 1.$$

Let $x = \frac{1}{2}$ and $y = \frac{1}{3}$.

$$\arctan\left(\frac{1}{2}\right) + \arctan\left(\frac{1}{3}\right) = \arctan\frac{\frac{1}{2} + \frac{1}{3}}{1 - \left(\frac{1}{2} \cdot \frac{1}{3}\right)} = \arctan\frac{\frac{5}{6}}{1 - \frac{1}{6}} = \arctan\frac{\frac{5}{6}}{\frac{5}{6}} = \arctan 1 = \frac{\pi}{4}$$

Section P.6 Exponential and Logarithmic Functions

1. (a) $25^{3/2} = 5^3 = 125$

(b) $81^{1/2} = 9$

(c) $3^{-2} = \dfrac{1}{3^2} = \dfrac{1}{9}$

(d) $27^{-1/3} = \dfrac{1}{27^{1/3}} = \dfrac{1}{3}$

3. (a) $(5^2)(5^3) = 5^{2+3} = 5^5 = 3125$

(b) $(5^2)(5^{-3}) = 5^{2-3} = 5^{-1} = \dfrac{1}{5}$

(c) $\dfrac{5^3}{25^2} = \dfrac{5^3}{5^4} = \dfrac{1}{5}$

(d) $\left(\dfrac{1}{4}\right)^2 2^6 = \dfrac{2^6}{2^4} = 2^2 = 4$

5. (a) $e^2(e^4) = e^6$

(b) $(e^3)^4 = e^{12}$

(c) $(e^3)^{-2} = e^{-6} = \dfrac{1}{e^6}$

(d) $\dfrac{e^5}{e^3} = e^2$

7. $3^x = 81 \Rightarrow x = 4$

9. $\left(\frac{1}{3}\right)^{x-1} = 27 \Rightarrow 3^{1-x} = 27 \Rightarrow 1 - x = 3 \Rightarrow x = -2$

11. $4^3 = (x+2)^3 \Rightarrow 4 = x + 2 \Rightarrow x = 2$

13. $x^{3/4} = 8 \Rightarrow x = 8^{4/3} = 2^4 = 16$

15. $e^{-2x} = e^5 \Rightarrow -2x = 5 \Rightarrow x = -\frac{5}{2}$

17. $\left(1 + \dfrac{1}{1{,}000{,}000}\right)^{1{,}000{,}000} \approx 2.718280469$

$e \approx 2.718281828$

$e > \left(1 + \dfrac{1}{1{,}000{,}000}\right)^{1{,}000{,}000}$

19. $y = 3^x$

x	-2	-1	0	1	2
y	$\frac{1}{9}$	$\frac{1}{3}$	1	3	9

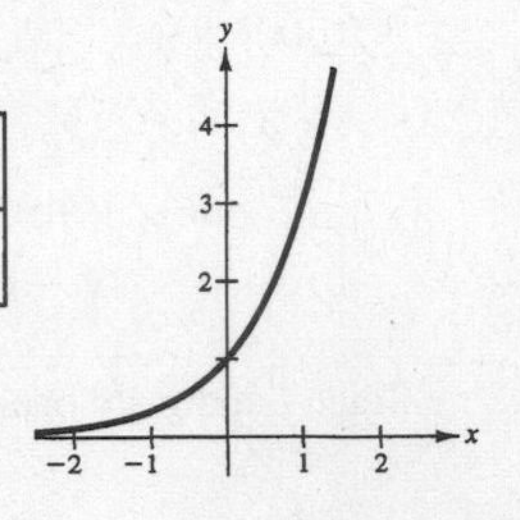

21. $y = \left(\frac{1}{3}\right)^x = 3^{-x}$

x	-2	-1	0	1	2
y	9	3	1	$\frac{1}{3}$	$\frac{1}{9}$

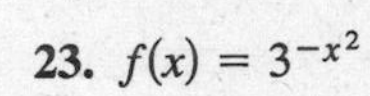

23. $f(x) = 3^{-x^2}$

x	0	± 1	± 2
y	1	$\frac{1}{3}$	0.0123

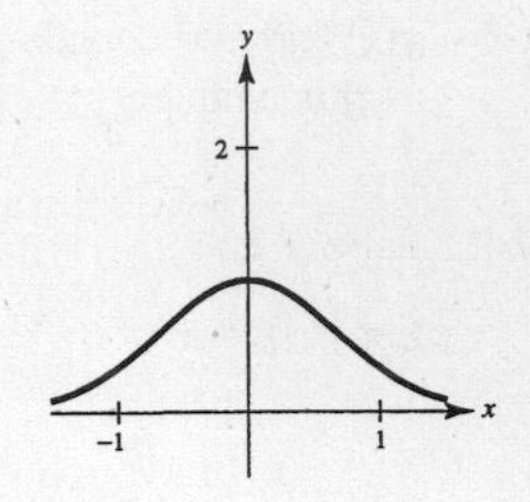

25. $h(x) = e^{x-2}$

x	0	1	2	3	4
y	e^{-2}	e^{-1}	1	e	e^2

27. $y = e^{-x^2}$

Symmetric with respect to the y-axis

Horizontal asymptote: $y = 0$

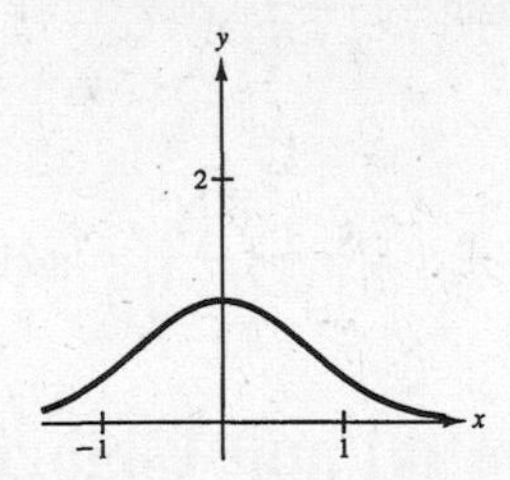

29. 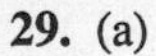 (a)

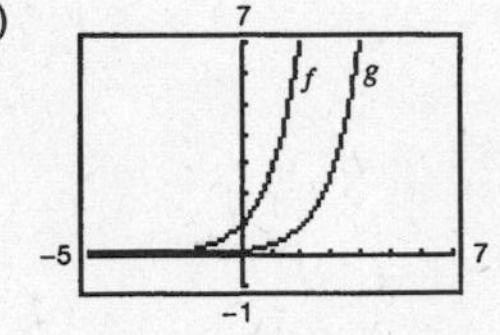

Horizontal shift 2 units to the right

 (b)

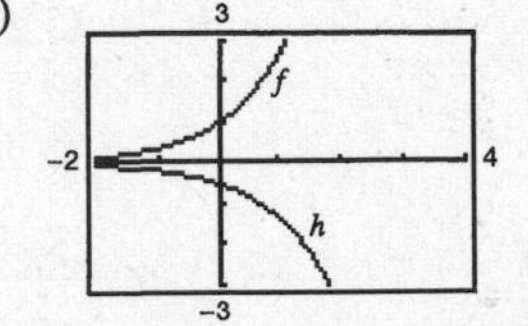

A reflection in the x-axis and a vertical shrink

(c)

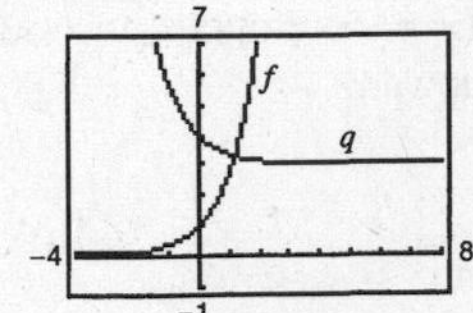

Vertical shift 3 units upward and a reflection in the y-axis

31. $y = Ce^{ax}$

Matches (c)

33. $y = C(1 - e^{-ax})$

Vertical shift C units

Reflection in both the x- and y-axes

Matches (a)

35. $e^0 = 1$

$\ln 1 = 0$

37. $\ln 2 = 0.6931\ldots$

$e^{0.6931\ldots} = 2$

39. $f(x) = 3 \ln x$

Domain: $x > 0$

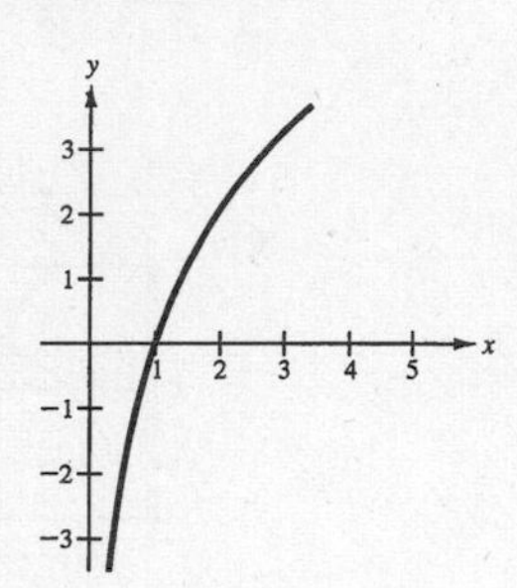

41. $f(x) = \ln 2x$

Domain: $x > 0$

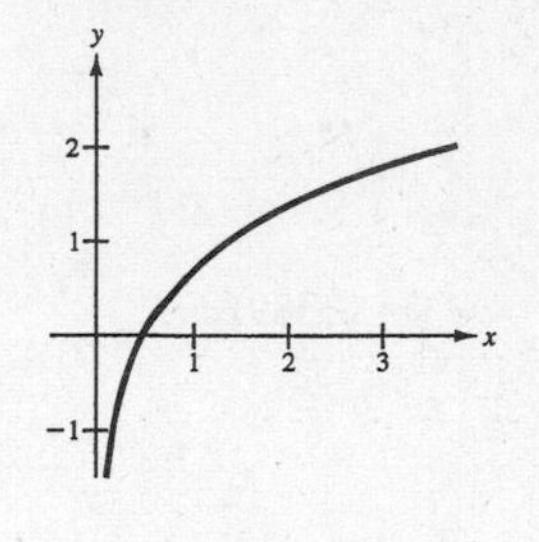

43. $f(x) = \ln(x - 1)$

Domain: $x > 1$

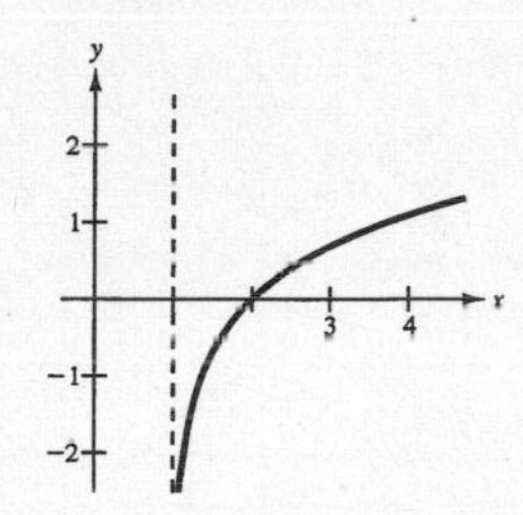

45. $f(x) = e^{2x}$

$g(x) = \ln\sqrt{x} = \frac{1}{2}\ln x$

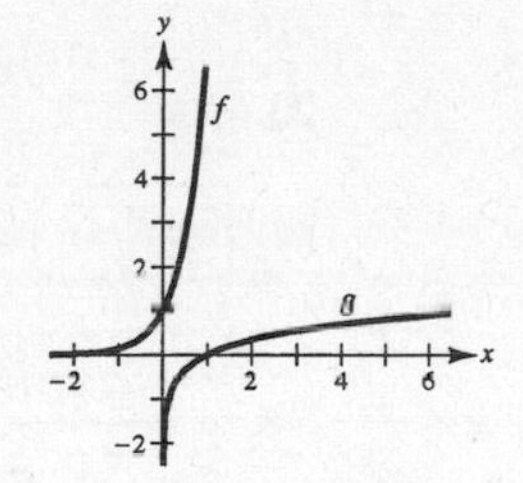

47. $f(x) = e^x - 1$

$g(x) = \ln(x + 1)$

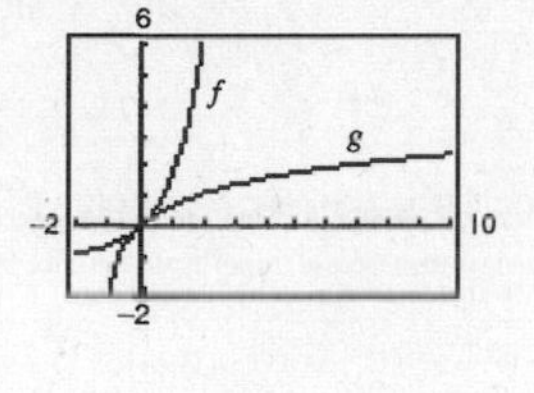

49. $\ln e^{x^2} = x^2$

51. $e^{\ln(5x+2)} = 5x + 2$

53. $e^{\ln\sqrt{x}} = \sqrt{x}$

55. (a) $\ln 6 = \ln 2 + \ln 3 \approx 1.7917$

(b) $\ln \frac{2}{3} = \ln 2 - \ln 3 \approx -0.4055$

(c) $\ln 81 = 4 \ln 3 \approx 4.3944$

(d) $\ln \sqrt{3} = \frac{1}{2}\ln 3 \approx 0.5493$

57. The domain of the natural logarithmic function is $(0, \infty)$ and the range is $(-\infty, \infty)$. The function is continuous, increasing, and one-to-one and its graph is convave downward. In addition, if a and b are positive numbers and n is rational, then $\ln(1) = 0$, $\ln(ab) = \ln a + \ln b$, $\ln(a^n) = n \ln a$ and

$$\ln\left(\frac{a}{b}\right) = \ln a - \ln b.$$

59. The domain of $f(x) = e^x$ is $(-\infty, \infty)$ and the range is $(0, \infty)$. $f(x)$ is continuous, increasing, one-to-one, and concave upward.

61. $\ln \frac{2}{3} = \ln 2 - \ln 3$

63. $\ln \frac{xy}{z} = \ln x + \ln y - \ln z$

65. $\ln \frac{1}{5} = \ln 1 - \ln 5 = -\ln 5$

67. $\ln\left(\frac{x^2 - 1}{x^3}\right)^3 = 3[\ln(x^2 - 1) - \ln x^3]$

$= 3[\ln(x + 1) + \ln(x - 1) - 3 \ln x]$

69. $\ln 3e^2 = \ln 3 + 2 \ln e = 2 + \ln 3$

71. $\ln(x - 2) - \ln(x + 2) = \ln \dfrac{x - 2}{x + 2}$

73. $\dfrac{1}{3}[2 \ln(x + 3) + \ln x - \ln(x^2 - 1)] = \dfrac{1}{3} \ln \dfrac{x(x + 3)^2}{x^2 - 1} = \ln \sqrt[3]{\dfrac{x(x + 3)^2}{x^2 - 1}}$

75. $2 \ln 3 - \dfrac{1}{2} \ln(x^2 + 1) = \ln \dfrac{9}{\sqrt{x^2 + 1}}$

77. (a) $e^{\ln x} = 4$

$x = 4$

(b) $\ln e^{2x} = 3$

$2x = 3$

$x = \frac{3}{2}$

79. (a) $\ln x = 2$

$x = e^2 \approx 7.3891$

(b) $e^x = 4$

$x = \ln 4 \approx 1.3863$

81.

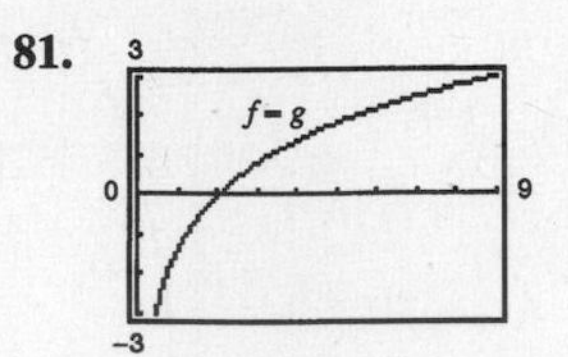

83. $\ln x = \ln \left[\left(\dfrac{x}{y}\right) y\right]$

$= \ln \dfrac{x}{y} + \ln y \Rightarrow \ln \dfrac{x}{y} = \ln x - \ln y$

Review Exercises for Chapter P

1. $y = 2x - 3$

$x = 0 \Rightarrow y = 2(0) - 3 = -3 \Rightarrow (0, -3)$ *y*-intercept

$y = 0 \Rightarrow 0 = 2x - 3 \Rightarrow x = \frac{3}{2} \Rightarrow \left(\frac{3}{2}, 0\right)$ *x*-intercept

3. $y = \dfrac{x - 1}{x - 2}$

$x = 0 \Rightarrow y = \dfrac{0 - 1}{0 - 2} = \dfrac{1}{2} \Rightarrow \left(0, \dfrac{1}{2}\right)$ *y*-intercept

$y = 0 \Rightarrow 0 = \dfrac{x - 1}{x - 2} \Rightarrow x = 1 \Rightarrow (1, 0)$ *x*-intercept

5. Symmetric with respect to *y*-axis since

$(-x)^2 y - (-x)^2 + 4y = 0$

$x^2 y - x^2 + 4y = 0.$

7. $y = -\frac{1}{2}x + \frac{3}{2}$

9. $-\frac{1}{3}x + \frac{5}{6}y = 1$

$-\frac{2}{5}x + y = \frac{6}{5}$

$y = \frac{2}{5}x + \frac{6}{5}$

Slope: $\frac{2}{5}$

y-intercept: $\frac{6}{5}$

11. $y = 7 - 6x - x^2$

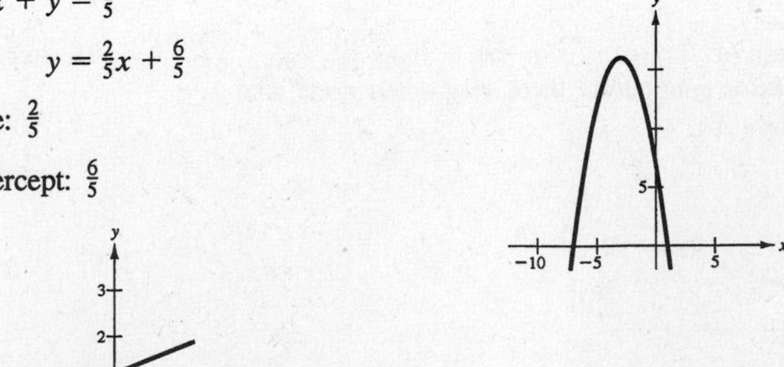

13. $y = \sqrt{5 - x}$

Domain: $(-\infty, 5]$

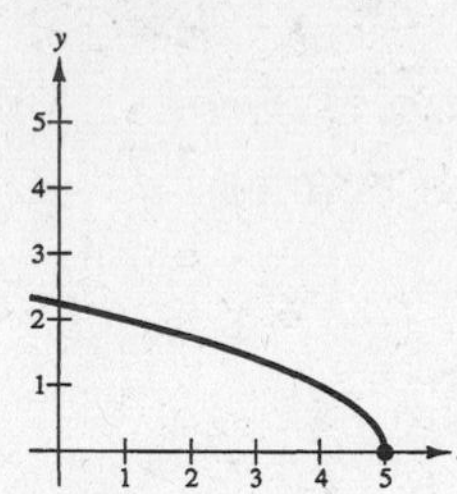

15.
$$\begin{aligned} 3x - 4y &= 8 \\ 4x + 4y &= 20 \\ \hline 7x \quad &= 28 \\ x &= 4 \\ y &= 1 \end{aligned}$$

Point: $(4, 1)$

17.

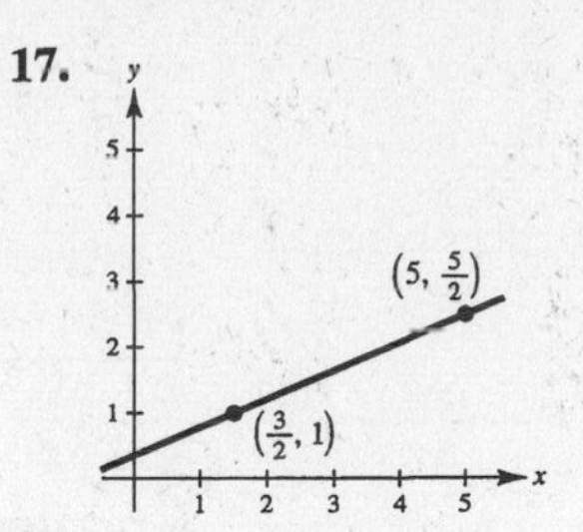

$$\text{Slope} = \frac{(5/2) - 1}{5 - (3/2)} = \frac{3/2}{7/2} = \frac{3}{7}$$

19. $\dfrac{1 - t}{1 - 0} = \dfrac{1 - 5}{1 - (-2)}$

$$1 - t = -\frac{4}{3}$$

$$t = \frac{7}{3}$$

21.
$$y - (-5) = \tfrac{3}{2}(x - 0)$$
$$y = \tfrac{3}{2}x - 5$$
$$2y - 3x + 10 = 0$$

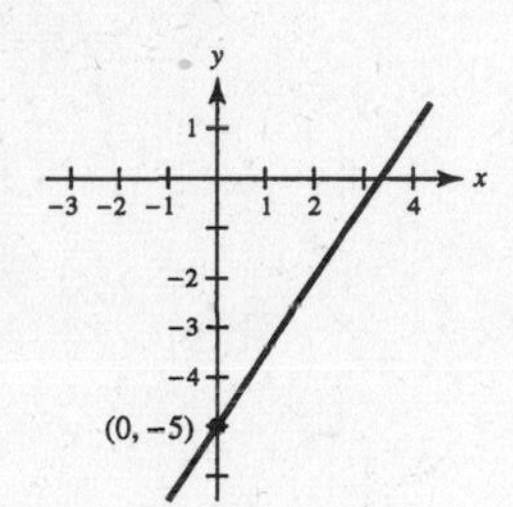

23.
$$y - 0 = -\tfrac{2}{3}(x - (-3))$$
$$y = -\tfrac{2}{3}x - 2$$
$$3y + 2x + 6 = 0$$

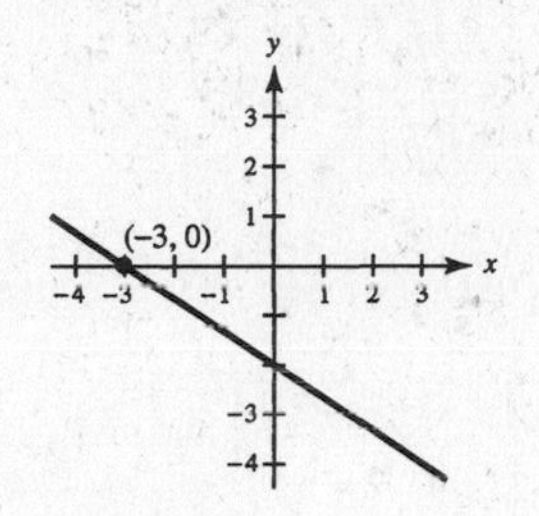

25. (a) $y - 4 = \dfrac{7}{16}(x + 2)$

$$16y - 64 = 7x + 14$$
$$0 = 7x - 16y + 78$$

(b) Slope of line is $\dfrac{5}{3}$.

$$y - 4 = \frac{5}{3}(x + 2)$$
$$3y - 12 = 5x + 10$$
$$0 = 5x - 3y + 22$$

(c) $m = \dfrac{4 - 0}{-2 - 0} = -2$

$$y = -2x$$
$$2x + y = 0$$

(d) $x = -2$

$$x + 2 = 0$$

27. The slope is -850. $V = -850t + 12{,}500$.

$V(3) = -850(3) + 12{,}500 = \$9950$

29. $x - y^2 = 0$

$$y = \pm\sqrt{x}$$

Not a function of x since there are two values of y for some x.

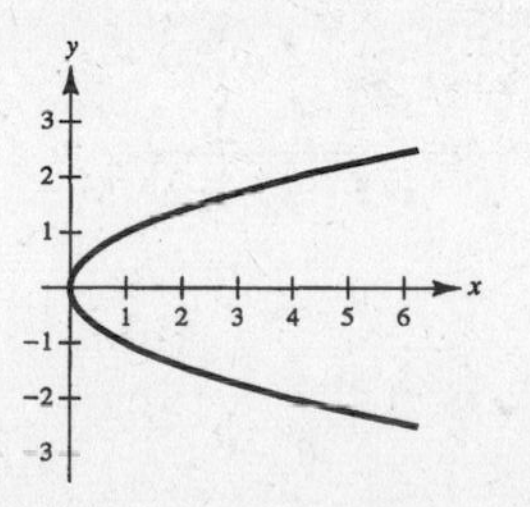

31. $y = x^2 - 2x$

Function of x since there is one value of y for each x.

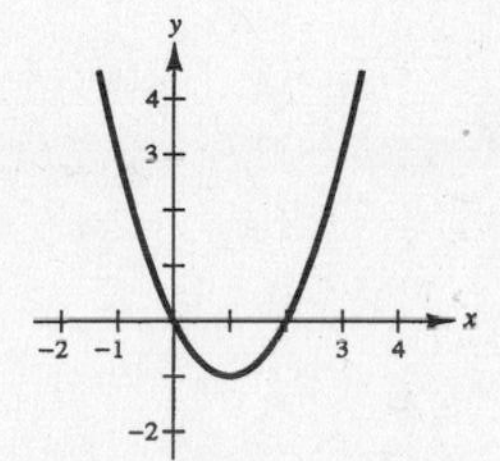

33. $f(x) = \dfrac{1}{x}$

(a) $f(0)$ does not exist.

(b) $$\frac{f(1+\Delta x) - f(1)}{\Delta x} = \frac{\dfrac{1}{1+\Delta x} - \dfrac{1}{1}}{\Delta x} = \frac{1 - 1 - \Delta x}{(1+\Delta x)\Delta x}$$
$$= \frac{-1}{1+\Delta x},\ \Delta x \neq -1, 0$$

35. (a) Domain: $36 - x^2 \geq 0 \Rightarrow -6 \leq x \leq 6$ or $[-6, 6]$

Range: $[0, 6]$

(b) Domain: all $x \neq 5$ or $(-\infty, 5), (5, \infty)$

Range: all $y \neq 0$ or $(-\infty, 0), (0, \infty)$

(c) Domain: all x or $(-\infty, \infty)$

Range: all y or $(-\infty, \infty)$

37. (a) $f(x) = x^3 + c, c = -2, 0, 2$

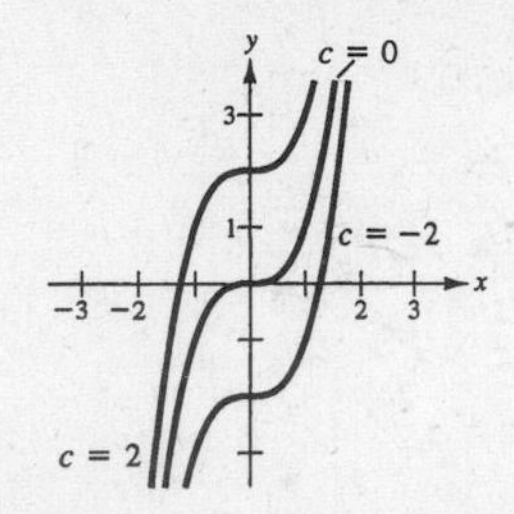

(b) $f(x) = (x - c)^3, c = -2, 0, 2$

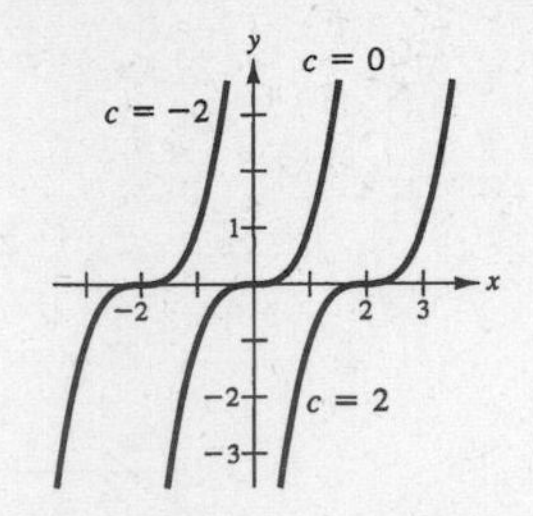

(c) $f(x) = (x - 2)^3 + c, c = -2, 0, 2$

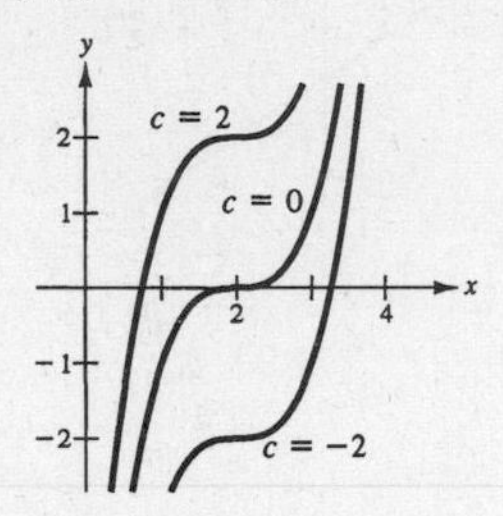

(d) $f(x) = cx^3, c = -2, 0, 2$

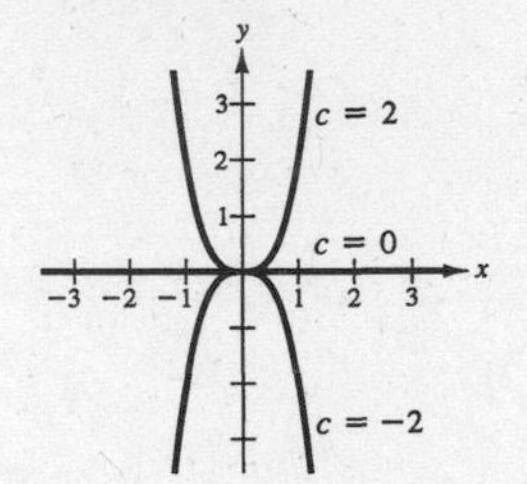

39. (a) 3 (cubic), negative leading coefficient

(b) 4 (quartic), positive leading coefficient

(c) 2 (quadratic), negative leading coefficient

(d) 5, positive leading coefficient

41. (a) Yes, y is a function of t. At each time t, there is one and only one displacement y.

(b) The amplitude is approximately

$$(0.25 - (-0.25))/2 = 0.25.$$

The period is approximately 1.1.

(c) One model is $y = \dfrac{1}{4}\cos\left(\dfrac{2\pi}{1.1}t\right) \approx \dfrac{1}{4}\cos(5.7t)$

(d)

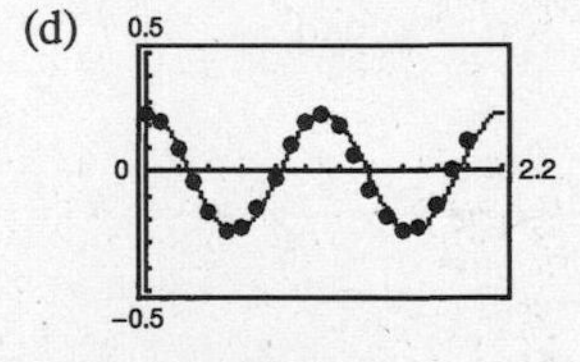

43. (a)
$$f(x) = \tfrac{1}{2}x - 3$$
$$y = \tfrac{1}{2}x - 3$$
$$2(y + 3) = x$$
$$2(x + 3) = y$$
$$f^{-1}(x) = 2x + 6$$

(b)

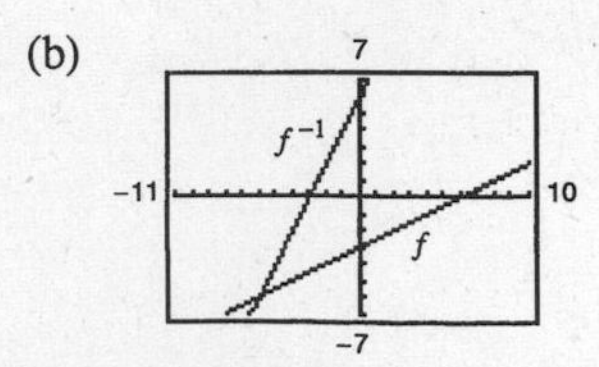

(c) $f^{-1}(f(x)) = f^{-1}\left(\tfrac{1}{2}x - 3\right) = 2\left(\tfrac{1}{2}x - 3\right) + 6 = x$

$f(f^{-1}(x)) = f(2x + 6) = \tfrac{1}{2}(2x + 6) - 3 = x$

45. (a) $f(x) = \sqrt{x+1}$

$y = \sqrt{x+1}$

$y^2 - 1 = x$

$x^2 - 1 = y$

$f^{-1}(x) = x^2 - 1, x \geq 0$

(b)

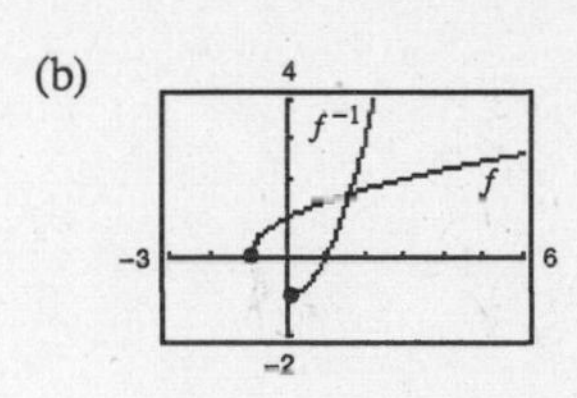

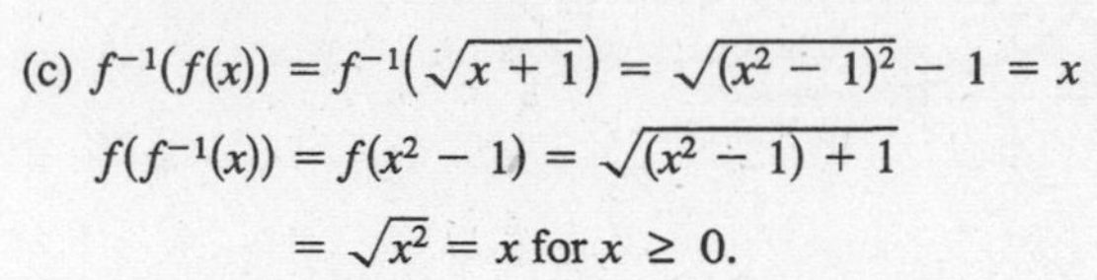

(c) $f^{-1}(f(x)) = f^{-1}\left(\sqrt{x+1}\right) = \sqrt{(x^2-1)^2} - 1 = x$

$f(f^{-1}(x)) = f(x^2 - 1) = \sqrt{(x^2-1)+1}$

$= \sqrt{x^2} = x$ for $x \geq 0$.

47. (a) $f(x) = \sqrt[3]{x+1}$

$y = \sqrt[3]{x+1}$

$y^3 - 1 = x$

$x^3 - 1 = y$

$f^{-1}(x) = x^3 - 1$

(b)

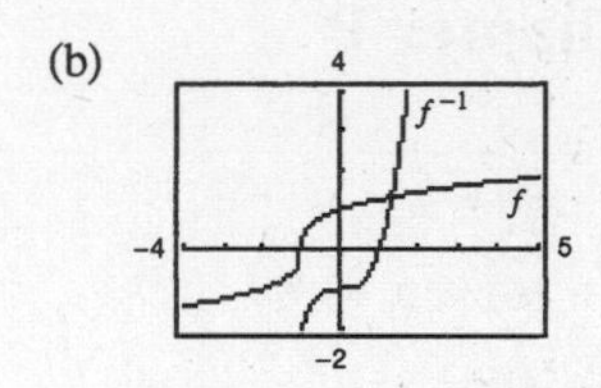

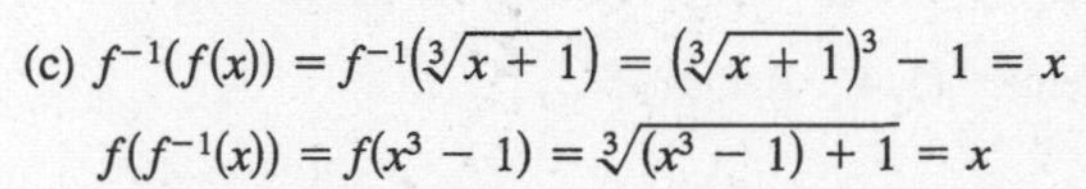

(c) $f^{-1}(f(x)) = f^{-1}\left(\sqrt[3]{x+1}\right) = \left(\sqrt[3]{x+1}\right)^3 - 1 = x$

$f(f^{-1}(x)) = f(x^3 - 1) = \sqrt[3]{(x^3-1)+1} = x$

49. $f(x) = 2\arctan(x+3)$

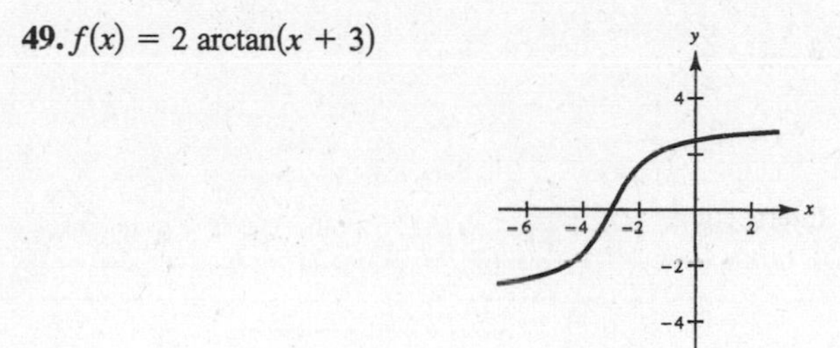

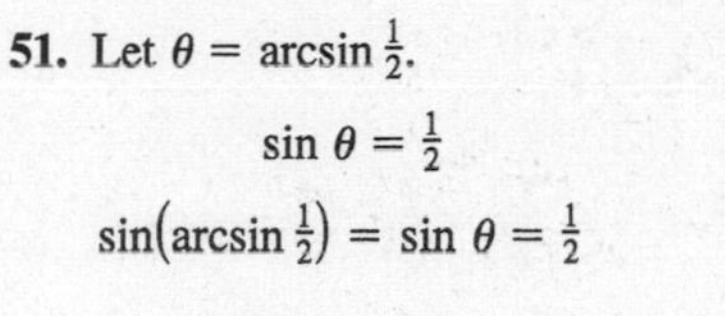

51. Let $\theta = \arcsin\frac{1}{2}$.

$\sin\theta = \frac{1}{2}$

$\sin\left(\arcsin\frac{1}{2}\right) = \sin\theta = \frac{1}{2}$

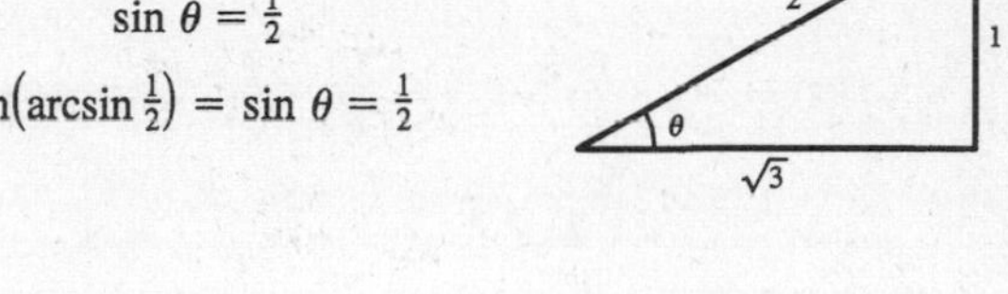

53. $f(x) = \ln x + 3$

Vertical shift 3 units upward

Vertical asymptote: $x = 0$

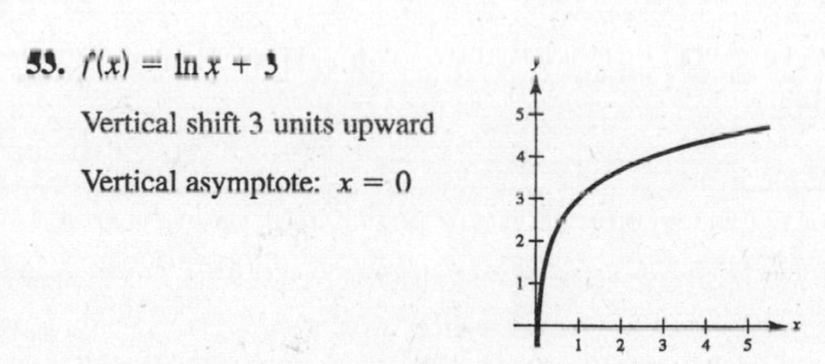

55. $\ln\sqrt[5]{\frac{4x^2-1}{4x^2+1}} = \frac{1}{5}\ln\frac{(2x-1)(2x+1)}{4x^2+1}$

$= \frac{1}{5}[\ln(2x-1) + \ln(2x+1) - \ln(4x^2+1)]$

57. $\ln 3 + \frac{1}{3}\ln(4 - x^2) - \ln x = \ln 3 + \ln\sqrt[3]{4-x^2} - \ln x$

$= \ln\left(\frac{3\sqrt[3]{4-x^2}}{x}\right)$

59. $\ln\sqrt{x+1} = 2$

$\sqrt{x+1} = e^2$

$x + 1 = e^4$

$x = e^4 - 1 \approx 53.598$

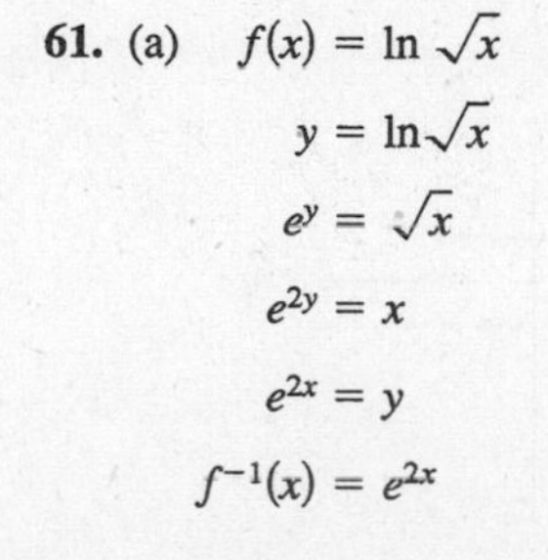

61. (a) $f(x) = \ln\sqrt{x}$

$y = \ln\sqrt{x}$

$e^y = \sqrt{x}$

$e^{2y} = x$

$e^{2x} = y$

$f^{-1}(x) = e^{2x}$

(b)

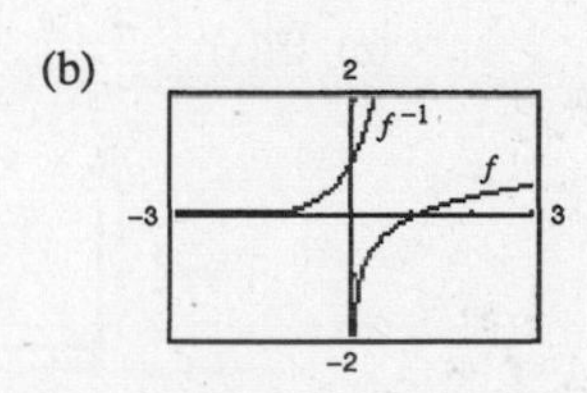

(c) $f^{-1}(f(x)) = f^{-1}\left(\ln\sqrt{x}\right) = e^{2\ln\sqrt{x}} = e^{\ln x} = x$

$f(f^{-1}(x)) = f(e^{2x}) = \ln\sqrt{e^{2x}} = \ln e^x = x$

63. $y = e^{-x/2}$

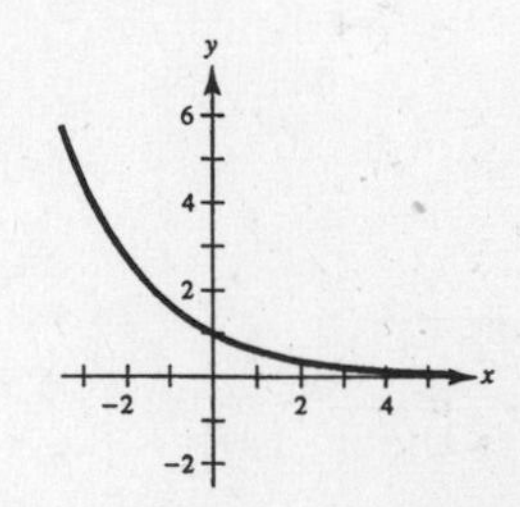

Problem Solving for Chapter P

1. (a)
$$x^2 - 6x + y^2 - 8y = 0$$
$$(x^2 - 6x + 9) + (y^2 - 8y + 16) = 9 + 16$$
$$(x - 3)^2 + (y - 4)^2 = 25$$

Center: $(3, 4)$ Radius: 5

(b) Slope of line from $(0, 0)$ to $(3, 4)$ is $\frac{4}{3}$. Slope of tangent line is $-\frac{3}{4}$. Hence,

$y - 0 = -\frac{3}{4}(x - 0) \Rightarrow y = -\frac{3}{4}x$ Tangent line

(c) Slope of line from $(6, 0)$ to $(3, 4)$ is $\frac{4 - 0}{3 - 6} = -\frac{4}{3}$.

Slope of tangent line is $\frac{3}{4}$. Hence,

$y - 0 = \frac{3}{4}(x - 6) \Rightarrow y = \frac{3}{4}x - \frac{9}{2}$ Tangent line

(d) $-\frac{3}{4}x = \frac{3}{4}x - \frac{9}{2}$

$\frac{3}{2}x = \frac{9}{2}$

$x = 3$

Intersection: $\left(3, -\frac{9}{4}\right)$

3. $H(x) = \begin{cases} 1 & x \geq 0 \\ 0 & x < 0 \end{cases}$

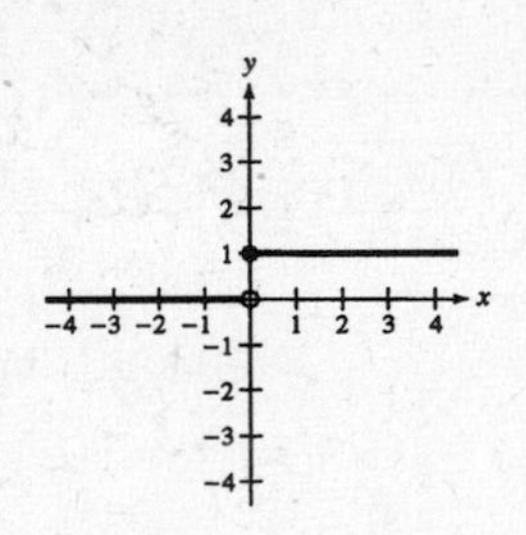

(a) $H(x) - 2$

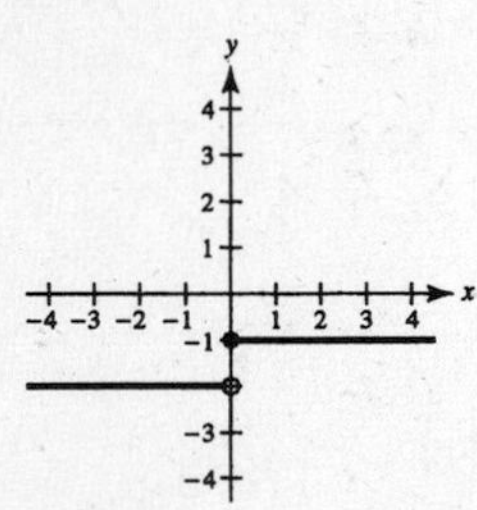

(b) $H(x - 2)$

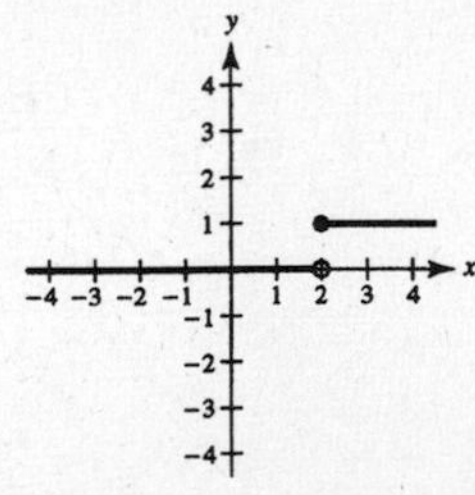

(c) $-H(x)$

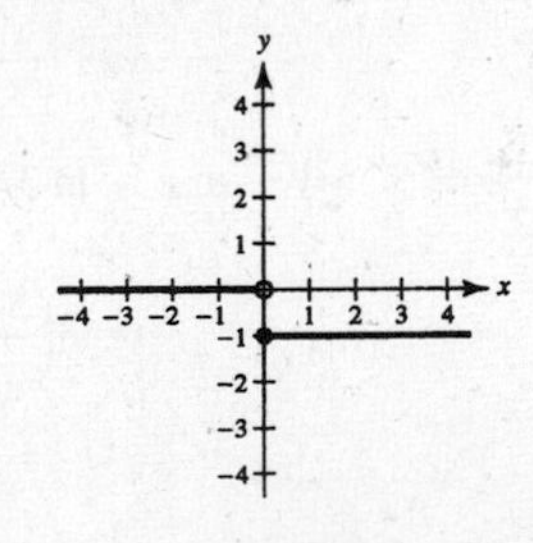

(d) $H(-x)$

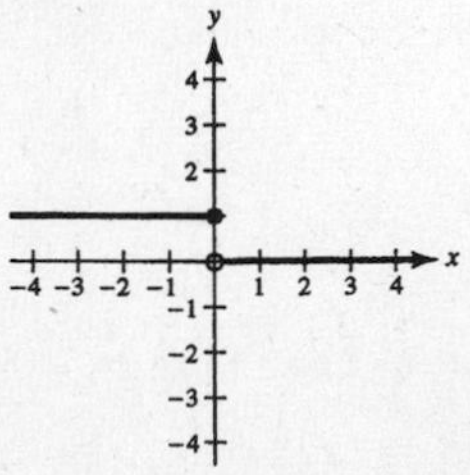

(e) $\frac{1}{2}H(x)$

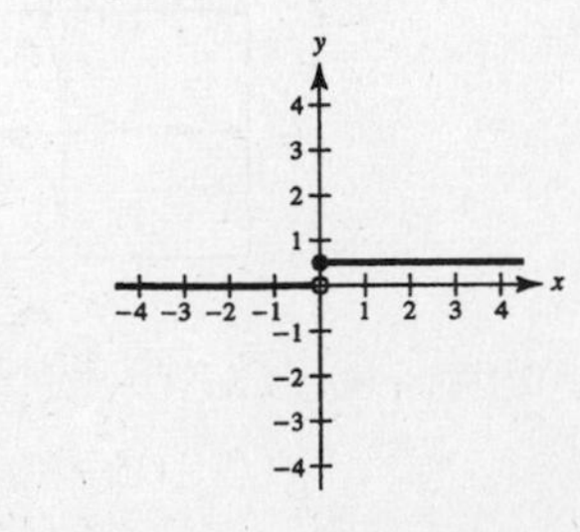

(f) $-H(x - 2) + 2$

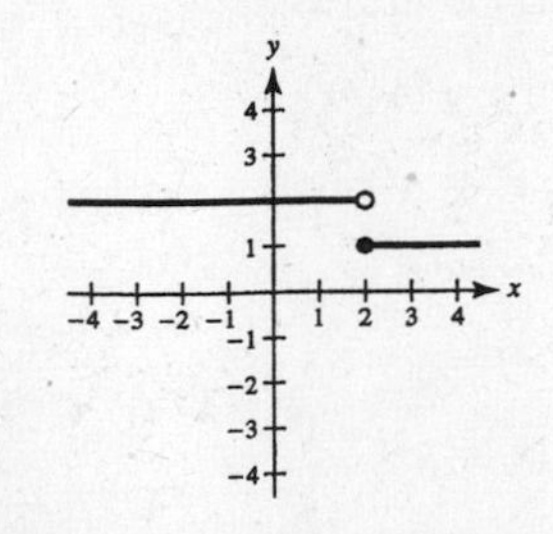

5. (a) $x + 2y = 100 \Rightarrow y = \dfrac{100 - x}{2}$

$$A(x) = xy = x\left(\frac{100 - x}{2}\right) = -\frac{x^2}{2} + 50x$$

Domain: $0 < x < 100$

(b)

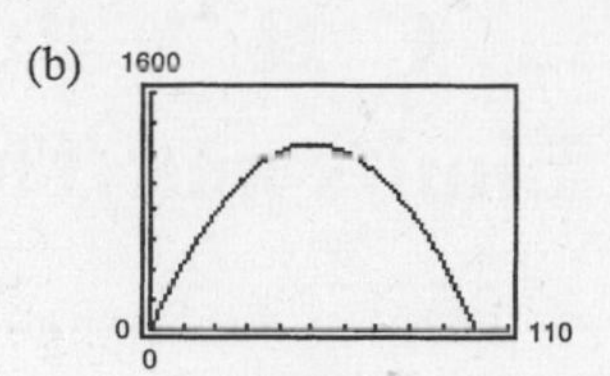

Maximum of 1250 m² at $x = 50$ m, $y = 25$ m.

(c)
$$A(x) = -\frac{1}{2}(x^2 - 100x)$$
$$= -\frac{1}{2}(x^2 - 100x + 2500) + 1250$$
$$= -\frac{1}{2}(x - 50)^2 + 1250$$

$A(50) = 1250$ m² is the maximum. $x = 50$ m, $y = 25$ m.

7. The length of the trip in the water is $\sqrt{2^2 + x^2}$, and the length of the trip over land is $\sqrt{1 + (3 - x)^2}$. Hence, the total time is

$$T = \frac{\sqrt{4 + x^2}}{2} + \frac{\sqrt{1 + (3 - x)^2}}{4} \text{ hours.}$$

9. (a) Slope $= \dfrac{9 - 4}{3 - 2} = 5.$

Slope of tangent line is less than 5.

(b) Slope $= \dfrac{4 - 1}{2 - 1} = 3.$

Slope of tangent line is greater than 3.

(c) Slope $= \dfrac{4.41 - 4}{2.1 - 2} = 4.1.$

Slope of tangent line is less than 4.1.

(d)
$$\text{Slope} = \frac{f(2 + h) - f(2)}{(2 + h) - 2}$$
$$= \frac{(2 + h)^2 - 4}{h}$$
$$= \frac{4h + h^2}{h}$$
$$= 4 + h, \quad h \neq 0$$

(e) Letting h get closer and closer to 0, the slope approaches 4. Hence, the slope at (2, 4) is 4.

11. (a)
$$\frac{I}{x^2} = \frac{2I}{(x - 3)^2}$$
$$x^2 - 6x + 9 = 2x^2$$
$$x^2 + 6x - 9 = 0$$
$$x = \frac{-6 \pm \sqrt{36 + 36}}{2}$$
$$= -3 \pm \sqrt{18} \approx 1.2426, -7.2426$$

(b)
$$\frac{I}{x^2 + y^2} = \frac{2I}{(x - 3)^2 + y^2}$$
$$(x - 3)^2 + y^2 = 2(x^2 + y^2)$$
$$x^2 - 6x + 9 + y^2 = 2x^2 + 2y^2$$
$$x^2 + y^2 + 6x - 9 = 0$$
$$(x + 3)^2 + y^2 = 18$$

Circle of radius $\sqrt{18}$ and center $(-3, 0)$.

13.
$$d_1 d_2 = 1$$
$$[(x + 1)^2 + y^2][(x - 1)^2 + y^2] = 1$$
$$(x + 1)^2(x - 1)^2 + y^2[(x + 1)^2 + (x - 1)^2] + y^4 = 1$$
$$(x^2 - 1)^2 + y^2[2x^2 + 2] + y^4 = 1$$
$$x^4 - 2x^2 + 1 + 2x^2y^2 + 2y^2 + y^4 = 1$$
$$(x^4 + 2x^2y^2 + y^4) - 2x^2 + 2y^2 = 0$$
$$(x^2 + y^2)^2 = 2(x^2 - y^2)$$

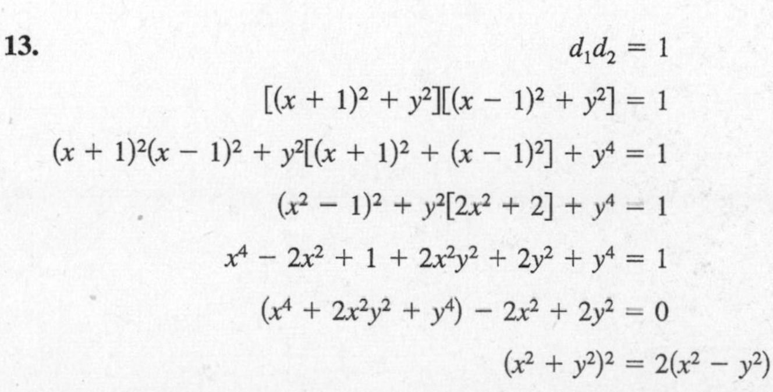

Let $y = 0$. Then $x^4 = 2x^2 \Rightarrow x = 0$ or $x^2 = 2$. Thus, $(0, 0)$, $(\sqrt{2}, 0)$ and $(-\sqrt{2}, 0)$ are on the curve.

CHAPTER 1
Limits and Their Properties

CHAPTER 1
Limits and Their Properties

Section 1.1 A Preview of Calculus

Solutions to Odd-Numbered Exercises

1. Precalculus: (20 ft/sec)(15 seconds) = 300 feet

3. Calculus required: slope of tangent line at $x = 2$ is rate of change, and equals about 0.16.

5. Precalculus: Area $= \frac{1}{2}bh = \frac{1}{2}(5)(3) = \frac{15}{2}$sq. units

7. Precalculus: Volume $= (2)(4)(3) = 24$ cubic units

9. (a)

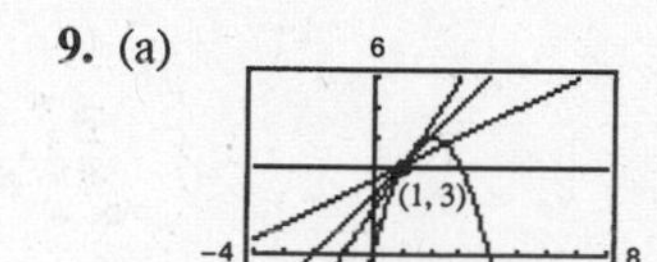

(b) The graphs of y_2 are approximations to the tangent line to y_1 at $x = 1$.

(c) The slope is approximately 2. For a better approximation make the list numbers smaller:

$$\{0.2, 0.1, 0.01, 0.001\}$$

11. (a) $D_1 = \sqrt{(5-1)^2 + (1-5)^2} = \sqrt{16+16} \approx 5.66$

(b) $D_2 = \sqrt{1 + \left(\frac{5}{2}\right)^2} + \sqrt{1 + \left(\frac{5}{2} - \frac{5}{3}\right)^2} + \sqrt{1 + \left(\frac{5}{3} - \frac{5}{4}\right)^2} + \sqrt{1 + \left(\frac{5}{4} - 1\right)^2}$

$\approx 2.693 + 1.302 + 1.083 + 1.031 \approx 6.11$

(c) Increase the number of line segments.

Section 1.2 Finding Limits Graphically and Numerically

1.

x	1.9	1.99	1.999	2.001	2.01	2.1
$f(x)$	0.3448	0.3344	0.3334	0.3332	0.3322	0.3226

$\lim_{x\to 2} \frac{x-2}{x^2-x-2} \approx 0.3333$ (Actual limit is $\frac{1}{3}$.)

3.

x	2.9	2.99	2.999	3.001	3.01	3.1
$f(x)$	−0.0641	−0.0627	−0.0625	−0.0625	−0.0623	−0.0610

$\lim_{x\to 3} \frac{[1/(x+1)] - (1/4)}{x-3} \approx -0.0625$ (Actual limit is $-\frac{1}{16}$.)

5.

x	−0.1	−0.01	−0.001	0.001	0.01	0.1
$f(x)$	0.9983	0.99998	1.0000	1.0000	0.99998	0.9983

$\lim_{x\to 0} \frac{\sin x}{x} \approx 1.0000$ (Actual limit is 1.) (Make sure you use radian mode.)

7.

x	-0.1	-0.01	-0.001	0.001	0.01	0.1
$f(x)$	0.9516	0.9950	0.9995	1.0005	1.0050	1.0517

$$\lim_{x\to 0} \frac{e^x - 1}{x} = 1$$

9.

x	-0.1	-0.01	-0.001	0.001	0.01	0.1
$f(x)$	1.0536	1.0050	1.0005	0.9995	0.9950	0.9531

$$\lim_{x\to 0} \frac{\ln(x+1)}{x} = 1$$

11. $\lim_{x\to 3} (4 - x) = 1$

13. $\lim_{x\to 3} \frac{|x-3|}{x-3}$ does not exist. For values of x to the left of 3, $|x-3|/(x-3)$ equals -1, whereas for values of x to the right of 3, $|x-3|/(x-3)$ equals 1.

15. $\lim_{x\to 1} \sqrt[3]{x}\ln|x-2| = 0$

17. $\lim_{x\to \pi/2} \tan x$ does not exist since the function increases and decreases without bound as x approaches $\pi/2$.

19. $C(t) = 0.75 - 0.50[\![-(t-1)]\!]$

(a)

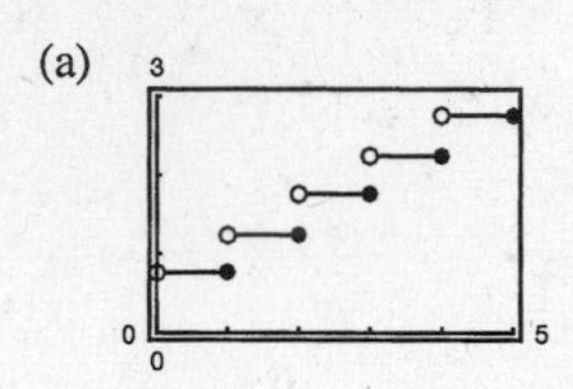

(b)

t	3	3.3	3.4	3.5	3.6	3.7	4
C	1.75	2.25	2.25	2.25	2.25	2.25	2.25

$\lim_{t\to 3.5} C(t) = 2.25$

(c)

t	2	2.5	2.9	3	3.1	3.5	4
C	1.25	1.75	1.75	1.75	2.25	2.25	2.25

$\lim_{t\to 3} C(t)$ does not exist. The values of C jump from 1.75 to 2.25 at $t = 3$.

21. You need to find δ such that $0 < |x-1| < \delta$ implies $|f(x) - 1| = \left|\frac{1}{x} - 1\right| < 0.1$. That is,

$$-0.1 < \frac{1}{x} - 1 < 0.1$$

$$1 - 0.1 < \frac{1}{x} < 1 + 0.1$$

$$\frac{9}{10} < \frac{1}{x} < \frac{11}{10}$$

$$\frac{10}{9} > x > \frac{10}{11}$$

$$\frac{10}{9} - 1 > x - 1 > \frac{10}{11} - 1$$

$$\frac{1}{9} > x - 1 > -\frac{1}{11}.$$

So take $\delta = \frac{1}{11}$. Then $0 < |x-1| < \delta$ implies

$$-\frac{1}{11} < x - 1 < \frac{1}{11}$$

$$-\frac{1}{11} < x - 1 < \frac{1}{9}.$$

Using the first series of equivalent inequalities, you obtain

$$|f(x) - 1| = \left|\frac{1}{x} - 1\right| < 0.1.$$

23. $\lim_{x\to 2} (3x + 2) = 8 = L$

$$|(3x + 2) - 8| < 0.01$$
$$|3x - 6| < 0.01$$
$$3|x - 2| < 0.01$$
$$0 < |x - 2| < \frac{0.01}{3} \approx 0.0033 = \delta$$

Hence, if $0 < |x - 2| < \delta = \frac{0.01}{3}$, you have:

$$3|x - 2| < 0.01$$
$$|3x - 6| < 0.01$$
$$|(3x + 2) - 8| < 0.01$$
$$|f(x) - L| < 0.01$$

25. $\lim_{x\to 2} (x^2 - 3) = 1 = L$

$$|(x^2 - 3) - 1| < 0.01$$
$$|x^2 - 4| < 0.01$$
$$|(x + 2)(x - 2)| < 0.01$$
$$|x + 2|\,|x - 2| < 0.01$$
$$|x - 2| < \frac{0.01}{|x + 2|}$$

If we assume $1 < x < 3$, then $\delta = 0.01/5 = 0.002$.

Hence, if $0 < |x - 2| < \delta = 0.002$, you have:

$$|x - 2| < 0.002 = \frac{1}{5}(0.01) < \frac{1}{|x + 2|}(0.01)$$
$$|x + 2||x - 2| < 0.01$$
$$|x^2 - 4| < 0.01$$
$$|(x^2 - 3) - 1| < 0.01$$
$$|f(x) - L| < 0.01$$

27. $\lim_{x\to 2} (x + 3) = 5$

Given $\epsilon > 0$:

$$|(x + 3) - 5| < \epsilon$$
$$|x - 2| < \epsilon = \delta$$

Hence, let $\delta = \epsilon$.

Hence, if $0 < |x - 2| < \delta = \epsilon$, you have:

$$|x - 2| < \epsilon$$
$$|(x + 3) - 5| < \epsilon$$
$$|f(x) - L| < \epsilon$$

29. $\lim_{x\to -4} \left(\frac{1}{2}x - 1\right) = \frac{1}{2}(-4) - 1 = -3$

Given $\epsilon > 0$:

$$\left|\left(\tfrac{1}{2}x - 1\right) - (-3)\right| < \epsilon$$
$$\left|\tfrac{1}{2}x + 2\right| < \epsilon$$
$$\tfrac{1}{2}|x - (-4)| < \epsilon$$
$$|x - (-4)| < 2\epsilon$$

Hence, let $\delta = 2\epsilon$.

Hence, if $0 < |x - (-4)| < \delta = 2\epsilon$, you have:

$$|x - (-4)| < 2\epsilon$$
$$\left|\tfrac{1}{2}x + 2\right| < \epsilon$$
$$\left|\left(\tfrac{1}{2}x - 1\right) + 3\right| < \epsilon$$
$$|f(x) - L| < \epsilon$$

31. $\lim_{x\to 6} 3 = 3$

Given $\epsilon > 0$:

$$|3 - 3| < \epsilon$$
$$0 < \epsilon$$

Hence, any $\delta > 0$ will work.

Hence, for any $\delta > 0$, you have:

$$|3 - 3| < \epsilon$$
$$|f(x) - L| < \epsilon$$

33. $\lim_{x\to 0} \sqrt[3]{x} = 0$

Given $\epsilon > 0$: $\left|\sqrt[3]{x} - 0\right| < \epsilon$

$$\left|\sqrt[3]{x}\right| < \epsilon$$
$$|x| < \epsilon^3 = \delta$$

Hence, let $\delta = \epsilon^3$.

Hence for $0 < |x - 0| < \delta = \epsilon^3$, you have:

$$|x| < \epsilon^3$$
$$|\sqrt[3]{x}| < \epsilon$$
$$|\sqrt[3]{x} - 0| < \epsilon$$
$$|f(x) - L| < \epsilon$$

35. $\lim_{x \to -2} |x - 2| = |(-2) - 2| = 4$

Given $\epsilon > 0$:

$$||x - 2| - 4| < \epsilon$$
$$|-(x - 2) - 4| < \epsilon \quad (x - 2 < 0)$$
$$|-x - 2| = |x + 2| = |x - (-2)| < \epsilon$$

Hence, $\delta = \epsilon$. Hence for $0 < |x - (-2)| < \delta = \epsilon$, you have:

$$|x + 2| < \epsilon$$
$$|-(x + 2)| < \epsilon$$
$$|-(x - 2) - 4| < \epsilon$$
$$||x - 2| - 4| < \epsilon \quad \text{(because } x - 2 < 0\text{)}$$
$$|f(x) - L| < \epsilon$$

37. $\lim_{x \to 1} (x^2 + 1) = 2$

Given $\epsilon > 0$:

$$|(x^2 + 1) - 2| < \epsilon$$
$$|x^2 - 1| < \epsilon$$
$$|(x + 1)(x - 1)| < \epsilon$$
$$|x - 1| < \frac{\epsilon}{|x + 1|}$$

If we assume $0 < x < 2$, then $\delta = \epsilon/3$. Hence for $0 < |x - 1| < \delta = \epsilon/3$, you have:

$$|x - 1| < \frac{1}{3}\epsilon < \frac{1}{|x + 1|}\epsilon$$
$$|x^2 - 1| < \epsilon$$
$$|(x^2 + 1) - 2| < \epsilon$$
$$|f(x) - 2| < \epsilon$$

39. $f(x) = \dfrac{\sqrt{x + 5} - 3}{x - 4}$

$\lim_{x \to 4} f(x) = \dfrac{1}{6}$

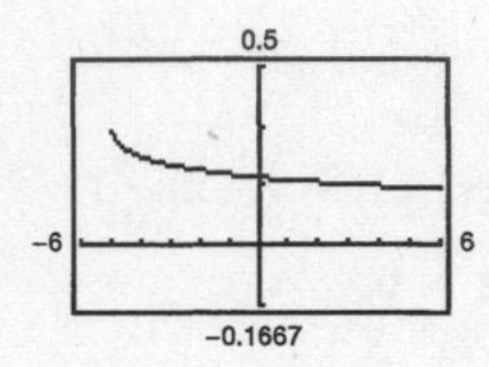

The domain is $[-5, 4) \cup (4, \infty)$.
The graphing utility does not show the hole at $\left(4, \frac{1}{6}\right)$.

41. $f(x) = \dfrac{x - 9}{\sqrt{x} - 3}$

$\lim_{x \to 9} f(x) = 6$

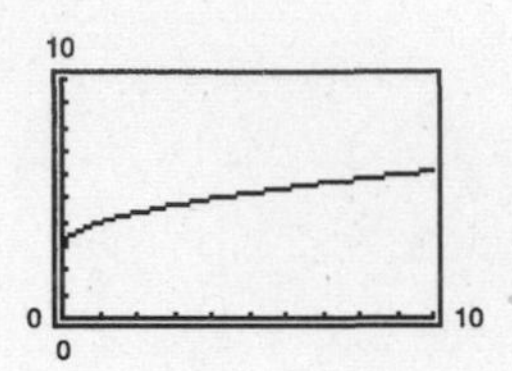

The domain is all $x \geq 0$ except $x = 9$. The graphing utility does not show the hole at $(9, 6)$.

43. $\lim_{x \to 8} f(x) = 25$ means that the values of f approach 25 as x gets closer and closer to 8.

45. (i) The values of f approach different numbers as x approaches c from different sides of c:

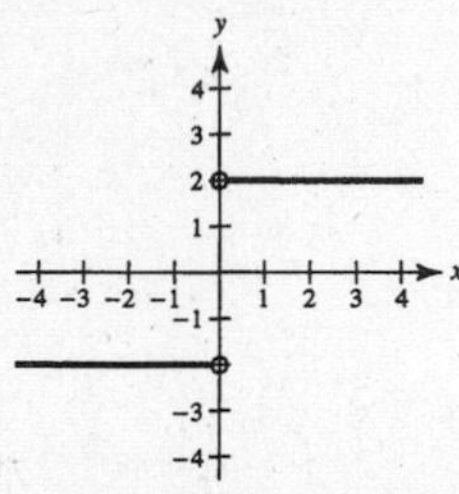

(ii) The values of f increase without bound as x approaches c:

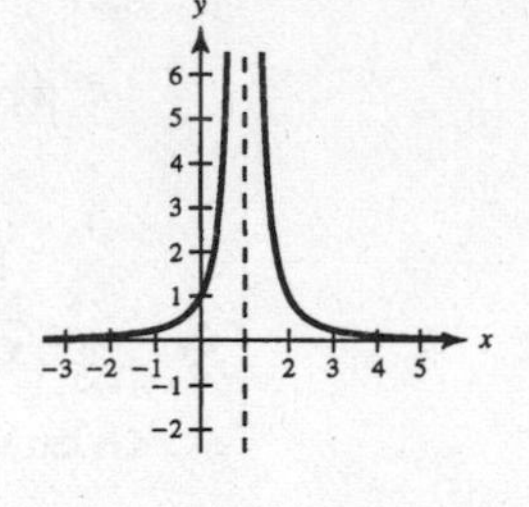

(iii) The values of f oscillate between two fixed numbers as x approaches c:

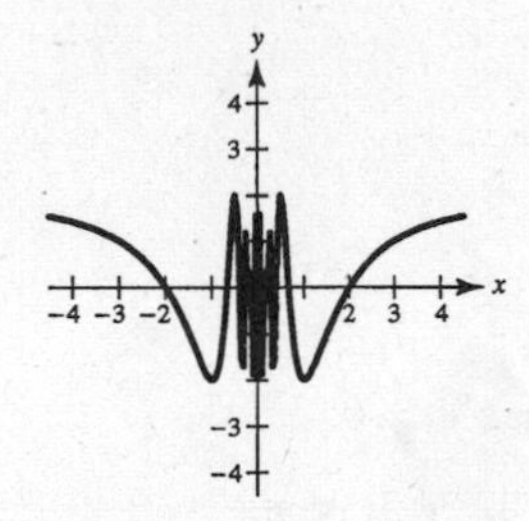

47. $f(x) = (1 + x)^{1/x}$

$\lim_{x \to 0} (1 + x)^{1/x} = e \approx 2.71828$

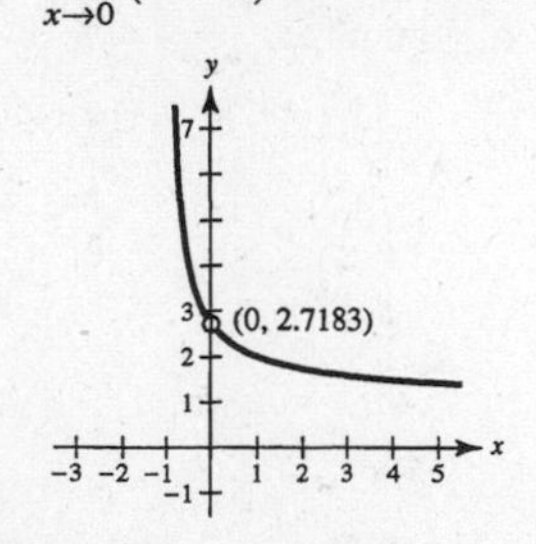

x	$f(x)$
-0.1	2.867972
-0.01	2.731999
-0.001	2.719642
-0.0001	2.718418
-0.00001	2.718295
-0.000001	2.718283

x	$f(x)$
0.1	2.593742
0.01	2.704814
0.001	2.716942
0.0001	2.718146
0.00001	2.718268
0.000001	2.718280

49. False; $f(x) = (\sin x)/x$ is undefined when $x = 0$. From Exercise 7, we have

$$\lim_{x \to 0} \frac{\sin x}{x} = 1.$$

51. False; let

$$f(x) = \begin{cases} x^2 - 4x, & x \neq 4 \\ 10, & x = 4 \end{cases}.$$

$$f(4) = 10$$

$$\lim_{x \to 4} f(x) = \lim_{x \to 4} (x^2 - 4x) = 0 \neq 10$$

53. If $\lim_{x \to c} f(x) = L_1$ and $\lim_{x \to c} f(x) = L_2$, then for every $\epsilon > 0$, there exists $\delta_1 > 0$ and $\delta_2 > 0$ such that $|x - c| < \delta_1 \Rightarrow |f(x) - L_1| < \epsilon$ and $|x - c| < \delta_2 \Rightarrow |f(x) - L_2| < \epsilon$. Let δ equal the smaller of δ_1 and δ_2. Then for $|x - c| < \delta$, we have $|L_1 - L_2| = |L_1 - f(x) + f(x) - L_2| \leq |L_1 - f(x)| + |f(x) - L_2| < \epsilon + \epsilon$. Therefore, $|L_1 - L_2| < 2\epsilon$. Since $\epsilon > 0$ is arbitrary, it follows that $L_1 = L_2$.

55. $\lim_{x \to c} [f(x) - L] = 0$ means that for every $\epsilon > 0$ there exists $\delta > 0$ such that if $0 < |x - c| < \delta$, then $|(f(x) - L) - 0| < \epsilon$. This means the same as $|f(x) - L| < \epsilon$ when $0 < |x - c| < \delta$. Thus, $\lim_{x \to c} f(x) = L$.

Section 1.3 Evaluating Limits Analytically

1.

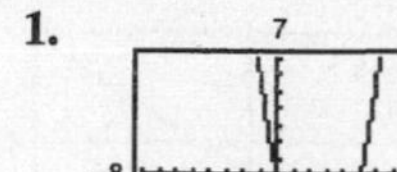
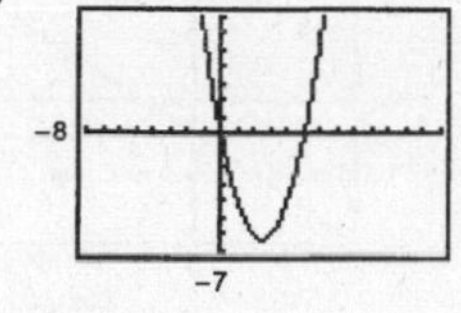

$h(x) = x^2 - 5x$

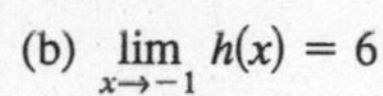

(a) $\lim_{x \to 5} h(x) = 0$

(b) $\lim_{x \to -1} h(x) = 6$

3.

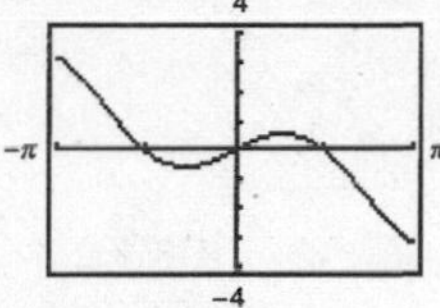

$f(x) = x \cos x$

(a) $\lim_{x \to 0} f(x) = 0$

(b) $\lim_{x \to \pi/3} f(x) \approx 0.524$ $\left(= \frac{\pi}{6}\right)$

5. $\lim_{x \to 2} x^4 = 2^4 = 16$

7. $\lim_{x \to 0} (2x - 1) = 2(0) - 1 = -1$

9. $\lim_{x \to -3} (2x^2 + 4x + 1) = 2(-3)^2 + 4(-3) + 1 = 18 - 12 + 1 = 7$

11. $\lim_{x \to 2} \frac{1}{x} = \frac{1}{2}$

13. $\lim_{x \to 1} \frac{x - 3}{x^2 + 4} = \frac{1 - 3}{1^2 + 4} = \frac{-2}{5} = -\frac{2}{5}$

15. $\lim_{x \to 7} \frac{5x}{\sqrt{x + 2}} = \frac{5(7)}{\sqrt{7 + 2}} = \frac{35}{\sqrt{9}} = \frac{35}{3}$

17. $\lim_{x \to 3} \sqrt{x + 1} = \sqrt{3 + 1} = 2$

19. $\lim_{x \to \pi/2} \sin x = \sin \frac{\pi}{2} = 1$

21. $\lim_{x \to 2} \cos \frac{\pi x}{3} = \cos \frac{\pi 2}{3} = -\frac{1}{2}$

23. $\lim_{x \to 0} \sec 2x = \sec 0 = 1$

25. $\lim_{x \to 5\pi/6} \sin x = \sin \frac{5\pi}{6} = \frac{1}{2}$

27. $\lim_{x \to 3} \tan\left(\frac{\pi x}{4}\right) = \tan \frac{3\pi}{4} = -1$

29. $\lim_{x \to 0} e^x \cos 2x = e^0 \cos 0 = 1$

31. $\lim_{x \to 1} (\ln 3x + e^x) = \ln 3 + e$

33. (a) $\lim_{x \to 1} f(x) = 5 - 1 = 4$

(b) $\lim_{x \to 4} g(x) = 4^3 = 64$

(c) $\lim_{x \to 1} g(f(x)) = g(f(1)) = g(4) = 64$

35. (a) $\lim_{x\to 1} f(x) = 4 - 1 = 3$

(b) $\lim_{x\to 3} g(x) = \sqrt{3+1} = 2$

(c) $\lim_{x\to 1} g(f(x)) = g(3) = 2$

37. (a) $\lim_{x\to c} [5g(x)] = 5 \lim_{x\to c} g(x) = 5(3) = 15$

(b) $\lim_{x\to c} [f(x) + g(x)] = \lim_{x\to c} f(x) + \lim_{x\to c} g(x) = 2 + 3 = 5$

(c) $\lim_{x\to c} [f(x)g(x)] = \left[\lim_{x\to c} f(x)\right]\left[\lim_{x\to c} g(x)\right] = (2)(3) = 6$

(d) $\lim_{x\to c} \dfrac{f(x)}{g(x)} = \dfrac{\lim_{x\to c} f(x)}{\lim_{x\to c} g(x)} = \dfrac{2}{3}$

39. (a) $\lim_{x\to c} [f(x)]^3 = \left[\lim_{x\to c} f(x)\right]^3 = (4)^3 = 64$

(b) $\lim_{x\to c} \sqrt{f(x)} = \sqrt{\lim_{x\to c} f(x)} = \sqrt{4} = 2$

(c) $\lim_{x\to c} [3f(x)] = 3 \lim_{x\to c} f(x) = 3(4) = 12$

(d) $\lim_{x\to c} [f(x)]^{3/2} = \left[\lim_{x\to c} f(x)\right]^{3/2} = (4)^{3/2} = 8$

41. $f(x) = -2x + 1$ and $g(x) = \dfrac{-2x^2 + x}{x}$ agree except at $x = 0$.

(a) $\lim_{x\to 0} g(x) = \lim_{x\to 0} f(x) = 1$

(b) $\lim_{x\to -1} g(x) = \lim_{x\to -1} f(x) = 3$

43. $f(x) = x(x+1)$ and $g(x) = \dfrac{x^3 - x}{x - 1}$ agree except at $x = 1$.

(a) $\lim_{x\to 1} g(x) = \lim_{x\to 1} f(x) = 2$

(b) $\lim_{x\to -1} g(x) = \lim_{x\to -1} f(x) = 0$

45. $f(x) = \dfrac{x^2 - 1}{x + 1}$ and $g(x) = x - 1$ agree except at $x = -1$.

$\lim_{x\to -1} f(x) = \lim_{x\to -1} g(x) = -2$

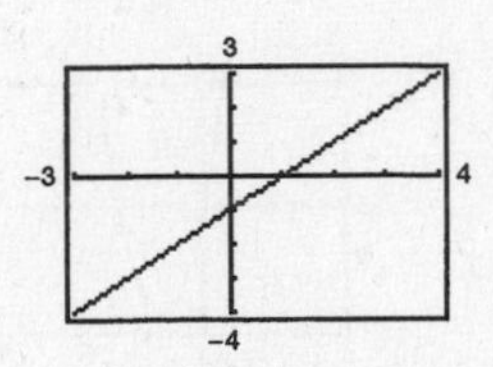

47. $f(x) = \dfrac{x^3 - 8}{x - 2}$ and $g(x) = x^2 + 2x + 4$ agree except at $x = 2$.

$\lim_{x\to 2} f(x) = \lim_{x\to 2} g(x) = 12$

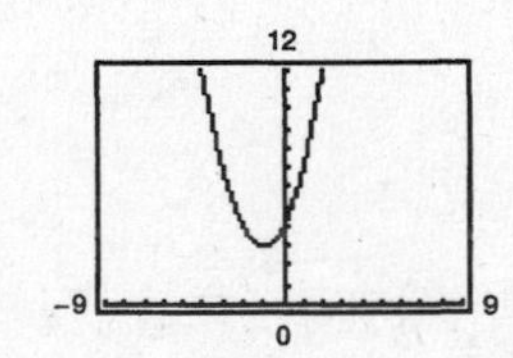

49. $f(x) = \dfrac{(x+4)\ln(x+6)}{x^2 - 16}$ and $g(x) = \dfrac{\ln(x+6)}{x - 4}$

$\lim_{x\to -4} f(x) = \lim_{x\to -4} g(x) = \dfrac{\ln 2}{-8} \approx -0.0866$

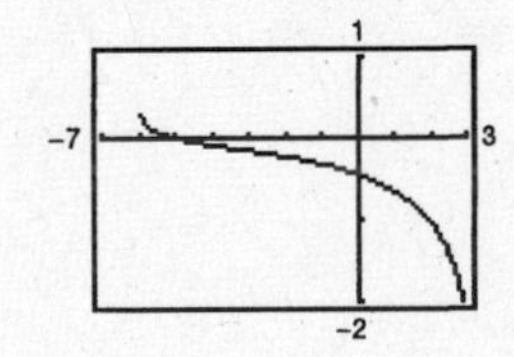

51. $\lim_{x\to 5} \dfrac{x - 5}{x^2 - 25} = \lim_{x\to 5} \dfrac{x - 5}{(x+5)(x-5)}$

$= \lim_{x\to 5} \dfrac{1}{x + 5} = \dfrac{1}{10}$

53. $\lim_{x\to -3} \dfrac{x^2 + x - 6}{x^2 - 9} = \lim_{x\to -3} \dfrac{(x+3)(x-2)}{(x+3)(x-3)}$

$= \lim_{x\to -3} \dfrac{x - 2}{x - 3} = \dfrac{-5}{-6} = \dfrac{5}{6}$

55. $\lim_{x\to 0} \dfrac{\sqrt{x+5} - \sqrt{5}}{x} = \lim_{x\to 0} \dfrac{\sqrt{x+5} - \sqrt{5}}{x} \cdot \dfrac{\sqrt{x+5} + \sqrt{5}}{\sqrt{x+5} + \sqrt{5}}$

$= \lim_{x\to 0} \dfrac{(x+5) - 5}{x\left(\sqrt{x+5} + \sqrt{5}\right)} = \lim_{x\to 0} \dfrac{1}{\sqrt{x+5} + \sqrt{5}} = \dfrac{1}{2\sqrt{5}} = \dfrac{\sqrt{5}}{10}$

57. $\lim_{x\to 4} \dfrac{\sqrt{x+5} - 3}{x - 4} = \lim_{x\to 4} \dfrac{\sqrt{x+5} - 3}{x - 4} \cdot \dfrac{\sqrt{x+5} + 3}{\sqrt{x+5} + 3}$

$= \lim_{x\to 4} \dfrac{(x+5) - 9}{(x-4)\left(\sqrt{x+5} + 3\right)} = \lim_{x\to 4} \dfrac{1}{\sqrt{x+5} + 3} = \dfrac{1}{\sqrt{9} + 3} = \dfrac{1}{6}$

59. $\lim_{x\to 0} \dfrac{\dfrac{1}{3+x}-\dfrac{1}{3}}{x} = \lim_{x\to 0} \dfrac{\dfrac{3-(3+x)}{3(3+x)}}{x}$

$= \lim_{x\to 0} \dfrac{-1}{3(3+x)} = -\dfrac{1}{9}$

61. $\lim_{\Delta x\to 0} \dfrac{2(x+\Delta x)-2x}{\Delta x} = \lim_{\Delta x\to 0} \dfrac{2x+2\Delta x-2x}{\Delta x}$

$= \lim_{\Delta x\to 0} 2 = 2$

63. $\lim_{\Delta x\to 0} \dfrac{(x+\Delta x)^2-2(x+\Delta x)+1-(x^2-2x+1)}{\Delta x} = \lim_{\Delta x\to 0} \dfrac{x^2+2x\Delta x+(\Delta x)^2-2x-2\Delta x+1-x^2+2x-1}{\Delta x}$

$= \lim_{\Delta x\to 0} (2x+\Delta x-2) = 2x-2$

65. $\lim_{x\to 0} \dfrac{\sqrt{x+2}-\sqrt{2}}{x} \approx 0.354$

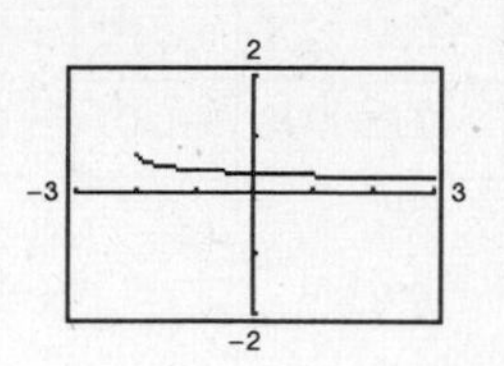

x	-0.1	-0.01	-0.001	0	0.001	0.01	0.1
$f(x)$	0.358	0.354	0.354	?	0.354	0.353	0.349

Analytically, $\lim_{x\to 0} \dfrac{\sqrt{x+2}-\sqrt{2}}{x} = \lim_{x\to 0} \dfrac{\sqrt{x+2}-\sqrt{2}}{x} \cdot \dfrac{\sqrt{x+2}+\sqrt{2}}{\sqrt{x+2}+\sqrt{2}}$

$= \lim_{x\to 0} \dfrac{x+2-2}{x\left(\sqrt{x+2}+\sqrt{2}\right)} = \lim_{x\to 0} \dfrac{1}{\sqrt{x+2}+\sqrt{2}} = \dfrac{1}{2\sqrt{2}} = \dfrac{\sqrt{2}}{4} \approx 0.354.$

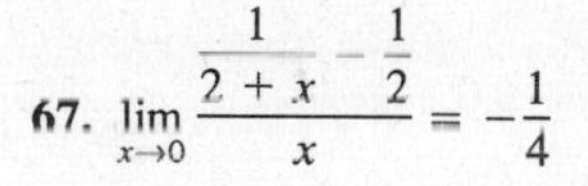

67. $\lim_{x\to 0} \dfrac{\dfrac{1}{2+x}-\dfrac{1}{2}}{x} = -\dfrac{1}{4}$

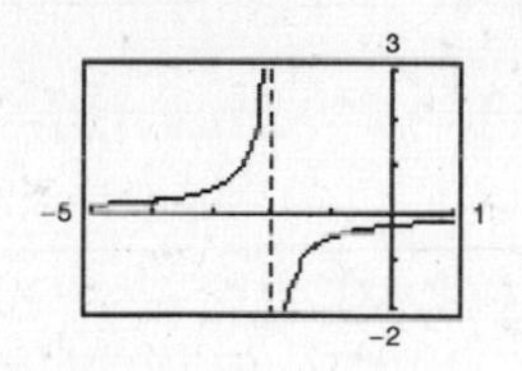

x	-0.1	-0.01	-0.001	0	0.001	0.01	0.1
$f(x)$	-0.263	-0.251	-0.250	?	-0.250	-0.249	-0.238

Analytically, $\lim_{x\to 0} \dfrac{\dfrac{1}{2+x}-\dfrac{1}{2}}{x} = \lim_{x\to 0} \dfrac{2-(2+x)}{2(2+x)} \cdot \dfrac{1}{x} = \lim_{x\to 0} \dfrac{-x}{2(2+x)} \cdot \dfrac{1}{x} = \lim_{x\to 0} \dfrac{-1}{2(2+x)} = -\dfrac{1}{4}.$

69. $\lim_{x\to 0} \dfrac{\sin x}{5x} = \lim_{x\to 0} \left[\left(\dfrac{\sin x}{x}\right)\left(\dfrac{1}{5}\right)\right] = (1)\left(\dfrac{1}{5}\right) = \dfrac{1}{5}$

71. $\lim_{x\to 0} \dfrac{\sin x(1-\cos x)}{2x^2} = \lim_{x\to 0} \left[\dfrac{1}{2} \cdot \dfrac{\sin x}{x} \cdot \dfrac{1-\cos x}{x}\right]$

$= \dfrac{1}{2}(1)(0) = 0$

73. $\lim_{x\to 0} \dfrac{\sin^2 x}{x} = \lim_{x\to 0} \left[\dfrac{\sin x}{x} \sin x\right] = (1) \sin 0 = 0$

75. $\lim_{h\to 0} \dfrac{(1-\cos h)^2}{h} = \lim_{h\to 0} \left[\dfrac{1-\cos h}{h}(1-\cos h)\right]$

$= (0)(0) = 0$

77. $\lim_{x\to \pi/2} \dfrac{\cos x}{\cot x} = \lim_{x\to \pi/2} \sin x = 1$

79. $\lim_{x\to 0} \dfrac{1-e^{-x}}{e^x-1} = \lim_{x\to 0} \dfrac{1-e^{-x}}{e^x-1} \cdot \dfrac{e^{-x}}{e^{-x}}$

$= \lim_{x\to 0} \dfrac{(1-e^{-x})e^{-x}}{1-e^{-x}} = \lim_{x\to 0} e^{-x} = 1$

81. $\lim_{t\to 0} \dfrac{\sin 3t}{2t} = \lim_{t\to 0} \left(\dfrac{\sin 3t}{3t}\right)\left(\dfrac{3}{2}\right) = (1)\left(\dfrac{3}{2}\right) = \dfrac{3}{2}$

83. $f(t) = \dfrac{\sin 3t}{t}$

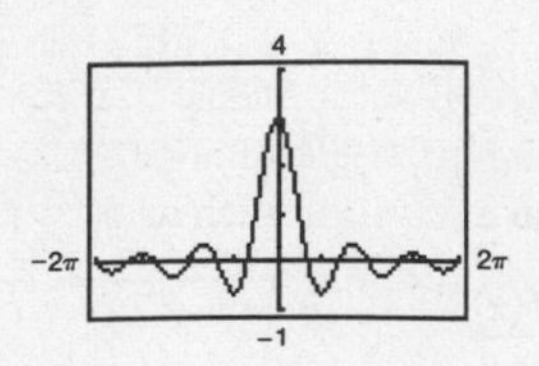

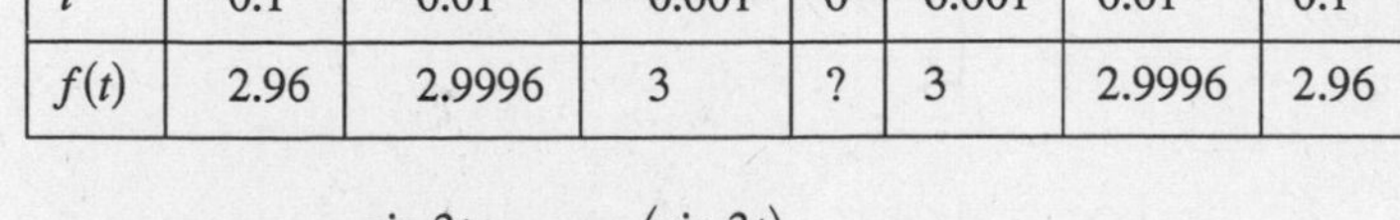

t	−0.1	−0.01	−0.001	0	0.001	0.01	0.1
$f(t)$	2.96	2.9996	3	?	3	2.9996	2.96

The limit appears to equal 3.

Analytically, $\displaystyle\lim_{t\to0} \frac{\sin 3t}{t} = \lim_{t\to0} 3\left(\frac{\sin 3t}{3t}\right) = 3(1) = 3.$

85. $f(x) = \dfrac{\sin x^2}{x}$

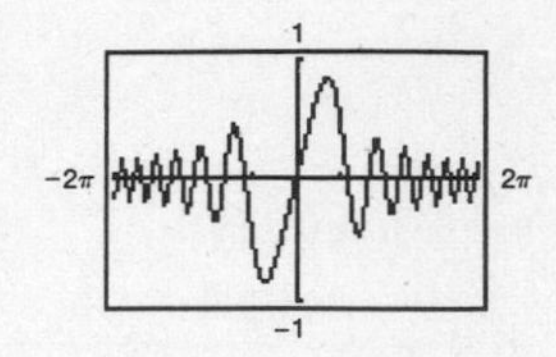

x	−0.1	−0.01	−0.001	0	0.001	0.01	0.1
$f(x)$	−0.099998	−0.01	−0.001	?	0.001	0.01	0.099998

Analytically, $\displaystyle\lim_{x\to0} \frac{\sin x^2}{x} = \lim_{x\to0} x\left(\frac{\sin x^2}{x^2}\right) = 0(1) = 0.$

87. $f(x) = \dfrac{\ln x}{x-1}$

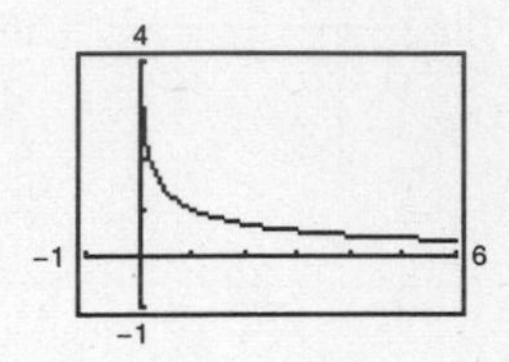

x	0.5	0.9	0.99	1.01	1.1	1.5
$f(x)$	1.3863	1.0536	1.0050	0.9950	0.9531	0.8109

$\displaystyle\lim_{x\to1} \frac{\ln x}{x-1} = 1$

89. $\displaystyle\lim_{\Delta x\to0} \frac{f(x+\Delta x) - f(x)}{\Delta x} = \lim_{\Delta x\to0} \frac{2(x+\Delta x) + 3 - (2x+3)}{\Delta x} = \lim_{\Delta x\to0} \frac{2x + 2\,\Delta x + 3 - 2x - 3}{\Delta x} = \lim_{\Delta x\to0} \frac{2\,\Delta x}{\Delta x} = 2$

91. $\displaystyle\lim_{\Delta x\to0} \frac{f(x+\Delta x) - f(x)}{\Delta x} = \lim_{\Delta x\to0} \frac{\dfrac{4}{x+\Delta x} - \dfrac{4}{x}}{\Delta x} = \lim_{\Delta x\to0} \frac{4x - 4(x+\Delta x)}{(x+\Delta x)x\,\Delta x} = \lim_{\Delta x\to0} \frac{-4}{(x+\Delta x)x} = \frac{-4}{x^2}$

93. $\displaystyle\lim_{x\to0} (4 - x^2) \le \lim_{x\to0} f(x) \le \lim_{x\to0} (4 + x^2)$

$\displaystyle 4 \le \lim_{x\to0} f(x) \le 4$

Therefore, $\displaystyle\lim_{x\to0} f(x) = 4.$

95. $f(x) = x \cos x$

$\displaystyle\lim_{x\to0} (x \cos x) = 0$

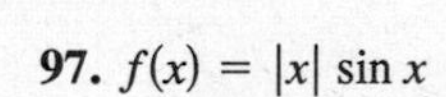

97. $f(x) = |x| \sin x$

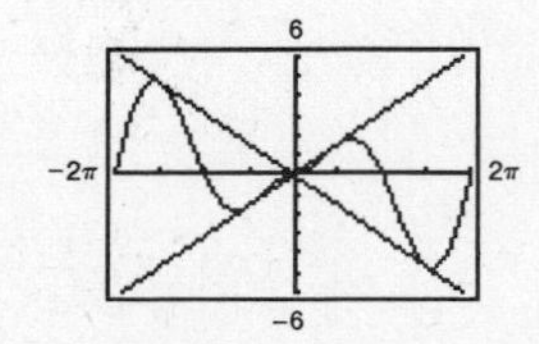

$\displaystyle\lim_{x\to0} |x| \sin x = 0$

99. $f(x) = x \sin \dfrac{1}{x}$

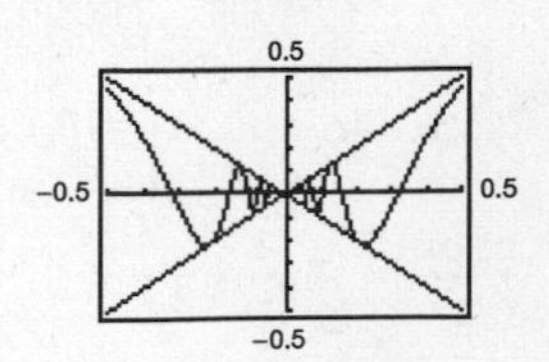

$\displaystyle\lim_{x\to0} \left(x \sin \frac{1}{x}\right) = 0$

101. We say that two functions f and g agree at all but one point (on an open interval) if $f(x) = g(x)$ for all x in the interval except for $x = c$, where c is in the interval.

103. An indeterminant form is obtained when evaluating a limit using direct substitution produces a meaningless fractional expression such as 0/0. That is,

$$\lim_{x\to c} \frac{f(x)}{g(x)}$$

for which $\lim_{x\to c} f(x) = \lim_{x\to c} g(x) = 0$

105. $f(x) = x,\ g(x) = \sin x,\ h(x) = \dfrac{\sin x}{x}$

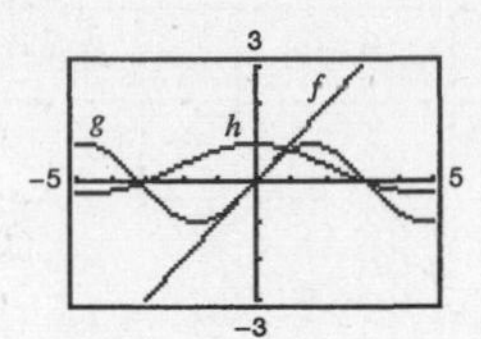

When you are "close to" 0 the magnitude of f is approximately equal to the magnitude of g. Thus, $|g|/|f| \approx 1$ when x is "close to" 0.

107. $s(t) = -16t^2 + 1000$

$$\lim_{t\to 5} \frac{s(5) - s(t)}{5 - t} = \lim_{t\to 5} \frac{600 - (-16t^2 + 1000)}{5 - t} = \lim_{t\to 5} \frac{16(t + 5)(t - 5)}{-(t - 5)} = \lim_{t\to 5} -16(t + 5) = -160 \text{ ft/sec}$$

Speed = 160 ft/sec

109. $s(t) = -4.9t^2 + 150$

$$\lim_{t\to 3} \frac{s(3) - s(t)}{3 - t} = \lim_{t\to 3} \frac{-4.9(3^2) + 150 - (-4.9t^2 + 150)}{3 - t} = \lim_{t\to 3} \frac{-4.9(9 - t^2)}{3 - t}$$

$$= \lim_{x\to 3} \frac{-4.9(3 - t)(3 + t)}{3 - t} = \lim_{x\to 3} -4.9(3 + t) = -29.4 \text{ m/sec}$$

111. Let $f(x) = 1/x$ and $g(x) = -1/x$. $\lim_{x\to 0} f(x)$ and $\lim_{x\to 0} g(x)$ do not exist.

$$\lim_{x\to 0} [f(x) + g(x)] = \lim_{x\to 0} \left[\frac{1}{x} + \left(-\frac{1}{x}\right)\right] = \lim_{x\to 0} [0] = 0$$

113. Given $f(x) = b$, show that for every $\epsilon > 0$ there exists a $\delta > 0$ such that $|f(x) - b| < \epsilon$ whenever $|x - c| < \delta$. Since $|f(x) - b| = |b - b| = 0 < \epsilon$ for any $\epsilon > 0$, then any value of $\delta > 0$ will work.

115. If $b = 0$, then the property is true because both sides are equal to 0. If $b \neq 0$, let $\epsilon > 0$ be given. Since $\lim_{x\to c} f(x) = L$, there exists $\delta > 0$ such that $|f(x) - L| < \epsilon/|b|$ whenever $0 < |x - c| < \delta$. Hence, wherever $0 < |x - c| < \delta$, we have $|b||f(x) - L| < \epsilon$ or $|bf(x) - bL| < \epsilon$ which implies that $\lim_{x\to c} [bf(x)] = bL$.

117.

$$-M|f(x)| \le f(x)g(x) \le M|f(x)|$$

$$\lim_{x\to c} (-M|f(x)|) \le \lim_{x\to c} f(x)g(x) \le \lim_{x\to c} (M|f(x)|)$$

$$-M(0) \le \lim_{x\to c} f(x)g(x) \le M(0)$$

$$0 \le \lim_{x\to c} f(x)g(x) \le 0$$

Therefore, $\lim_{x\to c} f(x)g(x) = 0$.

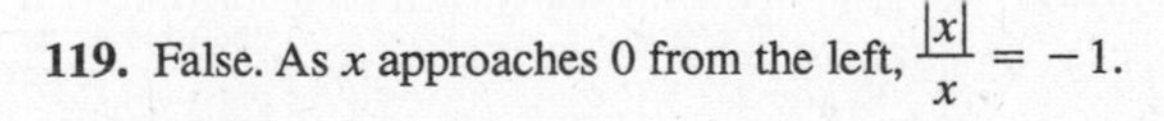

119. False. As x approaches 0 from the left, $\dfrac{|x|}{x} = -1$.

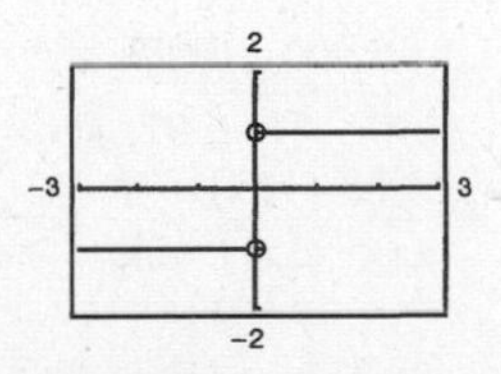

121. True.

123. False. The limit does not exist.

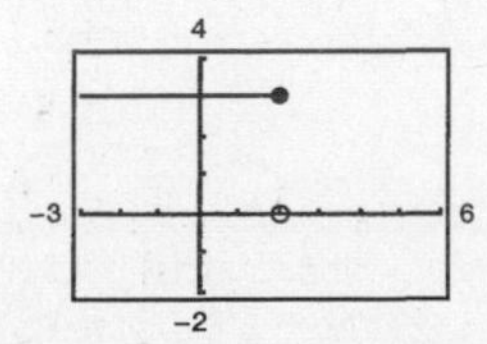

125. Let

$$f(x) = \begin{cases} 4, & \text{if } x \geq 0 \\ -4, & \text{if } x < 0 \end{cases}$$

$\lim_{x\to 0} |f(x)| = \lim_{x\to 0} 4 = 4.$

$\lim_{x\to 0} f(x)$ does not exist since for $x < 0, f(x) = -4$ and for $x \geq 0, f(x) = 4$.

127. $f(x) = \begin{cases} 0, & \text{if } x \text{ is rational} \\ 1, & \text{if } x \text{ is irrational} \end{cases}$

$$g(x) = \begin{cases} 0, & \text{if } x \text{ is rational} \\ x, & \text{if } x \text{ is irrational} \end{cases}$$

$\lim_{x\to 0} f(x)$ does not exist.

No matter how "close to" 0 x is, there are still an infinite number of rational and irrational numbers so that $\lim_{x\to 0} f(x)$ does not exist.

$\lim_{x\to 0} g(x) = 0.$

When x is "close to" 0, both parts of the function are "close to" 0.

129. (a) $\lim_{x\to 0} \frac{1 - \cos x}{x^2} = \lim_{x\to 0} \frac{1 - \cos x}{x^2} \cdot \frac{1 + \cos x}{1 + \cos x}$

$= \lim_{x\to 0} \frac{1 - \cos^2 x}{x^2(1 + \cos x)}$

$= \lim_{x\to 0} \frac{\sin^2 x}{x^2} \cdot \frac{1}{1 + \cos x}$

$= (1)\left(\frac{1}{2}\right) = \frac{1}{2}$

(b) Thus, $\frac{1 - \cos x}{x^2} \approx \frac{1}{2} \Rightarrow 1 - \cos x \approx \frac{1}{2}x^2$

$\Rightarrow \cos x \approx 1 - \frac{1}{2}x^2$ for $x \approx 0$.

(c) $\cos(0.1) \approx 1 - \frac{1}{2}(0.1)^2 = 0.995$

(d) $\cos(0.1) \approx 0.9950$, which agrees with part (c).

Section 1.4 Continuity and One-Sided Limits

1. (a) $\lim_{x\to 3^+} f(x) = 1$

(b) $\lim_{x\to 3^-} f(x) = 1$

(c) $\lim_{x\to 3} f(x) = 1$

The function is continuous at $x = 3$.

3. (a) $\lim_{x\to 3^+} f(x) = 0$

(b) $\lim_{x\to 3^-} f(x) = 0$

(c) $\lim_{x\to 3} f(x) = 0$

The function is NOT continuous at $x = 3$.

5. (a) $\lim_{x\to 4^+} f(x) = 2$

(b) $\lim_{x\to 4^-} f(x) = -2$

(c) $\lim_{x\to 4} f(x)$ does not exist

The function is NOT continuous at $x = 4$.

7. $\lim_{x\to 5^+} \frac{x - 5}{x^2 - 25} = \lim_{x\to 5^+} \frac{1}{x + 5} = \frac{1}{10}$

9. $\lim_{x\to -3^-} \frac{x}{\sqrt{x^2 - 9}}$ does not exist because $\frac{x}{\sqrt{x^2 - 9}}$ grows without bound as $x \to -3^-$.

11. $\lim_{x\to 0^-} \frac{|x|}{x} = \lim_{x\to 0^-} \frac{-x}{x} = -1$

13. $\lim_{\Delta x\to 0^-} \frac{\frac{1}{x + \Delta x} - \frac{1}{x}}{\Delta x} = \lim_{\Delta x\to 0^-} \frac{x - (x + \Delta x)}{x(x + \Delta x)} \cdot \frac{1}{\Delta x} = \lim_{\Delta x\to 0^-} \frac{-\Delta x}{x(x + \Delta x)} \cdot \frac{1}{\Delta x} = \lim_{\Delta x\to 0^-} \frac{-1}{x(x + \Delta x)} = \frac{-1}{x(x + 0)} = -\frac{1}{x^2}$

15. $\lim_{x\to 3^-} f(x) = \lim_{x\to 3^-} \frac{x + 2}{2} = \frac{5}{2}$

17. $\lim_{x\to 1^+} f(x) = \lim_{x\to 1^+} (x + 1) = 2$

$\lim_{x\to 1^-} f(x) = \lim_{x\to 1^-} (x^3 + 1) = 2$

$\lim_{x\to 1} f(x) = 2$

19. $\lim_{x\to \pi} \cot x$ does not exist since

$\lim_{x\to \pi^+} \cot x$ and $\lim_{x\to \pi^-} \cot x$

do not exist.

21. $\lim_{x\to 4^-} (3[\![x]\!] - 5) = 3(3) - 5 = 4$
$([\![x]\!] = 3$ for $3 < x < 4)$

23. $\lim_{x\to 3} (2 - [\![-x]\!])$ does not exist because

$\lim_{x\to 3^-} (2 - [\![-x]\!]) = 2 - (-3) = 5$

and

$\lim_{x\to 3^+} (2 - [\![-x]\!]) = 2 - (-4) = 6.$

25. $\lim_{x\to 3^+} \ln(x - 3) = \ln 0$
does not exist.

27. $\lim_{x\to 2^-} \ln[x^2(3 - x)] = \ln[4(1)] = \ln 4$

29. $f(x) = \dfrac{1}{x^2 - 4}$

has discontinuities at $x = -2$ and $x = 2$ since $f(-2)$ and $f(2)$ are not defined.

31. $f(x) = \dfrac{[\![x]\!]}{2} + x$

has discontinuities at each integer k since $\lim_{x\to k^-} f(x) \neq \lim_{x\to k^+} f(x)$.

33. $g(x) = \sqrt{25 - x^2}$ is continuous on $[-5, 5]$.

35. $\lim_{x\to 0^-} f(x) = 3 = \lim_{x\to 0^+} f(x)$.
f is continuous on $[-1, 4]$.

37. $f(x) = x^2 - 2x + 1$ is continuous for all real x.

39. $f(x) = 3x - \cos x$ is continuous for all real x.

41. $f(x) = \dfrac{x}{x^2 - x}$ is not continuous at $x = 0, 1$. Since

$\dfrac{x}{x^2 - x} = \dfrac{1}{x - 1}$ for $x \neq 0$, $x = 0$ is a removable discontinuity, whereas $x = 1$ is a nonremovable discontinuity.

43. $f(x) = \dfrac{x}{x^2 + 1}$ is continuous for all real x.

45. $f(x) = \dfrac{x + 2}{(x + 2)(x - 5)}$

has a nonremovable discontinuity at $x = 5$ since $\lim_{x\to 5} f(x)$ does not exist, and has a removable discontinuity at $x = -2$ since

$$\lim_{x\to -2} f(x) = \lim_{x\to -2} \frac{1}{x - 5} = -\frac{1}{7}.$$

47. $f(x) = \dfrac{|x + 2|}{x + 2}$ has a nonremovable discontinuity at $x = -2$ since $\lim_{x\to -2} f(x)$ does not exist.

49. $f(x) = \begin{cases} x, & x \le 1 \\ x^2, & x > 1 \end{cases}$

has a **possible** discontinuity at $x = 1$.

1. $f(1) = 1$

2. $\left.\begin{array}{l} \lim_{x\to 1^-} f(x) = \lim_{x\to 1^-} x = 1 \\ \lim_{x\to 1^+} f(x) = \lim_{x\to 1^+} x^2 = 1 \end{array}\right\} \lim_{x\to 1} f(x) = 1$

3. $f(1) = \lim_{x\to 1} f(x)$

f is continuous at $x = 1$, therefore, f is continuous for all real x.

51. $f(x) = \begin{cases} \dfrac{x}{2} + 1, & x \le 2 \\ 3 - x, & x > 2 \end{cases}$

has a possible discontinuity at $x = 2$.

1. $f(2) = \dfrac{2}{2} + 1 = 2$

2. $\left.\begin{array}{l} \lim_{x\to 2^-} f(x) = \lim_{x\to 2^-} \left(\dfrac{x}{2} + 1\right) = 2 \\ \lim_{x\to 2^+} f(x) = \lim_{x\to 2^+} (3 - x) = 1 \end{array}\right\} \lim_{x\to 2} f(x)$ does not exist.

Therefore, f has a nonremovable discontinuity at $x = 2$.

53. $f(x) = \begin{cases} \tan \frac{\pi x}{4}, & |x| < 1 \\ x, & |x| \geq 1 \end{cases} = \begin{cases} \tan \frac{\pi x}{4}, & -1 < x < 1 \\ x, & x \leq -1 \text{ or } x \geq 1 \end{cases}$ has **possible** discontinuities at $x = -1, x = 1$.

1. $f(-1) = -1$ $\qquad f(1) = 1$
2. $\lim_{x \to -1} f(x) = -1$ $\qquad \lim_{x \to 1} f(x) = 1$
3. $f(-1) = \lim_{x \to -1} f(x)$ $\qquad f(1) = \lim_{x \to 1} f(x)$

f is continuous at $x = \pm 1$, therefore, f is continuous for all real x.

55. $f(x) = \begin{cases} \ln(x + 1), & x \geq 0 \\ 1 - x^2, & x < 0 \end{cases}$ has a **possible** discontinuity at $x = 0$.

$f(0) = \ln(0 + 1) = \ln 1 = 0$

$\lim_{x \to 0^-} f(x) = 1 - 0 = 1; \quad \lim_{x \to 0^+} f(x) = 0$

Therefore, f has a nonremovable discontinuity at $x = 0$.

57. $f(x) = \csc 2x$ has nonremovable discontinuities at integer multiples of $\pi/2$.

59. $f(x) = [\![x - 1]\!]$ has nonremovable discontinuities at each integer k.

61. $\lim_{x \to 0^+} f(x) = 0$

$\lim_{x \to 0^-} f(x) = 0$

f is not continuous at $x = -2$.

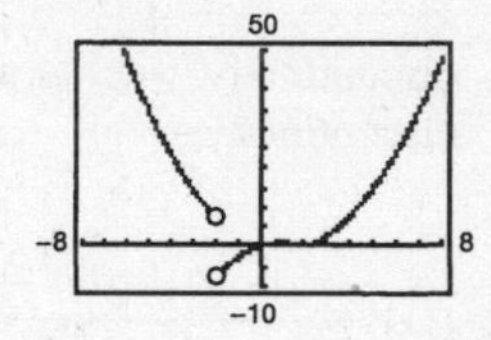

63. $f(2) = 8$

Find a so that $\lim_{x \to 2^+} ax^2 = 8 \implies a = \frac{8}{2^2} = 2$.

65. Find a and b such that $\lim_{x \to -1^+} (ax + b) = -a + b = 2$ and $\lim_{x \to 3^-} (ax + b) = 3a + b = -2$.

$$\begin{aligned} a - b &= -2 \\ (+)\ 3a + b &= -2 \\ \hline 4a \quad &= -4 \\ a &= -1 \\ b &= 2 + (-1) = 1 \end{aligned}$$

$$f(x) = \begin{cases} 2, & x \leq -1 \\ -x + 1, & -1 < x < 3 \\ -2, & x \geq 3 \end{cases}$$

67. $f(g(x)) = (x - 1)^2$

Continuous for all real x.

69. $f(g(x)) = \dfrac{1}{(x^2 + 5) - 6} = \dfrac{1}{x^2 - 1}$

Nonremovable discontinuities at $x = \pm 1$

71. $y = [\![x]\!] - x$

Nonremovable discontinuity at each integer

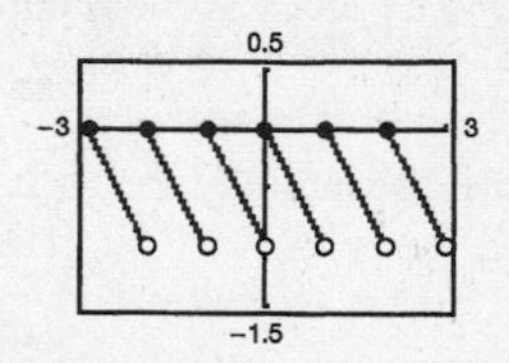

73. $f(x) = \begin{cases} 2x - 4, & x \leq 3 \\ x^2 - 2x, & x > 3 \end{cases}$

Nonremovable discontinuity at $x = 3$

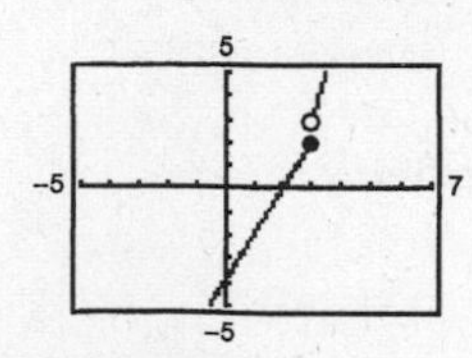

75. $f(x) = \dfrac{x}{x^2 + 1}$

Continuous on $(-\infty, \infty)$

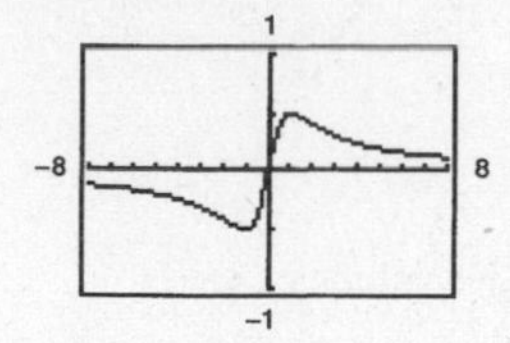

77. $f(x) = \csc \dfrac{x}{2}$

Continuous on: $\ldots, (-2\pi, 0), (0, 2\pi), (2\pi, 4\pi), \ldots$

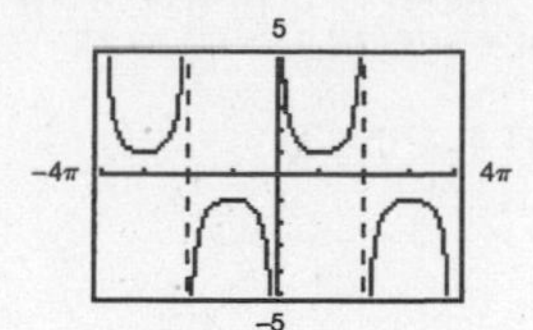

79. $f(x) = \dfrac{\sin x}{x}$

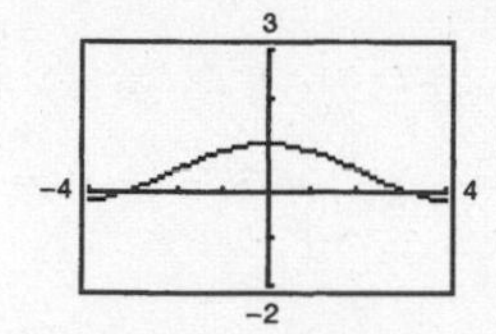

The graph **appears** to be continuous on the interval $[-4, 4]$. Since $f(0)$ is not defined, we know that f has a discontinuity at $x = 0$. This discontinuity is removable so it does not show up on the graph.

81. $f(x) = \dfrac{\ln(x^2 + 1)}{x}$

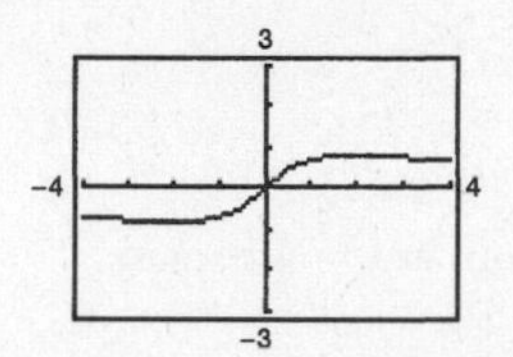

The graph **appears** to be continuous on the interval $[-4, 4]$. Since $f(0)$ is not defined, it is discontinuous there. Removable discontinuities do not always show up.

83. $f(x) = x^2 - 4x + 3$

$f(x)$ is continuous on $[2, 4]$.

$f(2) = -1$ and $f(4) = 3$

By the Intermediate Value Theorem, $f(c) = 0$ for at least one value of c between 2 and 4.

85. h is continuous on $\left[0, \dfrac{\pi}{2}\right]$.

$h(0) = -2 < 0$ and $h\left(\dfrac{\pi}{2}\right) \approx 0.91 > 0$

By the Intermediate Value Theorem, $h(c) = 0$ for at least one value of c between 0 and $\pi/2$.

87. $f(x) = x^3 + x - 1$

$f(x)$ is continuous on $[0, 1]$.

$f(0) = -1$ and $f(1) = 1$

By the Intermediate Value Theorem, $f(x) = 0$ for at least one value of c between 0 and 1. Using a graphing utility, we find that $x \approx 0.6823$.

89. $g(t) = 2 \cos t - 3t$

g is continuous on $[0, 1]$.

$g(0) = 2 > 0$ and $g(1) \approx -1.9 < 0$.

By the Intermediate Value Theorem, $g(t) = 0$ for at least one value c between 0 and 1. Using a graphing utility, we find that $t \approx 0.5636$.

91. $f(x) = x^2 + x - 1$

f is continuous on $[0, 5]$.

$f(0) = -1$ and $f(5) = 29$

$-1 < 11 < 29$

The Intermediate Value Theorem applies.

$$x^2 + x - 1 = 11$$

$$x^2 + x - 12 = 0$$

$$(x + 4)(x - 3) = 0$$

$$x = -4 \text{ or } x = 3$$

$c = 3$ ($x = -4$ is not in the interval.)

Thus, $f(3) = 11$.

93. $f(x) = x^3 - x^2 + x - 2$

f is continuous on $[0, 3]$.

$f(0) = -2$ and $f(3) = 19$

$-2 < 4 < 19$

The Intermediate Value Theorem applies.

$$x^3 - x^2 + x - 2 = 4$$

$$x^3 - x^2 + x - 6 = 0$$

$$(x - 2)(x^2 + x + 3) = 0$$

$$x = 2$$

($x^2 + x + 3$ has no real solution.)

$$c = 2$$

Thus, $f(2) = 4$.

95. (a) The limit does not exist at $x = c$.

(b) The function is not defined at $x = c$.

(c) The limit exists at $x = c$, but it is not equal to the value of the function at $x = c$.

(d) The limit does not exist at $x = c$.

97.

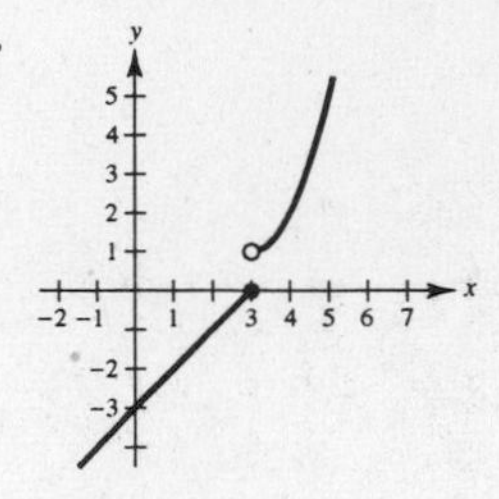

The function is not continuous at $x = 3$ because $\lim_{x\to 3^+} f(x) = 1 \neq 0 = \lim_{x\to 3^-} f(x)$.

99. The functions agree for integer values of x:

$$\left.\begin{aligned} g(x) &= 3 - [\![-x]\!] = 3 - (-x) = 3 + x \\ f(x) &= 3 + [\![x]\!] = 3 + x \end{aligned}\right\} \text{ for } x \text{ an integer}$$

However, for non-integer values of x, the functions differ by 1.

$f(x) = 3 + [\![x]\!] = g(x) - 1 = 2 - [\![-x]\!]$.

For example, $f\left(\frac{1}{2}\right) = 3 + 0 = 3$, $g\left(\frac{1}{2}\right) = 3 - (-1) = 4$.

101. $N(t) = 25\left(2\left[\!\!\left[\dfrac{t+2}{2}\right]\!\!\right] - t\right)$

t	0	1	1.8	2	3	3.8
$N(t)$	50	25	5	50	25	5

Discontinuous at every positive even integer. The company replenishes its inventory every two months.

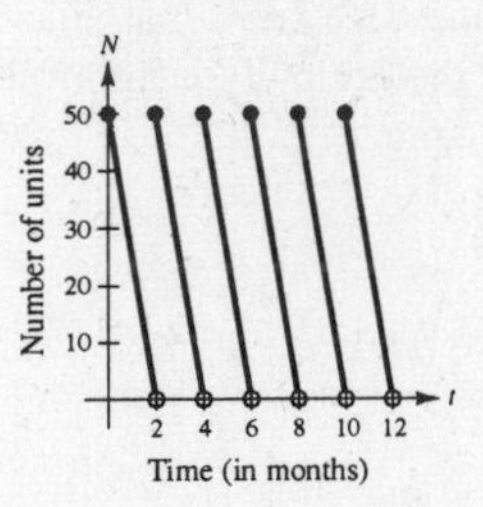

103. Let $V = \frac{4}{3}\pi r^3$ be the volume of a sphere of radius r. V is continuous on $[1, 5]$.

$V(1) = \frac{4}{3}\pi \approx 4.19$

$V(5) = \frac{4}{3}\pi(5^3) \approx 523.6$

Since $4.19 < 275 < 523.6$, the Intermediate Value Theorem implies that there is at least one value r between 1 and 5 such that $V(r) = 275$. (In fact, $r \approx 4.0341$.)

105. Let c be any real number. Then $\lim_{x\to c} f(x)$ does not exist since there are both rational and irrational numbers arbitrarily close to c. Therefore, f is not continuous at c.

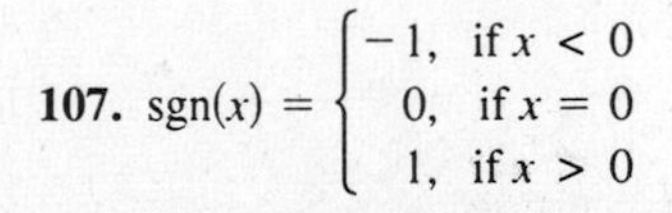

107. $\operatorname{sgn}(x) = \begin{cases} -1, & \text{if } x < 0 \\ 0, & \text{if } x = 0 \\ 1, & \text{if } x > 0 \end{cases}$

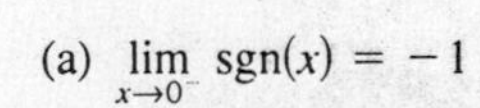

(a) $\lim_{x\to 0^-} \operatorname{sgn}(x) = -1$

(b) $\lim_{x\to 0^+} \operatorname{sgn}(x) = 1$

(c) $\lim_{x\to 0} \operatorname{sgn}(x)$ does not exist.

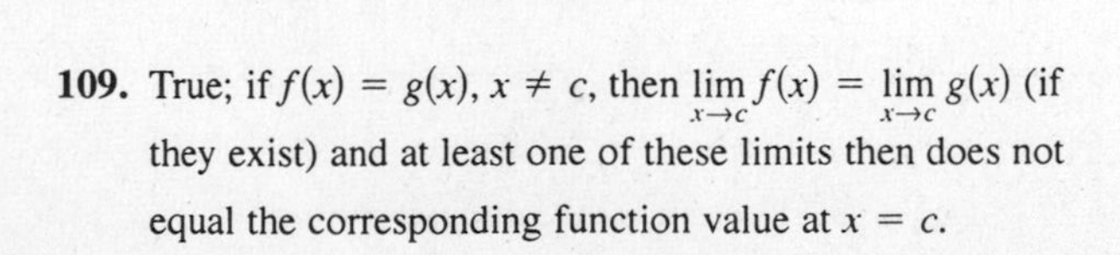

109. True; if $f(x) = g(x)$, $x \neq c$, then $\lim_{x\to c} f(x) = \lim_{x\to c} g(x)$ (if they exist) and at least one of these limits then does not equal the corresponding function value at $x = c$.

111. False; $f(1)$ is not defined and $\lim_{x\to 1} f(x)$ does not exist.

113. (a) 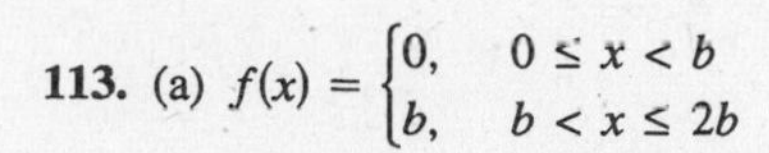
$f(x) = \begin{cases} 0, & 0 \le x < b \\ b, & b < x \le 2b \end{cases}$

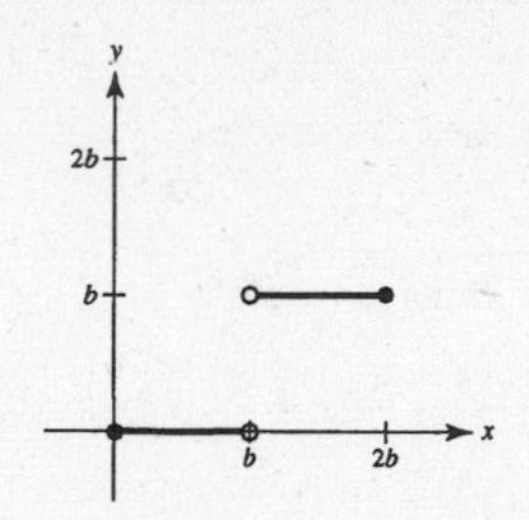

NOT continuous at $x = b$.

(b) $g(x) = \begin{cases} \dfrac{x}{2}, & 0 \le x \le b \\ b - \dfrac{x}{2}, & b < x \le 2b \end{cases}$

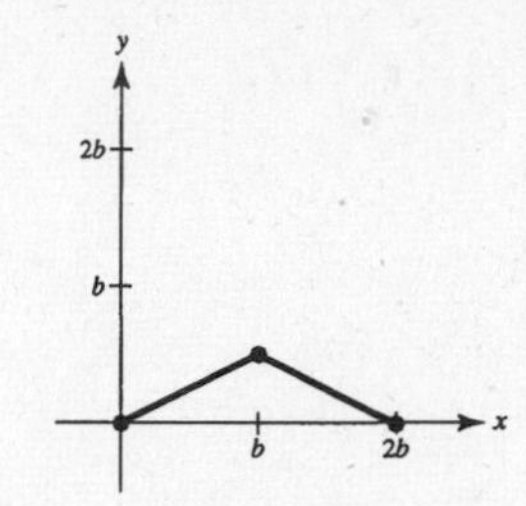

Continuous on $[0, 2b]$.

115. $f(x) = \dfrac{\sqrt{x + c^2} - c}{x}, \; c > 0$

Domain: $x + c^2 \ge 0 \Rightarrow x \ge -c^2$ and $x \ne 0$, $[-c^2, 0) \cup (0, \infty)$

$$\lim_{x \to 0} \frac{\sqrt{x + c^2} - c}{x} = \lim_{x \to 0} \frac{\sqrt{x + c^2} - c}{x} \cdot \frac{\sqrt{x + c^2} + c}{\sqrt{x + c^2} + c}$$

$$= \lim_{x \to 0} \frac{(x + c^2) - c^2}{x\left[\sqrt{x + c^2} + c\right]} = \lim_{x \to 0} \frac{1}{\sqrt{x + c^2} + c} = \frac{1}{2c}$$

Define $f(0) = 1/(2c)$ to make f continuous at $x = 0$.

117. $h(x) = x[\![x]\!]$

h has nonremovable discontinuities at

$x = \pm 1, \pm 2, \pm 3, \ldots.$

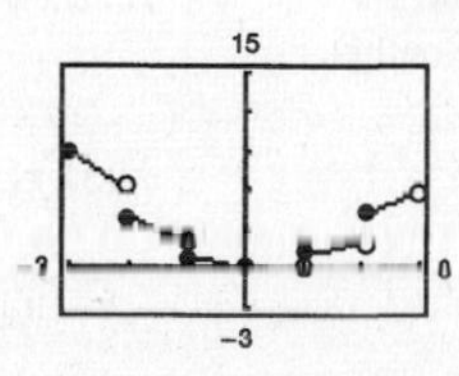

Section 1.5 Infinite Limits

1. $\displaystyle\lim_{x \to -2^+} 2\left|\frac{x}{x^2 - 4}\right| = \infty$

$\displaystyle\lim_{x \to -2^-} 2\left|\frac{x}{x^2 - 4}\right| = \infty$

3. $\displaystyle\lim_{x \to -2^+} \tan \frac{\pi x}{4} = -\infty$

$\displaystyle\lim_{x \to -2^-} \tan \frac{\pi x}{4} = \infty$

5. $f(x) = \dfrac{1}{x^2 - 9}$

x	-3.5	-3.1	-3.01	-3.001	-2.999	-2.99	-2.9	-2.5
$f(x)$	0.308	1.639	16.64	166.6	-166.7	-16.69	-1.695	-0.364

$\displaystyle\lim_{x \to -3^-} f(x) = \infty \qquad \lim_{x \to -3^+} f(x) = -\infty$

7. $f(x) = \dfrac{x^2}{x^2 - 9}$

x	-3.5	-3.1	-3.01	-3.001	-2.999	-2.99	-2.9	-2.5
$f(x)$	3.769	15.75	150.8	1501	-1499	-149.3	-14.25	-2.273

$\displaystyle\lim_{x \to -3^-} f(x) = \infty \qquad \lim_{x \to -3^+} f(x) = -\infty$

9. $\lim_{x \to 0^+} \frac{1}{x^2} = \infty = \lim_{x \to 0^-} \frac{1}{x^2}$

Therefore, $x = 0$ is a vertical asymptote.

11. $\lim_{x \to 2^+} \frac{x^2 - 2}{(x - 2)(x + 1)} = \infty$

$\lim_{x \to 2^-} \frac{x^2 - 2}{(x - 2)(x + 1)} = -\infty$

Therefore, $x = 2$ is a vertical asymptote.

$\lim_{x \to -1^+} \frac{x^2 - 2}{(x - 2)(x + 1)} = \infty$

$\lim_{x \to -1^-} \frac{x^2 - 2}{(x - 2)(x + 1)} = -\infty$

Therefore, $x = -1$ is a vertical asymptote.

13. $\lim_{x \to -2^-} \frac{x^2}{x^2 - 4} = \infty$ and $\lim_{x \to -2^+} \frac{x^2}{x^2 - 4} = -\infty$

Therefore, $x = -2$ is a vertical asymptote.

$\lim_{x \to 2^-} \frac{x^2}{x^2 - 4} = -\infty$ and $\lim_{x \to 2^+} \frac{x^2}{x^2 - 4} = \infty$

Therefore, $x = 2$ is a vertical asymptote.

15. No vertical asymptote since the denominator is never zero.

17. $f(x) = \tan 2x = \frac{\sin 2x}{\cos 2x}$ has vertical asymptotes at

$x = \frac{(2n + 1)\pi}{4} = \frac{\pi}{4} + \frac{n\pi}{2}$, n any integer.

19. $\lim_{t \to 0^+} \left(1 - \frac{4}{t^2}\right) = -\infty = \lim_{t \to 0^-} \left(1 - \frac{4}{t^2}\right)$

Therefore, $t = 0$ is a vertical asymptote.

21. $\lim_{x \to -2^+} \frac{x}{(x + 2)(x - 1)} = \infty$

$\lim_{x \to -2^-} \frac{x}{(x + 2)(x - 1)} = -\infty$

Therefore, $x = -2$ is a vertical asymptote.

$\lim_{x \to 1^+} \frac{x}{(x + 2)(x - 1)} = \infty$

$\lim_{x \to 1^-} \frac{x}{(x + 2)(x - 1)} = -\infty$

Therefore, $x = 1$ is a vertical asymptote.

23. $f(x) = \frac{x^3 + 1}{x + 1} = \frac{(x + 1)(x^2 - x + 1)}{x + 1}$

has no vertical asymptote since

$\lim_{x \to -1} f(x) = \lim_{x \to -1} (x^2 - x + 1) = 3$

25. $f(x) = \frac{e^{-2x}}{x - 1}$.

$x = 1$ is a vertical asymptote.

27. $h(t) = \frac{\ln(t^2 + 1)}{t + 2}$.

$t = -2$ is a vertical asymptote.

29. $f(x) = \frac{1}{e^x - 1}$.

$x = 0$ is a vertical asymptote.

31. $s(t) = \frac{t}{\sin t}$ has vertical asymptotes at $t = n\pi$, n a nonzero integer. There is no vertical asymptote at $t = 0$ since

$\lim_{t \to 0} \frac{t}{\sin t} = 1.$

33. $\lim_{x \to -1} \frac{x^2 - 1}{x + 1} = \lim_{x \to -1} (x - 1) = -2$

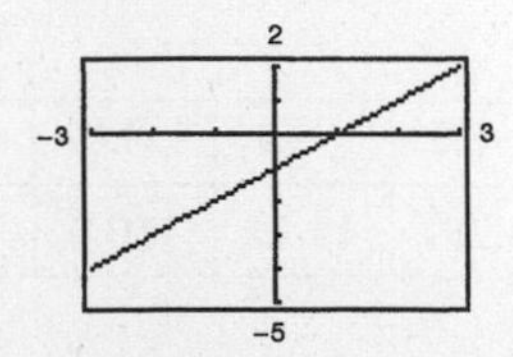

Removable discontinuity at $x = -1$

35. $\lim_{x \to -1^+} \frac{x^2 + 1}{x + 1} = \infty$

$\lim_{x \to -1^-} \frac{x^2 + 1}{x + 1} = -\infty$

Vertical asymptote at $x = -1$

37. $f(x) = \frac{e^{2(x+1)} - 1}{e^{x+1} - 1}$

$= \frac{(e^{x+1} - 1)(e^{x+1} + 1)}{e^{x+1} - 1} = e^{x+1}, \quad x \neq -1$

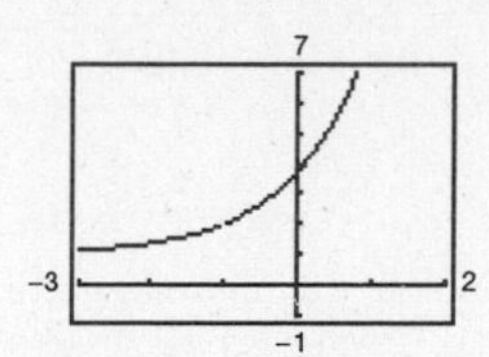

Removable discontinuity at $x = -1$

39. $\lim_{x \to 2^+} \frac{x - 3}{x - 2} = -\infty$

41. $\lim_{x \to 3^+} \frac{x^2}{(x - 3)(x + 3)} = \infty$

43. $\lim_{x \to -3^-} \frac{x^2 + 2x - 3}{x^2 + x - 6} = \lim_{x \to -3^-} \frac{(x - 1)(x + 3)}{(x - 2)(x + 3)}$

$= \lim_{x \to -3^-} \frac{x - 1}{x - 2}$

$= \frac{4}{5}$

45. $\lim_{x \to 1} \frac{x^2 - x}{(x^2 + 1)(x - 1)} = \lim_{x \to 1} \frac{x}{x^2 + 1}$

$= \frac{1}{2}$

47. $\lim_{x \to 0^-} \left(1 + \frac{1}{x}\right) = -\infty$

49. $\lim_{x \to 0^+} \frac{2}{\sin x} = \infty$

51. $\lim_{x \to (\pi/2)^-} \ln|\cos x| = \ln\left|\cos \frac{\pi}{2}\right| = \ln 0 = -\infty$

53. $\lim_{x \to (1/2)^-} x \sec(\pi x) = \infty$ and $\lim_{x \to (1/2)^+} x \sec(\pi x) = -\infty$.

Therefore, $\lim_{x \to (1/2)} x \sec(\pi x)$ does not exist.

55. $f(x) = \frac{x^2 + x + 1}{x^3 - 1}$

$\lim_{x \to 1^+} f(x) = \lim_{x \to 1^+} \frac{1}{x - 1} = \infty$

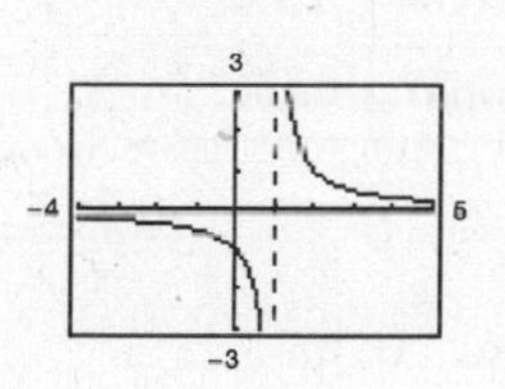

57. $f(x) = \frac{1}{x^2 - 25}$

$\lim_{x \to 5^-} f(x) = -\infty$

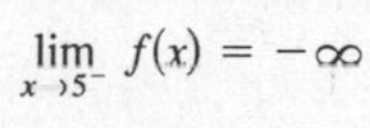

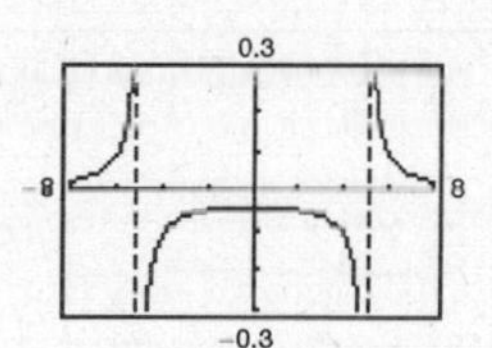

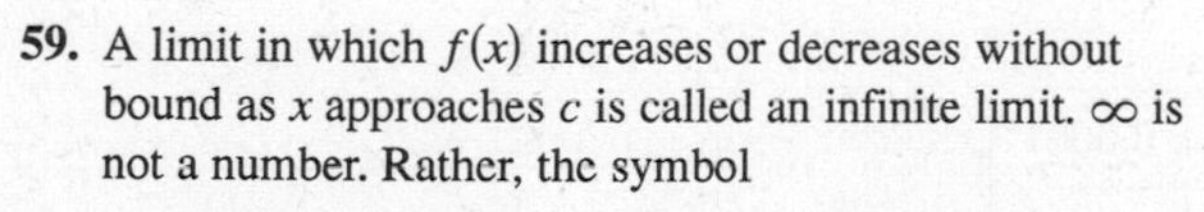

59. A limit in which $f(x)$ increases or decreases without bound as x approaches c is called an infinite limit. ∞ is not a number. Rather, the symbol

$\lim_{x \to c} f(x) = \infty$

says how the limit fails to exist.

61. One answer is $f(x) = \frac{x - 3}{(x - 6)(x + 2)} = \frac{x - 3}{x^2 - 4x - 12}$.

63.

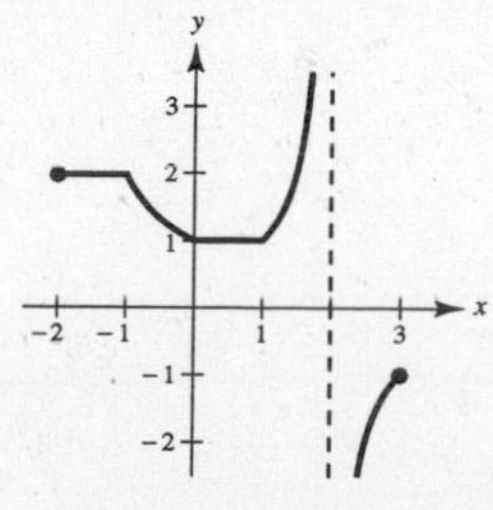

65. $S = \frac{k}{1 - r}$, $0 < |r| < 1$. Assume $k \neq 0$.

$\lim_{r \to 1^-} S = \lim_{r \to 1^-} \frac{k}{1 - r} = \infty \quad$ (or $-\infty$ if $k < 0$)

67. $C = \dfrac{528x}{100 - x}, \quad 0 \le x < 100$

(a) $C(25) = \$176$ million

(b) $C(50) = \$528$ million

(c) $C(75) = \$1584$ million

(d) $\displaystyle\lim_{x\to 100^-} \frac{528}{100 - x} = \infty;$

Thus, it is not possible.

69. (a) $r = \dfrac{2(7)}{\sqrt{625 - 49}} = \dfrac{7}{12}$ ft/sec

(b) $r = \dfrac{2(15)}{\sqrt{625 - 225}} = \dfrac{3}{2}$ ft/sec

(c) $\displaystyle\lim_{x\to 25^-} \frac{2x}{\sqrt{625 - x^2}} = \infty$

71. (a)

x	1	0.5	0.2	0.1	0.01	0.001	0.0001
$f(x)$	0.1585	0.0411	0.0067	0.0017	≈0	≈0	≈0

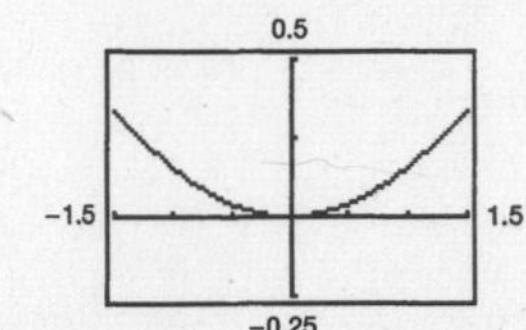

$\displaystyle\lim_{x\to 0^+} \frac{x - \sin x}{x} = 0$

(b)

x	1	0.5	0.2	0.1	0.01	0.001	0.0001
$f(x)$	0.1585	0.0823	0.0333	0.0167	0.0017	≈0	≈0

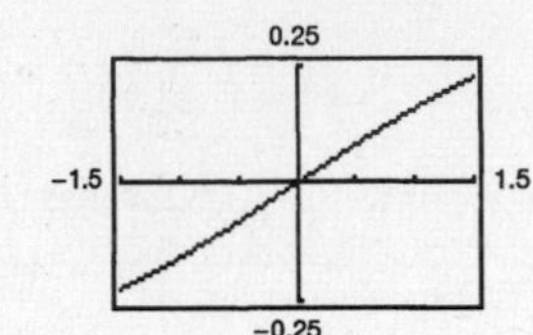

$\displaystyle\lim_{x\to 0^+} \frac{x - \sin x}{x^2} = 0$

(c)

x	1	0.5	0.2	0.1	0.01	0.001	0.0001
$f(x)$	0.1585	0.1646	0.1663	0.1666	0.1667	0.1667	0.1667

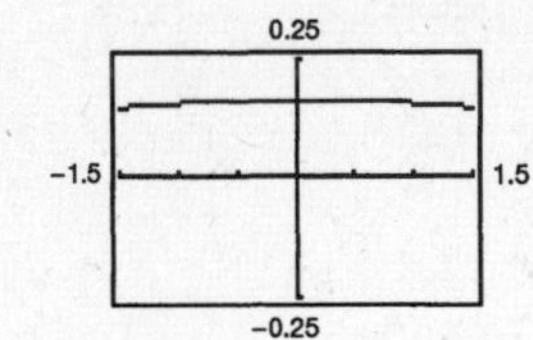

$\displaystyle\lim_{x\to 0^+} \frac{x - \sin x}{x^3} = 0.1667 \quad (1/6)$

(d)

x	1	0.5	0.2	0.1	0.01	0.001	0.0001
$f(x)$	0.1585	0.3292	0.8317	1.6658	16.67	166.7	1667.0

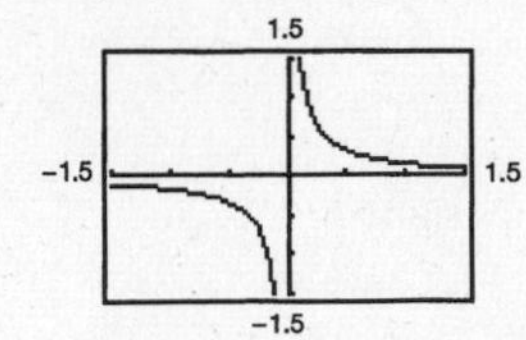

$\displaystyle\lim_{x\to 0^+} \frac{x - \sin x}{x^4} = \infty$

For $n \ge 3$, $\displaystyle\lim_{x\to 0^+} \frac{x - \sin x}{x^n} = \infty.$

73. (a) Because the circumference of the motor is half that of the saw arbor, the saw makes $1700/2 = 850$ revolutions per minute.

(b) The direction of rotation is reversed.

(c) $2(20 \cot \phi) + 2(10 \cot \phi)$: straight sections. The angle subtended in each circle is

$$2\pi - \left(2\left(\frac{\pi}{2} - \phi\right)\right) = \pi + 2\phi.$$

Thus, the length of the belt around the pulleys is

$$20(\pi + 2\phi) + 10(\pi + 2\phi) = 30(\pi + 2\phi).$$

Total length $= 60 \cot \phi + 30(\pi + 2\phi)$

Domain: $\left(0, \frac{\pi}{2}\right)$

(d)

ϕ	0.3	0.6	0.9	1.2	1.5
L	306.2	217.9	195.9	189.6	188.5

(e)

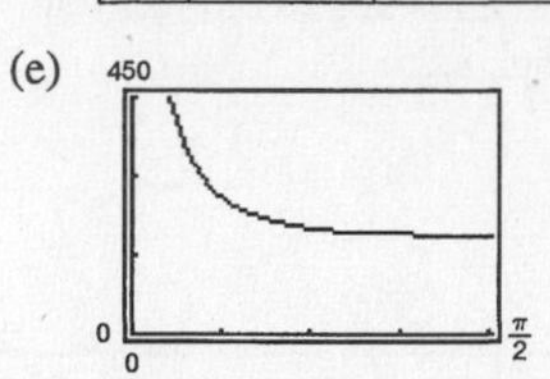

(f) $\lim_{\phi \to (\pi/2)^-} L = 60\pi \approx 188.5$ (All the belts are around pulleys.)

(g) $\lim_{\phi \to 0^+} L = \infty$

75. False; for instance, let

$$f(x) = \frac{x^2 - 1}{x - 1} \text{ or}$$

$$g(x) = \frac{x}{x^2 + 1}.$$

77. False; let

$$f(x) = \begin{cases} \frac{1}{x}, & x \neq 0 \\ 3, & x = 0. \end{cases}$$

The graph of f has a vertical asymptote at $x = 0$, but $f(0) = 3$.

79. Given $\lim_{x \to c} f(x) = \infty$ and $\lim_{x \to c} g(x) = L$:

(2) Product:

If $L > 0$, then for $\epsilon = L/2 > 0$ there exists $\delta_1 > 0$ such that $|g(x) - L| < L/2$ whenever $0 < |x - c| < \delta_1$. Thus, $L/2 < g(x) < 3L/2$. Since $\lim_{x \to c} f(x) = \infty$ then for $M > 0$, there exists $\delta_2 > 0$ such that $f(x) > M(2/L)$ whenever $|x - c| < \delta_2$. Let δ be the smaller of δ_1 and δ_2. Then for $0 < |x - c| < \delta$, we have $f(x)g(x) > M(2/L)(L/2) = M$. Therefore $\lim_{x \to c} f(x)g(x) = \infty$. The proof is similar for $L < 0$.

(3) Quotient: Let $\epsilon > 0$ be given.

There exists $\delta_1 > 0$ such that $f(x) > 3L/2\epsilon$ whenever $0 < |x - c| < \delta_1$ and there exists $\delta_2 > 0$ such that $|g(x) - L| < L/2$ whenever $0 < |x - c| < \delta_2$. This inequality gives us $L/2 < g(x) < 3L/2$. Let δ be the smaller of δ_1 and δ_2. Then for $0 < |x - c| < \delta$, we have

$$\left|\frac{g(x)}{f(x)}\right| < \frac{3L/2}{3L/2\epsilon} = \epsilon.$$

Therefore, $\lim_{x \to c} \frac{g(x)}{f(x)} = 0$.

81. Given $\lim_{x \to c} \frac{1}{f(x)} = 0$, suppose $\lim_{x \to c} f(x)$ exists and equals L. Then,

$$\lim_{x \to c} \frac{1}{f(x)} = \frac{\lim_{x \to c} 1}{\lim_{x \to c} f(x)} = \frac{1}{L} = 0.$$

This is not possible. Thus, $\lim_{x \to c} f(x)$ does not exist.

Review Exercises for Chapter 1

1. Calculus required. Using a graphing utility, you can estimate the length to be 8.3. Or, the length is slightly longer than the distance between the two points, 8.25.

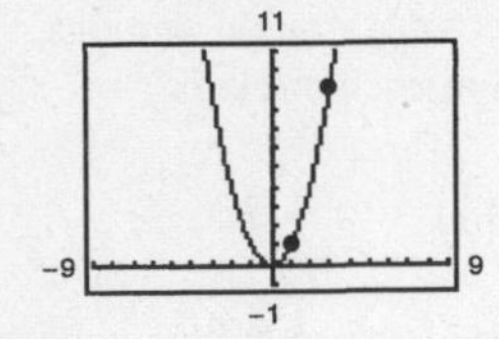

3.

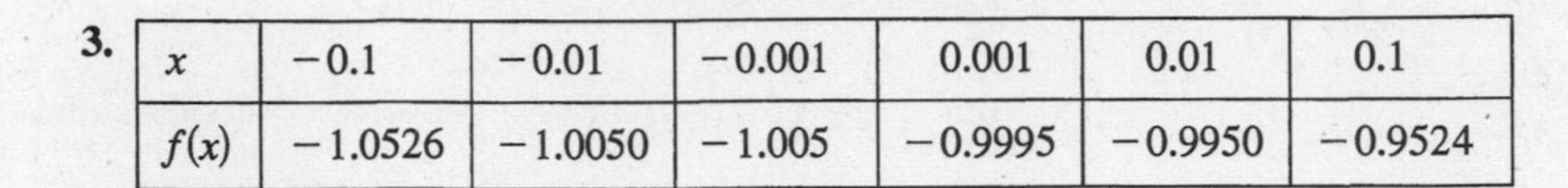

x	-0.1	-0.01	-0.001	0.001	0.01	0.1
$f(x)$	-1.0526	-1.0050	-1.005	-0.9995	-0.9950	-0.9524

$\lim_{x\to 0} f(x) \approx -1$

5.

x	-0.1	-0.01	-0.001	0.001	0.01	0.1
$f(x)$	0.8867	0.0988	0.0100	-0.0100	-0.1013	-1.1394

$\lim_{x\to 0} f(x) = 0$

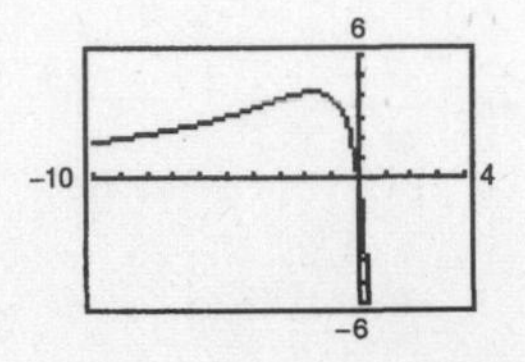

7. $h(x) = \dfrac{x^2 - 2x}{x}$

(a) $\lim_{x\to 0} h(x) = -2$

(b) $\lim_{x\to -1} h(x) = -3$

9. (a) $\lim_{t\to 0} f(t)$ does not exist.

(b) $\lim_{t\to -1} f(t) = 0$

11. $\lim_{x\to 1} (3 - x) = 3 - 1 = 2$

Let $\epsilon > 0$ be given. Choose $\delta = \epsilon$. Then for $0 < |x - 1| < \delta = \epsilon$, you have:

$$|x - 1| < \epsilon$$
$$|1 - x| < \epsilon$$
$$|(3 - x) - 2| < \epsilon$$
$$|f(x) - L| < \epsilon$$

13. $\lim_{x\to 2} (x^2 - 3) = 1$ Let $\epsilon > 0$ be given. We need

$$|x^2 - 3 - 1| < \epsilon \Rightarrow |x^2 - 4| = |(x - 2)(x + 2)| < \epsilon \Rightarrow |x - 2| < \frac{1}{|x + 2|}\epsilon.$$

Assuming, $1 < x < 3$, you can choose $\delta = \epsilon/5$. Hence, for $0 < |x - 2| < \delta = \epsilon/5$ you have:

$$|x - 2| < \frac{\epsilon}{5} < \frac{1}{|x + 2|}\epsilon$$
$$|x - 2||x + 2| < \epsilon$$
$$|x^2 - 4| < \epsilon$$
$$|(x^2 - 3) - 1| < \epsilon$$
$$|f(x) - L| < \epsilon$$

15. $\lim_{t\to 4} \sqrt{t + 2} = \sqrt{4 + 2} = \sqrt{6} \approx 2.45$

17. $\lim_{t\to -2} \dfrac{t + 2}{t^2 - 4} = \lim_{t\to -2} \dfrac{1}{t - 2} = -\dfrac{1}{4}$

19. $$\lim_{x\to 4} \frac{\sqrt{x} - 2}{x - 4} = \lim_{x\to 4} \frac{\sqrt{x} - 2}{(\sqrt{x} - 2)(\sqrt{x} + 2)}$$
$$= \lim_{x\to 4} \frac{1}{\sqrt{x} + 2} = \frac{1}{\sqrt{4} + 2} = \frac{1}{4}$$

21. $$\lim_{x\to 0} \frac{[1/(x + 1)] - 1}{x} = \lim_{x\to 0} \frac{1 - (x + 1)}{x(x + 1)}$$
$$= \lim_{x\to 0} \frac{-1}{x + 1} = -1$$

23. $\lim_{x\to-5}\dfrac{x^3+125}{x+5} = \lim_{x\to-5}\dfrac{(x+5)(x^2-5x+25)}{x+5}$

$= \lim_{x\to-5}(x^2-5x+25)$

$= 75$

25. $\lim_{x\to0}\dfrac{1-\cos x}{\sin x} = \lim_{x\to0}\left(\dfrac{x}{\sin x}\right)\left(\dfrac{1-\cos x}{x}\right) = (1)(0) = 0$

27. $\lim_{\Delta x\to0}\dfrac{\sin[(\pi/6)+\Delta x]-(1/2)}{\Delta x} = \lim_{\Delta x\to0}\dfrac{\sin(\pi/6)\cos\Delta x+\cos(\pi/6)\sin\Delta x-(1/2)}{\Delta x}$

$= \lim_{\Delta x\to0}\dfrac{1}{2}\cdot\dfrac{(\cos\Delta x-1)}{\Delta x} + \lim_{\Delta x\to0}\dfrac{\sqrt{3}}{2}\cdot\dfrac{\sin\Delta x}{\Delta x}$

$= 0 + \dfrac{\sqrt{3}}{2}(1) = \dfrac{\sqrt{3}}{2}$

29. $\lim_{x\to1} e^{x-1}\sin\dfrac{\pi x}{2} = e^0\sin\dfrac{\pi}{2} = 1$

31. $\lim_{x\to c}[f(x)\cdot g(x)] = \left(-\dfrac{3}{4}\right)\left(\dfrac{2}{3}\right) = -\dfrac{1}{2}$

33. $f(x) = \dfrac{\sqrt{2x+1}-\sqrt{3}}{x-1}$

(a)

x	1.1	1.01	1.001	1.0001
$f(x)$	0.5680	0.5764	0.5773	0.5773

$\lim_{x\to1^+}\dfrac{\sqrt{2x+1}-\sqrt{3}}{x-1} \approx 0.577$ (Actual limit is $\sqrt{3}/3$.)

(b)

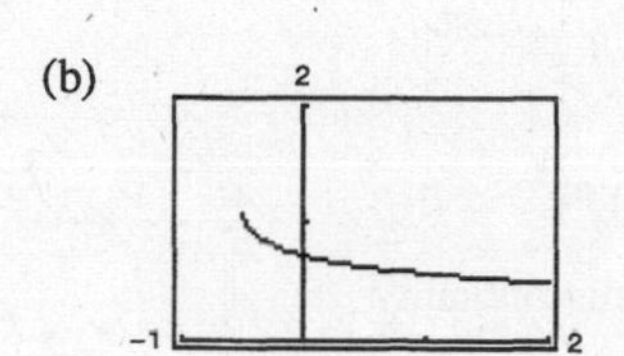

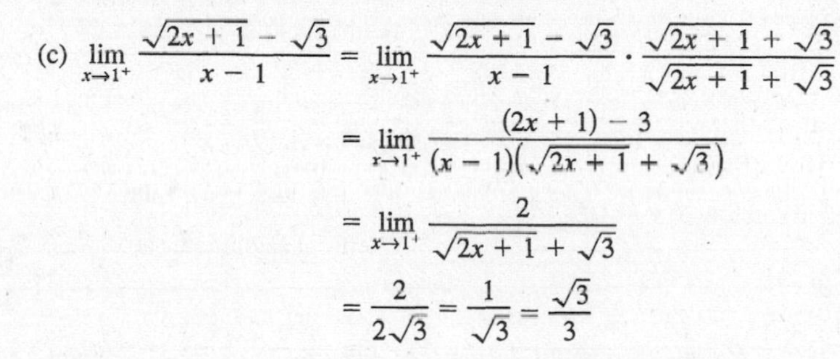

(c) $\lim_{x\to1^+}\dfrac{\sqrt{2x+1}-\sqrt{3}}{x-1} = \lim_{x\to1^+}\dfrac{\sqrt{2x+1}-\sqrt{3}}{x-1}\cdot\dfrac{\sqrt{2x+1}+\sqrt{3}}{\sqrt{2x+1}+\sqrt{3}}$

$= \lim_{x\to1^+}\dfrac{(2x+1)-3}{(x-1)\left(\sqrt{2x+1}+\sqrt{3}\right)}$

$= \lim_{x\to1^+}\dfrac{2}{\sqrt{2x+1}+\sqrt{3}}$

$= \dfrac{2}{2\sqrt{3}} = \dfrac{1}{\sqrt{3}} = \dfrac{\sqrt{3}}{3}$

35. $\lim_{t\to a}\dfrac{s(a)-s(t)}{a-t} = \lim_{t\to4}\dfrac{(-4.9(4)^2+200)-(-4.9t^2+200)}{4-t}$

$= \lim_{t\to4}\dfrac{4.9(t-4)(t+4)}{4-t}$

$= \lim_{t\to4} -4.9(t+4) = -39.2$ m/sec

37. $\lim_{x\to3^-}\dfrac{|x-3|}{x-3} = \lim_{x\to3^-}\dfrac{-(x-3)}{x-3}$

$= -1$

39. $\lim_{x\to2} f(x) = 0$

41. $\lim_{t\to1} h(t)$ does not exist because $\lim_{t\to1^-} h(t) = 1+1 = 2$ and $\lim_{t\to1^+} h(t) = \frac{1}{2}(1+1) = 1$.

43. $f(x) = [\![x+3]\!]$

$\lim_{x\to k^+}[\![x+3]\!] = k+3$ where k is an integer.

$\lim_{x\to k^-}[\![x+3]\!] = k+2$ where k is an integer.

Nonremovable discontinuity at each integer k

Continuous on $(k, k+1)$ for all integers k

45. $f(x) = \dfrac{3x^2-x-2}{x-1} = \dfrac{(3x+2)(x-1)}{x-1}$

$\lim_{x\to1} f(x) = \lim_{x\to1}(3x+2) = 5$

Removable discontinuity at $x = 1$

Continuous on $(-\infty, 1)\cup(1, \infty)$

47. $f(x) = \dfrac{1}{(x-2)^2}$

$$\lim_{x\to 2} \frac{1}{(x-2)^2} = \infty$$

Nonremovable discontinuity at $x = 2$
Continuous on $(-\infty, 2) \cup (2, \infty)$

49. $f(x) = \dfrac{3}{x+1}$

$$\lim_{x\to -1^-} f(x) = -\infty$$
$$\lim_{x\to -1^+} f(x) = \infty$$

Nonremovable discontinuity at $x = -1$
Continuous on $(-\infty, -1) \cup (-1, \infty)$

51. $f(x) = \csc \dfrac{\pi x}{2}$

Nonremovable discontinuities at each even integer.
Continuous on $(2k, 2k + 2)$ for all integers k.

53. $g(x) = 2e^{[\![x]\!]/4}$ is continuous on all intervals $(n, n + 1)$, where n is an integer. g has nonremovable discontinuities at each n.

55. $f(2) = 5$

Find c so that $\lim_{x\to 2^+} (cx + 6) = 5$.

$$c(2) + 6 = 5$$
$$2c = -1$$
$$c = -\tfrac{1}{2}$$

57. f is continuous on $[1, 2]$. $f(1) = -1 < 0$ and $f(2) = 13 > 0$. Therefore by the Intermediate Value Theorem, there is at least one value c in $(1, 2)$ such that $2c^3 - 3 = 0$.

59. $A = 5000(1.06)^{[\![2t]\!]}$

Nonremovable discontinuity every 6 months

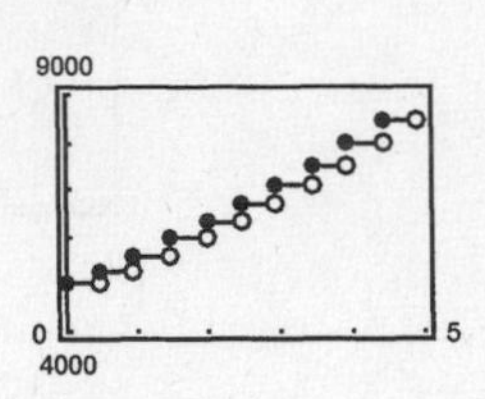

61. $g(x) = 1 + \dfrac{2}{x}$

Vertical asymptote at $x = 0$

63. $f(x) = \dfrac{8}{(x-10)^2}$

Vertical asymptote at $x = 10$

65. $g(x) = \ln(9 - x^2)$
$= \ln(3 - x) + \ln(3 + x)$

Vertical asymptotes at $x = \pm 3$

67. $\displaystyle\lim_{x\to -2^-} \frac{2x^2 + x + 1}{x + 2} = -\infty$

69. $\displaystyle\lim_{x\to -1^+} \frac{x+1}{x^3+1} = \lim_{x\to -1^+} \frac{1}{x^2 - x + 1} = \frac{1}{3}$

71. $\displaystyle\lim_{x\to 1^-} \frac{x^2 + 2x + 1}{x - 1} = -\infty$

73. $\displaystyle\lim_{x\to 0^+} \frac{\sin 4x}{5x} = \lim_{x\to 0^+} \left[\frac{4}{5}\left(\frac{\sin 4x}{4x}\right)\right] = \frac{4}{5}$

75. $\displaystyle\lim_{x\to 0^+} \frac{\csc 2x}{x} = \lim_{x\to 0^+} \frac{1}{x \sin 2x} = \infty$

77. $\displaystyle\lim_{x\to 0^+} \ln(\sin x) = -\infty$

79. $f(x) = \dfrac{\tan 2x}{x}$

(a) $\displaystyle\lim_{x\to 0} \frac{\tan 2x}{x} = 2$

x	-0.1	-0.01	-0.001	0.001	0.01	0.1
$f(x)$	2.0271	2.0003	2.0000	2.0000	2.0003	2.0271

Analytically, $\dfrac{\tan 2x}{x} = \dfrac{\sin 2x}{2x} \cdot \dfrac{2}{\cos 2x} \to 2$

(b) Yes, define:

$$f(x) = \begin{cases} \dfrac{\tan 2x}{x}, & x \neq 0 \\ 2, & x = 0 \end{cases}$$

Now $f(x)$ is continuous at $x = 0$.

Problem Solving for Chapter 1

1. (a) Perimeter $\triangle PAO = \sqrt{x^2 + (y-1)^2} + \sqrt{x^2 + y^2} + 1$

$$= \sqrt{x^2 + (x^2-1)^2} + \sqrt{x^2 + x^4} + 1$$

Perimeter $\triangle PBO = \sqrt{(x-1)^2 + y^2} + \sqrt{x^2 + y^2} + 1$

$$= \sqrt{(x-1)^2 + x^4} + \sqrt{x^2 + x^4} + 1$$

(b) $r(x) = \dfrac{\sqrt{x^2 + (x^2-1)^2} + \sqrt{x^2 + x^4} + 1}{\sqrt{(x-1)^2 + x^4} + \sqrt{x^2 + x^4} + 1}$

x	4	2	1	0.1	0.01
Perimeter $\triangle PAO$	33.02	9.08	3.41	2.10	2.01
Perimeter $\triangle PBO$	33.77	9.60	3.41	2.00	2.00
$r(x)$	0.98	0.95	1	1.05	1.005

(c) $\lim_{x\to 0^+} r(x) = \dfrac{1+0+1}{1+0+1} = \dfrac{2}{2} = 1$

3. (a) There are 6 triangles, each with a central angle of $60° = \pi/3$. Hence,

Area hexagon $= 6\left[\frac{1}{2}bh\right] = 6\left[\frac{1}{2}(1)\sin\frac{\pi}{3}\right]$

$$= \frac{3\sqrt{3}}{2} \approx 2.598.$$

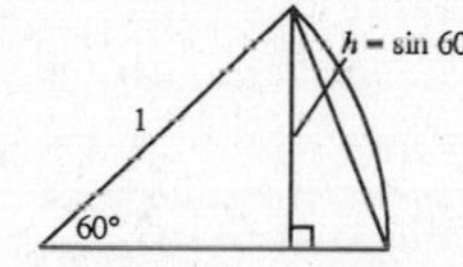

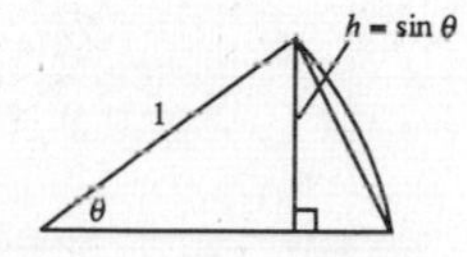

Error: $\pi - \dfrac{3\sqrt{3}}{2} \approx 0.5435.$

(b) There are n triangles, each with central angle of $\theta = 2\pi/n$. Hence,

$$A_n = n\left[\frac{1}{2}bh\right] = n\left[\frac{1}{2}(1)\sin\frac{2\pi}{n}\right] = \frac{n\sin(2\pi/n)}{2}.$$

(c)

n	6	12	24	48	96
A_n	2.598	3	3.106	3.133	3.139

(d) As n gets larger and larger, $2\pi/n$ approaches 0.

Letting $x = 2\pi/n$,

$$A_n = \frac{\sin(2\pi/n)}{2/n} = \frac{\sin(2\pi/n)}{(2\pi/n)}\pi = \frac{\sin x}{x}\pi$$

which approaches $(1)\pi = \pi$.

5. (a) Slope $= -\dfrac{12}{5}$

(b) Slope of tangent line is $\dfrac{5}{12}$.

$$y + 12 = \frac{5}{12}(x - 5)$$

$$y = \frac{5}{12}x - \frac{169}{12} \quad \text{Tangent line}$$

(c) $Q = (x, y) = \left(x, -\sqrt{169 - x^2}\right)$

$$m_x = \frac{-\sqrt{169 - x^2} + 12}{x - 5}$$

(d) $\lim_{x\to 5} m_x = \lim_{x\to 5} \dfrac{12 - \sqrt{169 - x^2}}{x - 5} \cdot \dfrac{12 + \sqrt{169 - x^2}}{12 + \sqrt{169 - x^2}}$

$$= \lim_{x\to 5} \frac{144 - (169 - x^2)}{(x-5)\left(12 + \sqrt{169 - x^2}\right)}$$

$$= \lim_{x\to 5} \frac{x^2 - 25}{(x-5)\left(12 + \sqrt{169 - x^2}\right)}$$

$$= \lim_{x\to 5} \frac{(x+5)}{12 + \sqrt{169 - x^2}}$$

$$= \frac{10}{12 + 12} = \frac{5}{12}$$

This is the same slope as part (b).

7. (a) $3 + x^{1/3} \geq 0$

$x^{1/3} \geq -3$

$x \geq -27$

Domain: $x \geq -27, x \neq 1$

(b)

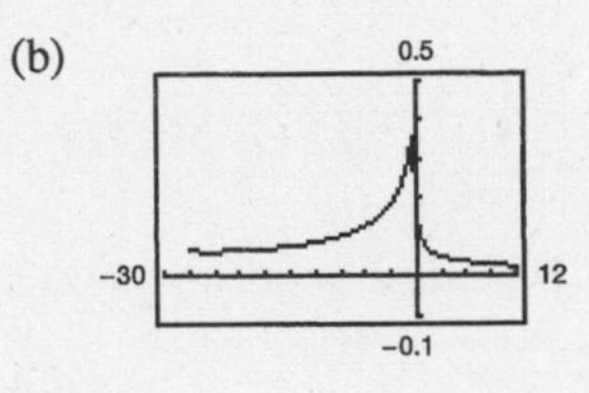

(c) $\lim_{x \to -27^+} f(x) = \dfrac{\sqrt{3 + (-27)^{1/3}} - 2}{-27 - 1}$

$= \dfrac{-2}{-28} = \dfrac{1}{14} \approx 0.0714$

(d) $\lim_{x \to 1} f(x) = \lim_{x \to 1} \dfrac{\sqrt{3 + x^{1/3}} - 2}{x - 1} \cdot \dfrac{\sqrt{3 + x^{1/3}} + 2}{\sqrt{3 + x^{1/3}} + 2}$

$= \lim_{x \to 1} \dfrac{3 + x^{1/3} - 4}{(x - 1)\left(\sqrt{3 + x^{1/3}} + 2\right)}$

$= \lim_{x \to 1} \dfrac{x^{1/3} - 1}{(x^{1/3} - 1)(x^{2/3} + x^{1/3} + 1)\left(\sqrt{3 + x^{1/3}} + 2\right)}$

$= \lim_{x \to 1} \dfrac{1}{(x^{2/3} + x^{1/3} + 1)\left(\sqrt{3 + x^{1/3}} + 2\right)}$

$= \dfrac{1}{(1 + 1 + 1)(2 + 2)} = \dfrac{1}{12}$

9. (a) $\lim_{x \to 2} f(x) = 3$: g_1, g_4

(b) f continuous at 2: g_1

(c) $\lim_{x \to 2^-} f(x) = 3$: g_1, g_3, g_4

11.

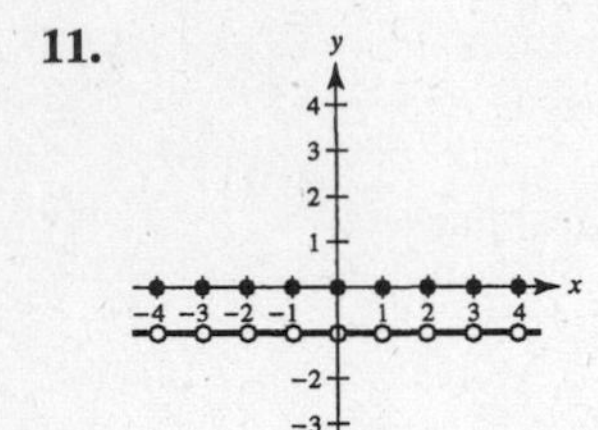

(a) $f(1) = [\![1]\!] + [\![-1]\!] = 1 + (-1) = 0$

$f(0) = 0$

$f\left(\frac{1}{2}\right) = 0 + (-1) = -1$

$f(-2.7) = -3 + 2 = -1$

(b) $\lim_{x \to 1^-} f(x) = -1$

$\lim_{x \to 1^+} f(x) = -1$

$\lim_{x \to 1/2} f(x) = -1$

(c) f is continuous for all real numbers except

$x = 0, \pm 1, \pm 2, \pm 3, \ldots$

13. (a)

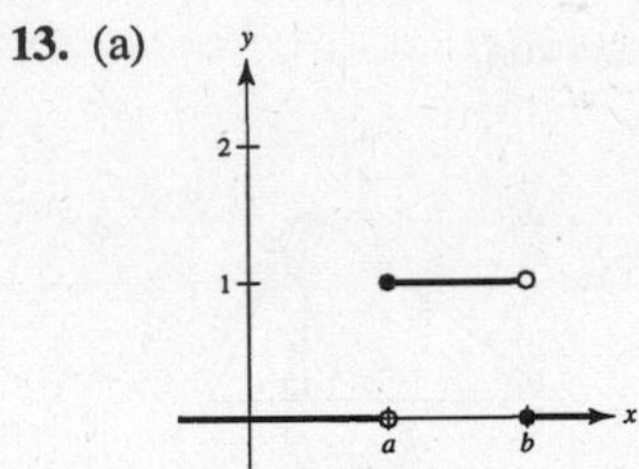

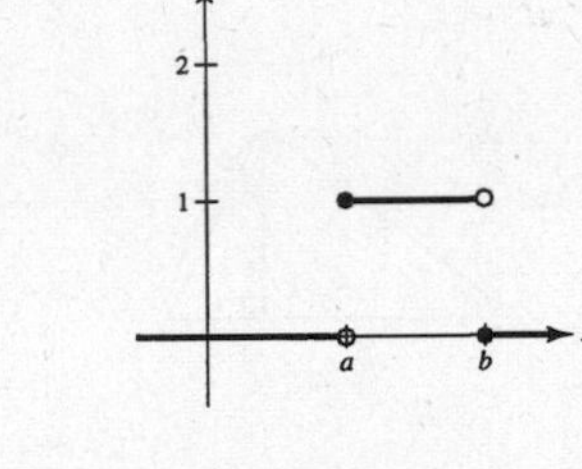

(b) (i) $\lim_{x \to a^+} P_{a,b}(x) = 1$

(ii) $\lim_{x \to a^-} P_{a,b}(x) = 0$

(iii) $\lim_{x \to b^+} P_{a,b}(x) = 0$

(iv) $\lim_{x \to b^-} P_{a,b}(x) = 1$

(c) $P_{a,b}$ is continuous for all positive real numbers except $x = a, b$.

(d) The area under the graph of u, and above the x-axis, is 1.

CHAPTER 2
Differentiation

CHAPTER 2
Differentiation

Section 2.1 The Derivative and the Tangent Line Problem

Solutions to Odd-Numbered Exercises

1. (a) $m = 0$

(b) $m = -3$

3. (a), (b)

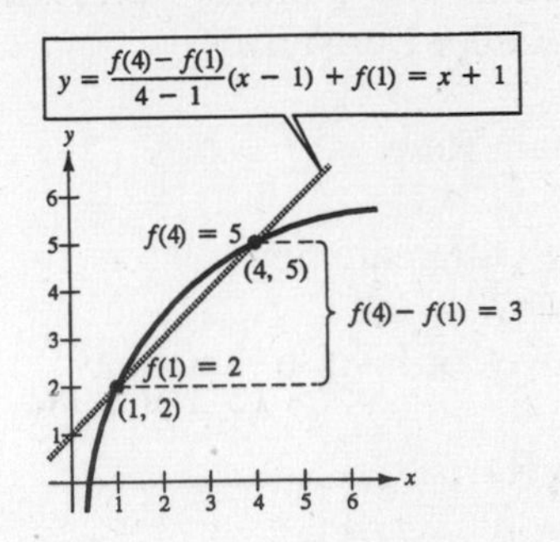

(c) $y = \dfrac{f(4) - f(1)}{4 - 1}(x - 1) + f(1)$

$= \dfrac{3}{3}(x - 1) + 2$

$= 1(x - 1) + 2$

$= x + 1$

5. $f(x) = 3 - 2x$ is a line.

Slope $= -2$

7. Slope at $(1, -3) = \lim\limits_{\Delta x \to 0} \dfrac{g(1 + \Delta x) - g(1)}{\Delta x}$

$= \lim\limits_{\Delta x \to 0} \dfrac{(1 + \Delta x)^2 - 4 - (-3)}{\Delta x}$

$= \lim\limits_{\Delta x \to 0} \dfrac{1 + 2(\Delta x) + (\Delta x)^2 - 1}{\Delta x}$

$= \lim\limits_{\Delta x \to 0} [2 + 2(\Delta x)] = 2$

9. Slope at $(0, 0) = \lim\limits_{\Delta t \to 0} \dfrac{f(0 + \Delta t) - f(0)}{\Delta t}$

$= \lim\limits_{\Delta t \to 0} \dfrac{3(\Delta t) - (\Delta t)^2 - 0}{\Delta t}$

$= \lim\limits_{\Delta t \to 0} (3 - \Delta t) = 3$

11. $f(x) = 3$

$f'(x) = \lim\limits_{\Delta x \to 0} \dfrac{f(x + \Delta x) - f(x)}{\Delta x}$

$= \lim\limits_{\Delta x \to 0} \dfrac{3 - 3}{\Delta x}$

$= \lim\limits_{\Delta x \to 0} 0 = 0$

13. $f(x) = -5x$

$f'(x) = \lim\limits_{\Delta x \to 0} \dfrac{f(x + \Delta x) - f(x)}{\Delta x}$

$= \lim\limits_{\Delta x \to 0} \dfrac{-5(x + \Delta x) - (-5x)}{\Delta x}$

$= \lim\limits_{\Delta x \to 0} -5 = -5$

15. $h(s) = 3 + \dfrac{2}{3}s$

$h'(s) = \lim\limits_{\Delta s \to 0} \dfrac{h(s + \Delta s) - h(s)}{\Delta s}$

$= \lim\limits_{\Delta s \to 0} \dfrac{3 + \frac{2}{3}(s + \Delta s) - \left(3 + \frac{2}{3}s\right)}{\Delta s}$

$= \lim\limits_{\Delta s \to 0} \dfrac{\frac{2}{3}\Delta s}{\Delta s} = \dfrac{2}{3}$

17. $f(x) = 2x^2 + x - 1$

$$f'(x) = \lim_{\Delta x \to 0} \frac{f(x + \Delta x) - f(x)}{\Delta x}$$

$$= \lim_{\Delta x \to 0} \frac{[2(x + \Delta x)^2 + (x + \Delta x) - 1] - [2x^2 + x - 1]}{\Delta x}$$

$$= \lim_{\Delta x \to 0} \frac{(2x^2 + 4x\,\Delta x + 2(\Delta x)^2 + x + \Delta x - 1) - (2x^2 + x - 1)}{\Delta x}$$

$$= \lim_{\Delta x \to 0} \frac{4x\,\Delta x + 2(\Delta x)^2 + \Delta x}{\Delta x} = \lim_{\Delta x \to 0} (4x + 2\Delta x + 1) = 4x + 1$$

19. $f(x) = x^3 - 12x$

$$f'(x) = \lim_{\Delta x \to 0} \frac{f(x + \Delta x) - f(x)}{\Delta x}$$

$$= \lim_{\Delta x \to 0} \frac{[(x + \Delta x)^3 - 12(x + \Delta x)] - [x^3 - 12x]}{\Delta x}$$

$$= \lim_{\Delta x \to 0} \frac{x^3 + 3x^2\,\Delta x + 3x(\Delta x)^2 + (\Delta x)^3 - 12x - 12\,\Delta x - x^3 + 12x}{\Delta x}$$

$$= \lim_{\Delta x \to 0} \frac{3x^2\,\Delta x + 3x(\Delta x)^2 + (\Delta x)^3 - 12\,\Delta x}{\Delta x}$$

$$= \lim_{\Delta x \to 0} (3x^2 + 3x\,\Delta x + (\Delta x)^2 - 12) = 3x^2 - 12$$

21. $f(x) = \dfrac{1}{x - 1}$

$$f'(x) = \lim_{\Delta x \to 0} \frac{f(x + \Delta x) - f(x)}{\Delta x}$$

$$= \lim_{\Delta x \to 0} \frac{\dfrac{1}{x + \Delta x - 1} - \dfrac{1}{x - 1}}{\Delta x}$$

$$= \lim_{\Delta x \to 0} \frac{(x - 1) - (x + \Delta x - 1)}{\Delta x(x + \Delta x - 1)(x - 1)}$$

$$= \lim_{\Delta x \to 0} \frac{-\Delta x}{\Delta x(x + \Delta x - 1)(x - 1)}$$

$$= \lim_{\Delta x \to 0} \frac{-1}{(x + \Delta x - 1)(x - 1)}$$

$$= -\frac{1}{(x - 1)^2}$$

23. $f(x) = \sqrt{x + 1}$

$$f'(x) = \lim_{\Delta x \to 0} \frac{f(x + \Delta x) - f(x)}{\Delta x}$$

$$= \lim_{\Delta x \to 0} \frac{\sqrt{x + \Delta x + 1} - \sqrt{x + 1}}{\Delta x} \cdot \left(\frac{\sqrt{x + \Delta x + 1} + \sqrt{x + 1}}{\sqrt{x + \Delta x + 1} + \sqrt{x + 1}}\right)$$

$$= \lim_{\Delta x \to 0} \frac{(x + \Delta x + 1) - (x + 1)}{\Delta x\left[\sqrt{x + \Delta x + 1} + \sqrt{x + 1}\right]}$$

$$= \lim_{\Delta x \to 0} \frac{1}{\sqrt{x + \Delta x + 1} + \sqrt{x + 1}}$$

$$= \frac{1}{\sqrt{x + 1} + \sqrt{x + 1}} = \frac{1}{2\sqrt{x + 1}}$$

25. (a) $f(x) = x^2 + 1$

$$\begin{aligned} f'(x) &= \lim_{\Delta x \to 0} \frac{f(x + \Delta x) - f(x)}{\Delta x} \\ &= \lim_{\Delta x \to 0} \frac{[(x + \Delta x)^2 + 1] - [x^2 + 1]}{\Delta x} \\ &= \lim_{\Delta x \to 0} \frac{2x\,\Delta x + (\Delta x)^2}{\Delta x} \\ &= \lim_{\Delta x \to 0} (2x + \Delta x) = 2x \end{aligned}$$

At (2, 5), the slope of the tangent line is $m = 2(2) = 4$. The equation of the tangent line is

$$\begin{aligned} y - 5 &= 4(x - 2) \\ y - 5 &= 4x - 8 \\ y &= 4x - 3. \end{aligned}$$

(b)

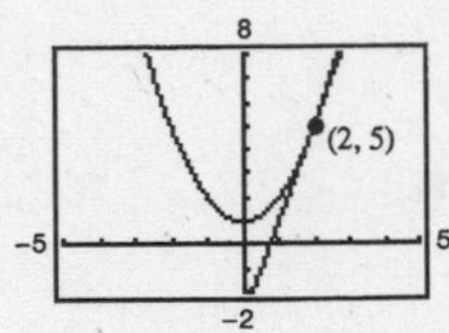

27. (a) $f(x) = x^3$

$$\begin{aligned} f'(x) &= \lim_{\Delta x \to 0} \frac{f(x + \Delta x) - f(x)}{\Delta x} \\ &= \lim_{\Delta x \to 0} \frac{(x + \Delta x)^3 - x^3}{\Delta x} \\ &= \lim_{\Delta x \to 0} \frac{3x^2\,\Delta x + 3x(\Delta x)^2 + (\Delta x)^3}{\Delta x} \\ &= \lim_{\Delta x \to 0} (3x^2 + 3x\,\Delta x + (\Delta x)^2) = 3x^2 \end{aligned}$$

At (2, 8), the slope of the tangent is $m = 3(2)^2 = 12$. The equation of the tangent line is

$$\begin{aligned} y - 8 &= 12(x - 2) \\ y &= 12x - 16. \end{aligned}$$

(b)

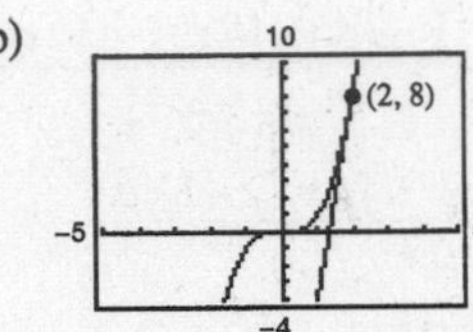

29. (a) $f(x) = \sqrt{x}$

$$\begin{aligned} f'(x) &= \lim_{\Delta x \to 0} \frac{f(x + \Delta x) - f(x)}{\Delta x} \\ &= \lim_{\Delta x \to 0} \frac{\sqrt{x + \Delta x} - \sqrt{x}}{\Delta x} \cdot \frac{\sqrt{x + \Delta x} + \sqrt{x}}{\sqrt{x + \Delta x} + \sqrt{x}} \\ &= \lim_{\Delta x \to 0} \frac{(x + \Delta x) - x}{\Delta x\left(\sqrt{x + \Delta x} + \sqrt{x}\right)} \\ &= \lim_{\Delta x \to 0} \frac{1}{\sqrt{x + \Delta x} + \sqrt{x}} = \frac{1}{2\sqrt{x}} \end{aligned}$$

At (1, 1), the slope of the tangent line is

$$m = \frac{1}{2\sqrt{1}} = \frac{1}{2}.$$

The equation of the tangent line is

$$\begin{aligned} y - 1 &= \frac{1}{2}(x - 1) \\ y &= \frac{1}{2}x + \frac{1}{2}. \end{aligned}$$

(b)

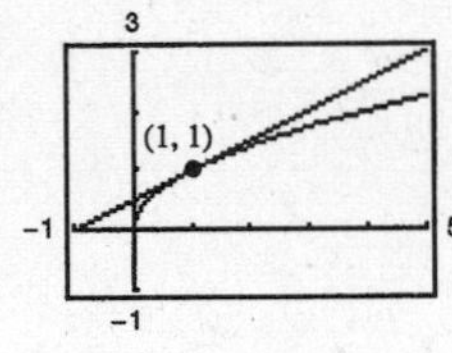

31. (a) $f(x) = x + \frac{4}{x}$

$$f'(x) = \lim_{\Delta x \to 0} \frac{f(x + \Delta x) - f(x)}{\Delta x}$$

$$= \lim_{\Delta x \to 0} \frac{(x + \Delta x) + \frac{4}{x + \Delta x} - \left(x + \frac{4}{x}\right)}{\Delta x}$$

$$= \lim_{\Delta x \to 0} \frac{x(x + \Delta x)(x + \Delta x) + 4x - x^2(x + \Delta x) - 4(x + \Delta x)}{x(\Delta x)(x + \Delta x)}$$

$$= \lim_{\Delta x \to 0} \frac{x^3 + 2x^2(\Delta x) + x(\Delta x)^2 - x^3 - x^2(\Delta x) - 4(\Delta x)}{x(\Delta x)(x + \Delta x)}$$

$$= \lim_{\Delta x \to 0} \frac{x^2(\Delta x) + x(\Delta x)^2 - 4(\Delta x)}{x(\Delta x)(x + \Delta x)}$$

$$= \lim_{\Delta x \to 0} \frac{x^2 + x(\Delta x) - 4}{x(x + \Delta x)}$$

$$= \frac{x^2 - 4}{x^2} = 1 - \frac{4}{x^2}$$

At (4, 5), the slope of the tangent line is:

$$m = 1 - \frac{4}{16} = \frac{3}{4}$$

The equation of the tangent line is:

$$y - 5 = \frac{3}{4}(x - 4)$$

$$y = \frac{3}{4}x + 2$$

(b)

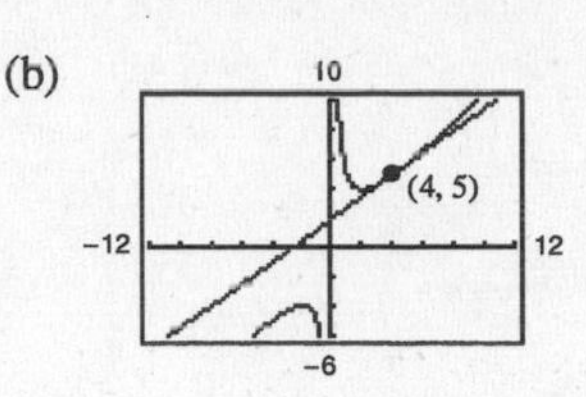

33. From Exercise 27 we know that $f'(x) = 3x^2$. Since the slope of the given line is 3, we have

$$3x^2 = 3$$

$$x = \pm 1.$$

Therefore, at the points (1, 1) and (−1, −1) the tangent lines are parallel to $3x - y + 1 = 0$. These lines have equations

$$y - 1 = 3(x - 1) \quad \text{and} \quad y + 1 = 3(x + 1)$$

$$y = 3x - 2 \qquad\qquad y = 3x + 2.$$

35. Using the limit definition of derivative,

$$f'(x) = \frac{-1}{2x\sqrt{x}}.$$

Since the slope of the given line is $-\frac{1}{2}$, we have

$$-\frac{1}{2x\sqrt{x}} = -\frac{1}{2}$$

$$x = 1.$$

Therefore, at the point (1, 1) the tangent line is parallel to $x + 2y - 6 = 0$. The equation of this line is

$$y - 1 = -\frac{1}{2}(x - 1)$$

$$y - 1 = -\frac{1}{2}x + \frac{1}{2}$$

$$y = -\frac{1}{2}x + \frac{3}{2}.$$

37. $g(5) = 2$ because the tangent line passes through (5, 2).

$$g'(5) = \frac{2 - 0}{5 - 9} = \frac{2}{-4} = -\frac{1}{2}$$

39. $f(x) = x \Rightarrow f'(x) = 1$ (b)

41. $f(x) = \sqrt{x} \Rightarrow f'(x)$ matches (a).

(decreasing slope as $x \to \infty$)

43.

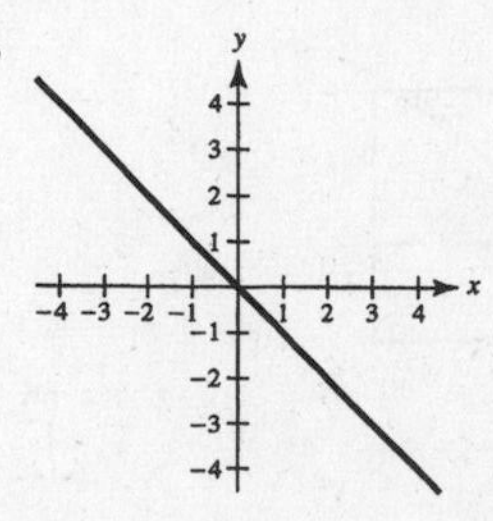

Answers will vary. Sample answer: $y = -x$

45. (a) If $f'(c) = 3$ and f is odd, then $f'(-c) = f'(c) = 3$.

(b) If $f'(c) = 3$ and f is even, then $f'(-c) = -f'(c) = -3$.

47. Let (x_0, y_0) be a point of tangency on the graph of f. By the limit definition for the derivative, $f'(x) = 4 - 2x$. The slope of the line through $(2, 5)$ and (x_0, y_0) equals the derivative of f at x_0:

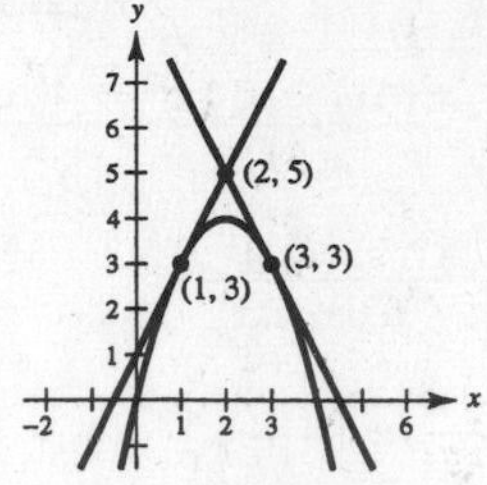

$$\frac{5 - y_0}{2 - x_0} = 4 - 2x_0$$

$$5 - y_0 = (2 - x_0)(4 - 2x_0)$$

$$5 - (4x_0 - x_0^2) = 8 - 8x_0 + 2x_0^2$$

$$0 = x_0^2 - 4x_0 + 3$$

$$0 = (x_0 - 1)(x_0 - 3) \Rightarrow x_0 = 1, 3$$

Therefore, the points of tangency are $(1, 3)$ and $(3, 3)$, and the corresponding slopes are 2 and -2. The equations of the tangent lines are:

$$y - 5 = 2(x - 2) \qquad y - 5 = -2(x - 2)$$

$$y = 2x + 1 \qquad y = -2x + 9$$

49. (a) $g'(0) = -3$

(b) $g'(3) = 0$

(c) Because $g'(1) = -\frac{8}{3}$, g is decreasing (falling) at $x = 1$.

(d) Because $g'(-4) = \frac{7}{3}$, g is increasing (rising) at $x = -4$.

(e) Because $g'(4)$ and $g'(6)$ are both positive, $g(6)$ is greater than $g(4)$, and $g(6) - g(4) > 0$.

(f) No, it is not possible. All you can say is that g is decreasing (falling) at $x = 2$.

51. $f(x) = \frac{1}{4}x^3$

By the limit definition of the derivative we have $f'(x) = \frac{3}{4}x^2$.

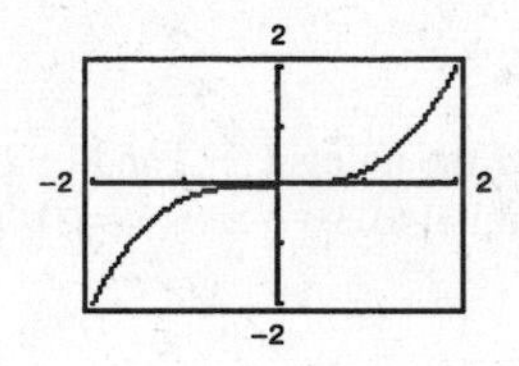

x	-2	-1.5	-1	-0.5	0	0.5	1	1.5	2
$f(x)$	-2	$-\frac{27}{32}$	$-\frac{1}{4}$	$-\frac{1}{32}$	0	$\frac{1}{32}$	$\frac{1}{4}$	$\frac{27}{32}$	2
$f'(x)$	3	$\frac{27}{16}$	$\frac{3}{4}$	$\frac{3}{16}$	0	$\frac{3}{16}$	$\frac{3}{4}$	$\frac{27}{16}$	3

53. $g(x) = \dfrac{f(x + 0.01) - f(x)}{0.01}$

$= [2(x + 0.01) - (x + 0.01)^2 - 2x + x^2] \cdot 100$

$= 2 - 2x - 0.01$

The graph of $g(x)$ is approximately the graph of $f'(x)$.

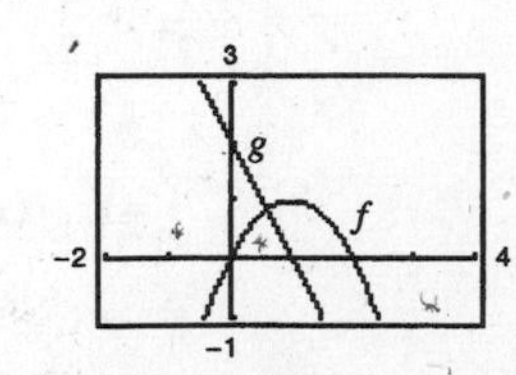

55. $f(2) = 2(4 - 2) = 4$, $f(2.1) = 2.1(4 - 2.1) = 3.99$

$$f'(2) \approx \frac{3.99 - 4}{2.1 - 2} = -0.1 \text{ [Exact: } f'(2) = 0]$$

57. $f(x) = \dfrac{1}{\sqrt{x}}$ and $f'(x) = \dfrac{-1}{2x^{3/2}}$

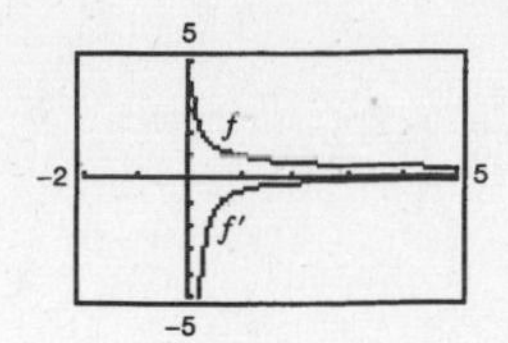

As $x \to \infty$, f is nearly horizontal and thus $f' \approx 0$.

59. $f(x) = 4 - (x - 3)^2$

$$S_{\Delta x}(x) = \frac{f(2 + \Delta x) - f(2)}{\Delta x}(x - 2) + f(2)$$

$$= \frac{4 - (2 + \Delta x - 3)^2 - 3}{\Delta x}(x - 2) + 3 = \frac{1 - (\Delta x - 1)^2}{\Delta x}(x - 2) + 3 = (-\Delta x + 2)(x - 2) + 3$$

(a) $\Delta x = 1$: $S_{\Delta x} = (x - 2) + 3 = x + 1$

$\Delta x = 0.5$: $S_{\Delta x} = \left(\dfrac{3}{2}\right)(x - 2) + 3 = \dfrac{3}{2}x$

$\Delta x = 0.1$: $S_{\Delta x} = \left(\dfrac{19}{10}\right)(x - 2) + 3 = \dfrac{19}{10}x - \dfrac{4}{5}$

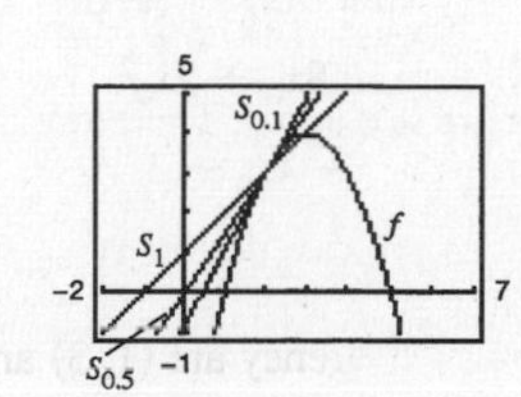

(b) As $\Delta x \to 0$, the line approaches the tangent line to f at $(2, 3)$.

61. $f(x) = x^2 - 1$, $c = 2$

$$f'(2) = \lim_{x \to 2} \frac{f(x) - f(2)}{x - 2} = \lim_{x \to 2} \frac{(x^2 - 1) - 3}{x - 2} = \lim_{x \to 2} \frac{(x - 2)(x + 2)}{x - 2} = \lim_{x \to 2}(x + 2) = 4$$

63. $f(x) = x^3 + 2x^2 + 1$, $c = -2$

$$f'(-2) = \lim_{x \to -2} \frac{f(x) - f(-2)}{x + 2} = \lim_{x \to -2} \frac{(x^3 + 2x^2 + 1) - 1}{x + 2} = \lim_{x \to -2} \frac{x^2(x + 2)}{x + 2} = \lim_{x \to -2} x^2 = 4$$

65. $g(x) = \sqrt{|x|}$, $c = 0$

$$g'(0) = \lim_{x \to 0} \frac{g(x) - g(0)}{x - 0} = \lim_{x \to 0} \frac{\sqrt{|x|}}{x}$$ Does not exist

As $x \to 0^-$, $\dfrac{\sqrt{|x|}}{x} = \dfrac{-1}{\sqrt{|x|}} \to -\infty$

As $x \to 0^+$, $\dfrac{\sqrt{|x|}}{x} = \dfrac{1}{\sqrt{x}} \to \infty$

67. $f(x) = (x - 6)^{2/3}$, $c = 6$

$$f'(6) = \lim_{x \to 6} \frac{f(x) - f(6)}{x - 6}$$

$$= \lim_{x \to 6} \frac{(x - 6)^{2/3} - 0}{x - 6}$$

$$= \lim_{x \to 6} \frac{1}{(x - 6)^{1/3}}$$

Does not exist

69. $h(x) = |x + 5|$, $c = -5$

$$h'(-5) = \lim_{x \to -5} \frac{h(x) - h(-5)}{x - (-5)}$$

$$= \lim_{x \to -5} \frac{|x + 5| - 0}{x + 5}$$

$$= \lim_{x \to -5} \frac{|x + 5|}{x + 5}$$

Does not exist

71. $f(x)$ is differentiable everywhere except at $x = -3$. (Sharp turn in the graph)

73. $f(x)$ is differentiable everywhere except at $x = -1$. (Discontinuity)

75. $f(x)$ is differentiable everywhere except at $x = 3$. (Sharp turn in the graph)

77. $f(x)$ is differentiable on the interval $(1, \infty)$. (At $x = 1$ the tangent line is vertical.)

79. $f(x)$ is differentiable everywhere except at $x = 0$. (Discontinuity)

81. $f(x) = |x - 1|$

The derivative from the left is

$$\lim_{x\to 1^-} \frac{f(x) - f(1)}{x - 1} = \lim_{x\to 1^-} \frac{|x - 1| - 0}{x - 1} = -1.$$

The derivative from the right is

$$\lim_{x\to 1^+} \frac{f(x) - f(1)}{x - 1} = \lim_{x\to 1^+} \frac{|x - 1| - 0}{x - 1} = 1.$$

The one-sided limits are not equal. Therefore, f is not differentiable at $x = 1$.

83. $f(x) = \begin{cases} (x-1)^3, & x \le 1 \\ (x-1)^2, & x > 1 \end{cases}$

The derivative from the left is

$$\lim_{x\to 1^-} \frac{f(x) - f(1)}{x - 1} = \lim_{x\to 1^-} \frac{(x - 1)^3 - 0}{x - 1} = \lim_{x\to 1^-} (x - 1)^2 = 0.$$

The derivative from the right is

$$\lim_{x\to 1^+} \frac{f(x) - f(1)}{x - 1} = \lim_{x\to 1^+} \frac{(x - 1)^2 - 0}{x - 1} = \lim_{x\to 1^+} (x - 1) = 0.$$

These one-sided limits are equal. Since f is continuous, f is differentiable at $x = 1$. $(f'(1) = 0)$

85. Note that f is continuous at $x = 2$. $f(x) = \begin{cases} x^2 + 1, & x \le 2 \\ 4x - 3, & x > 2 \end{cases}$

The derivative from the left is $\lim_{x\to 2^-} \frac{f(x) - f(2)}{x - 2} = \lim_{x\to 2^-} \frac{(x^2 + 1) - 5}{x - 2} = \lim_{x\to 2^-} (x + 2) = 4.$

The derivative from the right is $\lim_{x\to 2^+} \frac{f(x) - f(2)}{x - 2} = \lim_{x\to 2^+} \frac{(4x - 3) - 5}{x - 2} = \lim_{x\to 2^+} 4 = 4.$

The one-sided limits are equal. Therefore, f is differentiable at $x = 2$. $(f'(2) = 4)$

87. (a) The distance from $(3, 1)$ to the line $mx - y + 4 = 0$ is

$$d = \frac{|Ax_1 + By_1 + C|}{\sqrt{A^2 + B^2}} = \frac{|m(3) - 1(1) + 4|}{\sqrt{m^2 + 1}} = \frac{|3m + 3|}{\sqrt{m^2 + 1}}.$$

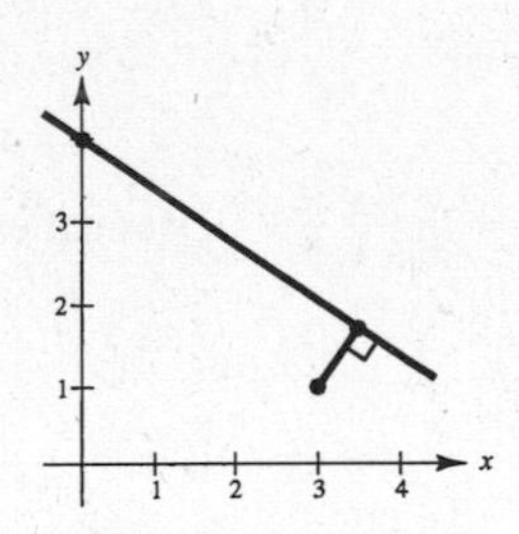

(b)

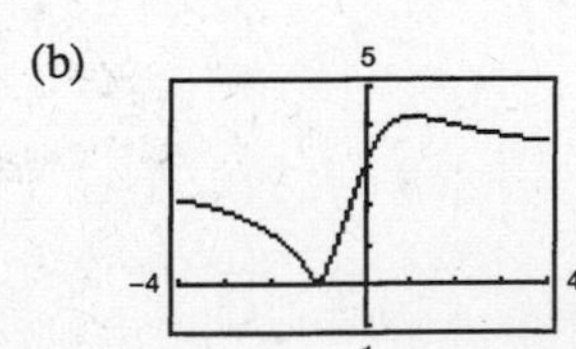

The function d is not differentiable at $m = -1$. This corresponds to the line $y = -x + 4$, which passes through the point $(3, 1)$.

89. False. the slope is $\lim_{\Delta x\to 0} \frac{f(2 + \Delta x) - f(2)}{\Delta x}$.

91. False. If the derivative from the left of a point does not equal the derivative from the right of a point, then the derivative does not exist at that point. For example, if $f(x) = |x|$, then the derivative from the left at $x = 0$ is -1 and the derivative from the right at $x = 0$ is 1. At $x = 0$, the derivative does not exist.

93. $f(x) = \begin{cases} x\sin(1/x), & x \neq 0 \\ 0, & x = 0 \end{cases}$

Using the Squeeze Theorem, we have $-|x| \leq x\sin(1/x) \leq |x|$, $x \neq 0$. Thus, $\lim_{x\to 0} x\sin(1/x) = 0 = f(0)$ and f is continuous at $x = 0$. Using the alternative form of the derivative we have

$$\lim_{x\to 0}\frac{f(x) - f(0)}{x - 0} = \lim_{x\to 0}\frac{x\sin(1/x) - 0}{x - 0} = \lim_{x\to 0}\left(\sin\frac{1}{x}\right).$$

Since this limit does not exist (it oscillates between -1 and 1), the function is not differentiable at $x = 0$.

$$g(x) = \begin{cases} x^2\sin(1/x), & x \neq 0 \\ 0, & x = 0 \end{cases}$$

Using the Squeeze Theorem again we have $-x^2 \leq x^2\sin(1/x) \leq x^2$, $x \neq 0$. Thus, $\lim_{x\to 0} x^2\sin(1/x) = 0 = f(0)$ and f is continuous at $x = 0$. Using the alternative form of the derivative again we have

$$\lim_{x\to 0}\frac{f(x) - f(0)}{x - 0} = \lim_{x\to 0}\frac{x^2\sin(1/x) - 0}{x - 0} = \lim_{x\to 0} x\sin\frac{1}{x} = 0.$$

Therefore, g is differentiable at $x = 0$, $g'(0) = 0$.

Section 2.2 Basic Differentiation Rules and Rates of Change

1. (a) $y = x^{1/2}$, $y' = \frac{1}{2}x^{-1/2}$, $y'(1) = \frac{1}{2}$

(b) $y = x^{3/2}$, $y' = \frac{3}{2}x^{1/2}$, $y'(1) = \frac{3}{2}$

(c) $y = x^2$, $y' = 2x$, $y'(1) = 2$

(d) $y = x^3$, $y' = 3x^2$, $y'(1) = 3$

3. $y = 8$
$y' = 0$

5. $y = x^6$
$y' = 6x^5$

7. $y = \sqrt[5]{x} = x^{1/5}$
$y' = \frac{1}{5}x^{-4/5} = \frac{1}{5x^{4/5}}$

9. $f(x) = x + 1$
$f'(x) = 1$

11. $f(t) = -2t^2 + 3t - 6$
$f'(x) = -4t + 3$

13. $g(x) = x^2 + 4x^3$
$g'(x) = 2x + 12x^2$

15. $s(t) = t^3 - 2t + 4$
$s'(t) = 3t^2 - 2$

17. $f(x) = 6x - 5e^x$
$f'(x) = 6 - 5e^x$

19. $y = \frac{\pi}{2}\sin\theta - \cos\theta$
$y' = \frac{\pi}{2}\cos\theta + \sin\theta$

21. $y = x^2 - \frac{1}{2}\cos x$
$y' = 2x + \frac{1}{2}\sin x$

23. $y = \frac{1}{2}e^x - 3\sin x$
$y' = \frac{1}{2}e^x - 3\cos x$

	Function	Rewrite	Derivative	Simplify
25.	$y = \frac{5}{2x^2}$	$y = \frac{5}{2}x^{-2}$	$y' = -5x^{-3}$	$y' = \frac{-5}{x^3}$
27.	$y = \frac{3}{(2x)^3}$	$y = \frac{3}{8}x^{-3}$	$y' = \frac{-9}{8}x^{-4}$	$y' = \frac{-9}{8x^4}$
29.	$y = \frac{\sqrt{x}}{x}$	$y = x^{-1/2}$	$y' = -\frac{1}{2}x^{-3/2}$	$y' = -\frac{1}{2x^{3/2}}$

31. $f(x) = \frac{3}{x^2} = 3x^{-2}, (1, 3)$

$f'(x) = -6x^{-3} = \frac{-6}{x^3}$

$f'(1) = -6$

33. $f(x) = -\frac{1}{2} + \frac{7}{5}x^3, \left(0, -\frac{1}{2}\right)$

$f'(x) = \frac{21}{5}x^2$

$f'(0) = 0$

35. $f(\theta) = 4 \sin \theta - \theta, (0, 0)$

$f'(\theta) = 4 \cos \theta - 1$

$f'(0) = 4(1) - 1 = 3$

37. $f(t) = \frac{3}{4}e^t$

$f'(t) = \frac{3}{4}e^t$

$f'(0) = \frac{3}{4}e^0 = \frac{3}{4}$

39. $g(t) = t^2 - \frac{4}{t^3} = t^2 - 4t^{-3}$

$g'(t) = 2t + 12t^{-4} = 2t + \frac{12}{t^4}$

41. $f(x) = \frac{x^3 - 3x^2 + 4}{x^2}$

$= x - 3 + 4x^{-2}$

$f'(x) = 1 - \frac{8}{x^3} = \frac{x^3 - 8}{x^3}$

43. $y = x(x^2 + 1) = x^3 + x$

$y' = 3x^2 + 1$

45. $f(x) = \sqrt{x} - 6\sqrt[3]{x}$

$= x^{1/2} - 6x^{1/3}$

$f'(x) = \frac{1}{2}x^{-1/2} - 2x^{-2/3}$

$= \frac{1}{2\sqrt{x}} - \frac{2}{x^{2/3}}$

47. $h(s) = s^{4/5} - s^{2/3}$

$h'(s) = \frac{4}{5}s^{-1/5} - \frac{2}{3}s^{-1/3}$

$= \frac{4}{5s^{1/5}} - \frac{2}{3s^{1/3}}$

49. $f(x) = 6\sqrt{x} + 5 \cos x = 6x^{1/2} + 5 \cos x$

$f'(x) = 3x^{-1/2} - 5 \sin x = \frac{3}{\sqrt{x}} - 5 \sin x$

51. $f(x) = x^{-2} - 2e^x$

$f'(x) = -2x^{-3} - 2e^x = \frac{-2}{x^3} - 2e^x$

53. (a) $y = x^4 - x$

$y' = 4x^3 - 1$

At $(-1, 2)$: $y' = 4(-1)^3 - 1 = -5.$

Tangent line: $y - 2 = -5(x + 1)$

$y = -5x - 3$

(b)

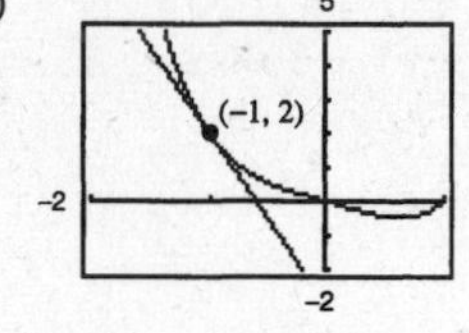

55. (a) $g(x) = x + e^x$

$g'(x) = 1 + e^x$

At $(0, 1)$: $g'(0) = 1 + 1 = 2$

Tangent line: $y - 1 = 2(x - 0)$

$y = 2x + 1$

(b)

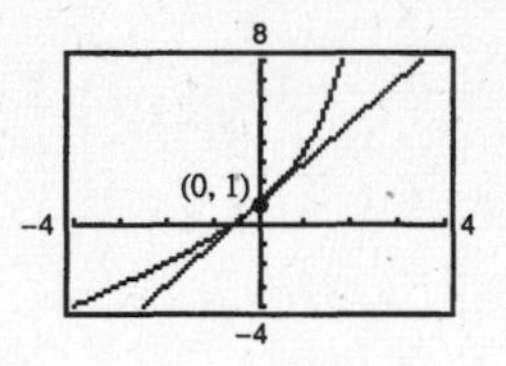

57. $y = x^4 - 8x^2 + 2$

$y' = 4x^3 - 16x$

$= 4x(x^2 - 4)$

$= 4x(x - 2)(x + 2)$

$y' = 0 \Rightarrow x = 0, \pm 2$

Horizontal tangents:

$(0, 2), (2, -14), (-2, -14)$

59. $y = x + \sin x, 0 \le x < 2\pi$

$y' = 1 + \cos x = 0$

$\cos x = -1 \Rightarrow x = \pi$

At $x = \pi, y = \pi$.

Horizontal tangent: (π, π)

61. $y = -4x + e^x$

$y' = -4 + e^x = 0$

$e^x = 4$

$x = \ln 4$

$(\ln 4, -4 \ln 4 + 4)$

63. $x^2 - kx = 4x - 9$ Equate functions

$2x - k = 4$ Equate derivatives

Hence, $k = 2x - 4$ and

$$x^2 - (2x - 4)x = 4x - 9 \Rightarrow -x^2 = -9 \Rightarrow x = \pm 3.$$

For $x = 3$, $k = 2$ and for $x = -3$, $k = -10$.

65. $\frac{k}{x} = -\frac{3}{4}x + 3$ Equate functions

$-\frac{k}{x^2} = -\frac{3}{4}$ Equate derivatives

Hence, $k = \frac{3}{4}x^2$ and

$$\frac{(3/4)x^2}{x} = \frac{-3}{4}x + 3 \Rightarrow \frac{3}{4}x = -\frac{3}{4}x + 3 \Rightarrow \frac{3}{2}x$$

$$= 3 \Rightarrow x = 2 \Rightarrow k = 3.$$

67. (a) The slope appears to be steepest between A and B.

(b) The average rate of change between A and B is **greater** than the instantaneous rate of change at B.

(c)

69. $g(x) = f(x) + 6 \Rightarrow g'(x) = f'(x)$

71.

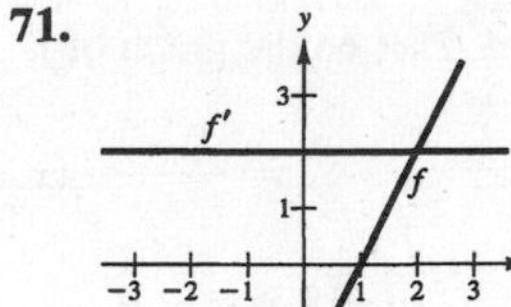
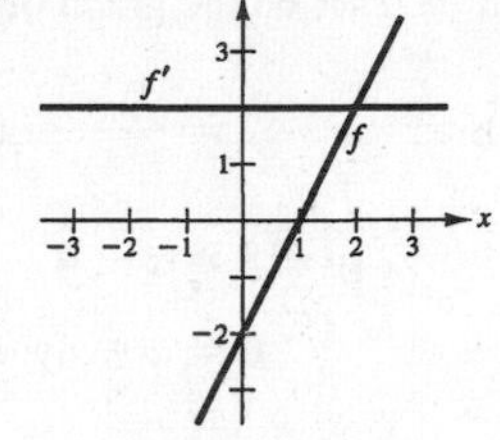

If f is linear then its derivative is a constant function.

$$f(x) = ax + b$$

$$f'(x) = a$$

73. Let (x_1, y_1) and (x_2, y_2) be the points of tangency on $y = x^2$ and $y = -x^2 + 6x - 5$, respectively. The derivatives of these functions are

$$y' = 2x \Rightarrow m = 2x_1 \quad \text{and} \quad y' = -2x + 6 \Rightarrow m = -2x_2 + 6.$$

$$m = 2x_1 = -2x_2 + 6$$

$$x_1 = -x_2 + 3$$

Since $y_1 = x_1^2$ and $y_2 = -x_2^2 + 6x_2 - 5$:

$$m = \frac{y_2 - y_1}{x_2 - x_1} = \frac{(-x_2^2 + 6x_2 - 5) - (x_1^2)}{x_2 - x_1} = -2x_2 + 6.$$

$$\frac{(-x_2^2 + 6x_2 - 5) - (-x_2 + 3)^2}{x_2 - (-x_2 + 3)} = -2x_2 + 6$$

$$(-x_2^2 + 6x_2 - 5) - (x_2^2 - 6x_2 + 9) = (-2x_2 + 6)(2x_2 - 3)$$

$$-2x_2^2 + 12x_2 - 14 = -4x_2^2 + 18x_2 - 18$$

$$2x_2^2 - 6x_2 + 4 = 0$$

$$2(x_2 - 2)(x_2 - 1) = 0$$

$$x_2 = 1 \text{ or } 2$$

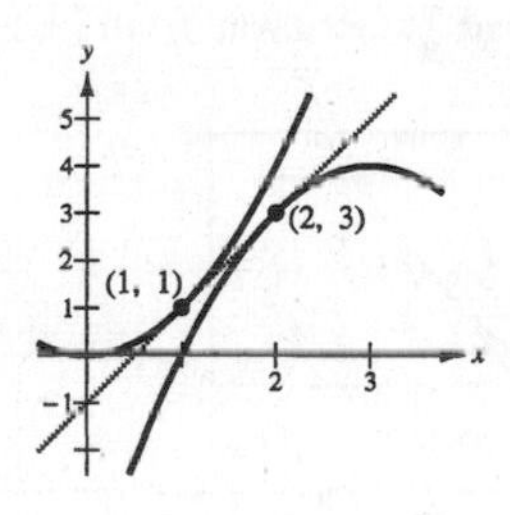

$x_2 = 1 \Rightarrow y_2 = 0, x_1 = 2$ and $y_1 = 4$

Thus, the tangent line through (1, 0) and (2, 4) is

$$y - 0 = \left(\frac{4 - 0}{2 - 1}\right)(x - 1) \Rightarrow y = 4x - 4.$$

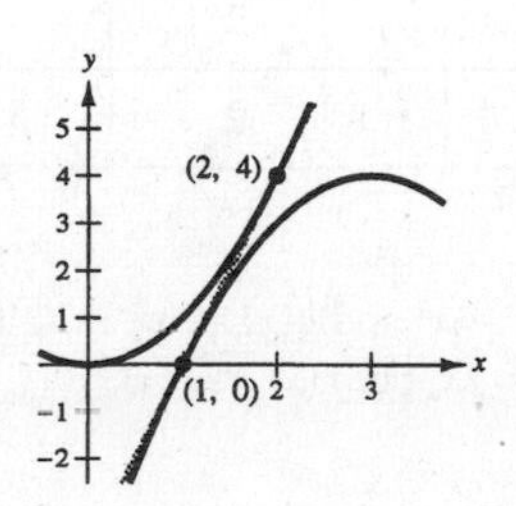

$x_2 = 2 \Rightarrow y_2 = 3, x_1 = 1$ and $y_1 = 1$

Thus, the tangent line through (2, 3) and (1, 1) is

$$y - 1 = \left(\frac{3 - 1}{2 - 1}\right)(x - 1) \Rightarrow y = 2x - 1.$$

75. $f(x) = \sqrt{x},\ (-4, 0)$

$$f'(x) = \frac{1}{2}x^{-1/2} = \frac{1}{2\sqrt{x}}$$

$$\frac{1}{2\sqrt{x}} = \frac{0 - y}{-4 - x}$$

$$4 + x = 2\sqrt{x}y$$

$$4 + x = 2\sqrt{x}\sqrt{x}$$

$$4 + x = 2x$$

$$x = 4,\ y = 2$$

The point (4, 2) is on the graph of f.

Tangent line: $y - 2 = \dfrac{0 - 2}{-4 - 4}(x - 4)$

$$4y - 8 = x - 4$$

$$0 = x - 4y + 4$$

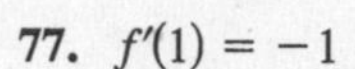

77. $f'(1) = -1$

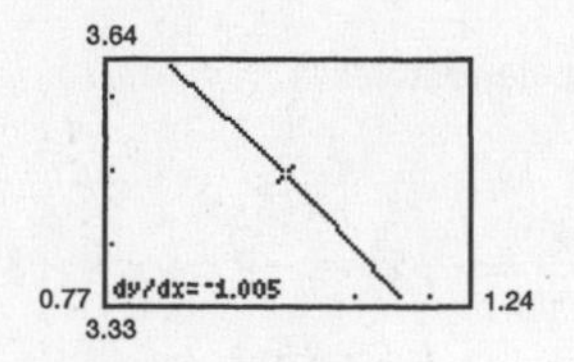

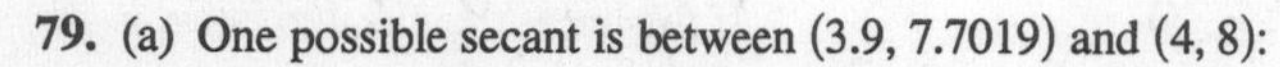

79. (a) One possible secant is between (3.9, 7.7019) and (4, 8):

$$y - 8 = \frac{8 - 7.7019}{4 - 3.9}(x - 4)$$

$$y - 8 = 2.981(x - 4)$$

$$y = S(x) = 2.981x - 3.924$$

(b) $f'(x) = \frac{3}{2}x^{1/2} \Rightarrow f'(4) = \frac{3}{2}(2) = 3$

$$T(x) = 3(x - 4) + 8 = 3x - 4$$

$S(x)$ is an approximation of the tangent line $T(x)$.

(c) As you move further away from (4, 8), the accuracy of the approximation T gets worse.

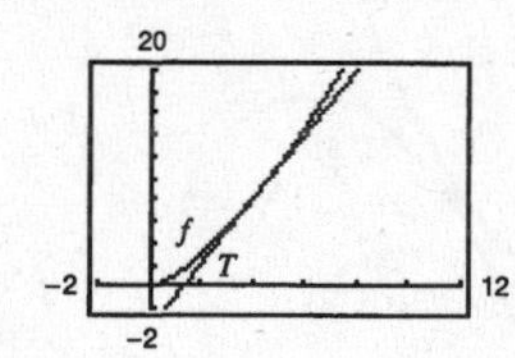

(d)

Δx	-3	-2	-1	-0.5	-0.1	0	0.1	0.5	1	2	3
$f(4 + \Delta x)$	1	2.828	5.196	6.548	7.702	8	8.302	9.546	11.180	14.697	18.520
$T(4 + \Delta x)$	-1	2	5	6.5	7.7	8	8.3	9.5	11	14	17

81. False. Let $f(x) = x^2$ and $g(x) = x^2 + 4$. Then $f'(x) = g'(x) = 2x$, but $f(x) \neq g(x)$.

83. False. If $y = \pi^2$, then $dy/dx = 0$. (π^2 is a constant.)

85. True. If $g(x) = 3f(x)$, then $g'(x) = 3f'(x)$.

87. $f(x) = -\dfrac{1}{x}, [1, 2]$

$f'(x) = \dfrac{1}{x^2}$

Instantaneous rate of change:

$(1, -1) \Rightarrow f'(1) = 1$

$\left(2, -\dfrac{1}{2}\right) \Rightarrow f'(2) = \dfrac{1}{4}$

Average rate of change:

$$\frac{f(2) - f(1)}{2 - 1} = \frac{(-1/2) - (-1)}{2 - 1} = \frac{1}{2}$$

89. $g(x) = x^2 + e^x, \quad [0, 1]$

$g'(x) = 2x + e^x$

Instantaneous rate of change:

$(0, 1)$: $g'(0) = 1$

$(1, 1 + e)$: $2 + e \approx 4.718$

Average rate of change:

$$\frac{g(1) - g(0)}{1 - 0} = \frac{(1 + e) - (1)}{1} = e \approx 2.718$$

91. (a) $s(t) = -16t^2 + 1362$

$v(t) = -32t$

(b) $\dfrac{s(2) - s(1)}{2 - 1} = 1298 - 1346 = -48$ ft/sec

(c) $v(t) = s'(t) = -32t$

When $t = 1$: $v(1) = -32$ ft/sec.

When $t = 2$: $v(2) = -64$ ft/sec.

(d) $-16t^2 + 1362 = 0$

$t^2 = \dfrac{1362}{16} \Rightarrow t = \dfrac{\sqrt{1362}}{4} \approx 9.226$ sec

(e) $v\left(\dfrac{\sqrt{1362}}{4}\right) = -32\left(\dfrac{\sqrt{1362}}{4}\right)$

$= -8\sqrt{1362} \approx -295.242$ ft/sec

93. $s(t) = -4.9t^2 + v_0t + s_0$

$= -4.9t^2 + 120t$

$v(t) = -9.8t + 120$

$v(5) = -9.8(5) + 120 = 71$ m/sec

$v(10) = -9.8(10) + 120 = 22$ m/sec

95.

(The velocity has been converted to miles per hour.)

97. 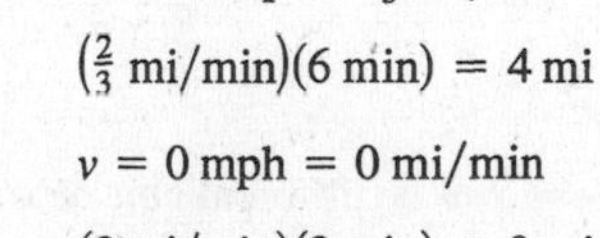

$v = 40 \text{ mph} = \frac{2}{3}$ mi/min

$(\frac{2}{3} \text{ mi/min})(6 \text{ min}) = 4$ mi

$v = 0 \text{ mph} = 0$ mi/min

$(0 \text{ mi/min})(2 \text{ min}) = 0$ mi

$v = 60 \text{ mph} = 1$ mi/min

$(1 \text{ mi/min})(2 \text{ min}) = 2$ mi

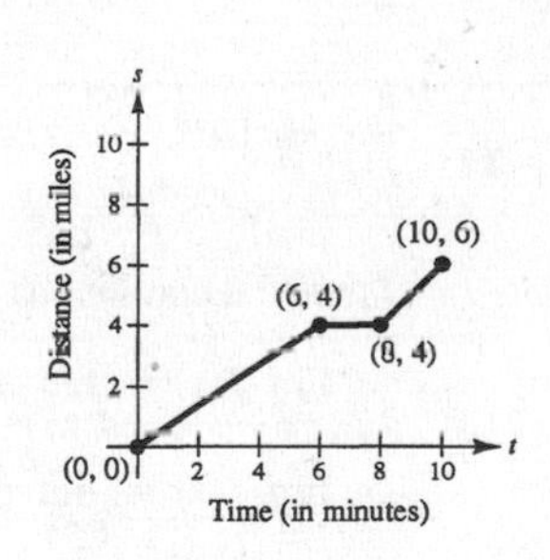

99. (a) Using a graphing utility, you obtain

$R = 0.167v - 0.02.$

(b) Using a graphing utility, you obtain

$B = 0.00586v^2 - 0.0239v + 0.46.$

(c) $T = R + B = 0.00586v^2 + 0.1431v + 0.44$

(d)

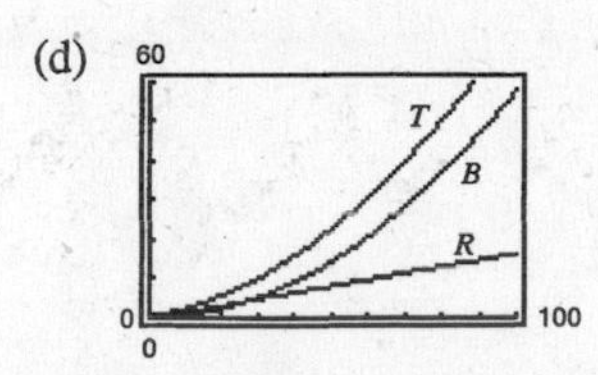

(e) $\dfrac{dT}{dv} = 0.01172v + 0.1431$

For $v = 40$, $T'(40) \approx 0.612$.

For $v = 80$, $T'(80) \approx 1.081$.

For $v = 100$, $T'(100) \approx 1.315$.

(f) For increasing speeds, the total stopping distance increases.

101. $A = s^2, \dfrac{dA}{ds} = 2s$

When $s = 4$ m,

$\dfrac{dA}{ds} = 8$ square meters per meter change in s.

103. $C = \dfrac{1{,}008{,}000}{Q} + 6.3Q$

$\dfrac{dC}{dQ} = -\dfrac{1{,}008{,}000}{Q^2} + 6.3$

$C(351) - C(350) \approx 5083.095 - 5085 \approx -\$1.91$

When $Q = 350$, $\dfrac{dC}{dQ} \approx -\$1.93$.

105. (a) $f'(1.47)$ is the rate of change of the amount of gasoline sold when the price is \$1.47 per gallon.

(b) $f'(1.47)$ is usually negative. As prices go up, sales go down.

107. $y = ax^2 + bx + c$

Since the parabola passes through (0, 1) and (1, 0), we have

$(0, 1)$: $1 = a(0)^2 + b(0) + c \Rightarrow c = 1$

$(1, 0)$: $0 = a(1)^2 + b(1) + 1 \Rightarrow b = -a - 1.$

Thus, $y = ax^2 + (-a - 1)x + 1$. From the tangent line $y = x - 1$, we know that the derivative is 1 at the point (1, 0).

$$y' = 2ax + (-a - 1)$$
$$1 = 2a(1) + (-a - 1)$$
$$1 = a - 1$$
$$a = 2$$
$$b = -a - 1 = -3$$

Therefore, $y = 2x^2 - 3x + 1$.

109. $y = x^3 - 9x$

$y' = 3x^2 - 9$

Tangent lines through $(1, -9)$:

$$y + 9 = (3x^2 - 9)(x - 1)$$
$$(x^3 - 9x) + 9 = 3x^3 - 3x^2 - 9x + 9$$
$$0 = 2x^3 - 3x^2 = x^2(2x - 3)$$
$$x = 0 \text{ or } x = \tfrac{3}{2}$$

The points of tangency are (0, 0) and $\left(\frac{3}{2}, -\frac{81}{8}\right)$. At (0, 0) the slope is $y'(0) = -9$. At $\left(\frac{3}{2}, -\frac{81}{8}\right)$ the slope is $y'\left(\frac{3}{2}\right) = -\frac{9}{4}$.

Tangent lines:

$y - 0 = -9(x - 0)$ and $y + \frac{81}{8} = -\frac{9}{4}\left(x - \frac{3}{2}\right)$

$y = -9x$ $\qquad$ $y = -\frac{9}{4}x - \frac{27}{4}$

$9x + y = 0$ $\qquad$ $9x + 4y + 27 = 0$

111. $f(x) = \begin{cases} ax^3, & x \le 2 \\ x^2 + b, & x > 2 \end{cases}$

f must be continuous at $x = 2$ to be differentiable at $x = 2$.

$$\left.\begin{aligned} \lim_{x\to 2^-} f(x) &= \lim_{x\to 2^-} ax^3 = 8a \\ \lim_{x\to 2^+} f(x) &= \lim_{x\to 2^+} (x^2 + b) = 4 + b \end{aligned}\right\} \quad \begin{aligned} 8a &= 4 + b \\ 8a - 4 &= b \end{aligned}$$

$$f'(x) = \begin{cases} 3ax^2, & x < 2 \\ 2x, & x > 2 \end{cases}$$

For f to be differentiable at $x = 2$, the left derivative must equal the right derivative.

$$3a(2)^2 = 2(2)$$
$$12a = 4$$
$$a = \tfrac{1}{3}$$
$$b = 8a - 4 = -\tfrac{4}{3}$$

113. Let $f(x) = \cos x$.

$$f'(x) = \lim_{\Delta x \to 0} \frac{f(x + \Delta x) - f(x)}{\Delta x}$$

$$= \lim_{\Delta x \to 0} \frac{\cos x \cos \Delta x - \sin x \sin \Delta x - \cos x}{\Delta x}$$

$$= \lim_{\Delta x \to 0} \frac{\cos x(\cos \Delta x - 1)}{\Delta x} - \lim_{\Delta x \to 0} \sin x\left(\frac{\sin \Delta x}{\Delta x}\right)$$

$$= 0 - \sin x(1) = -\sin x$$

Section 2.3 The Product and Quotient Rules and Higher-Order Derivatives

1. $g(x) = (x^2 + 2)(x^2 - 3x)$

$$g'(x) = (x^2 + 2)(2x - 3) + (x^2 - 3x)(2x)$$

$$= 2x^3 - 3x^2 + 4x - 6 + 2x^3 - 6x^2$$

$$= 4x^3 - 9x^2 + 4x - 6$$

3. $f(x) = x^3 \cos x$

$$f'(x) = x^3(-\sin x) + \cos x(3x^2)$$

$$= 3x^2 \cos x - x^3 \sin x$$

5. $f(x) = \dfrac{x}{x^2 + 1}$

$$f'(x) = \frac{(x^2 + 1)(1) - x(2x)}{(x^2 + 1)^2} = \frac{1 - x^2}{(x^2 + 1)^2}$$

7. $g(x) = \dfrac{\sin x}{x^2}$

$$g'(x) = \frac{x^2(\cos x) - \sin x(2x)}{(x^2)^2} = \frac{x \cos x - 2 \sin x}{x^3}$$

9. $f(x) = (x^3 - 3x)(2x^2 + 3x + 5)$

$$f'(x) = (x^3 - 3x)(4x + 3) + (2x^2 + 3x + 5)(3x^2 - 3)$$

$$= 10x^4 + 12x^3 - 3x^2 - 18x - 15$$

$$f'(0) = -15$$

11. $f(x) = \dfrac{x^2 - 4}{x - 3}$

$$f'(x) = \frac{(x - 3)(2x) - (x^2 - 4)(1)}{(x - 3)^2} = \frac{2x^2 - 6x - x^2 + 4}{(x - 3)^2}$$

$$= \frac{x^2 - 6x + 4}{(x - 3)^2}$$

$$f'(1) = \frac{1 - 6 + 4}{(1 - 3)^2} = -\frac{1}{4}$$

13. $f(x) = x \cos x$

$$f'(x) = (x)(-\sin x) + (\cos x)(1) = \cos x - x \sin x$$

$$f'\left(\frac{\pi}{4}\right) = \frac{\sqrt{2}}{2} - \frac{\pi}{4}\left(\frac{\sqrt{2}}{2}\right) = \frac{\sqrt{2}}{8}(4 - \pi)$$

15. $f(x) = e^x \sin x$

$$f'(x) = e^x \cos x + e^x \sin x = e^x(\cos x + \sin x)$$

$$f'(0) = 1$$

	Function	*Rewrite*	*Derivative*	*Simplify*
17.	$y = \dfrac{x^2 + 2x}{3}$	$y = \frac{1}{3}x^2 + \frac{2}{3}x$	$y' = \frac{2}{3}x + \frac{2}{3}$	$y' = \dfrac{2x + 2}{3}$
19.	$y = \dfrac{7}{3x^3}$	$y = \frac{7}{3}x^{-3}$	$y' = -7x^{-4}$	$y' = -\dfrac{7}{x^4}$
21.	$y = \dfrac{4x^{3/2}}{x}$	$y = 4\sqrt{x}, x > 0$	$y' = 2x^{-1/2}$	$y' = \dfrac{2}{\sqrt{x}}$

23. $f(x) = \dfrac{3 - 2x - x^2}{x^2 - 1}$

$$f'(x) = \frac{(x^2 - 1)(-2 - 2x) - (3 - 2x - x^2)(2x)}{(x^2 - 1)^2}$$

$$= \frac{2x^2 - 4x + 2}{(x^2 - 1)^2} = \frac{2(x - 1)^2}{(x^2 - 1)^2}$$

$$= \frac{2}{(x + 1)^2}, x \neq 1$$

25. $f(x) = x\left(1 - \dfrac{4}{x + 3}\right) = x - \dfrac{4x}{x + 3}$

$$f'(x) = 1 - \frac{(x + 3)4 - 4x(1)}{(x + 3)^2} = \frac{(x^2 + 6x + 9) - 12}{(x + 3)^2}$$

$$= \frac{x^2 + 6x - 3}{(x + 3)^2}$$

27. $f(x) = \dfrac{2x + 5}{\sqrt{x}} = 2x^{1/2} + 5x^{-1/2}$

$$f'(x) = x^{-1/2} - \frac{5}{2}x^{-3/2} = x^{-3/2}\left[x - \frac{5}{2}\right]$$

$$= \frac{2x - 5}{2x\sqrt{x}} = \frac{2x - 5}{2x^{3/2}}$$

29. $h(s) = (s^3 - 2)^2 = s^6 - 4s^3 + 4$

$$h'(s) = 6s^5 - 12s^2 = 6s^2(s^3 - 2)$$

31. $f(x) = \dfrac{2 - (1/x)}{x - 3} = \dfrac{2x - 1}{x(x - 3)} = \dfrac{2x - 1}{x^2 - 3x}$

$$f'(x) = \frac{(x^2 - 3x)2 - (2x - 1)(2x - 3)}{(x^2 - 3x)^2} = \frac{2x^2 - 6x - 4x^2 + 8x - 3}{(x^2 - 3x)^2}$$

$$= \frac{-2x^2 + 2x - 3}{(x^2 - 3x)^2} = -\frac{2x^2 - 2x + 3}{x^2(x - 3)^2}$$

33. $f(x) = (3x^3 + 4x)(x - 5)(x + 1)$

$$f'(x) = (9x^2 + 4)(x - 5)(x + 1) + (3x^3 + 4x)(1)(x + 1) + (3x^3 + 4x)(x - 5)(1)$$

$$= (9x^2 + 4)(x^2 - 4x - 5) + 3x^4 + 3x^3 + 4x^2 + 4x + 3x^4 - 15x^3 + 4x^2 - 20x$$

$$= 9x^4 - 36x^3 - 41x^2 - 16x - 20 + 6x^4 - 12x^3 + 8x^2 - 16x$$

$$= 15x^4 - 48x^3 - 33x^2 - 32x - 20$$

35. $f(x) = \dfrac{x^2 + c^2}{x^2 - c^2}$

$$f'(x) = \frac{(x^2 - c^2)(2x) - (x^2 + c^2)(2x)}{(x^2 - c^2)^2}$$

$$= \frac{-4xc^2}{(x^2 - c^2)^2}$$

37. $f(t) = t^2 \sin t$

$$f'(t) = t^2 \cos t + 2t \sin t$$

$$= t(t \cos t + 2 \sin t)$$

39. $f(t) = \dfrac{\cos t}{t}$

$$f'(t) = \frac{-t \sin t - \cos t}{t^2} = -\frac{t \sin t + \cos t}{t^2}$$

41. $f(x) = -e^x + \tan x$

$$f'(x) = -e^x + \sec^2 x$$

43. $g(t) = \sqrt[4]{t} + 8 \sec t = t^{1/4} + 8 \sec t$

$$g'(t) = \frac{1}{4}t^{-3/4} + 8 \sec t \tan t = \frac{1}{4t^{3/4}} + 8 \sec t \tan t$$

45. $y = \dfrac{3(1 - \sin x)}{2 \cos x} = \dfrac{3}{2}(\sec x - \tan x)$

$$y' = \frac{3}{2}(\sec x \tan x - \sec^2 x) = \frac{3}{2} \sec x(\tan x - \sec x)$$

$$= \frac{3}{2}(\sec x \tan x - \tan^2 x - 1)$$

47. $y = -\csc x - \sin x$

$$y' = \csc x \cot x - \cos x$$
$$= \frac{\cos x}{\sin^2 x} - \cos x$$
$$= \cos x(\csc^2 x - 1)$$
$$= \cos x \cot^2 x$$

49. $f(x) = x^2 \tan x$

$$f'(x) = x^2 \sec^2 x + 2x \tan x$$
$$= x(x \sec^2 x + 2 \tan x)$$

51. $y = 2x \sin x + x^2 e^x$

$$y' = 2x(\cos x) + 2 \sin x + x^2 e^x + 2xe^x$$
$$= 2x \cos x + 2 \sin x + xe^x(x + 2)$$

53. $y = \dfrac{e^x}{4\sqrt{x}}$

$$y' = \frac{4\sqrt{x}\,e^x - e^x \dfrac{4}{2\sqrt{x}}}{(4\sqrt{x})^2} = \frac{e^x\left(4\sqrt{x} - \dfrac{2}{\sqrt{x}}\right)}{16x} = \frac{e^x(4x - 2)}{16x^{3/2}}$$
$$= \frac{e^x(2x - 1)}{8x^{3/2}}$$

55. $g(x) = \left(\dfrac{x + 1}{x + 2}\right)(2x - 5)$

$$g'(x) = \frac{2x^2 + 8x - 1}{(x + 2)^2}$$ (form of answer may vary)

57. $g(\theta) = \dfrac{\theta}{1 - \sin \theta}$

$$g'(\theta) = \frac{1 - \sin \theta + \theta \cos \theta}{(\sin \theta - 1)^2}$$ (form of answer may vary)

59. $y = \dfrac{1 + \csc x}{1 - \csc x}$

$$y' = \frac{(1 - \csc x)(-\csc x \cot x) - (1 + \csc x)(\csc x \cot x)}{(1 - \csc x)^2} = \frac{-2 \csc x \cot x}{(1 - \csc x)^2}$$

$$y'\left(\frac{\pi}{6}\right) = \frac{-2(2)(\sqrt{3})}{(1 - 2)^2} = -4\sqrt{3}$$

61. $h(t) = \dfrac{\sec t}{t}$

$$h'(t) = \frac{t(\sec t \tan t) - (\sec t)(1)}{t^2}$$
$$= \frac{\sec t(t \tan t - 1)}{t^2}$$

$$h'(\pi) = \frac{\sec \pi(\pi \tan \pi - 1)}{\pi^2} = \frac{1}{\pi^2}$$

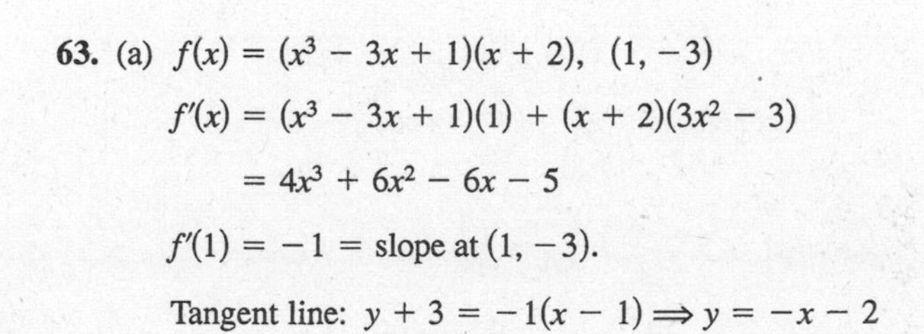

63. (a) $f(x) = (x^3 - 3x + 1)(x + 2), \ (1, -3)$

$$f'(x) = (x^3 - 3x + 1)(1) + (x + 2)(3x^2 - 3)$$
$$= 4x^3 + 6x^2 - 6x - 5$$

$f'(1) = -1 =$ slope at $(1, -3)$.

Tangent line: $y + 3 = -1(x - 1) \Rightarrow y = -x - 2$

(b)

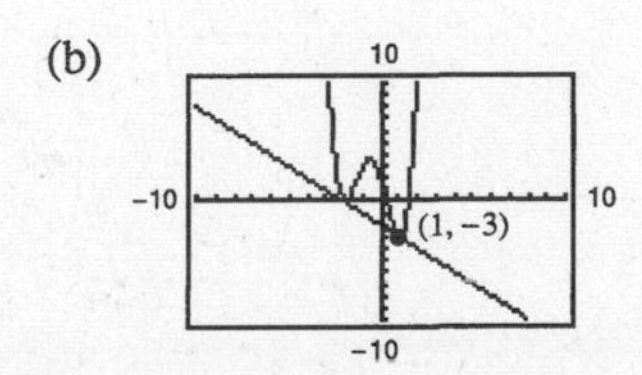

65. (a) $f(x) = \tan x, \left(\frac{\pi}{4}, 1\right)$

$f'(x) = \sec^2 x$

$f'\left(\frac{\pi}{4}\right) = 2 = \text{slope at } \left(\frac{\pi}{4}, 1\right).$

Tangent line:

$$y - 1 = 2\left(x - \frac{\pi}{4}\right)$$

$$y - 1 = 2x - \frac{\pi}{2}$$

$$4x - 2y - \pi + 2 = 0$$

(b)

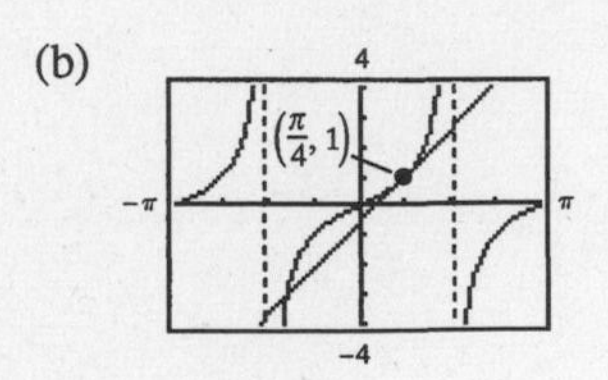

67. (a) $f(x) = (x - 1)e^x$

$f'(x) = (x - 1)e^x + e^x = e^x$

At $(1, 0)$, $f'(1) = e$

Tangent line: $y - 0 = e(x - 1)$

$y = e(x - 1)$

(b)

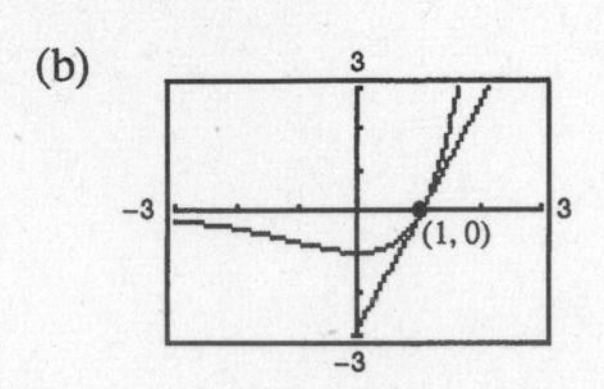

69. $f(x) = \dfrac{x^2}{x - 1}$

$$f'(x) = \frac{(x - 1)(2x) - x^2(1)}{(x - 1)^2}$$

$$= \frac{x^2 - 2x}{(x - 1)^2} = \frac{x(x - 2)}{(x - 1)^2}$$

$f'(x) = 0$ when $x = 0$ or $x = 2$.

Horizontal tangents are at $(0, 0)$ and $(2, 4)$.

71. $g(x) = \dfrac{8(x - 2)}{e^x}$

$$g'(x) = \frac{e^x(8) - 8(x - 2)e^x}{e^{2x}} = \frac{24 - 8x}{e^x}$$

$g'(x) = 0$ when $x = 3$.

Horizontal tangent at $(3, 8e^{-3})$

73. $f'(x) = \dfrac{(x + 2)3 - 3x(1)}{(x + 2)^2} = \dfrac{6}{(x + 2)^2}$

$$g'(x) = \frac{(x + 2)5 - (5x + 4)(1)}{(x + 2)^2} = \frac{6}{(x + 2)^2}$$

$$g(x) = \frac{5x + 4}{(x + 2)} = \frac{3x}{(x + 2)} + \frac{2x + 4}{(x + 2)} = f(x) + 2$$

f and g differ by a constant.

75. $f(x) = x^n \sin x$

$$f'(x) = x^n \cos x + nx^{n-1} \sin x$$

$$= x^{n-1}(x \cos x + n \sin x)$$

When $n = 1$: $f'(x) = x \cos x + \sin x$.

When $n = 2$: $f'(x) = x(x \cos x + 2 \sin x)$.

When $n = 3$: $f'(x) = x^2(x \cos x + 3 \sin x)$.

When $n = 4$: $f'(x) = x^3(x \cos x + 4 \sin x)$.

For general n, $f'(x) = x^{n-1}(x \cos x + n \sin x)$.

77. Area $= A(t) = (2t + 1)\sqrt{t} = 2t^{3/2} + t^{1/2}$

$$A'(t) = 2\left(\frac{3}{2}t^{1/2}\right) + \frac{1}{2}t^{-1/2}$$

$$= 3t^{1/2} + \frac{1}{2}t^{-1/2}$$

$$= \frac{6t + 1}{2\sqrt{t}} \text{ cm}^2/\text{sec}$$

79. $P(t) = 500\left[1 + \dfrac{4t}{50 + t^2}\right]$

$$P'(t) = 500\left[\frac{(50 + t^2)(4) - (4t)(2t)}{(50 + t^2)^2}\right]$$

$$= 500\left[\frac{200 - 4t^2}{(50 + t^2)^2}\right]$$

$$= 2000\left[\frac{50 - t^2}{(50 + t^2)^2}\right]$$

$P'(2) \approx 31.55$ bacteria per hour

81. (a) $\sec x = \dfrac{1}{\cos x}$

$$\frac{d}{dx}[\sec x] = \frac{d}{dx}\left[\frac{1}{\cos x}\right] = \frac{(\cos x)(0) - (1)(-\sin x)}{(\cos x)^2} = \frac{\sin x}{\cos x \cos x} = \frac{1}{\cos x} \cdot \frac{\sin x}{\cos x} = \sec x \tan x$$

(b) $\csc x = \dfrac{1}{\sin x}$

$$\frac{d}{dx}[\csc x] = \frac{d}{dx}\left[\frac{1}{\sin x}\right] = \frac{(\sin x)(0) - (1)(\cos x)}{(\sin x)^2} = -\frac{\cos x}{\sin x \sin x} = -\frac{1}{\sin x} \cdot \frac{\cos x}{\sin x} = -\csc x \cot x$$

(c) $\cot x = \dfrac{\cos x}{\sin x}$

$$\frac{d}{dx}[\cot x] = \frac{d}{dx}\left[\frac{\cos x}{\sin x}\right] = \frac{\sin x(-\sin x) - (\cos x)(\cos x)}{(\sin x)^2} = -\frac{\sin^2 x + \cos^2 x}{\sin^2 x} = -\frac{1}{\sin^2 x} = -\csc^2 x$$

83. $f(x) = 4x^{3/2}$

$f'(x) = 6x^{1/2}$

$f''(x) = 3x^{-1/2} = \dfrac{3}{\sqrt{x}}$

85. $f(x) = \dfrac{x}{x-1}$

$f'(x) = \dfrac{(x-1)(1) - x(1)}{(x-1)^2} = \dfrac{-1}{(x-1)^2}$

$f''(x) = \dfrac{2}{(x-1)^3}$

87. $f(x) = 3 \sin x$

$f'(x) = 3 \cos x$

$f''(x) = -3 \sin x$

89. $g(x) = \dfrac{e^x}{x}$

$g'(x) = \dfrac{xe^x - e^x}{x^2}$

$g''(x) = \dfrac{x^2(xe^x + e^x - e^x) - 2x(xe^x - e^x)}{x^4}$

$= \dfrac{e^x}{x^3}(x^2 - 2x + 2)$

91. $f'(x) = x^2$

$f''(x) = 2x$

93. $f'''(x) = 2\sqrt{x}$

$f^{(4)}(x) = \dfrac{1}{2}(2)x^{-1/2} = \dfrac{1}{\sqrt{x}}$

95.

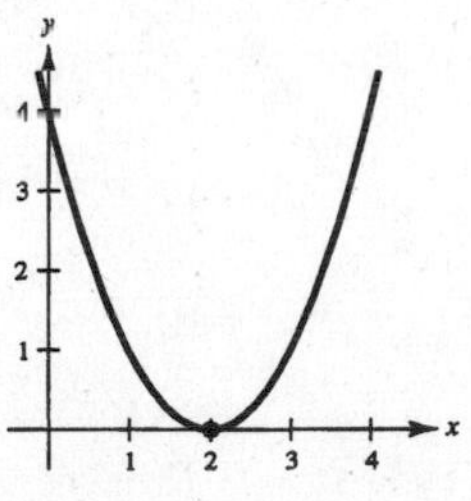

$f(2) = 0$

One such function is $f(x) = (x-2)^2$.

97. $f(x) = 2g(x) + h(x)$

$f'(x) = 2g'(x) + h'(x)$

$f'(2) = 2g'(2) + h'(2)$

$= 2(-2) + 4$

$= 0$

99. $f(x) = \dfrac{g(x)}{h(x)}$

$f'(x) = \dfrac{h(x)g'(x) - g(x)h'(x)}{[h(x)]^2}$

$f'(2) = \dfrac{h(2)g'(2) - g(2)h'(2)}{[h(2)]^2}$

$= \dfrac{(-1)(-2) - (3)(4)}{(-1)^2}$

$= -10$

101.

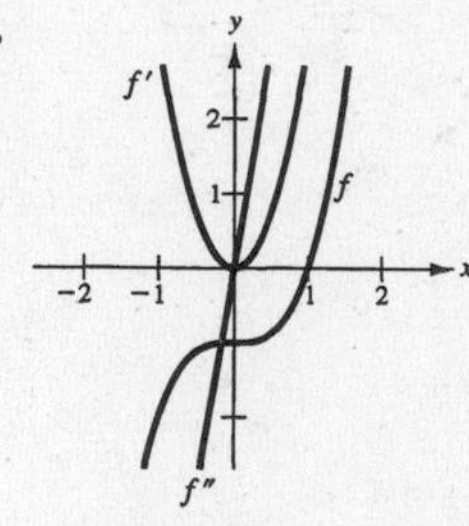

It appears that f is cubic; so f' would be quadratic and f'' would be linear.

103. $v(t) = 64 - t^2,\ 0 \le t \le 8$

$a(t) = -2t$

$v(3) = 55$ m/sec

$a(3) = -6$ m/sec²

The speed of the object is decreasing.

105. $v(t) = \dfrac{100t}{2t + 15}$

$$a(t) = \frac{(2t + 15)(100) - (100t)(2)}{(2t + 15)^2}$$

$$= \frac{1500}{(2t + 15)^2}$$

(a) $a(5) = \dfrac{1500}{[2(5) + 15]^2} = 2.4 \text{ ft/sec}^2$

(b) $a(10) = \dfrac{1500}{[2(10) + 15]^2} \approx 1.2 \text{ ft/sec}^2$

(c) $a(20) = \dfrac{1500}{[2(20) + 15]^2} \approx 0.5 \text{ ft/sec}^2$

107. $f(x) = g(x)h(x)$

(a) $f'(x) = g(x)h'(x) + h(x)g'(x)$

$$f''(x) = g(x)h''(x) + g'(x)h'(x) + h(x)g''(x) + h'(x)g'(x)$$
$$= g(x)h''(x) + 2g'(x)h'(x) + h(x)g''(x)$$
$$f'''(x) = g(x)h'''(x) + g'(x)h''(x) + 2g'(x)h''(x) + 2g''(x)h'(x) + h(x)g'''(x) + h'(x)g''(x)$$
$$= g(x)h'''(x) + 3g'(x)h''(x) + 3g''(x)h'(x) + g'''(x)h(x)$$
$$f^{(4)}(x) = g(x)h^{(4)}(x) + g'(x)h'''(x) + 3g'(x)h'''(x) + 3g''(x)h''(x) + 3g''(x)h''(x) + 3g'''(x)h'(x)$$
$$+ g'''(x)h'(x) + g^{(4)}(x)h(x)$$
$$= g(x)h^{(4)}(x) + 4g'(x)h'''(x) + 6g''(x)h''(x) + 4g'''(x)h'(x) + g^{(4)}(x)h(x)$$

(b) $$f^{(n)}(x) = g(x)h^{(n)}(x) + \frac{n(n-1)(n-2)\cdots(2)(1)}{1[(n-1)(n-2)\cdots(2)(1)]}g'(x)h^{(n-1)}(x) + \frac{n(n-1)(n-2)\cdots(2)(1)}{(2)(1)[(n-2)(n-3)\cdots(2)(1)]}g''(x)h^{(n-2)}(x)$$
$$+ \frac{n(n-1)(n-2)\cdots(2)(1)}{(3)(2)(1)[(n-3)(n-4)\cdots(2)(1)]}g'''(x)h^{(n-3)}(x) + \cdots$$
$$+ \frac{n(n-1)(n-2)\cdots(2)(1)}{[(n-1)(n-2)\cdots(2)(1)](1)}g^{(n-1)}(x)h'(x) + g^{(n)}(x)h(x)$$
$$= g(x)h^{(n)}(x) + \frac{n!}{1!(n-1)!}g'(x)h^{(n-1)}(x) + \frac{n!}{2!(n-2)!}g''(x)h^{(n-2)}(x) + \cdots$$
$$+ \frac{n!}{(n-1)!1!}g^{(n-1)}(x)h'(x) + g^{(n)}(x)h(x)$$

Note: $n! = n(n-1)\ldots 3 \cdot 2 \cdot 1$ (read "n factorial.")

109. $f(x) = \cos x$ $\qquad f\left(\frac{\pi}{3}\right) = \cos\frac{\pi}{3} = \frac{1}{2}$

$f'(x) = -\sin x$ $\qquad f'\left(\frac{\pi}{3}\right) = -\sin\frac{\pi}{3} = -\frac{\sqrt{3}}{2}$

$f''(x) = -\cos x$ $\qquad f''\left(\frac{\pi}{3}\right) = -\cos\frac{\pi}{3} = -\frac{1}{2}$

(a) $P_1(x) = f'(a)(x - a) + f(a) = -\frac{\sqrt{3}}{2}\left(x - \frac{\pi}{3}\right) + \frac{1}{2}$

$P_2(x) = \frac{1}{2}f''(a)(x - a)^2 + f'(a)(x - a) + f(a)$

$= -\frac{1}{4}\left(x - \frac{\pi}{3}\right)^2 - \frac{\sqrt{3}}{2}\left(x - \frac{\pi}{3}\right) + \frac{1}{2}$

(b)

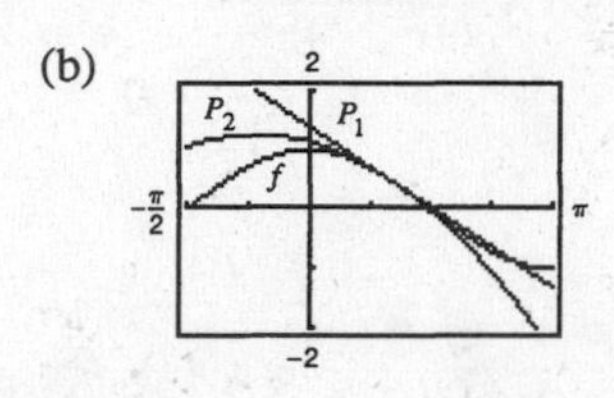

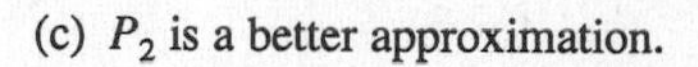

(c) P_2 is a better approximation.

(d) The accuracy worsens as you move farther away from $x = a = (\pi/3)$.

111. False. If $y = f(x)g(x)$, then

$\frac{dy}{dx} = f(x)g'(x) + g(x)f'(x).$

113. True

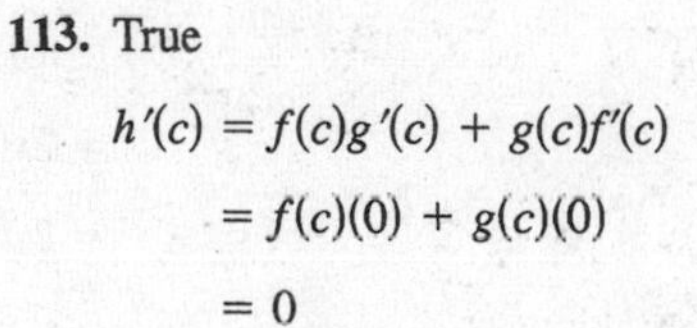

$h'(c) = f(c)g'(c) + g(c)f'(c)$

$= f(c)(0) + g(c)(0)$

$= 0$

115. True

117. $f(x) = x|x| = \begin{cases} x^2, & \text{if } x \geq 0 \\ -x^2, & \text{if } x < 0 \end{cases}$

$f'(x) = \begin{cases} 2x, & \text{if } x \geq 0 \\ -2x, & \text{if } x < 0 \end{cases} = 2|x|$

$f''(x) = \begin{cases} 2, & \text{if } x > 0 \\ -2, & \text{if } x < 0 \end{cases}$

$f''(0)$ does not exist since the left and right derivatives are not equal.

Section 2.4 The Chain Rule

$y = f(g(x))$	$u = g(x)$	$y = f(u)$
1. $y = (6x - 5)^4$	$u = 6x - 5$	$y = u^4$
3. $y = \sqrt{x^2 - 1}$	$u = x^2 - 1$	$y = \sqrt{u}$
5. $y = \csc^3 x$	$u = \csc x$	$y = u^3$
7. $y = e^{-2x}$	$u = -2x$	$y = e^u$

9. $y = (2x - 7)^3$

$y' = 3(2x - 7)^2(2) = 6(2x - 7)^2$

11. $g(x) = 3(4 - 9x)^4$

$g'(x) = 12(4 - 9x)^3(-9) = -108(4 - 9x)^3$

13. $f(x) = (9 - x^2)^{2/3}$

$$f'(x) = \frac{2}{3}(9 - x^2)^{-1/3}(-2x) = -\frac{4x}{3(9 - x^2)^{1/3}}$$

15. $f(t) = (1 - t)^{1/2}$

$$f'(t) = \frac{1}{2}(1 - t)^{-1/2}(-1) = -\frac{1}{2\sqrt{1 - t}}$$

17. $y = (9x^2 + 4)^{1/3}$

$$y' = \frac{1}{3}(9x^2 + 4)^{-2/3}(18x) = \frac{6x}{(9x^2 + 4)^{2/3}}$$

19. $y = 2(4 - x^2)^{1/4}$

$$y' = 2\left(\frac{1}{4}\right)(4 - x^2)^{-3/4}(-2x)$$

$$= \frac{-x}{\sqrt[4]{(4 - x^2)^3}}$$

21. $y = (x - 2)^{-1}$

$$y' = -1(x - 2)^{-2}(1) = \frac{-1}{(x - 2)^2}$$

23. $f(t) = (t - 3)^{-2}$

$$f'(t) = -2(t - 3)^{-3} = \frac{-2}{(t - 3)^3}$$

25. $y = (x + 2)^{-1/2}$

$$\frac{dy}{dx} = -\frac{1}{2}(x + 2)^{-3/2} = -\frac{1}{2(x + 2)^{3/2}}$$

27. $f(x) = x^2(x - 2)^4$

$$f'(x) = x^2[4(x - 2)^3(1)] + (x - 2)^4(2x)$$

$$= 2x(x - 2)^3[2x + (x - 2)]$$

$$= 2x(x - 2)^3(3x - 2)$$

29. $y = x\sqrt{1 - x^2} = x(1 - x^2)^{1/2}$

$$y' = x\left[\frac{1}{2}(1 - x^2)^{-1/2}(-2x)\right] + (1 - x^2)^{1/2}(1)$$

$$= -x^2(1 - x^2)^{-1/2} + (1 - x^2)^{1/2}$$

$$= (1 - x^2)^{-1/2}[-x^2 + (1 - x^2)]$$

$$= \frac{1 - 2x^2}{\sqrt{1 - x^2}}$$

31. $y = \dfrac{x}{\sqrt{x^2 + 1}} = x(x^2 + 1)^{-1/2}$

$$y' = x\left[-\frac{1}{2}(x^2 + 1)^{-3/2}(2x)\right] + (x^2 + 1)^{-1/2}(1)$$

$$= -x^2(x^2 + 1)^{-3/2} + (x^2 + 1)^{-1/2}$$

$$= (x^2 + 1)^{-3/2}[-x^2 + (x^2 + 1)]$$

$$= \frac{1}{(x^2 + 1)^{3/2}}$$

33. $g(x) = \left(\dfrac{x + 5}{x^2 + 2}\right)^2$

$$g'(x) = 2\left(\frac{x + 5}{x^2 + 2}\right)\left(\frac{(x^2 + 2) - (x + 5)(2x)}{(x^2 + 2)^2}\right)$$

$$= \frac{2(x + 5)(2 - 10x - x^2)}{(x^2 + 2)^3}$$

35. $f(v) = \left(\dfrac{1 - 2v}{1 + v}\right)^3$

$$f'(v) = 3\left(\frac{1 - 2v}{1 + v}\right)^2\left(\frac{(1 + v)(-2) - (1 - 2v)}{(1 + v)^2}\right)$$

$$= \frac{-9(1 - 2v)^2}{(1 + v)^4}$$

37. $y = \dfrac{\sqrt{x} + 1}{x^2 + 1}$

$$y' = \frac{1 - 3x^2 - 4x^{3/2}}{2\sqrt{x}(x^2 + 1)^2}$$

The zero of y' corresponds to the point on the graph of y where the tangent line is horizontal.

39. $g(t) = \dfrac{3t^2}{\sqrt{t^2 + 2t - 1}}$

$$g'(t) = \frac{3t(t^2 + 3t - 2)}{(t^2 + 2t - 1)^{3/2}}$$

The zeros of g' correspond to the points on the graph of g where the tangent lines are horizontal.

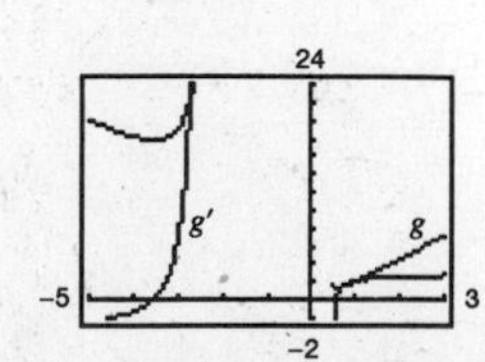

41. $y = \sqrt{\dfrac{x+1}{x}}$

$y' = -\dfrac{\sqrt{(x+1)/x}}{2x(x+1)}$

y' has no zeros.

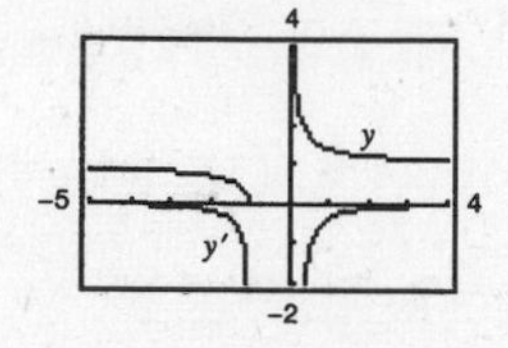

43. 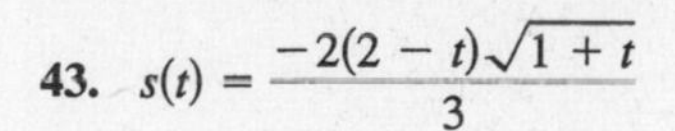 $s(t) = \dfrac{-2(2-t)\sqrt{1+t}}{3}$

$s'(t) = \dfrac{t}{\sqrt{1+t}}$

The zero of $s'(t)$ corresponds to the point on the graph of $s(t)$ where the tangent line is horizontal.

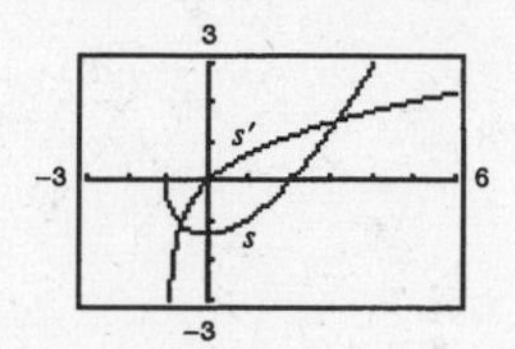

45. $y = \dfrac{\cos \pi x + 1}{x}$

$\dfrac{dy}{dx} = \dfrac{-\pi x \sin \pi x - \cos \pi x - 1}{x^2}$

$= -\dfrac{\pi x \sin \pi x + \cos \pi x + 1}{x^2}$

The zeros of y' correspond to the points on the graph of y where the tangent lines are horizontal.

47. (a) $y = \sin x$

$y' = \cos x$

$y'(0) = 1$

1 cycle in $[0, 2\pi]$

(b) $y = \sin 2x$

$y' = 2 \cos 2x$

$y'(0) = 2$

2 cycles in $[0, 2\pi]$

The slope of $\sin ax$ at the origin is a.

49. (a) $y = e^{3x}$

$y' = 3e^{3x}$

At $(0, 1)$, $y' = 3$

(b) $y = e^{-3x}$

$y' = -3e^{-3x}$

At $(0, 1)$, $y' = -3$

51. $y = \ln x^3 = 3 \ln x$

$y' = \dfrac{3}{x}$

At $(1, 0)$, $y' = 3$

53. $y = \ln x^2 = 2 \ln x$

$y' = \dfrac{2}{x}$

At $(1, 0)$, $y' = 2$

55. $y = \cos 3x$

$\dfrac{dy}{dx} = -3 \sin 3x$

57. $g(x) = 3 \tan 4x$

$g'(x) = 12 \sec^2 4x$

59. $f(\theta) = \dfrac{1}{4} \sin^2 2\theta = \dfrac{1}{4}(\sin 2\theta)^2$

$f'(\theta) = 2\left(\dfrac{1}{4}\right)(\sin 2\theta)(\cos 2\theta)(2)$

$= \sin 2\theta \cos 2\theta = \dfrac{1}{2} \sin 4\theta$

61. $y = \sqrt{x} + \dfrac{1}{4} \sin(2x)^2$

$= \sqrt{x} + \dfrac{1}{4} \sin(4x^2)$

$\dfrac{dy}{dx} = \dfrac{1}{2} x^{-1/2} + \dfrac{1}{4} \cos(4x^2)(8x)$

$= \dfrac{1}{2\sqrt{x}} + 2x \cos(2x)^2$

63. $y = \sin(\cos x)$

$$\frac{dy}{dx} = \cos(\cos x) \cdot (-\sin x)$$

$$= -\sin x \cos(\cos x)$$

65. $f(x) = e^{2x}$

$$f'(x) = 2e^{2x}$$

67. $y = e^{\sqrt{x}}$

$$\frac{dy}{dx} = \frac{e^{\sqrt{x}}}{2\sqrt{x}}$$

69. $g(t) = (e^{-t} + e^{t})^3$

$$g'(t) = 3(e^{-t} + e^{t})^2(e^{t} - e^{-t})$$

71. $y = \ln e^{x^2} = x^2$

$$\frac{dy}{dx} = 2x$$

73. $y = \dfrac{2}{e^x + e^{-x}} = 2(e^x + e^{-x})^{-1}$

$$\frac{dy}{dx} = -2(e^x + e^{-x})^{-2}(e^x - e^{-x})$$

$$= \frac{-2(e^x - e^{-x})}{(e^x + e^{-x})^2}$$

75. $y = x^2e^x - 2xe^x + 2e^x = e^x(x^2 - 2x + 2)$

$$\frac{dy}{dx} = e^x(2x - 2) + e^x(x^2 - 2x + 2) = x^2e^x$$

77. $f(x) = e^{-x} \ln x$

$$f'(x) = e^{-x}\left(\frac{1}{x}\right) - e^{-x}\ln x = e^{-x}\left(\frac{1}{x} - \ln x\right)$$

79. $y = e^x(\sin x + \cos x)$

$$\frac{dy}{dx} = e^x(\cos x - \sin x) + (\sin x + \cos x)(e^x)$$

$$= e^x(2\cos x) = 2e^x \cos x$$

81. $g(x) = \ln x^2 = 2\ln x$

$$g'(x) = \frac{2}{x}$$

83. $y = (\ln x)^4$

$$\frac{dy}{dx} = 4(\ln x)^3\left(\frac{1}{x}\right) = \frac{4(\ln x)^3}{x}$$

85. $y = \ln x\sqrt{x^2 - 1} = \ln x + \frac{1}{2}\ln(x^2 - 1)$

$$\frac{dy}{dx} = \frac{1}{x} + \frac{1}{2}\left(\frac{2x}{x^2 - 1}\right) = \frac{2x^2 - 1}{x(x^2 - 1)}$$

87. $f(x) = \ln\dfrac{x}{x^2 + 1} = \ln x - \ln(x^2 + 1)$

$$f'(x) = \frac{1}{x} - \frac{2x}{x^2 + 1} = \frac{1 - x^2}{x(x^2 + 1)}$$

89. $g(t) = \dfrac{\ln t}{t^2}$

$$g'(t) = \frac{t^2(1/t) - 2t\ln t}{t^4} = \frac{1 - 2\ln t}{t^3}$$

91. $y = \ln\sqrt{\dfrac{x + 1}{x - 1}} = \dfrac{1}{2}[\ln(x + 1) - \ln(x - 1)]$

$$\frac{dy}{dx} = \frac{1}{2}\left[\frac{1}{x + 1} - \frac{1}{x - 1}\right] = \frac{1}{1 - x^2}$$

93. $y = \dfrac{-\sqrt{x^2 + 1}}{x} + \ln\left(x + \sqrt{x^2 + 1}\right)$

$$\frac{dy}{dx} = \frac{-x\left(x/\sqrt{x^2 + 1}\right) + \sqrt{x^2 + 1}}{x^2} + \left(\frac{1}{x + \sqrt{x^2 + 1}}\right)\left(1 + \frac{x}{\sqrt{x^2 + 1}}\right)$$

$$= \frac{1}{x^2\sqrt{x^2 + 1}} + \left(\frac{1}{x + \sqrt{x^2 + 1}}\right)\left(\frac{\sqrt{x^2 + 1} + x}{\sqrt{x^2 + 1}}\right)$$

$$= \frac{1}{x^2\sqrt{x^2 + 1}} + \frac{1}{\sqrt{x^2 + 1}} = \frac{1 + x^2}{x^2\sqrt{x^2 + 1}} = \frac{\sqrt{x^2 + 1}}{x^2}$$

95. $y = \ln|\sin x|$

$$\frac{dy}{dx} = \frac{\cos x}{\sin x} = \cot x$$

97. $y = \ln\left|\frac{\cos x}{\cos x - 1}\right|$

$$= \ln|\cos x| - \ln|\cos x - 1|$$

$$\frac{dy}{dx} = \frac{-\sin x}{\cos x} - \frac{-\sin x}{\cos x - 1} = -\tan x + \frac{\sin x}{\cos x - 1}$$

99. $y = \ln\left|\frac{-1 + \sin x}{2 + \sin x}\right|$

$$= \ln|-1 + \sin x| - \ln|2 + \sin x|$$

$$\frac{dy}{dx} = \frac{\cos x}{-1 + \sin x} - \frac{\cos x}{2 + \sin x}$$

$$= \frac{3\cos x}{(\sin x - 1)(\sin x + 2)}$$

101. $f(x) = 2(x^2 - 1)^3$

$$f'(x) = 6(x^2 - 1)^2(2x)$$

$$= 12x(x^4 - 2x^2 + 1)$$

$$= 12x^5 - 24x^3 + 12x$$

$$f''(x) = 60x^4 - 72x^2 + 12$$

$$= 12(5x^2 - 1)(x^2 - 1)$$

103. $f(x) = \sin x^2$

$$f'(x) = 2x\cos x^2$$

$$f''(x) = 2x[2x(-\sin x^2)] + 2\cos x^2$$

$$= 2[\cos x^2 - 2x^2 \sin x^2]$$

105. $f(x) = (3 + 2x)e^{-3x}$

$$f'(x) = (3 + 2x)(-3e^{-3x}) + 2e^{-3x}$$

$$= (-7 - 6x)e^{-3x}$$

$$f''(x) = (-7 - 6x)(-3e^{-3x}) - 6e^{-3x}$$

$$= 3(6x + 5)e^{-3x}$$

107. $s(t) = (t^2 + 2t + 8)^{1/2}, \quad (2, 4)$

$$s'(t) = \frac{1}{2}(t^2 + 2t + 8)^{-1/2}(2t + 2)$$

$$= \frac{t + 1}{\sqrt{t^2 + 2t + 8}}$$

$$s'(2) = \frac{3}{4}$$

109. $f(x) = \frac{3}{x^3 - 4} = 3(x^3 - 4)^{-1}, \quad \left(-1, -\frac{3}{5}\right)$

$$f'(x) = -3(x^3 - 4)^{-2}(3x^2) = -\frac{9x^2}{(x^3 - 4)^2}$$

$$f'(-1) = -\frac{9}{25}$$

111. $f(t) = \frac{3t + 2}{t - 1}, \quad (0, -2)$

$$f'(t) = \frac{(t - 1)(3) - (3t + 2)(1)}{(t - 1)^2} = \frac{-5}{(t - 1)^2}$$

$$f'(0) = -5$$

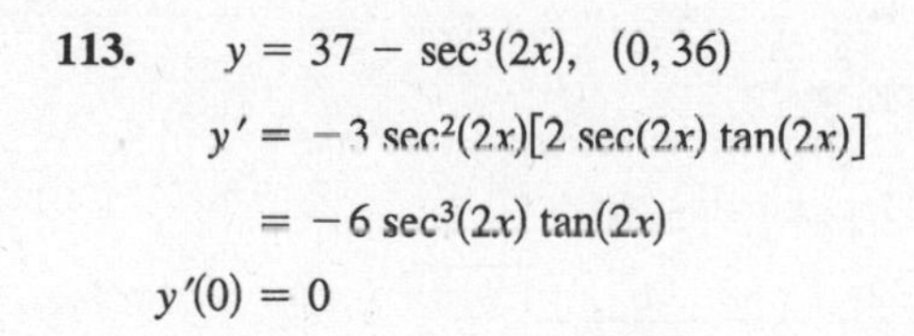

113. $y = 37 - \sec^3(2x), \quad (0, 36)$

$$y' = -3\sec^2(2x)[2\sec(2x)\tan(2x)]$$

$$= -6\sec^3(2x)\tan(2x)$$

$$y'(0) = 0$$

115. (a) $f(x) = \sqrt{3x^2 - 2}, \quad (3, 5)$

$$f'(x) = \frac{1}{2}(3x^2 - 2)^{-1/2}(6x)$$

$$= \frac{3x}{\sqrt{3x^2 - 2}}$$

$$f'(3) = \frac{9}{5}$$

Tangent line:

$$y - 5 = \frac{9}{5}(x - 3) \Rightarrow 9x - 5y - 2 = 0$$

(b)

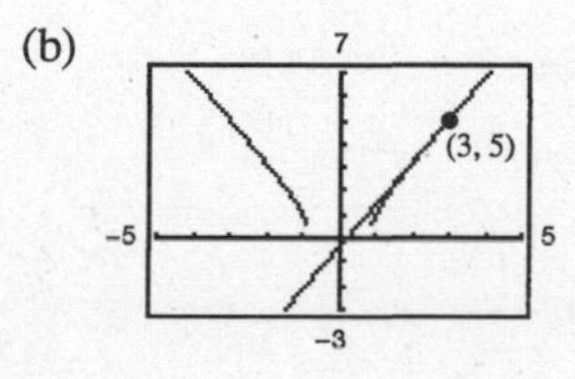

117. (a) $f(x) = \sin 2x, \quad (\pi, 0)$

$f'(x) = 2\cos 2x$

$f'(\pi) = 2$

Tangent line:

$y = 2(x - \pi) \Rightarrow 2x - y - 2\pi = 0$

(b)

119. (a) $y = 4 - x^2 - \ln\left(\frac{1}{2}x + 1\right), \quad (0, 4)$

$$\frac{dy}{dx} = -2x - \frac{1}{(1/2)x + 1}\left(\frac{1}{2}\right)$$

$$= -2x - \frac{1}{x + 2}$$

When $x = 0$, $\frac{dy}{dx} = -\frac{1}{2}$.

Tangent line: $y - 4 = -\frac{1}{2}(x - 0)$

$$y = -\frac{1}{2}x + 4$$

(b)

121. $f(x) = 4^x$

$f'(x) = (\ln 4)\, 4^x$

123. $y = 5^{x-2}$

$\frac{dy}{dx} = (\ln 5)\, 5^{x-2}$

125. $g(t) = t^2 2^t$

$g'(t) = t^2 (\ln 2)\, 2^t + (2t)\, 2^t$

$= t2^t (t \ln 2 + 2)$

$= 2^t t(2 + t \ln 2)$

127. $h(\theta) = 2^{-\theta} \cos \pi\theta$

$h'(\theta) = 2^{-\theta}(-\pi \sin \pi\theta) - (\ln 2)2^{-\theta} \cos \pi\theta$

$= -2^{-\theta}[(\ln 2) \cos \pi\theta + \pi \sin \pi\theta]$

129. $y = \log_3 x$

$\frac{dy}{dx} = \frac{1}{x \ln 3}$

131. $f(x) = \log_2 \frac{x^2}{x - 1}$

$= 2 \log_2 x - \log_2 (x - 1)$

$$f'(x) = \frac{2}{x \ln 2} - \frac{1}{(x - 1) \ln 2}$$

$$= \frac{x - 2}{(\ln 2)x(x - 1)}$$

133. $y = \log_5 \sqrt{x^2 - 1} = \frac{1}{2} \log_5 (x^2 - 1)$

$$\frac{dy}{dx} = \frac{1}{2} \cdot \frac{2x}{(x^2 - 1)\ln 5} = \frac{x}{(x^2 - 1)\ln 5}$$

135. $g(t) = \frac{10 \log_4 t}{t} = \frac{10}{\ln 4}\left(\frac{\ln t}{t}\right)$

$$g'(t) = \frac{10}{\ln 4}\left[\frac{t(1/t) - \ln t}{t^2}\right]$$

$$= \frac{10}{t^2 \ln 4}[1 - \ln t] = \frac{5}{t^2 \ln 2}(1 - \ln t)$$

137.

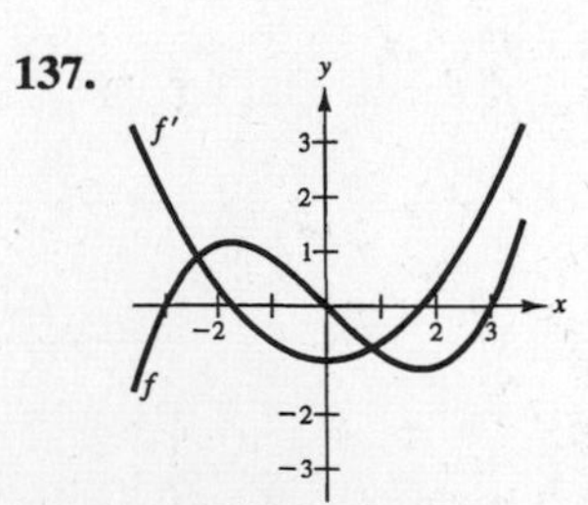

The zeros of f' correspond to the points where the graph of f has horizontal tangents.

139.

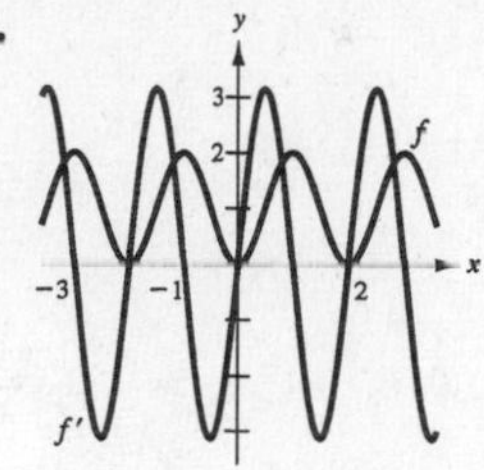

The zeros of f' correspond to the points where the graph of f has horizontal tangents.

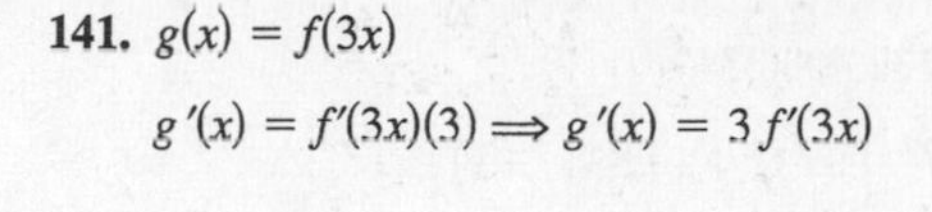

141. $g(x) = f(3x)$

$g'(x) = f'(3x)(3) \Rightarrow g'(x) = 3f'(3x)$

143. (a) $f(x) = g(x)h(x)$

$f'(x) = g(x)h'(x) + g'(x)h(x)$

$f'(5) = (-3)(-2) + (6)(3) = 24$

(b) $f(x) = g(h(x))$

$f'(x) = g'(h(x))h'(x)$

$f'(5) = g'(3)(-2) = -2g'(3)$

Need $g'(3) = g'(h(5))$ to find $f'(5)$.

(c) $f(x) = \dfrac{g(x)}{h(x)}$

$$f'(x) = \frac{h(x)g'(x) - g(x)h'(x)}{[h(x)]^2}$$

$$f'(5) = \frac{(3)(6) - (-3)(-2)}{(3)^2} = \frac{12}{9} = \frac{4}{3}$$

(d) $f(x) = [g(x)]^3$

$f'(x) = 3[g(x)]^2 g'(x)$

$f'(5) = 3(-3)^2(6) = 162$

145. $f(x) = e^{x/2}, f(0) = 1$

$f'(x) = \frac{1}{2}e^{x/2}, f'(0) = \frac{1}{2}$

$f''(x) = \frac{1}{4}e^{x/2}, f''(0) = \frac{1}{4}$

$P_1(x) = 1 + \frac{1}{2}(x - 0) = \frac{x}{2} + 1, P_1(0) = 1$

$P_1'(x) = \frac{1}{2}, P_1'(0) = \frac{1}{2}$

$P_2(x) = 1 + \frac{1}{2}(x - 0) + \frac{1}{8}(x - 0)^2 = \frac{x^2}{8} + \frac{x}{2} + 1, P_2(0) = 1$

$P_2'(x) = \frac{1}{4}x + \frac{1}{2}, P_2'(0) = \frac{1}{2}$

$P_2''(x) = \frac{1}{4}, P_2''(0) = \frac{1}{4}$

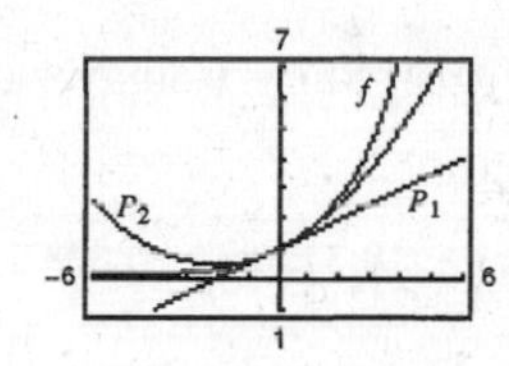

The values of f, P_1, P_2 and their first derivatives agree at $x = 0$. The values of the second derivatives of f and P_2 agree at $x = 0$.

147. (a) $y = e^x$

$y_1 = 1 + x$

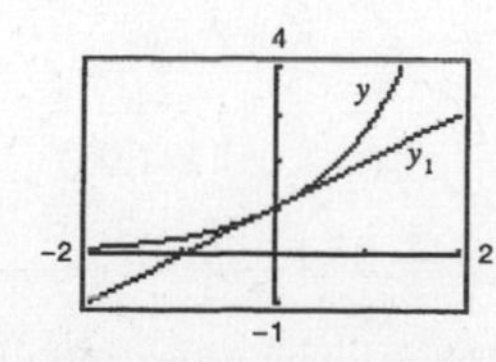

(b) $y = e^x$

$y_2 = 1 + x + \left(\dfrac{x^2}{2}\right)$

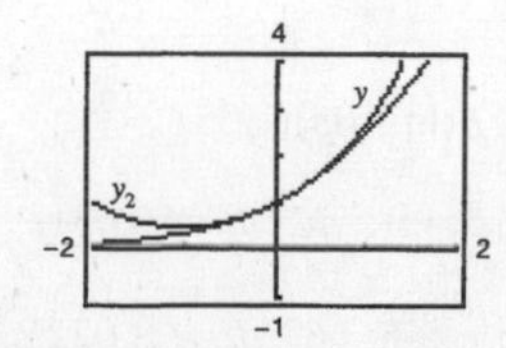

(c) $y = e^x$

$y_3 = 1 + x + \dfrac{x^2}{2} + \dfrac{x^3}{6}$

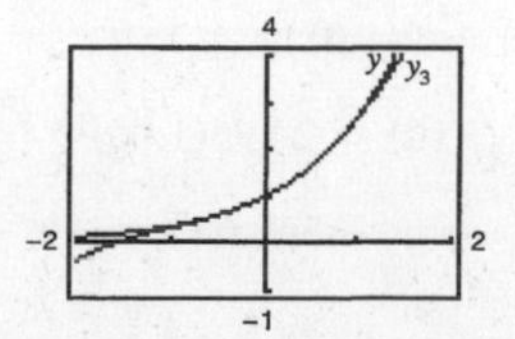

149. (a) $F = 132{,}400(331 - v)^{-1}$

$F' = (-1)(132{,}400)(331 - v)^{-2}(-1)$

$= \dfrac{132{,}400}{(331 - v)^2}$

When $v = 30$, $F' \approx 1.461$.

(b) $F = 132{,}400(331 + v)^{-1}$

$F' = (-1)(132{,}400)(331 + v)^{-2}(1)$

$= \dfrac{-132{,}400}{(331 + v)^2}$

When $v = 30$, $F' \approx -1.016$.

151. $\theta = 0.2 \cos 8t$

The maximum angular displacement is $\theta = 0.2$ (since $-1 \le \cos 8t \le 1$).

$\dfrac{d\theta}{dt} = 0.2[-8 \sin 8\mathrm{t}] = -1.6 \sin 8t$

When $t = 3$, $d\theta/dt = -1.6 \sin 24 \approx 1.4489$ radians per second.

153. $S = C(R^2 - r^2)$

$\dfrac{dS}{dt} = C\left(2R\dfrac{dR}{dt} - 2r\dfrac{dr}{dt}\right)$

Since r is constant, we have $dr/dt = 0$ and

$\dfrac{dS}{dt} = (1.76 \times 10^5)(2)(1.2 \times 10^{-2})(10^{-5})$

$= 4.224 \times 10^{-2} = 0.04224$.

155. $V = \dfrac{4}{3}\pi r^3, \quad \dfrac{dr}{dt} = 3.$

$\dfrac{dv}{dt} = 4\pi r^2 \dfrac{dr}{dt}$

$= 4\pi(8)^2(3)$

$= 768\pi$ cubic inches/second

157. (a) $x = -1.6372t^3 + 19.3120t^2 - 0.5082t - 0.6162$

(b) $C = 60x + 1350$

$= 60(-1.6372t^3 + 19.3120t^2 - 0.5082t - 0.6162) + 1350$

$\dfrac{dC}{dt} = 60(-4.9116t^2 + 38.624t - 0.5082)$

$= -294.696t^2 + 2317.44t - 30.492$

(c) The function dC/dt is quadratic, not linear. The cost function levels off at the end of the day, perhaps due to fatigue.

159. $V(t) = 20{,}000\left(\dfrac{3}{4}\right)^t$

(a)

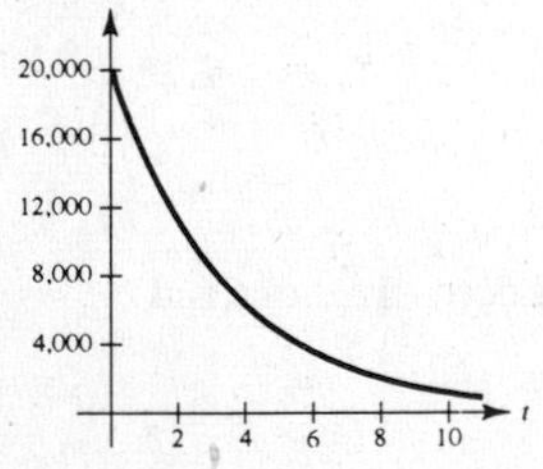

$V(2) = 20{,}000\left(\dfrac{3}{4}\right)^2 = \$11{,}250$

(b) $\dfrac{dV}{dt} = 20{,}000\left(\ln \dfrac{3}{4}\right)\left(\dfrac{3}{4}\right)^t$

When $t = 1$: $\dfrac{dV}{dt} \approx -4315.23$

When $t = 4$: $\dfrac{dV}{dt} \approx -1820.49$

161. $C(t) = P(1.05)^t$

(a) $C(10) = 24.95(1.05)^{10}$

$\approx \$40.64$

(b) $\dfrac{dC}{dt} = P(\ln 1.05)(1.05)^t$

When $t = 1$: $dC/dt \approx 0.051P$

When $t = 8$: $dC/dt \approx 0.072P$

(c) $\dfrac{dC}{dt} = (\ln 1.05)[P(1.05)^t]$

$= (\ln 1.05)C(t)$

The constant of proportionality is $\ln 1.05$.

163. $f(x + p) = f(x)$ for all x.

(a) Yes, $f'(x + p) = f'(x)$, which shows that f' is periodic as well.

(b) Yes, let $g(x) = f(2x)$, so $g'(x) = 2f'(2x)$. Since f' is periodic, so is g'.

165. $g(x) = |2x - 3|$

$$g'(x) = 2\left(\frac{2x - 3}{|2x - 3|}\right), \quad x \neq \frac{3}{2}$$

167. $h(x) = |x|\cos x$

$$h'(x) = -|x| \sin x + \frac{x}{|x|} \cos x, \quad x \neq 0$$

169. (a) $f(x) = \tan \dfrac{\pi x}{4}$ $\qquad f(1) = 1$

$$f'(x) = \frac{\pi}{4} \sec^2 \frac{\pi x}{4} \qquad f'(1) = \frac{\pi}{4}(2) = \frac{\pi}{2}$$

$$f''(x) = \frac{\pi}{2} \sec^2 \frac{\pi x}{4} \cdot \tan \frac{\pi x}{4}\left(\frac{\pi}{4}\right) \qquad f''(1) = \frac{\pi^2}{8}(2)(1) = \frac{\pi^2}{4}$$

$$P_1(x) = f'(1)(x - 1) + f(1) = \frac{\pi}{2}(x - 1) + 1$$

$$P_2(x) = \frac{1}{2}\left(\frac{\pi^2}{4}\right)(x - 1)^2 + f'(1)(x - 1) + f(1) = \frac{\pi^2}{8}(x - 1)^2 + \frac{\pi}{2}(x - 1) + 1$$

(b)

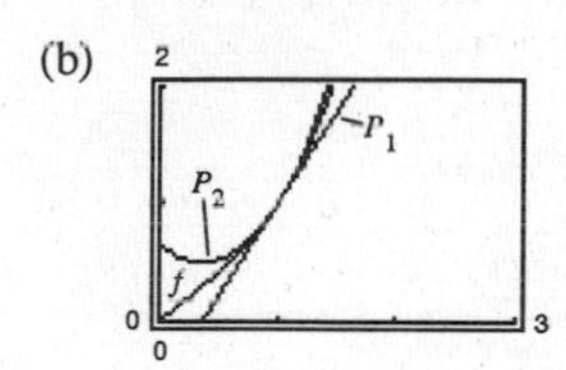

(c) P_2 is a better approximation than P_1

(d) The accuracy worsens as you move away from $x = c = 1$.

171. $f(x) = e^{-x^2/2}, f(0) = 1$

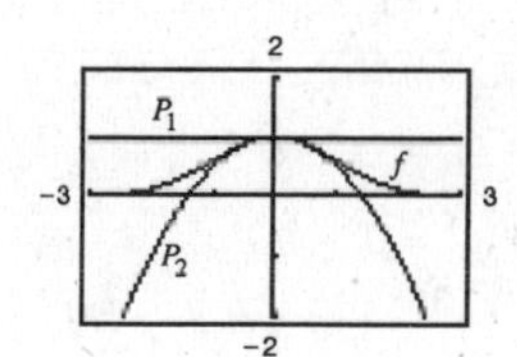

$f'(x) = -xe^{-x^2/2}, f'(0) = 0$

$f''(x) = x^2e^{-x^2/2} - e^{-x^2/2} = e^{-x^2/2}(x^2 - 1), f''(0) = -1$

$P_1(x) = 1 + 0(x - 0) = 1, P_1(0) = 1$

$P_1'(x) = 0,\ P_1'(0) = 0$

$P_2(x) = 1 + 0(x - 0) - \dfrac{1}{2}(x - 0)^2 = 1 - \dfrac{x^2}{2}, P_2(0) = 1$

$P_2'(x) = -x, P_2'(0) = 0$

$P_2''(x) = -1, P_2''(0) = -1$

The values of f, P_1, P_2 and their first derivatives agree at $x = 0$. The values of the second derivatives of f and P_2 agree at $x = 0$.

173. $g = \sqrt{x(x+n)}$

$= \sqrt{x^2 + nx}$

$\frac{dg}{dx} = \frac{1}{2}(x^2 + nx)^{-1/2}(2x + n)$

$= \frac{2x + n}{2\sqrt{x^2 + nx}}$

$= \frac{(2x + n)/2}{\sqrt{x(x+n)}}$

$= \frac{[x + (x + n)]/2}{\sqrt{x(x+n)}}$

$= \frac{a}{g}$

175. False. If $y = (1 - x)^{1/2}$, then

$y' = \frac{1}{2}(1 - x)^{-1/2}(-1)$.

177. True

Section 2.5 Implicit Differentiation

1. $x^2 + y^2 = 36$

$2x + 2yy' = 0$

$y' = \frac{-x}{y}$

3. $x^{1/2} + y^{1/2} = 9$

$\frac{1}{2}x^{-1/2} + \frac{1}{2}y^{-1/2}y' = 0$

$y' = -\frac{x^{-1/2}}{y^{-1/2}} = -\sqrt{\frac{y}{x}}$

5. $x^3 - xy + y^2 = 4$

$3x^2 - xy' - y + 2yy' = 0$

$(2y - x)y' = y - 3x^2$

$y' = \frac{y - 3x^2}{2y - x}$

7. $xe^y - 10x + 3y = 0$

$xe^y\frac{dy}{dx} + e^y - 10 + 3\frac{dy}{dx} = 0$

$\frac{dy}{dx}(xe^y + 3) = 10 - e^y$

$\frac{dy}{dx} = \frac{10 - e^y}{xe^y + 3}$

9. $x^3y^3 - y - x = 0$

$3x^3y^2y' + 3x^2y^3 - y' - 1 = 0$

$(3x^3y^2 - 1)y' = 1 - 3x^2y^3$

$y' = \frac{1 - 3x^2y^3}{3x^3y^2 - 1}$

11. $x^3 - 2x^2y + 3xy^2 = 38$

$3x^2 - 2x^2y' - 4xy + 6xyy' + 3y^2 = 0$

$2x(3y - x)y' = 4xy - 3x^2 - 3y^2$

$y' = \frac{4xy - 3x^2 - 3y^2}{2x(3y - x)}$

13. $\sin x + 2\cos 2y = 1$

$\cos x - 4(\sin 2y)y' = 0$

$y' = \frac{\cos x}{4\sin 2y}$

15. $\sin x = x(1 + \tan y)$

$\cos x = x(\sec^2 y)y' + (1 + \tan y)(1)$

$y' = \frac{\cos x - \tan y - 1}{x\sec^2 y}$

17. $y = \sin(xy)$

$y' = [xy' + y]\cos(xy)$

$y' - x\cos(xy)y' = y\cos(xy)$

$y' = \frac{y\cos(xy)}{1 - x\cos(xy)}$

19. $x^2 - 3\ln y + y^2 = 10$

$2x - \frac{3}{y}\frac{dy}{dx} + 2y\frac{dy}{dx} = 0$

$2x = \frac{dy}{dx}\left(\frac{3}{y} - 2y\right)$

$\frac{dy}{dx} = \frac{2x}{(3/y) - 2y} = \frac{2xy}{3 - 2y^2}$

21. (a) $x^2 + y^2 = 16$

$$y^2 = 16 - x^2$$

$$y = \pm\sqrt{16 - x^2}$$

(b)

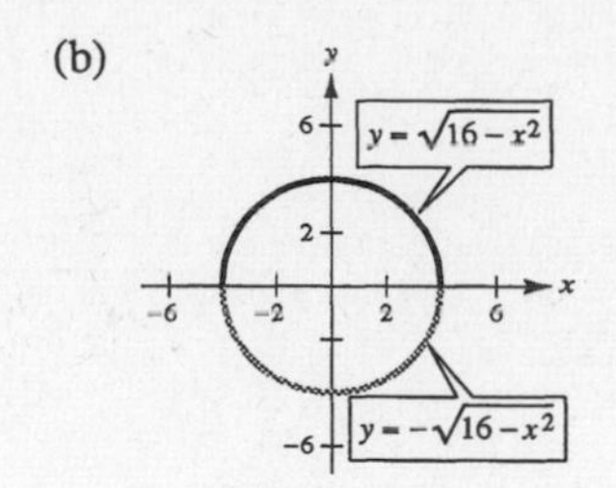

(c) Explicitly:

$$\frac{dy}{dx} = \pm\frac{1}{2}(16 - x^2)^{-1/2}(-2x)$$

$$= \frac{\mp x}{\sqrt{16 - x^2}} = \frac{-x}{\pm\sqrt{16 - x^2}} = \frac{-x}{y}$$

(d) Implicitly:

$$2x + 2yy' = 0$$

$$y' = -\frac{x}{y}$$

23. (a) $16y^2 = 144 - 9x^2$

$$y^2 = \frac{1}{16}(144 - 9x^2) = \frac{9}{16}(16 - x^2)$$

$$y = \pm\frac{3}{4}\sqrt{16 - x^2}$$

(b)

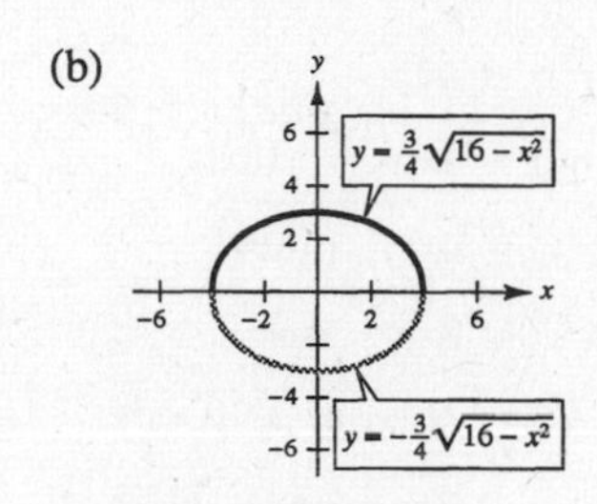

(c) Explicitly:

$$\frac{dy}{dx} = \pm\frac{3}{8}(16 - x^2)^{-1/2}(-2x)$$

$$= \mp\frac{3x}{4\sqrt{16 - x^2}} = \frac{-3x}{4(4/3)y} = \frac{-9x}{16y}$$

(d) Implicitly:

$$18x + 32yy' = 0$$

$$y' = \frac{-9x}{16y}$$

25. $xy = 4$

$$xy' + y(1) = 0$$

$$xy' = -y$$

$$y' = \frac{-y}{x}$$

At $(-4, -1)$: $y' = -\frac{1}{4}$

27. $y^2 = \dfrac{x^2 - 9}{x^2 + 9}$

$$2yy' = \frac{(x^2 + 9)(2x) - (x^2 - 9)2x}{(x^2 + 9)^2}$$

$$y' = \frac{18x}{(x^2 + 9)^2 y}$$

At $(3, 0)$: y' is undefined.

29. $x^{2/3} + y^{2/3} = 5$

$$\frac{2}{3}x^{-1/3} + \frac{2}{3}y^{-1/3}y' = 0$$

$$y' = \frac{-x^{-1/3}}{y^{-1/3}} = -\sqrt[3]{\frac{y}{x}}$$

At $(8, 1)$: $y' = -\frac{1}{2}$

31. $\tan(x + y) = x$

$$(1 + y')\sec^2(x + y) = 1$$

$$y' = \frac{1 - \sec^2(x + y)}{\sec^2(x + y)}$$

$$= \frac{-\tan^2(x + y)}{\tan^2(x + y) + 1} = -\sin^2(x + y)$$

$$= -\frac{x^2}{x^2 + 1}$$

33.

$$3e^{xy} - x = 0, (3, 0)$$

$$3e^{xy}[xy' + y] - 1 = 0$$

$$3e^{xy}xy' = 1 - 3ye^{xy}$$

$$y' = \frac{1 - 3ye^{xy}}{3xe^{xy}}$$

At (3, 0): $y' = \frac{1}{9}$

35.

$$(x^2 + 4)y = 8$$

$$(x^2 + 4)y' + y(2x) = 0$$

$$y' = \frac{-2xy}{x^2 + 4}$$

$$= \frac{-2x[8/(x^2 + 4)]}{x^2 + 4}$$

$$= \frac{-16x}{(x^2 + 4)^2}$$

At (2, 1): $y' = \frac{-32}{64} = -\frac{1}{2}$

$\left(\text{Or, you could just solve for } y\text{: } y = \frac{8}{x^2 + 4}\right)$

37.

$$(x^2 + y^2)^2 = 4x^2y$$

$$2(x^2 + y^2)(2x + 2yy') = 4x^2y' + y(8x)$$

$$4x^3 + 4x^2yy' + 4xy^2 + 4y^3y' = 4x^2y' + 8xy$$

$$4x^2yy' + 4y^3y' - 4x^2y' = 8xy - 4x^3 - 4xy^2$$

$$4y'(x^2y + y^3 - x^2) = 4(2xy - x^3 - xy^2)$$

$$y' = \frac{2xy - x^3 - xy^2}{x^2y + y^3 - x^2}$$

At (1, 1): $y' = 0$

39.

$$\tan y = x$$

$$y'\sec^2 y = 1$$

$$y' = \frac{1}{\sec^2 y} = \cos^2 y, \ -\frac{\pi}{2} < y < \frac{\pi}{2}$$

$$\sec^2 y = 1 + \tan^2 y = 1 + x^2$$

$$y' = \frac{1}{1 + x^2}$$

41.

$$x^2 + y^2 = 36$$

$$2x + 2yy' = 0$$

$$y' = \frac{-x}{y}$$

$$y'' = \frac{y(-1) + xy'}{y^2} = \frac{-y + x\left(-\frac{x}{y}\right)}{y^2} = \frac{-y^2 - x^2}{y^3} = \frac{-36}{y^3}$$

43.

$$x^2 - y^2 = 16$$

$$2x - 2yy' = 0$$

$$y' = \frac{x}{y}$$

$$x - yy' = 0$$

$$1 - yy'' - (y')^2 = 0$$

$$1 - yy'' - \left(\frac{x}{y}\right)^2 = 0$$

$$y^2 - y^3y'' = x^2$$

$$y'' = \frac{y^2 - x^2}{y^3} = \frac{-16}{y^3}$$

45.

$$y^2 = x^3$$

$$2yy' = 3x^2$$

$$y' = \frac{3x^2}{2y} = \frac{3x^2}{2y} \cdot \frac{xy}{xy} = \frac{3y}{2x} \cdot \frac{x^3}{y^2} = \frac{3y}{2x}$$

$$y'' = \frac{2x(3y') - 3y(2)}{4x^2}$$

$$= \frac{2x[3 \cdot (3y/2x)] - 6y}{4x^2}$$

$$= \frac{3y}{4x^2} = \frac{3x}{4y}$$

47.

$$\sqrt{x} + \sqrt{y} = 4$$

$$\frac{1}{2}x^{-1/2} + \frac{1}{2}y^{-1/2}y' = 0$$

$$y' = \frac{-\sqrt{y}}{\sqrt{x}}$$

At (9, 1), $y' = -\frac{1}{3}$

Tangent line: $y - 1 = -\frac{1}{3}(x - 9)$

$$y = -\frac{1}{3}x + 4$$

$$x + 3y - 12 = 0$$

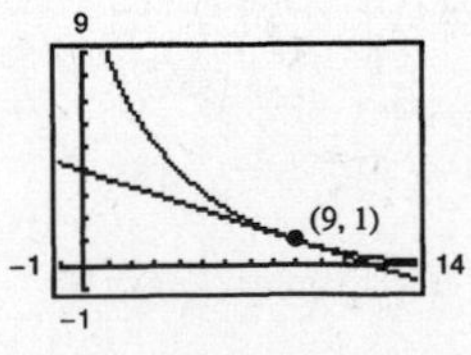

49. $x^2 + y^2 = 25$

$2x + 2yy' = 0$

$$y' = \frac{-x}{y}$$

At (4, 3):

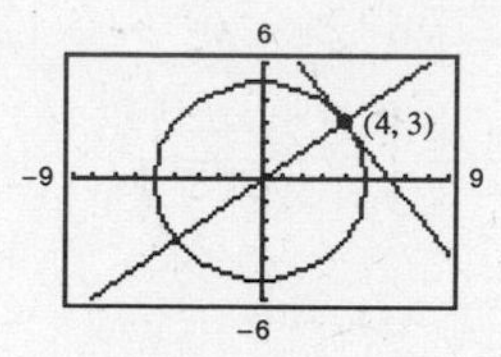

Tangent line: $y - 3 = \frac{-4}{3}(x - 4) \Longrightarrow 4x + 3y - 25 = 0$

Normal line: $y - 3 = \frac{3}{4}(x - 4) \Longrightarrow 3x - 4y = 0.$

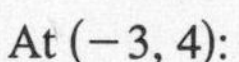

At $(-3, 4)$:

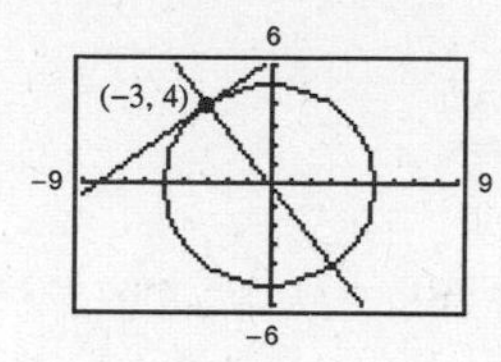

Tangent line: $y - 4 = \frac{3}{4}(x + 3) \Longrightarrow 3x - 4y + 25 = 0$

Normal line: $y - 4 = \frac{-4}{3}(x + 3) \Longrightarrow 4x + 3y = 0.$

51. $x^2 + y^2 = r^2$

$2x + 2yy' = 0$

$$y' = \frac{-x}{y} = \text{slope of tangent line}$$

$$\frac{y}{x} = \text{slope of normal line}$$

Let (x_0, y_0) be a point on the circle. If $x_0 = 0$, then the tangent line is horizontal, the normal line is vertical and, hence, passes through the origin. If $x_0 \neq 0$, then the equation of the normal line is

$$y - y_0 = \frac{y_0}{x_0}(x - x_0)$$

$$y = \frac{y_0}{x_0}x$$

which passes through the origin.

53. $25x^2 + 16y^2 + 200x - 160y + 400 = 0$

$$50x + 32yy' + 200 - 160y' = 0$$

$$y' = \frac{200 + 50x}{160 - 32y}$$

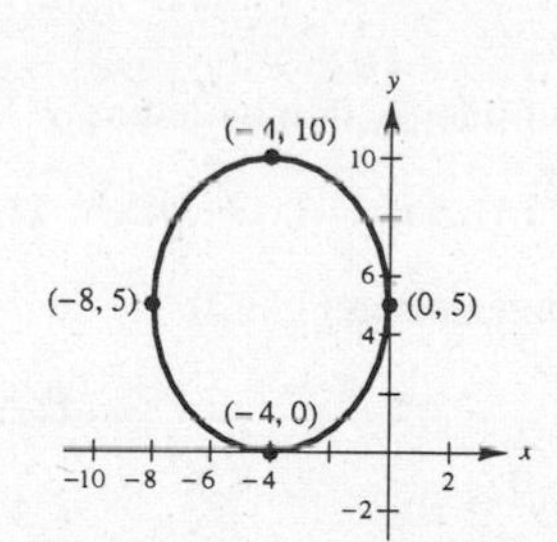

Horizontal tangents occur when $x = -4$:

$$25(16) + 16y^2 + 200(-4) - 160y + 400 = 0$$

$$y(y - 10) = 0 \Longrightarrow y = 0, 10$$

Horizontal tangents: $(-4, 0), (-4, 10)$.

Vertical tangents occur when $y = 5$:

$$25x^2 + 400 + 200x - 800 + 400 = 0$$

$$25x(x + 8) = 0 \Longrightarrow x = 0, -8$$

Vertical tangents: $(0, 5), (-8, 5)$.

55. $y = x\sqrt{x^2 - 1}$

$$\ln y = \ln x + \frac{1}{2}\ln(x^2 - 1)$$

$$\frac{1}{y}\left(\frac{dy}{dx}\right) = \frac{1}{x} + \frac{x}{x^2 - 1}$$

$$\frac{dy}{dx} = y\left[\frac{2x^2 - 1}{x(x^2 - 1)}\right] = \frac{2x^2 - 1}{\sqrt{x^2 - 1}}$$

57. $y = \dfrac{x^2\sqrt{3x - 2}}{(x - 1)^2}$

$$\ln y = 2\ln x + \frac{1}{2}\ln(3x - 2) - 2\ln(x - 1)$$

$$\frac{1}{y}\left(\frac{dy}{dx}\right) = \frac{2}{x} + \frac{3}{2(3x - 2)} - \frac{2}{x - 1}$$

$$\frac{dy}{dx} = y\left[\frac{3x^2 - 15x + 8}{2x(3x - 2)(x - 1)}\right]$$

$$= \frac{3x^3 - 15x^2 + 8x}{2(x - 1)^3\sqrt{3x - 2}}$$

59. $y = \dfrac{x(x - 1)^{3/2}}{\sqrt{x + 1}}$

$$\ln y = \ln x + \frac{3}{2}\ln(x - 1) - \frac{1}{2}\ln(x + 1)$$

$$\frac{1}{y}\left(\frac{dy}{dx}\right) = \frac{1}{x} + \frac{3}{2}\left(\frac{1}{x - 1}\right) - \frac{1}{2}\left(\frac{1}{x + 1}\right)$$

$$\frac{dy}{dx} = \frac{y}{2}\left[\frac{2}{x} + \frac{3}{x - 1} - \frac{1}{x + 1}\right]$$

$$= \frac{y}{2}\left[\frac{4x^2 + 4x - 2}{x(x^2 - 1)}\right] = \frac{(2x^2 + 2x - 1)\sqrt{x - 1}}{(x + 1)^{3/2}}$$

61. $y = x^{2/x}$

$$\ln y = \frac{2}{x}\ln x$$

$$\frac{1}{y}\left(\frac{dy}{dx}\right) = \frac{2}{x}\left(\frac{1}{x}\right) + \ln x\left(-\frac{2}{x^2}\right) = \frac{2}{x^2}(1 - \ln x)$$

$$\frac{dy}{dx} = \frac{2y}{x^2}(1 - \ln x) = 2x^{(2/x)-2}(1 - \ln x)$$

63. $y = (x - 2)^{x+1}$

$$\ln y = (x + 1)\ln(x - 2)$$

$$\frac{1}{y}\left(\frac{dy}{dx}\right) = (x + 1)\left(\frac{1}{x - 2}\right) + \ln(x - 2)$$

$$\frac{dy}{dx} = y\left[\frac{x + 1}{x - 2} + \ln(x - 2)\right]$$

$$= (x - 2)^{x+1}\left[\frac{x + 1}{x - 2} + \ln(x - 2)\right]$$

65. Find the points of intersection by letting $y^2 = 4x$ in the equation $2x^2 + y^2 = 6$.

$2x^2 + 4x = 6$ and $(x + 3)(x - 1) = 0$

The curves intersect at $(1, \pm 2)$.

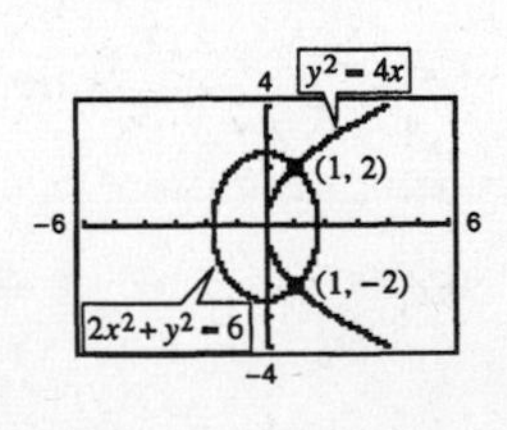

Ellipse:	Parabola:
$4x + 2yy' = 0$	$2yy' = 4$
$y' = -\dfrac{2x}{y}$	$y' = \dfrac{2}{y}$

At $(1, 2)$, the slopes are:

$y' = -1$	$y' = 1.$

At $(1, -2)$, the slopes are:

$y' = 1$	$y' = -1.$

Tangents are perpendicular.

67. $y = -x$ and $x = \sin y$

Point of intersection: $(0, 0)$

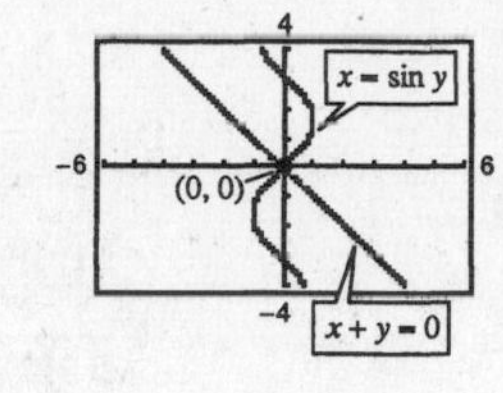

$\underline{y = -x:}$ $\qquad$ $\underline{x = \sin y:}$

$y' = -1$ $\qquad$ $1 = y' \cos y$

$$y' = \sec y$$

At $(0, 0)$, the slopes are:

$y' = -1 \quad y' = 1$

Tangents are perpendicular.

69.

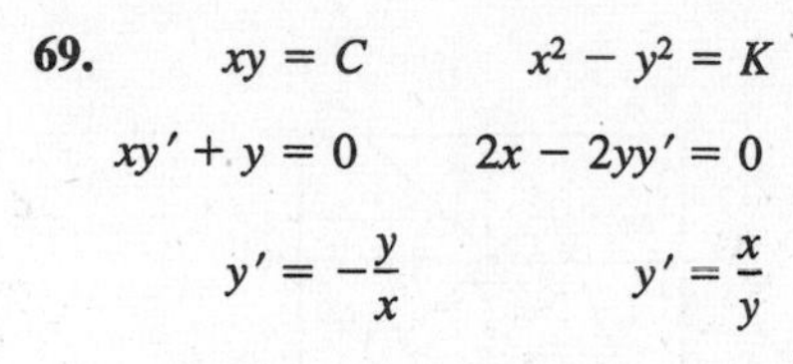

$xy = C$ $\qquad$ $x^2 - y^2 = K$

$xy' + y = 0$ $\qquad$ $2x - 2yy' = 0$

$y' = -\dfrac{y}{x}$ $\qquad$ $y' = \dfrac{x}{y}$

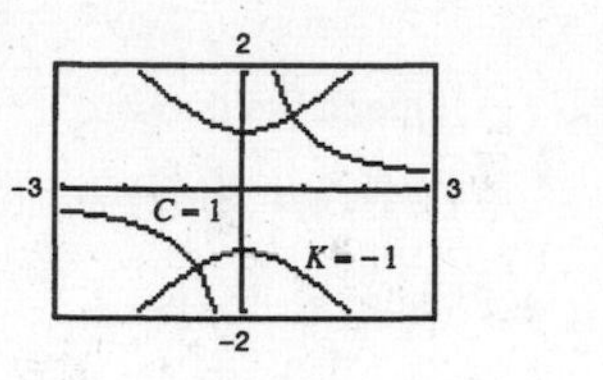

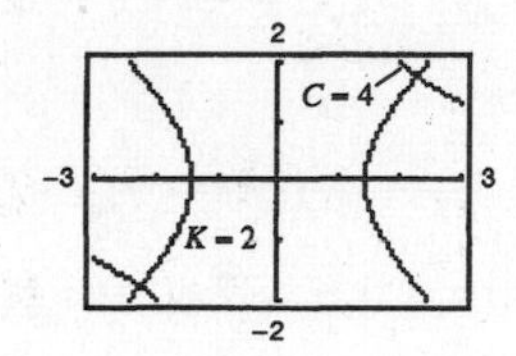

At any point of intersection (x, y) the product of the slopes is $(-y/x)(x/y) = -1$. The curves are orthogonal.

71. $2y^2 - 3x^4 = 0$

(a) $4yy' - 12x^3 = 0$

$$4yy' = 12x^3$$

$$y' = \frac{12x^3}{4y} = \frac{3x^3}{y}$$

(b) $4y\dfrac{dy}{dt} - 12x^3\dfrac{dx}{dt} = 0$

$$y\frac{dy}{dt} = 3x^3\frac{dx}{dt}$$

73. $\cos \pi y - 3 \sin \pi x = 1$

(a) $-\pi \sin(\pi y)y' - 3\pi \cos \pi x = 0$

$$y' = \frac{-3 \cos \pi x}{\sin \pi y}$$

(b) $-\pi \sin(\pi y)\dfrac{dy}{dt} - 3\pi \cos(\pi x)\dfrac{dx}{dt} = 0$

$$-\sin(\pi y)\frac{dy}{dt} = 3 \cos(\pi x)\frac{dx}{dt}$$

75. Answers will vary. A function is in explicit form if y is written as a function of x: $y = f(x)$. For example, $y = x^3$. An implicit equation is not in the form $y = f(x)$. For example, $x^2 + y^2 = 5$.

77. (a) $x^4 = 4(4x^2 - y^2)$

$$4y^2 = 16x^2 - x^4$$

$$y^2 = 4x^2 - \frac{1}{4}x^4$$

$$y = \pm\sqrt{4x^2 - \frac{1}{4}x^4}$$

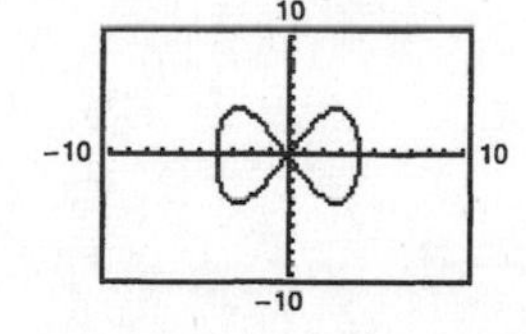

(b) $y = 3 \Longrightarrow 9 = 4x^2 - \dfrac{1}{4}x^4$

$$36 = 16x^2 - x^4$$

$$x^4 - 16x^2 + 36 = 0$$

$$x^2 = \frac{16 \pm \sqrt{256 - 144}}{2} = 8 \pm \sqrt{28}$$

Note that $x^2 = 8 \pm \sqrt{28} = 8 \pm 2\sqrt{7} = \left(1 \pm \sqrt{7}\right)^2$.

—CONTINUED—

77. —CONTINUED—

Hence, there are four values of x:

$$-1-\sqrt{7}, 1-\sqrt{7}, -1+\sqrt{7}, 1+\sqrt{7}$$

To find the slope, $2yy' = 8x - x^3 \Rightarrow y' = \dfrac{x(8-x^2)}{2(3)}$.

For $x = -1 - \sqrt{7}$, $y' = \frac{1}{3}(\sqrt{7}+7)$, and the line is

$$y_1 = \tfrac{1}{3}(\sqrt{7}+7)(x+1+\sqrt{7}) + 3 = \tfrac{1}{3}[(\sqrt{7}+7)x + 8\sqrt{7} + 23].$$

For $x = 1 - \sqrt{7}$, $y' = \frac{1}{3}(\sqrt{7}-7)$, and the line is

$$y_2 = \tfrac{1}{3}(\sqrt{7}-7)(x-1+\sqrt{7}) + 3 = \tfrac{1}{3}[(\sqrt{7}-7)x + 23 - 8\sqrt{7}].$$

For $x = -1 + \sqrt{7}$, $y' = -\frac{1}{3}(\sqrt{7}-7)$, and the line is

$$y_3 = -\tfrac{1}{3}(\sqrt{7}-7)(x+1-\sqrt{7}) + 3 = -\tfrac{1}{3}[(\sqrt{7}-7)x - (23 - 8\sqrt{7})].$$

For $x = 1 + \sqrt{7}$, $y' = -\frac{1}{3}(\sqrt{7}+7)$, and the line is

$$y_4 = -\tfrac{1}{3}(\sqrt{7}+7)(x-1-\sqrt{7}) + 3 = -\tfrac{1}{3}[(\sqrt{7}+7)x - (8\sqrt{7} + 23)].$$

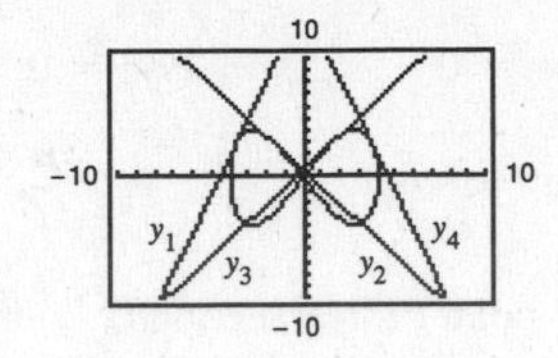

(c) Equating y_3 and y_4,

$$-\frac{1}{3}(\sqrt{7}-7)(x+1-\sqrt{7}) + 3 = -\frac{1}{3}(\sqrt{7}+7)(x-1-\sqrt{7}) + 3$$

$$(\sqrt{7}-7)(x+1-\sqrt{7}) = (\sqrt{7}+7)(x-1-\sqrt{7})$$

$$\sqrt{7}x + \sqrt{7} - 7 - 7x - 7 + 7\sqrt{7} = \sqrt{7}x - \sqrt{7} - 7 + 7x - 7 - 7\sqrt{7}$$

$$16\sqrt{7} = 14x$$

$$x = \frac{8\sqrt{7}}{7}$$

If $x = \dfrac{8\sqrt{7}}{7}$, then $y = 5$ and the lines intersect at $\left(\dfrac{8\sqrt{7}}{7}, 5\right)$.

79.

$$y = x^{p/q}$$

$$y^q = x^p$$

$$qy^{q-1}y' = px^{p-1}$$

$$y' = \frac{p}{q}\cdot\frac{x^{p-1}}{y^{q-1}} = \frac{p}{q}\cdot\frac{x^{p-1}y}{x^p} = \frac{p}{q}x^{p/q-1}$$

So, if $y = x^n = x^{p/q}$, $y' = nx^{n-1}$.

Section 2.6 Derivatives of Inverse Functions

1. $f(x) = x^3 + 2x - 1,\ f(1) = 2 = a$

$f'(x) = 3x^2 + 2$

$$(f^{-1})'(2) = \frac{1}{f'(f^{-1}(2))} = \frac{1}{f'(1)} = \frac{1}{3(1)^2 + 2} = \frac{1}{5}$$

3. $f(x) = \sin x,\ f\left(\frac{\pi}{6}\right) = \frac{1}{2} = a$

$f'(x) = \cos x$

$$(f^{-1})'\left(\frac{1}{2}\right) = \frac{1}{f'(f^{-1}(1/2))} = \frac{1}{f'(\pi/6)} = \frac{1}{\cos(\pi/6)}$$

$$= \frac{1}{\sqrt{3}/2} = \frac{2\sqrt{3}}{3}$$

5. $f(x) = x^3 - \frac{4}{x},\ f(2) = 6 = a$

$f'(x) = 3x^2 + \frac{4}{x^2}$

$$(f^{-1})'(6) = \frac{1}{f'(f^{-1}(6))} = \frac{1}{f'(2)} = \frac{1}{3(2)^2 + (4/2^2)} = \frac{1}{13}$$

7. $f(x) = x^3,\ \left(\frac{1}{2}, \frac{1}{8}\right)$

$f'(x) = 3x^2$

$f'\left(\frac{1}{2}\right) = \frac{3}{4}$

$f^{-1}(x) = \sqrt[3]{x},\ \left(\frac{1}{8}, \frac{1}{2}\right)$

$(f^{-1})'(x) = \frac{1}{3\sqrt[3]{x^2}}$

$(f^{-1})'\left(\frac{1}{8}\right) = \frac{4}{3}$

9. $f(x) = \sqrt{x-4},\ (5, 1)$

$f'(x) = \frac{1}{2\sqrt{x-4}}$

$f'(5) = \frac{1}{2}$

$f^{-1}(x) = x^2 + 4,\ (1, 5)$

$(f^{-1})'(x) = 2x$

$(f^{-1})'(1) = 2$

11. $x = y^3 - 7y^2 + 2$

$1 = 3y^2\frac{dy}{dx} - 14y\frac{dy}{dx}$

$\frac{dy}{dx} = \frac{1}{3y^2 - 14y}$

At $(-4, 1)$, $\frac{dy}{dx} = \frac{1}{3 - 14} = \frac{-1}{11}$.

Alternate solution: let $f(x) = x^3 - 7x^2 + 2$.

Then $f'(x) = 3x^2 - 14x$ and $f'(1) = -11$.

Hence, $\frac{dy}{dx} = \frac{1}{-11} = \frac{-1}{11}$.

13. $f(x) = 2\arcsin(x - 1)$

$$f'(x) = \frac{2}{\sqrt{1 - (x-1)^2}} = \frac{2}{\sqrt{2x - x^2}}$$

15. $g(x) = 3\arccos\frac{x}{2}$

$$g'(x) = \frac{-3(1/2)}{\sqrt{1 - (x^2/4)}} = \frac{-3}{\sqrt{4 - x^2}}$$

17. $f(x) = \arctan\frac{x}{a}$

$$f'(x) = \frac{1/a}{1 + (x^2/a^2)} = \frac{a}{a^2 + x^2}$$

19. $g(x) = \frac{\arcsin 3x}{x}$

$$g'(x) = \frac{x(3/\sqrt{1 - 9x^2}) - \arcsin 3x}{x^2}$$

$$= \frac{3x - \sqrt{1 - 9x^2}\arcsin 3x}{x^2\sqrt{1 - 9x^2}}$$

21. $g(x) = \dfrac{\arccos x}{x+1}$

$$g'(x) = \frac{(x+1)\dfrac{-1}{\sqrt{1-x^2}} - \arccos x}{(x+1)^2}$$

$$= -\frac{x+1+\sqrt{1-x^2}\arccos x}{(x+1)^2\sqrt{1-x^2}}$$

23. $h(x) = \operatorname{arccot} 6x$

$$h'(x) = \frac{-6}{1+36x^2}$$

25. $h(t) = \sin(\arccos t) = \sqrt{1-t^2}$

$$h'(t) = \frac{1}{2}(1-t^2)^{-1/2}(-2t)$$

$$= \frac{-t}{\sqrt{1-t^2}}$$

27. $y = x \arccos x - \sqrt{1-x^2}$

$$y' = \arccos x - \frac{x}{\sqrt{1-x^2}} - \frac{1}{2}(1-x^2)^{-1/2}(-2x)$$

$$= \arccos x$$

29. $y = \dfrac{1}{2}\left(\dfrac{1}{2}\ln\dfrac{x+1}{x-1} + \arctan x\right) = \dfrac{1}{4}[\ln(x+1) - \ln(x-1)] + \dfrac{1}{2}\arctan x$

$$\frac{dy}{dx} = \frac{1}{4}\left(\frac{1}{x+1} - \frac{1}{x-1}\right) + \frac{1/2}{1+x^2} = \frac{1}{1-x^4}$$

31. $g(t) = \tan(\arcsin t) = \dfrac{t}{\sqrt{1-t^2}}$

$$g'(t) = \frac{\sqrt{1-t^2} - t\left(-t/\sqrt{1-t^2}\right)}{1-t^2}$$

$$= \frac{1}{(1-t^2)^{3/2}}$$

33. $y = x \arcsin x + \sqrt{1-x^2}$

$$\frac{dy}{dx} = x\left(\frac{1}{\sqrt{1-x^2}}\right) + \arcsin x - \frac{x}{\sqrt{1-x^2}} = \arcsin x$$

35. $y = 8 \arcsin\dfrac{x}{4} - \dfrac{x\sqrt{16-x^2}}{2}$

$$y' = 2\frac{1}{\sqrt{1-(x/4)^2}} - \frac{\sqrt{16-x^2}}{2} - \frac{x}{4}(16-x^2)^{-1/2}(-2x)$$

$$= \frac{8}{\sqrt{16-x^2}} - \frac{\sqrt{16-x^2}}{2} + \frac{x^2}{2\sqrt{16-x^2}}$$

$$= \frac{16-(16-x^2)+x^2}{2\sqrt{16-x^2}} = \frac{x^2}{\sqrt{16-x^2}}$$

37. $y = \arctan x + \dfrac{x}{1+x^2}$

$$y' = \frac{1}{1+x^2} + \frac{(1+x^2) - x(2x)}{(1+x^2)^2}$$

$$= \frac{(1+x^2)+(1-x^2)}{(1+x^2)^2}$$

$$= \frac{2}{(1+x^2)^2}$$

39. $f(x) = \arccos x$

$$f'(x) = \frac{-1}{\sqrt{1-x^2}} = -2 \text{ when } x = \pm\frac{\sqrt{3}}{2}.$$

When $x = \sqrt{3}/2$, $f(\sqrt{3}/2) = \pi/6$. When $x = -\sqrt{3}/2$, $f(-\sqrt{3}/2) = 5\pi/6$.

Tangent lines: $y - \dfrac{\pi}{6} = -2\left(x - \dfrac{\sqrt{3}}{2}\right) \Rightarrow y = -2x + \left(\dfrac{\pi}{6} + \sqrt{3}\right)$

$y - \dfrac{5\pi}{6} = -2\left(x + \dfrac{\sqrt{3}}{2}\right) \Rightarrow y = -2x + \left(\dfrac{5\pi}{6} - \sqrt{3}\right)$

41. $f(x) = \arcsin x, \quad a = \dfrac{1}{2}$

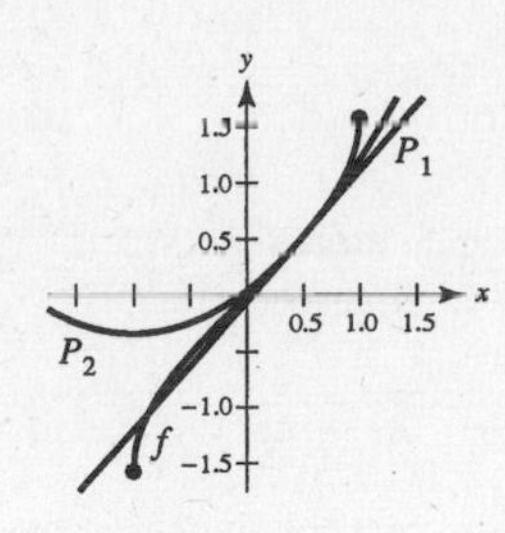

$$f'(x) = \frac{1}{\sqrt{1 - x^2}}$$

$$f''(x) = \frac{x}{(1 - x^2)^{3/2}}$$

$$P_1(x) = f\left(\frac{1}{2}\right) + f'\left(\frac{1}{2}\right)\left(x - \frac{1}{2}\right) = \frac{\pi}{6} + \frac{2\sqrt{3}}{3}\left(x - \frac{1}{2}\right)$$

$$P_2(x) = f\left(\frac{1}{2}\right) + f'\left(\frac{1}{2}\right)\left(x - \frac{1}{2}\right) + \frac{1}{2}f''\left(\frac{1}{2}\right)\left(x - \frac{1}{2}\right)^2 = \frac{\pi}{6} + \frac{2\sqrt{3}}{3}\left(x - \frac{1}{2}\right) + \frac{2\sqrt{3}}{9}\left(x - \frac{1}{2}\right)^2$$

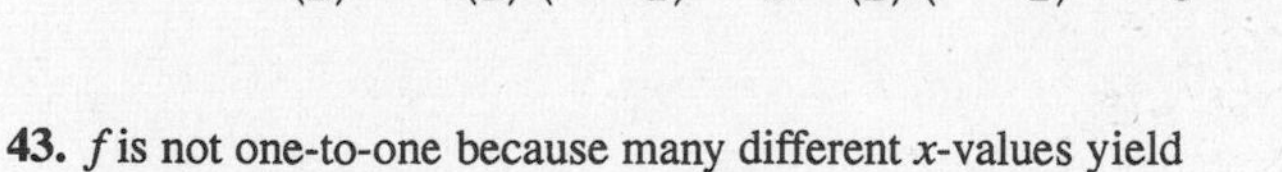

43. f is not one-to-one because many different x-values yield the same y-value.

Example: $f(0) = f(\pi) = 0$

Not continuous at $\dfrac{(2n - 1)\pi}{2}$, where n is an integer

45. Theorem 2.17: Let f be a function that is differentiable on an interval I. If f has an inverse function g, then g is differentiable at any x for which $f'(g(x)) \neq 0$. Moreover,

$$g'(x) = \frac{1}{f'(g)(x))}, \quad f'(g(x)) \neq 0$$

47. (a) $h(t) = -16t^2 + 256$

$-16t^2 + 256 = 0$ when $t = 4$ sec.

(b) $\tan\theta = \dfrac{h}{500} = \dfrac{-16t^2 + 256}{500}$

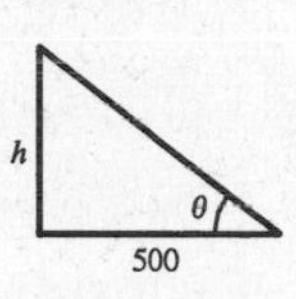

$$\theta = \arctan\left[\frac{16}{500}(-t^2 + 16)\right]$$

$$\frac{d\theta}{dt} = \frac{-8t/125}{1 + [(4/125)(-t^2 + 16)]^2} = \frac{-1000t}{15{,}625 + 16(16 - t^2)^2}$$

When $t = 1$, $d\theta/dt \approx -0.0520$ rad/sec.

When $t = 2$, $d\theta/dt \approx -0.1116$ rad/sec.

49. (a)

(b) $y' = -0.361 + 0.002t + \dfrac{79.564}{1 + t^2}$

$y'(20) \approx -0.123$ million/yr

$y'(60) \approx -0.219$ million/yr

51. $f(x) = kx + \sin x$

$f'(x) = k + \cos x \geq 0$ for $k \geq 1$

$f'(x) = k + \cos x \leq 0$ for $k \leq -1$

Therefore, $f(x) = kx + \sin x$ is strictly monotonic and has an inverse for $k \leq -1$ or $k \geq 1$.

53. True

$$\frac{d}{dx}[\arctan x] = \frac{1}{1 + x^2} > 0 \text{ for all } x$$

55. True

$$\frac{d}{dx}[\arctan(\tan x)] = \frac{\sec^2 x}{1 + \tan^2 x} = \frac{\sec^2 x}{\sec^2 x} = 1$$

57. $f(x) = \arcsin\left(\frac{x-2}{2}\right) - 2\arcsin\frac{\sqrt{x}}{2}, \quad 0 \le x \le 4$

$$f'(x) = \frac{1/2}{\sqrt{1-[(x-2)/2]^2}} - 2\left[\frac{1/(4\sqrt{x})}{\sqrt{1-(\sqrt{x}/2)^2}}\right]$$

$$= \frac{1}{2\sqrt{1-(1/4)(x^2-4x+4)}} - \frac{1}{2\sqrt{x}\sqrt{1-(x/4)}} = \frac{1}{2\sqrt{x-(x^2/4)}} - \frac{1}{2\sqrt{x-(x^2/4)}} = 0$$

Since the derivative is zero, we conclude that the function is constant. (By letting $x = 0$ in $f(x)$, you can see that the constant is $-\pi/2$.)

Section 2.7 Related Rates

1. $y = \sqrt{x}$

$$\frac{dy}{dt} = \left(\frac{1}{2\sqrt{x}}\right)\frac{dx}{dt}$$

$$\frac{dx}{dt} = 2\sqrt{x}\frac{dy}{dt}$$

(a) When $x = 4$ and $dx/dt = 3$,

$$\frac{dy}{dt} = \frac{1}{2\sqrt{4}}(3) = \frac{3}{4}.$$

(b) When $x = 25$ and $dy/dt = 2$,

$$\frac{dx}{dt} = 2\sqrt{25}\,(2) = 20.$$

3. $xy = 4$

$$x\frac{dy}{dt} + y\frac{dx}{dt} = 0$$

$$\frac{dy}{dt} = \left(-\frac{y}{x}\right)\frac{dx}{dt}$$

$$\frac{dx}{dt} = \left(-\frac{x}{y}\right)\frac{dy}{dt}$$

(a) When $x = 8$, $y = 1/2$, and $dx/dt = 10$,

$$\frac{dy}{dt} = -\frac{1/2}{8}(10) = -\frac{5}{8}.$$

(b) When $x = 1$, $y = 4$, and $dy/dt = -6$,

$$\frac{dx}{dt} = -\frac{1}{4}(-6) = \frac{3}{2}.$$

5. $y = x^2 + 1$

$$\frac{dx}{dt} = 2$$

$$\frac{dy}{dt} = 2x\,\frac{dx}{dt}$$

(a) When $x = -1$,

$$\frac{dy}{dt} = 2(-1)(2) = -4 \text{ cm/sec.}$$

(b) When $x = 0$,

$$\frac{dy}{dt} = 2(0)(2) = 0 \text{ cm/sec.}$$

(c) When $x = 1$,

$$\frac{dy}{dt} = 2(1)(2) = 4 \text{ cm/sec.}$$

7. $y = \tan x$

$$\frac{dx}{dt} = 2$$

$$\frac{dy}{dt} = \sec^2 x\,\frac{dx}{dt}$$

(a) When $x = -\pi/3$,

$$\frac{dy}{dt} = (2)^2(2) = 8 \text{ cm/sec.}$$

(b) When $x = -\pi/4$,

$$\frac{dy}{dt} = (\sqrt{2})^2(2) = 4 \text{ cm/sec.}$$

(c) When $x = 0$,

$$\frac{dy}{dt} = (1)^2(2) = 2 \text{ cm/sec.}$$

9. (a) $\frac{dx}{dt}$ negative $\Rightarrow \frac{dy}{dt}$ positive

(b) $\frac{dy}{dt}$ positive $\Rightarrow \frac{dx}{dt}$ negative

11. Yes, y changes at a constant rate: $\frac{dy}{dt} = a \cdot \frac{dx}{dt}$.

No, the rate $\frac{dy}{dt}$ is a multiple of $\frac{dx}{dt}$.

13. $D = \sqrt{x^2 + y^2}$

$= \sqrt{x^2 + (x^2 + 1)^2}$

$= \sqrt{x^4 + 3x^2 + 1}$

$\dfrac{dx}{dt} = 2$

$\dfrac{dD}{dt} = \dfrac{1}{2}(x^4 + 3x^2 + 1)^{-1/2}(4x^3 + 6x)\dfrac{dx}{dt}$

$= \dfrac{2x^3 + 3x}{\sqrt{x^4 + 3x^2 + 1}}\dfrac{dx}{dt}$

$= \dfrac{4x^3 + 6x}{\sqrt{x^4 + 3x^2 + 1}}$

15. $A = \pi r^2$

$\dfrac{dr}{dt} = 3$

$\dfrac{dA}{dt} = 2\pi r\dfrac{dr}{dt}$

(a) When $r = 6$,

$\dfrac{dA}{dt} = 2\pi(6)(3) = 36\pi \text{ cm}^2/\text{min}.$

(b) When $r = 24$,

$\dfrac{dA}{dt} = 2\pi(24)(3) = 144\pi \text{ cm}^2/\text{min}.$

17. (a) $\sin\dfrac{\theta}{2} = \dfrac{(1/2)b}{s} \Rightarrow b = 2s\sin\dfrac{\theta}{2}$

$\cos\dfrac{\theta}{2} = \dfrac{h}{s} \Rightarrow h = s\cos\dfrac{\theta}{2}$

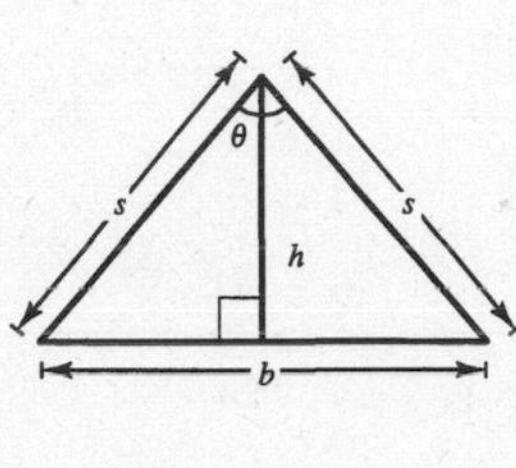

$A = \dfrac{1}{2}bh$

$= \dfrac{1}{2}\left(2s\sin\dfrac{\theta}{2}\right)\left(s\cos\dfrac{\theta}{2}\right)$

$= \dfrac{s^2}{2}\left(2\sin\dfrac{\theta}{2}\cos\dfrac{\theta}{2}\right)$

$= \dfrac{s^2}{2}\sin\theta$

(b) $\dfrac{dA}{dt} = \dfrac{s^2}{2}\cos\theta\dfrac{d\theta}{dt}$ where $\dfrac{d\theta}{dt} = \dfrac{1}{2}$ rad/min.

When $\theta = \dfrac{\pi}{6}, \dfrac{dA}{dt} = \dfrac{s^2}{2}\left(\dfrac{\sqrt{3}}{2}\right)\left(\dfrac{1}{2}\right) = \dfrac{\sqrt{3}s^2}{8}$

When $\theta = \dfrac{\pi}{3}, \dfrac{dA}{dt} = \dfrac{s^2}{2}\left(\dfrac{1}{2}\right)\left(\dfrac{1}{2}\right) = \dfrac{s^2}{8}$

(c) If $d\theta/dt$ is constant, dA/dt is proportional to $\cos\theta$.

19. $V = \dfrac{1}{3}\pi h(108 - h^2) = 36\pi h - \dfrac{1}{3}\pi h^3, \quad 0 < h < 6$

$\dfrac{dV}{dt} = 36\pi\dfrac{dh}{dt} - \pi h^2\dfrac{dh}{dt}$

When $\dfrac{dV}{dt} = 3$ and $h = 2$,

$3 = 36\pi\dfrac{dh}{dt} - 4\pi\dfrac{dh}{dt}$

$\Rightarrow \dfrac{dh}{dt} = \dfrac{3}{36\pi - 4\pi} = \dfrac{3}{32\pi} \approx 0.0298$ m/min.

21. $s = 6x^2$

$\dfrac{dx}{dt} = 3$

$\dfrac{ds}{dt} = 12x\dfrac{dx}{dt}$

(a) When $x = 1$, $\dfrac{ds}{dt} = 12(1)(3) = 36 \text{ cm}^2/\text{sec}.$

(b) When $x = 10$, $\dfrac{ds}{dt} = 12(10)(3) = 360 \text{ cm}^2/\text{sec}.$

23. $V = \dfrac{1}{3}\pi r^2 h = \dfrac{1}{3}\pi\left(\dfrac{9}{4}h^2\right)h \quad$ [since $2r = 3h$]

$= \dfrac{3\pi}{4}h^3$

$\dfrac{dV}{dt} = 10$

$\dfrac{dV}{dt} = \dfrac{9\pi}{4}h^2\dfrac{dh}{dt} \Rightarrow \dfrac{dh}{dt} = \dfrac{4(dV/dt)}{9\pi h^2}$

When $h = 15$, $\dfrac{dh}{dt} = \dfrac{4(10)}{9\pi(15)^2} = \dfrac{8}{405\pi}$ ft/min.

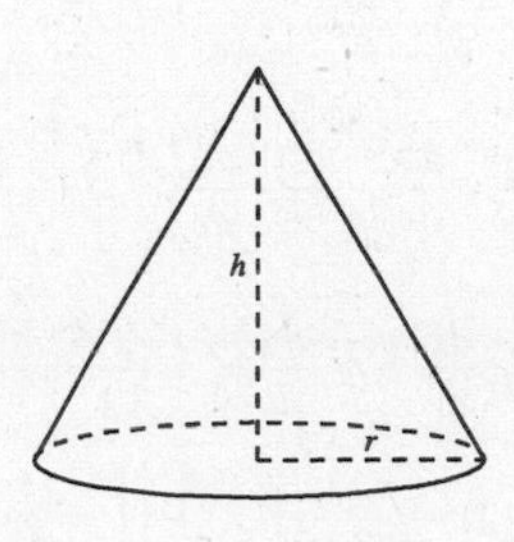

25.

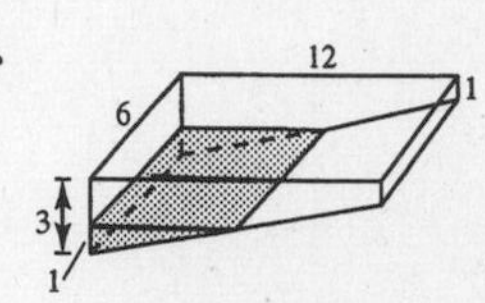

(a) Total volume of pool $= \frac{1}{2}(2)(12)(6) + (1)(6)(12) = 144 \text{ m}^3$

Volume of 1 m. of water $= \frac{1}{2}(1)(6)(6) = 18 \text{ m}^3$

(see similar triangle diagram)

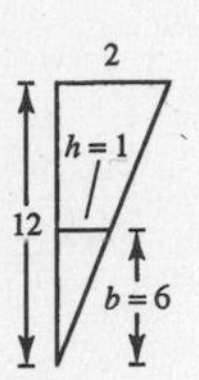

% pool filled $= \frac{18}{144}(100\%) = 12.5\%$

(b) Since for $0 \le h \le 2$, $b = 6h$, you have

$$V = \frac{1}{2}bh(6) = 3bh = 3(6h)h = 18h^2$$

$$\frac{dV}{dt} = 36h\frac{dh}{dt} = \frac{1}{4} \Rightarrow \frac{dh}{dt} = \frac{1}{144h} = \frac{1}{144(1)} = \frac{1}{144} \text{ m/min.}$$

27. $x^2 + y^2 = 25^2$

$$2x\frac{dx}{dt} + 2y\frac{dy}{dt} = 0$$

$$\frac{dy}{dt} = \frac{-x}{y} \cdot \frac{dx}{dt} = \frac{-2x}{y} \text{ since } \frac{dx}{dt} = 2.$$

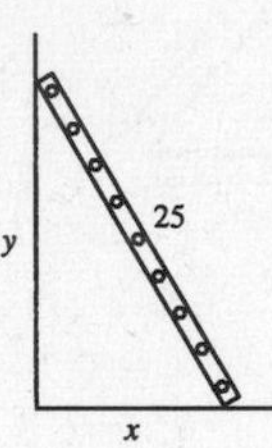

(a) When $x = 7$,

$$y = \sqrt{576} = 24, \frac{dy}{dt} = \frac{-2(7)}{24} = \frac{-7}{12} \text{ ft/sec.}$$

When $x = 15$,

$$y = \sqrt{400} = 20, \frac{dy}{dt} = \frac{-2(15)}{20} = \frac{-3}{2} \text{ ft/sec.}$$

When $x = 24$,

$$y = 7, \frac{dy}{dt} = \frac{-2(24)}{7} = \frac{-48}{7} \text{ ft/sec.}$$

(b) $A = \frac{1}{2}xy$

$$\frac{dA}{dt} = \frac{1}{2}\left(x\frac{dy}{dt} + y\frac{dx}{dt}\right)$$

From part (a) we have

$x = 7, y = 24, \frac{dx}{dt} = 2$, and $\frac{dy}{dt} = -\frac{7}{12}$. Thus,

$$\frac{dA}{dt} = \frac{1}{2}\left[7\left(-\frac{7}{12}\right) + 24(2)\right] = \frac{527}{24} \approx 21.96 \text{ ft}^2\text{/sec.}$$

(c) $\tan\theta = \frac{x}{y}$

$$\sec^2\theta\frac{d\theta}{dt} = \frac{1}{y} \cdot \frac{dx}{dt} - \frac{x}{y^2} \cdot \frac{dy}{dt}$$

$$\frac{d\theta}{dt} = \cos^2\theta\left[\frac{1}{y} \cdot \frac{dx}{dt} - \frac{x}{y^2} \cdot \frac{dy}{dt}\right]$$

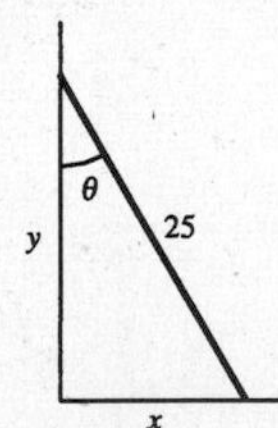

Using $x = 7, y = 24, \frac{dx}{dt} = 2, \frac{dy}{dt} = -\frac{7}{12}$ and $\cos\theta = \frac{24}{25}$, we have

$$\frac{d\theta}{dt} = \left(\frac{24}{25}\right)^2\left[\frac{1}{24}(2) - \frac{7}{(24)^2}\left(-\frac{7}{12}\right)\right] = \frac{1}{12} \text{ rad/sec.}$$

29. When $y = 6, x = \sqrt{12^2 - 6^2} = 6\sqrt{3}$, and

$$s = \sqrt{x^2 + (12 - y)^2}$$

$$= \sqrt{108 + 36} = 12.$$

$$x^2 + (12 - y)^2 = s^2$$

$$2x\frac{dx}{dt} + 2(12 - y)(-1)\frac{dy}{dt} = 2s\frac{ds}{dt}$$

$$x\frac{dx}{dt} + (y - 12)\frac{dy}{dt} = s\frac{ds}{dt}$$

Also, $x^2 + y^2 = 12^2$

$$2x\frac{dx}{dt} + 2y\frac{dy}{dt} = 0 \Rightarrow \frac{dy}{dt} = \frac{-x}{y}\frac{dx}{dt}.$$

Thus, $x\frac{dx}{dt} + (y - 12)\left(\frac{-x}{y}\frac{dx}{dt}\right) = s\frac{ds}{dt}$

$$\frac{dx}{dt}\left[x - x + \frac{12x}{y}\right] = s\frac{ds}{dt} \Rightarrow \frac{dx}{dt} = \frac{sy}{12x}\cdot\frac{ds}{dt} = \frac{(12)(6)}{(12)(6\sqrt{3})}(-0.2) = \frac{-1}{5\sqrt{3}} = \frac{-\sqrt{3}}{15} \text{ m/sec (horizontal)}$$

$$\frac{dy}{dt} = \frac{-x}{y}\frac{dx}{dt} = \frac{-6\sqrt{3}}{6}\cdot\frac{(-\sqrt{3})}{15} = \frac{1}{5} \text{ m/sec (vertical).}$$

31. (a) $s^2 = x^2 + y^2$

$$\frac{dx}{dt} = -450$$

$$\frac{dy}{dt} = -600$$

$$2s\frac{ds}{dt} = 2x\frac{dx}{dt} + 2y\frac{dy}{dt}$$

$$\frac{ds}{dt} = \frac{x(dx/dt) + y(dy/dt)}{s}$$

When $x = 150$ and $y = 200$, $s = 250$ and

$$\frac{ds}{dt} = \frac{150(-450) + 200(-600)}{250} = -750 \text{ mph.}$$

(b) $t = \frac{250}{750} = \frac{1}{3}$ hr $= 20$ min

33. $s^2 = 90^2 + x^2$

$$x = 30$$

$$\frac{dx}{dt} = -28$$

$$2s\frac{ds}{dt} = 2x\frac{dx}{dt} \Rightarrow \frac{ds}{dt} = \frac{x}{s}\cdot\frac{dx}{dt}$$

When $x = 30$,

$$s = \sqrt{90^2 + 30^2} = 30\sqrt{10}$$

$$\frac{ds}{dt} = \frac{30}{30\sqrt{10}}(-28) = \frac{-28}{\sqrt{10}} \approx -8.85 \text{ ft/sec.}$$

35. (a) $\frac{15}{6} = \frac{y}{y - x} \Rightarrow 15y - 15x = 6y$

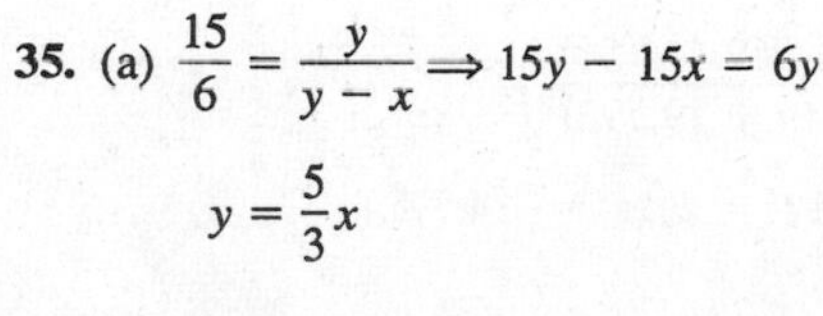

$$y = \frac{5}{3}x$$

$$\frac{dx}{dt} = 5$$

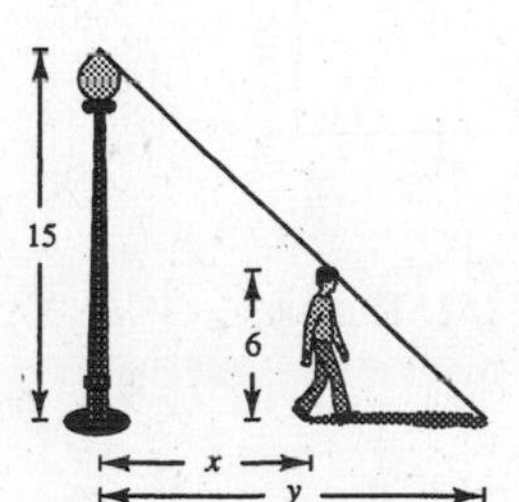

$$\frac{dy}{dt} = \frac{5}{3}\cdot\frac{dx}{dt} = \frac{5}{3}(5) = \frac{25}{3} \text{ ft/sec}$$

(b) $\frac{d(y - x)}{dt} = \frac{dy}{dt} - \frac{dx}{dt} = \frac{25}{3} - 5 = \frac{10}{3}$ ft/sec

37. $x(t) = \frac{1}{2}\sin\frac{\pi t}{6}, x^2 + y^2 = 1$

(a) Period: $\frac{2\pi}{\pi/6} = 12$ seconds

(b) When $x = \frac{1}{2}, y = \sqrt{1^2 - \left(\frac{1}{2}\right)^2} = \frac{\sqrt{3}}{2}$ m.

Lowest point: $\left(0, \frac{\sqrt{3}}{2}\right)$

(c) When $x = \frac{1}{4}, y = \sqrt{1 - \left(\frac{1}{4}\right)^2} = \frac{\sqrt{15}}{4}$ and $t = 1$

$$\frac{dx}{dt} = \frac{1}{2}\left(\frac{\pi}{6}\right)\cos\frac{\pi t}{6} = \frac{\pi}{12}\cos\frac{\pi t}{6}$$

$$x^2 + y^2 = 1$$

$$2x\frac{dx}{dt} + 2y\frac{dy}{dt} = 0 \Longrightarrow \frac{dy}{dt} = \frac{-x}{y}\frac{dx}{dt}.$$

Thus, $\frac{dy}{dt} = -\frac{1/4}{\sqrt{15}/4} \cdot \frac{\pi}{12}\cos\left(\frac{\pi}{6}\right)$

$$= \frac{-\pi}{\sqrt{15}}\left(\frac{1}{12}\right)\frac{\sqrt{3}}{2} = \frac{-\pi}{24}\frac{1}{\sqrt{5}} = \frac{-\sqrt{5}\pi}{120}.$$

Speed $= \left|\frac{-\sqrt{5}\pi}{120}\right| = \frac{\sqrt{5}\pi}{120}$ m/sec

39. Since the evaporation rate is proportional to the surface area, $dV/dt = k(4\pi r^2)$. However, since $V = (4/3)\pi r^3$, we have

$$\frac{dV}{dt} = 4\pi r^2\frac{dr}{dt}.$$

Therefore,

$$k(4\pi r^2) = 4\pi r^2\frac{dr}{dt} \Longrightarrow k = \frac{dr}{dt}.$$

41.

$$pV^{1.3} = k$$

$$1.3\,pV^{0.3}\frac{dV}{dt} + V^{1.3}\frac{dp}{dt} = 0$$

$$V^{0.3}\left(1.3p\frac{dV}{dt} + V\frac{dp}{dt}\right) = 0$$

$$1.3p\frac{dV}{dt} = -V\frac{dp}{dt}$$

43.

$$\tan\theta = \frac{y}{30}$$

$$\frac{dy}{dt} = 3 \text{ m/sec.}$$

$$\sec^2\theta \cdot \frac{d\theta}{dt} = \frac{1}{30}\frac{dy}{dt}$$

$$\frac{d\theta}{dt} = \frac{1}{30}\cos^2\theta \cdot \frac{dy}{dt}$$

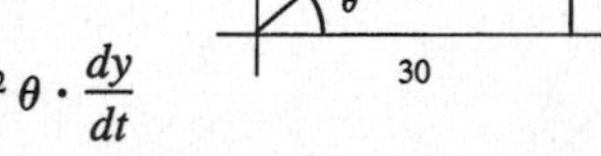

When $y = 30$, $\theta = \pi/4$ and $\cos\theta = \sqrt{2}/2$. Thus,

$$\frac{d\theta}{dt} = \frac{1}{30}\left(\frac{1}{2}\right)(3) = \frac{1}{20} \text{ rad/sec.}$$

45. $H = \frac{4347}{400,000,000}e^{369,444/(50t + 19,793)}$

(a) $t = 65° \Longrightarrow H \approx 99.79\%$

$t = 80° \Longrightarrow H \approx 60.20\%$

(b) $H' = H \cdot \left(\frac{-369,444(50)}{(50t + 19,793)^2}\right)t'$

At $t = 75$ and $t' = 2$, $H' \approx -4.7\%$.

47. $\dfrac{d\theta}{dt} = (10 \text{ rev/sec})(2\pi \text{ rad/rev}) = 20\pi \text{ rad/sec}$

(a) $\cos\theta = \dfrac{x}{20}$

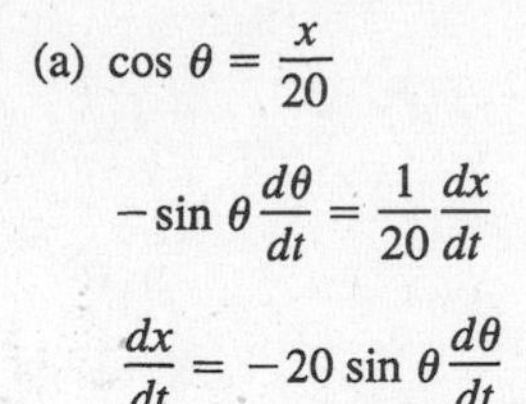

$$-\sin\theta \frac{d\theta}{dt} = \frac{1}{20}\frac{dx}{dt}$$

$$\frac{dx}{dt} = -20\sin\theta \frac{d\theta}{dt}$$

$$= (-20\sin\theta)(20\pi) = -400\pi\sin\theta$$

(b)

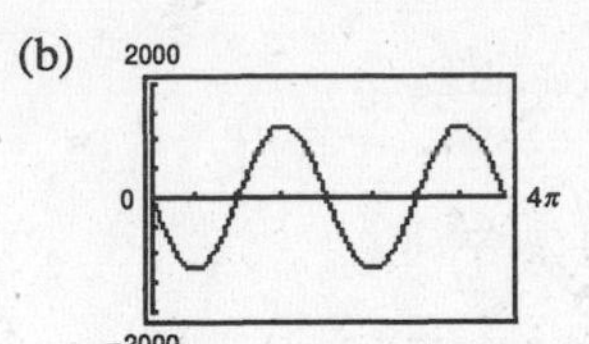

(c) $|dx/dt| = |-400\pi\sin\theta|$ is greatest when

$|\sin\theta| = 1 \Rightarrow \theta = (\pi/2) + n\pi.$

$|dx/dt|$ is least when $\theta = n\pi$.

(d) For $\theta = 30°$,

$$\frac{dx}{dt} = -400\pi\sin(30°) = -200\pi \text{ cm/sec.}$$

For $\theta = 60°$,

$$\frac{dx}{dt} = -400\pi\sin(60°) = -200\sqrt{3}\,\pi \text{ cm/sec.}$$

49. $\tan\theta = \dfrac{x}{50} \Rightarrow x = 50\tan\theta$

$$\frac{dx}{dt} = 50\sec^2\theta\frac{d\theta}{dt}$$

$$2 = 50\sec^2\theta\frac{d\theta}{dt}$$

$$\frac{d\theta}{dt} = \frac{1}{25}\cos^2\theta, \ -\frac{\pi}{4} \le \theta \le \frac{\pi}{4}$$

51. $\tan\theta = \dfrac{y}{x}, y = 5$

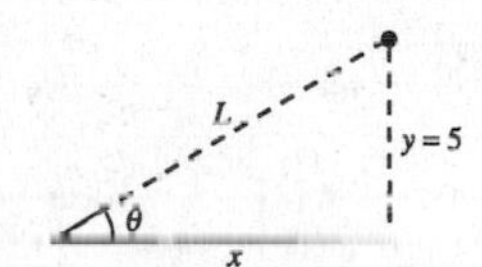

$$\frac{dx}{dt} = -600 \text{ mi/hr}$$

$$(\sec^2\theta)\frac{d\theta}{dt} = -\frac{5}{x^2}\cdot\frac{dx}{dt}$$

$$\frac{d\theta}{dt} = \cos^2\theta\left(-\frac{5}{x^2}\right)\frac{dx}{dt} = \frac{x^2}{L^2}\left(-\frac{5}{x^2}\right)\frac{dx}{dt}$$

$$= \left(-\frac{5^2}{L^2}\right)\left(\frac{1}{5}\right)\frac{dx}{dt} = (-\sin^2\theta)\left(\frac{1}{5}\right)(-600) = 120\sin^2\theta$$

(a) When $\theta = 30°$,

$$\frac{d\theta}{dt} = \frac{120}{4} = 30 \text{ rad/hr}$$

$$= 30 \text{ rad/hr} = \frac{1}{2} \text{ rad/min.}$$

(b) When $\theta = 60°$,

$$\frac{d\theta}{dt} = 120\left(\frac{3}{4}\right) = 90 \text{ rad/hr}$$

$$= \frac{3}{2} \text{ rad/min.}$$

(c) When $\theta = 75°$,

$$\frac{d\theta}{dt} = 120\sin^2 75° \approx 111.96 \text{ rad/hr}$$

$$\approx 1.87 \text{ rad/min.}$$

53. $x^2 + y^2 = 25$; acceleration of the top of the ladder $= \dfrac{d^2y}{dt^2}$

First derivative: $2x\dfrac{dx}{dt} + 2y\dfrac{dy}{dt} = 0$

$$x\frac{dx}{dt} + y\frac{dy}{dt} = 0$$

Second derivative: $x\dfrac{d^2x}{dt^2} + \dfrac{dx}{dt}\cdot\dfrac{dx}{dt} + y\dfrac{d^2y}{dt^2} + \dfrac{dy}{dt}\cdot\dfrac{dy}{dt} = 0$

$$\frac{d^2y}{dt^2} = \left(\frac{1}{y}\right)\left[-x\frac{d^2x}{dt^2} - \left(\frac{dx}{dt}\right)^2 - \left(\frac{dy}{dt}\right)^2\right]$$

When $x = 7, y = 24, \dfrac{dy}{dt} = -\dfrac{7}{12}$, and $\dfrac{dx}{dt} = 2$ (see Exercise 27). Since $\dfrac{dx}{dt}$ is constant, $\dfrac{d^2x}{dt^2} = 0$.

$$\frac{d^2y}{dt^2} = \frac{1}{24}\left[-7(0) - (2)^2 - \left(-\frac{7}{12}\right)^2\right] = \frac{1}{24}\left[-4 - \frac{49}{144}\right] = \frac{1}{24}\left[-\frac{625}{144}\right] \approx -0.1808 \text{ ft/sec}^2$$

55. (a) Using a graphing utility, you obtain $m(s) = -0.881s^2 + 29.10s - 206.2$

(b) $\dfrac{dm}{dt} = \dfrac{dm}{ds}\dfrac{ds}{dt} = (-1.762s + 29.10)\dfrac{ds}{dt}$

(c) If $t = 5$ (1995), then $s = 15.5$ and $\dfrac{ds}{dt} = 1.2$.

Thus, $\dfrac{dm}{dt} = (-1.762(15.5) + 29.10)(1.2) \approx 2.15$ million.

Section 2.8 Newton's Method

1. $f(x) = x^2 - 3$

$f'(x) = 2x$

$x_1 = 1.7$

n	x_n	$f(x_n)$	$f'(x_n)$	$\frac{f(x_n)}{f'(x_n)}$	$x_n - \frac{f(x_n)}{f'(x_n)}$
1	1.7000	−0.1100	3.4000	−0.0324	1.7324
2	1.7324	0.0012	3.4648	0.0003	1.7321

3. $f(x) = \sin x$

$f'(x) = \cos x$

$x_1 = 3$

n	x_n	$f(x_n)$	$f'(x_n)$	$\frac{f(x_n)}{f'(x_n)}$	$x_n - \frac{f(x_n)}{f'(x_n)}$
1	3.0000	0.1411	−0.9900	−0.1425	3.1425
2	3.1425	−0.0009	−1.0000	0.0009	3.1416

5. $f(x) = x^3 + x - 1$

$f'(x) = 3x^2 + 1$

Approximation of the zero of f is 0.682.

n	x_n	$f(x_n)$	$f'(x_n)$	$\frac{f(x_n)}{f'(x_n)}$	$x_n - \frac{f(x_n)}{f'(x_n)}$
1	0.5000	−0.3750	1.7500	−0.2143	0.7143
2	0.7143	0.0788	2.5307	0.0311	0.6832
3	0.6832	0.0021	2.4003	0.0009	0.6823

7. $f(x) = 3\sqrt{x-1} - x$

$f'(x) = \dfrac{3}{2\sqrt{x-1}} - 1$

Approximation of the zero of f is 1.146.

Similarly, the other zero is approximately 7.854.

n	x_n	$f(x_n)$	$f'(x_n)$	$\dfrac{f(x_n)}{f'(x_n)}$	$x_n - \dfrac{f(x_n)}{f'(x_n)}$
1	1.2000	0.1416	2.3541	0.0602	1.1398
2	1.1398	−0.0181	3.0118	−0.0060	1.1458
3	1.1458	−0.0003	2.9284	−0.0001	1.1459

9. $f(x) = x - e^{-x}$

$f'(x) = 1 + e^{-x}$

$x_1 = 0.5$

n	x_n
1	0.5
2	0.566311
3	0.567143
4	0.567143

Approximate zero: 0.567

11. $f(x) = x^3 + 3$

$f'(x) = 3x^2$

Approximation of the zero of f is −1.442.

n	x_n	$f(x_n)$	$f'(x_n)$	$\dfrac{f(x_n)}{f'(x_n)}$	$x_n - \dfrac{f(x_n)}{f'(x_n)}$
1	−1.5000	−0.3750	6.7500	−0.0556	−1.4444
2	−1.4444	−0.0134	6.2589	−0.0021	−1.4423
3	−1.4423	−0.0003	6.2407	−0.0001	−1.4422

13. $f(x) = x^3 - 3.9x^2 + 4.79x - 1.881$

$f'(x) = 3x^2 - 7.8x + 4.79$

n	x_n	$f(x_n)$	$f'(x_n)$	$\dfrac{f(x_n)}{f'(x_n)}$	$x_n - \dfrac{f(x_n)}{f'(x_n)}$
1	0.5000	−0.3360	1.6400	−0.2049	0.7049
2	0.7049	−0.0921	0.7824	−0.1177	0.8226
3	0.8226	−0.0231	0.4037	−0.0573	0.8799
4	0.8799	−0.0045	0.2495	−0.0181	0.8980
5	0.8980	−0.0004	0.2048	−0.0020	0.9000
6	0.9000	0.0000	0.2000	0.0000	0.9000

Approximation of the zero of f is 0.900.

n	x_n	$f(x_n)$	$f'(x_n)$	$\dfrac{f(x_n)}{f'(x_n)}$	$x_n - \dfrac{f(x_n)}{f'(x_n)}$
1	1.1	0.0000	−0.1600	−0.0000	1.1000

Approximation of the zero of f is 1.100.

n	x_n	$f(x_n)$	$f'(x_n)$	$\dfrac{f(x_n)}{f'(x_n)}$	$x_n - \dfrac{f(x_n)}{f'(x_n)}$
1	1.9	0.0000	0.8000	0.0000	1.9000

Approximation of the zero of f is 1.900.

15. $f(x) = x + \sin(x + 1)$

$f'(x) = 1 + \cos(x + 1)$

Approximation of the zero of f is -0.489.

n	x_n	$f(x_n)$	$f'(x_n)$	$\frac{f(x_n)}{f'(x_n)}$	$x_n - \frac{f(x_n)}{f'(x_n)}$
1	−0.5000	−0.0206	1.8776	−0.0110	−0.4890
2	−0.4890	0.0000	1.8723	0.0000	−0.4890

17. $h(x) = f(x) - g(x) = 2x + 1 - \sqrt{x + 4}$

$h'(x) = 2 - \dfrac{1}{2\sqrt{x + 4}}$

Point of intersection of the graphs of f and g occurs when $x \approx 0.569$.

n	x_n	$h(x_n)$	$h'(x_n)$	$\frac{h(x_n)}{h'(x_n)}$	$x_n - \frac{h(x_n)}{h'(x_n)}$
1	0.6000	0.0552	1.7669	0.0313	0.5687
2	0.5687	−0.0001	1.7661	0.0000	0.5687

19. $h(x) = f(x) - g(x) = x - \tan x$

$h'(x) = 1 - \sec^2 x$

Point of intersection of the graphs of f and g occurs when $x \approx 4.493$.

n	x_n	$h(x_n)$	$h'(x_n)$	$\frac{h(x_n)}{h'(x_n)}$	$x_n - \frac{h(x_n)}{h'(x_n)}$
1	4.5000	−0.1373	−21.5048	0.0064	4.4936
2	4.4936	−0.0039	−20.2271	0.0002	4.4934

21. $h(x) = \ln x + x$

$h'(x) = \dfrac{1}{x} + 1$

$x_1 = 0.5$

Approximate intersection at $x \approx 0.567$

n	x_n
1	0.5
2	0.564382
3	0.567139
4	0.567143

23. $h(x) = \arctan x - \arccos x$

$h'(x) = \dfrac{1}{1 + x^2} + \dfrac{1}{\sqrt{1 - x^2}}$

Approximate intersection at $x \approx 0.786$

n	x_n
1	0.5
2	0.798537
3	0.786251
4	0.786151

25. $f(x) = x^2 - a = 0$

$f'(x) = 2x$

$$x_{i+1} = x_i - \frac{x_i^2 - a}{2x_i}$$

$$= \frac{2x_i^2 - x_i^2 + a}{2x_i} = \frac{x_i^2 + a}{2x_i} = \frac{x_i}{2} + \frac{a}{2x_i}$$

27. $x_{i+1} = \dfrac{x_i^2 + 7}{2x_i}$

i	1	2	3	4	5
x_i	2.0000	2.7500	2.6477	2.6458	2.6458

$\sqrt{7} \approx 2.646$

29. $x_{i+1} = \dfrac{3x_i^4 + 6}{4x_i^3}$

$\sqrt[4]{6} \approx 1.565$

i	1	2	3	4
x_i	1.5000	1.5694	1.5651	1.5651

31. $f(x) = 1 + \cos x$

$f'(x) = -\sin x$

Approximation of the zero: 3.141

n	x_n	$f(x_n)$	$f'(x_n)$	$\frac{f(x_n)}{f'(x_n)}$	$x_n - \frac{f(x_n)}{f'(x_n)}$
1	3.0000	0.0100	−0.1411	−0.0709	3.0709
2	3.0709	0.0025	−0.0706	−0.0354	3.1063
3	3.1063	0.0006	−0.0353	−0.0176	3.1239
4	3.1239	0.0002	−0.0177	−0.0088	3.1327
5	3.1327	0.0000	−0.0089	−0.0044	3.1371
6	3.1371	0.0000	−0.0045	−0.0022	3.1393
7	3.1393	0.0000	−0.0023	−0.0011	3.1404
8	3.1404	0.0000	−0.0012	−0.0006	3.1410

33. $y = 2x^3 - 6x^2 + 6x - 1 = f(x)$

$y' = 6x^2 - 12x + 6 = f'(x)$

$x_1 = 1$

$f'(x) = 0$; therefore, the method fails.

n	x_n	$f(x_n)$	$f'(x_n)$
1	1	1	0

35. $y = -x^3 + 6x^2 - 10x + 6 = f(x)$

$y' = -3x^2 + 12x - 10 = f'(x)$

$x_1 = 2$

$x_2 = 1$

$x_3 = 2$

$x_4 = 1$ and so on.

Fails to converge

37. Answers will vary.

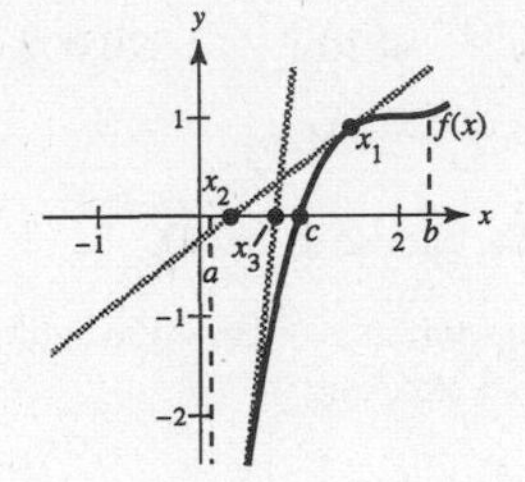

Newton's Method uses tangent lines to approximate c such that $f(c) = 0$.

First, estimate an initial x_1 close to c (see graph).

Then determine x_2 by $x_2 = x_1 - \frac{f(x_1)}{f'(x_1)}$.

Calculate a third estimate by $x_3 = x_2 - \frac{f(x_2)}{f'(x_2)}$.

Continue this process until $|x_n - x_{n+1}|$ is within the desired accuracy.

Let x_{n+1} be the final approximation of c.

39. Let $g(x) = f(x) - x = \cos x - x$

$g'(x) = -\sin x - 1.$

The fixed point is approximately 0.74.

n	x_n	$g(x_n)$	$g'(x_n)$	$\frac{g(x_n)}{g'(x_n)}$	$x_n - \frac{g(x_n)}{g'(x_n)}$
1	1.0000	−0.4597	−1.8415	0.2496	0.7504
2	0.7504	−0.0190	−1.6819	0.0113	0.7391
3	0.7391	0.0000	−1.6736	0.0000	0.7391

41. $g(x) = e^{x/10} - x$

n	x_n
1	1.0
2	1.118238
3	1.118326

Approximate fixed point: 1.12

43. $f(x) = x^3 - 3x^2 + 3,\ f'(x) = 3x^2 - 6x$

(a)

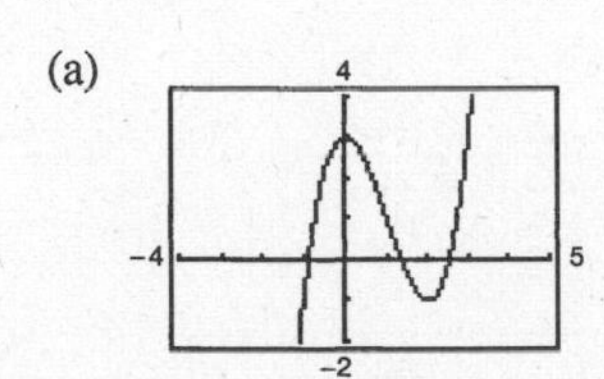

(b) $x_1 = 1$

$x_2 = x_1 - \frac{f(x_1)}{f'(x_1)} \approx 1.333$

Continuing, the zero is 1.347.

(c) $x_1 = \frac{1}{4}$

$x_2 = x_1 - \frac{f(x_1)}{f'(x_1)} \approx 2.405$

Continuing, the zero is 2.532.

(d)

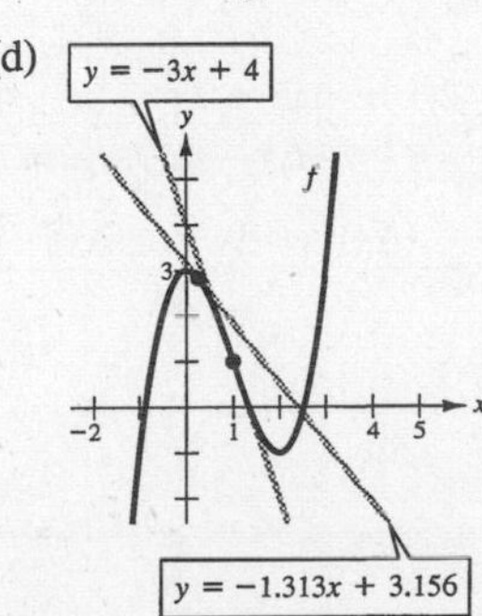

The x-intercepts correspond to the values resulting from the first iteration of Newton's Method.

(e) If the initial guess x_1 is not "close to" the desired zero of the function, the x-intercept of the tangent line may approximate another zero of the function.

45. $f(x) = \frac{1}{x} - a = 0$

$f'(x) = -\frac{1}{x^2}$

$x_{n+1} = x_n - \frac{(1/x_n) - a}{-1/x_n^2} = x_n + x_n^2\left(\frac{1}{x_n} - a\right) = x_n + x_n - x_n^2a = 2x_n - x_n^2a = x_n(2 - ax_n)$

47. $2{,}500{,}000 = -76x^3 + 4830x^2 - 320{,}000$

$76x^3 - 4830x^2 + 2{,}820{,}000 = 0$

Let $f(x) = 76x^3 - 4830x^2 + 2{,}820{,}000$

$f'(x) = 228x^2 - 9660x.$

From the graph, choose $x_1 = 40$.

The zero occurs when $x \approx 38.4356$ which corresponds to \$384,356.

n	x_n	$f(x_n)$	$f'(x_n)$	$\frac{f(x_n)}{f'(x_n)}$	$x_n - \frac{f(x_n)}{f'(x_n)}$
1	40.0000	−44000.0000	−21600.0000	2.0370	37.9630
2	37.9630	17157.6209	−38131.4039	−0.4500	38.4130
3	38.4130	780.0914	−34642.2263	−0.0225	38.4355
4	38.4355	2.6308	−34465.3435	−0.0001	38.4356

49. False. Let $f(x) = (x^2 - 1)/(x - 1)$. $x = 1$ is a discontinuity. It is not a zero of $f(x)$. This statement would be true if $f(x) = p(x)/q(x)$ is given in **reduced** form.

51. True

53. $f(x) = \frac{1}{4}x^3 - 3x^2 + \frac{3}{4}x - 2$

$f'(x) = \frac{3}{4}x^2 - 6x + \frac{3}{4}$

Let $x_1 = 12$.

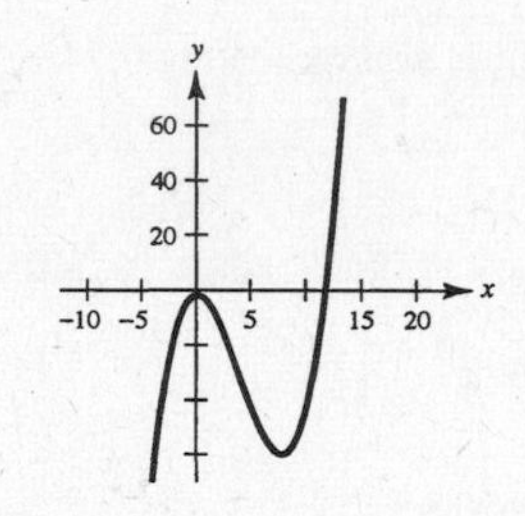

n	x_n	$f(x_n)$	$f'(x_n)$	$\frac{f(x_n)}{f'(x_n)}$	$x_n - \frac{f(x_n)}{f'(x_n)}$
1	12.0000	7.0000	36.7500	0.1905	11.8095
2	11.8095	0.2151	34.4912	0.0062	11.8033
3	11.8033	0.0015	34.4186	0.0000	11.8033

Approximation: $x \approx 11.803$

Review Exercises for Chapter 2

1. $f(x) = x^2 - 2x + 3$

$$f'(x) = \lim_{\Delta x \to 0} \frac{f(x + \Delta x) - f(x)}{\Delta x}$$

$$= \lim_{\Delta x \to 0} \frac{[(x + \Delta x)^2 - 2(x + \Delta x) + 3] - [x^2 - 2x + 3]}{\Delta x}$$

$$= \lim_{\Delta x \to 0} \frac{(x^2 + 2x(\Delta x) + (\Delta x)^2 - 2x - 2(\Delta x) + 3) - (x^2 - 2x + 3)}{\Delta x}$$

$$= \lim_{\Delta x \to 0} \frac{2x(\Delta x) + (\Delta x)^2 - 2(\Delta x)}{\Delta x} = \lim_{\Delta x \to 0} (2x + \Delta x - 2) = 2x - 2$$

3. $f(x) = \sqrt{x} + 1$

$$f'(x) = \lim_{\Delta x \to 0} \frac{f(x + \Delta x) - f(x)}{\Delta x}$$

$$= \lim_{\Delta x \to 0} \frac{(\sqrt{x + \Delta x} + 1) - (\sqrt{x} + 1)}{\Delta x}$$

$$= \lim_{\Delta x \to 0} \frac{\sqrt{x + \Delta x} - \sqrt{x}}{\Delta x} \cdot \frac{\sqrt{x + \Delta x} + \sqrt{x}}{\sqrt{x + \Delta x} + \sqrt{x}}$$

$$= \lim_{\Delta x \to 0} \frac{(x + \Delta x) - x}{\Delta x(\sqrt{x + \Delta x} + \sqrt{x})}$$

$$= \lim_{\Delta x \to 0} \frac{1}{\sqrt{x + \Delta x} + \sqrt{x}} = \frac{1}{2\sqrt{x}}$$

5. f is differentiable for all $x \neq -1$.

7. $f(x) = 4 - |x - 2|$

(a) Continuous at $x = 2$.

(b) Not differentiable at $x = 2$ because of the sharp turn in the graph.

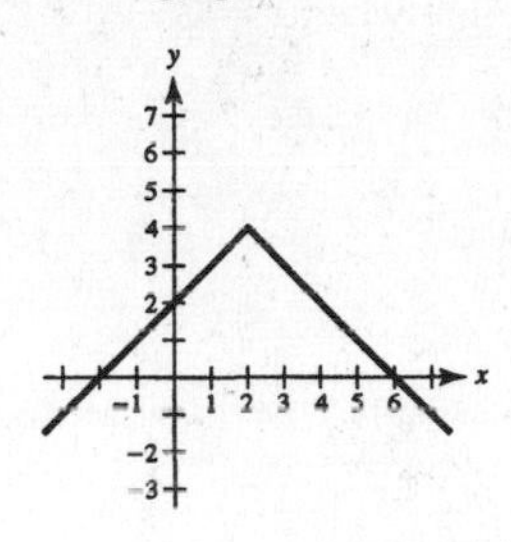

9. Using the limit definition, you obtain $g'(x) = \frac{4}{3}x - \frac{1}{6}$.

At $x = -1$, $g'(-1) = -\frac{4}{3} - \frac{1}{6} = \frac{-3}{2}$

11. (a) Using the limit defintion, $f'(x) = 3x^2$.

At $x = -1$, $f'(-1) = 3$. The tangent line is

$y - (-2) = 3(x - (-1))$

$y = 3x + 1$

(b)

13. $$g'(2) = \lim_{x \to 2} \frac{g(x) - g(2)}{x - 2}$$

$$= \lim_{x \to 2} \frac{x^2(x - 1) - 4}{x - 2}$$

$$= \lim_{x \to 2} \frac{x^3 - x^2 - 4}{x - 2}$$

$$= \lim_{x \to 2} \frac{(x - 2)(x^2 + x + 2)}{x - 2}$$

$$= \lim_{x \to 2} (x^2 + x + 2) = 8$$

15.

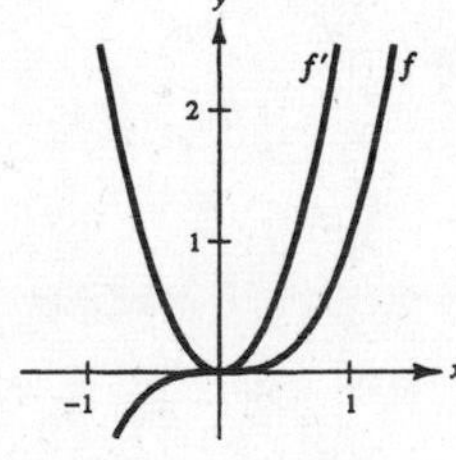

17. $y = 25$

$y' = 0$

19. $f(x) = x^8$

$f'(x) = 8x^7$

21. $h(t) = 3t^4$

$h'(t) = 12t^3$

23. $f(x) = x^3 - 3x^2$

$f'(x) = 3x^2 - 6x = 3x(x - 2)$

25. $h(x) = 6\sqrt{x} + 3\sqrt[3]{x} = 6x^{1/2} + 3x^{1/3}$

$h'(x) = 3x^{-1/2} + x^{-2/3} = \dfrac{3}{\sqrt{x}} + \dfrac{1}{\sqrt[3]{x^2}}$

27. $g(t) = \dfrac{2}{3t^2}$

$g(t) = \dfrac{2}{3}t^{-2}$

$g'(t) = \dfrac{-4}{3}t^{-3} = \dfrac{-4}{3t^3}$

29. $f(\theta) = 2\theta - 3\sin\theta$

$f'(\theta) = 2 - 3\cos\theta$

31. $f(t) = 3\cos t - 4e^t$

$f'(t) = -3\sin t - 4e^t$

33. $F = 200\sqrt{T}$

$F'(t) = \dfrac{100}{\sqrt{T}}$

(a) When $T = 4$, $F'(4) = 50$ vibrations/sec/lb.

(b) When $T = 9$, $F'(9) = 33\frac{1}{3}$ vibrations/sec/lb.

35. $s(t) = -16t^2 + s_0$

$s(9.2) = -16(9.2)^2 + s_0 = 0$

$s_0 = 1354.24$

The building is approximately 1354 feet high (or 415 m).

37. (a)

Total horizontal distance: 50

(b) $0 = x - 0.02x^2$

$0 = x\left(1 - \dfrac{x}{50}\right)$ implies $x = 50$.

(c) Ball reaches maximum height when $x = 25$.

(d) $y = x - 0.02x^2$

$y' = 1 - 0.04x$

$y'(0) = 1$

$y'(10) = 0.6$

$y'(25) = 0$

$y'(30) = -0.2$

$y'(50) = -1$

(e) $y'(25) = 0$

39. $x(t) = t^2 - 3t + 2 = (t - 2)(t - 1)$

(a) $v(t) = x'(t) = 2t - 3$

(b) $v(t) < 0$ for $t < \frac{3}{2}$.

(c) $v(t) = 0$ for $t = \frac{3}{2}$.

$x = \left(\frac{3}{2} - 2\right)\left(\frac{3}{2} - 1\right) = \left(-\frac{1}{2}\right)\left(\frac{1}{2}\right) = -\frac{1}{4}$

(d) $x(t) = 0$ for $t = 1, 2$.

$|v(1)| = |2(1) - 3| = 1$

$|v(2)| = |2(2) - 3| = 1$

The speed is 1 when the position is 0.

41. $f(x) = (3x^2 + 7)(x^2 - 2x + 3)$

$f'(x) = (3x^2 + 7)(2x - 2) + (x^2 - 2x + 3)(6x)$

$= 2(6x^3 - 9x^2 + 16x - 7)$

43. $h(x) = \sqrt{x}\sin x = x^{1/2}\sin x$

$h'(x) = \dfrac{1}{2\sqrt{x}}\sin x + \sqrt{x}\cos x$

45. $f(x) = \dfrac{2x^3 - 1}{x^2}$

$f(x) = 2x - x^{-2}$

$f'(x) = 2 + 2x^{-3} = 2\left(1 + \dfrac{1}{x^3}\right)$

$= \dfrac{2(x^3 + 1)}{x^3}$

47. $f(x) = \dfrac{x^2 + x - 1}{x^2 - 1}$

$f'(x) = \dfrac{(x^2 - 1)(2x + 1) - (x^2 + x - 1)(2x)}{(x^2 - 1)^2}$

$= \dfrac{-(x^2 + 1)}{(x^2 - 1)^2}$

49. $f(x) = \dfrac{1}{4 - 3x^2}$

$f(x) = (4 - 3x^2)^{-1}$

$f'(x) = -(4 - 3x^2)^{-2}(-6x) = \dfrac{6x}{(4 - 3x^2)^2}$

51. $y = \dfrac{x^2}{\cos x}$

$y' = \dfrac{\cos x\,(2x) - x^2(-\sin x)}{\cos^2 x} = \dfrac{2x \cos x + x^2 \sin x}{\cos^2 x}$

53. $y = 3x^2 \sec x$

$y' = 3x^2 \sec x \tan x + 6x \sec x$

55. $y = -x \tan x$

$y' = -x \sec^2 x - \tan x$

57. $y = 4xe^x$

$y' = 4xe^x + 4e^x = 4e^x(x + 1)$

59. $g(t) = t^3 - 3t + 2$

$g'(t) = 3t^2 - 3$

$g''(t) = 6t$

61. $f(\theta) = 3 \tan \theta$

$f'(\theta) = 3 \sec^2 \theta$

$f''(\theta) = 6 \sec \theta\,(\sec \theta \tan \theta) = 6 \sec^2 \theta \tan \theta$

63.

$$\begin{aligned} y &= 2 \sin x + 3 \cos x \\ y' &= 2 \cos x - 3 \sin x \\ y'' &= -2 \sin x - 3 \cos x \\ y'' + y &= -(2 \sin x + 3 \cos x) + (2 \sin x + 3 \cos x) \\ &= 0 \end{aligned}$$

65. $f(x) = \sec x,\ \ g(x) = \csc x,\ \ f'(x) = g'(x)\ \text{ on }\ [0, 2\pi)$:

$$\begin{aligned} \sec x \tan x &= -\csc x \cot x \\ \frac{\sec x \tan x}{\csc x \cot x} &= -1 \\ \frac{\frac{1}{\cos x} \cdot \frac{\sin x}{\cos x}}{\frac{1}{\sin x} \cdot \frac{\cos x}{\sin x}} &= -1 \\ \frac{\sin^3 x}{\cos^3 x} &= -1 \\ \tan^3 x &= -1 \\ \tan x &= -1 \\ x &= \frac{3\pi}{4}, \frac{7\pi}{4} \end{aligned}$$

67. $f(x) = \sqrt{1 - x^3}$

$f(x) = (1 - x^3)^{1/2}$

$f'(x) = \frac{1}{2}(1 - x^3)^{-1/2}(-3x^2)$

$= -\dfrac{3x^2}{2\sqrt{1 - x^3}}$

69. $h(x) = \left(\dfrac{x - 3}{x^2 + 1}\right)^2$

$h'(x) = 2\left(\dfrac{x - 3}{x^2 + 1}\right)\left(\dfrac{(x^2 + 1)(1) - (x - 3)(2x)}{(x^2 + 1)^2}\right)$

$= \dfrac{2(x - 3)(-x^2 + 6x + 1)}{(x^2 + 1)^3}$

71. $f(s) = (s^2 - 1)^{5/2}(s^3 + 5)$

$f'(s) = (s^2 - 1)^{5/2}(3s^2) + (s^3 + 5)\left(\dfrac{5}{2}\right)(s^2 - 1)^{3/2}(2s)$

$= s(s^2 - 1)^{3/2}[3s(s^2 - 1) + 5(s^3 + 5)]$

$= s(s^2 - 1)^{3/2}(8s^3 - 3s + 25)$

73. $y = 3 \cos(3x + 1)$

$y' = -3 \sin(3x + 1)(3)$

$y' = -9 \sin(3x + 1)$

75. $y = \dfrac{1}{2} \csc 2x$

$y' = \dfrac{1}{2}(-\csc 2x \cot 2x)(2)$

$= -\csc 2x \cot 2x$

77. $y = \frac{x}{2} - \frac{\sin 2x}{4}$

$y' = \frac{1}{2} - \frac{1}{4}\cos 2x(2)$

$= \frac{1}{2}(1 - \cos 2x) = \sin^2 x$

79. $y = \frac{2}{3}\sin^{3/2} x - \frac{2}{7}\sin^{7/2} x$

$y' = \sin^{1/2} x \cos x - \sin^{5/2} x \cos x$

$= (\cos x)\sqrt{\sin x}(1 - \sin^2 x)$

$= (\cos^3 x)\sqrt{\sin x}$

81. $y = \frac{\sin \pi x}{x + 2}$

$y' = \frac{(x + 2)\pi \cos \pi x - \sin \pi x}{(x + 2)^2}$

83. $g(t) = t^2 e^{t/4}$

$g'(t) = \frac{1}{4}t^2 e^{t/4} + 2te^{t/4}$

$= \frac{1}{4}te^{t/4}[t + 8]$

85. $y = \sqrt{e^{2x} + e^{-2x}} = (e^{2x} + e^{-2x})^{1/2}$

$y' = \frac{1}{2}(e^{2x} + e^{-2x})^{-1/2}(2e^{2x} - 2e^{-2x})$

$= \frac{e^{2x} - e^{-2x}}{\sqrt{e^{2x} + e^{-2x}}}$

87. $g(x) = \frac{x^2}{e^x}$

$g'(x) = \frac{e^x(2x) - x^2 e^x}{e^{2x}} = \frac{x(2 - x)}{e^x}$

89. $g(x) = \ln\sqrt{x} = \frac{1}{2}\ln x$

$g'(x) = \frac{1}{2x}$

91. $f(x) = x\sqrt{\ln x}$

$f'(x) = \left(\frac{x}{2}\right)(\ln x)^{-1/2}\left(\frac{1}{x}\right) + \sqrt{\ln x}$

$= \frac{1}{2\sqrt{\ln x}} + \sqrt{\ln x} = \frac{1 + 2\ln x}{2\sqrt{\ln x}}$

93. $y = \frac{1}{b^2}\left[\ln(a + bx) + \frac{a}{a + bx}\right]$

$\frac{dy}{dx} = \frac{1}{b^2}\left[\frac{b}{a + bx} - \frac{ab}{(a + bx)^2}\right] = \frac{x}{(a + bx)^2}$

95. $y = -\frac{1}{a}\ln\left(\frac{a + bx}{x}\right) = -\frac{1}{a}[\ln(a + bx) - \ln x]$

$\frac{dy}{dx} = -\frac{1}{a}\left(\frac{b}{a + bx} - \frac{1}{x}\right) = \frac{1}{x(a + bx)}$

97. $f(t) = t^2(t - 1)^5$

$f'(t) = t(t - 1)^4(7t - 2)$

The zeros of f' correspond to the points on the graph of f where the tangent line is horizontal.

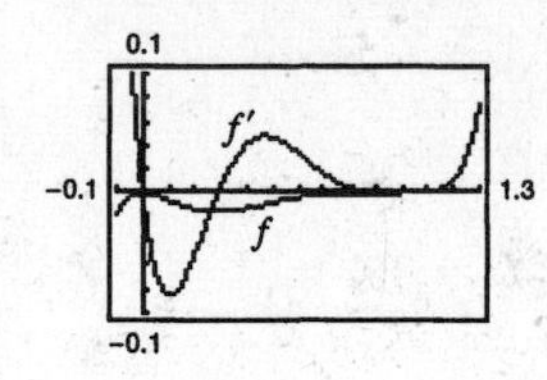

99. $g(x) = 2x(x + 1)^{-1/2}$

$g'(x) = \frac{x + 2}{(x + 1)^{3/2}}$

g' does not equal zero for any value of x in the domain. The graph of g has no horizontal tangent lines.

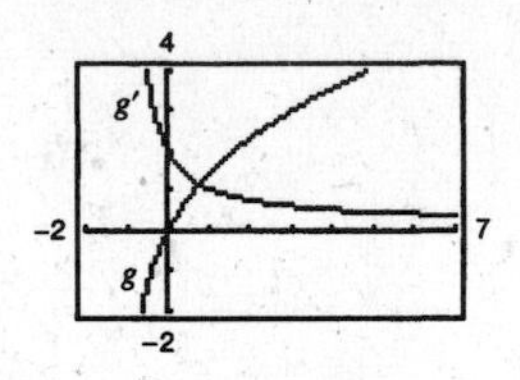

101. $f(t) = (t + 1)^{1/2}(t + 1)^{1/3} = (t + 1)^{5/6}$

$f'(t) = \frac{5}{6(t + 1)^{1/6}}$

f' does not equal zero for any x in the domain. The graph of f has no horizontal tangent lines.

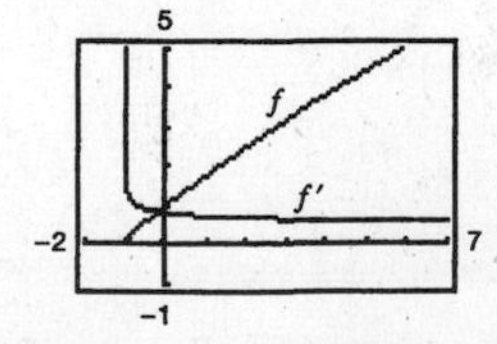

103. $y = \tan\sqrt{1 - x}$

$y' = -\frac{\sec^2\sqrt{1 - x}}{2\sqrt{1 - x}}$

y' does not equal zero for any x in the domain. The graph has no horizontal tangent lines.

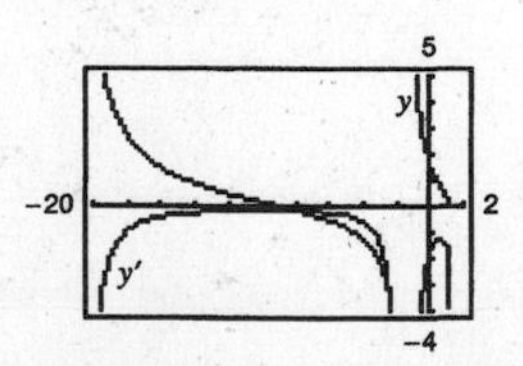

105. $y = 2x^2 + \sin 2x$

$y' = 4x + 2\cos 2x$

$y'' = 4 - 4\sin 2x$

107. $f(x) = \cot x$

$f'(x) = -\csc^2 x$

$f''(x) = -2\csc x(-\csc x \cdot \cot x)$

$= 2\csc^2 x \cot x$

109. $f(t) = \dfrac{t}{(1-t)^2}$

$f'(t) = \dfrac{t+1}{(1-t)^3}$

$f''(t) = \dfrac{2(t+2)}{(1-t)^4}$

111. $g(\theta) = \tan 3\theta - \sin(\theta - 1)$

$g'(\theta) = 3\sec^2 3\theta - \cos(\theta - 1)$

$g''(\theta) = 18\sec^2 3\theta \tan 3\theta + \sin(\theta - 1)$

113. $g(x) = x^3 \ln x$

$g'(x) = 3x^2 \ln x + x^3\left(\dfrac{1}{x}\right) = x^2(3\ln x + 1)$

$g''(x) = 2x(3\ln x + 1) + x^2\left(\dfrac{3}{x}\right)$

$= 6x\ln x + 5x$

115. $T = \dfrac{700}{t^2 + 4t + 10}$

$T = 700(t^2 + 4t + 10)^{-1}$

$T' = \dfrac{-1400(t+2)}{(t^2 + 4t + 10)^2}$

(a) When $t = 1$,

$T' = \dfrac{-1400(1+2)}{(1+4+10)^2} \approx -18.667$ deg/hr.

(b) When $t = 3$,

$T' = \dfrac{-1400(3+2)}{(9+12+10)^2} \approx -7.284$ deg/hr.

(c) When $t = 5$,

$T' = \dfrac{-1400(5+2)}{(25+20+10)^2} \approx -3.240$ deg/hr.

(d) When $t = 10$,

$T' = \dfrac{-1400(10+2)}{(100+40+10)^2} \approx -0.747$ deg/hr.

117. (a) You get an error message because $\ln h$ does not exist for $h = 0$.

(b) Reversing the data, you obtain

$h = 0.8627 - 6.4474 \ln p$.

(c)

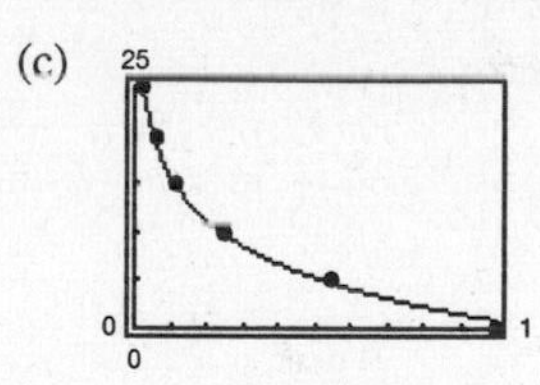

(d) If $p = 0.75$, $h \approx 2.72$ km.

(e) If $h = 13$ km, $p \approx 0.15$ atmosphere.

(f) $h = 0.8627 - 6.4474 \ln p$

$1 = -6.4474 \dfrac{1}{p}\dfrac{dp}{dh}$ (implicit differentiation)

$\dfrac{dp}{dh} = \dfrac{p}{-6.4474}$

For $h = 5$, $p = 0.55$ and $\dfrac{dp}{dh} = -0.0853$ atmos/km.

For $h = 20$, $p = 0.06$ and $\dfrac{dp}{dh} = -0.00931$ atmos/km.

As the altitude increases, the rate of change of pressure decreases.

119. $x^2 + 3xy + y^3 = 10$

$2x + 3xy' + 3y + 3y^2y' = 0$

$3(x + y^2)y' = -(2x + 3y)$

$y' = \dfrac{-(2x + 3y)}{3(x + y^2)}$

121. $\cos x^2 = xe^y$

$-\sin x^2(2x) = xe^y y' + e^y$

$xe^y y' = -e^y - 2x\sin x^2$

$y' = -\dfrac{e^y + 2x\sin x^2}{xe^y} = -\dfrac{1}{x} - 2e^{-y}\sin x^2$

123.
$$y\sqrt{x} - x\sqrt{y} = 16$$
$$y\left(\frac{1}{2}x^{-1/2}\right) + x^{1/2}y' - x\left(\frac{1}{2}y^{-1/2}y'\right) - y^{1/2} = 0$$
$$\left(\sqrt{x} - \frac{x}{2\sqrt{y}}\right)y' = \sqrt{y} - \frac{y}{2\sqrt{x}}$$
$$\frac{2\sqrt{xy} - x}{2\sqrt{y}}y' = \frac{2\sqrt{xy} - y}{2\sqrt{x}}$$
$$y' = \frac{2\sqrt{xy} - y}{2\sqrt{x}} \cdot \frac{2\sqrt{y}}{2\sqrt{xy} - x} = \frac{2y\sqrt{x} - y\sqrt{y}}{2x\sqrt{y} - x\sqrt{x}}$$

125.
$$x \sin y = y \cos x$$
$$(x \cos y)y' + \sin y = -y \sin x + y' \cos x$$
$$y'(x \cos y - \cos x) = -y \sin x - \sin y$$
$$y' = \frac{y \sin x + \sin y}{\cos x - x \cos y}$$

127. $x^2 + y^2 = 20$

$$2x + 2yy' = 0$$
$$y' = -\frac{x}{y}$$

At (2, 4): $y' = -\frac{1}{2}$

Tangent line: $y - 4 = -\frac{1}{2}(x - 2)$
$$x + 2y - 10 = 0$$

Normal line: $y - 4 = 2(x - 2)$
$$2x - y = 0$$

129. $y \ln x + y^2 = 0, \quad (e, -1)$

$$y' \ln x + \frac{y}{x} + 2yy' = 0$$
$$y'(\ln x + 2y) = \frac{-y}{x}$$
$$y' = \frac{-y}{x(\ln x + 2y)}$$

At $(e, -1)$: $y' = \frac{-1}{e}$

Tangent line: $y + 1 = \frac{-1}{e}(x - e)$
$$y = \frac{-1}{e}x$$

Normal line: $y + 1 = e(x - e)$
$$y = ex - e^2 - 1$$

131. $y = \dfrac{x\sqrt{x^2 + 1}}{x + 4}$

$$\ln y = \ln x + \frac{1}{2}\ln(x^2 + 1) - \ln(x + 4)$$
$$\frac{y'}{y} = \frac{1}{x} + \frac{x}{x^2 + 1} - \frac{1}{x + 4}$$
$$y' = \frac{x\sqrt{x^2 + 1}}{x + 4}\left(\frac{1}{x} + \frac{x}{x^2 + 1} - \frac{1}{x + 4}\right) = \frac{x^3 + 8x^2 + 4}{(x + 4)^2\sqrt{x^2 + 1}}$$

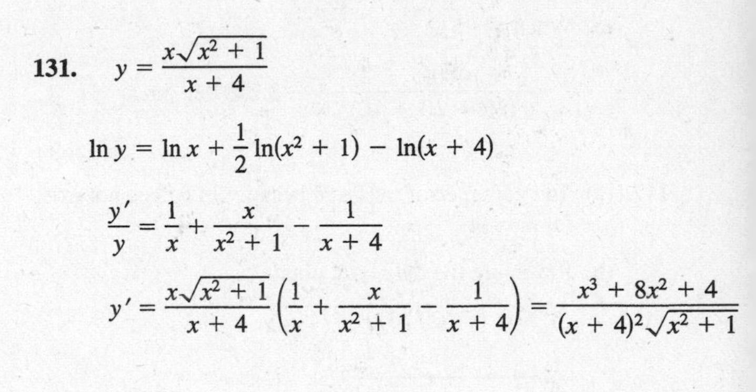

133. $f(x) = x^3 + 2$

$f^{-1}(x) = (x - 2)^{1/3}$

$(f^{-1})'(x) = \frac{1}{3}(x - 2)^{-2/3}$

$(f^{-1})'(-1) = \frac{1}{3}(-1 - 2)^{-2/3} = \frac{1}{3(-3)^{2/3}}$

$= \frac{1}{3^{5/3}} \approx 0.160$

135. $f(x) = \tan x$

$f\left(\frac{\pi}{6}\right) = \frac{\sqrt{3}}{3}$

$f'(x) = \sec^2 x$

$f'\left(\frac{\pi}{6}\right) = \frac{4}{3}$

$(f^{-1})'\left(\frac{\sqrt{3}}{3}\right) = \frac{1}{f'(\pi/6)} = \frac{3}{4}$

137. $y = \tan(\arcsin x) = \frac{x}{\sqrt{1 - x^2}}$

$y' = \frac{(1 - x^2)^{1/2} + x^2(1 - x^2)^{-1/2}}{1 - x^2} = (1 - x^2)^{-3/2}$

139. $y = x \operatorname{arcsec} x$

$y' = \frac{x}{|x|\sqrt{x^2 - 1}} + \operatorname{arcsec} x$

141. $y = x(\arcsin x)^2 - 2x + 2\sqrt{1 - x^2} \arcsin x$

$y' = \frac{2x \arcsin x}{\sqrt{1 - x^2}} + (\arcsin x)^2 - 2 + \frac{2\sqrt{1 - x^2}}{\sqrt{1 - x^2}} - \frac{2x}{\sqrt{1 - x^2}} \arcsin x = (\arcsin x)^2$

143. $y = \sqrt{x}$

$\frac{dy}{dt} = 2$ units/sec

$\frac{dy}{dt} = \frac{1}{2\sqrt{x}}\frac{dx}{dt} \Rightarrow \frac{dx}{dt} = 2\sqrt{x}\frac{dy}{dt} = 4\sqrt{x}$

(a) When $x = \frac{1}{2}$, $\frac{dx}{dt} = 2\sqrt{2}$ units/sec.

(b) When $x = 1$, $\frac{dx}{dt} = 4$ units/sec.

(c) When $x = 4$, $\frac{dx}{dt} = 8$ units/sec.

145. $\frac{s}{h} = \frac{1/2}{2}$

$s = \frac{1}{4}h$

$\frac{dV}{dt} = 1$

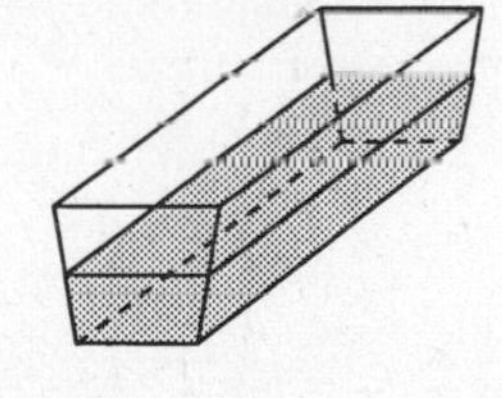

Width of water at depth h:

$w = 2 + 2s = 2 + 2\left(\frac{1}{4}h\right) = \frac{4 + h}{2}$

$V = \frac{5}{2}\left(2 + \frac{4 + h}{2}\right)h = \frac{5}{4}(8 + h)h$

$\frac{dV}{dt} = \frac{5}{2}(4 + h)\frac{dh}{dt}$

$\frac{dh}{dt} = \frac{2(dV/dt)}{5(4 + h)}$

When $h = 1$, $\frac{dh}{dt} = \frac{2}{25}$ m/min.

147. $s(t) = 60 - 4.9t^2$

$s'(t) = -9.8t$

$s = 35 = 60 - 4.9t^2$

$4.9t^2 = 25$

$t = \frac{5}{\sqrt{4.9}}$

$\tan 30° = \frac{1}{\sqrt{3}} = \frac{s(t)}{x(t)}$

$x(t) = \sqrt{3}\, s(t)$

$\frac{dx}{dt} = \sqrt{3}\frac{ds}{dt} = \sqrt{3}(-9.8)\frac{5}{\sqrt{4.9}}$

≈ -38.34 m/sec

149. $f(x) = x^3 - 3x - 1$

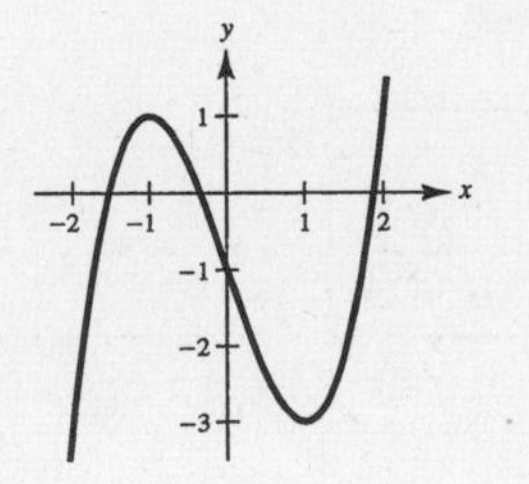

From the graph you can see that $f(x)$ has three real zeros.

$f'(x) = 3x^2 - 3$

n	x_n	$f(x_n)$	$f'(x_n)$	$\frac{f(x_n)}{f'(x_n)}$	$x_n - \frac{f(x_n)}{f'(x_n)}$
1	−1.5000	0.1250	3.7500	0.0333	−1.5333
2	−1.5333	−0.0049	4.0530	−0.0012	−1.5321

n	x_n	$f(x_n)$	$f'(x_n)$	$\frac{f(x_n)}{f'(x_n)}$	$x_n - \frac{f(x_n)}{f'(x_n)}$
1	−0.5000	0.3750	−2.2500	−0.1667	−0.3333
2	−0.3333	−0.0371	−2.6667	0.0139	−0.3472
3	−0.3472	−0.0003	−2.6384	0.0001	−0.3473

n	x_n	$f(x_n)$	$f'(x_n)$	$\frac{f(x_n)}{f'(x_n)}$	$x_n - \frac{f(x_n)}{f'(x_n)}$
1	−1.9000	0.1590	7.8300	0.0203	1.8797
2	1.8797	0.0024	7.5998	0.0003	1.8794

The three real zeros of $f(x)$ are $x \approx -1.532$, $x \approx -0.347$, and $x \approx 1.879$.

151. $g(x) = xe^x - 4$

$g'(x) = (x + 1)e^x$

From the graph, there is one zero near 1.
Newton's method gives

$x_1 = 1$

$x_2 = 1.23576$

$x_3 = 1.20297$

$x_4 = 1.20217$

To three decimal places, $x = 1.202$.

153. Find the zeros of $f(x) = x^4 - x - 3$.

$f'(x) = 4x^3 - 1$

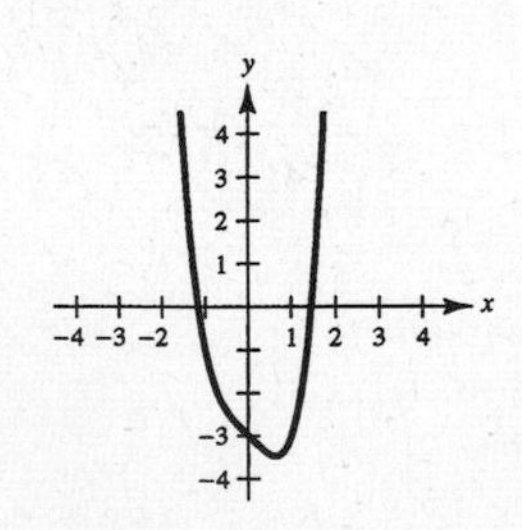

From the graph you can see that $f(x)$ has two real zeros.

f changes sign in $[-2, -1]$.

n	x_n	$f(x_n)$	$f'(x_n)$	$\frac{f(x_n)}{f'(x_n)}$	$x_n - \frac{f(x_n)}{f'(x_n)}$
1	−1.2000	0.2736	−7.9120	−0.0346	−1.1654
2	−1.1654	0.0100	−7.3312	−0.0014	−1.1640

On the interval $[-2, -1]$: $x \approx -1.164$.

f changes sign in $[1, 2]$.

—CONTINUED—

153. —CONTINUED—

n	x_n	$f(x_n)$	$f'(x_n)$	$\dfrac{f(x_n)}{f'(x_n)}$	$x_n - \dfrac{f(x_n)}{f'(x_n)}$
1	1.5000	0.5625	12.5000	0.0450	1.4550
2	1.4550	0.0268	11.3211	0.0024	1.4526
3	1.4526	−0.0003	11.2602	0.0000	1.4526

On the interval $[1, 2]$: $x \approx 1.453$.

Problem Solving for Chapter 2

1. (a) $x^2 + (y - r)^2 = r^2$ Circle

$x^2 = y$ Parabola

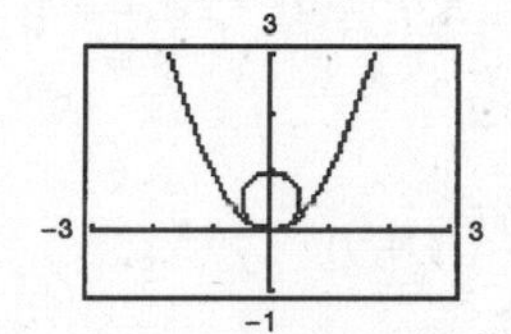

Substituting,

$$(y - r)^2 = r^2 - y$$

$$y^2 - 2ry + r^2 = r^2 - y$$

$$y^2 - 2ry + y = 0$$

$$y(y - 2r + 1) = 0$$

Since you want only one solution, let $1 - 2r = 0 \Rightarrow r = \frac{1}{2}$

Graph $y = x^2$ and $x^2 + (y - \frac{1}{2})^2 = \frac{1}{4}$

(b) Let (x, y) be a point of tangency: $x^2 + (y - b)^2 = 1 \Rightarrow 2x + 2(y - b)y' = 0 \Rightarrow y' = \dfrac{x}{b - y}$ (circle).

$y = x^2 \Rightarrow y' = 2x$ (parabola). Equating,

$$2x = \frac{x}{b - y}$$

$$2(b - y) = 1$$

$$b - y = \frac{1}{2} \Rightarrow b = y + \frac{1}{2}$$

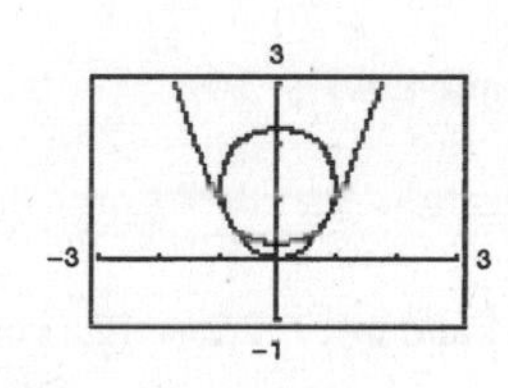

Also, $x^2 + (y - b)^2 = 1$ and $y = x^2$ imply

$$y + (y - b)^2 = 1 \Rightarrow y + \left[y - \left(y + \frac{1}{2}\right)\right]^2 = 1 \Rightarrow y + \frac{1}{4} = 1 \Rightarrow y = \frac{3}{4} \text{ and } b = \frac{5}{4}.$$

Center: $\left(0, \dfrac{5}{4}\right)$

Graph $y = x^2$ and $x^2 + \left(y - \dfrac{5}{4}\right)^2 = 1$

3. (a) $f(x) = \cos x$ $\quad P_1(x) = a_0 + a_1x$

$f(0) = 1$ $\quad P_1(0) = a_0 \Rightarrow a_0 = 1$

$f'(0) = 0$ $\quad P'_1(0) = a_1 \Rightarrow a_1 = 0$

$P_1(x) = 1$

(b) $f(x) = \cos x$ $\quad P_2(x) = a_0 + a_1x + a_2x^2$

$f(0) = 1$ $\quad P_2(0) = a_0 \Rightarrow a_0 = 1$

$f'(0) = 0$ $\quad P'_2(0) = a_1 \Rightarrow a_1 = 0$

$f''(0) = -1$ $\quad P''_2(0) = 2a_2 \Rightarrow a_2 = -\frac{1}{2}$

$P_2(x) = 1 - \frac{1}{2}x^2$

—CONTINUED—

3. —CONTINUED—

(c)

x	-1.0	-0.1	-0.001	0	0.001	0.1	1.0
$\cos x$	0.5403	0.9950	≈ 1	1	≈ 1	0.9950	0.5403
$P_2(x)$	0.5	0.9950	≈ 1	1	≈ 1	0.9950	0.5

$P_2(x)$ is a good approximation of $f(x) = \cos x$ when x is near 0.

(d) $f(x) = \sin x$

$f(0) = 0$

$f'(0) = 1$

$f''(0) = 0$

$f'''(0) = -1$

$P_3(x) = x - \frac{1}{6}x^3$

$P_3(x) = a_0 + a_1x + a_2x^2 + a_3x^3$

$P_3(0) = a_0 \Longrightarrow a_0 = 0$

$P'_3(0) = a_1 \Longrightarrow a_1 = 1$

$P''_3(0) = 2a_2 \Longrightarrow a_2 = 0$

$P'''_3(0) = 6a_3 \Longrightarrow a_3 = -\frac{1}{6}$

5. Let $p(x) = Ax^3 + Bx^2 + Cx + D$

$p'(x) = 3Ax^2 + 2Bx + C$

At $(1, 1)$: $A + B + C + D = 1$ Equation 1

$3A + 2B + C = 14$ Equation 2

At $(-1, -3)$: $-A + B - C + D = -3$ Equation 3

$3A - 2B + C = -2$ Equation 4

Adding Equations 1 and 3: $2B + 2D = -2$

Subtracting Equations 1 and 3: $2A + 2C = 4$

Adding Equations 2 and 4: $6A + 2C = 12$

Subtracting Equations 2 and 4: $4B = 16$

Hence, $B = 4$ and $D = \frac{1}{2}(-2 - 2B) = -5$

Subtracting $2A + 2C = 4$ and $6A + 2C = 12$, you obtain $4A = 8 \Longrightarrow A = 2$. Finally, $C = \frac{1}{2}(4 - 2A) = 0$

Thus, $p(x) = 2x^3 + 4x^2 - 5$.

7. (a) $x^4 = a^2x^2 - a^2y^2$

$a^2y^2 = a^2x^2 - x^4$

$$y = \frac{\pm\sqrt{a^2x^2 - x^4}}{a}$$

Graph: $y_1 = \dfrac{\sqrt{a^2x^2 - x^4}}{a}$ and $y_2 = -\dfrac{\sqrt{a^2x^2 - x^4}}{a}$

(b)

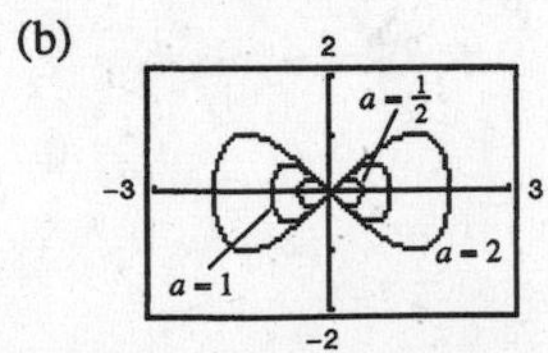

$(\pm a, 0)$ are the x-intercepts, along with $(0, 0)$.

—CONTINUED—

7. —CONTINUED—

(c) Differentiating implicitly,

$$4x^3 = 2a^2x - 2a^2yy'$$

$$y' = \frac{2a^2x - 4x^3}{2a^2y} = \frac{x(a^2 - 2x^2)}{a^2y} = 0 \Rightarrow 2x^2 = a^2 \Rightarrow x = \frac{\pm a}{\sqrt{2}}.$$

$$\left(\frac{a^2}{2}\right)^2 = a^2\left(\frac{a^2}{2}\right) - a^2y^2$$

$$\frac{a^4}{4} = \frac{a^4}{2} - a^2y^2$$

$$a^2y^2 = \frac{a^4}{4}$$

$$y^2 = \frac{a^2}{4}$$

$$y = \pm\frac{a}{2}$$

Four points: $\left(\frac{a}{\sqrt{2}}, \frac{a}{2}\right), \left(\frac{a}{\sqrt{2}}, -\frac{a}{2}\right), \left(-\frac{a}{\sqrt{2}}, \frac{a}{2}\right), \left(\frac{-a}{\sqrt{2}}, -\frac{a}{2}\right)$

9. (a)

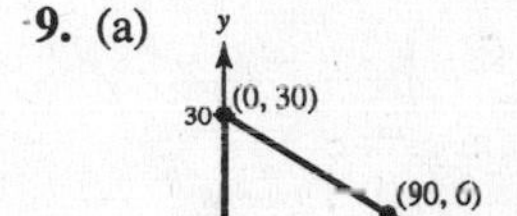

Not drawn to scale

Line determined by (0, 30) and (90, 6):

$$y - 30 = \frac{30 - 6}{0 - 90}(x - 0) = -\frac{24}{90}x = -\frac{4}{15}x \Rightarrow y = -\frac{4}{15}x + 30$$

When $x = 100$, $y = \frac{-4}{15}(100) + 30 = \frac{10}{3} > 3 \Rightarrow$ Shadow determined by man.

(b)

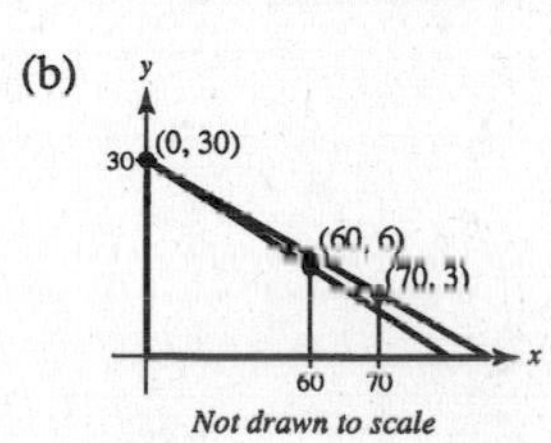

Not drawn to scale

Line determined by (0, 30) and (60, 6):

$$y - 30 = \frac{30 - 6}{0 - 60}(x - 0) = -\frac{2}{5}x \Rightarrow y = -\frac{2}{5}x + 30$$

When $x = 70$, $y = \frac{-2}{5}(70) + 30 = 2 < 3 \Rightarrow$ Shadow determined by child.

(c) Need (0, 30), $(d, 6)$, $(d + 10, 3)$ collinear.

$$\frac{30 - 6}{0 - d} = \frac{6 - 3}{d - (d + 10)} \Rightarrow \frac{24}{d} = \frac{3}{10} \Rightarrow d = 80 \text{ feet}$$

(d) Let y be the distance from the base of the street light to the tip of the shadow. We know that $\frac{dx}{dt} = -5$.

For $x > 80$, the shadow is determined by the man.

$$\frac{y}{30} = \frac{y - x}{6} \Rightarrow y = \frac{5}{4}x \text{ and } \frac{dy}{dt} = \frac{5}{4}\frac{dx}{dt} = \frac{-25}{4}.$$

For $x < 80$, the shadow is determined by the child.

$$\frac{y}{30} = \frac{y - x - 10}{3} \Rightarrow y = \frac{10}{9}x + \frac{100}{9} \text{ and } \frac{dy}{dt} = \frac{10}{9}\frac{dx}{dt} = \frac{-50}{9}.$$

Therefore,

$$\frac{dy}{dt} = \begin{cases} \frac{-25}{4} & x > 80 \\ \frac{-50}{9} & 0 < x < 80 \end{cases}$$

$\frac{dy}{dt}$ is not continuous at $x = 80$.

11. $y = \ln x$

$$y' = \frac{1}{x}$$

$$y - b = \frac{1}{a}(x - a)$$

$$y = \frac{1}{a}x + b - 1 \quad \text{Tangent line}$$

If $x = 0$, $c = b - 1$. Thus, $b - c = b - (b - 1) = 1$.

13. (a)

z (degrees)	0.1	0.01	0.0001
$\frac{\sin z}{z}$	0.0174524	0.0174533	0.0174533

(b) $\lim_{z \to 0} \frac{\sin z}{z} \approx 0.0174533$

In fact, $\lim_{z \to 0} \frac{\sin z}{z} = \frac{\pi}{180}$

(c)
$$\begin{aligned}\frac{d}{dz}(\sin z) &= \lim_{\Delta z \to 0} \frac{\sin(z + \Delta z) - \sin z}{\Delta z} \\ &= \lim_{\Delta z \to 0} \frac{\sin z \cdot \cos \Delta z + \sin \Delta z \cdot \cos z - \sin z}{\Delta z} \\ &= \lim_{\Delta z \to 0} \left[\sin z\left(\frac{\cos \Delta z - 1}{\Delta z}\right)\right] + \lim_{\Delta z \to 0} \left[\cos z\left(\frac{\sin \Delta z}{\Delta z}\right)\right] \\ &= (\sin z)(0) + (\cos z)\left(\frac{\pi}{180}\right) = \frac{\pi}{180}\cos z\end{aligned}$$

(d) $S(90) = \sin\left(\frac{\pi}{180}90\right) = \sin\frac{\pi}{2} = 1$; $C(180) = \cos\left(\frac{\pi}{180}180\right) = -1$

$$\frac{d}{dz}S(z) = \frac{d}{dz}\sin(cz) = c \cdot \cos(cz) = \frac{\pi}{180}C(z)$$

(e) The formulas for the derivatives are more complicated in degrees.

15. $j(t) = a'(t)$

(a) $j(t)$ is the rate of change of the acceleration.

(b) From Exercise 102 in Section 2.3,

$s(t) = -8.25t^2 + 66t$

$v(t) = -16.5t + 66$

$a(t) = -16.5$

$a'(t) = j(t) = 0$

The acceleration is constant.

CHAPTER 3
Applications of Differentiation

CHAPTER 3
Applications of Differentiation

Section 3.1 Extrema on an Interval

Solutions to Odd-Numbered Exercises

1. $f(x) = \dfrac{x^2}{x^2+4}$

$f'(x) = \dfrac{(x^2+4)(2x) - (x^2)(2x)}{(x^2+4)^2} = \dfrac{8x}{(x^2+4)^2}$

$f'(0) = 0$

3. $f(x) = x + \dfrac{27}{2x^2} = x + \dfrac{27}{2}x^{-2}$

$f'(x) = 1 - 27x^{-3} = 1 - \dfrac{27}{x^3}$

$f'(3) = 1 - \dfrac{27}{3^3} = 1 - 1 = 0$

5. $f(x) = (x+2)^{2/3}$

$f'(x) = \dfrac{2}{3}(x+2)^{-1/3}$

$f'(-2)$ is undefined.

7. Critical numbers: $x = 2$

$x = 2$: absolute maximum

9. Critical numbers: $x = 1, 2, 3$

$x = 1, 3$: absolute maximum

$x = 2$: absolute minimum

11. $f(x) = x^2(x-3) = x^3 - 3x^2$

$f'(x) = 3x^2 - 6x = 3x(x-2)$

Critical numbers: $x = 0, x = 2$

13. $g(t) = t\sqrt{4-t},\ t < 3$

$$g'(t) = t\left[\frac{1}{2}(4-t)^{-1/2}(-1)\right] + (4-t)^{1/2}$$

$$= \frac{1}{2}(4-t)^{-1/2}[-t + 2(4-t)]$$

$$= \frac{8-3t}{2\sqrt{4-t}}$$

Critical number is $t = \dfrac{8}{3}$.

15. $h(x) = \sin^2 x + \cos x,\ 0 < x < 2\pi$

$h'(x) = 2\sin x\cos x - \sin x = \sin x(2\cos x - 1)$

On $(0, 2\pi)$, critical numbers: $x = \dfrac{\pi}{3}, x = \pi, x = \dfrac{5\pi}{3}$

17. $f(x) = x^2\log_2(x^2+1) = x^2\dfrac{\ln(x^2+1)}{\ln 2}$

$$f'(x) = 2x\frac{\ln(x^2+1)}{\ln 2} + x^2\frac{2x}{\ln 2(x^2+1)}$$

$$= \frac{2x}{\ln 2}\left[\ln(x^2+1) + \frac{x^2}{x^2+1}\right] = 0 \Rightarrow x = 0$$

Critical number: $x = 0$

19. $f(x) = 2(3-x), [-1, 2]$

$f'(x) = -2 \Rightarrow$ No critical numbers

Left endpoint: $(-1, 8)$ Maximum

Right endpoint: $(2, 2)$ Minimum

21. $f(x) = -x^2 + 3x, [0, 3]$

$f'(x) = -2x + 3$

Left endpoint: $(0, 0)$ Minimum

Critical number: $\left(\frac{3}{2}, \frac{9}{4}\right)$ Maximum

Right endpoint: $(3, 0)$ Minimum

23. $f(x) = x^3 - \frac{3}{2}x^2, [-1, 2]$

$f'(x) = 3x^2 - 3x = 3x(x - 1)$

Left endpoint: $\left(-1, -\frac{5}{2}\right)$ Minimum

Right endpoint: $(2, 2)$ Maximum

Critical number: $(0, 0)$

Critical number: $\left(1, -\frac{1}{2}\right)$

25. $f(x) = 3x^{2/3} - 2x, [-1, 1]$

$f'(x) = 2x^{-1/3} - 2 = \dfrac{2(1 - \sqrt[3]{x})}{\sqrt[3]{x}}$

Left endpoint: $(-1, 5)$ Maximum

Critical number: $(0, 0)$ Minimum

Right endpoint: $(1, 1)$

27. $g(t) = \dfrac{t^2}{t^2 + 3}, [-1, 1]$

$g'(t) = \dfrac{6t}{(t^2 + 3)^2}$

Left endpoint: $\left(-1, \frac{1}{4}\right)$ Maximum

Critical number: $(0, 0)$ Minimum

Right endpoint: $\left(1, \frac{1}{4}\right)$ Maximum

29. $h(s) = \dfrac{1}{s - 2}, [0, 1]$

$h'(s) = \dfrac{-1}{(s - 2)^2}$

Left endpoint: $\left(0, -\frac{1}{2}\right)$ Maximum

Right endpoint: $(1, -1)$ Minimum

31. $y = e^x \sin x, \quad [0, \pi]$

$y' = e^x \sin x + e^x \cos x = e^x(\sin x + \cos x)$

Left endpoint: $(0, 0)$ Minimum

Critical number:

$\left(\frac{3\pi}{4}, \frac{\sqrt{2}}{2}e^{3\pi/4}\right) \approx \left(\frac{3\pi}{4}, 7.46\right)$ Maximum

Right endpoint: $(\pi, 0)$ Minimum

33. $f(x) = \cos \pi x, \left[0, \frac{1}{6}\right]$

$f'(x) = -\pi \sin \pi x$

Left endpoint: $(0, 1)$ Maximum

Right endpoint: $\left(\frac{1}{6}, \frac{\sqrt{3}}{2}\right)$ Minimum

35. $y = \dfrac{4}{x} + \tan\dfrac{\pi x}{8}, \ [1, 2]$

$y' = \dfrac{-4}{x^2} + \dfrac{\pi}{8}\sec^2\dfrac{\pi x}{8} = 0$

$\dfrac{\pi}{8}\sec^2\dfrac{\pi x}{8} = \dfrac{4}{x^2}$

On the interval $[1, 2]$, this equation has no solutions. Thus, there are no critical numbers.

Left endpoint: $(1, \sqrt{2} + 3) \approx (1, 4.4142)$ Maximum

Right endpoint: $(2, 3)$ Minimum

37. (a) Minimum: $(0, -3)$

Maximum: $(2, 1)$

(b) Minimum: $(0, -3)$

(c) Maximum: $(2, 1)$

(d) No extrema

39. $f(x) = \begin{cases} 2x + 2, & 0 \le x \le 1 \\ 4x^2, & 1 < x \le 3 \end{cases}$

Left endpoint: $(0, 2)$ Minimum

Right endpoint: $(3, 36)$ Maximum

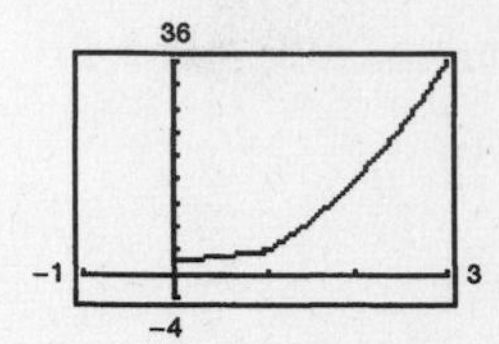

41. $f(x) = \dfrac{3}{x-1}, (1, 4]$

Right endpoint: $(4, 1)$ Minimum

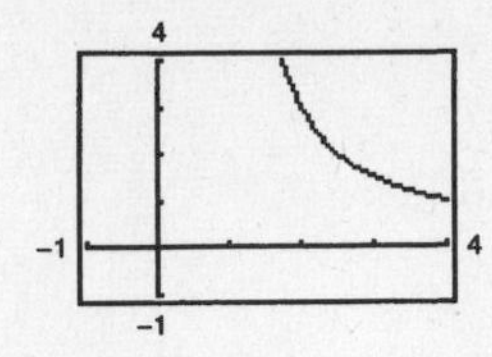

43. (a)

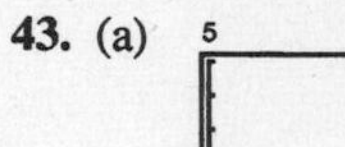

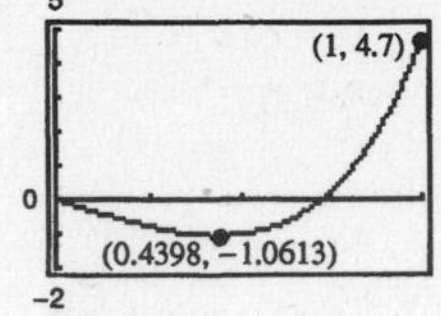

Maximum: $(1, 4.7)$ (endpoint)

Minimum: $(0.4398, -1.0613)$

(b)

$$f(x) = 3.2x^5 + 5x^3 - 3.5x, [0, 1]$$

$$f'(x) = 16x^4 + 15x^2 - 3.5$$

$$16x^4 + 15x^2 - 3.5 = 0$$

$$x^2 = \frac{-15 \pm \sqrt{(15)^2 - 4(16)(-3.5)}}{2(16)}$$

$$= \frac{-15 \pm \sqrt{449}}{32}$$

$$x = \sqrt{\frac{-15 + \sqrt{449}}{32}} \approx 0.4398$$

$$f(0) = 0$$

$$f(1) = 4.7 \text{ Maximum (endpoint)}$$

$$f\left(\sqrt{\frac{-15 + \sqrt{449}}{32}}\right) \approx -1.0613$$

Minimum: $(0.4398, -1.0613)$

45. (a)

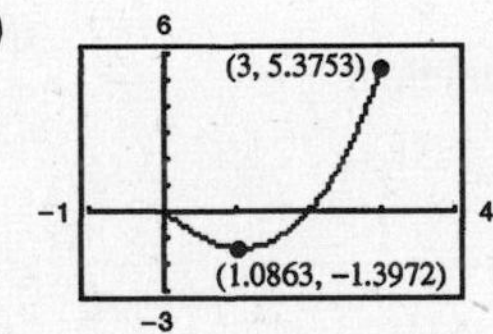

(b) $(1.0863, -1.3972)$ Minimum

47. (a)

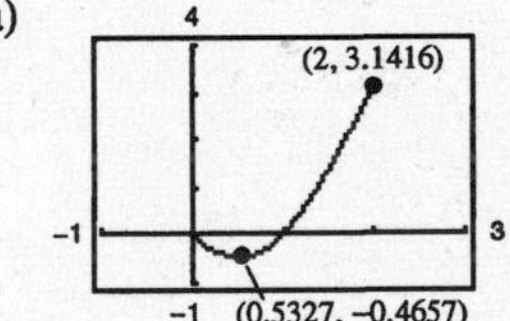

(b) $(0.5327, -0.4657)$ Minimum

49. $f(x) = (1 + x^3)^{1/2}, [0, 2]$

$f'(x) = \frac{3}{2}x^2(1 + x^3)^{-1/2}$

$f''(x) = \frac{3}{4}(x^4 + 4x)(1 + x^3)^{-3/2}$

$f'''(x) = -\frac{3}{8}(x^6 + 20x^3 - 8)(1 + x^3)^{-5/2}$

Setting $f''' = 0$, we have $x^6 + 20x^3 - 8 = 0$.

$x^3 = \frac{-20 \pm \sqrt{400 - 4(1)(-8)}}{2}$

$x = \sqrt[3]{-10 \pm \sqrt{108}} = \sqrt{3} - 1$

In the interval $[0, 2]$, choose

$x = \sqrt[3]{-10 + \sqrt{108}} = \sqrt{3} - 1 \approx 0.732.$

$\left|f''\left(\sqrt[3]{-10 + \sqrt{108}}\right)\right| \approx 1.47$ is the maximum value.

51. $f(x) = e^{-x^2/2}, \quad [0, 1]$

$f'(x) = -xe^{-x^2/2}$

$f''(x) = -x(-xe^{-x^2/2}) - e^{-x^2/2}$

$= e^{-x^2/2}(x^2 - 1)$

$f'''(x) = e^{-x^2/2}(2x) + (x^2 - 1)(-xe^{-x^2/2})$

$= xe^{-x^2/2}(3 - x^2)$

Left endpoint: $f''(0) = -1$

Right endpoint: $f''(1) = 0$

Maximum value of $|f''(x)|$ is 1 on $[0, 1]$.

53. $f(x) = (x + 1)^{2/3}, [0, 2]$

$f'(x) = \frac{2}{3}(x + 1)^{-1/3}$

$f''(x) = -\frac{2}{9}(x + 1)^{-4/3}$

$f'''(x) = \frac{8}{27}(x + 1)^{-7/3}$

$f^{(4)}(x) = -\frac{56}{81}(x + 1)^{-10/3}$

$f^{(5)}(x) = \frac{560}{243}(x + 1)^{-13/3}$

$|f^{(4)}(0)| = \frac{56}{81}$ is the maximum value.

55. $f(x) = \tan x$

f is continuous on $[0, \pi/4]$ but not on $[0, \pi]$.

$\lim_{x \to \pi/2^-} \tan x = \infty.$

57.

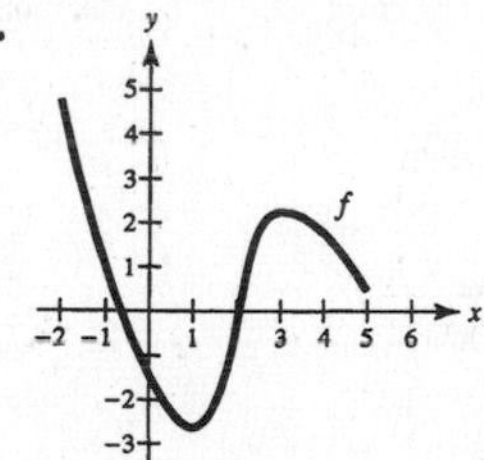

59. (a) Yes

(b) No

61. (a) No

(b) Yes

63. $P = VI - RI^2 = 12I - 0.5I^2, 0 \le I \le 15$

$P = 0$ when $I = 0$.

$P = 67.5$ when $I = 15$.

$P' = 12 - I = 0$

Critical number: $I = 12$ amps

When $I = 12$ amps, $P = 72$, the maximum output.

No, a 20-amp fuse would not increase the power output. P is decreasing for $I > 12$.

65. $$S = 6hs + \frac{3s^2}{2}\left(\frac{\sqrt{3} - \cos\theta}{\sin\theta}\right), \frac{\pi}{6} \le \theta \le \frac{\pi}{2}$$

$$\frac{dS}{d\theta} = \frac{3s^2}{2}(-\sqrt{3}\csc\theta\cot\theta + \csc^2\theta)$$

$$= \frac{3s^2}{2}\csc\theta(-\sqrt{3}\cot\theta + \csc\theta) = 0$$

$$\csc\theta = \sqrt{3}\cot\theta$$

$$\sec\theta = \sqrt{3}$$

$$\theta = \text{arcsec}\sqrt{3} \approx 0.9553 \text{ radians}$$

$$S\left(\frac{\pi}{6}\right) = 6hs + \frac{3s^2}{2}(\sqrt{3})$$

$$S\left(\frac{\pi}{2}\right) = 6hs + \frac{3s^2}{2}(\sqrt{3})$$

$$S(\text{arcsec}\sqrt{3}) = 6hs + \frac{3s^2}{2}(\sqrt{2})$$

S is minimum when $\theta = \text{arcsec}\sqrt{3} \approx 0.9553$ radians.

67. (a) $y = ax^2 + bx + c$

$y' = 2ax + b$

The coordinates of B are $(500, 30)$, and those of A are $(-500, 45)$. From the slopes at A and B,

$$-1000a + b = -0.09$$

$$1000a + b = 0.06.$$

Solving these two equations, you obtain $a = 3/40000$ and $b = -3/200$. From the points $(500, 30)$ and $(-500, 45)$, you obtain

$$30 = \frac{3}{40000}500^2 + 500\left(\frac{-3}{200}\right) + c$$

$$45 = \frac{3}{40000}500^2 - 500\left(\frac{-3}{200}\right) + c.$$

In both cases, $c = 18.75 = \frac{75}{4}$. Thus,

$$y = \frac{3}{40000}x^2 - \frac{3}{200}x + \frac{75}{4}.$$

(b)

x	-500	-400	-300	-200	-100	0	100	200	300	400	500
d	0	.75	3	6.75	12	18.75	12	6.75	3	.75	0

For $-500 \le x \le 0$, $d = (ax^2 + bx + c) - (-0.09x)$.

For $0 \le x \le 500$, $d = (ax^2 + bx + c) - (0.06x)$.

(c) The lowest point on the highway is $(100, 18)$, which is not directly over the point where the two hillsides come together.

69. True. See Exercise 27.

71. True.

Section 3.2 Rolle's Theorem and the Mean Value Theorem

1. Rolle's Theorem does not apply to $f(x) = 1 - |x - 1|$ over $[0, 2]$ since f is not differentiable at $x = 1$.

3. $f(x) = x^2 - x - 2 = (x - 2)(x + 1)$

x-intercepts: $(-1, 0), (2, 0)$

$f'(x) = 2x - 1 = 0$ at $x = \frac{1}{2}$.

5. $f(x) = x\sqrt{x + 4}$

x-intercepts: $(-4, 0), (0, 0)$

$$f'(x) = x\frac{1}{2}(x + 4)^{-1/2} + (x + 4)^{1/2}$$

$$= (x + 4)^{-1/2}\left(\frac{x}{2} + (x + 4)\right)$$

$$f'(x) = \left(\frac{3}{2}x + 4\right)(x + 4)^{-1/2} = 0 \text{ at } x = -\frac{8}{3}$$

7. $f(x) = x^2 - 2x, [0, 2]$

$f(0) = f(2) = 0$

f is continuous on $[0, 2]$. f is differentiable on $(0, 2)$. Rolle's Theorem applies.

$$f'(x) = 2x - 2$$

$$2x - 2 = 0 \Rightarrow x = 1$$

c value: 1

9. $f(x) = (x - 1)(x - 2)(x - 3), [1, 3]$

$f(1) = f(3) = 0$

f is continuous on $[1, 3]$. f is differentiable on $(1, 3)$. Rolle's Theorem applies.

$$f(x) = x^3 - 6x^2 + 11x - 6$$

$$f'(x) = 3x^2 - 12x + 11$$

$$3x^2 - 12x + 11 = 0 \Rightarrow x = \frac{6 \pm \sqrt{3}}{3}$$

$$c = \frac{6 - \sqrt{3}}{3}, c = \frac{6 + \sqrt{3}}{3}$$

11. $f(x) = x^{2/3} - 1, [-8, 8]$

$f(-8) = f(8) = 3$

f is continuous on $[-8, 8]$. f is not differentiable on $(-8, 8)$ since $f'(0)$ does not exist. Rolle's Theorem does not apply.

13. $f(x) = \dfrac{x^2 - 2x - 3}{x + 2}, [-1, 3]$

$f(-1) = f(3) = 0$

f is continuous on $[-1, 3]$. (**Note:** The discontinuity, $x = -2$, is not in the interval.) f is differentiable on $(-1, 3)$. Rolle's Theorem applies.

$$f'(x) = \frac{(x + 2)(2x - 2) - (x^2 - 2x - 3)(1)}{(x + 2)^2} = 0$$

$$\frac{x^2 + 4x - 1}{(x + 2)^2} = 0$$

$$x = \frac{-4 \pm 2\sqrt{5}}{2} = -2 \pm \sqrt{5}$$

c value: $-2 + \sqrt{5}$

15. $f(x) = (x^2 - 2x)e^x$, $[0, 2]$

$f(0) = f(2) = 0$

f is continuous on $[0, 2]$ and differentiable on $(0, 2)$, so Rolle's Theorem applies.

$$f'(x) = (x^2 - 2x)e^x + (2x - 2)e^x$$
$$= e^x(x^2 - 2)$$
$$= 0 \Rightarrow x = \sqrt{2}$$

c value: $\sqrt{2} \approx 1.414$

17. $f(x) = \sin x$, $[0, 2\pi]$

$f(0) = f(2\pi) = 0$

f is continuous on $[0, 2\pi]$. f is differentiable on $(0, 2\pi)$. Rolle's Theorem applies.

$$f'(x) = \cos x$$

c values: $\frac{\pi}{2}, \frac{3\pi}{2}$

19. $f(x) = \frac{6x}{\pi} - 4\sin^2 x$, $\left[0, \frac{\pi}{6}\right]$

$f(0) = f\left(\frac{\pi}{6}\right) = 0$

f is continuous on $[0, \pi/6]$. f is differentiable on $(0, \pi/6)$. Rolle's Theorem applies.

$$f'(x) = \frac{6}{\pi} - 8\sin x\cos x = 0$$
$$\frac{6}{\pi} = 8\sin x\cos x$$
$$\frac{3}{4\pi} = \frac{1}{2}\sin 2x$$
$$\frac{3}{2\pi} = \sin 2x$$
$$\frac{1}{2}\arcsin\left(\frac{3}{2\pi}\right) = x$$
$$x \approx 0.2489$$

c value: 0.2489

21. $f(x) = \tan x$, $[0, \pi]$

$f(0) = f(\pi) = 0$

f is not continuous on $[0, \pi]$ since $f(\pi/2)$ does not exist. Rolle's Theorem does not apply.

23. $f(x) = |x| - 1$, $[-1, 1]$

$f(-1) = f(1) = 0$

f is continuous on $[-1, 1]$. f is not differentiable on $(-1, 1)$ since $f'(0)$ does not exist. Rolle's Theorem does not apply.

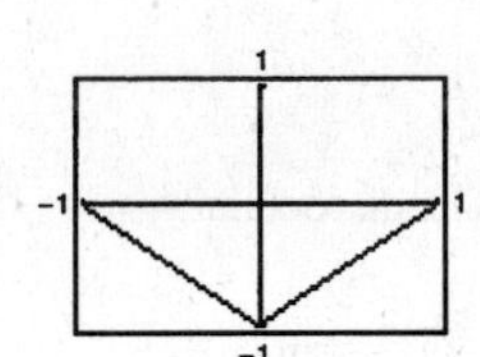

25. $f(x) = 4x - \tan \pi x, \left[-\frac{1}{4}, \frac{1}{4}\right]$

$$f\left(-\frac{1}{4}\right) = f\left(\frac{1}{4}\right) = 0$$

f is continuous on $[-1/4, 1/4]$. f is differentiable on $(-1/4, 1/4)$. Rolle's Theorem applies.

$$f'(x) = 4 - \pi \sec^2 \pi x = 0$$

$$\sec^2 \pi x = \frac{4}{\pi}$$

$$\sec \pi x = \pm\frac{2}{\sqrt{\pi}}$$

$$x = \pm\frac{1}{\pi} \operatorname{arcsec} \frac{2}{\sqrt{\pi}} = \pm\frac{1}{\pi} \arccos \frac{\sqrt{\pi}}{2}$$

$$\approx \pm 0.1533 \text{ radian}$$

c values: ± 0.1533 radian

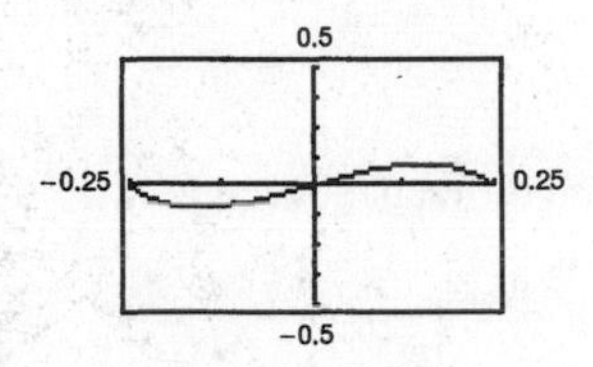

27. $f(x) = 2 + \arcsin(x^2 - 1), [-1, 1]$

$$f(-1) = f(1) = 2$$

$$f'(x) = \frac{2x}{\sqrt{1 - (x^2 - 1)^2}} = \frac{2x}{\sqrt{2x^2 + x^4}}$$

$f'(0)$ does not exist. Rolle's Theorem does not apply.

29. $f(t) = -16t^2 + 48t + 32$

(a) $f(1) = f(2) = 64$

(b) $v = f'(t)$ must be 0 at some time in $(1, 2)$.

$$f'(t) = -32t + 48 = 0$$

$$t = \frac{3}{2} \text{ seconds}$$

31.

tangent line
$(c_2, f(c_2))$
$(a, f(a))$
f
$(b, f(b))$
$(c_1, f(c_1))$
a
b
tangent line
secant line

33. $f(x) = \frac{1}{x - 3}, [0, 6]$

f has a discontinuity at $x = 3$.

35. $f(x) = x^2$ is continuous on $[-2, 1]$ and differentiable on $(-2, 1)$.

$$\frac{f(1) - f(-2)}{1 - (-2)} = \frac{1 - 4}{3} = -1$$

$f'(x) = 2x = -1$ when $x = -\frac{1}{2}$. Therefore,

$$c = -\frac{1}{2}.$$

37. $f(x) = x^{2/3}$ is continuous on $[0, 1]$ and differentiable on $(0, 1)$.

$$\frac{f(1) - f(0)}{1 - 0} = 1$$

$$f'(x) = \frac{2}{3}x^{-1/3} = 1$$

$$x = \left(\frac{2}{3}\right)^3 = \frac{8}{27}$$

$$c = \frac{8}{27}$$

39. $f(x) = \sqrt{2 - x}$ is continuous on $[-7, 2]$ and differentiable on $(-7, 2)$.

$$\frac{f(2) - f(-7)}{2 - (-7)} = \frac{0 - 3}{9} = -\frac{1}{3}$$

$$f'(x) = \frac{-1}{2\sqrt{2 - x}} = -\frac{1}{3}$$

$$2\sqrt{2 - x} = 3$$

$$\sqrt{2 - x} = \frac{3}{2}$$

$$2 - x = \frac{9}{4}$$

$$x = -\frac{1}{4}$$

$$c = -\frac{1}{4}$$

41. $f(x) = \sin x$ is continuous on $[0, \pi]$ and differentiable on $(0, \pi)$.

$$\frac{f(\pi) - f(0)}{\pi - 0} = \frac{0 - 0}{\pi} = 0$$

$$f'(x) = \cos x = 0$$

$$c = \frac{\pi}{2}$$

43. $f(x) = x \log_2 x = x \dfrac{\ln x}{\ln 2}$

f is continuous on $[1, 2]$ and differentiable on $(1, 2)$.

$$\frac{f(2) - f(1)}{2 - 1} = \frac{2 - 0}{2 - 1} = 2$$

$$f'(x) = x\frac{1}{x \ln 2} + \frac{\ln x}{\ln 2} = 2$$

$$1 + \ln x = 2 \ln 2 = \ln 4$$

$$xe = 4$$

$$x = \frac{4}{e} \Rightarrow c = \frac{4}{e}$$

45. $f(x) = \dfrac{x}{x + 1}$ on $\left[-\dfrac{1}{2}, 2\right]$.

(a)

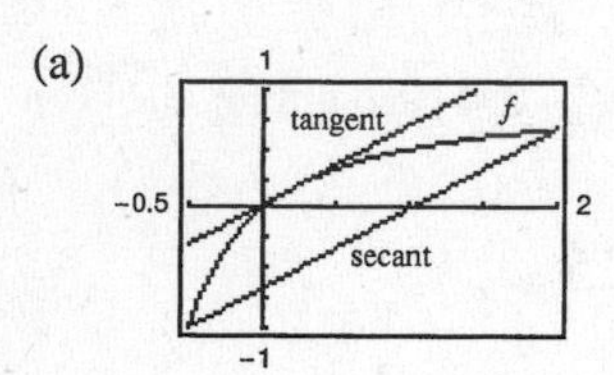

(b) Secant line:

$$\text{slope} = \frac{f(2) - f(-1/2)}{2 - (-1/2)} = \frac{2/3 - (-1)}{5/2} = \frac{2}{3}$$

$$y - \frac{2}{3} = \frac{2}{3}(x - 2)$$

$$3y - 2 = 2x - 4$$

$$3y - 2x + 2 = 0$$

(c) $f'(x) = \dfrac{1}{(x + 1)^2} = \dfrac{2}{3}$

$$(x + 1)^2 = \frac{3}{2}$$

$$x = -1 \pm \sqrt{\frac{3}{2}} = -1 \pm \frac{\sqrt{6}}{2}$$

In the interval $[-1/2, 2]$, $c = -1 + (\sqrt{6}/2)$.

$$f(c) = \frac{-1 + (\sqrt{6}/2)}{[-1 + (\sqrt{6}/2)] + 1} = \frac{-2 + \sqrt{6}}{\sqrt{6}} = \frac{-2}{\sqrt{6}} + 1$$

Tangent line: $y - 1 + \dfrac{2}{\sqrt{6}} = \dfrac{2}{3}\left(x - \dfrac{\sqrt{6}}{2} + 1\right)$

$$y - 1 + \frac{\sqrt{6}}{3} = \frac{2}{3}x - \frac{\sqrt{6}}{3} + \frac{2}{3}$$

$$3y - 2x - 5 + 2\sqrt{6} = 0$$

47. $f(x) = \sqrt{x}, \ [1, 9]$

$(1, 1), (9, 3)$

$$m = \frac{3 - 1}{9 - 1} = \frac{1}{4}$$

(a)

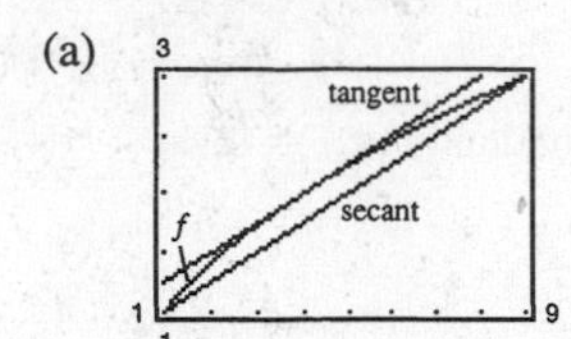

(b) Secant line: $y - 1 = \frac{1}{4}(x - 1)$

$$y = \frac{1}{4}x + \frac{3}{4}$$

$$0 = x - 4y + 3$$

(c) $f'(x) = \dfrac{1}{2\sqrt{x}}$

$$\frac{f(9) - f(1)}{9 - 1} = \frac{1}{4}$$

$$\frac{1}{2\sqrt{c}} = \frac{1}{4}$$

$$\sqrt{c} = 2$$

$$c = 4$$

$$(c, f(c)) = (4, 2)$$

$$m = f'(4) = \frac{1}{4}$$

Tangent line: $y - 2 = \frac{1}{4}(x - 4)$

$$y = \frac{1}{4}x + 1$$

$$0 = x - 4y + 4$$

49. $f(x) = 2e^{x/4} \cos \dfrac{\pi x}{4}, \quad 0 \le x \le 2$

$f(0) = 2, \ f(2) = 0$

$$m = \frac{0 - 2}{2 - 0} = -1$$

(a)

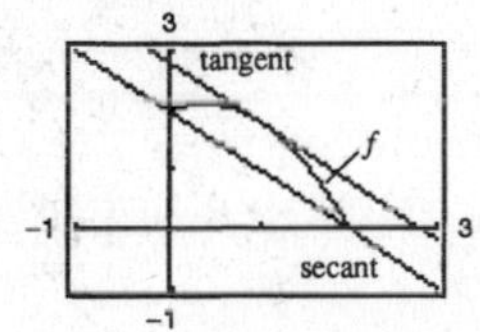

(b) Secant line: $y - 2 = -1(x - 0)$

$$y = -x + 2$$

$$f'(x) = 2\left(\frac{1}{4}e^{x/4} \cos \frac{\pi x}{4}\right) + 2e^{x/4}\left(-\sin \frac{\pi x}{4}\right)\frac{\pi}{4}$$

$$= e^{x/4}\left[\frac{1}{2} \cos \frac{\pi x}{4} - \frac{\pi}{2} \sin \frac{\pi x}{4}\right]$$

$$f'(c) = -1 \Rightarrow c \approx 1.0161, \ f(c) \approx 1.8$$

(c) Tangent line: $y - 1.8 = -1(x - 1.0161)$

$$y = -x + 2.8161$$

51. $s(t) = -4.9t^2 + 500$

(a) $V_{\text{avg}} = \dfrac{s(3) - s(0)}{3 - 0} = \dfrac{455.9 - 500}{3} = -14.7 \text{ m/sec}$

(b) $s(t)$ is continuous on $[0, 3]$ and differentiable on $(0, 3)$. Therefore, the Mean Value Theorem applies.

$$v(t) = s'(t) = -9.8t = -14.7 \text{ m/sec}$$

$$t = \frac{-14.7}{-9.8} = 1.5 \text{ seconds}$$

53. No. Let $f(x) = x^2$ on $[-1, 2]$.

$$f'(x) = 2x$$

$f'(0) = 0$ and zero is in the interval $(-1, 2)$ but $f(-1) \neq f(2)$.

55. Let $T(t)$ be the temperature of the object. Then $T(0) = 1500°$ and $T(5) = 390°$. The average temperature over the interval $[0, 5]$ is

$$\frac{390 - 1500}{5 - 0} = -222° \text{ F/hr}.$$

By the Mean Value Theorem, there exists a time t_0, $0 < t_0 < 5$, such that $T'(t_0) = -222$.

57. f is continuous on $[-5, 5]$ and does not satisfy the conditions of the Mean Value Theorem.
$\Rightarrow f$ is not differentiable on $(-5, 5)$.
Example: $f(x) = |x|$

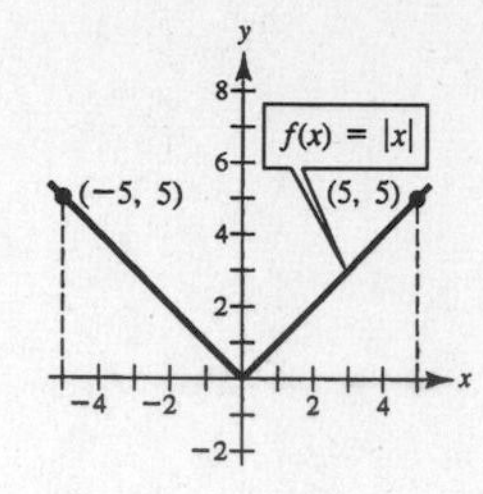

59. False. $f(x) = 1/x$ has a discontinuity at $x = 0$.

61. True. A polynomial is continuous and differentiable everywhere.

63. Suppose that $p(x) = x^{2n+1} + ax + b$ has two real roots x_1 and x_2. Then by Rolle's Theorem, since $p(x_1) = p(x_2) = 0$, there exists c in (x_1, x_2) such that $p'(c) = 0$. But $p'(x) = (2n + 1)x^{2n} + a \neq 0$, since $n > 0, a > 0$. Therefore, $p(x)$ cannot have two real roots.

65. If $p(x) = Ax^2 + Bx + C$, then

$$\begin{aligned} p'(x) = 2Ax + B = \frac{f(b) - f(a)}{b - a} &= \frac{(Ab^2 + Bb + C) - (Aa^2 + Ba + C)}{b - a} \\ &= \frac{A(b^2 - a^2) + B(b - a)}{b - a} \\ &= \frac{(b - a)[A(b + a) + B]}{b - a} \\ &= A(b + a) + B. \end{aligned}$$

Thus, $2Ax = A(b + a)$ and $x = (b + a)/2$ which is the midpoint of $[a, b]$.

67. $f(x) = \frac{1}{2}\cos x$ differentiable on $(-\infty, \infty)$.

$f'(x) = -\frac{1}{2}\sin x$

$-\frac{1}{2} \leq f'(x) \leq \frac{1}{2} \Rightarrow f'(x) < 1$ for all real numbers.

Thus, from Exercise 65, f has, at most, one fixed point. $(x \approx 0.4502)$

Section 3.3 Increasing and Decreasing Functions and the First Derivative Test

1. $f(x) = x^2 - 6x + 8$

Increasing on: $(3, \infty)$

Decreasing on: $(-\infty, 3)$

3. $y = \dfrac{x^3}{4} - 3x$

Increasing on: $(-\infty, -2), (2, \infty)$

Decreasing on: $(-2, 2)$

5. $f(x) = \dfrac{1}{x^2} = x^{-2}$

$f'(x) = \dfrac{-2}{x^3}$

Discontinuity: $x = 0$

Test intervals:	$-\infty < x < 0$	$0 < x < \infty$
Sign of $f'(x)$:	$f' > 0$	$f' < 0$
Conclusion:	Increasing	Decreasing

Increasing on $(-\infty, 0)$

Decreasing on $(0, \infty)$

7. $g(x) = x^2 - 2x - 8$

$g'(x) = 2x - 2$

Critical number: $x = 1$

Test intervals:	$-\infty < x < 1$	$1 < x < \infty$
Sign of $g'(x)$:	$g' < 0$	$g' > 0$
Conclusion:	Decreasing	Increasing

Increasing on: $(1, \infty)$

Decreasing on: $(-\infty, 1)$

9. $y = x\sqrt{16 - x^2}$ Domain: $[-4, 4]$

$y' = \dfrac{-2(x^2 - 8)}{\sqrt{16 - x^2}} = \dfrac{-2}{\sqrt{16 - x^2}}(x - 2\sqrt{2})(x + 2\sqrt{2})$

Critical numbers: $x = \pm 2\sqrt{2}$

Test intervals:	$-4 < x < -2\sqrt{2}$	$-2\sqrt{2} < x < 2\sqrt{2}$	$2\sqrt{2} < x < 4$
Sign of y':	$y' < 0$	$y' > 0$	$y' < 0$
Conclusion:	Decreasing	Increasing	Decreasing

Increasing on $(-2\sqrt{2}, 2\sqrt{2})$

Decreasing on $(-4, -2\sqrt{2}), (2\sqrt{2}, 4)$

11. $f(x) = x^2 - 6x$

$f'(x) = 2x - 6 = 0$

Critical number: $x = 3$

Test intervals:	$-\infty < x < 3$	$3 < x < \infty$
Sign of $f'(x)$:	$f' < 0$	$f' > 0$
Conclusion:	Decreasing	Increasing

Increasing on: $(3, \infty)$

Decreasing on: $(-\infty, 3)$

Relative minimum: $(3, -9)$

13. $f(x) = -2x^2 + 4x + 3$

$f'(x) = -4x + 4 = 0$

Critical number: $x = 1$

Test intervals:	$-\infty < x < 1$	$1 < x < \infty$
Sign of $f'(x)$:	$f' > 0$	$f' < 0$
Conclusion:	Increasing	Decreasing

Increasing on: $(-\infty, 1)$

Decreasing on: $(1, \infty)$

Relative maximum: $(1, 5)$

15. $f(x) = 2x^3 + 3x^2 - 12x$

$f'(x) = 6x^2 + 6x - 12 = 6(x + 2)(x - 1) = 0$

Critical numbers: $x = -2, 1$

Test intervals:	$-\infty < x < -2$	$-2 < x < 1$	$1 < x < \infty$
Sign of $f'(x)$:	$f' > 0$	$f' < 0$	$f' > 0$
Conclusion:	Increasing	Decreasing	Increasing

Increasing on: $(-\infty, -2), (1, \infty)$

Decreasing on: $(-2, 1)$

Relative maximum: $(-2, 20)$

Relative minimum: $(1, -7)$

17. $f(x) = x^2(3 - x) = 3x^2 - x^3$

$f'(x) = 6x - 3x^2 = 3x(2 - x)$

Critical numbers: $x = 0, 2$

Test intervals:	$-\infty < x < 0$	$0 < x < 2$	$2 < x < \infty$
Sign of $f'(x)$:	$f' < 0$	$f' > 0$	$f' < 0$
Conclusion:	Decreasing	Increasing	Decreasing

Increasing on: $(0, 2)$

Decreasing on: $(-\infty, 0), (2, \infty)$

Relative maximum: $(2, 4)$

Relative minimum: $(0, 0)$

19. $f(x) = \dfrac{x^5 - 5x}{5}$

$f'(x) = x^4 - 1$

Critical numbers: $x = -1, 1$

Test intervals:	$-\infty < x < -1$	$-1 < x < 1$	$1 < x < \infty$
Sign of $f'(x)$:	$f' > 0$	$f' < 0$	$f' > 0$
Conclusion:	Increasing	Decreasing	Increasing

Increasing on: $(-\infty, -1), (1, \infty)$

Decreasing on: $(-1, 1)$

Relative maximum: $\left(-1, \frac{4}{5}\right)$

Relative minimum: $\left(1, -\frac{4}{5}\right)$

21. $f(x) = x^{1/3} + 1$

$f'(x) = \frac{1}{3}x^{-2/3} = \frac{1}{3x^{2/3}}$

Critical number: $x = 0$

Test intervals:	$-\infty < x < 0$	$0 < x < \infty$
Sign of $f'(x)$:	$f' > 0$	$f' > 0$
Conclusion:	Increasing	Increasing

Increasing on: $(-\infty, \infty)$

No relative extrema

23. $f(x) = (x - 1)^{2/3}$

$f'(x) = \frac{2}{3(x - 1)^{1/3}}$

Critical number: $x = 1$

Test intervals:	$-\infty < x < 1$	$1 < x < \infty$
Sign of $f'(x)$:	$f' < 0$	$f' > 0$
Conclusion:	Decreasing	Increasing

Increasing on: $(1, \infty)$

Decreasing on: $(-\infty, 1)$

Relative minimum: $(1, 0)$

25. $f(x) = 5 - |x - 5|$

$f'(x) = -\frac{x - 5}{|x - 5|} = \begin{cases} 1, & x < 5 \\ -1, & x > 5 \end{cases}$

Critical number: $x = 5$

Test intervals:	$-\infty < x < 5$	$5 < x < \infty$
Sign of $f'(x)$:	$f' > 0$	$f' < 0$
Conclusion:	Increasing	Decreasing

Increasing on: $(-\infty, 5)$

Decreasing on: $(5, \infty)$

Relative maximum: $(5, 5)$

27. $f(x) = x + \frac{1}{x}$

$f'(x) = 1 - \frac{1}{x^2} = \frac{x^2 - 1}{x^2}$

Critical numbers: $x = -1, 1$

Discontinuity: $x = 0$

Test intervals:	$-\infty < x < -1$	$-1 < x < 0$	$0 < x < 1$	$1 < x < \infty$
Sign of $f'(x)$:	$f' > 0$	$f' < 0$	$f' < 0$	$f' > 0$
Conclusion:	Increasing	Decreasing	Decreasing	Increasing

Increasing on: $(-\infty, -1), (1, \infty)$

Decreasing on: $(-1, 0), (0, 1)$

Relative maximum: $(-1, -2)$

Relative minimum: $(1, 2)$

29. $f(x) = \dfrac{x^2}{x^2 - 9}$

$$f'(x) = \frac{(x^2 - 9)(2x) - (x^2)(2x)}{(x^2 - 9)^2} = \frac{-18x}{(x^2 - 9)^2}$$

Critical number: $x = 0$

Discontinuities: $x = -3, 3$

Test intervals:	$-\infty < x < -3$	$-3 < x < 0$	$0 < x < 3$	$3 < x < \infty$
Sign of $f'(x)$:	$f' > 0$	$f' > 0$	$f' < 0$	$f' < 0$
Conclusion:	Increasing	Increasing	Decreasing	Decreasing

Increasing on: $(-\infty, -3), (-3, 0)$

Decreasing on: $(0, 3), (3, \infty)$

Relative maximum: $(0, 0)$

31. $f(x) = \dfrac{x^2 - 2x + 1}{x + 1}$

$$f'(x) = \frac{(x + 1)(2x - 2) - (x^2 - 2x + 1)(1)}{(x + 1)^2} = \frac{x^2 + 2x - 3}{(x + 1)^2} = \frac{(x + 3)(x - 1)}{(x + 1)^2}$$

Critical numbers: $x = -3, 1$

Discontinuity: $x = -1$

Test intervals:	$-\infty < x < -3$	$-3 < x < -1$	$-1 < x < 1$	$1 < x < \infty$
Sign of $f'(x)$:	$f' > 0$	$f' < 0$	$f' < 0$	$f' > 0$
Conclusion:	Increasing	Decreasing	Decreasing	Increasing

Increasing on: $(-\infty, -3), (1, \infty)$

Decreasing on: $(-3, -1), (-1, 1)$

Relative maximum: $(-3, -8)$

Relative minimum: $(1, 0)$

33. $f(x) = (3 - x)e^{x-3}$

$$f'(x) = (3 - x)e^{x-3} - e^{x-3}$$
$$= e^{x-3}(2 - x)$$

Critical number: $x = 2$

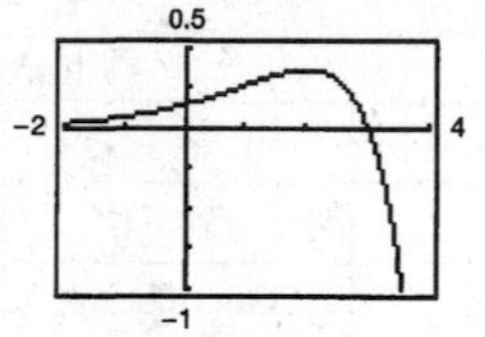

Test intervals:	$-\infty < x < 2$	$2 < x < \infty$
Sign of $f'(x)$:	$f' > 0$	$f' < 0$
Conclusion:	Increasing	Decreasing

Increasing on: $(-\infty, 2)$

Decreasing on: $(2, \infty)$

Relative maximum: $(2, e^{-1})$

35. $f(x) = 4(x - \arcsin x), \quad -1 \le x \le 1$

$$f'(x) = 4 - \frac{4}{\sqrt{1 - x^2}}$$

Critical number: $x = 0$

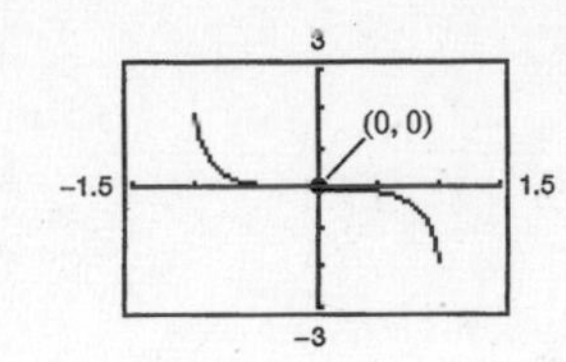

Test intervals:	$-1 \le x < 0$	$0 < x \le 1$
Sign of $f'(x)$:	$f' < 0$	$f' < 0$
Conclusion:	Decreasing	Decreasing

Decreasing on: $[-1, 1]$

No relative extrema

(Absolute maximum at $x = -1$, absolute minimum at $x = 1$)

37. $g(x) = (x)3^{-x}$

$g'(x) = (1 - x \ln 3)3^{-x}$

Critical number: $x = \dfrac{1}{\ln 3} \approx 0.9102$

Test intervals:	$-\infty < x < \dfrac{1}{\ln 3}$	$\dfrac{1}{\ln 3} < x < \infty$
Sign of $f'(x)$:	$f' > 0$	$f' < 0$
Conclusion:	Increasing	Decreasing

Increasing on: $\left(-\infty, \dfrac{1}{\ln 3}\right)$

Decreasing on: $\left(\dfrac{1}{\ln 3}, \infty\right)$

Relative minimum: $\left(\dfrac{1}{\ln 3}, \dfrac{1}{e \ln 3}\right) \approx (0.9102, 0.3349)$

39. $f(x) = x - \log_4 x = x - \dfrac{\ln x}{\ln 4}$

$$f'(x) = 1 - \frac{1}{x \ln 4} = 0 \Rightarrow x \ln 4 = 1 \Rightarrow x = \frac{1}{\ln 4} = \frac{1}{2 \ln 2}$$

Critical number: $x = \dfrac{1}{\ln 4}$

Test intervals:	$0 < x < \dfrac{1}{\ln 4}$	$\dfrac{1}{\ln 4} < x < \infty$
Sign of $f'(x)$:	$f' < 0$	$f' > 0$
Conclusion:	Decreasing	Increasing

Increasing on: $\left(\dfrac{1}{\ln 4}, \infty\right)$

Decreasing on: $\left(0, \dfrac{1}{\ln 4}\right)$

Relative minimum: $\left(\dfrac{1}{\ln 4}, \dfrac{1}{\ln 4} - \log_4\left(\dfrac{1}{\ln 4}\right)\right) = \left(\dfrac{1}{\ln 4}, \dfrac{\ln(\ln 4) + 1}{\ln 4}\right) \approx (0.7213, 0.9570)$

41. $f(x) = \frac{x}{2} + \cos x, 0 < x < 2\pi$

$f'(x) = \frac{1}{2} - \sin x = 0$

Critical numbers: $x = \frac{\pi}{6}, \frac{5\pi}{6}$

Test intervals:	$0 < x < \frac{\pi}{6}$	$\frac{\pi}{6} < x < \frac{5\pi}{6}$	$\frac{5\pi}{6} < x < 2\pi$
Sign of $f'(x)$:	$f' > 0$	$f' < 0$	$f' > 0$
Conclusion:	Increasing	Decreasing	Increasing

Increasing on: $\left(0, \frac{\pi}{6}\right), \left(\frac{5\pi}{6}, 2\pi\right)$

Decreasing on: $\left(\frac{\pi}{6}, \frac{5\pi}{6}\right)$

Relative maximum: $\left(\frac{\pi}{6}, \frac{\pi + 6\sqrt{3}}{12}\right)$

Relative minimum: $\left(\frac{5\pi}{6}, \frac{5\pi - 6\sqrt{3}}{12}\right)$

43. $f(x) = \sin^2 x + \sin x, 0 < x < 2\pi$

$f'(x) = 2 \sin x \cos x + \cos x = \cos x(2 \sin x + 1) = 0$

Critical numbers: $x = \frac{\pi}{2}, \frac{7\pi}{6}, \frac{3\pi}{2}, \frac{11\pi}{6}$

Test intervals:	$0 < x < \frac{\pi}{2}$	$\frac{\pi}{2} < x < \frac{7\pi}{6}$	$\frac{7\pi}{6} < x < \frac{3\pi}{2}$	$\frac{3\pi}{2} < x < \frac{11\pi}{6}$	$\frac{11\pi}{6} < x < 2\pi$
Sign of $f'(x)$:	$f' > 0$	$f' < 0$	$f' > 0$	$f' < 0$	$f' > 0$
Conclusion:	Increasing	Decreasing	Increasing	Decreasing	Increasing

Increasing on: $\left(0, \frac{\pi}{2}\right), \left(\frac{7\pi}{6}, \frac{3\pi}{2}\right), \left(\frac{11\pi}{6}, 2\pi\right)$

Decreasing on: $\left(\frac{\pi}{2}, \frac{7\pi}{6}\right), \left(\frac{3\pi}{2}, \frac{11\pi}{6}\right)$

Relative minima: $\left(\frac{7\pi}{6}, -\frac{1}{4}\right), \left(\frac{11\pi}{6}, -\frac{1}{4}\right)$

Relative maxima: $\left(\frac{\pi}{2}, 2\right), \left(\frac{3\pi}{2}, 0\right)$

45. $f(x) = 2x\sqrt{9 - x^2}, [-3, 3]$

(a) $f'(x) = \dfrac{2(9 - 2x^2)}{\sqrt{9 - x^2}}$

(b)

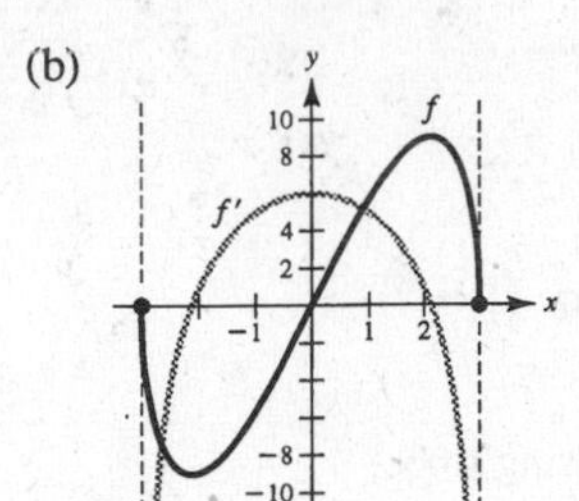

(c) $\dfrac{2(9 - 2x^2)}{\sqrt{9 - x^2}} = 0$

Critical numbers: $x = \pm\dfrac{3}{\sqrt{2}} = \pm\dfrac{3\sqrt{2}}{2}$

(d) Intervals:

$\left(-3, -\dfrac{3\sqrt{2}}{2}\right)$	$\left(-\dfrac{3\sqrt{2}}{2}, \dfrac{3\sqrt{2}}{2}\right)$	$\left(\dfrac{3\sqrt{2}}{2}, 3\right)$
$f'(x) < 0$	$f'(x) > 0$	$f'(x) < 0$
Decreasing	Increasing	Decreasing

f is increasing when f' is positive and decreasing when f' is negative.

47. $f(t) = t^2 \sin t, [0, 2\pi]$

(a) $f'(t) = t^2 \cos t + 2t \sin t$

$= t(t \cos t + 2 \sin t)$

(b)

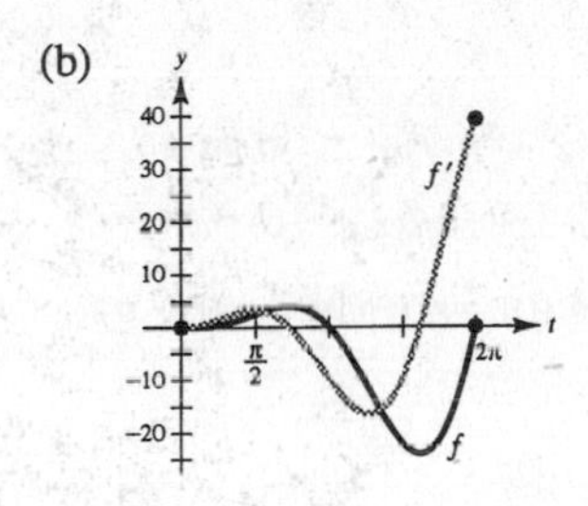

(c) $t(t \cos t + 2 \sin t) = 0$

$t = 0$ or $t = -2 \tan t$

$t \cot t = -2$

$t \approx 2.2889, 5.0870$ (graphing utility)

Critical numbers: $t = 2.2889, t = 5.0870$

(d) Intervals:

$(0, 2.2889)$	$(2.2889, 5.0870)$	$(5.0870, 2\pi)$
$f'(t) > 0$	$f'(t) < 0$	$f'(t) > 0$
Increasing	Decreasing	Increasing

f is increasing when f' is positive and decreasing when f' is negative.

49. (a) $f(x) = \dfrac{1}{2}(x^2 - \ln x), \quad (0, 3]$

$f'(x) = \dfrac{2x^2 - 1}{2x}$

(b)

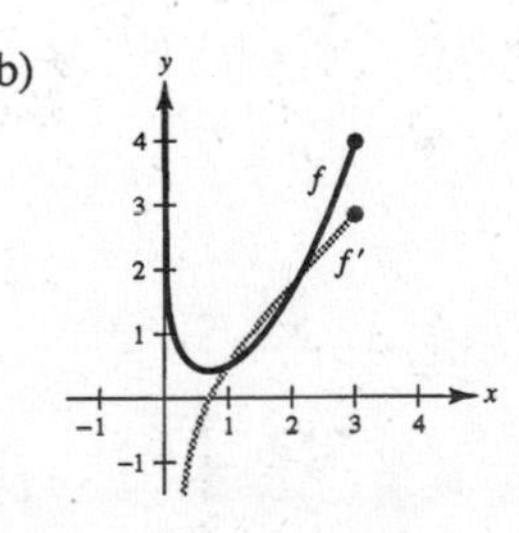

(c) $f'(x) = 0 \Rightarrow x = \dfrac{1}{\sqrt{2}} = \dfrac{\sqrt{2}}{2}$

(d) $f' > 0$ on $\left(\dfrac{\sqrt{2}}{2}, 3\right)$; $f' < 0$ on $\left(0, \dfrac{\sqrt{2}}{2}\right)$

51. $f(x) = \dfrac{x^5 - 4x^3 + 3x}{x^2 - 1} = \dfrac{(x^2 - 1)(x^3 - 3x)}{x^2 - 1} = x^3 - 3x,\ x \neq \pm 1$

$f(x) = g(x) = x^3 - 3x$ for all $x \neq \pm 1$.

$f'(x) = 3x^2 - 3 = 3(x^2 - 1), x \neq \pm 1 \quad f'(x) \neq 0$

f symmetric about origin

zeros of f: $(0, 0), (\pm\sqrt{3}, 0)$

No relative extrema

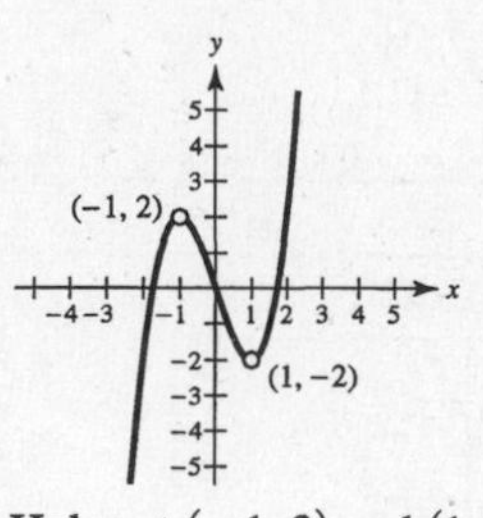

Holes at $(-1, 2)$ and $(1, -2)$

53. f is quadratic $\Rightarrow f'$ is a line.

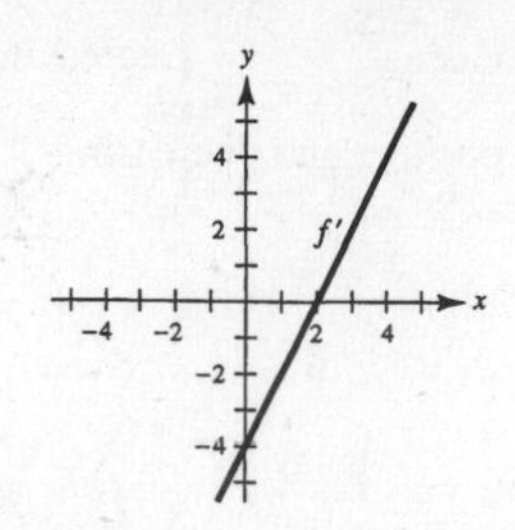

55. f has positive, but decreasing slope

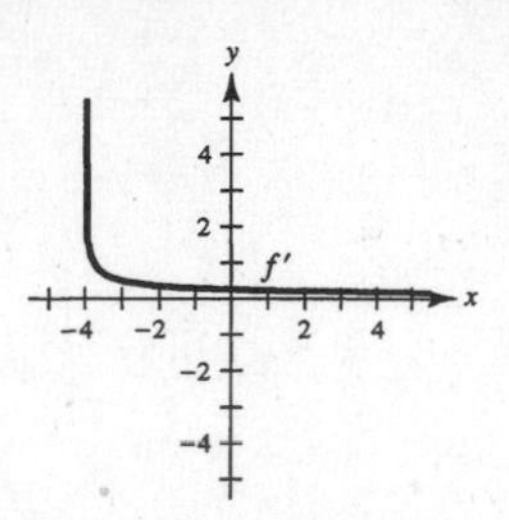

In Exercises 57–62, $f'(x) > 0$ on $(-\infty, -4)$, $f'(x) < 0$ on $(-4, 6)$ and $f'(x) > 0$ on $(6, \infty)$.

57. $g(x) = f(x) + 5$

$g'(x) = f'(x)$

$g'(0) = f'(0) < 0$

59. $g(x) = -f(x)$

$g'(x) = -f'(x)$

$g'(-6) = -f'(-6) < 0$

61. $g(x) = f(x - 10)$

$g'(x) = f'(x - 10)$

$g'(0) = f'(-10) > 0$

63. $f'(x) = \begin{cases} > 0, & x < 4 \Rightarrow f \text{ is increasing on } (-\infty, 4). \\ \text{undefined}, & x = 4 \\ < 0, & x > 4 \Rightarrow f \text{ is decreasing on } (4, \infty). \end{cases}$

Two possibilities for $f(x)$ are given below.

(a)

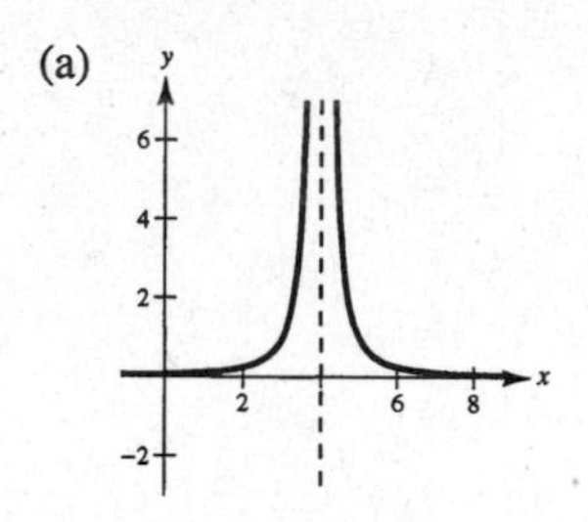

(b)

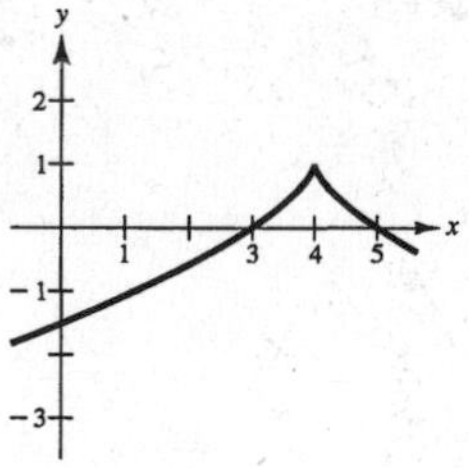

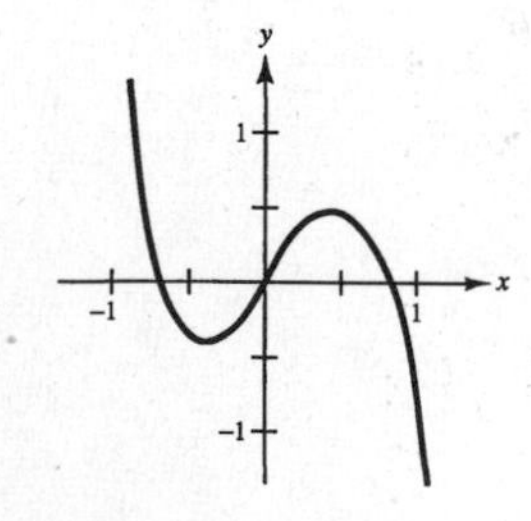

65. The critical numbers are in intervals $(-0.50, -0.25)$ and $(0.25, 0.50)$ since the sign of f' changes in these intervals. f is decreasing on approximately $(-1, -0.40)$, $(0.48, 1)$, and increasing on $(-0.40, 0.48)$.

Relative minimum when $x \approx -0.40$.

Relative maximum when $x \approx 0.48$.

67. $f(x) = x, g(x) = \sin x, 0 < x < \pi$

(a)

x	0.5	1	1.5	2	2.5	3
$f(x)$	0.5	1	1.5	2	2.5	3
$g(x)$	0.479	0.841	0.997	0.909	0.598	0.141

$f(x)$ seems greater than $g(x)$ on $(0, \pi)$.

(b)

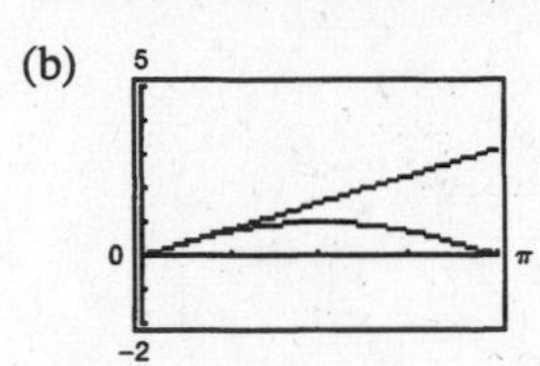

$x > \sin x$ on $(0, \pi)$

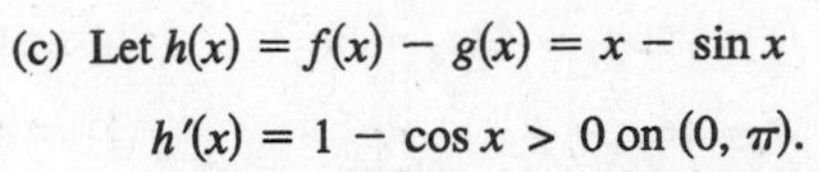

(c) Let $h(x) = f(x) - g(x) = x - \sin x$

$$h'(x) = 1 - \cos x > 0 \text{ on } (0, \pi).$$

Therefore, $h(x)$ is increasing on $(0, \pi)$. Since $h(0) = 0$, $h(x) > 0$ on $(0, \pi)$. Thus,

$$x - \sin x > 0$$

$$x > \sin x$$

$$f(x) > g(x) \text{ on } (0, \pi).$$

69. $v = k(R - r)r^2 = k(Rr^2 - r^3)$

$v' = k(2Rr - 3r^2)$

$= kr(2R - 3r) = 0$

$r = 0$ or $\frac{2}{3}R$

Maximum when $r = \frac{2}{3}R$.

71. $P = \dfrac{vR_1R_2}{(R_1 + R_2)^2}$, v and R_1 are constant

$$\frac{dP}{dR_2} = \frac{(R_1 + R_2)^2(vR_1) - vR_1R_2[2(R_1 + R_2)(1)]}{(R_1 + R_2)^4}$$

$$= \frac{vR_1(R_1 - R_2)}{(R_1 + R_2)^3} = 0 \Longrightarrow R_2 = R_1$$

Maximum when $R_1 = R_2$.

73. (a) $B = 0.1198t^4 - 4.4879t^3 + 56.9909t^2 - 223.0222t + 579.9541$

(b)

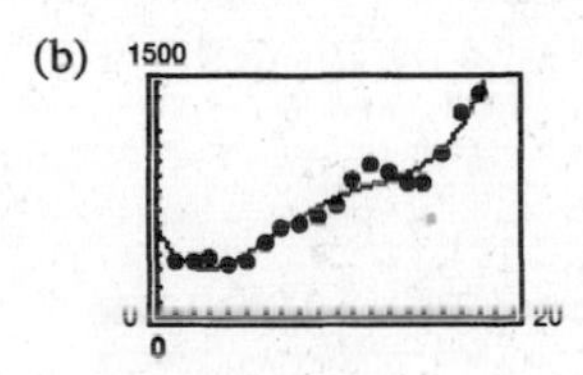

(c) $B' = 0$ for $t \approx 2.78$, or 1983, (311.1 thousand bankruptcies)
Actual minimum: 1984 (344.3 thousand bankruptcies)

75. (a) Use a cubic polynomial
$f(x) = a_3x^3 + a_2x^2 + a_1x + a_0$.

(b) $f'(x) = 3a_3x^2 + 2a_2x + a_1$.

(0, 0):	$0 = a_0$	$(f(0) = 0)$
	$0 = a_1$	$(f'(0) = 0)$
(2, 2):	$2 = 8a_3 + 4a_2$	$(f(2) = 2)$
	$0 = 12a_3 + 4a_2$	$(f'(2) = 0)$

(c) The solution is $a_0 = a_1 = 0, a_2 = \frac{3}{2}, a_3 = -\frac{1}{2}$:

$$f(x) = -\frac{1}{2}x^3 + \frac{3}{2}x^2.$$

(d)

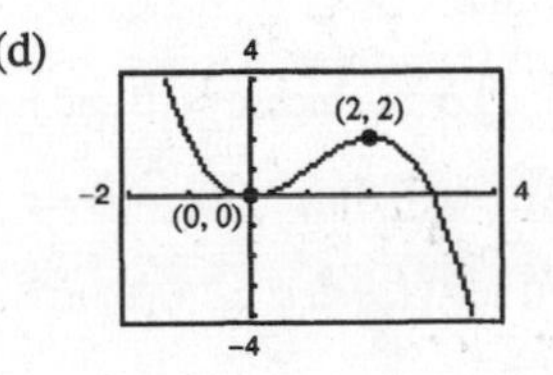

77. (a) Use a fourth degree polynomial $f(x) = a_4x^4 + a_3x^3 + a_2x^2 + a_1x + a_0$.

(b) $f'(x) = 4a_4x^3 + 3a_3x^2 + 2a_2x + a_1$

(0, 0):	$0 = a_0$	$(f(0) = 0)$
	$0 = a_1$	$(f'(0) = 0)$
(4, 0):	$0 = 256a_4 + 64a_3 + 16a_2$	$(f(4) = 0)$
	$0 = 256a_4 + 48a_3 + 8a_2$	$(f'(4) = 0)$
(2, 4):	$4 = 16a_4 + 8a_3 + 4a_2$	$(f(2) = 4)$
	$0 = 32a_4 + 12a_3 + 4a_2$	$(f'(2) = 0)$

(c) The solution is $a_0 = a_1 = 0, a_2 = 4, a_3 = -2, a_4 = \frac{1}{4}$.

$$f(x) = \frac{1}{4}x^4 - 2x^3 + 4x^2$$

(d)

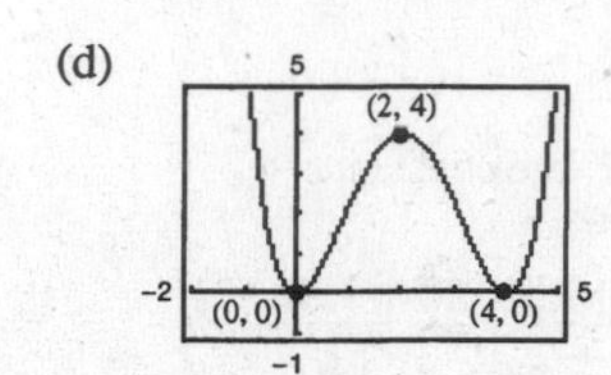

79. True

Let $h(x) = f(x) + g(x)$ where f and g are increasing. Then $h'(x) = f'(x) + g'(x) > 0$ since $f'(x) > 0$ and $g'(x) > 0$.

81. False

Let $f(x) = x^3$, then $f'(x) = 3x^2$ and f only has one critical number. Or, let $f(x) = x^3 + 3x + 1$, then $f'(x) = 3(x^2 + 1)$ has no critical numbers.

83. False. For example, $f(x) = x^3$ does not have a relative extrema at the critical number $x = 0$.

85. Assume that $f'(x) < 0$ for all x in the interval (a, b) and let $x_1 < x_2$ be any two points in the interval. By the Mean Value Theorem, we know there exists a number c such that $x_1 < c < x_2$, and

$$f'(c) = \frac{f(x_2) - f(x_1)}{x_2 - x_1}.$$

Since $f'(c) < 0$ and $x_2 - x_1 > 0$, then $f(x_2) - f(x_1) < 0$, which implies that $f(x_2) < f(x_1)$. Thus, f is decreasing on the interval.

87. Let $f(x) = (1 + x)^n - nx - 1$. Then

$$f'(x) = n(1 + x)^{n-1} - n$$
$$= n[(1 + x)^{n-1} - 1] > 0 \text{ since } x > 0 \text{ and } n > 1.$$

Thus, $f(x)$ is increasing on $(0, \infty)$. Since $f(0) = 0 \Rightarrow f(x) > 0$ on $(0, \infty)$

$$(1 + x)^n - nx - 1 > 0 \Rightarrow (1 + x)^n > 1 + nx.$$

Section 3.4 Concavity and the Second Derivative Test

1. $y = x^2 - x - 2,\ y'' = 2$

Concave upward: $(-\infty, \infty)$

3. $f(x) = \dfrac{24}{x^2 + 12},\ y'' = \dfrac{-144(4 - x^2)}{(x^2 + 12)^3}$

Concave upward: $(-\infty, -2), (2, \infty)$

Concave downward: $(-2, 2)$

5. $f(x) = \dfrac{x^2 + 1}{x^2 - 1},\ y'' = \dfrac{4(3x^2 + 1)}{(x^2 - 1)^3}$

Concave upward: $(-\infty, -1), (1, \infty)$

Concave downward: $(-1, 1)$

7. $f(x) = x^3 - 6x^2 + 12x$

$f'(x) = 3x^2 - 12x + 12$

$f''(x) = 6(x - 2) = 0$ when $x = 2$.

The concavity changes at $x = 2$. $(2, 8)$ is a point of inflection.

Concave upward: $(2, \infty)$

Concave downward: $(-\infty, 2)$

9. $f(x) = \dfrac{1}{4}x^4 - 2x^2$

$f'(x) = x^3 - 4x$

$f''(x) = 3x^2 - 4$

$f''(x) = 3x^2 - 4 = 0$ when $x = \pm\dfrac{2}{\sqrt{3}}$.

Test interval:	$-\infty < x < -\frac{2}{\sqrt{3}}$	$-\frac{2}{\sqrt{3}} < x < \frac{2}{\sqrt{3}}$	$\frac{2}{\sqrt{3}} < x < \infty$
Sign of $f''(x)$:	$f''(x) > 0$	$f''(x) < 0$	$f''(x) > 0$
Conclusion:	Concave upward	Concave downward	Concave upward

Points of inflection: $\left(\pm\dfrac{2}{\sqrt{3}}, -\dfrac{20}{9}\right)$

11. $f(x) = x(x - 4)^3$

$f'(x) = x[3(x - 4)^2] + (x - 4)^3$

$= (x - 4)^2(4x - 4)$

$f''(x) = 4(x - 1)[2(x - 4)] + 4(x - 4)^2$

$= 4(x - 4)[2(x - 1) + (x - 4)]$

$= 4(x - 4)(3x - 6) = 12(x - 4)(x - 2)$

$f''(x) = 12(x - 4)(x - 2) = 0$ when $x = 2, 4$.

Test interval:	$-\infty < x < 2$	$2 < x < 4$	$4 < x < \infty$
Sign of $f''(x)$:	$f''(x) > 0$	$f''(x) < 0$	$f''(x) > 0$
Conclusion:	Concave upward	Concave downward	Concave upward

Points of inflection: $(2, -16), (4, 0)$

13. $f(x) = x\sqrt{x+3}$, Domain: $[-3, \infty)$

$$f'(x) = x\left(\frac{1}{2}\right)(x+3)^{-1/2} + \sqrt{x+3} = \frac{3(x+2)}{2\sqrt{x+3}}$$

$$f''(x) = \frac{6\sqrt{x+3} - 3(x+2)(x+3)^{-1/2}}{4(x+3)} = \frac{3(x+4)}{4(x+3)^{3/2}}$$

$f''(x) > 0$ on the entire domain of f (except for $x = -3$, for which $f''(x)$ is undefined). There are no points of inflection.

Concave upward on $(-3, \infty)$

15. $f(x) = \dfrac{x}{x^2+1}$

$$f'(x) = \frac{1-x^2}{(x^2+1)^2}$$

$$f''(x) = \frac{2x(x^2-3)}{(x^2+1)^3} = 0 \text{ when } x = 0, \pm\sqrt{3}$$

Test intervals:	$-\infty < x < -\sqrt{3}$	$-\sqrt{3} < x < 0$	$0 < x < \sqrt{3}$	$\sqrt{3} < x < \infty$
Sign of $f''(x)$:	$f'' < 0$	$f'' > 0$	$f'' < 0$	$f'' > 0$
Conclusion:	Concave downward	Concave upward	Concave downward	Concave upward

Points of inflection: $\left(-\sqrt{3}, -\dfrac{\sqrt{3}}{4}\right), (0, 0), \left(\sqrt{3}, \dfrac{\sqrt{3}}{4}\right)$

17. $f(x) = \sin\left(\dfrac{x}{2}\right), 0 \le x \le 4\pi$

$$f'(x) = \frac{1}{2}\cos\left(\frac{x}{2}\right)$$

$$f''(x) = -\frac{1}{4}\sin\left(\frac{x}{2}\right)$$

$f''(x) = 0$ when $x = 0, 2\pi, 4\pi$.

Point of inflection: $(2\pi, 0)$

Test interval:	$0 < x < 2\pi$	$2\pi < x < 4\pi$
Sign of $f''(x)$:	$f'' < 0$	$f'' > 0$
Conclusion:	Concave downward	Concave upward

19. $f(x) = \sec\left(x - \dfrac{\pi}{2}\right), 0 < x < 4\pi$

$$f'(x) = \sec\left(x - \frac{\pi}{2}\right)\tan\left(x - \frac{\pi}{2}\right)$$

$$f''(x) = \sec^3\left(x - \frac{\pi}{2}\right) + \sec\left(x - \frac{\pi}{2}\right)\tan^2\left(x - \frac{\pi}{2}\right) \neq 0 \text{ for any } x \text{ in the domain of } f.$$

Concave upward: $(0, \pi), (2\pi, 3\pi)$

Concave downward: $(\pi, 2\pi), (3\pi, 4\pi)$

No points of inflection

21. $f(x) = 2\sin x + \sin 2x, 0 \le x \le 2\pi$

$f'(x) = 2\cos x + 2\cos 2x$

$f''(x) = -2\sin x - 4\sin 2x = -2\sin x(1 + 4\cos x)$

$f''(x) = 0$ when $x = 0, 1.823, \pi, 4.460$.

Test interval:	$0 < x < 1.823$	$1.823 < x < \pi$	$\pi < x < 4.460$	$4.460 < x < 2\pi$
Sign of $f''(x)$:	$f'' < 0$	$f'' > 0$	$f'' < 0$	$f'' > 0$
Conclusion:	Concave downward	Concave upward	Concave downward	Concave upward

Points of inflection: $(1.823, 1.452), (\pi, 0), (4.46, -1.452)$

23. $f(x) = x - \ln x$ Domain: $x > 0$

$f'(x) = 1 - \dfrac{1}{x}$

$f''(x) = \dfrac{1}{x^2}$ Concave upward $(0, \infty)$

25. $f(x) = x^4 - 4x^3 + 2$

$f'(x) = 4x^3 - 12x^2 = 4x^2(x - 3)$

$f''(x) = 12x^2 - 24x = 12x(x - 2)$

Critical numbers: $x = 0, x = 3$

However, $f''(0) = 0$, so we must use the First Derivative Test. $f'(x) < 0$ on the intervals $(-\infty, 0)$ and $(0, 3)$; hence, $(0, 2)$ is not an extremum. $f''(3) > 0$ so $(3, -25)$ is a relative minimum.

27. $f(x) = (x - 5)^2$

$f'(x) = 2(x - 5)$

$f''(x) = 2$

Critical number: $x = 5$

$f''(5) > 0$

Therefore, $(5, 0)$ is a relative minimum.

29. $f(x) = x^3 - 3x^2 + 3$

$f'(x) = 3x^2 - 6x = 3x(x - 2)$

$f''(x) = 6x - 6 = 6(x - 1)$

Critical numbers: $x = 0, x = 2$

$f''(0) = -6 < 0$

Therefore, $(0, 3)$ is a relative maximum.

$f''(2) = 6 > 0$

Therefore, $(2, -1)$ is a relative minimum.

31. $g(x) = x^2(6 - x)^3$

$g'(x) = x(x - 6)^2(12 - 5x)$

$g''(x) = 4(6 - x)(5x^2 - 24x + 18)$

Critical numbers: $x = 0, \frac{12}{5}, 6$

$g''(0) = 432 > 0$

Therefore, $(0, 0)$ is a relative minimum.

$g''\left(\frac{12}{5}\right) = -155.52 < 0$

Therefore, $\left(\frac{12}{5}, 268.7\right)$ is a relative minimum.

$g''(6) = 0$

Test fails by the First Derivative Test, $(6, 0)$ is not an extremum.

33. $f(x) = x^{2/3} - 3$

$f'(x) = \dfrac{2}{3x^{1/3}}$

$f''(x) = \dfrac{-2}{9x^{4/3}}$

Critical number: $x = 0$

However, $f''(0)$ is undefined, so we must use the First Derivative Test. Since $f'(x) < 0$ on $(-\infty, 0)$ and $f'(x) > 0$ on $(0, \infty)$, $(0, -3)$ is a relative minimum.

35. $f(x) = x + \dfrac{4}{x}$

$f'(x) = 1 - \dfrac{4}{x^2} = \dfrac{x^2 - 4}{x^2}$

$f''(x) = \dfrac{8}{x^3}$

Critical numbers: $x = \pm 2$

$f''(-2) < 0$

Therefore, $(-2, -4)$ is a relative maximum.

$f''(2) > 0$

Therefore, $(2, 4)$ is a relative minimum.

37. $f(x) = \cos x - x,\ 0 \le x \le 4\pi$

$f'(x) = -\sin x - 1 \le 0$

Therefore, f is non-increasing and there are no relative extrema.

39. $y = \dfrac{x^2}{2} - \ln x$

Domain: $x > 0$

$y' = x - \dfrac{1}{x} = \dfrac{(x + 1)(x - 1)}{x} = 0$ when $x = 1$.

$y'' = 1 + \dfrac{1}{x^2} > 0$

Relative minimum: $\left(1, \dfrac{1}{2}\right)$

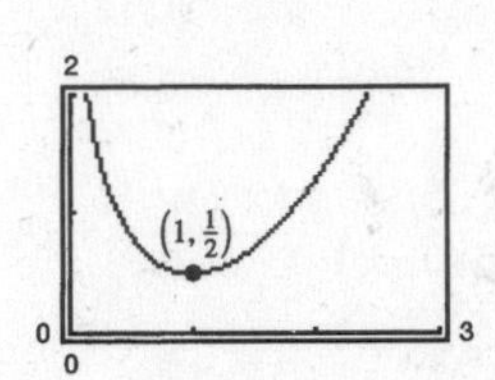

41. $y = \dfrac{x}{\ln x}$

Domain: $0 < x < 1,\ x > 1$

$y' = \dfrac{(\ln x)(1) - (x)(1/x)}{(\ln x)^2} = \dfrac{\ln x - 1}{(\ln x)^2} = 0$ when $x = e$.

$y'' = \dfrac{2 - \ln x}{x(\ln x)^3} = 0$ when $x = e^2$.

Relative minimum: (e, e)

Point of inflection: $\left(e^2, \dfrac{e^2}{2}\right)$

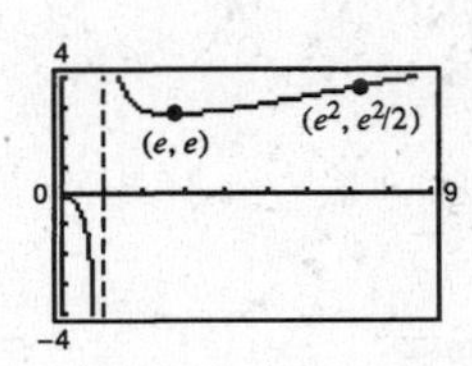

43. $f(x) = \dfrac{e^x + e^{-x}}{2}$

$f'(x) = \dfrac{e^x - e^{-x}}{2} = 0$ when $x = 0$.

$f''(x) = \dfrac{e^x + e^{-x}}{2} > 0$

Relative minimum: $(0, 1)$

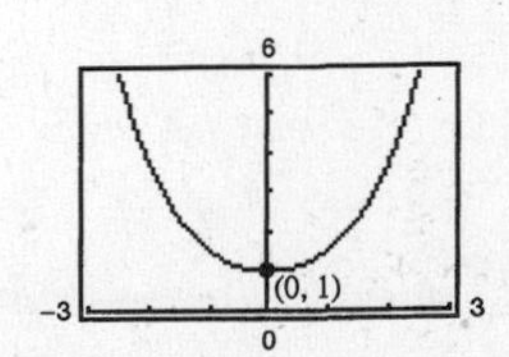

45. $f(x) = x^2e^{-x}$

$f'(x) = -x^2e^{-x} + 2xe^{-x} = xe^{-x}(2 - x) = 0$ when $x = 0, 2$.

$f''(x) = -e^{-x}(2x - x^2) + e^{-x}(2 - 2x) = e^{-x}(x^2 - 4x + 2) = 0$ when $x = 2 \pm \sqrt{2}$.

Relative minimum: $(0, 0)$

Relative maximum: $(2, 4e^{-2})$

$x = 2 \pm \sqrt{2}$

$y = \left(2 \pm \sqrt{2}\right)^2 e^{-(2\pm\sqrt{2})}$

Points of inflection: $(3.414, 0.384)$, $(0.586, 0.191)$

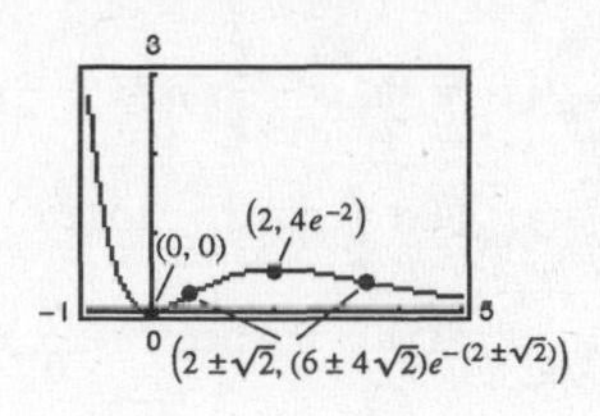

47. $f(x) = 8x(4^{-x})$

$f'(x) = -8(4^{-x})(x \ln 4 - 1)$

$f''(x) = 8(4^{-x}) \ln 4(x \ln 4 - 2)$

$f'(x) = 0 \Rightarrow x = \dfrac{1}{\ln 4} = \dfrac{1}{2 \ln 2}$

$f''\left(\dfrac{1}{2 \ln 2}\right) < 0 \Rightarrow$ relative maximum

$f''(x) = 0 \Rightarrow x = \dfrac{2}{\ln 4} = \dfrac{1}{\ln 2}$

Relative maximum: $\left(\dfrac{1}{2 \ln 2}, \dfrac{4e^{-1}}{\ln 2}\right)$

Point of inflection: $\left(\dfrac{1}{\ln 2}, \dfrac{8e^{-2}}{\ln 2}\right)$

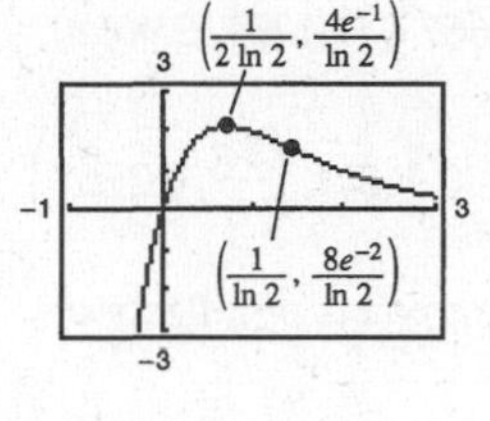

49. $f(x) = \operatorname{arcsec} x - x$

$f'(x) = \dfrac{1}{|x|\sqrt{x^2 - 1}} - 1 = 0$ when $|x|\sqrt{x^2 - 1} = 1$.

$x^2(x^2 - 1) = 1$

$x^4 - x^2 - 1 = 0$ when $x^2 = \dfrac{1 + \sqrt{5}}{2}$ or $x = \pm\sqrt{\dfrac{1 + \sqrt{5}}{2}} = \pm 1.272$.

Relative maximum: $(1.272, -0.606)$

Relative minimum: $(-1.272, 3.747)$

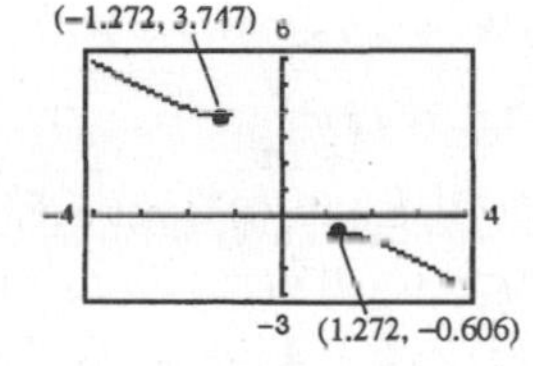

51. $f(x) = 0.2x^2(x - 3)^3, [-1, 4]$

(a) $f'(x) = 0.2x(5x - 6)(x - 3)^2$

$f''(x) = (x - 3)(4x^2 - 9.6x + 3.6)$

$= 0.4(x - 3)(10x^2 - 24x + 9)$

(b) $f''(0) < 0 \Rightarrow (0, 0)$ is a relative maximum.

$f''(\frac{6}{5}) > 0 \Rightarrow (1.2, -1.6796)$ is a relative minimum.

Points of inflection:

$(3, 0)$, $(0.4652, -0.7049)$, $(1.9348, -0.9049)$

(c)

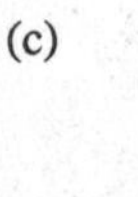
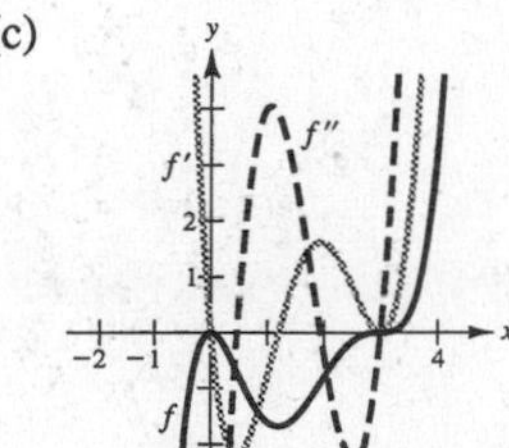

f is increasing when $f' > 0$ and decreasing when $f' < 0$. f is concave upward when $f'' > 0$ and concave downward when $f'' < 0$.

53. $f(x) = \sin x - \frac{1}{3}\sin 3x + \frac{1}{5}\sin 5x, [0, \pi]$

(a) $f'(x) = \cos x - \cos 3x + \cos 5x$

$f'(x) = 0$ when $x = \frac{\pi}{6}, x = \frac{\pi}{2}, x = \frac{5\pi}{6}$.

$f''(x) = -\sin x + 3\sin 3x - 5\sin 5x$

$f''(x) = 0$ when $x = \frac{\pi}{6}, x = \frac{5\pi}{6}, x \approx 1.1731, x \approx 1.9685$

(b) $f''\left(\frac{\pi}{2}\right) < 0 \Rightarrow \left(\frac{\pi}{2}, 1.53333\right)$ is a relative maximum.

Points of inflection: $\left(\frac{\pi}{6}, 0.2667\right)$, $(1.1731, 0.9638)$,

$(1.9685, 0.9637)$, $\left(\frac{5\pi}{6}, 0.2667\right)$

Note: $(0, 0)$ and $(\pi, 0)$ are not points of inflection since they are endpoints.

(c)

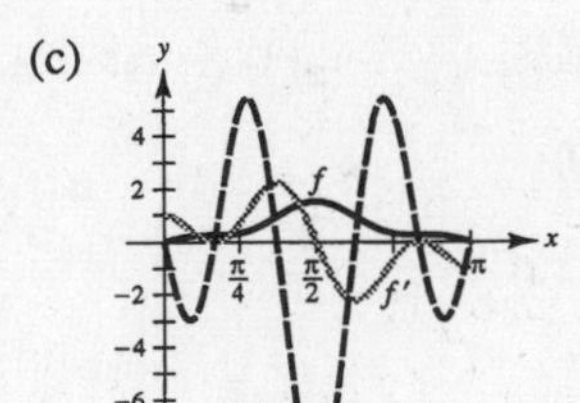

The graph of f is increasing when $f' > 0$ and decreasing when $f' < 0$. f is concave upward when $f'' > 0$ and concave downward when $f'' < 0$.

55. (a)

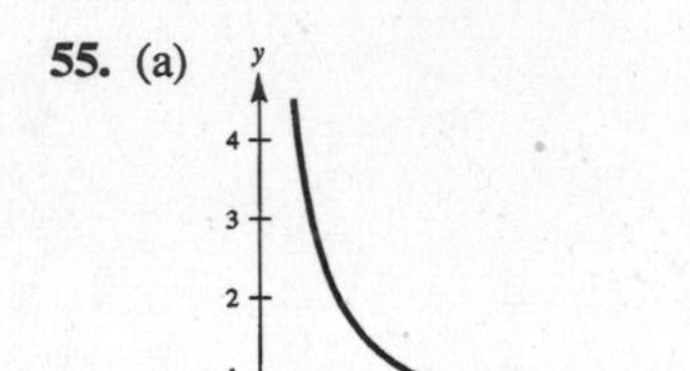

$f' < 0$ means f decreasing.

f' increasing means concave upward.

(b)

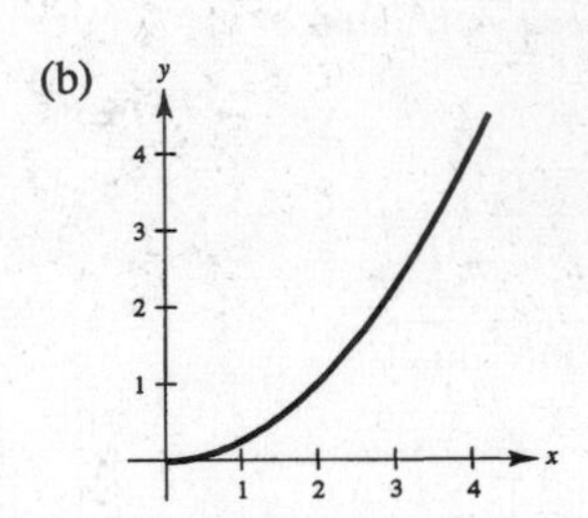

$f' > 0$ means f increasing.

f' increasing means concave upward.

57. Let $f(x) = x^4$.

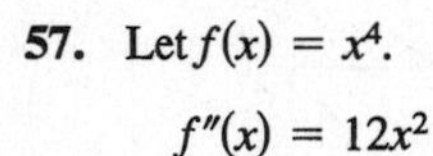

$f''(x) = 12x^2$

$f''(0) = 0$, but $(0, 0)$ is not a point of inflection.

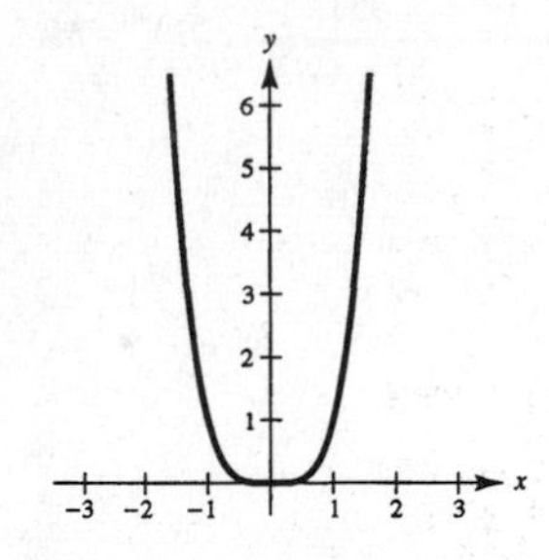

59.

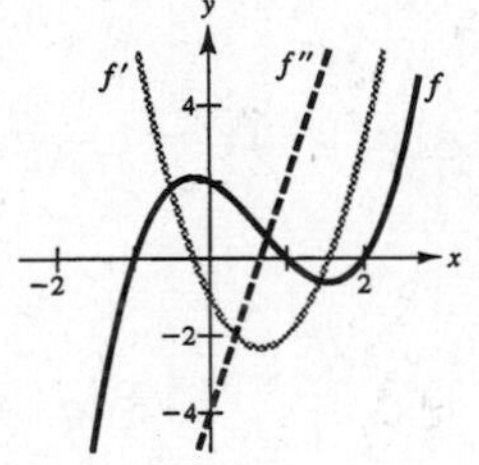

61.

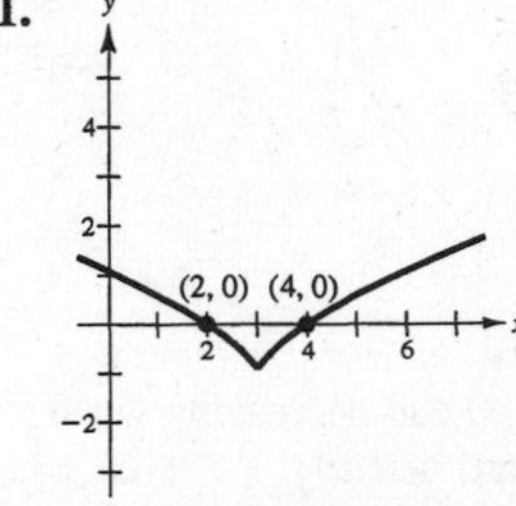

63.

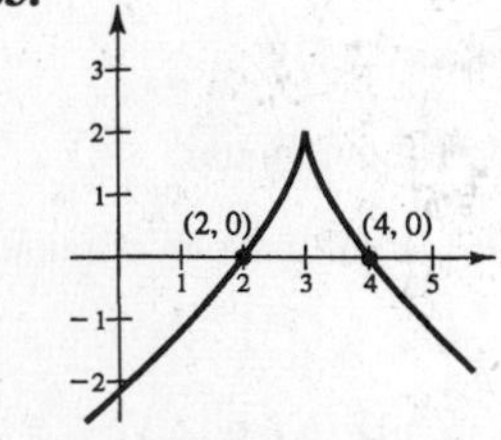

65.

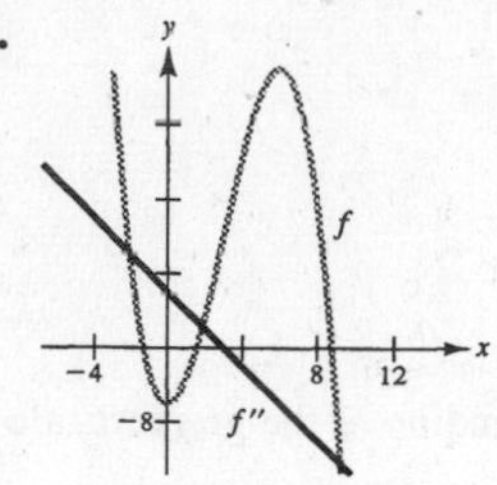

f'' is linear.

f' is quadratic.

f is cubic.

f concave upwards on $(-\infty, 3)$, downward on $(3, \infty)$.

67. (a) $n = 1$:

$f(x) = x - 2$

$f'(x) = 1$

$f''(x) = 0$

No inflection points

$n = 2$:

$f(x) = (x - 2)^2$

$f'(x) = 2(x - 2)$

$f''(x) = 2$

No inflection points

Relative minimum: $(2, 0)$

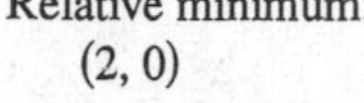

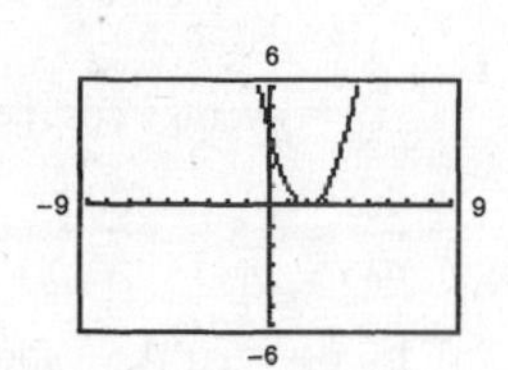

$n = 3$:

$f(x) = (x - 2)^3$

$f'(x) = 3(x - 2)^2$

$f''(x) = 6(x - 2)$

Inflection point: $(2, 0)$

$n = 4$:

$f(x) = (x - 2)^4$

$f'(x) = 4(x - 2)^3$

$f''(x) = 12(x - 2)^2$

No inflection points:

Relative minimum: $(2, 0)$

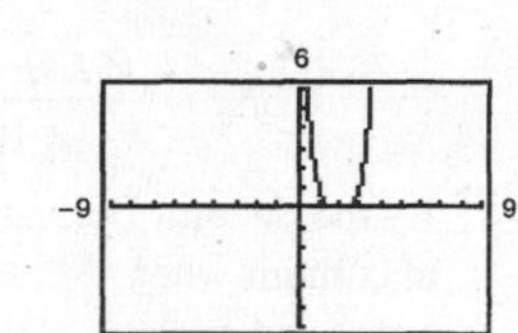

Conclusion: If $n \geq 3$ and n is odd, then $(2, 0)$ is an inflection point. If $n \geq 2$ and n is even, then $(2, 0)$ is a relative minimum.

(b) Let $f(x) = (x - 2)^n$, $f'(x) = n(x - 2)^{n-1}$, $f''(x) = n(n - 1)(x - 2)^{n-2}$.

For $n \geq 3$ and odd, $n - 2$ is also odd and the concavity changes at $x = 2$.

For $n \geq 4$ and even, $n - 2$ is also even and the concavity does not change at $x = 2$.

Thus, $x = 2$ is an inflection point if and only if $n \geq 3$ is odd.

69. $f(x) = ax^3 + bx^2 + cx + d$

Relative maximum: $(3, 3)$

Relative minimum: $(5, 1)$

Point of inflection: $(4, 2)$

$f'(x) = 3ax^2 + 2bx + c, f''(x) = 6ax + 2b$

$$\left.\begin{aligned} f(3) &= 27a + 9b + 3c + d = 3 \\ f(5) &= 125a + 25b + 5c + d = 1 \end{aligned}\right\} 98a + 16b + 2c = -2 \Rightarrow 49a + 8b + c = -1$$

$f'(3) = 27a + 6b + c = 0,\ f''(4) = 24a + 2b = 0$

$$\begin{array}{ll} 49a + 8b + c = -1 & 24a + 2b = 0 \\ \underline{27a + 6b + c = 0} & \underline{22a + 2b = -1} \\ 22a + 2b \quad = -1 & 2a \quad = 1 \end{array}$$

$a = \frac{1}{2}, b = -6, c = \frac{45}{2}, d = -24$

$f(x) = \frac{1}{2}x^3 - 6x^2 + \frac{45}{2}x - 24$

71. $f(x) = ax^3 + bx^2 + cx + d$

Maximum: $(-4, 1)$

Minimum: $(0, 0)$

(a) $f'(x) = 3ax^2 + 2bx + c, \quad f''(x) = 6ax + 2b$

$$f(0) = 0 \Rightarrow d = 0$$

$$f(-4) = 1 \Rightarrow -64a + 16b - 4c = 1$$

$$f'(-4) = 0 \Rightarrow 48a - 8b + c = 0$$

$$f'(0) = 0 \Rightarrow c = 0$$

Solving this system yields $a = \frac{1}{32}$ and $b = 6a = \frac{3}{16}$.

$$f(x) = \tfrac{1}{32}x^3 + \tfrac{3}{16}x^2$$

(b) The plane would be descending at the greatest rate at the point of inflection.

$$f''(x) = 6ax + 2b = \frac{3}{16}x + \frac{3}{8} = 0 \Rightarrow x = -2.$$

Two miles from touchdown.

73. $D = 2x^4 - 5Lx^3 + 3L^2x^2$

$$D' = 8x^3 - 15Lx^2 + 6L^2x = x(8x^2 - 15Lx + 6L^2) = 0$$

$$x = 0 \text{ or } x = \frac{15L \pm \sqrt{33}L}{16} = \left(\frac{15 \pm \sqrt{33}}{16}\right)L$$

By the Second Derivative Test, the deflection is maximum when

$$x = \left(\frac{15 - \sqrt{33}}{16}\right)L \approx 0.578L.$$

75. $C = 0.5x^2 + 15x + 5000$

$$\overline{C} = \frac{C}{x} = 0.5x + 15 + \frac{5000}{x}$$

$\overline{C}$ = average cost per unit

$$\frac{d\overline{C}}{dx} = 0.5 - \frac{5000}{x^2} = 0 \text{ when } x = 100$$

By the First Derivative Test, $\overline{C}$ is minimized when $x = 100$ units.

77. $S = \dfrac{5000t^2}{8 + t^2}$

$$S'(t) = \frac{80{,}000t}{(8 + t^2)^2}$$

$$S''(t) = \frac{80{,}000(8 - 3t^2)}{(8 + t^2)^3}$$

$S''(t) = 0$ for $t = \sqrt{8/3} \approx 1.633$.

Sales are increasing at the greatest rate at $t = 1.633$ years.

79. $f(x) = 2(\sin x + \cos x), \qquad f\left(\frac{\pi}{4}\right) = 2\sqrt{2}$

$f'(x) = 2(\cos x - \sin x), \qquad f'\left(\frac{\pi}{4}\right) = 0$

$f''(x) = 2(-\sin x - \cos x), \qquad f''\left(\frac{\pi}{4}\right) = -2\sqrt{2}$

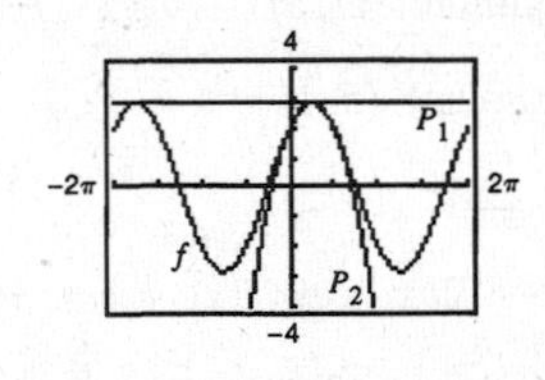

$$P_1(x) = 2\sqrt{2} + 0\left(x - \frac{\pi}{4}\right) = 2\sqrt{2}$$

$$P_1'(x) = 0$$

$$P_2(x) = 2\sqrt{2} + 0\left(x - \frac{\pi}{4}\right) + \frac{1}{2}\left(-2\sqrt{2}\right)\left(x - \frac{\pi}{4}\right)^2 = 2\sqrt{2} - \sqrt{2}\left(x - \frac{\pi}{4}\right)^2$$

$$P_2'(x) = -2\sqrt{2}\left(x - \frac{\pi}{4}\right)$$

$$P_2''(x) = -2\sqrt{2}$$

The values of f, P_1, P_2, and their first derivatives are equal at $x = \pi/4$. The values of the second derivatives of f and P_2 are equal at $x = \pi/4$. The approximations worsen as you move away from $x = \pi/4$.

81. $f(x) = \arctan x, \quad a = -1$

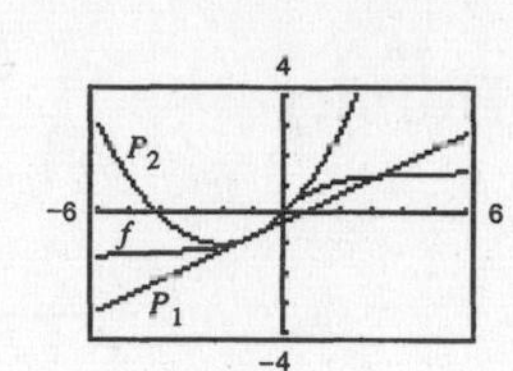

$$f'(x) = \frac{1}{1+x^2}$$

$$f''(x) = -\frac{2x}{(1+x^2)^2}$$

$$P_1(x) = f(-1) + f'(-1)(x+1) = -\frac{\pi}{4} + \frac{1}{2}(x+1)$$

$$P_2(x) = f(-1) + f'(-1)(x+1) + \frac{1}{2}f''(-1)(x+1)^2$$

$$= -\frac{\pi}{4} + \frac{1}{2}(x+1) + \frac{1}{4}(x+1)^2$$

83. $f(x) = x\sin\left(\frac{1}{x}\right)$

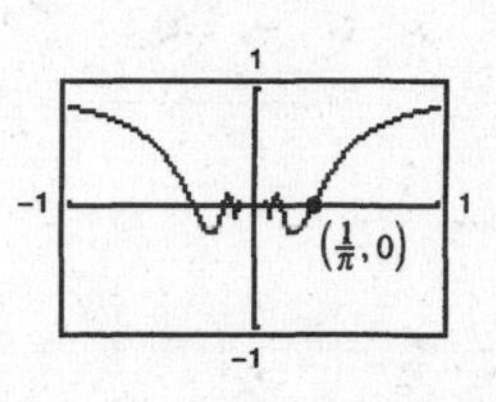

$$f'(x) = x\left[-\frac{1}{x^2}\cos\left(\frac{1}{x}\right)\right] + \sin\left(\frac{1}{x}\right) = -\frac{1}{x}\cos\left(\frac{1}{x}\right) + \sin\left(\frac{1}{x}\right)$$

$$f''(x) = -\frac{1}{x}\left[\frac{1}{x^2}\sin\left(\frac{1}{x}\right)\right] + \frac{1}{x^2}\cos\left(\frac{1}{x}\right) - \frac{1}{x^2}\cos\left(\frac{1}{x}\right) = -\frac{1}{x^3}\sin\left(\frac{1}{x}\right) = 0$$

$$x = \frac{1}{\pi}$$

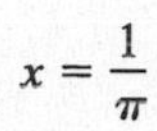

Point of inflection: $\left(\frac{1}{\pi}, 0\right)$

When $x > 1/\pi, f'' < 0$, so the graph is concave downward.

85. Assume the zeros of f are all real. Then express the function as $f(x) = a(x - r_1)(x - r_2)(x - r_3)$ where r_1, r_2, and r_3 are the distinct zeros of f. From the Product Rule for a function involving three factors, we have

$$f'(x) = a[(x - r_1)(x - r_2) + (x - r_1)(x - r_3) + (x - r_2)(x - r_3)]$$

$$f''(x) = a[(x - r_1) + (x - r_2) + (x - r_1) + (x - r_3) + (x - r_2) + (x - r_3)]$$

$$= a[6x - 2(r_1 + r_2 + r_3)].$$

Consequently, $f''(x) = 0$ if

$$x = \frac{2(r_1 + r_2 + r_3)}{6} = \frac{r_1 + r_2 + r_3}{3} = (\text{Average of } r_1, r_2, \text{and } r_3).$$

87. True. Let $y = ax^3 + bx^2 + cx + d, a \neq 0$. Then $y'' = 6ax + 2b = 0$ when $x = -(b/3a)$, and the concavity changes at this point.

89. False.

$$f(x) = 3\sin x + 2\cos x$$

$$f'(x) = 3\cos x - 2\sin x$$

$$3\cos x - 2\sin x = 0$$

$$3\cos x = 2\sin x$$

$$\tfrac{3}{2} = \tan x$$

Critical number: $x = \tan^{-1}\left(\frac{3}{2}\right)$

$f\left(\tan^{-1}\frac{3}{2}\right) \approx 3.60555$ is the maximum value of y.

91. False. Concavity is determined by f''.

Section 3.5 Limits at Infinity

1. $f(x) = \frac{3x^2}{x^2 + 2}$

No vertical asymptotes

Horizontal asymptote: $y = 3$

Matches (f)

3. $f(x) = \frac{x}{x^2 + 2}$

No vertical asymptotes

Horizontal asymptote: $y = 0$

Matches (d)

5. $f(x) = \frac{4\sin x}{x^2 + 1}$

No vertical asymptotes

Horizontal asymptotes: $y = 0$

Matches (b)

7. (a) $h(x) = \frac{f(x)}{x^2} = \frac{5x^3 - 3x^2 + 10}{x^2} = 5x - 3 + \frac{10}{x^2}$

$\lim_{x\to\infty} h(x) = \infty$ (Limit does not exist)

(b) $h(x) = \frac{f(x)}{x^3} = \frac{5x^3 - 3x^2 + 10}{x^3} = 5 - \frac{3}{x} + \frac{10}{x^3}$

$\lim_{x\to\infty} h(x) = 5$

(c) $h(x) = \frac{f(x)}{x^4} = \frac{5x^3 - 3x^2 + 10}{x^4} = \frac{5}{x} - \frac{3}{x^2} + \frac{10}{x^4}$

$\lim_{x\to\infty} h(x) = 0$

9. (a) $\lim_{x\to\infty} \frac{x^2 + 2}{x^3 - 1} = 0$

(b) $\lim_{x\to\infty} \frac{x^2 + 2}{x^2 - 1} = 1$

(c) $\lim_{x\to\infty} \frac{x^2 + 2}{x - 1} = \infty$ (Limit does not exist)

11. (a) $\lim_{x\to\infty} \frac{5 - 2x^{3/2}}{3x^2 - 4} = 0$

(b) $\lim_{x\to\infty} \frac{5 - 2x^{3/2}}{3x^{3/2} - 4} = -\frac{2}{3}$

(c) $\lim_{x\to\infty} \frac{5 - 2x^{3/2}}{3x - 4} = -\infty$ (Limit does not exist)

13. $\lim_{x\to\infty} \frac{2x - 1}{3x + 2} = \lim_{x\to\infty} \frac{2 - (1/x)}{3 + (2/x)} = \frac{2 - 0}{3 + 0} = \frac{2}{3}$

15. $\lim_{x\to\infty} \frac{x}{x^2 - 1} = \lim_{x\to\infty} \frac{1/x}{1 - (1/x^2)} = \frac{0}{1} = 0$

17. $\lim_{x\to-\infty} \frac{6x^2}{x + 3} = \lim_{x\to-\infty} \frac{6x}{1 + (3/x)} = -\infty$

Limit does not exist.

19. $\lim_{x\to-\infty} \frac{x}{\sqrt{x^2 - x}} = \lim_{x\to-\infty} \frac{1}{\frac{\sqrt{x^2 - x}}{-\sqrt{x^2}}}$, $\left(\text{for } x < 0 \text{ we have } x = -\sqrt{x^2}\right)$

$= \lim_{x\to-\infty} \frac{-1}{\sqrt{1 - (1/x)}} = -1$

21. $\lim_{x\to-\infty} \frac{2x + 1}{\sqrt{x^2 - x}} = \lim_{x\to-\infty} \frac{2 + \frac{1}{x}}{\left(\frac{\sqrt{x^2 - x}}{-\sqrt{x^2}}\right)}$ $\left(\text{for } x < 0, x = -\sqrt{x^2}\right)$

$= \lim_{x\to-\infty} \frac{-2 - (1/x)}{\sqrt{x + (1/x)}} = -2$

23. Since $(-1/x) \le (\sin(2x))/x \le (1/x)$ for all $x \ne 0$, we have by the Squeeze Theorem,

$$\lim_{x\to\infty} -\frac{1}{x} \le \lim_{x\to\infty} \frac{\sin(2x)}{x} \le \lim_{x\to\infty} \frac{1}{x}$$

$$0 \le \lim_{x\to\infty} \frac{\sin(2x)}{x} \le 0.$$

Therefore, $\lim_{x\to\infty} \frac{\sin(2x)}{x} = 0.$

25. $\lim_{x\to\infty} \frac{1}{2x + \sin x} = 0$

27. $\lim_{x\to\infty} (2 - 5e^{-x}) = 2$

29. $\lim_{x\to-\infty} \frac{3}{1 + 2e^x} = 3$

31. $\lim_{x\to\infty} \log_{10}(1 + 10^{-x}) = 0$

33. $\lim_{t\to\infty} \left(\frac{5}{t} - \arctan t\right) = 0 - \frac{\pi}{2} = -\frac{\pi}{2}$

35. (a) $f(x) = \frac{|x|}{x + 1}$

$$\lim_{x\to\infty} \frac{|x|}{x + 1} = 1$$

$$\lim_{x\to-\infty} \frac{|x|}{x + 1} = -1$$

Therefore, $y = 1$ and $y = -1$ are both horizontal asymptotes.

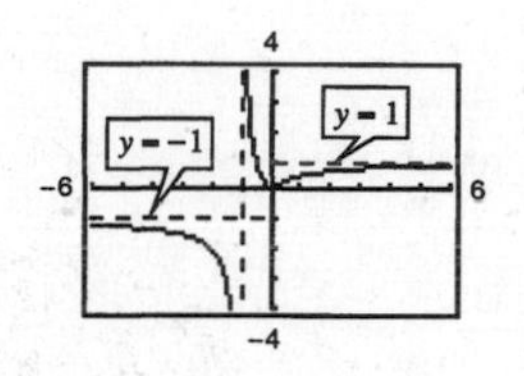

37. $\lim_{x\to\infty} x \sin\frac{1}{x} = \lim_{t\to 0^+} \frac{\sin t}{t} = 1$

(Let $x = 1/t$.)

39. $\lim_{x\to-\infty} \left(x + \sqrt{x^2 + 3}\right) = \lim_{x\to-\infty} \left[\left(x + \sqrt{x^2 + 3}\right) \cdot \frac{x - \sqrt{x^2 + 3}}{x - \sqrt{x^2 + 3}}\right] = \lim_{x\to-\infty} \frac{-3}{x - \sqrt{x^2 + 3}} = 0$

41. $\lim_{x\to\infty} \left(x - \sqrt{x^2 + x}\right) = \lim_{x\to\infty} \left[\left(x - \sqrt{x^2 + x}\right) \cdot \frac{x + \sqrt{x^2 + x}}{x + \sqrt{x^2 + x}}\right]$

$$= \lim_{x\to\infty} \frac{-x}{x + \sqrt{x^2 + x}} = \lim_{x\to\infty} \frac{-1}{1 + \sqrt{1 + (1/x)}} = -\frac{1}{2}$$

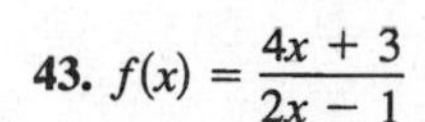

43. $f(x) = \frac{4x + 3}{2x - 1}$

x	10^0	10^1	10^2	10^3	10^4	10^5	10^6
$f(x)$	7	2.26	2.025	2.0025	2.0003	2	2

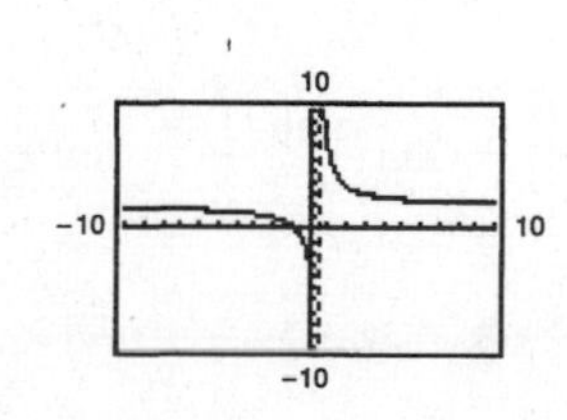

$\lim_{x\to\infty} f(x) = 2$

Analytically, $\lim_{x\to\infty} \frac{4x + 3}{2x - 1} = \lim_{x\to\infty} \frac{4 + \frac{3}{x}}{2 - \frac{1}{x}} = \frac{4}{2} = 2$

45. $f(x) = \dfrac{-6x}{\sqrt{4x^2 + 5}}$

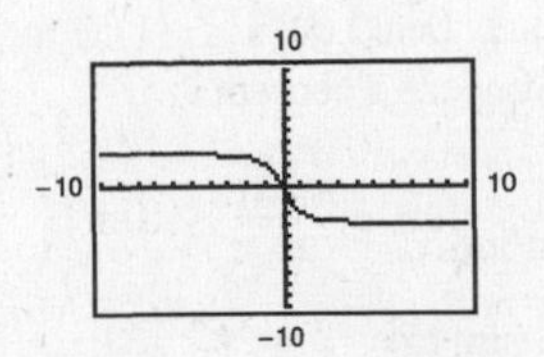

x	10^0	10^1	10^2	10^3	10^4	10^5	10^6
$f(x)$	-2	-2.98	-2.9998	-3	-3	-3	-3

$\lim_{x\to\infty} f(x) = -3$

Analytically, $\displaystyle\lim_{x\to\infty} \frac{-6x}{\sqrt{4x^2 + 5}} = \lim_{x\to\infty} \frac{-6}{\sqrt{4 + 5/x^2}} = \frac{-6}{2} = -3$

47. $f(x) = 5 - \dfrac{1}{x^2 + 1}$

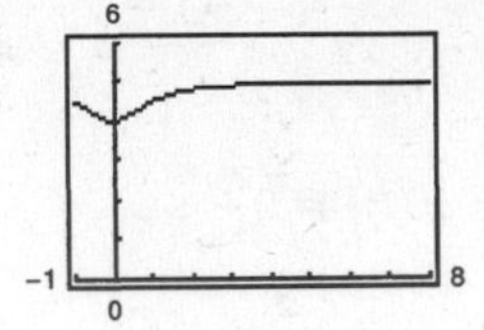

x	10^0	10^1	10^2	10^3	10^4	10^5	10^6
$f(x)$	4.5	4.99	4.9999	4.999999	5	5	5

$\lim_{x\to\infty} f(x) = 5$

Analytically, $\displaystyle\lim_{x\to\infty} \left(5 - \frac{1}{x^2 + 1}\right) = 5 - 0 = 5$

49.

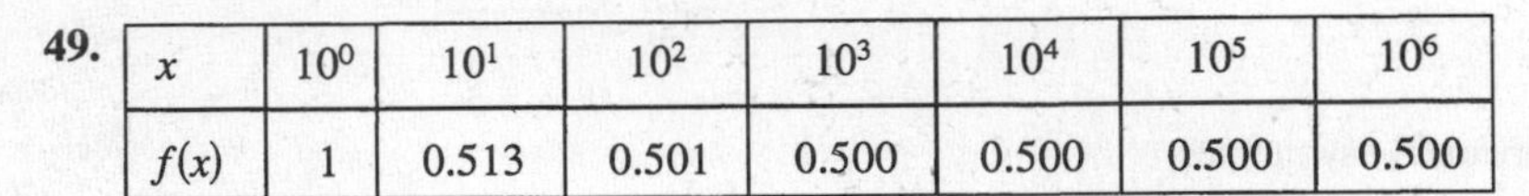

x	10^0	10^1	10^2	10^3	10^4	10^5	10^6
$f(x)$	1	0.513	0.501	0.500	0.500	0.500	0.500

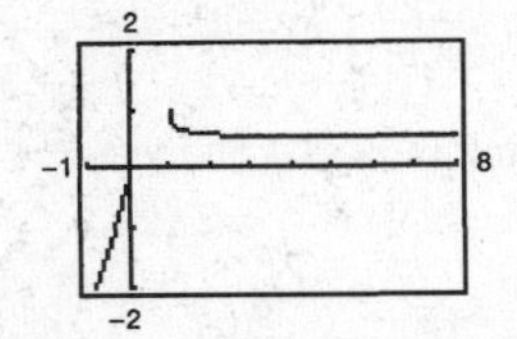

$$\lim_{x\to\infty} \left(x - \sqrt{x(x-1)}\right) = \lim_{x\to\infty} \frac{x - \sqrt{x^2 - x}}{1} \cdot \frac{x + \sqrt{x^2 - x}}{x + \sqrt{x^2 - x}}$$

$$= \lim_{x\to\infty} \frac{x}{x + \sqrt{x^2 - x}}$$

$$= \lim_{x\to\infty} \frac{1}{1 + \sqrt{1 - (1/x)}}$$

$$= \frac{1}{2}$$

51.

x	10^0	10^1	10^2	10^3	10^4	10^5	10^6
$f(x)$	0.479	0.500	0.500	0.500	0.500	0.500	0.500

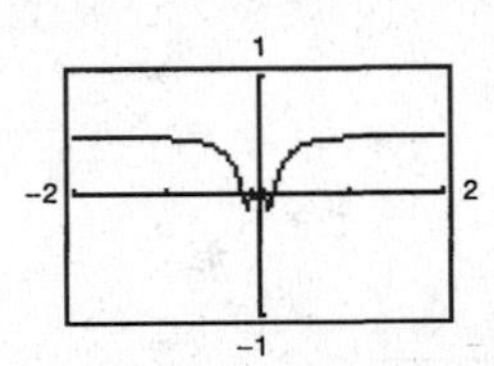

Let $x = 1/t$.

$$\lim_{x\to\infty} x \sin\left(\frac{1}{2x}\right) = \lim_{t\to 0^+} \frac{\sin(t/2)}{t} = \lim_{t\to 0^+} \frac{1}{2}\,\frac{\sin(t/2)}{t/2} = \frac{1}{2}$$

53. $y = \dfrac{2 + x}{1 - x}$

Intercepts: $(-2, 0), (0, 2)$

Symmetry: none

Horizontal asymptote: $y = -1$ since

$$\lim_{x \to -\infty} \frac{2 + x}{1 - x} = -1 = \lim_{x \to \infty} \frac{2 + x}{1 - x}.$$

Discontinuity: $x = 1$ (Vertical asymptote)

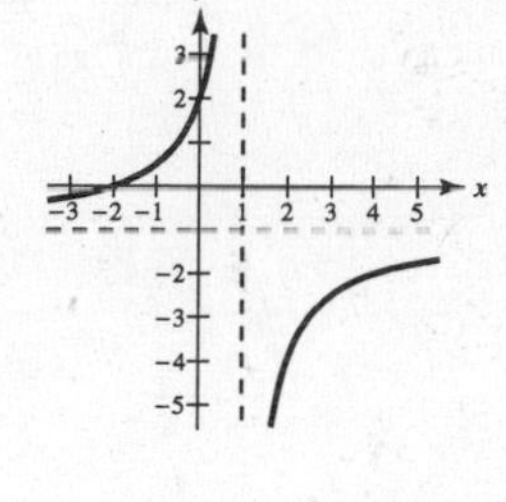

55. $y = \dfrac{x}{x^2 - 4}$

Intercept: $(0, 0)$

Symmetry: origin

Horizontal asymptote: $y = 0$

Vertical asymptote: $x = \pm 2$

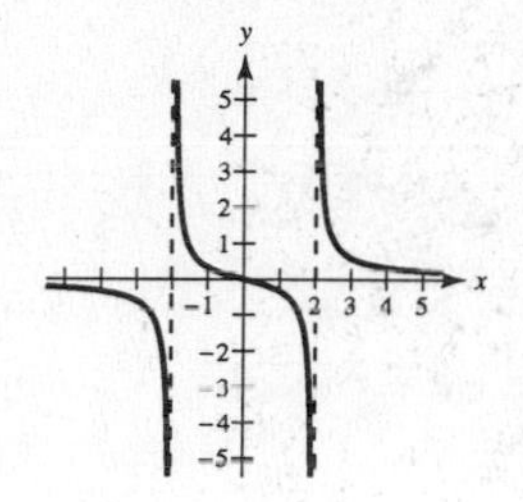

57. $y = \dfrac{x^2}{x^2 + 9}$

Intercept: $(0, 0)$

Symmetry: y-axis

Horizontal asymptote: $y = 1$ since

$$\lim_{x \to -\infty} \frac{x^2}{x^2 + 9} = 1 = \lim_{x \to \infty} \frac{x^2}{x^2 + 9}.$$

Relative minimum: $(0, 0)$

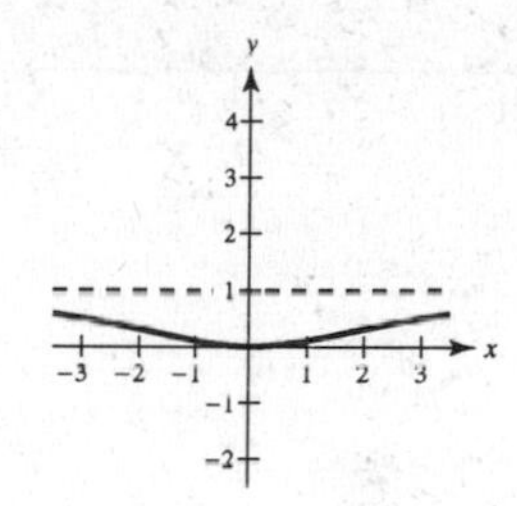

59. $y = \dfrac{2x^2}{x^2 - 4}$

Intercept: $(0, 0)$

Symmetry: y-axis

Horizontal asymptote: $y = 2$

Vertical asymptotes: $x = +2$

Relative maximum $= (0, 0)$

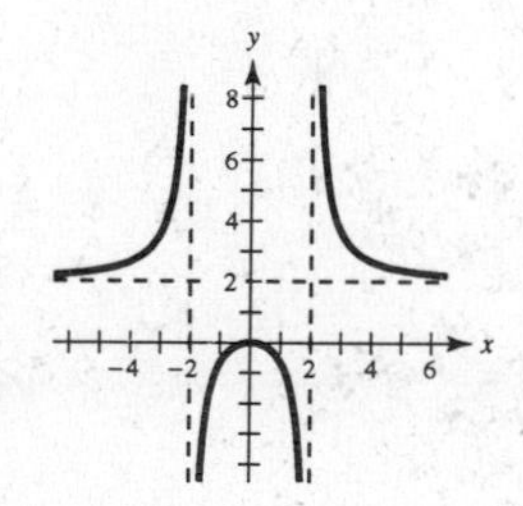

61. $xy^2 = 4$

Domain: $x > 0$

Intercepts: none

Symmetry: x-axis

Horizontal asymptote: $y = 0$ since

$$\lim_{x \to \infty} \frac{2}{\sqrt{x}} = 0 = \lim_{x \to \infty} -\frac{2}{\sqrt{x}}.$$

Discontinuity: $x = 0$ (Vertical asymptote)

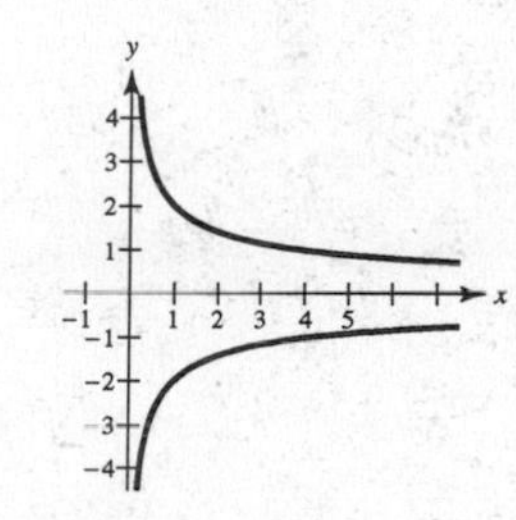

63. $y = \dfrac{2x}{1 - x}$

Intercept: $(0, 0)$

Symmetry: none

Horizontal asymptote: $y = -2$ since

$$\lim_{x \to -\infty} \frac{2x}{1 - x} = -2 = \lim_{x \to \infty} \frac{2x}{1 - x}.$$

Discontinuity: $x = 1$ (Vertical asymptote)

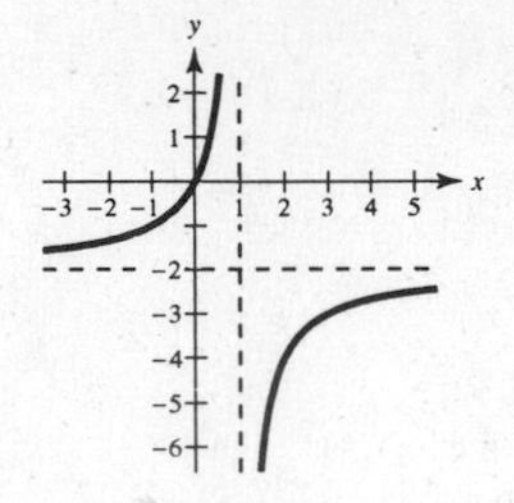

65. $y = 2 - \dfrac{3}{x^2}$

Intercepts: $\left(\pm\sqrt{3/2}, 0\right)$

Symmetry: y-axis

Horizontal asymptote: $y = 2$ since

$$\lim_{x \to -\infty} \left(2 - \frac{3}{x^2}\right) = 2 = \lim_{x \to \infty} \left(2 - \frac{3}{x^2}\right).$$

Discontinuity: $x = 0$ (Vertical asymptote)

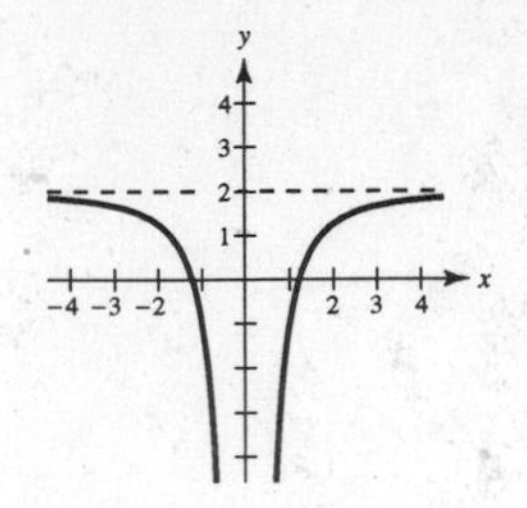

67. $y = 3 + \dfrac{2}{x}$

Intercept: $y = 0 = 3 + \dfrac{2}{x} \Rightarrow \dfrac{2}{x} = -3 \Rightarrow x = -\dfrac{2}{3};\ \left(-\dfrac{2}{3}, 0\right)$

Symmetry: none

Horizontal asymptote: $y = 3$

Vertical asymptote: $x = 0$

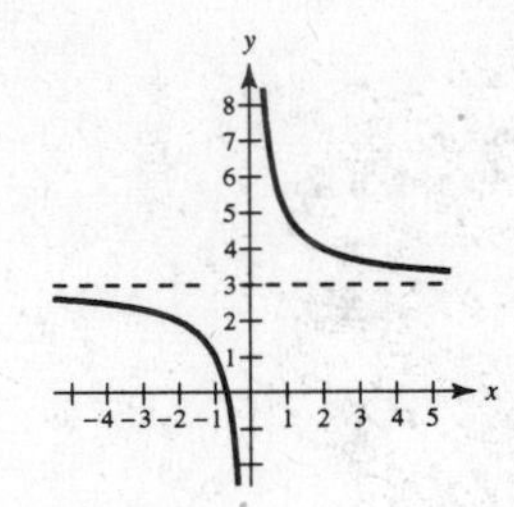

69. $y = \dfrac{x^3}{\sqrt{x^2 - 4}}$

Domain: $(-\infty, -2), (2, \infty)$

Intercepts: none

Symmetry: origin

Horizontal asymptote: none

Vertical asymptotes: $x = \pm 2$ (discontinuities)

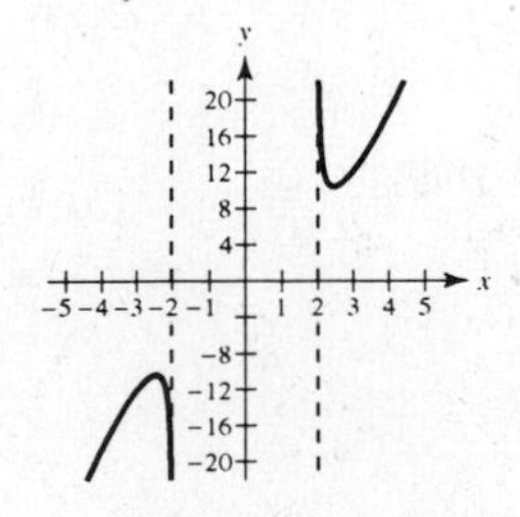

71. $f(x) = 5 - \dfrac{1}{x^2} = \dfrac{5x^2 - 1}{x^2}$

Domain: $(-\infty, 0), (0, \infty)$

$f'(x) = \dfrac{2}{x^3} \Rightarrow$ No relative extrema

$f''(x) = -\dfrac{6}{x^4} \Rightarrow$ No points of inflection

Vertical asymptote: $x = 0$

Horizontal asymptote: $y = 5$

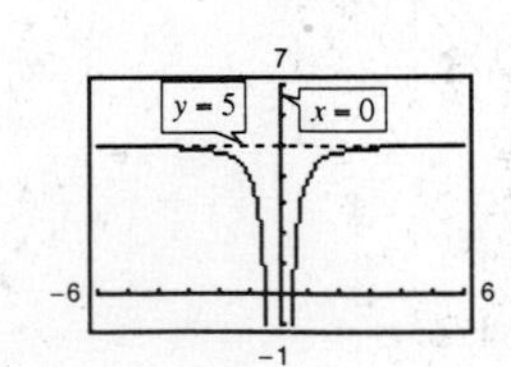

73. $f(x) = \dfrac{x}{x^2 - 4}$

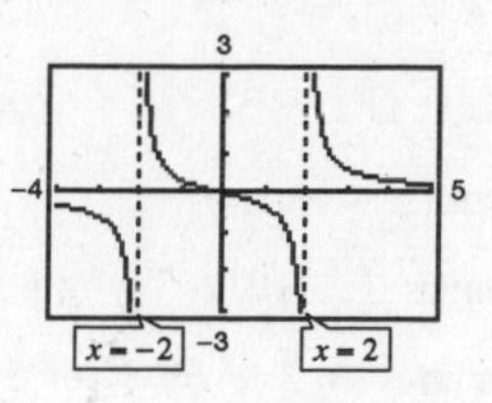

$$f'(x) = \frac{(x^2 - 4) - x(2x)}{(x^2 - 4)^2}$$

$$= \frac{-(x^2 + 4)}{(x^2 - 4)^2} \neq 0 \text{ for any } x \text{ in the domain of } f.$$

$$f''(x) = \frac{(x^2 - 4)^2(-2x) + (x^2 + 4)(2)(x^2 - 4)(2x)}{(x^2 - 4)^2}$$

$$= \frac{2x(x^2 + 12)}{(x^2 - 4)^3} = 0 \text{ when } x = 0.$$

Since $f''(x) > 0$ on $(-2, 0)$ and $f''(x) < 0$ on $(0, 2)$, then $(0, 0)$ is a point of inflection.

Vertical asymptotes: $x = \pm 2$

Horizontal asymptote: $y = 0$

75. $f(x) = \dfrac{x - 2}{x^2 - 4x + 3} = \dfrac{x - 2}{(x - 1)(x - 3)}$

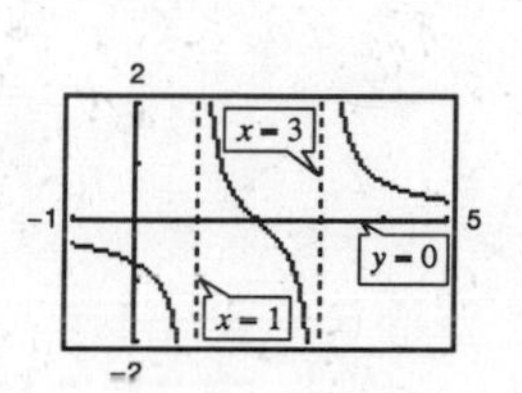

$$f'(x) = \frac{(x^2 - 4x + 3) - (x - 2)(2x - 4)}{(x^2 - 4x + 3)^2} = \frac{-x^2 + 4x - 5}{(x^2 - 4x + 3)^2} \neq 0$$

$$f''(x) = \frac{(x^2 - 4x + 3)^2(-2x + 4) - (-x^2 + 4x - 5)(2)(x^2 - 4x + 3)(2x - 4)}{(x^2 - 4x + 3)^4}$$

$$= \frac{2(x^3 - 6x^2 + 15x - 14)}{(x^2 - 4x + 3)^3} = \frac{2(x - 2)(x^2 - 4x + 7)}{(x^2 - 4x + 3)^3} = 0 \text{ when } x = 2.$$

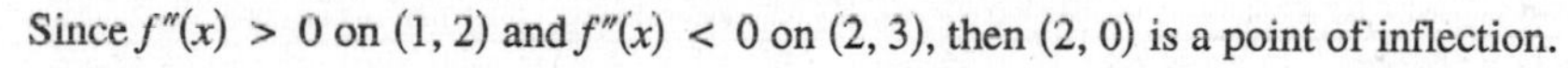

Since $f''(x) > 0$ on $(1, 2)$ and $f''(x) < 0$ on $(2, 3)$, then $(2, 0)$ is a point of inflection.

Vertical asymptotes: $x = 1, x = 3$

Horizontal asymptote: $y = 0$

77. $f(x) = \dfrac{3x}{\sqrt{4x^2 + 1}}$

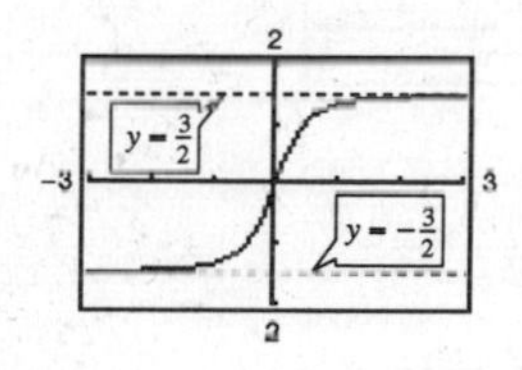

$$f'(x) = \frac{3}{(4x^2 + 1)^{3/2}} \Rightarrow \text{No relative extrema}$$

$$f''(x) = \frac{-36x}{(4x^2 + 1)^{5/2}} = 0 \text{ when } x = 0.$$

Point of inflection: $(0, 0)$

Horizontal asymptotes: $y = \pm\dfrac{3}{2}$

No vertical asymptotes

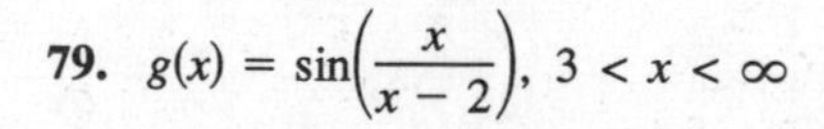

79. $g(x) = \sin\left(\dfrac{x}{x - 2}\right),\ 3 < x < \infty$

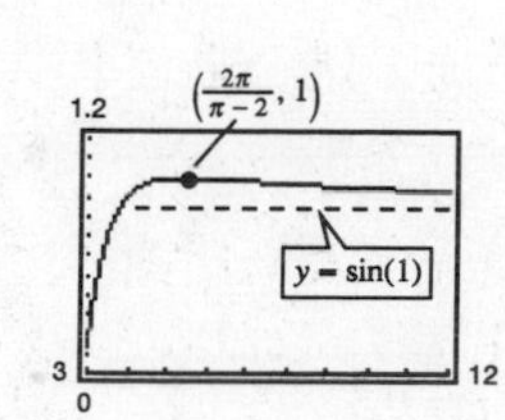

$$g'(x) = \frac{-2\cos\left(\dfrac{x}{x - 2}\right)}{(x - 2)^2}$$

Horizontal asymptote: $y = \sin(1)$

Relative maximum: $\dfrac{x}{x - 2} = \dfrac{\pi}{2} \Rightarrow x = \dfrac{2\pi}{\pi - 2} \approx 5.5039$

No vertical asymptotes

81. $f(x) = 2 + (x^2 - 3)e^{-x}$

$f'(x) = -e^{-x}(x + 1)(x - 3)$

Critical numbers: $x = -1, x = 3$

Relative minimum: $(-1, 2 - 2e) \approx (-1, -3.4366)$

Relative maximum: $(3, 2 + 6e^{-3}) \approx (3, 2.2987)$

Horizontal asymptote: $y = 2$

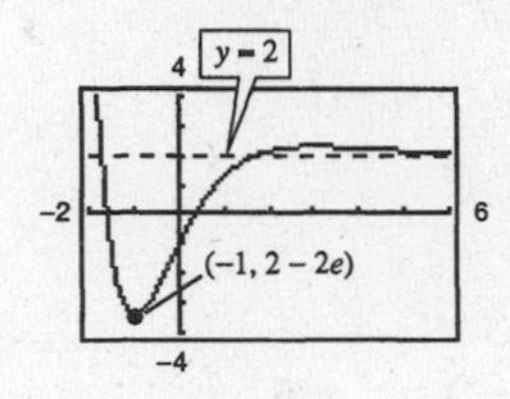

83. (a)

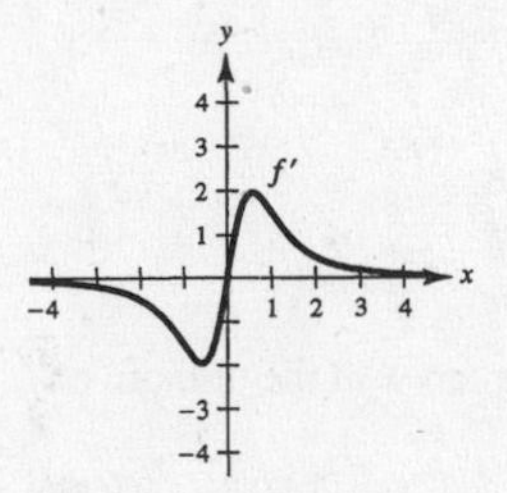

(b) $\lim_{x \to \infty} f(x) = 3$ $\quad \lim_{x \to \infty} f'(x) = 0$

(c) Since $\lim_{x \to \infty} f(x) = 3$, the graph approaches that of a horizontal line, $\lim_{x \to \infty} f'(x) = 0$.

85. Yes. For example, let $f(x) = \dfrac{6|x - 2|}{\sqrt{(x - 2)^2 + 1}}$.

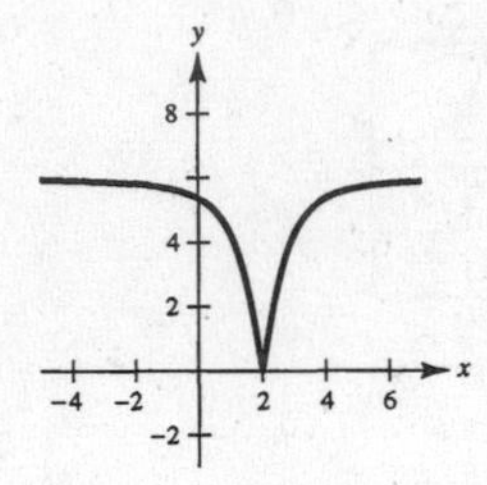

87. $f(x) = \dfrac{x^3 - 3x^2 + 2}{x(x - 3)}, g(x) = x + \dfrac{2}{x(x - 3)}$

(a)

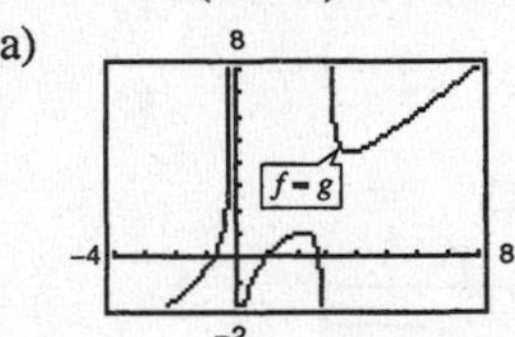

(b) $f(x) = \dfrac{x^3 - 3x^2 + 2}{x(x - 3)}$

$= \dfrac{x^2(x - 3)}{x(x - 3)} + \dfrac{2}{x(x - 3)}$

$= x + \dfrac{2}{x(x - 3)} = g(x)$

(c)

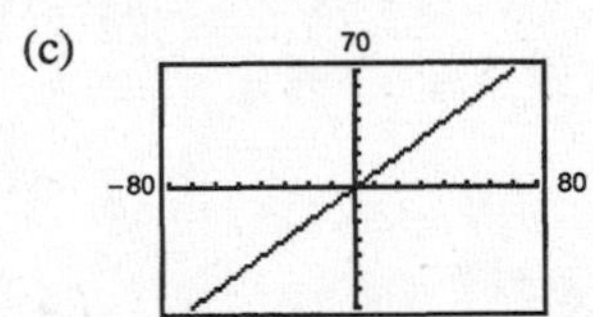

The graph appears as the slant asymptote $y = x$.

89. $C = 0.5x + 500$

$\overline{C} = \dfrac{C}{x}$

$\overline{C} = 0.5 + \dfrac{500}{x}$

$\lim_{x \to \infty} \left(0.5 + \dfrac{500}{x}\right) = 0.5$

91. line: $mx - y + 4 = 0$

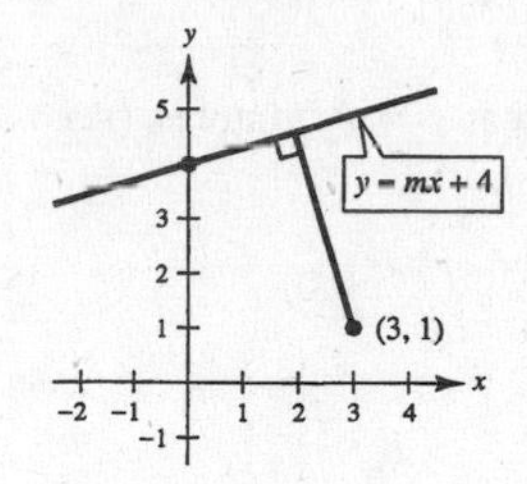

(a) $d = \dfrac{|Ax_1 + By_1 + C|}{\sqrt{A^2 + B^2}} = \dfrac{|m(3) - 1(1) + 4|}{\sqrt{m^2 + 1}}$

$= \dfrac{|3m + 3|}{\sqrt{m^2 + 1}}$

(b)

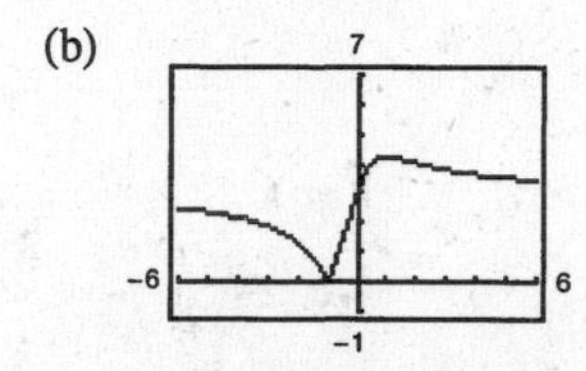

(c) $\lim_{m \to \infty} d(m) = 3 = \lim_{m \to -\infty} d(m)$

The line approaches the vertical line $x = 0$. Hence, the distance approaches 3.

93. (a) $\lim_{n \to \infty} \dfrac{0.83}{1 + e^{-0.2n}} = 0.83 = 83\%$

(b) $P' = \dfrac{0.166e^{-0.2n}}{(1 + e^{-0.2n})^2}$

$P'(3) \approx 0.038$

$P'(10) \approx 0.017$

95. (a) $T_1(t) = -0.003t^2 + 0.677t + 26.564$

(b)

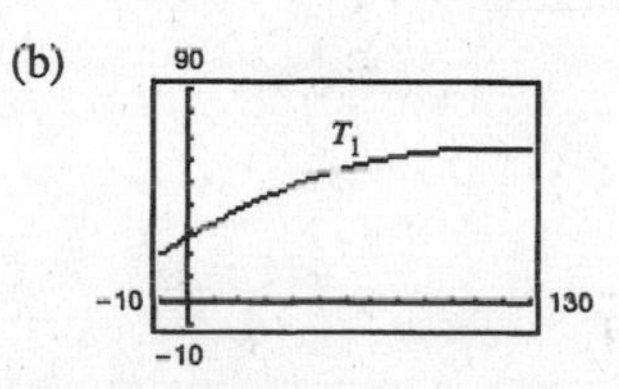

(c)

$T_2 = \dfrac{1451 + 86t}{58 + t}$

(d) $T_1(0) \approx 26.6$

$T_2(0) \approx 25.0$

(e) $\lim_{t \to \infty} T_2 = \dfrac{86}{1} = 86$

(f) The limiting temperature is 86. T_1 has no horizontal asymptote.

97. Answers will vary.

99. False. Let $f(x) = \dfrac{2x}{\sqrt{x^2 + 2}}$.

Section 3.6 A Summary of Curve Sketching

1. f has constant negative slope. Matches (D)

3. The slope is periodic, and zero at $x = 0$. Matches (A)

5. (a) $f'(x) = 0$ for $x = -2$ and $x = 2$

f' is negative for $-2 < x < 2$ (decreasing function).

f' is positive for $x > 2$ and $x < -2$ (increasing function).

(b) $f''(x) = 0$ at $x = 0$ (Inflection point).

f'' is positive for $x > 0$ (Concave upwards).

f'' is negative for $x < 0$ (Concave downward).

(c) f' is increasing on $(0, \infty)$. $(f'' > 0)$

(d) $f'(x)$ is minimum at $x = 0$. The rate of change of f at $x = 0$ is less than the rate of change of f for all other values of x.

7. $y = \dfrac{x^2}{x^2 + 3}$

$y' = \dfrac{6x}{(x^2 + 3)^2} = 0$ when $x = 0$.

$y'' = \dfrac{18(1 - x^2)}{(x^2 + 3)^3} = 0$ when $x = \pm 1$.

Horizontal asymptote: $y = 1$

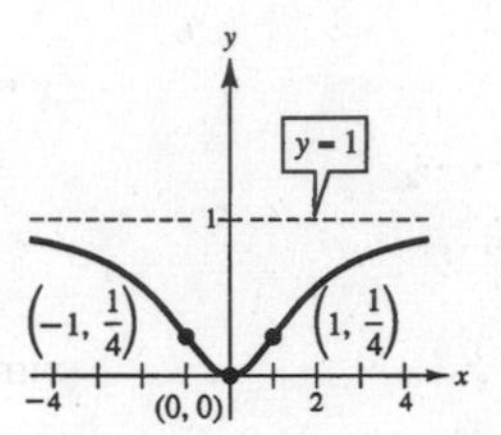

	y	y'	y''	Conclusion
$-\infty < x < -1$		$-$	$-$	Decreasing, concave down
$x = -1$	$\frac{1}{4}$	$-$	0	Point of inflection
$-1 < x < 0$		$-$	$+$	Decreasing, concave up
$x = 0$	0	0	$+$	Relative minimum
$0 < x < 1$		$+$	$+$	Increasing, concave up
$x = 1$	$\frac{1}{4}$	$+$	0	Point of inflection
$1 < x < \infty$		$+$	$-$	Increasing, concave down

9. $y = \dfrac{1}{x - 2} - 3$

$y' = -\dfrac{1}{(x - 2)^2} < 0$ when $x \neq 2$.

$y'' = \dfrac{2}{(x - 2)^3}$

No relative extrema, no points of inflection

Intercepts: $\left(\frac{7}{3}, 0\right), \left(0, -\frac{7}{2}\right)$

Vertical asymptote: $x = 2$

Horizontal asymptote: $y = -3$

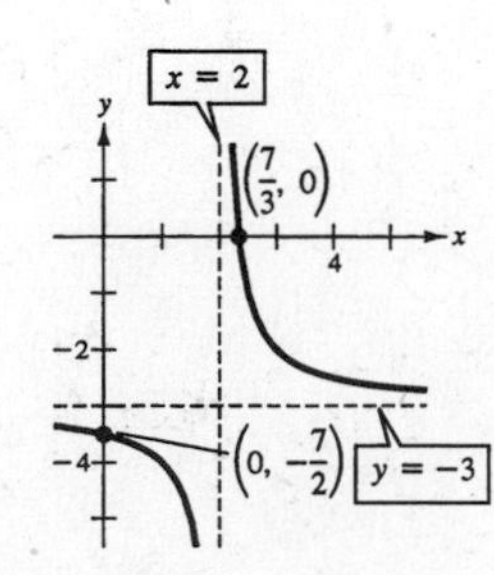

11. $y = \dfrac{2x}{x^2 - 1}$

$y' = \dfrac{-2(x^2 + 1)}{(x^2 - 1)^2} < 0$ if $x \neq \pm 1$.

$y'' = \dfrac{4x(x^2 + 3)}{(x^2 - 1)^3} = 0$ if $x = 0$.

Inflection point: $(0, 0)$

Intercept: $(0, 0)$

Vertical asymptote: $x = \pm 1$

Horizontal asymptote: $y = 0$

Symmetry with respect to the origin

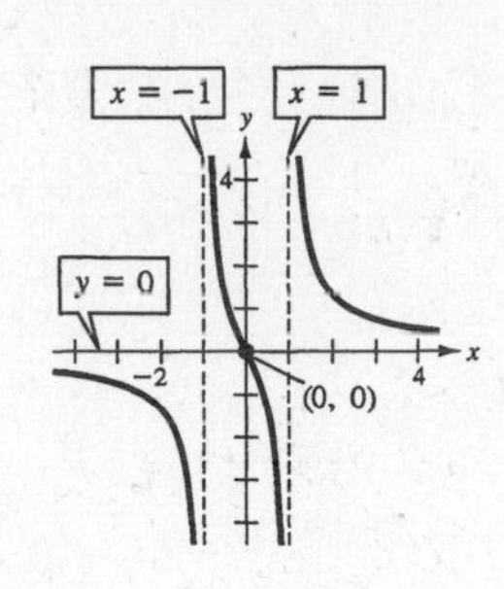

13. $g(x) = x + \dfrac{4}{x^2 + 1}$

$g'(x) = 1 - \dfrac{8x}{(x^2 + 1)^2} = \dfrac{x^4 + 2x^2 - 8x + 1}{(x^2 + 1)^2} = 0$ when $x \approx 0.1292, 1.6085$

$g''(x) = \dfrac{8(3x^2 - 1)}{(x^2 + 1)^3} = 0$ when $x = \pm\dfrac{\sqrt{3}}{3}$

$g''(0.1292) < 0$, therefore, $(0.1292, 4.064)$ is relative maximum.

$g''(1.6085) > 0$, therefore, $(1.6085, 2.724)$ is a relative minimum.

Points of inflection: $\left(-\dfrac{\sqrt{3}}{3}, 2.423\right), \left(\dfrac{\sqrt{3}}{3}, 3.577\right)$

Intercepts: $(0, 4), (-1.3788, 0)$

Slant asymptote: $y = x$

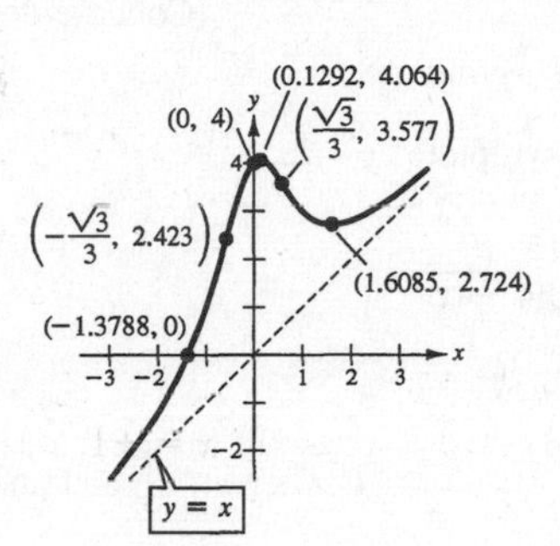

15. $f(x) = \dfrac{x^2 + 1}{x} = x + \dfrac{1}{x}$

$f'(x) = 1 - \dfrac{1}{x^2} = 0$ when $x = \pm 1$.

$f''(x) = \dfrac{2}{x^3} \neq 0$

Relative maximum: $(-1, -2)$

Relative minimum: $(1, 2)$

Vertical asymptote: $x = 0$

Slant asymptote: $y = x$

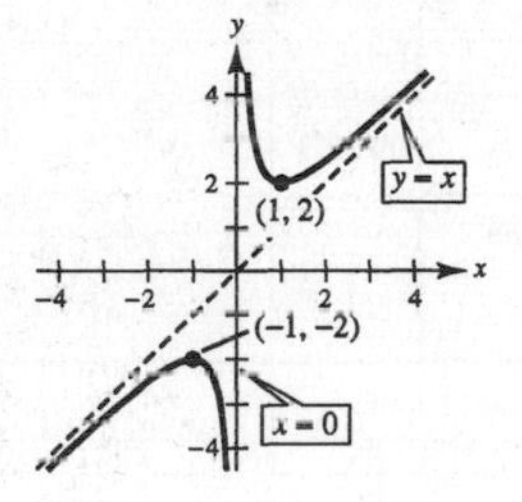

17. $y = \dfrac{x^2 - 6x + 12}{x - 4} = x - 2 + \dfrac{4}{x - 4}$

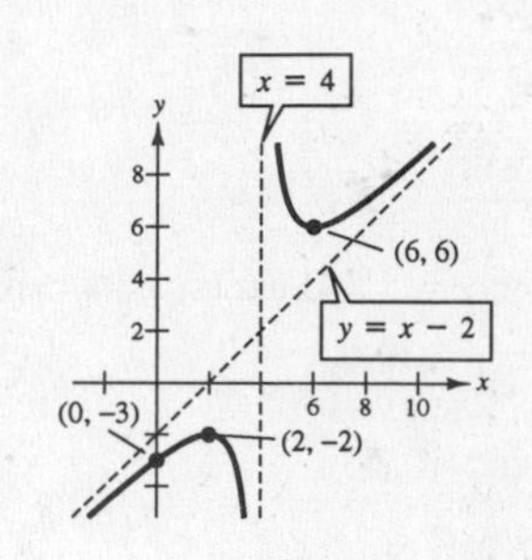

$y' = 1 - \dfrac{4}{(x - 4)^2}$

$= \dfrac{(x - 2)(x - 6)}{(x - 4)^2} = 0$ when $x = 2, 6$.

$y'' = \dfrac{8}{(x - 4)^3}$

$y'' < 0$ when $x = 2$.

Therefore, $(2, -2)$ is a relative maximum.

$y'' > 0$ when $x = 6$.

Therefore, $(6, 6)$ is a relative minimum.

Vertical asymptote: $x = 4$

Slant asymptote: $y = x - 2$

19. 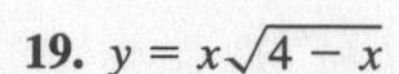 $y = x\sqrt{4 - x}$

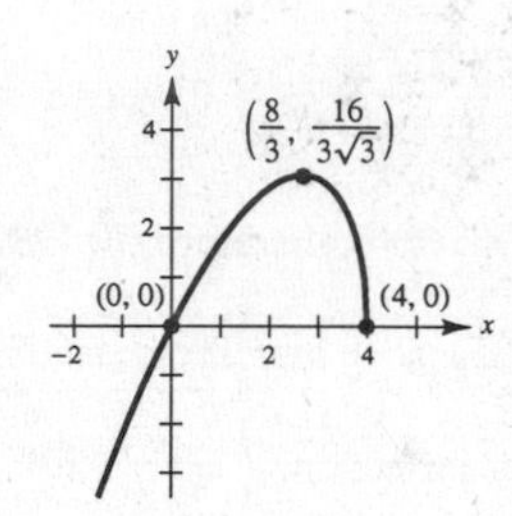

Domain: $(-\infty, 4]$

$y' = \dfrac{8 - 3x}{2\sqrt{4 - x}} = 0$ when $x = \dfrac{8}{3}$ and undefined when $x = 4$.

$y'' = \dfrac{3x - 16}{4(4 - x)^{3/2}} = 0$ when $x = \dfrac{16}{3}$ and undefined when $x = 4$.

Note: $x = \frac{16}{3}$ is not in the domain.

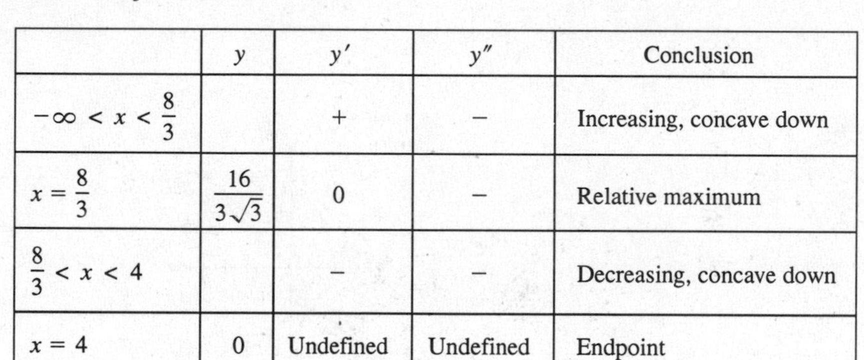

	y	y'	y''	Conclusion
$-\infty < x < \frac{8}{3}$		$+$	$-$	Increasing, concave down
$x = \frac{8}{3}$	$\frac{16}{3\sqrt{3}}$	0	$-$	Relative maximum
$\frac{8}{3} < x < 4$		$-$	$-$	Decreasing, concave down
$x = 4$	0	Undefined	Undefined	Endpoint

21. $h(x) = x\sqrt{9 - x^2}$ Domain: $-3 \le x \le 3$

$h'(x) = \dfrac{9 - 2x^2}{\sqrt{9 - x^2}} = 0$ when $x = \pm\dfrac{3}{\sqrt{2}} = \pm\dfrac{3\sqrt{2}}{2}$.

$h''(x) = \dfrac{x(2x^2 - 27)}{(9 - x^2)^{3/2}} = 0$ when $x = 0$.

Relative maximum: $\left(\dfrac{3\sqrt{2}}{2}, \dfrac{9}{2}\right)$

Relative minimum: $\left(-\dfrac{3\sqrt{2}}{2}, -\dfrac{9}{2}\right)$

Intercepts: $(0, 0)$, $(\pm 3, 0)$

Symmetric with respect to the origin

Point of inflection: $(0, 0)$

23. $y = 3x^{2/3} - 2x$

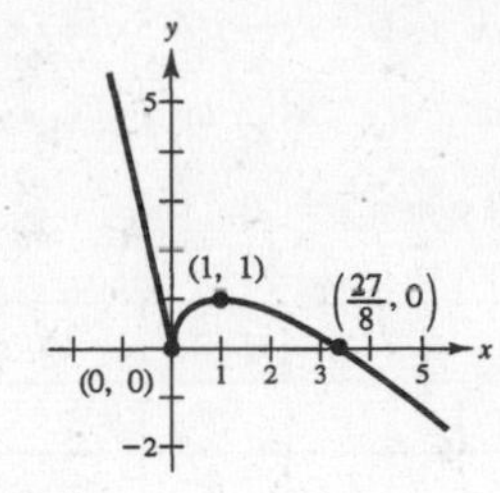

$$y' = 2x^{-1/3} - 2 = \frac{2(1 - x^{1/3})}{x^{1/3}}$$

$= 0$ when $x = 1$ and undefined when $x = 0$.

$$y'' = \frac{-2}{3x^{4/3}} < 0 \text{ when } x \neq 0.$$

	y	y'	y''	Conclusion
$-\infty < x < 0$		$-$	$-$	Decreasing, concave down
$x = 0$	0	Undefined	Undefined	Relative minimum
$0 < x < 1$		$+$	$-$	Increasing, concave down
$x = 1$	1	0	$-$	Relative maximum
$1 < x < \infty$		$-$	$-$	Decreasing, concave down

25. $y = x^3 - 3x^2 + 3$

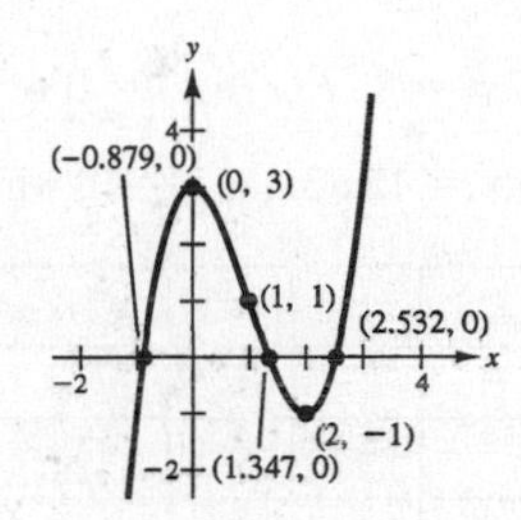

$y' = 3x^2 - 6x = 3x(x - 2) = 0$ when $x = 0, x = 2$.

$y'' = 6x - 6 = 6(x - 1) = 0$ when $x = 1$.

	y	y'	y''	Conclusion
$-\infty < x < 0$		$+$	$-$	Increasing, concave down
$x = 0$	3	0	$-$	Relative maximum
$0 < x < 1$		$-$	$-$	Decreasing, concave down
$x = 1$	1	$-$	0	Point of inflection
$1 < x < 2$		$-$	$+$	Decreasing, concave up
$x = 2$	-1	0	$+$	Relative minimum
$2 < x < \infty$		$+$	$+$	Increasing, concave up

27. $y = 2 - x - x^3$

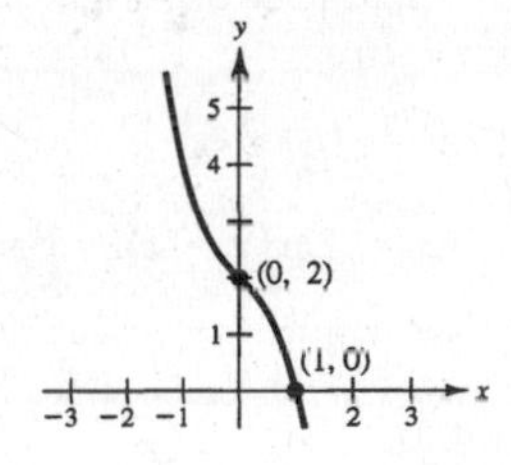

$y' = -1 - 3x^2$

No critical numbers

$y'' = -6x = 0$ when $x = 0$.

	y	y'	y''	Conclusion
$-\infty < x < 0$		$-$	$+$	Decreasing, concave up
$x = 0$	2	$-$	0	Point of inflection
$0 < x < \infty$		$-$	$-$	Decreasing, concave down

29. $f(x) = 3x^3 - 9x + 1$

$f'(x) = 9x^2 - 9 = 9(x^2 - 1) = 0$ when $x = \pm 1$.

$f''(x) = 18x = 0$ when $x = 0$

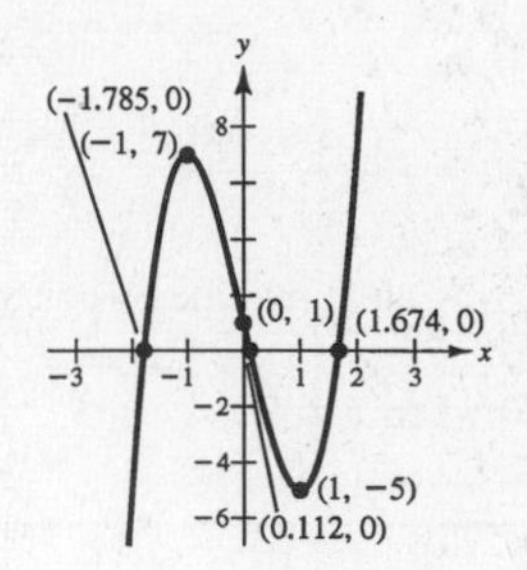

	$f(x)$	$f'(x)$	$f''(x)$	Conclusion
$-\infty < x < -1$		$+$	$-$	Increasing, concave down
$x = -1$	7	0	$-$	Relative maximum
$-1 < x < 0$		$-$	$-$	Decreasing, concave down
$x = 0$	1	$-$	0	Point of inflection
$0 < x < 1$		$-$	$+$	Decreasing, concave up
$x = 1$	-5	0	$+$	Relative minimum
$1 < x < \infty$		$+$	$+$	Increasing, concave up

31. $y = 3x^4 + 4x^3$

$y' = 12x^3 + 12x^2 = 12x^2(x + 1) = 0$ when $x = 0, x = -1$.

$y'' = 36x^2 + 24x = 12x(3x + 2) = 0$ when $x = 0, x = -\frac{2}{3}$.

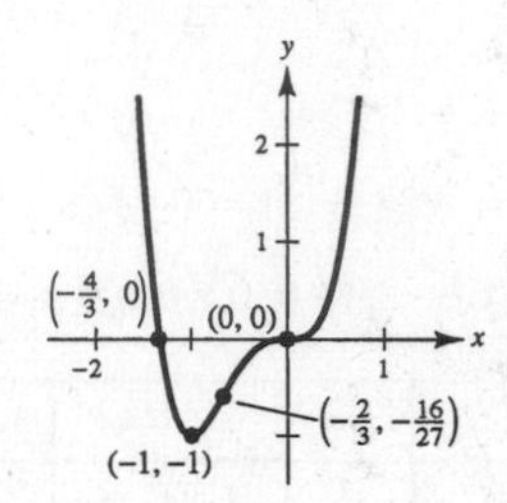

	y	y'	y''	Conclusion
$-\infty < x < -1$		$-$	$+$	Decreasing, concave up
$x = -1$	-1	0	$+$	Relative minimum
$-1 < x < -\frac{2}{3}$		$+$	$+$	Increasing, concave up
$x = -\frac{2}{3}$	$-\frac{16}{27}$	$+$	0	Point of inflection
$-\frac{2}{3} < x < 0$		$+$	$-$	Increasing, concave down
$x = 0$	0	0	0	Point of inflection
$0 < x < \infty$		$+$	$+$	Increasing, concave up

33. $f(x) = x^4 - 4x^3 + 16x$

$f'(x) = 4x^3 - 12x^2 + 16 = 4(x + 1)(x - 2)^2 = 0$ when $x = -1, x = 2$.

$f''(x) = 12x^2 - 24x = 12x(x - 2) = 0$ when $x = 0, x = 2$.

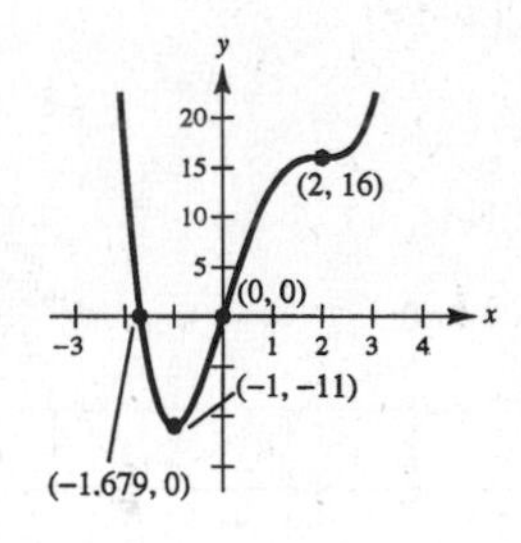

	$f(x)$	$f'(x)$	$f''(x)$	Conclusion
$-\infty < x < -1$		$-$	$+$	Decreasing, concave up
$x = -1$	-11	0	$+$	Relative minimum
$-1 < x < 0$		$+$	$+$	Increasing, concave up
$x = 0$	0	$+$	0	Point of inflection
$0 < x < 2$		$+$	$-$	Increasing, concave down
$x = 2$	16	0	0	Point of inflection
$2 < x < \infty$		$+$	$+$	Increasing, concave up

35. $y = x^5 - 5x$

$y' = 5x^4 - 5 = 5(x^4 - 1) = 0$ when $x = \pm 1$.

$y'' = 20x^3 = 0$ when $x = 0$.

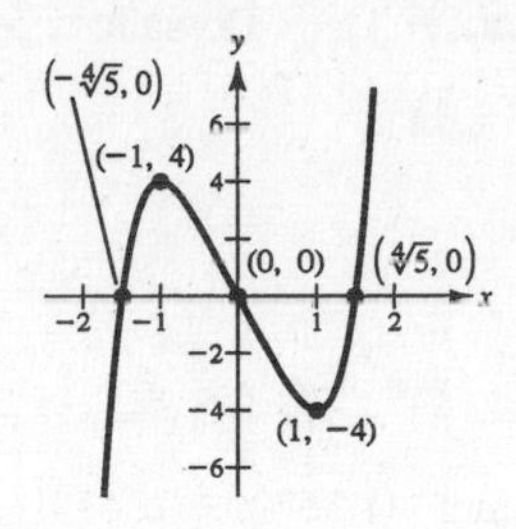

	y	y'	y''	Conclusion
$-\infty < x < -1$		$+$	$-$	Increasing, concave down
$x = -1$	4	0	$-$	Relative maximum
$-1 < x < 0$		$-$	$-$	Decreasing, concave down
$x = 0$	0	$-$	0	Point of inflection
$0 < x < 1$		$-$	$+$	Decreasing, concave up
$x = 1$	-4	0	$+$	Relative minimum
$1 < x < \infty$		$+$	$+$	Increasing, concave up

37. $y = |2x - 3|$

$y' = \dfrac{2(2x - 3)}{|2x - 3|}$ undefined at $x = \dfrac{3}{2}$.

$y'' = 0$

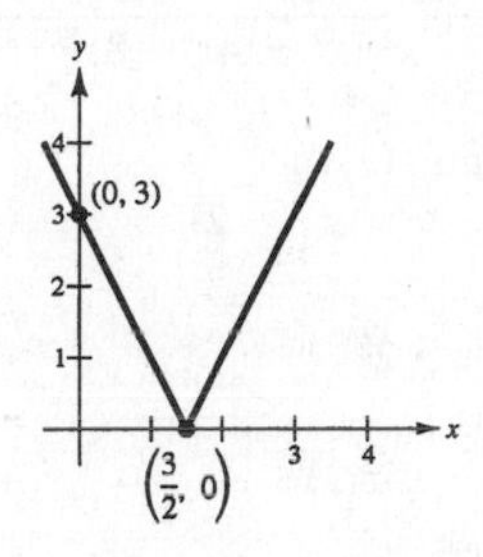

	y	y'	Conclusion
$-\infty < x < \frac{3}{2}$		$-$	Decreasing
$x = \frac{3}{2}$	0	Undefined	Relative minimum
$\frac{3}{2} < x < \infty$		$+$	Increasing

39. $f(x) = e^{3x}(2 - x)$

$f'(x) = -e^{3x} + 3(2 - x)e^{3x} = e^{3x}(5 - 3x)$

Critical point: $\left(\frac{5}{3}, 49.47\right)$

$f''(x) = -3e^{3x}(-4 + 3x) = 0$ when $x = \frac{4}{3}$.

Relative maximum: $\left(\frac{5}{3}, 49.47\right)$

Inflection point: $\left(\frac{4}{3}, 36.40\right)$

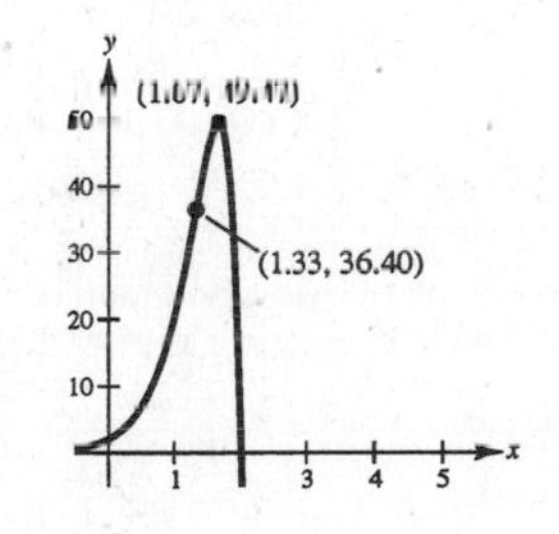

41. $g(t) = \dfrac{10}{1 + 4e^{-t}}$

$g'(t) = \dfrac{40e^{-t}}{(1 + 4e^{-t})^2} > 0$ for all t.

$g''(t) = \dfrac{40e^{-t}(4e^{-t} - 1)}{(1 + 4e^{-t})^3} = 0$ at $x \approx 1.3863$.

$\lim_{t \to \infty} g(t) = 10$

$\lim_{t \to -\infty} g(t) = 0$

Inflection point: $(1.3863, 5)$

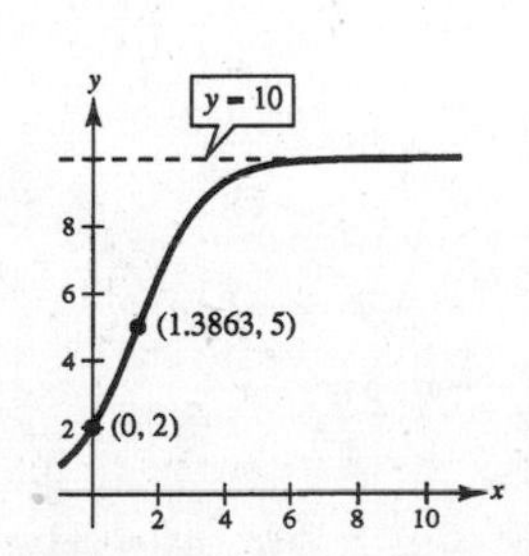

43. $y = (x - 1)\ln(x - 1)$, Domain: $x > 1$

$y' = 1 + \ln(x - 1)$

$y'' = \dfrac{1}{x - 1}$

Critical number:

$\ln(x - 1) = -1 \Rightarrow (x - 1) = e^{-1} \Rightarrow x = 1 + e^{-1}$

Relative minimum: $(1.3679, -0.3679)$

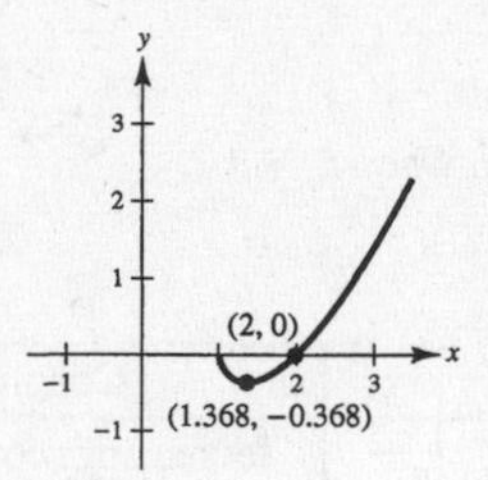

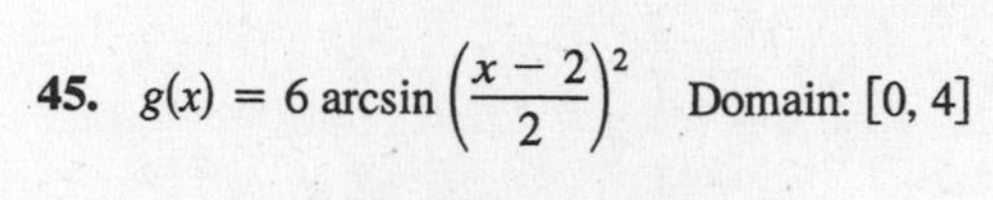

45. $g(x) = 6 \arcsin\left(\dfrac{x - 2}{2}\right)^2$ Domain: $[0, 4]$

$g'(x) = \dfrac{12(x - 2)}{\sqrt{(4x - x^2)(x^2 - 4x + 8)}}$

$g''(x) = \dfrac{12(x^4 - 8x^3 + 24x^2 - 32x + 32)}{[(4x - x^2)(x^2 - 4x + 8)]^{3/2}}$

Relative minimum: $(2, 0)$

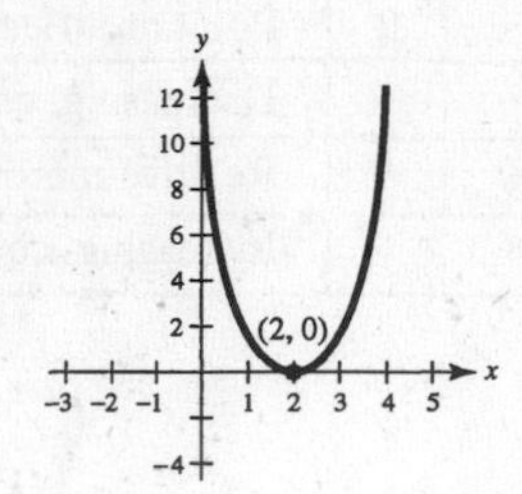

47. $f(x) = \dfrac{x}{2^{x-2}} = x \cdot 2^{2-x}$

$f'(x) = 2^{2-x}(1 - x\ln 2)$

$f''(x) = 2^{2-x}\ln 2(x\ln 2 - 2)$

Relative maximum: $\left(\dfrac{1}{\ln 2}, 2.1230\right) \approx (1.4427, 2.1230)$

Inflection point: $\left(\dfrac{2}{\ln 2}, 1.5620\right)$

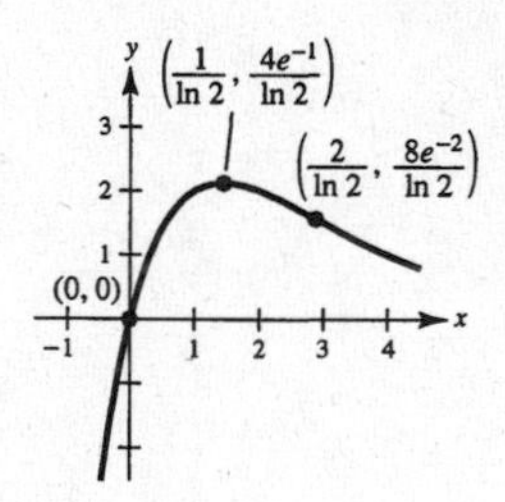

49. $g(x) = \log_4(x - x^2) = \dfrac{\ln(x - x^2)}{\ln 4}$, Domain: $0 < x < 1$

$g'(x) = \dfrac{2x - 1}{\ln 4 \cdot x(x - 1)}$

$g''(x) = \dfrac{-2x^2 + 2x - 1}{\ln 4 \cdot x^2(x - 1)^2}$

Relative maximum: $\left(\dfrac{1}{2}, -1\right)$

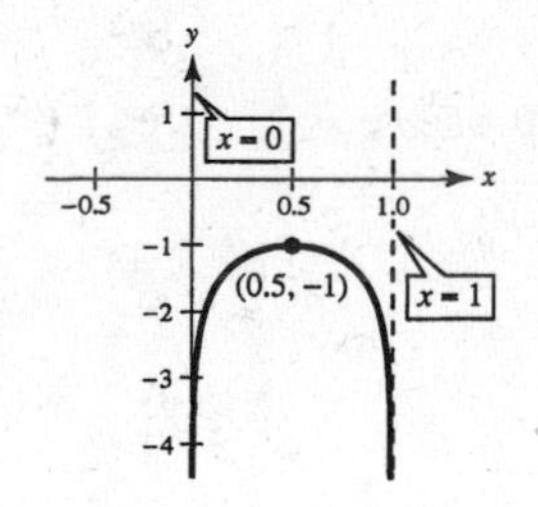

51. $y = \sin x - \dfrac{1}{18}\sin 3x, 0 \le x \le 2\pi$

$y' = \cos x - \dfrac{1}{6}\cos 3x = 0$ when $x = \dfrac{\pi}{2}, \dfrac{3\pi}{2}$.

$y'' = -\sin x + \dfrac{1}{2}\sin 3x = 0$ when $x = 0, \dfrac{\pi}{6}, \dfrac{5\pi}{6}, \pi, \dfrac{7\pi}{6}, \dfrac{11\pi}{6}$.

Relative maximum: $\left(\dfrac{\pi}{2}, \dfrac{19}{18}\right)$

Relative minimum: $\left(\dfrac{3\pi}{2}, -\dfrac{19}{18}\right)$

Inflection points: $\left(\dfrac{\pi}{6}, \dfrac{4}{9}\right), \left(\dfrac{5\pi}{6}, \dfrac{4}{9}\right), (\pi, 0), \left(\dfrac{7\pi}{6}, -\dfrac{4}{9}\right), \left(\dfrac{11\pi}{6}, -\dfrac{4}{9}\right)$

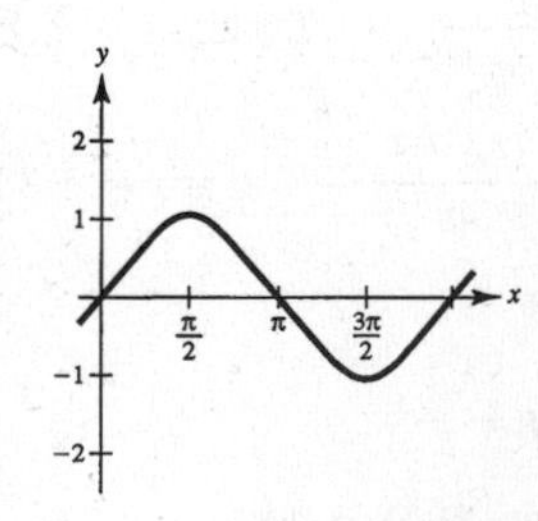

53. $y = 2x - \tan x, \; -\dfrac{\pi}{2} < x < \dfrac{\pi}{2}$

$y' = 2 - \sec^2 x = 0$ when $x = \pm\dfrac{\pi}{4}$.

$y'' = -2\sec^2 x \tan x = 0$ when $x = 0$.

Relative maximum: $\left(\dfrac{\pi}{4}, \dfrac{\pi}{2} - 1\right)$

Relative minimum: $\left(-\dfrac{\pi}{4}, 1 - \dfrac{\pi}{2}\right)$

Inflection point: $(0, 0)$

Vertical asymptotes: $x = \pm\dfrac{\pi}{2}$

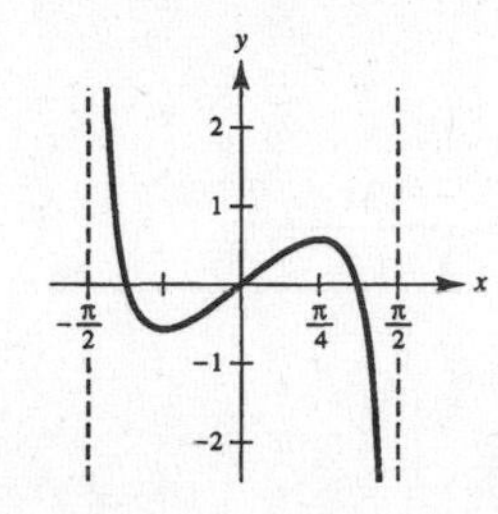

55. $y = 2(\csc x + \sec x), \; 0 < x < \dfrac{\pi}{2}$

$y' = 2(\sec x \tan x - \csc x \cot x) = 0 \Rightarrow x = \pi/4$

Relative minimum: $\left(\dfrac{\pi}{4}, 4\sqrt{2}\right)$

Vertical asymptotes: $x = 0, x = \dfrac{\pi}{2}$

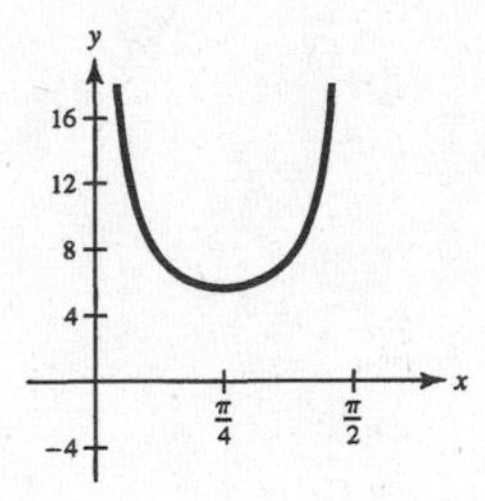

57. $g(x) = x \tan x, \; -\dfrac{3\pi}{2} < x < \dfrac{3\pi}{2}$

$$g'(x) = \frac{x + \sin x \cos x}{\cos^2 x} = 0 \text{ when } x = 0.$$

$$g''(x) = \frac{2(\cos x + x \sin x)}{\cos^3 x}$$

Vertical asymptotes: $x = -\dfrac{3\pi}{2}, -\dfrac{\pi}{2}, \dfrac{\pi}{2}, \dfrac{3\pi}{2}$

Intercepts: $(-\pi, 0), (0, 0), (\pi, 0)$

Symmetric with respect to y-axis.

Increasing on $\left(0, \dfrac{\pi}{2}\right)$ and $\left(\dfrac{\pi}{2}, \dfrac{3\pi}{2}\right)$

Points of inflection: $(\pm 2.80, 0)$

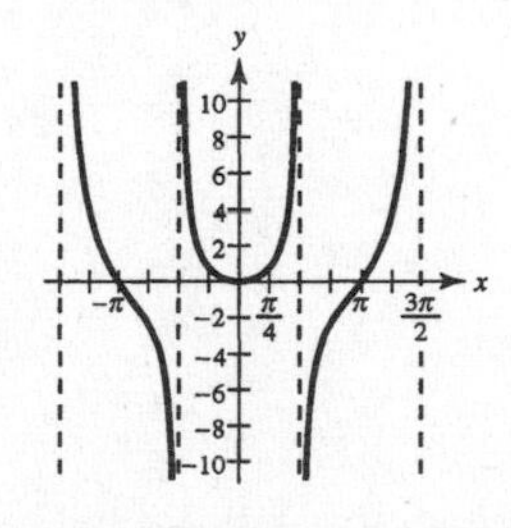

59. $f(x) = \dfrac{20x}{x^2 + 1} - \dfrac{1}{x} = \dfrac{19x^2 - 1}{x(x^2 + 1)}$

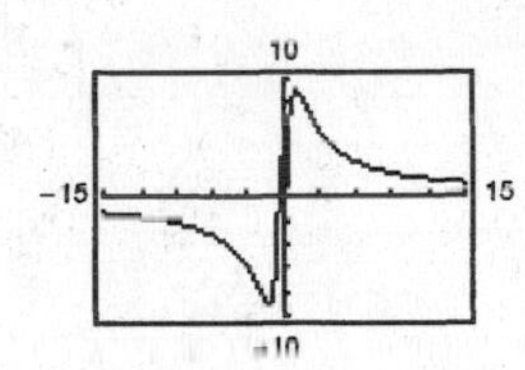

$x = 0$ vertical asymptote

$y = 0$ horizontal asymptote

Minimum: $(-1.10, -9.05)$

Maximum: $(1.10, 9.05)$

Points of inflection: $(-1.84, -7.86), (1.84, 7.86)$

61. $y = \dfrac{x}{\sqrt{x^2 + 7}}$

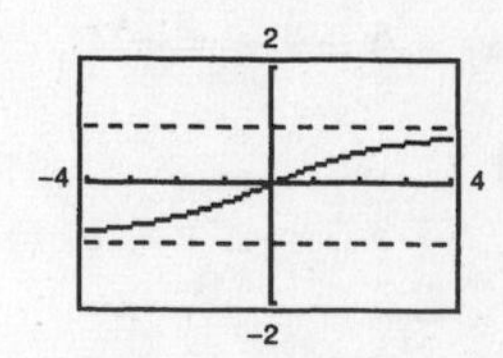

(0, 0) point of inflection

$y = \pm 1$ horizontal asymptotes

63. 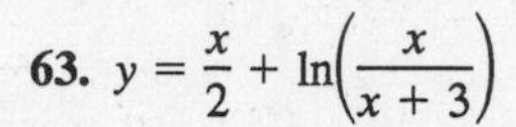 $y = \dfrac{x}{2} + \ln\left(\dfrac{x}{x+3}\right)$

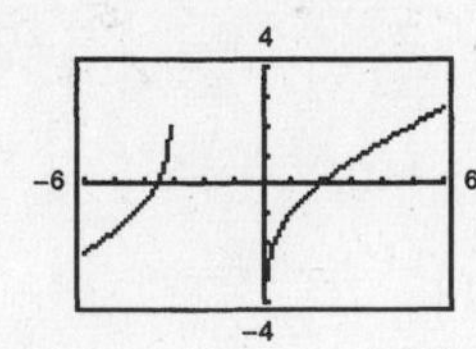

Vertical asymptotes: $x = -3, x = 0$

Slant asymptote: $y = \dfrac{x}{2}$

65. Vertical asymptote: $x = 5$

Horizontal asymptote: $y = 0$

$$y = \frac{1}{x - 5}$$

67. Vertical asymptote: $x = 5$

Slant asymptote: $y = 3x + 2$

$$y = 3x + 2 + \frac{1}{x - 5} = \frac{3x^2 - 13x - 9}{x - 5}$$

69. 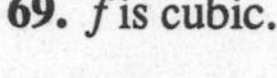 f is cubic.

f' is quadratic.

f'' is linear.

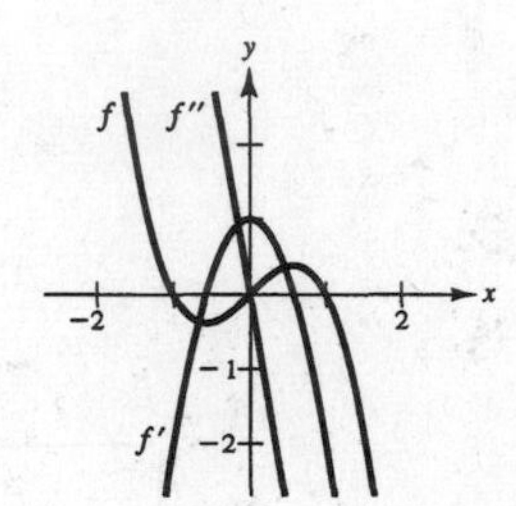

71.

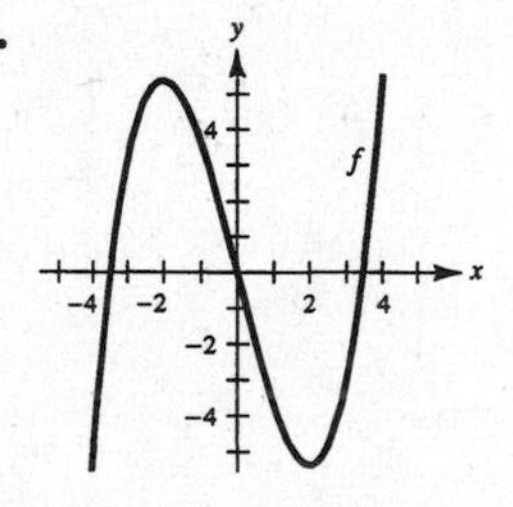

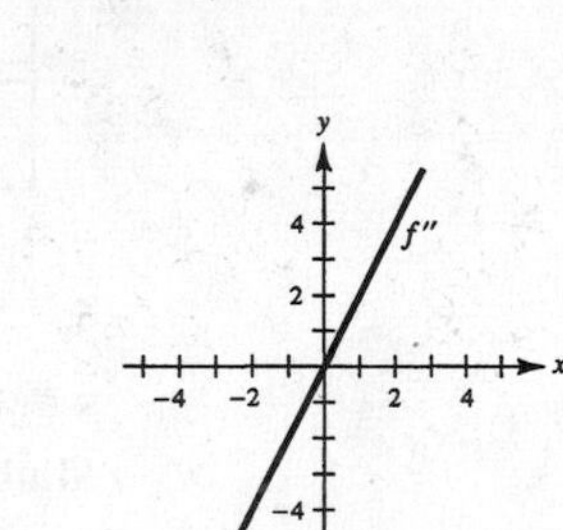

(any vertical translate of f will do)

73.

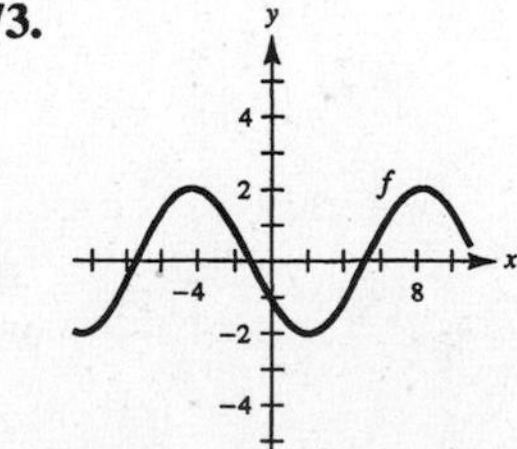

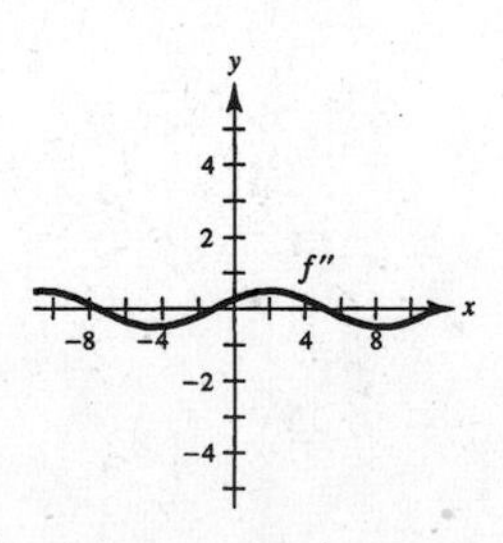

(any vertical translate of f will do)

75. Since the slope is negative, the function is decreasing on (2, 8), and hence $f(3) > f(5)$.

77. $f(x) = \dfrac{4(x-1)^2}{x^2-4x+5}$

Vertical asymptote: none

Horizontal asymptote: $y = 4$

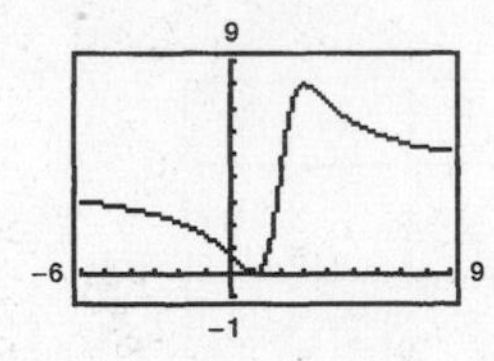

The graph crosses the horizontal asymptote $y = 4$. If a function has a vertical asymptote at $x = c$, the graph would not cross it since $f(c)$ is undefined.

79. $h(x) = \dfrac{6-2x}{3-x}$

$= \dfrac{2(3-x)}{3-x} = \begin{cases} 2, & \text{if } x \neq 3 \\ \text{Undefined}, & \text{if } x = 3 \end{cases}$

The rational function is not reduced to lowest terms.

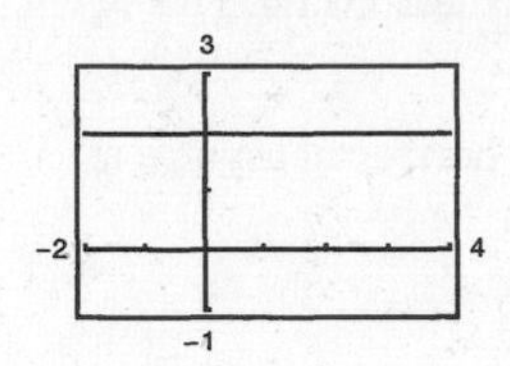

hole at $(3, 2)$

81. $f(x) = -\dfrac{x^2-3x-1}{x-2} = -x + 1 + \dfrac{3}{x-2}$

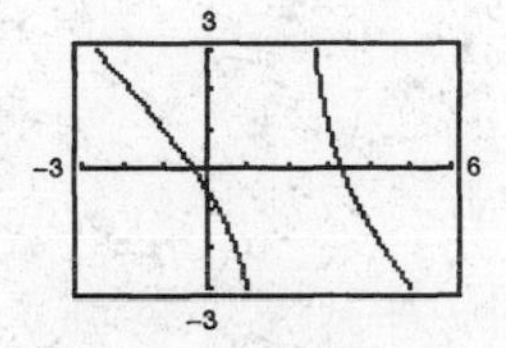

The graph appears to approach the slant asymptote $y = -x + 1$.

83. $f(x) = \dfrac{\cos^2 \pi x}{\sqrt{x^2+1}}, \ (0, 4)$

(a)

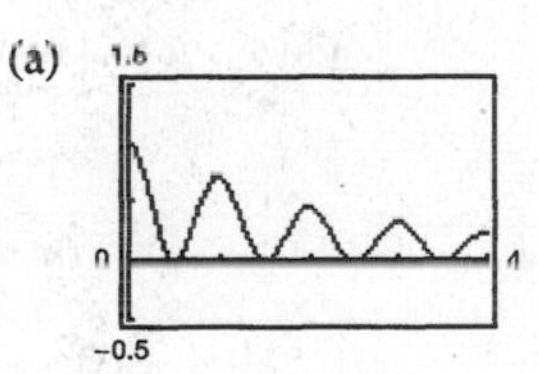

On $(0, 4)$ there seem to be 7 critical numbers:

0.5, 1.0, 1.5, 2.0, 2.5, 3.0, 3.5

(b) $f'(x) = \dfrac{-\cos \pi x(x \cos \pi x + 2\pi(x^2+1)\sin \pi x)}{(x^2+1)^{3/2}} = 0$

Critical numbers $\approx \dfrac{1}{2}, 0.97, \dfrac{3}{2}, 1.98, \dfrac{5}{2}, 2.98, \dfrac{7}{2}$.

The critical numbers where maxima occur appear to be integers in part (a), but approximating them using f' shows that they are not integers.

85. $f(x) = \dfrac{ax}{(x-b)^2}$

(a) The graph has a vertical asymptote at $x = b$. If $a > 0$, the graph approaches ∞ as $x \to b$. If $a < 0$, the graph approaches $-\infty$ as $x \to b$. The graph approaches its vertical asymptote faster as $|a| \to 0$.

(b) As b varies, the position of the vertical asymptote changes: $x = b$. Also, the coordinates of the minimum $(a > 0)$ or maximum $(a < 0)$ are changed.

87. $g(x) = \ln f(x), \ f(x) > 0$

$g'(x) = \dfrac{f'(x)}{f(x)}$

(a) Yes. If the graph of g is increasing, then $g'(x) > 0$. Since $f(x) > 0$, you know that $f'(x) = g'(x)f(x)$ and, thus, $f'(x) > 0$. Therefore, the graph of f is increasing.

(b) No. Let $f(x) = x^2 + 1$ (positive and concave up). $g(x) = \ln(x^2 + 1)$ is not concave up.

89. $f(x) = \dfrac{3x^n}{x^4 + 1}$

(a) For n even, f is symmetric about the y-axis. For n odd, f is symmetric about the origin.

(b) The x-axis will be the horizontal asymptote if the degree of the numerator is less than 4. That is, $n = 0, 1, 2, 3$.

(c) $n = 4$ gives $y = 3$ as the horizontal asymptote.

(d) There is a slant asymptote $y = 3x$ if $n = 5$:

$$\frac{3x^5}{x^4 + 1} = 3x - \frac{3x}{x^4 + 1}.$$

(e)

n	0	1	2	3	4	5
M	1	2	3	2	1	0
N	2	3	4	5	2	3

91. Tangent line at P: $y - y_0 = f'(x_0)(x - x_0)$

(a) Let $y = 0$: $-y_0 = f'(x_0)(x - x_0)$

$$f'(x_0)x = x_0 f'(x_0) - y_0$$

$$x = x_0 - \frac{y_0}{f'(x_0)} = x_0 - \frac{f(x_0)}{f'(x_0)}$$

x-intercept: $\left(x_0 - \dfrac{f(x_0)}{f'(x_0)}, 0\right)$

(b) Let $x = 0$: $y - y_0 = f'(x_0)(-x_0)$

$$y = y_0 - x_0 f'(x_0)$$

$$y = f(x_0) - x_0 f'(x_0)$$

y-intercept: $(0, f(x_0) - x_0 f'(x_0))$

(c) Normal line: $y - y_0 = -\dfrac{1}{f'(x_0)}(x - x_0)$

Let $y = 0$: $-y_0 = -\dfrac{1}{f'(x_0)}(x - x_0)$

$$-y_0 f'(x_0) = -x + x_0$$

$$x = x_0 + y_0 f'(x_0) = x_0 + f(x_0)f'(x_0)$$

x-intercept: $(x_0 + f(x_0)f'(x_0), 0)$

(d) Let $x = 0$: $y - y_0 = \dfrac{-1}{f'(x_0)}(-x_0)$

$$y = y_0 + \frac{x_0}{f'(x_0)}$$

y-intercept: $\left(0, y_0 + \dfrac{x_0}{f'(x_0)}\right)$

(e) $|BC| = \left|x_0 - \dfrac{f(x_0)}{f'(x_0)} - x_0\right| = \left|\dfrac{f(x_0)}{f'(x_0)}\right|$

(f) $|PC|^2 = y_0^2 + \left(\dfrac{f(x_0)}{f'(x_0)}\right)^2 = \dfrac{f(x_0)^2 f'(x_0)^2 + f(x_0)^2}{f'(x_0)^2}$

$$|PC| = \left|\frac{f(x_0)\sqrt{1 + [f'(x_0)]^2}}{f'(x_0)}\right|$$

(g) $|AB| = |x_0 - (x_0 + f(x_0)f'(x_0))| = |f(x_0)f'(x_0)|$

(h) $|AP|^2 = f(x_0)^2 f'(x_0)^2 + y_0^2$

$$|AP| = |f(x_0)|\sqrt{1 + [f'(x_0)]^2}$$

In Exercises 93-95

$$\mathbf{f(x) = ax^3 + bx^2 + cx + d}$$

$$\mathbf{f'(x) = 3ax^2 + 2bx + c}$$

$$\mathbf{f''(x) = 6ax + 2b}$$

$f'(x) = 0$ when $\mathbf{x = \dfrac{-2b \pm \sqrt{4b^2 - 12ac}}{6a} = \dfrac{-b \pm \sqrt{b^2 - 3ac}}{3a}}$.

Furthermore,

$\displaystyle\lim_{x\to\infty} f(x) = \infty$ **if and only if** $a > 0$.

$\displaystyle\lim_{x\to\infty} f(x) = -\infty$ **if and only if** $a < 0$.

93. Since $\displaystyle\lim_{x\to\infty} f(x) = \infty$, $a > 0$. Also, since there is only one critical point, the discriminant is zero.

$b^2 - 3ac = 0 \Rightarrow b^2 = 3ac$

95. Since $\displaystyle\lim_{x\to\infty} f(x) = \infty$, $a > 0$. Also, since there are two critical points, the discriminant is positive.

$b^2 - 3ac > 0 \Rightarrow b^2 > 3ac$

Section 3.7 Optimization Problems

1. (a)

First Number, x	Second Number	Product, P
10	$110 - 10$	$10(110 - 10) = 1000$
20	$110 - 20$	$20(110 - 20) = 1800$
30	$110 - 30$	$30(110 - 30) = 2400$
40	$110 - 40$	$40(110 - 40) = 2800$
50	$110 - 50$	$50(110 - 50) = 3000$
60	$110 - 60$	$60(110 - 60) = 3000$

(b)

First Number, x	Second Number	Product, P
10	$110 - 10$	$10(110 - 10) = 1000$
20	$110 - 20$	$20(110 - 20) = 1800$
30	$110 - 30$	$30(110 - 30) = 2400$
40	$110 - 40$	$40(110 - 40) = 2800$
50	$110 - 50$	$50(110 - 50) = 3000$
60	$110 - 60$	$60(110 - 60) = 3000$
70	$110 - 70$	$70(110 - 70) = 2800$
80	$110 - 80$	$80(110 - 80) = 2400$
90	$110 - 90$	$90(110 - 90) = 1800$
100	$110 - 100$	$100(110 - 100) = 1000$

The maximum is attained near $x = 50$ and 60.

(c) $P = x(110 - x) = 110x - x^2$

(d)

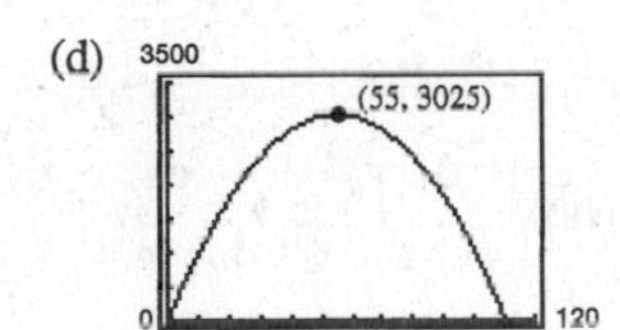

The solution appears to be $x = 55$.

(e) $\dfrac{dP}{dx} = 110 - 2x = 0$ when $x = 55$.

$\dfrac{d^2P}{dx^2} = -2 < 0$

P is a maximum when $x = 110 - x = 55$.

The two numbers are 55 and 55.

3. Let x and y be two positive numbers such that $xy = 192$.

$$S = x + y = x + \frac{192}{x}$$

$$\frac{dS}{dx} = 1 - \frac{192}{x^2} = 0 \text{ when } x = \sqrt{192}.$$

$$\frac{d^2S}{dx^2} = \frac{384}{x^3} > 0 \text{ when } x = \sqrt{192}.$$

S is a minimum when $x = y = \sqrt{192}$.

5. Let x be a positive number.

$$S = x + \frac{1}{x}$$

$$\frac{dS}{dx} = 1 - \frac{1}{x^2} = 0 \text{ when } x = 1.$$

$$\frac{d^2S}{dx^2} = \frac{2}{x^3} > 0 \text{ when } x = 1.$$

The sum is a minimum when $x = 1$ and $1/x = 1$.

7. Let x be the length and y the width of the rectangle.

$$2x + 2y = 100$$
$$y = 50 - x$$
$$A = xy = x(50 - x)$$
$$\frac{dA}{dx} = 50 - 2x = 0 \text{ when } x = 25.$$
$$\frac{d^2A}{dx^2} = -2 < 0 \text{ when } x = 25.$$

A is maximum when $x = y = 25$ meters.

9. Let x be the length and y the width of the rectangle.

$$xy = 64$$
$$y = \frac{64}{x}$$
$$P = 2x + 2y = 2x + 2\left(\frac{64}{x}\right) = 2x + \frac{128}{x}$$
$$\frac{dP}{dx} = 2 - \frac{128}{x^2} = 0 \text{ when } x = 8.$$
$$\frac{d^2P}{dx^2} = \frac{256}{x^3} > 0 \text{ when } x = 8.$$

P is minimum when $x = y = 8$ feet.

11. $d = \sqrt{(x-4)^2 + (\sqrt{x} - 0)^2}$

$= \sqrt{x^2 - 7x + 16}$

Since d is smallest when the expression inside the radical is smallest, you need only find the critical numbers of

$$f(x) = x^2 - 7x + 16.$$
$$f'(x) = 2x - 7 = 0$$
$$x = \tfrac{7}{2}$$

By the First Derivative Test, the point nearest to $(4, 0)$ is $\left(7/2, \sqrt{7/2}\right)$.

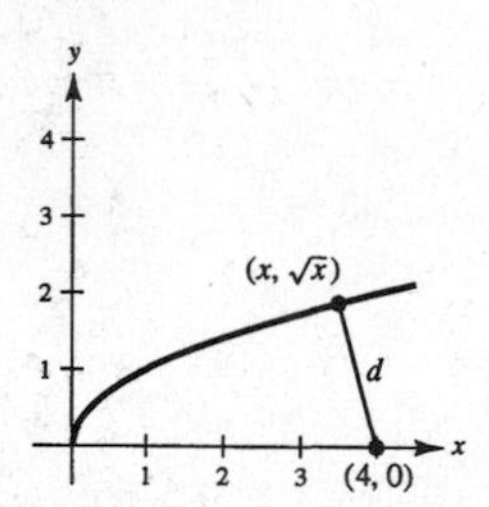

13. $d = \sqrt{(x-2)^2 + [x^2 - (1/2)]^2}$

$= \sqrt{x^4 - 4x + (17/4)}$

Since d is smallest when the expression inside the radical is smallest, you need only find the critical numbers of

$$f(x) = x^4 - 4x + \tfrac{17}{4}.$$
$$f'(x) = 4x^3 - 4 = 0$$
$$x = 1$$

By the First Derivative Test, the point nearest to $\left(2, \frac{1}{2}\right)$ is $(1, 1)$.

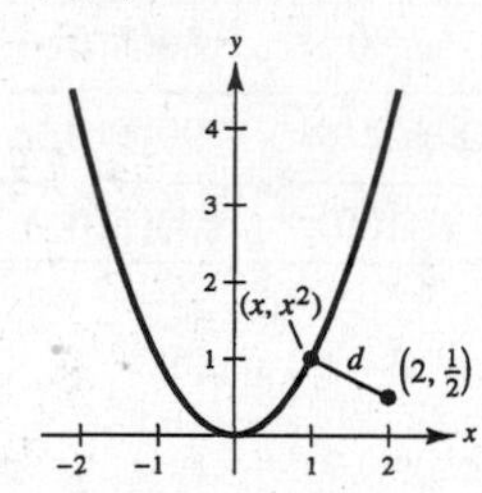

15. $\dfrac{dQ}{dx} = kx(Q_0 - x) = kQ_0x - kx^2$

$$\frac{d^2Q}{dx^2} = kQ_0 - 2kx$$
$$= k(Q_0 - 2x) = 0 \text{ when } x = \frac{Q_0}{2}.$$
$$\frac{d^3Q}{dx^3} = -2k < 0 \text{ when } x = \frac{Q_0}{2}.$$

dQ/dx is maximum when $x = Q_0/2$.

17. $xy = 180{,}000$ (see figure)

$S = x + 2y = \left(x + \dfrac{360{,}000}{x}\right)$ where S is the length of fence needed.

$$\frac{dS}{dx} = 1 - \frac{360{,}000}{x^2} = 0 \text{ when } x = 600.$$
$$\frac{d^2S}{dx^2} = \frac{720{,}000}{x^3} > 0 \text{ when } x = 600.$$

S is a minimum when $x = 600$ meters and $y = 300$ meters.

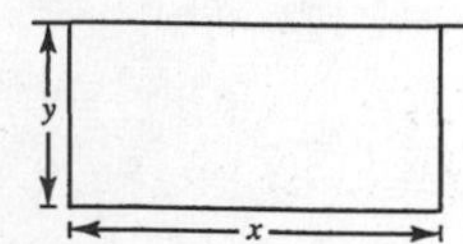

19. (a) $A = 4(\text{area of side}) + 2(\text{area of Top})$

(a) $A = 4(3)(11) + 2(3)(3) = 150$ square inches

(b) $A = 4(5)(5) + 2(5)(5) = 150$ square inches

(c) $A = 4(3.25)(6) + 2(6)(6) = 150$ square inches

(b) $V = (\text{length})(\text{width})(\text{height})$

(a) $V = (3)(3)(11) = 99$ cubic inches

(b) $V = (5)(5)(5) = 125$ cubic inches

(c) $V = (6)(6)(3.25) = 117$ cubic inches

(c) $S = 4xy + 2x^2 = 150 \Rightarrow y = \dfrac{150 - 2x^2}{4x}$

$$V = x^2y = x^2\left(\frac{150 - 2x^2}{4x}\right) = \frac{75}{2}x - \frac{1}{2}x^3$$

$$V' = \frac{75}{2} - \frac{3}{2}x^2 = 0 \Rightarrow x = \pm 5$$

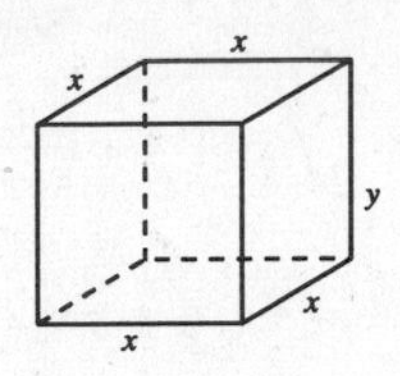

By the First Derivative Test, $x = 5$ yields the maximum volume. Dimensions: $5 \times 5 \times 5$. (A cube!)

21. (a) $V = x(s - 2x)^2, 0 < x < \dfrac{s}{2}$

$$\frac{dV}{dx} = 2x(s - 2x)(-2) + (s - 2x)^2$$

$$= (s - 2x)(s - 6x) = 0 \text{ when } x = \frac{s}{2}, \frac{s}{6} \ (s/2 \text{ is not in the domain}).$$

$$\frac{d^2V}{dx^2} = 24x - 8s$$

$$\frac{d^2V}{dx^2} < 0 \text{ when } x = \frac{s}{6}.$$

$$V = \frac{2s^3}{27} \text{ is maximum when } x = \frac{s}{6}.$$

(b) If the length is doubled, $V = \frac{2}{27}(2s)^3 = 8\left(\frac{2}{27}s^3\right)$. Volume is increased by a factor of 8.

23. $16 = 2y + x + \pi\left(\dfrac{x}{2}\right)$

$$32 = 4y + 2x + \pi x$$

$$y = \frac{32 - 2x - \pi x}{4}$$

$$A = xy + \frac{\pi}{2}\left(\frac{x}{2}\right)^2 = \left(\frac{32 - 2x - \pi x}{4}\right)x + \frac{\pi x^2}{8}$$

$$= 8x - \frac{1}{2}x^2 - \frac{\pi}{4}x^2 + \frac{\pi}{8}x^2$$

$$\frac{dA}{dx} = 8 - x - \frac{\pi}{2}x + \frac{\pi}{4}x = 8 - x\left(1 + \frac{\pi}{4}\right)$$

$$= 0 \text{ when } x = \frac{8}{1 + (\pi/4)} = \frac{32}{4 + \pi}.$$

$$\frac{d^2A}{dx^2} = -\left(1 + \frac{\pi}{4}\right) < 0 \text{ when } x = \frac{32}{4 + \pi}$$

$$y = \frac{32 - 2[32/(4 + \pi)] - \pi[32/(4 + \pi)]}{4} = \frac{16}{4 + \pi}$$

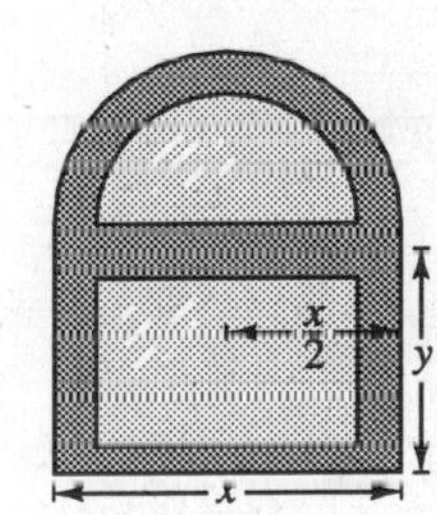

The area is maximum when $y = \dfrac{16}{4 + \pi}$ feet and $x = \dfrac{32}{4 + \pi}$ feet.

25. (a) $\dfrac{y-2}{0-1} = \dfrac{0-2}{x-1}$

$$y = 2 + \frac{2}{x-1}$$

$$L = \sqrt{x^2 + y^2} = \sqrt{x^2 + \left(2 + \frac{2}{x-1}\right)^2}$$

$$= \sqrt{x^2 + 4 + \frac{8}{x-1} + \frac{4}{(x-1)^2}}, \quad x > 1$$

(b)

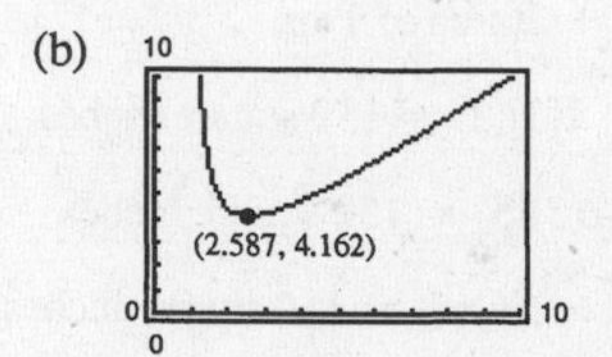

L is minimum when $x \approx 2.587$ and $L \approx 4.162$.

(c) Area $= A(x) = \dfrac{1}{2}xy = \dfrac{1}{2}x\left(2 + \dfrac{2}{x-1}\right) = x + \dfrac{x}{x-1}$

$$A'(x) = 1 + \frac{(x-1) - x}{(x-1)^2} = 1 - \frac{1}{(x-1)^2} = 0$$

$$(x-1)^2 = 1$$

$$x - 1 = \pm 1$$

$$x = 0, 2 \text{ (select } x = 2)$$

Then $y = 4$ and $A = 4$.

Vertices: $(0, 0)$, $(2, 0)$, $(0, 4)$

27. $A = 2xy = 2x\sqrt{25 - x^2}$ (see figure)

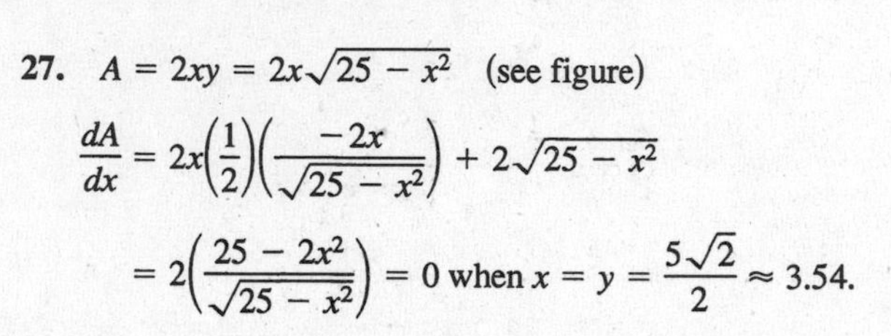

$$\frac{dA}{dx} = 2x\left(\frac{1}{2}\right)\left(\frac{-2x}{\sqrt{25 - x^2}}\right) + 2\sqrt{25 - x^2}$$

$$= 2\left(\frac{25 - 2x^2}{\sqrt{25 - x^2}}\right) = 0 \text{ when } x = y = \frac{5\sqrt{2}}{2} \approx 3.54.$$

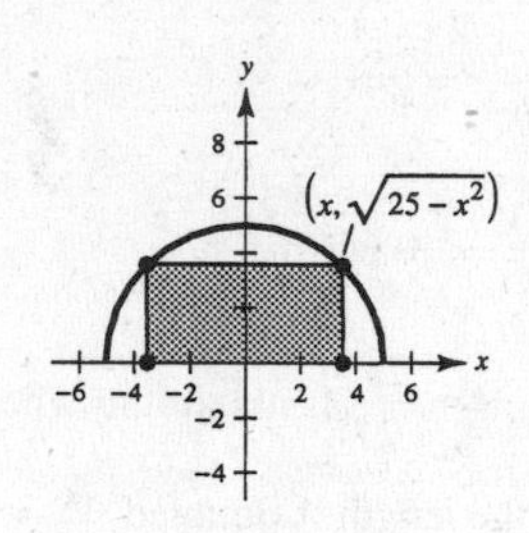

By the First Derivative Test, the inscribed rectangle of maximum area has vertices

$$\left(\pm\frac{5\sqrt{2}}{2}, 0\right), \left(\pm\frac{5\sqrt{2}}{2}, \frac{5\sqrt{2}}{2}\right).$$

Width: $\dfrac{5\sqrt{2}}{2}$; Length: $5\sqrt{2}$

29. $xy = 30 \Rightarrow y = \dfrac{30}{x}$

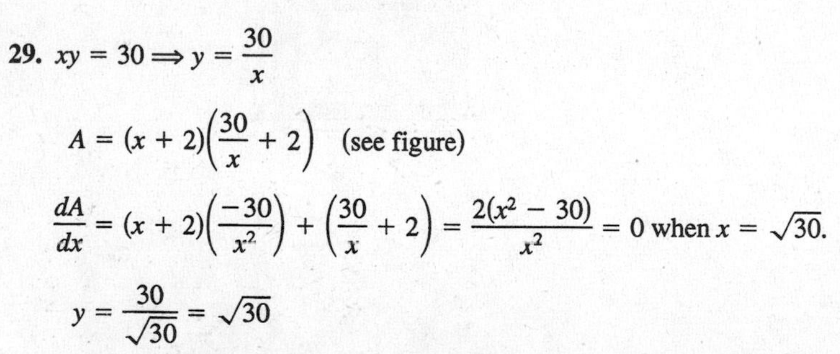

$$A = (x+2)\left(\frac{30}{x} + 2\right) \text{ (see figure)}$$

$$\frac{dA}{dx} = (x+2)\left(\frac{-30}{x^2}\right) + \left(\frac{30}{x} + 2\right) = \frac{2(x^2 - 30)}{x^2} = 0 \text{ when } x = \sqrt{30}.$$

$$y = \frac{30}{\sqrt{30}} = \sqrt{30}$$

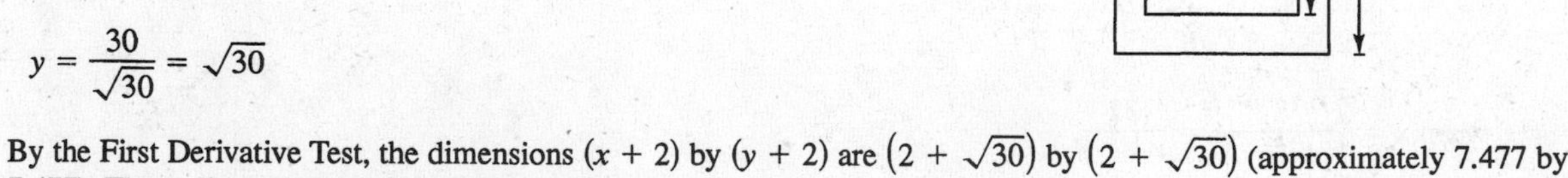

By the First Derivative Test, the dimensions $(x + 2)$ by $(y + 2)$ are $\left(2 + \sqrt{30}\right)$ by $\left(2 + \sqrt{30}\right)$ (approximately 7.477 by 7.477). These dimensions yield a minimum area.

31. $V = \pi r^2 h = 22$ cubic inches or $h = \dfrac{22}{\pi r^2}$

(a)

Radius, r	Height	Surface Area
0.2	$\dfrac{22}{\pi(0.2)^2}$	$2\pi(0.2)\left[0.2 + \dfrac{22}{\pi(0.2)^2}\right] \approx 220.3$
0.4	$\dfrac{22}{\pi(0.4)^2}$	$2\pi(0.4)\left[0.4 + \dfrac{22}{\pi(0.4)^2}\right] \approx 111.0$
0.6	$\dfrac{22}{\pi(0.6)^2}$	$2\pi(0.6)\left[0.6 + \dfrac{22}{\pi(0.6)^2}\right] \approx 75.6$
0.8	$\dfrac{22}{\pi(0.8)^2}$	$2\pi(0.8)\left[0.8 + \dfrac{22}{\pi(0.8)^2}\right] \approx 59.0$

(b)

Radius, r	Height	Surface Area
0.2	$\dfrac{22}{\pi(0.2)^2}$	$2\pi(0.2)\left[0.2 + \dfrac{22}{\pi(0.2)^2}\right] \approx 220.3$
0.4	$\dfrac{22}{\pi(0.4)^2}$	$2\pi(0.4)\left[0.4 + \dfrac{22}{\pi(0.4)^2}\right] \approx 111.0$
0.6	$\dfrac{22}{\pi(0.6)^2}$	$2\pi(0.6)\left[0.6 + \dfrac{22}{\pi(0.6)^2}\right] \approx 75.6$
0.8	$\dfrac{22}{\pi(0.8)^2}$	$2\pi(0.8)\left[0.8 + \dfrac{22}{\pi(0.8)^2}\right] \approx 59.0$
1.0	$\dfrac{22}{\pi(1.0)^2}$	$2\pi(1.0)\left[1.0 + \dfrac{22}{\pi(1.0)^2}\right] \approx 50.3$
1.2	$\dfrac{22}{\pi(1.2)^2}$	$2\pi(1.2)\left[1.2 + \dfrac{22}{\pi(1.2)^2}\right] \approx 45.7$
1.4	$\dfrac{22}{\pi(1.4)^2}$	$2\pi(1.4)\left[1.4 + \dfrac{22}{\pi(1.4)^2}\right] \approx 43.7$
1.6	$\dfrac{22}{\pi(1.6)^2}$	$2\pi(1.6)\left[1.6 + \dfrac{22}{\pi(1.6)^2}\right] \approx 43.6$
1.8	$\dfrac{22}{\pi(1.8)^2}$	$2\pi(1.8)\left[1.8 + \dfrac{22}{\pi(1.8)^2}\right] \approx 44.8$
2.0	$\dfrac{22}{\pi(2.0)^2}$	$2\pi(2.0)\left[2.0 + \dfrac{22}{\pi(2.0)^2}\right] \approx 47.1$

The minimum seems to be about 43.6 for $r = 1.6$.

(c) $S = 2\pi r^2 + 2\pi rh$

$$= 2\pi r(r + h) = 2\pi r\left[r + \frac{22}{\pi r^2}\right] = 2\pi r^2 + \frac{44}{r}$$

(d)

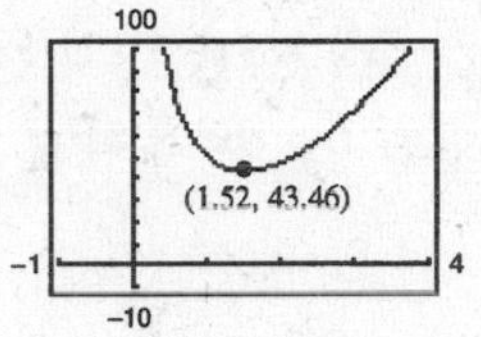

100

(1.52, 43.46)

−1 4

−10

The minimum seems to be 43.46 for $r \approx 1.52$.

(e) $\dfrac{dS}{dr} = 4\pi r - \dfrac{44}{r^2} = 0$ when $r = \sqrt[3]{11/\pi} \approx 1.52$ in.

$h = \dfrac{22}{\pi r^2} \approx 3.04$ in.

Note: Notice that

$$h = \frac{22}{\pi r^2} = \frac{22}{\pi(11/\pi)^{2/3}} = 2\left(\frac{11^{1/3}}{\pi^{1/3}}\right) = 2r.$$

33. Let x be the sides of the square ends and y the length of the package.

$$P = 4x + y = 108 \Rightarrow y = 108 - 4x$$
$$V = x^2y = x^2(108 - 4x) = 108x^2 - 4x^3$$
$$\frac{dV}{dx} = 216x - 12x^2$$
$$= 12x(18 - x) = 0 \text{ when } x = 18.$$
$$\frac{d^2V}{dx^2} = 216 - 24x = -216 < 0 \text{ when } x = 18.$$

The volume is maximum when $x = 18$ inches and $y = 108 - 4(18) = 36$ inches.

35. $V = \frac{1}{3}\pi x^2h = \frac{1}{3}\pi x^2\left(r + \sqrt{r^2 - x^2}\right)$ (see figure)

$$\frac{dV}{dx} = \frac{1}{3}\pi\left[\frac{-x^3}{\sqrt{r^2 - x^2}} + 2x\left(r + \sqrt{r^2 - x^2}\right)\right] = \frac{\pi x}{3\sqrt{r^2 - x^2}}\left(2r^2 + 2r\sqrt{r^2 - x^2} - 3x^2\right) = 0$$

$$2r^2 + 2r\sqrt{r^2 - x^2} - 3x^2 = 0$$
$$2r\sqrt{r^2 - x^2} = 3x^2 - 2r^2$$
$$4r^2(r^2 - x^2) = 9x^4 - 12x^2r^2 + 4r^4$$
$$0 = 9x^4 - 8x^2r^2 = x^2(9x^2 - 8r^2)$$
$$x = 0, \frac{2\sqrt{2}r}{3}$$

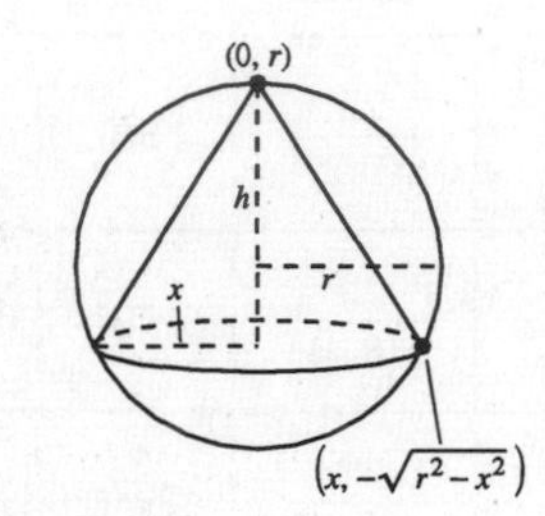

By the First Derivative Test, the volume is a maximum when

$$x = \frac{2\sqrt{2}r}{3} \text{ and } h = r + \sqrt{r^2 - x^2} = \frac{4r}{3}.$$

Thus, the maximum volume is

$$V = \frac{1}{3}\pi\left(\frac{8r^2}{9}\right)\left(\frac{4r}{3}\right) = \frac{32\pi r^3}{81} \text{ cubic units.}$$

37. No, there is no minimum area. If the sides are x and y, then $2x + 2y = 20 \Rightarrow y = 10 - x$. The area is $A(x) = x(10 - x) = 10x - x^2$. This can be made arbitrarily small by selecting $x \approx 0$.

39. $V = 12 = \frac{4}{3}\pi r^3 + \pi r^2h$

$$h = \frac{12 - (4/3)\pi r^3}{\pi r^2} = \frac{12}{\pi r^2} - \frac{4}{3}r$$

$$S = 4\pi r^2 + 2\pi rh = 4\pi r^2 + 2\pi r\left(\frac{12}{\pi r^2} - \frac{4}{3}r\right)$$
$$= 4\pi r^2 + \frac{24}{r} - \frac{8}{3}\pi r^2 = \frac{4}{3}\pi r^2 + \frac{24}{r}$$

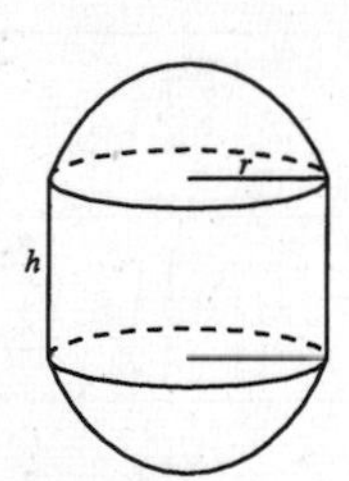

$$\frac{dS}{dr} = \frac{8}{3}\pi r - \frac{24}{r^2} = 0 \text{ when } r = \sqrt[3]{9/\pi} \approx 1.42 \text{ cm.}$$

$$\frac{d^2S}{dr^2} = \frac{8}{3}\pi + \frac{48}{r^3} > 0 \text{ when } r = \sqrt[3]{9/\pi} \text{ cm.}$$

The surface area is minimum when $r = \sqrt[3]{9/\pi}$ cm and $h = 0$. The resulting solid is a sphere of radius $r \approx 1.42$ cm.

41. Let x be the length of a side of the square and y the length of a side of the triangle.

$$4x + 3y = 10$$

$$A = x^2 + \frac{1}{2}y\left(\frac{\sqrt{3}}{2}y\right)$$

$$= \frac{(10 - 3y)^2}{16} + \frac{\sqrt{3}}{4}y^2$$

$$\frac{dA}{dy} = \frac{1}{8}(10 - 3y)(-3) + \frac{\sqrt{3}}{2}y = 0$$

$$-30 + 9y + 4\sqrt{3}y = 0$$

$$y = \frac{30}{9 + 4\sqrt{3}}$$

$$\frac{d^2A}{dy^2} = \frac{9 + 4\sqrt{3}}{8} > 0$$

A is minimum when

$$y = \frac{30}{9 + 4\sqrt{3}} \text{ and } x = \frac{10\sqrt{3}}{9 + 4\sqrt{3}}.$$

43. Let S be the strength and k the constant of proportionality. Given $h^2 + w^2 = 24^2$, $h^2 = 24^2 - w^2$,

$$S = kwh^2$$

$$S = kw(576 - w^2) = k(576w - w^3)$$

$$\frac{dS}{dw} = k(576 - 3w^2) = 0 \text{ when } w = 8\sqrt{3}, h = 8\sqrt{6}.$$

$$\frac{d^2S}{dw^2} = -6kw < 0 \text{ when } w = 8\sqrt{3}.$$

These values yield a maximum.

45. $$R = \frac{v_0^2}{g}\sin 2\theta$$

$$\frac{dR}{d\theta} = \frac{2v_0^2}{g}\cos 2\theta = 0 \text{ when } \theta = \frac{\pi}{4}, \frac{3\pi}{4}.$$

$$\frac{d^2R}{d\theta^2} = -\frac{4v_0^2}{g}\sin 2\theta < 0 \text{ when } \theta = \frac{\pi}{4}.$$

By the Second Derivative Test, R is maximum when $\theta = \pi/4$.

47. $$\sin\alpha = \frac{h}{s} \Longrightarrow s = \frac{h}{\sin\alpha},\ 0 < \alpha < \frac{\pi}{2}$$

$$\tan\alpha = \frac{h}{2} \Longrightarrow h = 2\tan\alpha \Longrightarrow s = \frac{2\tan\alpha}{\sin\alpha} = 2\sec\alpha$$

$$I = \frac{k\sin\alpha}{s^2} = \frac{k\sin\alpha}{4\sec^2\alpha} = \frac{k}{4}\sin\alpha\cos^2\alpha$$

$$\frac{dI}{d\alpha} = \frac{k}{4}[\sin\alpha(-2\sin\alpha\cos\alpha) + \cos^2\alpha(\cos\alpha)]$$

$$= \frac{k}{4}\cos\alpha[\cos^2\alpha - 2\sin^2\alpha]$$

$$= \frac{k}{4}\cos\alpha[1 - 3\sin^2\alpha]$$

$$= 0 \text{ when } \alpha = \frac{\pi}{2}, \frac{3\pi}{2}, \text{ or when } \sin\alpha = \pm\frac{1}{\sqrt{3}}.$$

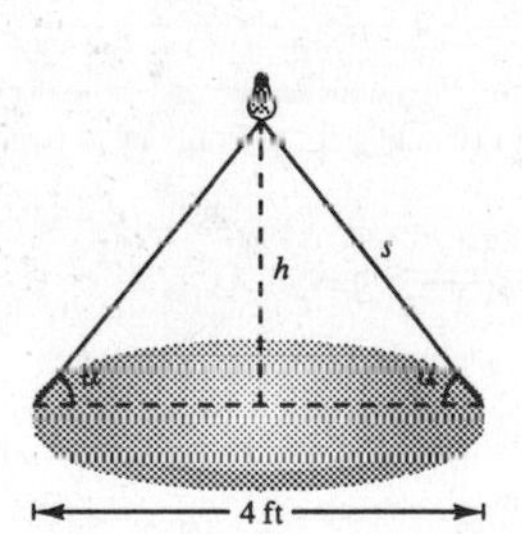

Since α is acute, we have

$$\sin\alpha = \frac{1}{\sqrt{3}} \Longrightarrow h = 2\tan\alpha = 2\left(\frac{1}{\sqrt{2}}\right) = \sqrt{2} \text{ feet.}$$

Since $(d^2I)/(d\alpha^2) = (k/4)\sin\alpha(9\sin^2\alpha - 7) < 0$ when $\sin\alpha = 1/\sqrt{3}$, this yields a maximum.

49.

$$S = \sqrt{x^2 + 4},\ L = \sqrt{1 + (3 - x)^2}$$

$$\text{Time} = T = \frac{\sqrt{x^2 + 4}}{2} + \frac{\sqrt{x^2 - 6x + 10}}{4}$$

$$\frac{dT}{dx} = \frac{x}{2\sqrt{x^2 + 4}} + \frac{x - 3}{4\sqrt{x^2 - 6x + 10}} = 0$$

$$\frac{x^2}{x^2 + 4} = \frac{9 - 6x + x^2}{4(x^2 - 6x + 10)}$$

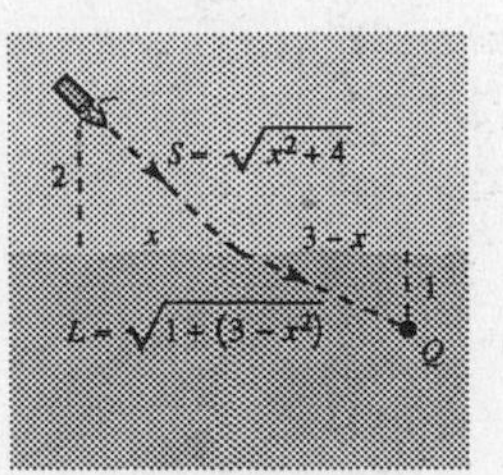

$$x^4 - 6x^3 + 9x^2 + 8x - 12 = 0$$

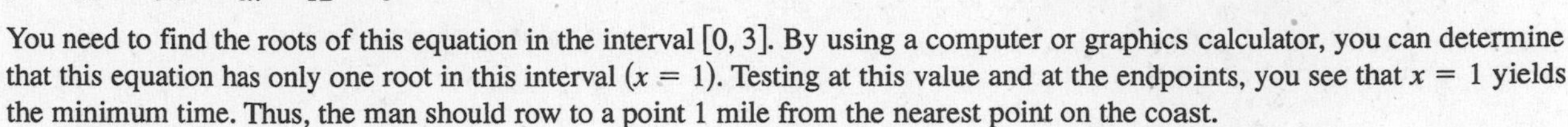

You need to find the roots of this equation in the interval [0, 3]. By using a computer or graphics calculator, you can determine that this equation has only one root in this interval ($x = 1$). Testing at this value and at the endpoints, you see that $x = 1$ yields the minimum time. Thus, the man should row to a point 1 mile from the nearest point on the coast.

51. $T = \dfrac{\sqrt{x^2 + 4}}{v_1} + \dfrac{\sqrt{x^2 - 6x + 10}}{v_2}$

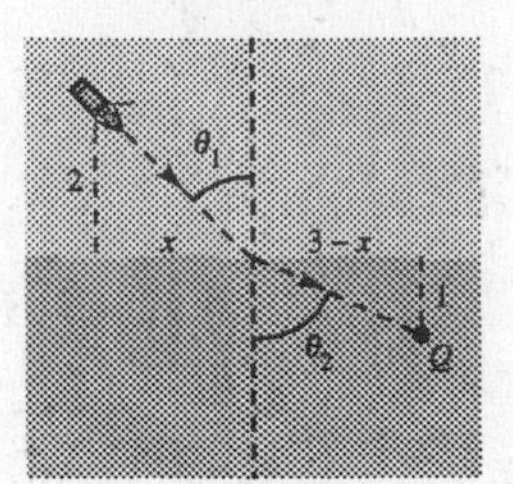

$$\frac{dT}{dx} = \frac{x}{v_1\sqrt{x^2 + 4}} + \frac{x - 3}{v_2\sqrt{x^2 - 6x + 10}} = 0$$

Since

$$\frac{x}{\sqrt{x^2 + 4}} = \sin \theta_1 \text{ and } \frac{x - 3}{\sqrt{x^2 - 6x + 10}} = -\sin \theta_2$$

we have

$$\frac{\sin \theta_1}{v_1} - \frac{\sin \theta_2}{v_2} = 0 \Longrightarrow \frac{\sin \theta_1}{v_1} = \frac{\sin \theta_2}{v_2}.$$

Since

$$\frac{d^2T}{dx^2} = \frac{4}{v_1(x^2 + 4)^{3/2}} + \frac{1}{v_2(x^2 - 6x + 10)^{3/2}} > 0$$

this condition yields a minimum time.

53. $C(x) = 2k\sqrt{x^2 + 4} + k(4 - x)$

$$C'(x) = \frac{2xk}{\sqrt{x^2 + 4}} - k = 0$$

$$2x = \sqrt{x^2 + 4}$$

$$4x^2 = x^2 + 4$$

$$3x^2 = 4$$

$$x = \frac{2}{\sqrt{3}}$$

Or, use Exercise 50(d): $\sin \theta = \dfrac{C_2}{C_1} = \dfrac{1}{2} \Rightarrow \theta = 30°.$

Thus, $x = \dfrac{2}{\sqrt{3}}$ kilometers

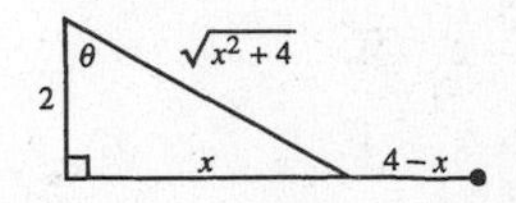

55. $x^2 + y^2 = (8.5 - x)^2 \Rightarrow y^2 = \dfrac{(289 - 68x)}{4}$

$$w^2 = 8.5^2 + (w - y)^2 \Rightarrow w = \frac{4y^2 + 289}{8y}$$

$$c^2 = w^2 + (8.5 - x)^2 = \left[\frac{4y^2 + 289}{8y}\right]^2 + (8.5 - x)^2$$

(a) Using a graphing utility, you find that c^2 is a minimum when

$$x = \frac{17}{8},\ y = \frac{17\sqrt{2}}{4} \text{ and } c = \frac{51\sqrt{3}}{8}.$$

(b) No, the answer would not change.

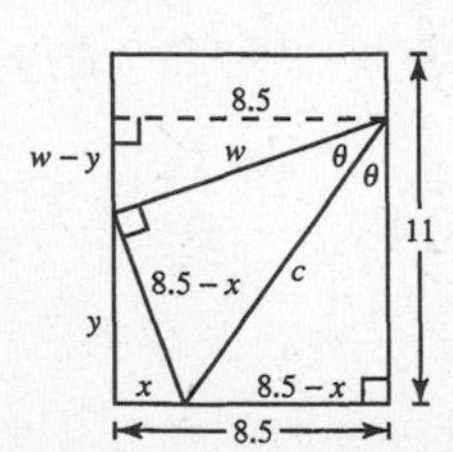

57. $p(t) = \dfrac{250}{1 + 4e^{-t/3}}$

$$p'(t) = \frac{1000}{3}\frac{e^{-t/3}}{(1 + 4e^{-t/3})^2}; \quad p'(2) \approx 18.35 \text{ elk/month}$$

$$p''(t) = \frac{1000}{9}\frac{e^{-t/3}(4e^{-t/3} - 1)}{1 + 4e^{-t/3}}$$

$$= 0 \text{ when } t \approx 4.16 \text{ months.}$$

59. $S_1 = (4m - 1)^2 + (5m - 6)^2 + (10m - 3)^2$

$$\frac{dS_1}{dm} = 2(4m - 1)(4) + 2(5m - 6)(5) + 2(10m - 3)(10) = 282m - 128 = 0 \text{ when } m = \frac{64}{141}.$$

Line: $y = \dfrac{64}{141}x$

$$S = \left|4\left(\frac{64}{141}\right) - 1\right| + \left|5\left(\frac{64}{141}\right) - 6\right| + \left|10\left(\frac{64}{141}\right) - 3\right|$$

$$= \left|\frac{256}{141} - 1\right| + \left|\frac{320}{141} - 6\right| + \left|\frac{640}{141} - 3\right| = \frac{858}{141} \approx 6.1 \text{ mi}$$

61. $C = 100\left(\dfrac{200}{x^2} + \dfrac{x}{x + 30}\right),\ 1 \le x$

$$C' = 100\left(-\frac{400}{x^3} + \frac{30}{(x + 30)^2}\right)$$

Approximation: $x \approx 40.45$ units, or 4045 units

63. $P = -\dfrac{1}{10}s^3 + 6s^2 + 400$

(a) $\dfrac{dP}{ds} = -\dfrac{3}{10}s^2 + 12s = -\dfrac{3}{10}s(s - 40) = 0$ when $x = 0,\ s = 40$.

$$\frac{d^2P}{ds^2} = -\frac{3}{5}s + 12$$

$\dfrac{d^2P}{ds^2}(0) > 0 \Rightarrow s = 0$ yields a minimum.

$\dfrac{d^2P}{ds^2}(40) < 0 \Rightarrow s = 40$ yields a maximum.

The maximum profit occurs when $s = 40$, which corresponds to \$40,000 ($P =$ \$3,600,000).

(b) $\dfrac{d^2P}{ds^2} = -\dfrac{3}{5}s + 12 = 0$ when $s = 20$.

The point of diminishing returns occurs when $s = 20$, which corresonds to \$20,000 being spent on advertising.

65. (a) $f(c) = f(c + x)$

$10ce^{-c} = 10(c + x)e^{-(c+x)}$

$\frac{c}{e^c} = \frac{c + x}{e^{c+x}}$

$ce^{c+x} = (c + x)e^c$

$ce^x = c + x$

$ce^x - c = x$

$c = \frac{x}{e^x - 1}$

(b) $A(x) = xf(c) = x\left[10\left(\frac{x}{e^x - 1}\right)e^{-x/(e^x-1)}\right]$

$= \frac{10x^2}{e^x - 1}e^{x/(1-e^x)}$

(c) $A(x) = \frac{10x^2}{e^x - 1}e^{x/(1-e^x)}$

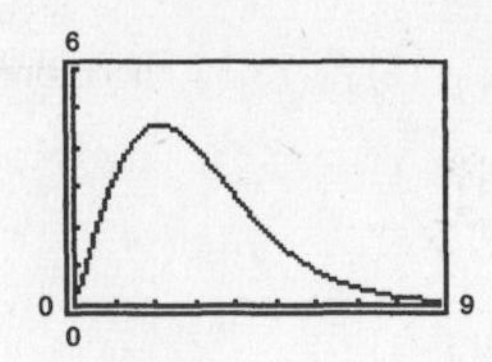

The maximum area is 4.591 for $x = 2.118$ and $f(x) = 2.547$.

(d) $c = \frac{x}{e^x - 1}$

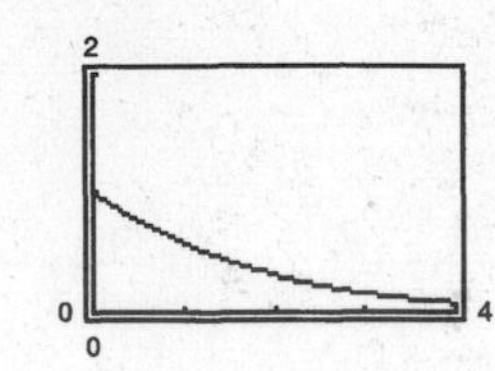

$\lim_{x \to 0^+} c = 1$

$\lim_{x \to \infty} c = 0$

Section 3.8 Differentials

1. $f(x) = x^2$

$f'(x) = 2x$

Tangent line at $(2, 4)$: $y - f(2) = f'(2)(x - 2)$

$y - 4 = 4(x - 2)$

$y = 4x - 4$

x	1.9	1.99	2	2.01	2.1
$f(x) = x^2$	3.6100	3.9601	4	4.0401	4.4100
$T(x) = 4x - 4$	3.6000	3.9600	4	4.0400	4.4000

3. $f(x) = x^5$

$f'(x) = 5x^4$

Tangent line at $(2, 32)$: $y - f(2) = f'(2)(x - 2)$

$y - 32 = 80(x - 2)$

$y = 80x - 128$

x	1.9	1.99	2	2.01	2.1
$f(x) = x^5$	24.7610	31.2080	32	32.8080	40.8410
$T(x) = 80x - 128$	24.0000	31.2000	32	32.8000	40.0000

5. $f(x) = \sin x$

$f'(x) = \cos x$

Tangent line at $(2, \sin 2)$:

$y - f(2) = f'(2)(x - 2)$

$y - \sin 2 = (\cos 2)(x - 2)$

$y = (\cos 2)(x - 2) + \sin 2$

x	1.9	1.99	2	2.01	2.1
$f(x) = \sin x$	0.9463	0.9134	0.9093	0.9051	0.8632
$T(x) = (\cos 2)(x - 2) + \sin 2$	0.9509	0.9135	0.9093	0.9051	0.8677

7. $y = f(x) = \frac{1}{2}x^3, f'(x) = \frac{3}{2}x^2, x = 2, \Delta x = dx = 0.1$

$\Delta y = f(x + \Delta x) - f(x)$

$= f(2.1) - f(2)$

$= 0.6305$

$dy = f'(x)dx$

$= f'(2)(0.1)$

$= 6(0.1) = 0.6$

9. $y = f(x) = x^4 + 1, f'(x) = 4x^3, x = -1, \Delta x = dx = 0.01$

$$\begin{aligned}\Delta y &= f(x + \Delta x) - f(x)\\ &= f(-0.99) - f(-1)\\ &= [(-0.99)^4 + 1] - [(-1)^4 + 1] \approx -0.0394\end{aligned}$$

$$\begin{aligned}dy &= f'(x)\,dx\\ &= f'(-1)(0.01)\\ &= (-4)(0.01) = -0.04\end{aligned}$$

11. $y = 3x^2 - 4$

$dy = 6x\,dx$

13. $y = \dfrac{x+1}{2x-1}$

$dy = \dfrac{-3}{(2x-1)^2}\,dx$

15. $y = x\sqrt{1-x^2}$

$dy = \left(x\dfrac{-x}{\sqrt{1-x^2}} + \sqrt{1-x^2}\right)dx = \dfrac{1-2x^2}{\sqrt{1-x^2}}\,dx$

17. $y = \ln\sqrt{4-x^2} = \dfrac{1}{2}\ln(4-x^2)$

$dy = \dfrac{1}{2}\dfrac{-2x}{4-x^2}\,dx = \dfrac{-x}{4-x^2}\,dx$

19. $y = 2x - \cot^2 x$

$dy = (2 + 2\cot x\csc^2 x)dx$

$= (2 + 2\cot x + 2\cot^3 x)dx$

21. $y = \dfrac{1}{3}\cos\left(\dfrac{6\pi x - 1}{2}\right)$

$dy = -\pi\sin\left(\dfrac{6\pi x - 1}{2}\right)dx$

23. $y = x\arcsin x$

$dy = \left(\dfrac{x}{\sqrt{1-x^2}} + \arcsin x\right)dx$

25. (a) $f(1.9) = f(2 - 0.1) \approx f(2) + f'(2)(-0.1)$

$\approx 1 + (1)(-0.1) = 0.9$

(b) $f(2.04) = f(2 + 0.04) \approx f(2) + f'(2)(0.04)$

$\approx 1 + (1)(0.04) = 1.04$

27. (a) $f(1.9) = f(2 - 0.1) \approx f(2) + f'(2)(-0.1)$

$\approx 1 + \left(-\frac{1}{2}\right)(-0.1) = 1.05$

(b) $f(2.04) = f(2 + 0.04) \approx f(2) + f'(2)(0.04)$

$\approx 1 + \left(-\frac{1}{2}\right)(0.04) = 0.98$

29. (a) $g(2.93) = g(3 - 0.07) \approx g(3) + g'(3)(-0.07)$

$\approx 8 + \left(-\frac{1}{2}\right)(-0.07) = 8.035$

(b) $g(3.1) = g(3 + 0.1) \approx g(3) + g'(3)(0.1)$

$\approx 8 + \left(-\frac{1}{2}\right)(0.1) = 7.95$

31. (a) $g(2.93) = g(3 - 0.07) \approx g(3) + g'(3)(-0.07)$

$\approx 8 + 0(-0.07) = 8$

(b) $g(3.1) = g(3 + 0.1) \approx g(3) + g'(3)(0.1)$

$\approx 8 + 0(0.1) = 8$

33. $A = x^2$

$x = 12$

$\Delta x = dx = \pm\frac{1}{64}$

$dA = 2x\,dx$

$\Delta A \approx dA = 2(12)\left(\pm\frac{1}{64}\right)$

$= \pm\frac{3}{8}$ square inches

35. $A = \pi r^2$

$r = 14$

$\Delta r = dr = \pm\frac{1}{4}$

$\Delta A \approx dA = 2\pi r\,dr = \pi(28)\left(\pm\frac{1}{4}\right)$

$= \pm 7\pi$ square inches

37. (a) $x = 15$ centimeter

$\Delta x = dx = \pm 0.05$ centimeters

$$A = x^2$$

$$dA = 2x\,dx = 2(15)(\pm 0.05)$$

$$= \pm 1.5 \text{ square centimeters}$$

Percentage error:

$$\frac{dA}{A} = \frac{1.5}{(15)^2} = 0.00666\ldots = \frac{2}{3}\%$$

(b) $$\frac{dA}{A} = \frac{2x\,dx}{x^2} = \frac{2\,dx}{x} \le 0.025$$

$$\frac{dx}{x} \le \frac{0.025}{2} = 0.0125 = 1.25\%$$

39. $r = 6$ inches

$\Delta r = dr = \pm 0.02$ inches

(a) $V = \dfrac{4}{3}\pi r^3$

$$dV = 4\pi r^2\,dr = 4\pi(6)^2(\pm 0.02) = \pm 2.88\pi \text{ cubic inches}$$

(b) $S = 4\pi r^2$

$$dS = 8\pi r\,dr = 8\pi(6)(\pm 0.02) = \pm 0.96\pi \text{ square inches}$$

(c) Relative error: $$\frac{dV}{V} = \frac{4\pi r^2\,dr}{(4/3)\pi r^3} = \frac{3dr}{r}$$

$$= \frac{3}{6}(0.02) = 0.01 = 1\%$$

Relative error: $$\frac{dS}{S} = \frac{8\pi r\,dr}{4\pi r^2} = \frac{2dr}{r}$$

$$= \frac{2(0.02)}{6} = 0.00666\ldots = \frac{2}{3}\%$$

41. $P = 100xe^{-x/400}$, x changes from 115 to 120.

$$dP = 100\left(e^{-x/400} - \frac{x}{400}e^{-x/400}\right)dx = e^{-115/400}\left(100 - \frac{115}{4}\right)(120 - 115) \approx 267.24$$

Approximate percentage change: $$\frac{dP}{P}(100) = \frac{267.24}{8626.57}(100) \approx 3.1\%$$

43. $V = \pi r^2 h = 40\pi r^2$, $r = 5$ cm, $h = 40$ cm, $dr = 0.2$ cm

$$\Delta V \approx dV = 80\pi r\,dr = 80\pi(5)(0.2) = 80\pi \text{ cm}^3$$

45. $\theta = 26°45' = 26.75°$

$d\theta = \pm 15' = \pm 0.25°$

(a) $h = 9.5 \csc\theta$

$$dh = -9.5 \csc\theta \cot\theta\,d\theta$$

$$\frac{dh}{h} = -\cot\theta\,d\theta$$

$$\left|\frac{dh}{h}\right| = (\cot 26.75°)(0.25°)$$

Converting to radians, $(\cot 0.4669)(0.0044)$
$\approx 0.0087 = 0.87\%$ (in radians).

(b) $$\left|\frac{dh}{h}\right| = \cot\theta\,d\theta \le 0.02$$

$$\frac{d\theta}{\theta} \le \frac{0.02}{\theta(\cot\theta)} = \frac{0.02\tan\theta}{\theta}$$

$$\frac{d\theta}{\theta} \le \frac{0.02\tan 26.75°}{26.75°} \approx \frac{0.02\tan 0.4669}{0.4669}$$

$$\approx 0.0216 = 2.16\% \text{ (in radians)}$$

47. $r = \frac{v_0^2}{32}(\sin 2\theta)$

$v_0 = 2200$ ft/sec

θ changes from 10° to 11°

$$dr = \frac{(2200)^2}{16}(\cos 2\theta)\,d\theta$$

$$\theta = 10\left(\frac{\pi}{180}\right)$$

$$d\theta = (11 - 10)\frac{\pi}{180}$$

$$\Delta r \approx dr$$

$$= \frac{(2200)^2}{16}\cos\left(\frac{20\pi}{180}\right)\left(\frac{\pi}{180}\right) \approx 4961 \text{ feet}$$

≈ 4961 feet

49. Let $f(x) = \sqrt{x}$, $x = 100$, $dx = -0.6$.

$$f(x + \Delta x) \approx f(x) + f'(x)\,dx$$

$$= \sqrt{x} + \frac{1}{2\sqrt{x}}\,dx$$

$$f(x + \Delta x) = \sqrt{99.4}$$

$$\approx \sqrt{100} + \frac{1}{2\sqrt{100}}(-0.6) = 9.97$$

Using a calculator: $\sqrt{99.4} \approx 9.96995$

51. Let $f(x) = \sqrt[4]{x}$, $x = 625$, $dx = -1$.

$$f(x + \Delta x) \approx f(x) + f'(x)\,dx = \sqrt[4]{x} + \frac{1}{4\sqrt[4]{x^3}}\,dx$$

$$f(x + \Delta x) = \sqrt[4]{624} \approx \sqrt[4]{625} + \frac{1}{4(\sqrt[4]{625})^3}(-1)$$

$$= 5 - \frac{1}{500} = 4.998$$

Using a calculator, $\sqrt[4]{624} \approx 4.9980$.

53. Let $f(x) = \sqrt{x}$, $x = 4$, $dx = 0.02$, $f'(x) = 1/(2\sqrt{x})$.

Then

$$f(4.02) \approx f(4) + f'(4)\,dx$$

$$\sqrt{4.02} \approx \sqrt{4} + \frac{1}{2\sqrt{4}}(0.02) = 2 + \frac{1}{4}(0.02).$$

55. In general, when $\Delta x \to 0$, dy approaches Δy.

57. True

59. True

Review Exercises for Chapter 3

1. A number c in the domain of f is a critical number if $f'(c) = 0$ or f' is undefined at c.

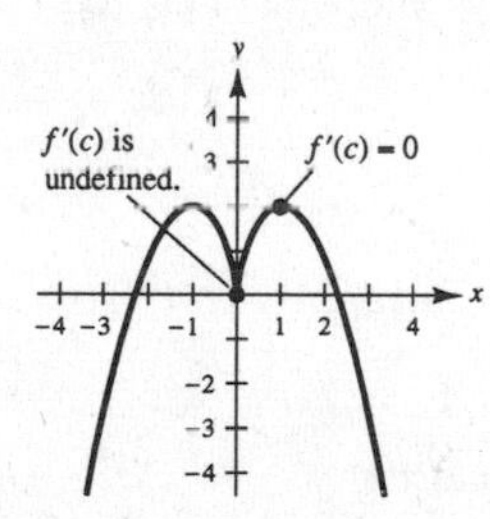

3. $g(x) = 2x + 5\cos x$, $[0, 2\pi]$

$g'(x) = 2 - 5\sin x$

$= 0$ when $\sin x = \frac{2}{5}$.

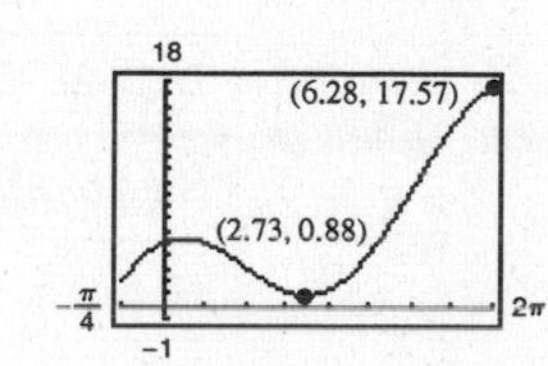

Critical numbers: $x \approx 0.41$, $x \approx 2.73$

Left endpoint: $(0, 5)$

Critical number: $(0.41, 5.41)$

Critical number: $(2.73, 0.88)$ Minimum

Right endpoint: $(2\pi, 17.57)$ Maximum

5. Yes. $f(-3) = f(2) = 0$. f is continuous on $[-3, 2]$, differentiable on $(-3, 2)$.

$f'(x) = (x + 3)(3x - 1) = 0$ for $x = \frac{1}{3}$.

$c = \frac{1}{3}$ satisfies $f'(c) = 0$.

7. $f(x) = 3 - |x - 4|$

(a)

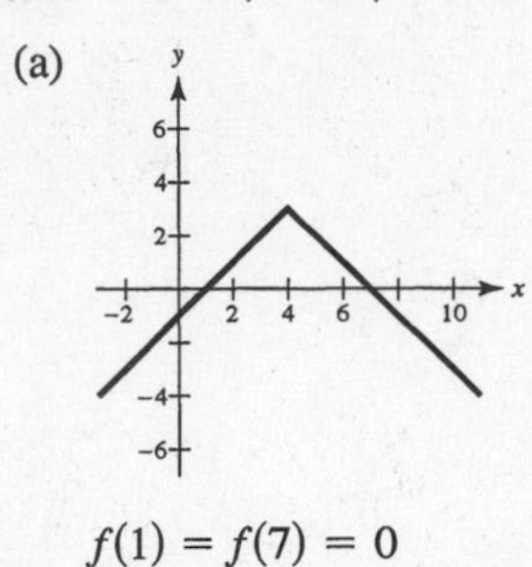

$f(1) = f(7) = 0$

(b) f is not differentiable at $x = 4$.

9.
$$f(x) = x^{2/3},\ 1 \le x \le 8$$
$$f'(x) = \frac{2}{3}x^{-1/3}$$
$$\frac{f(b) - f(a)}{b - a} = \frac{4 - 1}{8 - 1} = \frac{3}{7}$$
$$f'(c) = \frac{2}{3}c^{-1/3} = \frac{3}{7}$$
$$c = \left(\frac{14}{9}\right)^3 = \frac{2744}{729} \approx 3.764$$

11.
$$f(x) = x - \cos x,\ -\frac{\pi}{2} \le x \le \frac{\pi}{2}$$
$$f'(x) = 1 + \sin x$$
$$\frac{f(b) - f(a)}{b - a} = \frac{(\pi/2) - (-\pi/2)}{(\pi/2) - (-\pi/2)} = 1$$
$$f'(c) = 1 + \sin c = 1$$
$$c = 0$$

13.
$$f(x) = Ax^2 + Bx + C$$
$$f'(x) = 2Ax + B$$
$$\frac{f(x_2) - f(x_1)}{x_2 - x_1} = \frac{A(x_2^2 - x_1^2) + B(x_2 - x_1)}{x_2 - x_1}$$
$$= A(x_1 + x_2) + B$$
$$f'(c) = 2Ac + B = A(x_1 + x_2) + B$$
$$2Ac = A(x_1 + x_2)$$
$$c = \frac{x_1 + x_2}{2} = \text{Midpoint of } [x_1, x_2]$$

15. $f(x) = (x - 1)^2(x - 3)$

$f'(x) = (x - 1)^2(1) + (x - 3)(2)(x - 1)$

$= (x - 1)(3x - 7)$

Critical numbers: $x = 1$ and $x = \frac{7}{3}$

Interval:	$-\infty < x < 1$	$1 < x < \frac{7}{3}$	$\frac{7}{3} < x < \infty$
Sign of $f'(x)$:	$f'(x) > 0$	$f'(x) < 0$	$f'(x) > 0$
Conclusion:	Increasing	Decreasing	Increasing

17. $h(x) = \sqrt{x}(x - 3) = x^{3/2} - 3x^{1/2}$

Domain: $(0, \infty)$

$$h'(x) = \frac{3}{2}x^{1/2} - \frac{3}{2}x^{-1/2}$$
$$= \frac{3}{2}x^{-1/2}(x - 1) = \frac{3(x - 1)}{2\sqrt{x}}$$

Critical number: $x = 1$

Interval:	$0 < x < 1$	$1 < x < \infty$
Sign of $h'(x)$:	$h'(x) < 0$	$h'(x) > 0$
Conclusion:	Decreasing	Increasing

19. $f(t) = (2 - t)2^t$

$f'(t) = (2 - t)2^t \ln 2 - 2^t = 2^t[(2 - t)\ln 2 - 1]$

$f'(t) = 0$: $(2 - t)\ln 2 = 1$

$$2 - t = \frac{1}{\ln 2}$$

$$t = 2 - \frac{1}{\ln 2} \approx 0.5573 \quad \text{Critical number}$$

Interval:	$-\infty < t < 2 - \frac{1}{\ln 2}$	$2 - \frac{1}{\ln 2} < t < \infty$
Sign of $f'(t)$:	$f'(t) > 0$	$f'(t) < 0$
Conclusion:	Increasing	Decreasing

21. $h(t) = \frac{1}{4}t^4 - 8t$

$h'(t) = t^3 - 8 = 0$ when $t = 2$.

Relative minimum: $(2, -12)$

Test Interval:	$-\infty < t < 2$	$2 < t < \infty$
Sign of $h'(t)$:	$h'(t) < 0$	$h'(t) > 0$
Conclusion:	Decreasing	Increasing

23. $y = \frac{1}{3}\cos(12t) - \frac{1}{4}\sin(12t)$

$v = y' = -4\sin(12t) - 3\cos(12t)$

(a) When $t = \frac{\pi}{8}$, $y = \frac{1}{4}$ inch and $v = y' = 4$ inches/second.

(b) $y' = -4\sin(12t) - 3\cos(12t) = 0$ when $\frac{\sin(12t)}{\cos(12t)} = -\frac{3}{4} \Rightarrow \tan(12t) = -\frac{3}{4}$.

Therefore, $\sin(12t) = -\frac{3}{5}$ and $\cos(12t) = \frac{4}{5}$. The maximum displacement is

$$y = \left(\frac{1}{3}\right)\left(\frac{4}{5}\right) - \frac{1}{4}\left(-\frac{3}{5}\right) = \frac{5}{12} \text{ inch.}$$

(c) Period: $\frac{2\pi}{12} = \frac{\pi}{6}$

Frequency: $\frac{1}{\pi/6} = \frac{6}{\pi}$

25. $f(x) = x + \cos x$, $0 \le x \le 2\pi$

$f'(x) = 1 - \sin x$

$f''(x) = -\cos x = 0$ when $x = \frac{\pi}{2}, \frac{3\pi}{2}$.

Points of inflection: $\left(\frac{\pi}{2}, \frac{\pi}{2}\right), \left(\frac{3\pi}{2}, \frac{3\pi}{2}\right)$

Test Interval:	$0 < x < \frac{\pi}{2}$	$\frac{\pi}{2} < x < \frac{3\pi}{2}$	$\frac{3\pi}{2} < x < 2\pi$
Sign of $f''(x)$:	$f''(x) < 0$	$f''(x) > 0$	$f''(x) < 0$
Conclusion:	Concave downward	Concave upward	Concave downward

27. $g(x) = 2x^2(1 - x^2)$

$g'(x) = -4x(2x^2 - 1)$ Critical numbers: $x = 0, \pm\frac{1}{\sqrt{2}}$

$g''(x) = 4 - 24x^2$

$g''(0) = 4 > 0$ Relative minimum at $(0, 0)$

$g''\left(\pm\frac{1}{\sqrt{2}}\right) = -8 < 0$ Relative maximums at $\left(\pm\frac{1}{\sqrt{2}}, \frac{1}{2}\right)$

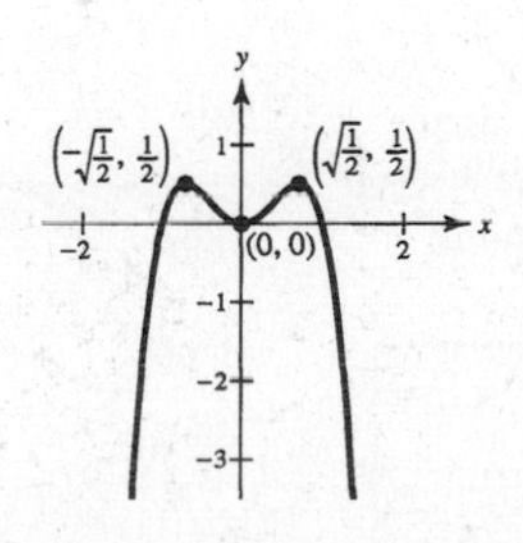

29.

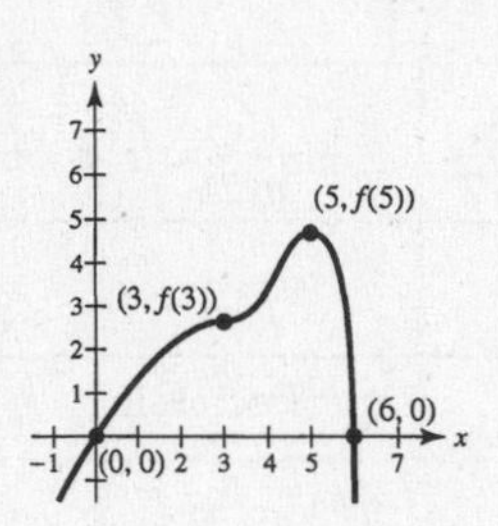

31. The first derivative is positive and the second derivative is negative. The graph is increasing and is concave down.

33. (a) $D = 0.0034t^4 - 0.2352t^3 + 4.9423t^2 - 20.8641t + 94.4025$

(b)

369

0

0 29

(c) Maximum at $(21.9, 319.5)$ (≈ 1992)

Minimum at $(2.6, 69.6)$ (≈ 1972)

(d) Outlays increasing at greatest rate at the point of inflection $(9.8, 173.7)$ (≈ 1979)

35. $\displaystyle\lim_{x\to\infty} \frac{2x^2}{3x^2+5} = \lim_{x\to\infty} \frac{2}{3+5/x^2} = \frac{2}{3}$

37. $\displaystyle\lim_{x\to\infty} \frac{5\cos x}{x} = 0$, since $|5\cos x| \le 5$.

39. $h(x) = \dfrac{2x+3}{x-4}$

Discontinuity: $x = 4$

$$\lim_{x\to\infty} \frac{2x+3}{x-4} = \lim_{x\to\infty} \frac{2+(3/x)}{1-(4/x)} = 2$$

Vertical asymptote: $x = 4$

Horizontal asymptote: $y = 2$

41. $f(x) = \dfrac{3}{x} - 2$

Discontinuity: $x = 0$

$$\lim_{x\to\infty}\left(\frac{3}{x} - 2\right) = -2$$

Vertical asymptote: $x = 0$

Horizontal asymptote: $y = -2$

43. $f(x) = \dfrac{5}{3+2e^{-x}}$

Horizontal asymptotes: $y = 0$ (to the left)

$y = \dfrac{5}{3}$ (to the right)

No vertical asymptotes

45. $g(x) = 3\ln(1 + e^{-x/4})$

Horizontal asymptote: $y = 0$ (to the right)

47. $f(x) = x^3 + \dfrac{243}{x}$

Relative minimum: $(3, 108)$

Relative maximum: $(-3, -108)$

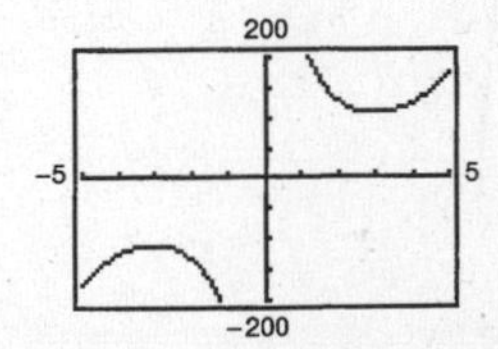

Vertical asymptote: $x = 0$

49. $f(x) = \dfrac{x-1}{1+3x^2}$

Relative minimum: $(-0.155, -1.077)$

Relative maximum: $(2.155, 0.077)$

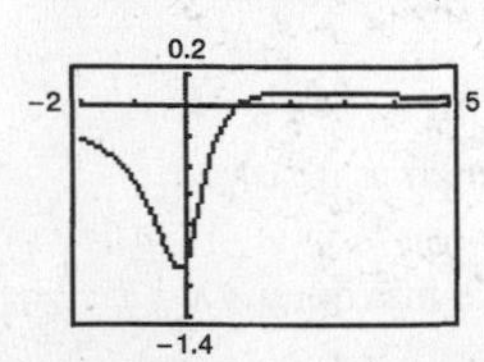

Horizontal asymptote: $y = 0$

51. $f(x) = 4x - x^2 = x(4 - x)$

Domain: $(-\infty, \infty)$; Range: $(-\infty, 4)$

$f'(x) = 4 - 2x = 0$ when $x = 2$.

$f''(x) = -2$

Therefore, $(2, 4)$ is a relative maximum.

Intercepts: $(0, 0), (4, 0)$

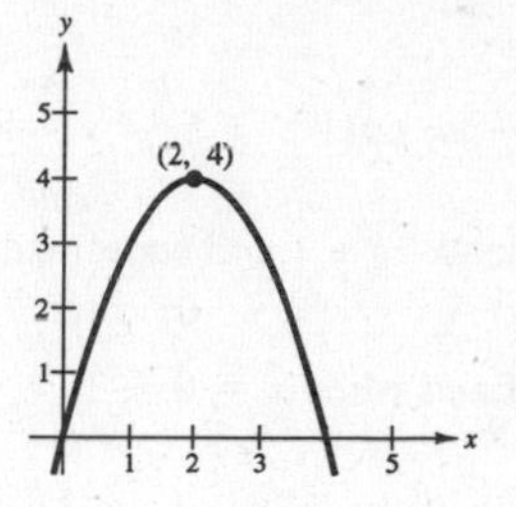

53. $f(x) = x\sqrt{16 - x^2}$, Domain: $[-4, 4]$, Range: $[-8, 8]$

Domain: $[-4, 4]$; Range: $[-8, 8]$

$$f'(x) = \frac{16 - 2x^2}{\sqrt{16 - x^2}} = 0 \text{ when } x = \pm 2\sqrt{2} \text{ and undefined when } x = \pm 4.$$

$$f''(x) = \frac{2x(x^2 - 24)}{(16 - x^2)^{3/2}}$$

$f''(-2\sqrt{2}) > 0$

Therefore, $(-2\sqrt{2}, -8)$ is a relative minimum.

$f''(2\sqrt{2}) < 0$

Therefore, $(2\sqrt{2}, 8)$ is a relative maximum.

Point of inflection: $(0, 0)$

Intercepts: $(-4, 0), (0, 0), (4, 0)$

Symmetry with respect to origin

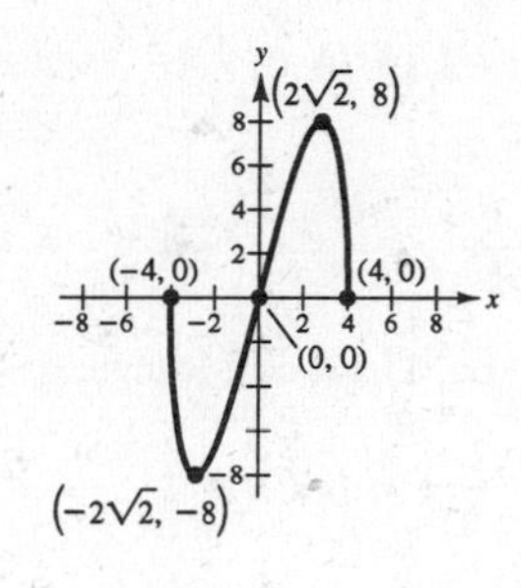

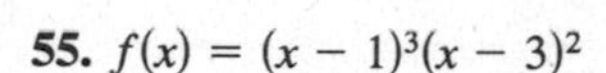

55. $f(x) = (x - 1)^3(x - 3)^2$

Domain: $(-\infty, \infty)$; Range: $(-\infty, \infty)$

$$f'(x) = (x - 1)^2(x - 3)(5x - 11) = 0 \text{ when } x = 1, \frac{11}{5}, 3.$$

$$f''(x) = 4(x - 1)(5x^2 - 22x + 23) = 0 \text{ when } x = 1, \frac{11 \pm \sqrt{6}}{5}.$$

$f''(3) > 0$

Therefore, $(3, 0)$ is a relative minimum.

$f''\left(\frac{11}{5}\right) < 0$

Therefore, $\left(\frac{11}{5}, \frac{3456}{3125}\right)$ is a relative maximum.

Points of inflection: $(1, 0), \left(\frac{11 - \sqrt{6}}{5}, 0.60\right), \left(\frac{11 + \sqrt{6}}{5}, 0.46\right)$

Intercepts: $(0, -9), (1, 0), (3, 0)$

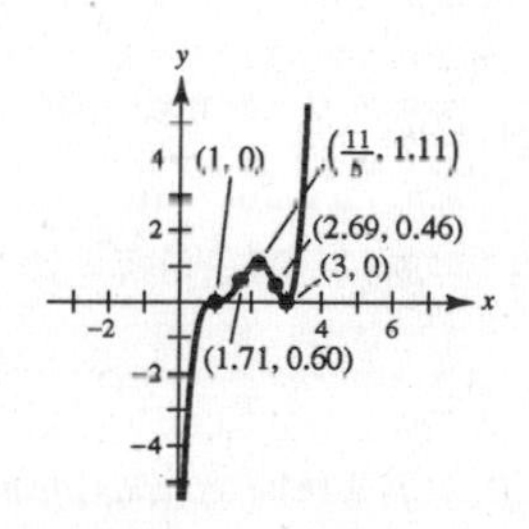

57. $f(x) = x^{1/3}(x + 3)^{2/3}$

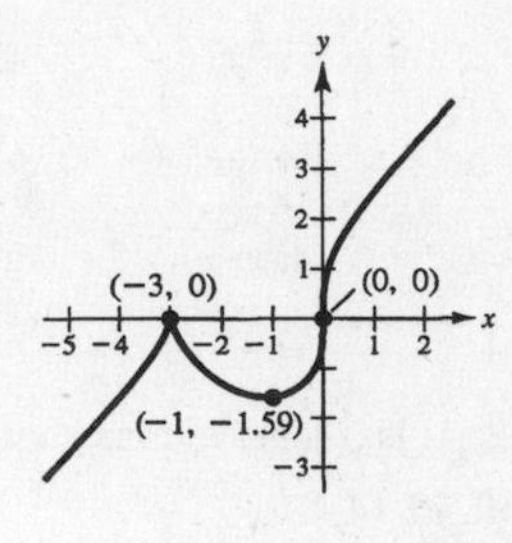

Domain: $(-\infty, \infty)$; Range: $(-\infty, \infty)$

$f'(x) = \dfrac{x + 1}{(x + 3)^{1/3}x^{2/3}} = 0$ when $x = -1$ and undefined when $x = -3, 0$.

$f''(x) = \dfrac{-2}{x^{5/3}(x + 3)^{4/3}}$ is undefined when $x = 0, -3$.

By the First Derivative Test $(-3, 0)$ is a relative maximum and $\left(-1, -\sqrt[3]{4}\right)$ is a relative minimum. $(0, 0)$ is a point of inflection.

Intercepts: $(-3, 0), (0, 0)$

59. $f(x) = \dfrac{x + 1}{x - 1}$

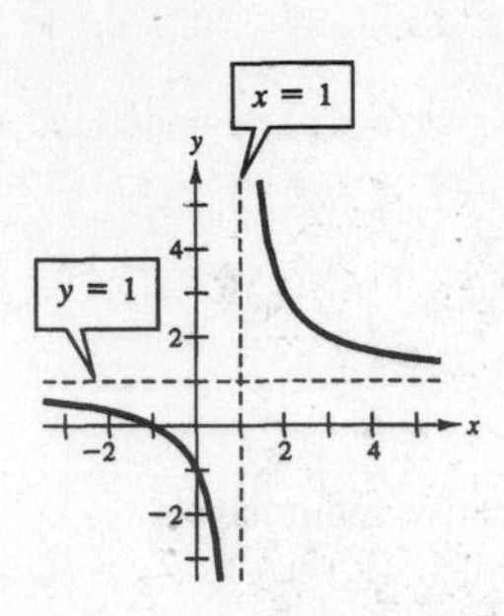

Domain: $(-\infty, 1), (1, \infty)$; Range: $(-\infty, 1), (1, \infty)$

$f'(x) = \dfrac{-2}{(x - 1)^2} < 0$ if $x \neq 1$.

$f''(x) = \dfrac{4}{(x - 1)^3}$

Horizontal asymptote: $y = 1$

Vertical asymptote: $x = 1$

Intercepts: $(-1, 0), (0, -1)$

61. $f(x) = \dfrac{4}{1 + x^2}$

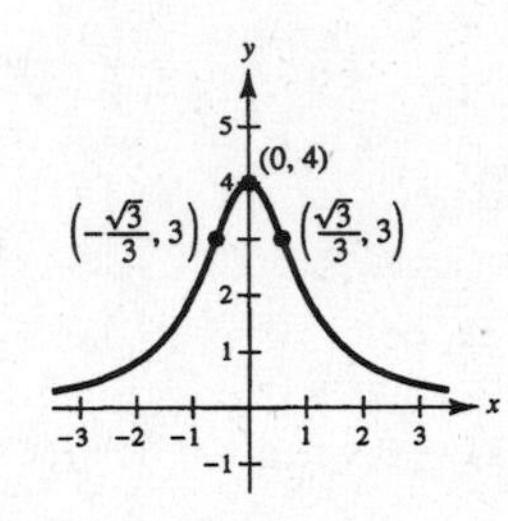

Domain: $(-\infty, \infty)$; Range: $(0, 4]$

$f'(x) = \dfrac{-8x}{(1 + x^2)^2} = 0$ when $x = 0$.

$f''(x) = \dfrac{-8(1 - 3x^2)}{(1 + x^2)^3} = 0$ when $x = \pm\dfrac{\sqrt{3}}{3}$.

$f''(0) < 0$

Therefore, $(0, 4)$ is a relative maximum.

Points of inflection: $\left(\pm\sqrt{3}/3, 3\right)$

Intercept: $(0, 4)$

Symmetric to the y-axis

Horizontal asymptote: $y = 0$

63. $f(x) = x^3 + x + \dfrac{4}{x}$

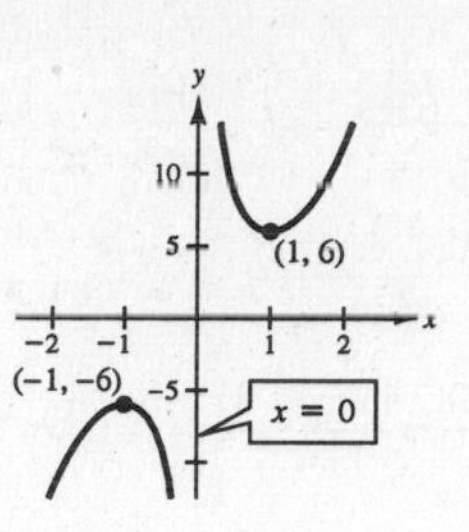

Domain: $(-\infty, 0), (0, \infty)$; Range: $(-\infty, -6], [6, \infty)$

$$f'(x) = 3x^2 + 1 - \frac{4}{x^2} = \frac{3x^4 + x^2 - 4}{x^2} = \frac{(3x^2 + 4)(x^2 - 1)}{x^2} = 0 \text{ when } x = \pm 1.$$

$$f''(x) = 6x + \frac{8}{x^3} = \frac{6x^4 + 8}{x^3} \neq 0$$

$f''(-1) < 0$

Therefore, $(-1, -6)$ is a relative maximum.

$f''(1) > 0$

Therefore, $(1, 6)$ is a relative minimum.

Vertical asymptote: $x = 0$

Symmetric with respect to origin

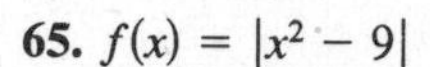

65. $f(x) = |x^2 - 9|$

Domain: $(-\infty, \infty)$; Range: $[0, \infty)$

$$f'(x) = \frac{2x(x^2 - 9)}{|x^2 - 9|} = 0 \text{ when } x = 0 \text{ and is undefined when } x = \pm 3.$$

$$f''(x) = \frac{2(x^2 - 9)}{|x^2 - 9|} \text{ is undefined at } x = \pm 3.$$

$f''(0) < 0$

Therefore, $(0, 9)$ is a relative maximum.

Relative minima: $(\pm 3, 0)$

Intercepts: $(\pm 3, 0), (0, 9)$

Symmetric to the y-axis

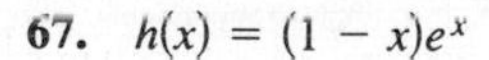

67. $h(x) = (1 - x)e^x$

$h'(x) = -xe^x$

$h''(x) = -(x + 1)e^x$

Horizontal asymptote: $y = 0$ (to the left)

Critical point: $(0, 1)$ (relative maximum)

Inflection point: $(-1, 2/e) \approx (-1, 0.736)$

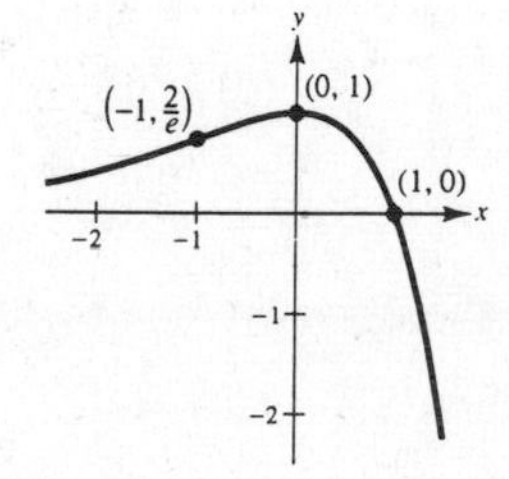

69. $g(x) = (x + 3)\ln(x + 3)$

$g'(x) = \ln(x + 3) + 1$

$$g''(x) = \frac{1}{x + 3}$$

Domain: $x > -3$

Relative minimum: $(-2.632, -0.368)$

Always concave upward

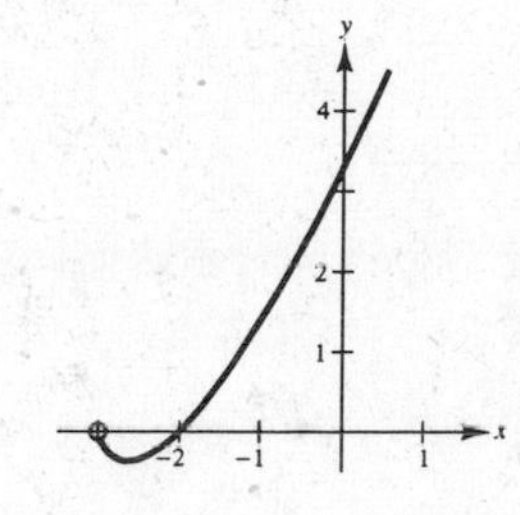

71. $f(x) = \dfrac{10 \log_4 x}{x} = \dfrac{10 \ln x}{x \ln 4}$

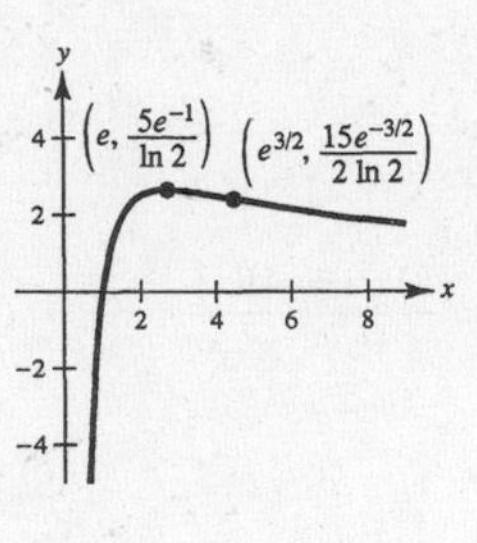

$f'(x) = \dfrac{10(1 - \ln x)}{x^2 \ln 4}$

$f''(x) = \dfrac{10(2 \ln x - 3)}{x^3 \ln 4}$

Relative maximum: $\left(e, \dfrac{5}{e \ln 2}\right) \approx (2.7183, 2.6537)$

Inflection point: $(e^{3/2}, 2.4143)$

73. $f(x) = x + \cos x$

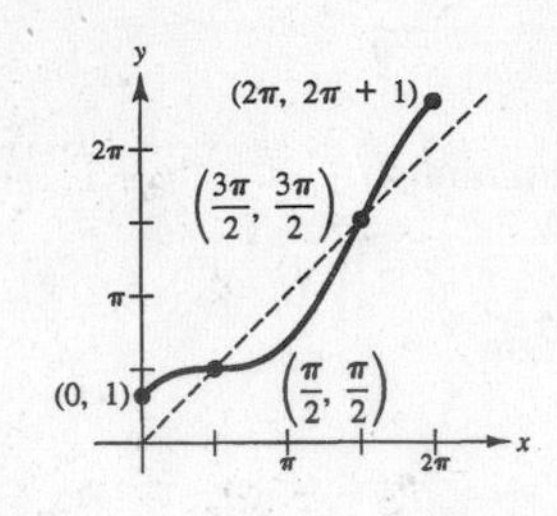

Domain: $[0, 2\pi]$; Range: $[1, 1 + 2\pi]$

$f'(x) = 1 - \sin x \geq 0$, f is increasing.

$f''(x) = -\cos x = 0$ when $x = \dfrac{\pi}{2}, \dfrac{3\pi}{2}$.

Points of inflection: $\left(\dfrac{\pi}{2}, \dfrac{\pi}{2}\right), \left(\dfrac{3\pi}{2}, \dfrac{3\pi}{2}\right)$

Intercept: $(0, 1)$

75. $y = 4x - 6 \arctan x$

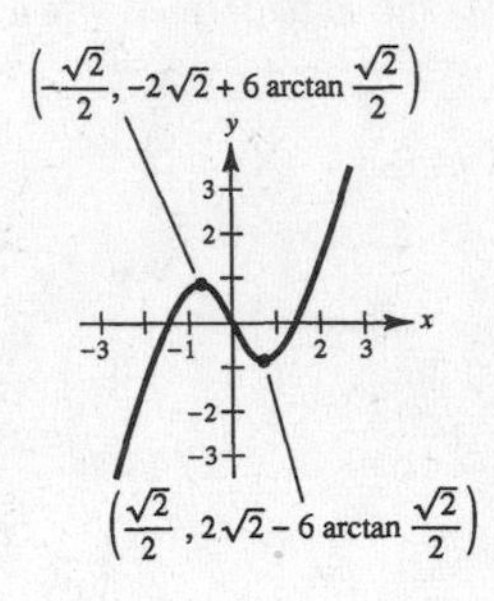

$y' = 4 - \dfrac{6}{1 + x^2} = \dfrac{4x^2 - 2}{1 + x^2}$

$y'' = \dfrac{12x}{(1 + x^2)^2}$

Relative maximum: $\left(-\dfrac{\sqrt{2}}{2}, -2\sqrt{2} + 6 \arctan \dfrac{\sqrt{2}}{2}\right)$

Relative minimum: $\left(\dfrac{\sqrt{2}}{2}, 2\sqrt{2} - 6 \arctan \dfrac{\sqrt{2}}{2}\right)$

Inflection point: $(0, 0)$

77. $x^2 + 4y^2 - 2x - 16y + 13 = 0$

(a) $(x^2 - 2x + 1) + 4(y^2 - 4y + 4) = -13 + 1 + 16$

$$(x - 1)^2 + 4(y - 2)^2 = 4$$

$$\frac{(x - 1)^2}{4} + \frac{(y - 2)^2}{1} = 1$$

The graph is an ellipse:

Maximum: $(1, 3)$

Minimum: $(1, 1)$

(b) $x^2 + 4y^2 - 2x - 16y + 13 = 0$

$$2x + 8y\frac{dy}{dx} - 2 - 16\frac{dy}{dx} = 0$$

$$\frac{dy}{dx}(8y - 16) = 2 - 2x$$

$$\frac{dy}{dx} = \frac{2 - 2x}{8y - 16} = \frac{1 - x}{4y - 8}$$

The critical numbers are $x = 1$ and $y = 2$. These correspond to the points $(1, 1)$, $(1, 3)$, $(2, -1)$, and $(2, 3)$. Hence, the maximum is $(1, 3)$ and the minimum is $(1, 1)$.

79. Let $t = 0$ at noon.

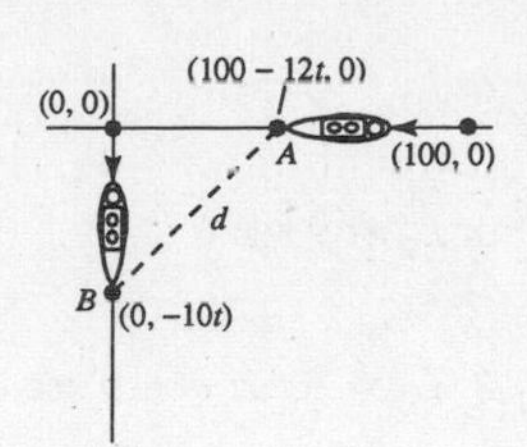

$$L = d^2 = (100 - 12t)^2 + (-10t)^2 = 10{,}000 - 2400t + 244t^2$$

$$\frac{dL}{dt} = -2400 + 488t = 0 \text{ when } t = \frac{300}{61} \approx 4.92 \text{ hr.}$$

Ship A at (40.98, 0); Ship B at (0, −49.18)

$$d^2 = 10{,}000 - 2400t + 244t^2$$

$$\approx 4098.36 \text{ when } t \approx 4.92 \approx 4{:}55 \text{ P.M.}.$$

$$d \approx 64 \text{ km}$$

81. We have points $(0, y)$, $(x, 0)$, and $(1, 8)$. Thus,

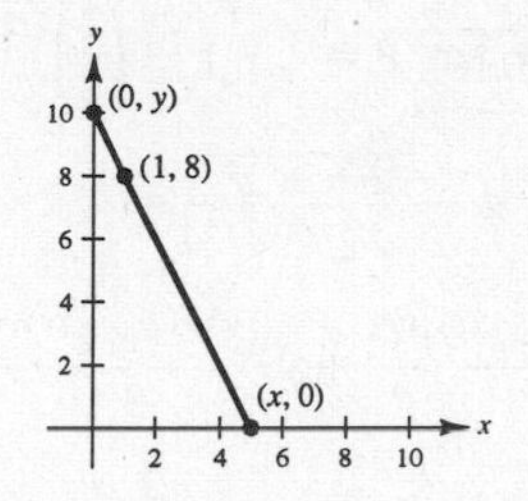

$$m = \frac{y - 8}{0 - 1} = \frac{0 - 8}{x - 1} \text{ or } y = \frac{8x}{x - 1}.$$

Let $f(x) = L^2 = x^2 + \left(\frac{8x}{x - 1}\right)^2$.

$$f'(x) = 2x + 128\left(\frac{x}{x - 1}\right)\left[\frac{(x - 1) - x}{(x - 1)^2}\right] = 0$$

$$x - \frac{64x}{(x - 1)^3} = 0$$

$$x[(x - 1)^3 - 64] = 0 \text{ when } x = 0, 5 \text{ (minimum).}$$

Vertices of triangle: (0, 0), (5, 0), (0, 10)

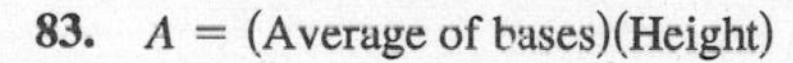

83. $A = \text{(Average of bases)(Height)}$

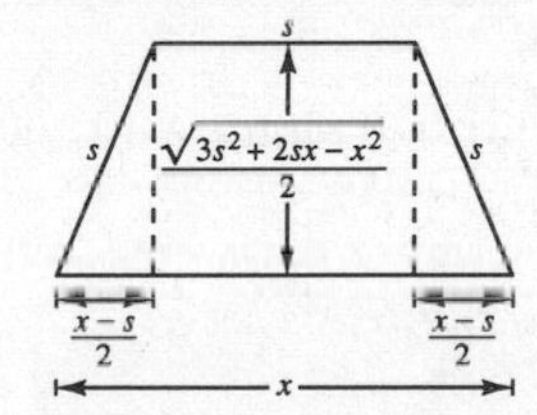

$$= \left(\frac{x + s}{2}\right)\frac{\sqrt{3s^2 + 2sx - x^2}}{2} \text{ (see figure)}$$

$$\frac{dA}{dx} = \frac{1}{4}\left[\frac{(s - x)(s + x)}{\sqrt{3s^2 + 2sx - x^2}} + \sqrt{3s^2 + 2sx - x^2}\right]$$

$$= \frac{2(2s - x)(s + x)}{4\sqrt{3s^2 + 2sx - x^2}} = 0 \text{ when } x = 2s.$$

A is a maximum when $x = 2s$.

85. You can form a right triangle with vertices $(0, 0)$, $(x, 0)$ and $(0, y)$. Assume that the hypotenuse of length L passes through $(4, 6)$.

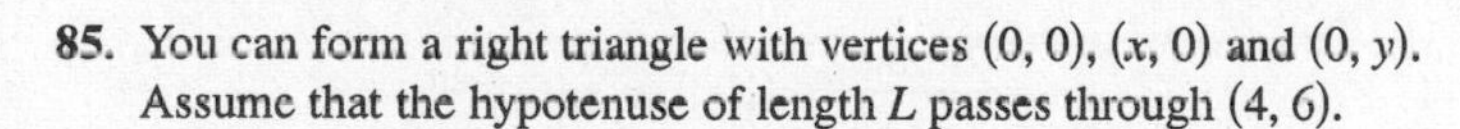

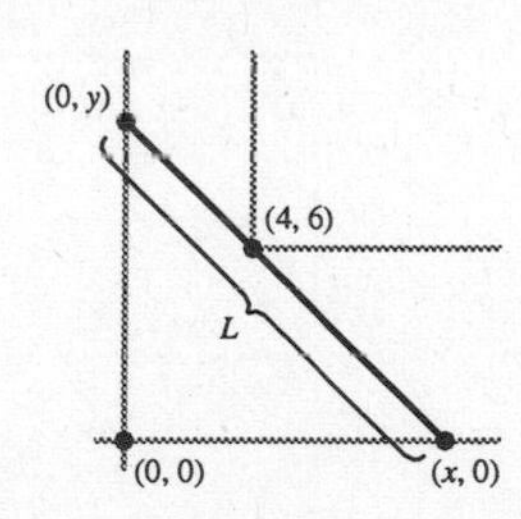

$$m = \frac{y - 6}{0 - 4} = \frac{6 - 0}{4 - x} \text{ or } y = \frac{6x}{x - 4}$$

Let $f(x) = L^2 = x^2 + y^2 = x^2 + \left(\frac{6x}{x - 4}\right)^2$.

$$f'(x) = 2x + 72\left(\frac{x}{x - 4}\right)\left[\frac{-4}{(x - 4)^2}\right] = 0$$

$$x[(x - 4)^3 - 144] = 0 \text{ when } x = 0 \text{ or } x = 4 + \sqrt[3]{144}.$$

$$L \approx 14.05 \text{ feet}$$

87. $\csc\theta = \dfrac{L_1}{6}$ or $L_1 = 6\csc\theta$ (see figure)

$\sec\theta = \dfrac{L_2}{9}$ or $L_2 = 9\sec\theta$

$$L = L_1 + L_2 = 6\csc\theta + 9\sec\theta$$

$$\frac{dL}{d\theta} = -6\csc\theta\cot\theta + 9\sec\theta\tan\theta = 0$$

$$\tan^3\theta = \frac{2}{3} \Longrightarrow \tan\theta = \frac{\sqrt[3]{2}}{\sqrt[3]{3}}$$

$$\sec\theta = \sqrt{1+\tan^2\theta} = \sqrt{1+\left(\frac{2}{3}\right)^{2/3}} = \frac{\sqrt{3^{2/3}+2^{2/3}}}{3^{1/3}}$$

$$\csc\theta = \frac{\sec\theta}{\tan\theta} = \frac{\sqrt{3^{2/3}+2^{2/3}}}{2^{1/3}}$$

$$L = 6\frac{(3^{2/3}+2^{2/3})^{1/2}}{2^{1/3}} + 9\frac{(3^{2/3}+2^{2/3})^{1/2}}{3^{1/3}} = 3(3^{2/3}+2^{2/3})^{3/2} \text{ ft} \approx 21.07 \text{ ft}$$ (Compare to Exercise 72 using $a = 9$ and $b = 6$.)

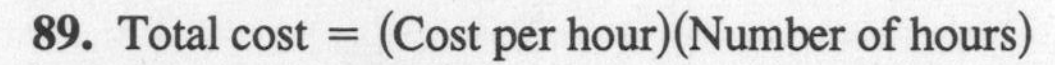

89. Total cost = (Cost per hour)(Number of hours)

$$T = \left(\frac{v^2}{600} + 5\right)\left(\frac{110}{v}\right) = \frac{11v}{60} + \frac{550}{v}$$

$$\frac{dT}{dv} = \frac{11}{60} - \frac{550}{v^2} = \frac{11v^2 - 33{,}000}{60v^2}$$

$$= 0 \text{ when } v = \sqrt{3000} = 10\sqrt{30} \approx 54.8 \text{ mph.}$$

$$\frac{d^2T}{dv^2} = \frac{1100}{v^3} > 0 \text{ when } v = 10\sqrt{30} \text{ so this value yields a minimum.}$$

91. $y = x(1 - \cos x) = x - x\cos x$

$$\frac{dy}{dx} = 1 + x\sin x - \cos x$$

$$dy = (1 + x\sin x - \cos x)\,dx$$

93. $S = 4\pi r^2.\ dr = \Delta r = \pm 0.025$

$$dS = 8\pi r\,dr = 8\pi(9)(\pm 0.025)$$

$$= \pm 1.8\pi \text{ square cm}$$

$$\frac{dS}{S}(100) = \frac{8\pi r\,dr}{4\pi r^2}(100) = \frac{2\,dr}{r}(100)$$

$$= \frac{2(\pm 0.025)}{9}(100) \approx \pm 0.56\%$$

$$V = \frac{4}{3}\pi r^3$$

$$dV = 4\pi r^2\,dr = 4\pi(9)^2(\pm 0.025)$$

$$= \pm 8.1\pi \text{ cubic cm}$$

$$\frac{dV}{V}(100) = \frac{4\pi r^2\,dr}{(4/3)\pi r^3}(100) = \frac{3\,dr}{r}(100)$$

$$= \frac{3(\pm 0.025)}{9}(100) \approx \pm 0.83\%$$

Problem Solving for Chapter 3

1. Assume $y_1 < d < y_2$. Let $g(x) = f(x) - d(x - a)$. g is continuous on $[a, b]$ and therefore has a minimum $(c, g(c))$ on $[a, b]$. The point c cannot be an endpoint of $[a, b]$ because

$$g'(a) = f'(a) - d = y_1 - d < 0$$

$$g'(b) = f'(b) - d = y_2 - d > 0$$

Hence, $a < c < b$ and $g'(c) = 0 \Rightarrow f'(c) = d$.

3. (a) For $a = -3, -2, -1, 0$, p has a relative maximum at $(0, 0)$.

For $a = 1, 2, 3$, p has a relative maximum at $(0, 0)$ and 2 relative minima.

(b) $p'(x) = 4ax^3 - 12x = 4x(ax^2 - 3) = 0 \Rightarrow x = 0, \pm\sqrt{\dfrac{3}{a}}$

$p''(x) = 12ax^2 - 12 = 12(ax^2 - 1)$

For $x = 0$, $p''(0) = -12 < 0 \Rightarrow p$ has a relative maximum at $(0, 0)$.

(c) If $a > 0$, $x = \pm\sqrt{\dfrac{3}{a}}$ are the remaining critical numbers.

$$p''\left(\pm\sqrt{\frac{3}{a}}\right) = 12a\left(\frac{3}{a}\right) - 12 = 24 > 0 \Rightarrow p \text{ has relative minima for } a > 0.$$

(d) $(0, 0)$ lies on $y = -3x^2$.

Let $x = \pm\sqrt{\dfrac{3}{a}}$. Then

$$p(x) = a\left(\frac{3}{a}\right)^2 - 6\left(\frac{3}{a}\right) = \frac{9}{a} - \frac{18}{a} = -\frac{9}{a}.$$

Thus, $y = -\dfrac{9}{a} = -3\left(\pm\sqrt{\dfrac{3}{a}}\right)^2 = -3x^2$ is satisfied by all the relative extrema of p.

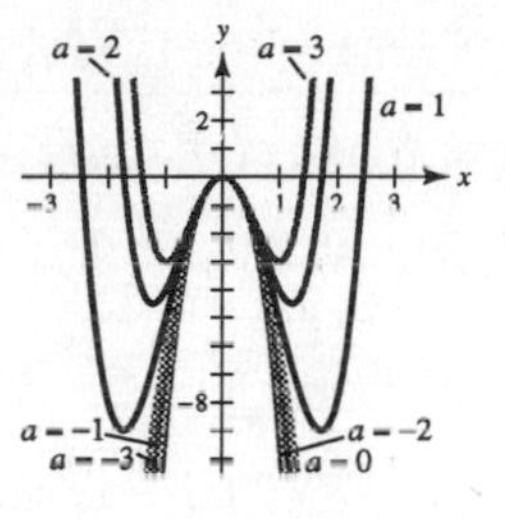

5. $p(x) = x^4 + ax^2 + 1$

(a) $p'(x) = 4x^3 + 2ax = 2x(2x^2 + a)$

$p''(x) = 12x^2 + 2a$

For $a \geq 0$, there is one relative minimum at $(0, 1)$.

(b) For $a < 0$, there is a relative maximum at $(0, 1)$.

(c) For $a < 0$, there are two relative minima at $x = \pm\sqrt{-\dfrac{a}{2}}$.

(d) There are either 1 or 3 critical points. The above analysis shows that there cannot be exactly two relative extrema.

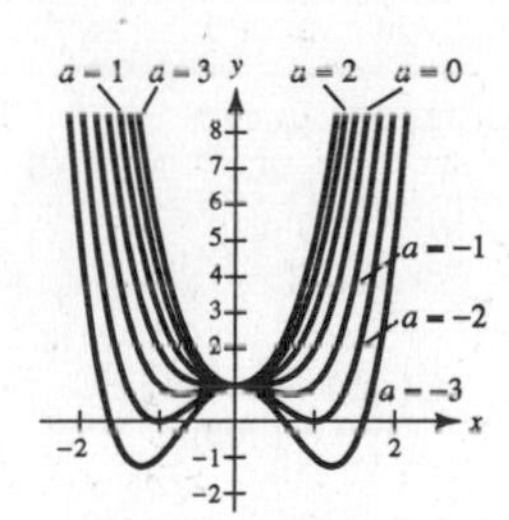

7. $f(x) = \frac{c}{x} + x^2$

$$f'(x) = -\frac{c}{x^2} + 2x = 0 \Rightarrow \frac{c}{x^2} = 2x \Rightarrow \quad x^3 = \frac{c}{2} \Rightarrow x = \sqrt[3]{\frac{c}{x}}$$

$$f''(x) = \frac{2c}{x^3} + 2$$

If $c = 0, f(x) = x^2$ has a relative minimum, but no relative maximum.

If $c > 0, x = \sqrt[3]{\frac{c}{2}}$ is a relative minimum, because $f''\left(\sqrt[3]{\frac{c}{2}}\right) > 0$.

If $c < 0, x = \sqrt[3]{\frac{c}{2}}$ is a relative minimum too.

Answer: all c.

9. Set $\frac{f(b) - f(a) - f'(a)(b - a)}{(b - a)^2} = k$.

Define $F(x) = f(x) - f(a) - f'(a)(x - a) - k(x - a)^2$.

$F(a) = 0, F(b) = f(b) - f(a) - f'(a)(b - a) - k(b - a)^2 = 0$

F is continuous on $[a, b]$ and differentiable on (a, b).

There exists $c_1, a < c_1 < b$, satisfying $F'(c_1) = 0$.

$F'(x) = f'(x) - f'(a) - 2k(x - a)$ satisfies the hypothesis of Rolle's Theorem on $[a, c_1]$:

$F'(a) = 0, F'(c_1) = 0$.

There exists $c_2, a < c_2 < c_1$ satisfying $F''(c_2) = 0$.

Finally, $F''(x) = f''(x) - 2k$ and $F''(c_2) = 0$ implies that

$$k = \frac{f''(c_2)}{2}.$$

Thus, $k = \frac{f(b) - f(a) - f'(a)(b - a)}{(b - a)^2} = \frac{f''(c_2)}{2} \Rightarrow f(b) = f(a) + f'(a)(b - a) + \frac{1}{2}f''(c_2)(b - a)^2$.

11. $E(\phi) = \frac{\tan \phi(1 - 0.1 \tan \phi)}{0.1 + \tan \phi} = \frac{10 \tan \phi - \tan^2 \phi}{1 + 10 \tan \phi}$

$$E'(\phi) = \frac{(1 + 10 \tan \phi)(10 \sec^2 \phi - 2 \tan \phi \sec^2 \phi) - (10 \tan \phi - \tan^2 \phi)10 \sec^2 \phi}{(1 + 10 \tan \phi)} = 0$$

$\Rightarrow (1 + 10 \tan \phi)(10 \sec^2 \phi - 2 \tan \phi \sec^2 \phi) = (10 \tan \phi - \tan^2 \phi)10 \sec^2 \phi$

$\Rightarrow 10 \sec^2 \phi - 2 \tan \phi \sec^2 \phi + 100 \tan \phi \sec^2 \phi - 20 \tan^2 \phi \sec^2 \phi$

$= 100 \tan \phi \sec^2 \phi - 10 \tan^2 \phi \sec^2 \phi$

$\Rightarrow 10 - 2 \tan \phi = 10 \tan^2 \phi$

$\Rightarrow 10 \tan^2 \phi + 2 \tan \phi - 10 = 0$

$$\tan \phi = \frac{-2 \pm \sqrt{4 + 400}}{20} \approx 0.90499, -1.10499$$

Using the positive value, $\phi \approx 0.7356$, or $42.14°$.

13. $v = -2400\pi \sin \theta$

$v' = -2400\pi \cos \theta = 0$

$\theta = \frac{\pi}{2} + 2n\pi, \frac{3\pi}{2} + 2n\pi$, n an integer

15. The line has equation $\frac{x}{3} + \frac{y}{4} = 1$ or $y = -\frac{4}{3}x + 4$.

Rectangle:

$$\text{Area} = A = xy = x\left(-\frac{4}{3}x + 4\right) = -\frac{4}{3}x^2 + 4x.$$

$$A'(x) = -\frac{8}{3}x + 4 = 0 \Rightarrow \frac{8}{3}x = 4 \Rightarrow x = \frac{3}{2}$$

Dimensions: $\frac{3}{2} \times 2$ Calculus was helpful.

Circle: The distance from the center (r, r) to the line $\frac{x}{3} + \frac{y}{4} - 1 = 0$ must be r:

$$r = \frac{\left|\frac{r}{3} + \frac{r}{4} - 1\right|}{\sqrt{\frac{1}{9} + \frac{1}{16}}} = \frac{12}{5}\left|\frac{7r - 12}{12}\right| = \frac{|7r - 12|}{5}$$

$$5r = |7r - 12| \Rightarrow r = 1 \text{ or } r = 6.$$

Clearly, $r = 1$.

Semicircle: The center lies on the line $\frac{x}{3} + \frac{y}{4} = 1$ and satisfies $x = y = r$.

Thus $\frac{r}{3} + \frac{r}{4} = 1 \Rightarrow \frac{7}{12}r = 1 \Rightarrow r = \frac{12}{7}$. No calculus necessary.

17. $y = (1 + x^2)^{-1}$

$$y' = \frac{-2x}{(1 + x^2)^2}$$

$$y'' = \frac{2(3x^2 - 1)}{(x^2 + 1)^3} = 0 \Rightarrow x = \pm\frac{1}{\sqrt{3}} = \pm\frac{\sqrt{3}}{3}$$

y'': +++ $-\frac{\sqrt{3}}{3}$ ---- 0 ---- $\frac{\sqrt{3}}{3}$ +++

The tangent line has greatest slope at $\left(-\frac{\sqrt{3}}{3}, \frac{3}{4}\right)$ and least slope at $\left(\frac{\sqrt{3}}{3}, \frac{3}{4}\right)$.

CHAPTER 4
Integration

CHAPTER 4
Integration

Section 4.1 Antiderivatives and Indefinite Integration

Solutions to Odd-Numbered Exercises

1. $\frac{d}{dx}\left(\frac{3}{x^3} + C\right) = \frac{d}{dx}(3x^{-3} + C) = -9x^{-4} = \frac{-9}{x^4}$

3. $\frac{d}{dx}\left(\frac{1}{3}x^3 - 4x + C\right) = x^2 - 4 = (x - 2)(x + 2)$

5. $\frac{dy}{dt} = 3t^2$

$y = t^3 + C$

Check: $\frac{d}{dt}[t^3 + C] = 3t^2$

7. $\frac{dy}{dx} = x^{3/2}$

$y = \frac{2}{5}x^{5/2} + C$

Check: $\frac{d}{dx}\left[\frac{2}{5}x^{5/2} + C\right] = x^{3/2}$

	Given	*Rewrite*	*Integrate*	*Simplify*
9.	$\int \sqrt[3]{x}\, dx$	$\int x^{1/3}\, dx$	$\frac{x^{4/3}}{4/3} + C$	$\frac{3}{4}x^{4/3} + C$
11.	$\int \frac{1}{x\sqrt{x}}\, dx$	$\int x^{-3/2}\, dx$	$\frac{x^{-1/2}}{-1/2} + C$	$-\frac{2}{\sqrt{x}} + C$
13.	$\int \frac{1}{2x^3}\, dx$	$\frac{1}{2}\int x^{-3}\, dx$	$\frac{1}{2}\left(\frac{x^{-2}}{-2}\right) + C$	$-\frac{1}{4x^2} + C$

15. $\int (x + 3)dx = \frac{x^2}{2} + 3x + C$

Check: $\frac{d}{dx}\left[\frac{x^2}{2} + 3x + C\right] = x + 3$

17. $\int (x^3 + 5)\, dx = \frac{1}{4}x^4 + 5x + C$

Check: $\frac{d}{dx}\left(\frac{1}{4}x^4 + 5x + C\right) = x^3 + 5$

19. $\int (x^{3/2} + 2x + 1)\, dx = \frac{2}{5}x^{5/2} + x^2 + x + C$

Check: $\frac{d}{dx}\left(\frac{2}{5}x^{5/2} + x^2 + x + C\right) = x^{3/2} + 2x + 1$

21. $\int \frac{1}{x^3}\, dx = \int x^{-3}\, dx = \frac{x^{-2}}{-2} + C = -\frac{1}{2x^2} + C$

Check: $\frac{d}{dx}\left(-\frac{1}{2x^2} + C\right) = \frac{1}{x^3}$

23. $\int \frac{x^2 + x + 1}{\sqrt{x}}\, dx = \int (x^{3/2} + x^{1/2} + x^{-1/2})\, dx = \frac{2}{5}x^{5/2} + \frac{2}{3}x^{3/2} + 2x^{1/2} + C = \frac{2}{15}x^{1/2}(3x^2 + 5x + 15) + C$

Check: $\frac{d}{dx}\left(\frac{2}{5}x^{5/2} + \frac{2}{3}x^{3/2} + 2x^{1/2} + C\right) = x^{3/2} + x^{1/2} + x^{-1/2} = \frac{x^2 + x + 1}{\sqrt{x}}$

25. $\int (x + 1)(3x - 2)\, dx = \int (3x^2 + x - 2)\, dx = x^3 + \frac{1}{2}x^2 - 2x + C$

Check: $\frac{d}{dx}\left(x^3 + \frac{1}{2}x^2 - 2x + C\right) = 3x^2 + x - 2 = (x + 1)(3x - 2)$

27. $\int y^2\sqrt{y}\,dy = \int y^{5/2}\,dy = \frac{2}{7}y^{7/2} + C$

Check: $\frac{d}{dy}\left(\frac{2}{7}y^{7/2} + C\right) = y^{5/2} = y^2\sqrt{y}$

29. $\int dx = \int 1\,dx = x + C$

Check: $\frac{d}{dx}(x + C) = 1$

31. $\int (2\sin x + 3\cos x)\,dx = -2\cos x + 3\sin x + C$

Check: $\frac{d}{dx}(-2\cos x + 3\sin x + C) = 2\sin x + 3\cos x$

33. $\int (1 - \csc t\cot t)\,dt = t + \csc t + C$

Check: $\frac{d}{dt}(t + \csc t + C) = 1 - \csc t\cot t$

35. $\int (2\sin x - 5e^x)\,dx = -2\cos x - 5e^x + C$

Check: $\frac{d}{dx}(-2\cos x - 5e^x + C) = 2\sin x - 5e^x$

37. $\int (\sec^2\theta - \sin\theta)\,d\theta = \tan\theta + \cos\theta + C$

Check: $\frac{d}{d\theta}(\tan\theta + \cos\theta + C) = \sec^2\theta - \sin\theta$

39. $\int (\tan^2 y + 1)\,dy = \int \sec^2 y\,dy = \tan y + C$

Check: $\frac{d}{dy}(\tan y + C) = \sec^2 y = \tan^2 y + 1$

41. $\int (2x - 4^x)\,dx = x^2 - \frac{4^x}{\ln 4} + C$

Check: $\frac{d}{dx}\left(x^2 - \frac{4^x}{\ln 4} + C\right) = 2x - 4^x$

43. $\int \left(x - \frac{5}{x}\right)dx = \frac{x^2}{2} - 5\ln|x| + C$

Check: $\frac{d}{dx}\left(\frac{x^2}{2} - 5\ln|x| + C\right) = x - \frac{5}{x}$

45. $f(x) = \cos x$

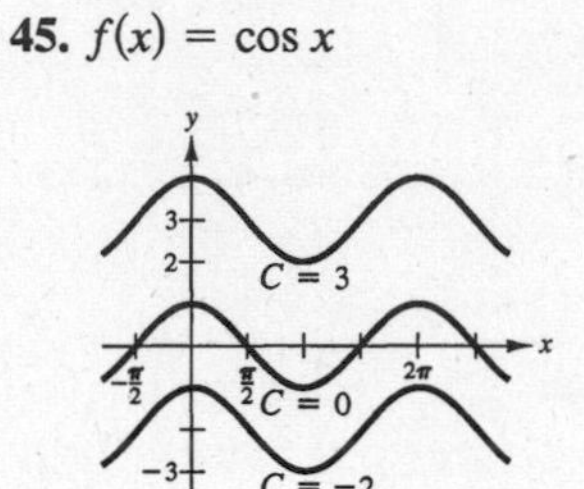

47. $f(x) = \ln x$

49. $f'(x) = 2$

$f(x) = 2x + C$

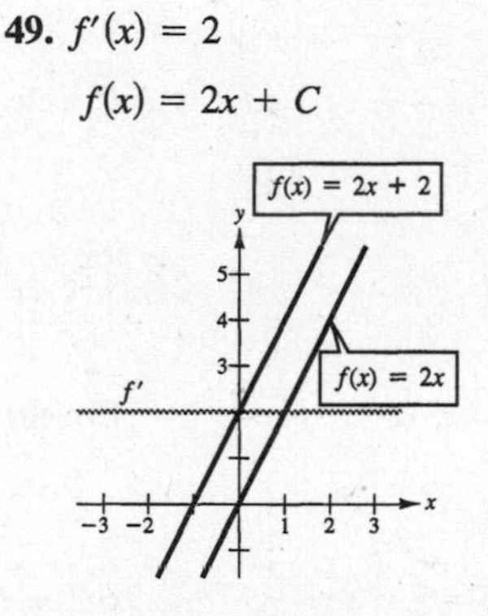

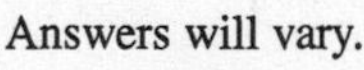
Answers will vary.

51. $f'(x) = 1 - x^2$

$f(x) = x - \frac{x^3}{3} + C$

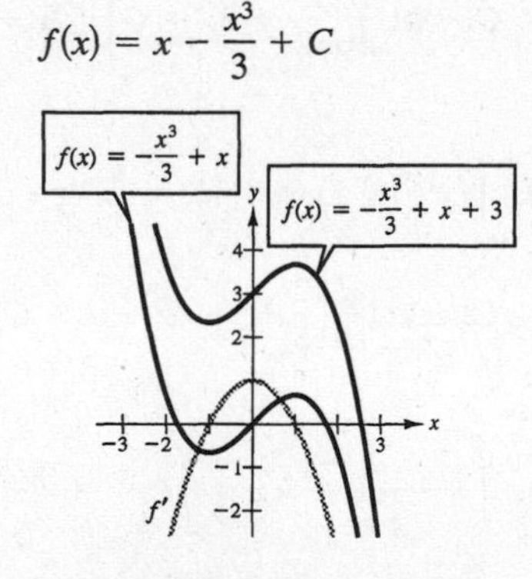

Answers will vary.

53. $\frac{dy}{dx} = 2x - 1,\ (1, 1)$

$y = \int (2x - 1)\,dx = x^2 - x + C$

$1 = (1)^2 - (1) + C \Rightarrow C = 1$

$y = x^2 - x + 1$

55. $\frac{dy}{dx} = \cos x,\ (0, 4)$

$y = \int \cos x\,dx = \sin x + C$

$4 = \sin 0 + C \Rightarrow C = 4$

$y = \sin x + 4$

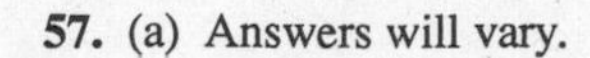

57. (a) Answers will vary.

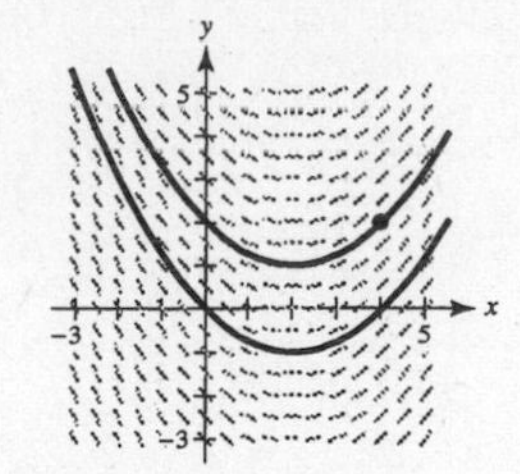

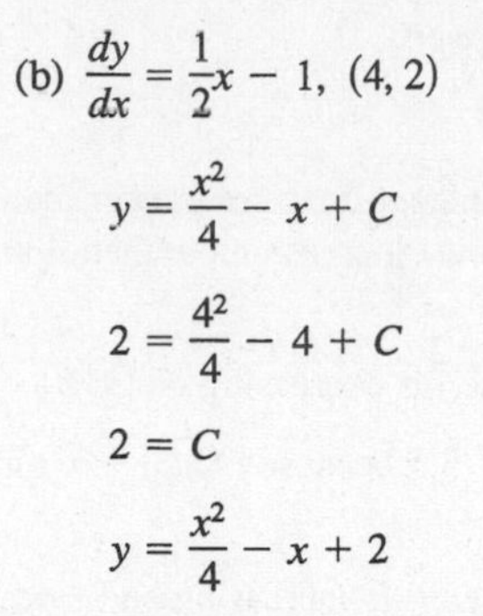

(b) $\dfrac{dy}{dx} = \dfrac{1}{2}x - 1, \ (4, 2)$

$$y = \frac{x^2}{4} - x + C$$

$$2 = \frac{4^2}{4} - 4 + C$$

$$2 = C$$

$$y = \frac{x^2}{4} - x + 2$$

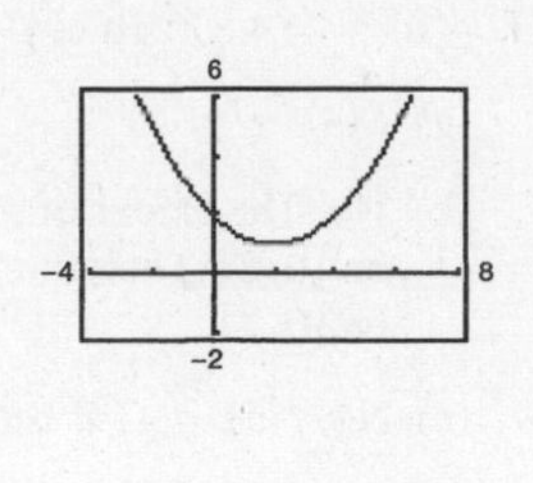

59. $f'(x) = 4x, f(0) = 6$

$$f(x) = \int 4x\,dx = 2x^2 + C$$

$$f(0) = 6 = 2(0)^2 + C \Rightarrow C = 6$$

$$f(x) = 2x^2 + 6$$

61. $h'(t) = 8t^3 + 5, h(1) = -4$

$$h(t) = \int (8t^3 + 5)dt = 2t^4 + 5t + C$$

$$h(1) = -4 = 2 + 5 + C \Rightarrow C = -11$$

$$h(t) = 2t^4 + 5t - 11$$

63. $f''(x) = 2$

$f'(2) = 5$

$f(2) = 10$

$$f'(x) = \int 2\,dx = 2x + C_1$$

$$f'(2) = 4 + C_1 = 5 \Rightarrow C_1 = 1$$

$$f'(x) - 2x + 1$$

$$f(x) = \int (2x + 1)\,dx = x^2 + x + C_2$$

$$f(2) = 6 + C_2 = 10 \Rightarrow C_2 = 4$$

$$f(x) = x^2 + x + 4$$

65. $f''(x) = x^{-3/2}$

$f'(4) = 2$

$f(0) = 0$

$$f'(x) = \int x^{-3/2}\,dx = -2x^{-1/2} + C_1 = -\frac{2}{\sqrt{x}} + C_1$$

$$f'(4) = -\frac{2}{2} + C_1 = 2 \Rightarrow C_1 = 3$$

$$f'(x) = -\frac{2}{\sqrt{x}} + 3$$

$$f(x) = \int (-2x^{-1/2} + 3)\,dx = -4x^{1/2} + 3x + C_2$$

$$f(0) = 0 + 0 + C_2 = 0 \Rightarrow C_2 = 0$$

$$f(x) = -4x^{1/2} + 3x = -4\sqrt{x} + 3x$$

67. $f''(x) = e^x$

$$f'(x) = \int e^x\,dx = e^x + C_1$$

$$f'(0) = 2 = e^0 + C_1 \Rightarrow C_1 = 1$$

$$f(x) = \int (e^x + 1)\,dx = e^x + x + C_2$$

$$f(0) = 5 = e^0 + 0 + C_2 \Rightarrow C_2 = 4$$

$$f(x) = e^x + x + 4$$

69. (a) $h(t) = \displaystyle\int (1.5t + 5)\,dt = 0.75t^2 + 5t + C$

$$h(0) = 0 + 0 + C = 12 \Rightarrow C = 12$$

$$h(t) = 0.75t^2 + 5t + 12$$

(b) $h(6) = 0.75(6)^2 + 5(6) + 12 = 69$ cm

71. $f(0) = -4$. Graph of f' is given.

(a) $f'(4) \approx -1.0$

(b) No. The slopes of the tangent lines are greater than 2 on $[0, 2]$. Therefore, f must increase more than 4 units on $[0, 4]$.

(c) No, $f(5) < f(4)$ because f is decreasing on $[4, 5]$.

(d) f is an maximum at $x = 3.5$ because $f'(3.5) \approx 0$ and the first derivative test.

(e) f is concave upward when f' is increasing on $(-\infty, 1)$ and $(5, \infty)$. f is concave downward on $(1, 5)$. Points of inflection at $x = 1, 5$.

(f) f'' is a minimum at $x = 3$.

(g)

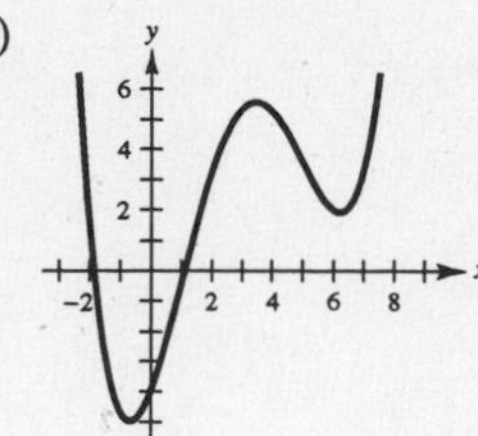

73. $a(t) = -32$ ft/sec^2

$v(t) = \int -32\,dt = -32t + C_1$

$v(0) = 60 = C_1$

$s(t) = \int (-32t + 60)dt = -16t^2 + 60t + C_2$

$s(0) = 6 = C_2$

$s(t) = -16t^2 + 60t + 6$ Position function

The ball reaches its maximim height when

$v(t) = -32t + 60 = 0$

$32t = 60$

$t = \frac{15}{8}$ seconds

$$s\left(\frac{15}{8}\right) = -16\left(\frac{15}{8}\right)^2 + 60\left(\frac{15}{8}\right) + 6 \approx 62.25 \text{ feet}$$

75. From Exercise 74, we have:

$s(t) = -16t^2 + v_0 t$

$s'(t) = -32t + v_0 = 0$ when $t = \frac{v_0}{32}$ = time to reach maximum height.

$$s\left(\frac{v_0}{32}\right) = -16\left(\frac{v_0}{32}\right)^2 + v_0\left(\frac{v_0}{32}\right) = 550$$

$$-\frac{v_0^2}{64} + \frac{v_0^2}{32} = 550$$

$v_0^2 = 35{,}200$

$v_0 \approx 187.617$ ft/sec

77. $a(t) = -9.8$

$v(t) = \int -9.8\,dt = -9.8t + C_1$

$v(0) = v_0 = C_1 \Rightarrow v(t) = -9.8t + v_0$

$f(t) = \int (-9.8t + v_0)\,dt = -4.9t^2 + v_0 t + C_2$

$f(0) = s_0 = C_2 \Rightarrow f(t) = -4.9t^2 + v_0 t + s_0$

79. $a = -1.6$

$v(t) = \int -1.6\,dt = -1.6t + v_0 = -1.6t,$

since the stone was dropped, $v_0 = 0$.

$s(t) = \int (-1.6t)\,dt = -0.8t^2 + s_0$

$s(20) = 0 \Rightarrow -0.8(20)^2 + s_0 = 0$

$s_0 = 320$

Thus, the height of the cliff is 320 meters.

$v(t) = -1.6t$

$v(20) = -32$ m/sec

81. $x(t) = t^3 - 6t^2 + 9t - 2 \qquad 0 \le t \le 5$

(a) $v(t) = x'(t) = 3t^2 - 12t + 9$

$= 3(t^2 - 4t + 3) = 3(t-1)(t-3)$

$a(t) = v'(t) = 6t - 12 = 6(t-2)$

(b) $v(t) > 0$ when $0 < t < 1$ or $3 < t < 5$.

(c) $a(t) = 6(t-2) = 0$ when $t = 2$.

$v(2) = 3(1)(-1) = -3$

83. $v(t) = \dfrac{1}{\sqrt{t}} = t^{-1/2} \qquad t > 0$

$x(t) = \int v(t)\,dt = 2t^{1/2} + C$

$x(1) = 4 = 2(1) + C \Rightarrow C = 2$

$x(t) = 2t^{1/2} + 2$ position function

$a(t) = v'(t) = -\dfrac{1}{2}t^{-3/2} = \dfrac{-1}{2t^{3/2}}$ acceleration

85. (a) $v(0) = 25$ km/hr $= 25 \cdot \dfrac{1000}{3600} = \dfrac{250}{36}$ m/sec

$v(13) = 80$ km/hr $= 80 \cdot \dfrac{1000}{3600} = \dfrac{800}{36}$ m/sec

$a(t) = a$ (constant acceleration)

$v(t) = at + C$

$v(0) = \dfrac{250}{36} \Rightarrow v(t) = at + \dfrac{250}{36}$

$v(13) = \dfrac{800}{36} = 13a + \dfrac{250}{36}$

$\dfrac{550}{36} = 13a$

$a = \dfrac{550}{468} = \dfrac{275}{234} \approx 1.175$ m/sec^2

(b) $s(t) = a\dfrac{t^2}{2} + \dfrac{250}{36}t \qquad (s(0) = 0)$

$s(13) = \dfrac{275}{234}\dfrac{(13)^2}{2} + \dfrac{250}{36}(13) \approx 189.58$ m

87. Truck: $v(t) = 30$

$s(t) = 30t$ (Let $s(0) = 0$.)

Automobile: $a(t) = 6$

$v(t) = 6t$ (Let $v(0) = 0$.)

$s(t) = 3t^2$ (Let $s(0) = 0$.)

At the point where the automobile overtakes the truck:

$30t = 3t^2$

$0 = 3t^2 - 30t$

$0 = 3t(t - 10)$ when $t = 10$ sec.

(a) $s(10) = 3(10)^2 = 300$ ft

(b) $v(10) = 6(10) = 60$ ft/sec ≈ 41 mph

89. $\dfrac{(1 \text{ mi/hr})(5280 \text{ ft/mi})}{(3600 \text{ sec/hr})} = \dfrac{22}{15}$ ft/sec

(a)

t	0	5	10	15	20	25	30
V_1(ft/sec)	0	3.67	10.27	23.47	42.53	66	95.33
V_2(ft/sec)	0	30.8	55.73	74.8	88	93.87	95.33

(b) $V_1(t) = 0.1068t^2 - 0.0416t + 0.3679$

$V_2(t) = -0.1208t^2 + 6.7991t - 0.0707$

(c) $S_1(t) = \int V_1(t)\,dt = \dfrac{0.1068}{3}t^3 - \dfrac{0.0416}{2}t^2 + 0.3679t$

$S_2(t) = \int V_2(t)\,dt = -\dfrac{0.1208t^3}{3} + \dfrac{6.7991t^2}{2} - 0.0707t$

[In both cases, the constant of integration is 0 because $S_1(0) = S_2(0) = 0$]

$S_1(30) \approx 953.5$ feet

$S_2(30) \approx 1970.3$ feet

The second car was going faster than the first until the end.

91. $a(t) = k$

$v(t) = kt$

$s(t) = \frac{k}{2}t^2$ since $v(0) = s(0) = 0$.

At the time of lift-off, $kt = 160$ and $(k/2)t^2 = 0.7$. Since $(k/2)t^2 = 0.7$,

$$t = \sqrt{\frac{1.4}{k}}$$

$$v\left(\sqrt{\frac{1.4}{k}}\right) = k\sqrt{\frac{1.4}{k}} = 160$$

$$1.4k = 160^2 \Rightarrow k = \frac{160^2}{1.4}$$

$$\approx 18{,}285.714 \text{ mi/hr}^2$$

$$\approx 7.45 \text{ ft/sec}^2.$$

93. True

95. True

97. False. For example, $\int x \cdot x\,dx \neq \int x\,dx \cdot \int x\,dx$ because $\frac{x^3}{3} + C \neq \left(\frac{x^2}{2} + C_1\right)\left(\frac{x^2}{2} + C_2\right)$

99. $f'(x) = \begin{cases} 1, & 0 \le x < 2 \\ 3x, & 2 \le x \le 5 \end{cases}$

$$f(x) = \begin{cases} x + C_1, & 0 \le x < 2 \\ \frac{3x^2}{2} + C_2, & 2 \le x \le 5 \end{cases}$$

$f(1) = 3 \Rightarrow 1 + C_1 = 3 \Rightarrow C_1 = 2$

f is continuous: Values must agree at $x = 2$:

$4 = 6 + C_2 \Rightarrow C_2 = -2$

$$f(x) = \begin{cases} x + 2, & 0 \le x < 2 \\ \frac{3x^2}{2} - 2, & 2 \le x \le 5 \end{cases}$$

The left and right hand derivatives at $x = 2$ do not agree.
Hence f is not differentiable at $x = 2$.

101. $\frac{d}{dx}(\ln|Cx|) = \frac{d}{dx}(\ln|C| + \ln|x|) = 0 + \frac{1}{x} = \frac{1}{x}$

Section 4.2 Area

1. $\sum_{i=1}^{5}(2i + 1) = 2\sum_{i=1}^{5} i + \sum_{i=1}^{5} 1 = 2(1 + 2 + 3 + 4 + 5) + 5 = 35$

3. $\sum_{k=0}^{4} \frac{1}{k^2 + 1} = 1 + \frac{1}{2} + \frac{1}{5} + \frac{1}{10} + \frac{1}{17} = \frac{158}{85}$

5. $\sum_{k=1}^{4} c = c + c + c + c = 4c$

7. $\sum_{i=1}^{9} \frac{1}{3i}$

9. $\sum_{j=1}^{8}\left[5\left(\frac{j}{8}\right) + 3\right]$

11. $\frac{2}{n}\sum_{i=1}^{n}\left[\left(\frac{2i}{n}\right)^3 - \left(\frac{2i}{n}\right)\right]$

13. $\frac{3}{n}\sum_{i=1}^{n}\left[2\left(1 + \frac{3i}{n}\right)^2\right]$

15. $\sum_{i=1}^{20} 2i = 2\sum_{i=1}^{20} i = 2\left[\frac{20(21)}{2}\right] = 420$

17. $\sum_{i=1}^{20}(i - 1)^2 = \sum_{i=1}^{19} i^2$

$$= \left[\frac{19(20)(39)}{6}\right] = 2470$$

19. $\sum_{i=1}^{15} i(i-1)^2 = \sum_{i=1}^{15} i^3 - 2\sum_{i=1}^{15} i^2 + \sum_{i=1}^{15} i$

$$= \frac{15^2(16)^2}{4} - 2\frac{15(16)(31)}{6} + \frac{15(16)}{2}$$

$$= 14{,}400 - 2{,}480 + 120$$

$$= 12{,}040$$

21. sum seq(x ^ 2 + 3, x, 1, 20, 1) = 2930 (TI-82)

$$\sum_{i=1}^{20} (i^2 + 3) = \frac{20(20+1)(2(20)+1)}{6} + 3(20)$$

$$= \frac{(20)(21)(41)}{6} + 60 = 2930$$

23. $S = \left[3 + 4 + \frac{9}{2} + 5\right](1) = \frac{33}{2} = 16.5$

$s = \left[1 + 3 + 4 + \frac{9}{2}\right](1) = \frac{25}{2} = 12.5$

25. $S = [3 + 3 + 5](1) = 11$

$s = [2 + 2 + 3](1) = 7$

27. $S(4) = \sqrt{\frac{1}{4}}\left(\frac{1}{4}\right) + \sqrt{\frac{1}{2}}\left(\frac{1}{4}\right) + \sqrt{\frac{3}{4}}\left(\frac{1}{4}\right) + \sqrt{1}\left(\frac{1}{4}\right) = \frac{1 + \sqrt{2} + \sqrt{3} + 2}{8} \approx 0.768$

$s(4) = 0\left(\frac{1}{4}\right) + \sqrt{\frac{1}{4}}\left(\frac{1}{4}\right) + \sqrt{\frac{1}{2}}\left(\frac{1}{4}\right) + \sqrt{\frac{3}{4}}\left(\frac{1}{4}\right) = \frac{1 + \sqrt{2} + \sqrt{3}}{8} \approx 0.518$

29. $S(5) = 1\left(\frac{1}{5}\right) + \frac{1}{6/5}\left(\frac{1}{5}\right) + \frac{1}{7/5}\left(\frac{1}{5}\right) + \frac{1}{8/5}\left(\frac{1}{5}\right) + \frac{1}{9/5}\left(\frac{1}{5}\right) = \frac{1}{5} + \frac{1}{6} + \frac{1}{7} + \frac{1}{8} + \frac{1}{9} \approx 0.746$

$s(5) = \frac{1}{6/5}\left(\frac{1}{5}\right) + \frac{1}{7/5}\left(\frac{1}{5}\right) + \frac{1}{8/5}\left(\frac{1}{5}\right) + \frac{1}{9/5}\left(\frac{1}{5}\right) + \frac{1}{2}\left(\frac{1}{5}\right) = \frac{1}{6} + \frac{1}{7} + \frac{1}{8} + \frac{1}{9} + \frac{1}{10} \approx 0.646$

31. $\lim_{n\to\infty} \left[\left(\frac{81}{n^4}\right)\frac{n^2(n+1)^2}{4}\right] = \frac{81}{4}\lim_{n\to\infty}\left[\frac{n^4 + 2n^3 + n^2}{n^4}\right] = \frac{81}{4}(1) = \frac{81}{4}$

33. $\lim_{n\to\infty} \left[\left(\frac{18}{n^2}\right)\frac{n(n+1)}{2}\right] = \frac{18}{2}\lim_{n\to\infty}\left[\frac{n^2 + n}{n^2}\right] = \frac{18}{2}(1) = 9$

35. $\sum_{i=1}^{n} \frac{2i+1}{n^2} = \frac{1}{n^2}\sum_{i=1}^{n}(2i+1) = \frac{1}{n^2}\left[2\frac{n(n+1)}{2} + n\right] = \frac{n+2}{n} = S(n)$

$S(10) = \frac{12}{10} = 1.2$

$S(100) = 1.02$

$S(1000) = 1.002$

$S(10{,}000) = 1.0002$

37. $\sum_{k=1}^{n} \frac{6k(k-1)}{n^3} = \frac{6}{n^3}\sum_{k=1}^{n}(k^2 - k) = \frac{6}{n^3}\left[\frac{n(n+1)(2n+1)}{6} - \frac{n(n+1)}{2}\right]$

$$= \frac{6}{n^2}\left[\frac{2n^2 + 3n + 1 - 3n - 3}{6}\right] = \frac{1}{n^2}[2n^2 - 2] = S(n)$$

$S(10) = 1.98$

$S(100) = 1.9998$

$S(1000) = 1.999998$

$S(10{,}000) = 1.99999998$

39. $\lim_{n\to\infty} \sum_{i=1}^{n}\left(\frac{16i}{n^2}\right) = \lim_{n\to\infty}\frac{16}{n^2}\sum_{i=1}^{n} i = \lim_{n\to\infty}\frac{16}{n^2}\left(\frac{n(n+1)}{2}\right) = \lim_{n\to\infty}\left[8\left(\frac{n^2+n}{n^2}\right)\right] = 8\lim_{n\to\infty}\left(1 + \frac{1}{n}\right) = 8$

41. $\displaystyle \lim_{n\to\infty} \sum_{i=1}^{n} \frac{1}{n^3}(i-1)^2 = \lim_{n\to\infty} \frac{1}{n^3}\sum_{i=1}^{n-1} i^2 = \lim_{n\to\infty} \frac{1}{n^3}\left[\frac{(n-1)(n)(2n-1)}{6}\right]$

$$= \lim_{n\to\infty} \frac{1}{6}\left[\frac{2n^3-3n^2+n}{n^3}\right] = \lim_{n\to\infty}\left[\frac{1}{6}\left(\frac{2-(3/n)+(1/n^2)}{1}\right)\right] = \frac{1}{3}$$

43. $\displaystyle \lim_{n\to\infty} \sum_{i=1}^{n}\left(1+\frac{i}{n}\right)\left(\frac{2}{n}\right) = 2\lim_{n\to\infty}\frac{1}{n}\left[\sum_{i=1}^{n} 1 + \frac{1}{n}\sum_{i=1}^{n} i\right] = 2\lim_{n\to\infty}\frac{1}{n}\left[n+\frac{1}{n}\left(\frac{n(n+1)}{2}\right)\right] = 2\lim_{n\to\infty}\left[1+\frac{n^2+n}{2n^2}\right] = 2\left(1+\frac{1}{2}\right) = 3$

45. (a)

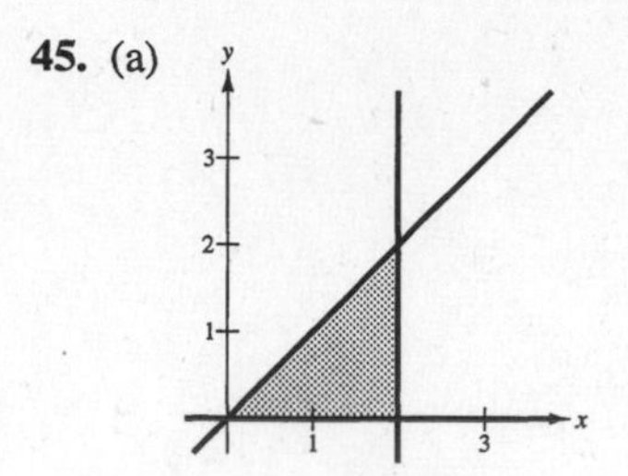

(b) $\Delta x = \dfrac{2-0}{n} = \dfrac{2}{n}$

Endpoints:

$$0 < 1\left(\frac{2}{n}\right) < 2\left(\frac{2}{n}\right) < \cdots < (n-1)\left(\frac{2}{n}\right) < n\left(\frac{2}{n}\right) = 2$$

(c) Since $y = x$ is increasing, $f(m_i) = f(x_{i-1})$ on $[x_{i-1}, x_i]$.

$$s(n) = \sum_{i=1}^{n} f(x_{i-1})\,\Delta x$$

$$= \sum_{i=1}^{n} f\left(\frac{2i-2}{n}\right)\left(\frac{2}{n}\right) = \sum_{i=1}^{n}\left[(i-1)\left(\frac{2}{n}\right)\right]\left(\frac{2}{n}\right)$$

(d) $f(M_i) = f(x_i)$ on $[x_{i-1}, x_i]$

$$S(n) = \sum_{i=1}^{n} f(x_i)\,\Delta x = \sum_{i=1}^{n} f\left(\frac{2i}{n}\right)\frac{2}{n} = \sum_{i=1}^{n}\left[i\left(\frac{2}{n}\right)\right]\left(\frac{2}{n}\right)$$

(e)

x	5	10	50	100
$s(n)$	1.6	1.8	1.96	1.98
$S(n)$	2.4	2.2	2.04	2.02

(f) $\displaystyle \lim_{n\to\infty}\sum_{i=1}^{n}\left[(i-1)\left(\frac{2}{n}\right)\right]\left(\frac{2}{n}\right) = \lim_{n\to\infty}\frac{4}{n^2}\sum_{i=1}^{n}(i-1)$

$$= \lim_{n\to\infty}\frac{4}{n^2}\left[\frac{n(n+1)}{2} - n\right]$$

$$= \lim_{n\to\infty}\left[\frac{2(n+1)}{n} - \frac{4}{n}\right] = 2$$

$$\lim_{n\to\infty}\sum_{i=1}^{n}\left[i\left(\frac{2}{n}\right)\right]\left(\frac{2}{n}\right) = \lim_{n\to\infty}\frac{4}{n^2}\sum_{i=1}^{n} i$$

$$= \lim_{n\to\infty}\left(\frac{4}{n^2}\right)\frac{n(n+1)}{2}$$

$$= \lim_{n\to\infty}\frac{2(n+1)}{n} = 2$$

47. $y = -2x + 3$ on $[0, 1]$. $\left(\textbf{Note: } \Delta x = \dfrac{1-0}{n} = \dfrac{1}{n}\right)$

$$s(n) = \sum_{i=1}^{n} f\left(\frac{i}{n}\right)\left(\frac{1}{n}\right) = \sum_{i=1}^{n}\left[-2\left(\frac{i}{n}\right)+3\right]\left(\frac{1}{n}\right)$$

$$= 3 - \frac{2}{n^2}\sum_{i=1}^{n} i = 3 - \frac{2(n+1)n}{2n^2} = 2 - \frac{1}{n}$$

Arca $= \displaystyle\lim_{n\to\infty} s(n) = 2$

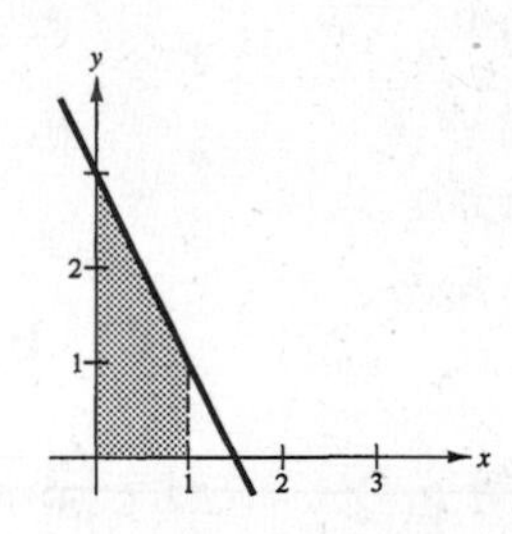

49. $y = x^2 + 2$ on $[0, 1]$. $\left(\textbf{Note: } \Delta x = \dfrac{1}{n}\right)$

$$S(n) = \sum_{i=1}^{n} f\left(\frac{i}{n}\right)\left(\frac{1}{n}\right) = \sum_{i=1}^{n}\left[\left(\frac{i}{n}\right)^2 + 2\right]\left(\frac{1}{n}\right)$$

$$= \left[\frac{1}{n^3}\sum_{i=1}^{n} i^2\right] + 2 = \frac{n(n+1)(2n+1)}{6n^3} + 2 = \frac{1}{6}\left(2+\frac{3}{n}+\frac{1}{n^2}\right) + 2$$

Area $= \displaystyle\lim_{n\to\infty} S(n) = \frac{7}{3}$

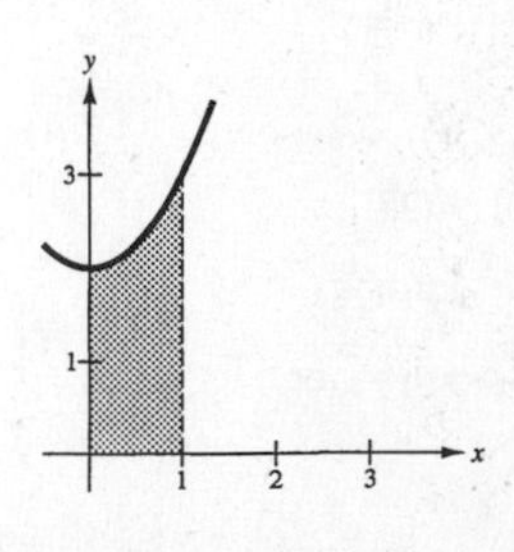

51. $y = 16 - x^2$ on $[1, 3]$. $\left(\textbf{Note: } \Delta x = \frac{2}{n}\right)$

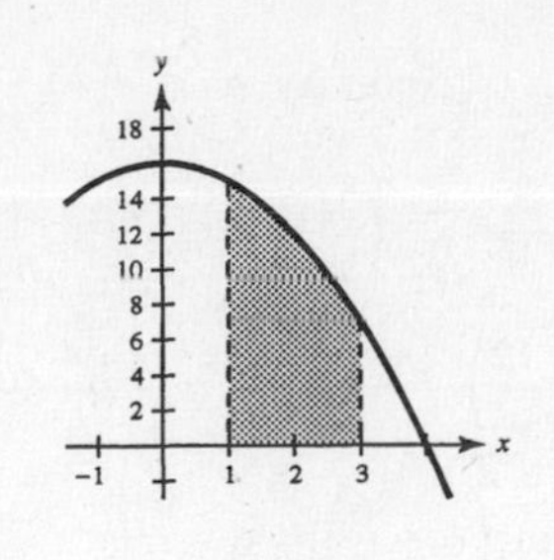

$$s(n) = \sum_{i=1}^{n} f\left(1 + \frac{2i}{n}\right)\left(\frac{2}{n}\right) = \sum_{i=1}^{n}\left[16 - \left(1 + \frac{2i}{n}\right)^2\right]\left(\frac{2}{n}\right)$$

$$= \frac{2}{n}\sum_{i=1}^{n}\left[15 - \frac{4i^2}{n^2} - \frac{4i}{n}\right]$$

$$= \frac{2}{n}\left[15n - \frac{4}{n^2}\frac{n(n+1)(2n+1)}{6} - \frac{4}{n}\frac{n(n+1)}{2}\right]$$

$$= 30 - \frac{8}{6n^2}(n+1)(2n+1) - \frac{4}{n}(n+1)$$

$$\text{Area} = \lim_{n\to\infty} s(n) = 30 - \frac{8}{3} - 4 = \frac{70}{3} = 23\frac{1}{3}$$

53. $y = 64 - x^3$ on $[1, 4]$. $\left(\textbf{Note: } \Delta x = \frac{4-1}{n} = \frac{3}{n}\right)$

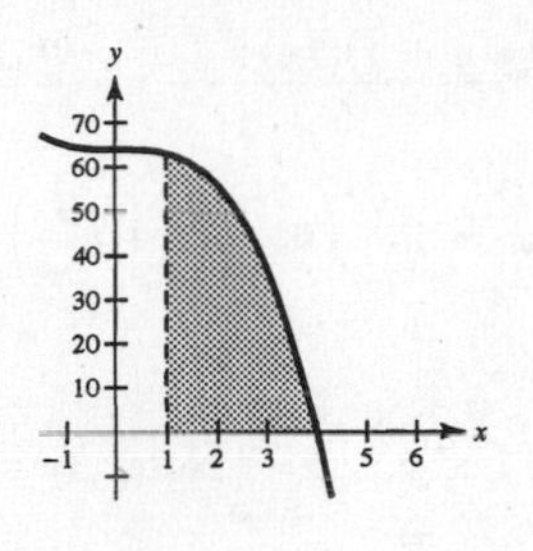

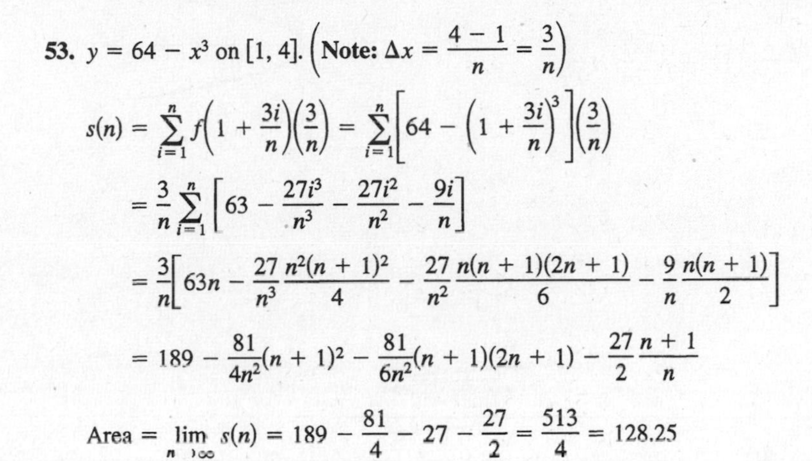

$$s(n) = \sum_{i=1}^{n} f\left(1 + \frac{3i}{n}\right)\left(\frac{3}{n}\right) = \sum_{i=1}^{n}\left[64 - \left(1 + \frac{3i}{n}\right)^3\right]\left(\frac{3}{n}\right)$$

$$= \frac{3}{n}\sum_{i=1}^{n}\left[63 - \frac{27i^3}{n^3} - \frac{27i^2}{n^2} - \frac{9i}{n}\right]$$

$$= \frac{3}{n}\left[63n - \frac{27}{n^3}\frac{n^2(n+1)^2}{4} - \frac{27}{n^2}\frac{n(n+1)(2n+1)}{6} - \frac{9}{n}\frac{n(n+1)}{2}\right]$$

$$= 189 - \frac{81}{4n^2}(n+1)^2 - \frac{81}{6n^2}(n+1)(2n+1) - \frac{27}{2}\frac{n+1}{n}$$

$$\text{Area} = \lim_{n\to\infty} s(n) = 189 - \frac{81}{4} - 27 - \frac{27}{2} = \frac{513}{4} = 128.25$$

55. $y = x^2 - x^3$ on $[-1, 1]$. $\left(\textbf{Note: } \Delta x = \frac{1-(-1)}{n} = \frac{2}{n}\right)$

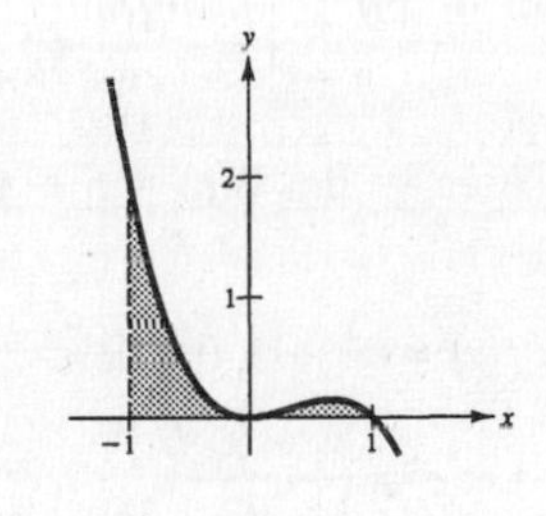

Again, $T(n)$ is neither an upper nor a lower sum.

$$T(n) = \sum_{i=1}^{n} f\left(-1 + \frac{2i}{n}\right)\left(\frac{2}{n}\right) = \sum_{i=1}^{n}\left[\left(-1 + \frac{2i}{n}\right)^2 - \left(-1 + \frac{2i}{n}\right)^3\right]\left(\frac{2}{n}\right)$$

$$= \sum_{i=1}^{n}\left[\left(1 - \frac{4i}{n} + \frac{4i^2}{n^2}\right) - \left(-1 + \frac{6i}{n} - \frac{12i^2}{n^2} + \frac{8i^3}{n^3}\right)\right]\left(\frac{2}{n}\right)$$

$$= \sum_{i=1}^{n}\left[2 - \frac{10i}{n} + \frac{16i^2}{n^2} - \frac{8i^3}{n^3}\right]\left(\frac{2}{n}\right) = \frac{4}{n}\sum_{i=1}^{n} 1 - \frac{20}{n^2}\sum_{i=1}^{n} i + \frac{32}{n^3}\sum_{i=1}^{n} i^2 - \frac{16}{n^4}\sum_{i=1}^{n} i^3$$

$$= \frac{4}{n}(n) - \frac{20}{n^2}\cdot\frac{n(n+1)}{2} + \frac{32}{n^3}\cdot\frac{n(n+1)(2n+1)}{6} - \frac{16}{n^4}\cdot\frac{n^2(n+1)^2}{4}$$

$$= 4 - 10\left(1 + \frac{1}{n}\right) + \frac{16}{3}\left(2 + \frac{3}{n} + \frac{1}{n^2}\right) - 4\left(1 + \frac{2}{n} + \frac{1}{n^2}\right)$$

$$\text{Area} = \lim_{n\to\infty} T(n) = 4 - 10 + \frac{32}{3} - 4 = \frac{2}{3}$$

57. $f(y) = 3y, 0 \le y \le 2 \quad \left(\textbf{Note: } \Delta y = \frac{2-0}{n} = \frac{2}{n}\right)$

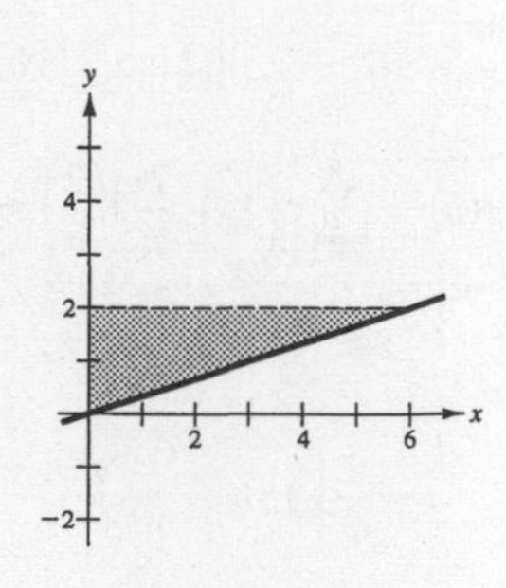

$$S(n) = \sum_{i=1}^{n} f(m_i)\,\Delta y = \sum_{i=1}^{n} f\left(\frac{2i}{n}\right)\left(\frac{2}{n}\right) = \sum_{i=1}^{n} 3\left(\frac{2i}{n}\right)\left(\frac{2}{n}\right)$$

$$= \frac{12}{n^2}\sum_{i=1}^{n} i = \left(\frac{12}{n^2}\right)\cdot\frac{n(n+1)}{2} = \frac{6(n+1)}{n} = 6 + \frac{6}{n}$$

$$\text{Area} = \lim_{n\to\infty} S(n) = \lim_{n\to\infty}\left(6 + \frac{6}{n}\right) = 6$$

59. $f(y) = y^2, 0 \le y \le 3 \quad \left(\textbf{Note: } \Delta y = \frac{3-0}{n} = \frac{3}{n}\right)$

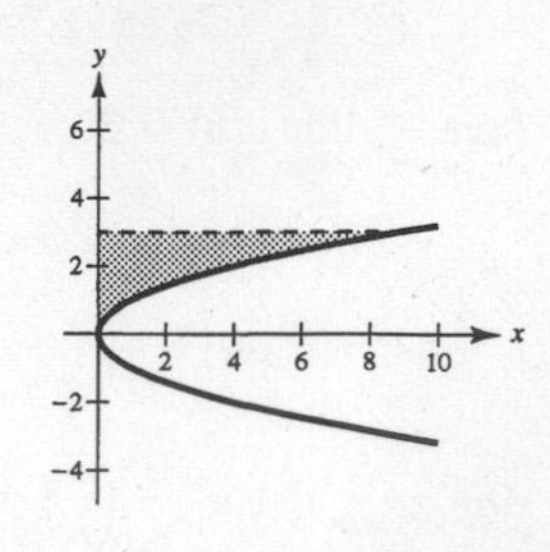

$$S(n) = \sum_{i=1}^{n} f\left(\frac{3i}{n}\right)\left(\frac{3}{n}\right) = \sum_{i=1}^{n}\left(\frac{3i}{n}\right)^2\left(\frac{3}{n}\right) = \frac{27}{n^3}\sum_{i=1}^{n} i^2$$

$$= \frac{27}{n^3}\cdot\frac{n(n+1)(2n+1)}{6} = \frac{9}{n^2}\left(\frac{2n^2+3n+1}{2}\right) = 9 + \frac{27}{2n} + \frac{9}{2n^2}$$

$$\text{Area} = \lim_{n\to\infty} S(n) = \lim_{n\to\infty}\left(9 + \frac{27}{2n} + \frac{9}{2n^2}\right) = 9$$

61. $g(y) = 4y^2 - y^3, 1 \le y \le 3. \quad \left(\textbf{Note: } \Delta y = \frac{3-1}{n} = \frac{2}{n}\right)$

$$S(n) = \sum_{i=1}^{n} g\left(1 + \frac{2i}{n}\right)\left(\frac{2}{n}\right)$$

$$= \sum_{i=1}^{n}\left[4\left(1+\frac{2i}{n}\right)^2 - \left(1+\frac{2i}{n}\right)^3\right]\frac{2}{n}$$

$$= \frac{2}{n}\sum_{i=1}^{n} 4\left[1 + \frac{4i}{n} + \frac{4i^2}{n^2}\right] - \left[1 + \frac{6i}{n} + \frac{12i^2}{n^2} + \frac{8i^3}{n^3}\right]$$

$$= \frac{2}{n}\sum_{i=1}^{n}\left[3 + \frac{10i}{n} + \frac{4i^2}{n^2} - \frac{8i^3}{n^3}\right] = \frac{2}{n}\left[3n + \frac{10}{n}\frac{n(n+1)}{2} + \frac{4}{n^2}\frac{n(n+1)(2n+1)}{6} - \frac{8}{n^2}\frac{n^2(n+1)^2}{4}\right]$$

$$\text{Area} = \lim_{n\to\infty} S(n) = 6 + 10 + \frac{8}{3} - 4 = \frac{44}{3}$$

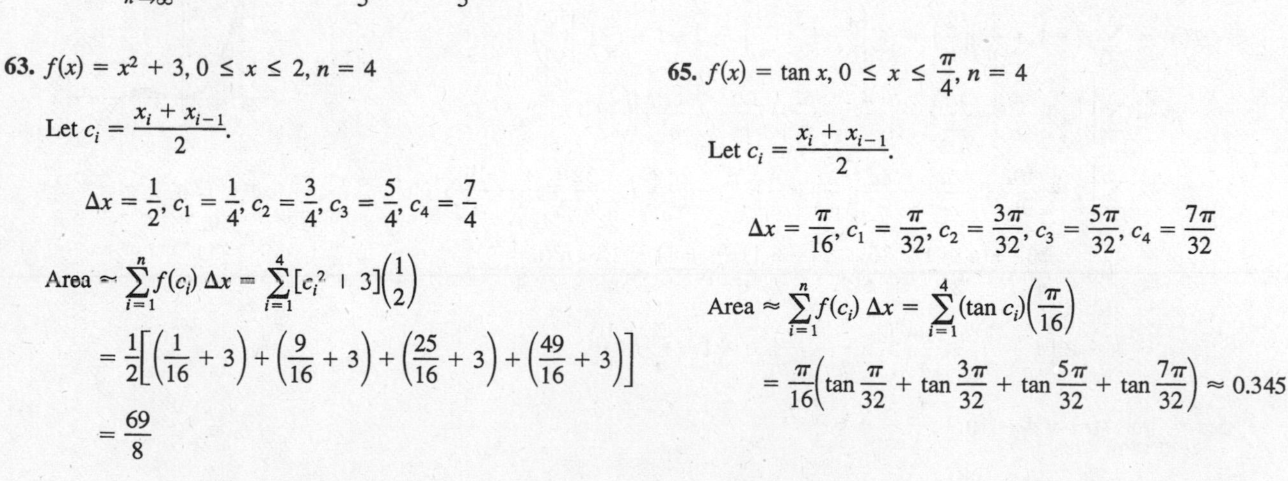

63. $f(x) = x^2 + 3, 0 \le x \le 2, n = 4$

Let $c_i = \dfrac{x_i + x_{i-1}}{2}$.

$$\Delta x = \frac{1}{2}, c_1 = \frac{1}{4}, c_2 = \frac{3}{4}, c_3 = \frac{5}{4}, c_4 = \frac{7}{4}$$

$$\text{Area} \approx \sum_{i=1}^{n} f(c_i)\,\Delta x = \sum_{i=1}^{4}[c_i^2 + 3]\left(\frac{1}{2}\right)$$

$$= \frac{1}{2}\left[\left(\frac{1}{16}+3\right) + \left(\frac{9}{16}+3\right) + \left(\frac{25}{16}+3\right) + \left(\frac{49}{16}+3\right)\right]$$

$$= \frac{69}{8}$$

65. $f(x) = \tan x, 0 \le x \le \frac{\pi}{4}, n = 4$

Let $c_i = \dfrac{x_i + x_{i-1}}{2}$.

$$\Delta x = \frac{\pi}{16}, c_1 = \frac{\pi}{32}, c_2 = \frac{3\pi}{32}, c_3 = \frac{5\pi}{32}, c_4 = \frac{7\pi}{32}$$

$$\text{Area} \approx \sum_{i=1}^{n} f(c_i)\,\Delta x = \sum_{i=1}^{4}(\tan c_i)\left(\frac{\pi}{16}\right)$$

$$= \frac{\pi}{16}\left(\tan\frac{\pi}{32} + \tan\frac{3\pi}{32} + \tan\frac{5\pi}{32} + \tan\frac{7\pi}{32}\right) \approx 0.345$$

67. $f(x) = \sqrt{x}$ on $[0, 4]$.

n	4	8	12	16	20
Approximate area	5.3838	5.3523	5.3439	5.3403	5.3384

(Exact value is 16/3)

69. $f(x) = \tan\left(\frac{\pi x}{8}\right)$ on $[1, 3]$.

n	4	8	12	16	20
Approximate area	2.2223	2.2387	2.2418	2.2430	2.2435

71. $f(x) = \ln x, \quad [1, 5]$

n	4	8	12	16	20
Approximate area	4.0786	4.0554	4.0509	4.0493	4.0485

73. We can use the line $y = x$ bounded by $x = a$ and $x = b$. The sum of the areas of these inscribed rectangles is the lower sum.

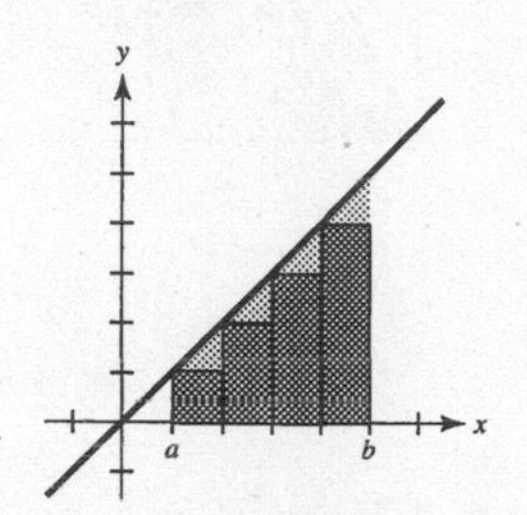

The sum of the areas of these circumscribed rectangles is the upper sum.

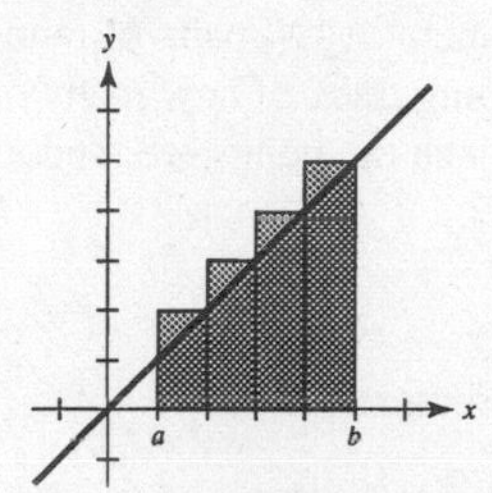

We can see that the rectangles do not contain all of the area in the first graph and the rectangles in the second graph cover more than the area of the region.

The exact value of the area lies between these two sums.

75. (a)

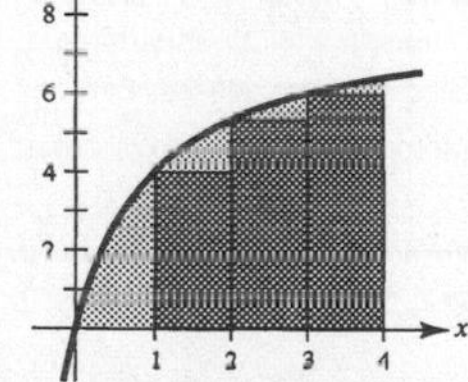

Lower sum:

$s(4) = 0 + 4 + 5\frac{1}{3} + 6 = 15\frac{1}{3} = \frac{46}{3} \approx 15.333$

(b)

Upper sum:

$S(4) = 4 + 5\frac{1}{3} + 6 + 6\frac{2}{5} = 21\frac{11}{15} = \frac{326}{15} \approx 21.733$

(c)

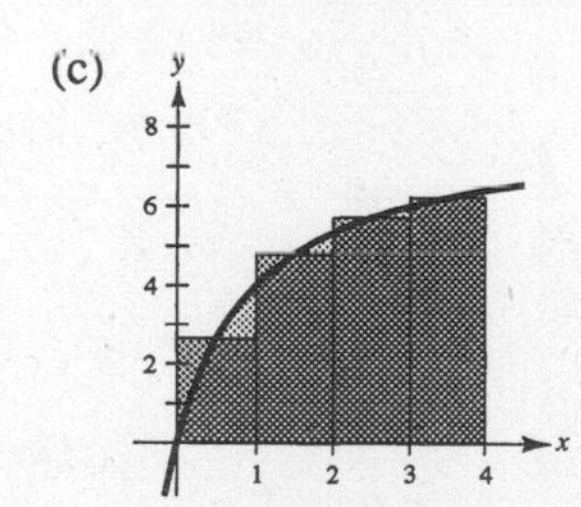

Midpoint Rule:

$M(4) = 2\frac{2}{3} + 4\frac{4}{5} + 5\frac{5}{7} + 6\frac{2}{9} = \frac{6112}{315} \approx 19.403$

(d) In each case, $\Delta x = 4/n$. The lower sum uses left endpoints, $(i - 1)(4/n)$. The upper sum uses right endpoints, $(i)(4/n)$. The Midpoint Rule uses midpoints, $\left(i - \frac{1}{2}\right)(4/n)$.

(e)

n	4	8	20	100	200
$s(n)$	15.333	17.368	18.459	18.995	19.06
$S(n)$	21.733	20.568	19.739	19.251	19.188
$M(n)$	19.403	19.201	19.137	19.125	19.125

(f) $s(n)$ increases because the lower sum approaches the exact value as n increases. $S(n)$ decreases because the upper sum approaches the exact value as n increases. Because of the shape of the graph, the lower sum is always smaller than the exact value, whereas the upper sum is always larger.

77.

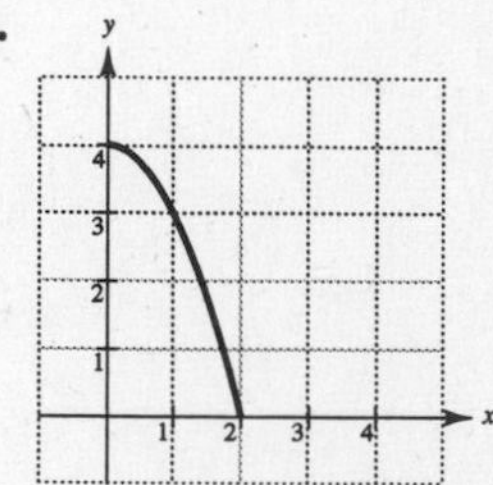

b. $A \approx 6$ square units

79. True. (Theorem 4.2 (2))

81. $f(x) = \sin x, \left[0, \frac{\pi}{2}\right]$

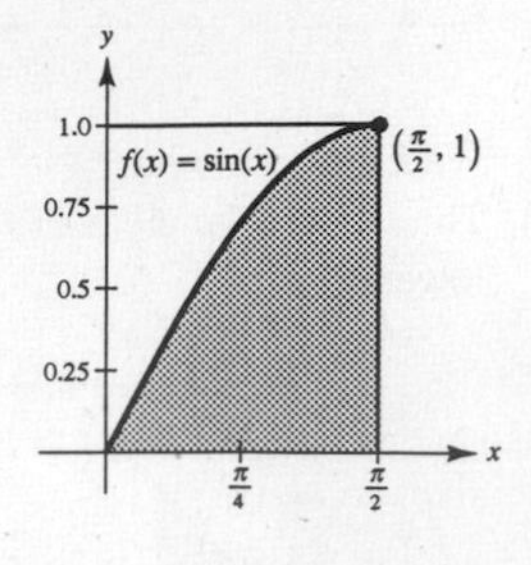

Let A_1 = area bounded by $f(x) = \sin x$, the x-axis, $x = 0$ and $x = \pi/2$. Let A_2 = area of the rectangle bounded by $y = 1, y = 0, x = 0$, and $x = \pi/2$. Thus, $A_2 = (\pi/2)(1) \approx 1.570796$. In this program, the computer is generating N_2 pairs of random points in the rectangle whose area is represented by A_2. It is keeping track of how many of these points, N_1, lie in the region whose area is represented by A_1. Since the points are randomly generated, we assume that

$$\frac{A_1}{A_2} \approx \frac{N_1}{N_2} \Rightarrow A_1 \approx \frac{N_1}{N_2}A_2.$$

The larger N_2 is the better the approximation to A_1.

83. Suppose there are n rows and $n + 1$ columns in the figure. The stars on the left total $1 + 2 + \cdots + n$, as do the stars on the right. There are $n(n + 1)$ stars in total, hence

$$2[1 + 2 + \cdots + n] = n(n + 1)$$

$$1 + 2 + \cdots + n = \tfrac{1}{2}(n)(n + 1).$$

85. (a) $y = (-4.09 \times 10^{-5})x^3 + 0.016x^2 - 2.67x + 452.9$

(b)

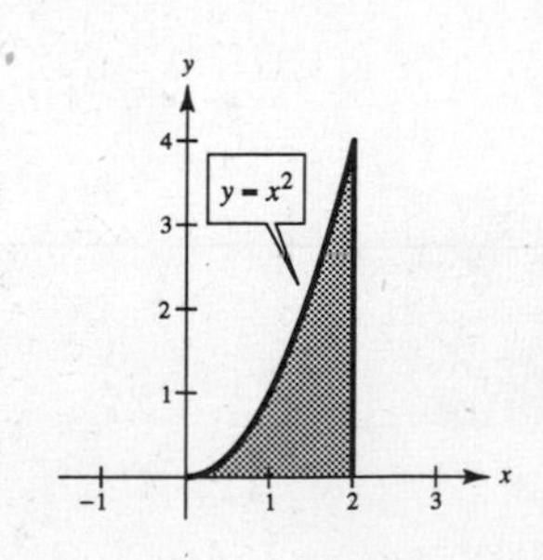

(c) Using the integration capability of a graphing utility, you obtain

$$A \approx 76{,}897.5 \text{ ft}^2.$$

Section 4.3 Reimann Sums and Definite Integrals

1. $f(x) = \sqrt{x}, y = 0, x = 0, x = 3, c_i = \dfrac{3i^2}{n^2}$

$$\Delta x_i = \frac{3i^2}{n^2} - \frac{3(i-1)^2}{n^2} = \frac{3}{n^2}(2i - 1)$$

$$\lim_{n\to\infty} \sum_{i=1}^{n} f(c_i)\Delta x_i = \lim_{n\to\infty} \sum_{i=1}^{n} \sqrt{\frac{3i^2}{n^2}}\,\frac{3}{n^2}(2i - 1)$$

$$= \lim_{n\to\infty} \frac{3\sqrt{3}}{n^3}\sum_{i=1}^{n}(2i^2 - i)$$

$$= \lim_{n\to\infty} \frac{3\sqrt{3}}{n^3}\left[2\frac{n(n+1)(2n+1)}{6} - \frac{n(n+1)}{2}\right]$$

$$= \lim_{n\to\infty} 3\sqrt{3}\left[\frac{(n+1)(2n+1)}{3n^2} - \frac{n+1}{2n^2}\right]$$

$$= 3\sqrt{3}\left[\frac{2}{3} - 0\right] = 2\sqrt{3} \approx 3.464$$

3. $y = 6$ on $[4, 10]$. $\left(\textbf{Note: } \Delta x = \dfrac{10-4}{n} = \dfrac{6}{n}, \|\Delta\| \to 0 \text{ as } n \to \infty\right)$

$$\sum_{i=1}^{n} f(c_i)\,\Delta x_i = \sum_{i=1}^{n} f\left(4 + \frac{6i}{n}\right)\left(\frac{6}{n}\right) = \sum_{i=1}^{n} 6\left(\frac{6}{n}\right) = \sum_{i=1}^{n} \frac{36}{n} = 36$$

$$\int_4^{10} 6\,dx = \lim_{n\to\infty} 36 = 36$$

5. $y = x^3$ on $[-1, 1]$. $\left(\textbf{Note: } \Delta x = \dfrac{1-(-1)}{n} = \dfrac{2}{n}, \|\Delta\| \to 0 \text{ as } n \to \infty\right)$

$$\sum_{i=1}^{n} f(c_i)\,\Delta x_i = \sum_{i=1}^{n} f\left(-1 + \frac{2i}{n}\right)\left(\frac{2}{n}\right) = \sum_{i=1}^{n}\left(-1 + \frac{2i}{n}\right)^3\left(\frac{2}{n}\right) = \sum_{i=1}^{n}\left[-1 + \frac{6i}{n} - \frac{12i^2}{n^2} + \frac{8i^3}{n^3}\right]\left(\frac{2}{n}\right)$$

$$= -2 + \frac{12}{n^2}\sum_{i=1}^{n} i - \frac{24}{n^3}\sum_{i=1}^{n} i^2 + \frac{16}{n^4}\sum_{i=1}^{n} i^3$$

$$= -2 + 6\left(1 + \frac{1}{n}\right) - 4\left(2 + \frac{3}{n} + \frac{1}{n^2}\right) + 4\left(1 + \frac{2}{n} + \frac{1}{n^2}\right) = \frac{2}{n}$$

$$\int_{-1}^{1} x^3\,dx = \lim_{n\to\infty} \frac{2}{n} = 0$$

7. $y = x^2 + 1$ on $[1, 2]$. $\left(\textbf{Note: } \Delta x = \dfrac{2-1}{n} = \dfrac{1}{n}, \|\Delta\| \to 0 \text{ as } n \to \infty\right)$

$$\sum_{i=1}^{n} f(c_i)\,\Delta x_i = \sum_{i=1}^{n} f\left(1 + \frac{i}{n}\right)\left(\frac{1}{n}\right) = \sum_{i=1}^{n}\left[\left(1 + \frac{i}{n}\right)^2 + 1\right]\left(\frac{1}{n}\right) = \sum_{i=1}^{n}\left[1 + \frac{2i}{n} + \frac{i^2}{n^2} + 1\right]\left(\frac{1}{n}\right)$$

$$= 2 + \frac{2}{n^2}\sum_{i=1}^{n} i + \frac{1}{n^3}\sum_{i=1}^{n} i^2 = 2 + \left(1 + \frac{1}{n}\right) + \frac{1}{6}\left(2 + \frac{3}{n} + \frac{1}{n^2}\right) = \frac{10}{3} + \frac{3}{2n} + \frac{1}{6n^2}$$

$$\int_1^2 (x^2 + 1)\,dx = \lim_{n\to\infty}\left(\frac{10}{3} + \frac{3}{2n} + \frac{1}{6n^2}\right) = \frac{10}{3}$$

9. $\displaystyle\lim_{\|\Delta\|\to 0} \sum_{i=1}^{n} (3c_i + 10)\,\Delta x_i = \int_{-1}^{5} (3x + 10)\,dx$

on the interval $[-1, 5]$.

11. $\displaystyle\lim_{\|\Delta\|\to 0} \sum_{i=1}^{n} \sqrt{c_i^2 + 4}\,\Delta x_i = \int_0^3 \sqrt{x^2 + 4}\,dx$

on the interval $[0, 3]$.

13. $\displaystyle\int_1^5 \left(1 + \frac{3}{x}\right) dx$

15. $\displaystyle\int_0^5 3\,dx$

17. $\displaystyle\int_1^4 \frac{2}{x}\,dx$

19. $\displaystyle\int_0^{\pi} \sin x\,dx$

21. $\displaystyle\int_0^2 y^3\,dy$

23. Rectangle

$A = bh = 3(4)$

$A = \displaystyle\int_0^3 4\,dx = 12$

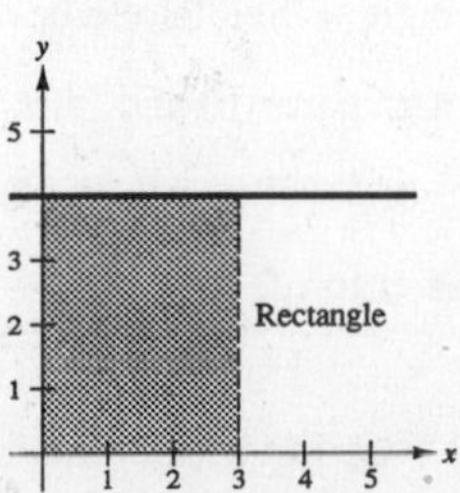

25. Triangle

$$A = \frac{1}{2}bh = \frac{1}{2}(4)(4)$$

$$A = \int_0^4 x\,dx = 8$$

27. Trapezoid

$$A = \frac{b_1 + b_2}{2}h = \left(\frac{5+9}{2}\right)2$$

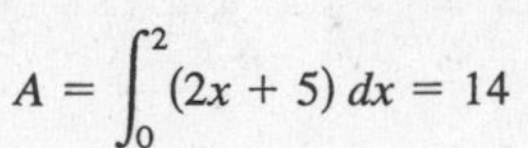

$$A = \int_0^2 (2x + 5)\,dx = 14$$

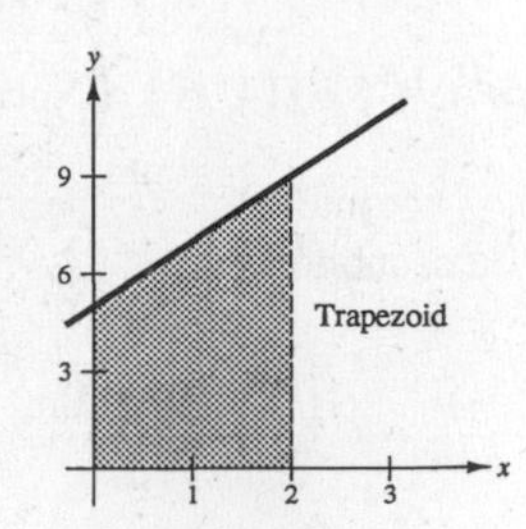

29. Triangle

$$A = \frac{1}{2}bh = \frac{1}{2}(4)(2) = 4$$

$$A = \int_{-2}^{2} (2 - |x|)\,dx = 4$$

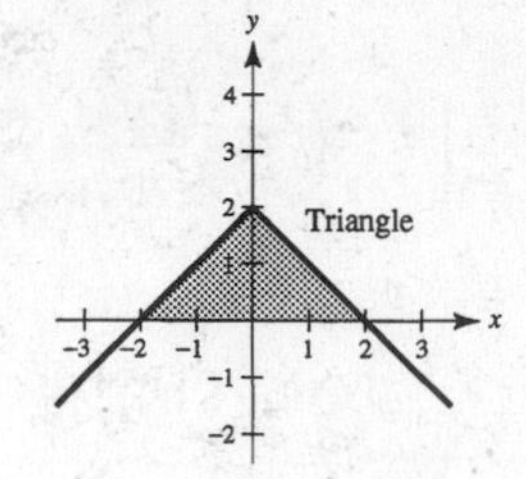

31. Semicircle

$$A = \frac{1}{2}\pi r^2 = \frac{1}{2}\pi(3)^2$$

$$A = \int_{-3}^{3} \sqrt{9 - x^2}\,dx = \frac{9\pi}{2}$$

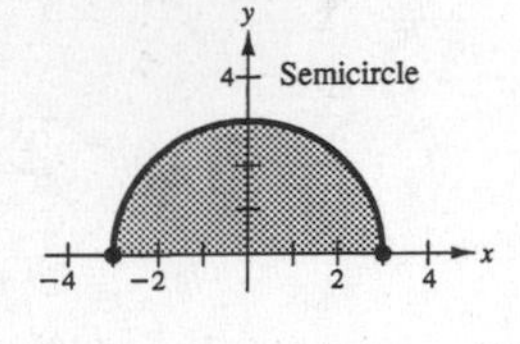

In Exercises 33–39, $\int_2^4 x^3\,dx = 60,\ \int_2^4 x\,dx = 6,\ \int_2^4 dx = 2$

33. $\int_4^2 x\,dx = -\int_2^4 x\,dx = -6$

35. $\int_2^4 4x\,dx = 4\int_2^4 x\,dx = 4(6) = 24$

37. $\int_2^4 (x - 8)\,dx = \int_2^4 x\,dx - 8\int_2^4 dx = 6 - 8(2) = -10$

39. $\int_2^4 \left(\frac{1}{2}x^3 - 3x + 2\right)dx = \frac{1}{2}\int_2^4 x^3\,dx - 3\int_2^4 x\,dx + 2\int_2^4 dx$

$= \frac{1}{2}(60) - 3(6) + 2(2) = 16$

41. (a) $\int_0^7 f(x)\,dx = \int_0^5 f(x)\,dx + \int_5^7 f(x)\,dx = 10 + 3 = 13$

(b) $\int_5^0 f(x)\,dx = -\int_0^5 f(x)\,dx = -10$

(c) $\int_5^5 f(x)\,dx = 0$

(d) $\int_0^5 3f(x)\,dx = 3\int_0^5 f(x)\,dx = 3(10) = 30$

43. (a) $\int_2^6 [f(x) + g(x)]\,dx = \int_2^6 f(x)\,dx + \int_2^6 g(x)\,dx$

$= 10 + (-2) = 8$

(b) $\int_2^6 [g(x) - f(x)]\,dx = \int_2^6 g(x)\,dx - \int_2^6 f(x)\,dx$

$= -2 - 10 = -12$

(c) $\int_2^6 2g(x)\,dx = 2\int_2^6 g(x)\,dx = 2(-2) = -4$

(d) $\int_2^6 3f(x)\,dx = 3\int_2^6 f(x)\,dx = 3(10) = 30$

45. (a) Quarter circle below x-axis: $-\frac{1}{4}\pi r^2 = -\frac{1}{4}\pi(2)^2 = -\pi$

(b) Triangle: $\frac{1}{2}bh = \frac{1}{2}(4)(2) = 4$

(c) Triangle + Semicircle below x-axis: $-\frac{1}{2}(2)(1) - \frac{1}{2}\pi(2)^2 = -(1 + 2\pi)$

(d) Sum of parts (b) and (c): $4 - (1 + 2\pi) = 3 - 2\pi$

(e) Sum of absolute values of (b) and (c): $4 + (1 + 2\pi) = 5 + 2\pi$

(f) Answer to (d) plus $2(10) = 20$: $(3 - 2\pi) + 20 = 23 - 2\pi$

47. The left endpoint approximation will be greater than the actual area: $>$

49. Because the curve is concave upward, the midpoint approximation will be less than the actual area: $<$

51. $f(x) = \dfrac{1}{x - 4}$

is not integrable on the interval $[3, 5]$ because f has a discontinuity at $x = 4$.

53.

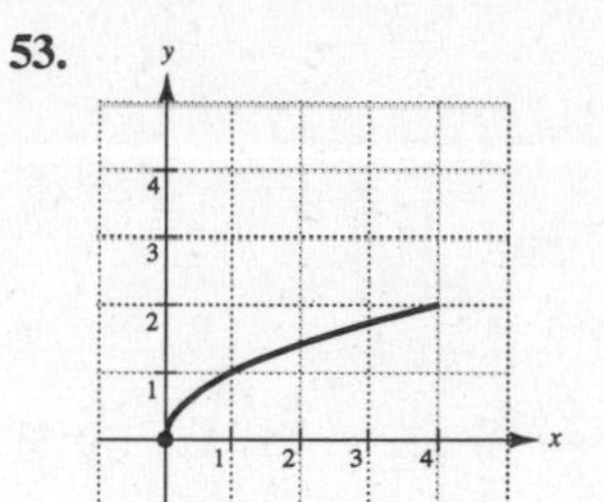

a. $A \approx 5$ square units

55. 

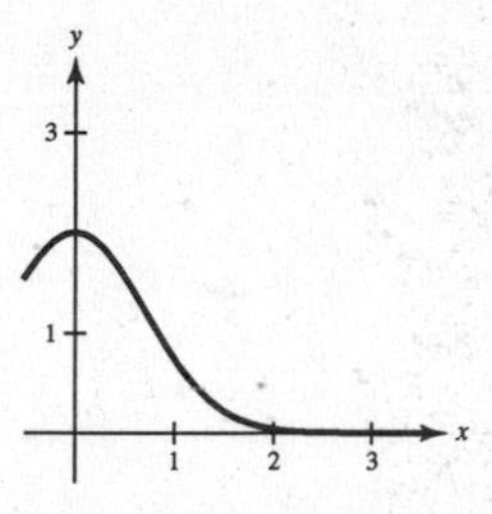

c. $\displaystyle\int_0^2 2e^{-x^2}\,dx \approx \frac{1}{2}(2)(2) = 2$

57. $\displaystyle\int_0^3 x\sqrt{3 - x}\,dx$

n	4	8	12	16	20
$L(n)$	3.6830	3.9956	4.0707	4.1016	4.1177
$M(n)$	4.3082	4.2076	4.1838	4.1740	4.1690
$R(n)$	3.6830	3.9956	4.0707	4.1016	4.1177

59. $\displaystyle\int_1^3 \frac{1}{x}\,dx$

n	4	8	12	16	20
$L(n)$	1.2833	1.1865	1.1562	1.1414	1.1327
$M(n)$	1.0898	1.0963	1.0976	1.0980	1.0982
$R(n)$	0.9500	1.0199	1.0451	1.0581	1.0660

61. $\displaystyle\int_0^{\pi/2} \sin^2 x\,dx$

n	4	8	12	16	20
$L(n)$	0.5890	0.6872	0.7199	0.7363	0.7461
$M(n)$	0.7854	0.7854	0.7854	0.7854	0.7854
$R(n)$	0.9817	0.8836	0.8508	0.8345	0.8247

63. True

65. True

67. False

$\displaystyle\int_0^2 (-x)\,dx = -2$

69. $f(x) = x^2 + 3x, [0, 8]$

$x_0 = 0, x_1 = 1, x_2 = 3, x_3 = 7, x_4 = 8$

$\Delta x_1 = 1, \Delta x_2 = 2, \Delta x_3 = 4, \Delta x_4 = 1$

$c_1 = 1, c_2 = 2, c_3 = 5, c_4 = 8$

$$\sum_{i=1}^{4} f(c_i)\,\Delta x = f(1)\,\Delta x_1 + f(2)\,\Delta x_2 + f(5)\,\Delta x_3 + f(8)\,\Delta x_4$$

$$= (4)(1) + (10)(2) + (40)(4) + (88)(1) = 272$$

71. $f(x) = \begin{cases} 1, & x \text{ is rational} \\ 0, & x \text{ is irrational} \end{cases}$

is not integrable on the interval $[0, 1]$. As $\|\Delta\| \to 0$, $f(c_i) = 1$ or $f(c_i) = 0$ in each subinterval since there are an infinite number of both rational and irrational numbers in any interval, no matter how small.

73. Let $f(x) = x^2, 0 \le x \le 1$, and $\Delta x_i = 1/n$. The appropriate Riemann Sum is

$$\sum_{i=1}^{n} f(c_i)\Delta x_i = \sum_{i=1}^{n} \left(\frac{i}{n}\right)^2 \frac{1}{n} = \frac{1}{n^3}\sum_{i=1}^{n} i^2.$$

$$\lim_{n\to\infty} \frac{1}{n^3}[1^2 + 2^2 + 3^2 + \cdots + n^2] = \lim_{n\to\infty} \frac{1}{n^3} \cdot \frac{n(2n+1)(n+1)}{6}$$

$$= \lim_{n\to\infty} \frac{2n^2 + 3n + 1}{6n^2} = \lim_{n\to\infty} \left(\frac{1}{3} + \frac{1}{2n} + \frac{1}{6n^2}\right) = \frac{1}{3}$$

Section 4.4 The Fundamental Theorem of Calculus

1. $f(x) = \dfrac{4}{x^2 + 1}$

$\displaystyle\int_0^{\pi} \frac{4}{x^2 + 1}\,dx$ is positive.

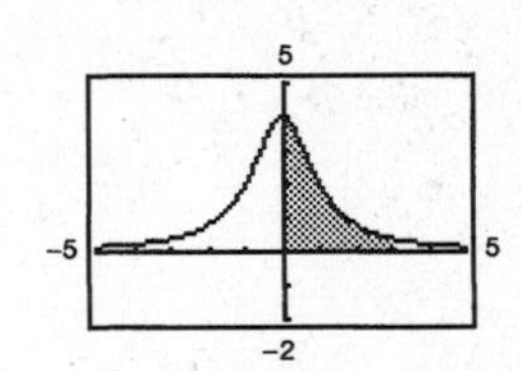

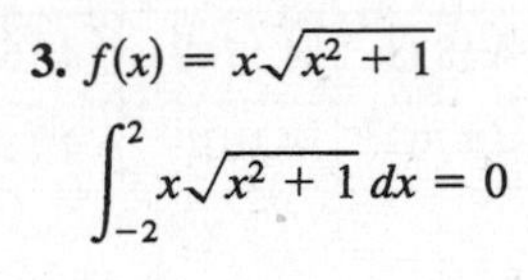

3. $f(x) = x\sqrt{x^2 + 1}$

$\displaystyle\int_{-2}^{2} x\sqrt{x^2 + 1}\,dx = 0$

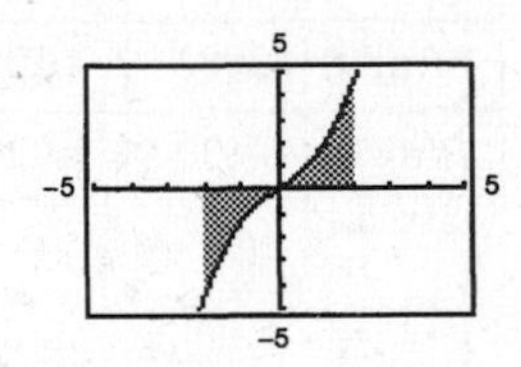

5. $\displaystyle\int_0^1 2x\,dx = \Big[x^2\Big]_0^1 = 1 - 0 = 1$

7. $\displaystyle\int_{-1}^{0} (x - 2)\,dx = \left[\frac{x^2}{2} - 2x\right]_{-1}^{0} = 0 - \left(\frac{1}{2} + 2\right) = -\frac{5}{2}$

9. $\displaystyle\int_{-1}^{1} (t^2 - 2)\,dt = \left[\frac{t^3}{3} - 2t\right]_{-1}^{1} = \left(\frac{1}{3} - 2\right) - \left(-\frac{1}{3} + 2\right) = -\frac{10}{3}$

11. $\displaystyle\int_0^1 (2t - 1)^2\,dt = \int_0^1 (4t^2 - 4t + 1)\,dt = \left[\frac{4}{3}t^3 - 2t^2 + t\right]_0^1 = \frac{4}{3} - 2 + 1 = \frac{1}{3}$

13. $\displaystyle\int_1^2 \left(\frac{3}{x^2} - 1\right)dx = \left[-\frac{3}{x} - x\right]_1^2 = \left(-\frac{3}{2} - 2\right) - (-3 - 1) = \frac{1}{2}$

15. $\displaystyle\int_1^4 \frac{u - 2}{\sqrt{u}}\,du = \int_1^4 (u^{1/2} - 2u^{-1/2})\,du = \left[\frac{2}{3}u^{3/2} - 4u^{1/2}\right]_1^4 = \left[\frac{2}{3}(\sqrt{4})^3 - 4\sqrt{4}\right] - \left[\frac{2}{3} - 4\right] = \frac{2}{3}$

17. $\displaystyle\int_{-1}^{1} (\sqrt[3]{t} - 2)\,dt = \left[\frac{3}{4}t^{4/3} - 2t\right]_{-1}^{1} = \left(\frac{3}{4} - 2\right) - \left(\frac{3}{4} + 2\right) = -4$

19. $\displaystyle\int_0^1 \frac{x - \sqrt{x}}{3}\,dx = \frac{1}{3}\int_0^1 (x - x^{1/2})\,dx = \frac{1}{3}\left[\frac{x^2}{2} - \frac{2}{3}x^{3/2}\right]_0^1 = \frac{1}{3}\left(\frac{1}{2} - \frac{2}{3}\right) = -\frac{1}{18}$

21. $\displaystyle\int_{-1}^0 (t^{1/3} - t^{2/3})\,dt = \left[\frac{3}{4}t^{4/3} - \frac{3}{5}t^{5/3}\right]_{-1}^0 = 0 - \left(\frac{3}{4} + \frac{3}{5}\right) = -\frac{27}{20}$

23. $\displaystyle\int_0^3 |2x - 3|\,dx = \int_0^{3/2} (3 - 2x)\,dx + \int_{3/2}^3 (2x - 3)\,dx \;\left(\text{split up the integral at the zero } x = \frac{3}{2}\right)$

$\displaystyle= \left[3x - x^2\right]_0^{3/2} + \left[x^2 - 3x\right]_{3/2}^3 = \left(\frac{9}{2} - \frac{9}{4}\right) - 0 + (9 - 9) - \left(\frac{9}{4} - \frac{9}{2}\right) = 2\left(\frac{9}{2} - \frac{9}{4}\right) = \frac{9}{2}$

25. $\displaystyle\int_0^3 |x^2 - 4|\,dx = \int_0^2 (4 - x^2)\,dx + \int_2^3 (x^2 - 4)\,dx$

$\displaystyle= \left[4x - \frac{x^3}{3}\right]_0^2 + \left[\frac{x^3}{3} - 4x\right]_2^3$

$\displaystyle= \left(8 - \frac{8}{3}\right) + (9 - 12) - \left(\frac{8}{3} - 8\right)$

$\displaystyle= \frac{23}{3}$

27. $\displaystyle\int_0^\pi (1 + \sin x)\,dx = \left[x - \cos x\right]_0^\pi = (\pi + 1) - (0 - 1) = 2 + \pi$

29. $\displaystyle\int_{-\pi/6}^{\pi/6} \sec^2 x\,dx = \left[\tan x\right]_{-\pi/6}^{\pi/6} = \frac{\sqrt{3}}{3} - \left(-\frac{\sqrt{3}}{3}\right) = \frac{2\sqrt{3}}{3}$

31. $\displaystyle\int_1^e \left(2x - \frac{1}{x}\right)dx = \left[x^2 - \ln x\right]_1^e = (e^2 - 1) - (1 - 0) = e^2 - 2$

33. $\displaystyle\int_{-\pi/3}^{\pi/3} 4\sec\theta\tan\theta\,d\theta = \left[4\sec\theta\right]_{-\pi/3}^{\pi/3} = 4(2) - 4(2) = 0$

35. $\displaystyle\int_0^2 (2^x + 6)\,dx = \left[\frac{2^x}{\ln 2} + 6x\right]_0^2 = \left(\frac{4}{\ln 2} + 12\right) - \left(\frac{1}{\ln 2} + 0\right) = \frac{3}{\ln 2} + 12$

37. $\displaystyle\int_{-1}^1 (e^\theta + \sin\theta)\,d\theta = \left[e^\theta - \cos\theta\right]_{-1}^1 = (e - \cos 1) - [e^{-1} - \cos(-1)] = e - \frac{1}{e}$

39. $\displaystyle\int_0^3 10{,}000(t - 6)\,dt = 10{,}000\left[\frac{t^2}{2} - 6t\right]_0^3 = -\$135{,}000$

41. $\displaystyle A = \int_0^1 (x - x^2)\,dx = \left[\frac{x^2}{2} - \frac{x^3}{3}\right]_0^1 = \frac{1}{6}$

43. $\displaystyle A = \int_0^3 (3 - x)\sqrt{x}\,dx = \int_0^3 (3x^{1/2} - x^{3/2})\,dx = \left[2x^{3/2} - \frac{2}{5}x^{5/2}\right]_0^3 = \left[\frac{x\sqrt{x}}{5}(10 - 2x)\right]_0^3 = \frac{12\sqrt{3}}{5}$

45. Since $y \geq 0$ on $[0, 2]$,

$$A = \int_0^2 (3x^2 + 1)\,dx = \Big[x^3 + x\Big]_0^2 = 8 + 2 = 10.$$

47. Since $y \geq 0$ on $[0, 2]$,

$$A = \int_0^2 (x^3 + x)\,dx = \left[\frac{x^4}{4} + \frac{x^2}{2}\right]_0^2 = 4 + 2 = 6.$$

49. $\displaystyle\int_1^e \frac{4}{x}\,dx = 4\ln x\Big]_1^e = 4\ln e - 4\ln 1 = 4$

51. $\displaystyle\int_0^2 \left(x - 2\sqrt{x}\right)dx = \left[\frac{x^2}{2} - \frac{4x^{3/2}}{3}\right]_0^2 = 2 - \frac{8\sqrt{2}}{3}$

$$f(c)(2 - 0) = \frac{6 - 8\sqrt{2}}{3}$$

$$c - 2\sqrt{c} = \frac{3 - 4\sqrt{2}}{3}$$

$$c - 2\sqrt{c} + 1 = \frac{3 - 4\sqrt{2}}{3} + 1$$

$$\left(\sqrt{c} - 1\right)^2 = \frac{6 - 4\sqrt{2}}{3}$$

$$\sqrt{c} - 1 = \pm\sqrt{\frac{6 - 4\sqrt{2}}{3}}$$

$$c = \left[1 \pm \sqrt{\frac{6 - 4\sqrt{2}}{3}}\right]^2$$

$$c \approx 0.4380 \text{ or } c \approx 1.7908$$

53. $\displaystyle\int_{-\pi/4}^{\pi/4} 2\sec^2 x\,dx = \Big[2\tan x\Big]_{-\pi/4}^{\pi/4} = 2(1) - 2(-1) = 4$

$$f(c)\left[\frac{\pi}{4} - \left(-\frac{\pi}{4}\right)\right] = 4$$

$$2\sec^2 c = \frac{8}{\pi}$$

$$\sec^2 c = \frac{4}{\pi}$$

$$\sec c = \pm\frac{2}{\sqrt{\pi}}$$

$$c = \pm\operatorname{arcsec}\left(\frac{2}{\sqrt{\pi}}\right)$$

$$= \pm\arccos\frac{\sqrt{\pi}}{2} \approx \pm 0.4817$$

55. $f(x) = 5 - \dfrac{1}{x}, \quad [1, 4]$

$$\int_1^4 \left(5 - \frac{1}{x}\right)dx = \Big[5x - \ln x\Big]_1^4 = (20 - \ln 4) - (5 - 0)$$

$$= 15 - \ln 4$$

$$f(c)(4 - 1) = 15 - \ln 4$$

$$\left(5 - \frac{1}{c}\right)(3) = 15 - \ln 4$$

$$15 - \frac{3}{c} = 15 - \ln 4$$

$$\frac{3}{c} = \ln 4$$

$$c = \frac{3}{\ln 4} \approx 2.1640$$

57. $\displaystyle\frac{1}{2 - (-2)}\int_{-2}^{2} (4 - x^2)\,dx = \frac{1}{4}\left[4x - \frac{1}{3}x^3\right]_{-2}^{2} = \frac{1}{4}\left[\left(8 - \frac{8}{3}\right) - \left(-8 + \frac{8}{3}\right)\right] = \frac{8}{3}$

Average value $= \dfrac{8}{3}$

$4 - x^2 = \dfrac{8}{3}$ when $x^2 = 4 - \dfrac{8}{3}$ or $x = \pm\dfrac{2\sqrt{3}}{3} \approx \pm 1.155.$

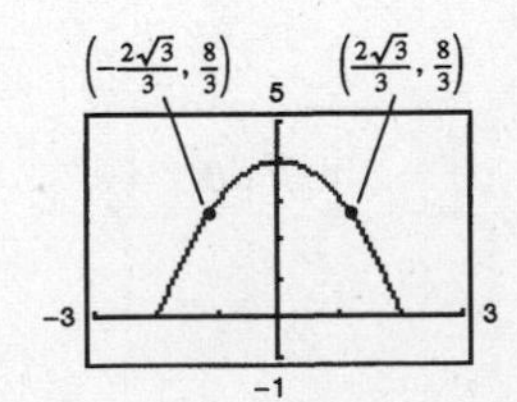

59. $f(x) = 2e^x, \quad [-1, 1]$

$$\text{Average value} = \frac{1}{1-(-1)}\int_{-1}^{1} 2e^x\,dx = \int_{-1}^{1} e^x\,dx$$

$$= e^x\Big]_{-1}^{1} = e - e^{-1} \approx 2.3504$$

$$2e^x = e - e^{-1}$$

$$e^x = \frac{1}{2}(e - e^{-1})$$

$$x = \ln\left(\frac{e - e^{-1}}{2}\right) \approx 0.1614$$

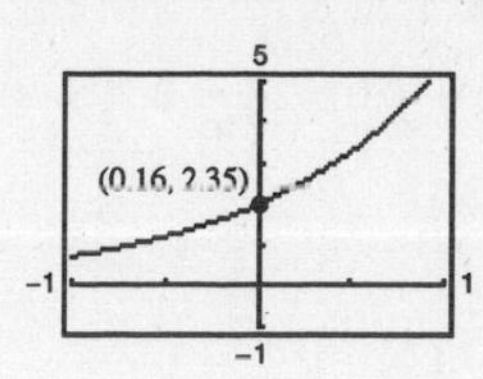

61. $\frac{1}{\pi - 0}\int_0^{\pi} \sin x\,dx = \left[-\frac{1}{\pi}\cos x\right]_0^{\pi} = \frac{2}{\pi}$

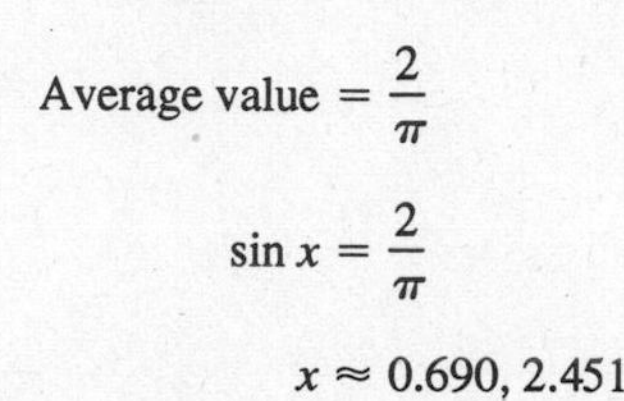

$$\text{Average value} = \frac{2}{\pi}$$

$$\sin x = \frac{2}{\pi}$$

$$x \approx 0.690, 2.451$$

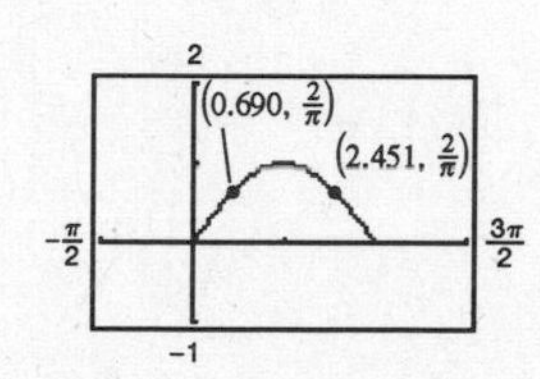

63. If f is continuous on $[a, b]$ and $F'(x) = f(x)$ on $[a, b]$, then $\int_a^b f(x)\,dx = F(b) - F(a)$.

65. $\int_0^2 f(x)\,dx = -(\text{area of region } A) = -1.5$

67. $\int_0^6 |f(x)|\,dx = -\int_0^2 f(x)\,dx + \int_2^6 f(x)\,dx = 1.5 + 5.0 = 6.5$

69. $\int_0^6 [2 + f(x)]\,dx = \int_0^6 2\,dx + \int_0^6 f(x)\,dx$

$$= 12 + 3.5 = 15.5$$

71. (a) $F(x) = k\sec^2 x$

$F(0) = k = 500$

$F(x) = 500\sec^2 x$

(b) $\frac{1}{\frac{\pi}{4} - 0}\int_0^{\pi/4} 500\sec^2 x\,dx = \frac{2000}{\pi}\Big[\tan x\Big]_0^{\pi/4}$

$$= \frac{2000}{\pi}(1 - 0)$$

$$= \frac{2000}{\pi} \approx 636.62 \text{ newtons}$$

73. $\frac{1}{5-0}\int_0^5 (0.1729t + 0.1522t^2 - 0.0374t^3)\,dt \approx \frac{1}{5}\Big[0.08645t^2 + 0.05073t^3 - 0.00935t^4\Big]_0^5 \approx 0.5318 \text{ liter}$

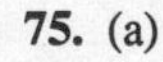

75. (a)

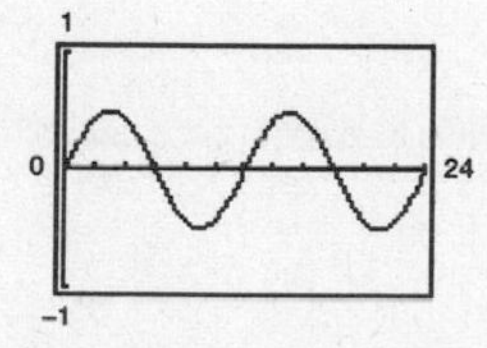

The area above the x-axis equals the area below the x-axis. Thus, the average value is zero.

(b)

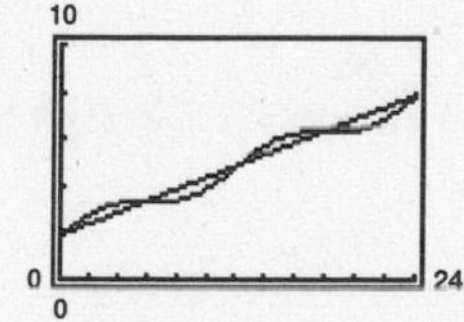

The average value of S appears to be g.

77. (a) $v = -8.61 \times 10^{-4}t^3 + 0.0782t^2 - 0.208t + 0.0952$

(b)

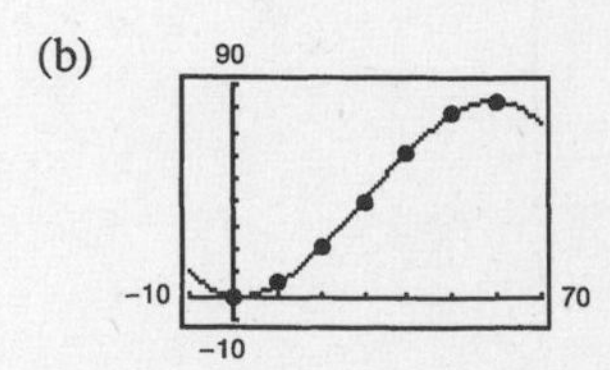

(c) $\displaystyle\int_0^{60} v(t)\,dt = \left[\frac{-8.61 \times 10^{-4}t^4}{4} + \frac{0.0782t^3}{3} - \frac{0.208t^2}{2} + 0.0952t\right]_0^{60} \approx 2476 \text{ meters}$

79. $\displaystyle F(x) = \int_0^x (t - 5)\,dt = \left[\frac{t^2}{2} - 5t\right]_0^x = \frac{x^2}{2} - 5x$

$\displaystyle F(2) = \frac{4}{2} - 5(2) = -8$

$\displaystyle F(5) = \frac{25}{2} - 5(5) = -\frac{25}{2}$

$\displaystyle F(8) = \frac{64}{2} - 5(8) = -8$

81. $\displaystyle F(x) = \int_1^x \frac{10}{v^2}\,dv = \int_1^x 10v^{-2}\,dv = \frac{-10}{v}\Bigg]_1^x$

$\displaystyle = -\frac{10}{x} + 10 = 10\left(1 - \frac{1}{x}\right)$

$\displaystyle F(2) = 10\left(\frac{1}{2}\right) = 5$

$\displaystyle F(5) = 10\left(\frac{4}{5}\right) = 8$

$\displaystyle F(8) = 10\left(\frac{7}{8}\right) = \frac{35}{4}$

83. $\displaystyle F(x) = \int_1^x \cos\theta\,d\theta = \sin\theta\Bigg]_1^x = \sin x - \sin 1$

$F(2) = \sin 2 - \sin 1 = 0.0678$

$F(5) = \sin 5 - \sin 1 \approx -1.8004$

$F(8) = \sin 8 - \sin 1 \approx 0.1479$

85. (a) $\displaystyle\int_0^x (t + 2)\,dt = \left[\frac{t^2}{2} + 2t\right]_0^x = \frac{1}{2}x^2 + 2x$

(b) $\displaystyle\frac{d}{dx}\left[\frac{1}{2}x^2 + 2x\right] = x + 2$

87. (a) $\displaystyle\int_8^x \sqrt[3]{t}\,dt = \left[\frac{3}{4}t^{4/3}\right]_8^x = \frac{3}{4}(x^{4/3} - 16) = \frac{3}{4}x^{4/3} - 12$

(b) $\displaystyle\frac{d}{dx}\left[\frac{3}{4}x^{4/3} - 12\right] = x^{1/3} = \sqrt[3]{x}$

89. (a) $\displaystyle\int_{\pi/4}^x \sec^2 t\,dt = \left[\tan t\right]_{\pi/4}^x = \tan x - 1$

(b) $\displaystyle\frac{d}{dx}[\tan x - 1] = \sec^2 x$

91. (a) $\displaystyle F(x) = \int_{-1}^x e^t\,dt = e^t\Bigg]_{-1}^x = e^x - e^{-1}$

(b) $\displaystyle\frac{d}{dx}(e^x - e^{-1}) = e^x$

93. $\displaystyle F(x) = \int_{-2}^x (t^2 - 2t)\,dt$

$F'(x) = x^2 - 2x$

95. $\displaystyle F(x) = \int_{-1}^x \sqrt{t^4 + 1}\,dt$

$F'(x) = \sqrt{x^4 + 1}$

97. $\displaystyle F(x) = \int_0^x t\cos t\,dt$

$F'(x) = x\cos x$

99. $F(x) = \int_x^{x+2} (4t + 1)\, dt$

$= \left[2t^2 + t\right]_x^{x+2}$

$= [2(x + 2)^2 + (x + 2)] - [2x^2 + x]$

$= 8x + 10$

$F'(x) = 8$

Alternate solution:

$F(x) = \int_x^{x+2} (4t + 1)\, dt$

$= \int_x^0 (4t + 1)\, dt + \int_0^{x+2} (4t + 1)\, dt$

$= -\int_0^x (4t + 1)\, dt + \int_0^{x+2} (4t + 1)\, dt$

$F'(x) = -(4x + 1) + 4(x + 2) + 1 = 8$

101. $F(x) = \int_0^{\sin x} \sqrt{t}\, dt = \left[\frac{2}{3}t^{3/2}\right]_0^{\sin x} = \frac{2}{3}(\sin x)^{3/2}$

$F'(x) = (\sin x)^{1/2} \cos x = \cos x\sqrt{\sin x}$

Alternate solution

$F(x) = \int_0^{\sin x} \sqrt{t}\, dt$

$F'(x) = \sqrt{\sin x}\frac{d}{dx}(\sin x) = \sqrt{\sin x}(\cos x)$

103. $F(x) = \int_0^{x^3} \sin t^2\, dt$

$F'(x) = \sin(x^3)^2 \cdot 3x^2 = 3x^2 \sin x^6$

105. $g(x) = \int_0^x f(t)\, dt$

$g(0) = 0, g(1) \approx \frac{1}{2}, g(2) \approx 1, g(3) \approx \frac{1}{2}, g(4) = 0$

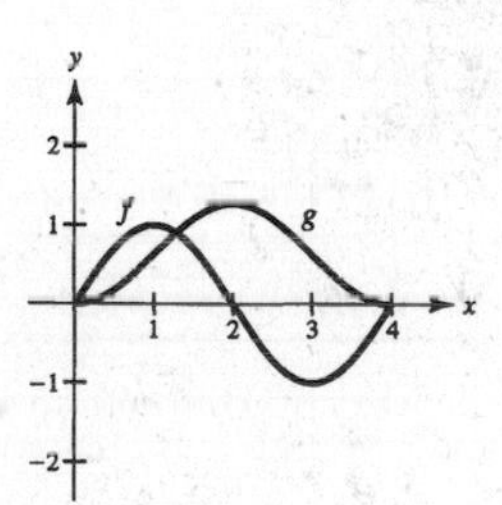

g has a relative maximum at $x = 2$.

107. (a) $C(x) = 5000\left(25 + 3\int_0^x t^{1/4}\, dt\right)$

$= 5000\left(25 + 3\left[\frac{4}{5}t^{5/4}\right]_0^x\right)$

$= 5000\left(25 + \frac{12}{5}x^{5/4}\right) = 1000(125 + 12x^{5/4})$

(b) $C(1) = 1000(125 + 12(1)) = \$137{,}000$

$C(5) = 1000(125 + 12(5)^{5/4}) \approx \$214{,}721$

$C(10) = 1000(125 + 12(10)^{5/4}) \approx \$338{,}394$

109. True

111. False; $\int_{-1}^1 x^{-2} dx = \int_{-1}^0 x^{-2} dx + \int_0^1 x^{-2} dx$

Each of these integrals is infinite. $f(x) = x^{-2}$ has a nonremovable discontinuity at $x = 0$.

113. $f(x) = \int_0^{1/x} \frac{1}{t^2 + 1}\, dt + \int_0^x \frac{1}{t^2 + 1}\, dt$

By the Second Fundamental Theorem of Calculus, we have

$f'(x) = \frac{1}{(1/x)^2 + 1}\left(-\frac{1}{x^2}\right) + \frac{1}{x^2 + 1}$

$= -\frac{1}{1 + x^2} + \frac{1}{x^2 + 1} = 0.$

Since $f'(x) = 0$, $f(x)$ must be constant.

115. $x(t) = t^3 - 6t^2 + 9t - 2$

$$x'(t) = 3t^2 - 12t + 9$$
$$= 3(t^2 - 4t + 3)$$
$$= 3(t - 3)(t - 1)$$

$$\text{Total distance} = \int_0^5 |x'(t)|\,dt$$
$$= \int_0^5 3|(t-3)(t-1)|\,dt$$
$$= 3\int_0^1 (t^2 - 4t + 3)\,dt - 3\int_1^3 (t^2 - 4t + 3)\,dt + 3\int_3^5 (t^2 - 4t + 3)\,dt$$
$$= 4 + 4 + 20$$
$$= 28 \text{ units}$$

117. $$\text{Total distance} = \int_1^4 |x'(t)|\,dt$$
$$= \int_1^4 |v(t)|\,dt$$
$$= \int_1^4 \frac{1}{\sqrt{t}}\,dt$$
$$= 2t^{1/2}\Big]_1^4$$
$$= 2(2 - 1) = 2 \text{ units}$$

Section 4.5 Integration by Substitution

$\int f(g(x))g'(x)\,dx$	$u = g(x)$	$du = g'(x)\,dx$
1. $\int (5x^2 + 1)^2(10x)\,dx$	$5x^2 + 1$	$10x\,dx$
3. $\int \frac{x}{\sqrt{x^2 + 1}}\,dx$	$x^2 + 1$	$2x\,dx$
5. $\int \tan^2 x \sec^2 x\,dx$	$\tan x$	$\sec^2 x\,dx$

7. $\int (1 + 2x)^4\, 2\,dx = \frac{(1 + 2x)^5}{5} + C$

Check: $\frac{d}{dx}\left[\frac{(1 + 2x)^5}{5} + C\right] = 2(1 + 2x)^4$

9. $\int (9 - x^2)^{1/2}(-2x)\,dx = \frac{(9 - x^2)^{3/2}}{3/2} + C = \frac{2}{3}(9 - x^2)^{3/2} + C$

Check: $\frac{d}{dx}\left[\frac{2}{3}(9 - x^2)^{3/2} + C\right] = \frac{2}{3} \cdot \frac{3}{2}(9 - x^2)^{1/2}(-2x) = \sqrt{9 - x^2}(-2x)$

11. $\int x^3(x^4+3)^2\,dx = \frac{1}{4}\int (x^4+3)^2(4x^3)\,dx = \frac{1}{4}\frac{(x^4+3)^3}{3} + C = \frac{(x^4+3)^3}{12} + C$

Check: $\frac{d}{dx}\left[\frac{(x^4+3)^3}{12} + C\right] = \frac{3(x^4+3)^2}{12}(4x^3) = (x^4+3)^2(x^3)$

13. $\int x^2(x^3-1)^4\,dx = \frac{1}{3}\int (x^3-1)^4(3x^2)\,dx = \frac{1}{3}\left[\frac{(x^3-1)^5}{5}\right] + C = \frac{(x^3-1)^5}{15} + C$

Check: $\frac{d}{dx}\left[\frac{(x^3-1)^5}{15} + C\right] = \frac{5(x^3-1)^4(3x^2)}{15} = x^2(x^3-1)^4$

15. $\int t\sqrt{t^2+2}\,dt = \frac{1}{2}\int (t^2+2)^{1/2}(2t)\,dt = \frac{1}{2}\frac{(t^2+2)^{3/2}}{3/2} + C = \frac{(t^2+2)^{3/2}}{3} + C$

Check: $\frac{d}{dt}\left[\frac{(t^2+2)^{3/2}}{3} + C\right] = \frac{3/2(t^2+2)^{1/2}(2t)}{3} = (t^2+2)^{1/2}t$

17. $\int 5x(1-x^2)^{1/3}\,dx = -\frac{5}{2}\int (1-x^2)^{1/3}(-2x)\,dx = -\frac{5}{2}\cdot\frac{(1-x^2)^{4/3}}{4/3} + C = -\frac{15}{8}(1-x^2)^{4/3} + C$

Check: $\frac{d}{dx}\left[-\frac{15}{8}(1-x^2)^{4/3} + C\right] = -\frac{15}{8}\cdot\frac{4}{3}(1-x^2)^{1/3}(-2x) = 5x(1-x^2)^{1/3} = 5x\sqrt[3]{1-x^2}$

19. $\int \frac{x}{(1-x^2)^3}\,dx = -\frac{1}{2}\int (1-x^2)^{-3}(-2x)\,dx = -\frac{1}{2}\frac{(1-x^2)^{-2}}{-2} + C = \frac{1}{4(1-x^2)^2} + C$

Check: $\frac{d}{dx}\left[\frac{1}{4(1-x^2)^2} + C\right] = \frac{1}{4}(-2)(1-x^2)^{-3}(-2x) = \frac{x}{(1-x^2)^3}$

21. $\int \frac{x^2}{(1+x^3)^2}\,dx = \frac{1}{3}\int (1+x^3)^{-2}(3x^2)\,dx = \frac{1}{3}\left[\frac{(1+x^3)^{-1}}{-1}\right] + C = -\frac{1}{3(1+x^3)} + C$

Check: $\frac{d}{dx}\left[-\frac{1}{3(1+x^3)} + C\right] = -\frac{1}{3}(-1)(1+x^3)^{-2}(3x^2) = \frac{x^2}{(1+x^3)^2}$

23. $\int \frac{x}{\sqrt{1-x^2}}\,dx = -\frac{1}{2}\int (1-x^2)^{-1/2}(-2x)\,dx = -\frac{1}{2}\frac{(1-x^2)^{1/2}}{1/2} + C = -\sqrt{1-x^2} + C$

Check: $\frac{d}{dx}[-(1-x^2)^{1/2} + C] = -\frac{1}{2}(1-x^2)^{-1/2}(-2x) = \frac{x}{\sqrt{1-x^2}}$

25. $\int \left(1+\frac{1}{t}\right)^3\left(\frac{1}{t^2}\right)dt = -\int \left(1+\frac{1}{t}\right)^3\left(-\frac{1}{t^2}\right)dt = -\frac{[1+(1/t)]^4}{4} + C$

Check: $\frac{d}{dt}\left[-\frac{[1+(1/t)]^4}{4} + C\right] = -\frac{1}{4}(4)\left(1+\frac{1}{t}\right)^3\left(-\frac{1}{t^2}\right) = \frac{1}{t^2}\left(1+\frac{1}{t}\right)^3$

27. $\int \frac{1}{\sqrt{2x}}\,dx = \frac{1}{2}\int (2x)^{-1/2}\,2\,dx = \frac{1}{2}\left[\frac{(2x)^{1/2}}{1/2}\right] + C = \sqrt{2x} + C$

Check: $\frac{d}{dx}\left[\sqrt{2x} + C\right] = \frac{1}{2}(2x)^{-1/2}(2) = \frac{1}{\sqrt{2x}}$

29. $\displaystyle\int \frac{x^2+3x+7}{\sqrt{x}}\,dx = \int (x^{3/2}+3x^{1/2}+7x^{-1/2})\,dx = \frac{2}{5}x^{5/2}+2x^{3/2}+14x^{1/2}+C = \frac{2}{5}\sqrt{x}(x^2+5x+35)+C$

Check: $\displaystyle\frac{d}{dx}\left[\frac{2}{5}x^{5/2}+2x^{3/2}+14x^{1/2}+C\right] = \frac{x^2+3x+7}{\sqrt{x}}$

31. $\displaystyle\int t^2\left(t-\frac{2}{t}\right)dt = \int (t^3-2t)\,dt = \frac{1}{4}t^4 - t^2 + C$

Check: $\displaystyle\frac{d}{dt}\left[\frac{1}{4}t^4 - t^2 + C\right] = t^3 - 2t = t^2\left(t-\frac{2}{t}\right)$

33. $\displaystyle\int (9-y)\sqrt{y}\,dy = \int (9y^{1/2}-y^{3/2})\,dy = 9\left(\frac{2}{3}y^{3/2}\right) - \frac{2}{5}y^{5/2} + C = \frac{2}{5}y^{3/2}(15-y)+C$

Check: $\displaystyle\frac{d}{dy}\left[\frac{2}{5}y^{3/2}(15-y)+C\right] = \frac{d}{dy}\left[6y^{3/2}-\frac{2}{5}y^{5/2}+C\right] = 9y^{1/2}-y^{3/2} = (9-y)\sqrt{y}$

35.
$$y = \int\left[4x + \frac{4x}{\sqrt{16-x^2}}\right]dx$$
$$= 4\int x\,dx - 2\int (16-x^2)^{-1/2}(-2x)\,dx$$
$$= 4\left(\frac{x^2}{2}\right) - 2\left[\frac{(16-x^2)^{1/2}}{1/2}\right] + C$$
$$= 2x^2 - 4\sqrt{16-x^2} + C$$

37.
$$y = \int \frac{x+1}{(x^2+2x-3)^2}\,dx$$
$$= \frac{1}{2}\int (x^2+2x-3)^{-2}(2x+2)\,dx$$
$$= \frac{1}{2}\left[\frac{(x^2+2x-3)^{-1}}{-1}\right] + C$$
$$= -\frac{1}{2(x^2+2x-3)} + C$$

39. (a)

(b) $\displaystyle\frac{dy}{dx} = x\sqrt{4-x^2},\ (2, 2)$

$$y = \int x\sqrt{4-x^2}\,dx = -\frac{1}{2}\int (4-x^2)^{1/2}(-2x\,dx)$$
$$= -\frac{1}{2}\cdot\frac{2}{3}(4-x^2)^{3/2} + C = -\frac{1}{3}(4-x^2)^{3/2} + C$$

$(2, 2)$: $2 = -\frac{1}{3}(4-2^2)^{3/2} + C \Rightarrow C = 2$

$$y = -\frac{1}{3}(4-x^2)^{3/2} + 2$$

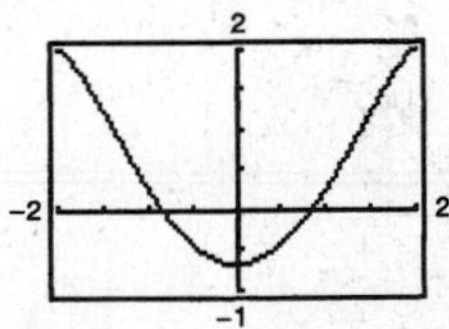

41. (a)

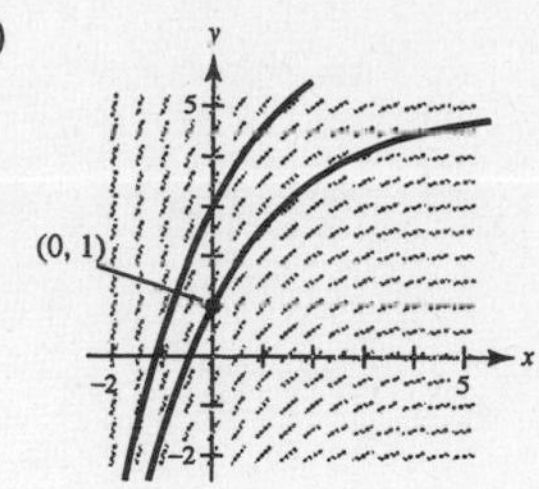

(b) $\dfrac{dy}{dx} = 2e^{-x/2}, \quad (0, 1)$

$$y = \int 2e^{-x/2}\,dx = -4\int e^{-x/2}\left(-\frac{1}{2}\,dx\right)$$

$$= -4e^{-x/2} + C$$

$$(0, 1):\ 1 = -4e^0 + C = -4 + C \Rightarrow C = 5$$

$$y = -4e^{-x/2} + 5$$

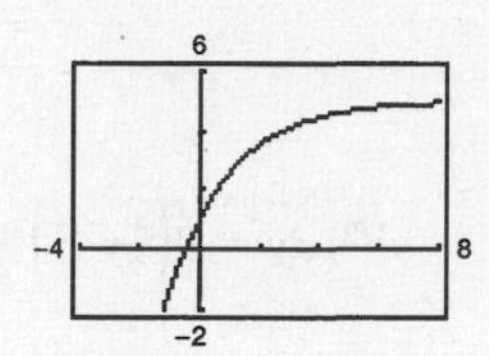

43. (a)

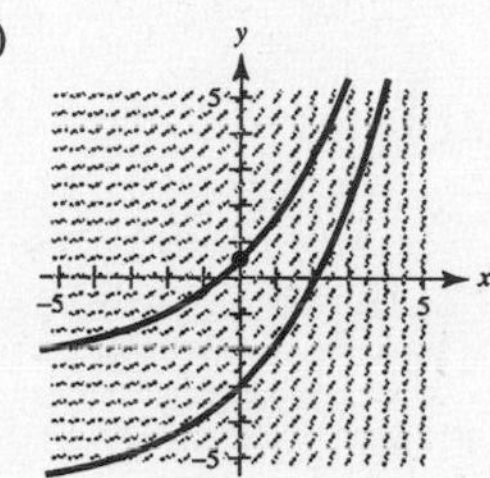

(b) $\dfrac{dy}{dx} = e^{x/3}, \quad \left(0, \dfrac{1}{2}\right)$

$$y = \int e^{x/3}\,dx = 3e^{x/3} + C$$

$$\left(0, \frac{1}{2}\right):\ \frac{1}{2} = 3e^{0/3} + C \Rightarrow C = -\frac{5}{2}$$

$$y = 3e^{x/3} - \frac{5}{2}$$

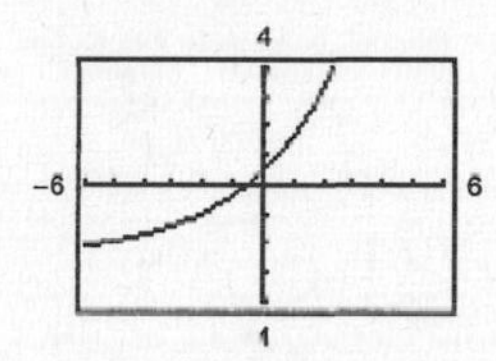

45. $\displaystyle\int \pi \sin \pi x\,dx = -\cos \pi x + C$

47. $\displaystyle\int \sin 2x\,dx = \frac{1}{2}\int (\sin 2x)(2x)\,dx = -\frac{1}{2}\cos 2x + C$

49. $\displaystyle\int \frac{1}{\theta^2}\cos\frac{1}{\theta}\,d\theta = -\int \cos\frac{1}{\theta}\left(-\frac{1}{\theta^2}\right)d\theta = -\sin\frac{1}{\theta} + C$

51. Let $u = 5x$, $du = 5\,dx$.

$$\int e^{5x}5\,dx = e^{5x} + C$$

53. $\displaystyle\int x^2e^{-x^3}\,dx = -\frac{1}{3}\int e^{-x^3}(-3x^2)\,dx$

$$= \frac{-1}{3}e^{-x^3} + C$$

55. $\displaystyle\int \sin 2x \cos 2x\,dx = \frac{1}{2}\int (\sin 2x)(2\cos 2x)\,dx = \frac{1}{2}\frac{(\sin 2x)^2}{2} + C = \frac{1}{4}\sin^2 2x + C$ OR

$\displaystyle\int \sin 2x \cos 2x\,dx = -\frac{1}{2}\int (\cos 2x)(-2\sin 2x)\,dx = -\frac{1}{2}\frac{(\cos 2x)^2}{2} + C_1 = -\frac{1}{4}\cos^2 2x + C_1$ OR

$\displaystyle\int \sin 2x \cos 2x\,dx = \frac{1}{2}\int 2\sin 2x \cos 2x\,dx = \frac{1}{2}\int \sin 4x\,dx = -\frac{1}{8}\cos 4x + C_2$

57. $\displaystyle\int \tan^4 x \sec^2 x\,dx = \frac{\tan^5 x}{5} + C = \frac{1}{5}\tan^5 x + C$

59. $\displaystyle\int \frac{\csc^2 x}{\cot^3 x}\,dx = -\int (\cot x)^{-3}(-\csc^2 x)\,dx$

$$= -\frac{(\cot x)^{-2}}{-2} + C = \frac{1}{2\cot^2 x} + C = \frac{1}{2}\tan^2 x + C = \frac{1}{2}(\sec^2 x - 1) + C = \frac{1}{2}\sec^2 x + C_1$$

61. $\displaystyle\int \cot^2 x\,dx = \int (\csc^2 x - 1)\,dx = -\cot x - x + C$

63. $\displaystyle\int e^x(e^x + 1)^2\,dx = \frac{(e^x + 1)^3}{3} + C$

65. Let $u = 1 - e^x$, $du = -e^x\,dx$.

$$\int e^x\sqrt{1 - e^x}\,dx = -\int (1 - e^x)^{1/2}(-e^x)\,dx = -\frac{2}{3}(1 - e^x)^{3/2} + C$$

67. $\displaystyle\int \frac{5 - e^x}{e^{2x}}\,dx = \int 5e^{-2x}\,dx - \int e^{-x}\,dx$

$$= -\frac{5}{2}e^{-2x} + e^{-x} + C$$

69. $\displaystyle\int e^{\sin \pi x}\cos \pi x\,dx = \frac{1}{\pi}\int e^{\sin \pi x}(\pi \cos \pi x)\,dx$

$$= \frac{1}{\pi}e^{\sin \pi x} + C$$

71. $\displaystyle\int e^{-x}\sec^2(e^{-x})dx = -\int \sec^2(e^{-x})\,(-e^{-x})dx$

$$= -\tan(e^{-x}) + C$$

73. $\displaystyle\int 3^{x/2}\,dx = 2\int 3^{x/2}\left(\frac{1}{2}\right)dx = 2\frac{3^{x/2}}{\ln 3} + C = \frac{2}{\ln 3}3^{x/2} + C$

75. $\displaystyle\int x5^{-x^2}\,dx = -\frac{1}{2}\int 5^{-x^2}(-2x)\,dx$

$$= -\left(\frac{1}{2}\right)\frac{5^{-x^2}}{\ln 5} + C$$

$$= \frac{-1}{2\ln 5}(5^{-x^2}) + C$$

77. $\displaystyle f(x) = \int x\sqrt{4 - x^2}\,dx = -\frac{1}{2}\int (4 - x^2)^{1/2}(-2x)\,dx$

$$= -\frac{1}{2}(4 - x^2)^{3/2}\left(\frac{2}{3}\right) + C = -\frac{1}{3}(4 - x^2)^{3/2} + C$$

$$f(2) = 2 = -\frac{1}{3}(0) + C \Rightarrow C = 2$$

$$f(x) = -\frac{1}{3}(4 - x^2)^{3/2} + 2$$

79. $\displaystyle f(x) = \int \cos\frac{x}{2}\,dx = 2\sin\frac{x}{2} + C$

Since $f(0) = 3 = 2\sin 0 + C$, $C = 3$. Thus,

$$f(x) = 2\sin\frac{x}{2} + 3.$$

81. $\displaystyle f(x) = \int 2e^{-x/4}\,dx = -8\int e^{-x/4}\left(-\frac{1}{4}\right)dx$

$$= -8e^{-x/4} + C$$

$$f(0) = 1 = -8 + C \Rightarrow C = 9$$

$$f(x) = -8e^{-x/4} + 9$$

83. $\displaystyle f'(x) = \int \frac{1}{2}(e^x + e^{-x})\,dx = \frac{1}{2}(e^x - e^{-x}) + C_1$

$$f'(0) = C_1 = 0$$

$$f(x) = \int \frac{1}{2}(e^x - e^{-x})\,dx = \frac{1}{2}(e^x + e^{-x}) + C_2$$

$$f(0) = 1 + C_2 = 1 \Rightarrow C_2 = 0$$

$$f(x) = \frac{1}{2}(e^x + e^{-x})$$

85. $u = x + 2, x = u - 2, dx = du$

$$\int x\sqrt{x+2}\,dx = \int (u-2)\sqrt{u}\,du$$
$$= \int (u^{3/2} - 2u^{1/2})\,du$$
$$= \frac{2}{5}u^{5/2} - \frac{4}{3}u^{3/2} + C$$
$$= \frac{2u^{3/2}}{15}(3u - 10) + C$$
$$= \frac{2}{15}(x+2)^{3/2}[3(x+2) - 10] + C$$
$$= \frac{2}{15}(x+2)^{3/2}(3x - 4) + C$$

87. $u = 1 - x, x = 1 - u, dx = -du$

$$\int x^2\sqrt{1-x}\,dx = -\int (1-u)^2\sqrt{u}\,du$$
$$= -\int (u^{1/2} - 2u^{3/2} + u^{5/2})\,du$$
$$= -\left(\frac{2}{3}u^{3/2} - \frac{4}{5}u^{5/2} + \frac{2}{7}u^{7/2}\right) + C$$
$$= -\frac{2u^{3/2}}{105}(35 - 42u + 15u^2) + C$$
$$= -\frac{2}{105}(1-x)^{3/2}[35 - 42(1-x) + 15(1-x)^2] + C$$
$$= -\frac{2}{105}(1-x)^{3/2}(15x^2 + 12x + 8) + C$$

89. $u = 2x - 1, x = \frac{1}{2}(u+1), dx = \frac{1}{2}\,du$

$$\int \frac{x^2 - 1}{\sqrt{2x-1}}\,dx = \int \frac{[(1/2)(u+1)]^2 - 1}{\sqrt{u}}\frac{1}{2}\,du$$
$$= \frac{1}{8}\int u^{-1/2}[(u^2 + 2u + 1) - 4]\,du$$
$$= \frac{1}{8}\int (u^{3/2} + 2u^{1/2} - 3u^{-1/2})\,du$$
$$= \frac{1}{8}\left(\frac{2}{5}u^{5/2} + \frac{4}{3}u^{3/2} - 6u^{1/2}\right) + C$$
$$= \frac{u^{1/2}}{60}(3u^2 + 10u - 45) + C$$
$$= \frac{\sqrt{2x-1}}{60}[3(2x-1)^2 + 10(2x-1) - 45] + C$$
$$= \frac{1}{60}\sqrt{2x-1}(12x^2 + 8x - 52) + C$$
$$= \frac{1}{15}\sqrt{2x-1}(3x^2 + 2x - 13) + C$$

91. $u = x + 1, x = u - 1, dx = du$

$$\int \frac{-x}{(x+1) - \sqrt{x+1}}\,dx = \int \frac{-(u-1)}{u - \sqrt{u}}\,du$$

$$= -\int \frac{(\sqrt{u}+1)(\sqrt{u}-1)}{\sqrt{u}(\sqrt{u}-1)}\,du$$

$$= -\int (1 + u^{-1/2})\,du$$

$$= -(u + 2u^{1/2}) + C$$

$$= -u - 2\sqrt{u} + C$$

$$= -(x+1) - 2\sqrt{x+1} + C$$

$$= -x - 2\sqrt{x+1} - 1 + C$$

$$= -(x + 2\sqrt{x+1}) + C_1$$

where $C_1 = -1 + C$.

93. Let $u = x^2 + 1, du = 2x\,dx$.

$$\int_{-1}^{1} x(x^2+1)^3\,dx = \frac{1}{2}\int_{-1}^{1} (x^2+1)^3(2x)\,dx$$

$$= \left[\frac{1}{8}(x^2+1)^4\right]_{-1}^{1} = 0$$

95. Let $u = x^3 + 1, du = 3x^2\,dx$

$$\int_1^2 2x^2\sqrt{x^3+1}\,dx = 2\cdot\frac{1}{3}\int_1^2 (x^3+1)^{1/2}(3x^2)\,dx$$

$$= \left[\frac{2}{3}\frac{(x^3+1)^{3/2}}{3/2}\right]_1^2$$

$$= \frac{4}{9}\left[(x^3+1)^{3/2}\right]_1^2$$

$$= \frac{4}{9}\left[27 - 2\sqrt{2}\right] = 12 - \frac{8}{9}\sqrt{2}$$

97. Let $u = 2x + 1, du = 2\,dx$.

$$\int_0^4 \frac{1}{\sqrt{2x+1}}\,dx = \frac{1}{2}\int_0^4 (2x+1)^{-1/2}(2)\,dx = \left[\sqrt{2x+1}\right]_0^4 = \sqrt{9} - \sqrt{1} = 2$$

99. $$\int_0^1 e^{-2x}\,dx = -\frac{1}{2}\int_0^1 e^{-2x}(-2)\,dx = -\frac{1}{2}e^{-2x}\Big]_0^1 = -\frac{1}{2}e^{-2} + \frac{1}{2}$$

101. $$\int_1^3 \frac{e^{3/x}}{x^2}\,dx = -\frac{1}{3}\int_1^3 e^{3/x}\left(-\frac{3}{x^2}\right)dx = \left[-\frac{1}{3}e^{3/x}\right]_1^3 = -\frac{1}{3}(e - e^3) = \frac{e}{3}(e^2 - 1)$$

103. Let $u = 1 + \sqrt{x}, du = \dfrac{1}{2\sqrt{x}}\,dx$.

$$\int_1^9 \frac{1}{\sqrt{x}(1+\sqrt{x})^2}\,dx = 2\int_1^9 (1+\sqrt{x})^{-2}\left(\frac{1}{2\sqrt{x}}\right)dx = \left[-\frac{2}{1+\sqrt{x}}\right]_1^9 = -\frac{1}{2} + 1 = \frac{1}{2}$$

105. $u = 2 - x, x = 2 - u, dx = -du$

When $x = 1, u = 1$. When $x = 2, u = 0$.

$$\int_1^2 (x-1)\sqrt{2-x}\,dx = \int_1^0 -[(2-u) - 1]\sqrt{u}\,du = \int_1^0 (u^{3/2} - u^{1/2})\,du = \left[\frac{2}{5}u^{5/2} - \frac{2}{3}u^{3/2}\right]_1^0 = -\left[\frac{2}{5} - \frac{2}{3}\right] = \frac{4}{15}$$

107. $\displaystyle\int_0^{\pi/2} \cos\left(\frac{2}{3}x\right) dx = \left[\frac{3}{2}\sin\left(\frac{2}{3}x\right)\right]_0^{\pi/2} = \frac{3}{2}\left(\frac{\sqrt{3}}{2}\right) = \frac{3\sqrt{3}}{4}$

109. $\displaystyle\int_{-1}^{2} 2^x\, dx = \left[\frac{2^x}{\ln 2}\right]_{-1}^{2} = \frac{1}{\ln 2}\left[4 - \frac{1}{2}\right] = \frac{7}{2\ln 2} = \frac{7}{\ln 4}$

111. $u = x + 1, x = u - 1, dx = du$

When $x = 0$, $u = 1$. When $x = 7$, $u = 8$.

$$\text{Area} = \int_0^7 x\sqrt[3]{x+1}\, dx = \int_1^8 (u-1)\sqrt[3]{u}\, du$$
$$= \int_1^8 (u^{4/3} - u^{1/3})\, du = \left[\frac{3}{7}u^{7/3} - \frac{3}{4}u^{4/3}\right]_1^8 = \left(\frac{384}{7} - 12\right) - \left(\frac{3}{7} - \frac{3}{4}\right) = \frac{1209}{28}$$

113. $\displaystyle A = \int_0^{\pi} (2\sin x + \sin 2x)\, dx = -\left[2\cos x + \frac{1}{2}\cos 2x\right]_0^{\pi} = 4$

115. $\displaystyle\text{Area} = \int_{\pi/2}^{2\pi/3} \sec^2\left(\frac{x}{2}\right) dx = 2\int_{\pi/2}^{2\pi/3} \sec^2\left(\frac{x}{2}\right)\left(\frac{1}{2}\right) dx = \left[2\tan\left(\frac{x}{2}\right)\right]_{\pi/2}^{2\pi/3} = 2(\sqrt{3} - 1)$

117. $\displaystyle\int_0^5 e^x\, dx = \left[e^x\right]_0^5 = e^5 - 1 \approx 147.413$

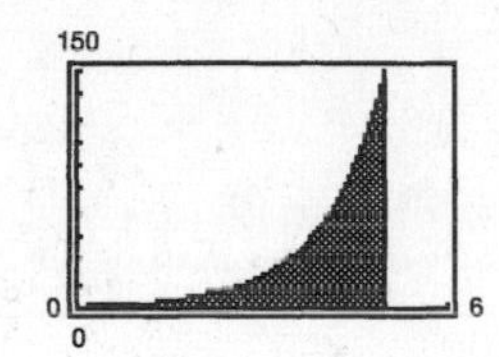

119. $\displaystyle\int_0^{\sqrt{6}} xe^{-x^2/4} dx = \left[-2e^{-x^2/4}\right]_0^{\sqrt{6}}$

$= -2e^{-3/2} + 2 \approx 1.554$

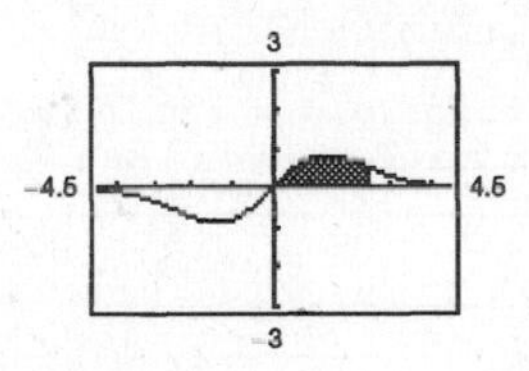

121. $\displaystyle\int_0^4 \frac{x}{\sqrt{2x+1}}\, dx \approx 3.333 = \frac{10}{3}$

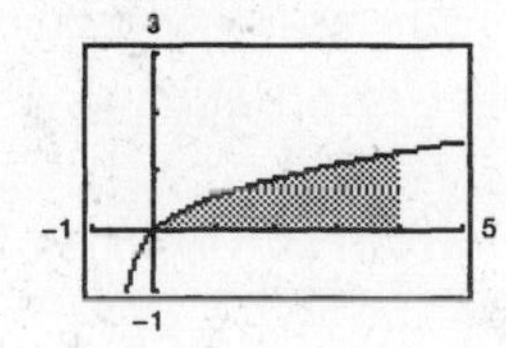

123. $\displaystyle\int_3^7 x\sqrt{x-3}\, dx \approx 28.8 = \frac{144}{5}$

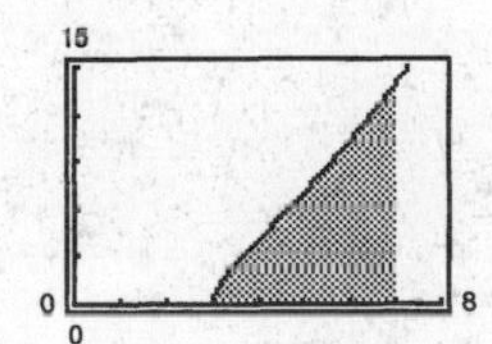

125. $\displaystyle\int_0^3 \left(\theta + \cos\frac{\theta}{6}\right) d\theta \approx 7.377$

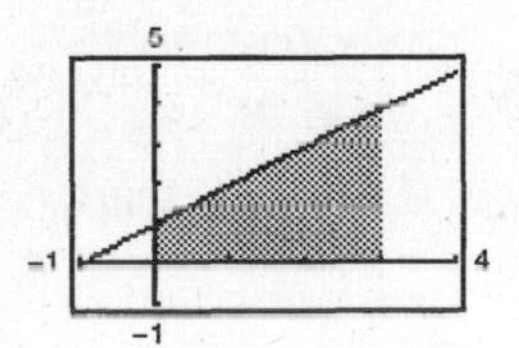

127. $\displaystyle\int_0^{\sqrt{2}} xe^{-(x^2/2)} dx = \left[-e^{-(x^2/2)}\right]_0^{\sqrt{2}}$

$= -e^{-1} + 1 \approx 0.632$

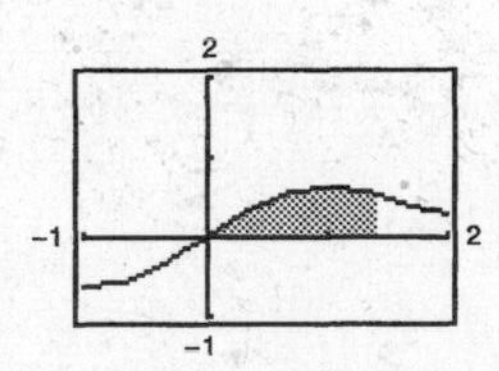

129. $\displaystyle\int (2x-1)^2\, dx = \frac{1}{2}\int (2x-1)^2\, 2\, dx = \frac{1}{6}(2x-1)^3 + C_1 = \frac{4}{3}x^3 - 2x^2 + x - \frac{1}{6} + C_1$

$\displaystyle\int (2x-1)^2\, dx = \int (4x^2 - 4x + 1)\, dx = \frac{4}{3}x^3 - 2x^2 + x + C_2$

They differ by a constant: $C_2 = C_1 - \frac{1}{6}$.

131. $f(x) = x^2(x^2 + 1)$ is even.

$$\int_{-2}^{2} x^2(x^2 + 1)\,dx = 2\int_0^2 (x^4 + x^2)\,dx = 2\left[\frac{x^5}{5} + \frac{x^3}{3}\right]_0^2$$

$$= 2\left[\frac{32}{5} + \frac{8}{3}\right] = \frac{272}{15}$$

133. $f(x) = x(x^2 + 1)^3$ is odd.

$$\int_{-2}^{2} x(x^2 + 1)^3\,dx = 0$$

135. $\displaystyle\int_0^2 x^2\,dx = \left[\frac{x^3}{3}\right]_0^2 = \frac{8}{3}$; the function x^2 is an even function.

(a) $\displaystyle\int_{-2}^{0} x^2\,dx = \int_0^2 x^2\,dx = \frac{8}{3}$

(b) $\displaystyle\int_{-2}^{2} x^2\,dx = 2\int_0^2 x^2\,dx = \frac{16}{3}$

(c) $\displaystyle\int_0^2 (-x^2)\,dx = -\int_0^2 x^2\,dx = -\frac{8}{3}$

(d) $\displaystyle\int_{-2}^{0} 3x^2\,dx = 3\int_0^2 x^2\,dx = 8$

137. $\displaystyle\int_{-4}^{4} (x^3 + 6x^2 - 2x - 3)\,dx = \int_{-4}^{4} (x^3 - 2x)\,dx + \int_{-4}^{4} (6x^2 - 3)\,dx = 0 + 2\int_0^4 (6x^2 - 3)\,dx = 2\left[2x^3 - 3x\right]_0^4 = 232$

139. Answers will vary.
See "Guidelines for Making a Change of Variables".

141. $f(x) = x(x^2 + 1)^2$ is odd. Hence, $\displaystyle\int_{-2}^{2} x(x^2 + 1)^2\,dx = 0.$

143. $\displaystyle\frac{dV}{dt} = \frac{k}{(t + 1)^2}$

$$V(t) = \int \frac{k}{(t + 1)^2}\,dt = -\frac{k}{t + 1} + C$$

$$V(0) = -k + C = 500{,}000$$

$$V(1) = -\frac{1}{2}k + C = 400{,}000$$

Solving this system yields $k = -200{,}000$ and $C = 300{,}000$. Thus,

$$V(t) = \frac{200{,}000}{t + 1} + 300{,}000.$$

When $t = 4$, $V(4) = \$340{,}000$.

145. $\displaystyle\frac{1}{b - a}\int_a^b \left[74.50 + 43.75\sin\frac{\pi t}{6}\right] dt = \frac{1}{b - a}\left[74.50t - \frac{262.5}{\pi}\cos\frac{\pi t}{6}\right]_a^b$

(a) $\displaystyle\frac{1}{3}\left[74.50t - \frac{262.5}{\pi}\cos\frac{\pi t}{6}\right]_0^3 = \frac{1}{3}\left(223.5 + \frac{262.5}{\pi}\right) \approx 102.352$ thousand units

(b) $\displaystyle\frac{1}{3}\left[74.50t - \frac{262.5}{\pi}\cos\frac{\pi t}{6}\right]_3^6 = \frac{1}{3}\left(447 + \frac{262.5}{\pi} - 223.5\right) \approx 102.352$ thousand units

(c) $\displaystyle\frac{1}{12}\left[74.50t - \frac{262.5}{\pi}\cos\frac{\pi t}{6}\right]_0^{12} = \frac{1}{12}\left(894 - \frac{262.5}{\pi} + \frac{262.5}{\pi}\right) = 74.5$ thousand units

147. $\dfrac{1}{b-a}\displaystyle\int_a^b [2\sin(60\pi t) + \cos(120\pi t)]\,dt = \dfrac{1}{b-a}\left[-\dfrac{1}{30\pi}\cos(60\pi t) + \dfrac{1}{120\,\pi}\sin(120\pi t)\right]_a^b$

(a) $\dfrac{1}{(1/60)-0}\left[-\dfrac{1}{30\pi}\cos(60\pi t) + \dfrac{1}{120\pi}\sin(120\pi t)\right]_0^{1/60} = 60\left[\left(\dfrac{1}{30\pi}+0\right) - \left(-\dfrac{1}{30\pi}\right)\right] = \dfrac{4}{\pi} \approx 1.273$ amps

(b) $\dfrac{1}{(1/240)-0}\left[-\dfrac{1}{30\pi}\cos(60\pi t) + \dfrac{1}{120\pi}\sin(120\pi t)\right]_0^{1/240} = 240\left[\left(-\dfrac{1}{30\sqrt{2}\pi} + \dfrac{1}{120\pi}\right) - \left(-\dfrac{1}{30\pi}\right)\right]$

$= \dfrac{2}{\pi}(5 - 2\sqrt{2}) \approx 1.382$ amps

(c) $\dfrac{1}{(1/30)-0}\left[-\dfrac{1}{30\pi}\cos(60\pi t) + \dfrac{1}{120\pi}\sin(120\pi t)\right]_0^{1/30} = 30\left[\left(\dfrac{1}{30\pi}\right) - \left(-\dfrac{1}{30\pi}\right)\right] = 0$ amps

149. $0.0665\displaystyle\int_{48}^{60} e^{-0.0139(t-48)^2}\,dt$

Graphing Utility: $0.4772 = 47.72\%$

151. False

$\displaystyle\int (2x+1)^2\,dx = \frac{1}{2}\int (2x+1)^2\,2\,dx = \frac{1}{6}(2x+1)^3 + C$

153. True

$$\int_{-10}^{10} (ax^3 + bx^2 + cx + d)\,dx = \underbrace{\int_{-10}^{10} (ax^3 + cx)\,dx}_{\text{Odd}} + \underbrace{\int_{-10}^{10} (bx^2 + d)\,dx}_{\text{Even}} = 0 + 2\int_0^{10} (bx^2 + d)\,dx$$

155. True

$4\displaystyle\int \sin x\cos x\,dx = 2\int \sin 2x\,dx = -\cos 2x + C$

157. Let $u = x + h$, then $du = dx$. When $x = a$, $u = a + h$. When $x = b$, $u = b + h$. Thus,

$$\int_a^b f(x+h)\,dx = \int_{a+h}^{b+h} f(u)\,du = \int_{a+h}^{b+h} f(x)\,dx.$$

Section 4.6 Numerical Integration

1. Exact: $\displaystyle\int_0^2 x^2\,dx = \left[\frac{1}{3}x^3\right]_0^2 = \frac{8}{3} \approx 2.6667$

Trapezoidal: $\displaystyle\int_0^2 x^2\,dx \approx \frac{1}{4}\left[0 + 2\left(\frac{1}{2}\right)^2 + 2(1)^2 + 2\left(\frac{3}{2}\right)^2 + (2)^2\right] = \frac{11}{4} = 2.7500$

Simpson's: $\displaystyle\int_0^2 x^2\,dx \approx \frac{1}{6}\left[0 + 4\left(\frac{1}{2}\right)^2 + 2(1)^2 + 4\left(\frac{3}{2}\right)^2 + (2)^2\right] = \frac{8}{3} \approx 2.6667$

3. Exact: $\displaystyle\int_0^2 x^3\,dx = \left[\frac{x^4}{4}\right]_0^2 = 4.000$

Trapezoidal: $\displaystyle\int_0^2 x^3\,dx \approx \frac{1}{4}\left[0 + 2\left(\frac{1}{2}\right)^3 + 2(1)^3 + 2\left(\frac{3}{2}\right)^3 + (2)^3\right] = \frac{17}{4} = 4.2500$

Simpson's: $\displaystyle\int_0^2 x^3\,dx \approx \frac{1}{6}\left[0 + 4\left(\frac{1}{2}\right)^3 + 2(1)^3 + 4\left(\frac{3}{2}\right)^3 + (2)^3\right] = \frac{24}{6} = 4.0000$

5. Exact: $\displaystyle\int_0^2 x^3\,dx = \left[\frac{1}{4}x^4\right]_0^2 = 4.0000$

Trapezoidal: $\displaystyle\int_0^2 x^3\,dx \approx \frac{1}{8}\left[0 + 2\left(\frac{1}{4}\right)^3 + 2\left(\frac{2}{4}\right)^3 + 2\left(\frac{3}{4}\right)^3 + 2(1)^3 + 2\left(\frac{5}{4}\right)^3 + 2\left(\frac{6}{4}\right)^3 + 2\left(\frac{7}{4}\right)^3 + 8\right] = 4.0625$

Simpson's: $\displaystyle\int_0^2 x^3\,dx \approx \frac{1}{12}\left[0 + 4\left(\frac{1}{4}\right)^3 + 2\left(\frac{2}{4}\right)^3 + 4\left(\frac{3}{4}\right)^3 + 2(1)^3 + 4\left(\frac{5}{4}\right)^3 + 2\left(\frac{6}{4}\right)^3 + 4\left(\frac{7}{4}\right)^3 + 8\right] = 4.0000$

7. Exact: $\displaystyle\int_4^9 \sqrt{x}\,dx = \left[\frac{2}{3}x^{3/2}\right]_4^9 = 18 - \frac{16}{3} = \frac{38}{3} \approx 12.6667$

Trapezoidal: $\displaystyle\int_4^9 \sqrt{x}\,dx \approx \frac{5}{16}\left[2 + 2\sqrt{\frac{37}{8}} + 2\sqrt{\frac{21}{4}} + 2\sqrt{\frac{47}{8}} + 2\sqrt{\frac{26}{4}} + 2\sqrt{\frac{57}{8}} + 2\sqrt{\frac{31}{4}} + 2\sqrt{\frac{67}{8}} + 3\right]$

≈ 12.6640

Simpson's: $\displaystyle\int_4^9 \sqrt{x}\,dx \approx \frac{5}{24}\left[2 + 4\sqrt{\frac{37}{8}} + \sqrt{21} + 4\sqrt{\frac{47}{8}} + \sqrt{26} + 4\sqrt{\frac{57}{8}} + \sqrt{31} + 4\sqrt{\frac{67}{8}} + 3\right] \approx 12.6667$

9. Exact: $\displaystyle\int_1^2 \frac{1}{(x+1)^2}\,dx = \left[-\frac{1}{x+1}\right]_1^2 = -\frac{1}{3} + \frac{1}{2} = \frac{1}{6} \approx 0.1667$

Trapezoidal: $\displaystyle\int_1^2 \frac{1}{(x+1)^2}\,dx \approx \frac{1}{8}\left[\frac{1}{4} + 2\left(\frac{1}{((5/4)+1)^2}\right) + 2\left(\frac{1}{((3/2)+1)^2}\right) + 2\left(\frac{1}{((7/4)+1)^2}\right) + \frac{1}{9}\right]$

$\displaystyle= \frac{1}{8}\left(\frac{1}{4} + \frac{32}{81} + \frac{8}{25} + \frac{32}{121} + \frac{1}{9}\right) \approx 0.1676$

Simpson's: $\displaystyle\int_1^2 \frac{1}{(x+1)^2}\,dx \approx \frac{1}{12}\left[\frac{1}{4} + 4\left(\frac{1}{((5/4)+1)^2}\right) + 2\left(\frac{1}{((3/2)+1)^2}\right) + 4\left(\frac{1}{((7/4)+1)^2}\right) + \frac{1}{9}\right]$

$\displaystyle= \frac{1}{12}\left(\frac{1}{4} + \frac{64}{81} + \frac{8}{25} + \frac{64}{121} + \frac{1}{9}\right) \approx 0.1667$

11. Trapezoidal: $\displaystyle\int_0^2 \sqrt{1+x^3}\,dx \approx \frac{1}{4}[1 + 2\sqrt{1+(1/8)} + 2\sqrt{2} + 2\sqrt{1+(27/8)} + 3] \approx 3.283$

Simpson's: $\displaystyle\int_0^2 \sqrt{1+x^3}\,dx \approx \frac{1}{6}[1 + 4\sqrt{1+(1/8)} + 2\sqrt{2} + 4\sqrt{1+(27/8)} + 3] \approx 3.240$

Graphing utility: 3.241

13. $\displaystyle\int_0^1 \sqrt{x}\sqrt{1-x}\,dx = \int_0^1 \sqrt{x(1-x)}\,dx$

Trapezoidal: $\displaystyle\int_0^1 \sqrt{x(1-x)}\,dx \approx \frac{1}{8}\left[0 + 2\sqrt{\frac{1}{4}\left(1-\frac{1}{4}\right)} + 2\sqrt{\frac{1}{2}\left(1-\frac{1}{2}\right)} + 2\sqrt{\frac{3}{4}\left(1-\frac{3}{4}\right)}\right] \approx 0.342$

Simpson's: $\displaystyle\int_0^1 \sqrt{x(1-x)}\,dx \approx \frac{1}{12}\left[0 + 4\sqrt{\frac{1}{4}\left(1-\frac{1}{4}\right)} + 2\sqrt{\frac{1}{2}\left(1-\frac{1}{2}\right)} + 4\sqrt{\frac{3}{4}\left(1-\frac{3}{4}\right)}\right] \approx 0.372$

Graphing utility: 0.393

15. Trapezoidal: $\displaystyle\int_0^{\sqrt{\pi/2}} \cos(x^2)\,dx \approx \frac{\sqrt{\pi/2}}{8}\left[\cos 0 + 2\cos\left(\frac{\sqrt{\pi/2}}{4}\right)^2 + 2\cos\left(\frac{\sqrt{\pi/2}}{2}\right)^2 + 2\cos\left(\frac{\sqrt{\pi/2}}{4}\right)^2 + \cos\left(\sqrt{\frac{\pi}{2}}\right)^2\right]$

≈ 0.957

Simpson's: $\displaystyle\int_0^{\sqrt{\pi/2}} \cos(x^2)\,dx \approx \frac{\sqrt{\pi/2}}{12}\left[\cos 0 + 4\cos\left(\frac{\sqrt{\pi/2}}{4}\right)^2 + 2\cos\left(\frac{\sqrt{\pi/2}}{2}\right)^2 + 4\cos\left(\frac{\sqrt{\pi/2}}{4}\right)^2 + \cos\left(\sqrt{\frac{\pi}{2}}\right)^2\right]$

≈ 0.978

Graphing utility: 0.977

17. Trapezoidal: $\displaystyle\int_1^{1.1} \sin x^2\,dx \approx \frac{1}{80}[\sin(1) + 2\sin(1.025)^2 + 2\sin(1.05)^2 + 2\sin(1.075)^2 + \sin(1.1)^2] \approx 0.089$

Simpson's: $\displaystyle\int_1^{1.1} \sin x^2\,dx \approx \frac{1}{120}[\sin(1) + 4\sin(1.025)^2 + 2\sin(1.05)^2 + 4\sin(1.075)^2 + \sin(1.1)^2] \approx 0.089$

Graphing utility: 0.089

19. $\displaystyle\int_0^2 x\ln(x+1)\,dx$

Trapezoidal: 1.684

Simpson's: 1.649

Graphing utility: 1.648

21. Trapezoidal: $\displaystyle\int_0^{\pi/4} x\tan x\,dx \approx \frac{\pi}{32}\left[0 + 2\left(\frac{\pi}{16}\right)\tan\left(\frac{\pi}{16}\right) + 2\left(\frac{2\pi}{16}\right)\tan\left(\frac{2\pi}{16}\right) + 2\left(\frac{3\pi}{16}\right)\tan\left(\frac{3\pi}{16}\right) + \frac{\pi}{4}\right] \approx 0.194$

Simpson's: $\displaystyle\int_0^{\pi/4} x\tan x\,dx \approx \frac{\pi}{48}\left[0 + 4\left(\frac{\pi}{16}\right)\tan\left(\frac{\pi}{16}\right) + 2\left(\frac{2\pi}{16}\right)\tan\left(\frac{2\pi}{16}\right) + 4\left(\frac{3\pi}{16}\right)\tan\left(\frac{3\pi}{16}\right) + \frac{\pi}{4}\right] \approx 0.186$

Graphing utility: 0.186

23. Trapezoidal: $\displaystyle\int_0^4 \sqrt{x}e^x\,dx \approx \frac{1}{2}[0 + 2e^1 + 2\sqrt{2}e^2 + 2\sqrt{3}e^3 + 2e^4] \approx \frac{205.1106}{2} \approx 102.555$

Simpson's: $\displaystyle\int_0^4 \sqrt{x}e^x\,dx \approx \frac{1}{3}[0 + 4e^1 + 2\sqrt{2}e^2 + 4\sqrt{3}e^3 + 2e^4] \approx \frac{208.1255}{3} \approx 93.375$

Graphing utility: 92.744

25. (a)

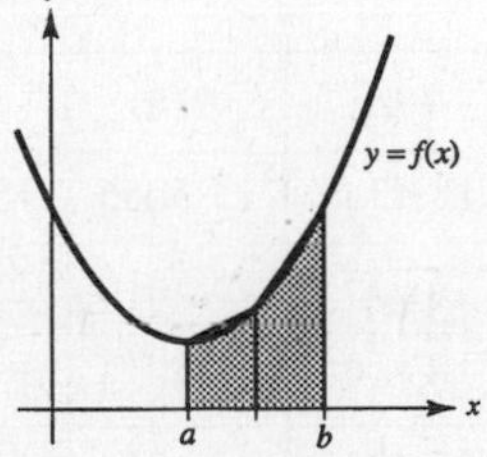

The Trapezoidal Rule overestimates the area if the graph of the integrand is concave up.

27. $f(x) = x^3$

$f'(x) = 3x^2$

$f''(x) = 6x$

$f'''(x) = 6$

$f^{(4)}(x) = 0$

(a) Trapezoidal: Error $\le \dfrac{(2-0)^3}{12(4^2)}(12) = 0.5$ since

$f''(x)$ is maximum in $[0, 2]$ when $x = 2$.

(b) Simpson's: Error $\le \dfrac{(2-0)^5}{180(4^4)}(0) = 0$ since

$f^{(4)}(x) = 0$.

29. $f''(x) = \frac{2}{x^3}$ in $[1, 3]$.

(a) $|f''(x)|$ is maximum when $x = 1$ and $|f''(1)| = 2$.

Trapezoidal: Error $\leq \frac{2^3}{12n^2}(2) < 0.00001$, $n^2 > 133{,}333.33$, $n > 365.15$; let $n = 366$.

$f^{(4)}(x) = \frac{24}{x^5}$ in $[1, 3]$

(b) $|f^{(4)}(x)|$ is maximum when $x = 1$ and $|f^{(4)}(1)| = 24$.

Simpson's: Error $\leq \frac{2^5}{180n^4}(24) < 0.00001$, $n^4 > 426{,}666.67$, $n > 25.56$; let $n = 26$.

31. $f(x) = \sqrt{1 + x}$

(a) $f''(x) = -\frac{1}{4(1 + x)^{3/2}}$ in $[0, 2]$.

$|f''(x)|$ is maximum when $x = 0$ and $|f''(0)| = \frac{1}{4}$.

Trapezoidal: Error $\leq \frac{8}{12n^2}\left(\frac{1}{4}\right) < 0.00001$, $n^2 > 16{,}666.67$, $n > 129.10$; let $n = 130$.

(b) $f^{(4)}(x) = \frac{-15}{16(1 + x)^{7/2}}$ in $[0, 2]$

$|f^{(4)}(x)|$ is maximum when $x = 0$ and $|f^{(4)}(0)| = \frac{15}{16}$.

Simpson's: Error $\leq \frac{32}{180n^4}\left(\frac{15}{16}\right) < 0.00001$, $n^4 > 16{,}666.67$, $n > 11.36$; let $n = 12$.

33. $f(x) = \tan(x^2)$

(a) $f''(x) = 2\sec^2(x^2)[1 + 4x^2\tan(x^2)]$ in $[0, 1]$.

$|f''(x)|$ is maximum when $x = 1$ and $|f''(1)| \approx 49.5305$.

Trapezoidal: Error $\leq \frac{(1 - 0)^3}{12n^2}(49.5305) < 0.00001$, $n^2 > 412{,}754.17$, $n > 642.46$; let $n = 643$.

(b) $f^{(4)}(x) = 8\sec^2(x^2)[12x^2 + (3 + 32x^4)\tan(x^2) + 36x^2\tan^2(x^2) + 48x^4\tan^3(x^2)]$ in $[0, 1]$

$|f^{(4)}(x)|$ is maximum when $x = 1$ and $|f^{(4)}(1)| \approx 9184.4734$.

Simpson's: Error $\leq \frac{(1 - 0)^5}{180n^4}(9184.4734) < 0.00001$, $n^4 > 5{,}102{,}485.22$, $n > 47.53$; let $n = 48$.

35. Let $f(x) = Ax^3 + Bx^2 + Cx + D$. Then $f^{(4)}(x) = 0$.

Simpson's: Error $\leq \frac{(b - a)^5}{180n^4}(0) = 0$

Therefore, Simpson's Rule is exact when approximating the integral of a cubic polynomial.

Example: $\int_0^1 x^3\,dx = \frac{1}{6}\left[0 + 4\left(\frac{1}{2}\right)^3 + 1\right] = \frac{1}{4}$

This is the exact value of the integral.

37. $f(x) = \sqrt{2 + 3x^2}$ on $[0, 4]$.

n	$L(n)$	$M(n)$	$R(n)$	$T(n)$	$S(n)$
4	12.7771	15.3965	18.4340	15.6055	15.4845
8	14.0868	15.4480	16.9152	15.5010	15.4662
10	14.3569	15.4544	16.6197	15.4883	15.4658
12	14.5386	15.4578	16.4242	15.4814	15.4657
16	14.7674	15.4613	16.1816	15.4745	15.4657
20	14.9056	15.4628	16.0370	15.4713	15.4657

39. $f(x) = \sin\sqrt{x}$ on $[0, 4]$.

n	$L(n)$	$M(n)$	$R(n)$	$T(n)$	$S(n)$
4	2.8163	3.5456	3.7256	3.2709	3.3996
8	3.1809	3.5053	3.6356	3.4083	3.4541
10	3.2478	3.4990	3.6115	3.4296	3.4624
12	3.2909	3.4952	3.5940	3.4425	3.4674
16	3.3431	3.4910	3.5704	3.4568	3.4730
20	3.3734	3.4888	3.5552	3.4643	3.4759

41. $f(x) = 6e^{-x^2/2}$ on $[0, 2]$.

n	$L(n)$	$M(n)$	$R(n)$	$T(n)$	$S(n)$
4	8.4410	7.1945	5.8470	7.1440	7.1770
8	7.8178	7.1820	6.5208	7.1693	7.1777
10	7.6911	7.1804	6.6535	7.1723	7.1777
12	7.6063	7.1796	6.7416	7.1740	7.1777
16	7.4999	7.1788	6.8514	7.1756	7.1777
20	7.4358	7.1784	6.9170	7.1764	7.1777

43. $A = \int_0^{\pi/2} \sqrt{x}\cos x\,dx$

Simpson's Rule: $n = 14$

$$\int_0^{\pi/2} \sqrt{x}\cos x\,dx \approx \frac{\pi}{84}\left[\sqrt{0}\cos 0 + 4\sqrt{\frac{\pi}{28}}\cos\frac{\pi}{28} + 2\sqrt{\frac{\pi}{14}}\cos\frac{\pi}{14} + 4\sqrt{\frac{3\pi}{28}}\cos\frac{3\pi}{28} + \cdots + \sqrt{\frac{\pi}{2}}\cos\frac{\pi}{2}\right]$$

$$\approx 0.701$$

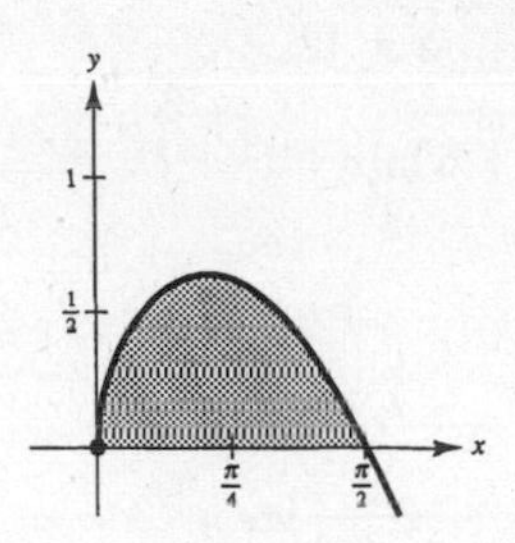

45. $W = \int_0^5 100x\sqrt{125 - x^3}\,dx$

Simpson's Rule: $n = 12$

$$\int_0^5 100x\sqrt{125 - x^3}\,dx \approx \frac{5}{3(12)}\left[0 + 400\left(\frac{5}{12}\right)\sqrt{125 - \left(\frac{5}{12}\right)^3} + 200\left(\frac{10}{12}\right)\sqrt{125 - \left(\frac{10}{12}\right)^3}\right.$$

$$\left. + 400\left(\frac{15}{12}\right)\sqrt{125 - \left(\frac{15}{12}\right)^3} + \cdots + 0\right] \sim 10{,}233.58 \text{ ft} \cdot \text{lb}$$

47. $\int_0^{1/2} \frac{6}{\sqrt{1 - x^2}}\,dx$ Simpson's Rule, $n = 6$

$$\pi \approx \frac{\left(\frac{1}{2} - 0\right)}{3(6)}[6 + 4(6.0209) + 2(6.0851) + 4(6.1968) + 2(6.3640) + 4(6.6002) + 6.9282]$$

$$\approx \frac{1}{36}[113.098] \approx 3.1416$$

49. Area $\approx \frac{1000}{2(10)}[125 + 2(125) + 2(120) + 2(112) + 2(90) + 2(90) + 2(95) + 2(88) + 2(75) + 2(35)] = 89{,}250$ sq m

51. $\int_0^t \sin\sqrt{x}\,dx = 2,\ n = 10$

By trial and error, we obtain $t \approx 2.477$.

Section 4.7 The Natural Logarithmic Function: Integration

1. $u = x + 1,\ du = dx$

$$\int \frac{1}{x+1}\,dx = \ln|x+1| + C$$

3. $u = 3 - 2x,\ du = -2\,dx$

$$\int \frac{1}{3-2x}\,dx = -\frac{1}{2}\int \frac{1}{3-2x}(-2)\,dx$$
$$= -\frac{1}{2}\ln|3-2x| + C$$

5. $u = x^2 + 1,\ du = 2x\,dx$

$$\int \frac{x}{x^2+1}\,dx = \frac{1}{2}\int \frac{1}{x^2+1}(2x)\,dx$$
$$= \frac{1}{2}\ln(x^2+1) + C$$
$$= \ln\sqrt{x^2+1} + C$$

7. $$\int \frac{x^2-4}{x}\,dx = \int \left(x - \frac{4}{x}\right)dx$$
$$= \frac{x^2}{2} - 4\ln|x| + C$$

9. $u = x^3 + 3x^2 + 9x,\ du = 3(x^2 + 2x + 3)\,dx$

$$\int \frac{x^2+2x+3}{x^3+3x^2+9x}\,dx = \frac{1}{3}\int \frac{3(x^2+2x+3)}{x^3+3x^2+9x}\,dx$$
$$= \frac{1}{3}\ln|x^3+3x^2+9x| + C$$

11. $$\int \frac{x^2-3x+2}{x+1}\,dx = \int \left(x - 4 + \frac{6}{x+1}\right)dx$$
$$= \frac{x^2}{2} - 4x + 6\ln|x+1| + C$$

13. $$\int \frac{x^3-3x^2+5}{x-3}\,dx = \int \left(x^2 + \frac{5}{x-3}\right)dx$$
$$= \frac{x^3}{3} + 5\ln|x-3| + C$$

15. $$\int \frac{x^4+x-4}{x^2+2}\,dx = \int \left(x^2 - 2 + \frac{x}{x^2+2}\right)dx$$
$$= \frac{x^3}{3} - 2x + \frac{1}{2}\ln(x^2+2) + C$$

17. $u = \ln x,\ du = \frac{1}{x}\,dx$

$$\int \frac{(\ln x)^2}{x}\,dx = \frac{1}{3}(\ln x)^3 + C$$

19. $u = x + 1,\ du = dx$

$$\int \frac{1}{\sqrt{x+1}}\,dx = \int (x+1)^{-1/2}\,dx$$
$$= 2(x+1)^{1/2} + C$$
$$= 2\sqrt{x+1} + C$$

21. $$\int \frac{2x}{(x-1)^2}\,dx = \int \frac{2x-2+2}{(x-1)^2}\,dx$$
$$= \int \frac{2(x-1)}{(x-1)^2}\,dx + 2\int \frac{1}{(x-1)^2}\,dx$$
$$= 2\int \frac{1}{x-1}\,dx + 2\int \frac{1}{(x-1)^2}\,dx$$
$$= 2\ln|x-1| - \frac{2}{(x-1)} + C$$

23. $$\int \frac{\cos\theta}{\sin\theta}\,d\theta = \ln|\sin\theta| + C$$

$(u = \sin\theta,\ du = \cos\theta\,d\theta)$

25. $$\int \csc 2x\,dx = \frac{1}{2}\int (\csc 2x)(2)\,dx$$
$$= -\frac{1}{2}\ln|\csc 2x + \cot 2x| + C$$

27. $$\int \frac{\cos t}{1+\sin t}\,dt = \ln|1+\sin t| + C$$

29. $\displaystyle\int \frac{\sec x \tan x}{\sec x - 1}\,dx = \ln|\sec x - 1| + C$

31. $\displaystyle\int e^{-x}\tan(e^{-x})\,dx = -\int \tan(e^{-x})\,(-e^{-x})dx$

$$= -(-\ln|\cos(e^{-x})|) + C$$

$$= \ln|\cos(e^{-x})| + C$$

33. $u = 1 + \sqrt{2x},\ du = \dfrac{1}{\sqrt{2x}}\,dx \Rightarrow (u - 1)\,du = dx$

$$\int \frac{1}{1 + \sqrt{2x}}\,dx = \int \frac{(u - 1)}{u}\,du = \int \left(1 - \frac{1}{u}\right) du$$

$$= u - \ln|u| + C_1$$

$$= \left(1 + \sqrt{2x}\right) - \ln\left|1 + \sqrt{2x}\right| + C_1$$

$$= \sqrt{2x} - \ln\left(1 + \sqrt{2x}\right) + C$$

where $C = C_1 + 1$.

35. $u = \sqrt{x} - 3,\ du = \dfrac{1}{2\sqrt{x}}\,dx \Rightarrow 2(u + 3)\,du = dx$

$$\int \frac{\sqrt{x}}{\sqrt{x} - 3}\,dx = 2\int \frac{(u + 3)^2}{u}\,du = 2\int \frac{u^2 + 6u + 9}{u}\,du = 2\int \left(u + 6 + \frac{9}{u}\right) du$$

$$= 2\left[\frac{u^2}{2} + 6u + 9\ln|u|\right] + C_1 = u^2 + 12u + 18\ln|u| + C_1$$

$$= \left(\sqrt{x} - 3\right)^2 + 12\left(\sqrt{x} - 3\right) + 18\ln\left|\sqrt{x} - 3\right| + C_1$$

$$= x + 6\sqrt{x} + 18\ln\left|\sqrt{x} - 3\right| + C \text{ where } C = C_1 - 27.$$

37. $\displaystyle y = \int \frac{3}{2 - x}\,dx$

$$= -3\int \frac{1}{x - 2}\,dx$$

$$= -3\ln|x - 2| + C$$

$(1, 0)$: $0 = -3\ln|1 - 2| + C \Rightarrow C = 0$

$y = -3\ln|x - 2|$

39. $\displaystyle s = \int \tan(2\theta)\,d\theta$

$$= \frac{1}{2}\int \tan(2\theta)(2\,d\theta)$$

$$= -\frac{1}{2}\ln|\cos 2\theta| + C$$

$(0, 2)$: $2 = -\dfrac{1}{2}\ln|\cos(0)| + C \Rightarrow C = 2$

$$s = -\frac{1}{2}\ln|\cos 2\theta| + 2$$

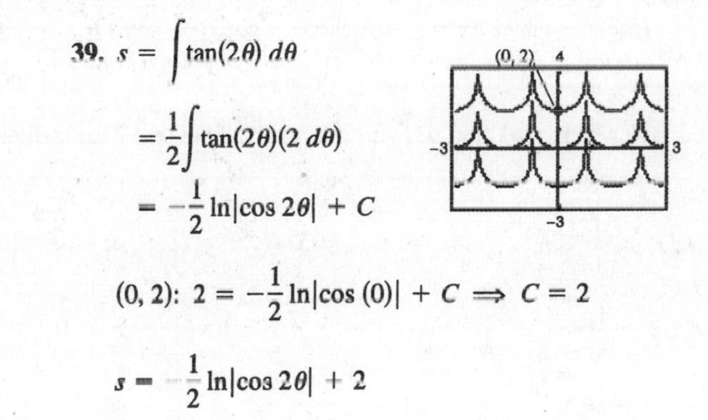

41. $\dfrac{dy}{dx} = \dfrac{1}{x + 2}$, $(0, 1)$

(a)

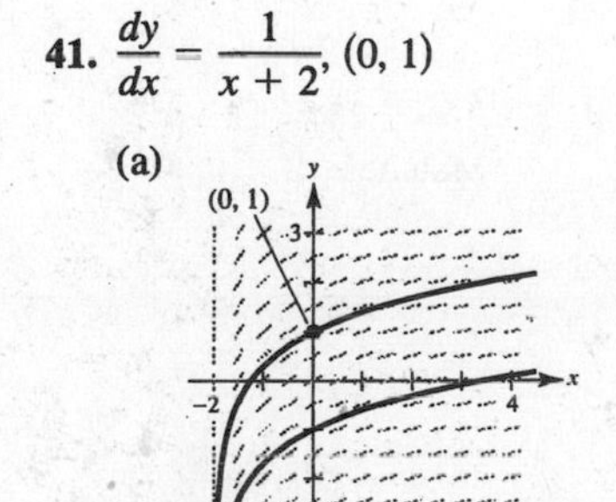

(b)

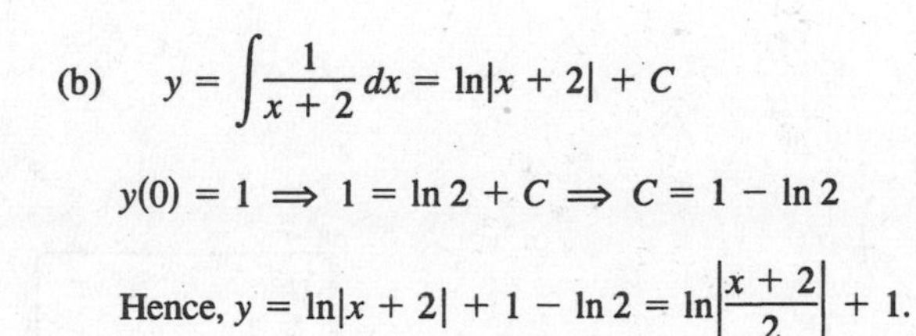

$$y = \int \frac{1}{x + 2}\,dx = \ln|x + 2| + C$$

$y(0) = 1 \Rightarrow 1 = \ln 2 + C \Rightarrow C = 1 - \ln 2$

Hence, $y = \ln|x + 2| + 1 - \ln 2 = \ln\left|\dfrac{x + 2}{2}\right| + 1.$

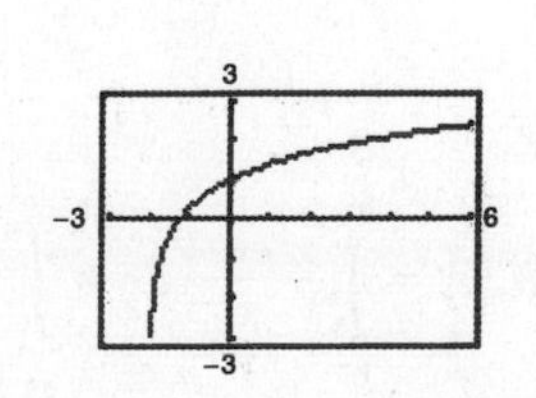

43. $\displaystyle\int_0^4 \frac{5}{3x+1}\,dx = \left[\frac{5}{3}\ln|3x+1|\right]_0^4$

$\displaystyle = \frac{5}{3}\ln 13 \approx 4.275$

45. $u = 1 + \ln x,\ du = \frac{1}{x}\,dx$

$\displaystyle\int_1^e \frac{(1+\ln x)^2}{x}\,dx = \left[\frac{1}{3}(1+\ln x)^3\right]_1^e = \frac{7}{3}$

47. $\displaystyle\int_0^2 \frac{x^2-2}{x+1}\,dx = \int_0^2 \left(x - 1 - \frac{1}{x+1}\right)dx$

$\displaystyle = \left[\frac{1}{2}x^2 - x - \ln|x+1|\right]_0^2 = -\ln 3$

49. $\displaystyle\int_1^2 \frac{1-\cos\theta}{\theta - \sin\theta}\,d\theta = \Big[\ln|\theta - \sin\theta|\Big]_1^2$

$\displaystyle = \ln\left|\frac{2-\sin 2}{1-\sin 1}\right| \approx 1.929$

51. $\displaystyle -\ln|\cos x| + C = \ln\left|\frac{1}{\cos x}\right| + C = \ln|\sec x| + C$

53. $\displaystyle \ln|\sec x + \tan x| + C = \ln\left|\frac{(\sec x + \tan x)(\sec x - \tan x)}{(\sec x - \tan x)}\right| + C = \ln\left|\frac{\sec^2 x - \tan^2 x}{\sec x - \tan x}\right| + C$

$\displaystyle = \ln\left|\frac{1}{\sec x - \tan x}\right| + C = -\ln|\sec x - \tan x| + C$

55. $\displaystyle\int \frac{1}{1+\sqrt{x}}\,dx = 2\left(1+\sqrt{x}\right) - 2\ln\left(1+\sqrt{x}\right) + C_1$

$= 2\left[\sqrt{x} - \ln\left(1+\sqrt{x}\right)\right] + C$ where $C = C_1 + 2$.

57. $\displaystyle\int \cos(1-x)\,dx = -\sin(1-x) + C$

59. $\displaystyle\int_{\pi/4}^{\pi/2} (\csc x - \sin x)\,dx = \Big[-\ln|\csc x + \cot x| + \cos x\Big]_{\pi/4}^{\pi/2} = \ln\left(\sqrt{2}+1\right) - \frac{\sqrt{2}}{2} \approx 0.174$

Note: In Exercises 61 and 63, you can use the Second Fundamental Theorem of Calculus or integrate the function.

61. $\displaystyle F(x) = \int_1^x \frac{1}{t}\,dt$

$\displaystyle F'(x) = \frac{1}{x}$

63. $\displaystyle F(x) = \int_x^{3x} \frac{1}{t}\,dt = \int_1^{3x} \frac{1}{t}\,dt - \int_1^x \frac{1}{t}\,dt$

$\displaystyle F'(x) = \frac{3}{3x} - \frac{1}{x} = 0$

65.

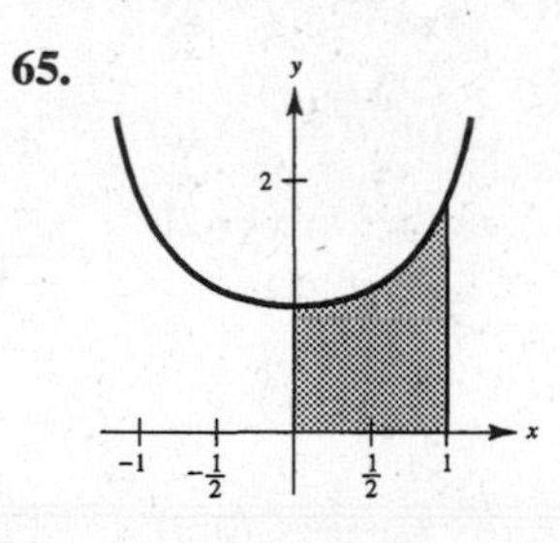

$A \approx 1.25$
Matches (d)

67. $\displaystyle A = \int_1^4 \frac{x^2+4}{x}\,dx = \int_1^4 \left(x + \frac{4}{x}\right)dx$

$\displaystyle = \left[\frac{x^2}{2} + 4\ln x\right]_1^4 = (8 + 4\ln 4) - \frac{1}{2}$

$\displaystyle = \frac{15}{2} + 8\ln 2 \approx 13.045$ square units

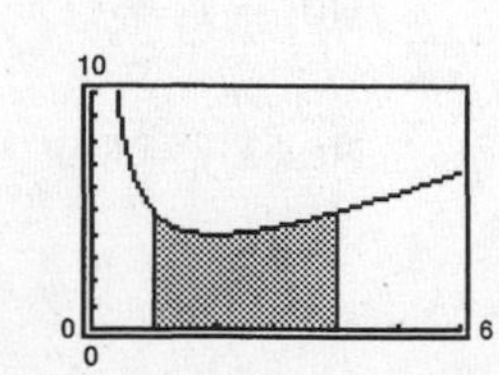

69. $\displaystyle\int_0^2 2\sec\frac{\pi x}{6}\,dx = \frac{12}{\pi}\int_0^2 \sec\left(\frac{\pi x}{6}\right)\frac{\pi}{6}\,dx$

$\displaystyle = \left[\frac{12}{\pi}\ln\left|\sec\frac{\pi x}{6} + \tan\frac{\pi x}{6}\right|\right]_0^2$

$\displaystyle = \frac{12}{\pi}\ln\left|\sec\frac{\pi}{3} + \tan\frac{\pi}{3}\right| - \frac{12}{\pi}\ln|1 + 0|$

$\displaystyle = \frac{12}{\pi}\ln\left(2 + \sqrt{3}\right) \approx 5.03041$

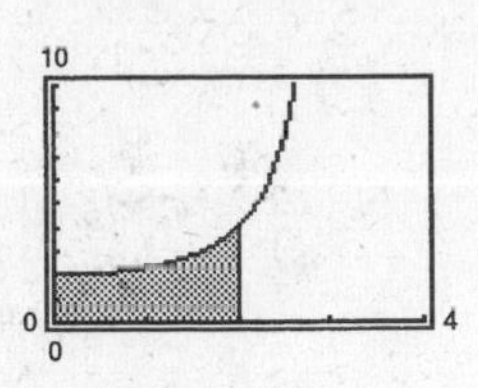

71. Power Rule

73. Substitution: $(u = x^2 + 4)$ and Log Rule

75. Divide the polynomials:

$$\frac{x^2}{x+1} = x - 1 + \frac{1}{x+1}$$

77. Average value $\displaystyle = \frac{1}{4-2}\int_2^4 \frac{8}{x^2}\,dx = 4\int_2^4 x^{-2}\,dx$

$\displaystyle = \left[-4\frac{1}{x}\right]_2^4$

$\displaystyle = -4\left(\frac{1}{4} - \frac{1}{2}\right) = 1$

79. Average value $\displaystyle = \frac{1}{e-1}\int_1^e \frac{\ln x}{x}\,dx = \frac{1}{e-1}\left[\frac{(\ln x)^2}{2}\right]_1^e$

$\displaystyle = \frac{1}{e-1}\left(\frac{1}{2}\right)$

$\displaystyle = \frac{1}{2e-2} \approx 0.291$

81. $\displaystyle P(t) = \int \frac{3000}{1 + 0.25t}\,dt = (3000)(4)\int \frac{0.25}{1 + 0.25t}\,dt = 12{,}000\ln|1 + 0.25t| + C$

$P(0) = 12{,}000\ln|1 + 0.25(0)| + C = 1000$

$C = 1000$

$P(t) = 12{,}000\ln|1 + 0.25t| + 1000 = 1000[12\ln|1 + 0.25t| + 1]$

$P(3) = 1000[12(\ln 1.75) + 1] \approx 7715$

83. $\displaystyle \frac{1}{50-40}\int_{40}^{50} \frac{90{,}000}{400 + 3x}\,dx = \Big[3000\ln|400 + 3x|\Big]_{40}^{50} \approx \$168.27$

85. (a) $2x^2 - y^2 = 8$

$y^2 = 2x^2 - 8$

$y_1 = \sqrt{2x^2 - 8}$

$y_2 = -\sqrt{2x^2 - 8}$

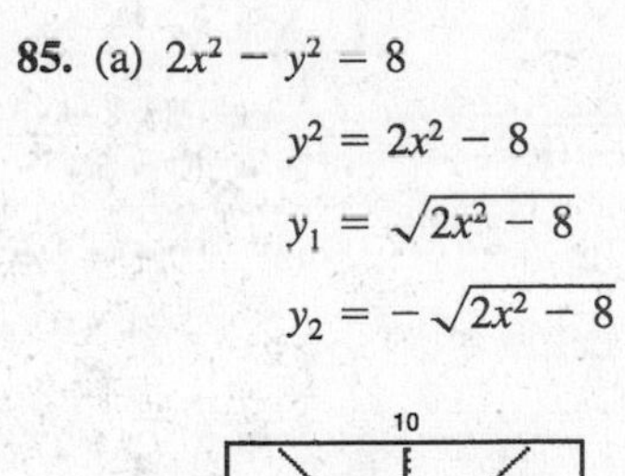

(b) $y^2 = e^{-\int(1/x)dx} = e^{-\ln x + C} = e^{\ln(1/x)}(e^C) = \dfrac{1}{x}k$

Let $k = 4$ and graph $y^2 = \dfrac{4}{x}$ $\left(\begin{array}{l} y_1 = 2/\sqrt{x} \\ y_2 = -2/\sqrt{x}\end{array}\right)$

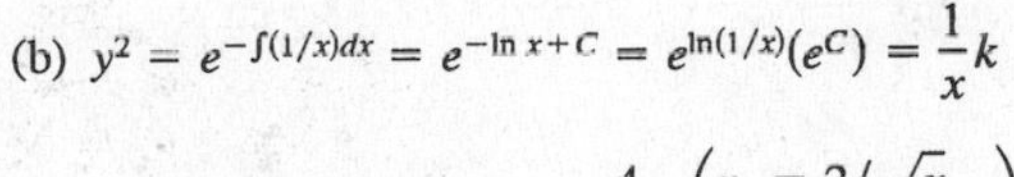

(c) In part (a), $2x^2 - y^2 = 8$

$4x - 2yy' = 0$

$y' = \dfrac{2x}{y}.$

In part (b), $y^2 = \dfrac{4}{x} = 4x^{-1}$

$2yy' = \dfrac{-4}{x^2}$

$y' = \dfrac{-2}{yx^2} = \dfrac{-2y}{y^2x^2} = \dfrac{-2y}{4x} = \dfrac{-y}{2x}.$

Using a graphing utility the graphs intersect at (2.214, 1.344). The slopes are 3.295 and $-0.304 = (-1)/3.295$, respectively.

87. False

$$\frac{1}{2}(\ln x) = \ln(x^{1/2}) \neq (\ln x)^{1/2}$$

89. True

$$\int \frac{1}{x}\,dx = \ln|x| + C_1$$
$$= \ln|x| + \ln|C| = \ln|Cx|,\; C \neq 0$$

91. $\dfrac{d}{dx}[\ln|x| + C] = \dfrac{1}{x}$

$$\frac{d}{dx}[\ln|u| + C] = \frac{1}{u}u'$$

Section 4.8 Inverse Trigonometric Functions: Integration

1. $\displaystyle\int \frac{5}{\sqrt{9 - x^2}}\,dx = 5\arcsin\left(\frac{x}{3}\right) + C$

3. Let $u = 3x$, $du = 3\,dx$.

$$\int_0^{1/6} \frac{1}{\sqrt{1 - 9x^2}}\,dx = \frac{1}{3}\int_0^{1/6} \frac{1}{\sqrt{1 - (3x)^2}}(3)\,dx = \left[\frac{1}{3}\arcsin(3x)\right]_0^{1/6} = \frac{\pi}{18}$$

5. $\displaystyle\int \frac{7}{16 + x^2}\,dx = \frac{7}{4}\arctan\left(\frac{x}{4}\right) + C$

7. Let $u = 2x$, $du = 2\,dx$.

$$\int_0^{\sqrt{3}/2} \frac{1}{1 + 4x^2}\,dx = \frac{1}{2}\int_0^{\sqrt{3}/2} \frac{2}{1 + (2x)^2}\,dx = \left[\frac{1}{2}\arctan(2x)\right]_0^{\sqrt{3}/2} = \frac{\pi}{6}$$

9. $\displaystyle\int \frac{1}{x\sqrt{4x^2 - 1}}\,dx = \int \frac{2}{2x\sqrt{(2x)^2 - 1}}\,dx = \operatorname{arcsec}|2x| + C$

11. $\displaystyle\int \frac{x^3}{x^2 + 1}\,dx = \int\left[x - \frac{x}{x^2 + 1}\right]dx = \int x\,dx - \frac{1}{2}\int \frac{2x}{x^2 + 1}\,dx = \frac{1}{2}x^2 - \frac{1}{2}\ln(x^2 + 1) + C$ (Use long division.)

13. $\displaystyle\int \frac{1}{\sqrt{1 - (x + 1)^2}}\,dx = \arcsin(x + 1) + C$

15. Let $u = t^2$, $du = 2t\,dt$.

$$\int \frac{t}{\sqrt{1 - t^4}}\,dt = \frac{1}{2}\int \frac{1}{\sqrt{1 - (t^2)^2}}(2t)\,dt = \frac{1}{2}\arcsin(t^2) + C$$

17. Let $u = \arcsin x$, $du = \dfrac{1}{\sqrt{1 - x^2}}\,dx$.

$$\int_0^{1/\sqrt{2}} \frac{\arcsin}{\sqrt{1 - x^2}}\,dx = \left[\frac{1}{2}\arcsin^2 x\right]_0^{1/\sqrt{2}} = \frac{\pi^2}{32} \approx 0.308$$

19. Let $u = 1 - x^2$, $du = -2x\,dx$.

$$\int_{-1/2}^{0} \frac{x}{\sqrt{1 - x^2}}\,dx = -\frac{1}{2}\int_{-1/2}^{0} (1 - x^2)^{-1/2}(-2x)\,dx$$
$$= \left[-\sqrt{1 - x^2}\right]_{-1/2}^{0} = \frac{\sqrt{3} - 2}{2}$$
$$\approx -0.134$$

21. Let $u = e^{2x}$, $du = 2e^{2x}\,dx$.

$$\int \frac{e^{2x}}{4 + e^{4x}}\,dx = \frac{1}{2}\int \frac{2e^{2x}}{4 + (e^{2x})^2}\,dx = \frac{1}{4}\arctan\frac{e^{2x}}{2} + C$$

23. Let $u = \cos x$, $du = -\sin x\,dx$.

$$\int_{\pi/2}^{\pi} \frac{\sin x}{1 + \cos^2 x}\,dx = -\int_{\pi/2}^{\pi} \frac{-\sin x}{1 + \cos^2 x}\,dx$$
$$= \left[-\arctan(\cos x)\right]_{\pi/2}^{\pi} = \frac{\pi}{4}$$

25. $\displaystyle\int \frac{1}{\sqrt{x}\sqrt{1-x}}\,dx.\ u = \sqrt{x}, x - u^2, dx = 2u\,du$

$$\int \frac{1}{u\sqrt{1-u^2}}(2u\,du) = 2\int \frac{du}{\sqrt{1-u^2}} = 2\arcsin u + C$$

$$= 2\arcsin\sqrt{x} + C$$

27. $\displaystyle\int \frac{x-3}{x^2+1}\,dx = \frac{1}{2}\int \frac{2x}{x^2+1}\,dx - 3\int \frac{1}{x^2+1}\,dx$

$$= \frac{1}{2}\ln(x^2+1) - 3\arctan x + C$$

29. $\displaystyle\int \frac{x+5}{\sqrt{9-(x-3)^2}}\,dx = \int \frac{(x-3)}{\sqrt{9-(x-3)^2}}\,dx + \int \frac{8}{\sqrt{9-(x-3)^2}}\,dx$

$$= -\sqrt{9-(x-3)^2} - 8\arcsin\left(\frac{x-3}{3}\right) + C$$

$$= -\sqrt{6x-x^2} + 8\arcsin\left(\frac{x}{3} - 1\right) + C$$

31. $\displaystyle\int_0^2 \frac{1}{x^2-2x+2}\,dx = \int_0^2 \frac{1}{1+(x-1)^2}\,dx = \Big[\arctan(x-1)\Big]_0^2 = \frac{\pi}{2}$

33. $\displaystyle\int \frac{2x}{x^2+6x+13}\,dx = \int \frac{2x+6}{x^2+6x+13}dx - 6\int \frac{1}{x^2+6x+13}\,dx = \int \frac{2x+6}{x^2+6x+13}\,dx - 6\int \frac{1}{4+(x+3)^2}\,dx$

$$= \ln|x^2+6x+13| - 3\arctan\left(\frac{x+3}{2}\right) + C$$

35. $\displaystyle\int \frac{1}{\sqrt{-x^2-4x}}\,dx = \int \frac{1}{\sqrt{4-(x+2)^2}}\,dx = \arcsin\left(\frac{x+2}{2}\right) + C$

37. Let $u = -x^2 - 4x$, $du = (-2x-4)\,dx$.

$$\int \frac{x+2}{\sqrt{-x^2-4x}}\,dx = -\frac{1}{2}\int (-x^2-4x)^{-1/2}(-2x-4)\,dx = -\sqrt{-x^2-4x} + C$$

39. $\displaystyle\int_2^3 \frac{2x-3}{\sqrt{4x-x^2}}\,dx = \int_2^3 \frac{2x-4}{\sqrt{4x-x^2}}\,dx + \int_2^3 \frac{1}{\sqrt{4x-x^2}}\,dx = -\int_2^3 (4x-x^2)^{-1/2}(4-2x)\,dx + \int_2^3 \frac{1}{\sqrt{4-(x-2)^2}}\,dx$

$$= \left[-2\sqrt{4x-x^2} + \arcsin\left(\frac{x-2}{2}\right)\right]_2^3 = 4 - 2\sqrt{3} + \frac{\pi}{6} \approx 1.059$$

41. Let $u = x^2 + 1$, $du = 2x\,dx$.

$$\int \frac{x}{x^4+2x^2+2}\,dx = \frac{1}{2}\int \frac{2x}{(x^2+1)^2+1}\,dx = \frac{1}{2}\arctan(x^2+1) + C$$

43. Let $u = \sqrt{e^t - 3}$. Then $u^2 + 3 = e^t$, $2u\,du = e^t\,dt$, and $\dfrac{2u\,du}{u^2+3} = dt$.

$$\int \sqrt{e^t-3}\,dt = \int \frac{2u^2}{u^2+3}\,du = \int 2\,du - \int 6\frac{1}{u^2+3}\,du$$

$$= 2u - 2\sqrt{3}\arctan\frac{u}{\sqrt{3}} + C = 2\sqrt{e^t-3} - 2\sqrt{3}\arctan\sqrt{\frac{e^t-3}{3}} + C$$

45. A perfect square trinomial is an expression in x with three terms that factor as a perfect square.

Example: $x^2 + 6x + 9 = (x+3)^2$

47. (a) $\int \frac{1}{\sqrt{1-x^2}}\,dx = \arcsin x + C,\ u = x$

(b) $\int \frac{x}{\sqrt{1-x^2}}\,dx = -\sqrt{1-x^2} + C,\ u = 1 - x^2$

(c) $\int \frac{1}{x\sqrt{1-x^2}}\,dx$ cannot be evaluated using the basic integration rules.

49. (a) $\int \sqrt{x-1}\,dx = \frac{2}{3}(x-1)^{3/2} + C,\ u = x - 1$

(b) Let $u = \sqrt{x-1}$. Then $x = u^2 + 1$ and $dx = 2u\,du$.

$$\int x\sqrt{x-1}\,dx = \int (u^2+1)(u)(2u)\,du = 2\int (u^4 + u^2)\,du = 2\left(\frac{u^5}{5} + \frac{u^3}{3}\right) + C$$

$$= \frac{2}{15}u^3(3u^2+5) + C = \frac{2}{15}(x-1)^{3/2}[3(x-1)+5] + C = \frac{2}{15}(x-1)^{3/2}(3x+2) + C$$

(c) Let $u = \sqrt{x-1}$. Then $x = u^2 + 1$ and $dx = 2u\,du$.

$$\int \frac{x}{\sqrt{x-1}}\,dx = \int \frac{u^2+1}{u}(2u)\,du = 2\int (u^2+1)\,du = 2\left(\frac{u^3}{3} + u\right) + C = \frac{2}{3}u(u^2+3) + C = \frac{2}{3}\sqrt{x-1}(x+2) + C$$

Note: In (b) and (c), substitution was necessary *before* the basic integration rules could be used.

51. (a)

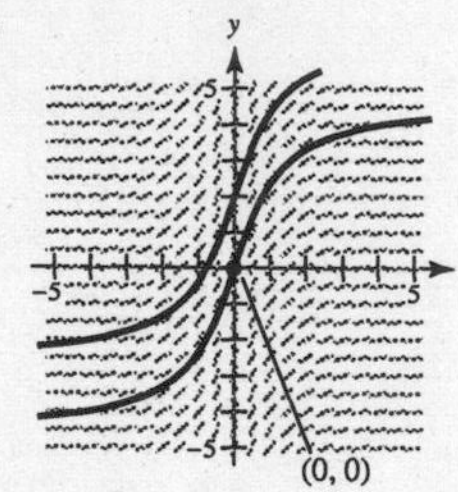

(b) $\frac{dy}{dx} = \frac{3}{1+x^2},\ (0, 0)$

$$y = 3\int \frac{dx}{1+x^2} = 3\arctan x + C$$

$(0, 0)$: $0 = 3\arctan(0) + C \Rightarrow C = 0$

$y = 3\arctan x$

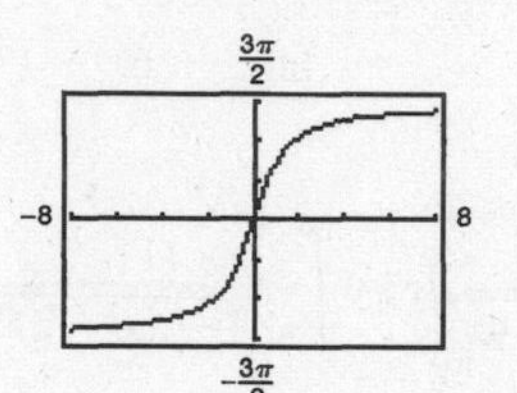

53. $\frac{dy}{dx} = \frac{10}{x\sqrt{x^2-1}},\ y(3) = 0$

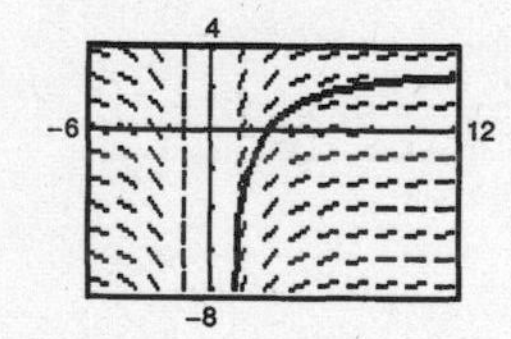

55. $A = \int_1^3 \frac{1}{x^2 - 2x + 1 + 4}\,dx = \int_1^3 \frac{1}{(x-1)^2 + 2^2}\,dx$

$$= \left[\frac{1}{2}\arctan\left(\frac{x-1}{2}\right)\right]_1^3 = \frac{1}{2}\arctan(1) = \frac{\pi}{8} \approx 0.3927$$

57. Area $\approx (1)(1) = 1$

Matches (c)

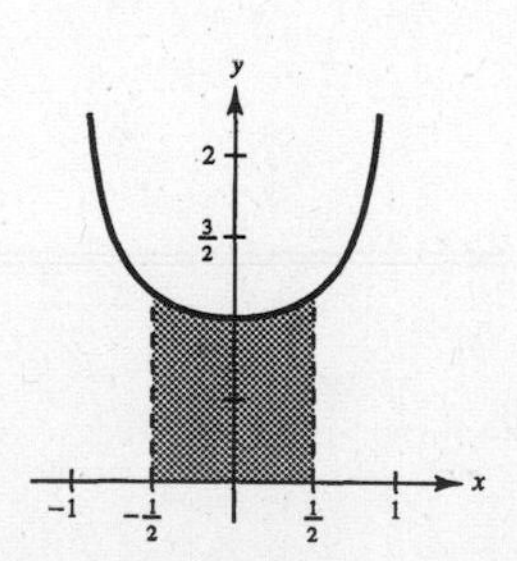

59. (a) $\int_0^1 \frac{4}{1+x^2}\,dx = \left[4\arctan x\right]_0^1 = 4\arctan 1 - 4\arctan 0 = 4\left(\frac{\pi}{4}\right) - 4(0) = \pi$

(b) Let $n = 6$.

$$4\int_0^1 \frac{4}{1+x^2}\,dx \approx 4\left(\frac{1}{36}\right)\left[1 + \frac{4}{1+(1/36)} + \frac{2}{1+(1/9)} + \frac{4}{1+(1/4)} + \frac{2}{1+(4/9)} + \frac{4}{1+(25/36)} + \frac{1}{2}\right] \approx 3.1415918$$

(c) 3.1415927

61. (a) $\frac{d}{dx}\left[\arcsin\left(\frac{u}{a}\right) + C\right] = \frac{1}{\sqrt{1 - (u^2/a^2)}}\left(\frac{u'}{a}\right) = \frac{u'}{\sqrt{a^2 - u^2}}$

Thus, $\int \frac{du}{\sqrt{a^2 - u^2}} = \arcsin\left(\frac{u}{a}\right) + C.$

(b) $\frac{d}{dx}\left[\frac{1}{a}\arctan\frac{u}{a} + C\right] = \frac{1}{a}\left[\frac{u'/a}{1 + (u/a)^2}\right] = \frac{1}{a^2}\left[\frac{u'}{(a^2 + u^2)/a^2}\right] = \frac{u'}{a^2 + u^2}$

Thus, $\int \frac{du}{a^2 + u^2} = \int \frac{u'}{a^2 + u^2}\,dx = \frac{1}{a}\arctan\frac{u}{a} + C.$

(c) Assume $u > 0$.

$\frac{d}{dx}\left[\frac{1}{a}\operatorname{arcsec}\frac{u}{a} + C\right] = \frac{1}{a}\left[\frac{u'/a}{(u/a)\sqrt{(u/a)^2 - 1}}\right] = \frac{1}{a}\left[\frac{u'}{u\sqrt{(u^2 - a^2)/a^2}}\right] = \frac{u'}{u\sqrt{u^2 - a^2}}$. The case $u < 0$ is handled in a similar manner.

Thus, $\int \frac{du}{u\sqrt{u^2 - a^2}} = \int \frac{u'}{u\sqrt{u^2 - a^2}}\,dx = \frac{1}{a}\operatorname{arcsec}\frac{|u|}{a} + C.$

63. (a) $v(t) = -32t + 500$

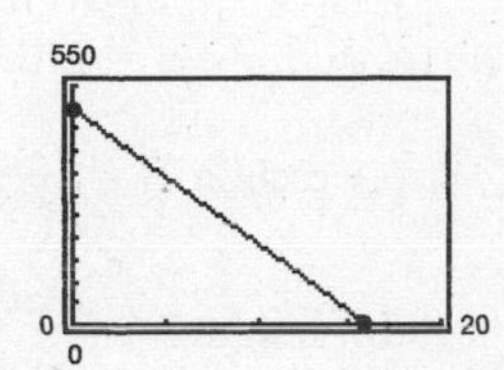

(b) $s(t) = \int v(t)\,dt = \int (-32t + 500)\,dt$

$= -16t^2 + 500t + C$

$s(0) = -16(0) + 500(0) + C = 0 \Rightarrow C = 0$

$s(t) = -16t^2 + 500t$

When the object reaches its maximum height, $v(t) = 0$.

$v(t) = -32t + 500 = 0$

$-32t = -500$

$t = 15.625$

$s(15.625) = -16(15.625)^2 + 500(15.625)$

$= 3906.25$ ft (Maximum height)

(c)
$$\int \frac{1}{32 + kv^2}\,dv = -\int dt$$

$$\frac{1}{\sqrt{32k}}\arctan\left(\sqrt{\frac{k}{32}}v\right) = -t + C_1$$

$$\arctan\left(\sqrt{\frac{k}{32}}v\right) = -\sqrt{32k}t + C$$

$$\sqrt{\frac{k}{32}}v = \tan\left(C - \sqrt{32k}\,t\right)$$

$$v = \sqrt{\frac{32}{k}}\tan\left(C - \sqrt{32k}\,t\right)$$

When $t = 0$, $v = 500$, $C = \arctan\left(500\sqrt{k/32}\right)$, and we have

$$v(t) = \sqrt{\frac{32}{k}}\tan\left[\arctan\left(500\sqrt{\frac{k}{32}}\right) - \sqrt{32k}\,t\right].$$

—CONTINUED—

63. —CONTINUED—

(d) When $k = 0.001$, $v(t) = \sqrt{32{,}000}\tan\left[\arctan\left(500\sqrt{0.00003125}\right) - \sqrt{0.032}\,t\right]$.

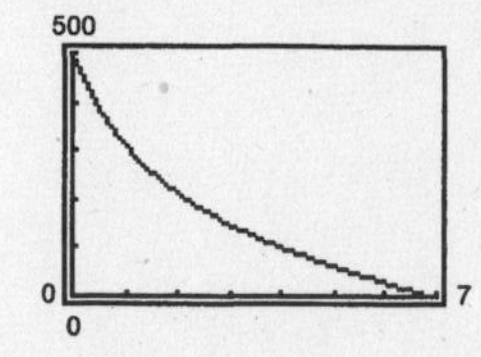

$v(t) = 0$ when $t_0 \approx 6.86$ sec.

(e) $h = \displaystyle\int_0^{6.86} \sqrt{32{,}000}\tan\left[\arctan\left(500\sqrt{0.00003125}\right) - \sqrt{0.032}\,t\right] dt$

Simpson's Rule: $n = 10$; $h \approx 1088$ feet

(f) Air resistance lowers the maximum height.

Section 4.9 Hyperbolic Functions

1. (a) $\sinh 3 = \dfrac{e^3 - e^{-3}}{2} \approx 10.018$

(b) $\tanh(-2) = \dfrac{\sinh(-2)}{\cosh(-2)} = \dfrac{e^{-2} - e^2}{e^{-2} + e^2} \approx -0.964$

3. (a) $\operatorname{csch}(\ln 2) = \dfrac{2}{e^{\ln 2} - e^{-\ln 2}} = \dfrac{2}{2 - (1/2)} = \dfrac{4}{3}$

(b) $\coth(\ln 5) = \dfrac{\cosh(\ln 5)}{\sinh(\ln 5)} = \dfrac{e^{\ln 5} + e^{-\ln 5}}{e^{\ln 5} - e^{-\ln 5}}$

$= \dfrac{5 + (1/5)}{5 - (1/5)} = \dfrac{13}{12}$

5. (a) $\cosh^{-1}(2) = \ln\left(2 + \sqrt{3}\right) \approx 1.317$

(b) $\operatorname{sech}^{-1}\left(\dfrac{2}{3}\right) = \ln\left(\dfrac{1 + \sqrt{1 - (4/9)}}{2/3}\right) \approx 0.962$

7. $\tanh^2 x + \operatorname{sech}^2 x = \left(\dfrac{e^x - e^{-x}}{e^x + e^{-x}}\right)^2 + \left(\dfrac{2}{e^x + e^{-x}}\right)^2 = \dfrac{e^{2x} - 2 + e^{-2x} + 4}{(e^x + e^{-x})^2} = \dfrac{e^{2x} + 2 + e^{-2x}}{e^{2x} + 2 + e^{-2x}} = 1$

9. $\sinh x \cosh y + \cosh x \sinh y = \left(\dfrac{e^x - e^{-x}}{2}\right)\left(\dfrac{e^y + e^{-y}}{2}\right) + \left(\dfrac{e^x + e^{-x}}{2}\right)\left(\dfrac{e^y - e^{-y}}{2}\right)$

$= \dfrac{1}{4}\left[e^{x+y} - e^{-x+y} + e^{x-y} - e^{-(x+y)} + e^{x+y} + e^{-x+y} - e^{x-y} - e^{-(x+y)}\right]$

$= \dfrac{1}{4}\left[2(e^{x+y} - e^{-(x+y)})\right] = \dfrac{e^{(x+y)} - e^{-(x+y)}}{2} = \sinh(x + y)$

11. $3\sinh x + 4\sinh^3 x = \sinh x(3 + 4\sinh^2 x) = \left(\dfrac{e^x - e^{-x}}{2}\right)\left[3 + 4\left(\dfrac{e^x - e^{-x}}{2}\right)^2\right]$

$= \left(\dfrac{e^x - e^{-x}}{2}\right)\left[3 + e^{2x} - 2 + e^{-2x}\right] = \dfrac{1}{2}(e^x - e^{-x})(e^{2x} + e^{-2x} + 1)$

$= \dfrac{1}{2}\left[e^{3x} + e^{-x} + e^x - e^x - e^{-3x} - e^{-x}\right] = \dfrac{e^{3x} - e^{-3x}}{2} = \sinh(3x)$

13. $\sinh x = \dfrac{3}{2}$

$\cosh^2 x - \left(\dfrac{3}{2}\right)^2 = 1 \Rightarrow \cosh^2 x = \dfrac{13}{4} \Rightarrow \cosh x = \dfrac{\sqrt{13}}{2}$

$\tanh x = \dfrac{3/2}{\sqrt{13}/2} = \dfrac{3\sqrt{13}}{13}$

$\operatorname{csch} x = \dfrac{1}{3/2} = \dfrac{2}{3}$

$\operatorname{sech} x = \dfrac{1}{\sqrt{13}/2} = \dfrac{2\sqrt{13}}{13}$

$\coth x = \dfrac{1}{3/\sqrt{13}} = \dfrac{\sqrt{13}}{3}$

15. $y = \sinh(1 - x^2)$

$y' = -2x \cosh(1 - x^2)$

17. $f(x) = \ln(\sinh x)$

$f'(x) = \dfrac{1}{\sinh x}(\cosh x) = \coth x$

19. $y = \ln\left(\tanh \dfrac{x}{2}\right)$

$y' = \dfrac{1/2}{\tanh(x/2)} \operatorname{sech}^2\left(\dfrac{x}{2}\right) = \dfrac{1}{2 \sinh(x/2) \cosh(x/2)}$

$= \dfrac{1}{\sinh x} = \operatorname{csch} x$

21. $h(x) = \dfrac{1}{4} \sinh 2x - \dfrac{x}{2}$

$h'(x) = \dfrac{1}{2} \cosh 2x - \dfrac{1}{2} = \dfrac{\cosh 2x - 1}{2} = \sinh^2 x$

23. $f(t) = \arctan(\sinh t)$

$f'(t) = \dfrac{1}{1 + \sinh^2 t}(\cosh t)$

$= \dfrac{\cosh t}{\cosh^2 t} = \operatorname{sech} t$

25. Let $y = g(x)$.

$y = x^{\cosh x}$

$\ln y = \cosh x \ln x$

$\dfrac{1}{y}\left(\dfrac{dy}{dx}\right) = \dfrac{\cosh x}{x} + \sinh x \ln x$

$\dfrac{dy}{dx} = \dfrac{y}{x}[\cosh x + x(\sinh x) \ln x]$

$= \dfrac{x^{\cosh x}}{x}[\cosh x + x(\sinh x) \ln x]$

27. $y = (\cosh x - \sinh x)^2$

$y' = 2(\cosh x - \sinh x)(\sinh x - \cosh x)$

$= -2(\cosh x - \sinh x)^2 = -2e^{-2x}$

29. $f(x) = \sin x \sinh x - \cos x \cosh x, \ -4 \le x \le 4$

$f'(x) = \sin x \cosh x + \cos x \sinh x - \cos x \sinh x + \sin x \cosh x$

$= 2 \sin x \cosh x = 0$ when $x = 0, \pm\pi$.

Relative maxima: $(\pm\pi, \cosh \pi)$

Relative minimum: $(0, -1)$

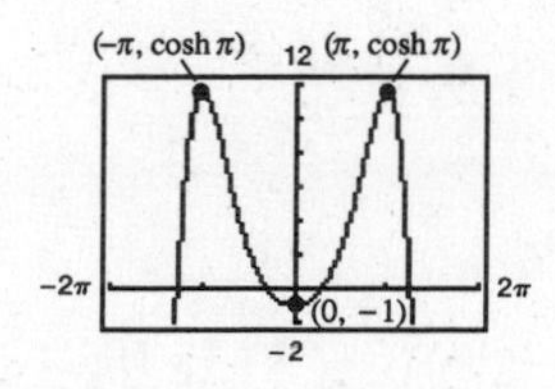

31. $g(x) = x \operatorname{sech} x = \dfrac{x}{\cosh x}$

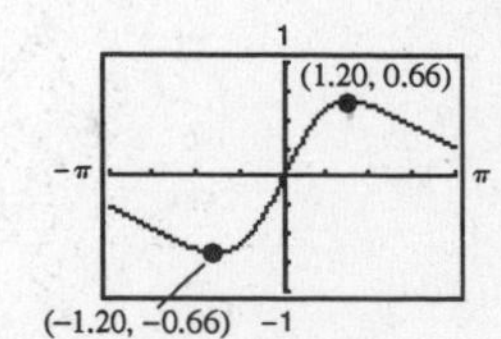

Relative maximum: $(1.20, 0.66)$

Relative minimum: $(-1.20, -0.66)$

33. $y = a \sinh x$

$y' = a \cosh x$

$y'' = a \sinh x$

$y''' = a \cosh x$

Therefore, $y''' - y' = 0$.

35. $f(x) = \tanh x$ $\qquad f(1) = \tanh(1) \approx 0.7616$

$f'(x) = \operatorname{sech}^2 x$ $\qquad f'(1) = \dfrac{1}{\cosh^2(1)} \approx 0.4200$

$f''(x) = -2 \operatorname{sech}^2 x \cdot \tanh x$ $\qquad f''(1) \approx -0.6397$

$P_1(x) = f(1) + f'(1)(x - 1) = 0.7616 + 0.42(x - 1)$

$P_2(x) = 0.7616 + 0.42(x - 1) - \dfrac{0.6397}{2}(x - 1)^2$

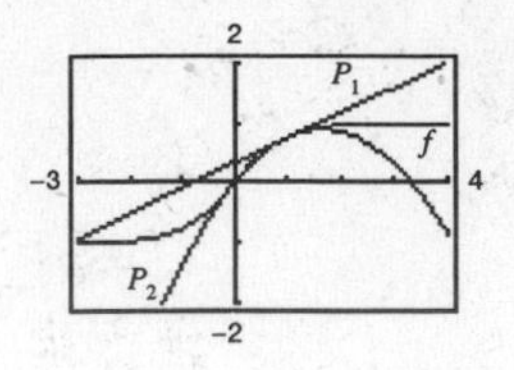

37. (a) $y = 10 + 15 \cosh \dfrac{x}{15}, \; -15 \le x \le 15$

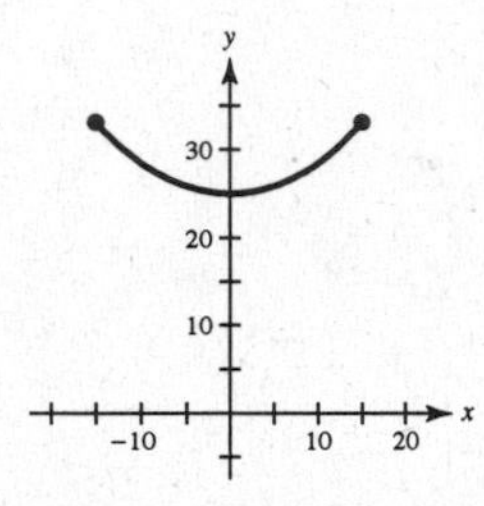

(b) At $x = \pm 15$, $y = 10 + 15 \cosh(1) \approx 33.146$.

At $x = 0$, $y = 10 + 15 \cosh(0) = 25$.

(c) $y' = \sinh \dfrac{x}{15}$. At $x = 15$, $y' = \sinh(1) \approx 1.175$

39. Let $u = 1 - 2x$, $du = -2\,dx$.

$$\int \sinh(1 - 2x)\,dx = -\frac{1}{2}\int \sinh(1 - 2x)(-2)\,dx = -\frac{1}{2}\cosh(1 - 2x) + C$$

41. Let $u = \cosh(x - 1)$, $du = \sinh(x - 1)\,dx$.

$$\int \cosh^2(x - 1)\sinh(x - 1)\,dx = \frac{1}{3}\cosh^3(x - 1) + C$$

43. Let $u = \sinh x$, $du = \cosh x\,dx$.

$$\int \frac{\cosh x}{\sinh x}\,dx = \ln|\sinh x| + C$$

45. Let $u = \dfrac{x^2}{2}$, $du = x\,dx$.

$$\int x \operatorname{csch}^2 \frac{x^2}{2}\,dx = \int \left(\operatorname{csch}^2 \frac{x^2}{2}\right) x\,dx = -\coth \frac{x^2}{2} + C$$

47. Let $u = \dfrac{1}{x}$, $du = -\dfrac{1}{x^2}\,dx$.

$$\int \frac{\operatorname{csch}(1/x)\coth(1/x)}{x^2}\,dx = -\int \operatorname{csch}\frac{1}{x}\coth\frac{1}{x}\left(-\frac{1}{x^2}\right)dx = \operatorname{csch}\frac{1}{x} + C$$

49. $$\int_0^4 \frac{1}{25 - x^2}\,dx = \frac{1}{10}\int_0^4 \frac{1}{5 - x}\,dx + \frac{1}{10}\int_0^4 \frac{1}{5 + x}\,dx = \left[\frac{1}{10}\ln\left|\frac{5 + x}{5 - x}\right|\right]_0^4 = \frac{1}{10}\ln 9 = \frac{1}{5}\ln 3$$

51. Let $u = 2x$, $du = 2\,dx$.

$$\int_0^{\sqrt{2}/4} \frac{2}{\sqrt{1-4x^2}}\,dx = \int_0^{\sqrt{2}/4} \frac{1}{\sqrt{1-(2x)^2}}(2)\,dx = \Big[\arcsin(2x)\Big]_0^{\sqrt{2}/4} = \frac{\pi}{4}$$

53. Let $u = x^2$, $du = 2x\,dx$.

$$\int \frac{x}{x^4+1}\,dx = \frac{1}{2}\int \frac{2x}{(x^2)^2+1}\,dx = \frac{1}{2}\arctan(x^2) + C$$

55. $y = \cosh^{-1}(3x)$

$$y' = \frac{3}{\sqrt{9x^2-1}}$$

57. $y = \sinh^{-1}(\tan x)$

$$y' = \frac{1}{\sqrt{\tan^2 x + 1}}(\sec^2 x) = |\sec x|$$

59. $y = \tanh^{-1}(\sin 2x)$

$$y' = \frac{1}{1-\sin^2 2x}(2\cos 2x) = 2\sec 2x$$

61. $y = 2x\sinh^{-1}(2x) - \sqrt{1+4x^2}$

$$y' = 2x\left(\frac{2}{\sqrt{1+4x^2}}\right) + 2\sinh^{-1}(2x) - \frac{4x}{\sqrt{1+4x^2}} = 2\sinh^{-1}(2x)$$

63. See the definitions

65. $y = a\,\text{sech}^{-1}\dfrac{x}{a} - \sqrt{a^2-x^2},\ a > 0$

$$\frac{dy}{dx} = \frac{-1}{(x/a)\sqrt{1-(x^2/a^2)}} + \frac{x}{\sqrt{a^2-x^2}} = \frac{-a^2}{x\sqrt{a^2-x^2}} + \frac{x}{\sqrt{a^2-x^2}} = \frac{x^2-a^2}{x\sqrt{a^2-x^2}} = \frac{-\sqrt{a^2-x^2}}{x}$$

67. $$\int \frac{1}{\sqrt{1+e^{2x}}}\,dx = \int \frac{e^x}{e^x\sqrt{1+(e^x)^2}}\,dx = -\text{csch}^{-1}(e^x) + C = -\ln\left(\frac{1+\sqrt{1+e^{2x}}}{e^x}\right) + C$$

69. Let $u = \sqrt{x}$, $du = \dfrac{1}{2\sqrt{x}}\,dx$.

$$\int \frac{1}{\sqrt{x}\sqrt{1+x}}\,dx = 2\int \frac{1}{\sqrt{1+(\sqrt{x})^2}}\left(\frac{1}{2\sqrt{x}}\right)dx = 2\sinh^{-1}\sqrt{x} + C = 2\ln\left(\sqrt{x}+\sqrt{1+x}\right) + C$$

71. $$\int \frac{-1}{4x-x^2}\,dx = \int \frac{1}{(x-2)^2-4}\,dx = \frac{1}{4}\ln\left|\frac{(x-2)-2}{(x-2)+2}\right| = \frac{1}{4}\ln\left|\frac{x-4}{x}\right| + C$$

73. $$\int \frac{1}{1-4x-2x^2}\,dx = \int \frac{1}{3-2(x+1)^2}\,dx = \frac{-1}{\sqrt{2}}\int \frac{\sqrt{2}}{\left[\sqrt{2}(x+1)\right]^2 - \left(\sqrt{3}\right)^2}\,dx$$

$$= \frac{-1}{2\sqrt{6}}\ln\left|\frac{\sqrt{2}(x+1)-\sqrt{3}}{\sqrt{2}(x+1)+\sqrt{3}}\right| + C = \frac{1}{2\sqrt{6}}\ln\left|\frac{\sqrt{2}(x+1)+\sqrt{3}}{\sqrt{2}(x+1)-\sqrt{3}}\right| + C$$

75. Let $u = 4x - 1$, $du = 4\,dx$.

$$y = \int \frac{1}{\sqrt{80+8x-16x^2}}\,dx = \frac{1}{4}\int \frac{4}{\sqrt{81-(4x-1)^2}}\,dx = \frac{1}{4}\arcsin\left(\frac{4x-1}{9}\right) + C$$

77. $$y = \int \frac{x^3-21x}{5+4x-x^2}\,dx = \int\left(-x-4+\frac{20}{5+4x-x^2}\right)dx = \int(-x-4)\,dx + 20\int \frac{1}{3^2-(x-2)^2}\,dx$$

$$= -\frac{x^2}{2} - 4x + \frac{20}{6}\ln\left|\frac{3+(x-2)}{3-(x-2)}\right| + C = -\frac{x^2}{2} - 4x + \frac{10}{3}\ln\left|\frac{1+x}{5-x}\right| + C = \frac{-x^2}{2} - 4x - \frac{10}{3}\ln\left|\frac{5-x}{x+1}\right| + C$$

79. $A = 2\int_0^4 \text{sech}\,\frac{x}{2}\,dx$

$= 2\int_0^4 \frac{2}{e^{x/2} + e^{-x/2}}\,dx$

$= 4\int_0^4 \frac{e^{x/2}}{(e^{x/2})^2 + 1}\,dx$

$= \left[8\arctan(e^{x/2})\right]_0^4$

$= 8\arctan(e^2) - 2\pi \approx 5.207$

81. $A = \int_0^2 \frac{5x}{\sqrt{x^4 + 1}}\,dx$

$= \frac{5}{2}\int_0^2 \frac{2x}{\sqrt{(x^2)^2 + 1}}\,dx$

$= \left[\frac{5}{2}\ln\left(x^2 + \sqrt{x^4 + 1}\right)\right]_0^2$

$= \frac{5}{2}\ln\left(4 + \sqrt{17}\right) \approx 5.237$

83. $\int \frac{3k}{16}\,dt = \int \frac{1}{x^2 - 12x + 32}\,dx$

$\frac{3kt}{16} = \int \frac{1}{(x-6)^2 - 4}\,dx = \frac{1}{2(2)}\ln\left|\frac{(x-6)-2}{(x-6)+2}\right| + C = \frac{1}{4}\ln\left|\frac{x-8}{x-4}\right| + C$

When $x = 0$: $t = 0$

$C = -\frac{1}{4}\ln(2)$

When $x = 1$: $t = 10$

$\frac{30k}{16} = \frac{1}{4}\ln\left|\frac{-7}{-3}\right| - \frac{1}{4}\ln(2) = \frac{1}{4}\ln\left(\frac{7}{6}\right)$

$k = \frac{2}{15}\ln\left(\frac{7}{6}\right)$

When $t = 20$: $\left(\frac{3}{16}\right)\left(\frac{2}{15}\right)\ln\left(\frac{7}{6}\right)(20) = \frac{1}{4}\ln\frac{x-8}{2x-8}$

$\ln\left(\frac{7}{6}\right)^2 = \ln\frac{x-8}{2x-8}$

$\frac{49}{36} = \frac{x-8}{2x-8}$

$62x = 104$

$x = \frac{104}{62} = \frac{52}{31} \approx 1.677 \text{ kg}$

85. $y = \cosh x = \frac{e^x + e^{-x}}{2}$

$y' = \frac{e^x - e^{-x}}{2} = \sinh x$

87. $y = \cosh^{-1} x$

$\cosh y = x$

$(\sinh y)(y') = 1$

$y' = \frac{1}{\sinh y} = \frac{1}{\sqrt{\cosh^2 y - 1}} = \frac{1}{\sqrt{x^2 - 1}}$

89. $y = \text{sech}\,x = \frac{2}{e^x + e^{-x}}$

$y' = -2(e^x + e^{-x})^{-2}(e^x - e^{-x}) = \left(\frac{-2}{e^x + e^{-x}}\right)\left(\frac{e^x - e^{-x}}{e^x + e^{-x}}\right) = -\text{sech}\,x\tanh x$

Review Exercises for Chapter 4

1.

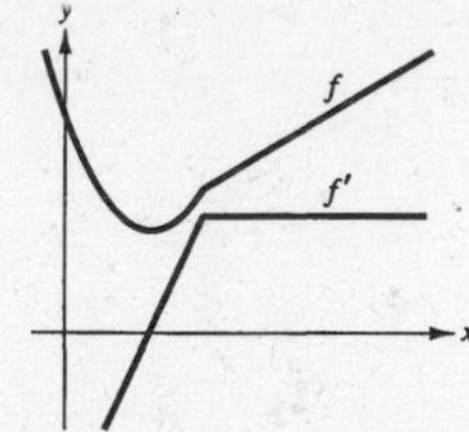

3. $\displaystyle\int (2x^2 + x - 1)\, dx = \frac{2}{3}x^3 + \frac{1}{2}x^2 - x + C$

5. $\displaystyle\int \frac{x^3 + 1}{x^2}\, dx = \int \left(x + \frac{1}{x^2}\right) dx = \frac{1}{2}x^2 - \frac{1}{x} + C$

7. $\displaystyle\int (4x - 3 \sin x)\, dx = 2x^2 + 3 \cos x + C$

9. $\displaystyle\int (5 - e^x)\, dx = 5x - e^x + C$

11. $\displaystyle\int \frac{5}{x}\, dx = 5 \ln|x| + C$

13. $f'(x) = -2x,\ (-1, 1)$

$\displaystyle f(x) = \int -2x\, dx = -x^2 + C$

When $x = -1$:

$y = -1 + C = 1$

$C = 2$

$y = 2 - x^2$

15. $a(t) = a$

$\displaystyle v(t) = \int a\, dt = at + C_1$

$v(0) = 0 + C_1 = 0$ when $C_1 = 0$.

$v(t) = at$

$\displaystyle s(t) = \int at\, dt = \frac{a}{2}t^2 + C_2$

$s(0) = 0 + C_2 = 0$ when $C_2 = 0$.

$s(t) = \dfrac{a}{2}t^2$

$s(30) = \dfrac{a}{2}(30)^2 = 3600$ or

$a = \dfrac{2(3600)}{(30)^2} = 8 \text{ ft/sec}^2.$

$v(30) = 8(30) = 240$ ft/sec

17. $a(t) = -32$

$v(t) = -32t + 96$

$s(t) = -16t^2 + 96t$

(a) $v(t) = -32t + 96 = 0$ when $t = 3$ sec.

(b) $s(3) = -144 + 288 = 144$ ft

(c) $v(t) = -32t + 96 = \dfrac{96}{2}$ when $t = \dfrac{3}{2}$ sec.

(d) $s\left(\dfrac{3}{2}\right) = -16\left(\dfrac{9}{4}\right) + 96\left(\dfrac{3}{2}\right) = 108$ ft

19. (a) $\displaystyle\sum_{i=1}^{10} (2i - 1)$

(b) $\displaystyle\sum_{i=1}^{n} i^3$

(c) $\displaystyle\sum_{i=1}^{10} (4i + 2)$

21. $y = \dfrac{10}{x^2+1}, \Delta x = \dfrac{1}{2}, n = 4$

$$S(n) = S(4) = \frac{1}{2}\left[\frac{10}{1} + \frac{10}{(1/2)^2+1} + \frac{10}{(1)^2+1} + \frac{10}{(3/2)^2+1}\right]$$

$$\approx 13.0385$$

$$s(n) = s(4) = \frac{1}{2}\left[\frac{10}{(1/2)^2+1} + \frac{10}{1+1} + \frac{10}{(3/2)^2+1} + \frac{10}{2^2+1}\right]$$

$$\approx 9.0385$$

$9.0385 <$ Area of Region < 13.0385

23. $y = 6 - x, \Delta x = \dfrac{4}{n}$, right endpoints

$$\text{Area} = \lim_{n\to\infty} \sum_{i=1}^{n} f(c_i)\,\Delta x$$

$$= \lim_{n\to\infty} \sum_{i=1}^{n} \left(6 - \frac{4i}{n}\right)\frac{4}{n}$$

$$= \lim_{n\to\infty} \frac{4}{n}\left[6n - \frac{4}{n}\frac{n(n+1)}{2}\right]$$

$$= \lim_{n\to\infty} \left[24 - 8\frac{n+1}{n}\right] = 24 - 8 = 16$$

25. $y = 5 - x^2, \Delta x = \dfrac{3}{n}$

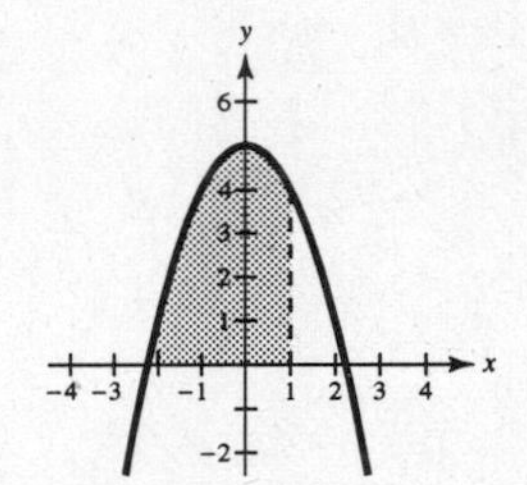

$$\text{Area} = \lim_{n\to\infty} \sum_{i=1}^{n} f(c_i)\,\Delta x$$

$$= \lim_{n\to\infty} \sum_{i=1}^{n} \left[5 - \left(-2 + \frac{3i}{n}\right)^2\right]\left(\frac{3}{n}\right)$$

$$= \lim_{n\to\infty} \frac{3}{n} \sum_{i=1}^{n} \left[1 + \frac{12i}{n} - \frac{9i^2}{n^2}\right]$$

$$= \lim_{n\to\infty} \frac{3}{n}\left[n + \frac{12}{n}\frac{n(n+1)}{2} - \frac{9}{n^2}\frac{n(n+1)(2n+1)}{6}\right]$$

$$= \lim_{n\to\infty} \left[3 + 18\frac{n+1}{n} - \frac{9}{2}\frac{(n+1)(2n+1)}{n^2}\right]$$

$$= 3 + 18 - 9 = 12$$

27. $x = 5y - y^2, 2 \le y \le 5, \Delta y = \dfrac{3}{n}$

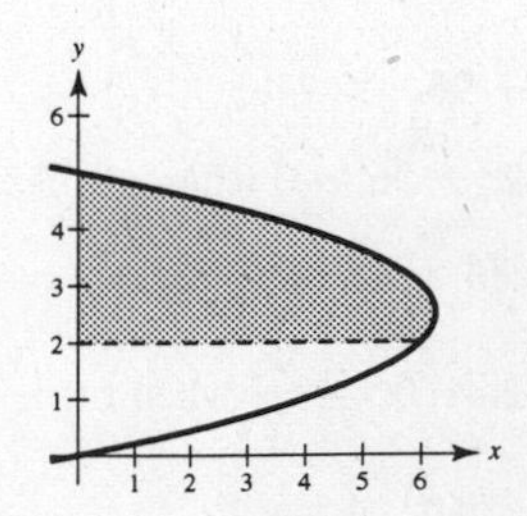

$$\text{Area} = \lim_{n\to\infty} \sum_{i=1}^{n} \left[5\left(2 + \frac{3i}{n}\right) - \left(2 + \frac{3i}{n}\right)^2\right]\left(\frac{3}{n}\right)$$

$$= \lim_{n\to\infty} \frac{3}{n} \sum_{i=1}^{n} \left[10 + \frac{15i}{n} - 4 - 12\frac{i}{n} - \frac{9i^2}{n^2}\right]$$

$$= \lim_{n\to\infty} \frac{3}{n} \sum_{i=1}^{n} \left[6 + \frac{3i}{n} - \frac{9i^2}{n^2}\right]$$

$$= \lim_{n\to\infty} \frac{3}{n}\left[6n + \frac{3}{n}\frac{n(n+1)}{2} - \frac{9}{n^2}\frac{n(n+1)(2n+1)}{6}\right]$$

$$= \left[18 + \frac{9}{2} - 9\right] = \frac{27}{2}$$

29. $\lim_{\|\Delta\| \to \infty} \sum_{i=1}^{n} (2c_i - 3)\, \Delta xi = \int_4^6 (2x - 3)\, dx$

31.

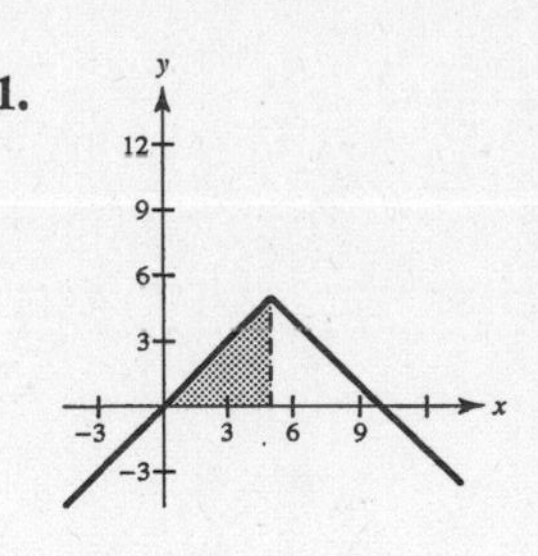

$$\int_0^5 (5 - |x - 5|)dx = \int_0^5 (5 - (5 - x))\, dx = \int_0^5 x\, dx = \frac{25}{2}$$

(triangle)

33. (a) $\int_2^6 [f(x) + g(x)]\, dx = \int_2^6 f(x)\, dx + \int_2^6 g(x)\, dx = 10 + 3 = 13$

(b) $\int_2^6 [f(x) - g(x)]\, dx = \int_2^6 f(x)\, dx - \int_2^6 g(x)\, dx = 10 - 3 = 7$

(c) $\int_2^6 [2f(x) - 3g(x)]\, dx = 2\int_2^6 f(x)\, dx - 3\int_2^6 g(x)\, dx = 2(10) - 3(3) = 11$

(d) $\int_2^6 5f(x)\, dx = 5\int_2^6 f(x)\, dx = 5(10) = 50$

35. $\int_1^8 (\sqrt[3]{x} + 1)\, dx = \left[\frac{3}{4}x^{4/3} + x\right]_1^8 = \left[\frac{3}{4}(16) + 8\right] - \left[\frac{3}{4} + 1\right] = \frac{73}{4}$ (c)

37. $\int_0^4 (2 + x)\, dx = \left[2x + \frac{x^2}{2}\right]_0^4 = 8 + \frac{16}{2} = 16$

39. $\int_{-1}^1 (4t^3 - 2t)\, dt = \left[t^4 - t^2\right]_{-1}^1 = 0$

41. $\int_4^9 x\sqrt{x}\, dx = \int_4^9 x^{3/2}\, dx = \left[\frac{2}{5}x^{5/2}\right]_4^9 = \frac{2}{5}\left[(\sqrt{9})^5 - (\sqrt{4})^5\right] = \frac{2}{5}(243 - 32) = \frac{422}{5}$

43. $\int_0^{3\pi/4} \sin\theta\, d\theta = \left[-\cos\theta\right]_0^{3\pi/4} = -\left(-\frac{\sqrt{2}}{2}\right) + 1 = 1 + \frac{\sqrt{2}}{2} = \frac{\sqrt{2} + 2}{2}$

45. $\int_0^2 (x + e^x)dx = \left[\frac{x^2}{2} + e^x\right]_0^2 = 2 + e^2 - 1 = 1 + e^2$

47. $\int_1^3 (2x - 1)\, dx = \left[x^2 - x\right]_1^3 = 6$

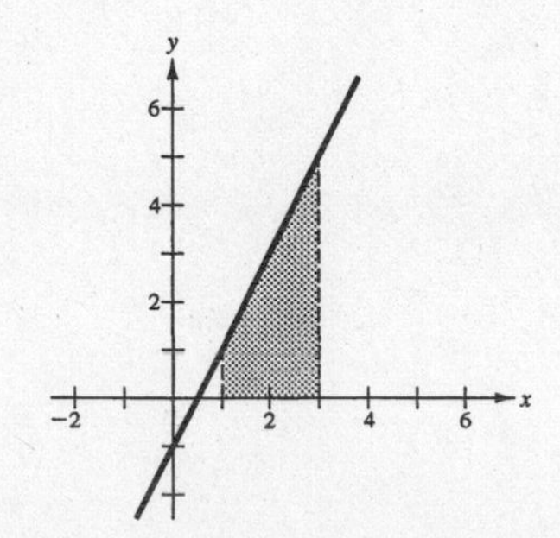

49. $\int_3^4 (x^2 - 9)\, dx = \left[\frac{x^3}{3} - 9x\right]_3^4$

$= \left(\frac{64}{3} - 36\right) - (9 - 27)$

$= \frac{64}{3} - \frac{54}{3} = \frac{10}{3}$

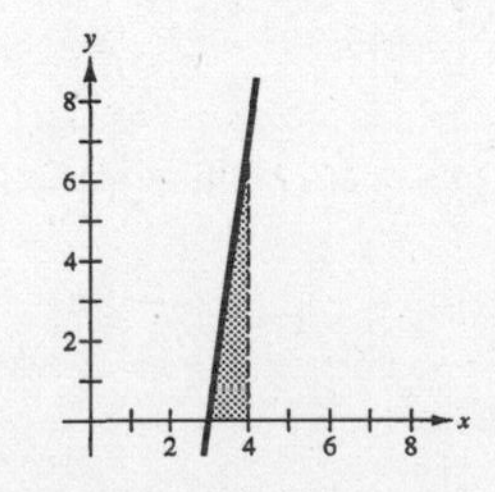

51. $\displaystyle\int_0^1 (x - x^3)\,dx = \left[\frac{x^2}{2} - \frac{x^4}{4}\right]_0^1 = \frac{1}{2} - \frac{1}{4} = \frac{1}{4}$

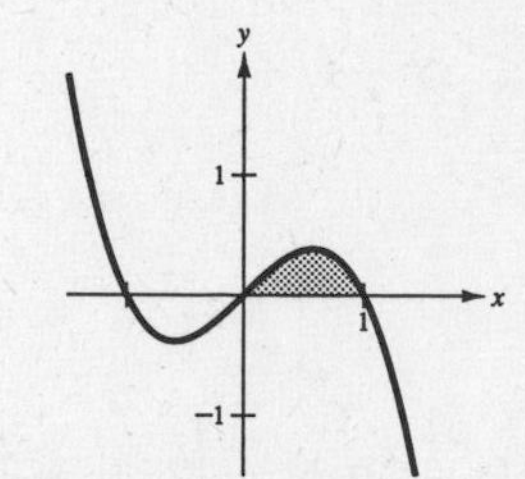

53. $\displaystyle \text{Area} = \int_1^9 \frac{4}{\sqrt{x}}\,dx = \left[\frac{4x^{1/2}}{(1/2)}\right]_1^9 = 8(3 - 1) = 16$

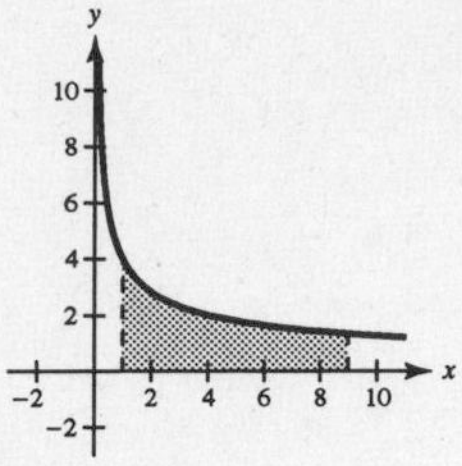

55. $y = \dfrac{2}{x}$

$\displaystyle \text{Area} = \int_1^3 \frac{2}{x}\,dx = \left[2 \ln x\right]_1^3 = 2 \ln 3 - 2 \ln 1 = \ln 9$

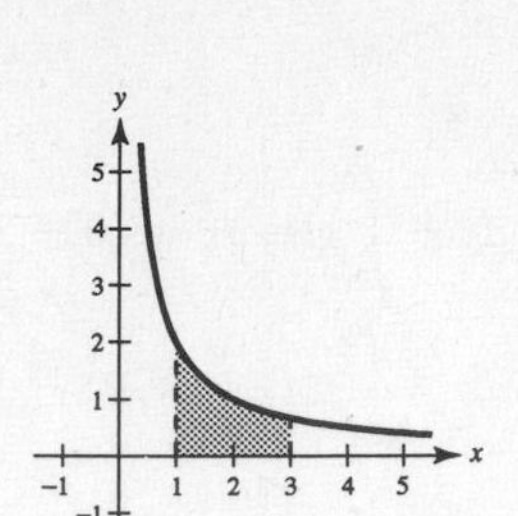

57. $\displaystyle \frac{1}{9-4}\int_4^9 \frac{1}{\sqrt{x}}\,dx = \left[\frac{1}{5} 2\sqrt{x}\right]_4^9 = \frac{2}{5}(3 - 2) = \frac{2}{5}$ Average value

$\dfrac{2}{5} = \dfrac{1}{\sqrt{x}}$

$\sqrt{x} = \dfrac{5}{2}$

$x = \dfrac{25}{4}$

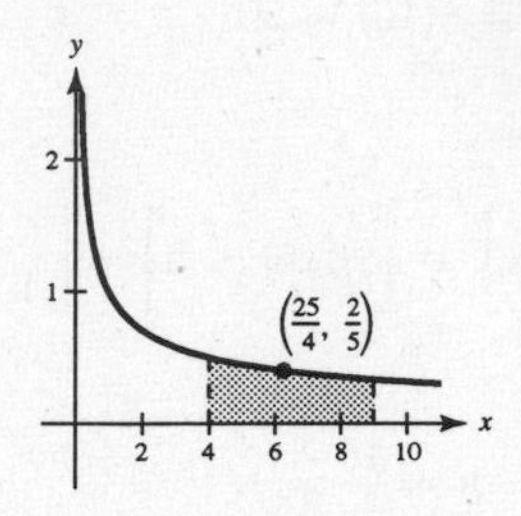

59. $F'(x) = x^2\sqrt{1 + x^3}$

61. $F'(x) = x^2 + 3x + 2$

63. $\displaystyle\int (x^2 + 1)^3\,dx = \int (x^6 + 3x^4 + 3x^2 + 1)\,dx = \frac{x^7}{7} + \frac{3}{5}x^5 + x^3 + x + C$

65. $u = x^3 + 3,\ du = 3x^2\,dx$

$$\int \frac{x^2}{\sqrt{x^3 + 3}}\,dx = \int (x^3 + 3)^{-1/2}\,x^2\,dx = \frac{1}{3}\int (x^3 + 3)^{-1/2}\,3x^2\,dx = \frac{2}{3}(x^3 + 3)^{1/2} + C$$

67. $u = 1 - 3x^2,\ du = -6x\,dx$

$$\int x(1 - 3x^2)^4\,dx = -\frac{1}{6}\int (1 - 3x^2)^4(-6x\,dx) = -\frac{1}{30}(1 - 3x^2)^5 + C = \frac{1}{30}(3x^2 - 1)^5 + C$$

69. $\displaystyle\int \sin^3 x \cos x\,dx = \frac{1}{4}\sin^4 x + C$

71. $\displaystyle\int \frac{\sin\theta}{\sqrt{1 - \cos\theta}}\,d\theta = \int (1 - \cos\theta)^{-1/2} \sin\theta\,d\theta = 2(1 - \cos\theta)^{1/2} + C = 2\sqrt{1 - \cos\theta} + C$

73. $\displaystyle\int \tan^n x \sec^2 x\,dx = \frac{\tan^{n+1} x}{n + 1} + C,\ n \neq -1$

75. $\int (1 + \sec \pi x)^2 \sec \pi x \tan \pi x \, dx = \frac{1}{\pi}\int (1 + \sec \pi x)^2(\pi \sec \pi x \tan \pi x)\, dx = \frac{1}{3\pi}(1 + \sec \pi x)^3 + C$

77. $\int xe^{-3x^2}\, dx = -\frac{1}{6}\int e^{-3x^2}(-6x)dx = -\frac{1}{6}e^{-3x^2} + C$

79. $\int (x + 1)5^{(x+1)^2}dx = \frac{1}{2}\int 5^{(x+1)^2}2(x + 1)dx$

$= \frac{1}{2 \ln 5}5^{(x+1)^2} + C$

81. $\int_{-1}^{2} x(x^2 - 4)\, dx = \frac{1}{2}\int_{-1}^{2} (x^2 - 4)(2x)\, dx = \frac{1}{2}\frac{(x^2 - 4)^2}{2}\Big]_{-1}^{2} = \frac{1}{4}[0 - 9] = -\frac{9}{4}$

83. $\int_0^3 \frac{1}{\sqrt{1 + x}}\, dx = \int_0^3 (1 + x)^{-1/2}\, dx = \left[2(1 + x)^{1/2}\right]_0^3 = 4 - 2 = 2$

85. $u = 1 - y, y = 1 - u, dy = -du$

When $y = 0$, $u = 1$. When $y = 1$, $u = 0$.

$$2\pi\int_0^1 (y + 1)\sqrt{1 - y}\, dy = 2\pi\int_1^0 -[(1 - u) + 1]\sqrt{u}\, du$$

$$= 2\pi\int_1^0 (u^{3/2} - 2u^{1/2})\, du = 2\pi\left[\frac{2}{5}u^{5/2} - \frac{4}{3}u^{3/2}\right]_1^0 = \frac{28\pi}{15}$$

87. $\int_0^{\pi} \cos\left(\frac{x}{2}\right) dx = 2\int_0^{\pi} \cos\left(\frac{x}{2}\right)\frac{1}{2}\, dx = \left[2 \sin\left(\frac{x}{2}\right)\right]_0^{\pi} = 2$

89. $p = 1.20 + 0.04t$

$$C = \frac{15,000}{M}\int_t^{t+1} p\, ds = \frac{15,000}{m}\int_t^{t+1} (1.20 + 0.04s)\, ds$$

(a) 2000 corresponds to $t = 10$.

$$C = \frac{15,000}{M}\int_{10}^{11} \lfloor 1.20 + 0.04t \rfloor\, dt$$

$$= \frac{15,000}{M}\left[1.20t + 0.02t^2\right]_{10}^{11} = \frac{24,300}{M}$$

(b) 2005 corresponds to $t = 15$.

$$C = \frac{15,000}{M}\left[1.20t + 0.02t^2\right]_{15}^{16} = \frac{27,300}{M}$$

91. Trapezoidal Rule ($n = 4$): $\int_1^2 \frac{1}{1 + x^3}\, dx \approx \frac{1}{8}\left[\frac{1}{1 + 1^3} + \frac{2}{1 + (1.25)^3} + \frac{2}{1 + (1.5)^3} + \frac{2}{1 + (1.75)^3} + \frac{1}{1 + 2^3}\right] \approx 0.257$

Simpson's Rule ($n = 4$): $\int_1^2 \frac{1}{1 + x^3}\, dx \approx \frac{1}{12}\left[\frac{1}{1 + 1^3} + \frac{4}{1 + (1.25)^3} + \frac{2}{1 + (1.5)^3} + \frac{4}{1 + (1.75)^3} + \frac{1}{1 + 2^3}\right] \approx 0.254$

Graphing utility: 0.254

93. Trapezoidal Rule ($n = 4$): $\int_0^{\pi/2} \sqrt{x} \cos x\, dx \approx 0.637$

Simpson's Rule ($n = 4$): 0.685

Graphing Utility: 0.704

95. Trapezoidal Rule ($n = 4$): $\int_{-1}^{1} e^{-x^2}\, dx \approx 1.463$

Simpson's Rule ($n = 4$) ≈ 1.494

Graphing Utility: 1.494

97. $u = 7x - 2$, $du = 7dx$

$$\int \frac{1}{7x - 2}\, dx = \frac{1}{7}\int \frac{1}{7x - 2}(7)\, dx = \frac{1}{7}\ln|7x - 2| + C$$

99. $\int \frac{\sin x}{1 + \cos x}\, dx = -\int \frac{-\sin x}{1 + \cos x}\, dx$

$= -\ln|1 + \cos x| + C$

101. $\displaystyle\int_1^4 \frac{x+1}{x}\,dx = \int_1^4\left(1+\frac{1}{x}\right)dx = \Big[x + \ln|x|\Big]_1^4 = 3 + \ln 4$

103. $\displaystyle\int_0^{\pi/3} \sec\theta\,d\theta = \Big[\ln|\sec\theta + \tan\theta|\Big]_0^{\pi/3} = \ln\left(2+\sqrt{3}\right)$

105. $\displaystyle\int \frac{e^{2x}-e^{-2x}}{e^{2x}+e^{-2x}}\,dx$. Let $u = e^{2x} + e^{-2x}$. Then $du = 2(e^{2x} - e^{-2x})dx$. $\displaystyle\int \frac{e^{2x}-e^{-2x}}{e^{2x}+e^{-2x}}\,dx = \frac{1}{2}\ln(e^{2x}+e^{-2x}) + C$

107. Let $u = e^{2x}$, $du = 2e^{2x}\,dx$.

$$\int \frac{1}{e^{2x}+e^{-2x}}\,dx = \int \frac{e^{2x}}{1+e^{4x}}\,dx = \frac{1}{2}\int \frac{1}{1+(e^{2x})^2}(2e^{2x})\,dx = \frac{1}{2}\arctan(e^{2x}) + C$$

109. Let $u = x^2$, $du = 2x\,dx$.

$$\int \frac{x}{\sqrt{1-x^4}}\,dx = \frac{1}{2}\int \frac{1}{\sqrt{1-(x^2)^2}}(2x)\,dx = \frac{1}{2}\arcsin x^2 + C$$

111. Let $u = 16 + x^2$, $du = 2x\,dx$.

$$\int \frac{x}{16+x^2}\,dx = \frac{1}{2}\int \frac{1}{16+x^2}(2x)\,dx = \frac{1}{2}\ln(16+x^2) + C$$

113. Let $u = \arctan\left(\frac{x}{2}\right)$, $du = \frac{2}{4+x^2}\,dx$.

$$\int \frac{\arctan(x/2)}{4+x^2}\,dx = \frac{1}{2}\int\left(\arctan\frac{x}{2}\right)\left(\frac{2}{4+x^2}\right)dx = \frac{1}{4}\left(\arctan\frac{x}{2}\right)^2 + C$$

115. $\displaystyle\int \frac{dy}{\sqrt{A^2-y^2}} = \int \sqrt{\frac{k}{m}}\,dt$

$$\arcsin\left(\frac{y}{A}\right) = \sqrt{\frac{k}{m}}\,t + C$$

Since $y = 0$ when $t = 0$, you have $C = 0$. Thus,

$$\sin\left(\sqrt{\frac{k}{m}}\,t\right) = \frac{y}{A}$$

$$y = A\sin\left(\sqrt{\frac{k}{m}}\,t\right)$$

117. $y = 2x - \cosh\sqrt{x}$

$$y' = 2 - \frac{1}{2\sqrt{x}}(\sinh\sqrt{x}) = 2 - \frac{\sinh\sqrt{x}}{2\sqrt{x}}$$

119. Let $u = x^2$, $du = 2x\,dx$.

$$\int \frac{x}{\sqrt{x^4-1}}\,dx = \frac{1}{2}\int \frac{1}{\sqrt{(x^2)^2-1}}(2x)\,dx = \frac{1}{2}\ln\left(x^2 + \sqrt{x^4-1}\right) + C$$

Problem Solving for Chapter 4

1. (a) $L(1) = \displaystyle\int_1^1 \frac{1}{t}\,dt = 0$

(b) $L'(x) = \dfrac{1}{x}$ by the Second Fundamental Theorem of Calculus.

$L'(1) = 1$

(c) $L(x) = 1 = \displaystyle\int_1^x \frac{1}{t}\,dt$ for $x \approx 2.718$

$\displaystyle\int_1^{2.718} \frac{1}{t}\,dt = 0.999896$ (**Note:** The exact value of x is e, the base of the natural logarithm function.)

—CONTINUED—

1. —CONTINUED—

(d) We first show that $\displaystyle\int_1^{x_1} \frac{1}{t}\,dt = \int_{1/x_1}^{1} \frac{1}{t}\,dt.$

To see this, let $u = \dfrac{t}{x_1}$ and $du = \dfrac{1}{x_1}\,dt.$

Then $\displaystyle\int_1^{x_1} \frac{1}{t}\,dt = \int_{1/x_1}^{1} \frac{1}{ux_1}(x_1\,du) = \int_{1/x_1}^{1} \frac{1}{u}\,du = \int_{1/x_1}^{1} \frac{1}{t}\,dt.$

Now,
$$
\begin{aligned}
L(x_1x_2) &= \int_1^{x_1x_2} \frac{1}{t}\,dt = \int_{1/x_1}^{x_2} \frac{1}{u}\,du\left(\text{using } u = \frac{t}{x_1}\right)\\
&= \int_{1/x_1}^{1} \frac{1}{u}\,du + \int_1^{x_2} \frac{1}{u}\,du\\
&= \int_1^{x_1} \frac{1}{u}\,du + \int_1^{x_2} \frac{1}{u}\,du\\
&= L(x_1) + L(x_2).
\end{aligned}
$$

3. $S(x) = \displaystyle\int_0^x \sin\left(\frac{\pi t^2}{2}\right) dt$

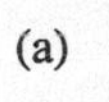
(a)

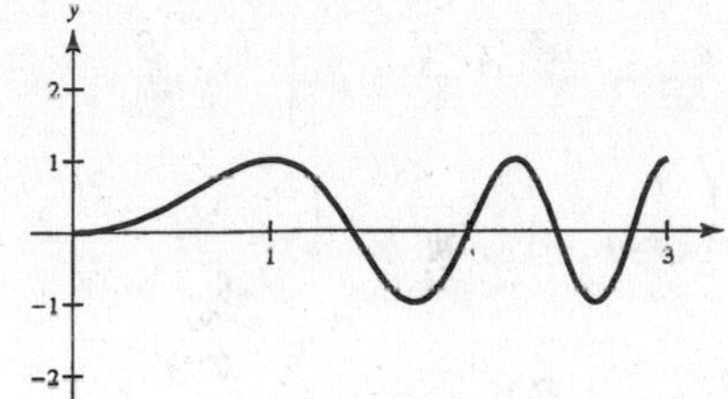

(b)

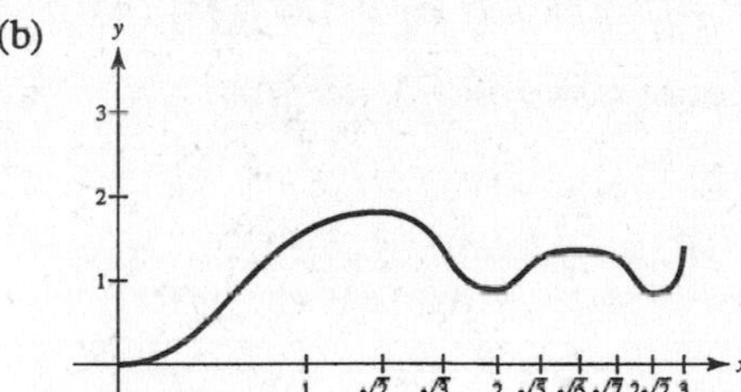

The zeros of $y = \sin\dfrac{\pi x^2}{2}$ correspond to the relative extrema of $S(x)$.

(c) $S'(x) = \sin\dfrac{\pi x^2}{2} = 0 \Rightarrow \dfrac{\pi x^2}{2} = n\pi \Rightarrow x^2 = 2n \Rightarrow x = \sqrt{2n}$, n integer.

Relative maximum at $x = \sqrt{2} \approx 1.4142$ and $x = \sqrt{6} \approx 2.4495$

Relative minimum at $x = 2$ and $x = \sqrt{8} \approx 2.8284$

(d) $S''(x) = \cos\left(\dfrac{\pi x^2}{2}\right)(\pi x) = 0 \Rightarrow \dfrac{\pi x^2}{2} = \dfrac{\pi}{2} + n\pi \Rightarrow x^2 = 1 + 2n \Rightarrow x = \sqrt{1 + 2n}$, n integer

Points of inflection at $x = 1, \sqrt{3}, \sqrt{5}$, and $\sqrt{7}$.

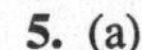
5. (a)

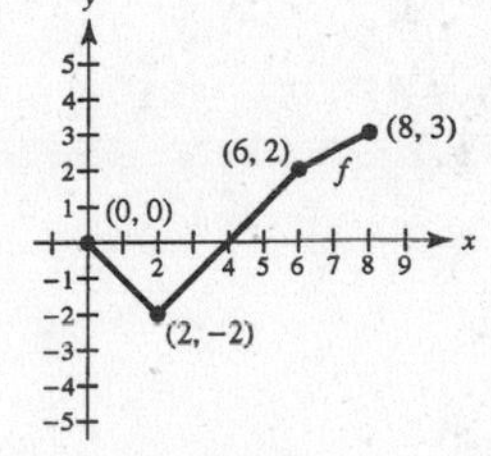

(b)

x	0	1	2	3	4	5	6	7	8
$F(x)$	0	$-\frac{1}{2}$	-2	$-\frac{7}{2}$	-4	$-\frac{7}{2}$	-2	$\frac{1}{4}$	3

—CONTINUED—

5. —CONTINUED—

(c) $f(x) = \begin{cases} -x, & 0 \le x < 2 \\ x - 4, & 2 \le x < 6 \\ \frac{1}{2}x - 1, & 6 \le x \le 8 \end{cases}$

$$F(x) = \int_0^x f(t)\,dt = \begin{cases} (-x^2/2), & 0 \le x < 2 \\ (x^2/2) - 4x + 4, & 2 \le x < 6 \\ (1/4)x^2 - x - 5, & 6 \le x \le 8 \end{cases}$$

$F'(x) = f(x)$. F is decreasing on $(0, 4)$ and increasing on $(4, 8)$. Therefore, the minimum is -4 at $x = 4$, and the maximum is 3 at $x = 8$.

(d) $F''(x) = f'(x) = \begin{cases} -1, & 0 < x < 2 \\ 1, & 2 < x < 6 \\ \frac{1}{2}, & 6 < x < 8 \end{cases}$

$x = 2$ is a point of inflection, whereas $x = 6$ is not.

7. (a) $\displaystyle\int_{-1}^{1} \cos x\,dx \approx \cos\left(-\frac{1}{\sqrt{3}}\right) + \cos\left(\frac{1}{\sqrt{3}}\right) = 2\cos\left(\frac{1}{\sqrt{3}}\right) \approx 1.6758$

$\displaystyle\int_{-1}^{1} \cos x\,dx = \sin x\Big]_{-1}^{1} = 2\sin(1) \approx 1.6829$

Error. $|1.6829 - 1.6758| = 0.0071$

(b) $\displaystyle\int_{-1}^{1} \frac{1}{1+x^2}\,dx \approx \frac{1}{1+(1/3)} + \frac{1}{1+(1/3)} = \frac{3}{2}$

(Note: exact answer is $\pi/2 \approx 1.5708$)

(c) Let $p(x) = ax^3 + bx^2 + cx + d$.

$$\int_{-1}^{1} p(x)\,dx = \left[\frac{ax^4}{4} + \frac{bx^3}{3} + \frac{cx^2}{2} + dx\right]_{-1}^{1} = \frac{2b}{3} + 2d$$

$$p\left(-\frac{1}{\sqrt{3}}\right) + p\left(\frac{1}{\sqrt{3}}\right) = \left(\frac{b}{3} + d\right) + \left(\frac{b}{3} + d\right) = \frac{2b}{3} + 2d$$

9. Consider $F(x) = [f(x)]^2 \Rightarrow F'(x) = 2f(x)f'(x)$. Thus,

$$\begin{aligned}\int_a^b f(x)f'(x)\,dx &= \int_a^b \frac{1}{2}F'(x)\,dx \\ &= \left[\frac{1}{2}F(x)\right]_a^b \\ &= \frac{1}{2}[F(b) - F(a)] \\ &= \frac{1}{2}[f(b)^2 - f(a)^2]\end{aligned}$$

11. Consider $\displaystyle\int_0^1 x^5\,dx = \frac{x^6}{6}\Big]_0^1 = \frac{1}{6}$.

The corresponding Riemann Sum using right endpoints is

$$\begin{aligned}S(n) &= \frac{1}{n}\left[\left(\frac{1}{n}\right)^5 + \left(\frac{2}{n}\right)^5 + \cdots + \left(\frac{n}{n}\right)^5\right] \\ &= \frac{1}{n^6}[1^5 + 2^5 + \cdots + n^5]\end{aligned}$$

Thus, $\displaystyle\lim_{n\to\infty} S(n) = \lim_{n\to\infty} \frac{1^5 + 2^5 + \cdots + n^5}{n^6} = \frac{1}{6}$.

13. By Theorem 4.8, $0 < f(x) \le M \Rightarrow \displaystyle\int_a^b f(x)\,dx \le \int_a^b M\,dx = M(b-a)$.

Similarly, $m \le f(x) \Rightarrow m(b-a) = \displaystyle\int_a^b m\,dx \le \int_a^b f(x)\,dx$.

Thus, $m(b-a) \le \displaystyle\int_a^b f(x)\,dx \le M(b-a)$. On the interval $[0, 1]$, $1 \le \sqrt{1+x^4} \le \sqrt{2}$ and $b - a = 1$.

Thus, $1 \le \displaystyle\int_0^1 \sqrt{1+x^4}\,dx \le \sqrt{2}$. $\left(\textbf{Note: } \displaystyle\int_0^1 \sqrt{1+x^4}\,dx \approx 1.0894\right)$

15. Since $-|f(x)| \le f(x) \le |f(x)|$,

$$-\int_a^b |f(x)|\,dx \le \int_a^b f(x)\,dx \le \int_a^b |f(x)|\,dx \Rightarrow \left|\int_a^b f(x)\,dx\right| \le \int_a^b |f(x)|\,dx.$$

17. Let $u = 1 + \sqrt{x}$, $\sqrt{x} = u - 1$, $x = u^2 - 2u + 1$,

$dx = (2u - 2)du.$

$$\text{Area} = \int_1^4 \frac{1}{\sqrt{x} + x}\,dx = \int_2^3 \frac{2u - 2}{(u - 1) + (u^2 - 2u + 1)}\,du$$

$$= \int_2^3 \frac{2(u - 1)}{u^2 - u}\,du$$

$$= \int_2^3 \frac{2}{u}\,du$$

$$= \Big[2 \ln u\Big]_2^3$$

$$= 2 \ln 3 - 2 \ln 2 = 2 \ln\left(\frac{3}{2}\right)$$

$$\approx 0.8109$$

19. $$\frac{1}{365}\int_0^{365} 100{,}000\left[1 + \sin\frac{2\pi(t - 60)}{365}\right]dt = \frac{100{,}000}{365}\left[t - \frac{365}{2\pi}\cos\frac{2\pi(t - 60)}{365}\right]_0^{365} = 100{,}000 \text{ lbs.}$$

CHAPTER 5
Differential Equations

CHAPTER 5
Differential Equations

Section 5.1 Differential Equations: Growth and Decay

Solutions to Odd-Numbered Exercises

1. $\dfrac{dy}{dx} = x + 2$

$$y = \int (x+2)\,dx = \frac{x^2}{2} + 2x + C$$

3. $\dfrac{dy}{dx} = y + 2$

$$\frac{dy}{y+2} = dx$$

$$\int \frac{1}{y+2}\,dy = \int dx$$

$$\ln|y+2| = x + C_1$$

$$y + 2 = e^{x+C_1} = Ce^x$$

$$y = Ce^x - 2$$

5. $y' = \dfrac{5x}{y}$

$$yy' = 5x$$

$$\int yy'\,dx = \int 5x\,dx$$

$$\int y\,dy = \int 5x\,dx$$

$$\frac{1}{2}y^2 = \frac{5}{2}x^2 + C_1$$

$$y^2 - 5x^2 = C$$

7. $y' = \sqrt{x}\,y$

$$\frac{y'}{y} = \sqrt{x}$$

$$\int \frac{y'}{y}\,dx = \int \sqrt{x}\,dx$$

$$\int \frac{dy}{y} = \int \sqrt{x}\,dx$$

$$\ln|y| = \frac{2}{3}x^{3/2} + C_1$$

$$y = e^{(2/3)x^{3/2} + C_1}$$

$$= e^{C_1}e^{(2/3)x^{3/2}}$$

$$= Ce^{(2/3)x^{3/2}}$$

9. $(1 + x^2)y' - 2xy = 0$

$$y' = \frac{2xy}{1+x^2}$$

$$\frac{y'}{y} = \frac{2x}{1+x^2}$$

$$\int \frac{y'}{y}\,dx = \int \frac{2x}{1+x^2}\,dx$$

$$\int \frac{dy}{y} = \int \frac{2x}{1+x^2}\,dx$$

$$\ln|y| = \ln(1+x^2) + C_1$$

$$\ln|y| = \ln(1+x^2) + \ln C$$

$$\ln|y| = \ln\left[C(1+x^2)\right]$$

$$y = C(1+x^2)$$

11. $\dfrac{dQ}{dt} = \dfrac{k}{t^2}$

$$\int \frac{dQ}{dt}\,dt = \int \frac{k}{t^2}\,dt$$

$$\int dQ = -\frac{k}{t} + C$$

$$Q = -\frac{k}{t} + C$$

13. $\dfrac{dN}{ds} = k(250 - s)$

$$\int \frac{dN}{ds}\,ds = \int k(250 - s)\,ds$$

$$\int dN = -\frac{k}{2}(250 - s)^2 + C$$

$$N = -\frac{k}{2}(250 - s)^2 + C$$

15. (a)

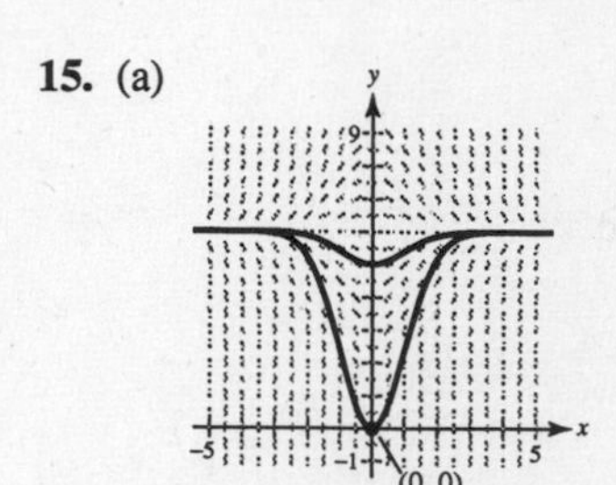

(b) $\dfrac{dy}{dx} = x(6 - y),\ (0, 0)$

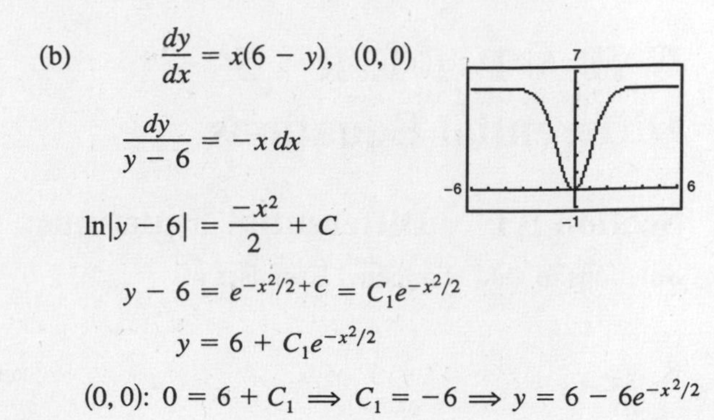

$$\frac{dy}{y-6} = -x\,dx$$

$$\ln|y - 6| = \frac{-x^2}{2} + C$$

$$y - 6 = e^{-x^2/2+C} = C_1e^{-x^2/2}$$

$$y = 6 + C_1e^{-x^2/2}$$

$$(0, 0):\ 0 = 6 + C_1 \Rightarrow C_1 = -6 \Rightarrow y = 6 - 6e^{-x^2/2}$$

17. $\dfrac{dy}{dt} = \dfrac{1}{2}t,\ (0, 10)$

$$\int dy = \int \frac{1}{2}t\,dt$$

$$y = \frac{1}{4}t^2 + C$$

$$10 = \frac{1}{4}(0)^2 + C \Rightarrow C = 10$$

$$y = \frac{1}{4}t^2 + 10$$

19. $\dfrac{dy}{dt} = -\dfrac{1}{2}y,\ (0, 10)$

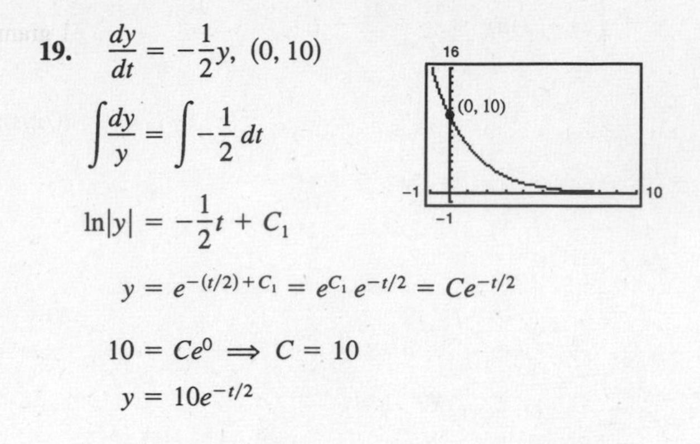

$$\int \frac{dy}{y} = \int -\frac{1}{2}\,dt$$

$$\ln|y| = -\frac{1}{2}t + C_1$$

$$y = e^{-(t/2)+C_1} = e^{C_1}e^{-t/2} = Ce^{-t/2}$$

$$10 = Ce^0 \Rightarrow C = 10$$

$$y = 10e^{-t/2}$$

21. $\dfrac{dy}{dx} = ky$

$y = Ce^{kx}$ (Theorem 5.16)

$(0, 4)$: $4 = Ce^0 = C$

$(3, 10)$: $10 = 4e^{3k} \Rightarrow k = \dfrac{1}{3}\ln\left(\dfrac{5}{2}\right) \approx 0.3054$

$y = 4e^{0.3054x}$

When $x = 6$, $y = 4e^{1/3\ln(5/2)(6)} = 4e^{\ln(5/2)^2}$

$$= 4\left(\frac{5}{2}\right)^2 = 25$$

23. $\dfrac{dV}{dt} = kV$

$V = Ce^{kt}$ (Theorem 5.16)

$(0, 20{,}000)$: $C = 20{,}000$

$(4, 12{,}500)$: $12{,}500 = 20{,}000e^{4k} \Rightarrow k = \dfrac{1}{4}\ln\left(\dfrac{5}{8}\right) \approx -0.1175$

$y = 20{,}000e^{-0.1175t}$

When $t = 6$, $V = 20{,}000e^{1/4\ln(5/8)(6)} = 20{,}000e^{\ln(5/8)^{3/2}}$

$$= 20{,}000\left(\frac{5}{8}\right)^{3/2} \approx 9882.118$$

25. $y = Ce^{kt},\ \left(0, \dfrac{1}{2}\right), (5, 5)$

$$C = \frac{1}{2}$$

$$y = \frac{1}{2}e^{kt}$$

$$5 = \frac{1}{2}e^{5k}$$

$$k = \frac{\ln 10}{5} \approx 0.4605$$

$$\frac{1}{2}e^{(\ln 10/5)t} = \frac{1}{2}(10^{t/5}) \quad \text{or} \quad y \approx \frac{1}{2}e^{0.4605t}$$

27. $y = Ce^{kt},\ (1, 1),\ (5, 5)$

$$1 = Ce^k$$

$$5 = Ce^{5k}$$

$$5Ce^k = Ce^{5k}$$

$$5e^k = e^{5k}$$

$$5 = e^{4k}$$

$$k = \frac{\ln 5}{4} \approx 0.4024$$

$$y = Ce^{0.4024t}$$

$$1 = Ce^{0.4024}$$

$$C \approx 0.6687 \qquad (C = 5^{-1/4})$$

$$y \approx 0.6687e^{0.4024t}$$

29. A differential equation in x and y is an equation that involves x, y and derivatives of y.

31. $\dfrac{dy}{dx} = \dfrac{1}{2}xy$

$\dfrac{dy}{dx} > 0$ when $xy > 0$. Quadrants I and III.

33. Since the initial quantity is 10 grams, $y = 10e^{[\ln(1/2)/1620]t}$. When $t = 1000$, $y = 10e^{[\ln(1/2)/1620](1000)} \approx 6.52$ grams. When $t = 10{,}000$, $y = 10e^{[\ln(1/2)/1620](10{,}000)} \approx 0.14$ gram.

35. Since $y = Ce^{[\ln(1/2)/1620]t}$, we have $0.5 = Ce^{[\ln(1/2)/1620](10{,}000)} \Rightarrow C \approx 36.07$.

Initial quantity: 36.07 grams.

When $t = 1000$, we have $y = Ce^{[\ln(1/2)/1620](1000)} \approx 23.51$ grams.

37. Since the initial quantity is 5 grams, we have $y = 5.0e^{[\ln(1/2)/5730]t}$.

When $t = 1000$, $y \approx 4.43$ g.

When $t = 10{,}000$, $y \approx 1.49$ g.

39. Since $y = Ce^{[\ln(1/2)/24{,}360]t}$, we have $2.1 = Ce^{[\ln(1/2)/24{,}360](1000)} \Rightarrow C \approx 2.16$. Thus, the initial quantity is 2.16 grams. When $t = 10{,}000$, $y = 2.16e^{[\ln(1/2)/24{,}360](10{,}000)} \approx 1.63$ grams.

41. Since $\dfrac{dy}{dx} = ky$, $y = Ce^{kt}$ or $y = y_0e^{kt}$.

$$\frac{1}{2}y_0 = y_0e^{1620k}$$

$$k = \frac{-\ln 2}{1620}$$

$$y = y_0e^{-(\ln 2)t/1620}.$$

When $t = 100$, $y = y_0e^{-(\ln 2)/16.2} \approx y_0(0.9581)$.
Therefore, 95.81% of the present amount still exists.

43. Since $A = 1000e^{0.06t}$, the time to double is given by $2000 = 1000e^{0.06t}$ and we have

$$2 = e^{0.06t}$$

$$\ln 2 = 0.06t$$

$$t = \frac{\ln 2}{0.06} \approx 11.55 \text{ years.}$$

Amount after 10 years: $A = 1000e^{(0.06)(10)} \approx \$1822.12$

45. Since $A = 750e^{rt}$ and $A = 1500$ when $t = 7.75$, we have the following.

$$1500 = 750e^{7.75r}$$

$$r = \frac{\ln 2}{7.75} \approx 0.0894 = 8.94\%$$

Amount after 10 years: $A = 750e^{0.0894(10)} \approx \$1833.67$

47. Since $A = 500e^{rt}$ and $A = 1292.85$ when $t = 10$, we have the following.

$$1292.85 = 500e^{10r}$$

$$r = \frac{\ln(1292.85/500)}{10} \approx 0.0950 = 9.50\%$$

The time to double is given by

$$1000 = 500e^{0.0950t}$$

$$t = \frac{\ln 2}{0.095} \approx 7.30 \text{ years.}$$

49. $500{,}000 = P\left(1 + \dfrac{0.075}{12}\right)^{(12)(20)}$

$$P = 500{,}000\left(1 + \frac{0.075}{12}\right)^{-240}$$

$$\approx \$112{,}087.09$$

51. $500{,}000 = P\left(1 + \dfrac{0.08}{12}\right)^{(12)(35)}$

$$P = 500{,}000\left(1 + \frac{0.08}{12}\right)^{-420}$$

$$= \$30{,}688.87$$

53. (a) $2000 = 1000(1 + 0.07)^t$

$$2 = 1.07^t$$

$$\ln 2 = t \ln 1.07$$

$$t = \frac{\ln 2}{\ln 1.07} \approx 10.24 \text{ years}$$

(b) $2000 = 1000\left(1 + \frac{0.07}{12}\right)^{12t}$

$$2 = \left(1 + \frac{0.007}{12}\right)^{12t}$$

$$\ln 2 = 12t \ln\left(1 + \frac{0.07}{12}\right)$$

$$t = \frac{\ln 2}{12 \ln(1 + (0.07/12))} \approx 9.93 \text{ years}$$

(c) $2000 = 1000\left(1 + \frac{0.07}{365}\right)^{365t}$

$$2 = \left(1 + \frac{0.07}{365}\right)^{365t}$$

$$\ln 2 = 365t \ln\left(1 + \frac{0.07}{365}\right)$$

$$t = \frac{\ln 2}{365 \ln(1 + (0.07/365))} \approx 9.90 \text{ years}$$

(d) $2000 = 1000e^{(0.07)t}$

$$2 = e^{0.07t}$$

$$\ln 2 = 0.07t$$

$$t = \frac{\ln 2}{0.07} \approx 9.90 \text{ years}$$

55. (a) $2000 = 1000(1 + 0.085)^t$

$$2 = 1.085^t$$

$$\ln 2 = t \ln 1.085$$

$$t = \frac{\ln 2}{\ln 1.085} \approx 8.50 \text{ years}$$

(b) $2000 = 1000\left(1 + \frac{0.085}{12}\right)^{12t}$

$$2 = \left(1 + \frac{0.085}{12}\right)^{12t}$$

$$\ln 2 = 12t \ln\left(1 + \frac{0.085}{12}\right)$$

$$t = \frac{1}{12}\frac{\ln 2}{\ln\left(1 + \frac{0.085}{12}\right)} \approx 8.18 \text{ years}$$

(c) $2000 = 1000\left(1 + \frac{0.085}{365}\right)^{365t}$

$$2 = \left(1 + \frac{0.085}{365}\right)^{365t}$$

$$\ln 2 = 365t \ln\left(1 + \frac{0.085}{365}\right)$$

$$t = \frac{1}{365}\frac{\ln 2}{\ln\left(1 + \frac{0.085}{365}\right)} \approx 8.16 \text{ years}$$

(d) $2000 = 1000e^{0.085t}$

$$2 = e^{0.085t}$$

$$\ln 2 = 0.085t$$

$$t = \frac{\ln 2}{0.085} \approx 8.15 \text{ years}$$

57. $P = Ce^{kt} = Ce^{-0.009t}$

$P(-1) = 8.2 = Ce^{-0.009(-1)} \Rightarrow C = 8.1265$

$P = 8.1265e^{-0.009t}$

$P(10) \approx 7.43$ or 7,430,000 people in 2010

59. $P = Ce^{kt} = Ce^{0.036t}$

$P(-1) = 4.6 = Ce^{0.036(-1)} \Rightarrow C = 4.7686$

$P = 4.7686e^{0.036t}$

$P(10) \approx 6.83$ or 6,830,000 people in 2010

61. If $k < 0$, the population decreases.

If $k > 0$, the population increases.

63. $P = Ce^{kx}$, $(0, 760)$, $(1000, 672.71)$

$C = 760$

$672.71 = 760e^{1000x}$

$$x = \frac{\ln(672.71/760)}{1000} \approx -0.000122$$

$P \approx 760e^{-0.000122x}$

When $x = 3000$, $P \approx 527.06$ mm Hg.

65. (a) $19 = 30(1 - e^{20k})$

$30e^{20k} = 11$

$k = \dfrac{\ln(11/30)}{20} \approx -0.0502$

$N \approx 30(1 - e^{-0.0502t})$

(b) $25 = 30(1 - e^{-0.0502t})$

$e^{-0.0502t} = \dfrac{1}{6}$

$t = \dfrac{-\ln 6}{-0.0502} \approx 36$ days

67. $S = Ce^{k/t}$

(a) $S = 5$ when $t = 1$

$5 = Ce^k$

$\lim\limits_{t\to\infty} Ce^{k/t} = C = 30$

$5 = 30e^k$

$k = \ln\frac{1}{6} \approx -1.7918$

$S = 30\left(\frac{1}{6}\right)^{1/t} \approx 30e^{-1.7918/t}$

(b) When $t = 5$, $S \approx 20.9646$ which is 20,965 units.

(c)

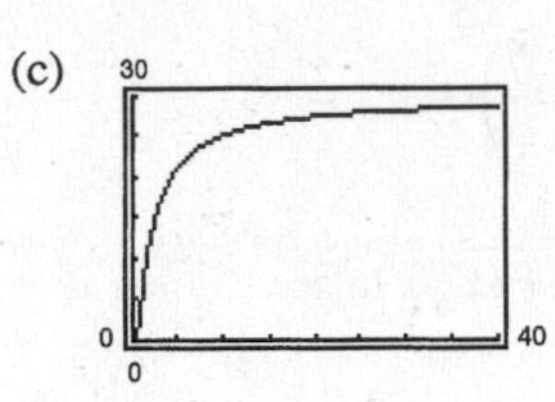

69. $A(t) = V(t)e^{-0.10t} = 100{,}000e^{0.8\sqrt{t}}\,e^{-0.10t} = 100{,}000e^{0.8\sqrt{t}-0.10t}$

$\dfrac{dA}{dt} = 100{,}000\left(\dfrac{0.4}{\sqrt{t}} - 0.10\right)e^{0.8\sqrt{t}-0.10t} = 0$ when $t = 16$.

The timber should be harvested in the year 2014, (1998 + 16). **Note:** You could also use a graphing utility to graph $A(t)$ and find the maximum of $A(t)$. Use the viewing rectangle $0 \le x \le 30$ and $0 \le y \le 600{,}000$.

71. $\beta(I) = 10\log_{10}\dfrac{I}{I_0}$, $I_0 = 10^{-16}$

(a) $\beta(10^{-14}) = 10\log_{10}\dfrac{10^{-14}}{10^{-16}} = 20$ decibels

(b) $\beta(10^{-9}) = 10\log_{10}\dfrac{10^{-9}}{10^{-16}} = 70$ decibels

(c) $\beta(10^{-6.5}) = 10\log_{10}\dfrac{10^{-6.5}}{10^{-16}} = 95$ decibels

(d) $\beta(10^{-4}) = 10\log_{10}\dfrac{10^{-4}}{10^{-16}} = 120$ decibels

73. $R = \dfrac{\ln I - \ln I_0}{\ln 10} = \dfrac{\ln I - 0}{\ln 10}$, $I = e^{R\ln 10} = 10^R$

(a) $8.3 = \dfrac{\ln I - 0}{\ln 10}$

$I = 10^{8.3} \approx 199{,}526{,}231.5$

(b) $2R = \dfrac{\ln I - 0}{\ln 10}$

$I = e^{2R\ln 10} = e^{2R\ln 10} = (e^{R\ln 10})^2 = (10^R)^2$

Increases by a factor of $e^{2R\ln 10}$ or 10^R.

(c) $\dfrac{dR}{dI} = \dfrac{1}{I\ln 10}$

75. False. If $y = Ce^{kt}$, $y' = Cke^{kt} \neq$ constant.

77. True

Section 5.2 Differential Equations: Separation of Variables

1. Differential equation: $y' = 4y$

Solution: $y = Ce^{4x}$

Check:

$y' = 4Ce^{4x} = 4y$

3. Differential equation: $y'' + y = 0$

Solution: $y = C_1\cos x + C_2\sin x$

Check:

$y' = -C_1\sin x + C_2\cos x$

$y'' = -C_1\cos x - C_2\sin x$

$y'' + y = -C_1\cos x - C_2\sin x + C_1\cos x + C_2\sin x = 0$

5. Differential equation: $y'' + y = \tan x$

$$y = -\cos x \ln|\sec x + \tan x|$$

$$y' = (-\cos x)\frac{1}{\sec x + \tan x}(\sec x \cdot \tan x + \sec^2 x) + \sin x \ln|\sec x + \tan x|$$

$$= \frac{(-\cos x)}{\sec x + \tan x}(\sec x)(\tan x + \sec x) + \sin x \ln|\sec x + \tan x|$$

$$= -1 + \sin x \ln|\sec x + \tan x|$$

$$y'' = (\sin x)\frac{1}{\sec x + \tan x}(\sec x \cdot \tan x + \sec^2 x) + \cos x \ln|\sec x + \tan x|$$

$$= (\sin x)(\sec x) + \cos x \ln|\sec x + \tan x|$$

Substituting,

$$y'' + y = (\sin x)(\sec x) + \cos x \ln|\sec x + \tan x| - \cos x \ln|\sec x + \tan x|$$

$$= \tan x.$$

In Exercises 7–11, the differential equation is $y^{(4)} - 16y = 0$.

7.

$$y = 3\cos x$$

$$y^{(4)} = 3\cos x$$

$$y^{(4)} - 16y = -45\cos x \neq 0,$$

No

9.

$$y = e^{-2x}$$

$$y^{(4)} = 16e^{-2x}$$

$$y^{(4)} - 16y = 16e^{-2x} - 16e^{-2x} = 0,$$

Yes

11.

$$y = C_1e^{2x} + C_2e^{-2x} + C_3 \sin 2x + C_4 \cos 2x$$

$$y^{(4)} = 16C_1e^{2x} + 16C_2e^{-2x} + 16C_3 \sin 2x + 16C_4 \cos 2x$$

$$y^{(4)} - 16y = 0,$$

Yes

In 13–17, the differential equation is $xy' - 2y = x^3e^x$.

13. $y = x^2,\ y' = 2x$

$$xy' - 2y = x(2x) - 2(x^2) = 0 \neq x^3e^x$$

No

15. $y = x^2(2 + e^x),\ \ y' = x^2(e^x) + 2x(2 + e^x)$

$$xy' - 2y = x[x^2e^x + 2xe^x + 4x] - 2[x^2e^x + 2x^2] = x^3e^x,$$

Yes

17. $y = \ln x,\ y' = \dfrac{1}{x}$

$$xy' - 2y = x\left(\frac{1}{x}\right) - 2\ln x \neq x^3e^x$$

No

19. $y = Ce^{kx}$

$$\frac{dy}{dx} = Cke^{kx}$$

Since $dy/dx = 0.07y$, we have $Cke^{kx} = 0.07Ce^{kx}$. Thus, $k = 0.07$.

C cannot be determined.

21. $y = Ce^{-x/2}$ passes through $(0, 3)$.

$3 = Ce^0 = C \Rightarrow C = 3$

Particular solution: $y = 3e^{-x/2}$

23. $y^2 = Cx^3$ passes through $(4, 4)$

$16 = C(64) \Rightarrow C = \frac{1}{4}$

Particular solution: $y^2 = \frac{1}{4}x^3$ or $4y^2 = x^3$

25. Differential equation: $4yy' - x = 0$

General solution: $4y^2 - x^2 = C$

Particular solutions: $C = 0$, Two intersecting lines
$C = \pm 1$, $C = \pm 4$, Hyperbolas

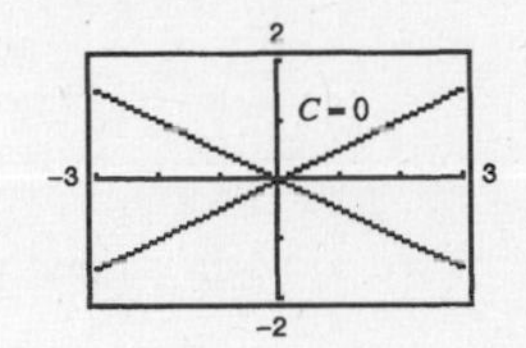

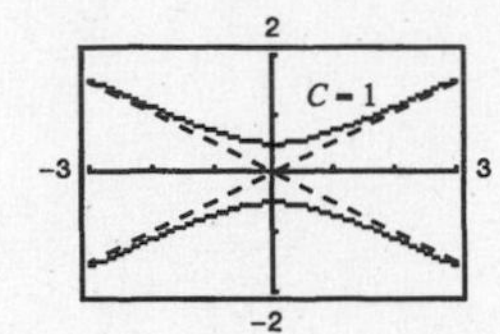

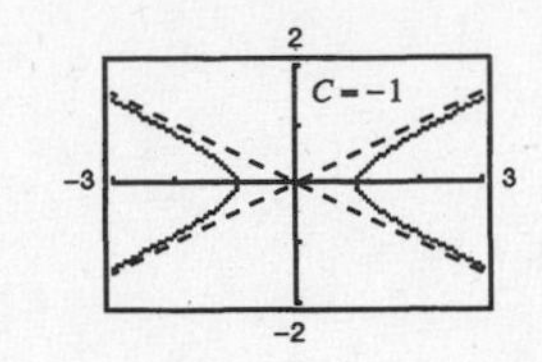

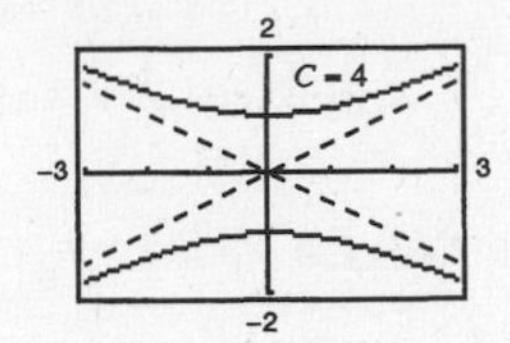

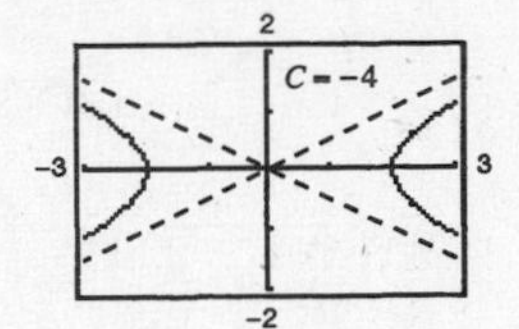

27. Differential equation: $y' + 2y = 0$

General Solution: $y = Ce^{-2x}$

$y' + 2y = C(-2)e^{-2x} + 2(Ce^{-2x}) = 0$

Initial condition: $y(0) = 3$, $3 = Ce^0 = C$

Particular solution: $y = 3e^{-2x}$

29. Differential equation: $y'' + 9y = 0$

General solution: $y = C_1 \sin 3x + C_2 \cos 3x$

$y' = 3C_1 \cos 3x - 3C_2 \sin 3x,$

$y'' = -9C_1 \sin 3x - 9C_2 \cos 3x$

$y'' + 9y = (-9C_1 \sin 3x - 9C_2 \cos 3x) +$

$9(C_1 \sin 3x + C_2 \cos 3x) = 0$

Initial conditions: $y\left(\frac{\pi}{6}\right) = 2,\ y'\left(\frac{\pi}{6}\right) = 1$

$$2 = C_1 \sin\left(\frac{\pi}{2}\right) + C_2 \cos\left(\frac{\pi}{2}\right) \Rightarrow C_1 = 2$$

$$y' = 3C_1 \cos 3x - 3C_2 \sin 3x$$

$$1 = 3C_1 \cos\left(\frac{\pi}{2}\right) - 3C_2 \sin\left(\frac{\pi}{2}\right)$$

$$= -3C_2 \Rightarrow C_2 = -\frac{1}{3}$$

Particular solution: $y = 2 \sin 3x - \frac{1}{3} \cos 3x$

31. Differential equation: $x^2y'' - 3xy' + 3y = 0$

General solution: $y = C_1x + C_2x^3$

$y' = C_1 + 3C_2x^2,\ y'' = 6C_2x$

$x^2y'' - 3xy' + 3y = x^2(6C_2x) - 3x(C_1 + 3C_2x^2) +$

$3(C_1x + C_2x^3) = 0$

Initial conditions: $y(2) = 0,\ y'(2) = 4$

$$0 = 2C_1 + 8C_2$$

$$y' = C_1 + 3C_2x^2$$

$$4 = C_1 + 12C_2$$

$$\left.\begin{array}{l} C_1 + 4C_2 = 0 \\ C_1 + 12C_2 = 4 \end{array}\right\} \quad C_2 = \frac{1}{2},\ C_1 = -2$$

Particular solution: $y = -2x + \frac{1}{2}x^3$

33. $\dfrac{dy}{dx} = 3x^2$

$$y = \int 3x^2\,dx = x^3 + C$$

35. $\dfrac{dy}{dx} = \dfrac{x}{1 + x^2}$

$$y = \int \frac{x}{1 + x^2}\,dx = \frac{1}{2}\ln(1 + x^2) + C$$

$(u = 1 + x^2,\ du = 2x\,dx)$

37. $\dfrac{dy}{dx} = \dfrac{x-2}{x} = 1 - \dfrac{2}{x}$

$$y = \int \left[1 - \frac{2}{x}\right] dx$$

$$= x - 2\ln|x| + C = x - \ln x^2 + C$$

39. $\dfrac{dy}{dx} = \sin 2x$

$$y = \int \sin 2x\, dx = -\frac{1}{2}\cos 2x + C$$

$(u = 2x,\ du = 2\,dx)$

41. $\dfrac{dy}{dx} = x\sqrt{x-3}$ Let $u = \sqrt{x-3}$, then $x = u^2 + 3$ and $dx = 2u\,du$.

$$y = \int x\sqrt{x-3}\,dx = \int (u^2+3)(u)(2u)\,du$$

$$= 2\int (u^4 + 3u^2)\,du = 2\left(\frac{u^5}{5} + u^3\right) + C = \frac{2}{5}(x-3)^{5/2} + 2(x-3)^{3/2} + C$$

43. $\dfrac{dy}{dx} = xe^{x^2}$

$$y = \int xe^{x^2}\,dx = \frac{1}{2}e^{x^2} + C$$

$(u = x^2,\ du = 2x\,dx)$

45. $\dfrac{dy}{dx} = \dfrac{x}{y}$

$$\int y\,dy = \int x\,dx$$

$$\frac{y^2}{2} = \frac{x^2}{2} + C_1$$

$$y^2 - x^2 = C$$

47. $\dfrac{dr}{ds} = 0.05r$

$$\int \frac{dr}{r} = \int 0.05\,ds$$

$$\ln|r| = 0.05s + C_1$$

$$r = e^{0.05s + C_1} = Ce^{0.05s}$$

49. $(2 + x)y' = 3y$

$$\int \frac{dy}{y} = \int \frac{3}{2+x}\,dx$$

$$\ln|y| = 3\ln|2+x| + \ln C = \ln|C(2+x)^3|$$

$$y = C(x+2)^3$$

51. $yy' = \sin x$

$$\int y\,dy = \int \sin x\,dx$$

$$\frac{y^2}{2} = -\cos x + C_1$$

$$y^2 = -2\cos x + C$$

53. $\sqrt{1-4x^2}\,\dfrac{dy}{dx} = x$

$$dy = \frac{x}{\sqrt{1-4x^2}}\,dx$$

$$\int dy = \int \frac{x}{\sqrt{1-4x^2}}\,dx$$

$$= -\frac{1}{8}\int (1-4x^2)^{-1/2}(-8x\,dx)$$

$$y = -\frac{1}{4}(1-4x^2)^{1/2} + C$$

55. $y\ln x - xy' = 0$

$$\int \frac{dy}{y} = \int \frac{\ln x}{x}\,dx\left(u = \ln x,\ du = \frac{dx}{x}\right)$$

$$\ln|y| = \frac{1}{2}(\ln x)^2 + C_1$$

$$y = e^{(1/2)(\ln x)^2 + C_1} = Ce^{(\ln x)^2/2}$$

57. $\sqrt{1-x^2}\dfrac{dy}{dx} = \sqrt{1-y^2}$

$$\int \frac{1}{\sqrt{1-y^2}}\,dy = \int \frac{1}{\sqrt{1-x^2}}\,dx$$

$\arcsin y = \arcsin x + C$ or $\arcsin x - \arcsin y = C_1$

Taking the sine of both sides:

$$\sin[\arcsin x - \arcsin y] = \sin C_1 = C_2$$

$$x[\cos(\arcsin y)] - y[\cos(\arcsin x)] = C_2$$

$$x\sqrt{1-y^2} - y\sqrt{1-x^2} = C_2$$

59. $yy' - e^x = 0$

$$\int y\,dy = \int e^x\,dx$$

$$\frac{y^2}{2} = e^x + C_1$$

$$y^2 = 2e^x + C$$

Initial condition: $y(0) = 4,\ 16 = 2 + C,\ C = 14$

Particular solution: $y^2 = 2e^x + 14$

61. $y(x+1) + y' = 0$

$$\int \frac{dy}{y} = -\int (x+1)\,dx$$

$$\ln y = -\frac{(x+1)^2}{2} + C_1$$

$$y = Ce^{-(x+1)^2/2}$$

Initial condition: $y(-2) = 1,\ 1 = Ce^{-1/2},\ C = e^{1/2}$

Particular solution: $y = e^{[1-(x+1)^2]/2} = e^{-(x^2+2x)/2}$

63. $y(1+x^2)y' = x(1+y^2)$

$$\frac{y}{1+y^2}\,dy = \frac{x}{1+x^2}\,dx$$

$$\frac{1}{2}\ln(1+y^2) = \frac{1}{2}\ln(1+x^2) + C_1$$

$$\ln(1+y^2) = \ln(1+x^2) + \ln C = \ln[C(1+x^2)]$$

$$1 + y^2 = C(1+x^2)$$

$y(0) = \sqrt{3}$: $1 + 3 = C \Rightarrow C = 4$

$$1 + y^2 = 4(1+x^2)$$

$$y^2 = 3 + 4x^2$$

65. $\dfrac{du}{dv} = uv\sin v^2$

$$\int \frac{du}{u} = \int v\sin v^2\,dv$$

$$\ln|u| = -\frac{1}{2}\cos v^2 + C_1$$

$$u = Ce^{-(\cos v^2)/2}$$

Initial condition: $u(0) = 1,\ C = \dfrac{1}{e^{-1/2}} = e^{1/2}$

Particular solution: $u = e^{(1-\cos v^2)/2}$

67. $dP - kP\,dt = 0$

$$\int \frac{dP}{P} = k\int dt$$

$$\ln|P| = kt + C_1$$

$$P = Ce^{kt}$$

Initial condition: $P(0) = P_0,\ P_0 = Ce^0 = C$

Particular solution: $P = P_0e^{kt}$

69. $\dfrac{dy}{dx} = x$

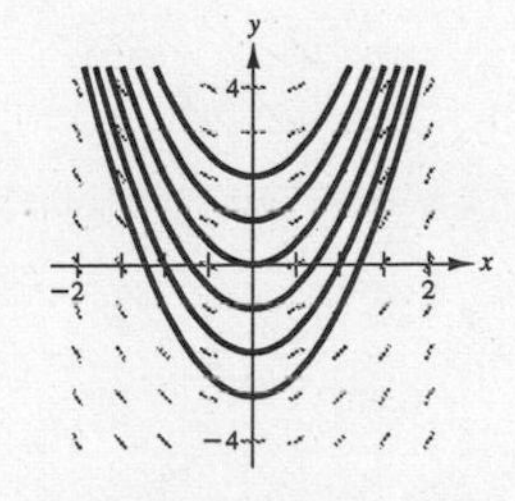

$$y = \int x\,dx = \frac{1}{2}x^2 + C$$

71. $\dfrac{dy}{dx} = \dfrac{-y}{x}$

$$\frac{dy}{y} = -\frac{dx}{x}$$

$$\ln|y| = -\ln|x| + C = \ln|x|^{-1} + \ln C = \ln|Cx^{-1}|$$

$$y = \frac{C}{x}$$

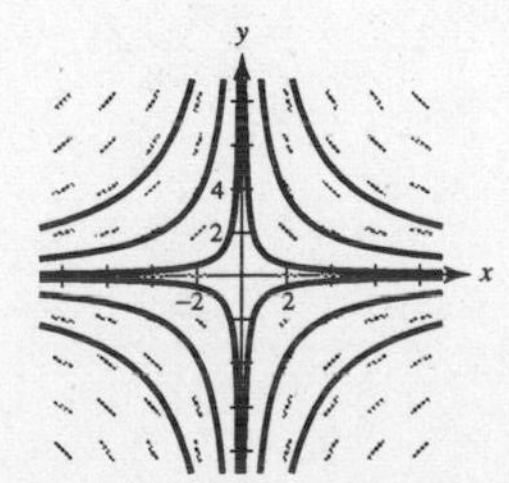

73. $\dfrac{dy}{dx} = 4 - y$

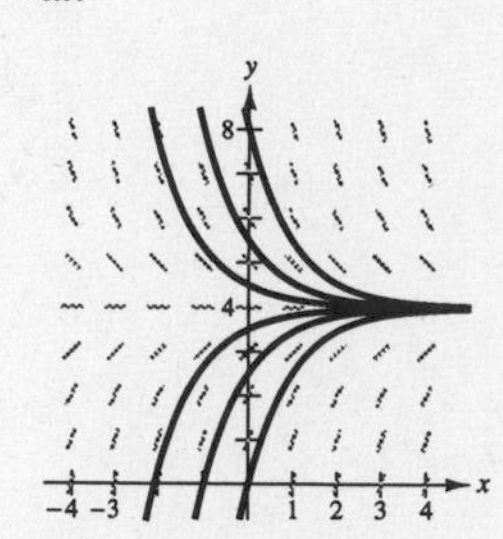

$$\int \frac{dy}{4-y} = \int dx$$

$$\ln|4 - y| = -x + C_1$$

$$4 - y = e^{-x+C_1}$$

$$y = 4 + Ce^{-x}$$

75. $\dfrac{dy}{dx} = 0.5y,\ y(0) = 6$

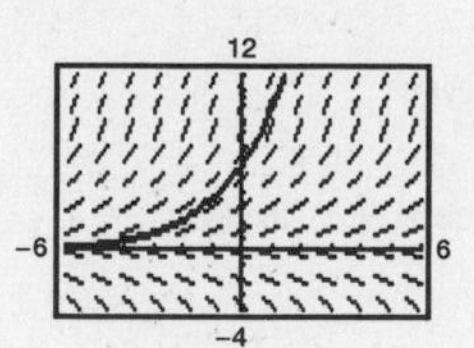

77. $\dfrac{dy}{dx} = 0.02y(10 - y),\ y(0) = 2$

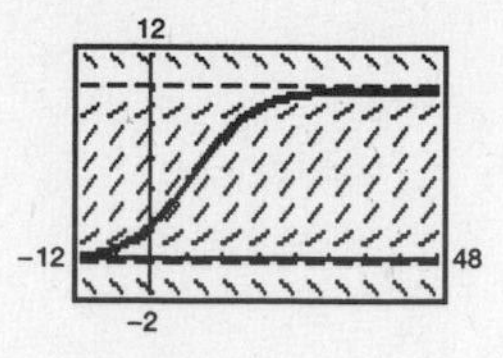

79. $\dfrac{dy}{dx} = \dfrac{-9x}{16y}$

$$\int 16y\,dy = -\int 9x\,dx$$

$$8y^2 = \frac{-9}{2}x^2 + C$$

Initial condition: $y(1) = 1,\ 8 = -\dfrac{9}{2} + C,\ C = \dfrac{25}{2}$

Particular solution: $8y^2 = \dfrac{-9}{2}x^2 + \dfrac{25}{2},$

$$16y^2 + 9x^2 = 25$$

81. $\dfrac{dy}{dx} = \dfrac{1 + y^2}{e^{0.5x}};\quad (0, 0)$

$$\int \frac{1}{1+y^2}\,dy = \int e^{-0.5x}\,dx$$

$$\arctan y = -2e^{-0.5x} + C$$

$$\arctan(0) = -2e^0 + C \Longrightarrow C = 2$$

$$y = \tan[2 - 2e^{-0.5x}]$$

83. $m = \dfrac{dy}{dx} = \dfrac{0 - y}{(x + 2) - x} = -\dfrac{y}{2}$

$$\int \frac{dy}{y} = \int -\frac{1}{2}\,dx$$

$$\ln|y| = -\frac{1}{2}x + C_1$$

$$y = Ce^{-x/2}$$

85. $f(x, y) = x^3 - 4xy^2 + y^3$

$$f(tx, ty) = t^3x^3 - 4t\,xt^2y^2 + t^3y^3$$

$$= t^3(x^3 - 4xy^2 + y^3)$$

Homogeneous of degree 3

87. $f(x, y) = \dfrac{x^2y^2}{\sqrt{x^2 + y^2}}$

$$f(tx, ty) = \frac{t^4x^2y^2}{\sqrt{t^2x^2 + t^2y^2}} = t^3\frac{x^2y^2}{\sqrt{x^2 + y^2}}$$

Homogeneous of degree 3

89. $f(x, y) = 2\ln xy$

$$f(tx, ty) = 2\ln[tx\,ty]$$

$$= 2\ln[t^2xy] = 2(\ln t^2 + \ln xy)$$

Not homogeneous

91. $f(x, y) = 2\ln\dfrac{x}{y}$

$$f(tx, ty) = 2\ln\frac{tx}{ty} = 2\ln\frac{x}{y}$$

Homogeneous degree 0

93.

$$y' = \frac{x+y}{2x}, y = vx$$

$$v + x\frac{dv}{dx} = \frac{x+vx}{2x}$$

$$x\frac{dv}{dx} = \frac{1+v}{2} - v = \frac{1-v}{2}$$

$$2\int\frac{dv}{1-v} = \int\frac{dx}{x}$$

$$-\ln(1-v)^2 = \ln|x| + \ln C = \ln|Cx|$$

$$\frac{1}{(1-v)^2} = |Cx|$$

$$\frac{1}{[1-(y/x)]^2} = |Cx|$$

$$\frac{x^2}{(x-y)^2} = |Cx|$$

$$|x| = C(x-y)^2$$

95.

$$y' = \frac{x-y}{x+y}, y = vx$$

$$v + x\frac{dv}{dx} = \frac{x-xv}{x+xv}$$

$$v\,dx + x\,dv = \frac{1-v}{1+v}dx$$

$$x\,dv = \left(\frac{1-v}{1+v} - v\right)dx = \frac{1-2v-v^2}{1+v}dx$$

$$\int\frac{v+1}{v^2+2v-1}dv = -\int\frac{dx}{x}$$

$$\frac{1}{2}\ln|v^2+2v-1| = -\ln|x| + \ln C_1 = \ln\left|\frac{C_1}{x}\right|$$

$$|v^2+2v-1| = \frac{C}{x^2}$$

$$\left|\frac{y^2}{x^2} + 2\frac{y}{x} - 1\right| = \frac{C}{x^2}$$

$$|y^2+2xy-x^2| = C$$

97.

$$y' = \frac{xy}{x^2-y^2}, y = vx$$

$$v + x\frac{dv}{dx} = \frac{x^2v}{x^2-x^2v^2}$$

$$v\,dx + x\,dv = \frac{v}{1-v^2}dx$$

$$x\,dv = \left(\frac{v}{1-v^2} - v\right)dx = \left(\frac{v^3}{1-v^2}\right)dx$$

$$\int\frac{1-v^2}{v^3}dv = \int\frac{dx}{x}$$

$$-\frac{1}{2v^2} - \ln|v| = \ln|x| + \ln C_1 = \ln|C_1x|$$

$$\frac{-1}{2v^2} = \ln|C_1xv|$$

$$\frac{-x^2}{2y^2} = \ln|C_1y|$$

$$y = Ce^{-x^2/2y^2}$$

99.

$$x\,dy - (2xe^{-y/x} + y)\,dx = 0, y = vx$$

$$x(v\,dx + x\,dv) - (2xe^{-v} + vx)\,dx = 0$$

$$\int e^v\,dv = \int\frac{2}{x}dx$$

$$e^v = \ln C_1x^2$$

$$e^{y/x} = \ln C_1 + \ln x^2$$

$$e^{y/x} = C + \ln x^2$$

Initial condition: $y(1) = 0, 1 = C$

Particular solution: $e^{y/x} = 1 + \ln x^2$

101.

$$\left(x\sec\frac{y}{x} + y\right)dx - x\,dy = 0, y = vx$$

$$(x\sec v + xv)dx - x(v\,dx + x\,dv) = 0$$

$$(\sec v + v)\,dx = v\,dx + x\,dv$$

$$\int\cos v\,dv = \int\frac{dx}{x}$$

$$\sin v = \ln x + \ln C_1$$

$$x = Ce^{\sin v}$$

$$= Ce^{\sin(y/x)}$$

Initial condition: $y(1) = 0, 1 = Ce^0 = C$

Particular solution: $x = e^{\sin(y/x)}$

103. $\dfrac{dy}{dx} = k(y-4)$

The direction field satisfies $(dy/dx) = 0$ along $y = 4$; but not along $y = 0$. Matches (a).

105. $\dfrac{dy}{dx} = ky(y-4)$

The direction field satisfies $(dy/dx) = 0$ along $y = 0$ and $y = 4$. Matches (c).

107. $\dfrac{dy}{dt} = ky,\ y = Ce^{kt}$

Initial conditions: $y(0) = y_0$

$$y(1620) = \frac{y_0}{2}$$

$$C = y_0$$

$$\frac{y_0}{2} = y_0 e^{1620k}$$

$$k = \frac{\ln(1/2)}{1620}$$

Particular solution: $y = y_0 e^{-t(\ln 2)/1620}$

When $t = 25$, $y \approx 0.989y_0$, $y = 98.9\%$ of y_0.

109.

$$\frac{dw}{dt} = k(1200 - w)$$

$$\int \frac{dw}{1200 - w} = \int k\,dt$$

$$\ln|1200 - w| = -kt + C_1$$

$$1200 - w = e^{-kt + C_1} = Ce^{-kt}$$

$$w = 1200 - Ce^{-kt}$$

$$w(0) = 60 = 1200 - C \Rightarrow C = 1200 - 60 = 1140$$

$$w = 1200 - 1140e^{-kt}$$

(a)

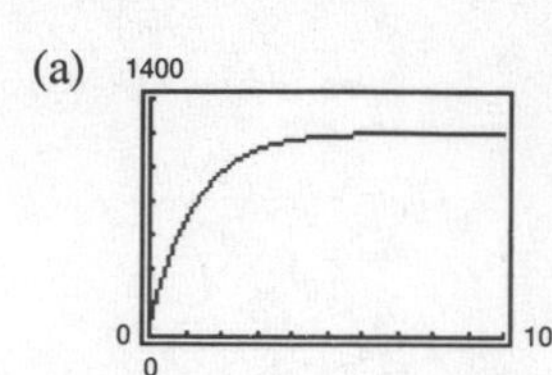

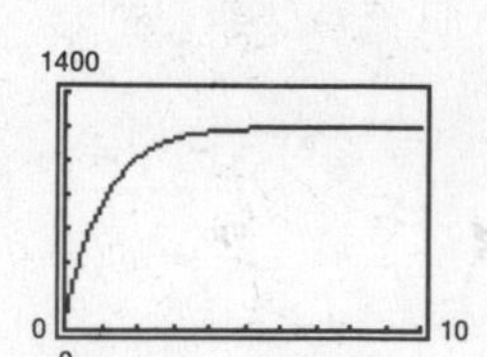

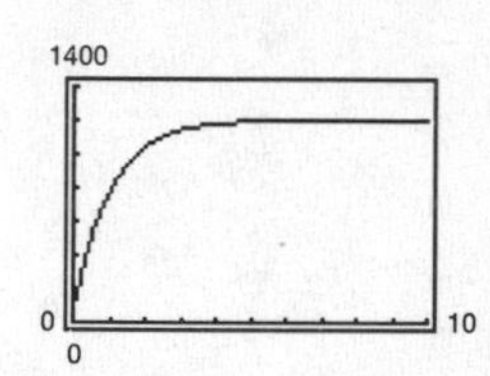

(b) $k = 0.8$: $t = 1.31$ years

$k = 0.9$: $t = 1.16$ years

$k = 1.0$: $t = 1.05$ years

(c) Maximum weight: 1200 pounds

$\lim_{t\to\infty} w = 1200$

111. (a) $\dfrac{dS}{dt} = k_1 S(L - S)$

$S = \dfrac{L}{1 + Ce^{-kt}}$ is a solution because

$$\frac{dS}{dt} = -L(1 + Ce^{-kt})^{-2}(-Cke^{-kt})$$

$$= \frac{LCke^{-kt}}{(1 + Ce^{-kt})^2}$$

$$= \left(\frac{k}{L}\right)\frac{L}{1 + Ce^{-kt}} \cdot \frac{CLe^{-kt}}{1 + Ce^{-kt}}$$

$$= \left(\frac{k}{L}\right)\frac{L}{1 + Ce^{-kt}} \cdot \left(L - \frac{L}{1 + Ce^{-kt}}\right)$$

$$= k_1 S(L - S), \text{ where } k_1 = \frac{k}{L}.$$

$L = 100$. Also, $S = 10$ when $t = 0 \Rightarrow C = 9$.
And, $S = 20$ when $t = 1 \Rightarrow k = -\ln(4/9)$.

Particular solution: $S = \dfrac{100}{1 + 9e^{\ln(4/9)t}} = \dfrac{100}{1 + 9e^{-0.8109t}}$

(b) $\dfrac{dS}{dt} = \ln\left(\dfrac{4}{9}\right)S(100 - S)$

$$\frac{d^2S}{dt^2} = \ln\left(\frac{4}{9}\right)\left[S\left(-\frac{dS}{dt}\right) + (100 - S)\frac{dS}{dt}\right]$$

$$= \ln\left(\frac{4}{9}\right)(100 - 2S)\frac{dS}{dt}$$

$$= 0 \text{ when } S = 50 \text{ or } \frac{dS}{dt} = 0.$$

Choosing $S = 50$, we have:

$$50 = \frac{100}{1 + 9e^{\ln(4/9)t}}$$

$$2 = 1 + 9e^{\ln(4/9)t}$$

$$\frac{\ln(1/9)}{\ln(4/9)} = t$$

$$t \approx 2.7 \text{ months}$$

—CONTINUED—

111. —CONTINIUED—

(c)

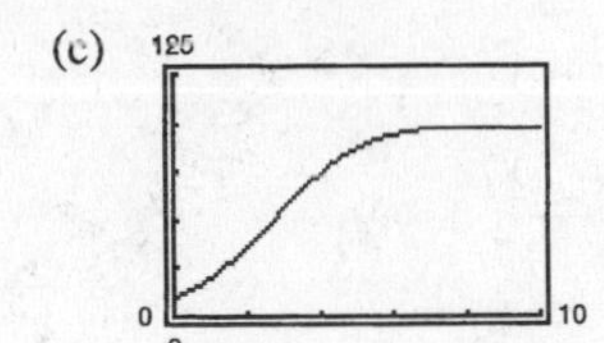

(d)

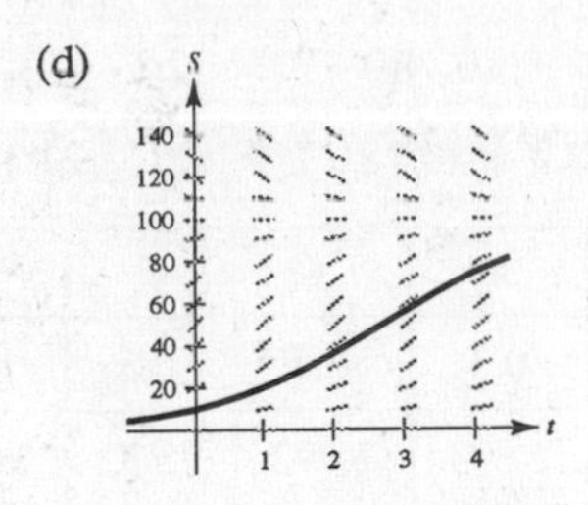

(e) Sales will decrease toward the line $S = L$.

113. (a) $\frac{dx}{dt} = k(m - x)(n - x), \quad m \neq n$

$$k\,dt = \frac{dx}{(x - m)(x - n)} = \left[\frac{1}{x - m} - \frac{1}{x - n}\right]\frac{1}{m - n}\,dx$$

$$\int (m - n)k\,dt = \int \left(\frac{1}{x - m} - \frac{1}{x - n}\right) dx$$

$$(m - n)k\,t + C_1 = \ln|x - m| - \ln|x - n| = \ln\left|\frac{x - m}{x - n}\right|$$

$$\frac{x - m}{x - n} = Ce^{(m-n)kt}$$

$$x - m = Cxe^{(m-n)kt} - Cne^{(m-n)kt}$$

$$xe^{nkt} - me^{nkt} = Cxe^{mkt} - Cne^{mkt}$$

$$x = \frac{me^{nkt} - Cne^{mkt}}{e^{nkt} - Ce^{mkt}}$$

(b) $\frac{dx}{dt} = k(m - x)^2, \quad m = n.$

$$\int (x - m)^{-2}\,dx = \int k\,dt$$

$$-(x - m)^{-1} = kt + C$$

$$\frac{1}{m - x} = kt + C$$

$$m - x = \frac{1}{kt + C}$$

$$x = m - \frac{1}{kt + C}$$

115. (a) $\frac{dv}{dt} = k(W - v)$

$$\int \frac{dv}{W - v} = \int k\,dt$$

$$-\ln(W - v) = kt + C_1$$

$$v = W - Ce^{-kt}$$

Initial conditions:

$W = 20$, $v = 0$ when $t = 0$, and

$v = 5$ when $t = 1$.

$C = 20$, $k = -\ln(3/4)$

Particular solution:

$$v = 20(1 - e^{\ln(3/4)t}) \approx 20(1 - e^{-0.2877t})$$

(b) $s = \int 20(1 - e^{-0.2877t})\,dt$

$$\approx 20[t + 3.4761e^{-0.2877t}] + C$$

Since $S(0) = 0$, $C \approx -69.5$ and we have $s \approx 20t + 69.5(e^{-0.2877t} - 1)$.

117. $y' = 0.5xy$, $y(0) = 3$

(a) $\frac{dy}{dx} = \frac{1}{2}xy$

$$\int \frac{2\,dy}{y} = \int x\,dx$$

$$2 \ln y = \frac{x^2}{2} + C_1$$

$$\ln y = \frac{x^2}{4} + C_2$$

$$y = e^{x^2/4 + C_2} = Ce^{x^2/4}$$

$$3 = Ce^0 \Longrightarrow C = 3 \Longrightarrow y = 3e^{x^2/4}$$

$$y(1) \approx 3.8521$$

(b) For $\Delta x = 0.25$, $y(1) \approx y_4 \approx 3.5953$.

For $\Delta x = 0.1$, $y(1) \approx y_{10} \approx 3.7439$.

The accuracy improves as Δx decreases.

119. $y' = x + y, \quad y(0) = 2, \quad n = 10, \quad h = 0.1$

$y_1 = y_0 + hF(x_0, y_0) = 2 + (0.1)(0 + 2) = 2.2$

$y_2 = y_1 + hF(x_1, y_1) = 2.2 + (0.1)(0.1 + 2.2) = 2.43$, etc.

n	0	1	2	3	4	5	6	7	8	9	10
x_n	0	0.1	0.2	0.3	0.4	0.5	0.6	0.7	0.8	0.9	1.0
y_n	2	2.2	2.43	2.693	2.992	3.332	3.715	4.146	4.631	5.174	5.781

121. $y' = 3x - 2y, \quad y(0) = 3, \quad n = 10, \quad h = 0.05$

$y_1 = y_0 + hF(x_0, y_0) = 3 + (0.05)(3(0) - 2(3)) = 2.7$

$y_2 = y_1 + hF(x_1, y_1) = 2.7 + (0.05)(3(0.05) - 2(2.7)) = 2.4375$, etc.

n	0	1	2	3	4	5	6	7	8	9	10
x_n	0	0.05	0.1	0.15	0.2	0.25	0.3	0.35	0.4	0.45	0.5
y_n	3	2.7	2.438	2.209	2.010	1.839	1.693	1.569	1.464	1.378	1.308

123. $y' = e^{xy}, \quad y(0) = 1, \quad n = 10, \quad h = 0.1$

$y_1 = y_0 + hF(x_0, y_0) = 1 + (0.1)e^{0(1)} = 1.1$

$y_2 = y_1 + hF(x_1, y_1) = 1.1 + (0.1)e^{(0.1)(1.1)} \approx 1.2116$, etc.

n	0	1	2	3	4	5	6	7	8	9	10
x_n	0	0.1	0.2	0.3	0.4	0.5	0.6	0.7	0.8	0.9	1.0
y_n	1	1.1	1.212	1.339	1.488	1.670	1.900	2.213	2.684	3.540	5.958

125. Given family (circles): $x^2 + y^2 = C$

$$2x + 2yy' = 0$$

$$y' = -\frac{x}{y}$$

Orthogonal trajectory (lines): $y' = \frac{y}{x}$

$$\int \frac{dy}{y} = \int \frac{dx}{x}$$

$$\ln|y| = \ln|x| + \ln K$$

$$y = Kx$$

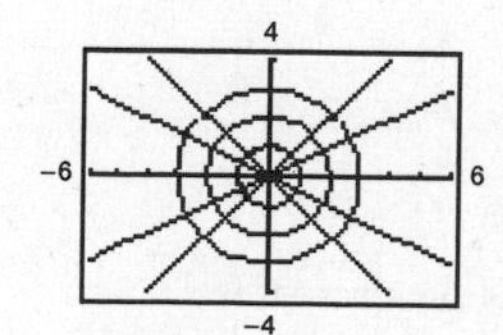

127. Given family (parabolas): $x^2 = Cy$

$$2x = Cy'$$

$$y' = \frac{2x}{C} = \frac{2x}{x^2/y} = \frac{2y}{x}$$

Orthogonal trajectory (ellipses): $y' = -\frac{x}{2y}$

$$2\int y\,dy = -\int x\,dx$$

$$y^2 = -\frac{x^2}{2} + K_1$$

$$x^2 + 2y^2 = K$$

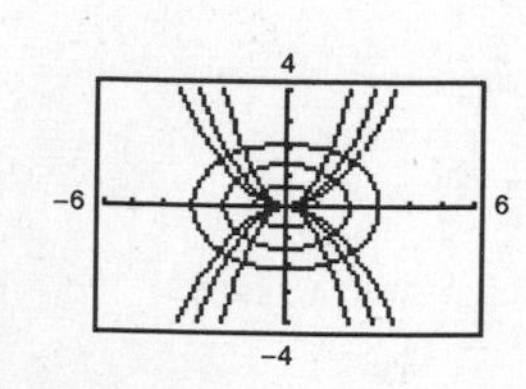

129. Given family: $y^2 = Cx^3$

$$2yy' = 3Cx^2$$

$$y' = \frac{3Cx^2}{2y} = \frac{3x^2}{2y}\left(\frac{y^2}{x^3}\right) = \frac{3y}{2x}$$

Orthogonal trajectory (ellipses): $y' = -\dfrac{2x}{3y}$

$$3\int y\,dy = -2\int x\,dx$$

$$\frac{3y^2}{2} = -x^2 + K_1$$

$$3y^2 + 2x^2 = K$$

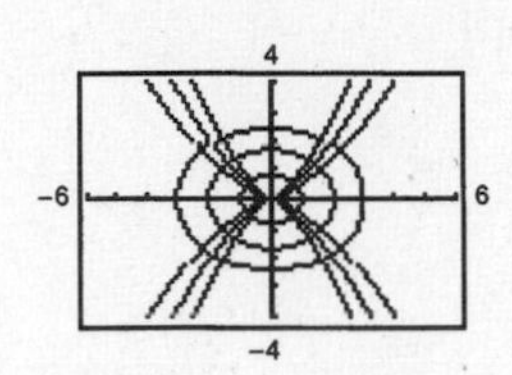

131. A general solution of order n has n arbitrary constants while in a particular solution initial conditions are given in order to solve for all these constants.

133. $M(x, y)dx + N(x, y)dy = 0$, where M and N are homogeneous functions of the same degree.

135. False. Consider Example 2. $y = x^3$ is a solution to $xy' - 3y = 0$, but $y = x^3 + 1$ is not a solution.

137. False

$$f(tx, ty) = t^2x^2 + t^2xy + 2$$

$$\neq t^2 f(x, y)$$

Section 5.3 First-Order Linear Differential Equations

1. (a),(c)

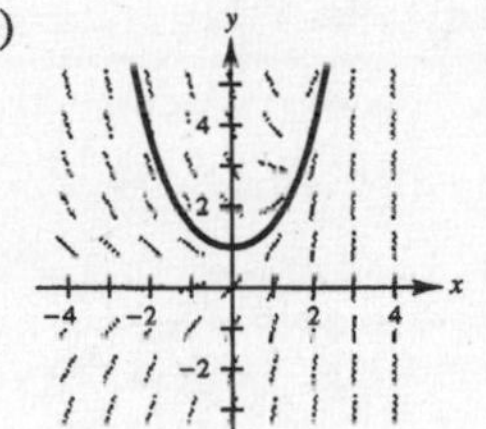

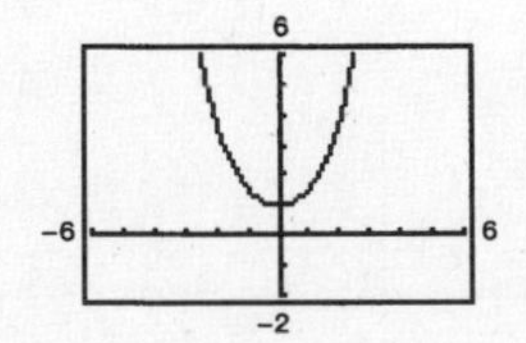

(b) $\dfrac{dy}{dx} = e^x - y$

$\dfrac{dy}{dx} + y = e^x$ Integrating factor: $e^{\int dx} = e^x$

$$e^x y' + e^x y = e^{2x}$$

$$(ye^x) = \int e^{2x}\,dx$$

$$ye^x = \frac{1}{2}e^{2x} + C$$

$$y(0) = 1 \Rightarrow 1 = \frac{1}{2} + C \Rightarrow C = \frac{1}{2}$$

$$ye^x = \frac{1}{2}e^{2x} + \frac{1}{2}$$

$$y = \frac{1}{2}e^x + \frac{1}{2}e^{-x} = \frac{1}{2}(e^x + e^{-x})$$

3. $\dfrac{dy}{dx} + \left(\dfrac{1}{x}\right)y = 3x + 4$

Integrating factor: $e^{\int (1/x)\,dx} = e^{\ln x} = x$

$$xy = \int x(3x + 4)\,dx = x^3 + 2x^2 + C$$

$$y = x^2 + 2x + \frac{C}{x}$$

5. $y' - y = 10$

Integrating factor: $e^{\int -1\,dx} = e^{-x}$

$$e^{-x}y' - e^{-x}y = 10e^{-x}$$

$$ye^{-x} = \int 10e^{-x}\,dx = -10e^{-x} + C$$

$$y = -10 + Ce^{x}$$

7. $(y + 1)\cos x\,dx = dy$

$y' = (y + 1)\cos x = y\cos x + \cos x$

$y' - (\cos x)y = \cos x$

Integrating factor: $e^{\int -\cos x\,dx} = e^{-\sin x}$

$$y'e^{-\sin x} - (\cos x)e^{-\sin x}y = (\cos x)e^{-\sin x}$$

$$ye^{-\sin x} = \int (\cos x)e^{-\sin x}\,dx$$

$$= -e^{-\sin x} + C$$

$$y = -1 + Ce^{\sin x}$$

9. $(x - 1)y' + y = x^2 - 1$

$y' + \left(\dfrac{1}{x - 1}\right)y = x + 1$

Integrating factor: $e^{\int [1/(x-1)]\,dx} = e^{\ln|x-1|} = x - 1$

$$y(x - 1) = \int (x^2 - 1)\,dx = \frac{1}{3}x^3 - x + C_1$$

$$y = \frac{x^3 - 3x + C}{3(x - 1)}$$

11. $y' - 3x^2y = e^{x^3}$

Integrating factor: $e^{-\int 3x^2\,dx} = e^{-x^3}$

$$ye^{-x^3} = \int e^{x^3}e^{-x^3}\,dx = \int dx = x + C$$

$$y = (x + C)e^{x^3}$$

13. $y'\cos^2 x + y - 1 = 0$

$y' + (\sec^2 x)y = \sec^2 x$

Integrating factor: $e^{\int \sec^2 x\,dx} = e^{\tan x}$

$$ye^{\tan x} = \int \sec^2 x e^{\tan x}\,dx = e^{\tan x} + C$$

$$y = 1 + Ce^{-\tan x}$$

Boundary condition: $y(0) = 5, C = 4$

Particular solution: $y = 1 + 4e^{-\tan x}$

15. $y' + y\tan x = \sec x + \cos x$

Integrating factor: $e^{\int \tan x\,dx} = e^{\ln|\sec x|} = \sec x$

$$y\sec x = \int \sec x(\sec x + \cos x)\,dx = \tan x + x + C$$

$$y = \sin x + x\cos x + C\cos x$$

Boundary condition: $y(0) = 1, 1 = C$

Particular solution: $y = \sin x + (x + 1)\cos x$

17. $y' + \left(\dfrac{1}{x}\right)y = 0$

Integrating factor: $e^{\int (1/x)\,dx} = e^{\ln|x|} = x$

Separation of variables:

$$\frac{dy}{dx} = -\frac{y}{x}$$

$$\int \frac{1}{y}\,dy = \int -\frac{1}{x}\,dx$$

$$\ln y = -\ln x + \ln C$$

$$\ln xy = \ln C$$

$$xy = C$$

Boundary condition: $y(2) = 2, C = 4$

Particular solution: $xy = 4$

19. $x\,dy = (x + y + 2)\,dx, \quad y(1) = 10$

$$\frac{dy}{dx} = \frac{x + y + 2}{x} = 1 + \frac{y}{x} + \frac{2}{x}$$

$$y' - \frac{1}{x}y = 1 + \frac{2}{x}$$

Integrating factor: $e^{\int -1/x\,dx} = e^{-\ln x} = \dfrac{1}{x}$

$$y = x\int\left(1 + \frac{2}{x}\right)\frac{1}{x}\,dx$$

$$= x\int\left(\frac{1}{x} + \frac{2}{x^2}\right)dx$$

$$= x\left[\ln|x| - \frac{2}{x} + C\right.$$

$y(1) = 10{:}\quad 10 = 1[-2] + C \Rightarrow C = 12$

$y = x\ln x - 2 + 12x$

21. $y' + 3x^2y = x^2y^3$

$n - 3, Q - x^2, P = 3x^2$

$$y^{-2}e^{\int(-2)3x^2\,dx} = \int(-2)x^2e^{\int(-2)3x^2\,dx}\,dx$$

$$y^{-2}e^{-2x^3} = -\int 2x^2e^{-2x^3}\,dx$$

$$y^{-2}e^{-2x^3} = \frac{1}{3}e^{-2x^3} + C$$

$$y^{-2} = \frac{1}{3} + Ce^{2x^3}$$

$$\frac{1}{y^2} = Ce^{2x^3} + \frac{1}{3}$$

23. $y' + \left(\dfrac{1}{x}\right)y = xy^2$

$n = 2, Q = x, P = x^{-1}$

$e^{\int-(1/x)\,dx} = e^{-\ln|x|} = x^{-1}$

$$y^{-1}x^{-1} = \int -x(x^{-1})\,dx = -x + C$$

$$\frac{1}{y} = -x^2 + Cx$$

$$y = \frac{1}{Cx - x^2}$$

25. $y' - y = e^x\sqrt[3]{y},\ n = \dfrac{1}{3},\ Q = e^x,\ P = -1$

$e^{\int-(2/3)\,dx} = e^{-(2/3)x}$

$$y^{2/3}e^{-(2/3)x} = \int\frac{2}{3}e^xe^{-(2/3)x}\,dx = \int\frac{2}{3}e^{(1/3)x}\,dx$$

$$y^{2/3}e^{-(2/3)x} = 2e^{(1/3)x} + C$$

$$y^{2/3} = 2e^x + Ce^{2x/3}$$

27. (a)

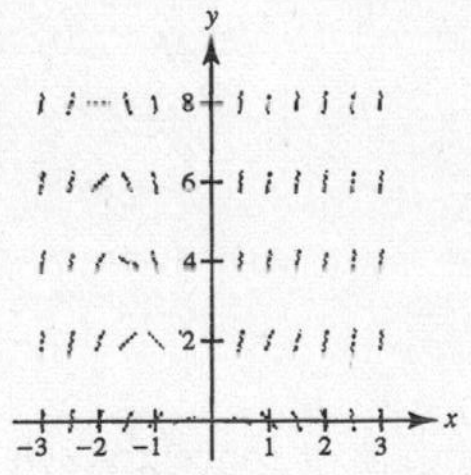

(b) $\dfrac{dy}{dx} - \dfrac{1}{x}y = x^2$

Integrating factor $e^{\int -1/x\,dx} = e^{-\ln x} = \dfrac{1}{x}$

$$\frac{1}{x}y' - \frac{1}{x^2}y = x$$

$$\left(\frac{1}{x}y\right) = \int x\,dx = \frac{x^2}{2} + C$$

$$y = \frac{x^3}{2} + Cx$$

$(-2, 4){:}\ 4 = \dfrac{-8}{2} - 2C \Rightarrow C = -4 \Rightarrow y = \dfrac{x^3}{2} - 4x = \dfrac{1}{2}x(x^2 - 8)$

$(2, 8){:}\ 8 = \dfrac{8}{2} + 2C \Rightarrow C = 2 \Rightarrow y = \dfrac{x^3}{2} + 2x = \dfrac{1}{2}x(x^2 + 4)$

(c)

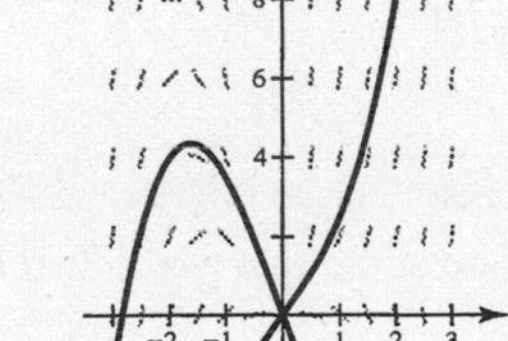

29. (a)

(b) $y' + (\cot x)y = 2$

Integrating factor: $e^{\int \cot x\, dx} = e^{\ln|\sin x|} = \sin x$

$y' \sin x + (\cos x)y = 2 \sin x$

$$y \sin x = \int 2 \sin x\, dx = -2 \cos x + C$$

$$y = -2 \cot x + C \csc x$$

$(1, 1)$: $1 = -2 \cot 1 + C \csc 1 \Rightarrow C = \dfrac{1 + 2 \cot 1}{\csc 1} = \sin 1 + 2 \cos 1$

$$y = -2 \cot x + (\sin 1 + 2 \cos 1) \csc x$$

$(3, -1)$: $-1 = -2 \cot 3 + C \csc 3 \Rightarrow C = \dfrac{2 \cot 3 - 1}{\csc 3} = 2 \cos 3 - \sin 3$

$$y = -2 \cot x + (2 \cos 3 - \sin 3) \csc x$$

(c)

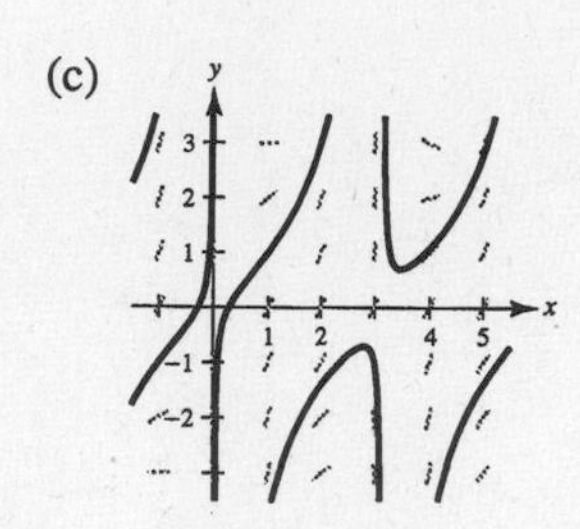

31.
$$\frac{dP}{dt} = kP + N,\ N \text{ constant}$$

$$\frac{dP}{kP + N} = dt$$

$$\int \frac{1}{kP + N}\, dP = \int dt$$

$$\frac{1}{k}\ln(kP + N) = t + C_1$$

$$\ln(kP + N) = kt + C_2$$

$$kP + N = e^{kt + C_2}$$

$$P = \frac{C_3 e^{kt} - N}{k}$$

$$P = Ce^{kt} - \frac{N}{k}$$

When $t = 0$: $P = P_0$

$$P_0 = C - \frac{N}{k} \Rightarrow C = P_0 + \frac{N}{k}$$

$$P = \left(P_0 + \frac{N}{k}\right)e^{kt} - \frac{N}{k}$$

33. (a) $A = \dfrac{P}{r}(e^{rt} - 1)$

$$A = \frac{100{,}000}{0.06}(e^{0.06(5)} - 1) \approx 583{,}098.01$$

(b) $A = \dfrac{250{,}000}{0.05}(e^{0.05(10)} - 1) \approx 3{,}243{,}606.35$

35. (a) $\dfrac{dQ}{dt} = q - kQ$, q constant

(b) $Q' + kQ = q$

Let $P(t) = k$, $Q(t) = q$, then the integrating factor is $u(t) = e^{kt}$.

$$Q = e^{-kt}\int qe^{kt}\, dt = e^{-kt}\left(\frac{q}{k}e^{kt} + C\right) = \frac{q}{k} + Ce^{-kt}$$

When $t = 0$: $Q = Q_0$

$$Q_0 = \frac{q}{k} + C \Rightarrow C = Q_0 - \frac{q}{k}$$

$$Q = \frac{q}{k} + \left(Q_0 - \frac{q}{k}\right)e^{-kt}$$

(c) $\displaystyle\lim_{t\to\infty} Q = \frac{q}{k}$

37. Let Q be the number of pounds of concentrate in the solution at any time t. Since the number of gallons of solution in the tank at any time t is $v_0 + (r_1 - r_2)t$ and since the tank loses r_2 gallons of solution per minute, it must lose concentrate at the rate

$$\left[\frac{Q}{v_0 + (r_1 - r_2)t}\right]r_2.$$

The solution gains concentrate at the rate r_1q_1. Therefore, the net rate of change is

$$\frac{dQ}{dt} = q_1r_1 - \left[\frac{Q}{v_0 + (r_1 - r_2)t}\right]r_2 \quad \text{or}$$

$$\frac{dQ}{dt} + \frac{r_2Q}{v_0 + (r_1 - r_2)t} = q_1r_1.$$

39. (a) $Q' + \dfrac{r^2 Q}{v_0 + (r_1 - r_2)t} = q_1 r_1$

$Q(0) = q_0,\ q_0 = 25,\ q_1 = 0,\ v_0 = 200,$

$r_1 = 10,\ r_2 = 10,\ Q' + \dfrac{1}{20}Q = 0$

$$\int \frac{1}{Q}\,dQ = \int -\frac{1}{20}\,dt$$

$$\ln Q = -\frac{1}{20}t + \ln C_1$$

$$Q = Ce^{-(1/20)t}$$

Initial condition: $Q(0) = 25,\ C = 25$

Particular solution: $Q = 25e^{-(1/20)t}$

(b) $15 = 25e^{-(1/20)t}$

$$\ln\left(\frac{3}{5}\right) = -\frac{1}{20}t$$

$$t = -20\ln\left(\frac{3}{5}\right) \approx 10.2 \text{ min}$$

(c) $\lim_{t\to\infty} 25e^{-(1/20)t} = 0$

41. (a) The volume of the solution in the tank is given by $v_0 + (r_1 - r_2)t$. Therefore, $100 + (5 - 3)t = 200$ or $t = 50$ minutes.

(b) $Q' + \dfrac{r_2 Q}{v_0 + (r_1 - r_2)t} = q_1 r_1$

$Q(0) = q_0,\ q_0 = 0,\ q_1 = 0.5,\ v_0 = 100,\ r_1 = 5,\ r_2 = 3,\ Q' + \dfrac{3}{100 + 2t}Q = 2.5$

Integrating factor: $e^{\int [3/(100+2t)]\,dt} = (50 + t)^{3/2}$

$$Q(50 + t)^{3/2} = \int 2.5(50 + t)^{3/2}\,dt = (50 + t)^{5/2} + C$$

$$Q = (50 + t) + C(50 + t)^{-3/2}$$

Initial condition: $Q(0) = 0,\ 0 = 50 + C(50^{-3/2}),\ C = -50^{5/2}$

Particular solution: $Q = (50 + t) - 50^{-5/2}(50 + t)^{-3/2}$

$$Q(50) = 100 - 50^{5/2}(100)^{-3/2} = 100 - \frac{25}{\sqrt{2}} \approx 82.32 \text{ lbs}$$

43. $mg = 8 \Rightarrow m = \dfrac{8}{g}$

$$\frac{dv}{dt} + \frac{kv}{m} = g$$

By Example 6, $v = \dfrac{mg}{k}(1 - e^{-kt/m})$

$$= \frac{8}{k}(1 - e^{-ktg/8})$$

$v(5) = -101 = \dfrac{8}{k}(1 - e^{k(5)(32)/8}) = \dfrac{8}{k}(1 - e^{20k})$

Solving for k using a graphing utility, $k \approx -0.0502$

Thus $v(t) = -159.36(1 - e^{-0.2008t})$

Limiting velocity is -159.36 ft/sec.

45. $L\dfrac{dI}{dt} + RI = E_0,\ I' + \dfrac{R}{L}I = \dfrac{E_0}{L}$

Integrating factor: $e^{\int (R/L)\,dt} = e^{Rt/L}$

$$I e^{Rt/L} = \int \frac{E_0}{L}e^{Rt/L}\,dt = \frac{E_0}{R}e^{Rt/L} + C$$

$$I = \frac{E_0}{R} + Ce^{-Rt/L}$$

47. $y' + P(x)y = Q(x)$

Integrating factor: $e^{\int P(x)dx}$

49. $y' - 2x = 0$

$$\int dy = \int 2x\,dx$$

$$y = x^2 + C$$

Matches c.

51. $y' - 2xy = 0$

$$\int \frac{dy}{y} = \int 2x\,dx$$

$$\ln y = x^2 + C_1$$

$$y = Ce^{x^2}$$

Matches a.

53. $e^{2x+y}\,dx - e^{x-y}\,dy = 0$

Separation of variables:

$$e^{2x}e^{y}\,dx = e^{x}e^{-y}\,dy$$

$$\int e^{x}\,dx = \int e^{-2y}\,dy$$

$$e^{x} = -\frac{1}{2}e^{-2y} + C_1$$

$$2e^{x} + e^{-2y} = C$$

55. $(y\cos x - \cos x)\,dx + dy = 0$

Separation of variables:

$$\int \cos x\,dx = \int \frac{-1}{y-1}\,dy$$

$$\sin x = -\ln(y-1) + \ln C$$

$$\ln(y-1) = -\sin x + \ln C$$

$$y = Ce^{-\sin x} + 1$$

57. $(3y^2 + 4xy)\,dx + (2xy + x^2)\,dy = 0$

Homogeneous: $y = vx$, $dy = v\,dx + x\,dv$

$$(3v^2x^2 + 4vx^2)\,dx + (2vx^2 + x^2)(v\,dx + x\,dv) = 0$$

$$\int \frac{5}{x}\,dx + \int \left(\frac{2v+1}{v^2+v}\right)dv = 0$$

$$\ln x^5 + \ln|v^2 + v| = \ln C$$

$$x^5(v^2 + v) = C$$

$$x^3y^2 + x^4y = C$$

59. $(2y - e^x)\,dx + x\,dy = 0$

Linear: $y' + \left(\frac{2}{x}\right)y = \frac{1}{x}e^x$

Integrating factor: $e^{\int (2/x)\,dx} = e^{\ln x^2} = x^2$

$$yx^2 = \int x^2\frac{1}{x}e^x\,dx = e^x(x-1) + C$$

$$y = \frac{e^x}{x^2}(x-1) + \frac{C}{x^2}$$

61. $(x^2y^4 - 1)\,dx + x^3y^3\,dy = 0$

$$y' + \left(\frac{1}{x}\right)y = x^{-3}y^{-3}$$

Bernoulli: $n = -3$, $Q = x^{-3}$, $P = x^{-1}$,

$$e^{\int (4/x)\,dx} = e^{\ln x^4} = x^4$$

$$y^4x^4 = \int 4(x^{-3})(x^4)\,dx = 2x^2 + C$$

$$x^4y^4 - 2x^2 = C$$

63. $3(y - 4x^2)\,dx = -x\,dy$

$$x\frac{dy}{dx} = -3y + 12x^2$$

$$y' + \frac{3}{x}y = 12x$$

Integrating factor: $e^{\int (3/x)\,dx} = e^{3\ln x} = x^3$

$$y'x^3 + \frac{3}{x}x^3y = 12x(x^3) = 12x^4$$

$$yx^3 = \int 12x^4\,dx = \frac{12}{5}x^5 + C$$

$$y = \frac{12}{5}x^2 + \frac{C}{x^3}$$

65. False

$y' + xy = x^2$ is first-order linear.

Review Exercises for Chapter 5

1. $y = Ce^{-2x}$

$y' = -2Ce^{-2x} = -2y$

3. $y = C(\sec x + \tan x)$

$y' = C(\sec x \cdot \tan x + \sec^2 x)$

$= C(\tan x + \sec x)\sec x$

$= y \cdot \sec x$

5. $P(h) = 30e^{kh}$

$P(18{,}000) = 30e^{18{,}000k} = 15$

$k = \dfrac{\ln(1/2)}{18{,}000} = \dfrac{-\ln 2}{18{,}000}$

$P(h) = 30e^{-(h \ln 2)/18{,}000}$

$P(35{,}000) = 30e^{-(35{,}000 \ln 2)/18{,}000} \approx 7.79$ inches

7. $P = Ce^{0.015t}$

$2C = Ce^{0.015t}$

$2 = e^{0.015t}$

$\ln 2 = 0.015t$

$t = \dfrac{\ln 2}{0.015} \approx 46.21$ years

9. (a)

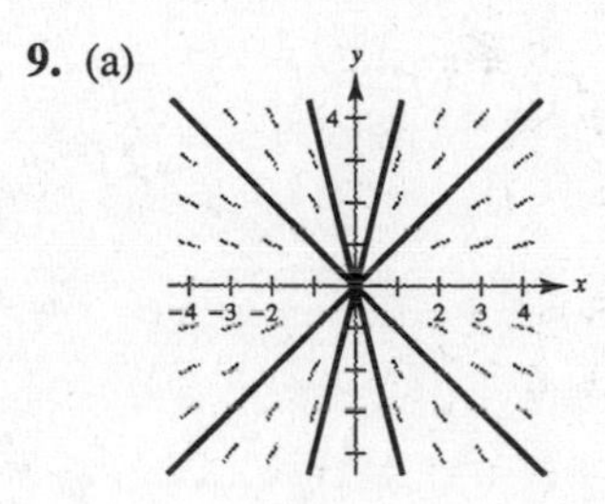

(b) $\dfrac{dy}{dx} = \dfrac{y}{x}$

$\displaystyle\int \frac{dy}{y} = \int \frac{dx}{x}$

$\ln y = \ln x + \ln C$

$y = Cx$

11. $\dfrac{dy}{dx} = \dfrac{x^2 + 3}{x}$

$\displaystyle\int dy = \int \left(x + \frac{3}{x}\right) dx$

$y = \dfrac{x^2}{2} + 3\ln|x| + C$

13. $y' - 2xy = 0$

$\dfrac{dy}{dx} = 2xy$

$\displaystyle\int \frac{1}{y}\,dy = \int 2x\,dx$

$\ln|y| = x^2 + C_1$

$e^{x^2 + C_1} = y$

$y = Ce^{x^2}$

15. $\dfrac{dy}{dx} = 4x(1 + y^2)$

$\displaystyle\int \frac{1}{1 + y^2}\,dy = \int 4x\,dx$

$\arctan y = 2x^2 + C$

$y = \tan(2x^2 + C)$

17. $\dfrac{dy}{dx} = \dfrac{e^x - e^{-x}}{e^x + e^{-x}}$

$y = \displaystyle\int \frac{e^x - e^{-x}}{e^x + e^{-x}}\,dx = \ln(e^x + e^{-x}) + C$

19. $y' - \dfrac{2y}{x} = \dfrac{1}{x}y'$

$(x - 1)\dfrac{dy}{dx} = 2y$

$\displaystyle\int \frac{1}{y}\,dy = \int \frac{2}{x - 1}\,dx$

$\ln|y| = \ln(x - 1)^2 + \ln C$

$y = C(x - 1)^2$

21. $3x^2y^2\,dx + (2x^3y + x^3y^4)\,dy = 0$

$$3x^2y^2\,dx + x^3(2y + y^4)\,dy = 0$$

$$\int \frac{3}{x}\,dx + \int \left(\frac{2}{y} + y^2\right)dy = 0$$

$$\ln x^3 + \ln y^2 + \frac{1}{3}y^3 = C_1$$

$$3\ln x^3y^2 + y^3 = C$$

23. $\dfrac{dy}{dx} = \dfrac{x^2 + y^2}{2xy}$ (homogeneous differential equation)

$$(x^2 + y^2)\,dx - 2xy\,dy = 0$$

Let $y = vx$, $dy = x\,dv + v\,dx$.

$$(x^2 + v^2x^2)\,dx - 2x(vx)(x\,dv + v\,dx) = 0$$

$$(x^2 + v^2x^2 - 2x^2v^2)\,dx - 2x^3v\,dv = 0$$

$$(x^2 - x^2v^2)\,dx = 2x^3v\,dv$$

$$(1 - v^2)\,dx = 2xv\,dv$$

$$\int \frac{dx}{x} = \int \frac{2v}{1 - v^2}\,dv$$

$$\ln|x| = -\ln|1 - v^2| + C_1 = -\ln|1 - v^2| + \ln C$$

$$x = \frac{C}{1 - v^2} = \frac{C}{1 - (y/x)^2} = \frac{Cx^2}{x^2 - y^2}$$

$$1 = \frac{Cx}{x^2 - y^2} \quad \text{or} \quad C_1 = \frac{x}{x^2 - y^2}$$

25. $xy\,dy = 2(2x^2 + y^2)\,dx$

Homogeneous: Let $y = vx$, $dy = x\,dv + v\,dx$.

$$x(vx)(x\,dv + v\,dx) - 2(2x^2 + v^2x^2)\,dx = 0$$

$$[v^2x^2 - 4x^2 - 2v^2x^2]dx + x^3v\,dv = 0$$

$$(-4 - v^2)dx + x\,v\,dv = 0$$

$$xv\,dv = (4 + v^2)dx$$

$$\int \frac{v}{4 + v^2}\,dv = \int \frac{dx}{x}$$

$$\frac{1}{2}\ln|4 + v^2| = \ln|x| + C_1$$

$$\ln|4 + v^2| = 2\ln|x| + \ln C_2$$

$$4 + v^2 = Cx^2$$

$$4 + \left(\frac{y}{x}\right)^2 = Cx^2$$

$$4x^2 + y^2 = Cx^4$$

$$\frac{x^4}{4x^2 + y^2} = C_2$$

27. $y = \dfrac{2}{x^2 + C}$; $\left(1, \dfrac{1}{2}\right)$

$$\frac{1}{2} = \frac{2}{1 + C} \Rightarrow C = 3 \Rightarrow y = \frac{2}{x^2 + 3}$$

29.

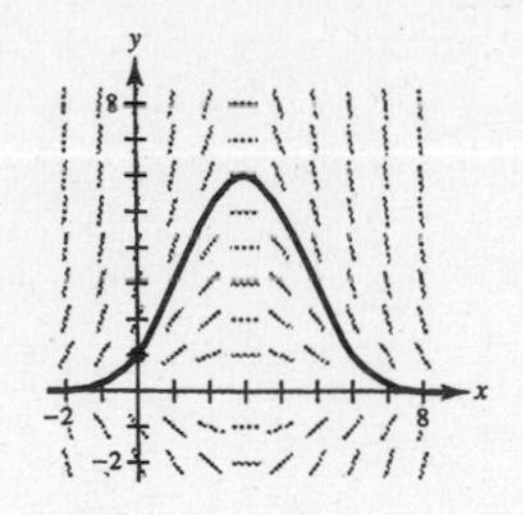

31. $\ln(1 + y)\,dx + \left(\dfrac{1}{1 + y}\right) dy = 0$

$$\int dx + \int \frac{1}{(1 + y)\ln(1 + y)}\,dy = C_1$$

$$x + \ln|\ln(1 + y)| = C_1$$

$$\ln|\ln(1 + y)| = C_1 - x$$

$$\ln|1 + y| = e^{C_1 - x} = Ce^{-x}$$

Initial condition: $y(0) = 2$

$$\ln 3 = C$$

Particular solution: $\ln|1 + y| = (\ln 3)e^{-x}$

33. $y' = x^2 - y^2,\ y(0) = 1,\ \Delta x = 0.05,$

$y(0.8) \approx y_{16} \approx 0.6724$

35. $y' = x + 0.4y^2,\ y(0) = 1,\ \Delta x = 0.1,$

$y(1.5) \approx y_{15} \approx 4.7400$

37. $y' = (x + 1)e^{-y^2/100},\ y(0) = 4,\ \Delta x = 0.2$

$y(2) \approx y_{10} \approx 6.8492$

39.

$$(x - C)^2 + y^2 = C^2$$

$$x^2 - 2Cx + C^2 + y^2 = C^2$$

$$\frac{x^2 + y^2}{x} = 2C$$

$$\frac{x(2x + 2yy') - (x^2 + y^2)}{x^2} = 0$$

$$2x^2 + 2xyy' - x^2 - y^2 = 0$$

$$y' = \frac{y^2 - x^2}{2xy}$$

The negative reciprocal of y' is the slope of the orthogonal trajectories.

$$\frac{dy}{dx} = \frac{2xy}{x^2 - y^2}$$

$2xy\,dx + (y^2 - x^2)dy = 0$

Homogeneous

$x = vy,\ dx = v\,dy + y\,dv$

$$2vy^2(v\,dy + y\,dv) + (y^2 - v^2y^2)dy = 0$$

$$\int \frac{2v}{1 + v^2}\,dv + \int \frac{1}{y}\,dy = 0$$

$$\ln(1 + v^2) + \ln|y| = \ln K_1$$

$$y^2 + x^2 = K_1 y$$

Circles: $x^2 + (y - K)^2 = K^2$

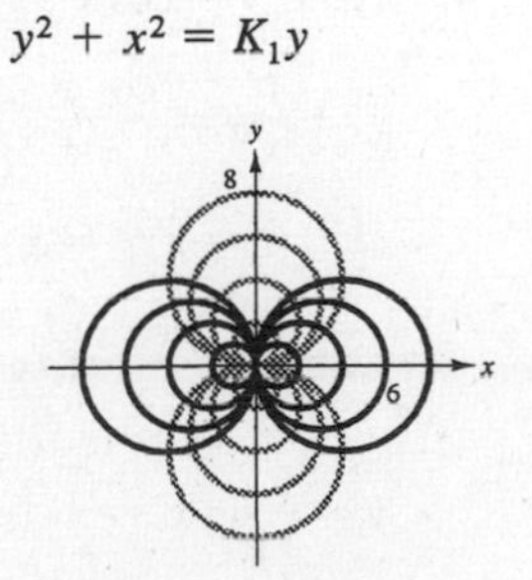

41. $2x^3 + 3y^2 - 6y = C$

$6x^2 + 6yy' - 6y' = 0$

$y'(y - 1) = -x^2$

$(1 - y)y' = x^2$

$x = 2, y = 0$: $16 = C$

Particular Solution: $2x^3 + 3y^2 - 6y = 16$

43. $\dfrac{dv}{dt} = kv - 9.8$

(a) $$\int \frac{dv}{kv - 9.8} = \int dt$$

$$\frac{1}{k}\ln|kv - 9.8| = t + C_1$$

$$\ln|kv - 9.8| = kt + C_2$$

$$kv - 9.8 = e^{kt + C_2} = C_3 e^{kt}$$

$$v = \frac{1}{k}\left[9.8 + C_3 e^{kt}\right]$$

At $t = 0$, $v_0 = \dfrac{1}{k}(9.8 + C_3) \Rightarrow C_3 = kv_0 - 9.8$

$$v = \frac{1}{k}[9.8 + (kv_0 - 9.8)e^{kt}]$$

Note that $k < 0$ since the object is moving downward.

(b) $\displaystyle\lim_{t\to\infty} v(t) = \frac{9.8}{k}$

(c) $$s(t) = \int \frac{1}{k}[9.8 + (kv_0 - 9.8)e^{kt}]\,dt$$

$$= \frac{1}{k}\left[9.8t + \frac{1}{k}(kv_0 - 9.8)e^{kt}\right] + C$$

$$= \frac{9.8t}{k} + \frac{1}{k^2}(kv_0 - 9.8)e^{kt} + C$$

$$s(0) = \frac{1}{k^2}(kv_0 - 9.8) + C \Rightarrow C = s_0 - \frac{1}{k^2}(kv_0 - 9.8)$$

$$s(t) = \frac{9.8t}{k} + \frac{1}{k^2}(kv_0 - 9.8)e^{kt} + s_0 - \frac{1}{k^2}(kv_0 - 9.8)$$

$$= \frac{9.8t}{k} + \frac{1}{k^2}(kv_0 - 9.8)(e^{kt} - 1) + s_0$$

45. $\dfrac{dy}{dx} - \dfrac{y}{x} = 2 + \sqrt{x}$

Integrating factor: $e^{-\int(1/x)\,dx} = e^{-\ln|x|} = \dfrac{1}{x}$

$$y\left(\frac{1}{x}\right) = \int \frac{1}{x}(2 + \sqrt{x})dx = \ln x^2 + 2\sqrt{x} + C$$

$$y = x\ln x^2 + 2x^{3/2} + Cx$$

47. $\dfrac{dy}{dx} - \dfrac{y}{x} = \dfrac{x}{y}$, Bernoulli

$n = -1$, $P = -\dfrac{1}{x}$, $Q = x$,

$$e^{\int(-2/x)\,dx} = e^{\ln x^{-2}} = x^{-2}$$

$$y^2x^{-2} = \int 2(x)x^{-2}\,dx = \ln x^2 + C$$

$$y^2 = x^2\ln x^2 + Cx^2$$

49. $xy' + y = \cos x$

$y' + \dfrac{1}{x}y = \dfrac{\cos x}{x}$ linear

$$e^{\int 1/x\,dx} = e^{\ln x} = x$$

$$yx = \int x\left(\frac{\cos x}{x}\right)dx = \int \cos x\,dx = \sin x + C$$

$$y = \frac{1}{x}\sin x + \frac{C}{x}$$

$$yx - \sin x = C$$

51. $y' - \left(\dfrac{a}{x}\right)y = bx^3$

Integrating factor: $e^{-\int(a/x)dx} = e^{-a\ln x} = x^{-a}$

$$yx^{-a} = \int bx^3(x^{-a})\,dx = \frac{b}{4 - a}x^{4-a} + C$$

$$y = \frac{bx^4}{4 - a} + Cx^a$$

53. $y' - y = -xe^{-2x}y^3$ Bernoulli, $n = 3$

$e^{\int -2(-1)\,dx} = e^{2x}$

$$y^{-2}e^{2x} = \int (-2)(-xe^{-2x})(e^{2x})\,dx = \int 2x\,dx = x^2 + C$$

$$\frac{1}{y^2} = (x^2 + C)e^{-2x} \quad \text{or} \quad \frac{e^{2x} - x^2y^2}{y^2} = C$$

55. $y' - 2y = e^x$

Integrating factor: $e^{\int -2\,dx} = e^{-2x}$

$$ye^{-2x} = \int e^{-2x}e^x\,dx + C = -e^{-x} + C$$

$$y = Ce^{2x} - e^x$$

Initial condition: $y(0) = 4$

$$4 = C - 1 \Rightarrow C = 5$$

Particular solution: $y = 5e^{2x} - e^x$

57. (a)

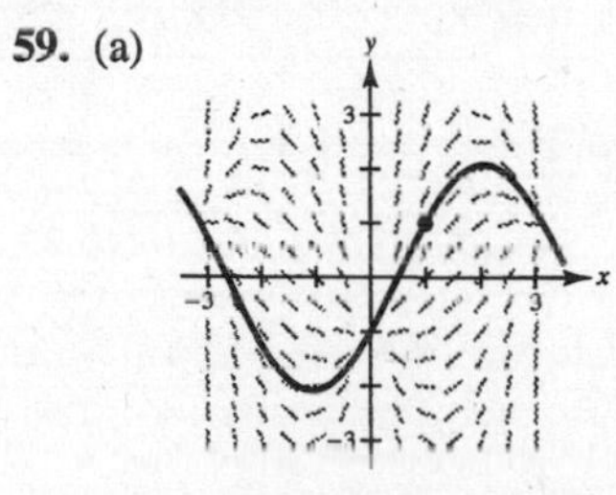

(c)

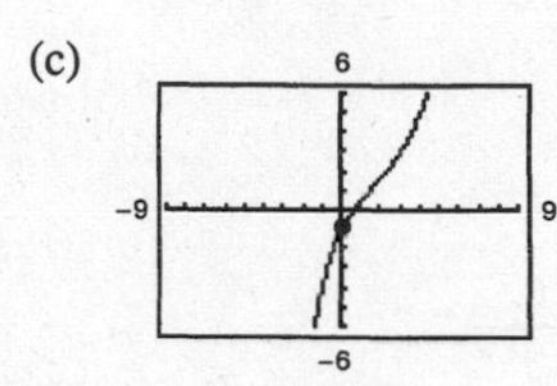

(b) $y' = e^{x/2} - y$

$y' + y = e^{x/2}$. Integrating factor: $e^{\int dx} = e^x$

$$ye^x = \int e^{x/2}e^x\,dx = \int e^{(3/2)x}\,dx = \frac{2}{3}e^{(3/2)x} + C$$

$$y = \frac{2}{3}e^{x/2} + Ce^{-x}$$

$$y(0) = -1 = \frac{2}{3} + C \Rightarrow C = -\frac{5}{3}$$

$$y = \frac{2}{3}e^{x/2} - \frac{5}{3}e^{-x} = \frac{1}{3}[2e^{x/2} - 5e^{-x}]$$

59. (a)

(c)

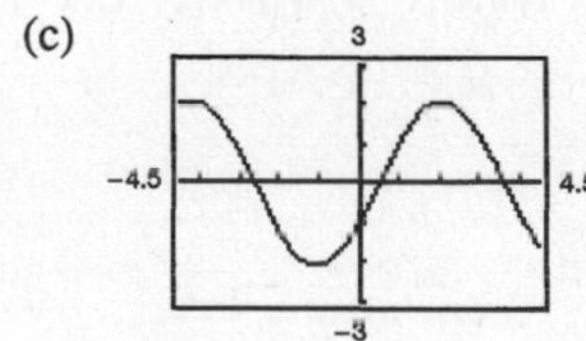

(b) $\dfrac{dy}{dx} = \csc x + y\cot x$

$$\frac{dy}{dx} - (\cot x)y = \csc x$$

Integrating factor: $e^{\int -\cot x\,dx} = e^{-\ln|\sin x|} = \csc x$

$$\csc x \cdot y' - \csc x \cot x \cdot y = \csc^2 x$$

$$(y \csc x)' = \csc^2 x$$

$$y \csc x = \int \csc^2 x\,dx = -\cot x + C$$

$$y = -\cos x + C\sin x$$

$$y(1) = 1 \Rightarrow 1 = -\cos 1 + C\sin 1 \Rightarrow C = \frac{1 + \cos 1}{\sin 1} \approx 1.83$$

61. (a) $\dfrac{ds}{dh} = \dfrac{k}{h}$

$$\int ds = \int \frac{k}{h}\,dh$$

$$s = k\ln h + C_1$$

$$= k\ln Ch$$

$$= k\ln h + C_1$$

(b) Since $s = 25$ when $h = 2$ and $s = 12$ when $h = 10$, it follows that $25 = k\ln 2C$ and $12 = k\ln 10C$, which implies

$$C = \frac{1}{2}e^{-(25/13)\ln 5} \approx 0.0605 \text{ and } k = \frac{25}{\ln 2C} = \frac{-13}{\ln 5} \approx -8.0774.$$

Therefore, s is given by the following.

$$s = -\frac{13}{\ln 5}\ln\left[\frac{h}{2}e^{-(25/13)\ln 5}\right]$$

$$= -\frac{13}{\ln 5}\left[\ln\frac{h}{2} - \frac{25}{13}\ln 5\right] = -\frac{1}{\ln 5}\left[13\ln\frac{h}{2} - 25\ln 5\right]$$

$$= 25 - \frac{13\ln(h/2)}{\ln 5}, \quad 2 \le h \le 15$$

63. $\dfrac{dA}{dt} - rA = -P$

For this linear differential equation, we have $P(t) = -r$ and $Q(t) = -P$. Therefore, the integrating factor is

$u(x) = e^{\int -r\,dt} = e^{-rt}$ and the solution is

$$A = e^{rt}\int -Pe^{-rt}\,dt = e^{rt}\left(\frac{P}{r}e^{-rt} + C\right) = \frac{P}{r} + Ce^{rt}.$$

Since $A = A_0$ when $t = 0$, we have $C = A_0 - (P/r)$ which implies that

$$A = \frac{P}{r} + \left(A_0 - \frac{P}{r}\right)e^{rt}.$$

65. $$A = \frac{200{,}000}{0.14} + \left(1{,}000{,}000 - \frac{200{,}000}{0.14}\right)e^{0.14t}$$

$$0 = 200{,}000\left[\frac{50}{7} + \left(5 - \frac{50}{7}\right)e^{0.14t}\right]$$

$$e^{0.14t} = \frac{10}{3}$$

$$t = \frac{\ln(10/3)}{0.14} \approx 8.6 \text{ years}$$

Problem Solving for Chapter 5

1. (a) $$\frac{dy}{dt} = y^{1.01}$$

$$\int y^{-1.01}\,dy = \int dt$$

$$\frac{y^{-0.01}}{-0.01} = t + C_1$$

$$\frac{1}{y^{0.01}} = -0.01t + C$$

$$y^{0.01} = \frac{1}{C - 0.01t}$$

$$y = \frac{1}{(C - 0.01t)^{100}}$$

$$y(0) = 1:\ 1 = \frac{1}{C^{100}} \Rightarrow C = 1$$

Hence, $y = \dfrac{1}{(1 - 0.01t)^{100}}$.

For $T = 100$, $\lim\limits_{t\to T^-} y = \infty$.

(b) $$\int y^{-(1+\varepsilon)}\,dy = \int k\,dt$$

$$\frac{y^{-\varepsilon}}{-\varepsilon} = kt + C_1$$

$$y^{-\varepsilon} = -\varepsilon kt + C$$

$$y = \frac{1}{(C - \varepsilon kt)^{1/\varepsilon}}$$

$$y(0) = y_0 = \frac{1}{C^{1/\varepsilon}} \Rightarrow C^{1/\varepsilon} = \frac{1}{y_0} \Rightarrow C = \left(\frac{1}{y_0}\right)^{\varepsilon}$$

Hence, $y = \dfrac{1}{\left(\dfrac{1}{y_0^{\varepsilon}} - \varepsilon kt\right)^{1/\varepsilon}}$.

For $t \to \dfrac{1}{y_0^{\varepsilon}\varepsilon k}$, $y \to \infty$.

3. Since $\dfrac{dy}{dt} = k(y - 20)$,

$$\int \frac{1}{y - 20}\,dy = \int k\,dt$$

$$\ln|y - 20| = kt + C$$

$$y = Ce^{kt} + 20.$$

When $t = 0$, $y = 72$. Therefore, $C = 52$.

When $t = 1$, $y = 48$. Therefore, $48 = 52e^k + 20$, $e^k = (28/52) = (7/13)$, and $k = \ln(7/13)$.
Thus, $y = 52e^{[\ln(7/13)]t} + 20$.

When $t = 5$, $y = 52e^{5\ln(7/13)} + 20 \approx 22.35°$.

5. $u = 1 - x, x = 1 - u, dx = -du$

When $x = a, u = 1 - a$. When $x = b, u = 1 - b$.

$$P_{a,b} = \int_a^b \frac{15}{4}x\sqrt{1-x}\,dx = \frac{15}{4}\int_{1-a}^{1-b} -(1-u)\sqrt{u}\,du$$

$$= \frac{15}{4}\int_{1-a}^{1-b}(u^{3/2} - u^{1/2})\,du = \frac{15}{4}\left[\frac{2}{5}u^{5/2} - \frac{2}{3}u^{3/2}\right]_{1-a}^{1-b} = \frac{15}{4}\left[\frac{2u^{3/2}}{15}(3u-5)\right]_{1-a}^{1-b} = \left[-\frac{(1-x)^{3/2}}{2}(3x+2)\right]_a^b$$

(a) $P_{0.50,\,0.75} = \left[-\frac{(1-x)^{3/2}}{2}(3x+2)\right]_{0.50}^{0.75} = 0.353 = 35.3\%$

(b) $P_{0,b} = \left[-\frac{(1-x)^{3/2}}{2}(3x+2)\right]_0^b = -\frac{(1-b)^{3/2}}{2}(3b+2) + 1 = 0.5$

$(1-b)^{3/2}(3b+2) = 1$

$b \approx 0.586 = 58.6\%$

7. (a) $\int_0^x e^t\,dt > \int_0^x dt$

$e^t\Big]_0^x > t\Big]_0^x$

$e^x - 1 > x - 0$

$e^x > x + 1$

(b) From Part (a),

$\int_0^x e^t\,dt > \int_0^x (t+1)\,dt$

$e^t\Big]_0^x > \left[\frac{t^2}{t} + t\right]_0^x$

$e^x - 1 > \frac{x^2}{2} + x$

$e^x > \frac{x^2}{2} + x + 1$

(c) Using mathematical induction, assuming $e^x \geq 1 + x + \cdots + \frac{x^k}{k!}$,

$\int_0^x e^t\,dt > \int_0^x \left[1 + t + \cdots + \frac{t^k}{k!}\right] dt$

$e^x - 1 > x + \frac{x^2}{2} + \cdots + \frac{x^{k+1}}{(k+1)!}$

$e^x > 1 + x + \frac{x^2}{2} + \cdots + \frac{x^{k+1}}{(k+1)!}$

9. $\tan\theta_1 = \dfrac{3}{x}$ $\qquad \tan\theta_2 = \dfrac{6}{10-x}$

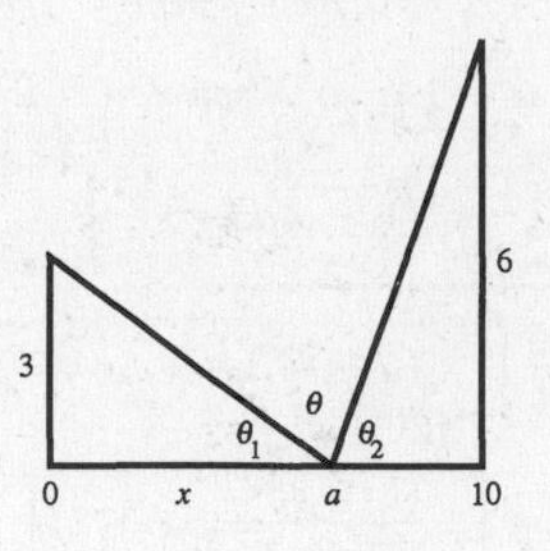

Minimize $\theta_1 + \theta_2$:

$$f(x) = \theta_1 + \theta_2 = \arctan\left(\frac{3}{x}\right) + \arctan\left(\frac{6}{10-x}\right)$$

$$f'(x) = \frac{1}{1+\dfrac{9}{x^2}}\left(\frac{-3}{x^2}\right) + \frac{1}{1+\dfrac{36}{(10-x)^2}}\left(\frac{6}{(10-x)^2}\right) = 0$$

$$\frac{3}{x^2+9} = \frac{6}{(10-x)^2+36}$$

$$(10-x)^2 + 36 = 2(x^2+9)$$

$$100 - 20x + x^2 + 36 = 2x^2 + 18$$

$$x^2 + 20x - 118 = 0$$

$$x = \frac{-20 \pm \sqrt{20^2 - 4(-118)}}{2} = -10 \pm \sqrt{218}$$

$a = -10 + \sqrt{218} \approx 4.7648 \qquad f(a) \approx 1.4153$

$\theta = \pi - (\theta_1 + \theta_2) \approx 1.7263$ or $98.9°$

Endpoints: $a = 0$: $\theta \approx 1.0304$

$a = 10$: $\theta \approx 1.2793$

Maximum is 1.7263 at $a = -10 + \sqrt{218} \approx 4.7648$.

11. $L'(x) = \displaystyle\lim_{\Delta x\to 0} \frac{L(x+\Delta x) - L(x)}{\Delta x}$

$= \displaystyle\lim_{\Delta x\to 0} \frac{L(x) + L(\Delta x) - L(x)}{\Delta x}$

$= \displaystyle\lim_{\Delta x\to 0} \frac{L(\Delta x)}{\Delta x}$

Also, $L'(0) = \displaystyle\lim_{\Delta x\to 0} \frac{L(\Delta x) - L(0)}{\Delta x}$

But, $L(0) = 0$ because $L(0) = L(0+0) = L(0) + L(0) \Rightarrow L(0) = 0$.

Thus, $L'(x) = L'(0)$, for all x.

The graph of L is a line through the origin of slope $L'(0)$.

13. (a) $\dfrac{\text{Area sector}}{\text{Area circle}} = \dfrac{t}{2\pi} \Rightarrow \text{Area sector} = \dfrac{t}{2\pi}(\pi) = \dfrac{t}{2}$

(b) Area $AOP = \dfrac{1}{2}(\text{base})(\text{height}) - \displaystyle\int_1^{\cosh t} \sqrt{x^2-1}\,dx$

$A(t) = \dfrac{1}{2}\cosh t \cdot \sinh t - \displaystyle\int_1^{\cosh t} \sqrt{x^2-1}\,dx$

$A'(t) = \frac{1}{2}[\cosh^2 t + \sinh^2 t] - \sqrt{\cosh^2 t - 1}\,\sinh t$

$= \frac{1}{2}[\cosh^2 t + \sinh^2 t] - \sinh^2 t$

$= \frac{1}{2}[\cosh^2 t - \sinh^2 t] = \frac{1}{2}$

$A(t) = \frac{1}{2}t + C$. But, $A(0) = C = 0 \Rightarrow C = 0$

Thus, $A(t) = \frac{1}{2}t$ or $t = 2A(t)$.

CHAPTER 6
Applications of Integration

CHAPTER 6
Applications of Integration

Section 6.1 Area of a Region Between Two Curves

Solutions to Odd-Numbered Exercises

1. $A = \int_0^6 [0 - (x^2 - 6x)]\, dx = -\int_0^6 (x^2 - 6x)\, dx$

3. $A = \int_0^3 [(-x^2 + 2x + 3) - (x^2 - 4x + 3)]\, dx = \int_0^3 (-2x^2 + 6x)\, dx$

5. $A = 2\int_{-1}^0 3(x^3 - x)\, dx = 6\int_{-1}^0 (x^3 - x)\, dx \quad \text{or} \quad -6\int_0^1 (x^3 - x)\, dx$

7. $\int_0^4 \left[(x + 1) - \frac{x}{2}\right] dx$

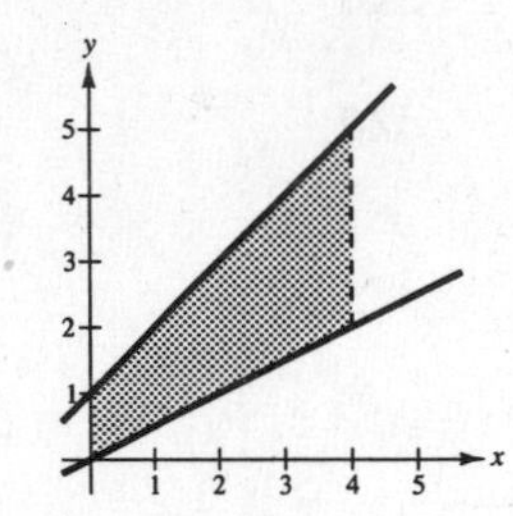

9. $\int_0^6 \left[4(2^{-x/3}) - \frac{x}{6}\right] dx$

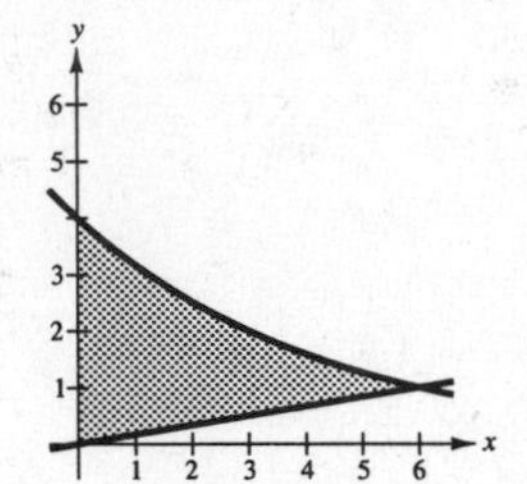

11. $\int_{-\pi/3}^{\pi/3} [2 - \sec x]\, dx$

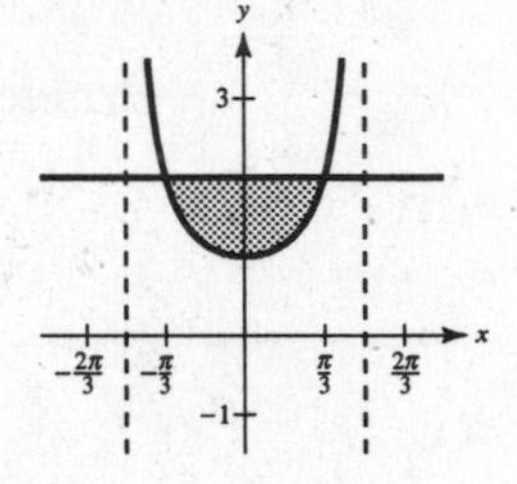

13. $f(x) = x + 1$

$g(x) = (x - 1)^2$

$A \approx 4$

Matches (d)

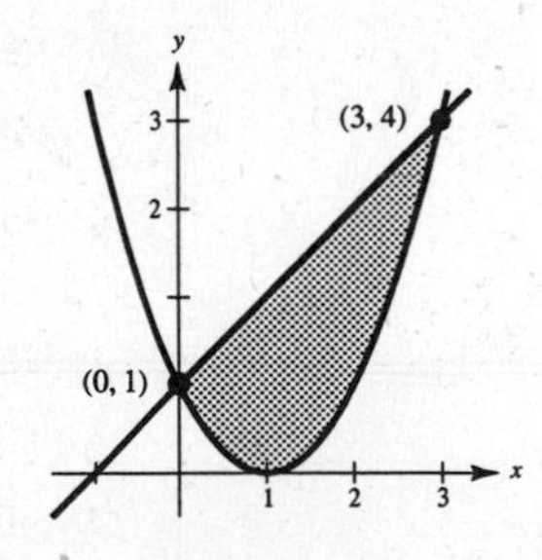

15.
$$\begin{aligned} A &= \int_0^2 \left[\left(\frac{1}{2}x^3 + 2\right) - (x + 1)\right] dx \\ &= \int_0^2 \left(\frac{1}{2}x^3 - x + 1\right) dx \\ &= \left[\frac{x^4}{8} - \frac{x^2}{2} + x\right]_0^2 \\ &= \left(\frac{16}{8} - \frac{4}{2} + 2\right) - 0 = 2 \end{aligned}$$

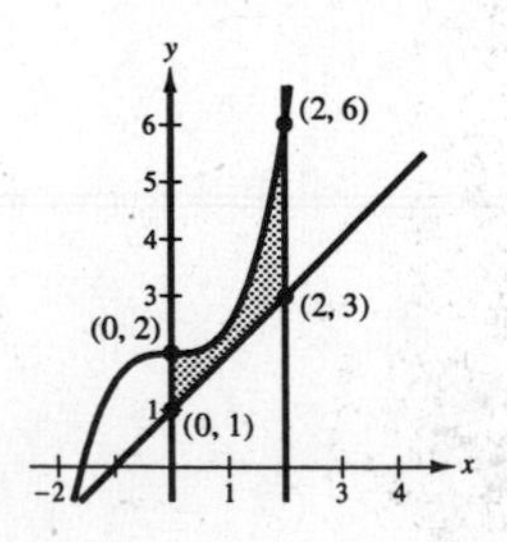

17. The points of intersection are given by:

$$x^2 - 4x = 0$$

$$x(x - 4) = 0 \quad \text{when} \quad x = 0, 4$$

$$A = \int_0^4 [g(x) - f(x)]\,dx$$

$$= -\int_0^4 (x^2 - 4x)\,dx$$

$$= -\left[\frac{x^3}{3} - 2x^2\right]_0^4$$

$$= \frac{32}{3}$$

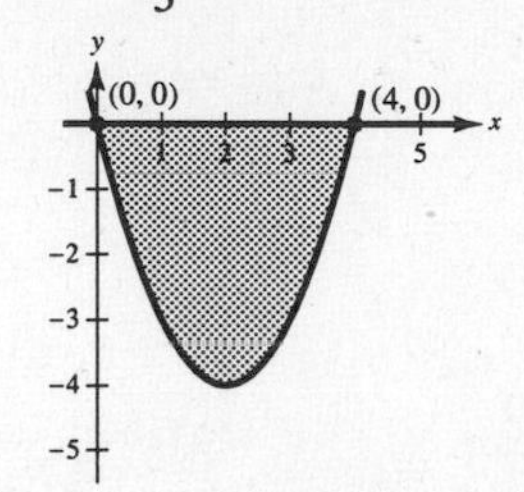

19. The points of intersection are given by:

$$x^2 + 2x + 1 = 3x + 3$$

$$(x - 2)(x + 1) = 0 \quad \text{when} \quad x = -1, 2$$

$$A = \int_{-1}^2 [g(x) - f(x)]\,dx$$

$$= \int_{-1}^2 [(3x + 3) - (x^2 + 2x + 1)]\,dx$$

$$= \int_{-1}^2 (2 + x - x^2)\,dx$$

$$= \left[2x + \frac{x^2}{2} - \frac{x^3}{3}\right]_{-1}^2 = \frac{9}{2}$$

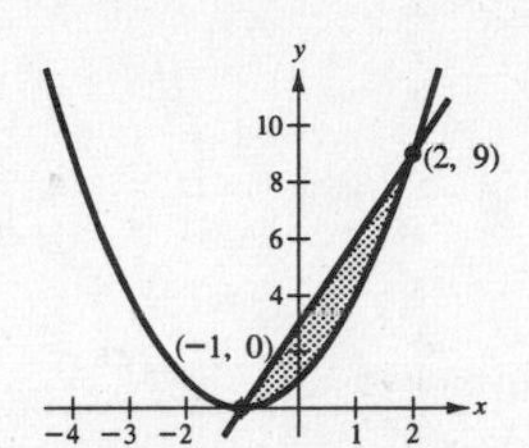

21. The points of intersection are given by:

$$x = 2 - x \quad \text{and} \quad x = 0 \quad \text{and} \quad 2 - x = 0$$

$$x = 1 \qquad\qquad x = 0 \qquad\qquad x = 2$$

$$A = \int_0^1 [(2 - y) - (y)]\,dy = \Big[2y - y^2\Big]_0^1 = 1$$

Note that if we integrate with respect to x, we need two integrals. Also, note that the region is a triangle.

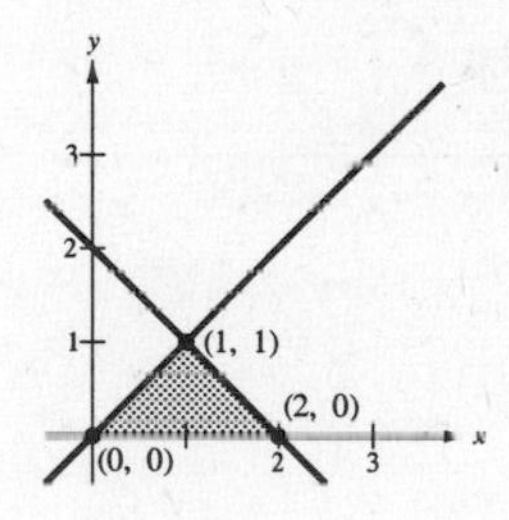

23. The points of intersection are given by:

$$\sqrt{3x} + 1 = x + 1$$

$$\sqrt{3x} = x \quad \text{when} \quad x = 0, 3$$

$$A = \int_0^3 [f(x) - g(x)]\,dx$$

$$= \int_0^3 \left[\left(\sqrt{3x} + 1\right) - (x + 1)\right]dx$$

$$= \int_0^3 [(3x)^{1/2} - x]\,dx$$

$$= \left[\frac{2}{9}(3x)^{3/2} - \frac{x^2}{2}\right]_0^3 = \frac{3}{2}$$

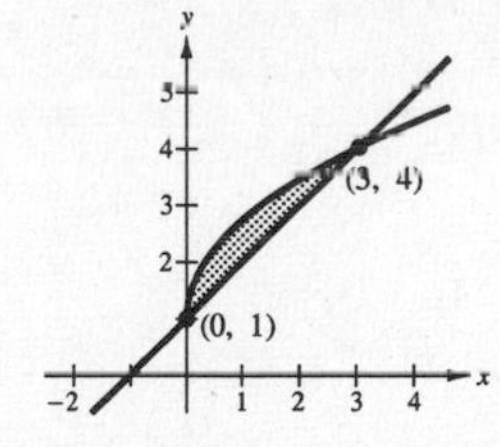

25. The points of intersection are given by:

$$y^2 = y + 2$$

$$(y - 2)(y + 1) = 0 \quad \text{when} \quad y = -1, 2$$

$$A = \int_{-1}^2 [g(y) - f(y)]\,dy$$

$$= \int_{-1}^2 [(y + 2) - y^2]\,dy$$

$$= \left[2y + \frac{y^2}{2} - \frac{y^3}{3}\right]_{-1}^2 = \frac{9}{2}$$

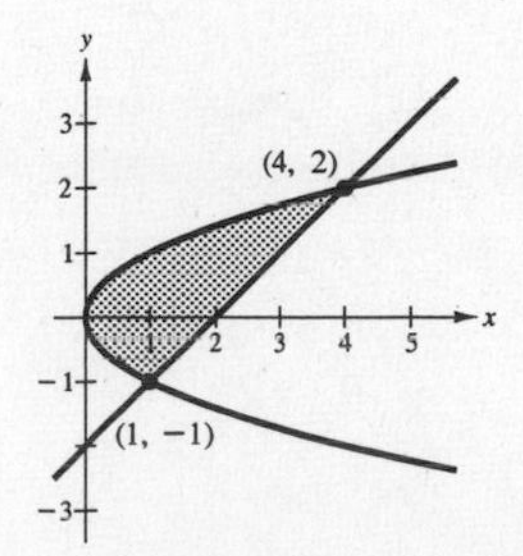

27. $A = \int_{-1}^{2} [f(y) - g(y)]\,dy$

$= \int_{-1}^{2} [(y^2 + 1) - 0]\,dy$

$= \left[\frac{y^3}{3} + y\right]_{-1}^{2} = 6$

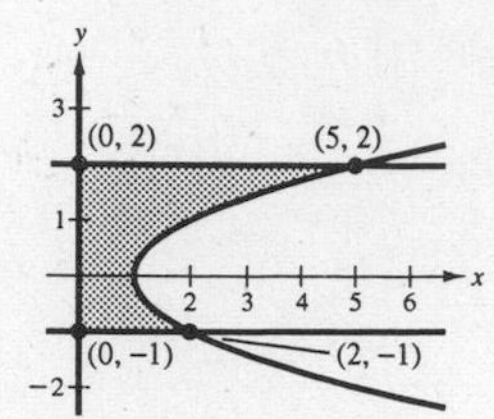

29. $y = \frac{10}{x} \Rightarrow x = \frac{10}{y}$

$A = \int_{2}^{10} \frac{10}{y}\,dy$

$= \left[10 \ln y\right]_{2}^{10}$

$= 10(\ln 10 - \ln 2)$

$= 10 \ln 5 \approx 16.0944$

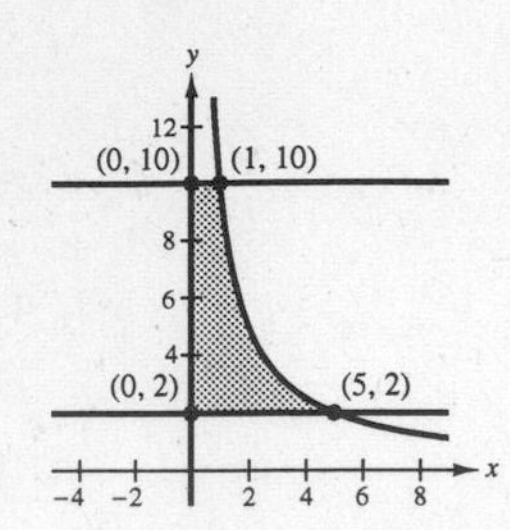

31. The points of intersection are given by:

$x^3 - 3x^2 + 3x = x^2$

$x(x - 1)(x - 3) = 0$ when $x = 0, 1, 3$

$A = \int_{0}^{1} [f(x) - g(x)]\,dx + \int_{1}^{3} [g(x) - f(x)]\,dx$

$= \int_{0}^{1} [(x^3 - 3x^2 + 3x) - x^2]\,dx + \int_{1}^{3} [x^2 - (x^3 - 3x^2 + 3x)]\,dx$

$= \int_{0}^{1} (x^3 - 4x^2 + 3x)\,dx + \int_{1}^{3} (-x^3 + 4x^2 - 3x)\,dx$

$= \left[\frac{x^4}{4} - \frac{4}{3}x^3 + \frac{3}{2}x^2\right]_{0}^{1} + \left[\frac{-x^4}{4} + \frac{4}{3}x^3 - \frac{3}{2}x^2\right]_{1}^{3} = \frac{5}{12} + \frac{8}{3} = \frac{37}{12}$

Numerical Approximation: $0.417 + 2.667 \approx 3.083$

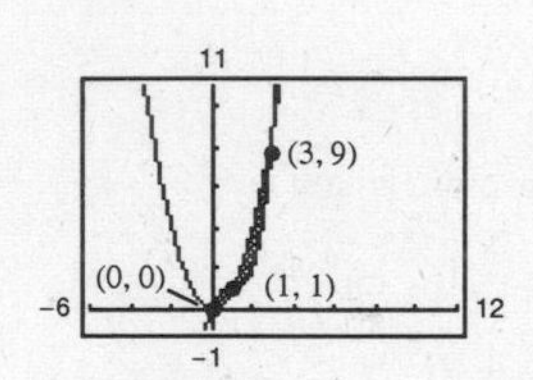

33. The points of intersection are given by:

$x^2 - 4x + 3 = 3 + 4x - x^2$

$2x(x - 4) = 0$ when $x = 0, 4$

$A = \int_{0}^{4} [(3 + 4x - x^2) - (x^2 - 4x + 3)]\,dx$

$= \int_{0}^{4} (-2x^2 + 8x)\,dx$

$= \left[-\frac{2x^3}{3} + 4x^2\right]_{0}^{4} = \frac{64}{3}$

Numerical Approximation: 21.333

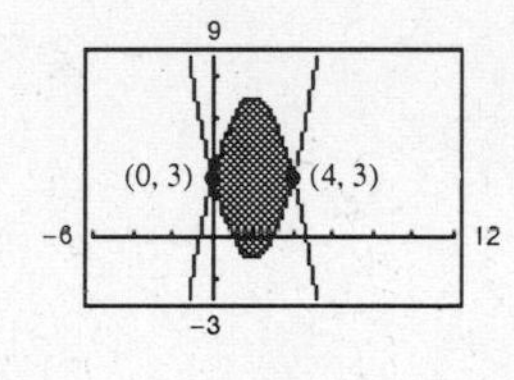

35. $f(x) = x^4 - 4x^2,\ g(x) = x^2 - 4$

The points of intersection are given by:

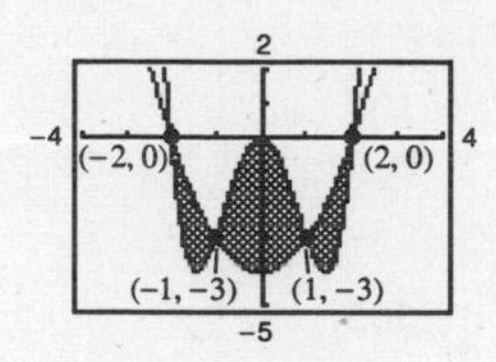

$$x^4 - 4x^2 = x^2 - 4$$

$$x^4 - 5x^2 + 4 = 0$$

$$(x^2 - 4)(x^2 - 1) = 0 \text{ when } x = \pm 2, \pm 1$$

By symmetry,

$$A = 2\int_0^1 [(x^4 - 4x^2) - (x^2 - 4)]\,dx + 2\int_1^2 [(x^2 - 4) - (x^4 - 4x^2)]\,dx$$

$$= 2\int_0^1 (x^4 - 5x^2 + 4)\,dx + 2\int_1^2 (-x^4 + 5x^2 - 4)\,dx$$

$$= 2\left[\frac{x^5}{5} - \frac{5x^3}{3} + 4x\right]_0^1 + 2\left[-\frac{x^5}{5} + \frac{5x^3}{3} - 4x\right]_1^2$$

$$= 2\left[\frac{1}{5} - \frac{5}{3} + 4\right] + 2\left[\left(-\frac{32}{5} + \frac{40}{3} - 8\right) - \left(-\frac{1}{5} + \frac{5}{3} - 4\right)\right] = 8.$$

Numerical Approximation: $5.067 + 2.933 = 8.0$

37. The points of intersection are given by:

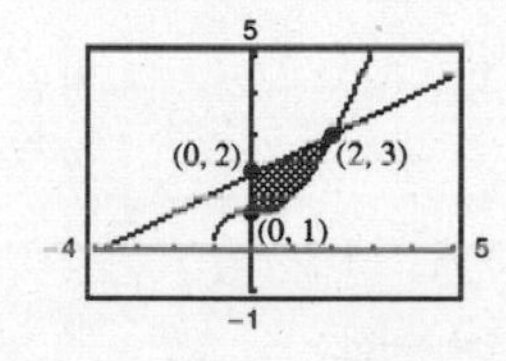

$$\frac{1}{1 + x^2} = \frac{x^2}{2}$$

$$x^4 + x^2 - 2 = 0$$

$$(x^2 + 2)(x^2 - 1) = 0$$

$$x = \pm 1$$

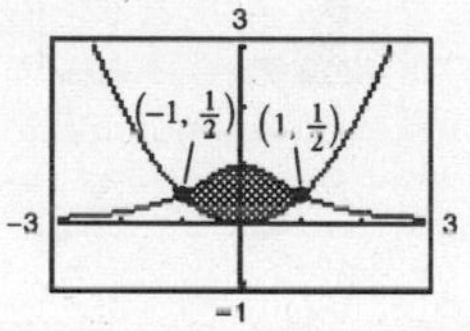

$$A = 2\int_0^1 [f(x) - g(x)]dx$$

$$= 2\int_0^1 \left[\frac{1}{1 + x^2} - \frac{x^2}{2}\right]dx$$

$$= 2\left[\arctan x - \frac{x^3}{6}\right]_0^1$$

$$= 2\left(\frac{\pi}{4} - \frac{1}{6}\right) = \frac{\pi}{2} - \frac{1}{3} \approx 1.237$$

Numerical Approximation: 1.237

39. $\sqrt{1 + x^3} \le \frac{1}{2}x + 2$ on $[0, 2]$

Numerical approximation: 1.759

$$A = \int_0^2 \left[\frac{1}{2}x + 2 - \sqrt{1 + x^3}\right] dx \approx 1.759$$

5
(0, 2)
(2, 3)
(0, 1)
-4
5
-1

41. $A = 2\displaystyle\int_0^{\pi/3} [f(x) - g(x)]\,dx$

$$= 2\int_0^{\pi/3} (2\sin x - \tan x)\,dx$$

$$= 2\Big[-2\cos x + \ln|\cos x|\Big]_0^{\pi/3}$$

$$= 2(1 - \ln 2) \approx 0.614$$

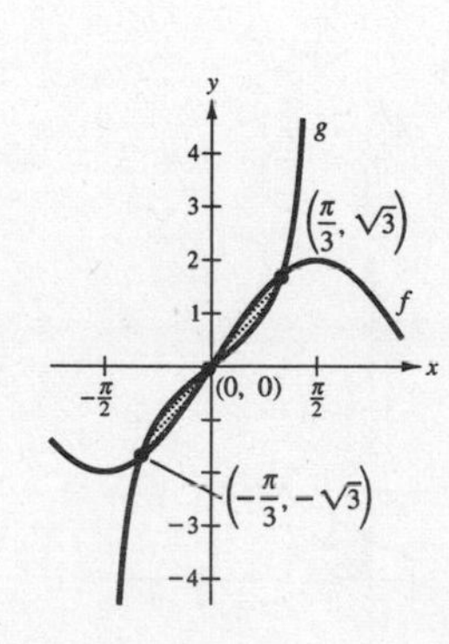

43. $A = \int_0^{2\pi} [(2 - \cos x) - \cos x]\, dx$

$= 2\int_0^{2\pi} (1 - \cos x)\, dx$

$= 2\Big[x - \sin x\Big]_0^{2\pi} = 4\pi \approx 12.566$

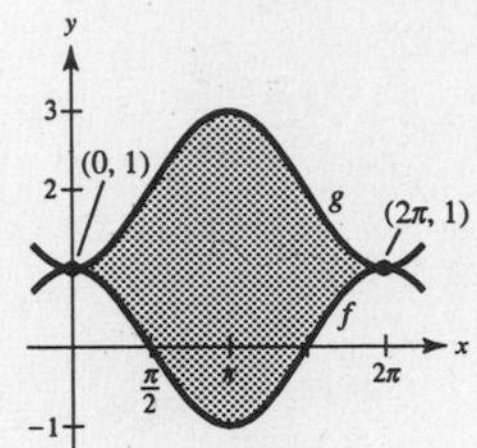

45. $A = \int_0^1 [xe^{-x^2} - 0]\, dx$

$= \left[-\frac{1}{2}e^{-x^2}\right]_0^1 = \frac{1}{2}\left(1 - \frac{1}{e}\right) \approx 0.316$

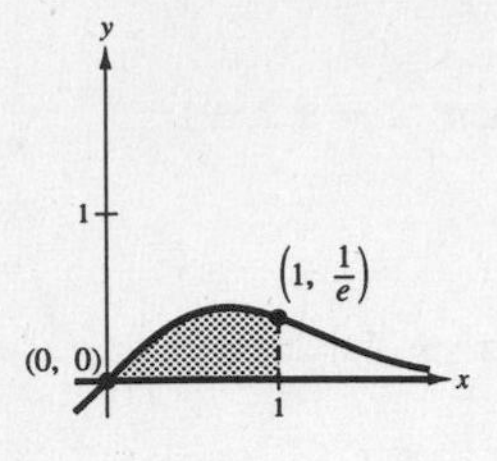

47. $A = \int_0^{\pi} [(2\sin x + \sin 2x) - 0]\, dx$

$= \left[-2\cos x - \frac{1}{2}\cos 2x\right]_0^{\pi} = 4.0$

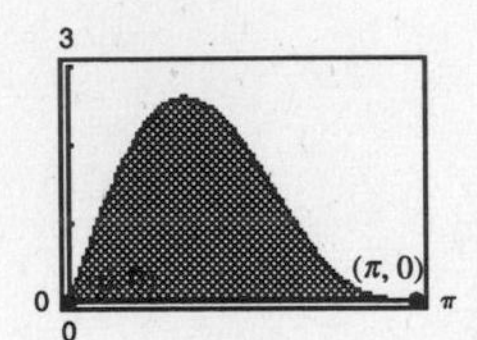

49. $A = \int_1^3 \left[\frac{1}{x^2}e^{1/x} - 0\right] dx$

$= \left[-e^{1/x}\right]_1^3 = e - e^{1/3} \approx 1.323$

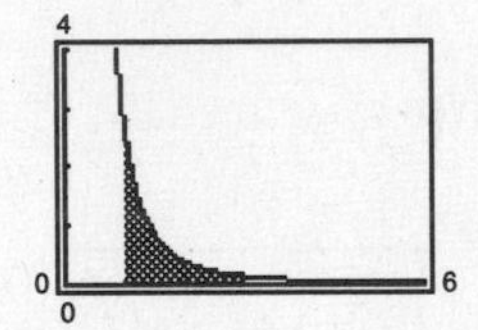

51. (a) $y = \sqrt{\frac{x^3}{4 - x}}$, $y = 0$, $x = 3$

(b) $A = \int_0^3 \sqrt{\frac{x^3}{4 - x}}\, dx$,

No, it cannot be evaluated by hand.

(c) 4.7721

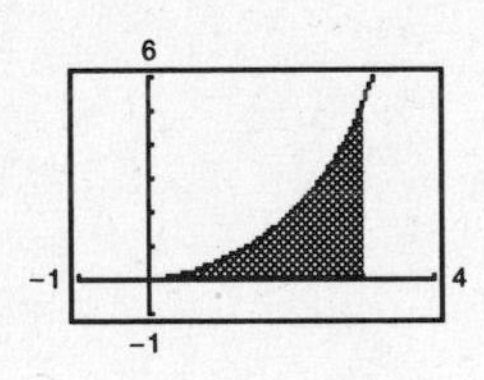

53. $F(x) = \int_0^x \left(\frac{1}{2}t + 1\right) dt = \left[\frac{t^2}{4} + t\right]_0^x = \frac{x^2}{4} + x$

(a) $F(0) = 0$

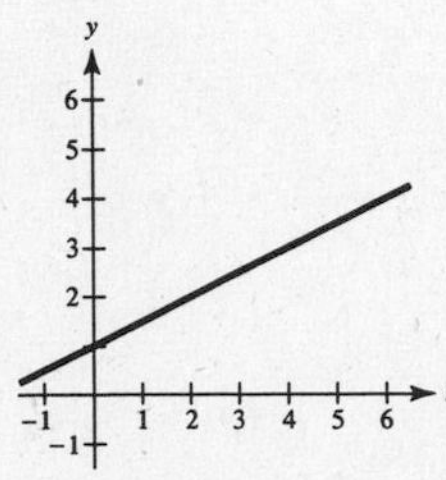

(b) $F(2) = \frac{2^2}{4} + 2 = 3$

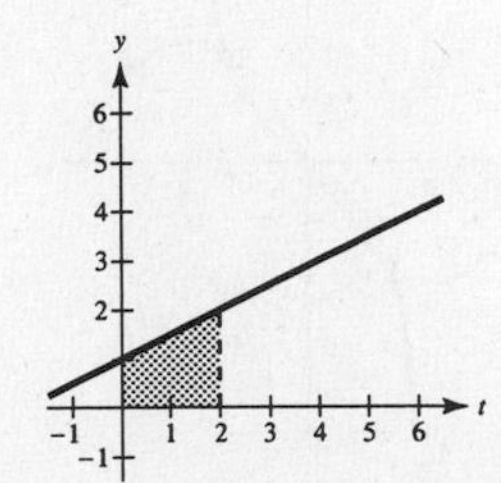

(c) $F(6) = \frac{6^2}{4} + 6 = 15$

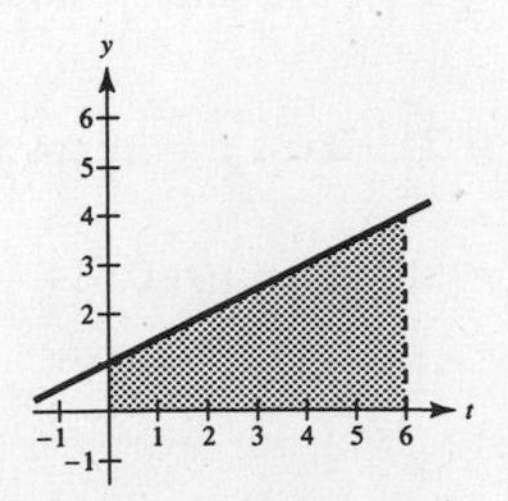

55. $F(\alpha) = \int_{1}^{\alpha} \cos \frac{\pi\theta}{2}\, d\theta = \left[\frac{2}{\pi}\sin\frac{\pi\theta}{2}\right]_{-1}^{\alpha} = \frac{2}{\pi}\sin\frac{\pi\alpha}{2} + \frac{2}{\pi}$

(a) $F(-1) = 0$

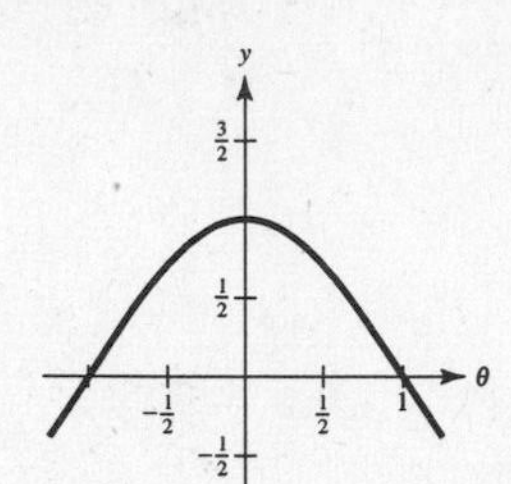

(b) $F(0) = \frac{2}{\pi} \approx 0.6366$

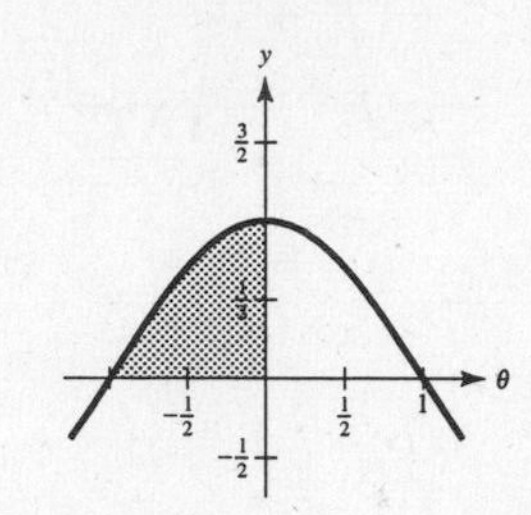

(c) $F\left(\frac{1}{2}\right) = \frac{2 + \sqrt{2}}{\pi} \approx 1.0868$

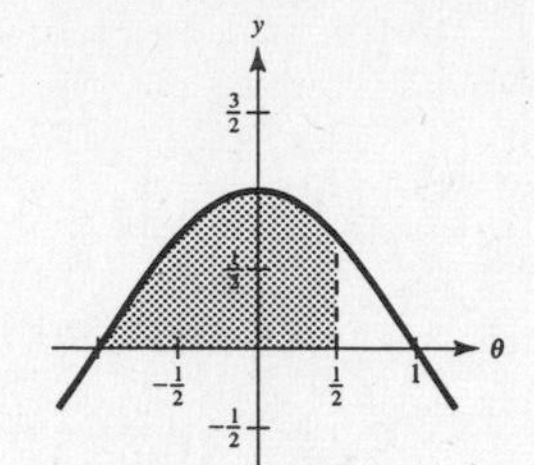

57. $A = \int_{0}^{c} \left[\left(\frac{b-a}{c}y + a\right) - \frac{b}{c}y\right] dy$

$= \int_{0}^{c} \left(-\frac{a}{c}y + a\right) dy$

$= \left[-\frac{a}{2c}y^2 + ay\right]_{0}^{c}$

$= -\frac{ac}{2} + ac = \frac{ac}{2} \quad \left(= \frac{1}{2}\text{(base)(height)}\right)$

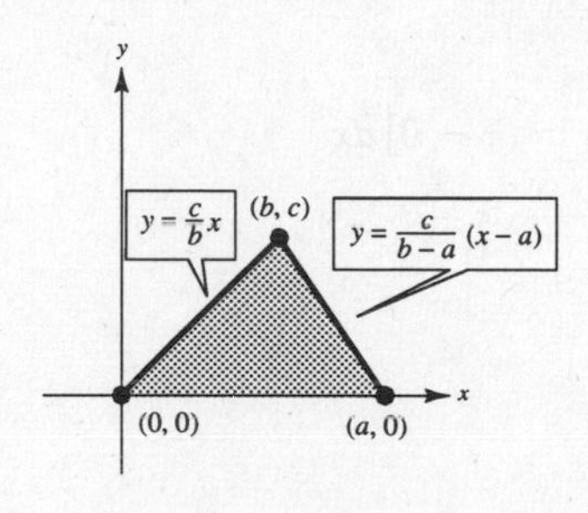

59. $f(x) = x^3$

$f'(x) = 3x^2$

At $(1, 1)$, $f'(1) = 3$.

Tangent line:

$y - 1 = 3(x - 1)$ or $y = 3x - 2$

The tangent line intersects $f(x) = x^3$ at $x = -2$.

$A = \int_{-2}^{1} [x^3 - (3x - 2)]\, dx = \left[\frac{x^4}{4} - \frac{3x^2}{2} + 2x\right]_{-2}^{1} = \frac{27}{4}$

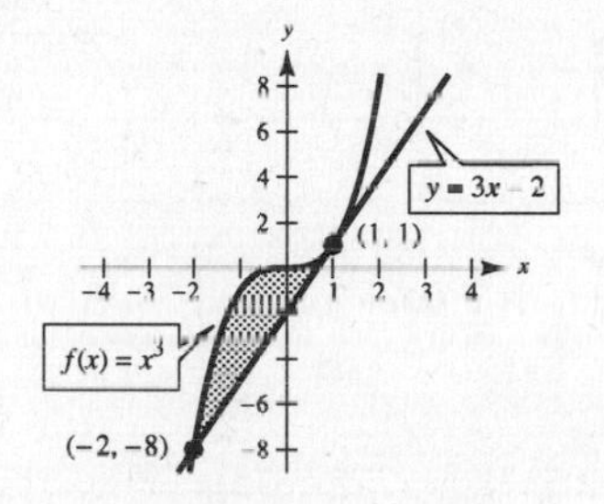

61. The variable is y.

63. $x^4 - 2x^2 + 1 \le 1 - x^2$ on $[-1, 1]$

$A = \int_{-1}^{1} [(1 - x^2) - (x^4 - 2x^2 + 1)]\, dx$

$= \int_{-1}^{1} (x^2 - x^4)\, dx$

$= \left[\frac{x^3}{3} - \frac{x^5}{5}\right]_{-1}^{1} = \frac{4}{15}$

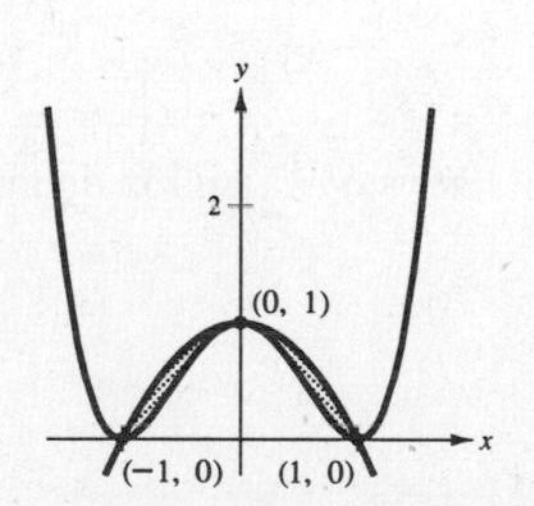

You can use a single integral because $x^4 - 2x^2 + 1 \le 1 - x^2$ on $[-1, 1]$.

65. Offer 2 is better because the accumulated salary (area under the curve) is larger.

67. $A = \int_{-3}^{3} (9 - x^2)\,dx = 36$

$$\int_{-\sqrt{9-b}}^{\sqrt{9-b}} [(9 - x^2) - b]\,dx = 18$$

$$\int_{0}^{\sqrt{9-b}} [(9 - b) - x^2]\,dx = 9$$

$$\left[(9 - b)x - \frac{x^3}{3}\right]_0^{\sqrt{9-b}} = 9$$

$$\frac{2}{3}(9 - b)^{3/2} = 9$$

$$(9 - b)^{3/2} = \frac{27}{2}$$

$$9 - b = \frac{9}{\sqrt[3]{4}}$$

$$b = 9 - \frac{9}{\sqrt[3]{4}} \approx 3.330$$

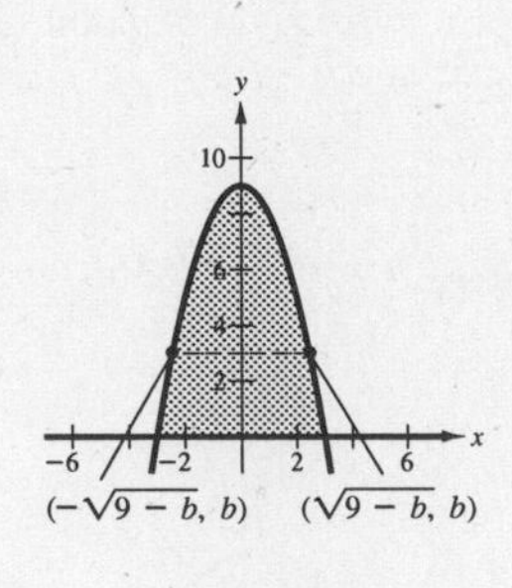

69. $\lim_{\|\Delta\| \to 0} \sum_{i=1}^{n} (x_i - x_i^2)\,\Delta x$

where $x_i = \dfrac{i}{n}$ and $\Delta x = \dfrac{1}{n}$ is the same as

$$\int_0^1 (x - x^2)\,dx = \left[\frac{x^2}{2} - \frac{x^3}{3}\right]_0^1 = \frac{1}{6}.$$

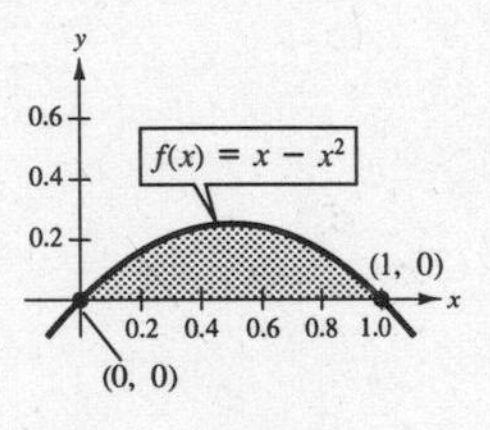

71. $\int_0^5 [(7.21 + 0.58t) - (7.21 + 0.45t)]\,dt = \int_0^5 0.13t\,dt = \left[\dfrac{0.13t^2}{2}\right]_0^5 = \$1.625$ billion

73. (a) $y_1 = (275.0675)(1.0537)^t = (275.0675)e^{0.0523t}]$

(b) $y_2 = (239.9407)(1.0417)^t = (239.9407)e^{0.0408t}$

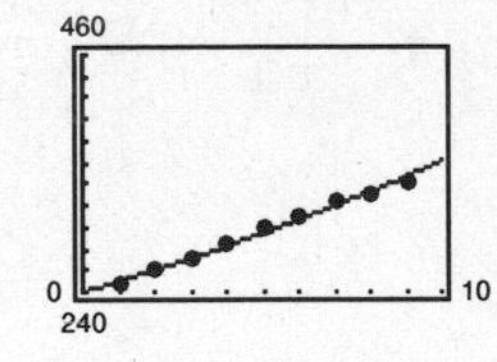

(c) $\int_{10}^{15} (y_1 - y_2)\,dt \approx 649.5$ billion dollars

(d) No, model $y_1 > y_2$ forever because $1.0537 > 1.0417$.

No, these models are not accurate. According to news reports, $E > R$ eventually.

75. The total area is 8 times the area of the shaded region to the right. A point (x, y) is on the upper boundary of the region if

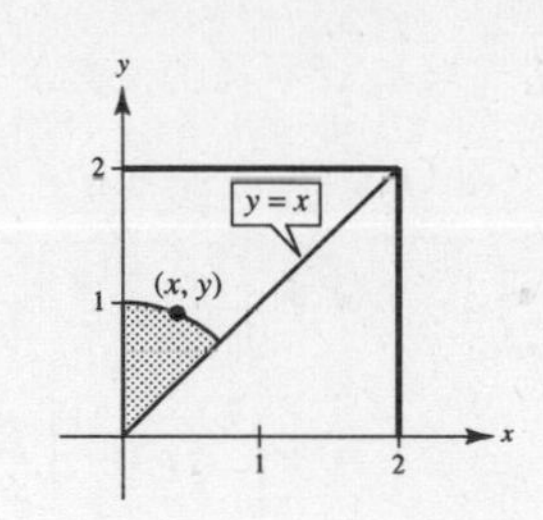

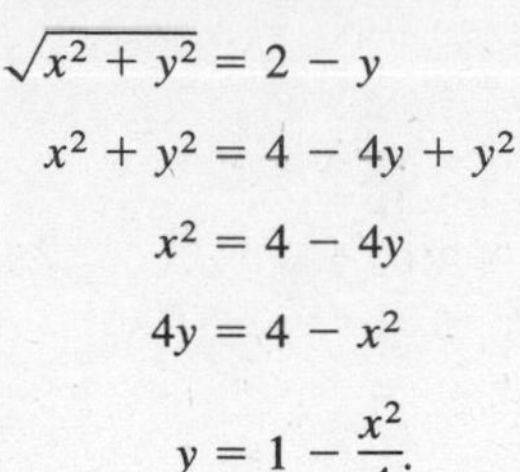

$$\sqrt{x^2 + y^2} = 2 - y$$
$$x^2 + y^2 = 4 - 4y + y^2$$
$$x^2 = 4 - 4y$$
$$4y = 4 - x^2$$
$$y = 1 - \frac{x^2}{4}.$$

We now determine where this curve intersects the line $y = x$.

$$x = 1 - \frac{x^2}{4}$$
$$x^2 + 4x - 4 = 0$$
$$x = \frac{-4 \pm \sqrt{16 + 16}}{2} = -2 \pm 2\sqrt{2} \Rightarrow x = -2 + 2\sqrt{2}$$

$$\text{Total area} = 8\int_0^{-2+2\sqrt{2}} \left(1 - \frac{x^2}{4} - x\right) dx$$
$$= 8\left[x - \frac{x^3}{12} - \frac{x^2}{2}\right]_0^{-2+2\sqrt{2}} = \frac{16}{3}(4\sqrt{2} - 5) \approx 8(0.4379) = 3.503$$

77. (a) $A = 2\left[\int_0^5 \left(1 - \frac{1}{3}\sqrt{5 - x}\right) dx + \int_5^{5.5} (1 - 0)\, dx\right]$

$$= 2\left(\left[x + \frac{2}{9}(5 - x)^{3/2}\right]_0^5 + \left[x\right]_5^{5.5}\right) = 2\left(5 - \frac{10\sqrt{5}}{9} + 5.5 - 5\right) \approx 6.031 \text{ m}^2$$

(b) $V = 2A \approx 2(6.031) \approx 12.062 \text{ m}^3$

(c) $5000\, V \approx 5000(12.062) = 60{,}310$ pounds

79. True

81. False. Let $f(x) = x$ and $g(x) = 2x - x^2$. f and g intersect at $(1, 1)$, the midpoint of $[0, 2]$. But

$$\int_a^b [f(x) - g(x)]\, dx = \int_0^2 [x - (2x - x^2)]\, dx = \frac{2}{3} \neq 0.$$

Section 6.2 Volume: The Disk Method

1. $V = \pi\int_0^1 (-x + 1)^2\, dx = \pi\int_0^1 (x^2 - 2x + 1)\, dx = \pi\left[\frac{x^3}{3} - x^2 + x\right]_0^1 = \frac{\pi}{3}$

3. $V = \pi\int_1^4 (\sqrt{x})^2\, dx = \pi\int_1^4 x\, dx = \pi\left[\frac{x^2}{2}\right]_1^4 = \frac{15\pi}{2}$

5. $V = \pi\int_0^1 [(x^2)^2 - (x^3)^2]\, dx = \pi\int_0^1 (x^4 - x^6)\, dx = \pi\left[\frac{x^5}{5} - \frac{x^7}{7}\right]_0^1 = \frac{2\pi}{35}$

7. $y = x^2 \Rightarrow x = \sqrt{y}$

$$V = \pi\int_0^4 (\sqrt{y})^2\, dy = \pi\int_0^4 y\, dy$$
$$= \pi\left[\frac{y^2}{2}\right]_0^4 = 8\pi$$

9. $y = x^{2/3} \Rightarrow x = y^{3/2}$

$$V = \pi\int_0^1 (y^{3/2})^2\, dy = \pi\int_0^1 y^3\, dy = \pi\left[\frac{y^4}{4}\right]_0^1 = \frac{\pi}{4}$$

11. $y = \sqrt{x},\ y = 0,\ x = 4$

(a) $R(x) = \sqrt{x},\ r(x) = 0$

$$V = \pi \int_0^4 \left(\sqrt{x}\right)^2 dx$$

$$= \pi \int_0^4 x\, dx = \left[\frac{\pi}{2}x^2\right]_0^4 = 8\pi$$

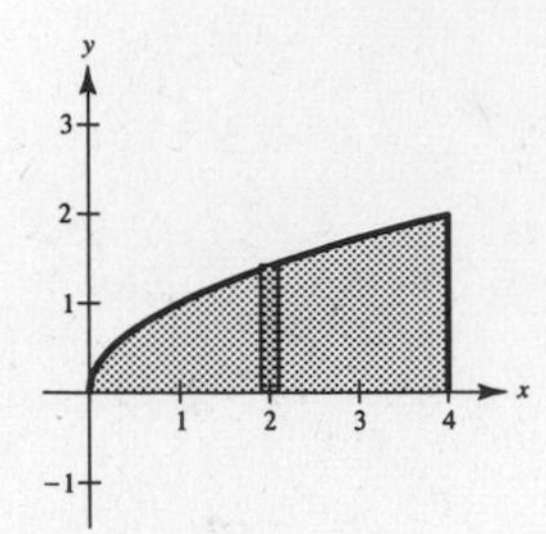

(b) $R(y) = 4,\ r(y) = y^2$

$$V = \pi \int_0^2 (16 - y^4)\, dy$$

$$= \pi\left[16y - \frac{1}{5}y^5\right]_0^2 = \frac{128\pi}{5}$$

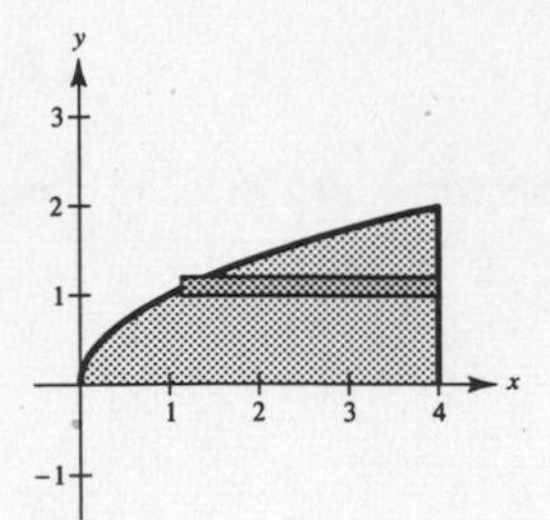

(c) $R(y) = 4 - y^2,\ r(y) = 0$

$$V = \pi \int_0^2 (4 - y^2)^2\, dy$$

$$= \pi \int_0^2 (16 - 8y^2 + y^4)\, dy$$

$$= \pi\left[16y - \frac{8}{3}y^3 + \frac{1}{5}y^5\right]_0^2 = \frac{256\pi}{15}$$

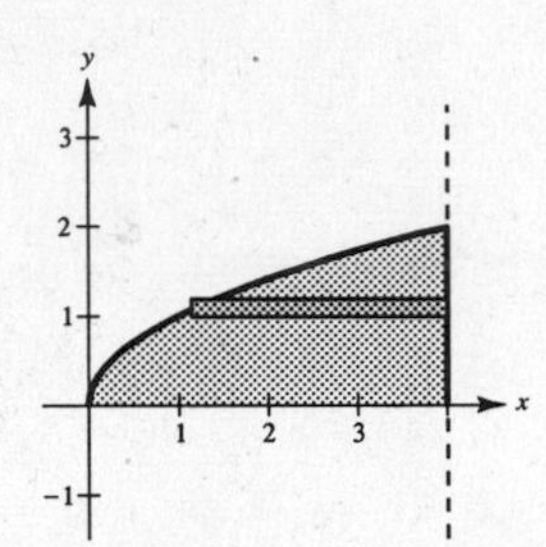

(d) $R(y) = 6 - y^2,\ r(y) = 2$

$$V = \pi \int_0^2 [(6 - y^2)^2 - 4]\, dy$$

$$= \pi \int_0^2 (32 - 12y^2 + y^4)\, dy$$

$$= \pi\left[32y - 4y^3 + \frac{1}{5}y^5\right]_0^2 = \frac{192\pi}{5}$$

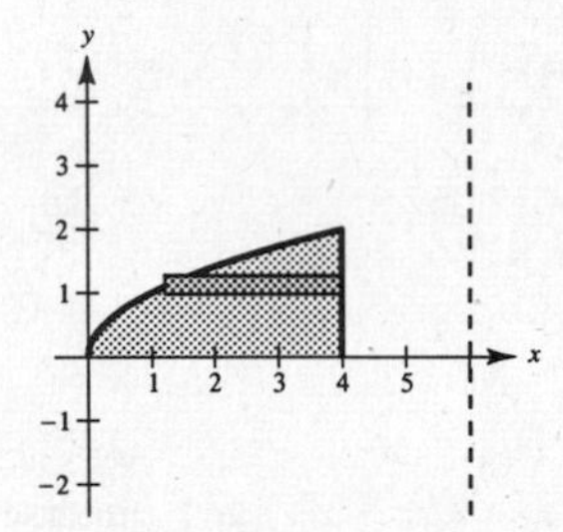

13. $y = x^2,\ y = 4x - x^2$ intersect at $(0, 0)$ and $(2, 4)$.

(a) $R(x) = 4x - x^2\ \ r(x) = x^2$

$$V = \pi \int_0^2 [(4x - x^2)^2 - x^4]\, dx$$

$$= \pi \int_0^2 (16x^2 - 8x^3)\, dx$$

$$= \pi\left[\frac{16}{3}x^3 - 2x^4\right]_0^2 = \frac{32\pi}{3}$$

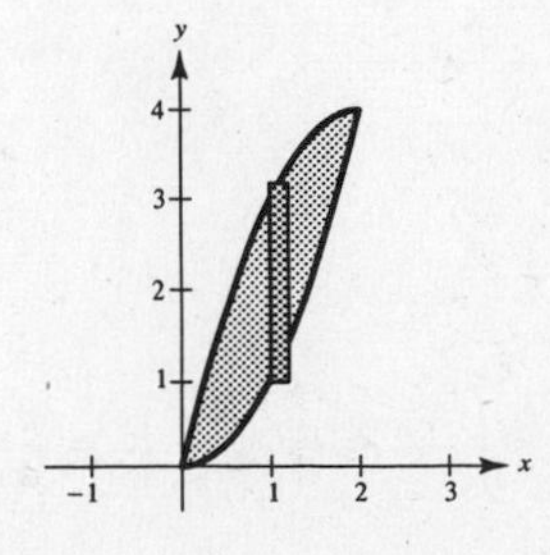

(b) $R(x) = 6 - x^2,\ r(x) = 6 - (4x - x^2)$

$$V = \pi \int_0^2 [(6 - x^2)^2 - (6 - 4x + x^2)^2]\, dx$$

$$= 8\pi \int_0^2 (x^3 - 5x^2 + 6x)\, dx$$

$$= 8\pi\left[\frac{x^4}{4} - \frac{5}{3}x^3 + 3x^2\right]_0^2 = \frac{64\pi}{3}$$

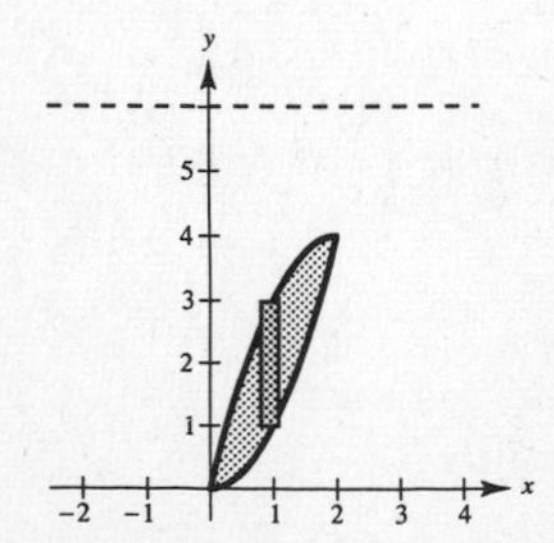

15. $R(x) = 4 - x,\ r(x) = 1$

$$V = \pi \int_0^3 [(4 - x)^2 - (1)^2]\, dx$$

$$= \pi \int_0^3 (x^2 - 8x + 15)\, dx$$

$$= \pi \left[\frac{x^3}{3} - 4x^2 + 15x\right]_0^3 = 18\pi$$

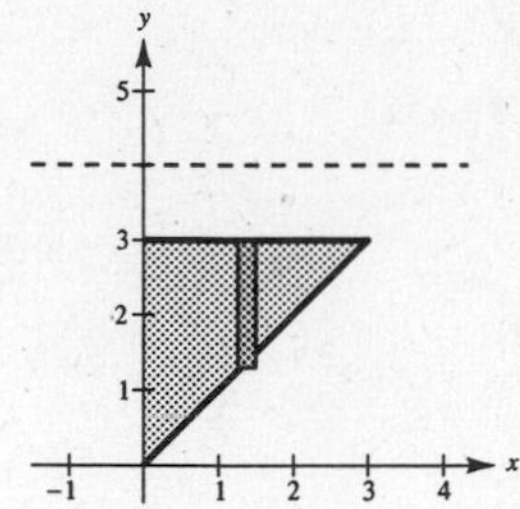

17. $R(x) = 4,\ r(x) = 4 - \dfrac{1}{1 + x}$

$$V = \pi \int_0^3 \left[4^2 - \left(4 - \frac{1}{1 + x}\right)^2\right] dx$$

$$= \pi \int_0^3 \left[\frac{8}{1 + x} - \frac{1}{(1 + x)^2}\right] dx$$

$$= \pi \left[8 \ln(1 + x) + \frac{1}{1 + x}\right]_0^3$$

$$= \pi \left[8 \ln 4 + \frac{1}{4} - 1\right]$$

$$= \left(8 \ln 4 - \frac{3}{4}\right)\pi \approx 32.485$$

19. $R(y) = 6 - y,\ r(y) = 0$

$$V = \pi \int_0^4 (6 - y)^2\, dy$$

$$= \pi \int_0^4 (y^2 - 12y + 36)\, dy$$

$$= \pi \left[\frac{y^3}{3} - 6y^2 + 36y\right]_0^4$$

$$= \frac{208\pi}{3}$$

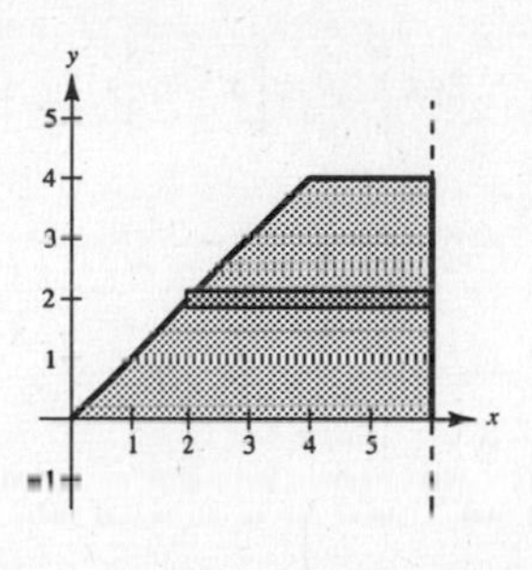

21. $R(y) = 6 - y^2,\ r(y) = 2$

$$V = \pi \int_{-2}^2 [(6 - y^2)^2 - (2)^2]\, dy$$

$$= 2\pi \int_0^2 (y^4 - 12y^2 + 32)\, dy$$

$$= 2\pi \left[\frac{y^5}{5} - 4y^3 + 32y\right]_0^2$$

$$= \frac{384\pi}{5}$$

23. $R(x) = \dfrac{1}{\sqrt{x + 1}},\ r(x) = 0$

$$V = \pi \int_0^3 \left(\frac{1}{\sqrt{x + 1}}\right)^2 dx$$

$$= \pi \int_0^3 \frac{1}{x + 1}\, dx$$

$$= \Big[\pi \ln|x + 1|\Big]_0^3 = \pi \ln 4$$

25. $R(x) = \dfrac{1}{x},\ r(x) = 0$

$$V = \pi\int_1^4 \left(\frac{1}{x}\right)^2 dx$$

$$= \pi\left[-\frac{1}{x}\right]_1^4$$

$$= \frac{3\pi}{4}$$

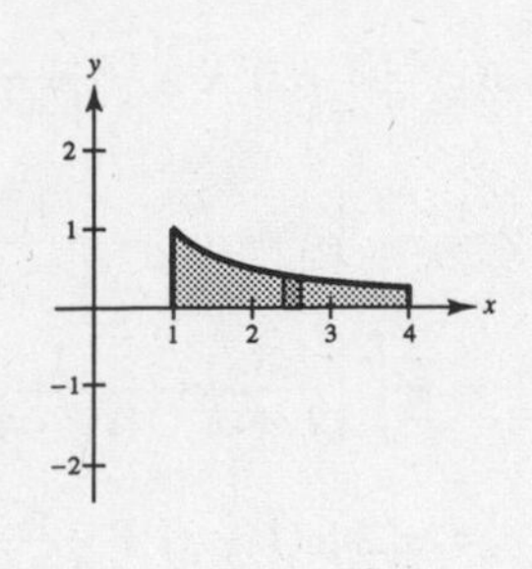

27. $R(x) = e^{-x},\ r(x) = 0$

$$V = \pi\int_0^1 (e^{-x})^2\, dx$$

$$= \pi\int_0^1 e^{-2x}\, dx$$

$$= \left[-\frac{\pi}{2}e^{-2x}\right]_0^1$$

$$= \frac{\pi}{2}(1 - e^{-2}) \approx 1.358$$

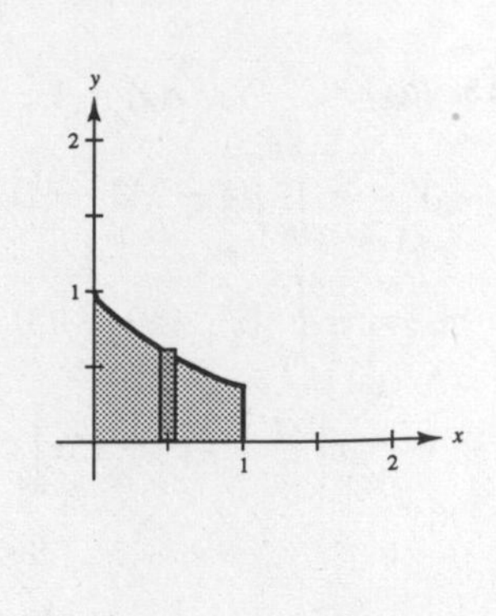

29.

$$x^2 + 1 = -x^2 + 2x + 5$$

$$x^2 - x - 2 = 0$$

$$(x - 2)(x + 1) = 0$$

$(-1, 2)$, $(2, 5)$ are points of intersection.

$$V = \pi\int_0^2 [(5 + 2x - x^2)^2 - (x^2 + 1)^2)]\, dx + \pi\int_2^3 [(x^2 + 1)^2 - (5 + 2x - x^2)^2]\, dx$$

$$= \pi\int_0^2 (-4x^3 - 8x^2 + 20x + 24)\, dx + \pi\int_2^3 (4x^3 + 8x^2 - 20x - 24)\, dx$$

$$= \pi\left[-x^4 - \frac{8}{3}x^3 + 10x^2 + 24x\right]_0^2 + \pi\left[x^4 + \frac{8}{3}x^3 - 10x^2 - 24x\right]_2^3$$

$$= \pi\frac{152}{3} + \pi\frac{125}{3} = \frac{277\pi}{3}$$

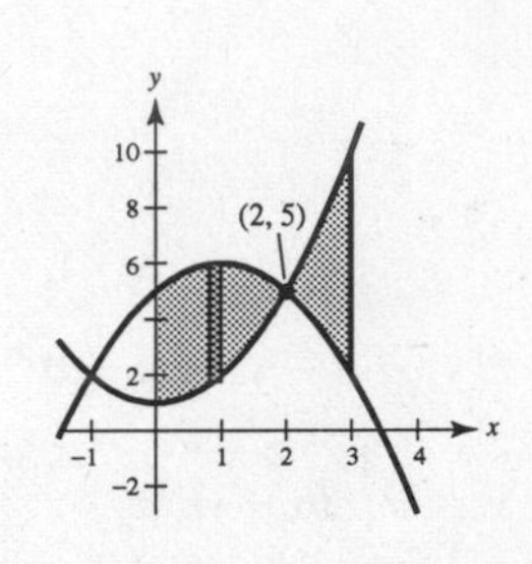

31. $y = 6 - 3x \Rightarrow x = \dfrac{1}{3}(6 - y)$

$$V = \pi\int_0^6 \left[\frac{1}{3}(6 - y)\right]^2 dy$$

$$= \frac{\pi}{9}\int_0^6 [36 - 12y + y^2]\, dy$$

$$= \frac{\pi}{9}\left[36y - 6y^2 + \frac{y^3}{3}\right]_0^6$$

$$= \frac{\pi}{9}\left[216 - 216 + \frac{216}{3}\right]$$

$$= 8\pi\left(= \frac{1}{3}\pi r^2 h - \text{volume at core}\right)$$

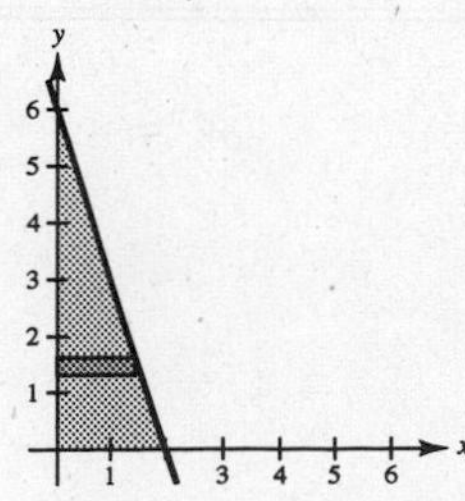

33. $V = \pi\displaystyle\int_0^{\pi} [\sin x]^2\, dx \approx 4.9348$

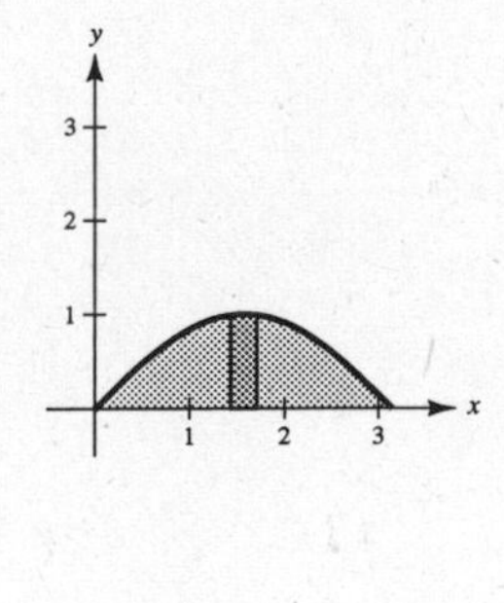

35. $V = \pi \int_0^2 [e^{-x^2}]^2 \, dx \approx 1.9686$

37. $V = \pi \int_{-1}^2 [e^{x/2} + e^{-x/2}]^2 \, dx \approx 49.0218$

39. $A \approx 3$

Matches (a)

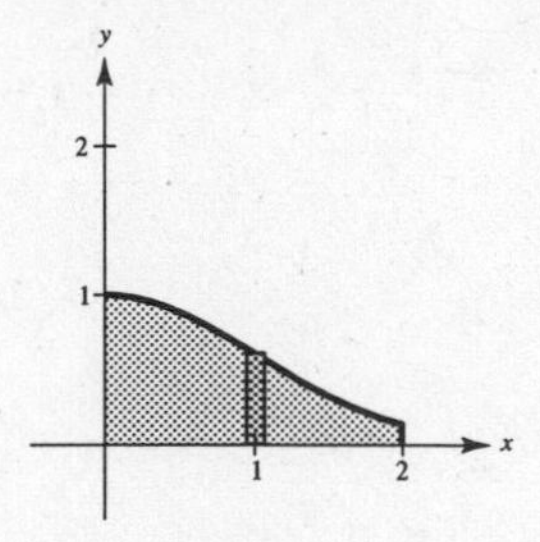

41. Disk Method:

$$V = \pi \int_a^b [R(x)]^2 \, dx \quad \text{or} \quad V = \pi \int_c^d [R(y)]^2 \, dy$$

Washer Method:

$$V = \pi \int_a^b ([R(x)]^2 - [r(x)]^2) \, dx \quad \text{or}$$

$$V = \pi \int_c^d ([R(y)]^2 - [r(y)]^2) \, dy$$

43.

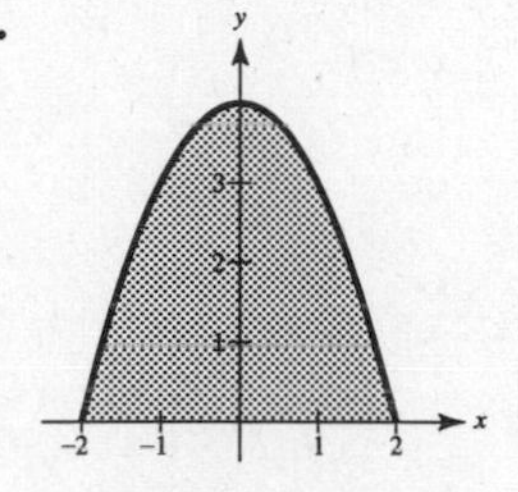

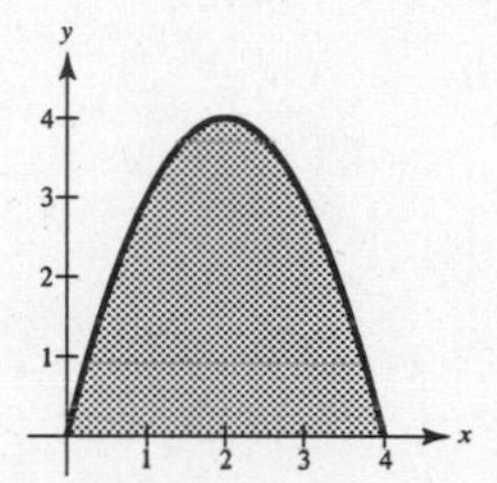

The volumes are the same because the solid has been translated horizontally.

45. $R(x) = \frac{1}{2}x, \; r(x) = 0$

$$V = \pi \int_0^6 \frac{1}{4}x^2 \, dx$$

$$= \left[\frac{\pi}{12}x^3\right]_0^6 = 18\pi$$

Note: $V = \frac{1}{3}\pi r^2 h$

$$= \frac{1}{3}\pi(3^2)6$$

$$= 18\pi$$

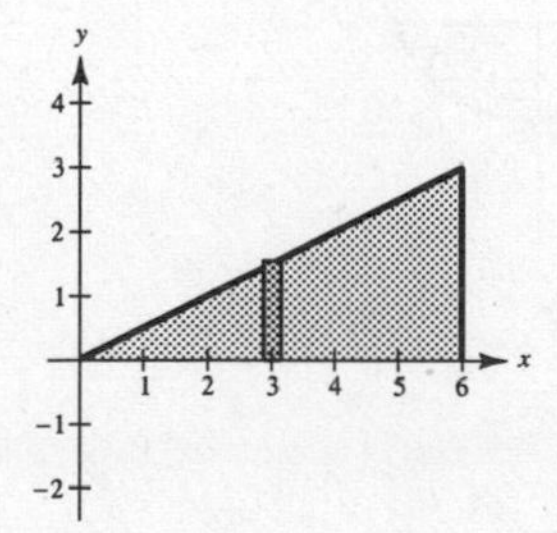

47. $R(x) = \sqrt{r^2 - x^2}, \; r(x) = 0$

$$V = \pi \int_{-r}^r (r^2 - x^2) \, dx$$

$$= 2\pi \int_0^r (r^2 - x^2) \, dx$$

$$= 2\pi \left[r^2 x - \frac{1}{3}x^3\right]_0^r$$

$$= 2\pi\left(r^3 - \frac{1}{3}r^3\right) = \frac{4}{3}\pi r^3$$

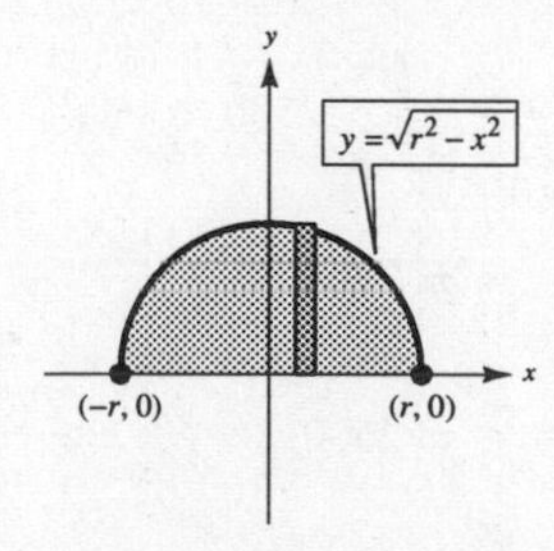

49. $x = r - \frac{r}{H}y = r\left(1 - \frac{y}{H}\right)$, $R(y) = r\left(1 - \frac{y}{H}\right)$, $r(y) = 0$

$$V = \pi\int_0^h \left[r\left(1 - \frac{y}{H}\right)\right]^2 dy = \pi r^2 \int_0^h \left(1 - \frac{2}{H}y + \frac{1}{H^2}y^2\right) dy$$

$$= \pi r^2\left[y - \frac{1}{H}y^2 + \frac{1}{3H^2}y^3\right]_0^h$$

$$= \pi r^2\left(h - \frac{h^2}{H} + \frac{h^3}{3H^2}\right)$$

$$= \pi r^2 h\left(1 - \frac{h}{H} + \frac{h^2}{3H^2}\right)$$

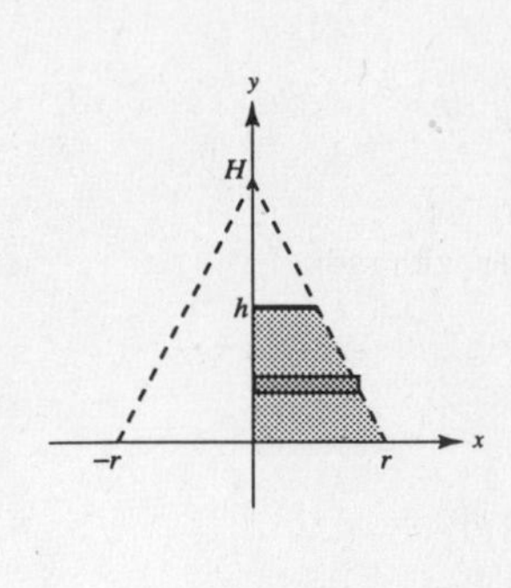

51. $V = \pi\int_0^2 \left(\frac{1}{8}x^2\sqrt{2 - x}\right)^2 dx = \frac{\pi}{64}\int_0^2 x^4(2 - x)\, dx = \frac{\pi}{64}\left[\frac{2x^5}{5} - \frac{x^6}{6}\right]_0^2 = \frac{\pi}{30}$

53. (a) $R(x) = \frac{3}{5}\sqrt{25 - x^2}$, $r(x) = 0$

$$V = \frac{9\pi}{25}\int_{-5}^{5} (25 - x^2)\, dx$$

$$= \frac{18\pi}{25}\int_0^5 (25 - x^2)\, dx$$

$$= \frac{18\pi}{25}\left[25x - \frac{x^3}{3}\right]_0^5 = 60\pi$$

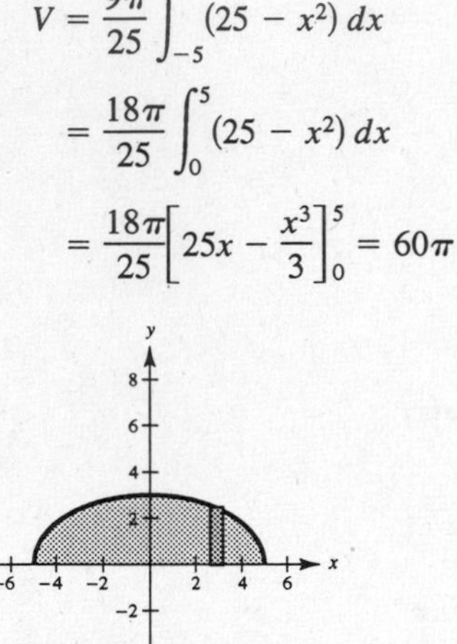

(b) $R(y) = \frac{5}{3}\sqrt{9 - y^2}$, $r(y) = 0$, $x \geq 0$

$$V = \frac{25\pi}{9}\int_0^3 (9 - y^2)\, dy$$

$$= \frac{25\pi}{9}\left[9y - \frac{y^3}{3}\right]_0^3 = 50\pi$$

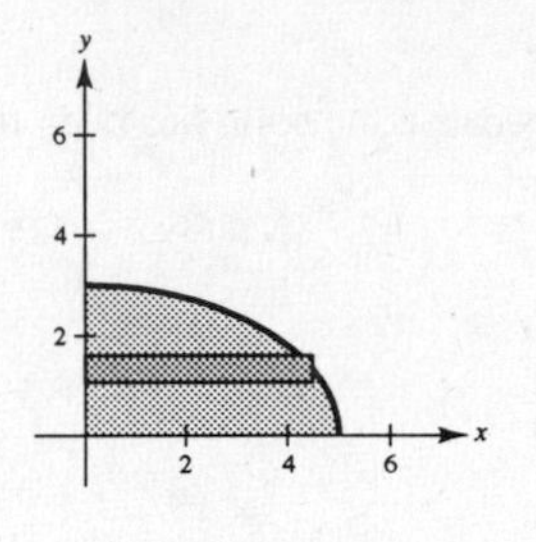

55. Total volume: $V = \frac{4\pi(50)^3}{3} = \frac{500{,}000\pi}{3}$ ft^3

Volume of water in the tank:

$$\pi\int_{-50}^{y_0} (\sqrt{2500 - y^2})^2\, dy = \pi\int_{-50}^{y_0} (2500 - y^2)\, dy$$

$$= \pi\left[2500y - \frac{y^3}{3}\right]_{-50}^{y_0}$$

$$= \pi\left(2500y_0 - \frac{y_0^3}{3} + \frac{250{,}000}{3}\right)$$

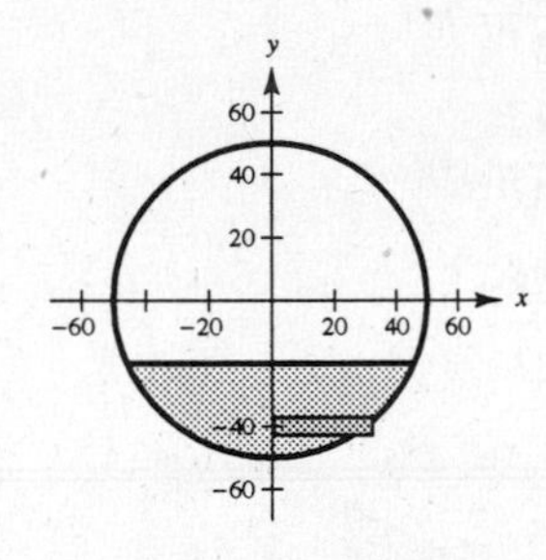

When the tank is one-fourth of its capacity:

$$\frac{1}{4}\left(\frac{500{,}000\pi}{3}\right) = \pi\left(2500y_0 - \frac{y_0^3}{3} + \frac{250{,}000}{3}\right)$$

$$125{,}000 = 7500y_0 - y_0^3 + 250{,}000$$

$$y_0^3 - 7500y_0 - 125{,}000 = 0$$

$$y_0 \approx -17.36$$

Depth: $-17.36 - (-50) = 32.64$ feet

When the tank is three-fourths of its capacity the depth is $100 - 32.64 = 67.36$ feet.

57. (a) $\pi \int_0^h r^2\,dx$ (ii)

is the volume of a right circular cylinder with radius r and height h.

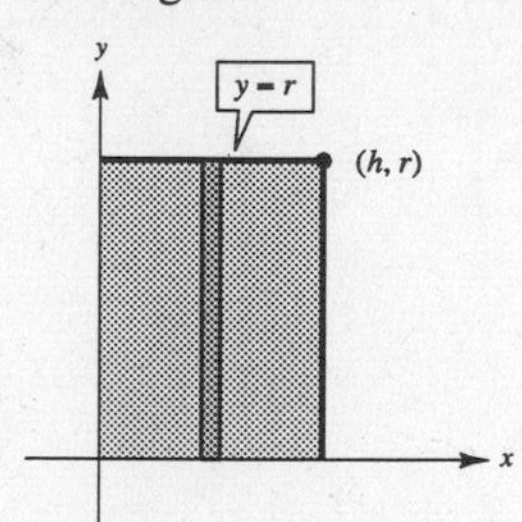

(b) $\pi \int_{-b}^{b} \left(a\sqrt{1 - \frac{x^2}{b^2}}\right)^2 dx$ (iv)

is the volume of an ellipsoid with axes $2a$ and $2b$.

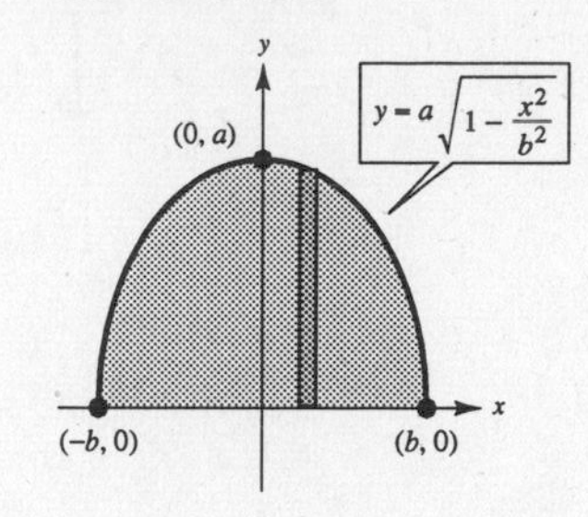

(c) $\pi \int_{-r}^{r} \left(\sqrt{r^2 - x^2}\right)^2 dx$ (iii)

is the volume of a sphere with radius r.

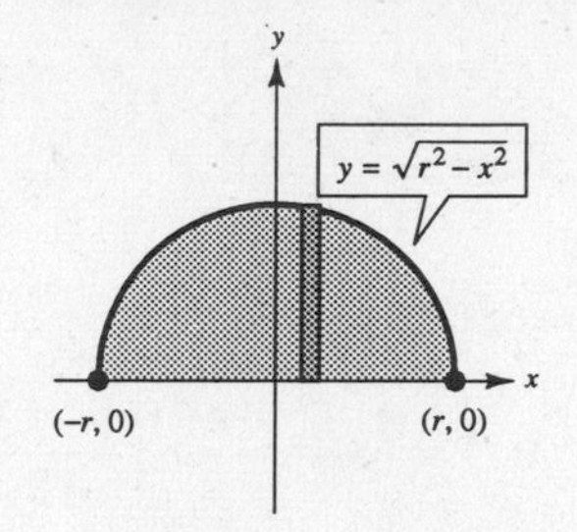

(d) $\pi \int_0^h \left(\frac{rx}{h}\right)^2 dx$ (i)

is the volume of a right circular cone with the radius of the base as r and height h.

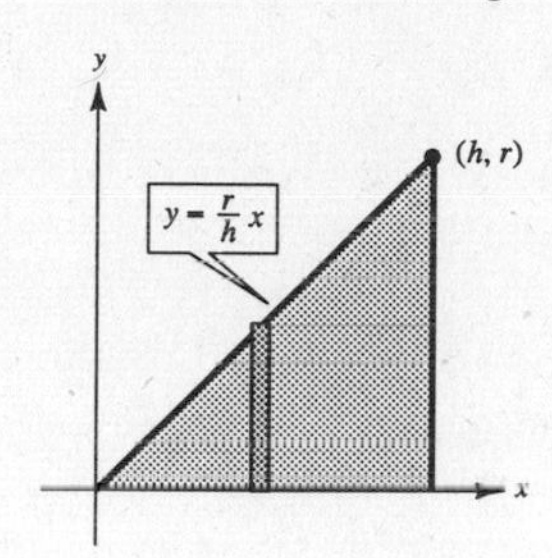

(e) $\pi \int_{-r}^{r} \left[\left(R + \sqrt{r^2 - x^2}\right)^2 - \left(R - \sqrt{r^2 - x^2}\right)^2\right] dx$ (v)

is the volume of a torus with the radius of its circular cross section as r and the distance from the axis of the torus to the center of its cross section as R.

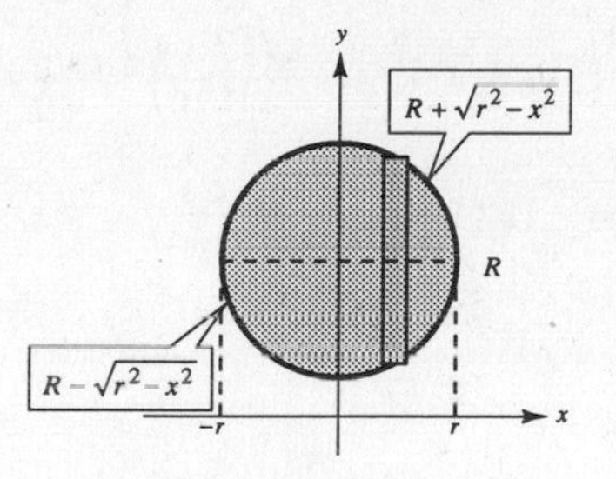

59.

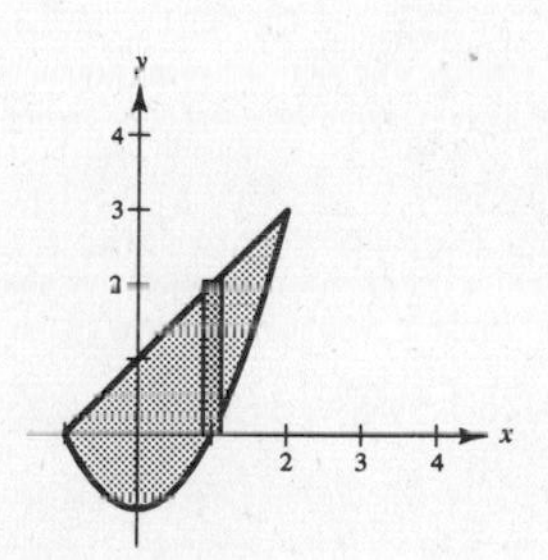

Base of Cross Section $= (x + 1) - (x^2 - 1) = 2 + x - x^2$

(a) $A(x) = b^2 = (2 + x - x^2)^2$

$$= 4 + 4x - 3x^2 - 2x^3 + x^4$$

$$V = \int_{-1}^{2} (4 + 4x - 3x^2 - 2x^3 + x^4)\,dx$$

$$= \left[4x + 2x^2 - x^3 - \frac{1}{2}x^4 + \frac{1}{5}x^5\right]_{-1}^{2} = \frac{81}{10}$$

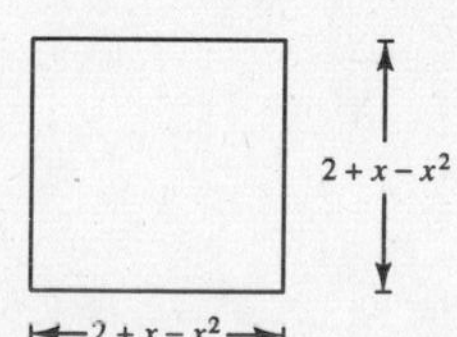

(b) $A(x) = bh = (2 + x - x^2)1$

$$V = \int_{-1}^{2} (2 + x - x^2)\,dx = \left[2x + \frac{x^2}{2} - \frac{x^3}{3}\right]_{-1}^{2} = \frac{9}{2}$$

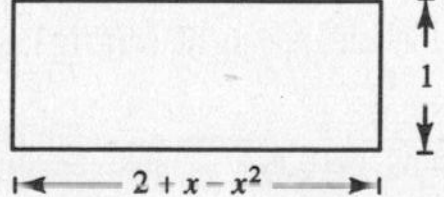

61.

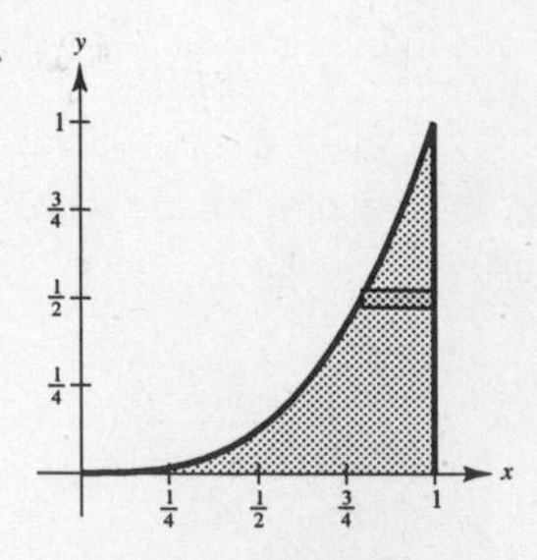

(a) $A(y) = b^2 = \left(1 - \sqrt[3]{y}\right)^2$

$$V = \int_0^1 \left(1 - \sqrt[3]{y}\right)^2 dy$$

$$= \int_0^1 (1 - 2y^{1/3} + y^{2/3})\, dy$$

$$= \left[y - \frac{3}{2}y^{4/3} + \frac{3}{5}y^{5/3}\right]_0^1 = \frac{1}{10}$$

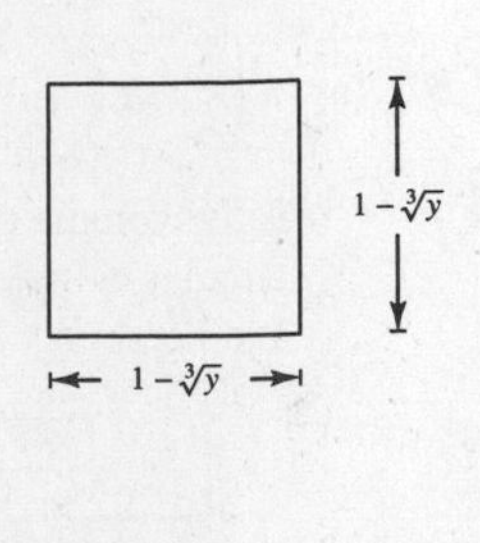

Base of Cross Section $= 1 - \sqrt[3]{y}$

(b) $A(y) = \frac{1}{2}\pi r^2 = \frac{1}{2}\pi\left(\frac{1 - \sqrt[3]{y}}{2}\right)^2 = \frac{1}{8}\pi\left(1 - \sqrt[3]{y}\right)^2$

$$V = \frac{1}{8}\pi \int_0^1 \left(1 - \sqrt[3]{y}\right)^2 dy = \frac{\pi}{8}\left(\frac{1}{10}\right) = \frac{\pi}{80}$$

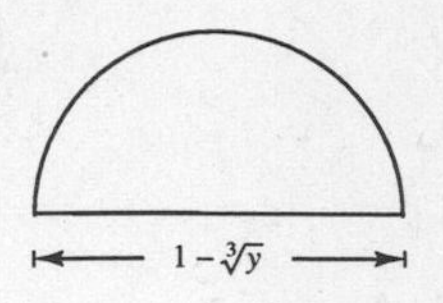

(c) $A(y) = \frac{1}{2}bh = \frac{1}{2}\left(1 - \sqrt[3]{y}\right)\left(\frac{\sqrt{3}}{2}\right)\left(1 - \sqrt[3]{y}\right)$

$$= \frac{\sqrt{3}}{4}\left(1 - \sqrt[3]{y}\right)^2$$

$$V = \frac{\sqrt{3}}{4}\int_0^1 \left(1 - \sqrt[3]{y}\right)^2 dy = \frac{\sqrt{3}}{4}\left(\frac{1}{10}\right) = \frac{\sqrt{3}}{40}$$

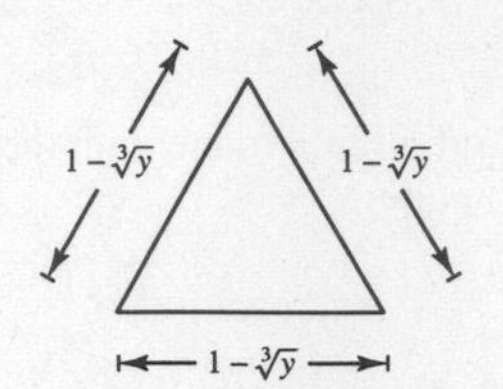

(d) $A(y) = \frac{1}{2}\pi ab = \frac{\pi}{2}(2)\left(1 - \sqrt[3]{y}\right)\frac{1 - \sqrt[3]{y}}{2}$

$$= \frac{\pi}{2}\left(1 - \sqrt[3]{y}\right)^2$$

$$V = \frac{\pi}{2}\int_0^1 \left(1 - \sqrt[3]{y}\right)^2 dy = \frac{\pi}{2}\left(\frac{1}{10}\right) = \frac{\pi}{20}$$

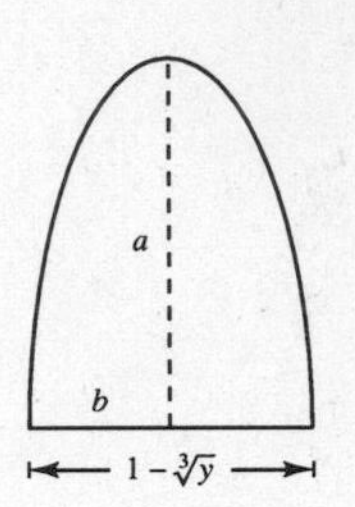

63. Let $A_1(x)$ and $A_2(x)$ equal the areas of the cross sections of the two solids for $a \le x \le b$. Since $A_1(x) = A_2(x)$, we have

$$V_1 = \int_a^b A_1(x)\, dx = \int_a^b A_2(x)\, dx = V_2$$

Thus, the volumes are the same.

65. $\frac{4}{3}\pi(25 - r^2)^{3/2} = \frac{1}{2}\left(\frac{4}{3}\right)\pi(125)$

$$(25 - r^2)^{3/2} = \frac{125}{2}$$

$$25 - r^2 = \left(\frac{125}{2}\right)^{2/3}$$

$$25 - \frac{25}{(2^{2/3})} = r^2$$

$$25(1 - 2^{-2/3}) = r^2$$

$$r = 5\sqrt{1 - 2^{-2/3}} \approx 3.0415$$

67. (a) Since the cross sections are isosceles right triangles:

$$A(x) = \frac{1}{2}bh = \frac{1}{2}\left(\sqrt{r^2 - y^2}\right)\left(\sqrt{r^2 - y^2}\right) = \frac{1}{2}(r^2 - y^2)$$

$$V = \frac{1}{2}\int_{-r}^{r} (r^2 - y^2)\, dy = \int_0^r (r^2 - y^2)\, dy = \left[r^2y - \frac{y^3}{3}\right]_0^r = \frac{2}{3}r^3$$

(b) $A(x) = \frac{1}{2}bh = \frac{1}{2}\sqrt{r^2 - y^2}\left(\sqrt{r^2 - y^2}\tan\theta\right) = \frac{\tan\theta}{2}(r^2 - y^2)$

$$V = \frac{\tan\theta}{2}\int_{-r}^{r} (r^2 - y^2)\, dy = \tan\theta \int_0^r (r^2 - y^2)\, dy = \tan\theta\left[r^2y - \frac{y^3}{3}\right]_0^r = \frac{2}{3}r^3\tan\theta$$

As $\theta \to 90°$, $V \to \infty$.

Section 6.3 Volume: The Shell Method

1. $p(x) = x$

$h(x) = x$

$$V = 2\pi\int_0^2 x(x)\,dx = \left[\frac{2\pi x^3}{3}\right]_0^2 = \frac{16\pi}{3}$$

3. $p(x) = x$

$h(x) = \sqrt{x}$

$$V = 2\pi\int_0^4 x\sqrt{x}\,dx$$

$$= 2\pi\int_0^4 x^{3/2}\,dx$$

$$= \left[\frac{4\pi}{5}x^{5/2}\right]_0^4 = \frac{128\pi}{5}$$

5. $p(x) = x$

$h(x) = x^2$

$$V = 2\pi\int_0^2 x^3\,dx$$

$$= \left[\frac{\pi}{2}x^4\right]_0^2 = 8\pi$$

7. $p(x) = x$

$h(x) = (4x - x^2) - x^2 = 4x - 2x^2$

$$V = 2\pi\int_0^2 x(4x - 2x^2)\,dx$$

$$= 4\pi\int_0^2 (2x^2 - x^3)\,dx$$

$$= 4\pi\left[\frac{2}{3}x^3 - \frac{1}{4}x^4\right]_0^2 = \frac{16\pi}{3}$$

9. $p(x) = x$

$h(x) = 4 - (4x - x^2) = x^2 - 4x + 4$

$$V = 2\pi\int_0^2 (x^3 - 4x^2 + 4x)\,dx$$

$$= 2\pi\left[\frac{x^4}{4} - \frac{4}{3}x^3 + 2x^2\right]_0^2 = \frac{8\pi}{3}$$

11. $p(x) = x$

$h(x) = \dfrac{1}{\sqrt{2\pi}}e^{-x^2/2}$

$$V = 2\pi\int_0^1 x\left(\frac{1}{\sqrt{2\pi}}e^{-x^2/2}\right)dx$$

$$= \sqrt{2\pi}\int_0^1 e^{-x^2/2}x\,dx$$

$$= \left[-\sqrt{2\pi}\,e^{-x^2/2}\right]_0^1 = \sqrt{2\pi}\left(1 - \frac{1}{\sqrt{e}}\right) \approx 0.986$$

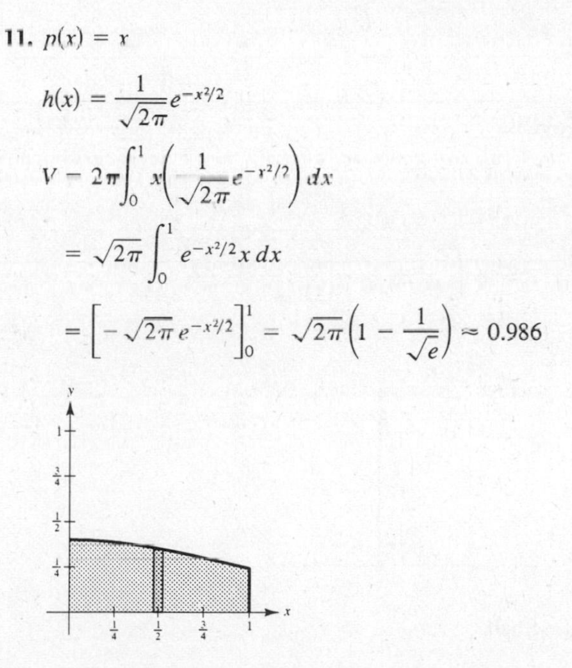

13. $p(y) = y$

$h(y) = 2 - y$

$$V = 2\pi\int_0^2 y(2 - y)\,dy$$

$$= 2\pi\int_0^2 (2y - y^2)\,dy$$

$$= 2\pi\left[y^2 - \frac{y^3}{3}\right]_0^2 = \frac{8\pi}{3}$$

15. $p(y) = y$ and $h(y) = 1$ if $0 \le y < \dfrac{1}{2}$.

$p(y) = y$ and $h(y) = \dfrac{1}{y} - 1$ if $\dfrac{1}{2} \le y \le 1$.

$$V = 2\pi \int_0^{1/2} y\,dy + 2\pi \int_{1/2}^1 (1 - y)\,dy$$

$$= 2\pi\left[\frac{y^2}{2}\right]_0^{1/2} + 2\pi\left[y - \frac{y^2}{2}\right]_{1/2}^1 = \frac{\pi}{4} + \frac{\pi}{4} = \frac{\pi}{2}$$

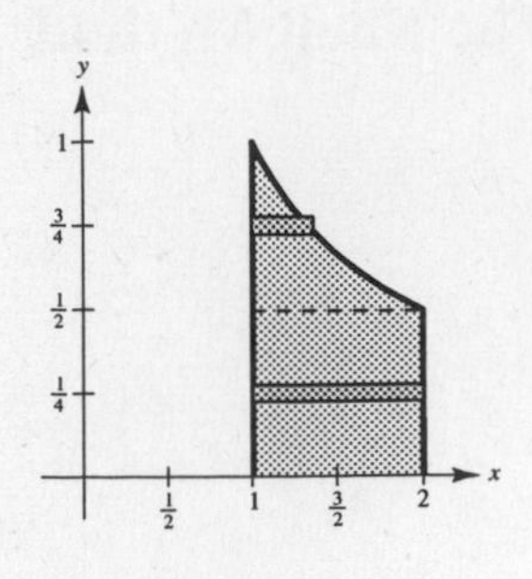

17. $p(x) = 4 - x$

$h(x) = 4x - x^2 - x^2 = 4x - 2x^2$

$$V = 2\pi \int_0^2 (4 - x)(4x - 2x^2)\,dx$$

$$= 2\pi(2) \int_0^2 (x^3 - 6x^2 + 8x)\,dx$$

$$= 4\pi\left[\frac{x^4}{4} - 2x^3 + 4x^2\right]_0^2 = 16\pi$$

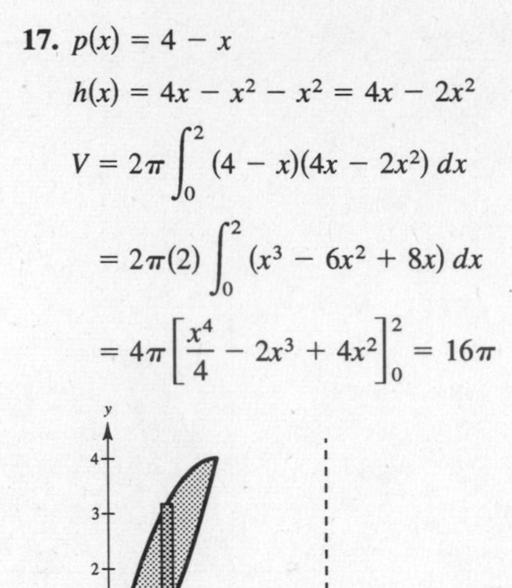

19. $p(x) = 5 - x$

$h(x) = 4x - x^2$

$$V = 2\pi \int_0^4 (5 - x)(4x - x^2)\,dx$$

$$= 2\pi \int_0^4 (x^3 - 9x^2 + 20x)\,dx$$

$$= 2\pi\left[\frac{x^4}{4} - 3x^3 + 10x^2\right]_0^4 = 64\pi$$

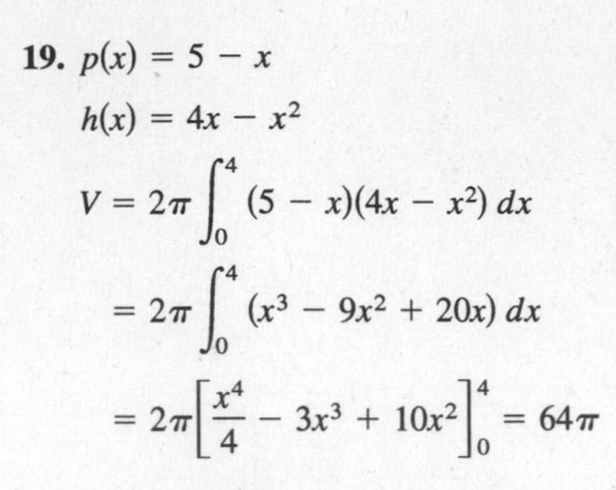

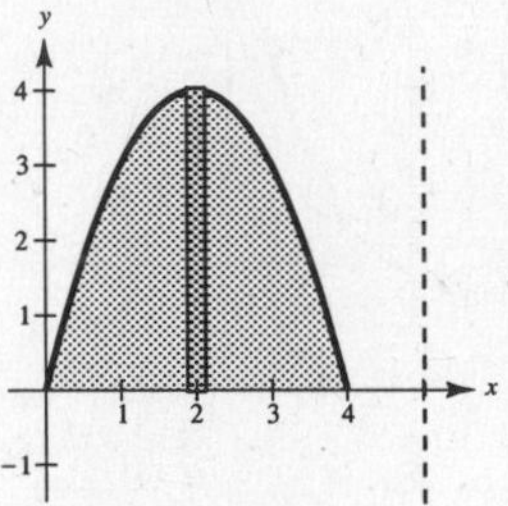

21. (a) **Disk**

$R(x) = x^3$

$r(x) = 0$

$$V = \pi \int_0^2 x^6\,dx = \pi\left[\frac{x^7}{7}\right]_0^2 = \frac{128\pi}{7}$$

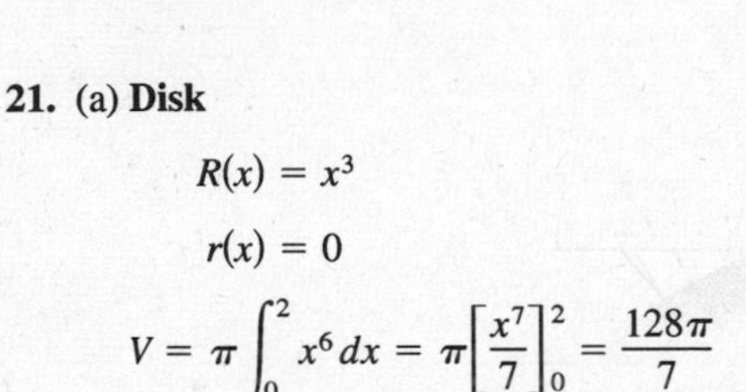

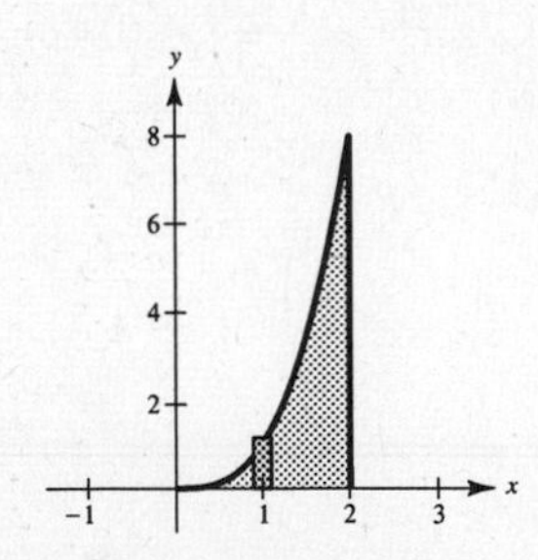

(b) **Shell**

$p(x) = x$

$h(x) = x^3$

$$V = 2\pi \int_0^2 x^4\,dx = 2\pi\left[\frac{x^5}{5}\right]_0^2 = \frac{64\pi}{5}$$

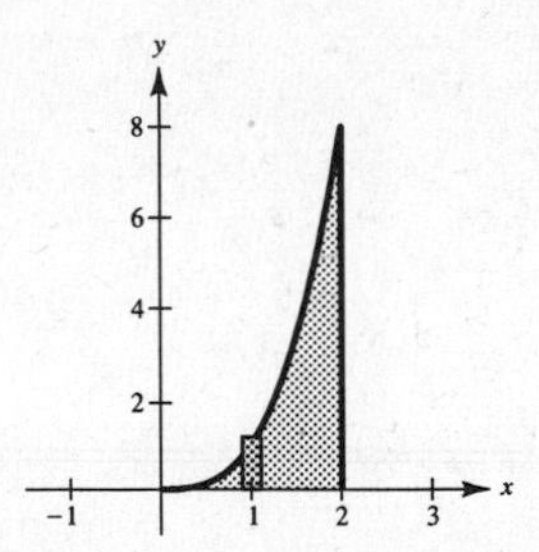

(c) **Shell**

$p(x) = 4 - x$

$h(x) = x^3$

$$V = 2\pi \int_0^2 (4 - x)x^3\,dx$$

$$= 2\pi \int_0^2 (4x^3 - x^4)\,dx$$

$$= 2\pi\left[x^4 - \frac{1}{5}x^5\right]_0^2 = \frac{96\pi}{5}$$

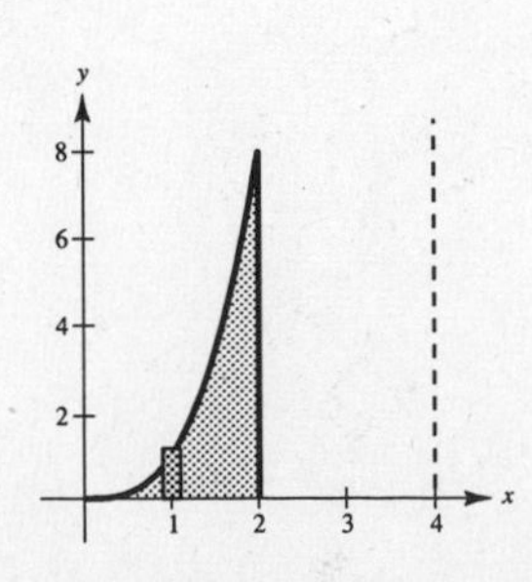

23. (a) **Shell**

$p(y) = y$

$h(y) = (a^{1/2} - y^{1/2})^2$

$$V = 2\pi\int_0^a y(a - 2a^{1/2}y^{1/2} + y)\,dy$$

$$= 2\pi\int_0^a (ay - 2a^{1/2}y^{3/2} + y^2)\,dy$$

$$= 2\pi\left[\frac{a}{2}y^2 - \frac{4a^{1/2}}{5}y^{5/2} + \frac{y^3}{3}\right]_0^a$$

$$= 2\pi\left[\frac{a^3}{2} - \frac{4a^3}{5} + \frac{a^3}{3}\right] = \frac{\pi a^3}{15}$$

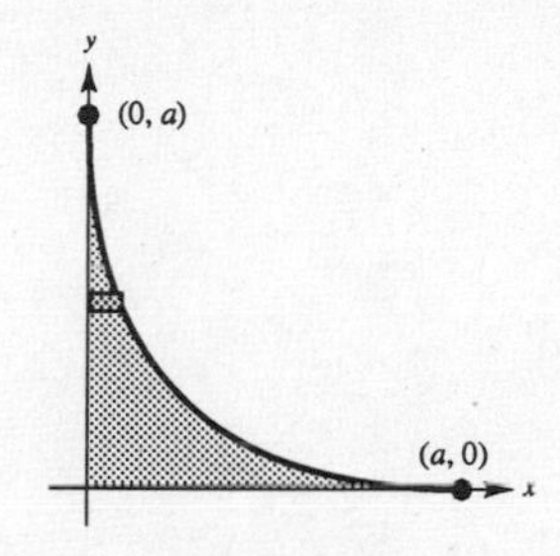

(b) Same as part (a) by symmetry

(c) **Shell**

$p(x) = a - x$

$h(x) = (a^{1/2} - x^{1/2})^2$

$$V = 2\pi\int_0^a (a - x)(a^{1/2} - x^{1/2})^2\,dx$$

$$= 2\pi\int_0^a (a^2 - 2a^{3/2}x^{1/2} + 2a^{1/2}x^{3/2} - x^2)\,dx$$

$$= 2\pi\left[a^2x - \frac{4}{3}a^{3/2}x^{3/2} + \frac{4}{5}a^{1/2}x^{5/2} - \frac{1}{3}x^3\right]_0^a = \frac{4\pi a^3}{15}$$

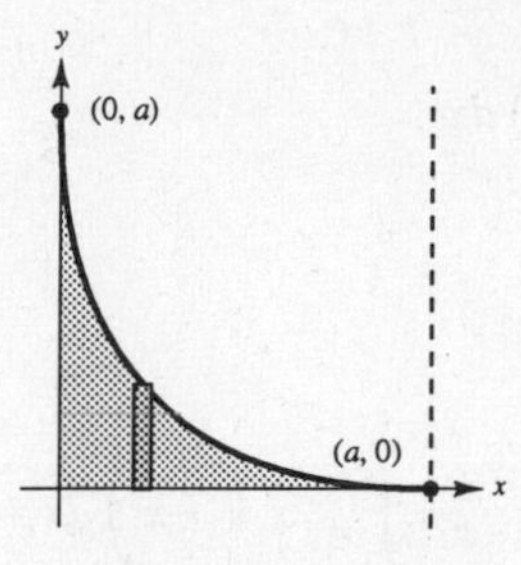

25. $V = 2\pi\int_c^d p(y)h(y)\,dy$ or $V = 2\pi\int_a^b p(x)h(x)\,dx$

27. $\pi\int_1^5 (x - 1)\,dx = \pi\int_1^5 \left(\sqrt{x - 1}\right)^2 dx$

This integral represents the volume of the solid generated by revolving the region bounded by $y = \sqrt{x - 1}$, $y = 0$, and $x = 5$ about the x-axis by using the Disk Method.

$$2\pi\int_0^2 y[5 - (y^2 + 1)]\,dy$$

represents this same volume by using the Shell Method.

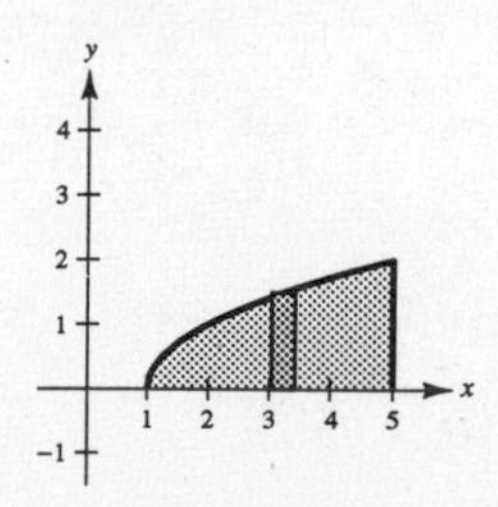

Disk Method

29. (a)

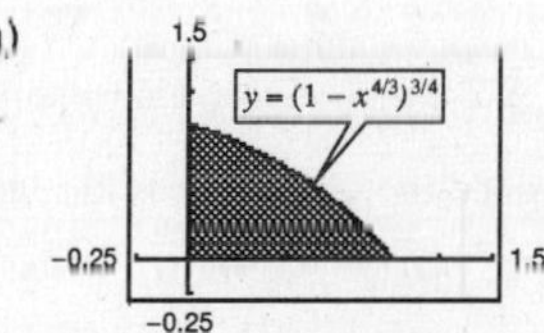

(b) $x^{4/3} + y^{4/3} = 1,\ x = 0,\ y = 0$

$y = (1 - x^{4/3})^{3/4}$

$$V = 2\pi\int_0^1 x(1 - x^{4/3})^{3/4}\,dx \approx 1.5056$$

31. (a)

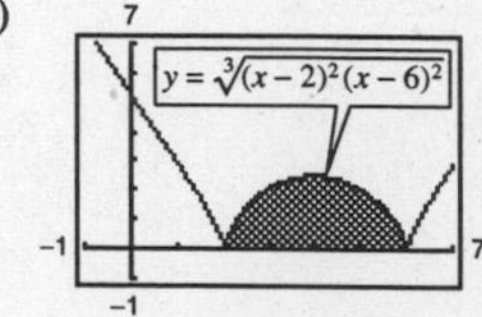

(b) $V = 2\pi \int_2^6 x\sqrt[3]{(x-2)^2(x-6)^2}\,dx \approx 187.249$

33. $y = 2e^{-x}$, $y = 0$, $x = 0$, $x = 2$

Volume ≈ 7.5

Matches (d)

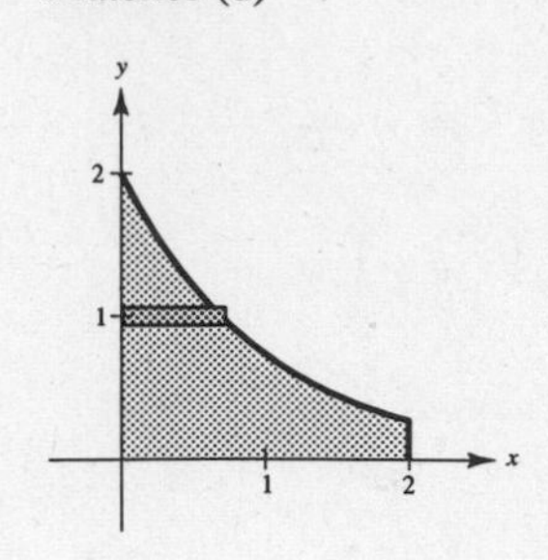

35. $p(x) = x$

$h(x) = 2 - \frac{1}{2}x^2$

$$V = 2\pi \int_0^2 x\left(2 - \frac{1}{2}x^2\right) dx = 2\pi \int_0^2 \left(2x - \frac{1}{2}x^3\right) dx = 2\pi\left[x^2 - \frac{1}{8}x^4\right]_0^2 = 4\pi \text{ (total volume)}$$

Now find x_0 such that

$$\pi = 2\pi \int_0^{x_0} \left(2x - \frac{1}{2}x^3\right) dx$$

$$1 = 2\left[x^2 - \frac{1}{8}x^4\right]_0^{x_0}$$

$$1 = 2x_0^2 - \frac{1}{4}x_0^4$$

$$x_0^4 - 8x_0^2 + 4 = 0$$

$$x_0^2 = 4 \pm 2\sqrt{3} \quad \text{(Quadratic Formula)}$$

Take $x_0 = \sqrt{4 - 2\sqrt{3}}$ since the other root is too large.

Diameter: $2\sqrt{4 - 2\sqrt{3}} \approx 1.464$

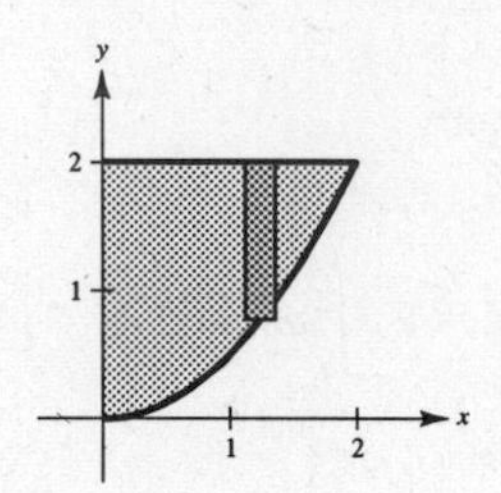

37. $V = 4\pi \int_{-1}^{1} (2 - x)\sqrt{1 - x^2}\,dx$

$$= 8\pi \int_{-1}^{1} \sqrt{1 - x^2}\,dx - 4\pi \int_{-1}^{1} x\sqrt{1 - x^2}\,dx$$

$$= 8\pi\left(\frac{\pi}{2}\right) + 2\pi \int_{-1}^{1} x(1 - x^2)^{1/2}(-2)\,dx$$

$$= 4\pi^2 + \left[2\pi\left(\frac{2}{3}\right)(1 - x^2)^{3/2}\right]_{-1}^{1} = 4\pi^2$$

39. Disk Method

$R(y) = \sqrt{r^2 - y^2}$

$r(y) = 0$

$$V = \pi \int_{r-h}^{r} (r^2 - y^2)\,dy$$

$$= \pi\left[r^2y - \frac{y^3}{3}\right]_{r-h}^{r} = \frac{1}{3}\pi h^2(3r - h)$$

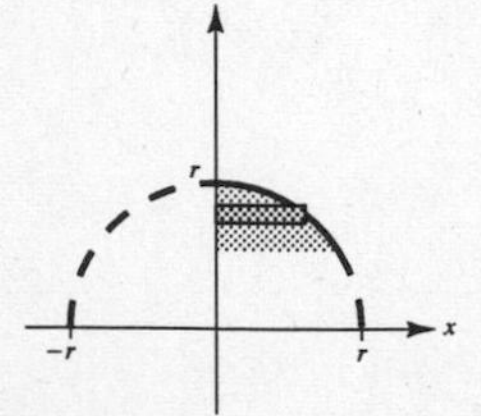

41. (a) $2\pi \int_0^r hx\left(1 - \frac{x}{r}\right) dx$ (ii)

is the volume of a right circular cone with the radius of the base as r and height h.

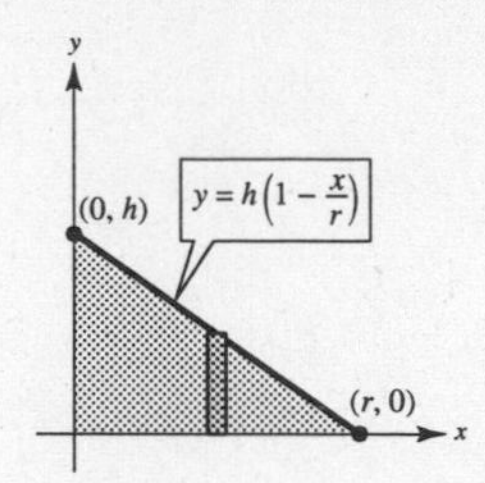

(b) $2\pi \int_{-r}^r (R - x)\left(2\sqrt{r^2 - x^2}\right) dx$ (v)

is the volume of a torus with the radius of its circular cross section as r and the distance from the axis of the torus to the center of its cross section as R.

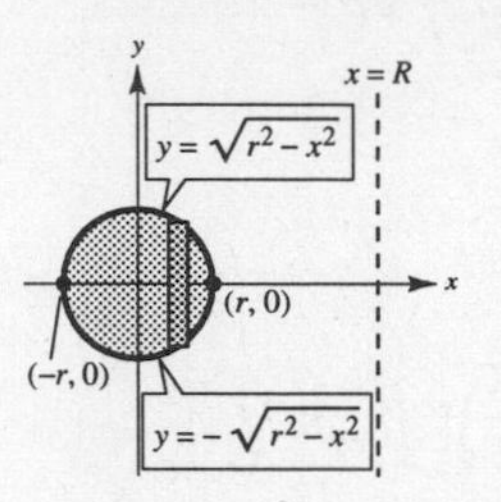

(c) $2\pi \int_0^r 2x\sqrt{r^2 - x^2}\, dx$ (iii) is the volume of a sphere with radius r.

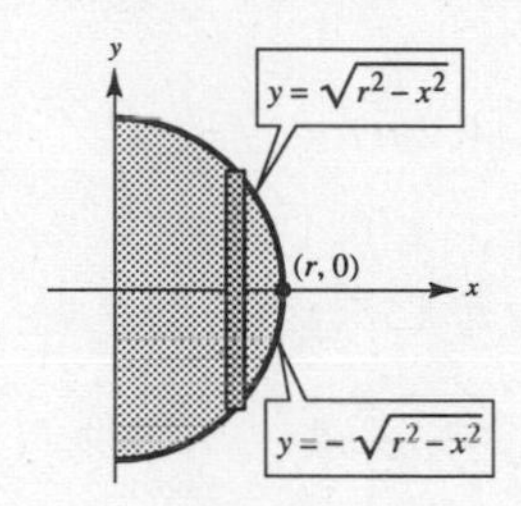

(d) $2\pi \int_0^r hx\, dx$ (i) is the volume of a right circular cylinder with a radius of r and a height of h.

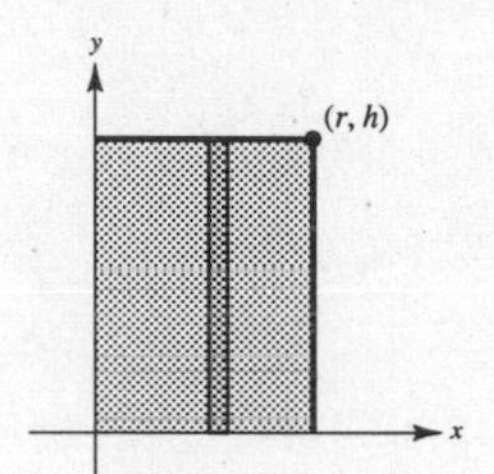

(e) $2\pi \int_0^b 2ax\sqrt{1 - (x^2/b^2)}\, dx$ (iv)

is the volume of an ellipsoid with axes $2a$ and $2b$.

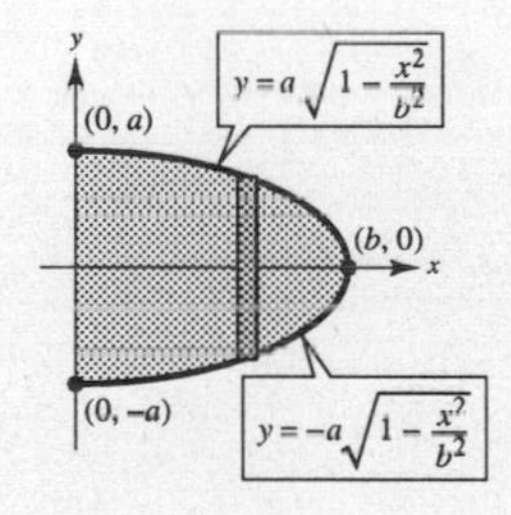

43. (a) $V = 2\pi \int_0^{200} xf(x)\, dx$

$$\approx \frac{2\pi(200)}{3(8)}[0 + 4(25)(19) + 2(50)(19) + 4(75)(17) + 2(100)15 + 4(125)(14) + 2(150)(10) + 4(175)(6) + 0]$$

$\approx$ 1,366,593 cubic feet

(b) $d = -0.000561x^2 + 0.0189x + 19.39$

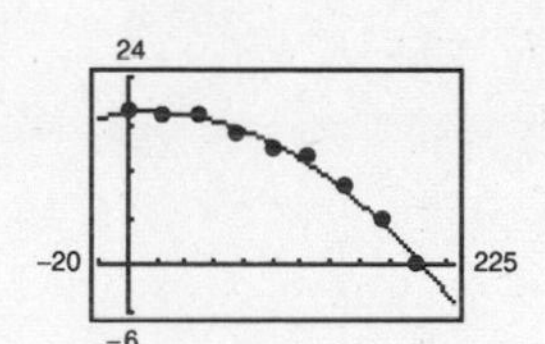

(c) $V \approx 2\pi \int_0^{200} xd(x)\, dx \approx 2\pi(213{,}800) = 1{,}343{,}345$ cubic feet

(d) Number gallons $\approx V(7.48) = 10{,}048{,}221$ gallons

Section 6.4 Arc Length and Surfaces of Revolution

1. $(0, 0), (5, 12)$

(a) $d = \sqrt{(5-0)^2 + (12-0)^2} = 13$

(b) $y = \frac{12}{5}x$

$y' = \frac{12}{5}$

$$s = \int_0^5 \sqrt{1 + \left(\frac{12}{5}\right)^2}\,dx = \left[\frac{13}{5}x\right]_0^5 = 13$$

3. $y = \frac{2}{3}x^{3/2} + 1$

$y' = x^{1/2}, [0, 1]$

$$s = \int_0^1 \sqrt{1+x}\,dx$$

$$= \left[\frac{2}{3}(1+x)^{3/2}\right]_0^1$$

$$= \frac{2}{3}\left(\sqrt{8} - 1\right) \approx 1.219$$

5. $y = \frac{3}{2}x^{2/3}$

$y' = \frac{1}{x^{1/3}}, [1, 8]$

$$s = \int_1^8 \sqrt{1 + \left(\frac{1}{x^{1/3}}\right)^2}\,dx$$

$$= \int_1^8 \sqrt{\frac{x^{2/3}+1}{x^{2/3}}}\,dx$$

$$= \frac{3}{2}\int_1^8 \sqrt{x^{2/3}+1}\left(\frac{2}{3x^{1/3}}\right)dx$$

$$= \frac{3}{2}\left[\frac{2}{3}(x^{2/3}+1)^{3/2}\right]_1^8$$

$$= 5\sqrt{5} - 2\sqrt{2} \approx 8.352$$

7. $y = \frac{x^4}{8} + \frac{1}{4x^2}$

$y' = \frac{1}{2}x^3 - \frac{1}{2x^3}, [1, 2]$

$$1 + (y')^2 = \left(\frac{1}{2}x^3 + \frac{1}{2x^3}\right)^2, [1, 2]$$

$$s = \int_a^b \sqrt{1 + (y')^2}\,dx$$

$$= \int_1^2 \left(\frac{1}{2}x^3 + \frac{1}{2x^3}\right)dx$$

$$= \left[\frac{1}{8}x^4 - \frac{1}{4x^2}\right]_1^2 = \frac{33}{16} \approx 2.063$$

9. $y = \ln(\sin x), \left[\frac{\pi}{4}, \frac{3\pi}{4}\right]$

$y' = \frac{1}{\sin x}\cos x = \cot x$

$1 + (y')^2 = 1 + \cot^2 x = \csc^2 x$

$$s = \int_{\pi/4}^{3\pi/4} \csc x\,dx$$

$$= \Big[\ln\left|\csc x - \cot x\right|\Big]_{\pi/4}^{3\pi/4}$$

$$= \ln\left(\sqrt{2} + 1\right) - \ln\left(\sqrt{2} - 1\right) \approx 1.763$$

11. (a) $y = 4 - x^2, 0 \le x \le 2$

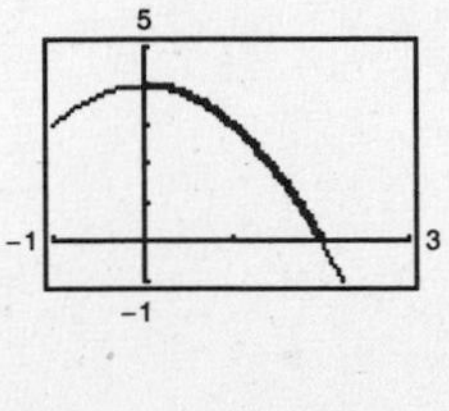

(b) $y' = -2x$

$1 + (y')^2 = 1 + 4x^2$

$$L = \int_0^2 \sqrt{1 + 4x^2}\,dx$$

(c) $L \approx 4.647$

13. (a) $y = \dfrac{1}{x}, 1 \le x \le 3$

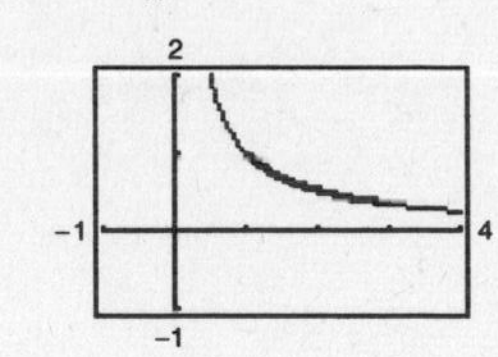

(b) $$y' = -\frac{1}{x^2}$$
$$1 + (y')^2 = 1 + \frac{1}{x^4}$$
$$L = \int_1^3 \sqrt{1 + \frac{1}{x^4}}\, dx$$

(c) $L \approx 2.147$

15. (a) $y = \sin x, 0 \le x \le \pi$

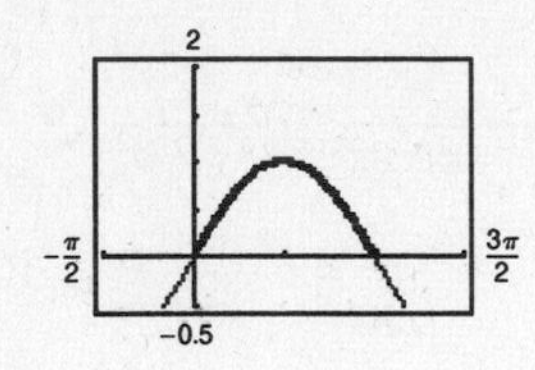

(b) $$y' = \cos x$$
$$1 + (y')^2 = 1 + \cos^2 x$$
$$L = \int_0^{\pi} \sqrt{1 + \cos^2 x}\, dx$$

(c) $L \approx 3.820$

17. (a) $x = e^{-y}, 0 \le y \le 2$

$y = -\ln x$

$1 \ge x \ge e^{-2} \approx 0.135$

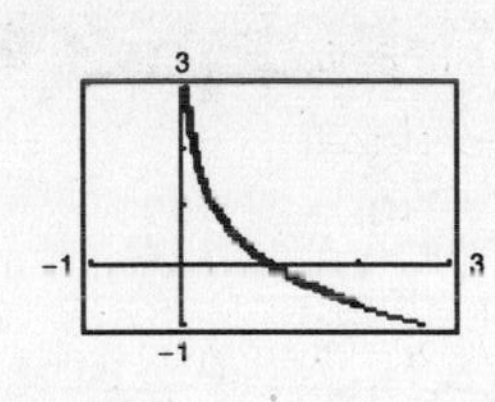

(b) $$y' = -\frac{1}{x}$$
$$1 + (y')^2 = 1 + \frac{1}{x^2}$$
$$L = \int_{e^{-2}}^1 \sqrt{1 + \frac{1}{x^2}}\, dx$$

(c) $L \approx 2.221$

Alternatively, you can do all the computations with respect to y.

(a) $x = e^{-y} \quad 0 \le y \le 2$

(b)

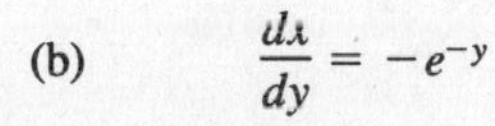

$$\frac{dx}{dy} = -e^{-y}$$
$$1 + \left(\frac{dx}{dy}\right)^2 = 1 + e^{-2y}$$
$$L = \int_0^2 \sqrt{1 + e^{-2y}}\, dy$$

(c) $L \approx 2.221$

19. (a) $y = 2 \arctan x, 0 \le x \le 1$

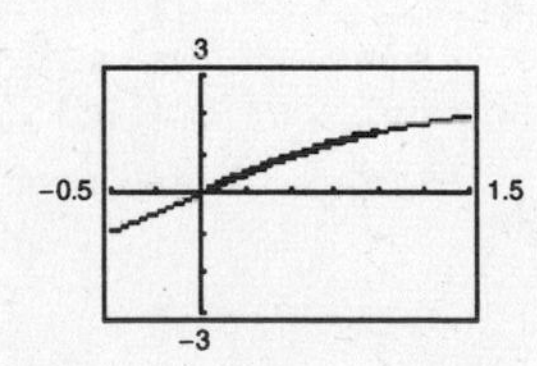

(b) $$y' = \frac{2}{1 + x^2}$$
$$L = \int_0^1 \sqrt{1 + \frac{4}{(1 + x^2)^2}}\, dx$$

(c) $L \approx 1.871$

21. $\displaystyle\int_0^2 \sqrt{1 + \left[\frac{d}{dx}\left(\frac{5}{x^2+1}\right)\right]^2}\,dx$

$s \approx 5$

Matches (b)

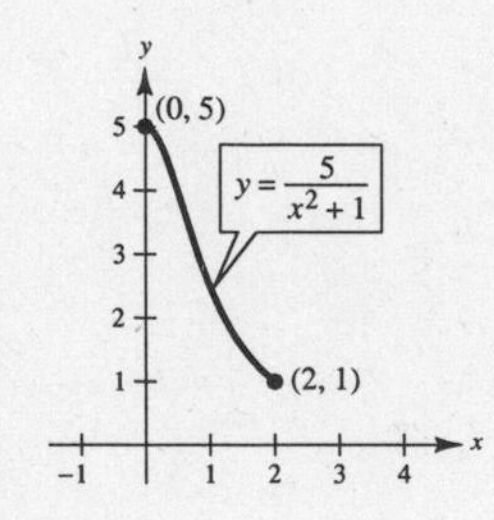

23. $y = x^3$, $[0, 4]$

(a) $d = \sqrt{(4-0)^2 + (64-0)^2} \approx 64.125$

(b) $d = \sqrt{(1-0)^2 + (1-0)^2} + \sqrt{(2-1)^2 + (8-1)^2} + \sqrt{(3-2)^2 + (27-8)^2} + \sqrt{(4-3)^2 + (64-27)^2}$

≈ 64.525

(c) $s = \displaystyle\int_0^4 \sqrt{1 + (3x^2)^2}\,dx = \int_0^4 \sqrt{1 + 9x^4}\,dx \approx 64.666$

(d) 64.672

25. (a)

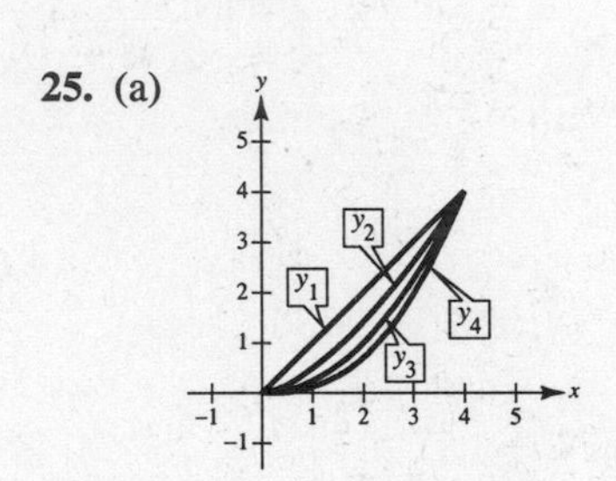

(b) y_1, y_2, y_3, y_4

(c) $y_1' = 1$, $L_1 = \displaystyle\int_0^4 \sqrt{2}\,dx \approx 5.657$

$y_2' = \dfrac{3}{4}x^{1/2}$, $L_2 = \displaystyle\int_0^4 \sqrt{1 + \frac{9x}{16}}\,dx \approx 5.759$

$y_3' = \dfrac{1}{2}x$, $L_3 = \displaystyle\int_0^4 \sqrt{1 + \frac{x^2}{4}}\,dx \approx 5.916$

$y_4' = \dfrac{5}{16}x^{3/2}$, $L_4 = \displaystyle\int_0^4 \sqrt{1 + \frac{25}{256}x^3}\,dx \approx 6.063$

27. $y = \dfrac{1}{3}[x^{3/2} - 3x^{1/2} + 2]$

When $x = 0$, $y = \frac{2}{3}$. Thus, the fleeting object has traveled $\frac{2}{3}$ units when it is caught.

$$y' = \frac{1}{3}\left[\frac{3}{2}x^{1/2} - \frac{3}{2}x^{-1/2}\right] = \left(\frac{1}{2}\right)\frac{x-1}{x^{1/2}}$$

$$1 + (y')^2 = 1 + \frac{(x-1)^2}{4x} = \frac{(x+1)^2}{4x}$$

$$s = \int_0^1 \frac{x+1}{2x^{1/2}}\,dx = \frac{1}{2}\int_0^1 (x^{1/2} + x^{-1/2})\,dx = \frac{1}{2}\left[\frac{2}{3}x^{3/2} + 2x^{1/2}\right]_0^1 = \frac{4}{3} = 2\left(\frac{2}{3}\right)$$

The pursuer has traveled twice the distance that the fleeing object has traveled when it is caught.

29. $y = 20\cosh\dfrac{x}{20}$, $-20 \le x \le 20$

$$y' = \sinh\frac{x}{20}$$

$$1 + (y')^2 = 1 + \sinh^2\frac{x}{20} = \cosh^2\frac{x}{20}$$

$$L = \int_{-20}^{20} \cosh\frac{x}{20}\,dx = 2\int_0^{20} \cosh\frac{x}{20}\,dx = 2(20)\sinh\frac{x}{20}\bigg]_0^{20}$$

$$= 40\sinh(1) \approx 47.008 \text{ m.}$$

31.

$$y = \sqrt{9 - x^2}$$

$$y' = \frac{-x}{\sqrt{9 - x^2}}$$

$$1 + (y')^2 = \frac{9}{9 - x^2}$$

$$s = \int_0^2 \sqrt{\frac{9}{9 - x^2}}\,dx$$

$$= \int_0^2 \frac{3}{\sqrt{9 - x^2}}\,dx$$

$$= \left[3 \arcsin \frac{x}{3}\right]_0^2$$

$$= 3\left(\arcsin \frac{2}{3} - \arcsin 0\right)$$

$$= 3 \arcsin \frac{2}{3} \approx 2.1892$$

33. $y = \dfrac{x^3}{3}$

$$y' = x^2, [0, 3]$$

$$S = 2\pi \int_0^3 \frac{x^3}{3}\sqrt{1 + x^4}\,dx$$

$$= \frac{\pi}{6} \int_0^3 (1 + x^4)^{1/2}(4x^3)\,dx$$

$$= \left[\frac{\pi}{9}(1 + x^4)^{3/2}\right]_0^3$$

$$= \frac{\pi}{9}\left(82\sqrt{82} - 1\right) \approx 258.85$$

35.

$$y = \frac{x^3}{6} + \frac{1}{2x}$$

$$y' = \frac{x^2}{2} - \frac{1}{2x^2}$$

$$1 + (y')^2 = \left(\frac{x^2}{2} + \frac{1}{2x^2}\right)^2, [1, 2]$$

$$S = 2\pi \int_1^2 \left(\frac{x^3}{6} + \frac{1}{2x}\right)\left(\frac{x^2}{2} + \frac{1}{2x^2}\right) dx$$

$$= 2\pi \int_1^2 \left(\frac{x^5}{12} + \frac{x}{3} + \frac{1}{4x^3}\right) dx$$

$$= 2\pi\left[\frac{x^6}{72} + \frac{x^2}{6} - \frac{1}{8x^2}\right]_1^2 = \frac{47\pi}{16}$$

37. $y = \sqrt[3]{x} + 2$

$$y' = \frac{1}{3x^{2/3}}, [1, 8]$$

$$S = 2\pi \int_1^8 x\sqrt{1 + \frac{1}{9x^{4/3}}}\,dx$$

$$= \frac{2\pi}{3} \int_1^8 x^{1/3}\sqrt{9x^{4/3} + 1}\,dx$$

$$= \frac{\pi}{18} \int_1^8 (9x^{4/3} + 1)^{1/2}(12x^{1/3})\,dx$$

$$= \left[\frac{\pi}{27}(9x^{4/3} + 1)^{3/2}\right]_1^8$$

$$= \frac{\pi}{27}\left(145\sqrt{145} - 10\sqrt{10}\right) \approx 199.48$$

39. $y = \sin x$

$$y' = \cos x, [0, \pi]$$

$$S = 2\pi \int_0^\pi \sin x\sqrt{1 + \cos^2 x}\,dx$$

$$\approx 14.4236$$

41. A rectifiable curve is one that has a finite arc length.

43. The precalculus formula is the surface area formula for the lateral surface of the frustum of a right circular cone. The representative element is

$$2\pi f(d_i)\sqrt{\Delta x_i^2 + \Delta y_i^2} = 2\pi f(d_i)\sqrt{1 + \left(\frac{\Delta y_i}{\Delta x_i}\right)^2}\,\Delta x_i.$$

45.

$$y = \frac{hx}{r}$$

$$y' = \frac{h}{r}$$

$$1 + (y')^2 = \frac{r^2 + h^2}{r^2}$$

$$S = 2\pi \int_0^r x\sqrt{\frac{r^2 + h^2}{r^2}}\,dx$$

$$= \left[\frac{2\pi\sqrt{r^2 + h^2}}{r}\left(\frac{x^2}{2}\right)\right]_0^r = \pi r\sqrt{r^2 + h^2}$$

47.

$$y = \sqrt{9 - x^2}$$

$$y' = \frac{-x}{\sqrt{9 - x^2}}$$

$$\sqrt{1 + (y')^2} = \frac{3}{\sqrt{9 - x^2}}$$

$$S = 2\pi \int_0^2 \frac{3x}{\sqrt{9 - x^2}}\,dx$$

$$= -3\pi \int_0^2 \frac{-2x}{\sqrt{9 - x^2}}\,dx$$

$$= \left[-6\pi\sqrt{9 - x^2}\right]_0^2$$

$$= 6\pi\left(3 - \sqrt{5}\right) \approx 14.40$$

See figure in Exercise 48.

49.

$$y = \frac{1}{3}x^{1/2} - x^{3/2}$$

$$y' = \frac{1}{6}x^{-1/2} - \frac{3}{2}x^{1/2} = \frac{1}{6}(x^{-1/2} - 9x^{1/2})$$

$$1 + (y')^2 = 1 + \frac{1}{36}(x^{-1} - 18 + 81x) = \frac{1}{36}(x^{-1/2} + 9x^{1/2})^2$$

$$S = 2\pi \int_0^{1/3}\left(\frac{1}{3}x^{1/2} - x^{3/2}\right)\sqrt{\frac{1}{36}(x^{-1/2} + 9x^{1/2})^2}\,dx = \frac{2\pi}{6}\int_0^{1/3}\left(\frac{1}{3}x^{1/2} - x^{3/2}\right)(x^{-1/2} + 9x^{1/2})\,dx$$

$$= \frac{\pi}{3}\int_0^{1/3}\left(\frac{1}{3} + 2x - 9x^2\right)dx = \frac{\pi}{3}\left[\frac{1}{3}x + x^2 - 3x^3\right]_0^{1/3} = \frac{\pi}{27}\text{ ft}^2 \approx 0.1164\text{ ft}^2 \approx 16.8\text{ in}^2$$

Amount of glass needed: $V = \frac{\pi}{27}\left(\frac{0.015}{12}\right) \approx 0.00015\text{ ft}^3 \approx 0.25\text{ in}^3$

51. (a) $y = f(x) = 0.0000001953x^4 - 0.0001804x^3 + 0.0496x^2 - 4.8323x + 536.9270$

(b) Area $= \int_0^{400} f(x)\,dx \approx 131{,}734.5$ square feet

≈ 3.0 acres (1 acre $= 43{,}560$ square feet)

(Answers will vary.)

(c) $L = \int_0^{400} \sqrt{1 + f'(x)^2}\,dx \approx 794.9$ feet

(Answers will vary.)

53. (a) $V = \pi \int_1^b \frac{1}{x^2}\,dx = \left[-\frac{\pi}{x}\right]_1^b = \pi\left(1 - \frac{1}{b}\right)$

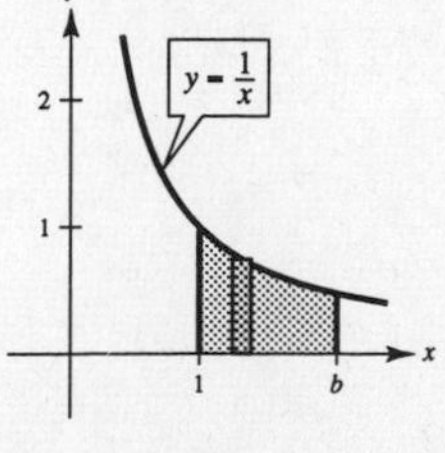

(b)
$$S = 2\pi \int_1^b \frac{1}{x}\sqrt{1 + \left(-\frac{1}{x^2}\right)^2}\,dx$$

$$= 2\pi \int_1^b \frac{1}{x}\sqrt{1 + \frac{1}{x^4}}\,dx$$

$$= 2\pi \int_1^b \frac{\sqrt{x^4 + 1}}{x^3}\,dx$$

—CONTINUED—

53. —CONTINUED—

(c) $\lim_{b\to\infty} V = \lim_{b\to\infty} \pi\left(1 - \frac{1}{b}\right) = \pi$

(d) Since

$$\frac{\sqrt{x^4+1}}{x^3} > \frac{\sqrt{x^4}}{x^3} = \frac{1}{x} > 0 \text{ on } [1, b]$$

we have

$$\int_1^b \frac{\sqrt{x^4+1}}{x^3}\,dx > \int_1^b \frac{1}{x}\,dx = \Big[\ln x\Big]_1^b = \ln b$$

and $\lim_{b\to\infty} \ln b \to \infty$. Thus,

$$\lim_{b\to\infty} 2\pi \int_1^b \frac{\sqrt{x^4+1}}{x^3}\,dx = \infty.$$

55. (a) Area of circle with radius L: $A = \pi L^2$

Area of sector with central angle θ (in radians)

$$S = \frac{\theta}{2\pi}A = \frac{\theta}{2\pi}(\pi L^2) = \frac{1}{2}L^2\theta$$

(b) Let s be the arc length of the sector, which is the circumference of the base of the cone. Here, $s = L\theta = 2\pi r$, and you have

$$S = \frac{1}{2}L^2\theta = \frac{1}{2}L^2\left(\frac{s}{L}\right) = \frac{1}{2}Ls = \frac{1}{2}L(2\pi r) = \pi rL$$

(c) The lateral surface area of the frustum is the difference of the large cone and the small one.

$$S = \pi r_2(L + L_1) - \pi r_1 L_1$$
$$= \pi r_2 L + \pi L_1(r_2 - r_1)$$

By similar triangles, $\frac{L + L_1}{r_2} = \frac{L_1}{r_1} \Rightarrow Lr_1 = L_1(r_2 - r_1)$

Hence,

$$S = \pi r_2 L + \pi L_1(r_2 - r_1) = \pi r_2 L + \pi L r_1$$
$$= \pi L(r_1 + r_2).$$

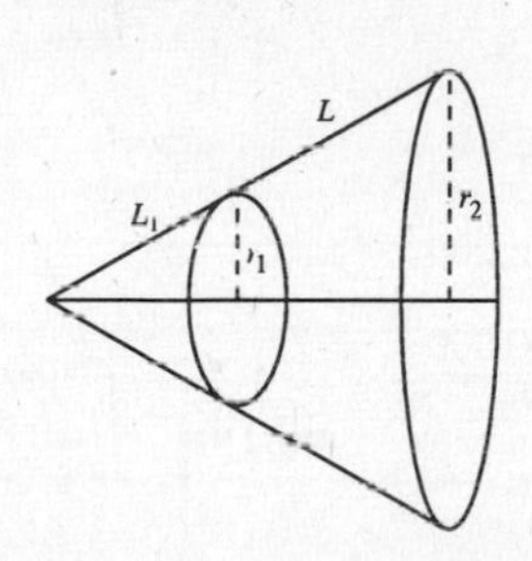

Section 6.5 Work

1. $W = Fd = (100)(10) = 1000$ ft · lb

3. $W = Fd = (112)(4) = 448$ joules (newton-meters)

5. Work equals force times distance, $W = FD$.

7. Since the work equals the area under the force function, you have $(c) < (d) < (a) < (b)$.

9. $F(x) = kx$

$5 = k(4)$

$k = \frac{5}{4}$

$$W = \int_0^7 \frac{5}{4}x\,dx = \left[\frac{5}{8}x^2\right]_0^7$$
$$= \frac{245}{8} \text{ in} \cdot \text{lb}$$
$$= 30.625 \text{ in} \cdot \text{lb} \approx 2.55 \text{ ft} \cdot \text{lb}$$

11. $F(x) = kx$

$$250 = k(30) \Rightarrow k = \frac{25}{3}$$

$$W = \int_{20}^{50} F(x)\,dx = \int_{20}^{50} \frac{25}{3}x\,dx = \left.\frac{25x^2}{6}\right]_{20}^{50}$$
$$= 8750 \text{ n} \cdot \text{cm} = 87.5 \text{ joules or Nm}$$

13. $F(x) = kx$

$20 = k(9)$

$k = \frac{20}{9}$

$W = \int_0^{12} \frac{20}{9}x\,dx = \left[\frac{10}{9}x^2\right]_0^{12} = 160 \text{ in} \cdot \text{lb} = \frac{40}{3} \text{ ft} \cdot \text{lb}$

15. $W = 18 = \int_0^{1/3} kx\,dx = \frac{kx^2}{2}\Big]_0^{1/3} = \frac{k}{18} \Rightarrow k = 324$

$W = \int_{1/3}^{7/12} 324x\,dx = 162x^2\Big]_{1/3}^{7/12} = 37.125 \text{ ft} \cdot \text{lbs}$

[**Note:** 4 inches $= \frac{1}{3}$ foot]

17. Assume that Earth has a radius of 4000 miles.

$F(x) = \frac{k}{x^2}$

$5 = \frac{k}{(4000)^2}$

$k = 80{,}000{,}000$

$F(x) = \frac{80{,}000{,}000}{x^2}$

(a) $W = \int_{4000}^{4100} \frac{80{,}000{,}000}{x^2}\,dx = \left[\frac{-80{,}000{,}000}{x}\right]_{4000}^{4100} \approx 487.8 \text{ mi} \cdot \text{tons}$

$\approx 5.15 \times 10^9 \text{ ft} \cdot \text{lb}$

(b) $W = \int_{4000}^{4300} \frac{80{,}000{,}000}{x^2}\,dx \approx 1395.3 \text{ mi} \cdot \text{ton}$

$\approx 1.47 \times 10^{10} \text{ ft} \cdot \text{lb}$

19. Assume that the earth has a radius of 4000 miles.

$F(x) = \frac{k}{x^2}$

$10 = \frac{k}{(4000)^2}$

$k = 160{,}000{,}000$

$F(x) = \frac{160{,}000{,}000}{x^2}$

(a) $W = \int_{4000}^{15{,}000} \frac{160{,}000{,}000}{x^2}\,dx = \left[-\frac{160{,}000{,}000}{x}\right]_{4000}^{15{,}000} \approx -10{,}666.667 + 40{,}000$

$= 29{,}333.333 \text{ mi} \cdot \text{ton}$

$\approx 2.93 \times 10^4 \text{ mi} \cdot \text{ton}$

$\approx 3.10 \times 10^{11} \text{ ft} \cdot \text{lb}$

(b) $W = \int_{4000}^{26{,}000} \frac{160{,}000{,}000}{x^2}\,dx = \left[-\frac{160{,}000{,}000}{x}\right]_{4000}^{26{,}000} \approx -6{,}153.846 + 40{,}000$

$= 33{,}846.154 \text{ mi} \cdot \text{ton}$

$\approx 3.38 \times 10^4 \text{ mi} \cdot \text{ton}$

$\approx 3.57 \times 10^{11} \text{ ft} \cdot \text{lb}$

21. Weight of each layer: $62.4(20)\,\Delta y$

Distance: $4 - y$

(a) $W = \int_2^4 62.4(20)(4 - y)\,dy = \left[4992y - 624y^2\right]_2^4 = 2496 \text{ ft} \cdot \text{lb}$

(b) $W = \int_0^4 62.4(20)(4 - y)\,dy = \left[4992y - 624y^2\right]_0^4 = 9984 \text{ ft} \cdot \text{lb}$

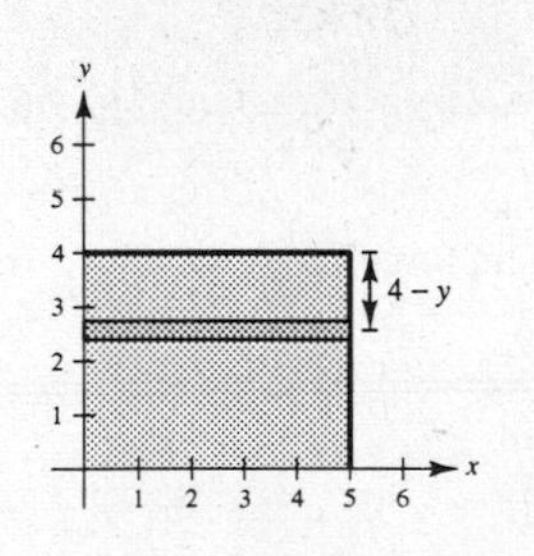

23. Volume of disk: $\pi(2)^2\,\Delta y = 4\pi\,\Delta y$

Weight of disk of water: $9800(4\pi)\,\Delta y$

Distance the disk of water is moved: $5 - y$

$W = \int_0^4 (5 - y)(9800)4\pi\,dy = 39{,}200\pi \int_0^4 (5 - y)\,dy$

$= 39{,}200\pi\left[5y - \frac{y^2}{2}\right]_0^4$

$= 39{,}200\pi(12) = 470{,}400\pi$ newton–meters

25. Volume of disk: $\pi\left(\frac{2}{3}y\right)^2 \Delta y$

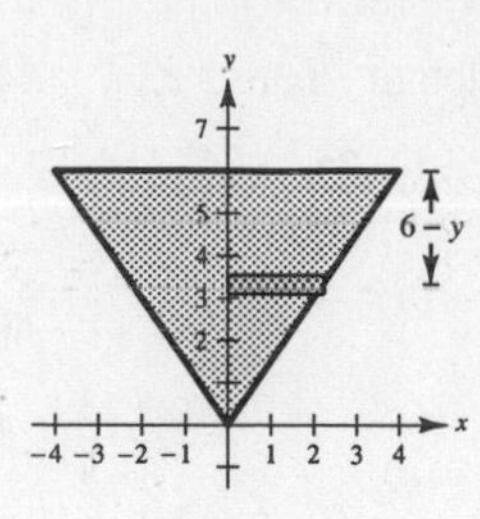

Weight of disk: $62.4\pi\left(\frac{2}{3}y\right)^2 \Delta y$

Distance: $6 - y$

$$W = \frac{4(62.4)\pi}{9}\int_0^6 (6-y)y^2\,dy = \frac{4}{9}(62.4)\pi\left[2y^3 - \frac{1}{4}y^4\right]_0^6 = 2995.2\pi \text{ ft} \cdot \text{lb}$$

27. Volume of disk: $\pi\left(\sqrt{36-y^2}\right)^2 \Delta y$

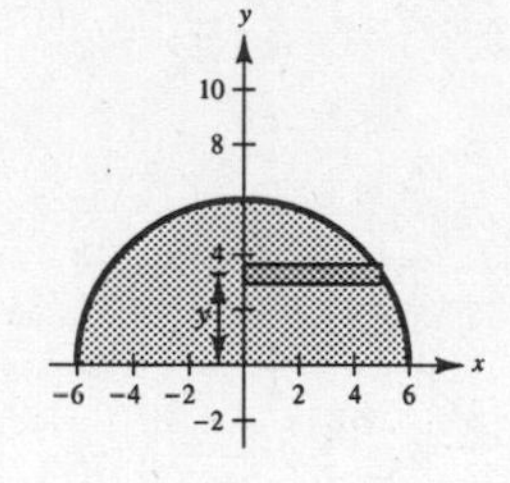

Weight of disk: $62.4\pi(36-y^2)\,\Delta y$

Distance: y

$$W = 62.4\pi\int_0^6 y(36-y^2)\,dy$$

$$= 62.4\pi\int_0^6 (36y - y^3)\,dy = 62.4\pi\left[18y^2 - \frac{1}{4}y^4\right]_0^6$$

$$= 20{,}217.6\pi \text{ ft} \cdot \text{lb}$$

29. Volume of layer: $V = lwh = 4(2)\sqrt{(9/4) - y^2}\,\Delta y$

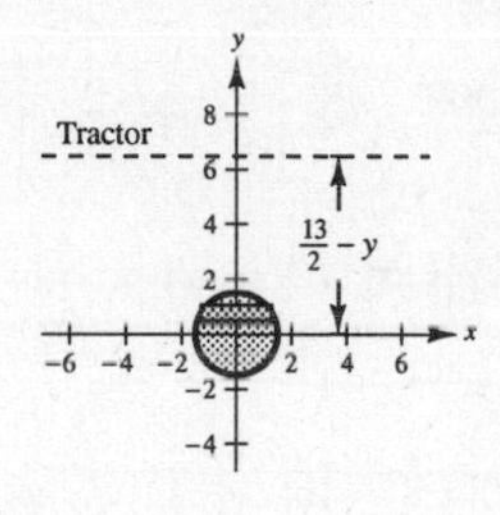

Weight of layer: $W = 42(8)\sqrt{(9/4) - y^2}\,\Delta y$

Distance: $\frac{13}{2} - y$

$$W = \int_{-1.5}^{1.5} 42(8)\sqrt{(9/4) - y^2}\left(\frac{13}{2} - y\right)dy$$

$$= 336\left[\frac{13}{2}\int_{-1.5}^{1.5}\sqrt{(9/4) - y^2}\,dy - \int_{-1.5}^{1.5}\sqrt{(9/4) - y^2}\,y\,dy\right]$$

The second integral is zero since the integrand is odd and the limits of integration are symmetric to the origin. The first integral represents the area of a semicircle of radius $\frac{3}{2}$. Thus, the work is

$$W = 336\left(\frac{13}{2}\right)\pi\left(\frac{3}{2}\right)^2\left(\frac{1}{2}\right) = 2457\pi \text{ ft} \cdot \text{lb}$$

31. Weight of section of chain: $3\,\Delta y$

Distance: $15 - y$

$$W = 3\int_0^{15}(15-y)\,dy$$

$$= \left[-\frac{3}{2}(15-y)^2\right]_0^{15}$$

$$= 337.5 \text{ ft} \cdot \text{lb}$$

33. The lower 5 feet of chain are raised 10 feet with a constant force.

$$W_1 = 3(5)(10) = 150 \text{ ft} \cdot \text{lb}$$

The top 10 feet of chain are raised with a variable force.

Weight per section: $3\,\Delta y$

Distance: $10 - y$

$$W_2 = 3\int_0^{10}(10-y)\,dy = \left[-\frac{3}{2}(10-y)^2\right]_0^{10}$$

$$= 150 \text{ ft} \cdot \text{lb}$$

$$W = W_1 + W_2 = 300 \text{ ft} \cdot \text{lb}$$

35. Weight of section of chain: $3\,\Delta y$

Distance: $15 - 2y$

$$W = 3\int_0^{7.5}(15-2y)\,dy = \left[-\frac{3}{4}(15-2y)^2\right]_0^{7.5}$$

$$= \frac{3}{4}(15)^2 = 168.75 \text{ ft}\cdot\text{lb}$$

37. Work to pull up the ball: $W_1 = 500(15) = 7500$ ft · lb

Work to wind up the top 15 feet of cable: force is variable

Weight per section: $1\,\Delta y$

Distance: $15 - x$

$$W_2 = \int_0^{15}(15-x)\,dx = \left[-\frac{1}{2}(15-x)^2\right]_0^{15}$$

$$= 112.5 \text{ ft}\cdot\text{lb}$$

Work to lift the lower 25 feet of cable with a constant force:

$$W_3 = (1)(25)(15) = 375 \text{ ft}\cdot\text{lb}$$

$$W = W_1 + W_2 + W_3 = 7500 + 112.5 + 375$$

$$= 7987.5 \text{ ft}\cdot\text{lb}$$

39. $p = \dfrac{k}{V}$

$$1000 = \frac{k}{2}$$

$$k = 2000$$

$$W = \int_2^3 \frac{2000}{V}\,dV = \Big[2000\ln|V|\Big]_2^3$$

$$= 2000\ln\left(\frac{3}{2}\right) \approx 810.93 \text{ ft}\cdot\text{lb}$$

41. $F(x) = \dfrac{k}{(2-x)^2}$

$$W = \int_{-2}^{1}\frac{k}{(2-x)^2}\,dx = \left[\frac{k}{2-x}\right]_{-2}^{1} = k\left(1-\frac{1}{4}\right)$$

$$= \frac{3k}{4}\text{(units of work)}$$

43. $W = \displaystyle\int_0^5 1000[1.8 - \ln(x+1)]\,dx \approx 3249.44$ ft · lb

45. $W = \displaystyle\int_0^5 100x\sqrt{125 - x^3}\,dx \approx 10{,}330.3$ ft · lb

Section 6.6 Moments, Centers of Mass, and Centroids

1. $\bar{x} = \dfrac{6(-5) + 3(1) + 5(3)}{6+3+5} = -\dfrac{6}{7}$

3. $\bar{x} = \dfrac{1(7) + 1(8) + 1(12) + 1(15) + 1(18)}{1+1+1+1+1} = 12$

5. (a) $\bar{x} = \dfrac{(7+5) + (8+5) + (12+5) + (15+5) + (18+5)}{5} = 17 = 12 + 5$

(b) $\bar{x} = \dfrac{12(-6-3) + 1(-4-3) + 6(-2-3) + 3(0-3) + 11(8-3)}{12+1+6+3+11} = \dfrac{-99}{33} = -3$

7. $50x = 75(L - x) = 75(10 - x)$

$$50x = 750 - 75x$$

$$125x = 750$$

$$x = 6 \text{ feet}$$

9. $\bar{x} = \dfrac{5(2) + 1(-3) + 3(1)}{5+1+3} = \dfrac{10}{9}$

$$\bar{y} = \frac{5(2) + 1(1) + 3(-4)}{5+1+3} = -\frac{1}{9}$$

$$(\bar{x}, \bar{y}) = \left(\frac{10}{9}, -\frac{1}{9}\right)$$

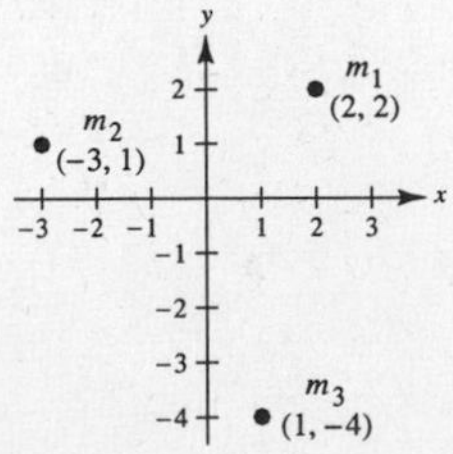

11. $\bar{x} = \dfrac{3(-2) + 4(-1) + 2(7) + 1(0) + 6(-3)}{3 + 4 + 2 + 1 + 6} = -\dfrac{7}{8}$

$\bar{y} = \dfrac{3(-3) + 4(0) + 2(1) + 1(0) + 6(0)}{3 + 4 + 2 + 1 + 6} = -\dfrac{7}{16}$

$(\bar{x}, \bar{y}) = \left(-\dfrac{7}{8}, -\dfrac{7}{16}\right)$

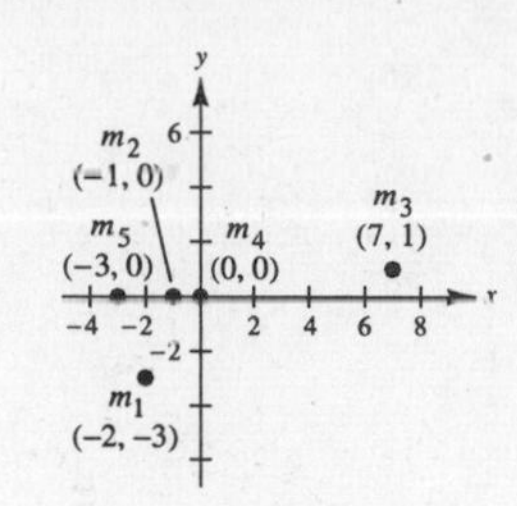

13. $m = \rho \displaystyle\int_0^4 \sqrt{x}\, dx = \left[\frac{2\rho}{3} x^{3/2}\right]_0^4 = \frac{16\rho}{3}$

$M_x = \rho \displaystyle\int_0^4 \frac{\sqrt{x}}{2}(\sqrt{x})\, dx = \left[\rho \frac{x^2}{4}\right]_0^4 = 4\rho$

$\bar{y} = \dfrac{M_x}{m} = 4\rho\left(\dfrac{3}{16\rho}\right) = \dfrac{3}{4}$

$M_y = \rho \displaystyle\int_0^4 x\sqrt{x}\, dx = \left[\rho \frac{2}{5} x^{5/2}\right]_0^4 = \frac{64\rho}{5}$

$\bar{x} = \dfrac{M_y}{m} = \dfrac{64\rho}{5}\left(\dfrac{3}{16\rho}\right) = \dfrac{12}{5}$

$(\bar{x}, \bar{y}) = \left(\dfrac{12}{5}, \dfrac{3}{4}\right)$

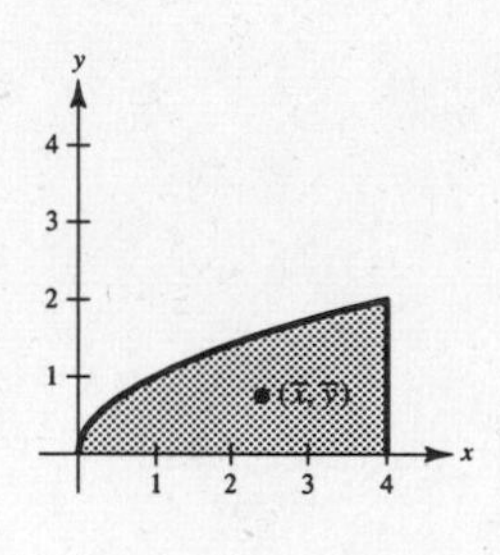

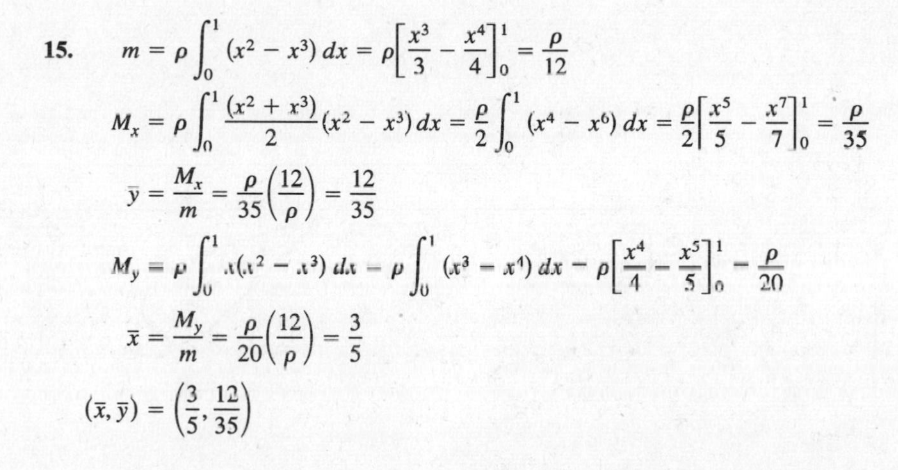

15. $m = \rho \displaystyle\int_0^1 (x^2 - x^3)\, dx = \rho\left[\frac{x^3}{3} - \frac{x^4}{4}\right]_0^1 = \frac{\rho}{12}$

$M_x = \rho \displaystyle\int_0^1 \frac{(x^2 + x^3)}{2}(x^2 - x^3)\, dx = \frac{\rho}{2}\int_0^1 (x^4 - x^6)\, dx = \frac{\rho}{2}\left[\frac{x^5}{5} - \frac{x^7}{7}\right]_0^1 = \frac{\rho}{35}$

$\bar{y} = \dfrac{M_x}{m} = \dfrac{\rho}{35}\left(\dfrac{12}{\rho}\right) = \dfrac{12}{35}$

$M_y = \rho \displaystyle\int_0^1 x(x^2 - x^3)\, dx = \rho \int_0^1 (x^3 - x^4)\, dx = \rho\left[\frac{x^4}{4} - \frac{x^5}{5}\right]_0^1 = \frac{\rho}{20}$

$\bar{x} = \dfrac{M_y}{m} = \dfrac{\rho}{20}\left(\dfrac{12}{\rho}\right) = \dfrac{3}{5}$

$(\bar{x}, \bar{y}) = \left(\dfrac{3}{5}, \dfrac{12}{35}\right)$

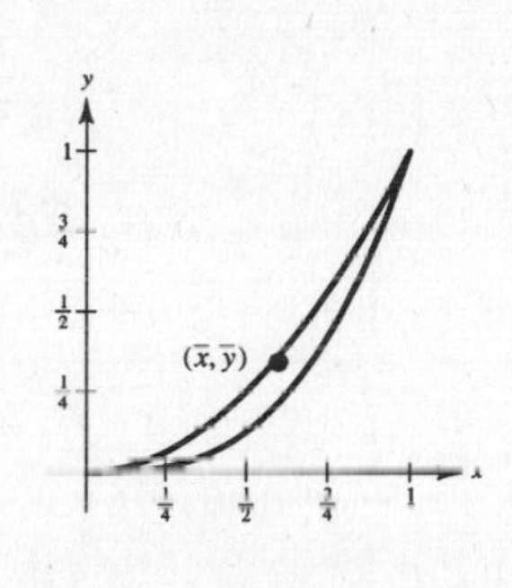

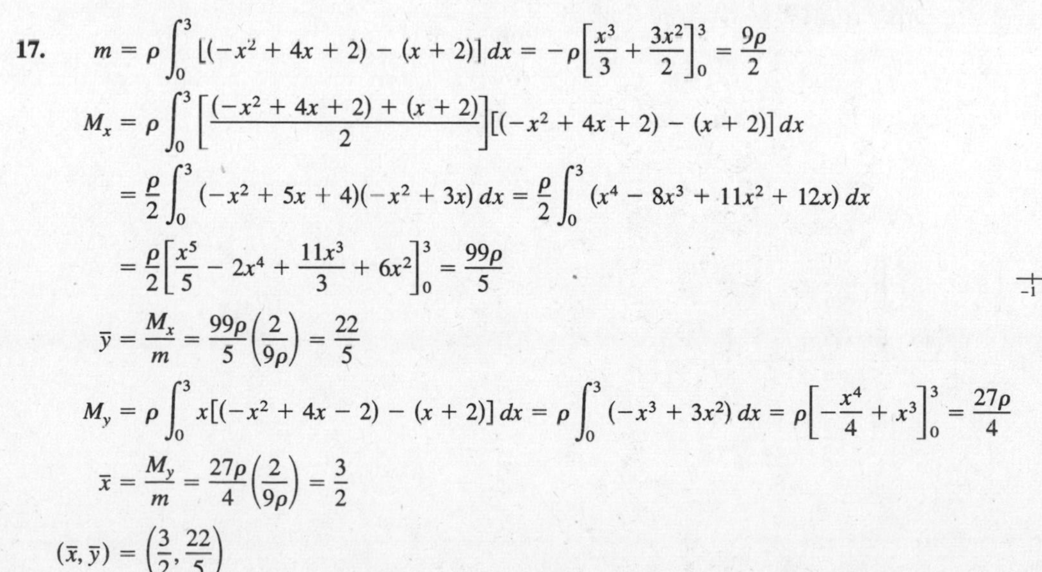

17. $m = \rho \displaystyle\int_0^3 [(-x^2 + 4x + 2) - (x + 2)]\, dx = -\rho\left[\frac{x^3}{3} + \frac{3x^2}{2}\right]_0^3 = \frac{9\rho}{2}$

$M_x = \rho \displaystyle\int_0^3 \left[\frac{(-x^2 + 4x + 2) + (x + 2)}{2}\right][(-x^2 + 4x + 2) - (x + 2)]\, dx$

$= \dfrac{\rho}{2} \displaystyle\int_0^3 (-x^2 + 5x + 4)(-x^2 + 3x)\, dx = \frac{\rho}{2}\int_0^3 (x^4 - 8x^3 + 11x^2 + 12x)\, dx$

$= \dfrac{\rho}{2}\left[\dfrac{x^5}{5} - 2x^4 + \dfrac{11x^3}{3} + 6x^2\right]_0^3 = \dfrac{99\rho}{5}$

$\bar{y} = \dfrac{M_x}{m} = \dfrac{99\rho}{5}\left(\dfrac{2}{9\rho}\right) = \dfrac{22}{5}$

$M_y = \rho \displaystyle\int_0^3 x[(-x^2 + 4x - 2) - (x + 2)]\, dx = \rho \int_0^3 (-x^3 + 3x^2)\, dx = \rho\left[-\frac{x^4}{4} + x^3\right]_0^3 = \frac{27\rho}{4}$

$\bar{x} = \dfrac{M_y}{m} = \dfrac{27\rho}{4}\left(\dfrac{2}{9\rho}\right) = \dfrac{3}{2}$

$(\bar{x}, \bar{y}) = \left(\dfrac{3}{2}, \dfrac{22}{5}\right)$

19. $$m = \rho\int_0^8 x^{2/3}\,dx = \rho\left[\frac{3}{5}x^{5/3}\right]_0^8 = \frac{96\rho}{5}$$

$$M_x = \rho\int_0^8 \frac{x^{2/3}}{2}(x^{2/3})\,dx = \frac{\rho}{2}\left[\frac{3}{7}x^{7/3}\right]_0^8 = \frac{192\rho}{7}$$

$$\bar{y} = \frac{M_x}{m} = \frac{192\rho}{7}\left(\frac{5}{96\rho}\right) = \frac{10}{7}$$

$$M_y = \rho\int_0^8 x(x^{2/3})\,dx = \rho\left[\frac{3}{8}x^{8/3}\right]_0^8 = 96\rho$$

$$\bar{x} = \frac{M_y}{m} = 96\rho\left(\frac{5}{96\rho}\right) = 5$$

$$(\bar{x}, \bar{y}) = \left(5, \frac{10}{7}\right)$$

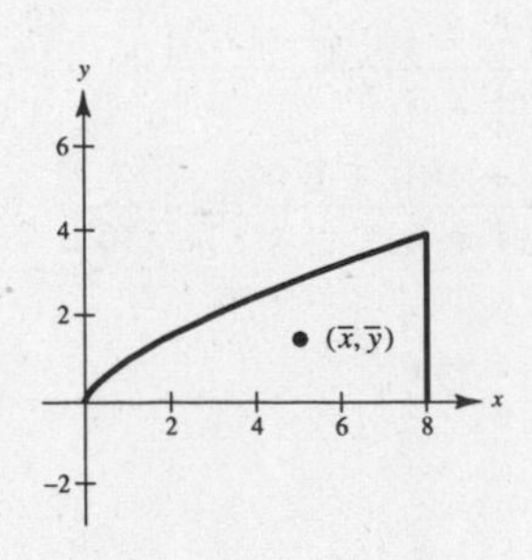

21. $$m = 2\rho\int_0^2 (4 - y^2)\,dy = 2\rho\left[4y - \frac{y^3}{3}\right]_0^2 = \frac{32\rho}{3}$$

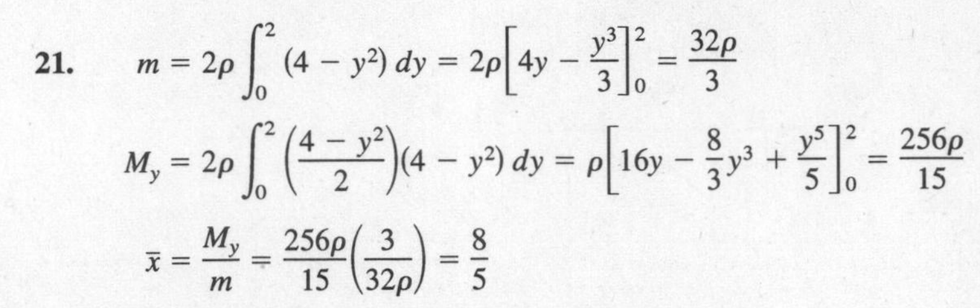

$$M_y = 2\rho\int_0^2 \left(\frac{4 - y^2}{2}\right)(4 - y^2)\,dy = \rho\left[16y - \frac{8}{3}y^3 + \frac{y^5}{5}\right]_0^2 = \frac{256\rho}{15}$$

$$\bar{x} = \frac{M_y}{m} = \frac{256\rho}{15}\left(\frac{3}{32\rho}\right) = \frac{8}{5}$$

By symmetry, M_x and $\bar{y} = 0$.

$$(\bar{x}, \bar{y}) = \left(\frac{8}{5}, 0\right)$$

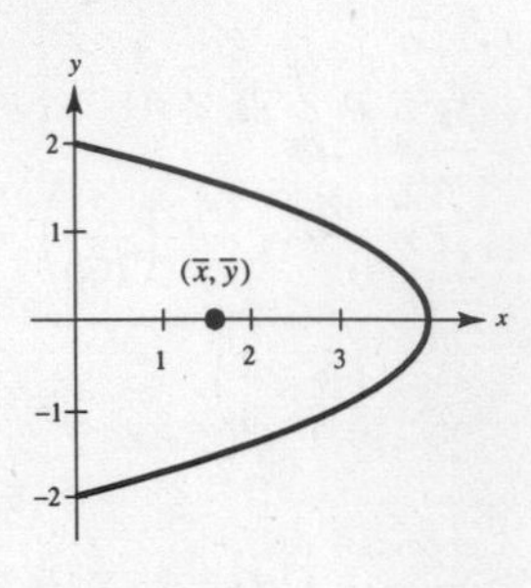

23. $$m = \rho\int_0^3 [(2y - y^2) - (-y)]\,dy = \rho\left[\frac{3y^2}{2} - \frac{y^3}{3}\right]_0^3 = \frac{9\rho}{2}$$

$$M_y = \rho\int_0^3 \frac{[(2y - y^2) + (-y)]}{2}[(2y - y^2) - (-y)]\,dy = \frac{\rho}{2}\int_0^3 (y - y^2)(3y - y^2)\,dy$$

$$= \frac{\rho}{2}\int_0^3 (y^4 - 4y^3 + 3y^2)\,dy = \frac{\rho}{2}\left[\frac{y^5}{5} - y^4 + y^3\right]_0^3 = -\frac{27\rho}{10}$$

$$\bar{x} = \frac{M_y}{m} = -\frac{27\rho}{10}\left(\frac{2}{9\rho}\right) = -\frac{3}{5}$$

$$M_x = \rho\int_0^3 y[(2y - y^2) - (-y)]\,dy = \rho\int_0^3 (3y^2 - y^3)\,dy = \rho\left[y^3 - \frac{y^4}{4}\right]_0^3 = \frac{27\rho}{4}$$

$$\bar{y} = \frac{M_x}{m} = \frac{27\rho}{4}\left(\frac{2}{9\rho}\right) = \frac{3}{2}$$

$$(\bar{x}, \bar{y}) = \left(-\frac{3}{5}, \frac{3}{2}\right)$$

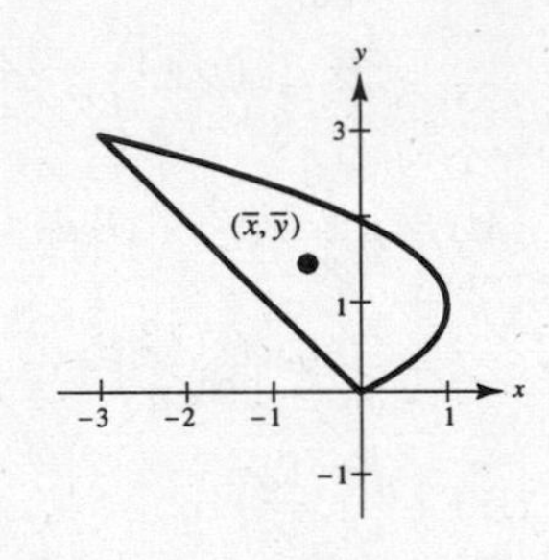

25. $$A = \int_0^1 (x - x^2)\,dx = \left[\frac{1}{2}x^2 - \frac{x^3}{3}\right]_0^1 = \frac{1}{6}$$

$$M_x = \frac{1}{2}\int_0^1 (x^2 - x^4)\,dx = \frac{1}{2}\left[\frac{x^3}{3} - \frac{x^5}{5}\right]_0^1 = \frac{1}{2}\left(\frac{1}{3} - \frac{1}{5}\right) = \frac{1}{15}$$

$$M_y = \int_0^1 (x^2 - x^3)\,dx = \left[\frac{x^3}{3} - \frac{x^4}{4}\right]_0^1 = \left(\frac{1}{3} - \frac{1}{4}\right) = \frac{1}{12}$$

27. $A = \int_0^3 (2x + 4)\,dx = \left[x^2 + 4x\right]_0^3 = 9 + 12 = 21$

$$M_x = \frac{1}{2}\int_0^3 (2x + 4)^2\,dx = \int_0^3 (2x^2 + 8x + 8)\,dx = \left[\frac{2x^3}{3} + 4x^2 + 8x\right]_0^3 = 18 + 36 + 24 = 78$$

$$M_y = \int_0^3 (2x^2 + 4x)\,dx = \left[\frac{2x^3}{3} + 2x^2\right]_0^3 = 18 + 18 = 36$$

29. $m = \rho\int_0^5 10x\sqrt{125 - x^3}\,dx \approx 1033.0\rho$

$$M_x = \rho\int_0^5 \left(\frac{10x\sqrt{125 - x^3}}{2}\right)\left(10x\sqrt{125 - x^3}\right)dx = 50\rho\int_0^5 x^2(125 - x^3)\,dx = \frac{3{,}124{,}375\rho}{24} \approx 130{,}208\rho$$

$$M_y = \rho\int_0^5 10x^2\sqrt{125 - x^3}\,dx = -\frac{10\rho}{3}\int_0^5 \sqrt{125 - x^3}(-3x^2)\,dx = \frac{12{,}500\sqrt{5}\rho}{9} \approx 3105.6\rho$$

$$\bar{x} = \frac{M_y}{m} \approx 3.0$$

$$\bar{y} = \frac{M_x}{m} \approx 126.0$$

Therefore, the centroid is (3.0, 126.0).

31. $m = \rho\int_{-20}^{20} 5\sqrt[3]{400 - x^2}\,dx \approx 1239.76\rho$

$$M_x = \rho\int_{-20}^{20} \frac{5\sqrt[3]{400 - x^2}}{2}\left(5\sqrt[3]{400 - x^2}\right)dx$$

$$= \frac{25\rho}{2}\int_{-20}^{20} (400 - x^2)^{2/3}\,dx \approx 20064.27$$

$$\bar{y} = \frac{M_x}{m} \approx 16.18$$

$\bar{x} = 0$ by symmetry. Therefore, the centroid is (0, 16.2).

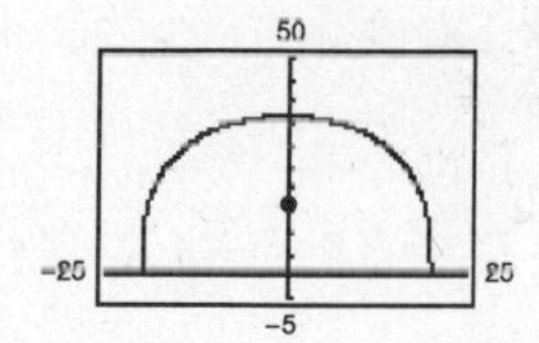

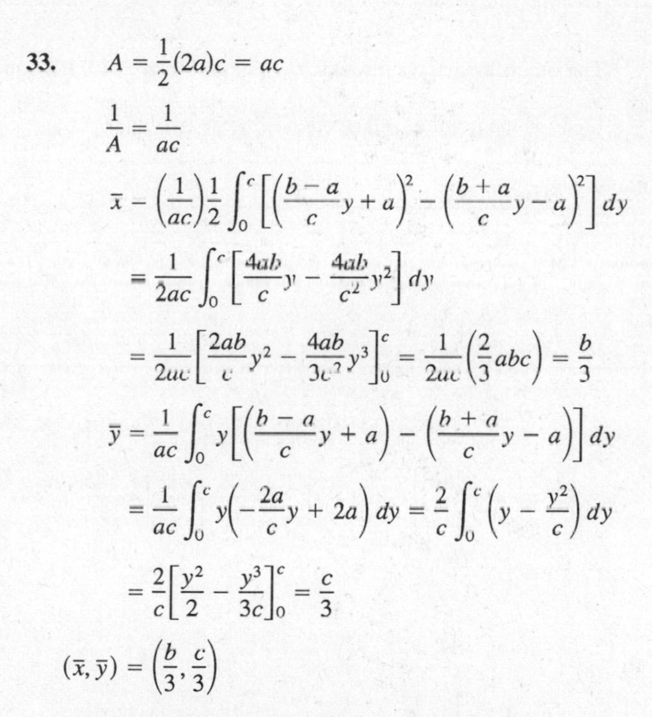

33. $A = \frac{1}{2}(2a)c = ac$

$$\frac{1}{A} = \frac{1}{ac}$$

$$\bar{x} = \left(\frac{1}{ac}\right)\frac{1}{2}\int_0^c \left[\left(\frac{b-a}{c}y + a\right)^2 - \left(\frac{b+a}{c}y - a\right)^2\right]dy$$

$$= \frac{1}{2ac}\int_0^c \left[\frac{4ab}{c}y - \frac{4ab}{c^2}y^2\right]dy$$

$$= \frac{1}{2ac}\left[\frac{2ab}{c}y^2 - \frac{4ab}{3c^2}y^3\right]_0^c = \frac{1}{2ac}\left(\frac{2}{3}abc\right) = \frac{b}{3}$$

$$\bar{y} = \frac{1}{ac}\int_0^c y\left[\left(\frac{b-a}{c}y + a\right) - \left(\frac{b+a}{c}y - a\right)\right]dy$$

$$= \frac{1}{ac}\int_0^c y\left(-\frac{2a}{c}y + 2a\right)dy = \frac{2}{c}\int_0^c \left(y - \frac{y^2}{c}\right)dy$$

$$= \frac{2}{c}\left[\frac{y^2}{2} - \frac{y^3}{3c}\right]_0^c = \frac{c}{3}$$

$$(\bar{x}, \bar{y}) = \left(\frac{b}{3}, \frac{c}{3}\right)$$

In Exercise 73 of Section P.2, you found that $(b/3, c/3)$ is the point of intersection of the medians.

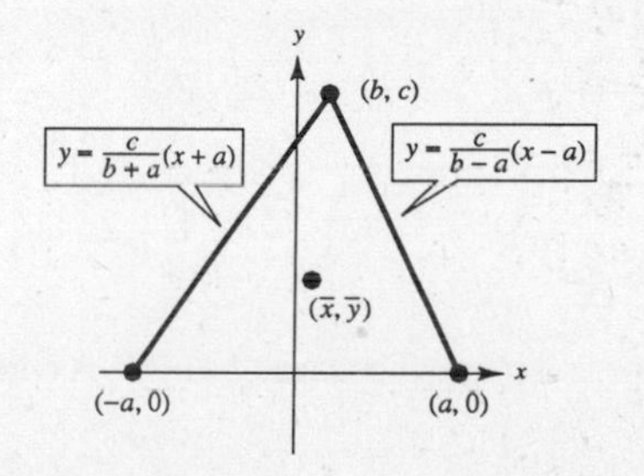

35. $A = \frac{c}{2}(a + b)$

$$\frac{1}{A} = \frac{2}{c(a+b)}$$

$$\bar{x} = \frac{2}{c(a+b)}\int_0^c x\left(\frac{b-a}{c}x + a\right)dx = \frac{2}{c(a+b)}\int_0^c \left(\frac{b-a}{c}x^2 + ax\right)dx = \frac{2}{c(a+b)}\left[\frac{b-a}{c}\frac{x^3}{3} + \frac{ax^2}{2}\right]_0^c$$

$$= \frac{2}{c(a+b)}\left[\frac{(b-a)c^2}{3} + \frac{ac^2}{2}\right] = \frac{2}{c(a+b)}\left[\frac{2bc^2 - 2ac^2 + 3ac^2}{6}\right] = \frac{c(2b+a)}{3(a+b)} = \frac{(a+2b)c}{3(a+b)}$$

$$\bar{y} = \frac{2}{c(a+b)}\frac{1}{2}\int_0^c \left(\frac{b-a}{c}x + a\right)^2 dx = \frac{1}{c(a+b)}\int_0^c \left[\left(\frac{b-a}{c}\right)^2 x^2 + \frac{2a(b-a)}{c}x + a^2\right]dx$$

$$= \frac{1}{c(a+b)}\left[\left(\frac{b-a}{c}\right)^2\frac{x^3}{3} + \frac{2a(b-a)}{c}\frac{x^2}{2} + a^2x\right]_0^c = \frac{1}{c(a+b)}\left[\frac{(b-a)^2c}{3} + ac(b-a) + a^2c\right]$$

$$= \frac{1}{3c(a+b)}[(b^2 - 2ab + a^2)c + 3ac(b-a) + 3a^2c]$$

$$= \frac{1}{3(a+b)}[b^2 - 2ab + a^2 + 3ab - 3a^2 + 3a^2] = \frac{a^2 + ab + b^2}{3(a+b)}$$

Thus, $(\bar{x}, \bar{y}) = \left(\frac{(a+2b)c}{3(a+b)}, \frac{a^2+ab+b^2}{3(a+b)}\right)$.

The one line passes through $(0, a/2)$ and $(c, b/2)$. It's equation is $y = \frac{b-a}{2c}x + \frac{a}{2}$.

The other line passes through $(0, -b)$ and $(c, a + b)$. It's equation is $y = \frac{a+2b}{c}x - b$.

$(\bar{x}, \bar{y})$ is the point of intersection of these two lines.

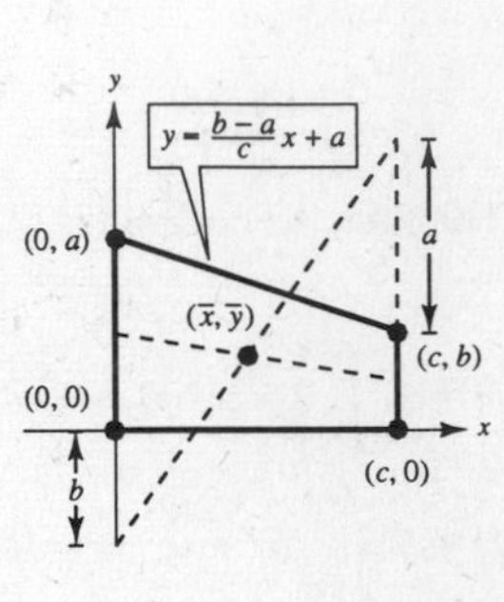

37. $\bar{x} = 0$ by symmetry

$$A = \frac{1}{2}\pi ab$$

$$\frac{1}{A} = \frac{2}{\pi ab}$$

$$\bar{y} = \frac{2}{\pi ab}\frac{1}{2}\int_{-a}^{a}\left(\frac{b}{a}\sqrt{a^2 - x^2}\right)^2 dx$$

$$= \frac{1}{\pi ab}\left(\frac{b^2}{a^2}\right)\left[a^2x - \frac{x^3}{3}\right]_{-a}^{a} = \frac{b}{\pi a^3}\left[\frac{4a^3}{3}\right] = \frac{4b}{3\pi}$$

$$(\bar{x}, \bar{y}) = \left(0, \frac{4b}{3\pi}\right)$$

39. (a)

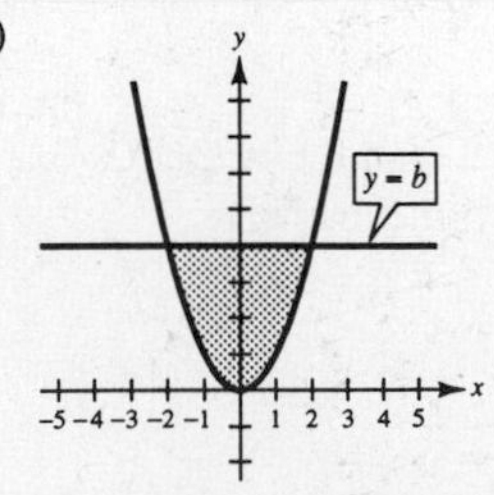

(b) $\bar{x} = 0$ by symmetry

(c) $M_y = \int_{-\sqrt{b}}^{\sqrt{b}} x(b - x^2)\,dx = 0$ because $bx - x^3$ is odd

(d) $\bar{y} > \frac{b}{2}$ since there is more area above $y = \frac{b}{2}$ than below

(e) $M_x = \int_{-\sqrt{b}}^{\sqrt{b}} \frac{(b + x^2)(b - x^2)}{2}\,dx$

$$= \int_{-\sqrt{b}}^{\sqrt{b}} \frac{b^2 - x^4}{2}\,dx = \frac{1}{2}\left[b^2x - \frac{x^5}{5}\right]_{-\sqrt{b}}^{\sqrt{b}}$$

$$= b^2\sqrt{b} - \frac{b^2\sqrt{b}}{5} = \frac{4b^2\sqrt{b}}{5}$$

$$A = \int_{-\sqrt{b}}^{\sqrt{b}} (b - x^2)\,dx = \left[bx - \frac{x^3}{3}\right]_{-\sqrt{b}}^{\sqrt{b}}$$

$$= \left(b\sqrt{b} - \frac{b\sqrt{b}}{3}\right)2 = 4\frac{b\sqrt{b}}{3}$$

$$\bar{y} = \frac{M_x}{A} = \frac{4b^2\sqrt{b}/5}{4b\sqrt{b}/3} = \frac{3}{5}b.$$

41. (a) $\bar{x} = 0$ by symmetry

$$A = 2\int_0^{40} f(x)\,dx = \frac{2(40)}{3(4)}[30 + 4(29) + 2(26) + 4(20) + 0] = \frac{20}{3}(278) = \frac{5560}{3}$$

$$M_x = \int_{-40}^{40} \frac{f(x)^2}{2}\,dx = \frac{40}{3(4)}[30^2 + 4(29)^2 + 2(26)^2 + 4(20)^2 + 0] = \frac{10}{3}(7216) = \frac{72160}{3}$$

$$\bar{y} = \frac{M_x}{A} = \frac{72160/3}{5560/3} = \frac{72160}{5560} \approx 12.98$$

$(\bar{x}, \bar{y}) = (0, 12.98)$

(b) $y = (-1.02 \times 10^{-5})x^4 - 0.0019x^2 + 29.28$

(c) $\bar{y} = \frac{M_x}{A} \approx \frac{23697.68}{1843.54} \approx 12.85$

$(\bar{x}, \bar{y}) = (0, 12.85)$

43. Centroids of the given regions: (1, 0) and (3, 0)

Area: $A = 4 + \pi$

$$\bar{x} = \frac{4(1) + \pi(3)}{4 + \pi} = \frac{4 + 3\pi}{4 + \pi}$$

$$\bar{y} = \frac{4(0) + \pi(0)}{4 + \pi} = 0$$

$$(\bar{x}, \bar{y}) = \left(\frac{4 + 3\pi}{4 + \pi}, 0\right) \approx (1.88, 0)$$

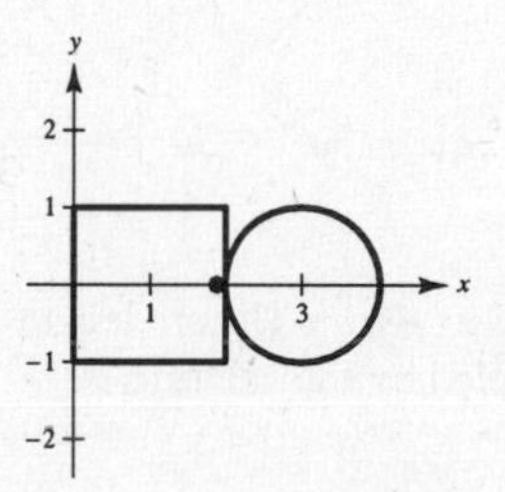

45. Centroids of the given regions: $\left(0, \frac{3}{2}\right)$, $(0, 5)$, and $\left(0, \frac{15}{2}\right)$

Area: $A = 15 + 12 + 7 = 34$

$$\bar{x} = \frac{15(0) + 12(0) + 7(0)}{34} = 0$$

$$\bar{y} = \frac{15(3/2) + 12(5) + 7(15/2)}{34} = \frac{135}{34}$$

$$(\bar{x}, \bar{y}) = \left(0, \frac{135}{34}\right)$$

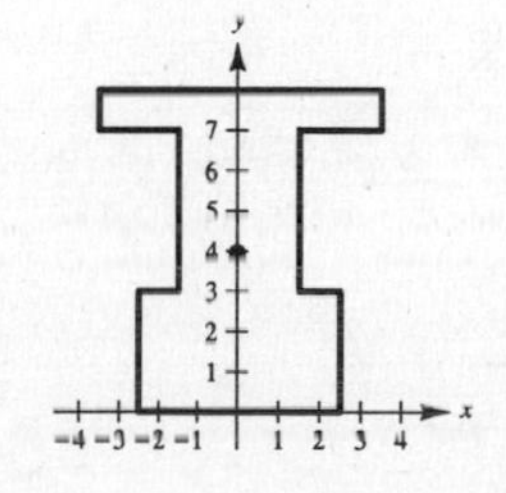

47. Centroids of the given regions: (1, 0) and (3, 0)

Mass: $4 + 2\pi$

$$\bar{x} = \frac{4(1) + 2\pi(3)}{4 + 2\pi} = \frac{2 + 3\pi}{2 + \pi}$$

$$\bar{y} = 0$$

$$(\bar{x}, \bar{y}) = \left(\frac{2 + 3\pi}{2 + \pi}, 0\right) \approx (2.22, 0)$$

49. $V = 2\pi rA = 2\pi(5)(16\pi) = 160\pi^2 \approx 1579.14$

51. $A = \frac{1}{2}(4)(4) = 8$

$$\bar{y} = \left(\frac{1}{8}\right)\frac{1}{2}\int_0^4 (4 + x)(4 - x)\,dx = \frac{1}{16}\left[16x - \frac{x^3}{3}\right]_0^4 = \frac{8}{3}$$

$$r = \bar{y} = \frac{8}{3}$$

$$V = 2\pi rA = 2\pi\left(\frac{8}{3}\right)(8) = \frac{128\pi}{3} \approx 134.04$$

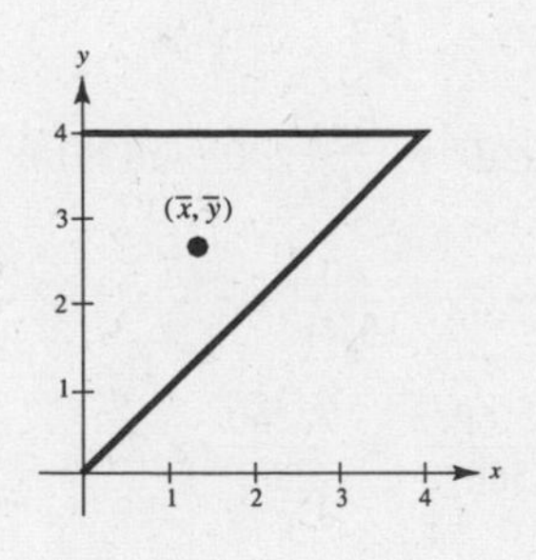

53. $m = m_1 + \cdots + m_n$

$M_y = m_1x_1 + \cdots + m_nx_n$

$M_x = m_1y_1 + \cdots + m_ny_n$

$\bar{x} = \frac{M_y}{m}, \bar{y} = \frac{M_x}{m}$

55. (a) Yes. $(\bar{x}, \bar{y}) = \left(\frac{5}{6}, \frac{5}{18} + 2\right) = \left(\frac{5}{6}, \frac{41}{18}\right)$

(b) Yes. $(\bar{x}, \bar{y}) = \left(\frac{5}{6} + 2, \frac{5}{18}\right) = \left(\frac{17}{6}, \frac{5}{18}\right)$

(c) Yes. $(\bar{x}, \bar{y}) = \left(\frac{5}{6}, -\frac{5}{18}\right)$

(d) No.

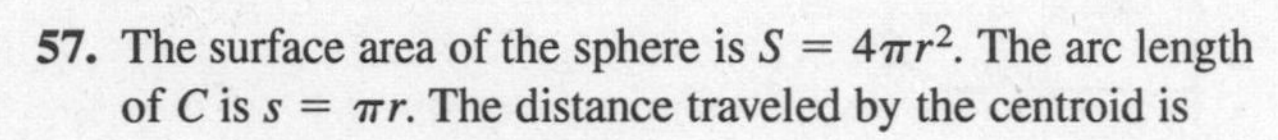

57. The surface area of the sphere is $S = 4\pi r^2$. The arc length of C is $s = \pi r$. The distance traveled by the centroid is

$$d = \frac{S}{s} = \frac{4\pi r^2}{\pi r} = 4r.$$

This distance is also the circumference of the circle of radius y.

$$d = 2\pi y$$

Thus, $2\pi y = 4r$ and we have $y = 2r/\pi$. Therefore, the centroid of the semicircle $y = \sqrt{r^2 - x^2}$ is $(0, 2r/\pi)$.

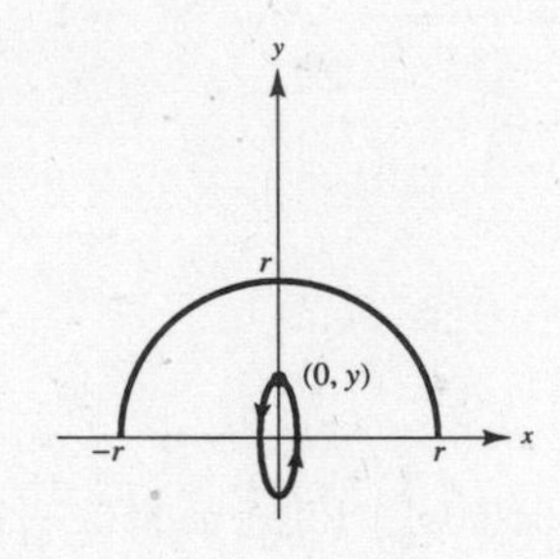

59. $A = \int_0^1 x^n\,dx = \left[\frac{x^{n+1}}{n+1}\right]_0^1 = \frac{1}{n+1}$

$$m = \rho A = \frac{\rho}{n+1}$$

$$M_x = \frac{\rho}{2}\int_0^1 (x^n)^2\,dx = \left[\frac{\rho}{2}\cdot\frac{x^{2n+1}}{2n+1}\right]_0^1 = \frac{\rho}{2(2n+1)}$$

$$M_y = \rho\int_0^1 x(x^n)\,dx = \left[\rho\cdot\frac{x^{n+2}}{n+2}\right]_0^1 = \frac{\rho}{n+2}$$

$$\bar{x} = \frac{M_y}{m} = \frac{n+1}{n+2}$$

$$\bar{y} = \frac{M_x}{m} = \frac{n+1}{2(2n+1)} = \frac{n+1}{4n+2}$$

Centroid: $\left(\frac{n+1}{n+2}, \frac{n+1}{4n+2}\right)$

As $n \to \infty$, $(\bar{x}, \bar{y}) \to \left(1, \frac{1}{4}\right)$.

The graph approaches the x-axis and the line $x = 1$ as $n \to \infty$.

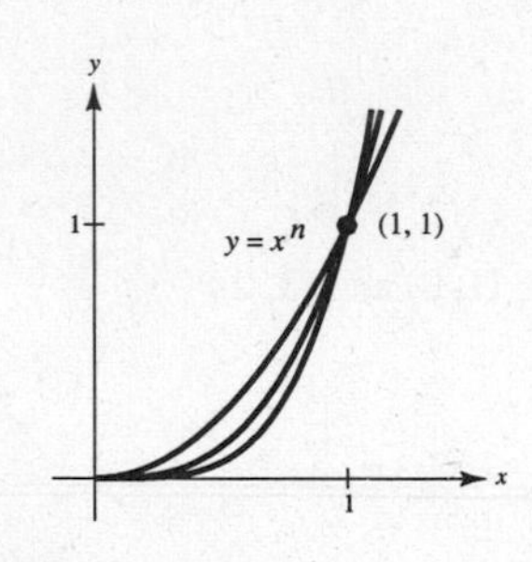

Section 6.7 Fluid Pressure and Fluid Force

1. $F = PA = [62.4(5)](3) = 936 \text{ lb}$

3. $F = 62.4(h + 2)(6) - (62.4)(h)(6)$

$= 62.4(2)(6) = 748.8 \text{ lb}$

5. $h(y) = 3 - y$

$L(y) = 4$

$$F = 62.4 \int_0^3 (3 - y)(4)\, dy$$

$$= 249.6 \int_0^3 (3 - y)\, dy$$

$$= 249.6\left[3y - \frac{y^2}{2}\right]_0^3 = 1123.2 \text{ lb}$$

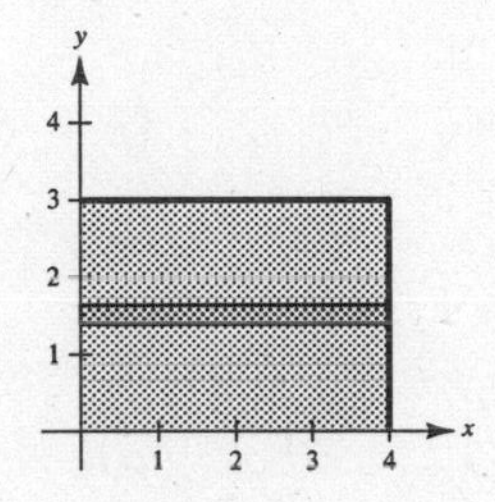

7. $h(y) = 3 - y$

$$L(y) = 2\left(\frac{y}{3} + 1\right)$$

$$F = 2(62.4) \int_0^3 (3 - y)\left(\frac{y}{3} + 1\right) dy$$

$$= 124.8 \int_0^3 \left(3 - \frac{y^2}{3}\right) dy$$

$$= 124.8\left[3y - \frac{y^3}{9}\right]_0^3 = 748.8 \text{ lb}$$

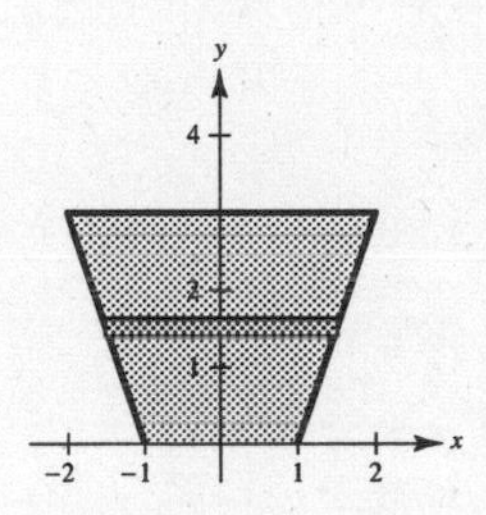

9. $h(y) = 4 - y$

$L(y) = 2\sqrt{y}$

$$F = 2(62.4) \int_0^4 (4 - y)\sqrt{y}\, dy$$

$$= 124.8 \int_0^4 (4y^{1/2} - y^{3/2})\, dy$$

$$= 124.8\left[\frac{8y^{3/2}}{3} - \frac{2y^{5/2}}{5}\right]_0^4 = 1064.96 \text{ lb}$$

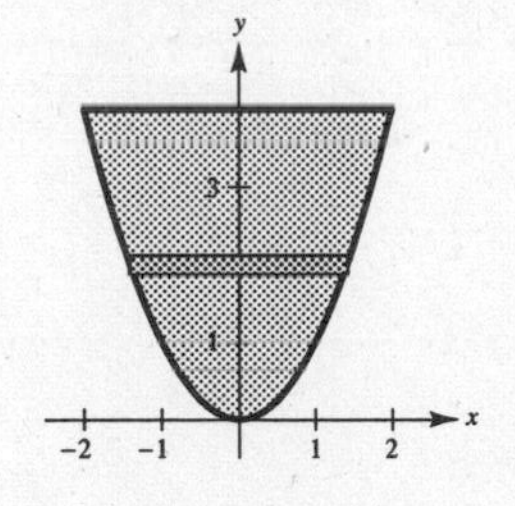

11. $h(y) = 4 - y$

$L(y) = 2$

$$F = 9800 \int_0^2 2(4 - y)\, dy$$

$$= 9800\left[8y - y^2\right]_0^2 = 117{,}600 \text{ Newtons}$$

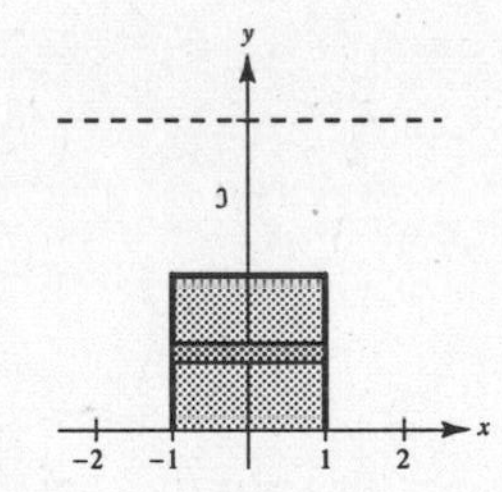

13. $h(y) = 12 - y$

$L(y) = 6 - \dfrac{2y}{3}$

$$F = 9800 \int_0^9 (12 - y)\left(6 - \frac{2y}{3}\right) dy$$

$$= 9800\left[72y - 7y^2 + \frac{2y^3}{9}\right]_0^9 = 2{,}381{,}400 \text{ Newtons}$$

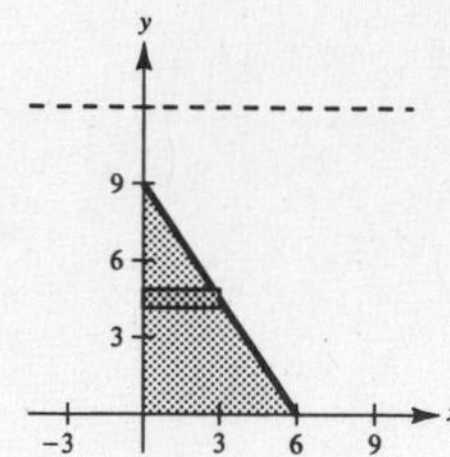

15. $h(y) = 2 - y$

$L(y) = 10$

$$F = 140.7 \int_0^2 (2 - y)(10)\, dy$$

$$= 1407 \int_0^2 (2 - y)\, dy$$

$$= 1407\left[2y - \frac{y^2}{2}\right]_0^2 = 2814 \text{ lb}$$

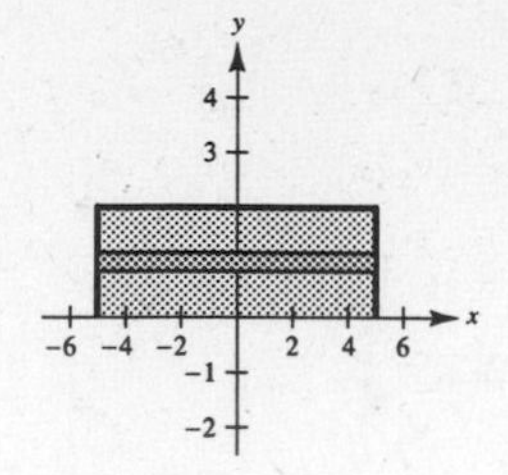

17. $h(y) = 4 - y$

$L(y) = 6$

$$F = 140.7 \int_0^4 (4 - y)(6)\, dy$$

$$= 844.2 \int_0^4 (4 - y)\, dy$$

$$= 844.2\left[4y - \frac{y^2}{2}\right]_0^4 = 6753.6 \text{ lb}$$

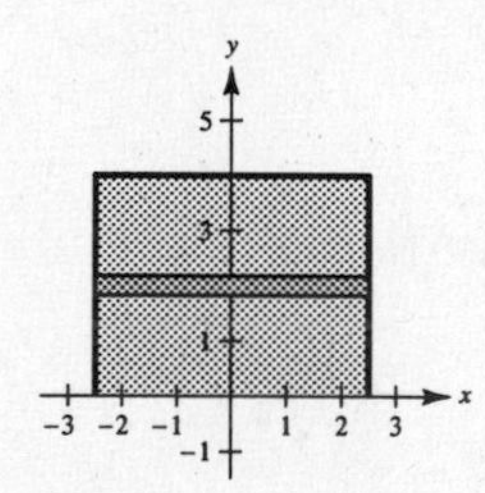

19. $h(y) = -y$

$L(y) = 2\left(\dfrac{1}{2}\right)\sqrt{9 - 4y^2}$

$$F = 42 \int_{-3/2}^{0} (-y)\sqrt{9 - 4y^2}\, dy$$

$$= \frac{42}{8} \int_{-3/2}^{0} (9 - 4y^2)^{1/2}(-8y)\, dy$$

$$= \left[\left(\frac{21}{4}\right)\left(\frac{2}{3}\right)(9 - 4y^2)^{3/2}\right]_{-3/2}^{0} = 94.5 \text{ lb}$$

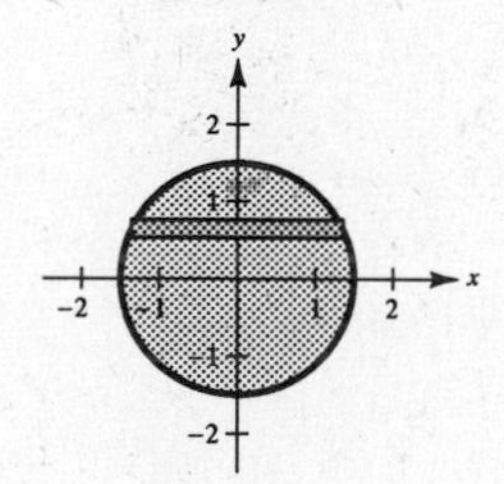

21. $h(y) = k - y$

$L(y) = 2\sqrt{r^2 - y^2}$

$$F = w \int_{-r}^{r} (k - y)\sqrt{r^2 - y^2}\,(2)\, dy$$

$$= w\left[2k \int_{-r}^{r} \sqrt{r^2 - y^2}\, dy + \int_{-r}^{r} \sqrt{r^2 - y^2}\,(-2y)\, dy\right]$$

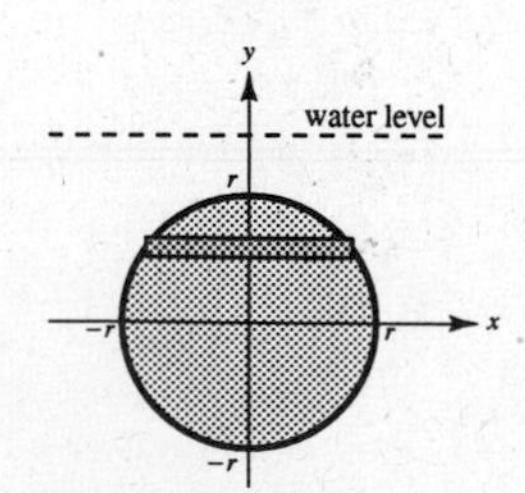

The second integral is zero since its integrand is odd and the limits of integration are symmetric to the origin. The first integral is the area of a semicircle with radius r.

$$F = w\left[(2k)\frac{\pi r^2}{2} + 0\right] = wk\pi r^2$$

23. $h(y) = k - y$

$L(y) = b$

$$F = w\int_{-h/2}^{h/2} (k - y)b\,dy$$

$$= wb\left[ky - \frac{y^2}{2}\right]_{-h/2}^{h/2} = wb(hk) = wkhb$$

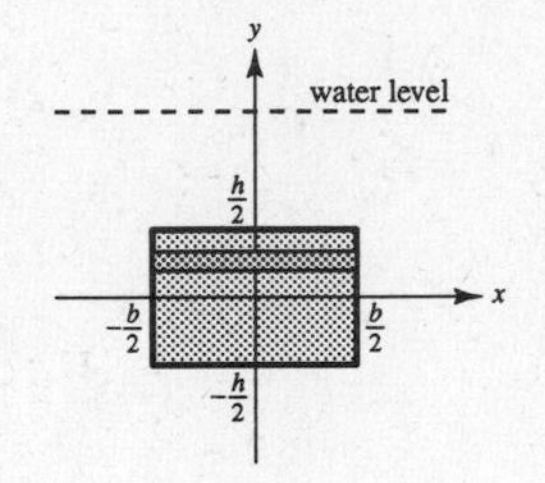

25. From Exercise 23:

$$F = 64(15)(1)(1) = 960 \text{ lb}$$

27. $h(y) = 4 - y$

$$F = 62.4\int_0^4 (4 - y)L(y)\,dy$$

Using Simpson's Rule with $n = 8$ we have:

$$F \approx 62.4\left(\frac{4 - 0}{3(8)}\right)[0 + 4(3.5)(3) + 2(3)(5) + 4(2.5)(8) + 2(2)(9) + 4(1.5)(10) + 2(1)(10.25) + 4(0.5)(10.5) + 0]$$

$$= 3010.8 \text{ lb}$$

29. $h(y) = 12 - y$

$L(y) = 2(4^{2/3} - y^{2/3})^{3/2}$

$$F = 62.4\int_0^4 2(12 - y)(4^{2/3} - y^{2/3})^{3/2}\,dy$$

$$\approx 6448.73 \text{ lb}$$

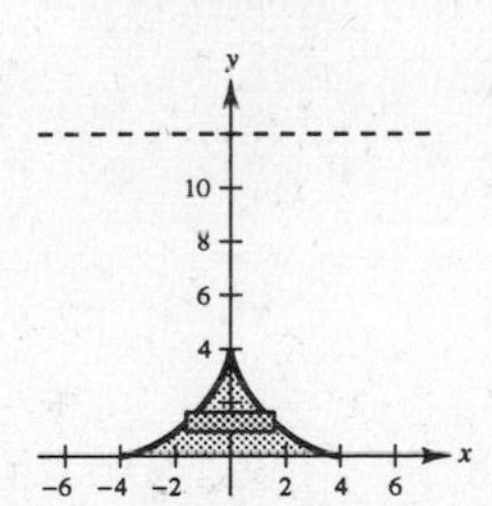

31. (a) If the fluid force is one half of 1123.2 lb, and the height of the water is b, then

$h(y) = b - y$

$L(y) = 4$

$$F = 62.4\int_0^b (b - y)(4)\,dy = \frac{1}{2}(1123.2)$$

$$\int_0^b (b - y)\,dy = 2.25$$

$$\left[by - \frac{y^2}{2}\right]_0^b = 2.25$$

$$b^2 - \frac{b^2}{2} = 2.25$$

$$b^2 = 4.5 \Rightarrow b \approx 2.12 \text{ ft.}$$

(b) The pressure increases with increasing depth.

33. $F = Fw = w\int_c^d h(y)L(y)\,dy$, see page 471.

Review Exercises for Chapter 6

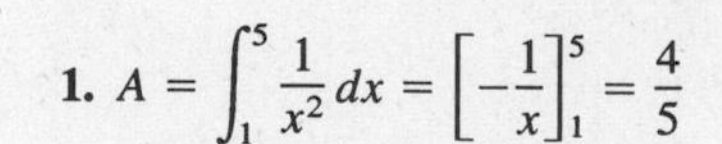

1. $A = \int_1^5 \frac{1}{x^2}\,dx = \left[-\frac{1}{x}\right]_1^5 = \frac{4}{5}$

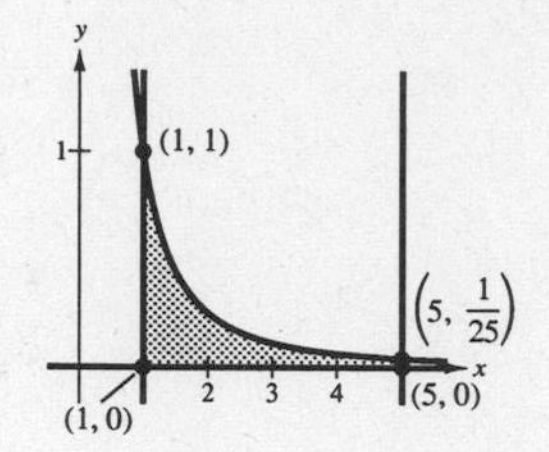

3. $A = \int_{-1}^1 \frac{1}{x^2+1}\,dx$

$= \left[\arctan x\right]_{-1}^1$

$= \frac{\pi}{4} - \left(-\frac{\pi}{4}\right) = \frac{\pi}{2}$

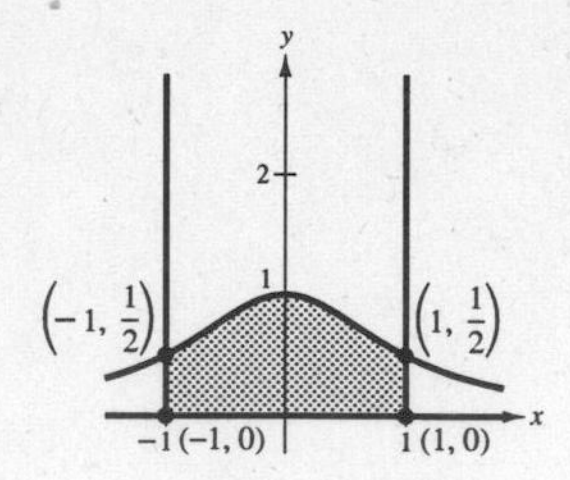

5. $A = 2\int_0^1 (x - x^3)\,dx$

$= 2\left[\frac{1}{2}x^2 - \frac{1}{4}x^4\right]_0^1$

$= \frac{1}{2}$

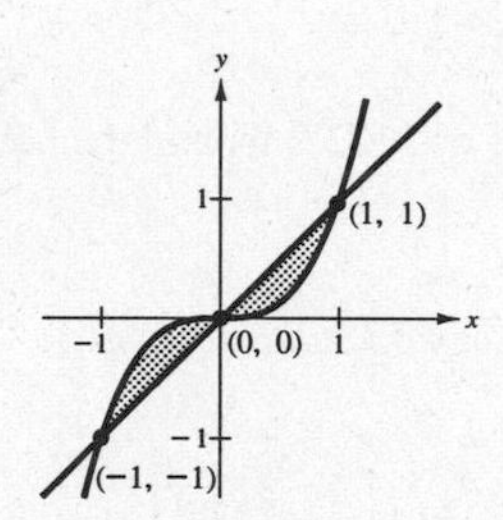

7. $A = \int_0^2 (e^2 - e^x)\,dx$

$= \left[xe^2 - e^x\right]_0^2$

$= e^2 + 1$

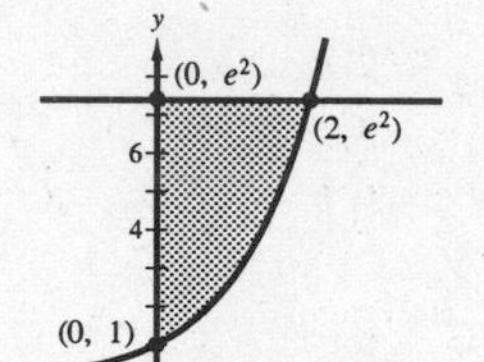

9. $A = \int_{\pi/4}^{5\pi/4} (\sin x - \cos x)\,dx$

$= \left[-\cos x - \sin x\right]_{\pi/4}^{5\pi/4}$

$= \left(\frac{1}{\sqrt{2}} + \frac{1}{\sqrt{2}}\right) - \left(-\frac{1}{\sqrt{2}} - \frac{1}{\sqrt{2}}\right)$

$= \frac{4}{\sqrt{2}} = 2\sqrt{2}$

11. $A = \int_0^8 [(3 + 8x - x^2) - (x^2 - 8x + 3)]\,dx$

$= \int_0^8 (16x - 2x^2)\,dx$

$= \left[8x^2 - \frac{2}{3}x^3\right]_0^8 = \frac{512}{3} \approx 170.667$

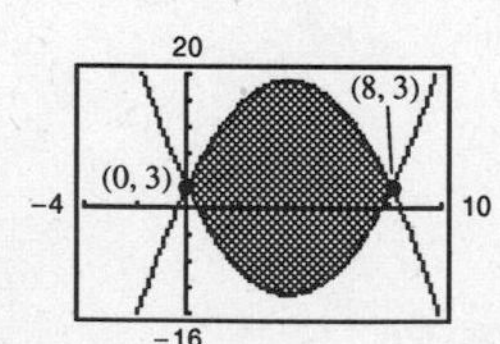

13. $y = (1 - \sqrt{x})^2$

$$A = \int_0^1 (1 - \sqrt{x})^2\, dx$$

$$= \int_0^1 (1 - 2x^{1/2} + x)\, dx$$

$$= \left[x - \frac{4}{3}x^{3/2} + \frac{1}{2}x^2\right]_0^1 = \frac{1}{6} \approx 0.1667$$

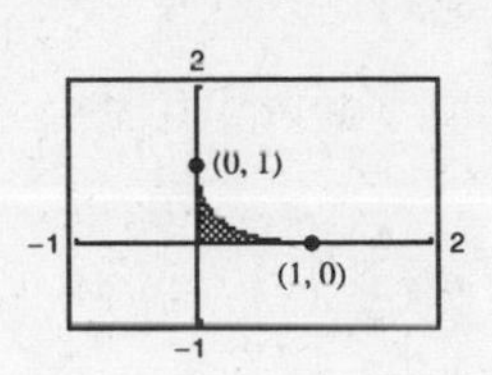

15. $x = y^2 - 2y \Rightarrow x + 1 = (y - 1)^2 \Rightarrow y = 1 \pm \sqrt{x + 1}$

$$A = \int_{-1}^0 \left[\left(1 + \sqrt{x+1}\right) - \left(1 - \sqrt{x+1}\right)\right] dx = \int_{-1}^0 2\sqrt{x+1}\, dx$$

$$A = \int_0^2 [0 - (y^2 - 2y)]\, dy = \int_0^2 (2y - y^2)\, dy = \left[y^2 - \frac{1}{3}y^3\right]_0^2 = \frac{4}{3}$$

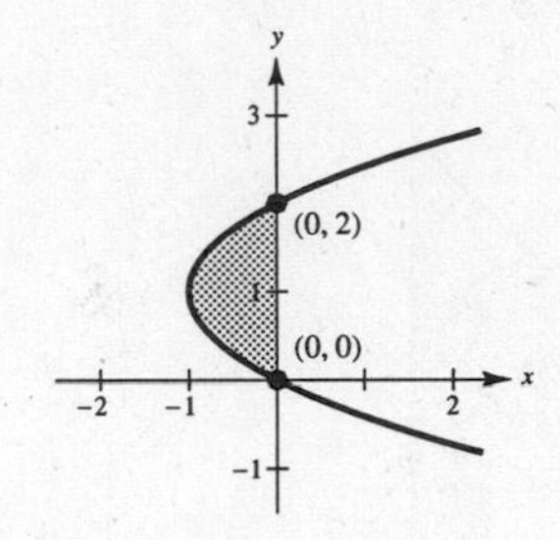

17. $A = \int_0^2 \left[1 - \left(1 - \frac{x}{2}\right)\right] dx + \int_2^3 [1 - (x - 2)]\, dx$

$$= \int_0^2 \frac{x}{2}\, dx + \int_2^3 (3 - x)\, dx$$

$$y = 1 - \frac{x}{2} \Rightarrow x = 2 - 2y$$

$$y = x - 2 \Rightarrow x = y + 2,\ y = 1$$

$$A = \int_0^1 [(y + 2) - (2 - 2y)]\, dy$$

$$= \int_0^1 3y\, dy = \left[\frac{3}{2}y^2\right]_0^1 = \frac{3}{2}$$

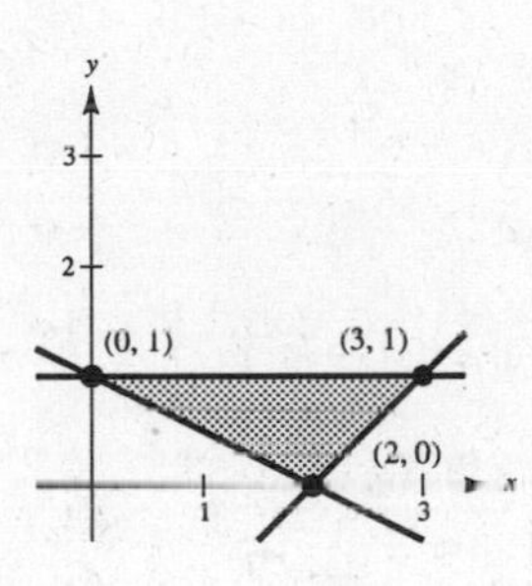

19. Job 1 is better. The salary for Job 1 is greater than the salary for Job 2 for all the years except the first and 10th years.

21. (a) **Disk**

$$V = \pi \int_0^4 x^2\, dx = \left[\frac{\pi x^3}{3}\right]_0^4 = \frac{64\pi}{3}$$

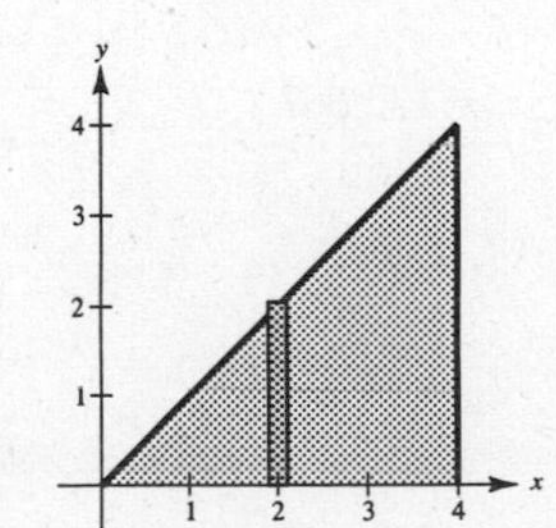

(b) **Shell**

$$V = 2\pi \int_0^4 x^2\, dx = \left[\frac{2\pi}{3}x^3\right]_0^4 = \frac{128\pi}{3}$$

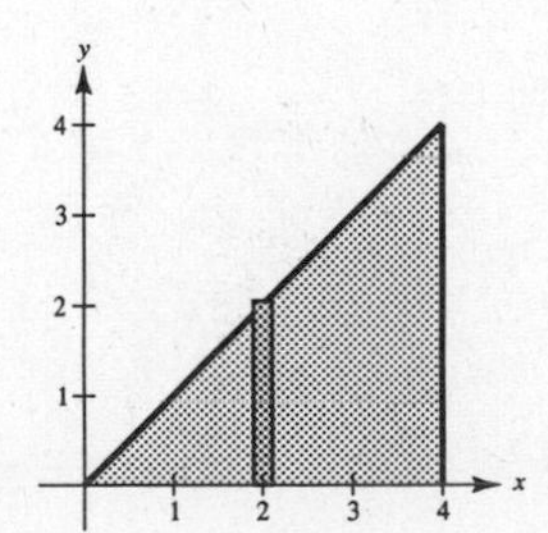

—CONTINUED—

21. —CONTINUED—

(c) **Shell**

$$V = 2\pi \int_0^4 (4 - x)x\,dx$$
$$= 2\pi \int_0^4 (4x - x^2)\,dx$$
$$= 2\pi\left[2x^2 - \frac{x^3}{3}\right]_0^4 = \frac{64\pi}{3}$$

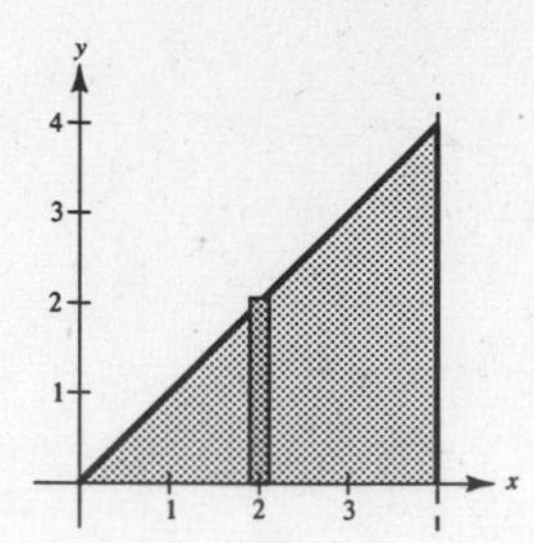

(d) **Shell**

$$V = 2\pi \int_0^4 (6 - x)x\,dx$$
$$= 2\pi \int_0^4 (6x - x^2)\,dx$$
$$= 2\pi\left[3x^2 - \frac{1}{3}x^3\right]_0^4 = \frac{160\pi}{3}$$

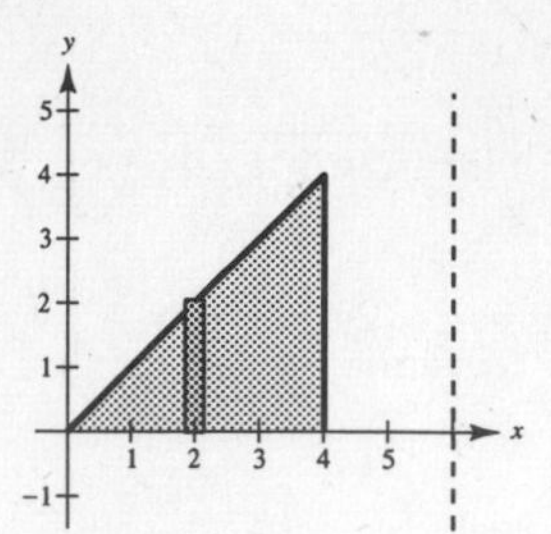

23. (a) **Shell**

$$V = 4\pi \int_0^4 x\left(\frac{3}{4}\right)\sqrt{16 - x^2}\,dx$$
$$= \left[3\pi\left(-\frac{1}{2}\right)\left(\frac{2}{3}\right)(16 - x^2)^{3/2}\right]_0^4 = 64\pi$$

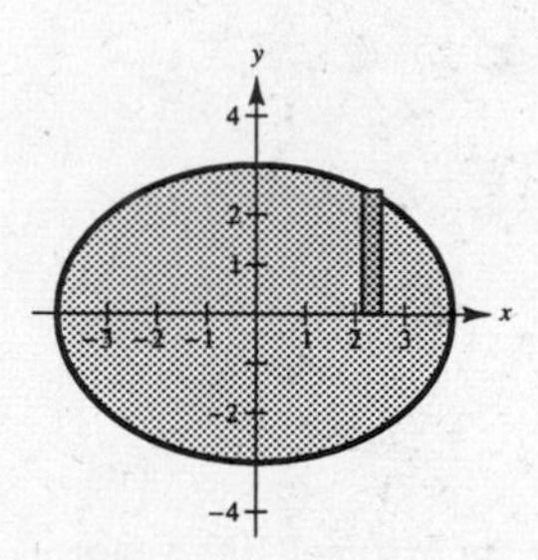

(b) **Disk**

$$V = 2\pi \int_0^4 \left[\frac{3}{4}\sqrt{16 - x^2}\right]^2 dx$$
$$= \frac{9\pi}{8}\left[16x - \frac{x^3}{3}\right]_0^4 = 48\pi$$

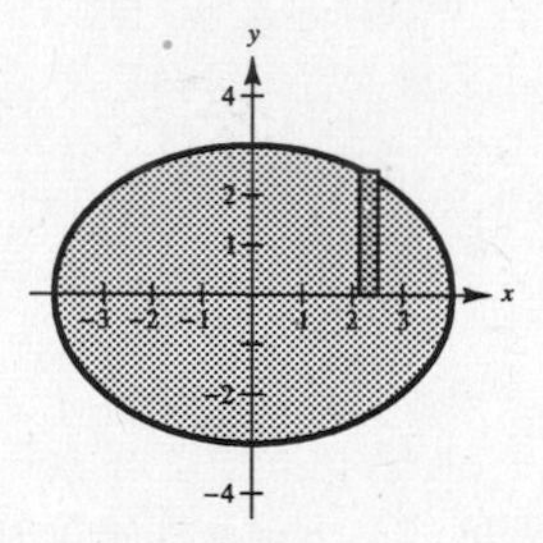

25. Shell

$$V = 2\pi \int_0^1 \frac{x}{x^4 + 1}\,dx$$
$$= \pi \int_0^1 \frac{(2x)}{(x^2)^2 + 1}\,dx$$
$$= \left[\pi \arctan(x^2)\right]_0^1$$
$$= \pi\left[\frac{\pi}{4} - 0\right] = \frac{\pi^2}{4}$$

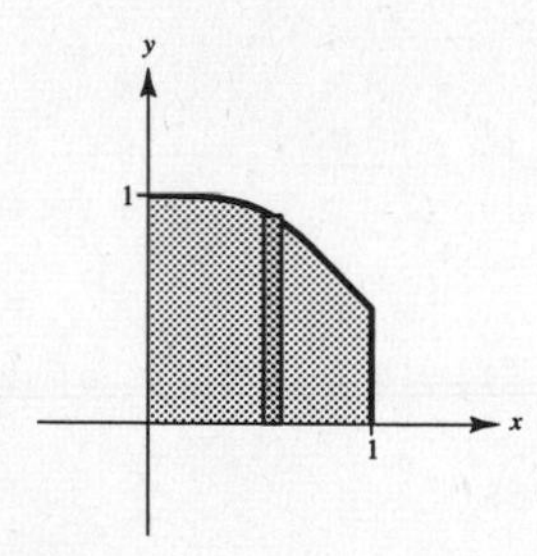

27. **Shell**

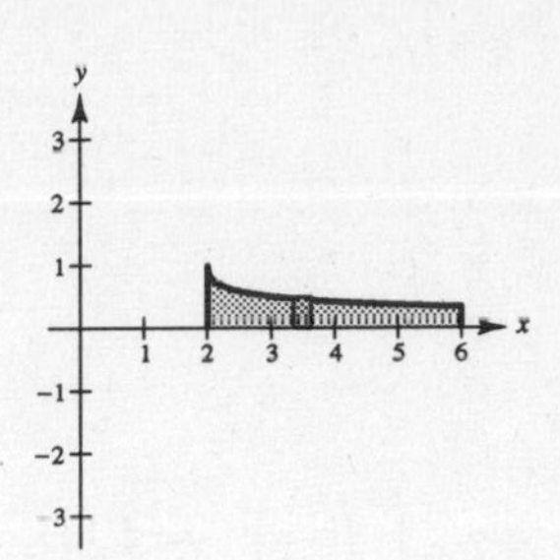

$u = \sqrt{x-2}$

$x = u^2 + 2$

$dx = 2u\,du$

$$V = 2\pi \int_2^6 \frac{x}{1+\sqrt{x-2}}\,dx = 4\pi \int_0^2 \frac{(u^2+2)u}{1+u}\,du$$

$$= 4\pi \int_0^2 \frac{u^3+2u}{1+u}\,du = 4\pi \int_0^2 \left(u^2 - u + 3 - \frac{3}{1+u}\right) du$$

$$= 4\pi \left[\frac{1}{3}u^3 - \frac{1}{2}u^2 + 3u - 3\ln(1+u)\right]_0^2 = \frac{4\pi}{3}(20 - 9\ln 3) \approx 42.359$$

29. Since $y \le 0$, $A = -\int_{-1}^{0} x\sqrt{x+1}\,dx$.

$u = x + 1$

$x = u - 1$

$dx = du$

$$A = -\int_0^1 (u-1)\sqrt{u}\,du = -\int_0^1 (u^{3/2} - u^{1/2})\,du$$

$$= -\left[\frac{2}{5}u^{5/2} - \frac{2}{3}u^{3/2}\right]_0^1 = \frac{4}{15}$$

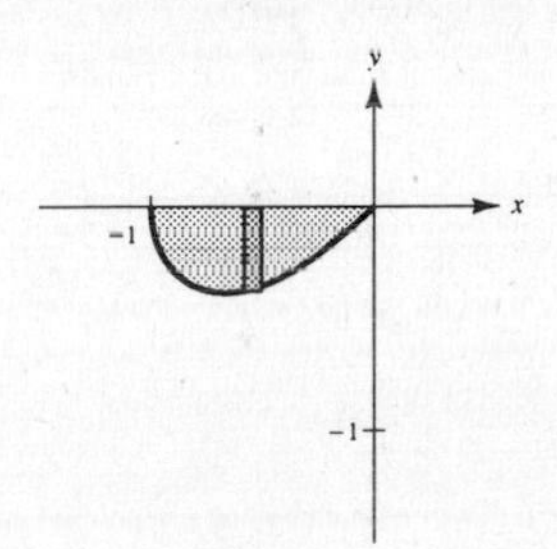

31. From Exercise 23(a) we have: $V = 64\pi \text{ ft}^3$

$$\frac{1}{4}V = 16\pi$$

Disk:

$$\pi \int_{-3}^{y_0} \frac{16}{9}(9 - y^2)\,dy = 16\pi$$

$$\frac{1}{9}\int_{-3}^{y_0} (9 - y^2)\,dy = 1$$

$$\left[9y - \frac{1}{3}y^3\right]_{-3}^{y_0} = 9$$

$$\left(9y_0 - \frac{1}{3}y_0^3\right) - (-27 + 9) - 9$$

$$y_0^3 - 27y_0 - 27 = 0$$

By Newton's Method, $y_0 \approx -1.042$ and the depth of the gasoline is $3 - 1.042 = 1.958$ ft.

33.

$$f(x) = \frac{4}{5}x^{5/4}$$

$$f'(x) = x^{1/4}$$

$$1 + [f'(x)]^2 = 1 + \sqrt{x}$$

$$u = 1 + \sqrt{x}$$

$$x = (u-1)^2$$

$$dx = 2(u-1)\,du$$

$$s = \int_0^4 \sqrt{1+\sqrt{x}}\,dx = 2\int_1^3 \sqrt{u}(u-1)\,du$$

$$= 2\int_1^3 (u^{3/2} - u^{1/2})\,du$$

$$= 2\left[\frac{2}{5}u^{5/2} - \frac{2}{3}u^{3/2}\right]_1^3 = \frac{4}{15}\left[u^{3/2}(3u-5)\right]_1^3$$

$$= \frac{8}{15}(1 + 6\sqrt{3}) \approx 6.076$$

35. $y = 300\cosh\left(\frac{x}{2000}\right) - 280,\ -2000 \le x \le 2000$

$$y' = \frac{3}{20}\sinh\left(\frac{x}{2000}\right)$$

$$s = \int_{-2000}^{2000} \sqrt{1 + \left[\frac{3}{20}\sinh\left(\frac{x}{2000}\right)\right]^2}\,dx$$

$$= \frac{1}{20}\int_{-2000}^{2000} \sqrt{400 + 9\sinh^2\left(\frac{x}{2000}\right)}\,dx$$

≈ 4018.2 ft (by Simpson's Rule or graphing utility)

37. $y = \frac{3}{4}x$

$$y' = \frac{3}{4}$$

$$1 + (y')^2 = \frac{25}{16}$$

$$S = 2\pi \int_0^4 \left(\frac{3}{4}x\right)\sqrt{\frac{25}{16}}\,dx = \left[\left(\frac{15\pi}{8}\right)\frac{x^2}{2}\right]_0^4 = 15\pi$$

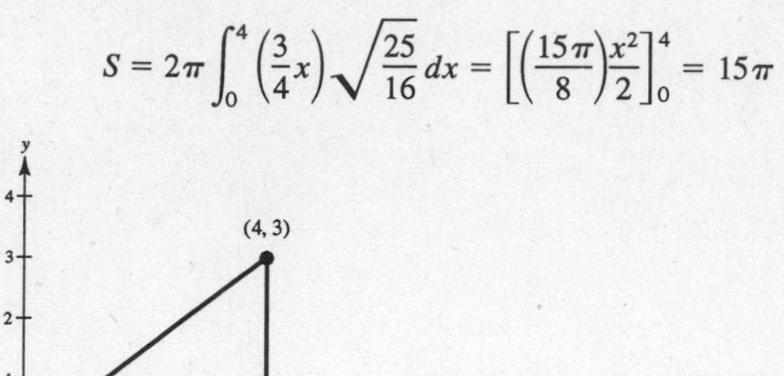

39. $F = kx$

$$4 = k(1)$$

$$F = 4x$$

$$W = \int_0^5 4x\,dx = \Big[2x^2\Big]_0^5$$

$$= 50 \text{ in} \cdot \text{lb} \approx 4.167 \text{ ft} \cdot \text{lb}$$

41. Volume of disk: $\pi\left(\frac{1}{3}\right)^2 \Delta y$

Weight of disk: $62.4\pi\left(\frac{1}{3}\right)^2 \Delta y$

Distance: $175 - y$

$$W = \frac{62.4\pi}{9}\int_0^{150} (175 - y)\,dy = \frac{62.4\pi}{9}\left[175y - \frac{y^2}{2}\right]_0^{150}$$

$$= 104{,}000\pi \text{ ft} \cdot \text{lb} \approx 163.4 \text{ ft} \cdot \text{ton}$$

43. Weight of section of chain: $5\,\Delta x$

Distance moved: $10 - x$

$$W = 5\int_0^{10} (10 - x)\,dx = \left[-\frac{5}{2}(10 - x)^2\right]_0^{10}$$

$$= 250 \text{ ft} \cdot \text{lb}$$

45. $W = \int_a^b F(x)\,dx$

$$80 = \int_0^4 ax^2 dx = \frac{ax^3}{3}\Big]_0^4 = \frac{64}{3}a$$

$$a = \frac{3(80)}{64} = \frac{15}{4} = 3.75$$

47. $A = \int_0^a (\sqrt{a} - \sqrt{x})^2\,dx = \int_0^a (a - 2\sqrt{a}\,x^{1/2} + x)\,dx = \left[ax - \frac{4}{3}\sqrt{a}\,x^{3/2} + \frac{1}{2}x^2\right]_0^a = \frac{a^2}{6}$

$$\frac{1}{A} = \frac{6}{a^2}$$

$$\bar{x} = \frac{6}{a^2}\int_0^a x(\sqrt{a} - \sqrt{x})^2\,dx = \frac{6}{a^2}\int_0^a (ax - 2\sqrt{a}\,x^{3/2} + x^2)\,dx = \frac{a}{5}$$

$$\bar{y} = \left(\frac{6}{a^2}\right)\frac{1}{2}\int_0^a (\sqrt{a} - \sqrt{x})^4\,dx$$

$$= \frac{3}{a^2}\int_0^a (a^2 - 4a^{3/2}x^{1/2} + 6ax - 4a^{1/2}x^{3/2} + x^2)\,dx$$

$$= \frac{3}{a^2}\left[a^2x - \frac{8}{3}a^{3/2}x^{3/2} + 3ax^2 - \frac{8}{5}a^{1/2}x^{5/2} + \frac{1}{3}x^3\right]_0^a = \frac{a}{5}$$

$$(\bar{x}, \bar{y}) = \left(\frac{a}{5}, \frac{a}{5}\right)$$

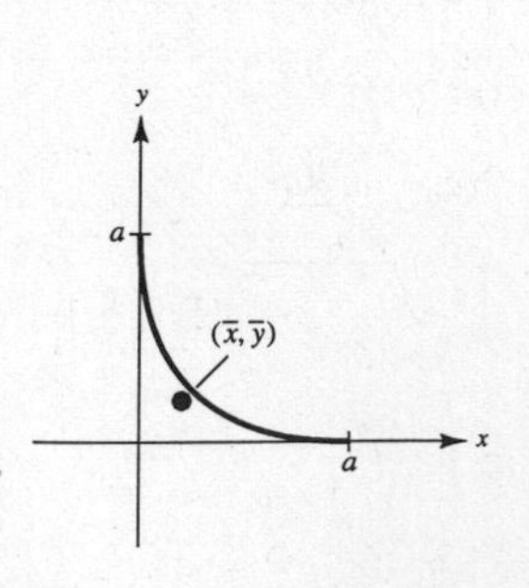

49. By symmetry, $x = 0$.

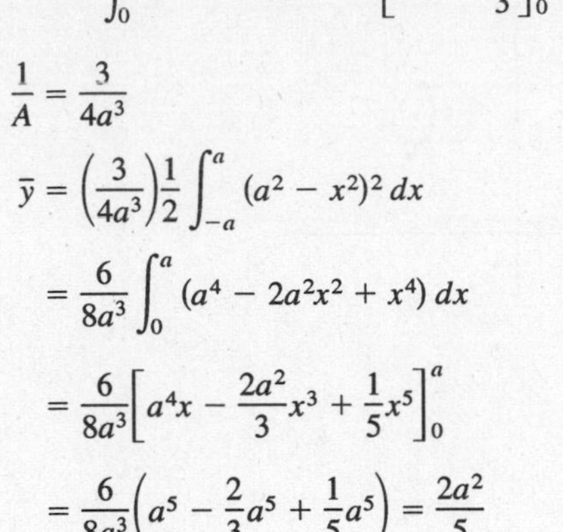

$$A = 2\int_0^1 (a^2 - x^2)\,dx = 2\left[a^2x - \frac{x^3}{3}\right]_0^a = \frac{4a^3}{3}$$

$$\frac{1}{A} = \frac{3}{4a^3}$$

$$\bar{y} = \left(\frac{3}{4a^3}\right)\frac{1}{2}\int_{-a}^{a} (a^2 - x^2)^2\,dx$$

$$= \frac{6}{8a^3}\int_0^a (a^4 - 2a^2x^2 + x^4)\,dx$$

$$= \frac{6}{8a^3}\left[a^4x - \frac{2a^2}{3}x^3 + \frac{1}{5}x^5\right]_0^a$$

$$= \frac{6}{8a^3}\left(a^5 - \frac{2}{3}a^5 + \frac{1}{5}a^5\right) = \frac{2a^2}{5}$$

$$(\bar{x}, \bar{y}) = \left(0, \frac{2a^2}{5}\right)$$

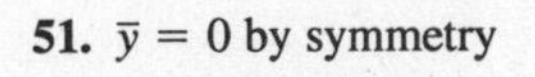

51. $\bar{y} = 0$ by symmetry

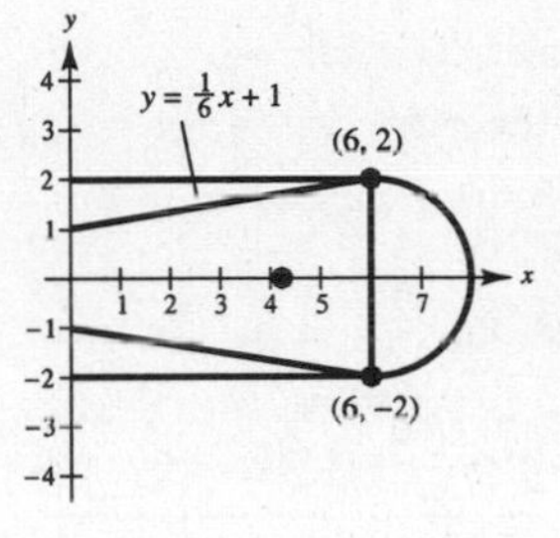

For the trapezoid:

$$m = [(4)(6) - (1)(6)]\rho = 18\rho$$

$$M_y = \rho\int_0^6 x\left[\left(\frac{1}{6}x + 1\right) - \left(-\frac{1}{6}x - 1\right)\right]dx$$

$$= \rho\int_0^6 \left(\frac{1}{3}x^2 + 2x\right)dx = \rho\left[\frac{x^3}{9} + x^2\right]_0^6 = 60\rho$$

For the semicircle:

$$m = \left(\frac{1}{2}\right)(\pi)(2)^2\rho = 2\pi\rho$$

$$M_y = \rho\int_6^8 x\left[\sqrt{4 - (x-6)^2} - \left(-\sqrt{4 - (x-6)^2}\right)\right]dx = 2\rho\int_6^8 x\sqrt{4 - (x-6)^2}\,dx$$

Let $u = x - 6$, then $x = u + 6$ and $dx = du$. When $x = 6$, $u = 0$. When $x = 8$, $u = 2$.

$$M_y = 2\rho\int_0^2 (u + 6)\sqrt{4 - u^2}\,du = 2\rho\int_0^2 u\sqrt{4 - u^2}\,du + 12\rho\int_0^2 \sqrt{4 - u^2}\,du$$

$$= 2\rho\left[\left(-\frac{1}{2}\right)\left(\frac{2}{3}\right)(4 - u^2)^{3/2}\right]_0^2 + 12\rho\left[\frac{\pi(2)^2}{4}\right] = \frac{16\rho}{3} + 12\pi\rho = \frac{4\rho(4 + 9\pi)}{3}$$

Thus, we have:

$$\bar{x}(18\rho + 2\pi\rho) = 60\rho + \frac{4\rho(4 + 9\pi)}{3}$$

$$\bar{x} = \frac{180\rho + 4\rho(4 + 9\pi)}{3} \cdot \frac{1}{2\rho(9 + \pi)} = \frac{2(9\pi + 49)}{3(\pi + 9)}$$

The centroid of the blade is $\left(\dfrac{2(9\pi + 49)}{3(\pi + 9)}, 0\right)$.

53. Let D = surface of liquid; ρ = weight per cubic volume.

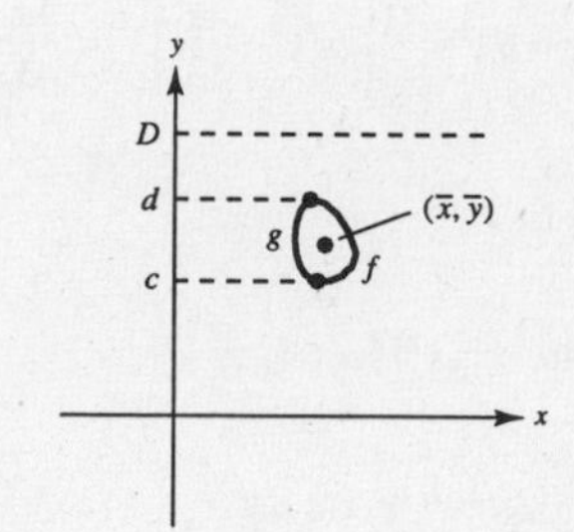

$$F = \rho \int_c^d (D - y)[f(y) - g(y)]\, dy$$

$$= \rho\left[\int_c^d D[f(y) - g(y)]\, dy - \int_c^d y[f(y) - g(y)]\, dy\right]$$

$$= \rho\left[\int_c^d [f(y) - g(y)]\, dy\right]\left[D - \frac{\int_c^d y[f(y) - g(y)]\, dy}{\int_c^d [f(y) - g(y)]\, dy}\right]$$

$$= \rho(\text{Area})(D - \bar{y})$$

$$= \rho(\text{Area})(\text{depth of centroid})$$

Problem Solving for Chapter 6

1. $T = \frac{1}{2}c(c^2) = \frac{1}{2}c^3$

$$R = \int_0^c (cx - x^2)\, dx = \left[\frac{cx^2}{2} - \frac{x^3}{3}\right]_0^c = \frac{c^3}{2} - \frac{c^3}{3} = \frac{c^3}{6}$$

$$\lim_{c\to 0^+} \frac{T}{R} = \lim_{c\to 0^+} \frac{\frac{1}{2}c^3}{\frac{1}{6}c^3} = 3$$

3. (a) $\frac{1}{2}V = \int_0^1 \left[\pi\left(2 + \sqrt{1 - y^2}\right)^2 - \pi\left(2 - \sqrt{1 - y^2}\right)^2\right] dy$

$$= \pi\int_0^1 \left[\left(4 + 4\sqrt{1 - y^2} + (1 - y^2)\right) - \left(4 - 4\sqrt{1 - y^2} + (1 - y^2)\right)\right] dy$$

$$= 8\pi\int_0^1 \sqrt{1 - y^2}\, dy \quad \text{(Integral represents 1/4 (area of circle))}$$

$$= 8\pi\left(\frac{\pi}{4}\right) = 2\pi^2 \Rightarrow V = 4\pi^2$$

(b) $(x - R)^2 + y^2 = r^2 \Rightarrow x = R \pm \sqrt{r^2 - y^2}$

$$\frac{1}{2}V = \int_0^r \left[\pi\left(R + \sqrt{r^2 - y^2}\right)^2 - \pi\left(R - \sqrt{r^2 - y^2}\right)^2\right] dy$$

$$= \pi\int_0^r 4R\sqrt{r^2 - y^2}\, dy$$

$$= \pi(4R)\frac{1}{4}\pi r^2 = \pi^2 r^2 R$$

$$V = 2\pi^2 r^2 R$$

5. $V = 2(2\pi)\int_{\sqrt{r^2 - (h^2/4)}}^r x\sqrt{r^2 - x^2}\, dx$

$$= -2\pi\left[\frac{2}{3}(r^2 - x^2)^{3/2}\right]_{\sqrt{r^2 - (h^2/4)}}^r$$

$$= \frac{-4\pi}{3}\left[-\frac{h^3}{8}\right] = \frac{\pi h^3}{6} \text{ which does not depend on } r!$$

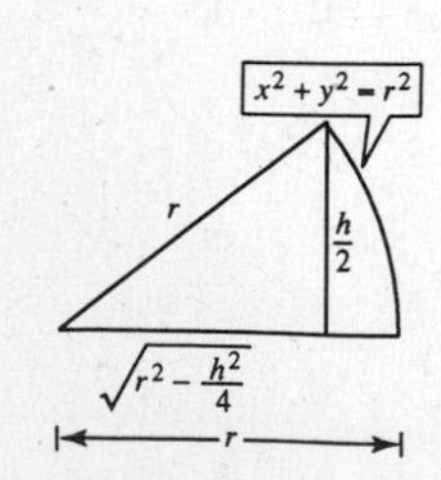

7. (a) Tangent at A: $y = x^3, y' = 3x^2$

$$y - 1 = 3(x - 1)$$
$$y = 3x - 2$$

To find point B: $x^3 = 3x - 2$

$$x^3 - 3x + 2 = 0$$
$$(x - 1)^2(x + 2) = 0 \Rightarrow B = (-2, -8)$$

Tangent at B: $y = x^3, y' = 3x^2$

$$y + 8 = 12(x + 2)$$
$$y = 12x + 16$$

To find point C: $x^3 = 12x + 16$

$$x^3 - 12x - 16 = 0$$
$$(x + 2)^2(x - 4) = 0 \Rightarrow C = (4, 64)$$

$$\text{Area of } R = \int_{-2}^{1} (x^3 - 3x + 2)\,dx = \frac{27}{4}$$

$$\text{Area of } S = \int_{-2}^{4} (12x + 16 - x^3)\,dx = 108$$

$$\text{Area of } S = 16(\text{area of } R) \quad \left[\frac{\text{area } S}{\text{area } R} = 16\right]$$

(b) Tangent at $A(a, a^3)$: $y - a^3 = 3a^2(x - a)$

$$y = 3a^2x - 2a^3$$

To find point B: $x^3 - 3a^2x + 2a^3 = 0$

$$(x - a)^2(x + 2a) = 0 \Rightarrow$$
$$B = (-2a, -8a^3)$$

Tangent at B: $y + 8a^3 = 12a^2(x + 2a)$

$$y = 12a^2x + 16a^3$$

To find point C: $x^3 - 12a^2x - 16a^3 = 0$

$$(x + 2a)^2(x - 4a) = 0 \Rightarrow$$
$$C = (4a, 64a^3)$$

$$\text{Area of } R = \int_{-2a}^{a} [x^3 - 3a^2x + 2a^3]\,dx = \frac{27}{4}a^4$$

$$\text{Area of } S = \int_{-2a}^{4a} [12a^2x + 16a^3 - x^3]\,dx = 108a^4$$

Area of S = 16(area of R)

9. $s(x) = \displaystyle\int_a^x \sqrt{1 + f'(t)^2}\,dt$

(a) $s'(x) = \dfrac{ds}{dx} = \sqrt{1 + f'(x)^2}$

(b) $ds = \sqrt{1 + f'(x)^2}\,dx$

$$(ds)^2 = [1 + f'(x)^2](dx)^2 = \left[1 + \left(\frac{dy}{dx}\right)^2\right](dx)^2 = (dx)^2 + (dy)^2$$

(c) $s(x) = \displaystyle\int_1^x \sqrt{1 + \left(\frac{3}{2}t^{1/2}\right)^2}\,dt = \int_1^x \sqrt{1 + \frac{9}{4}t}\,dt$

(d) $s(2) = \displaystyle\int_1^2 \sqrt{1 + \frac{9}{4}t}\,dt = \left[\frac{8}{27}\left(1 + \frac{9}{4}t\right)^{3/2}\right]_1^2 = \frac{22}{27}\sqrt{22} - \frac{13}{27}\sqrt{13} \approx 2.0858$

This is the length of the curve $y = x^{3/2}$ from $x = 1$ to $x = 2$.

11. (a) $\bar{y} = 0$ by symmetry

$$M_y = \int_1^6 x\left(\frac{1}{x^3} - \left(-\frac{1}{x^3}\right)\right)dx = \int_1^6 \frac{2}{x^2}\,dx = \left[-\frac{2}{x}\right]_1^6 = \frac{5}{3}$$

$$m = 2\int_1^6 \frac{1}{x^3}\,dx = \left[-\frac{1}{x^2}\right]_1^6 = \frac{35}{36}$$

$$\bar{x} = \frac{5/3}{35/36} = \frac{12}{7} \qquad (\bar{x}, \bar{y}) = \left(\frac{12}{7}, 0\right)$$

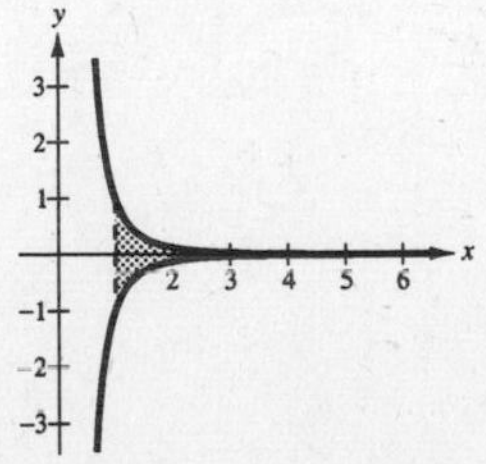

(b) $m = 2\displaystyle\int_1^b \frac{1}{x^3}\,dx = \frac{b^2 - 1}{b^2}$

$$M_y = 2\int_1^b \frac{1}{x^2}\,dx = \frac{2(b - 1)}{b}$$

$$\bar{x} = \frac{2(b - 1)/b}{(b^2 - 1)/b^2} = \frac{2b}{b + 1} \qquad (\bar{x}, \bar{y}) = \left(\frac{2b}{b + 1}, 0\right)$$

(c) $\displaystyle\lim_{b\to\infty} \bar{x} = \lim_{b\to\infty} \frac{2b}{b + 1} = 2 \qquad (\bar{x}, \bar{y}) = (2, 0)$

13. (a) $W = \text{area} = 2 + 4 + 6 = 12$

(b) $W = \text{area} = 3 + (1 + 1) + 2 + \frac{1}{2} = 7\frac{1}{2}$

15. Point of equilibrium: $50 - 0.5x = 0.125x$

$x = 80, p = 10$

$(P_0, x_0) = (10, 80)$

$$\text{Consumer surplus} = \int_0^{80} [(50 - 0.5x) - 10]\, dx = 1600$$

$$\text{Producer surplus} = \int_0^{80} [10 - 0.125x]\, dx = 400$$

17. We use Exercise 23 in Section 6.7:

(a) Wall at shallow end

From Exercise 23: $F = 62.4(2)(4)(20) = 9984$ lb

(b) Wall at deep end

From Exercise 23: $F = 62.4(4)(8)(20) = 39{,}936$ lb

(c) Side wall

From Exercise 23: $F_1 = 62.4(2)(4)(40) = 19{,}968$ lb

$$F_2 = 62.4 \int_0^4 (8 - y)(10y)\, dy$$

$$= 624 \int_0^4 (8y - y^2)\, dy = 624\left[4y^2 - \frac{y^3}{3}\right]_0^4$$

$$= 26{,}624 \text{ lb}$$

Total force: $F_1 + F_2 = 46{,}592$ lb

CHAPTER 7
Integration Techniques, L'Hôpital's Rule, and Improper Integrals

CHAPTER 7
Integration Techniques, L'Hôpital's Rule, and Improper Integrals

Section 7.1 Basic Integration Rules

Solutions to Odd-Numbered Exercises

1. (a) $\frac{d}{dx}\left[2\sqrt{x^2+1}+C\right] = 2\left(\frac{1}{2}\right)(x^2+1)^{-1/2}(2x) = \frac{2x}{\sqrt{x^2+1}}$

(b) $\frac{d}{dx}\left[\sqrt{x^2+1}+C\right] = \frac{1}{2}(x^2+1)^{-1/2}(2x) = \frac{x}{\sqrt{x^2+1}}$

(c) $\frac{d}{dx}\left[\frac{1}{2}\sqrt{x^2+1}+C\right] = \frac{1}{2}\left(\frac{1}{2}\right)(x^2+1)^{-1/2}(2x) = \frac{x}{2\sqrt{x^2+1}}$

(d) $\frac{d}{dx}\left[\ln(x^2+1)+C\right] = \frac{2x}{x^2+1}$

$\int \frac{x}{\sqrt{x^2+1}}\,dx$ matches (b).

3. (a) $\frac{d}{dx}\left[\ln\sqrt{x^2+1}+C\right] = \frac{1}{2}\left(\frac{2x}{x^2+1}\right) = \frac{x}{x^2+1}$

(b) $\frac{d}{dx}\left[\frac{2x}{(x^2+1)^2}+C\right] = \frac{(x^2+1)^2(2)-(2x)(2)(x^2+1)(2x)}{(x^2+1)^4} = \frac{2(1-3x^2)}{(x^2+1)^3}$

(c) $\frac{d}{dx}\left[\arctan x + C\right] = \frac{1}{1+x^2}$

(d) $\frac{d}{dx}\left[\ln(x^2+1)+C\right] = \frac{2x}{x^2+1}$

$\int \frac{1}{x^2+1}\,dx$ matches (c).

5. $\int (3x-2)^4\,dx$

$u = 3x-2,\ du = 3\,dx,\ n = 4$

Use $\int u^n\,du$.

7. $\int \frac{1}{\sqrt{x}(1-2\sqrt{x})}\,dx$

$u = 1-2\sqrt{x},\ du = -\frac{1}{\sqrt{x}}\,dx$

Use $\int \frac{du}{u}$.

9. $\int \frac{3}{\sqrt{1-t^2}}\,dt$

$u = t,\ du = dt,\ a = 1$

Use $\int \frac{du}{\sqrt{a^2-u^2}}$

11. $\int t\sin t^2\,dt$

$u = t^2,\ du = 2t\,dt$

Use $\int \sin u\,du$.

13. $\int (\cos x)e^{\sin x}\,dx$

$u = \sin x,\ du = \cos x\,dx$

Use $\int e^u\,du$.

15. Let $u = -2x + 5$, $du = -2\,dx$.

$$\int (-2x+5)^{3/2}\,dx = -\frac{1}{2}\int (-2x+5)^{3/2}(-2)\,dx$$
$$= -\frac{1}{5}(-2x+5)^{5/2} + C$$

17. Let $u = z - 4$, $du = dz$

$$\int \frac{5}{(z-4)^5}\,dz = 5\int (z-4)^{-5}\,dz = 5\frac{(z-4)^{-4}}{-4} + C$$
$$= \frac{-5}{4(z-4)^4} + C$$

19. Let $u = t^3 - 1$, $du = 3t^2\,dt$.

$$\int t^2\sqrt[3]{t^3-1}\,dt = \frac{1}{3}\int (t^3-1)^{1/3}(3t^2)\,dt$$
$$= \frac{1}{3}\frac{(t^3-1)^{4/3}}{4/3} + C$$
$$= \frac{(t^3-1)^{4/3}}{4} + C$$

21. $$\int\left[v + \frac{1}{(3v-1)^3}\right]dv = \int v\,dv + \frac{1}{3}\int (3v-1)^{-3}(3)dv$$
$$= \frac{1}{2}v^2 - \frac{1}{6(3v-1)^2} + C$$

23. Let $u = -t^3 + 9t + 1$, $du = (-3t^2 + 9)\,dt = -3(t^2 - 3)\,dt$.

$$\int \frac{t^2-3}{-t^3+9t+1}\,dt = -\frac{1}{3}\int \frac{-3(t^2-3)}{-t^3+9t+1}\,dt = -\frac{1}{3}\ln|-t^3+9t+1| + C$$

25. $$\int \frac{x^2}{x-1}\,dx = \int (x+1)\,dx + \int \frac{1}{x-1}\,dx$$
$$= \frac{1}{2}x^2 + x + \ln|x-1| + C$$

27. Let $u = 1 + e^x$, $du = e^x\,dx$.

$$\int \frac{e^x}{1+e^x}\,dx = \ln(1+e^x) + C$$

29. $$\int (1+2x^2)^2\,dx = \int (4x^4 + 4x^2 + 1)dx = \frac{4}{5}x^5 + \frac{4}{3}x^3 + x + C = \frac{x}{15}(12x^4 + 20x^2 + 15) + C$$

31. Let $u = 2\pi x^2$, $du = 4\pi x\,dx$.

$$\int x(\cos 2\pi x^2)\,dx = \frac{1}{4\pi}\int (\cos 2\pi x^2)(4\pi x)\,dx$$
$$= \frac{1}{4\pi}\sin 2\pi x^2 + C$$

33. Let $u = \pi x$, $du = \pi\,dx$.

$$\int \csc(\pi x)\cot(\pi x)\,dx = \frac{1}{\pi}\int \csc(\pi x)\cot(\pi x)\pi\,dx = -\frac{1}{\pi}\csc(\pi x) + C$$

35. Let $u = 5x$, $du = 5\,dx$.

$$\int e^{5x}\,dx = \frac{1}{5}\int e^{5x}(5)\,dx = \frac{1}{5}e^{5x} + C$$

37. Let $u = 1 + e^x$, $du = e^x\,dx$.

$$\int \frac{2}{e^{-x}+1}\,dx = 2\int\left(\frac{1}{e^{-x}+1}\right)\left(\frac{e^x}{e^x}\right)dx$$
$$= 2\int \frac{e^x}{1+e^x}\,dx = 2\ln(1+e^x) + C$$

39. $\displaystyle\int \frac{\ln x^2}{x}\,dx = 2\int (\ln x)\frac{1}{x}\,dx = 2\frac{(\ln x)^2}{2} + C = (\ln x)^2 + C$

41. $\displaystyle\int \frac{1 + \sin x}{\cos x}\,dx = \int (\sec x + \tan x)\,dx = \ln|\sec x + \tan x| + \ln|\sec x| + C = \ln|\sec x(\sec x + \tan x)| + C$

43. $$\frac{1}{\cos\theta - 1} = \frac{1}{\cos\theta - 1}\cdot\frac{\cos\theta + 1}{\cos\theta + 1} = \frac{\cos\theta + 1}{\cos^2\theta - 1} = \frac{\cos\theta + 1}{-\sin^2\theta}$$

$$= -\csc\theta\cdot\cot\theta - \csc^2\theta$$

$$\int \frac{1}{\cos\theta - 1}\,d\theta = \int(-\csc\theta\cot\theta - \csc^2\theta)\,d\theta$$

$$= \csc\theta + \cot\theta + C$$

$$= \frac{1}{\sin\theta} + \frac{\cos\theta}{\sin\theta} + C$$

$$= \frac{1 + \cos\theta}{\sin\theta} + C$$

45. $$\int \frac{3z + 2}{z^2 + 9}\,dz = \frac{3}{2}\int \frac{2z}{z^2 + 9}\,dz + 2\int \frac{dz}{z^2 + 9}$$

$$= \frac{3}{2}\ln(z^2 + 9) + \frac{2}{3}\arctan\left(\frac{z}{3}\right) + C$$

47. Let $u = 2t - 1$, $du = 2\,dt$.

$$\int \frac{-1}{\sqrt{1 - (2t - 1)^2}}\,dt = -\frac{1}{2}\int \frac{2}{\sqrt{1 - (2t - 1)^2}}\,dt$$

$$= -\frac{1}{2}\arcsin(2t - 1) + C$$

49. Let $u = \cos\left(\frac{2}{t}\right)$, $du = \frac{2\sin(2/t)}{t^2}\,dt$.

$$\int \frac{\tan(2/t)}{t^2}\,dt = \frac{1}{2}\int \frac{1}{\cos(2/t)}\left[\frac{2\sin(2/t)}{t^2}\right]dt$$

$$= \frac{1}{2}\ln\left|\cos\left(\frac{2}{t}\right)\right| + C$$

51. $\displaystyle\int \frac{3}{\sqrt{6x - x^2}}\,dx = 3\int \frac{1}{\sqrt{9 - (x - 3)^2}}\,dx = 3\arcsin\left(\frac{x - 3}{3}\right) + C$

53. $\displaystyle\int \frac{4}{4x^2 + 4x + 65}\,dx = \int \frac{1}{[x + (1/2)]^2 + 16}\,dx = \frac{1}{4}\arctan\left[\frac{x + (1/2)}{4}\right] + C = \frac{1}{4}\arctan\left(\frac{2x + 1}{8}\right) + C$

55. $\displaystyle\frac{ds}{dt} = \frac{t}{\sqrt{1 - t^4}},\ \left(0, -\frac{1}{2}\right)$

(a)

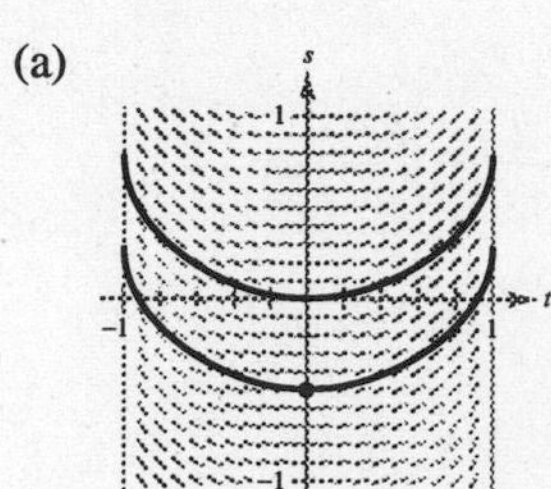

(b) $u = t^2$, $du = 2t\,dt$

$$\int \frac{t}{\sqrt{1 - t^4}}\,dt = \frac{1}{2}\int \frac{2t}{\sqrt{1 - (t^2)^2}}\,dt = \frac{1}{2}\arcsin t^2 + C$$

$\left(0, -\frac{1}{2}\right)$: $-\frac{1}{2} = \frac{1}{2}\arcsin 0 + C \Rightarrow C = -\frac{1}{2}$

$$s = \frac{1}{2}\arcsin t^2 - \frac{1}{2}$$

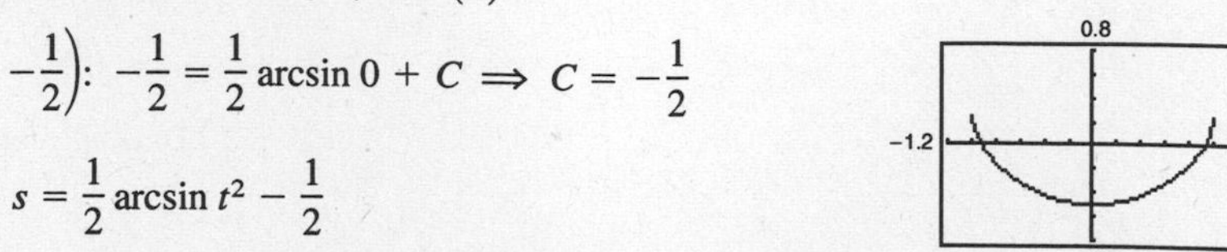

57.

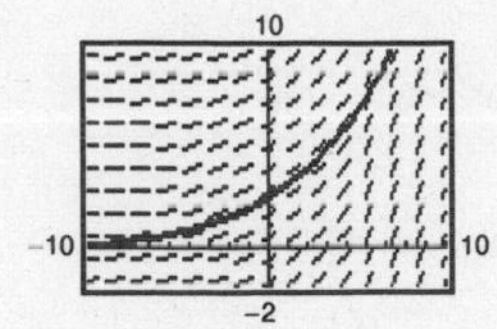

$y = 3e^{0.2x}$

59. $y = \int (1 + e^x)^2\, dx = \int (e^{2x} + 2e^x + 1)\, dx$

$= \frac{1}{2}e^{2x} + 2e^x + x + C$

61. $\frac{dy}{dx} = \frac{\sec^2 x}{4 + \tan^2 x}$

Let $u = \tan x$, $du = \sec^2 x\, dx$.

$y = \int \frac{\sec^2 x}{4 + \tan^2 x}\, dx = \frac{1}{2}\arctan\left(\frac{\tan x}{2}\right) + C$

63. Let $u = 2x$, $du = 2\, dx$.

$\int_0^{\pi/4} \cos 2x\, dx = \frac{1}{2}\int_0^{\pi/4} \cos 2x(2)\, dx$

$= \left[\frac{1}{2}\sin 2x\right]_0^{\pi/4} = \frac{1}{2}$

65. Let $u = -x^2$, $du = -2x\, dx$.

$\int_0^1 xe^{-x^2}\, dx = -\frac{1}{2}\int_0^1 e^{-x^2}(-2x)\, dx = \left[-\frac{1}{2}e^{-x^2}\right]_0^1$

$= \frac{1}{2}(1 - e^{-1}) \approx 0.316$

67. Let $u = x^2 + 9$, $du = 2x\, dx$.

$\int_0^4 \frac{2x}{\sqrt{x^2 + 9}}\, dx = \int_0^4 (x^2 + 9)^{-1/2}(2x)\, dx$

$= \left[2\sqrt{x^2 + 9}\right]_0^4 = 4$

69. Let $u = 3x$, $du = 3\, dx$.

$\int_0^{2/\sqrt{3}} \frac{1}{4 + 9x^2}\, dx = \frac{1}{3}\int_0^{2/\sqrt{3}} \frac{3}{4 + (3x)^2}\, dx$

$= \left[\frac{1}{6}\arctan\left(\frac{3x}{2}\right)\right]_0^{2/\sqrt{3}}$

$= \frac{\pi}{18} \approx 0.175$

71. $\int \frac{1}{x^2 + 4x + 13}\, dx = \frac{1}{3}\arctan\left(\frac{x + 2}{3}\right) + C$

The antiderivatives are vertical translations of each other.

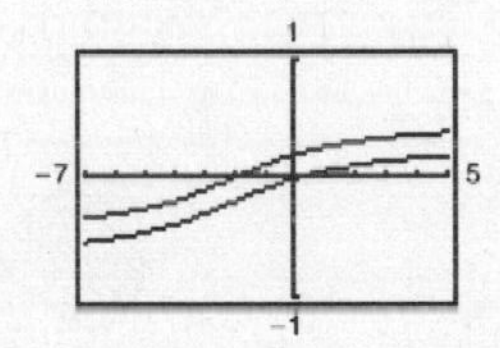

73. $\int \frac{1}{1 + \sin\theta}\, d\theta = \tan\theta - \sec\theta + C \left(\text{or } \frac{-2}{1 + \tan(\theta/2)}\right)$

The antiderivatives are vertical translations of each other.

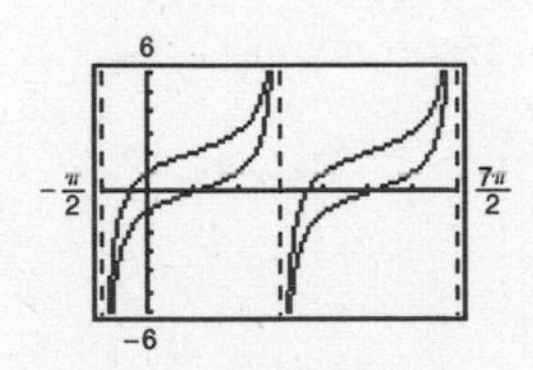

75. Power Rule: $\int u^n\, du = \frac{u^{n+1}}{n + 1} + C, n \neq -1.$

$u = x^2 + 1, n = 3$

77. Log Rule: $\int \frac{du}{u} = \ln|u| + C, u = x^2 + 1.$

79. The are equivalent because

$e^{x + C_1} = e^x \cdot e^{C_1} = Ce^x, C = e^{C_1}$

81. $\sin x + \cos x = a \sin(x + b)$

$\sin x + \cos x = a \sin x \cos b + a \cos x \sin b$

$\sin x + \cos x = (a \cos b) \sin x + (a \sin b) \cos x$

Equate coefficients of like terms to obtain the following.

$1 = a \cos b$ and $1 = a \sin b$

Thus, $a = 1/\cos b$. Now, substitute for a in $1 = a \sin b$.

$$1 = \left(\frac{1}{\cos b}\right)\sin b$$

$$1 = \tan b \implies b = \frac{\pi}{4}$$

Since $b = \dfrac{\pi}{4}$, $a = \dfrac{1}{\cos(\pi/4)} = \sqrt{2}$. Thus, $\sin x + \cos x = \sqrt{2}\sin\left(x + \dfrac{\pi}{4}\right)$.

$$\int \frac{dx}{\sin x + \cos x} = \int \frac{dx}{\sqrt{2}\sin(x + (\pi/4))} = \frac{1}{\sqrt{2}}\int \csc\left(x + \frac{\pi}{4}\right) dx = -\frac{1}{\sqrt{2}} \ln\left|\csc\left(x + \frac{\pi}{4}\right) + \cot\left(x + \frac{\pi}{4}\right)\right| + C$$

83. $\displaystyle\int_0^2 \frac{4x}{x^2 + 1}\, dx \approx 3$

Matches (a).

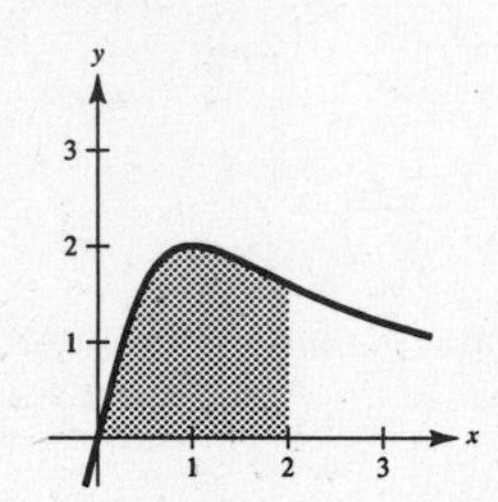

85. Let $u = 1 - x^2$, $du = -2x\, dx$.

$$A = 4\int_0^1 x\sqrt{1 - x^2}\, dx$$

$$= -2\int_0^1 (1 - x^2)^{1/2}(-2x)\, dx$$

$$= \left[-\frac{4}{3}(1 - x^2)^{3/2}\right]_0^1 = \frac{4}{3}$$

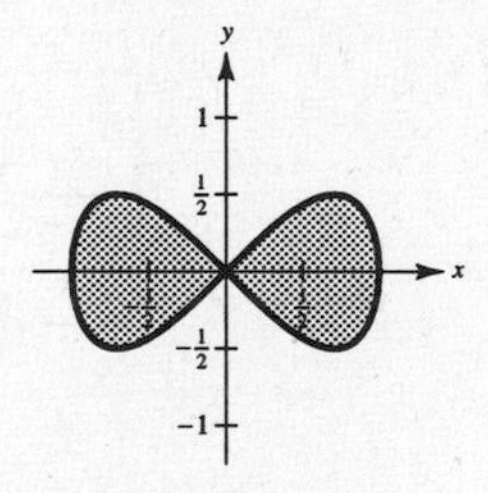

87. $\displaystyle\int_0^{1/a} (x - ax^2)\, dx = \left[\frac{1}{2}x^2 - \frac{a}{3}x^3\right]_0^{1/a}$

$$= \frac{1}{6a^2}$$

Let $\dfrac{1}{6a^2} = \dfrac{2}{3}$, $12a^2 = 3$, $a = \dfrac{1}{2}$.

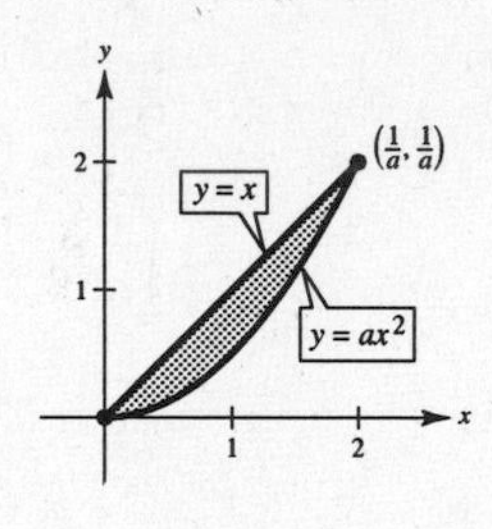

89. (a) **Shell Method:**

Let $u = -x^2$, $du = -2x\, dx$.

$$V = 2\pi\int_0^1 xe^{-x^2}\, dx$$

$$= -\pi\int_0^1 e^{-x^2}(-2x)\, dx$$

$$= \left[-\pi e^{-x^2}\right]_0^1$$

$$= \pi(1 - e^{-1}) \approx 1.986$$

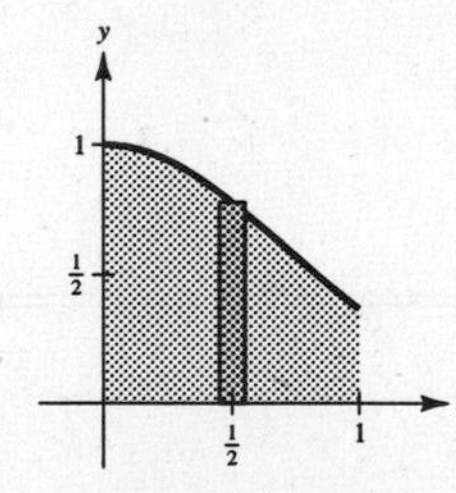

(b) **Shell Method:**

$$V = 2\pi\int_0^b xe^{-x^2}\, dx$$

$$= \left[-\pi e^{-x^2}\right]_0^b$$

$$= \pi(1 - e^{-b^2}) = \frac{4}{3}$$

$$e^{-b^2} = \frac{3\pi - 4}{3\pi}$$

$$b = \sqrt{\ln\left(\frac{3\pi}{3\pi - 4}\right)}$$

$$\approx 0.743$$

91. $A = \int_0^4 \frac{5}{\sqrt{25 - x^2}}\, dx = \left[5 \arcsin \frac{x}{5}\right]_0^4 = 5 \arcsin \frac{4}{5}$

$$\bar{x} = \frac{1}{A}\int_0^4 x\left(\frac{5}{\sqrt{25 - x^2}}\right) dx$$

$$= \frac{1}{5\arcsin(4/5)}\left(-\frac{5}{2}\right)\int_0^4 (25 - x^2)^{-1/2}(-2x)\, dx$$

$$= \frac{1}{5\arcsin(4/5)}(-5)\left[(25 - x^2)^{1/2}\right]_0^4$$

$$= -\frac{1}{\arcsin(4/5)}[3 - 5]$$

$$= \frac{2}{\arcsin(4/5)} \approx 2.157$$

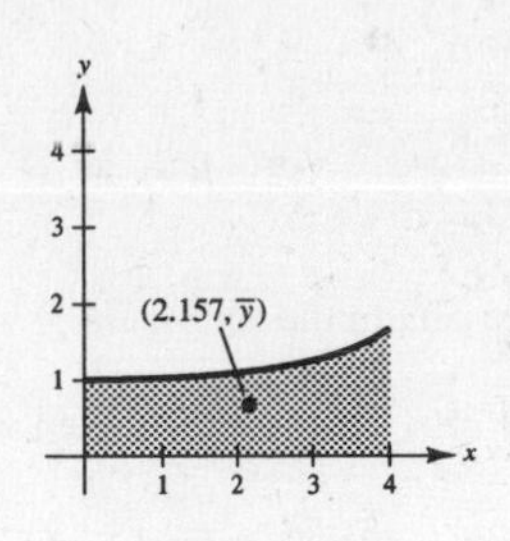

93.

$$y = \tan(\pi x)$$

$$y' = \pi \sec^2(\pi x)$$

$$1 + (y')^2 = 1 + \pi^2 \sec^4(\pi x)$$

$$s = \int_0^{1/4} \sqrt{1 + \pi^2 \sec^4(\pi x)}\, dx$$

$$\approx 1.0320$$

Section 7.2 Integration by Parts

1. $\frac{d}{dx}[\sin x - x\cos x] = \cos x - (-x \sin x + \cos x) = x \sin x$. Matches (b)

3. $\frac{d}{dx}[x^2e^x - 2xe^x + 2e^x] = x^2e^x + 2xe^x - 2xe^x - 2e^x + 2e^x = x^2e^x$. Matches (c)

5. $\int xe^{2x}\, dx$

$u = x,\ dv = e^{2x}\, dx$

7. $\int (\ln x)^2\, dx$

$u = (\ln x)^2,\ dv = dx$

9. $\int x \sec^2 x\, dx$

$u = x,\ dv = \sec^2 x\, dx$

11. $dv = e^{-2x}\, dx \implies v = \int e^{-2x}\, dx = -\frac{1}{2}e^{-2x}$

$u = x \implies du = dx$

$$\int xe^{-2x}\, dx = -\frac{1}{2}xe^{-2x} - \int -\frac{1}{2}e^{-2x}\, dx$$

$$= -\frac{1}{2}xe^{-2x} - \frac{1}{4}e^{-2x} + C = \frac{-1}{4e^{2x}}(2x + 1) + C$$

13. Use integration by parts three times.

(1) $dv = e^x\,dx \Rightarrow v = \int e^x\,dx = e^x$ $\quad u = x^3 \Rightarrow du = 3x^2\,dx$

(2) $dv = e^x\,dx \Rightarrow v = \int e^x\,dx = e^x$ $\quad u = x^2 \Rightarrow du = 2x\,dx$

(3) $dv = e^x\,dx \Rightarrow v = \int e^x\,dx = e^x$ $\quad u = x \Rightarrow du = dx$

$$\int x^3 e^x\,dx = x^3e^x - 3\int x^2e^x\,dx = x^3e^x - 3x^2e^x + 6\int xe^x\,dx$$

$$= x^3e^x - 3x^2e^x + 6xe^x - 6e^x + C = e^x(x^3 - 3x^2 + 6x - 6) + C$$

15. $\int x^2 e^{x^3}\,dx = \frac{1}{3}\int e^{x^3}(3x^2)dx = \frac{1}{3}e^{x^3} + C$

17. $dv = t\,dt \Rightarrow v = \int t\,dt = \frac{t^2}{2}$

$u = \ln(t+1) \Rightarrow du = \frac{1}{t+1}\,dt$

$$\int t\ln(t+1)\,dt = \frac{t^2}{2}\ln(t+1) - \frac{1}{2}\int \frac{t^2}{t+1}\,dt$$

$$= \frac{t^2}{2}\ln(t+1) - \frac{1}{2}\int\left(t - 1 + \frac{1}{t+1}\right)dt$$

$$= \frac{t^2}{2}\ln(t+1) - \frac{1}{2}\left[\frac{t^2}{2} - t + \ln(t+1)\right] + C$$

$$= \frac{1}{4}[2(t^2 - 1)\ln|t+1| - t^2 + 2t] + C$$

19. Let $u = \ln x$, $du = \frac{1}{x}\,dx$.

$$\int \frac{(\ln x)^2}{x}\,dx = \int (\ln x)^2\left(\frac{1}{x}\right)dx = \frac{(\ln x)^3}{3} + C$$

21. $dv = \frac{1}{(2x+1)^2}\,dx \Rightarrow v = \int (2x+1)^{-2}\,dx = -\frac{1}{2(2x+1)}$

$u = xe^{2x} \Rightarrow du = (2xe^{2x} + e^{2x})\,dx = e^{2x}(2x+1)\,dx$

$$\int \frac{xe^{2x}}{(2x+1)^2}\,dx = -\frac{xe^{2x}}{2(2x+1)} + \int \frac{e^{2x}}{2}\,dx$$

$$= \frac{-xe^{2x}}{2(2x+1)} + \frac{e^{2x}}{4} + C$$

$$= \frac{e^{2x}}{4(2x+1)} + C$$

23. Use integration by parts twice.

(1) $dv = e^x\,dx \Rightarrow v = \int e^x\,dx = e^x$ $\quad u = x^2 \Rightarrow du = 2x\,dx$

(2) $dv = e^x\,dx \Rightarrow v = \int e^x\,dx = e^x$ $\quad u = x \Rightarrow du = dx$

$$\int (x^2 - 1)e^x\,dx = \int x^2e^x\,dx - \int e^x dx = x^2e^x - 2\int xe^x\,dx - e^x$$

$$= x^2e^x - 2\left[xe^x - \int e^x\,dx\right] - e^x = x^2e^x - 2xe^x + e^x + C = (x-1)^2e^x + C$$

25. $dv = \sqrt{x-1}\,dx \Rightarrow \quad v = \int (x-1)^{1/2} dx = \frac{2}{3}(x-1)^{3/2}$

$u = x \quad \Rightarrow du = dx$

$$\int x\sqrt{x-1}\,dx = \frac{2}{3}x(x-1)^{3/2} - \frac{2}{3}\int (x-1)^{3/2}\,dx$$

$$= \frac{2}{3}x(x-1)^{3/2} - \frac{4}{15}(x-1)^{5/2} + C$$

$$= \frac{2(x-1)^{3/2}}{15}(3x+2) + C$$

27. $dv = \cos x\,dx \Rightarrow \quad v = \int \cos x\,dx = \sin x$

$u = x \quad \Rightarrow du = dx$

$$\int x\cos x\,dx = x\sin x - \int \sin x\,dx = x\sin x + \cos x + C$$

29. Use integration by parts three times.

(1) $u = x^3,\ du = 3x^2,\ dv = \sin x\,dx,\ v = -\cos x$

$$\int x^3 \sin dx = -x^3\cos x + 3\int x^2\cos x\,dx$$

(2) $u = x^2,\ du = 2x\,dx,\ dv = \cos x\,dx,\ v = \sin x$

$$\int x^3\sin x\,dx = -x^3\cos x + 3\left[x^2\sin x - 2\int x\sin x\,dx\right]$$

$$= -x^3\cos x + 3x^2\sin x - 6\int x\sin x\,dx$$

(3) $u = x,\ du = dx,\ dv = \sin x\,dx,\ v = -\cos x$

$$\int x^3\sin x\,dx = -x^3\cos x + 3x^2\sin x - 6\left[-x\cos x + \int \cos x\,dx\right]$$

$$= -x^3\cos x + 3x^2\sin x + 6x\cos x - 6\sin x + C$$

31. $u = t,\ du = dt,\ dv = \csc t\cot t\,dt,\ v = -\csc t$

$$\int t\csc t\cot t\,dt = -t\csc t + \int \csc t\,dt$$

$$= -t\csc t - \ln|\csc t + \cot t| + C$$

33. $dv = dx \quad \Rightarrow \quad v = \int dx = x$

$u = \arctan x \Rightarrow du = \frac{1}{1+x^2}\,dx$

$$\int \arctan x\,dx = x\arctan x - \int \frac{x}{1+x^2}\,dx$$

$$= x\arctan x - \frac{1}{2}\ln(1+x^2) + C$$

35. Use integration by parts twice.

(1) $dv = e^{2x}dx \Rightarrow \quad v = \int e^{2x}\,dx = \frac{1}{2}e^{2x}$

$u = \sin x \Rightarrow du = \cos x\,dx$

(2) $dv = e^{2x}\,dx \Rightarrow \quad v = \int e^{2x}\,dx = \frac{1}{2}e^{2x}$

$u = \cos x \Rightarrow du = -\sin x\,dx$

$$\int e^{2x}\sin x\,dx = \frac{1}{2}e^{2x}\sin x - \frac{1}{2}\int e^{2x}\cos x\,dx = \frac{1}{2}e^{2x}\sin x - \frac{1}{2}\left(\frac{1}{2}e^{2x}\cos x + \frac{1}{2}\int e^{2x}\sin x\,dx\right)$$

$$\frac{5}{4}\int e^{2x}\sin x\,dx = \frac{1}{2}e^{2x}\sin x - \frac{1}{4}e^{2x}\cos x$$

$$\int e^{2x}\sin x\,dx = \frac{1}{5}e^{2x}(2\sin x - \cos x) + C$$

37. $y' = xe^{x^2}$

$$y = \int xe^{x^2}\,dx = \frac{1}{2}e^{x^2} + C$$

39. Use integration by parts twice.

(1) $dv = \dfrac{1}{\sqrt{2+3t}}\,dt \Rightarrow v = \displaystyle\int (2+3t)^{-1/2}\,dt = \frac{2}{3}\sqrt{2+3t}$

$u = t^2 \Rightarrow du = 2t\,dt$

(2) $dv = \sqrt{2+3t}\,dt \Rightarrow v = \displaystyle\int (2+3t)^{1/2}\,dt = \frac{2}{9}(2+3t)^{3/2}$

$u = t \Rightarrow du = dt$

$$\begin{aligned} y = \int \frac{t^2}{\sqrt{2+3t}}\,dt &= \frac{2t^2\sqrt{2+3t}}{3} - \frac{4}{3}\int t\sqrt{2+3t}\,dt \\ &= \frac{2t^2\sqrt{2+3t}}{3} - \frac{4}{3}\left[\frac{2t}{9}(2+3t)^{3/2} - \frac{2}{9}\int (2+3t)^{3/2}\,dt\right] \\ &= \frac{2t^2\sqrt{2+3t}}{3} - \frac{8t}{27}(2+3t)^{3/2} + \frac{16}{405}(2+3t)^{5/2} + C \\ &= \frac{2\sqrt{2+3t}}{405}(27t^2 - 24t + 32) + C \end{aligned}$$

41. $(\cos y)y' = 2x$

$$\int \cos y\,dy = \int 2x\,dx$$

$$\sin y = x^2 + C$$

43. (a)

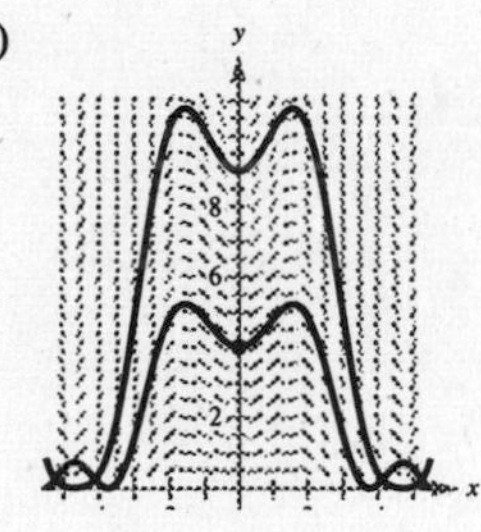

(b) $\dfrac{dy}{dx} = x\sqrt{y}\cos x,\ (0, 4)$

$$\int \frac{dy}{\sqrt{y}} = \int x\cos x\,dx$$

$$\int y^{-1/2}\,dy = \int x\cos x\,dx \qquad (u = x,\ du = dx,\ dv = \cos x\,dx,\ v = \sin x)$$

$$2y^{1/2} = x\sin x - \int \sin x\,dx$$

$$= x\sin x + \cos x + C$$

$(0, 4)$: $2(4)^{1/2} = 0 + 1 + C \Rightarrow C = 3$

$2\sqrt{y} = x\sin x + \cos x + 3$

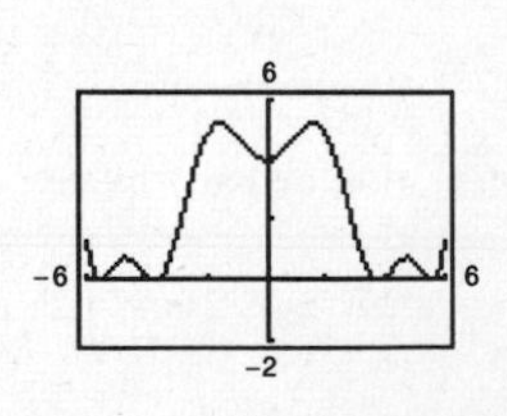

45. $\dfrac{dy}{dx} = \dfrac{x}{y}e^{x/8},\ y(0) = 2$

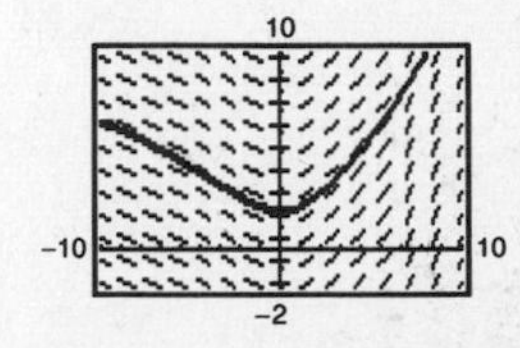

47. $u = x,\ du = dx,\ dv = e^{-x/2}\,dx,\ v = -2e^{-x/2}$

$$\int xe^{-x/2}\,dx = -2xe^{-x/2} + \int 2e^{-x/2}\,dx = -2xe^{-x/2} - 4e^{-x/2} + C$$

$$\text{Thus, } \int_0^4 xe^{-x/2}\,dx = \Big[-2xe^{-x/2} - 4e^{-x/2}\Big]_0^4$$
$$= -8e^{-2} - 4e^{-2} + 4$$
$$= -12e^{-2} + 4 \approx 2.376.$$

49. See Exercise 27.

$$\int_0^{\pi/2} x\cos x\,dx = \Big[x\sin x + \cos x\Big]_0^{\pi/2} = \frac{\pi}{2} - 1$$

51. $u = \arccos x,\ du = -\dfrac{1}{\sqrt{1-x^2}}\,dx,\ dv = dx,\ v = x$

$$\int \arccos x\,dx = x\arccos x + \int \frac{x}{\sqrt{1-x^2}}\,dx = x\arccos x - \sqrt{1-x^2} + C$$

$$\text{Thus, } \int_0^{1/2} \arccos x = \Big[x\arccos x - \sqrt{1-x^2}\Big]_0^{1/2}$$
$$= \frac{1}{2}\arccos\left(\frac{1}{2}\right) - \sqrt{\frac{3}{4}} + 1$$
$$= \frac{\pi}{6} - \frac{\sqrt{3}}{2} + 1 \approx 0.658.$$

53. Use integration by parts twice.

(1) $dv = e^x\,dx \implies v = \int e^x\,dx = e^x$

$u = \sin x \implies du = \cos x\,dx$

(2) $dv = e^x\,dx \implies v = \int e^x\,dx = e^x$

$u = \cos x \implies du = -\sin x\,dx$

$$\int e^x\sin x\,dx = e^x\sin x - \int e^x\cos x\,dx = e^x\sin x - e^x\cos x - \int e^x\sin x\,dx$$

$$2\int e^x\sin x\,dx = e^x(\sin x - \cos x)$$

$$\int e^x\sin x\,dx = \frac{e^x}{2}(\sin x - \cos x) + C$$

$$\text{Thus, } \int_0^1 e^x\sin x\,dx = \left[\frac{e^x}{2}(\sin x - \cos x)\right]_0^1 = \frac{e}{2}(\sin 1 - \cos 1) + \frac{1}{2} = \frac{e(\sin 1 - \cos 1) + 1}{2} \approx 0.909.$$

55. $dv = x^2\,dx,\ v = \dfrac{x^3}{3},\ u = \ln x,\ du = \dfrac{1}{x}\,dx$

$$\int x^2\ln x\,dx = \frac{x^3}{3}\ln x - \int \frac{x^3}{3}\left(\frac{1}{x}\right)dx$$
$$= \frac{x^3}{3}\ln x - \frac{1}{3}\int x^2\,dx$$

$$\text{Hence, } \int_1^2 x^2\ln x\,dx = \left[\frac{x^3}{3}\ln x - \frac{1}{9}x^3\right]_1^2$$
$$= \frac{8}{3}\ln 2 - \frac{8}{9} + \frac{1}{9} = \frac{8}{3}\ln 2 - \frac{7}{2} \approx 1.071.$$

57. $dv = x\,dx,\ v = \dfrac{x^2}{2},\ u = \operatorname{arcsec} x,\ du = \dfrac{1}{x\sqrt{x^2-1}}\,dx$

$$\int x \operatorname{arcsec} x\,dx = \frac{x^2}{2}\operatorname{arcsec} x - \int \frac{x^2/2}{x\sqrt{x^2-1}}\,dx$$
$$= \frac{x^2}{2}\operatorname{arcsec} x - \frac{1}{4}\int \frac{2x}{\sqrt{x^2-1}}\,dx$$
$$= \frac{x^2}{2}\operatorname{arcsec} x - \frac{1}{2}\sqrt{x^2-1} + C$$

Hence,

$$\int_2^4 x \operatorname{arcsec} x\,dx = \left[\frac{x^2}{2}\operatorname{arcsec} x - \frac{1}{2}\sqrt{x^2-1}\right]_2^4$$
$$= \left(8 \operatorname{arcsec} 4 - \frac{\sqrt{15}}{2}\right) - \left(\frac{2\pi}{3} - \frac{\sqrt{3}}{2}\right)$$
$$= 8 \operatorname{arcsec} 4 - \frac{\sqrt{15}}{2} + \frac{\sqrt{3}}{2} - \frac{2\pi}{3}$$
$$\approx 7.380.$$

59. $$\int x^2 e^{2x}\,dx = x^2\left(\frac{1}{2}e^{2x}\right) - (2x)\left(\frac{1}{4}e^{2x}\right) + 2\left(\frac{1}{8}e^{2x}\right) + C$$
$$= \frac{1}{2}x^2e^{2x} - \frac{1}{2}xe^{2x} + \frac{1}{4}e^{2x} + C$$
$$= \frac{1}{4}e^{2x}(2x^2 - 2x + 1) + C$$

Alternate signs	u and its derivatives	v' and its antiderivatives
$+$	x^2	e^{2x}
$-$	$2x$	$\frac{1}{2}e^{2x}$
$+$	2	$\frac{1}{4}e^{2x}$
$-$	0	$\frac{1}{8}e^{2x}$

61. $$\int x^3 \sin x\,dx = x^3(-\cos x) - 3x^2(-\sin x) + 6x\cos x - 6\sin x + C$$
$$= -x^3\cos x + 3x^2\sin x + 6x\cos x - 6\sin x + C$$
$$= (3x^2 - 6)\sin x - (x^3 - 6x)\cos x + C$$

Alternate signs	u and its derivatives	v' and its antiderivatives
$+$	x^3	$\sin x$
$-$	$3x^2$	$-\cos x$
$+$	$6x$	$-\sin x$
$-$	6	$\cos x$
$+$	0	$\sin x$

63. $$\int x\sec^2 x\,dx = x\tan x + \ln|\cos x| + C$$

Alternate signs	u and its derivatives	v' and its antiderivatives
$+$	x	$\sec^2 x$
$-$	1	$\tan x$
$+$	0	$-\ln\lvert\cos x\rvert$

65. Integration by parts is based on the product rule.

67. No. Substitution.

69. Yes. $u = x^2,\ dv = e^{2x}\,dx$

71. Yes. Let $u = x$ and $du = \dfrac{1}{\sqrt{x+1}}\,dx$.

(Substitution also works. Let $u = \sqrt{x+1}$)

73. $$\int t^3e^{-4t}\,dt = -\frac{e^{-4t}}{128}(32t^3 + 24t^2 + 12t + 3) + C$$

75. $$\int_0^{\pi/2} e^{-2x}\sin 3x\,dx = \left[\frac{e^{-2x}(-2\sin 3x - 3\cos 3x)}{13}\right]_0^{\pi/2} = \frac{1}{13}(2e^{-\pi} + 3) \approx 0.2374$$

77. (a) $dv = \sqrt{2x-3}\,dx \Rightarrow v = \int (2x-3)^{1/2}\,dx = \frac{1}{3}(2x-3)^{3/2}$

$u = 2x \Rightarrow du = 2\,dx$

$$\int 2x\sqrt{2x-3}\,dx = \frac{2}{3}x(2x-3)^{3/2} - \frac{2}{3}\int (2x-3)^{3/2}\,dx$$

$$= \frac{2}{3}x(2x-3)^{3/2} - \frac{2}{15}(2x-3)^{5/2} + C$$

$$= \frac{2}{15}(2x-3)^{3/2}(3x+3) + C = \frac{2}{5}(2x-3)^{3/2}(x+1) + C$$

(b) $u = 2x - 3 \Rightarrow x = \dfrac{u+3}{2}$ and $dx = \dfrac{1}{2}\,du$

$$\int 2x\sqrt{2x-3}\,dx = \int 2\left(\frac{u+3}{2}\right)u^{1/2}\left(\frac{1}{2}\right)du = \frac{1}{2}\int (u^{3/2} + 3u^{1/2})\,du = \frac{1}{2}\left[\frac{2}{5}u^{5/2} + 2u^{3/2}\right] + C$$

$$= \frac{1}{5}u^{3/2}(u+5) + C = \frac{1}{5}(2x-3)^{3/2}[(2x-3)+5] + C = \frac{2}{5}(2x-3)^{3/2}(x+1) + C$$

79. (a) $dv = \dfrac{x}{\sqrt{4+x^2}}\,dx \Rightarrow v = \int (4+x^2)^{-1/2}x\,dx = \sqrt{4+x^2}$

$u = x^2 \Rightarrow du = 2x\,dx$

$$\int \frac{x^3}{\sqrt{4+x^2}}\,dx = x^2\sqrt{4+x^2} - 2\int x\sqrt{4+x^2}\,dx$$

$$= x^2\sqrt{4+x^2} - \frac{2}{3}(4+x^2)^{3/2} + C = \frac{1}{3}\sqrt{4+x^2}(x^2-8) + C$$

(b) $u = 4 + x^2 \Rightarrow x^2 = u - 4$ and $2x\,dx = du \Rightarrow x\,dx = \dfrac{1}{2}\,du$

$$\int \frac{x^3}{\sqrt{4+x^2}}\,dx = \int \frac{x^2}{\sqrt{4+x^2}}\,x\,dx = \int \frac{u-4}{\sqrt{u}}\,\frac{1}{2}\,du$$

$$= \frac{1}{2}\int (u^{1/2} - 4u^{-1/2})\,du = \frac{1}{2}\left(\frac{2}{3}u^{3/2} - 8u^{1/2}\right) + C$$

$$= \frac{1}{3}u^{1/2}(u-12) + C = \frac{1}{3}\sqrt{4+x^2}\,[(4+x^2) - 12] + C = \frac{1}{3}\sqrt{4+x^2}\,(x^2-8) + C$$

81. $n = 0$: $\int \ln x\,dx = x(\ln x - 1) + C$

$n = 1$: $\int x\ln x\,dx = \dfrac{x^2}{4}(2\ln x - 1) + C$

$n = 2$: $\int x^2\ln x\,dx = \dfrac{x^3}{9}(3\ln x - 1) + C$

$n = 3$: $\int x^3\ln x\,dx = \dfrac{x^4}{16}(4\ln x - 1) + C$

$n = 4$: $\int x^4\ln x\,dx = \dfrac{x^5}{25}(5\ln x - 1) + C$

In general, $\int x^n\ln x\,dx = \dfrac{x^{n+1}}{(n+1)^2}[(n+1)\ln x - 1] + C$. (See Exercise 85.)

83. $dv = \sin x\,dx \Rightarrow \quad v = -\cos x$

$u = x^n \quad \Rightarrow du = nx^{n-1}\,dx$

$$\int x^n \sin x\,dx = -x^n \cos x + n\int x^{n-1}\cos x\,dx$$

85. $dv = x^n\,dx \Rightarrow \quad v = \dfrac{x^{n+1}}{n+1}$

$u = \ln x \quad \Rightarrow du = \dfrac{1}{x}\,dx$

$$\int x^n \ln x\,dx = \frac{x^{n+1}}{n+1}\ln x - \int \frac{x^n}{n+1}\,dx$$

$$= \frac{x^{n+1}}{n+1}\ln x - \frac{x^{n+1}}{(n+1)^2} + C$$

$$= \frac{x^{n+1}}{(n+1)^2}[(n+1)\ln x - 1] + C$$

87. Use integration by parts twice.

(1) $dv = e^{ax}\,dx \Rightarrow \quad v = \dfrac{1}{a}e^{ax}$

$u = \sin bx \Rightarrow du = b\cos bx\,dx$

(2) $dv = e^{ax}\,dx \Rightarrow \quad v = \dfrac{1}{a}e^{ax}$

$u = \cos bx \Rightarrow du = -b\sin bx\,dx$

$$\int e^{ax}\sin bx\,dx = \frac{e^{ax}\sin bx}{a} - \frac{b}{a}\int e^{ax}\cos bx\,dx$$

$$= \frac{e^{ax}\sin bx}{a} - \frac{b}{a}\left[\frac{e^{ax}\cos bx}{a} + \frac{b}{a}\int e^{ax}\sin bx\,dx\right] = \frac{e^{ax}\sin bx}{a} - \frac{b}{a^2}e^{ax}\cos bx - \frac{b^2}{a^2}\int e^{ax}\sin bx\,dx$$

Therefore, $\left(1 + \dfrac{b^2}{a^2}\right)\displaystyle\int e^{ax}\sin bx\,dx = \dfrac{e^{ax}(a\sin bx - b\cos bx)}{a^2}$

$$\int e^{ax}\sin bx\,dx = \frac{e^{ax}(a\sin bx - b\cos bx)}{a^2 + b^2} + C.$$

89. $n = 3$ (Use formula in Exercise 85.)

$$\int x^3 \ln x\,dx = \frac{x^4}{16}[4\ln x - 1] + C$$

91. $a = 2, b = 3$ (Use formula in Exercise 88.)

$$\int e^{2x}\cos 3x\,dx = \frac{e^{2x}(2\cos 3x + 3\sin 3x)}{13} + C$$

93. $dv = e^{-x}\,dx \Rightarrow \quad v = -e^{-x}$

$u = x \quad \Rightarrow du = dx$

$$A = \int_0^4 xe^{-x}\,dx = \Big[-xe^{-x}\Big]_0^4 + \int_0^4 e^{-x}\,dx = \frac{-4}{e^4} - \Big[e^{-x}\Big]_0^4$$

$$= 1 - \frac{5}{e^4} \approx 0.908$$

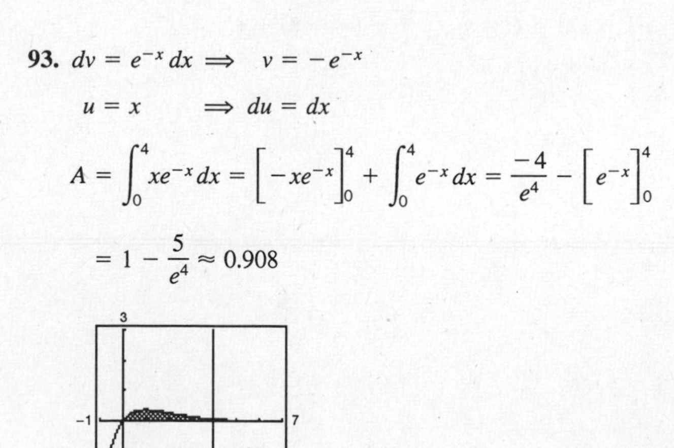

95. $A = \displaystyle\int_0^1 e^{-x}\sin(\pi x)\,dx$

$$= \left[\frac{e^{-x}(-\sin \pi x - \pi\cos \pi x)}{1 + \pi^2}\right]_0^1$$

$$= \frac{1}{1+\pi^2}\left(\frac{\pi}{e} + \pi\right) = \frac{\pi}{1+\pi^2}\left(\frac{1}{e} + 1\right)$$

≈ 0.395 (See Exercise 87.)

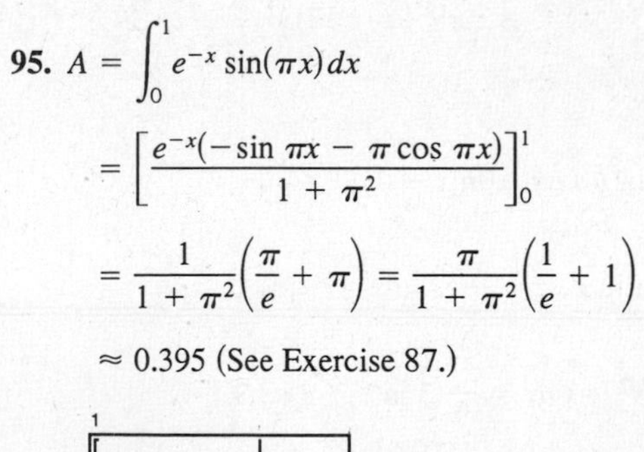

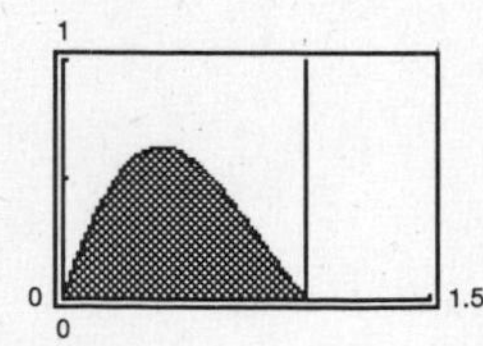

97. (a) $A = \int_1^e \ln x\, dx = \Big[-x + x \ln x\Big]_1^e = 1$ (See Exercise 4.)

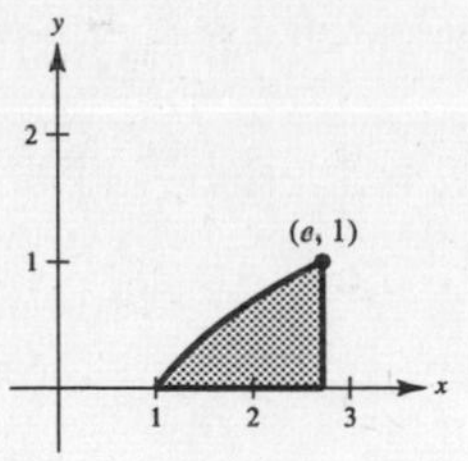

(b) $R(x) = \ln x,\ r(x) = 0$

$$V = \pi \int_1^e (\ln x)^2\, dx$$

$$= \pi\Big[x(\ln x)^2 - 2x \ln x + 2x\Big]_1^e \text{ (Use integration by parts twice, see Exercise 7.)}$$

$$= \pi(e - 2) \approx 2.257$$

(c) $p(x) = x,\ h(x) = \ln x$

$$V = 2\pi \int_1^e x \ln x\, dx = 2\pi\left[\frac{x^2}{4}(-1 + 2\ln x)\right]_1^e$$

$$= \frac{(e^2 + 1)\pi}{2} \approx 13.177 \text{ (See Exercise 85.)}$$

(d) $$\bar{x} = \frac{\int_1^e x \ln x\, dx}{1} = \frac{e^2 + 1}{4} \approx 2.097$$

$$\bar{y} = \frac{\frac{1}{2}\int_1^e (\ln x)^2 dx}{1} = \frac{e - 2}{2} \approx 0.359$$

$$(\bar{x}, \bar{y}) = \left(\frac{e^2 + 1}{4}, \frac{e - 2}{2}\right) \approx (2.097, 0.359)$$

99. $$\text{Average value} = \frac{1}{\pi}\int_0^{\pi} e^{-4t}(\cos 2t + 5 \sin 2t)\, dt$$

$$= \frac{1}{\pi}\left[e^{-4t}\left(\frac{-4\cos 2t + 2\sin 2t}{20}\right) + 5e^{-4t}\left(\frac{-4\sin 2t - 2\cos 2t}{20}\right)\right]_0^{\pi} \text{ (From Exercises 87 and 88)}$$

$$= \frac{7}{10\pi}(1 - e^{-4\pi}) \approx 0.223$$

101. $c(t) = 100{,}000 + 4000t,\ r = 5\%,\ t_1 = 10$

$$P = \int_0^{10} (100{,}000 + 4000t)e^{-0.05t}\, dt = 4000\int_0^{10} (25 + t)e^{-0.05t}\, dt$$

Let $u = 25 + t,\ dv = e^{-0.05t}dt,\ du = dt,\ v = -\frac{100}{5}e^{-0.05t}$

$$P = 4000\left\{\left[(25 + t)\left(-\frac{100}{5}e^{-0.05t}\right)\right]_0^{10} + \frac{100}{5}\int_0^{10} e^{-0.05t}\, dt\right\}$$

$$= 4000\left\{\left[(25 + t)\left(-\frac{100}{5}e^{-0.05t}\right)\right]_0^{10} - \left[\frac{10{,}000}{25}e^{-0.05t}\right]_0^{10}\right\} \approx \$931{,}265$$

103. $$\int_{-\pi}^{\pi} x \sin nx\, dx = \left[-\frac{x}{n}\cos nx + \frac{1}{n^2}\sin nx\right]_{-\pi}^{\pi}$$

$$= -\frac{\pi}{n}\cos \pi n - \frac{\pi}{n}\cos(-\pi n)$$

$$= -\frac{2\pi}{n}\cos \pi n$$

$$= \begin{cases} -(2\pi/n), & \text{if } n \text{ is even} \\ (2\pi/n), & \text{if } n \text{ is odd} \end{cases}$$

105. Let $u = x$, $dv = \sin\left(\frac{n\pi}{2}x\right)dx$, $du = dx$, $v = -\frac{2}{n\pi}\cos\left(\frac{n\pi}{2}x\right)$.

$$I_1 = \int_0^1 x\sin\left(\frac{n\pi}{2}x\right)dx = \left[\frac{-2x}{n\pi}\cos\left(\frac{n\pi}{2}x\right)\right]_0^1 + \frac{2}{n\pi}\int_0^1 \cos\left(\frac{n\pi}{2}x\right)dx$$

$$= -\frac{2}{n\pi}\cos\left(\frac{n\pi}{2}\right) + \left[\left(\frac{2}{n\pi}\right)^2\sin\left(\frac{n\pi}{2}x\right)\right]_0^1$$

$$= -\frac{2}{n\pi}\cos\left(\frac{n\pi}{2}\right) + \left(\frac{2}{n\pi}\right)^2\sin\left(\frac{n\pi}{2}\right)$$

Let $u = (-x + 2)$, $dv = \sin\left(\frac{n\pi}{2}x\right)dx$, $du = -dx$, $v = -\frac{2}{n\pi}\cos\left(\frac{n\pi}{2}x\right)$.

$$I_2 = \int_1^2 (-x+2)\sin\left(\frac{n\pi}{2}x\right)dx = \left[\frac{-2(-x+2)}{n\pi}\cos\left(\frac{n\pi}{2}x\right)\right]_1^2 - \frac{2}{n\pi}\int_1^2 \cos\left(\frac{n\pi}{2}x\right)dx$$

$$= \frac{2}{n\pi}\cos\left(\frac{n\pi}{2}\right) - \left[\left(\frac{2}{n\pi}\right)^2\sin\left(\frac{n\pi}{2}x\right)\right]_1^2$$

$$= \frac{2}{n\pi}\cos\left(\frac{n\pi}{2}\right) + \left(\frac{2}{n\pi}\right)^2\sin\left(\frac{n\pi}{2}\right)$$

$$h(I_1 + I_2) = b_n = h\left[\left(\frac{2}{n\pi}\right)^2\sin\left(\frac{n\pi}{2}\right) + \left(\frac{2}{n\pi}\right)^2\sin\left(\frac{n\pi}{2}\right)\right] = \frac{8h}{(n\pi)^2}\sin\left(\frac{n\pi}{2}\right)$$

107. Shell Method:

$$V = 2\pi\int_a^b xf(x)\,dx$$

$$dv = x\,dx \implies v = \frac{x^2}{2}$$

$$u = f(x) \implies du = f'(x)\,dx$$

$$V = 2\pi\left[\frac{x^2}{2}f(x) - \int\frac{x^2}{2}f'(x)dx\right]_a^b$$

$$= \pi\left[(b^2f(b) - a^2f(a)) - \int_a^b x^2f'(x)\,dx\right]$$

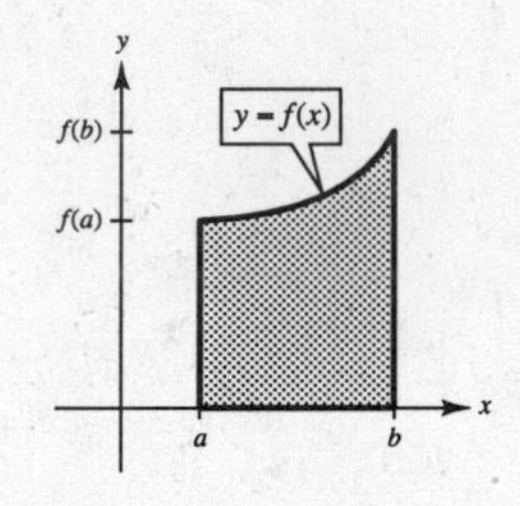

Disk Method:

$$V = \pi\int_0^{f(a)} (b^2 - a^2)\,dy + \pi\int_{f(a)}^{f(b)} [b^2 - [f^{-1}(y)]^2]\,dy$$

$$= \pi(b^2 - a^2)f(a) + \pi b^2(f(b) - f(a)) - \pi\int_{f(a)}^{f(b)} [f^{-1}(y)]^2\,dy$$

$$= \pi\left[(b^2f(b) - a^2f(a)) - \int_{f(a)}^{f(b)} [f^{-1}(y)]^2\,dy\right]$$

Since $x = f^{-1}(y)$, we have $f(x) = y$ and $f'(x)dx = dy$. When $y = f(a)$, $x = a$. When $y = f(b)$, $x = b$. Thus,

$$\int_{f(a)}^{f(b)} [f^{-1}(y)]^2\,dy = \int_a^b x^2f'(x)\,dx$$

and the volumes are the same.

109. $f'(x) = xe^{-x}$

(a) $f(x) = \int xe^{-x}\,dx = -xe^{-x} - e^{-x} + C$

(Parts: $u = x$, $dv = e^{-x}\,dx$)

$f(0) = 0 = -1 + C \Rightarrow C = 1$

$f(x) = -xe^{-x} - e^{-x} + 1$

(b)

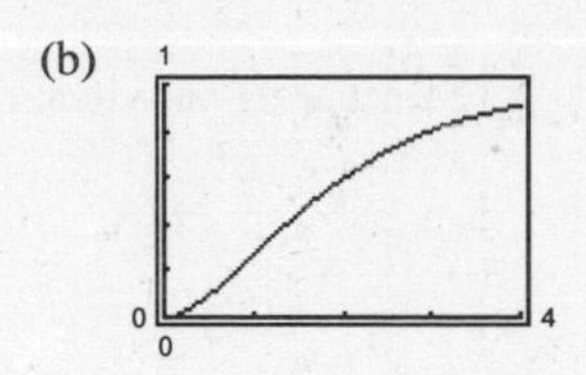

(c) You obtain the points

n	x_n	y_n
0	0	0
1	0.05	0
2	0.10	2.378×10^{-3}
3	0.15	0.0069
4	0.20	0.0134
⋮	⋮	⋮
80	4.0	0.9064

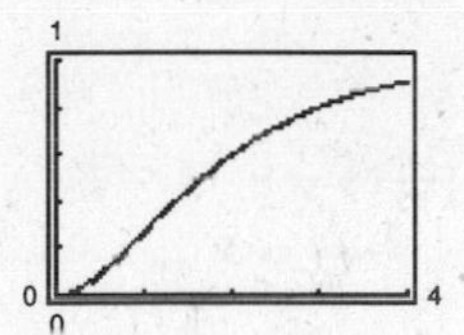

(d) You obtain the points

n	x_n	y_n
0	0	0
1	0.1	0
2	0.2	0.0090484
3	0.3	0.025423
4	0.4	0.047648
⋮	⋮	⋮
40	4.0	0.9039

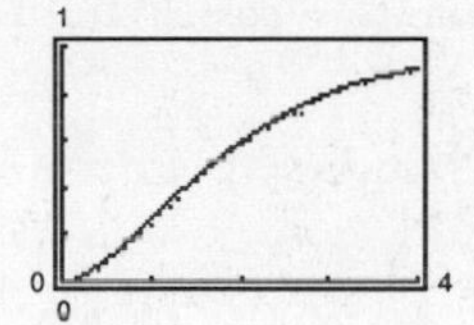

(e) $f(4) = 0.9084$

The approximations are tangent line approximations. The results in (c) are better because Δx is smaller.

Section 7.3 Trigonometric Integrals

1. $f(x) = \sin^4 x + \cos^4 x$

(a)
$$\begin{aligned}\sin^4 x + \cos^4 x &= \left(\frac{1 - \cos 2x}{2}\right)^2 + \left(\frac{1 + \cos 2x}{2}\right)^2 \\ &= \frac{1}{4}[1 - 2\cos 2x + \cos^2 2x + 1 + 2\cos 2x + \cos^2 2x] \\ &= \frac{1}{4}\left[2 + 2\frac{1 + \cos 4x}{2}\right] \\ &= \frac{1}{4}[3 + \cos 4x]\end{aligned}$$

(b)
$$\begin{aligned}\sin^4 x + \cos^4 x &= (\sin^2 x)^2 + \cos^4 x \\ &= (1 - \cos^2 x)^2 + \cos^4 x \\ &= 1 - 2\cos^2 x + 2\cos^4 x\end{aligned}$$

(c)
$$\begin{aligned}\sin^4 x + \cos^4 x &= \sin^4 x + 2\sin^2 x\cos^2 x + \cos^4 x - 2\sin^2 x\cos^2 x \\ &= (\sin^2 x + \cos^2 x)^2 - 2\sin^2 x\cos^2 x \\ &= 1 - 2\sin^2 x\cos^2 x\end{aligned}$$

—CONTINUED—

1. —CONTINUED—

(d) $1 - 2\sin^2 x\cos^2 x = 1 - (2\sin x\cos x)(\sin x\cos x)$

$$= 1 - (\sin 2x)\left(\frac{1}{2}\sin 2x\right)$$

$$= 1 - \frac{1}{2}\sin^2(2x)$$

(e) Four ways. There is often more than one way to rewrite a trigonometric expression.

3. Let $u = \cos x$, $du = -\sin x\,dx$.

$$\int \cos^3 x\sin x\,dx = -\int \cos^3 x(-\sin x)\,dx$$

$$= -\frac{1}{4}\cos^4 x + C$$

5. Let $u = \sin 2x$, $du = 2\cos 2x\,dx$.

$$\int \sin^5 2x\cos 2x\,dx = \frac{1}{2}\int \sin^5 2x(2\cos 2x)dx$$

$$= \frac{1}{12}\sin^6 2x + C$$

7. Let $u = \cos x$, $du = -\sin x\,dx$.

$$\int \sin^5 x\cos^2 x\,dx = \int \sin x(1 - \cos^2 x)^2\cos^2 x\,dx$$

$$= -\int (\cos^2 x - 2\cos^4 x + \cos^6 x)(-\sin x)\,dx = \frac{-1}{3}\cos^3 x + \frac{2}{5}\cos^5 x - \frac{1}{7}\cos^7 x + C$$

9. $$\int \cos^3\theta\sqrt{\sin\theta}\,d\theta = \int \cos\theta(1 - \sin^2\theta)(\sin\theta)^{1/2}\,d\theta$$

$$= \int [(\sin\theta)^{1/2} - (\sin\theta)^{5/2}]\cos\theta\,d\theta$$

$$= \frac{2}{3}(\sin\theta)^{3/2} - \frac{2}{7}(\sin\theta)^{7/2} + C$$

11. $$\int \cos^2 3x\,dx = \int \frac{1 + \cos 6x}{2}\,dx$$

$$= \frac{1}{2}\left(x + \frac{1}{6}\sin 6x\right) + C$$

$$= \frac{1}{12}(6x + \sin 6x) + C$$

13. $$\int \sin^2\alpha \cdot \cos^2\alpha\,d\alpha = \int \frac{1 - \cos 2\alpha}{2} \cdot \frac{1 + \cos 2\alpha}{2}\,d\alpha$$

$$= \frac{1}{4}\int (1 - \cos^2 2\alpha)\,d\alpha$$

$$= \frac{1}{4}\int \left(1 - \frac{1 + \cos 4\alpha}{2}\right)d\alpha$$

$$= \frac{1}{8}\int (1 - \cos 4\alpha)\,d\alpha$$

$$= \frac{1}{8}\left[\alpha - \frac{1}{4}\sin 4\alpha\right] + C$$

$$= \frac{1}{32}[4\alpha - \sin 4\alpha] + C$$

15. Integration by parts.

$$dv = \sin^2 x\,dx = \frac{1-\cos 2x}{2} \Rightarrow v = \frac{x}{2} - \frac{\sin 2x}{4} = \frac{1}{4}(2x - \sin 2x)$$

$$u = x \Rightarrow du = dx$$

$$\int x\sin^2 x\,dx = \frac{1}{4}x(2x - \sin 2x) - \frac{1}{4}\int (2x - \sin 2x)\,dx$$

$$= \frac{1}{4}x(2x - \sin 2x) - \frac{1}{4}\left(x^2 + \frac{1}{2}\cos 2x\right) + C = \frac{1}{8}(2x^2 - 2x\sin 2x - \cos 2x) + C$$

17. Let $u = \sin x$, $du = \cos x\,dx$.

$$\int_0^{\pi/2} \cos^3 x\,dx = \int_0^{\pi/2} (1 - \sin^2 x)\cos x\,dx$$

$$= \left[\sin x - \frac{1}{3}\sin^3 x\right]_0^{\pi/2} = \frac{2}{3}$$

19. Let $u = \sin x$, $du = \cos x\,dx$.

$$\int_0^{\pi/2} \cos^7 x\,dx = \int_0^{\pi/2} (1 - \sin^2 x)^3 \cos x\,dx = \int_0^{\pi/2} (1 - 3\sin^2 x + 3\sin^4 x - \sin^6 x)\cos x\,dx$$

$$= \left[\sin x - \sin^3 x + \frac{3}{5}\sin^5 x - \frac{1}{7}\sin^7 x\right]_0^{\pi/2} = \frac{16}{35}$$

21. $\displaystyle\int \sec(3x)\,dx = \frac{1}{3}\ln|\sec 3x + \tan 3x| + C$

23. $\displaystyle\int \sec^4 5x\,dx = \int (1 + \tan^2 5x)\sec^2 5x\,dx$

$$= \frac{1}{5}\left(\tan 5x + \frac{\tan^3 5x}{3}\right) + C$$

$$= \frac{\tan 5x}{15}(3 + \tan^2 5x) + C$$

25. $dv = \sec^2 \pi x\,dx \Rightarrow v = \dfrac{1}{\pi}\tan \pi x$

$u = \sec \pi x \Rightarrow du = \pi \sec \pi x \tan \pi x\,dx$

$$\int \sec^3 \pi x\,dx = \frac{1}{\pi}\sec \pi x \tan \pi x - \int \sec \pi x \tan^2 \pi x\,dx = \frac{1}{\pi}\sec \pi x \tan \pi x - \int \sec \pi x(\sec^2 \pi x - 1)\,dx$$

$$2\int \sec^3 \pi x\,dx = \frac{1}{\pi}(\sec \pi x \tan \pi x + \ln|\sec \pi x + \tan \pi x|) + C_1$$

$$\int \sec^3 \pi x\,dx = \frac{1}{2\pi}(\sec \pi x \tan \pi x + \ln|\sec \pi x + \tan \pi x|) + C$$

27. $\displaystyle\int \tan^5 \frac{x}{4}\,dx = \int \left(\sec^2 \frac{x}{4} - 1\right)\tan^3 \frac{x}{4}\,dx$

$$= \int \tan^3 \frac{x}{4}\sec^2 \frac{x}{4}\,dx - \int \tan^3 \frac{x}{4}\,dx$$

$$= \tan^4 \frac{x}{4} - \int \left(\sec^2 \frac{x}{4} - 1\right)\tan \frac{x}{4}\,dx$$

$$= \tan^4 \frac{x}{4} - 2\tan^2 \frac{x}{4} - 4\ln\left|\cos \frac{x}{4}\right| + C$$

29. $u = \tan x$, $du = \sec^2 x\,dx$

$$\int \sec^2 x \tan x\,dx = \frac{1}{2}\tan^2 x + C$$

31. $\int \tan^2 x \sec^2 x \, dx = \dfrac{\tan^3 x}{3} + C$

33. $\int \sec^6 4x \tan 4x \, dx = \dfrac{1}{4}\int \sec^5 4x(4 \sec 4x \tan 4x) \, dx$

$= \dfrac{\sec^6 4x}{24} + C$

35. Let $u = \sec x$, $du = \sec x \tan x \, dx$.

$\int \sec^3 x \tan x \, dx = \int \sec^2 x(\sec x \tan x) \, dx$

$= \dfrac{1}{3}\sec^3 x + C$

37. $\int \dfrac{\tan^2 x}{\sec x} \, dx = \int \dfrac{(\sec^2 x - 1)}{\sec x} \, dx$

$= \int (\sec x - \cos x) \, dx$

$= \ln|\sec x + \tan x| - \sin x + C$

39. $r = \int \sin^4(\pi\theta) \, d\theta = \dfrac{1}{4}\int [1 - \cos(2\pi\theta)]^2 \, d\theta$

$= \dfrac{1}{4}\int [1 - 2\cos(2\pi\theta) + \cos^2(2\pi\theta)] \, d\theta$

$= \dfrac{1}{4}\int \left[1 - 2\cos(2\pi\theta) + \dfrac{1 + \cos(4\pi\theta)}{2}\right] d\theta$

$= \dfrac{1}{4}\left[\theta - \dfrac{1}{\pi}\sin(2\pi\theta) + \dfrac{\theta}{2} + \dfrac{1}{8\pi}\sin(4\pi\theta)\right] + C$

$= \dfrac{1}{32\pi}[12\pi\theta - 8\sin(2\pi\theta) + \sin(4\pi\theta)] + C$

41. $y = \int \tan^3 3x \sec 3x \, dx$

$= \int (\sec^2 3x - 1)\sec 3x \tan 3x \, dx$

$= \dfrac{1}{3}\int \sec^2 3x(3 \sec 3x \tan 3x) \, dx - \dfrac{1}{3}\int 3 \sec 3x \tan 3x \, dx$

$= \dfrac{1}{9}\sec^3 3x - \dfrac{1}{3}\sec 3x + C$

43. (a)

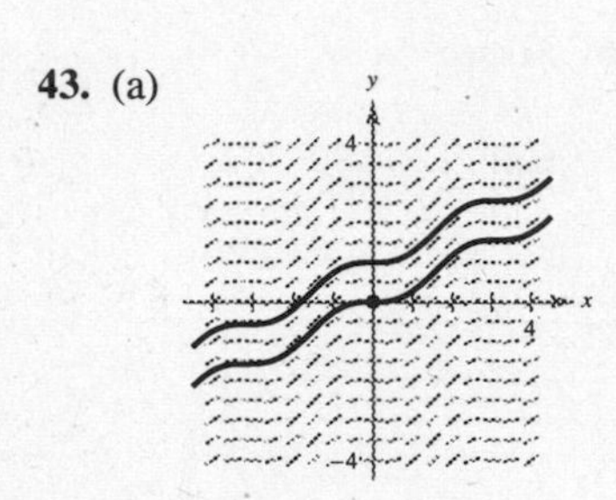

(b) $\dfrac{dy}{dx} = \sin^2 x$, $(0, 0)$

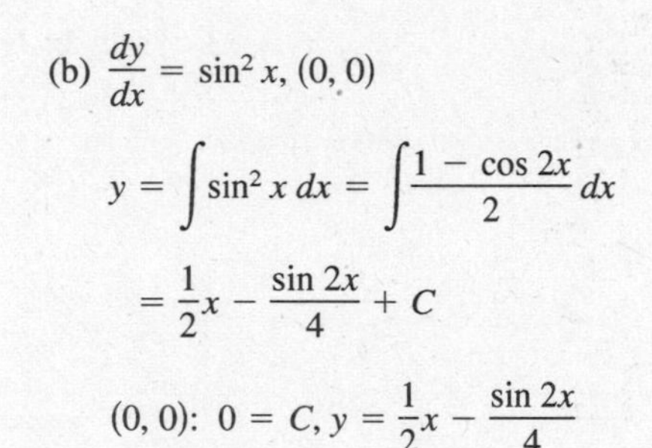

$y = \int \sin^2 x \, dx = \int \dfrac{1 - \cos 2x}{2} \, dx$

$= \dfrac{1}{2}x - \dfrac{\sin 2x}{4} + C$

$(0, 0)$: $0 = C$, $y = \dfrac{1}{2}x - \dfrac{\sin 2x}{4}$

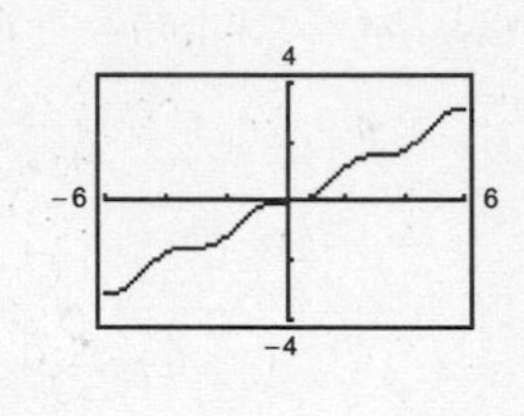

45. $\dfrac{dy}{dx} = \dfrac{3 \sin x}{y}$, $y(0) = 2$

8
-9 9
-4

47. $\int \sin 3x \cos 2x \, dx = \dfrac{1}{2}\int (\sin 5x + \sin x) \, dx$

$= \dfrac{-1}{2}\left(\dfrac{1}{5}\cos 5x + \cos x\right) + C$

$= \dfrac{-1}{10}(\cos 5x + 5 \cos x) + C$

49. $\int \sin\theta \sin 3\theta \, d\theta = \dfrac{1}{2}\int (\cos 2\theta - \cos 4\theta) \, d\theta$

$= \dfrac{1}{2}\left(\dfrac{1}{2}\sin 2\theta - \dfrac{1}{4}\sin 4\theta\right) + C$

$= \dfrac{1}{8}(2 \sin 2\theta - \sin 4\theta) + C$

51. $\int \cot^3 2x \, dx = \int (\csc^2 2x - 1)\cot 2x \, dx$

$= -\dfrac{1}{2}\int \cot 2x(-2\csc^2 2x) \, dx - \dfrac{1}{2}\int \dfrac{2 \cos 2x}{\sin 2x} \, dx$

$= -\dfrac{1}{4}\cot^2 2x - \dfrac{1}{2}\ln|\sin 2x| + C$

$= \dfrac{1}{4}(\ln|\csc^2 2x| - \cot^2 2x) + C$

53. Let $u = \cot\theta$, $du = -\csc^2\theta\, d\theta$.

$$\int \csc^4\theta\, d\theta = \int \csc^2\theta(1 + \cot^2\theta)\, d\theta$$
$$= \int \csc^2\theta\, d\theta + \int \csc^2\theta \cot^2\theta\, d\theta$$
$$= -\cot\theta - \frac{1}{3}\cot^3\theta + C$$

55. $$\int \frac{\cot^2 t}{\csc t}\, dt = \int \frac{\csc^2 t - 1}{\csc t}\, dt$$
$$= \int (\csc t - \sin t) dt$$
$$= \ln|\csc t - \cot t| + \cos t + C$$

57. $$\int \frac{1}{\sec x \tan x}\, dx = \int \frac{\cos^2 x}{\sin x}\, dx = \int \frac{1 - \sin^2 x}{\sin x}\, dx$$
$$= \int (\csc x - \sin x)\, dx$$
$$= \ln|\csc x - \cot x| + \cos x + C$$

59. $$\int (\tan^4 t - \sec^4 t)\, dt = \int (\tan^2 t + \sec^2 t)(\tan^2 t - \sec^2 t)\, dt \qquad (\tan^2 t - \sec^2 t = -1)$$
$$= -\int (\tan^2 t + \sec^2 t)\, dt = -\int (2\sec^2 t - 1)\, dt = -2\tan t + t + C$$

61. $$\int_{-\pi}^{\pi} \sin^2 x\, dx = 2\int_0^{\pi} \frac{1 - \cos 2x}{2}\, dx$$
$$= \left[x - \frac{1}{2}\sin 2x\right]_0^{\pi} = \pi$$

63. $$\int_0^{\pi/4} \tan^3 x\, dx = \int_0^{\pi/4} (\sec^2 x - 1)\tan x\, dx$$
$$= \int_0^{\pi/4} \sec^2 x \tan x\, dx - \int_0^{\pi/4} \frac{\sin x}{\cos x}\, dx$$
$$= \left[\frac{1}{2}\tan^2 x + \ln|\cos x|\right]_0^{\pi/4}$$
$$= \frac{1}{2}(1 - \ln 2)$$

65. Let $u = 1 + \sin t$, $du = \cos t\, dt$.

$$\int_0^{\pi/2} \frac{\cos t}{1 + \sin t}\, dt = \left[\ln|1 + \sin t|\right]_0^{\pi/2} = \ln 2$$

67. Let $u = \sin x$, $du = \cos x\, dx$.

$$\int_{-\pi/2}^{\pi/2} \cos^3 x\, dx = 2\int_0^{\pi/2} (1 - \sin^2 x)\cos x\, dx$$
$$= 2\left[\sin x - \frac{1}{3}\sin^3 x\right]_0^{\pi/2} = \frac{4}{3}$$

69. $$\int \cos^4 \frac{x}{2} dx = \frac{1}{16}[6x + 8\sin x + \sin 2x] + C$$

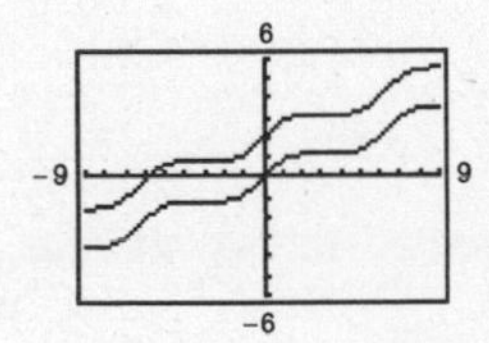

71. $$\int \sec^5 \pi x\, dx = \frac{1}{4\pi}\left\{\sec^3 \pi x \tan \pi x + \frac{3}{2}[\sec \pi x \tan \pi x + \ln|\sec \pi x + \tan \pi x|]\right\} + C$$

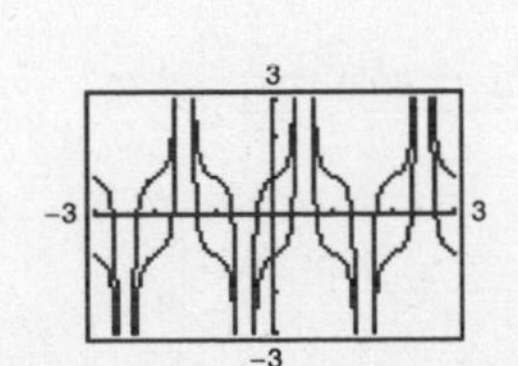

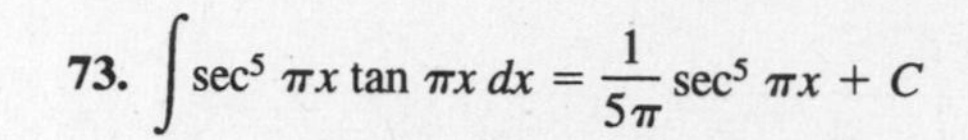

73. $\displaystyle\int \sec^5 \pi x \tan \pi x \, dx = \frac{1}{5\pi} \sec^5 \pi x + C$

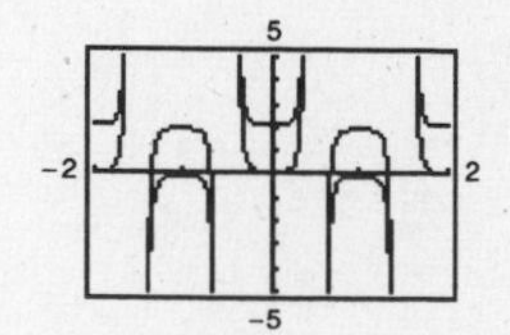

75. $\displaystyle\int_0^{\pi/4} \sin 2\theta \sin 3\theta \, d\theta = \frac{1}{2}\left[\sin \theta - \frac{1}{5}\sin 5\theta\right]_0^{\pi/4} = \frac{3\sqrt{2}}{10}$

77. $\displaystyle\int_0^{\pi/2} \sin^4 x \, dx = \frac{1}{4}\left[\frac{3x}{2} - \sin 2x + \frac{1}{8}\sin 4x\right]_0^{\pi/2}$

$\displaystyle= \frac{3\pi}{16}$

79. (a) Save one sine factor and convert the remaining sine factors to cosine. Then expand and integrate.

(b) Save one cosine factor and convert the remaining cosine factors to sine. Then expand and integrate.

(c) Make repeated use of the power reducing formula to convert the integrand to odd powers of the cosine.

81. (a) Let $u = \tan 3x$, $du = 3 \sec^2 3x \, dx$.

$$\int \sec^4 3x \tan^3 3x \, dx = \int \sec^2 3x \tan^3 3x \sec^2 3x \, dx$$

$$= \frac{1}{3}\int (\tan^2 3x + 1) \tan^3 3x (3 \sec^2 3x) \, dx$$

$$= \frac{1}{3}\int (\tan^5 3x + \tan^3 3x)(3 \sec^2 3x) dx$$

$$= \frac{\tan^6 3x}{18} + \frac{\tan^4 3x}{12} + C_1$$

Or let $u = \sec 3x$, $du = 3 \sec 3x \tan 3x \, dx$.

$$\int \sec^4 3x \tan^3 3x \, dx = \int \sec^3 3x \tan^2 3x \sec 3x \tan 3x \, dx$$

$$= \frac{1}{3}\int \sec^3 3x(\sec^2 3x - 1)(3 \sec 3x \tan 3x) \, dx$$

$$= \frac{\sec^6 3x}{18} - \frac{\sec^4 3x}{12} + C$$

(b)

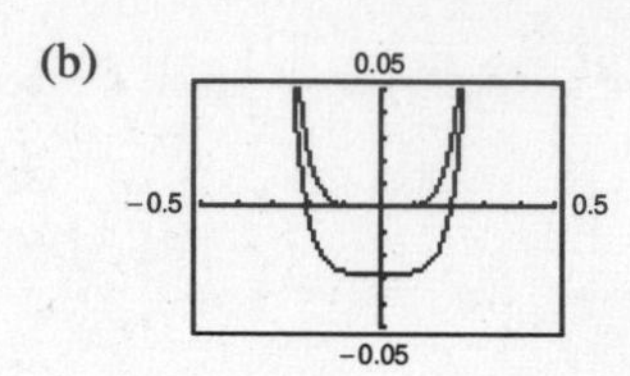

(c) $\displaystyle\frac{\sec^6 3x}{18} - \frac{\sec^4 3x}{12} + C = \frac{(1 + \tan^2 3x)^3}{18} - \frac{(1 + \tan^2 3x)^2}{12} + C$

$$= \frac{1}{18}\tan^6 3x + \frac{1}{6}\tan^4 3x + \frac{1}{6}\tan^2 3x + \frac{1}{18} - \frac{1}{12}\tan^4 3x - \frac{1}{6}\tan^2 3x - \frac{1}{12} + C$$

$$= \frac{\tan^6 3x}{18} + \frac{\tan^4 3x}{12} + \left(\frac{1}{18} - \frac{1}{12}\right) + C$$

$$= \frac{\tan^6 3x}{18} + \frac{\tan^4 3x}{12} + C_2$$

83. $\displaystyle A = \int_0^1 \sin^2(\pi x) \, dx$

$\displaystyle= \int_0^1 \frac{1 - \cos(2\pi x)}{2} \, dx$

$\displaystyle= \left[\frac{x}{2} - \frac{1}{4\pi}\sin(2\pi x)\right]_0^1$

$\displaystyle= \frac{1}{2}$

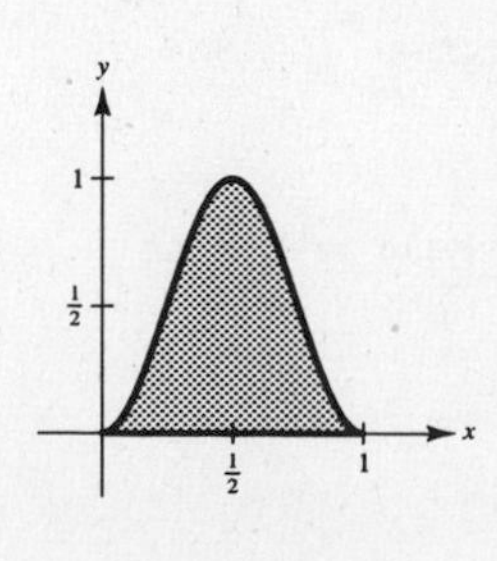

85. (a) $V = \pi\int_0^{\pi} \sin^2 x\,dx = \frac{\pi}{2}\int_0^{\pi}(1 - \cos 2x)\,dx = \frac{\pi}{2}\left[x - \frac{1}{2}\sin 2x\right]_0^{\pi} = \frac{\pi^2}{2}$

(b) $A = \int_0^{\pi} \sin x\,dx = \left[-\cos x\right]_0^{\pi} = 1 + 1 = 2$

Let $u = x$, $dv = \sin x\,dx$, $du = dx$, $v = -\cos x$.

$$\bar{x} = \frac{1}{A}\int_0^{\pi} x\sin x\,dx = \frac{1}{2}\left[\left[-x\cos x\right]_0^{\pi} + \int_0^{\pi}\cos x\,dx\right] = \frac{1}{2}\left[-x\cos x + \sin x\right]_0^{\pi} = \frac{\pi}{2}$$

$$\bar{y} = \frac{1}{2A}\int_0^{\pi}\sin^2 x\,dx$$

$$= \frac{1}{8}\int_0^{\pi}(1 - \cos 2x)\,dx$$

$$= \frac{1}{8}\left[x - \frac{1}{2}\sin 2x\right]_0^{\pi} = \frac{\pi}{8}$$

$$(\bar{x}, \bar{y}) = \left(\frac{\pi}{2}, \frac{\pi}{8}\right)$$

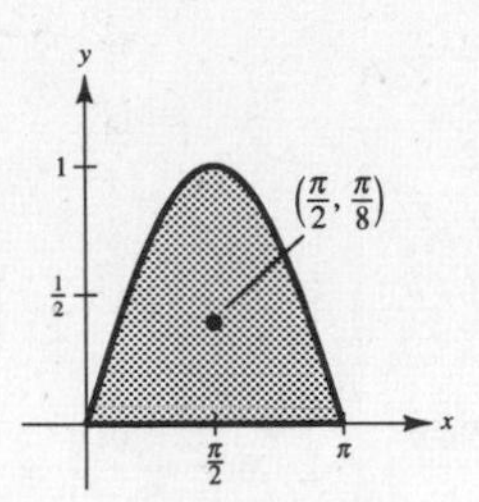

87. $dv = \sin x\,dx \implies v = -\cos x$

$u = \sin^{n-1} x \implies du = (n - 1)\sin^{n-2} x\cos x\,dx$

$$\int \sin^n x\,dx = -\sin^{n-1} x\cos x + (n - 1)\int \sin^{n-2} x\cos^2 x\,dx$$

$$= -\sin^{n-1} x\cos x + (n - 1)\int \sin^{n-2} x(1 - \sin^2 x)\,dx$$

$$= -\sin^{n-1} x\cos x + (n - 1)\int \sin^{n-2} x\,dx - (n - 1)\int \sin^n x\,dx$$

Therefore, $n\int \sin^n x\,dx = -\sin^{n-1} x\cos x + (n - 1)\int \sin^{n-2} x\,dx$

$$\int \sin^n x\,dx = \frac{-\sin^{n-1} x\cos x}{n} + \frac{n - 1}{n}\int \sin^{n-2} x\,dx.$$

89. Let $u = \sin^{n-1} x$, $du = (n - 1)\sin^{n-2} x\cos x\,dx$, $dv = \cos^m x\sin x\,dx$, $v = \dfrac{-\cos^{m+1} x}{m + 1}$.

$$\int \cos^m x\sin^n x\,dx = \frac{-\sin^{n-1} x\cos^{m+1} x}{m + 1} + \frac{n - 1}{m + 1}\int \sin^{n-2} x\cos^{m+2} x\,dx$$

$$= \frac{-\sin^{n-1} x\cos^{m+1} x}{m + 1} + \frac{n - 1}{m + 1}\int \sin^{n-2} x\cos^m x(1 - \sin^2 x)\,dx$$

$$= \frac{-\sin^{n-1} x\cos^{m+1} x}{m + 1} + \frac{n - 1}{m + 1}\int \sin^{n-2} x\cos^m x\,dx - \frac{n - 1}{m + 1}\int \sin^n x\cos^m x\,dx$$

$$\frac{m + n}{m + 1}\int \cos^m x\sin^n x\,dx = \frac{-\sin^{n-1} x\cos^{m+1} x}{m + 1} + \frac{n - 1}{m + 1}\int \sin^{n-2} x\cos^m x\,dx$$

$$\int \cos^m x\sin^n x\,dx = \frac{-\cos^{m+1} x\sin^{n-1}}{m + n} + \frac{n - 1}{m + n}\int \cos^m x\sin^{n-2} x\,dx$$

91. $$\int \sin^5 x\,dx = -\frac{\sin^4 x \cos x}{5} + \frac{4}{5}\int \sin^3 x\,dx$$

$$= -\frac{\sin^4 x \cos x}{5} + \frac{4}{5}\left[-\frac{\sin^2 x \cos x}{3} + \frac{2}{3}\int \sin x\,dx\right]$$

$$= -\frac{1}{5}\sin^4 x \cos x - \frac{4}{15}\sin^2 x \cos x - \frac{8}{15}\cos x + C$$

$$= -\frac{\cos x}{15}[3\sin^4 x + 4\sin^2 x + 8] + C$$

93. $$\int \sec^4\left(\frac{2\pi x}{5}\right) dx = \frac{5}{2\pi}\int \sec^4\left(\frac{2\pi x}{5}\right)\frac{2\pi}{5}\,dx$$

$$= \frac{5}{2\pi}\left[\frac{1}{3}\sec^2\left(\frac{2\pi x}{5}\right)\tan\left(\frac{2\pi x}{5}\right) + \frac{2}{3}\int \sec^2\left(\frac{2\pi x}{5}\right)\frac{2\pi}{5}\,dx\right]$$

$$= \frac{5}{6\pi}\left[\sec^2\left(\frac{2\pi x}{5}\right)\tan\left(\frac{2\pi x}{5}\right) + 2\tan\left(\frac{2\pi x}{5}\right)\right] + C$$

$$= \frac{5}{6\pi}\tan\left(\frac{2\pi x}{5}\right)\left[\sec^2\left(\frac{2\pi x}{5}\right) + 2\right] + C$$

95. (a) $f(t) = a_0 + a_1 \cos\frac{\pi t}{6} + b_1 \sin\frac{\pi t}{6}$ where:

$$a_0 = \frac{1}{12}\int_0^{12} f(t)\,dt$$

$$a_1 = \frac{1}{6}\int_0^{12} f(t)\cos\frac{\pi t}{6}\,dt$$

$$b_1 = \frac{1}{6}\int_0^{12} f(t)\sin\frac{\pi t}{6}\,dt$$

$$a_0 \approx \frac{12-0}{3(12)^2}[30.9 + 4(32.2) + 2(41.1) + 4(53.7) + 2(64.6) + 4(74.0) + 2(78.2) + 4(77.0) + 2(71.0) + 4(60.1) + 2(47.1) + 4(35.7) + 30.9] \approx 55.46$$

$$a_1 \approx \frac{12-0}{6(3)(12)}\left[30.9\cos 0 + 4\left(32.2\cos\frac{\pi}{6}\right) + 2\left(41.1\cos\frac{\pi}{3}\right) + 4\left(53.7\cos\frac{\pi}{2}\right) + 2\left(64.6\cos\frac{2\pi}{3}\right) + 4\left(74.0\cos\frac{5\pi}{6}\right) + 2(78.2\cos\pi) + 4\left(77.0\cos\frac{7\pi}{6}\right) + 2\left(71.0\cos\frac{4\pi}{3}\right) + 4\left(60.1\cos\frac{3\pi}{2}\right) + 2\left(47.1\cos\frac{5\pi}{3}\right) + 4\left(35.7\cos\frac{11\pi}{6}\right) + 30.9\cos 2\pi\right] \approx -23.88$$

$$b_1 \approx \frac{12-0}{6(3)(12)}\left[30.9\sin 0 + 4\left(32.2\sin\frac{\pi}{6}\right) + 2\left(41.1\sin\frac{\pi}{3}\right) + 4\left(53.7\sin\frac{\pi}{2}\right) + 2\left(64.6\sin\frac{2\pi}{3}\right) + 4\left(74.0\sin\frac{5\pi}{6}\right) + 2(78.2\sin\pi) + 4\left(77.0\sin\frac{7\pi}{6}\right) + 2\left(71.0\sin\frac{4\pi}{3}\right) + 4\left(60.1\sin\frac{3\pi}{2}\right) + 2\left(47.1\sin\frac{5\pi}{3}\right) + 4\left(35.7\sin\frac{11\pi}{6}\right) + 30.9\sin 2\pi\right] \approx -3.34$$

$$H(t) \approx 55.46 - 23.88\cos\frac{\pi t}{6} - 3.34\sin\frac{\pi t}{6}$$

—CONTINUED—

95. —CONTINUED—

(b) $a_0 \approx \frac{12-0}{3(12)^2}[18.0 + 4(17.7) + 2(25.8) + 4(36.1) + 2(45.4) + 4(55.2) + 2(59.9) + 4(59.4) + 2(53.1) +$

$4(43.2) + 2(34.3) + 4(24.2) + 18.0] \approx 39.34$

$$a_1 \approx \frac{12-0}{6(3)(12)}\Bigg[18.0\cos 0 + 4\left(17.7\cos\frac{\pi}{6}\right) + 2\left(25.8\cos\frac{\pi}{3}\right) + 4\left(36.1\cos\frac{\pi}{2}\right) + 2\left(45.4\cos\frac{2\pi}{3}\right) +$$

$$4\left(55.2\cos\frac{5\pi}{6}\right) + 2(59.9\cos\pi) + 4\left(59.4\cos\frac{7\pi}{6}\right) + 2\left(53.1\cos\frac{4\pi}{3}\right) +$$

$$4\left(43.2\cos\frac{3\pi}{2}\right) + 2\left(34.3\cos\frac{5\pi}{3}\right) + 4\left(24.2\cos\frac{11\pi}{6}\right) + 18\cos 2\pi\Bigg] \approx -20.78$$

$$b_1 \approx \frac{12-0}{6(3)(12)}\Bigg[18.0\sin 0 + 4\left(17.7\sin\frac{\pi}{6}\right) + 2\left(25.8\sin\frac{\pi}{3}\right) + 4\left(36.1\sin\frac{\pi}{2}\right) + 2\left(45.4\sin\frac{2\pi}{3}\right) +$$

$$4\left(55.2\sin\frac{5\pi}{6}\right) + 2(59.9\sin\pi) + 4\left(59.4\sin\frac{7\pi}{6}\right) + 2\left(53.1\sin\frac{4\pi}{3}\right) +$$

$$4\left(43.2\sin\frac{3\pi}{2}\right) + 2\left(34.3\sin\frac{5\pi}{3}\right) + 4\left(24.2\sin\frac{11\pi}{6}\right) + 18\sin 2\pi\Bigg] \approx -4.33$$

$$L(t) \approx 39.34 - 20.78\cos\frac{\pi t}{6} - 4.33\sin\frac{\pi t}{6}$$

(c) The difference between the maximum and minimum temperatures is greatest in the summer.

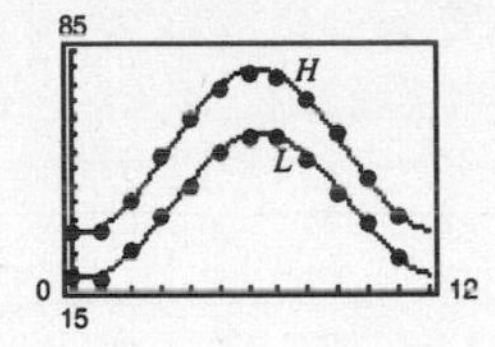

97. $\displaystyle\int_{-\pi}^{\pi}\cos(mx)\cos(nx)\,dx = \frac{1}{2}\left[\frac{\sin(m+n)x}{m+n} + \frac{\sin(m-n)x}{m-n}\right]_{-\pi}^{\pi} = 0,\ (m \neq n)$

$$\int_{-\pi}^{\pi}\sin(mx)\sin(nx)\,dx = \frac{1}{2}\int_{-\pi}^{\pi}[\cos(m-n)x - \cos(m+n)x]\,dx$$

$$= \frac{1}{2}\left[\frac{\sin(m-n)x}{m-n} - \frac{\sin(m+n)x}{m+n}\right]_{-\pi}^{\pi} = 0,\ (m \neq n)$$

$$\int_{-\pi}^{\pi}\sin(mx)\cos(nx)\,dx = \frac{1}{2}\int_{-\pi}^{\pi}[\sin(m+n)x + \sin(m-n)x]\,dx$$

$$= -\frac{1}{2}\left[\frac{\cos(m+n)x}{m+n} + \frac{\cos(m-n)x}{m-n}\right]_{-\pi}^{\pi},\ (m \neq n)$$

$$= -\frac{1}{2}\left[\left(\frac{\cos(m+n)\pi}{m+n} + \frac{\cos(m-n)\pi}{m-n}\right) - \left(\frac{\cos(m+n)(-\pi)}{m+n} + \frac{\cos(m-n)(-\pi)}{m-n}\right)\right]$$

$= 0$, since $\cos(-\theta) = \cos\theta$.

$$\int_{-\pi}^{\pi}\sin(mx)\cos(mx)\,dx = \frac{1}{m}\left[\frac{\sin^2(mx)}{2}\right]_{-\pi}^{\pi} = 0$$

Section 7.4 Trigonometric Substitution

1. $$\frac{d}{dx}\left[4\ln\left|\frac{\sqrt{x^2+16}-4}{x}\right| + \sqrt{x^2+16} + C\right] = \frac{d}{dx}\left[4\ln\left|\sqrt{x^2+16}-4\right| - 4\ln|x| + \sqrt{x^2+16} + C\right]$$

$$= 4\left[\frac{x/\sqrt{x^2+16}}{\sqrt{x^2+16}-4}\right] - \frac{4}{x} + \frac{x}{\sqrt{x^2+16}}$$

$$= \frac{4x}{\sqrt{x^2+16}\left(\sqrt{x^2+16}-4\right)} - \frac{4}{x} + \frac{x}{\sqrt{x^2+16}}$$

$$= \frac{4x^2 - 4\sqrt{x^2+16}\left(\sqrt{x^2+16}-4\right) + x^2\left(\sqrt{x^2+16}-4\right)}{x\sqrt{x^2+16}\left(\sqrt{x^2+16}-4\right)}$$

$$= \frac{4x^2 - 4(x^2+16) + 16\sqrt{x^2+16} + x^2\sqrt{x^2+16} - 4x^2}{x\sqrt{x^2+16}\left(\sqrt{x^2+16}-4\right)}$$

$$= \frac{\sqrt{x^2+16}(x^2+16) - 4(x^2+16)}{x\sqrt{x^2+16}\left(\sqrt{x^2+16}-4\right)}$$

$$= \frac{(x^2+16)\left(\sqrt{x^2+16}-4\right)}{x\sqrt{x^2+16}\left(\sqrt{x^2+16}-4\right)} = \frac{\sqrt{x^2+16}}{x}$$

Indefinite integral: $\displaystyle\int \frac{\sqrt{x^2+16}}{x}\,dx$ Matches (b)

3. $$\frac{d}{dx}\left[8\arcsin\frac{x}{4} - \frac{x\sqrt{16-x^2}}{2} + C\right] = 8\frac{1/4}{\sqrt{1-(x/4)^2}} - \frac{x(1/2)(16-x^2)^{-1/2}(-2x) + \sqrt{16-x^2}}{2}$$

$$= \frac{8}{\sqrt{16-x^2}} + \frac{x^2}{2\sqrt{16-x^2}} - \frac{\sqrt{16-x^2}}{2}$$

$$= \frac{16}{2\sqrt{16-x^2}} + \frac{x^2}{2\sqrt{16-x^2}} - \frac{(16-x^2)}{2\sqrt{16-x^2}} = \frac{x^2}{\sqrt{16-x^2}}$$

Matches (a)

5. Let $x = 5\sin\theta$, $dx = 5\cos\theta\,d\theta$, $\sqrt{25-x^2} = 5\cos\theta$.

$$\int \frac{1}{(25-x^2)^{3/2}}\,dx = \int \frac{5\cos\theta}{(5\cos\theta)^3}\,d\theta$$

$$= \frac{1}{25}\int \sec^2\theta\,d\theta$$

$$= \frac{1}{25}\tan\theta + C$$

$$= \frac{x}{25\sqrt{25-x^2}} + C$$

5
x
θ
$\sqrt{25-x^2}$

7. Same substitution as in Exercise 5

$$\int \frac{\sqrt{25-x^2}}{x}\,dx = \int \frac{25\cos^2\theta\,d\theta}{5\sin\theta} = 5\int \frac{1-\sin^2\theta}{\sin\theta}\,d\theta = 5\int (\csc\theta - \sin\theta)\,d\theta$$

$$= 5[\ln|\csc\theta - \cot\theta| + \cos\theta] + C = 5\ln\left|\frac{5-\sqrt{25-x^2}}{x}\right| + \sqrt{25-x^2} + C$$

9. Let $x = 2 \sec \theta$, $dx = 2 \sec \theta \tan \theta \, d\theta$, $\sqrt{x^2 - 4} = 2 \tan \theta$.

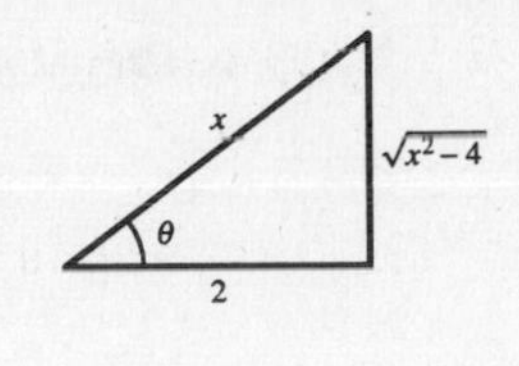

$$\begin{aligned}\int \frac{1}{\sqrt{x^2 - 4}}\,dx &= \int \frac{2 \sec \theta \tan \theta \, d\theta}{2 \tan \theta} = \int \sec \theta \, d\theta = \ln|\sec \theta + \tan\theta| + C_1 \\ &= \ln\left|\frac{x}{2} + \frac{\sqrt{x^2 - 4}}{2}\right| + C_1 \\ &= \ln\left|x + \sqrt{x^2 - 4}\right| - \ln 2 + C_1 = \ln\left|x + \sqrt{x^2 - 4}\right| + C\end{aligned}$$

11. Same substitution as in Exercise 9

$$\begin{aligned}\int x^3 \sqrt{x^2 - 4}\,dx &= \int (8 \sec^3 \theta)(2 \tan \theta)(2 \sec \theta \tan \theta)\,d\theta = 32 \int \tan^2 \theta \sec^4 \theta \, d\theta \\ &= 32 \int \tan^2 \theta (1 + \tan^2 \theta) \sec^2 \theta \, d\theta = 32\left(\frac{\tan^3 \theta}{3} + \frac{\tan^5 \theta}{5}\right) + C \\ &= \frac{32}{15} \tan^3 \theta [5 + 3 \tan^2 \theta] + C = \frac{32}{15} \frac{(x^2 - 4)^{3/2}}{8}\left[5 + 3\frac{(x^2 - 4)}{4}\right] + C \\ &= \frac{1}{15}(x^2 - 4)^{3/2}[20 + 3(x^2 - 4)] + C = \frac{1}{15}(x^2 - 4)^{3/2}(3x^2 + 8) + C\end{aligned}$$

13. Let $x = \tan \theta$, $dx = \sec^2 \theta \, d\theta$, $\sqrt{1 + x^2} = \sec \theta$.

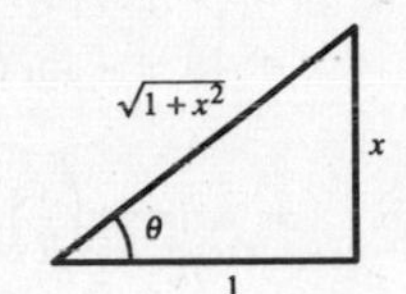

$$\int x\sqrt{1 + x^2}\,dx = \int \tan \theta (\sec \theta) \sec^2 \theta \, d\theta = \frac{\sec^3 \theta}{3} + C = \frac{1}{3}(1 + x^2)^{3/2} + C$$

Note: This integral could have been evaluated with the Power Rule.

15. Same substitution as in Exercise 13

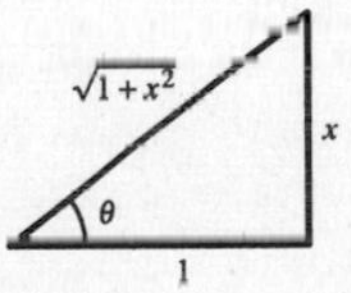

$$\begin{aligned}\int \frac{1}{(1 + x^2)^2}\,dx &= \int \frac{1}{(\sqrt{1 + x^2})^4}\,dx \\ &= \int \frac{\sec^2 \theta \, d\theta}{\sec^4 \theta} \\ &= \int \cos^2 \theta \, d\theta = \frac{1}{2}\int (1 + \cos 2\theta)\,d\theta \\ &= \frac{1}{2}\left[\theta + \frac{\sin 2\theta}{2}\right] \\ &= \frac{1}{2}[\theta + \sin \theta \cos \theta] + C \\ &= \frac{1}{2}\left[\arctan x + \left(\frac{x}{\sqrt{1 + x^2}}\right)\left(\frac{1}{\sqrt{1 + x^2}}\right)\right] + C \\ &= \frac{1}{2}\left[\arctan x + \frac{x}{1 + x^2}\right] + C\end{aligned}$$

17. Let $u = 3x$, $a = 2$, and $du = 3\,dx$.

$$\begin{aligned}\int \sqrt{4 + 9x^2}\,dx &= \frac{1}{3}\int \sqrt{(2)^2 + (3x)^2}\,3\,dx \\ &= \frac{1}{3}\left(\frac{1}{2}\right)\left(3x\sqrt{4 + 9x^2} + 4 \ln\left|3x + \sqrt{4 + 9x^2}\right|\right) + C \\ &= \frac{1}{2}x\sqrt{4 + 9x^2} + \frac{2}{3}\ln\left|3x + \sqrt{4 + 9x^2}\right| + C\end{aligned}$$

19. $$\int \frac{x}{\sqrt{x^2+9}}\,dx = \frac{1}{2}\int (x^2+9)^{-1/2}(2x)\,dx$$
$$= \sqrt{x^2+9} + C$$

(Power Rule)

21. $$\int \frac{1}{\sqrt{16-x^2}}\,dx = \arcsin\left(\frac{x}{4}\right) + C$$

23. Let $x = 2\sin\theta$, $dx = 2\cos\theta\,d\theta$, $\sqrt{4-x^2} = 2\cos\theta$.

$$\int \sqrt{16-4x^2}\,dx = 2\int \sqrt{4-x^2}\,dx$$
$$= 2\int 2\cos\theta(2\cos\theta\,d\theta)$$
$$= 8\int \cos^2\theta\,d\theta$$
$$= 4\int (1+\cos 2\theta)\,d\theta$$
$$= 4\left[\theta + \frac{1}{2}\sin 2\theta\right] + C$$
$$= 4\theta + 4\sin\theta\cos\theta + C$$
$$= 4\arcsin\left(\frac{x}{2}\right) + x\sqrt{4-x^2} + C$$

25. Let $x = 3\sec\theta$, $dx = 3\sec\theta\tan\theta\,d\theta$,
$\sqrt{x^2-9} = 3\tan\theta$.

$$\int \frac{1}{\sqrt{x^2-9}}\,dx = \int \frac{3\sec\theta\tan\theta\,d\theta}{3\tan\theta}$$
$$= \int \sec\theta\,d\theta$$
$$= \ln|\sec\theta + \tan\theta| + C_1$$
$$= \ln\left|\frac{x}{3} + \frac{\sqrt{x^2-9}}{3}\right| + C_1$$
$$= \ln\left|x + \sqrt{x^2-9}\right| + C$$

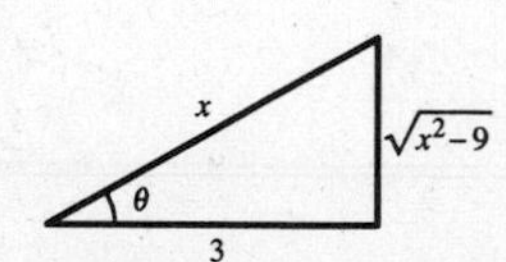

27. Let $x = \sin\theta$, $dx = \cos\theta\,d\theta$, $\sqrt{1-x^2} = \cos\theta$.

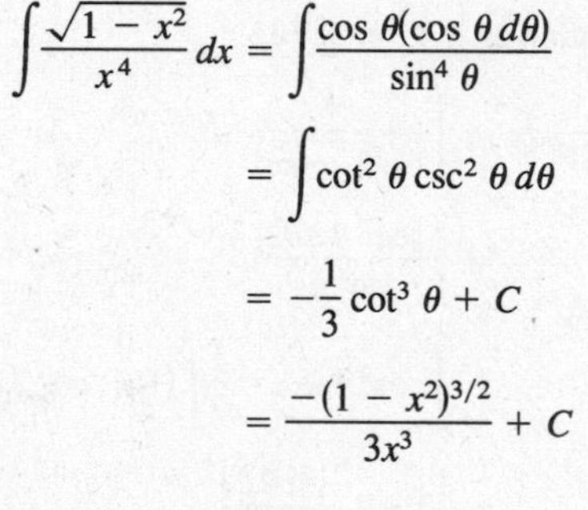

$$\int \frac{\sqrt{1-x^2}}{x^4}\,dx = \int \frac{\cos\theta(\cos\theta\,d\theta)}{\sin^4\theta}$$
$$= \int \cot^2\theta\csc^2\theta\,d\theta$$
$$= -\frac{1}{3}\cot^3\theta + C$$
$$= \frac{-(1-x^2)^{3/2}}{3x^3} + C$$

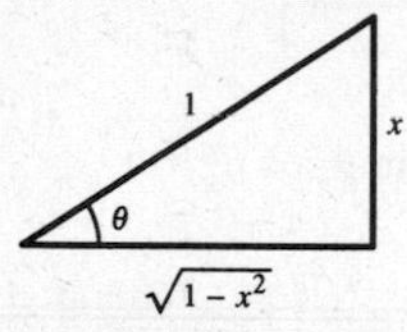

29. Same substitutions as in Exercise 28

$$\int \frac{1}{x\sqrt{4x^2+9}}\,dx = \int \frac{(3/2)\sec^2\theta\,d\theta}{(3/2)\tan\theta\,3\sec\theta}$$
$$= \frac{1}{3}\int \csc\theta\,d\theta = -\frac{1}{3}\ln|\csc\theta + \cot\theta| + C = -\frac{1}{3}\ln\left|\frac{\sqrt{4x^2+9}+3}{2x}\right| + C$$

31. Let $x = \sqrt{5}\tan\theta$, $dx = \sqrt{5}\sec^2\theta\,d\theta$, $x^2 + 5 = 5\sec^2\theta$.

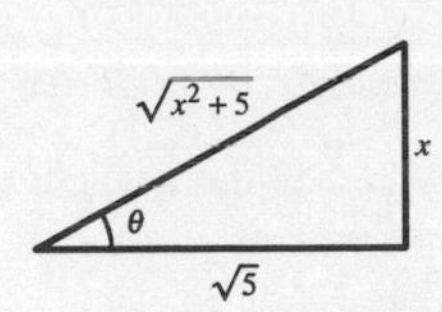

$$\int \frac{-5x}{(x^2+5)^{3/2}}\,dx = \int \frac{-5\sqrt{5}\tan\theta}{(5\sec^2\theta)^{3/2}}\sqrt{5}\sec^2\theta\,d\theta$$

$$= -\sqrt{5}\int \frac{\tan\theta}{\sec\theta}\,d\theta$$

$$= -\sqrt{5}\int \sin\theta\,d\theta$$

$$= \sqrt{5}\cos\theta + C$$

$$= \sqrt{5}\,\frac{\sqrt{5}}{\sqrt{x^2+5}} + C$$

$$= \frac{5}{\sqrt{x^2+5}} + C$$

33. Let $u = 1 + e^{2x}$, $du = 2e^{2x}\,dx$.

$$\int e^{2x}\sqrt{1+e^{2x}}\,dx = \frac{1}{2}\int (1+e^{2x})^{1/2}(2e^{2x})\,dx = \frac{1}{3}(1+e^{2x})^{3/2} + C$$

35. Let $e^x = \sin\theta$, $e^x\,dx = \cos\theta\,d\theta$, $\sqrt{1-e^{2x}} = \cos\theta$.

$$\int e^x\sqrt{1-e^{2x}}\,dx = \int \cos^2\theta\,d\theta$$

$$= \frac{1}{2}\int (1+\cos 2\theta)\,d\theta$$

$$= \frac{1}{2}\left[\theta + \frac{\sin 2\theta}{2}\right]$$

$$= \frac{1}{2}(\theta + \sin\theta\cos\theta) + C = \frac{1}{2}\left(\arcsin e^x + e^x\sqrt{1-e^{2x}}\right) + C$$

37. Let $x = \sqrt{2}\tan\theta$, $dx = \sqrt{2}\sec^2\theta\,d\theta$, $x^2 + 2 = 2\sec^2\theta$.

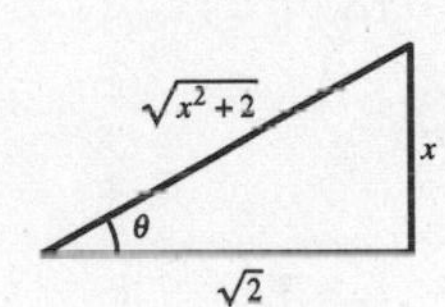

$$\int \frac{1}{4+4x^2+x^4}\,dx = \int \frac{1}{(x^2+2)^2}\,dx$$

$$= \int \frac{\sqrt{2}\sec^2\theta\,d\theta}{4\sec^4\theta}$$

$$= \frac{\sqrt{2}}{4}\int \cos^2\theta\,d\theta$$

$$= \frac{\sqrt{2}}{4}\left(\frac{1}{2}\right)\int (1+\cos 2\theta)\,d\theta$$

$$= \frac{\sqrt{2}}{8}\left(\theta + \frac{1}{2}\sin 2\theta\right) + C$$

$$= \frac{\sqrt{2}}{8}(\theta + \sin\theta\cos\theta) + C$$

$$= \frac{1}{4}\left[\frac{x}{x^2+2} + \frac{1}{\sqrt{2}}\arctan\frac{x}{\sqrt{2}}\right] + C$$

39. Since $x > \frac{1}{2}$,

$$u = \operatorname{arcsec} 2x, \Rightarrow du = \frac{1}{x\sqrt{4x^2 - 1}}\,dx,\ dv = dx \Rightarrow v = x$$

$$\int \operatorname{arcsec} 2x\,dx = x \operatorname{arcsec} 2x - \int \frac{1}{\sqrt{4x^2 - 1}}\,dx$$

$$2x = \sec\theta,\ dx = \frac{1}{2}\sec\theta\tan\theta\,d\theta,\ \sqrt{4x^2 - 1} = \tan\theta$$

$$\int \operatorname{arcsec} 2x\,dx = x \operatorname{arcsec} 2x - \int \frac{(1/2)\sec\theta\tan\theta\,d\theta}{\tan\theta} = x \operatorname{arcsec} 2x - \frac{1}{2}\int \sec\theta\,d\theta$$

$$= x \operatorname{arcsec} 2x - \frac{1}{2}\ln\left|\sec\theta + \tan\theta\right| + C = x \operatorname{arcsec} 2x - \frac{1}{2}\ln\left|2x + \sqrt{4x^2 - 1}\right| + C.$$

41. $$\int \frac{1}{\sqrt{4x - x^2}}\,dx = \int \frac{1}{\sqrt{4 - (x - 2)^2}}\,dx = \arcsin\left(\frac{x - 2}{2}\right) + C$$

43. Let $x + 2 = 2\tan\theta$, $dx = 2\sec^2\theta\,d\theta$, $\sqrt{(x + 2)^2 + 4} = 2\sec\theta$.

$$\int \frac{x}{\sqrt{x^2 + 4x + 8}}\,dx = \int \frac{x}{\sqrt{(x + 2)^2 + 4}}\,dx = \int \frac{(2\tan\theta - 2)(2\sec^2\theta)\,d\theta}{2\sec\theta}$$

$$= 2\int (\tan\theta - 1)(\sec\theta)\,d\theta$$

$$= 2[\sec\theta - \ln|\sec\theta + \tan\theta|] + C_1$$

$$= 2\left[\frac{\sqrt{(x + 2)^2 + 4}}{2} - \ln\left|\frac{\sqrt{(x + 2)^2 + 4}}{2} + \frac{x + 2}{2}\right|\right] + C_1$$

$$= \sqrt{x^2 + 4x + 8} - 2\left[\ln\left|\sqrt{x^2 + 4x + 8} + (x + 2)\right| - \ln 2\right] + C_1$$

$$= \sqrt{x^2 + 4x + 8} - 2\ln\left|\sqrt{x^2 + 4x + 8} + (x + 2)\right| + C$$

45. Let $t = \sin\theta$, $dt = \cos\theta\,d\theta$, $1 - t^2 = \cos^2\theta$.

(a) $$\int \frac{t^2}{(1 - t^2)^{3/2}}\,dt = \int \frac{\sin^2\theta\cos\theta\,d\theta}{\cos^3\theta}$$

$$= \int \tan^2\theta\,d\theta$$

$$= \int (\sec^2\theta - 1)\,d\theta$$

$$= \tan\theta - \theta + C$$

$$= \frac{t}{\sqrt{1 - t^2}} - \arcsin t + C$$

Thus, $$\int_0^{\sqrt{3}/2} \frac{t^2}{(1 - t^2)^{3/2}}\,dt = \left[\frac{t}{\sqrt{1 - t^2}} - \arcsin t\right]_0^{\sqrt{3}/2} = \frac{\sqrt{3}/2}{\sqrt{1/4}} - \arcsin\frac{\sqrt{3}}{2} = \sqrt{3} - \frac{\pi}{3} \approx 0.685.$$

(b) When $t = 0$, $\theta = 0$. When $t = \sqrt{3}/2$, $\theta = \pi/3$. Thus,

$$\int_0^{\sqrt{3}/2} \frac{t^2}{(1 - t^2)^{3/2}}\,dt = \Big[\tan\theta - \theta\Big]_0^{\pi/3} = \sqrt{3} - \frac{\pi}{3} \approx 0.685.$$

47. (a) Let $x = 3\tan\theta$, $dx = 3\sec^2\theta\,d\theta$, $\sqrt{x^2+9} = 3\sec\theta$.

$$\int \frac{x^3}{\sqrt{x^2+9}}\,dx = \int \frac{(27\tan^3\theta)(3\sec^2\theta\,d\theta)}{3\sec\theta}$$

$$= 27\int (\sec^2\theta - 1)\sec\theta\tan\theta\,d\theta$$

$$= 27\left[\frac{1}{3}\sec^3\theta - \sec\theta\right] + C = 9[\sec^3\theta - 3\sec\theta] + C$$

$$= 9\left[\left(\frac{\sqrt{x^2+9}}{3}\right)^3 - 3\left(\frac{\sqrt{x^2+9}}{3}\right)\right] + C = \frac{1}{3}(x^2+9)^{3/2} - 9\sqrt{x^2+9} + C$$

Thus, $\displaystyle\int_0^3 \frac{x^3}{\sqrt{x^2+9}}\,dx = \left[\frac{1}{3}(x^2+9)^{3/2} - 9\sqrt{x^2+9}\right]_0^3$

$$= \left(\frac{1}{3}(54\sqrt{2}) - 27\sqrt{2}\right) - (9 - 27)$$

$$= 18 - 9\sqrt{2} = 9\left(2 - \sqrt{2}\right) \approx 5.272.$$

(b) When $x = 0$, $\theta = 0$. When $x = 3$, $\theta = \pi/4$. Thus,

$$\int_0^3 \frac{x^3}{\sqrt{x^2+9}}\,dx = 9\left[\sec^3\theta - 3\sec\theta\right]_0^{\pi/4} = 9\left(2\sqrt{2} - 3\sqrt{2}\right) - 9(1 - 3) = 9\left(2 - \sqrt{2}\right) \approx 5.272.$$

49. (a) Let $x = 3\sec\theta$, $dx = 3\sec\theta\tan\theta\,d\theta$, $\sqrt{x^2-9} = 3\tan\theta$.

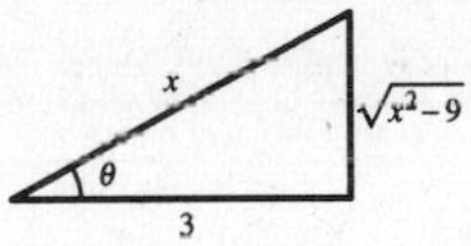

$$\int \frac{x^2}{\sqrt{x^2-9}}\,dx = \int \frac{9\sec^2\theta}{3\tan\theta}\,3\sec\theta\tan\theta\,d\theta$$

$$= 9\int \sec^3\theta\,d\theta$$

$$= 9\left[\frac{1}{2}\sec\theta\tan\theta + \frac{1}{2}\int \sec\theta\,d\theta\right] \qquad \text{(7.3 Exercise 90)}$$

$$= \frac{9}{2}[\sec\theta\tan\theta + \ln|\sec\theta + \tan\theta|]$$

$$= \frac{9}{2}\left[\frac{x}{3}\cdot\frac{\sqrt{x^2-9}}{3} + \ln\left|\frac{x}{3} + \frac{\sqrt{x^2-9}}{3}\right|\right]$$

Hence,

$$\int_4^6 \frac{x^2}{\sqrt{x^2-9}}\,dx = \frac{9}{2}\left[\frac{x\sqrt{x^2-9}}{9} + \ln\left|\frac{x}{3} + \frac{\sqrt{x^2-9}}{3}\right|\right]_4^6$$

$$= \frac{9}{2}\left[\left(\frac{6\sqrt{27}}{9} + \ln\left|2 + \frac{\sqrt{27}}{3}\right|\right) - \left(\frac{4\sqrt{7}}{9} + \ln\left|\frac{4}{3} + \frac{\sqrt{7}}{3}\right|\right)\right]$$

$$= 9\sqrt{3} - 2\sqrt{7} + \frac{9}{2}\left(\ln\left(\frac{6+\sqrt{27}}{3}\right) - \ln\left(\frac{4+\sqrt{7}}{3}\right)\right)$$

$$= 9\sqrt{3} - 2\sqrt{7} + \frac{9}{2}\ln\left(\frac{6+3\sqrt{3}}{4+\sqrt{7}}\right)$$

$$= 9\sqrt{3} - 2\sqrt{7} + \frac{9}{2}\ln\left(\frac{\left(4-\sqrt{7}\right)\left(2+\sqrt{3}\right)}{3}\right) \approx 12.644.$$

—CONTINUED—

49. —CONTINUED—

(b) When $x = 4$, $\theta = \operatorname{arcsec}\left(\frac{4}{3}\right)$.

When $x = 6$, $\theta = \operatorname{arcsec}(2) = \frac{\pi}{3}$.

$$\int_4^6 \frac{x^2}{\sqrt{x^2-9}}\,dx = \frac{9}{2}\Big[\sec\theta\tan\theta + \ln|\sec\theta + \tan\theta|\Big]_{\operatorname{arcsec}(4/3)}^{\pi/3}$$

$$= \frac{9}{2}\left[2\cdot\sqrt{3} + \ln\left|2+\sqrt{3}\right|\right] - \frac{9}{2}\left[\frac{4}{3}\frac{\sqrt{7}}{3} + \ln\left|\frac{4}{3} + \frac{\sqrt{7}}{3}\right|\right]$$

$$= 9\sqrt{3} - 2\sqrt{7} + \frac{9}{2}\ln\left(\frac{6+3\sqrt{3}}{4+\sqrt{7}}\right) \approx 12.644$$

51. $\displaystyle\int \frac{x^2}{\sqrt{x^2+10x+9}}\,dx = \frac{1}{2}\sqrt{x^2+10x+9}\,(x-15) + 33\ln\left|(x+5) + \sqrt{x^2+10x+9}\right| + C$

53. $\displaystyle\int \frac{x^2}{\sqrt{x^2-1}}\,dx = \frac{1}{2}\left(x\sqrt{x^2-1} + \ln\left|x + \sqrt{x^2-1}\right|\right) + C$

55. (a) $u = a\sin\theta$ (b) $u = a\tan\theta$ (c) $u = a\sec\theta$

57. $\displaystyle A = 4\int_0^a \frac{b}{a}\sqrt{a^2-x^2}\,dx$

$$= \frac{4b}{a}\int_0^a \sqrt{a^2-x^2}\,dx$$

$$= \left[\frac{4b}{a}\left(\frac{1}{2}\right)\left(a^2\arcsin\frac{x}{a} + x\sqrt{a^2-x^2}\right)\right]_0^a$$

$$= \frac{2b}{a}\left(a^2\left(\frac{\pi}{2}\right)\right)$$

$$= \pi ab$$

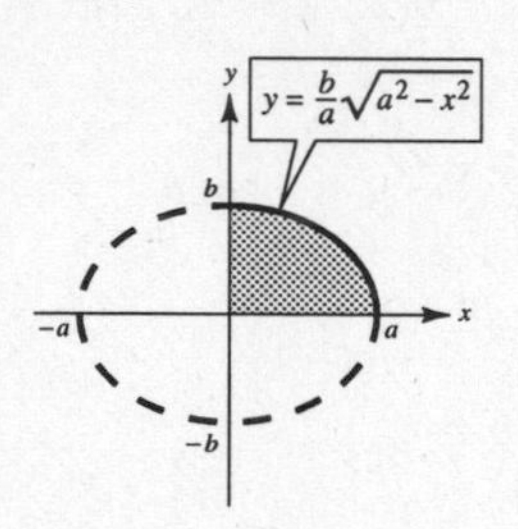

Note: See Theorem 7.2 for $\int\sqrt{a^2-x^2}\,dx$.

59. $x^2 + y^2 = a^2$

$$x = \pm\sqrt{a^2-y^2}$$

$$A = 2\int_h^a \sqrt{a^2-y^2}\,dy = \left[a^2\arcsin\left(\frac{y}{a}\right) + y\sqrt{a^2-y^2}\right]_h^a \qquad \text{(Theorem 7.2)}$$

$$= \left(a^2\frac{\pi}{2}\right) - \left(a^2\arcsin\left(\frac{h}{a}\right) + h\sqrt{a^2-h^2}\right)$$

$$= \frac{a^2\pi}{2} - a^2\arcsin\left(\frac{h}{a}\right) - h\sqrt{a^2-h^2}$$

61. Let $x - 3 = \sin\theta$, $dx = \cos\theta\,d\theta$, $\sqrt{1-(x-3)^2} = \cos\theta$.

Shell Method:

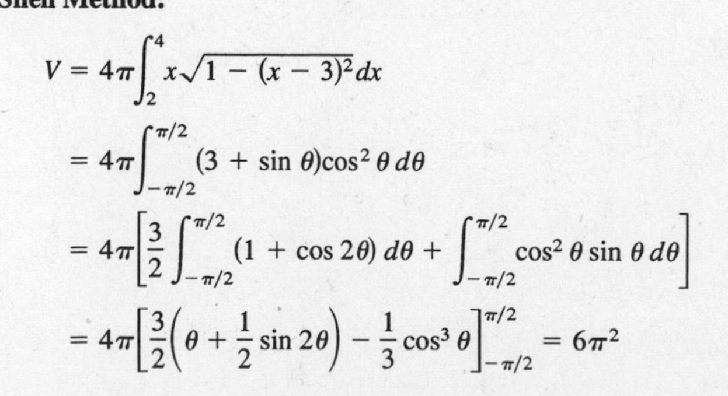

$$V = 4\pi\int_2^4 x\sqrt{1-(x-3)^2}\,dx$$

$$= 4\pi\int_{-\pi/2}^{\pi/2} (3+\sin\theta)\cos^2\theta\,d\theta$$

$$= 4\pi\left[\frac{3}{2}\int_{-\pi/2}^{\pi/2}(1+\cos 2\theta)\,d\theta + \int_{-\pi/2}^{\pi/2}\cos^2\theta\sin\theta\,d\theta\right]$$

$$= 4\pi\left[\frac{3}{2}\left(\theta + \frac{1}{2}\sin 2\theta\right) - \frac{1}{3}\cos^3\theta\right]_{-\pi/2}^{\pi/2} = 6\pi^2$$

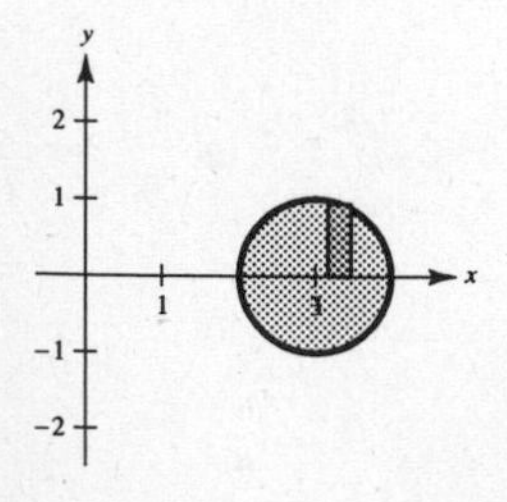

63. $y = \ln x,\ y' = \dfrac{1}{x},\ 1 + (y')^2 = 1 + \dfrac{1}{x^2} = \dfrac{x^2+1}{x^2}$

Let $x = \tan\theta,\ dx = \sec^2\theta\,d\theta,\ \sqrt{x^2+1} = \sec\theta.$

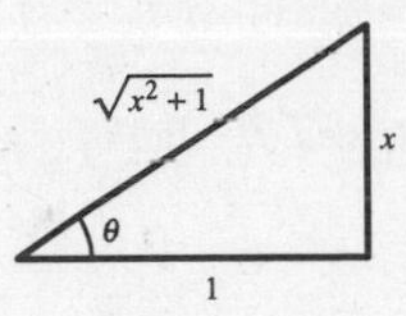

$$
\begin{aligned}
s &= \int_1^5 \sqrt{\frac{x^2+1}{x^2}}\,dx = \int_1^5 \frac{\sqrt{x^2+1}}{x}\,dx\\
&= \int_a^b \frac{\sec\theta}{\tan\theta}\sec^2\theta\,d\theta = \int_a^b \frac{\sec\theta}{\tan\theta}(1+\tan^2\theta)\,d\theta\\
&= \int_a^b (\csc\theta + \sec\theta\tan\theta)\,d\theta\\
&= \Big[-\ln|\csc\theta + \cot\theta| + \sec\theta\Big]_a^b\\
&= \left[-\ln\left|\frac{\sqrt{x^2+1}}{x} + \frac{1}{x}\right| + \sqrt{x^2+1}\right]_1^5\\
&= \left[-\ln\left(\frac{\sqrt{26}+1}{5}\right) + \sqrt{26}\right] - \left[-\ln(\sqrt{2}+1) + \sqrt{2}\right]\\
&= \ln\left[\frac{5(\sqrt{2}+1)}{\sqrt{26}+1}\right] + \sqrt{26} - \sqrt{2} \approx 4.367 \text{ or } \ln\left[\frac{\sqrt{26}-1}{5(\sqrt{2}-1)}\right] + \sqrt{26} - \sqrt{2}
\end{aligned}
$$

65. Length of one arch of sine curve: $y = \sin x,\ y' = \cos x$

$$L_1 = \int_0^{\pi} \sqrt{1+\cos^2 x}\,dx$$

Length of one arch of cosine curve: $y = \cos x,\ y' = -\sin x$

$$
\begin{aligned}
L_2 &= \int_{-\pi/2}^{\pi/2} \sqrt{1+\sin^2 x}\,dx\\
&= \int_{-\pi/2}^{\pi/2} \sqrt{1+\cos^2\left(x - \frac{\pi}{2}\right)}\,dx, \qquad u = x - \frac{\pi}{2},\ du = dx\\
&= \int_{-\pi}^{0} \sqrt{1+\cos^2 u}\,du\\
&= \int_0^{\pi} \sqrt{1+\cos^2 u}\,du = L_1
\end{aligned}
$$

67. (a)

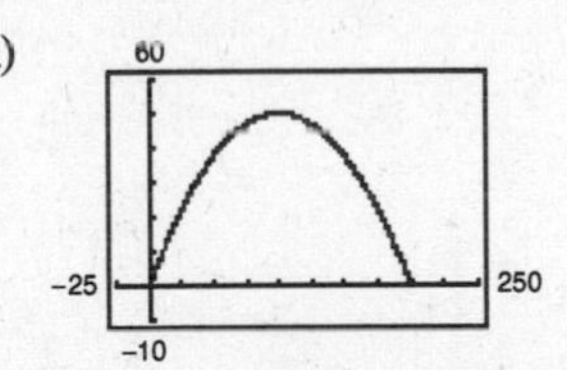

(b) $y = 0$ for $x = 200$ (range)

(c) $y = x - 0.005x^2,\ y' = 1 - 0.01x,\ 1 + (y')^2 = 1 + (1 - 0.01x)^2$

Let $u = 1 - 0.01x,\ du = -0.01\,dx,\ a = 1.$ (See Theorem 7.2.)

$$
\begin{aligned}
s &= \int_0^{200} \sqrt{1 + (1-0.01x)^2}\,dx = -100\int_0^{200} \sqrt{(1-0.01x)^2 + 1}\,(-0.01)\,dx\\
&= -50\left[(1-0.01x)\sqrt{(1-0.01x)^2+1} + \ln\left|(1-0.01x) + \sqrt{(1-0.01x)^2+1}\right|\right]_0^{200}\\
&= -50\left[\left(-\sqrt{2} + \ln\left|-1+\sqrt{2}\right|\right) - \left(\sqrt{2} + \ln\left|1+\sqrt{2}\right|\right)\right]\\
&= 100\sqrt{2} + 50\ln\left(\frac{\sqrt{2}+1}{\sqrt{2}-1}\right) \approx 229.559
\end{aligned}
$$

69. Let $x = 3\tan\theta$, $dx = 3\sec^2\theta\,d\theta$, $\sqrt{x^2+9} = 3\sec\theta$.

$$A = 2\int_0^4 \frac{3}{\sqrt{x^2+9}}\,dx = 6\int_0^4 \frac{dx}{\sqrt{x^2+9}} = 6\int_a^b \frac{3\sec^2\theta\,d\theta}{3\sec\theta}$$

$$= 6\int_a^b \sec\theta\,d\theta = \Big[6\ln|\sec\theta + \tan\theta|\Big]_a^b = \left[6\ln\left|\frac{\sqrt{x^2+9}+x}{3}\right|\right]_0^4 = 6\ln 3$$

$\bar{x} = 0$ (by symmetry)

$$\bar{y} = \frac{1}{2}\left(\frac{1}{A}\right)\int_{-4}^4 \left(\frac{3}{\sqrt{x^2+9}}\right)^2 dx$$

$$= \frac{9}{12\ln 3}\int_{-4}^4 \frac{1}{x^2+9}\,dx$$

$$= \frac{3}{4\ln 3}\left[\frac{1}{3}\arctan\frac{x}{3}\right]_{-4}^4$$

$$= \frac{2}{4\ln 3}\arctan\frac{4}{3} \approx 0.422$$

$$(\bar{x}, \bar{y}) = \left(0, \frac{1}{2\ln 3}\arctan\frac{4}{3}\right) \approx (0, 0.422)$$

71. $y = x^2$, $\quad y' = 2x$, $\quad 1 + (y') = 1 + 4x^2$

$2x = \tan\theta$, $dx = \frac{1}{2}\sec^2\theta\,d\theta$, $\sqrt{1+4x^2} = \sec\theta$

(For $\int\sec^5\theta\,d\theta$ and $\int\sec^3\theta\,d\theta$, see Exercise 90 in Section 7.3)

$$S = 2\pi\int_0^{\sqrt{2}} x^2\sqrt{1+4x^2}\,dx = 2\pi\int_a^b \left(\frac{\tan\theta}{2}\right)^2(\sec\theta)\left(\frac{1}{2}\sec^2\theta\right)d\theta$$

$$= \frac{\pi}{4}\int_a^b \sec^3\theta\tan^2\theta\,d\theta = \frac{\pi}{4}\left[\int_a^b \sec^5\theta\,d\theta - \int_a^b \sec^3\theta\,d\theta\right]$$

$$= \frac{\pi}{4}\left\{\frac{1}{4}\left[\sec^3\theta\tan\theta + \frac{3}{2}(\sec\theta\tan\theta + \ln|\sec\theta + \tan\theta|)\right] - \frac{1}{2}(\sec\theta\tan\theta + \ln|\sec\theta + \tan\theta|)\right\}\Bigg]_a^b$$

$$= \frac{\pi}{4}\left[\frac{1}{4}[(1+4x^2)^{3/2}(2x)] - \frac{1}{8}[(1+4x^2)^{1/2}(2x) + \ln\left|\sqrt{1+4x^2}+2x\right|]\right]_0^{\sqrt{2}}$$

$$= \frac{\pi}{4}\left[\frac{54\sqrt{2}}{4} - \frac{6\sqrt{2}}{8} - \frac{1}{8}\ln\left(3+2\sqrt{2}\right)\right]$$

$$= \frac{\pi}{4}\left(\frac{51\sqrt{2}}{4} - \frac{\ln\left(3+2\sqrt{2}\right)}{8}\right) = \frac{\pi}{32}\left[102\sqrt{2} - \ln\left(3+2\sqrt{2}\right)\right] \approx 13.989$$

73. (a) Area of representative rectangle: $2\sqrt{1-y^2}\,\Delta y$

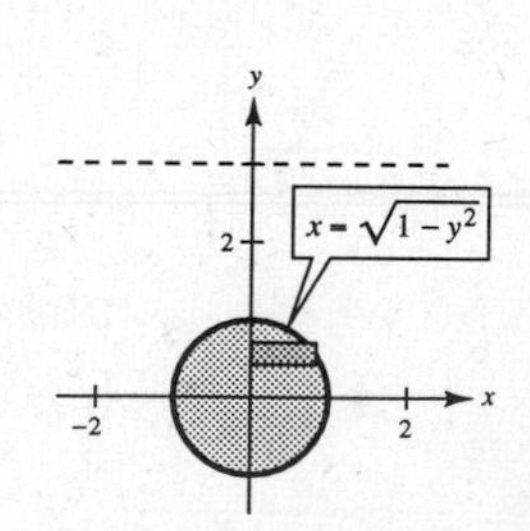

Pressure: $2(62.4)(3-y)\sqrt{1-y^2}\,\Delta y$

$$F = 124.8\int_{-1}^1 (3-y)\sqrt{1-y^2}\,dy$$

$$= 124.8\left[3\int_{-1}^1 \sqrt{1-y^2}\,dy - \int_{-1}^1 y\sqrt{1-y^2}\,dy\right]$$

$$= 124.8\left[\frac{3}{2}\left(\arcsin y + y\sqrt{1-y^2}\right) + \frac{1}{2}\left(\frac{2}{3}\right)(1-y^2)^{3/2}\right]_{-1}^1$$

$$= (62.4)3[\arcsin 1 - \arcsin(-1)] = 187.2\pi \text{ lb}$$

(b) $$F = 124.8\int_{-1}^1 (d-y)\sqrt{1-y^2}\,dy = 124.8d\int_{-1}^1 \sqrt{1-y^2}\,dy - 124.8\int_{-1}^1 y\sqrt{1-y^2}\,dy$$

$$= 124.8\left(\frac{d}{2}\right)\left[\arcsin y + y\sqrt{1-y^2}\right]_{-1}^1 - 124.8(0) = 62.4\pi d \text{ lb}$$

75. (a) $m = \dfrac{dy}{dx} = \dfrac{y - \left(y + \sqrt{144 - x^2}\right)}{x - 0}$

$= -\dfrac{\sqrt{144 - x^2}}{x}$

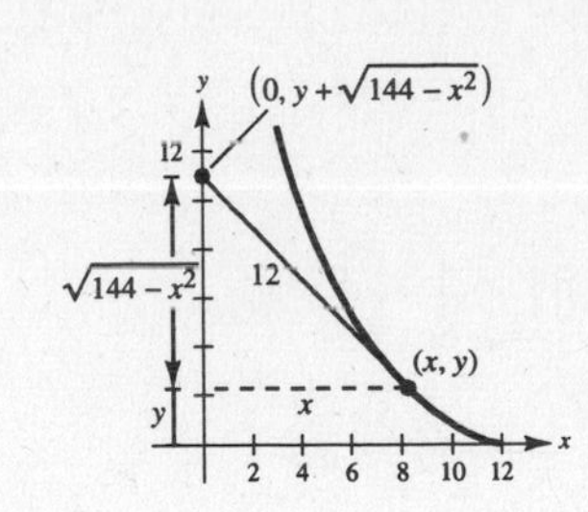

(b) $y = -\displaystyle\int \frac{\sqrt{144 - x^2}}{x}\,dx$

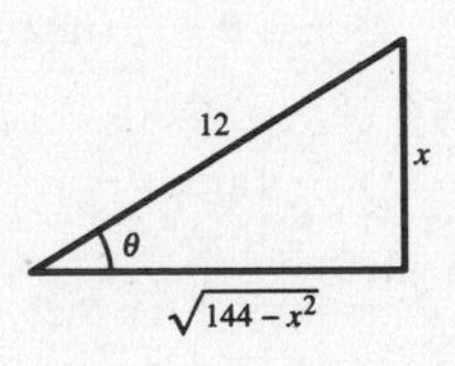

Let $x = 12\sin\theta$, $dx = 12\cos\theta\,d\theta$, $\sqrt{144 - x^2} = 12\cos\theta$.

$$y = -\int \frac{12\cos\theta}{12\sin\theta}\,12\cos\theta\,d\theta = -12\int \frac{1 - \sin^2\theta}{\sin\theta}\,d\theta$$

$$= -12\int(\csc\theta - \sin\theta)\,d\theta = -12\ln|\csc\theta - \cot\theta| - 12\cos\theta + C$$

$$= -12\ln\left|\frac{12}{x} - \frac{\sqrt{144 - x^2}}{x}\right| - 12\left(\frac{\sqrt{144 - x^2}}{12}\right) + C$$

$$= -12\ln\left|\frac{12 - \sqrt{144 - x^2}}{x}\right| - \sqrt{144 - x^2} + C$$

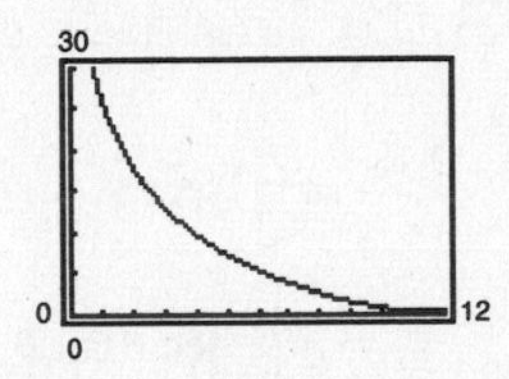

When $x = 12$, $y = 0 \Rightarrow C = 0$. Thus, $y = -12\ln\left(\dfrac{12 - \sqrt{144 - x^2}}{x}\right) - \sqrt{144 - x^2}$.

Note: $\dfrac{12 - \sqrt{144 - x^2}}{x} > 0$ for $0 < x \le 12$

(c) Vertical asymptote: $x = 0$

(d) $y + \sqrt{144 - x^2} = 12 \Rightarrow y = 12 - \sqrt{144 - x^2}$

Thus,

$$12 - \sqrt{144 - x^2} = -12\ln\left(\frac{12 - \sqrt{144 - x^2}}{x}\right) - \sqrt{144 - x^2}$$

$$1 = \ln\left(\frac{12 - \sqrt{144 - x^2}}{x}\right)$$

$$xe^{-1} = 12 - \sqrt{144 - x^2}$$

$$(xe^{-1} - 12)^2 = \left(-\sqrt{144 - x^2}\right)^2$$

$$x^2e^{-2} - 24xe^{-1} + 144 = 144 - x^2$$

$$x^2(e^{-2} + 1) - 24xe^{-1} = 0$$

$$x[x(e^{-2} + 1) - 24e^{-1}] = 0$$

$$x = 0 \text{ or } x = \frac{24e^{-1}}{e^{-2} + 1} \approx 7.77665.$$

Therefore,

$$s = \int_{7.77665}^{12} \sqrt{1 + \left(-\frac{\sqrt{144 - x^2}}{x}\right)^2}\,dx = \int_{7.77665}^{12} \sqrt{\frac{x^2 + (144 - x^2)}{x^2}}\,dx$$

$$= \int_{7.77665}^{12} \frac{12}{x}\,dx = \Big[12\ln|x|\Big]_{7.77665}^{12} = 12(\ln 12 - \ln 7.77665) \approx 5.2 \text{ meters.}$$

77. True

$$\int \frac{dx}{\sqrt{1-x^2}} = \int \frac{\cos\theta\,d\theta}{\cos\theta} = \int d\theta$$

79. False

$$\int_0^{\sqrt{3}} \frac{dx}{(\sqrt{1+x^2})^3} = \int_0^{\pi/3} \frac{\sec^2\theta\,d\theta}{\sec^3\theta} = \int_0^{\pi/3} \cos\theta\,d\theta$$

81. Let $u = a\sin\theta$, $du = a\cos\theta\,d\theta$, $\sqrt{a^2-u^2} = a\cos\theta$.

$$\int \sqrt{a^2-u^2}\,du = \int a^2\cos^2\theta\,d\theta = a^2\int \frac{1+\cos 2\theta}{2}\,d\theta$$

$$= \frac{a^2}{2}\left(\theta + \frac{1}{2}\sin 2\theta\right) + C = \frac{a^2}{2}(\theta + \sin\theta\cos\theta) + C$$

$$= \frac{a^2}{2}\left[\arcsin\frac{u}{a} + \left(\frac{u}{a}\right)\left(\frac{\sqrt{a^2+u^2}}{a}\right)\right] + C = \frac{1}{2}\left[a^2\arcsin\frac{u}{a} + u\sqrt{a^2-u^2}\right] + C$$

Let $u = a\sec\theta$, $du = a\sec\theta\tan\theta\,d\theta$, $\sqrt{u^2-a^2} = a\tan\theta$.

$$\int \sqrt{u^2-a^2}\,du = \int a\tan\theta(a\sec\theta\tan\theta)\,d\theta = a^2\int \tan^2\theta\sec\theta\,d\theta$$

$$= a^2\int(\sec^2\theta - 1)\sec\theta\,d\theta = a^2\int(\sec^3\theta - \sec\theta)\,d\theta$$

$$= a^2\left[\frac{1}{2}\sec\theta\tan\theta + \frac{1}{2}\int\sec\theta\,d\theta\right] - a^2\int\sec\theta\,d\theta = a^2\left[\frac{1}{2}\sec\theta\tan\theta - \frac{1}{2}\ln|\sec\theta + \tan\theta|\right]$$

$$= \frac{a^2}{2}\left[\frac{u}{a}\cdot\frac{\sqrt{u^2-a^2}}{a} - \ln\left|\frac{u}{a} + \frac{\sqrt{u^2-a^2}}{a}\right|\right] + C_1$$

$$= \frac{1}{2}\left[u\sqrt{u^2-a^2} - a^2\ln|u + \sqrt{u^2-a^2}|\right] + C$$

Let $u = a\tan\theta$, $du = a\sec^2\theta\,d\theta$, $\sqrt{u^2+a^2} = a\sec\theta$.

$$\int \sqrt{u^2+a^2}\,du = \int (a\sec\theta)(a\sec^2\theta)\,d\theta$$

$$= a^2\int\sec^3\theta\,d\theta = a^2\left[\frac{1}{2}\sec\theta\tan\theta + \frac{1}{2}\ln|\sec\theta + \tan\theta|\right] + C_1$$

$$= \frac{a^2}{2}\left[\frac{\sqrt{u^2+a^2}}{a}\cdot\frac{u}{a} + \ln\left|\frac{\sqrt{u^2+a^2}}{a} + \frac{u}{a}\right|\right] + C_1 = \frac{1}{2}\left[u\sqrt{u^2+a^2} + a^2\ln|u + \sqrt{u^2+a^2}|\right] + C$$

Section 7.5 Partial Fractions

1. $\dfrac{5}{x^2-10x} = \dfrac{5}{x(x-10)} = \dfrac{A}{x} + \dfrac{B}{x-10}$

3. $\dfrac{2x-3}{x^3+10x} = \dfrac{2x-3}{x(x^2+10)} = \dfrac{A}{x} + \dfrac{Bx+C}{x^2+10}$

5. $\dfrac{16x}{x^3-10x^2} = \dfrac{16x}{x^2(x-10)} = \dfrac{A}{x} + \dfrac{B}{x^2} + \dfrac{C}{x-10}$

7. $\dfrac{1}{x^2-1} = \dfrac{1}{(x+1)(x-1)} = \dfrac{A}{x+1} + \dfrac{B}{x-1}$

$$1 = A(x-1) + B(x+1)$$

When $x = -1$, $1 = -2A$, $A = -\frac{1}{2}$.

When $x = 1$, $1 = 2B$, $B = \frac{1}{2}$.

$$\int \frac{1}{x^2-1}\,dx = -\frac{1}{2}\int\frac{1}{x+1}\,dx + \frac{1}{2}\int\frac{1}{x-1}\,dx$$

$$= -\frac{1}{2}\ln|x+1| + \frac{1}{2}\ln|x-1| + C$$

$$= \frac{1}{2}\ln\left|\frac{x-1}{x+1}\right| + C$$

9. $\dfrac{3}{x^2+x-2} = \dfrac{3}{(x-1)(x+2)} = \dfrac{A}{x-1} + \dfrac{B}{x+2}$

$$3 = A(x+2) + B(x-1)$$

When $x = 1$, $3 = 3A$, $A = 1$.
When $x = -2$, $3 = -3B$, $B = -1$.

$$\int \frac{3}{x^2+x-2}\,dx = \int \frac{1}{x-1}\,dx - \int \frac{1}{x+2}\,dx$$
$$= \ln|x-1| - \ln|x+2| + C$$
$$= \ln\left|\frac{x-1}{x+2}\right| + C$$

11. $\dfrac{5-x}{2x^2+x-1} = \dfrac{5-x}{(2x-1)(x+1)} = \dfrac{A}{2x-1} + \dfrac{B}{x+1}$

$$5 - x = A(x+1) + B(2x-1)$$

When $x = \frac{1}{2}$, $\frac{9}{2} = \frac{3}{2}A$, $A = 3$.

When $x = -1$, $6 = -3B$, $B = -2$.

$$\int \frac{5-x}{2x^2+x-1}\,dx = 3\int \frac{1}{2x-1}\,dx - 2\int \frac{1}{x+1}\,dx$$
$$= \frac{3}{2}\ln|2x-1| - 2\ln|x+1| + C$$

13. $\dfrac{x^2+12x+12}{x(x+2)(x-2)} = \dfrac{A}{x} + \dfrac{B}{x+2} + \dfrac{C}{x-2}$

$$x^2 + 12x + 12 = A(x+2)(x-2) + Bx(x-2) + Cx(x+2)$$

When $x = 0$, $12 = -4A$, $A = -3$. When $x = -2$, $-8 = 8B$, $B = -1$. When $x = 2$, $40 = 8C$, $C = 5$.

$$\int \frac{x^2+12x+12}{x^3-4x}\,dx = 5\int \frac{1}{x-2}\,dx - \int \frac{1}{x+2}\,dx - 3\int \frac{1}{x}\,dx$$
$$= 5\ln|x-2| - \ln|x+2| - 3\ln|x| + C$$

15. $\dfrac{2x^3-4x^2-15x+5}{x^2-2x-8} = 2x + \dfrac{x+5}{(x-4)(x+2)} = 2x + \dfrac{A}{x-4} + \dfrac{B}{x+2}$

$$x + 5 = A(x+2) + B(x-4)$$

When $x = 4$, $9 = 6A$, $A = \frac{3}{2}$. When $x = -2$, $3 = -6B$, $B = -\frac{1}{2}$.

$$\int \frac{2x^3-4x^2-15x+5}{x^2-2x-8}\,dx = \int \left[2x + \frac{3/2}{x-4} - \frac{1/2}{x+2}\right] dx$$
$$= x^2 + \frac{3}{2}\ln|x-4| - \frac{1}{2}\ln|x+2| + C$$

17. $\dfrac{4x^2+2x-1}{x^2(x+1)} = \dfrac{A}{x} + \dfrac{B}{x^2} + \dfrac{C}{x+1}$

$$4x^2 + 2x - 1 = Ax(x+1) + B(x+1) + Cx^2$$

When $x = 0$, $B = -1$. When $x = -1$, $C = 1$. When $x = 1$, $A = 3$.

$$\int \frac{4x^2+2x-1}{x^3+x^2}\,dx = \int \left[\frac{3}{x} - \frac{1}{x^2} + \frac{1}{x+1}\right] dx = 3\ln|x| + \frac{1}{x} + \ln|x+1| + C$$
$$= \frac{1}{x} + \ln|x^4 + x^3| + C$$

19. $\dfrac{x^2+3x-4}{x^3-4x^2+4x} = \dfrac{x^2+3x-4}{x(x-2)^2} = \dfrac{A}{x} + \dfrac{B}{(x-2)} + \dfrac{C}{(x-2)^2}$

$$x^2 + 3x - 4 = A(x-2)^2 + Bx(x-2) + Cx$$

When $x = 0$, $-4 = 4A \implies A = -1$. When $x = 2$, $6 = 2C \implies C = 3$. When $x = 1$, $0 = -1 - B + 3 \implies B = 2$.

$$\int \frac{x^2+3x-4}{x^3-4x^2+4x}\,dx = \int \frac{-1}{x}\,dx + \int \frac{2}{(x-2)}\,dx + \int \frac{3}{(x-2)^2}\,dx$$
$$= -\ln|x| + 2\ln|x-2| - \frac{3}{(x-2)} + C$$

21. $\dfrac{x^2 - 1}{x(x^2 + 1)} = \dfrac{A}{x} + \dfrac{Bx + C}{x^2 + 1}$

$$x^2 - 1 = A(x^2 + 1) + (Bx + C)x$$

When $x = 0, A = -1$. When $x = 1, 0 = -2 + B + C$. When $x = -1, 0 = -2 + B - C$. Solving these equations we have $A = -1, B = 2, C = 0$.

$$\int \frac{x^2 - 1}{x^3 + x}\,dx = -\int \frac{1}{x}\,dx + \int \frac{2x}{x^2 + 1}\,dx$$

$$= \ln|x^2 + 1| - \ln|x| + C$$

$$= \ln\left|\frac{x^2 + 1}{x}\right| + C$$

23. $\dfrac{x^2}{x^4 - 2x^2 - 8} = \dfrac{A}{x - 2} + \dfrac{B}{x + 2} + \dfrac{Cx + D}{x^2 + 2}$

$$x^2 = A(x + 2)(x^2 + 2) + B(x - 2)(x^2 + 2) + (Cx + D)(x + 2)(x - 2)$$

When $x = 2, 4 = 24A$. When $x = -2, 4 = -24B$. When $x = 0, 0 = 4A - 4B - 4D$, and when $x = 1$, $1 = 9A - 3B - 3C - 3D$. Solving these equations we have $A = \frac{1}{6}, B = -\frac{1}{6}, C = 0, D = \frac{1}{3}$.

$$\int \frac{x^2}{x^4 - 2x^2 - 8}\,dx = \frac{1}{6}\left[\int \frac{1}{x - 2}\,dx - \int \frac{1}{x + 2}\,dx + 2\int \frac{1}{x^2 + 2}\,dx\right]$$

$$= \frac{1}{6}\left[\ln\left|\frac{x - 2}{x + 2}\right| + \sqrt{2}\arctan\frac{x}{\sqrt{2}}\right] + C$$

25. $\dfrac{x}{(2x - 1)(2x + 1)(4x^2 + 1)} = \dfrac{A}{2x - 1} + \dfrac{B}{2x + 1} + \dfrac{Cx + D}{4x^2 + 1}$

$$x = A(2x + 1)(4x^2 + 1) + B(2x - 1)(4x^2 + 1) + (Cx + D)(2x - 1)(2x + 1)$$

When $x = \frac{1}{2}, \frac{1}{2} = 4A$. When $x = -\frac{1}{2}, -\frac{1}{2} = -4B$. When $x = 0, 0 = A - B - D$, and when $x = 1$, $1 = 15A + 5B + 3C + 3D$. Solving these equations we have $A = \frac{1}{8}, B = \frac{1}{8}, C = -\frac{1}{2}, D = 0$.

$$\int \frac{x}{16x^4 - 1}\,dx = \frac{1}{8}\left[\int \frac{1}{2x - 1}\,dx + \int \frac{1}{2x + 1}\,dx - 4\int \frac{x}{4x^2 + 1}\,dx\right]$$

$$= \frac{1}{16}\ln\left|\frac{4x^2 - 1}{4x^2 + 1}\right| + C$$

27. $\dfrac{x^2 + 5}{(x + 1)(x^2 - 2x + 3)} = \dfrac{A}{x + 1} + \dfrac{Bx + C}{x^2 - 2x + 3}$

$$x^2 + 5 = A(x^2 - 2x + 3) + (Bx + C)(x + 1)$$

$$= (A + B)x^2 + (-2A + B + C)x + (3A + C)$$

When $x = -1, A = 1$. By equating coefficients of like terms, we have $A + B = 1, -2A + B + C = 0$, $3A + C = 5$. Solving these equations we have $A = 1, B = 0, C = 2$.

$$\int \frac{x^2 + 5}{x^3 - x^2 + x + 3}\,dx = \int \frac{1}{x + 1}\,dx + 2\int \frac{1}{(x - 1)^2 + 2}\,dx$$

$$= \ln|x + 1| + \sqrt{2}\arctan\left(\frac{x - 1}{\sqrt{2}}\right) + C$$

29. $\dfrac{3}{(2x+1)(x+2)} = \dfrac{A}{2x+1} + \dfrac{B}{x+2}$

$$3 = A(x+2) + B(2x+1)$$

When $x = -\frac{1}{2}, A = 2$. When $x = -2, B = -1$.

$$\begin{aligned}\int_0^1 \frac{3}{2x^2+5x+2}\,dx &= \int_0^1 \frac{2}{2x+1}\,dx - \int_0^1 \frac{1}{x+2}\,dx \\ &= \Big[\ln|2x+1| - \ln|x+2|\Big]_0^1 \\ &= \ln 2\end{aligned}$$

31. $\dfrac{x+1}{x(x^2+1)} = \dfrac{A}{x} + \dfrac{Bx+C}{x^2+1}$

$$x + 1 = A(x^2+1) + (Bx+C)x$$

When $x = 0, A = 1$. When $x = 1, 2 = 2A + B + C$. When $x = -1, 0 = 2A + B - C$. Solving these equations we have $A = 1, B = -1, C = 1$.

$$\begin{aligned}\int_1^2 \frac{x+1}{x(x^2+1)}\,dx &= \int_1^2 \frac{1}{x}\,dx - \int_1^2 \frac{x}{x^2+1}\,dx + \int_1^2 \frac{1}{x^2+1}\,dx \\ &= \left[\ln|x| - \frac{1}{2}\ln(x^2+1) + \arctan x\right]_1^2 \\ &= \frac{1}{2}\ln\frac{8}{5} - \frac{\pi}{4} + \arctan 2 \\ &\approx 0.557\end{aligned}$$

33. $\displaystyle\int \frac{3x\,dx}{x^2-6x+9} = 3\ln|x-3| - \frac{9}{x-3} + C$

$(4, 0)$: $3\ln|4-3| - \dfrac{9}{4-3} + C = 0 \Longrightarrow C = 9$

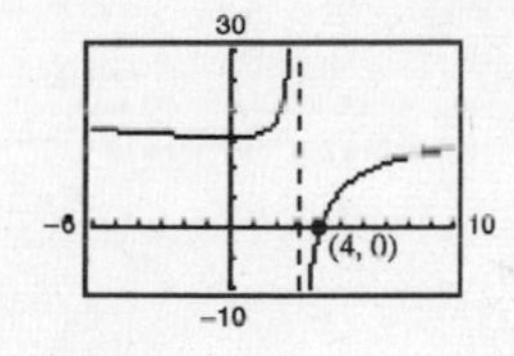

35. $\displaystyle\int \frac{x^2+x+2}{(x^2+2)^2}\,dx = \frac{\sqrt{2}}{2}\arctan\frac{x}{\sqrt{2}} - \frac{1}{2(x^2+2)} + C$

$(0, 1)$: $0 - \dfrac{1}{4} + C = 1 \Longrightarrow C = \dfrac{5}{4}$

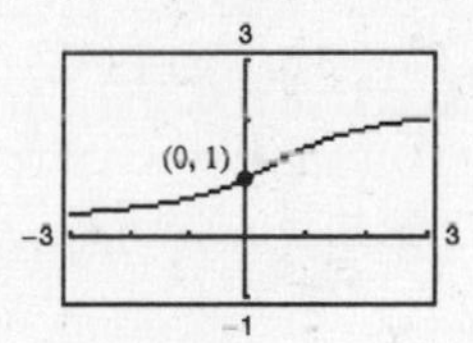

37. $\displaystyle\int \frac{2x^2-2x+3}{x^3-x^2-x-2}\,dx = \ln|x-2| + \frac{1}{2}\ln|x^2+x+1| - \sqrt{3}\arctan\left(\frac{2x+1}{\sqrt{3}}\right) + C$

$(3, 10)$: $0 + \dfrac{1}{2}\ln 13 - \sqrt{3}\arctan\dfrac{7}{\sqrt{3}} + C = 10 \Longrightarrow C = 10 - \dfrac{1}{2}\ln 13 + \sqrt{3}\arctan\dfrac{7}{\sqrt{3}}$

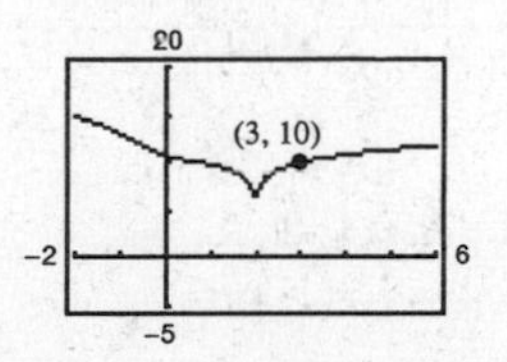

39. $\displaystyle\int \frac{1}{x^2-4}\,dx = \frac{1}{4}\ln\left|\frac{x-2}{x+2}\right| + C$

$(6, 4)$: $\dfrac{1}{4}\ln\left|\dfrac{4}{8}\right| + C = 4 \Longrightarrow C = 4 - \dfrac{1}{4}\ln\dfrac{1}{2} = 4 + \dfrac{1}{4}\ln 2$

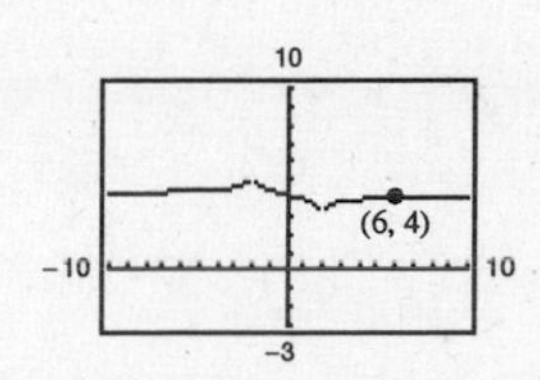

41. Let $u = \cos x$ $du = -\sin x\,dx$.

$$\frac{1}{u(u-1)} = \frac{A}{u} + \frac{B}{u-1}$$

$$1 = A(u-1) + Bu$$

When $u = 0$, $A = -1$. When $u = 1$, $B = 1$, $u = \cos x$, $du = -\sin x\,dx$.

$$\int \frac{\sin x}{\cos x(\cos x - 1)}\,dx = -\int \frac{1}{u(u-1)}\,du$$

$$= \int \frac{1}{u}\,du - \int \frac{1}{u-1}\,du$$

$$= \ln|u| - \ln|u-1| + C$$

$$= \ln\left|\frac{u}{u-1}\right| + C$$

$$= \ln\left|\frac{\cos x}{\cos x - 1}\right| + C$$

43. $$\int \frac{3\cos x}{\sin^2 x + \sin x - 2}\,dx = 3\int \frac{1}{u^2 + u - 2}\,du$$

$$= \ln\left|\frac{u-1}{u+2}\right| + C$$

$$= \ln\left|\frac{-1 + \sin x}{2 + \sin x}\right| + C$$

(From Exercise 9 with $u = \sin x$, $du = \cos x\,dx$)

45. Let $u = e^x$, $du = e^x\,dx$.

$$\frac{1}{(u-1)(u+4)} = \frac{A}{u-1} + \frac{B}{u+4}$$

$$1 = A(u+4) + B(u-1)$$

When $u = 1$, $A = \frac{1}{5}$. When $u = -4$, $B = -\frac{1}{5}$, $u = e^x$, $du = e^x\,dx$.

$$\int \frac{e^x}{(e^x - 1)(e^x + 4)}\,dx = \int \frac{1}{(u-1)(u+4)}\,du$$

$$= \frac{1}{5}\left(\int \frac{1}{u-1}\,du - \int \frac{1}{u+4}\,du\right)$$

$$= \frac{1}{5}\ln\left|\frac{u-1}{u+4}\right| + C$$

$$= \frac{1}{5}\ln\left|\frac{e^x - 1}{e^x + 4}\right| + C$$

47. $$\frac{1}{x(a+bx)} = \frac{A}{x} + \frac{B}{a+bx}$$

$$1 = A(a+bx) + Bx$$

When $x = 0$, $1 = aA \Rightarrow A = 1/a$.
When $x = -a/b$, $1 = -(a/b)B \Rightarrow B = -b/a$.

$$\int \frac{1}{x(a+bx)}\,dx = \frac{1}{a}\int\left(\frac{1}{x} - \frac{b}{a+bx}\right)dx$$

$$= \frac{1}{a}(\ln|x| - \ln|a+bx|) + C$$

$$= \frac{1}{a}\ln\left|\frac{x}{a+bx}\right| + C$$

49. $$\frac{x}{(a+bx)^2} = \frac{A}{a+bx} + \frac{B}{(a+bx)^2}$$

$$x = A(a+bx) + B$$

When $x = -a/b$, $B = -a/b$.
When $x = 0$, $0 = aA + B \Rightarrow A = 1/b$.

$$\int \frac{x}{(a+bx)^2}\,dx = \int\left(\frac{1/b}{a+bx} + \frac{-a/b}{(a+bx)^2}\right)dx$$

$$= \frac{1}{b}\int \frac{1}{a+bx}\,dx - \frac{a}{b}\int \frac{1}{(a+bx)^2}\,dx$$

$$= \frac{1}{b^2}\ln|a+bx| + \frac{a}{b^2}\left(\frac{1}{a+bx}\right) + C$$

$$= \frac{1}{b^2}\left(\frac{a}{a+bx} + \ln|a+bx|\right) + C$$

51. $\dfrac{dy}{dx} = \dfrac{6}{4 - x^2}$, $y(0) = 3$

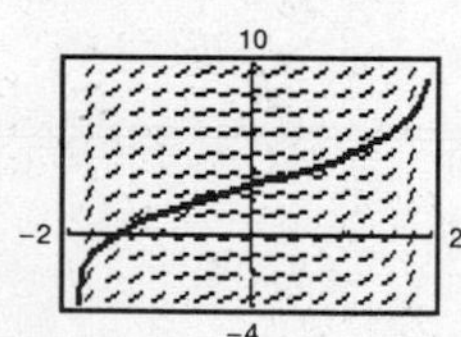

53. Dividing x^3 by $x - 5$.

55. (a) Substitution: $u = x^2 + 2x - 8$

(b) Partial fractions

(c) Trigonometric substitution (tan) or inverse tangent rule

57. Average Cost $= \dfrac{1}{80 - 75}\displaystyle\int_{75}^{80} \frac{124p}{(10 + p)(100 - p)}\, dp$

$$= \frac{1}{5}\int_{75}^{80}\left(\frac{-124}{(10 + p)11} + \frac{1240}{(100 - p)11}\right) dp$$

$$= \frac{1}{5}\left[\frac{-124}{11}\ln(10 + p) - \frac{1240}{11}\ln(100 - p)\right]_{75}^{80}$$

$$\approx \frac{1}{5}(24.51) = 4.9$$

Approximately $490,000.

59. $A = \displaystyle\int_1^3 \frac{10}{x(x^2 + 1)}\, dx \approx 3$

Matches (c)

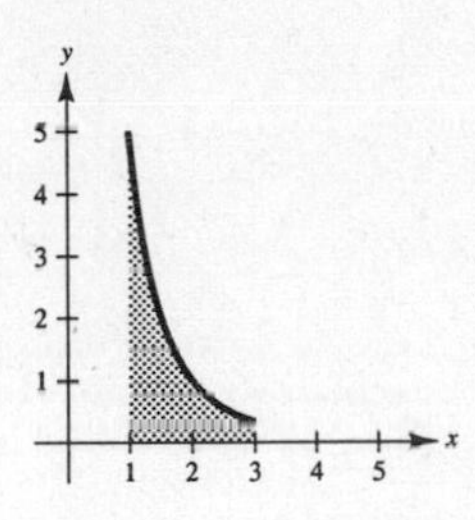

61.
$$\frac{1}{(x + 1)(n - x)} = \frac{A}{x + 1} + \frac{B}{n - x}, A = B = \frac{1}{n + 1}$$

$$\frac{1}{n + 1}\int\left(\frac{1}{x + 1} + \frac{1}{n - x}\right) dx = kt + C$$

$$\frac{1}{n + 1}\ln\left|\frac{x + 1}{n - x}\right| = kt + C$$

When $t = 0, x = 0, C = \dfrac{1}{n + 1}\ln\dfrac{1}{n}$.

$$\frac{1}{n + 1}\ln\left|\frac{x + 1}{n - x}\right| = kt + \frac{1}{n + 1}\ln\frac{1}{n}$$

$$\frac{1}{n + 1}\left[\ln\left|\frac{x + 1}{n - x}\right| - \ln\frac{1}{n}\right] = kt$$

$$\ln\frac{nx + n}{n - x} = (n + 1)kt$$

$$\frac{nx + n}{n - x} = e^{(n + 1)kt}$$

$$x = \frac{n[e^{(n + 1)kt} - 1]}{n + e^{(n + 1)kt}}$$

Note: $\displaystyle\lim_{t \to \infty} x = n$

63. $\dfrac{x}{1+x^4} = \dfrac{Ax+B}{x^2+\sqrt{2}x+1} + \dfrac{Cx+D}{x^2-\sqrt{2}x+1}$

$$x = (Ax+B)(x^2-\sqrt{2}x+1) + (Cx+D)(x^2+\sqrt{2}x+1)$$
$$= (A+C)x^3 + (B+D-\sqrt{2}A+\sqrt{2}C)x^2 + (A+C-\sqrt{2}B+\sqrt{2}D)x + (B+D)$$

$0 = A + C \Rightarrow C = -A$

$0 = B + D - \sqrt{2}A + \sqrt{2}C$ $\quad -2\sqrt{2}A = 0 \Rightarrow A = 0$ and $C = 0$

$1 = A + C - \sqrt{2}B + \sqrt{2}D$ $\quad -2\sqrt{2}B = 1 \Rightarrow B = -\dfrac{\sqrt{2}}{4}$ and $D = \dfrac{\sqrt{2}}{4}$

$0 = B + D \Rightarrow D = -B$

Thus,

$$\int_0^1 \frac{x}{1+x^4}\,dx = \int_0^1 \left[\frac{-\sqrt{2}/4}{x^2+\sqrt{2}x+1} + \frac{\sqrt{2}/4}{x^2-\sqrt{2}x+1}\right]dx$$
$$= \frac{\sqrt{2}}{4}\int_0^1 \left[\frac{-1}{\left[x+(\sqrt{2}/2)\right]^2+(1/2)} + \frac{1}{\left[x-(\sqrt{2}/2)\right]^2+(1/2)}\right]dx$$
$$= \frac{\sqrt{2}}{4}\cdot\frac{1}{1/\sqrt{2}}\left[-\arctan\left(\frac{x+(\sqrt{2}/2)}{1/\sqrt{2}}\right) + \arctan\left(\frac{x-(\sqrt{2}/2)}{1/\sqrt{2}}\right)\right]_0^1$$
$$= \frac{1}{2}\left[-\arctan(\sqrt{2}x+1) + \arctan(\sqrt{2}x-1)\right]_0^1$$
$$= \frac{1}{2}\left[\left(-\arctan(\sqrt{2}+1) + \arctan(\sqrt{2}-1)\right) - \left(-\arctan 1 + \arctan(-1)\right)\right]$$
$$= \frac{1}{2}\left[\arctan(\sqrt{2}-1) - \arctan(\sqrt{2}+1) + \frac{\pi}{4} + \frac{\pi}{4}\right].$$

Since $\arctan x - \arctan y = \arctan[(x-y)/(1+xy)]$, we have:

$$\int_0^1 \frac{x}{1+x^4}\,dx = \frac{1}{2}\left[\arctan\left(\frac{(\sqrt{2}-1)-(\sqrt{2}+1)}{1+(\sqrt{2}-1)(\sqrt{2}+1)}\right) + \frac{\pi}{2}\right] = \frac{1}{2}\left[\arctan\left(\frac{-2}{2}\right) + \frac{\pi}{2}\right] = \frac{1}{2}\left[-\frac{\pi}{4} + \frac{\pi}{2}\right] = \frac{\pi}{8}$$

Section 7.6 Integration by Tables and Other Integration Techniques

1. By Formula 6: $\displaystyle\int \frac{x^2}{1+x}\,dx = -\frac{x}{2}(2-x) + \ln|1+x| + C$

3. By Formula 26: $\displaystyle\int e^x\sqrt{1+e^{2x}}\,dx = \frac{1}{2}\left[e^x\sqrt{e^{2x}+1} + \ln\left|e^x + \sqrt{e^{2x}+1}\right|\right] + C$

$u = e^x,\ du = e^x\,dx$

5. By Formula 44: $\displaystyle\int \frac{1}{x^2\sqrt{1-x^2}}\,dx = -\frac{\sqrt{1-x^2}}{x} + C$

7. By Formulas 50 and 48: $\int \sin^4(2x)\,dx = \frac{1}{2}\int \sin^4(2x)(2)\,dx$

$$= \frac{1}{2}\left[\frac{-\sin^3(2x)\cos(2x)}{4} + \frac{3}{4}\int \sin^2(2x)(2)\,dx\right]$$

$$= \frac{1}{2}\left[\frac{-\sin^3(2x)\cos(2x)}{4} + \frac{3}{8}(2x - \sin 2x \cos 2x)\right] + C$$

$$= \frac{1}{16}(6x - 3\sin 2x \cos 2x - 2\sin^3 2x \cos 2x) + C$$

9. By Formula 57: $\int \frac{1}{\sqrt{x}(1 - \cos\sqrt{x})}\,dx = 2\int \frac{1}{1 - \cos\sqrt{x}}\left(\frac{1}{2\sqrt{x}}\right)dx$

$$= -2\left(\cot\sqrt{x} + \csc\sqrt{x}\right) + C$$

$u = \sqrt{x},\ du = \frac{1}{2\sqrt{x}}\,dx$

11. By Formula 84:

$$\int \frac{1}{1 + e^{2x}}\,dx = x - \frac{1}{2}\ln(1 + e^{2x}) + C$$

13. By Formula 89:

$$\int x^3 \ln x\,dx = \frac{x^4}{16}(4\ln|x| - 1) + C$$

15. (a) By Formulas 83 and 82: $\int x^2e^x\,dx = x^2e^x - 2\int xe^x\,dx$

$$= x^2e^x - 2[(x - 1)e^x + C_1]$$

$$= x^2e^x - 2xe^x + 2e^x + C$$

(b) Integration by parts: $u = x^2,\ du = 2x\,dx,\ dv = e^x\,dx,\ v = e^x$

$$\int x^2e^x\,dx = x^2e^x - \int 2xe^x\,dx$$

Parts again: $u = 2x,\ du = 2\,dx,\ dv = e^x\,dx,\ v = e^x$

$$\int x^2e^x\,dx = x^2e^x - \left[2xe^x - \int 2e^x\,dx\right] = x^2e^x - 2xe^x + 2e^x + C$$

17. (a) By Formula: 12, $a = b = 1,\ u = x$, and

$$\int \frac{1}{x^2(x + 1)}\,dx = \frac{-1}{1}\left(\frac{1}{x} + \frac{1}{1}\ln\left|\frac{x}{1 + x}\right|\right) + C$$

$$= \frac{-1}{x} - \ln\left|\frac{x}{1 + x}\right| + C$$

$$= \frac{-1}{x} + \ln\left|\frac{x + 1}{x}\right| + C$$

(b) Partial fractions:

$$\frac{1}{x^2(x + 1)} = \frac{A}{x} + \frac{B}{x^2} + \frac{C}{x + 1}$$

$$1 = Ax(x + 1) + B(x + 1) + Cx^2$$

$x = 0$: $1 = B$

$x = -1$: $1 = C$

$x = 1$: $1 = 2A + 2 + 1 \Rightarrow A = -1$

$$\int \frac{1}{x^2(x + 1)}\,dx = \int\left[\frac{-1}{x} + \frac{1}{x^2} + \frac{1}{x + 1}\right]dx$$

$$= -\ln|x| - \frac{1}{x} + \ln|x + 1| + C$$

$$= -\frac{1}{x} - \ln\left|\frac{x}{x + 1}\right| + C$$

19. By Formula 81: $\int xe^{x^2} = \frac{1}{2}e^{x^2} + C$

21. By Formula 79: $\int x \operatorname{arcsec}(x^2 + 1)\, dx = \frac{1}{2}\int \operatorname{arcsec}(x^2 + 1)(2x)\, dx$

$$= \frac{1}{2}\left[(x^2 + 1)\operatorname{arcsec}(x^2 + 1) - \ln\left((x^2 + 1) + \sqrt{x^4 + 2x^2}\right)\right] + C$$

$u = x^2 + 1,\ du = 2x\, dx$

23. By Formula 89: $\int x^2 \ln x\, dx = \frac{x^3}{9}\left(-1 + 3\ln|x|\right) + C$

25. By Formula 35: $\int \frac{1}{x^2\sqrt{x^2 - 4}}\, dx = \frac{\sqrt{x^2 - 4}}{4x} + C$

27. By Formula 4: $\int \frac{2x}{(1 - 3x)^2}\, dx = 2\int \frac{x}{(1 - 3x)^2}\, dx = \frac{2}{9}\left(\ln|1 - 3x| + \frac{1}{1 - 3x}\right) + C$

29. By Formula 76:

$$\int e^x \arccos e^x\, dx = e^x \arccos e^x - \sqrt{1 - e^{2x}} + C$$

$u = e^x,\ du = e^x\, dx$

31. By Formula 73:

$$\int \frac{x}{1 - \sec x^2}\, dx = \frac{1}{2}\int \frac{2x}{1 - \sec x^2}\, dx$$

$$= \frac{1}{2}(x^2 + \cot x^2 + \csc x^2) + C$$

33. By Formula 23: $\int \frac{\cos x}{1 + \sin^2 x}\, dx = \arctan(\sin x) + C$

$u = \sin x,\ du = \cos x\, dx$

35. By Formula 14: $\int \frac{\cos\theta}{3 + 2\sin\theta + \sin^2\theta}\, d\theta = \frac{\sqrt{2}}{2}\arctan\left(\frac{1 + \sin\theta}{\sqrt{2}}\right) + C$

$u = \sin\theta,\ du = \cos\theta\, d\theta$

37. By Formula 35: $\int \frac{1}{x^2\sqrt{2 + 9x^2}}\, dx = 3\int \frac{3}{(3x)^2\sqrt{\left(\sqrt{2}\right)^2 + (3x)^2}}\, dx$

$$= -\frac{3\sqrt{2 + 9x^2}}{6x} + C$$

$$= -\frac{\sqrt{2 + 9x^2}}{2x} + C$$

39. By Formulas 54 and 55:

$$\int t^3 \cos t\, dt = t^3 \sin t - 3\int t^2 \sin t\, dt$$

$$= t^3 \sin t - 3\left[-t^2 \cos t + 2\int t \cos t\, dt\right]$$

$$= t^3 \sin t + 3t^2 \cos t - 6\left[t \sin t - \int \sin t\, dt\right]$$

$$= t^3 \sin t + 3t^2 \cos t - 6t \sin t - 6\cos t + C$$

41. By Formula 3: $\displaystyle\int \frac{\ln x}{x(3 + 2\ln x)}\,dx = \frac{1}{4}\left(2\ln|x| - 3\ln|3 + 2\ln|x||\right) + C$

$u = \ln x,\ du = \frac{1}{x}\,dx$

43. By Formulas 1, 25, and 33:
$$\int \frac{x}{(x^2 - 6x + 10)^2}\,dx = \frac{1}{2}\int \frac{2x - 6 + 6}{(x^2 - 6x + 10)^2}\,dx$$
$$= \frac{1}{2}\int (x^2 - 6x + 10)^{-2}(2x - 6)\,dx + 3\int \frac{1}{[(x - 3)^2 + 1]^2}\,dx$$
$$= -\frac{1}{2(x^2 - 6x + 10)} + \frac{3}{2}\left[\frac{x - 3}{x^2 - 6x + 10} + \arctan(x - 3)\right] + C$$
$$= \frac{3x - 10}{2(x^2 - 6x + 10)} + \frac{3}{2}\arctan(x - 3) + C$$

45. By Formula 31:
$$\int \frac{x}{\sqrt{x^4 - 6x^2 + 5}}\,dx = \frac{1}{2}\int \frac{2x}{\sqrt{(x^2 - 3)^2 - 4}}\,dx$$
$$= \frac{1}{2}\ln\left|x^2 - 3 + \sqrt{x^4 - 6x^2 + 5}\right| + C$$

$u = x^2 - 3,\ du = 2x\,dx$

47.
$$\int \frac{x^3}{\sqrt{4 - x^2}}\,dx = \int \frac{8\sin^3\theta(2\cos\theta\,d\theta)}{2\cos\theta}$$ 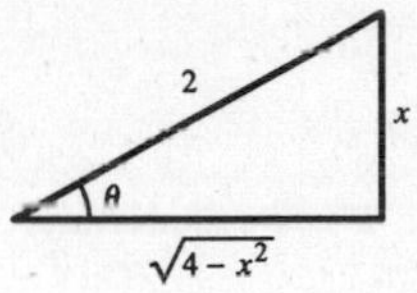
$$= 8\int (1 - \cos^2\theta)\sin\theta\,d\theta$$
$$= 8\int [\sin\theta - \cos^2\theta(\sin\theta)]\,d\theta$$
$$= -8\cos\theta + \frac{8\cos^3\theta}{3} + C$$
$$= \frac{-\sqrt{4 - x^2}}{3}(x^2 + 8) + C$$

$x = 2\sin\theta,\ dx = 2\cos\theta\,d\theta,\ \sqrt{4 - x^2} = 2\cos\theta$

49. By Formula 8:
$$\int \frac{e^{3x}}{(1 + e^x)^3}\,dx = \int \frac{(e^x)^2}{(1 + e^x)^3}(e^x)\,dx$$
$$= \frac{2}{1 + e^x} - \frac{1}{2(1 + e^x)^2} + \ln|1 + e^x| + C$$

$u = e^x,\ du = e^x\,dx$

51.
$$\frac{u^2}{(a + bu)^2} = \frac{1}{b^2} - \frac{(2a/b)u + (a^2/b^2)}{(a + bu)^2} = \frac{1}{b^2} + \frac{A}{a + bu} + \frac{B}{(a + bu)^2}$$
$$-\frac{2a}{b}u - \frac{a^2}{b^2} = A(a + bu) + B = (aA + B) + bAu$$

Equating the coefficients of like terms we have $aA + B = -a^2/b^2$ and $bA = -2a/b$. Solving these equations we have $A = -2a/b^2$ and $B = a^2/b^2$.

$$\int \frac{u^2}{(a + bu)^2}\,du = \frac{1}{b^2}\int du - \frac{2a}{b^2}\left(\frac{1}{b}\right)\int \frac{1}{a + bu}b\,du + \frac{a^2}{b^2}\left(\frac{1}{b}\right)\int \frac{1}{(a + bu)^2}b\,du = \frac{1}{b^2}u - \frac{2a}{b^3}\ln|a + bu| - \frac{a^2}{b^3}\left(\frac{1}{a + bu}\right) + C$$
$$= \frac{1}{b^3}\left(bu - \frac{a^2}{a + bu} - 2a\ln|a + bu|\right) + C$$

53. When we have $u^2 + a^2$:

$$u = a \tan\theta$$
$$du = a\sec^2\theta\, d\theta$$
$$u^2 + a^2 = a^2\sec^2\theta$$
$$\int \frac{1}{(u^2+a^2)^{3/2}}\,du = \int \frac{a\sec^2\theta\,d\theta}{a^3\sec^3\theta}$$
$$= \frac{1}{a^2}\int \cos\theta\,d\theta$$
$$= \frac{1}{a^2}\sin\theta + C$$
$$= \frac{u}{a^2\sqrt{u^2+a^2}} + C$$

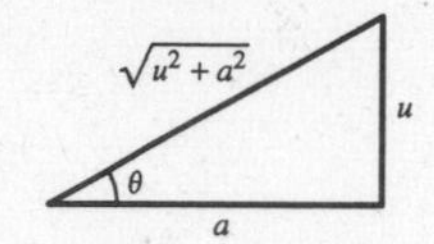

When we have $u^2 - a^2$:

$$u = a\sec\theta$$
$$du = a\sec\theta\tan\theta\,d\theta$$
$$u^2 - a^2 = a^2\tan^2\theta$$
$$\int \frac{1}{(u^2-a^2)^{3/2}}\,du = \int \frac{a\sec\theta\tan\theta\,d\theta}{a^3\tan^3\theta}$$
$$= \frac{1}{a^2}\int \frac{\cos\theta}{\sin^2\theta}\,d\theta$$
$$= -\frac{1}{a^2}\csc\theta + C$$
$$= \frac{-u}{a^2\sqrt{u^2-a^2}} + C$$

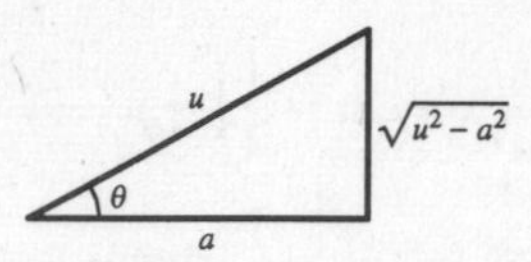

55. $$\int (\arctan u)\,du = u\arctan u - \frac{1}{2}\int \frac{2u}{1+u^2}\,du$$
$$= u\arctan u - \frac{1}{2}\ln(1+u^2) + C$$
$$= u\arctan u - \ln\sqrt{1+u^2} + C$$

$w = \arctan u,\ dv = du,\ dw = \dfrac{du}{1+u^2},\ v = u$

57. $$\int \frac{1}{x^{3/2}\sqrt{1-x}}\,dx = \frac{-2\sqrt{1-x}}{\sqrt{x}} + C$$

$\left(\frac{1}{2}, 5\right)$: $\dfrac{-2\sqrt{1/2}}{\sqrt{1/2}} + C = 5 \Rightarrow C = 7$

$$y = \frac{-2\sqrt{1-x}}{\sqrt{x}} + 7$$

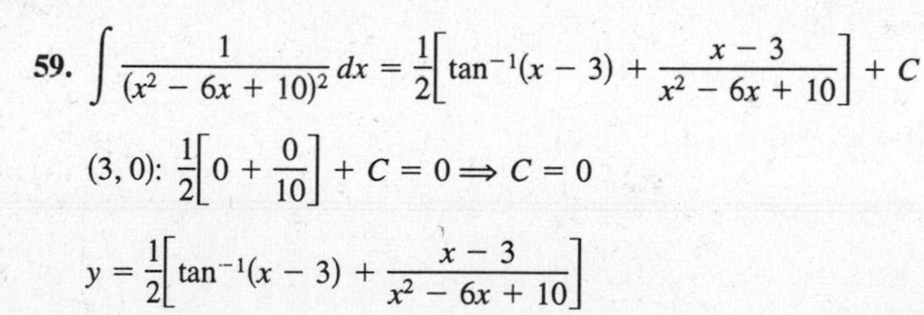

59. $$\int \frac{1}{(x^2-6x+10)^2}\,dx = \frac{1}{2}\left[\tan^{-1}(x-3) + \frac{x-3}{x^2-6x+10}\right] + C$$

$(3, 0)$: $\dfrac{1}{2}\left[0 + \dfrac{0}{10}\right] + C = 0 \Rightarrow C = 0$

$$y = \frac{1}{2}\left[\tan^{-1}(x-3) + \frac{x-3}{x^2-6x+10}\right]$$

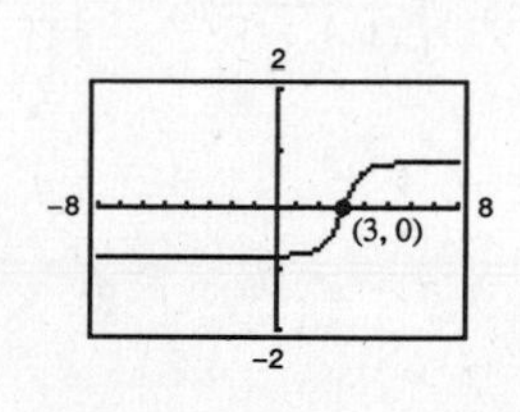

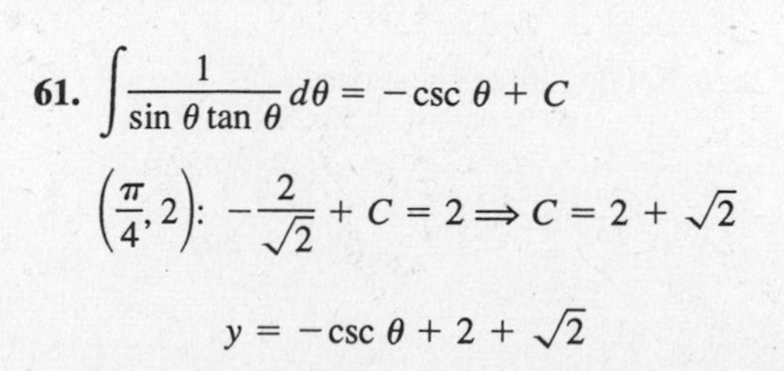

61. $$\int \frac{1}{\sin\theta\tan\theta}\,d\theta = -\csc\theta + C$$

$\left(\frac{\pi}{4}, 2\right)$: $-\dfrac{2}{\sqrt{2}} + C = 2 \Rightarrow C = 2 + \sqrt{2}$

$$y = -\csc\theta + 2 + \sqrt{2}$$

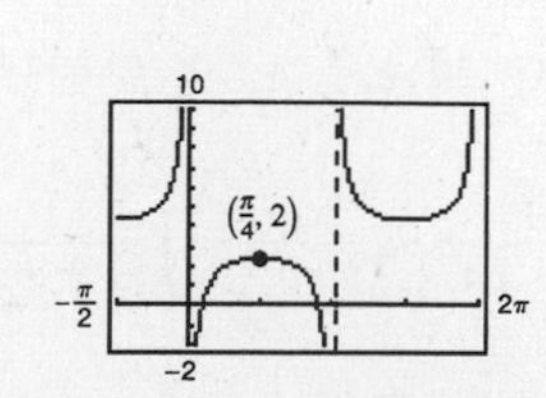

63. $\displaystyle\int \frac{1}{2 - 3\sin\theta}\,d\theta = \int\left[\frac{\dfrac{2\,du}{1+u^2}}{2 - 3\left(\dfrac{2u}{1+u^2}\right)}\right]$

$\displaystyle = \int \frac{2}{2(1+u^2) - 6u}\,du$

$\displaystyle = \int \frac{1}{u^2 - 3u + 1}\,du$

$\displaystyle = \int \frac{1}{\left(u - \frac{3}{2}\right)^2 - \frac{5}{4}}\,du$

$\displaystyle = \frac{1}{\sqrt{5}}\ln\left|\frac{\left(u - \frac{3}{2}\right) - \frac{\sqrt{5}}{2}}{\left(u - \frac{3}{2}\right) + \frac{\sqrt{5}}{2}}\right| + C$

$\displaystyle = \frac{1}{\sqrt{5}}\ln\left|\frac{2u - 3 - \sqrt{5}}{2u - 3 + \sqrt{5}}\right| + C$

$\displaystyle = \frac{1}{\sqrt{5}}\ln\left|\frac{2\tan\left(\frac{\theta}{2}\right) - 3 - \sqrt{5}}{2\tan\left(\frac{\theta}{2}\right) - 3 + \sqrt{5}}\right| + C$

$u = \tan\dfrac{\theta}{2}$

65. $\displaystyle\int_0^{\pi/2} \frac{1}{1 + \sin\theta + \cos\theta}\,d\theta = \int_0^1\left[\frac{\dfrac{2\,du}{1+u^2}}{1 + \dfrac{2u}{1+u^2} + \dfrac{1-u^2}{1+u^2}}\right]$

$\displaystyle = \int_0^1 \frac{1}{1+u}\,du$

$\displaystyle = \Big[\ln|1+u|\Big]_0^1$

$= \ln 2$

$u = \tan\dfrac{\theta}{2}$

67. $\displaystyle\int \frac{\sin\theta}{3 - 2\cos\theta}\,d\theta = \frac{1}{2}\int \frac{2\sin\theta}{3 - 2\cos\theta}\,d\theta$

$\displaystyle = \frac{1}{2}\ln|u| + C$

$\displaystyle = \frac{1}{2}\ln(3 - 2\cos\theta) + C$

$u = 3 - 2\cos\theta,\ du = 2\sin\theta\,d\theta$

69. $\displaystyle\int \frac{\cos\sqrt{\theta}}{\sqrt{\theta}}\,d\theta = 2\int \cos\sqrt{\theta}\left(\frac{1}{2\sqrt{\theta}}\right)d\theta$

$= 2\sin\sqrt{\theta} + C$

$u = \sqrt{\theta},\ du = \dfrac{1}{2\sqrt{\theta}}\,d\theta$

71. $\displaystyle A = \int_0^8 \frac{x}{\sqrt{x+1}}\,dx$

$\displaystyle = \left[\frac{-2(2-x)}{3}\sqrt{x+1}\right]_0^8$

$\displaystyle = 12 - \left(-\frac{4}{3}\right)$

$\displaystyle = \frac{40}{3} \approx 13.333$ square units

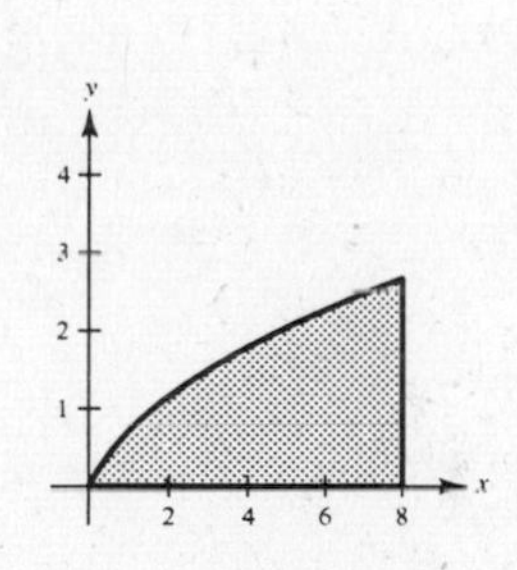

73. Arctangent Formula, Formula 23,

$\displaystyle\int \frac{1}{u^2 + 1}\,du,\ u = e^x$

75. Substitution: $u = x^2,\ du = 2x\,dx$

Then Formula 81.

77. Cannot be integrated.

79. Answers will vary. For example,

$\displaystyle\int (2x)e^{2x}\,dx$

can be integrated by first letting $u = 2x$ and then using Formula 82.

81. $W = \int_0^5 2000xe^{-x}\,dx$

$= -2000\int_0^5 -xe^{-x}\,dx$

$= 2000\int_0^5 (-x)e^{-x}(-1)\,dx$

$= 2000\Big[(-x)e^{-x} - e^{-x}\Big]_0^5$

$= 2000\left(-\dfrac{6}{e^5} + 1\right)$

≈ 1919.145 ft · lbs

83. (a) $V = 20(2)\int_0^3 \dfrac{2}{\sqrt{1+y^2}}\,dy$

$= \left[80 \ln\left|y + \sqrt{1+y^2}\right|\right]_0^3$

$= 80 \ln(3 + \sqrt{10})$

≈ 145.5 cubic feet

$W = 148(80 \ln(3 + \sqrt{10}))$

$= 11{,}840 \ln(3 + \sqrt{10})$

$\approx 21{,}530.4$ lb

(b) By symmetry, $\bar{x} = 0$.

$$M = \rho(2)\int_0^3 \frac{2}{\sqrt{1+y^2}}\,dy = \left[4\rho \ln\left|y + \sqrt{1+y^2}\right|\right]_0^3 = 4\rho \ln(3 + \sqrt{10})$$

$$M_x = 2\rho\int_0^3 \frac{2y}{\sqrt{1+y^2}}\,dy = \left[4\rho\sqrt{1+y^2}\right]_0^3 = 4\rho(\sqrt{10} - 1)$$

$$\bar{y} = \frac{M_x}{M} = \frac{4\rho(\sqrt{10}-1)}{4\rho \ln(3+\sqrt{10})} \approx 1.19$$

Centroid: $(\bar{x}, \bar{y}) \approx (0, 1.19)$

85. (a) $\int_0^4 \dfrac{k}{2+3x}\,dx = 10$

$$k = \frac{10}{\int_0^4 \frac{1}{2+3x}\,dx} \approx \frac{10}{0.6486}$$

$$= 15.417 \left(= \frac{30}{\ln 7}\right)$$

(b) $\int_0^4 \dfrac{15.417}{2+3x}\,dx$

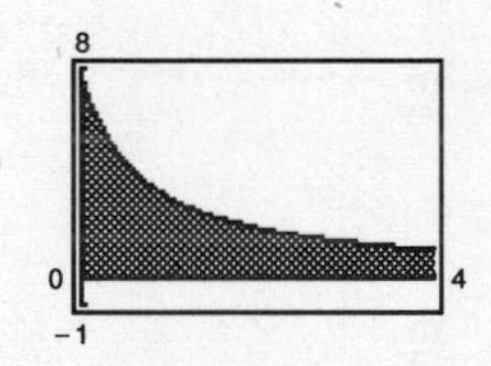

87. False. You might need to convert your integral using substitution or algebra.

Section 7.7 Indeterminate Forms and L'Hôpital's Rule

1. $\lim_{x\to 0} \dfrac{\sin 5x}{\sin 2x} \approx 2.5\left(\text{exact: } \dfrac{5}{2}\right)$

x	-0.1	-0.01	-0.001	0.001	0.01	0.1
$f(x)$	2.4132	2.4991	2.500	2.500	2.4991	2.4132

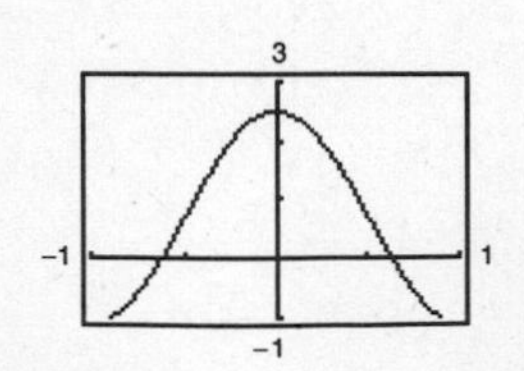

3. $\lim_{x\to\infty} x^5 e^{-x/100} \approx 0$

x	1	10	10^2	10^3	10^4	10^5
$f(x)$	0.9900	90,484	3.7×10^9	4.5×10^{10}	0	0

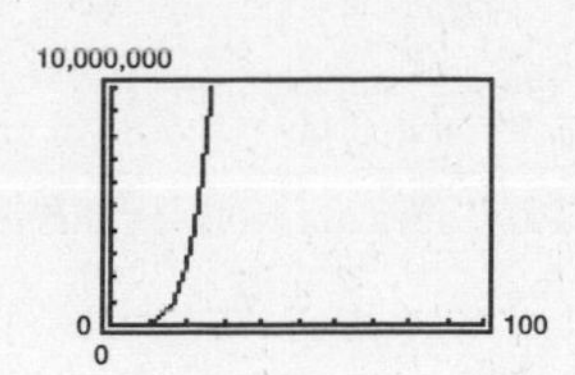

5. (a) $\lim_{x\to3} \frac{2(x-3)}{x^2-9} = \lim_{x\to3} \frac{2(x-3)}{(x+3)(x-3)} = \lim_{x\to3} \frac{2}{x+3} = \frac{1}{3}$

(b) $\lim_{x\to3} \frac{2(x-3)}{x^2-9} = \lim_{x\to3} \frac{(d/dx)[2(x-3)]}{(d/dx)[x^2-9]} = \lim_{x\to3} \frac{2}{2x} = \frac{2}{6} = \frac{1}{3}$

7. (a) $\lim_{x\to3} \frac{\sqrt{x+1}-2}{x-3} = \lim_{x\to3} \frac{\sqrt{x+1}-2}{x-3} \cdot \frac{\sqrt{x+1}+2}{\sqrt{x+1}+2} = \lim_{x\to3} \frac{(x+1)-4}{(x-3)\left[\sqrt{x+1}+2\right]} = \lim_{x\to3} \frac{1}{\sqrt{x+1}+2} = \frac{1}{4}$

(b) $\lim_{x\to3} \frac{\sqrt{x+1}-2}{x-3} = \lim_{x\to3} \frac{(d/dx)\left[\sqrt{x+1}-2\right]}{(d/dx)[x-3]} = \lim_{x\to3} \frac{1/\left(2\sqrt{x+1}\right)}{1} = \frac{1}{4}$

9. (a) $\lim_{x\to\infty} \frac{5x^2-3x+1}{3x^2-5} = \lim_{x\to\infty} \frac{5-(3/x)+(1/x^2)}{3-(5/x^2)} = \frac{5}{3}$

(b) $\lim_{x\to\infty} \frac{5x^2-3x+1}{3x^2-5} = \lim_{x\to\infty} \frac{(d/dx)[5x^2-3x+1]}{(d/dx)[3x^2-5]} = \lim_{x\to\infty} \frac{10x-3}{6x} = \lim_{x\to\infty} \frac{(d/dx)[10x-3]}{(d/dx)[6x]} = \lim_{x\to\infty} \frac{10}{6} = \frac{5}{3}$

11. $\lim_{x\to2} \frac{x^2-x-2}{x-2} = \lim_{x\to2} \frac{2x-1}{1} = 3$

13. $\lim_{x\to0} \frac{\sqrt{4-x^2}-2}{x} = \lim_{x\to0} \frac{-x/\sqrt{4-x^2}}{1} = 0$

15. $\lim_{x\to0} \frac{e^x-(1-x)}{x} = \lim_{x\to0} \frac{e^x+1}{1} = 2$

17. Case 1: $n = 1$

$$\lim_{x\to0^+} \frac{e^x-(1+x)}{x} = \lim_{x\to0^+} \frac{e^x-1}{1} = 0$$

Case 2: $n = 2$

$$\lim_{x\to0^+} \frac{e^x-(1+x)}{x^2} = \lim_{x\to0^+} \frac{e^x-1}{2x} = \lim_{x\to0^+} \frac{e^x}{2} = \frac{1}{2}$$

Case 3: $n \geq 3$

$$\lim_{x\to0^+} \frac{e^x-(1+x)}{x^n} = \lim_{x\to0^+} \frac{e^x-1}{nx^{n-1}} = \lim_{x\to0^+} \frac{e^x}{n(n-1)x^{n-2}} = \infty$$

19. $\lim_{x\to0} \frac{\sin 2x}{\sin 3x} = \lim_{x\to0} \frac{2\cos 2x}{3\cos 3x} = \frac{2}{3}$

21. $\lim_{x\to0} \frac{\arcsin x}{x} = \lim_{x\to0} \frac{1/\sqrt{1-x^2}}{1} = 1$

23. $\lim_{x\to\infty} \frac{3x^2-2x+1}{2x^2+3} = \lim_{x\to\infty} \frac{6x-2}{4x}$

$= \lim_{x\to\infty} \frac{6}{4} = \frac{3}{2}$

25. $\lim_{x\to\infty} \frac{x^2+2x+3}{x-1} = \lim_{x\to\infty} \frac{2x+2}{1} = \infty$

27. $\lim_{x\to\infty} \frac{x^3}{e^{x/2}} = \lim_{x\to\infty} \frac{3x^2}{(1/2)e^{x/2}}$

$= \lim_{x\to\infty} \frac{6x}{(1/4)e^{x/2}} = \lim_{x\to\infty} \frac{6}{(1/8)e^{x/2}} = 0$

29. $\lim_{x\to\infty} \dfrac{x}{\sqrt{x^2+1}} = \lim_{x\to\infty} \dfrac{1}{\sqrt{1+(1/x^2)}} = 1$

Note: L'Hôpital's Rule does not work on this limit.
See Exercise 79.

31. $\lim_{x\to\infty} \dfrac{\cos x}{x} = 0$ by Squeeze Theorem

$\left(\dfrac{\cos x}{x} \le \dfrac{1}{x}, \text{ for } x > 0\right)$

33. $\lim_{x\to\infty} \dfrac{\ln x}{x^2} = \lim_{x\to\infty} \dfrac{1/x}{2x} = \lim_{x\to\infty} \dfrac{1}{2x^2} = 0$

35. $\lim_{x\to\infty} \dfrac{e^x}{x^2} = \lim_{x\to\infty} \dfrac{e^x}{2x} = \lim_{x\to\infty} \dfrac{e^x}{2} = \infty$

37. (a) $\lim_{x\to 0^+} (-x\ln x) = (-0)(-\infty) = (0)(\infty)$

(b) $\lim_{x\to 0^+} (-x\ln x) = \lim_{x\to 0^+} \dfrac{\ln x}{-1/x}$

$= \lim_{x\to 0^+} \dfrac{1/x}{1/x^2}$

$= \lim_{x\to 0^+} x = 0$

(c)

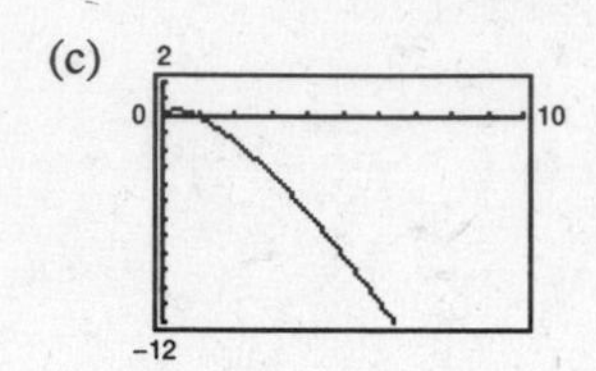

39. (a) $\lim_{x\to\infty} \left(x \sin\dfrac{1}{x}\right) = (\infty)(0)$

(b) $\lim_{x\to\infty} x\sin\dfrac{1}{x} = \lim_{x\to\infty} \dfrac{\sin(1/x)}{1/x}$

$= \lim_{x\to\infty} \dfrac{(-1/x^2)\cos(1/x)}{-1/x^2}$

$= \lim_{x\to\infty} \cos\left(\dfrac{1}{x}\right) = 1$

(c)

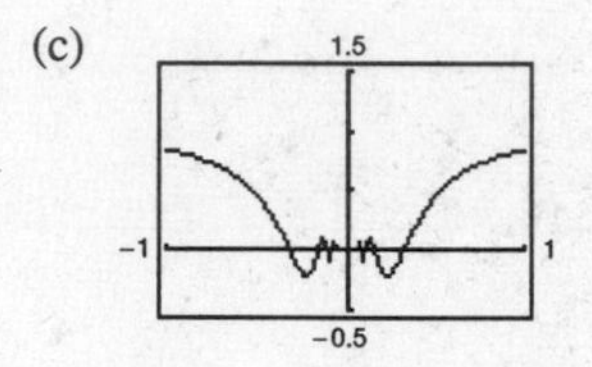

41. (a) $\lim_{x\to 0^+} x^{1/x} = 0^\infty = 0$, not indeterminant
(See Exercise 95)

(b) Let $y = x^{1/x}$

$\ln y = \ln x^{1/x} = \dfrac{1}{x}\ln x.$

Since $x \to 0^+$, $\dfrac{1}{x}\ln x \to (\infty)(-\infty) = -\infty$. Hence,

$\ln y \to -\infty \implies y \to 0^+.$

Therefore, $\lim_{x\to 0^+} x^{1/x} = 0.$

(c)

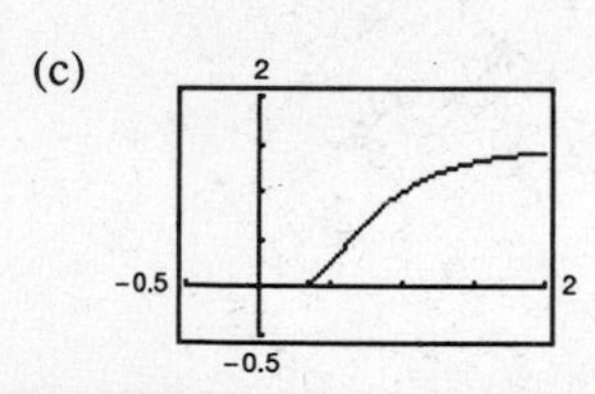

43. (a) $\lim_{x\to\infty} x^{1/x} = \infty^0$

(b) Let $y = \lim_{x\to\infty} x^{1/x}$.

$\ln y = \lim_{x\to\infty} \dfrac{\ln x}{x} = \lim_{x\to\infty} \left(\dfrac{1/x}{1}\right) = 0$

Thus, $\ln y = 0 \implies y = e^0 = 1$. Therefore,

$\lim_{x\to\infty} x^{1/x} = 1.$

(c)

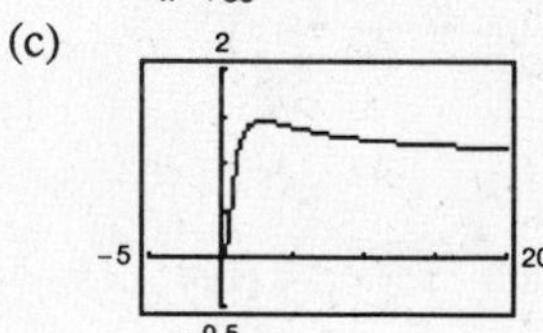

45. (a) $\lim_{x\to 0^+} (1+x)^{1/x} = 1^\infty$

(b) Let $y = \lim_{x\to 0^+} (1+x)^{1/x}$.

$\ln y = \lim_{x\to 0^+} \dfrac{\ln(1+x)}{x}$

$= \lim_{x\to 0^+} \left(\dfrac{1/(1+x)}{1}\right) = 1$

Thus, $\ln y = 1 \implies y = e^1 = e$.

Therefore, $\lim_{x\to 0^+} (1+x)^{1/x} = e$.

(c)

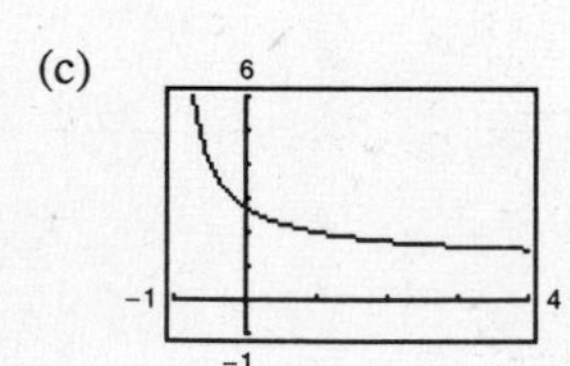

47. (a) $\lim_{x\to0^+} [3(x)^{x/2}] = 0^0$

(b) Let $y = \lim_{x\to0^+} 3(x)^{x/2}$.

$$\ln y = \lim_{x\to0^+}\left[\ln 3 + \frac{x}{2}\ln x\right]$$

$$= \lim_{x\to0^+}\left[\ln 3 + \frac{\ln x}{2/x}\right]$$

$$= \lim_{x\to0^+} \ln 3 + \lim_{x\to0^+}\frac{1/x}{-2/x^2}$$

$$= \lim_{x\to0^+} \ln 3 - \lim_{x\to0^+}\frac{x}{2}$$

$$= \ln 3$$

Hence, $\lim_{x\to0^+} 3(x)^{x/2} = 3$.

(c)

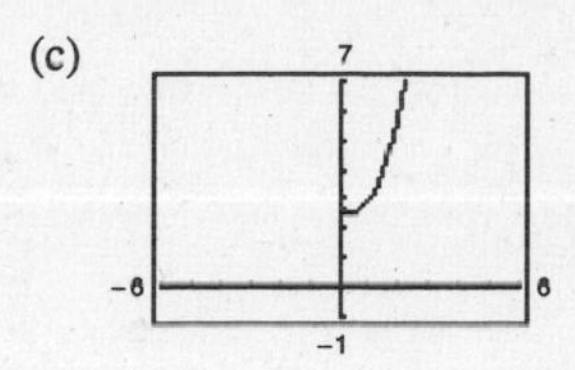

49. (a) $\lim_{x\to1^+} (\ln x)^{x-1} = 0^0$

(b) Let $y = \lim_{x\to1^+} (\ln x)^{x-1}$

$$= \lim_{x\to1^+} (x-1)\ln x = 0$$

Hence, $\lim_{x\to1^+} (\ln x)^{x-1} = 1$

(c)

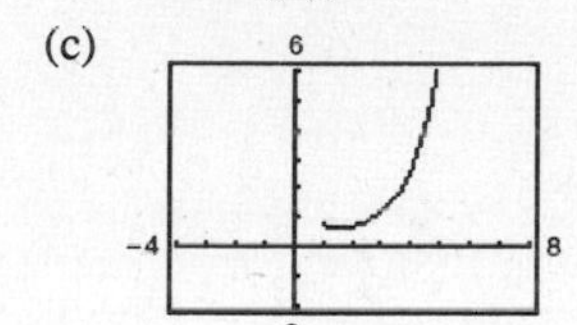

51. (a) $\lim_{x\to2^+}\left(\frac{8}{x^2-4} - \frac{x}{x-2}\right) = \infty - \infty$

(b) $\lim_{x\to2^+}\left(\frac{8}{x^2-4} - \frac{x}{x-2}\right) = \lim_{x\to2^+}\frac{8 - x(x+2)}{x^2-4}$

$$= \lim_{x\to2^+}\frac{(2-x)(4+x)}{(x+2)(x-2)}$$

$$= \lim_{x\to2^+}\frac{-(x+4)}{x+2} = \frac{-3}{2}$$

(c)

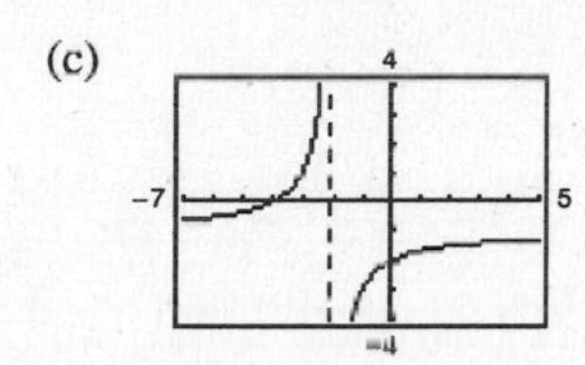

53. (a) $\lim_{x\to1^+}\left(\frac{3}{\ln x} - \frac{2}{x-1}\right) = \infty - \infty$

(b) $\lim_{x\to1^+}\left(\frac{3}{\ln x} - \frac{2}{x-1}\right) = \lim_{x\to1^+}\frac{3x - 3 - 2\ln x}{(x-1)\ln x}$

$$= \lim_{x\to1^+}\frac{3 - (2/x)}{[(x-1)/x] + \ln x} = \infty$$

(c)

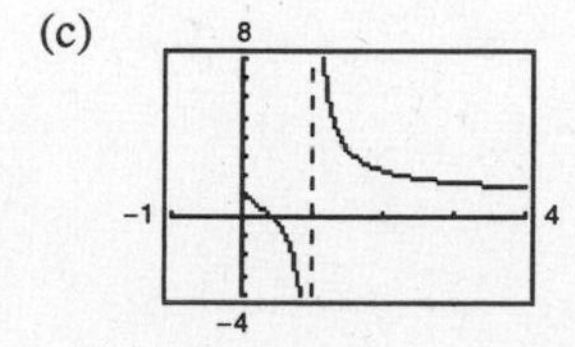

55. (a)

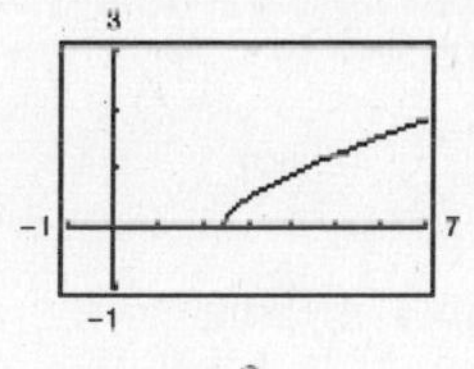

(b) $\lim_{x\to3}\frac{x-3}{\ln(2x-5)} = \lim_{x\to3}\frac{1}{2/(2x-5)}$

$$= \lim_{x\to3}\frac{2x-5}{2} = \frac{1}{2}$$

57. (a)

(b) $\lim_{x\to\infty}\left(\sqrt{x^2+5x+2} - x\right) = \lim_{x\to\infty}\left(\sqrt{x^2+5x+2} - x\right)\frac{\left(\sqrt{x^2+5x+2}+x\right)}{\left(\sqrt{x^2+5x+2}+x\right)}$

$$= \lim_{x\to\infty}\frac{(x^2+5x+2) - x^2}{\sqrt{x^2+5x+2}+x}$$

$$= \lim_{x\to\infty}\frac{5x+2}{\sqrt{x^2+5x+2}+x}$$

$$= \lim_{x\to\infty}\frac{5 + (2/x)}{\sqrt{1 + (5/x) + (2/x^2)} + 1} = \frac{5}{2}$$

59. $\dfrac{0}{0}, \dfrac{\infty}{\infty}, 0 \cdot \infty, 1^{\infty}, 0^0, \infty - \infty, \infty^0$

61. (a) Let $f(x) = x^2 - 25$ and $g(x) = x - 5$.

(b) Let $f(x) = (x - 5)^2$ and $g(x) = x^2 - 25$.

(c) Let $f(x) = x^2 - 25$ and $g(x) = (x - 5)^3$.

63. $\displaystyle \lim_{x\to\infty} \frac{x^2}{e^{5x}} = \lim_{x\to\infty} \frac{2x}{5e^{5x}} = \lim_{x\to\infty} \frac{2}{25e^{5x}} = 0$

65.
$$\begin{aligned}
\lim_{x\to\infty} \frac{(\ln x)^3}{x} &= \lim_{x\to\infty} \frac{3(\ln x)^2(1/x)}{1} \\
&= \lim_{x\to\infty} \frac{3(\ln x)^2}{x} \\
&= \lim_{x\to\infty} \frac{6(\ln x)(1/x)}{1} \\
&= \lim_{x\to\infty} \frac{6(\ln x)}{x} = \lim_{x\to\infty} \frac{6}{x} = 0
\end{aligned}$$

67.
$$\begin{aligned}
\lim_{x\to\infty} \frac{(\ln x)^n}{x^m} &= \lim_{x\to\infty} \frac{n(\ln x)^{n-1}/x}{mx^{m-1}} \\
&= \lim_{x\to\infty} \frac{n(\ln x)^{n-1}}{mx^m} \\
&= \lim_{x\to\infty} \frac{n(n-1)(\ln x)^{n-2}}{m^2x^m} \\
&= \cdots = \lim_{x\to\infty} \frac{n!}{m^n x^m} = 0
\end{aligned}$$

69.

x	10	10^2	10^4	10^6	10^8	10^{10}
$\dfrac{(\ln x)^4}{x}$	2.811	4.498	0.720	0.036	0.001	0.000

71. $y = x^{1/x}, x > 0$

Horizontal asymptote: $y = 1$ (See Exercise 43)

$$\ln y = \frac{1}{x} \ln x$$

$$\frac{1}{y}\frac{dy}{dx} = \frac{1}{x}\left(\frac{1}{x}\right) + (\ln x)\left(-\frac{1}{x^2}\right)$$

$$\frac{dy}{dx} = x^{1/x}\left(\frac{1}{x^2}\right)(1 - \ln x) = x^{(1/x)-2}(1 - \ln x) = 0$$

Critical number: $x = e$

Intervals:	$(0, e)$	(e, ∞)
Sign of dy/dx:	$+$	$-$
$y = f(x)$:	Increasing	Decreasing

Relative maximum: $(e, e^{1/e})$

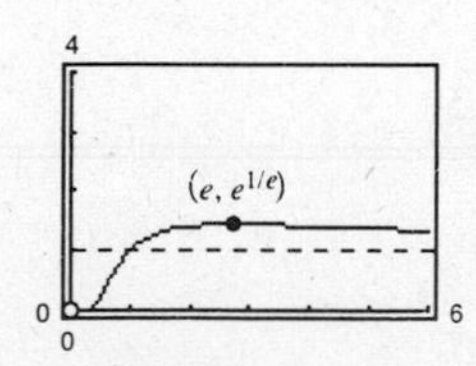

73. $y = 2xe^{-x}$

$$\lim_{x\to\infty} \frac{2x}{e^x} = \lim_{x\to\infty} \frac{2}{e^x} = 0$$

Horizontal asymptote: $y = 0$

$$\frac{dy}{dx} = 2x(-e^{-x}) + 2e^{-x}$$
$$= 2e^{-x}(1 - x) = 0$$

Critical number: $x = 1$

Intervals:	$(-\infty, 1)$	$(1, \infty)$
Sign of dy/dx:	$+$	$-$
$y = f(x)$:	Increasing	Decreasing

Relative maximum: $\left(1, \dfrac{2}{e}\right)$

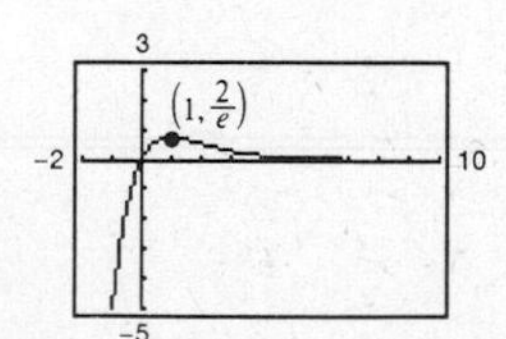

75. $\displaystyle \lim_{x\to 0} \frac{e^{2x} - 1}{e^x} = \frac{0}{1} = 0$

Limit is not of the form $0/0$ or ∞/∞.
L'Hôpital's Rule does not apply.

77. $\displaystyle \lim_{x\to\infty} x \cos \frac{1}{x} = \infty(1) = \infty$

Limit is not of the form $0/0$ or ∞/∞.
L'Hôpital's Rule does not apply.

79. (a) $\lim_{x\to\infty} \frac{x}{\sqrt{x^2+1}} = \lim_{x\to\infty} \frac{x/x}{\sqrt{x^2+1}/x}$

$= \lim_{x\to\infty} \frac{1}{\sqrt{x^2+1}/\sqrt{x^2}}$

$= \lim_{x\to\infty} \frac{1}{\sqrt{1+(1/x^2)}}$

$= \frac{1}{\sqrt{1+0}} = 1$

(b) $\lim_{x\to\infty} \frac{x}{\sqrt{x^2+1}} = \lim_{x\to\infty} \frac{1}{x/\sqrt{x^2+1}}$

$= \lim_{x\to\infty} \frac{\sqrt{x^2+1}}{x} = \lim_{x\to\infty} \frac{x/\sqrt{x^2+1}}{1}$

$= \lim_{x\to\infty} \frac{x}{\sqrt{x^2+1}}$

Applying L'Hôpital's rule twice results in the original limit, so L'Hôpital's rule fails.

(c)

1.5

−6 6

−1.5

81. $\lim_{k\to 0} \frac{32\left(1 - e^{-kt} + \frac{v_0 ke^{-kt}}{32}\right)}{k} = \lim_{k\to 0} \frac{32(1-e^{-kt})}{k} + \lim_{k\to 0} (v_0 e^{-kt})$

$= \lim_{k\to 0} \frac{32(0+te^{-kt})}{1} + \lim_{k\to 0} \left(\frac{v_0}{e^{kt}}\right) = 32t + v_0$

83. Area of triangle: $\frac{1}{2}(2x)(1-\cos x) = x - x\cos x$

Shaded area: Area of rectangle − Area under curve

$$2x(1-\cos x) - 2\int_0^x (1-\cos t)\,dt = 2x(1-\cos x) - 2\Big[t - \sin t\Big]_0^x$$

$$= 2x(1-\cos x) - 2(x - \sin x) = 2\sin x - 2x\cos x$$

Ratio: $\lim_{x\to 0} \frac{x - x\cos x}{2\sin x - 2x\cos x} = \lim_{x\to 0} \frac{1 + x\sin x - \cos x}{2\cos x + 2x\sin x - 2\cos x}$

$= \lim_{x\to 0} \frac{1 + x\sin x - \cos x}{2x\sin x}$

$= \lim_{x\to 0} \frac{x\cos x + \sin x + \sin x}{2x\cos x + 2\sin x}$

$= \lim_{x\to 0} \frac{x\cos x + 2\sin x}{2x\cos x + 2\sin x} \cdot \frac{1/\cos x}{1/\cos x}$

$= \lim_{x\to 0} \frac{x + 2\tan x}{2x + 2\tan x}$

$= \lim_{x\to 0} \frac{1 + 2\sec^2 x}{2 + 2\sec^2 x} = \frac{3}{4}$

85. $f(x) = x^3,\ g(x) = x^2 + 1,\ [0, 1]$

$\frac{f(b) - f(a)}{g(b) - g(a)} = \frac{f'(c)}{g'(c)}$

$\frac{f(1) - f(0)}{g(1) - g(0)} = \frac{3c^2}{2c}$

$\frac{1}{1} = \frac{3c}{2}$

$c = \frac{2}{3}$

87. $f(x) = \sin x,\ g(x) = \cos x,\ \left[0, \frac{\pi}{2}\right]$

$\frac{f(\pi/2) - f(0)}{g(\pi/2) - g(0)} = \frac{f'(c)}{g'(c)}$

$\frac{1}{-1} = \frac{\cos c}{-\sin c}$

$-1 = -\cot c$

$c = \frac{\pi}{4}$

89. False. L'Hôpital's Rule does not apply since

$\lim_{x\to 0}(x^2+x+1)\neq 0.$

$$\lim_{x\to 0^+}\frac{x^2+x+1}{x}=\lim_{x\to 0^+}\left(x+1+\frac{1}{x}\right)=1+\infty=\infty$$

91. True

93. (a) $\sin\theta = BD$

$\cos\theta = DO \implies AD = 1-\cos\theta$

$$\text{Area }\triangle ABD=\frac{1}{2}bh=\frac{1}{2}(1-\cos\theta)\sin\theta=\frac{1}{2}\sin\theta-\frac{1}{2}\sin\theta\cos\theta$$

(b) Area of sector: $\frac{1}{2}\theta$

$$\text{Shaded area: }\frac{1}{2}\theta-\text{Area }\triangle OBD=\frac{1}{2}\theta-\frac{1}{2}(\cos\theta)(\sin\theta)=\frac{1}{2}\theta-\frac{1}{2}\sin\theta\cos\theta$$

(c) $$R=\frac{(1/2)\sin\theta-(1/2)\sin\theta\cos\theta}{(1/2)\theta-(1/2)\sin\theta\cos\theta}=\frac{\sin\theta-\sin\theta\cos\theta}{\theta-\sin\theta\cos\theta}$$

(d) $$\lim_{\theta\to 0}R=\lim_{\theta\to 0}\frac{\sin\theta-(1/2)\sin 2\theta}{\theta-(1/2)\sin 2\theta}$$

$$=\lim_{\theta\to 0}\frac{\cos\theta-\cos 2\theta}{1-\cos 2\theta}=\lim_{\theta\to 0}\frac{-\sin\theta+2\sin 2\theta}{2\sin 2\theta}=\lim_{\theta\to 0}\frac{-\cos\theta+4\cos 2\theta}{4\cos 2\theta}=\frac{3}{4}$$

95. $\lim_{x\to a} f(x)^{g(x)}$

$y = f(x)^{g(x)}$

$\ln y = g(x)\ln f(x)$

$\lim_{x\to a} g(x)\ln f(x) = (\infty)(-\infty) = -\infty$

As $x\to a$, $\ln y \implies -\infty$, and hence $y = 0$. Thus,

$\lim_{x\to a} f(x)^{g(x)} = 0.$

97. $$f'(a)(b-a)-\int_a^b f''(t)(t-b)\,dt=f'(a)(b-a)-\left\{\Big[f'(t)(t-b)\Big]_a^b-\int_a^b f'(t)\,dt\right\}$$

$$=f'(a)(b-a)+f'(a)(a-b)+\Big[f(t)\Big]_a^b=f(b)-f(a)$$

$dv = f''(t)dt \implies v = f'(t)$

$u = t-b \implies du = dt$

Section 7.8 Improper Integrals

1. Infinite discontinuity at $x = 0$.

$$\int_0^4 \frac{1}{\sqrt{x}}\,dx=\lim_{b\to 0^+}\int_b^4\frac{1}{\sqrt{x}}\,dx$$

$$=\lim_{b\to 0^+}\Big[2\sqrt{x}\Big]_b^4$$

$$=\lim_{b\to 0^+}\left(4-2\sqrt{b}\right)=4$$

Converges

3. Infinite discontinuity at $x = 1$.

$$\int_0^2 \frac{1}{(x-1)^2}\,dx = \int_0^1 \frac{1}{(x-1)^2}\,dx + \int_1^2 \frac{1}{(x-1)^2}\,dx$$

$$= \lim_{b\to 1^-} \int_0^b \frac{1}{(x-1)^2}\,dx + \lim_{c\to 1^+} \int_c^2 \frac{1}{(x-1)^2}\,dx$$

$$= \lim_{b\to 1^-} \left[-\frac{1}{x-1}\right]_0^b + \lim_{c\to 1^+} \left[-\frac{1}{x-1}\right]_c^2 = (\infty - 1) + (-1 + \infty)$$

Diverges

5. Infinite limit of integration.

$$\int_0^\infty e^{-x}\,dx = \lim_{b\to\infty} \int_0^b e^{-x}\,dx$$

$$= \lim_{b\to\infty} \Big[-e^{-x}\Big]_0^b = 0 + 1 = 1$$

Converges

7. $\displaystyle\int_{-1}^1 \frac{1}{x^2}\,dx \neq -2$

because the integrand is not defined at $x = 0$.
Diverges

9. $\displaystyle\int_1^\infty \frac{1}{x^2}\,dx = \lim_{b\to\infty} \int_1^b \frac{1}{x^2}\,dx$

$$= \lim_{b\to\infty} \left[-\frac{1}{x}\right]_1^b = 1$$

11. $\displaystyle\int_1^\infty \frac{3}{\sqrt[3]{x}}\,dx = \lim_{b\to\infty} \int_1^b 3x^{-1/3}\,dx$

$$= \lim_{b\to\infty} \left[\frac{9}{2}x^{2/3}\right]_1^b = \infty$$

Diverges

13. $\displaystyle\int_{-\infty}^0 xe^{-2x}\,dx = \lim_{b\to-\infty} \int_b^0 xe^{-2x}\,dx = \lim_{b\to-\infty} \frac{1}{4}\Big[(-2x-1)e^{-2x}\Big]_b^0 = \lim_{b\to-\infty} \frac{1}{4}[-1 + (2b+1)e^{-2b}] = -\infty$ (Integration by parts)

Diverges

15. $\displaystyle\int_0^\infty x^2e^{-x}\,dx = \lim_{b\to\infty} \int_0^b x^2e^{-x}\,dx = \lim_{b\to\infty} \Big[-e^{-x}(x^2+2x+2)\Big]_0^b = \lim_{b\to\infty} \left(-\frac{b^2+2b+2}{e^b} + 2\right) = 2$

Since $\displaystyle\lim_{b\to\infty} \left(-\frac{b^2+2b+2}{e^b}\right) = 0$ by L'Hôpital's Rule.

17. $\displaystyle\int_0^\infty e^{-x}\cos x\,dx = \lim_{b\to\infty} \frac{1}{2}\Big[e^{-x}(-\cos x + \sin x)\Big]_0^b$

$$= \frac{1}{2}[0 - (-1)] = \frac{1}{2}$$

19. $\displaystyle\int_4^\infty \frac{1}{x(\ln x)^3}\,dx = \lim_{b\to\infty} \int_4^b (\ln x)^{-3}\frac{1}{x}\,dx$

$$= \lim_{b\to\infty} \left[-\frac{1}{2}(\ln x)^{-2}\right]_4^b$$

$$= -\frac{1}{2}(\ln b)^{-2} + \frac{1}{2}(\ln 4)^{-2}$$

$$= \frac{1}{2}\frac{1}{(2\ln 2)^2} = \frac{1}{8(\ln 2)^2}$$

21. $\displaystyle\int_{-\infty}^\infty \frac{2}{4+x^2}\,dx = \int_{-\infty}^0 \frac{2}{4+x^2}\,dx + \int_0^\infty \frac{2}{4+x^2}\,dx$

$$= \lim_{b\to-\infty} \int_b^0 \frac{2}{4+x^2}\,dx + \lim_{c\to\infty} \int_0^c \frac{2}{4+x^2}\,dx$$

$$= \lim_{b\to-\infty} \left[\arctan\left(\frac{x}{2}\right)\right]_b^0 + \lim_{c\to\infty} \left[\arctan\left(\frac{x}{2}\right)\right]_0^c$$

$$= \left(0 - \left(-\frac{\pi}{2}\right)\right) + \left(\frac{\pi}{2} - 0\right) = \pi$$

23. $\displaystyle\int_0^\infty \frac{1}{e^x + e^{-x}}\,dx = \lim_{b\to\infty}\int_0^b \frac{e^x}{1+e^{2x}}\,dx$

$\displaystyle = \lim_{b\to\infty}\Big[\arctan(e^x)\Big]_0^b$

$\displaystyle = \frac{\pi}{2} - \frac{\pi}{4} = \frac{\pi}{4}$

25. $\displaystyle\int_0^\infty \cos \pi x\,dx = \lim_{b\to\infty}\left[\frac{1}{\pi}\sin \pi x\right]_0^b$

Diverges since $\sin \pi x$ does not approach a limit as $x\to\infty$.

27. $\displaystyle\int_0^1 \frac{1}{x^2}\,dx = \lim_{b\to 0^+}\left[\frac{-1}{x}\right]_b^1 = -1 + \infty$

Diverges

29. $\displaystyle\int_0^8 \frac{1}{\sqrt[3]{8-x}}\,dx = \lim_{b\to 8^-}\int_0^b \frac{1}{\sqrt[3]{8-x}}\,dx = \lim_{b\to 8^-}\left[\frac{-3}{2}(8-x)^{2/3}\right]_0^b = 6$

31. $\displaystyle\int_0^1 x\ln x\,dx = \lim_{b\to 0^+}\left[\frac{x^2}{2}\ln|x| - \frac{x^2}{4}\right]_b^1 = \lim_{b\to 0^+}\left[\frac{-1}{4} - \frac{b^2\ln b}{2} + \frac{b^2}{4}\right] = \frac{-1}{4}$ since $\displaystyle\lim_{b\to 0^+}(b^2\ln b) = 0$ by L'Hôpital's Rule.

33. $\displaystyle\int_0^{\pi/2} \tan\theta\,d\theta = \lim_{b\to(\pi/2)^-}\Big[\ln|\sec\theta|\Big]_0^b = \infty$

Diverges

35. $\displaystyle\int_2^4 \frac{2}{x\sqrt{x^2-4}}\,dx = \lim_{b\to 2^+}\int_b^4 \frac{2}{x\sqrt{x^2-4}}\,dx$

$\displaystyle = \lim_{b\to 2^+}\left[\operatorname{arcsec}\left|\frac{x}{2}\right|\right]_b^4$

$\displaystyle = \lim_{b\to 2^+}\left(\operatorname{arcsec} 2 - \operatorname{arcsec}\left(\frac{b}{2}\right)\right)$

$\displaystyle = \frac{\pi}{3} - 0 = \frac{\pi}{3}$

37. $\displaystyle\int_2^4 \frac{1}{\sqrt{x^2-4}} = \lim_{b\to 2^+}\left[\ln\left|x + \sqrt{x^2-4}\right|\right]_b^4$

$= \ln\left(4 + 2\sqrt{3}\right) - \ln 2$

$= \ln\left(2+\sqrt{3}\right) \approx 1.317$

39. $\displaystyle\int_0^2 \frac{1}{\sqrt[3]{x-1}}\,dx = \int_0^1 \frac{1}{\sqrt[3]{x-1}}\,dx + \int_1^2 \frac{1}{\sqrt[3]{x-1}}\,dx$

$\displaystyle = \lim_{b\to 1^-}\left[\frac{3}{2}(x-1)^{2/3}\right]_0^b + \lim_{c\to 1^+}\left[\frac{3}{2}(x-1)^{2/3}\right]_c^2 = \frac{-3}{2} + \frac{3}{2} = 0$

41. $\displaystyle\int_0^\infty \frac{4}{\sqrt{x}(x+6)}\,dx = \int_0^1 \frac{4}{\sqrt{x}(x+6)}\,dx + \int_1^\infty \frac{4}{\sqrt{x}(x+6)}\,dx$

Let $u = \sqrt{x}$, $u^2 = x$, $2u\,du = dx$.

$\displaystyle\int \frac{4}{\sqrt{x}(x+6)}\,dx = \int \frac{4(2u\,du)}{u(u^2+6)} = 8\int \frac{du}{u^2+6} = \frac{8}{\sqrt{6}}\arctan\left(\frac{u}{\sqrt{6}}\right) + C = \frac{8}{\sqrt{6}}\arctan\left(\frac{\sqrt{x}}{\sqrt{6}}\right) + C$

Thus, $\displaystyle\int_0^\infty \frac{4}{\sqrt{x}(x+6)}\,dx = \lim_{b\to 0^+}\left[\frac{8}{\sqrt{6}}\arctan\left(\frac{\sqrt{x}}{\sqrt{6}}\right)\right]_b^1 + \lim_{c\to\infty}\left[\frac{8}{\sqrt{6}}\arctan\left(\frac{\sqrt{x}}{\sqrt{6}}\right)\right]_1^c$

$\displaystyle = \left(\frac{8}{\sqrt{6}}\arctan\left(\frac{1}{\sqrt{6}}\right) - \frac{8}{\sqrt{6}}0\right) + \left(\frac{8}{\sqrt{6}}\frac{\pi}{2} - \frac{8}{\sqrt{6}}\arctan\left(\frac{1}{\sqrt{6}}\right)\right)$

$\displaystyle = \frac{8\pi}{2\sqrt{6}} = \frac{2\pi\sqrt{6}}{3}.$

43. If $p = 1$, $\int_1^{\infty} \frac{1}{x}\,dx = \lim_{b\to\infty} \int_1^b \frac{1}{x}\,dx = \lim_{b\to\infty} \ln x \Big]_1^b$.

Diverges. For $p \neq 1$,

$$\int_1^{\infty} \frac{1}{x^p}\,dx = \lim_{b\to\infty} \left[\frac{x^{1-p}}{1-p}\right]_1^b = \lim_{b\to\infty} \left[\frac{b^{1-p}}{1-p} - \frac{1}{1-p}\right].$$

This converges to $\frac{1}{p-1}$ if $1 - p < 0$ or $p > 1$.

45. For $n = 1$ we have

$$\begin{aligned}\int_0^{\infty} xe^{-x}\,dx &= \lim_{b\to\infty} \int_0^b xe^{-x}\,dx \\ &= \lim_{b\to\infty} \Big[-e^{-x}x - e^{-x}\Big]_0^b \qquad \text{(Parts: } u = x,\ dv = e^{-x}\,dx) \\ &= \lim_{b\to\infty} \left[-e^{-b}b - e^{-b} + 1\right] \\ &= \lim_{b\to\infty} \left[\frac{-b}{e^b} - \frac{1}{e^b} + 1\right] = 1 \qquad \text{(L'Hôpital's Rule)}\end{aligned}$$

Assume that $\int_0^{\infty} x^n e^{-x}\,dx$ converges. Then for $n + 1$ we have

$$\int x^{n+1}e^{-x}\,dx = -x^{n+1}e^{-x} + (n+1)\int x^n e^{-x}\,dx$$

by parts ($u = x^{n+1}$, $du = (n+1)x^n\,dx$, $dv = e^{-x}\,dx$, $v = -e^{-x}$).

Thus,

$$\int_0^{\infty} x^{n+1}e^{-x}\,dx = \lim_{b\to\infty} \Big[-x^{n+1}e^{-x}\Big]_0^b + (n+1)\int_0^{\infty} x^n e^{-x}\,dx = 0 + (n+1)\int_0^{\infty} x^n e^{-x}\,dx, \text{ which converges.}$$

47. $\int_0^1 \frac{1}{x^3}\,dx$ diverges

(See Exercise 44, $p = 3 \nless 1$.)

49. $\int_1^{\infty} \frac{1}{x^3}\,dx = \frac{1}{3-1} = \frac{1}{2}$ converges.

(See Exercise 43, $p = 3$.)

51. Since $\frac{1}{x^2+5} \le \frac{1}{x^2}$ on $[1, \infty)$ and $\int_1^{\infty} \frac{1}{x^2}\,dx$ converges by Exercise 43, $\int_1^{\infty} \frac{1}{x^2+5}\,dx$ converges.

53. Since $\frac{1}{\sqrt[3]{x(x-1)}} \ge \frac{1}{\sqrt[3]{x^2}}$ on $[2, \infty)$ and $\int_2^{\infty} \frac{1}{\sqrt[3]{x^2}}\,dx$ diverges by Exercise 43, $\int_2^{\infty} \frac{1}{\sqrt[3]{x(x-1)}}\,dx$ diverges.

55. Since $e^{-x^2} \le e^{-x}$ on $[1, \infty)$ and $\int_0^{\infty} e^{-x}\,dx$ converges (see Exercise 5), $\int_0^{\infty} e^{-x^2}\,dx$ converges.

57. Answers will vary. See pages 540, 543.

59. $\int_{-1}^1 \frac{1}{x^3}\,dx = \int_{-1}^0 \frac{1}{x^3}\,dx + \int_0^1 \frac{1}{x^3}\,dx$

These two integrals diverge by Exercise 44.

61. $f(t) = 1$

$$F(s) = \int_0^{\infty} e^{-st}\,dt = \lim_{b\to\infty} \left[-\frac{1}{s}e^{-st}\right]_0^b = \frac{1}{s},\ s > 0$$

63. $f(t) = t^2$

$$\begin{aligned}F(s) &= \int_0^{\infty} t^2 e^{-st}\,dt = \lim_{b\to\infty} \left[\frac{1}{s^3}(-s^2t^2 - 2st - 2)e^{-st}\right]_0^b \\ &= \frac{2}{s^3},\ s > 0\end{aligned}$$

65. $f(t) = \cos at$

$$F(s) = \int_0^{\infty} e^{-st}\cos at\,dt$$

$$= \lim_{b\to\infty}\left[\frac{e^{-st}}{s^2+a^2}(-s\cos at + a\sin at)\right]_0^b$$

$$= 0 + \frac{s}{s^2+a^2} = \frac{s}{s^2+a^2},\ s > 0$$

67. $f(t) = \cosh at$

$$F(s) = \int_0^{\infty} e^{-st}\cosh at\,dt = \int_0^{\infty} e^{-st}\left(\frac{e^{at}+e^{-at}}{2}\right)dt = \frac{1}{2}\int_0^{\infty}\left[e^{t(-s+a)} + e^{t(-s-a)}\right]dt$$

$$= \lim_{b\to\infty}\frac{1}{2}\left[\frac{1}{(-s+a)}e^{t(-s+a)} + \frac{1}{(-s-a)}e^{t(-s-a)}\right]_0^b = 0 - \frac{1}{2}\left[\frac{1}{(-s+a)} + \frac{1}{(-s-a)}\right]$$

$$= \frac{-1}{2}\left[\frac{1}{(-s+a)} + \frac{1}{(-s-a)}\right] = \frac{s}{s^2-a^2},\ s > |a|$$

69. (a) $A = \int_0^{\infty} e^{-x}\,dx$

$$= \lim_{b\to\infty}\left[-e^{-x}\right]_0^b = 0 - (-1) = 1$$

(b) **Disk:**

$$V = \pi\int_0^{\infty}(e^{-x})^2\,dx$$

$$= \lim_{b\to\infty}\pi\left[-\frac{1}{2}e^{-2x}\right]_0^b = \frac{\pi}{2}$$

(c) **Shell:**

$$V = 2\pi\int_0^{\infty} xe^{-x}\,dx$$

$$= \lim_{b\to\infty}\left\{2\pi\left[-e^{-x}(x+1)\right]_0^b\right\} = 2\pi$$

71. $x^{2/3} + y^{2/3} = 4$

$$\frac{2}{3}x^{-1/3} + \frac{2}{3}y^{-1/3}y' = 0$$

$$y' = \frac{-y^{1/3}}{x^{1/3}}$$

$$\sqrt{1+(y')^2} = \sqrt{1+\frac{y^{2/3}}{x^{2/3}}} = \sqrt{\frac{x^{2/3}+y^{2/3}}{x^{2/3}}} = \sqrt{\frac{4}{x^{2/3}}} = \frac{2}{x^{1/3}},\ (x > 0)$$

$$s = 4\int_0^8 \frac{2}{x^{1/3}}\,dx = \lim_{b\to 0^+}\left[8\cdot\frac{3}{2}x^{2/3}\right]_b^8 = 48$$

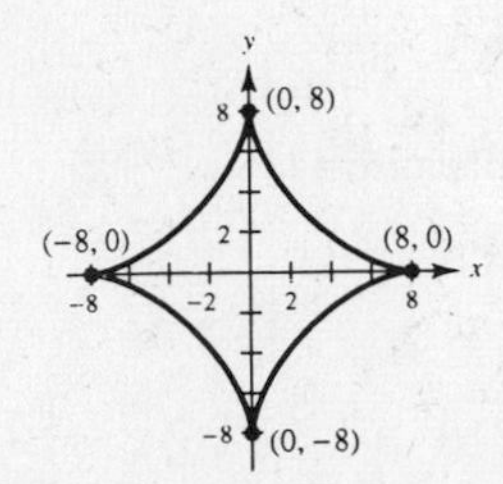

73. $\Gamma(n) = \int_0^{\infty} x^{n-1}e^{-x}\,dx$

(a) $\Gamma(1) = \int_0^{\infty} e^{-x}\,dx = \lim_{b\to\infty}\Big[-e^{-x}\Big]_0^b = 1$

$\Gamma(2) = \int_0^{\infty} xe^{-x}\,dx = \lim_{b\to\infty}\Big[-e^{-x}(x+1)\Big]_0^b = 1$

$\Gamma(3) = \int_0^{\infty} x^2e^{-x}\,dx = \lim_{b\to\infty}\Big[-x^2e^{-x} - 2xe^{-x} - 2e^{-x}\Big]_0^b = 2$

(b) $\Gamma(n+1) = \int_0^{\infty} x^n e^{-x}\,dx = \lim_{b\to\infty}\Big[-x^n e^{-x}\Big]_0^b + \lim_{b\to\infty} n\int_0^b x^{n-1}e^{-x}\,dx = 0 + n\Gamma(n) \qquad (u = x^n,\ dv = e^{-x}\,dx)$

(c) $\Gamma(n) = (n-1)!$

75. (a) $\int_{-\infty}^{\infty} \frac{1}{7}e^{-t/7}\,dt = \int_0^{\infty} \frac{1}{7}e^{-t/7}\,dt = \lim_{b\to\infty}\Big[-e^{-t/7}\Big]_0^b = 1$

(b) $\int_0^4 \frac{1}{7}e^{-t/7}\,dt = \Big[-e^{-t/7}\Big]_0^4 = -e^{-4/7} + 1$

$\approx 0.4353 = 43.53\%$

(c) $\int_0^{\infty} t\Big[\frac{1}{7}e^{-t/7}\Big]\,dt = \lim_{b\to\infty}\Big[-te^{-t/7} - 7e^{-t/7}\Big]_0^b$

$= 0 + 7 = 7$

77. (a) $C = 650{,}000 + \int_0^5 25{,}000\,e^{-0.06t}\,dt = 650{,}000 - \Big[\frac{25{,}000}{0.06}e^{-0.06t}\Big]_0^5 \approx \$757{,}992.41$

(b) $C = 650{,}000 + \int_0^{10} 25{,}000e^{-0.06t}\,dt \approx \$837{,}995.15$

(c) $C = 650{,}000 + \int_0^{\infty} 25{,}000e^{-0.06t}\,dt = 650{,}000 - \lim_{b\to\infty}\Big[\frac{25{,}000}{0.06}e^{-0.06t}\Big]_0^b \approx \$1{,}066{,}666.67$

79. Let $x = a\tan\theta$, $dx = a\sec^2\theta\,d\theta$, $\sqrt{a^2+x^2} = a\sec\theta$.

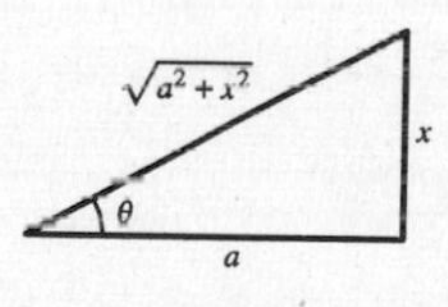

$\int \frac{1}{(a^2+x^2)^{3/2}}\,dx = \int \frac{a\sec^2\theta\,d\theta}{a^3\sec^3\theta} = \frac{1}{a^2}\int \cos\theta\,d\theta$

$= \frac{1}{a^2}\sin\theta = \frac{1}{a^2}\frac{x}{\sqrt{a^2+x^2}}$

Hence,

$P = k\int_1^{\infty} \frac{1}{(a^2+x^2)^{3/2}}\,dx = \frac{k}{a^2}\lim_{b\to\infty}\Big[\frac{x}{\sqrt{a^2+x^2}}\Big]_1^b$

$= \frac{k}{a^2}\Big[1 - \frac{1}{\sqrt{a^2+1}}\Big] = \frac{k\left(\sqrt{a^2+1}-1\right)}{a^2\sqrt{a^2+1}}.$

81. $\frac{10}{x^2-2x} = \frac{10}{x(x-2)} \Longrightarrow x = 0, 2.$

You must analyze three improper integrals, and each must converge in order for the original integral to converge.

$\int_0^3 f(x)\,dx = \int_0^1 f(x)\,dx + \int_1^2 f(x)\,dx + \int_2^3 f(x)\,dx$

83. For $n = 1$,

$$I_1 = \int_0^\infty \frac{x}{(x^2+1)^4}\,dx = \lim_{b\to\infty} \frac{1}{2}\int_0^b (x^2+1)^{-4}(2x\,dx) = \lim_{b\to\infty}\left[-\frac{1}{6}\frac{1}{(x^2+1)^3}\right]_0^b = \frac{1}{6}.$$

For $n > 1$,

$$I_n = \int_0^\infty \frac{x^{2n-1}}{(x^2+1)^{n+3}}\,dx = \lim_{b\to\infty}\left[\frac{-x^{2n-2}}{2(n+2)(x^2+1)^{n+2}}\right]_0^b + \frac{n-1}{n+2}\int_0^\infty \frac{x^{2n-3}}{(x^2+1)^{n+2}}\,dx = 0 + \frac{n-1}{n+2}(I_{n-1})$$

$$u = x^{2n-2},\ du = (2n-2)x^{2n-3}\,dx,\ dv = \frac{x}{(x^2+1)^{n+3}}\,dx,\ v = \frac{-1}{2(n+2)(x^2+1)^{n+2}}$$

(a) $\displaystyle\int_0^\infty \frac{x}{(x^2+1)^4}\,dx = \lim_{b\to\infty}\left[-\frac{1}{6(x^2+1)^3}\right]_0^b = \frac{1}{6}$

(b) $\displaystyle\int_0^\infty \frac{x^3}{(x^2+1)^5}\,dx = \frac{1}{4}\int_0^\infty \frac{x}{(x^2+1)^4}\,dx = \frac{1}{4}\left(\frac{1}{6}\right) = \frac{1}{24}$

(c) $\displaystyle\int_0^\infty \frac{x^5}{(x^2+1)^6} = \frac{2}{5}\int_0^\infty \frac{x^3}{(x^2+1)^5}\,dx = \frac{2}{5}\left(\frac{1}{24}\right) = \frac{1}{60}$

85. False. $f(x) = 1/(x+1)$ is continuous on $[0, \infty)$, $\lim_{x\to\infty} 1/(x+1) = 0$, but $\displaystyle\int_0^\infty \frac{1}{x+1}\,dx = \lim_{b\to\infty}\Big[\ln|x+1|\Big]_0^b = \infty.$

Diverges

87. True

Review Exercises for Chapter 7

1. $\displaystyle\int x\sqrt{x^2-1}\,dx = \frac{1}{2}\int (x^2-1)^{1/2}(2x)\,dx$

$\displaystyle= \frac{1}{2}\frac{(x^2-1)^{3/2}}{3/2} + C$

$\displaystyle= \frac{1}{3}(x^2-1)^{3/2} + C$

3. $\displaystyle\int \frac{x}{x^2-1}\,dx = \frac{1}{2}\int \frac{2x}{x^2-1}\,dx$

$\displaystyle= \frac{1}{2}\ln|x^2-1| + C$

5. $\displaystyle\int \frac{\ln(2x)}{x}\,dx = \frac{(\ln 2x)^2}{2} + C$

7. $\displaystyle\int \frac{16}{\sqrt{16-x^2}}\,dx = 16\arcsin\left(\frac{x}{4}\right) + C$

9. $\displaystyle\int e^{2x}\sin 3x\,dx = -\frac{1}{3}e^{2x}\cos 3x + \frac{2}{3}\int e^{2x}\cos 3x\,dx$

$\displaystyle= -\frac{1}{3}e^{2x}\cos 3x + \frac{2}{3}\left(\frac{1}{3}e^{2x}\sin 3x - \frac{2}{3}\int e^{2x}\sin 3x\,dx\right)$

$\displaystyle\frac{13}{9}\int e^{2x}\sin 3x\,dx = -\frac{1}{3}e^{2x}\cos 3x + \frac{2}{9}e^{2x}\sin 3x$

$\displaystyle\int e^{2x}\sin 3x\,dx = \frac{e^{2x}}{13}(2\sin 3x - 3\cos 3x) + C$

(1) $dv = \sin 3x\,dx \Rightarrow v = -\frac{1}{3}\cos 3x$

$u = e^{2x} \Rightarrow du = 2e^{2x}\,dx$

(2) $dv = \cos 3x\,dx \Rightarrow v = \frac{1}{3}\sin 3x$

$u = e^{2x} \Rightarrow du = 2e^{2x}\,dx$

11. $u = x,\ du = dx,\ dv = (x-5)^{1/2}\,dx,\ v = \frac{2}{3}(x-5)^{3/2}$

$$\int x\sqrt{x-5}\,dx = \frac{2}{3}x(x-5)^{3/2} - \int \frac{2}{3}(x-5)^{3/2}\,dx$$

$$= \frac{2}{3}x(x-5)^{3/2} - \frac{4}{15}(x-5)^{5/2} + C$$

$$= (x-5)^{3/2}\left[\frac{2}{3}x - \frac{4}{15}(x-5)\right] + C$$

$$= (x-5)^{3/2}\left[\frac{6}{15}x + \frac{4}{3}\right] + C$$

$$= \frac{2}{15}(x-5)^{3/2}[3x+10] + C$$

13. $$\int x^2 \sin 2x\,dx = -\frac{1}{2}x^2\cos 2x + \int x\cos 2x\,dx$$

$$= -\frac{1}{2}x^2\cos 2x + \frac{1}{2}x\sin 2x - \frac{1}{2}\int \sin 2x\,dx$$

$$= -\frac{1}{2}x^2\cos 2x + \frac{x}{2}\sin 2x + \frac{1}{4}\cos 2x + C$$

(1) $dv = \sin 2x\,dx \Rightarrow v = -\frac{1}{2}\cos 2x$

$u = x^2 \Rightarrow du = 2x\,dx$

(2) $dv = \cos 2x\,dx \Rightarrow v = \frac{1}{2}\sin 2x$

$u = x \Rightarrow du = dx$

15. $$\int x \arcsin 2x\,dx = \frac{x^2}{2}\arcsin 2x - \int \frac{x^2}{\sqrt{1-4x^2}}\,dx$$

$$= \frac{x^2}{2}\arcsin 2x - \frac{1}{8}\int \frac{2(2x)^2}{\sqrt{1-(2x)^2}}\,dx$$

$$= \frac{x^2}{2}\arcsin 2x - \frac{1}{8}\left(\frac{1}{2}\right)\left[-(2x)\sqrt{1-4x^2} + \arcsin 2x\right] + C$$ (by Formula 43 of Integration Tables)

$$= \frac{1}{16}\left[(8x^2-1)\arcsin 2x + 2x\sqrt{1-4x^2}\right] + C$$

$dv = x\,dx \Rightarrow v = \frac{x^2}{2}$

$u = \arcsin 2x \Rightarrow du = \frac{2}{\sqrt{1-4x^2}}\,dx$

17. $$\int \cos^3(\pi x - 1)\,dx = \int [1 - \sin^2(\pi x - 1)]\cos(\pi x - 1)\,dx$$

$$= \frac{1}{\pi}\left[\sin(\pi x - 1) - \frac{1}{3}\sin^3(\pi x - 1)\right] + C$$

$$= \frac{1}{3\pi}\sin(\pi x - 1)[3 - \sin^2(\pi x - 1)] + C$$

$$= \frac{1}{3\pi}\sin(\pi x - 1)[3 - (1 - \cos^2(\pi x - 1))] + C$$

$$= \frac{1}{3\pi}\sin(\pi x - 1)[2 + \cos^2(\pi x - 1)] + C$$

19. $$\int \sec^4\left(\frac{x}{2}\right)dx = \int \left[\tan^2\left(\frac{x}{2}\right) + 1\right]\sec^2\left(\frac{x}{2}\right)dx$$

$$= \int \tan^2\left(\frac{x}{2}\right)\sec^2\left(\frac{x}{2}\right)dx + \int \sec^2\left(\frac{x}{2}\right)dx$$

$$= \frac{2}{3}\tan^3\left(\frac{x}{2}\right) + 2\tan\left(\frac{x}{2}\right) + C = \frac{2}{3}\left[\tan^3\left(\frac{x}{2}\right) + 3\tan\left(\frac{x}{2}\right)\right] + C$$

21. $$\int \frac{1}{1-\sin\theta}\,d\theta = \int \frac{1+\sin\theta}{\cos^2\theta}\,d\theta = \int (\sec^2\theta + \sec\theta\tan\theta)\,d\theta = \tan\theta + \sec\theta + C$$

23. $\displaystyle\int \frac{-12}{x^2\sqrt{4-x^2}}\,dx = \int \frac{-24\cos\theta\,d\theta}{(4\sin^2\theta)(2\cos\theta)}$

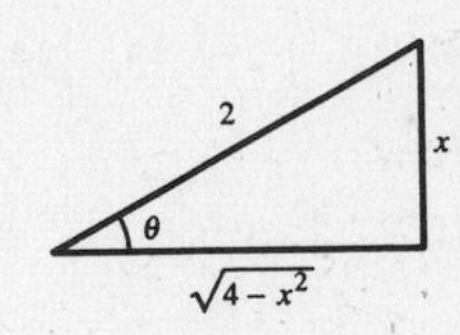

$$= -3\int \csc^2\theta\,d\theta$$

$$= 3\cot\theta + C$$

$$= \frac{3\sqrt{4-x^2}}{x} + C$$

$x = 2\sin\theta,\ dx = 2\cos\theta\,d\theta,\ \sqrt{4-x^2} = 2\cos\theta$

25. $x = 2\tan\theta$

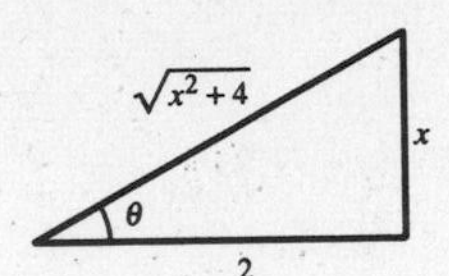

$dx = 2\sec^2\theta\,d\theta$

$4 + x^2 = 4\sec^2\theta$

$$\int \frac{x^3}{\sqrt{4+x^2}}\,dx = \int \frac{8\tan^3\theta}{2\sec\theta}\,2\sec^2\theta\,d\theta$$

$$= 8\int \tan^3\theta\sec\theta\,d\theta$$

$$= 8\int (\sec^2\theta - 1)\tan\theta\sec\theta\,d\theta$$

$$= 8\left[\frac{\sec^3\theta}{3} - \sec\theta\right] + C$$

$$= 8\left[\frac{(x^2+4)^{3/2}}{24} - \frac{\sqrt{x^2+4}}{2}\right] + C$$

$$= \sqrt{x^2+4}\left[\frac{1}{3}(x^2+4) - 4\right] + C$$

$$= \frac{1}{3}x^2\sqrt{x^2+4} - \frac{8}{3}\sqrt{x^2+4} + C$$

$$= \frac{1}{3}(x^2+4)^{1/2}(x^2-8) + C$$

27. $\displaystyle\int \sqrt{4-x^2}\,dx = \int (2\cos\theta)(2\cos\theta)\,d\theta$

$$= 2\int (1 + \cos 2\theta)\,d\theta$$

$$= 2\left(\theta + \frac{1}{2}\sin 2\theta\right) + C$$

$$= 2(\theta + \sin\theta\cos\theta) + C$$

$$= 2\left[\arcsin\left(\frac{x}{2}\right) + \frac{x}{2}\left(\frac{\sqrt{4-x^2}}{2}\right)\right] + C$$

$$= \frac{1}{2}\left[4\arcsin\left(\frac{x}{2}\right) + x\sqrt{4-x^2}\right] + C$$

$x = 2\sin\theta,\ dx = 2\cos\theta\,d\theta,\ \sqrt{4-x^2} = 2\cos\theta$

29. (a) $x = 2\tan\theta,$

$dx = 2\sec^2\theta\,d\theta$

$\sqrt{4+x^2} = 2\sec\theta$

$$\int \frac{x^3}{\sqrt{4+x^2}}\,dx = \int \frac{8\tan^3\theta}{2\sec\theta} 2\sec^2\theta\,d\theta$$

$$= 8\int \frac{\sin^3\theta}{\cos^4\theta}\,d\theta$$

$$= 8\int (1-\cos^2\theta)\cos^{-4}\theta\sin\theta\,d\theta$$

$$= 8\int (\cos^{-4}\theta - \cos^{-2}\theta)\sin\theta\,d\theta$$

$$= 8\left[\frac{\cos^{-3}\theta}{3} + \frac{\cos^{-1}\theta}{-1}\right] + C$$

$$= \frac{8}{3}\sec\theta(\sec^2\theta - 3) + C$$

$$= \frac{8}{3}\frac{\sqrt{4+x^2}}{2}\left(\frac{4+x^2}{4} - 3\right) + C$$

$$= \frac{1}{3}\sqrt{4+x^2}(x^2-8) + C$$

(b) $$\int \frac{x^3}{\sqrt{4+x^2}}\,dx = \int \frac{x^2}{\sqrt{4+x^2}}x\,dx$$

$$= \int \frac{(u^2-4)}{u}u\,du$$

$$= \int (u^2-4)\,du$$

$$= \frac{1}{3}u^3 - 4u + C$$

$$= \frac{u}{3}(u^2-12) + C$$

$$= \frac{\sqrt{4+x^2}}{3}(x^2-8) + C$$

$u^2 = 4 + x^2,\ 2u\,du = 2x\,dx$

(c) $$\int \frac{x^3}{\sqrt{4+x^2}}\,dx = x^2\sqrt{4+x^2} - \int 2x\sqrt{4+x^2}\,dx$$

$$= x^2\sqrt{4+x^2} - \frac{2}{3}(4+x^2)^{3/2} + C$$

$$= \frac{\sqrt{4+x^2}}{3}(x^2-8) + C$$

$$dv = \frac{x}{\sqrt{4+x^2}}\,dx \implies v = \sqrt{4+x^2}$$

$$u = x^2 \implies du = 2x\,dx$$

31. $$\frac{x-28}{x^2-x-6} = \frac{A}{x-3} + \frac{B}{x+2}$$

$$x - 28 = A(x+2) + B(x-3)$$

$$x = -2 \implies -30 = B(-5) \implies B = 6$$

$$x = 3 \implies -25 = A(5) \implies A = -5$$

$$\int \frac{x-28}{x^2-x-6}\,dx = \int\left(\frac{-5}{x-3} + \frac{6}{x+2}\right)dx = -5\ln|x-3| + 6\ln|x+2| + C$$

33. $$\frac{x^2+2x}{(x-1)(x^2+1)} = \frac{A}{x-1} + \frac{Bx+C}{x^2+1}$$

$$x^2 + 2x = A(x^2+1) + (Bx+C)(x-1)$$

Let $x = 1$: $3 = 2A \implies A = \frac{3}{2}$

Let $x = 0$: $0 = A - C \implies C = \frac{3}{2}$

Let $x = 2$: $8 = 5A + 2B + C \implies B = -\frac{1}{2}$

$$\int \frac{x^2+2x}{x^3-x^2+x-1}\,dx = \frac{3}{2}\int\frac{1}{x-1}\,dx - \frac{1}{2}\int\frac{x-3}{x^2+1}\,dx$$

$$= \frac{3}{2}\int\frac{1}{x-1}\,dx - \frac{1}{4}\int\frac{2x}{x^2+1}\,dx + \frac{3}{2}\int\frac{1}{x^2+1}\,dx$$

$$= \frac{3}{2}\ln|x-1| - \frac{1}{4}\ln|x^2+1| + \frac{3}{2}\arctan x + C$$

$$= \frac{1}{4}[6\ln|x-1| - \ln(x^2+1) + 6\arctan x] + C$$

35. $\dfrac{x^2}{x^2+2x-15} = 1 + \dfrac{15-2x}{x^2+2x-15}$

$$\frac{15-2x}{(x-3)(x+5)} = \frac{A}{x-3} + \frac{B}{x+5}$$

$$15 - 2x = A(x+5) + B(x-3)$$

Let $x = 3$: $9 = 8A \Rightarrow A = \dfrac{9}{8}$

Let $x = -5$: $25 = -8B \Rightarrow B = -\dfrac{25}{8}$

$$\int \frac{x^2}{x^2+2x-15}\,dx = \int dx + \frac{9}{8}\int \frac{1}{x-3}\,dx - \frac{25}{8}\int \frac{1}{x+5}\,dx$$

$$= x + \frac{9}{8}\ln|x-3| - \frac{25}{8}\ln|x+5| + C$$

37. $\displaystyle\int \frac{x}{(2+3x)^2}\,dx = \frac{1}{9}\left[\frac{2}{2+3x} + \ln|2+3x|\right] + C$

(Formula 4)

39. $\displaystyle\int \frac{x}{1+\sin x^2}\,dx = \frac{1}{2}\int \frac{1}{1+\sin u}\,du$ $\quad (u = x^2)$

$$= \frac{1}{2}[\tan u - \sec u] + C \quad \text{(Formula 56)}$$

$$= \frac{1}{2}[\tan x^2 - \sec x^2] + C$$

41. $\displaystyle\int \frac{x}{x^2+4x+8}\,dx = \frac{1}{2}\left[\ln|x^2+4x+8| - 4\int \frac{1}{x^2+4x+8}\,dx\right]$ (Formula 15)

$$= \frac{1}{2}[\ln|x^2+4x+8|] - 2\left[\frac{2}{\sqrt{32-16}}\arctan\left(\frac{2x+4}{\sqrt{32-16}}\right)\right] + C \quad \text{(Formula 14)}$$

$$= \frac{1}{2}\ln|x^2+4x+8| - \arctan\left(1 + \frac{x}{2}\right) + C$$

43. $\displaystyle\int \frac{1}{\sin \pi x \cos \pi x}\,dx = \frac{1}{\pi}\int \frac{1}{\sin \pi x \cos \pi x}(\pi)\,dx \quad (u = \pi x)$

$$= \frac{1}{\pi}\ln|\tan \pi x| + C \quad \text{(Formula 58)}$$

45. $dv = dx \Rightarrow v = x$

$u = (\ln x)^n \Rightarrow du = n(\ln x)^{n-1}\dfrac{1}{x}dx$

$$\int (\ln x)^n\,dx = x(\ln x)^n - n\int (\ln x)^{n-1}\,dx$$

47. $\displaystyle\int \theta \sin\theta \cos\theta\,d\theta = \frac{1}{2}\int \theta \sin 2\theta\,d\theta$

$$= -\frac{1}{4}\theta\cos 2\theta + \frac{1}{4}\int \cos 2\theta\,d\theta = -\frac{1}{4}\theta\cos 2\theta + \frac{1}{8}\sin 2\theta + C = \frac{1}{8}(\sin 2\theta - 2\theta\cos 2\theta) + C$$

$dv = \sin 2\theta\,d\theta \Rightarrow v = -\dfrac{1}{2}\cos 2\theta$

$u = \theta \Rightarrow du = d\theta$

49. $\int \frac{x^{1/4}}{1+x^{1/2}}\,dx = 4\int \frac{u(u^3)}{1+u^2}\,du$

$= 4\int \left(u^2 - 1 + \frac{1}{u^2+1}\right) du$

$= 4\left(\frac{1}{3}u^3 - u + \arctan u\right) + C$

$= \frac{4}{3}[x^{3/4} - 3x^{1/4} + 3\arctan(x^{1/4})] + C$

$u = \sqrt[4]{x}, x = u^4, dx = 4u^3\,du$

51. $\int \sqrt{1+\cos x}\,dx = \int \frac{\sin x}{\sqrt{1-\cos x}}\,dx$

$= \int (1-\cos x)^{-1/2}(\sin x)\,dx$

$= 2\sqrt{1-\cos x} + C$

$u = 1 - \cos x, du = \sin x\,dx$

53. $\int \cos x \ln(\sin x)\,dx = \sin x \ln(\sin x) - \int \cos x\,dx$

$= \sin x \ln(\sin x) - \sin x + C$

$dv = \cos x\,dx \implies v = \sin x$

$u = \ln(\sin x) \implies du = \frac{\cos x}{\sin x}\,dx$

55. $y = \int \frac{9}{x^2-9}\,dx = \frac{3}{2}\ln\left|\frac{x-3}{x+3}\right| + C$

(by Formula 24 of Integration Tables)

57. $y = \int \ln(x^2+x)dx = x\ln|x^2+x| - \int \frac{2x^2+x}{x^2+x}\,dx$

$= x\ln|x^2+x| - \int \frac{2x+1}{x+1}\,dx$

$= x\ln|x^2+x| - \int 2dx + \int \frac{1}{x+1}\,dx$

$= x\ln|x^2+x| - 2x + \ln|x+1| + C$

$dv = dx \implies v = x$

$u = \ln(x^2+x) \implies du = \frac{2x+1}{x^2+x}\,dx$

59. $\int_2^{\sqrt{5}} x(x^2-4)^{3/2}\,dx = \left[\frac{1}{5}(x^2-4)^{5/2}\right]_2^{\sqrt{5}} = \frac{1}{5}$

61. $\int_1^4 \frac{\ln x}{x}\,dx = \left[\frac{1}{2}(\ln x)^2\right]_1^4 = \frac{1}{2}(\ln 4)^2 = 2(\ln 2)^2 \approx 0.961$

63. $\int_0^{\pi} x\sin x\,dx = \left[-x\cos x + \sin x\right]_0^{\pi} = \pi$

65. $A = \int_0^4 x\sqrt{4-x}\,dx = \int_2^0 (4-u^2)u(-2u)\,du$

$= \int_2^0 2(u^4 - 4u^2)\,du$

$= \left[2\left(\frac{u^5}{5} - \frac{4u^3}{3}\right)\right]_2^0 = \frac{128}{15}$

$u = \sqrt{4-x}, x = 4 - u^2, dx = -2u\,du$

67. By symmetry, $\bar{x} = 0, A = \frac{1}{2}\pi$.

$\bar{y} = \frac{2}{\pi}\left(\frac{1}{2}\right)\int_{-1}^{1} (\sqrt{1-x^2})^2\,dx = \frac{1}{\pi}\left[x - \frac{1}{3}x^3\right]_{-1}^{1} = \frac{4}{3\pi}$

$(\bar{x}, \bar{y}) = \left(0, \frac{4}{3\pi}\right)$

69. $s = \int_0^{\pi} \sqrt{1+\cos^2 x}\,dx \approx 3.82$

71. $\lim_{x\to 1}\left[\frac{(\ln x)^2}{x-1}\right] = \lim_{x\to 1}\left[\frac{2(1/x)\ln x}{1}\right] = 0$

73. $\lim_{x\to\infty}\frac{e^{2x}}{x^2} = \lim_{x\to\infty}\frac{2e^{2x}}{2x} = \lim_{x\to\infty}\frac{4e^{2x}}{2} = \infty$

75. $y = \lim_{x\to\infty}(\ln x)^{2/x}$

$\ln y = \lim_{x\to\infty}\frac{2\ln(\ln x)}{x} = \lim_{x\to\infty}\left[\frac{2/(x\ln x)}{1}\right] = 0$

Since $\ln y = 0, y = 1$.

77. $\lim_{n\to\infty} 1000\left(1 + \frac{0.09}{n}\right)^n = 1000 \lim_{n\to\infty}\left(1 + \frac{0.09}{n}\right)^n$

Let $y = \lim_{n\to\infty}\left(1 + \frac{0.09}{n}\right)^n$.

$$\ln y = \lim_{n\to\infty} n \ln\left(1 + \frac{0.09}{n}\right) = \lim_{n\to\infty} \frac{\ln\left(1 + \frac{0.09}{n}\right)}{\frac{1}{n}} = \lim_{n\to\infty}\left(\frac{\frac{-0.09/n^2}{1 + (0.09/n)}}{-\frac{1}{n^2}}\right) = \lim_{n\to\infty} \frac{0.09}{1 + \left(\frac{0.09}{n}\right)} = 0.09$$

Thus, $\ln y = 0.09 \Rightarrow y = e^{0.09}$ and $\lim_{n\to\infty} 1000\left(1 + \frac{0.09}{n}\right)^n = 1000e^{0.09} \approx 1094.17$.

79. $\int_0^{16} \frac{1}{\sqrt[4]{x}}\,dx = \lim_{b\to 0^+}\left[\frac{4}{3}x^{3/4}\right]_b^{16} = \frac{32}{3}$

Converges

81. $\int_1^{\infty} x^2 \ln x\,dx = \lim_{b\to\infty}\left[\frac{x^3}{9}(-1 + 3\ln x)\right]_1^b = \infty$

Diverges

83.
$$\int_0^{t_0} 500{,}000e^{-0.05t}\,dt = \left[\frac{500{,}000}{-0.05}e^{-0.05t}\right]_0^{t_0}$$
$$= \frac{-500{,}000}{0.05}(e^{-0.05t_0} - 1)$$
$$= 10{,}000{,}000(1 - e^{-0.05t_0})$$

(a) $t_0 = 20$: \$6,321,205.59

(b) $t_0 \to \infty$: \$10,000,000

85. (a) $P(13 \le x < \infty) = \frac{1}{0.95\sqrt{2\pi}}\int_{13}^{\infty} e^{-(x-12.9)^2/2(0.95)^2}\,dx \approx 0.4581$

(b) $P(15 \le x < \infty) = \frac{1}{0.95\sqrt{2\pi}}\int_{15}^{\infty} e^{-(x-12.9)^2/2(0.95)^2}\,dx \approx 0.0135$

Problem Solving for Chapter 7

1. (a) $\int_{-1}^{1} (1 - x^2)\,dx = \left[x - \frac{x^3}{3}\right]_{-1}^{1} = 2\left(1 - \frac{1}{3}\right) = \frac{4}{3}$

$$\int_{-1}^{1} (1 - x^2)^2\,dx = \int_{-1}^{1} (1 - 2x^2 + x^4)\,dx = \left[x - \frac{2x^3}{3} + \frac{x^5}{5}\right]_{-1}^{1} = 2\left(1 - \frac{2}{3} + \frac{1}{5}\right) = \frac{16}{15}$$

(b) Let $x = \sin u$, $dx = \cos u\,du$, $1 - x^2 = 1 - \sin^2 u = \cos^2 u$.

$$\int_{-1}^{1} (1 - x^2)^n\,dx = \int_{-\pi/2}^{\pi/2} (\cos^2 u)^n \cos u\,du$$
$$= \int_{-\pi/2}^{\pi/2} \cos^{2n+1} u\,du$$
$$= 2\left[\frac{2}{3}\cdot\frac{4}{5}\cdot\frac{6}{7}\cdot\cdot\cdot\frac{(2n)}{(2n + 1)}\right] \quad \text{(Wallis's Formula)}$$
$$= 2\left[\frac{2^2\cdot 4^2\cdot 6^2\cdot\cdot\cdot(2n)^2}{2\cdot 3\cdot 4\cdot 5\cdot\cdot\cdot(2n)(2n + 1)}\right]$$
$$= \frac{2(2^{2n})(n!)^2}{(2n + 1)!} = \frac{2^{2n+1}(n!)^2}{(2n + 1)!}$$

3.

$$\lim_{x\to\infty}\left(\frac{x+c}{x-c}\right)^x = 9$$

$$\lim_{x\to\infty} x \ln\left(\frac{x+c}{x-c}\right) = \ln 9$$

$$\lim_{x\to\infty} \frac{\ln(x+c) - \ln(x-c)}{1/x} = \ln 9$$

$$\lim_{x\to\infty} \frac{\dfrac{1}{x+c} - \dfrac{1}{x-c}}{-\dfrac{1}{x^2}} = \ln 9$$

$$\lim_{x\to\infty} \frac{-2c}{(x+c)(x-c)}(-x^2) = \ln 9$$

$$\lim_{x\to\infty}\left(\frac{2cx^2}{x^2-c^2}\right) = \ln 9$$

$$2c = \ln 9$$

$$2c = 2\ln 3$$

$$c = \ln 3$$

5. $\sin\theta = \dfrac{PB}{OP} = PB$, $\cos\theta = OB$

$AQ = \widehat{AP} = \theta$

$BR = OR + OB = OR + \cos\theta$

The triangles $\triangle AQR$ and $\triangle BPR$ are similar:

$$\frac{AR}{AQ} = \frac{BR}{BP} \Rightarrow \frac{OR+1}{\theta} = \frac{OR+\cos\theta}{\sin\theta}$$

$$\sin\theta(OR) + \sin\theta = (OR)\theta + \theta\cos\theta$$

$$OR = \frac{\theta\cos\theta - \sin\theta}{\sin\theta - \theta}$$

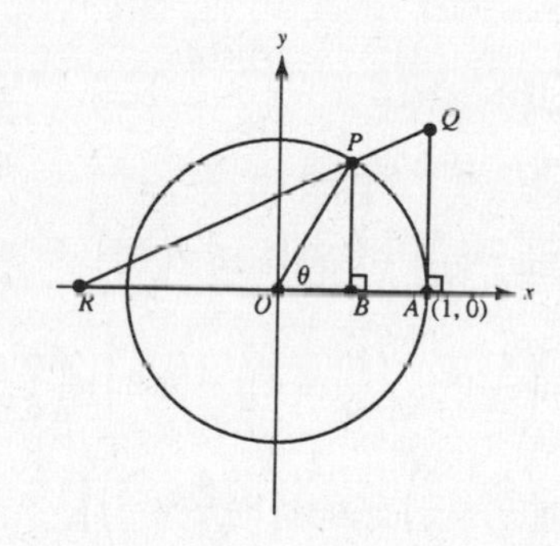

$$\lim_{\theta\to 0^+} OR = \lim_{\theta\to 0^+} \frac{\theta\cos\theta - \sin\theta}{\sin\theta - \theta}$$

$$= \lim_{\theta\to 0^+} \frac{-\theta\sin\theta + \cos\theta - \cos\theta}{\cos\theta - 1}$$

$$= \lim_{\theta\to 0^+} \frac{-\theta\sin\theta}{\cos\theta - 1}$$

$$= \lim_{\theta\to 0^+} \frac{-\sin\theta - \theta\cos\theta}{-\sin\theta}$$

$$= \lim_{\theta\to 0^+} \frac{\cos\theta + \cos\theta - \theta\sin\theta}{\cos\theta}$$

$$= 2$$

7. (a)

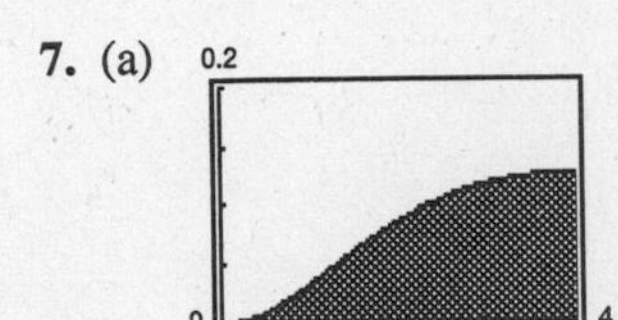

Area ≈ 0.2986

(b) Let $x = 3\tan\theta$, $dx = 3\sec^2\theta\,d\theta$, $x^2 + 9 = 9\sec^2\theta$.

$$\int \frac{x^2}{(x^2+9)^{3/2}}\,dx = \int \frac{9\tan^2\theta}{(9\sec^2\theta)^{3/2}}(3\sec^2\theta\,d\theta)$$
$$= \int \frac{\tan^2\theta}{\sec\theta}\,d\theta$$
$$= \int \frac{\sin^2\theta}{\cos\theta}\,d\theta$$
$$= \int \frac{1-\cos^2\theta}{\cos\theta}\,d\theta$$
$$= \ln|\sec\theta + \tan\theta| - \sin\theta + C$$

$$\text{Area} = \int_0^4 \frac{x^2}{(x^2+9)^{3/2}}\,dx = \Big[\ln|\sec\theta + \tan\theta| - \sin\theta\Big]_0^{\tan^{-1}(4/3)}$$
$$= \left[\ln\left(\frac{\sqrt{x^2+9}}{3} + \frac{x}{3}\right) - \frac{x}{\sqrt{x^2+9}}\right]_0^4$$
$$= \ln\left(\frac{5}{3} + \frac{4}{3}\right) - \frac{4}{5} = \ln 3 - \frac{4}{5}$$

(c) $x = 3\sinh u$, $dx = 3\cosh u\,du$, $x^2 + 9 = 9\sinh^2 u + 9 = 9\cosh^2 u$

$$A = \int_0^4 \frac{x^2}{(x^2+9)^{3/2}}\,dx = \int_0^{\sinh^{-1}(4/3)} \frac{9\sinh^2 u}{(9\cosh^2 u)^{3/2}}(3\cosh u\,du)$$
$$= \int_0^{\sinh^{-1}(4/3)} \tanh^2 u\,du$$
$$= \int_0^{\sinh^{-1}(4/3)} (1 - \operatorname{sech}^2 u)\,du$$
$$= \Big[u - \tanh u\Big]_0^{\sinh^{-1}(4/3)}$$
$$= \sinh^{-1}\left(\frac{4}{3}\right) - \tanh\left(\sinh^{-1}\left(\frac{4}{3}\right)\right)$$
$$= \ln\left(\frac{4}{3} + \sqrt{\frac{16}{9} + 1}\right) - \tanh\left[\ln\left(\frac{4}{3} + \sqrt{\frac{16}{9} + 1}\right)\right]$$
$$= \ln\left(\frac{4}{3} + \frac{5}{3}\right) - \tanh\left(\ln\left(\frac{4}{3} + \frac{5}{3}\right)\right)$$
$$= \ln 3 - \tanh(\ln 3)$$
$$= \ln 3 - \frac{3 - (1/3)}{3 + (1/3)}$$
$$= \ln 3 - \frac{4}{5}$$

9. $y = \ln(1 - x^2),\ y' = \dfrac{-2x}{1 - x^2}$

$$1 + (y')^2 = 1 + \frac{4x^2}{(1 - x^2)^2} = \frac{1 - 2x^2 + x^4 + 4x^2}{(1 - x^2)^2} = \left(\frac{1 + x^2}{1 - x^2}\right)^2$$

$$\begin{aligned}
\text{Arc length} &= \int_0^{1/2} \sqrt{1 + (y')^2}\, dx \\
&= \int_0^{1/2} \left(\frac{1 + x^2}{1 - x^2}\right) dx \\
&= \int_0^{1/2} \left(-1 + \frac{2}{1 - x^2}\right) dx \\
&= \int_0^{1/2} \left(-1 + \frac{1}{x + 1} + \frac{1}{1 - x}\right) dx \\
&= \Big[-x + \ln(1 + x) - \ln(1 - x)\Big]_0^{1/2} \\
&= \left(-\frac{1}{2} + \ln\frac{3}{2} - \ln\frac{1}{2}\right) \\
&= -\frac{1}{2} + \ln 3 - \ln 2 + \ln 2 \\
&= \ln 3 - \frac{1}{2} \approx 0.5986
\end{aligned}$$

11. Consider $\displaystyle\int \frac{1}{\ln x}\, dx$.

Let $u = \ln x,\ du = \dfrac{1}{x}\, dx,\ x = e^u$. Then $\displaystyle\int \frac{1}{\ln x}\, dx = \int \frac{1}{u} e^u\, du = \int \frac{e^u}{u}\, du.$

If $\displaystyle\int \frac{1}{\ln x}\, dx$ were elementary, then $\displaystyle\int \frac{e^u}{u}\, du$ would be too, which is false.

Hence, $\displaystyle\int \frac{1}{\ln x}\, dx$ is not elementary.

13. $x^4 + 1 = (x^2 + ax + b)(x^2 + cx + d)$

$\qquad = x^4 + (a + c)x^3 + (ac + b + d)x^2 + (ad + bc)x + bd$

$a = -c,\ b = d = 1,\ a = \sqrt{2}$

$\qquad x^4 + 1 = \left(x^2 + \sqrt{2}x + 1\right)\left(x^2 - \sqrt{2}x + 1\right)$

$$\begin{aligned}
\int_0^1 \frac{1}{x^4 + 1}\, dx &= \int_0^1 \frac{Ax + B}{x^2 + \sqrt{2}x + 1}\, dx + \int_0^1 \frac{Cx + D}{x^2 - \sqrt{2}x + 1}\, dx \\
&= \int_0^1 \frac{\frac{1}{2} + \frac{\sqrt{2}}{4}x}{x^2 + \sqrt{2}x + 1}\, dx - \int_0^1 \frac{-\frac{1}{2} + \frac{\sqrt{2}}{4}x}{x^2 + \sqrt{2}x + 1}\, dx \\
&= \frac{\sqrt{2}}{4}\Big[\arctan\left(\sqrt{2}x + 1\right) + \arctan\left(\sqrt{2}x - 1\right)\Big]_0^1 + \frac{\sqrt{2}}{8}\Big[\ln\left(x^2 + \sqrt{2}x + 1\right) - \ln\left(x^2 - \sqrt{2}x + 1\right)\Big]_0^1 \\
&= \frac{\sqrt{2}}{4}\left[\arctan\left(\sqrt{2} + 1\right) + \arctan\left(\sqrt{2} - 1\right)\right] + \frac{\sqrt{2}}{8}\left[\ln\left(2 + \sqrt{2}\right) - \ln\left(2 - \sqrt{2}\right)\right] - \frac{\sqrt{2}}{4}\left[\frac{\pi}{4} - \frac{\pi}{4}\right] - \frac{\sqrt{2}}{8}[0] \\
&\approx 0.5554 + 0.3116 \\
&\approx 0.8670
\end{aligned}$$

15. Using a graphing utility,

(a) $\lim_{x\to 0^+}\left(\cot x + \frac{1}{x}\right) = \infty$

(b) $\lim_{x\to 0^+}\left(\cot x - \frac{1}{x}\right) = 0$

(c) $\lim_{x\to 0^+}\left(\cot x + \frac{1}{x}\right)\left(\cot x - \frac{1}{x}\right) \approx -\frac{2}{3}.$

Analytically,

(a) $\lim_{x\to 0^+}\left(\cot x + \frac{1}{x}\right) = \infty + \infty = \infty$

(b) $$\lim_{x\to 0^+}\left(\cot x - \frac{1}{x}\right) = \lim_{x\to 0^+}\frac{x\cot x - 1}{x} = \lim_{x\to 0^+}\frac{x\cos x - \sin x}{x\sin x}$$

$$= \lim_{x\to 0^+}\frac{\cos x - x\sin x - \cos x}{\sin x + x\cos x} = \lim_{x\to 0^+}\frac{-x\sin x}{\sin x + x\cos x}$$

$$= \lim_{x\to 0^+}\frac{-\sin x - x\cos x}{\cos x + \cos x - x\sin x} = 0.$$

(c) $$\left(\cot x + \frac{1}{x}\right)\left(\cot x - \frac{1}{x}\right) = \cot^2 x - \frac{1}{x^2}$$

$$= \frac{x^2\cot^2 x - 1}{x^2}$$

$$\lim_{x\to 0^+}\frac{x^2\cot^2 x - 1}{x^2} = \lim_{x\to 0^+}\frac{2x\cot^2 x - 2x^2\cot x\csc^2 x}{2x}$$

$$= \lim_{x\to 0^+}\frac{\cot^2 x - x\cot x\csc^2 x}{1}$$

$$= \lim_{x\to 0^+}\frac{\cos^2 x\sin x - x\cos x}{\sin^3 x}$$

$$= \lim_{x\to 0^+}\frac{(1 - \sin^2 x)\sin x - x\cos x}{\sin^3 x}$$

$$= \lim_{x\to 0^+}\frac{\sin x - x\cos x}{\sin^3 x} - 1$$

Now, $$\lim_{x\to 0^+}\frac{\sin x - x\cos x}{\sin^3 x} = \lim_{x\to 0^+}\frac{\cos x - \cos x + x\sin x}{3\sin^2 x\cos x}$$

$$= \lim_{x\to 0^+}\frac{x}{3\sin x \cdot \cos x}$$

$$= \lim_{x\to 0^+}\left(\frac{x}{\sin x}\right)\frac{1}{3\cos x} = \frac{1}{3}.$$

Thus, $\lim_{x\to 0^+}\left(\cot x + \frac{1}{x}\right)\left(\cot x - \frac{1}{x}\right) = \frac{1}{3} - 1 = -\frac{2}{3}.$

The form $0 \cdot \infty$ is indeterminant.

17. $\dfrac{x^3 - 3x^2 + 1}{x^4 - 13x^2 + 12x} = \dfrac{P_1}{x} + \dfrac{P_2}{x-1} + \dfrac{P_3}{x+4} + \dfrac{P_4}{x-3} \implies c_1 = 0,\ c_2 = 1,\ c_3 = -4,\ c_4 = 3$

$N(x) = x^3 - 3x^2 + 1$

$D'(x) = 4x^3 - 26x + 12$

$$P_1 = \frac{N(0)}{D'(0)} = \frac{1}{12}$$

$$P_2 = \frac{N(1)}{D'(1)} = \frac{-1}{-10} = \frac{1}{10}$$

$$P_3 = \frac{N(-4)}{D'(-4)} = \frac{-111}{-140} = \frac{111}{140}$$

$$P_4 = \frac{N(3)}{D'(3)} = \frac{1}{42}$$

Thus, $\dfrac{x^3 - 3x^2 + 1}{x^4 - 13x^2 + 12x} = \dfrac{1/12}{x} + \dfrac{1/10}{x-1} + \dfrac{111/140}{x+4} + \dfrac{1/42}{x-3}.$

19. By parts,

$$\int_a^b f(x)g''(x)\,dx = \Big[f(x)g'(x)\Big]_a^b - \int_a^b f'(x)g'(x)\,dx$$

$$= -\int_a^b f'(x)g'(x)\,dx \quad 8$$

$$= \Big[-f'(x)g(x)\Big]_a^b + \int_a^b g(x)f''(x)\,dx$$

$$= \int_a^b f''(x)g(x)\,dx.$$

CHAPTER 8
Infinite Series

CHAPTER 8
Infinite Series

Section 8.1 Sequences

Solutions to Odd-Numbered Exercises

1. $a_n = 2^n$

$a_1 = 2^1 = 2$

$a_2 = 2^2 = 4$

$a_3 = 2^3 = 8$

$a_4 = 2^4 = 16$

$a_5 = 2^5 = 32$

3. $a_n = \left(-\frac{1}{2}\right)^n$

$a_1 = \left(-\frac{1}{2}\right)^1 = -\frac{1}{2}$

$a_2 = \left(-\frac{1}{2}\right)^2 = \frac{1}{4}$

$a_3 = \left(-\frac{1}{2}\right)^3 = -\frac{1}{8}$

$a_4 = \left(-\frac{1}{2}\right)^4 = \frac{1}{16}$

$a_5 = \left(-\frac{1}{2}\right)^5 = -\frac{1}{32}$

5. $a_n = \sin \frac{n\pi}{2}$

$a_1 = \sin \frac{\pi}{2} = 1$

$a_2 = \sin \pi = 0$

$a_3 = \sin \frac{3\pi}{2} = -1$

$a_4 = \sin 2\pi = 0$

$a_5 = \sin \frac{5\pi}{2} = 1$

7. $a_n = \frac{(-1)^{n(n+1)/2}}{n^2}$

$a_1 = \frac{(-1)^1}{1^2} = -1$

$a_2 = \frac{(-1)^3}{2^2} = -\frac{1}{4}$

$a_3 = \frac{(-1)^6}{3^2} = \frac{1}{9}$

$a_4 = \frac{(-1)^{10}}{4^2} = \frac{1}{16}$

$a_5 = \frac{(-1)^{15}}{5^2} = -\frac{1}{25}$

9. $a_n = 5 - \frac{1}{n} + \frac{1}{n^2}$

$a_1 = 5 - 1 + 1 = 5$

$a_2 = 5 - \frac{1}{2} + \frac{1}{4} = \frac{19}{4}$

$a_3 = 5 - \frac{1}{3} + \frac{1}{9} = \frac{43}{9}$

$a_4 = 5 - \frac{1}{4} + \frac{1}{16} = \frac{77}{16}$

$a_5 = 5 - \frac{1}{5} + \frac{1}{25} = \frac{121}{25}$

11. $a_n = \frac{3^n}{n!}$

$a_1 = \frac{3}{1!} = 3$

$a_2 = \frac{3^2}{2!} = \frac{9}{2}$

$a_3 = \frac{3^3}{3!} = \frac{27}{6}$

$a_4 = \frac{3^4}{4!} = \frac{81}{24}$

$a_5 = \frac{3^5}{5!} = \frac{243}{120}$

13. $a_1 = 3, a_{k+1} = 2(a_k - 1)$

$a_2 = 2(a_1 - 1)$

$= 2(3 - 1) = 4$

$a_3 = 2(a_2 - 1)$

$= 2(4 - 1) = 6$

$a_4 = 2(a_3 - 1)$

$= 2(6 - 1) = 10$

$a_5 = 2(a_4 - 1)$

$= 2(10 - 1) = 18$

15. $a_1 = 32, a_{k+1} = \frac{1}{2}a_k$

$a_2 = \frac{1}{2}a_1 = \frac{1}{2}(32) = 16$

$a_3 = \frac{1}{2}a_2 = \frac{1}{2}(16) = 8$

$a_4 = \frac{1}{2}a_3 = \frac{1}{2}(8) = 4$

$a_5 = \frac{1}{2}a_4 = \frac{1}{2}(4) = 2$

17. Because $a_1 = 8/(1 + 1) = 4$ and $a_2 = 8/(2 + 1) = \frac{8}{3}$, the sequence matches graph (d).

19. This sequence decreases and $a_1 = 4$, $a_2 = 4(0.5) = 2$. Matches (c).

21.

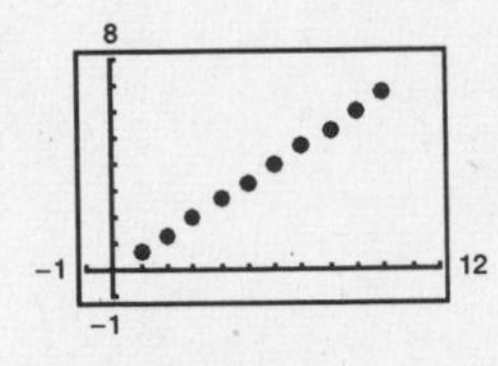

$a_n = \frac{2}{3}n, n = 1, \ldots, 10$

23.

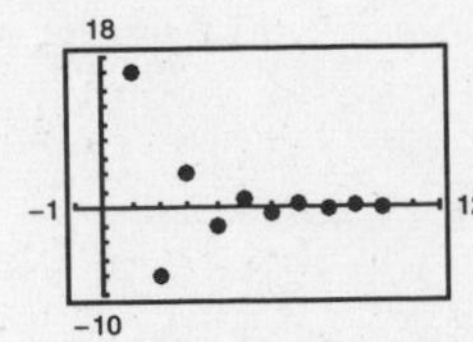

$a_n = 16(-0.5)^{n-1}, n = 1, \ldots, 10$

25.

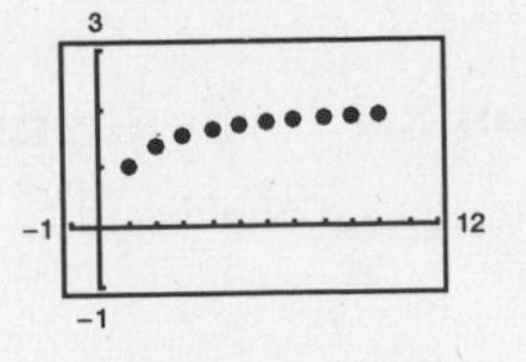

$a_n = \frac{2n}{n+1}, n = 1, 2, \ldots, 10$

27. $a_n = 3n - 1$

$a_5 = 3(5) - 1 = 14$

$a_6 = 3(6) - 1 = 17$

Add 3 to preceeding term.

29. 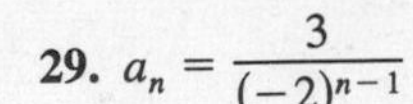 $a_n = \frac{3}{(-2)^{n-1}}$

$a_5 = \frac{3}{(-2)^4} = \frac{3}{16}$

$a_6 = \frac{3}{(-2)^5} = -\frac{3}{32}$

Multiply the preceeding term by $-\frac{1}{2}$.

31. $\frac{10!}{8!} = \frac{8!(9)(10)}{8!}$

$= (9)(10) = 90$

33. $\frac{(n+1)!}{n!} = \frac{n!(n+1)}{n!}$

$= n + 1$

35. $\frac{(2n-1)!}{(2n+1)!} = \frac{(2n-1)!}{(2n-1)!(2n)(2n+1)}$

$= \frac{1}{2n(2n+1)}$

37. $\lim_{n\to\infty} \frac{5n^2}{n^2+2} = 5$

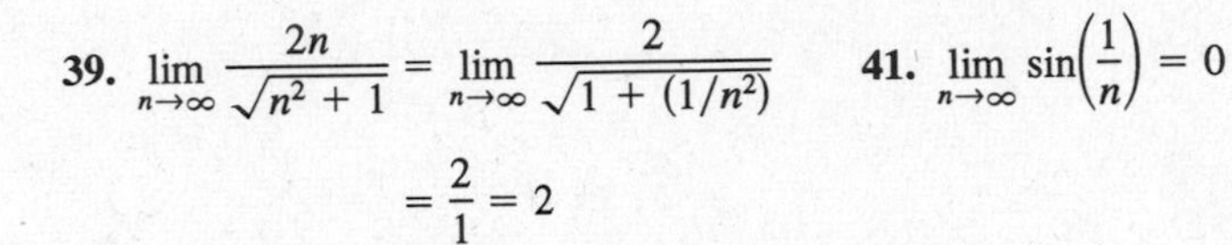

39. $\lim_{n\to\infty} \frac{2n}{\sqrt{n^2+1}} = \lim_{n\to\infty} \frac{2}{\sqrt{1 + (1/n^2)}}$

$= \frac{2}{1} = 2$

41. $\lim_{n\to\infty} \sin\left(\frac{1}{n}\right) = 0$

43.

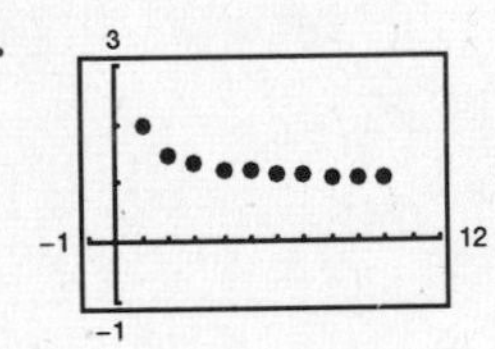

The graph seems to indicate that the sequence converges to 1. Analytically,

$$\lim_{n\to\infty} a_n = \lim_{n\to\infty} \frac{n+1}{n} = \lim_{x\to\infty} \frac{x+1}{x} = \lim_{x\to\infty} 1 = 1.$$

45.

The graph seems to indicate that the sequence diverges. Analytically, the sequence is

$$\{a_n\} = \{0, -1, 0, 1, 0, -1, \ldots\}.$$

Hence, $\lim_{n\to\infty} a_n$ does not exist.

47. $\lim_{n\to\infty} (-1)^n\left(\frac{n}{n+1}\right)$

does not exist (oscillates between -1 and 1), diverges.

49. $\lim_{n\to\infty} \frac{3n^2 - n + 4}{2n^2 + 1} = \frac{3}{2}$, converges

51. $\lim_{n\to\infty} \frac{1 + (-1)^n}{n} = 0$, converges

53. $\lim_{n\to\infty} \frac{\ln(n^3)}{2n} = \lim_{n\to\infty} \frac{3}{2}\frac{\ln(n)}{n}$

$= \lim_{n\to\infty} \frac{3}{2}\left(\frac{1}{n}\right) = 0$, converges

(L'Hôpital's Rule)

55. $\lim_{n\to\infty} \left(\frac{3}{4}\right)^n = 0$, converges

57. $\lim_{n\to\infty} \frac{(n+1)!}{n!} = \lim_{n\to\infty} (n+1) = \infty$, diverges

59. $\lim_{n\to\infty} \left(\frac{n-1}{n} - \frac{n}{n-1}\right) = \lim_{n\to\infty} \frac{(n-1)^2 - n^2}{n(n-1)}$

$= \lim_{n\to\infty} \frac{1-2n}{n^2-n} = 0$, converges

61. $\lim_{n\to\infty} \frac{n^p}{e^n} = 0$, converges

$(p > 0, n \geq 2)$

63. $a_n = \left(1 + \frac{k}{n}\right)^n$

$\lim_{n\to\infty} \left(1 + \frac{k}{n}\right)^n = \lim_{u\to 0} [(1+u)^{1/u}]^k = e^k$

where $u = \frac{k}{n}$, converges

65. $\lim_{n\to\infty} \frac{\sin n}{n} = \lim_{n\to\infty} (\sin n)\frac{1}{n} = 0$, converges

67. $a_n = 3n - 2$

69. $a_n = n^2 - 2$

71. $a_n = \frac{n+1}{n+2}$

73. $a_n = \frac{(-1)^{n-1}}{2^{n-2}}$

75. $a_n = 1 + \frac{1}{n} = \frac{n+1}{n}$

77. $a_n = \frac{n}{(n+1)(n+2)}$

79. $a_n = \frac{(-1)^{n-1}}{1 \cdot 3 \cdot 5 \cdots (2n-1)} = \frac{(-1)^{n-1} 2^n n!}{(2n)!}$

81. $a_n = 4 - \frac{1}{n} < 4 - \frac{1}{n+1} = a_{n+1}$,

monotonic; $|a_n| < 4$ bounded.

83. $\frac{n}{2^{n+2}} \overset{?}{\geq} \frac{n+1}{2^{(n+1)+2}}$

$2^{n+3} n \overset{?}{\geq} 2^{n+2}(n+1)$

$2n \overset{?}{\geq} n+1$

$n \geq 1$

Hence, $n \geq 1$

$2n \geq n+1$

$2^{n+3} n \geq 2^{n+2}(n+1)$

$\frac{n}{2^{n+2}} \geq \frac{n+1}{2^{(n+1)+2}}$

$a_n \geq a_{n+1}$

True; monotonic; $|a_n| \leq \frac{1}{8}$, bounded

85. $a_n = (-1)^n\left(\frac{1}{n}\right)$

$a_1 = -1$

$a_2 = \frac{1}{2}$

$a_3 = -\frac{1}{3}$

Not monotonic; $|a_n| \leq 1$, bounded

87. $a_n = \left(\frac{2}{3}\right)^n > \left(\frac{2}{3}\right)^{n+1} = a_{n+1}$

Monotonic; $|a_n| \leq \frac{2}{3}$, bounded

89. $a_n = \sin\left(\frac{n\pi}{6}\right)$

$a_1 = 0.500$

$a_2 = 0.8660$

$a_3 = 1.000$

$a_4 = 0.8660$

Not monotonic; $|a_n| \leq 1$, bounded

91. (a) $a_n = 5 + \frac{1}{n}$

$$\left|5 + \frac{1}{n}\right| \le 6 \Rightarrow \{a_n\} \text{ bounded}$$

$$a_n = 5 + \frac{1}{n} > 5 + \frac{1}{n+1}$$

$$= a_{n+1} \Rightarrow \{a_n\} \text{ monotonic}$$

Therefore, $\{a_n\}$ converges.

(b)

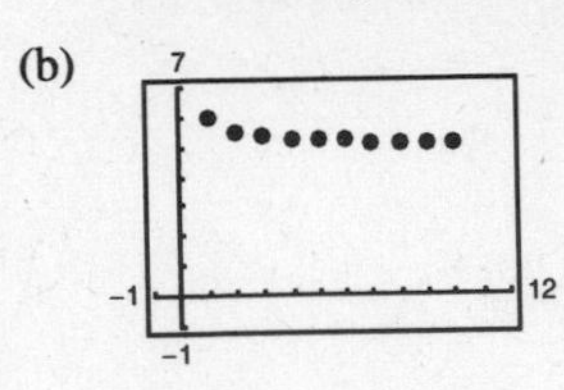

$$\lim_{n\to\infty}\left(5 + \frac{1}{n}\right) = 5$$

93. (a) $a_n = \frac{1}{3}\left(1 - \frac{1}{3^n}\right)$

$$\left|\frac{1}{3}\left(1 - \frac{1}{3^n}\right)\right| < \frac{1}{3} \Rightarrow \{a_n\} \text{ bounded}$$

$$a_n = \frac{1}{3}\left(1 - \frac{1}{3^n}\right) < \frac{1}{3}\left(1 - \frac{1}{3^{n+1}}\right)$$

$$= a_{n+1} \Rightarrow \{a_n\} \text{ monotonic}$$

Therefore, $\{a_n\}$ converges.

(b)

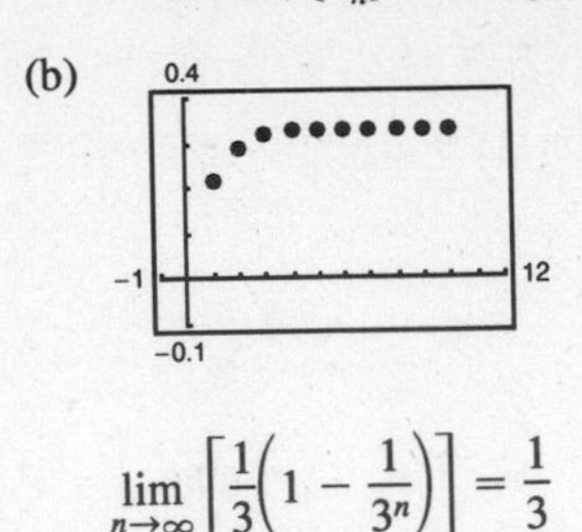

$$\lim_{n\to\infty}\left[\frac{1}{3}\left(1 - \frac{1}{3^n}\right)\right] = \frac{1}{3}$$

95. $A_n = P\left[1 + \frac{r}{12}\right]^n$

(a) $\lim_{n\to\infty} A_n = \infty$, divergent. The amount will grow arbitrarily large over time.

(b) $A_n = 9000\left[1 + \frac{0.115}{12}\right]^n$

$A_1 = \$9086.25$	$A_6 = \$9530.06$
$A_2 = \$9173.33$	$A_7 = \$9621.39$
$A_3 = \$9261.24$	$A_8 = \$9713.59$
$A_4 = \$9349.99$	$A_9 = \$9806.68$
$A_5 = \$9439.60$	$A_{10} = \$9900.66$

97. (a) A sequence is a function whose domain is the set of positive integers.

(b) A sequence converges if it has a limit.

(c) A bounded monotonic sequence is a sequence that has nondecreasing or nonincreasing terms, and an upper and lower bound.

99. $a_n = 10 - \frac{1}{n}$

101. $a_n = \frac{3n}{4n + 1}$

103. (a) $A_n = (0.8)^n (2.5)$ billion

(b) $A_1 = \$2$ billion

$A_2 = \$1.6$ billion

$A_3 = \$1.28$ billion

$A_4 = \$1.024$ billion

(c) $\lim_{n\to\infty} (0.8)^n (2.5) = 0$

105. (a) $a_n = -3.7262n^2 + 75.9167n + 684.25$

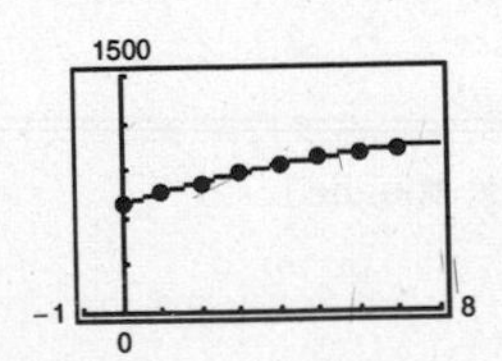

(b) For 2004, $n = 14$ and $a_{14} \approx 1017$, or $1017.

107. $a_n = \dfrac{10^n}{n!}$

(a) $a_9 = a_{10} = \dfrac{10^9}{9!}$

$= \dfrac{1{,}000{,}000{,}000}{362{,}880}$

$= \dfrac{1{,}562{,}500}{567}$

(b) Decreasing

(c) Factorials increase more rapidly than exponentials.

109. $\{a_n\} = \{\sqrt[n]{n}\} = \{n^{1/n}\}$

$a_1 = 1^{1/1} = 1$

$a_2 = \sqrt{2} \approx 1.4142$

$a_3 = \sqrt[3]{3} \approx 1.4422$

$a_4 = \sqrt[4]{4} \approx 1.4142$

$a_5 = \sqrt[5]{5} \approx 1.3797$

$a_6 = \sqrt[6]{6} \approx 1.3480$

Let $y = \lim_{n\to\infty} n^{1/n}$.

$$\ln y = \lim_{n\to\infty}\left(\frac{1}{n}\ln n\right)$$

$$= \lim_{n\to\infty}\frac{\ln n}{n} = \lim_{n\to\infty}\frac{1/n}{1} = 0$$

Since $\ln y = 0$, we have $y = e^0 = 1$. Therefore, $\lim_{n\to\infty}\sqrt[n]{n} = 1$.

111. $a_{n+2} = a_n + a_{n+1}$

(a) $a_1 = 1$ $\quad a_7 = 8 + 5 = 13$

$a_2 = 1$ $\quad a_8 = 13 + 8 = 21$

$a_3 = 1 + 1 = 2$ $\quad a_9 = 21 + 13 = 34$

$a_4 = 2 + 1 = 3$ $\quad a_{10} = 34 + 21 = 55$

$a_5 = 3 + 2 = 5$ $\quad a_{11} = 55 + 34 = 89$

$a_6 = 5 + 3 = 8$ $\quad a_{12} = 89 + 55 = 144$

(b) $b_n = \dfrac{a_{n+1}}{a_n}, n > 1$

$b_1 = \dfrac{1}{1} = 1$ $\quad b_6 = \dfrac{13}{8}$

$b_2 = \dfrac{2}{1} = 2$ $\quad b_7 = \dfrac{21}{13}$

$b_3 = \dfrac{3}{2}$ $\quad b_8 = \dfrac{34}{21}$

$b_4 = \dfrac{5}{3}$ $\quad b_9 = \dfrac{55}{34}$

$b_5 = \dfrac{8}{5}$ $\quad b_{10} = \dfrac{89}{55}$

(c) $1 + \dfrac{1}{b_{n-1}} = 1 + \dfrac{1}{a_n/a_{n-1}}$

$= 1 + \dfrac{a_{n-1}}{a_n}$

$= \dfrac{a_n + a_{n-1}}{a_n} = \dfrac{a_{n+1}}{a_n} = b_n$

(d) If $\lim_{n\to\infty} b_n = \rho$, then $\lim_{n\to\infty}\left(1 + \dfrac{1}{b_{n-1}}\right) = \rho$.

Since $\lim_{n\to\infty} b_n = \lim_{n\to\infty} b_{n-1}$ we have,

$1 + (1/\rho) = \rho$.

$\rho + 1 = \rho^2$

$0 = \rho^2 - \rho - 1$

$\rho = \dfrac{1 \pm \sqrt{1+4}}{2} = \dfrac{1 \pm \sqrt{5}}{2}$

Since a_n, and thus b_n, is positive,

$\rho = (1 + \sqrt{5})/2 \approx 1.6180$.

113. True

115. True

117. $a_1 = \sqrt{2} \approx 1.4142$

$a_2 = \sqrt{2 + \sqrt{2}} \approx 1.8478$

$a_3 = \sqrt{2 + \sqrt{2 + \sqrt{2}}} \approx 1.9616$

$a_4 = \sqrt{2 + \sqrt{2 + \sqrt{2 + \sqrt{2}}}} \approx 1.9904$

$a_5 = \sqrt{2 + \sqrt{2 + \sqrt{2 + \sqrt{2 + \sqrt{2}}}}} \approx 1.9976$

$\{a_n\}$ is increasing and bounded by 2, and hence converges to L. Letting $\lim_{n\to\infty} a_n = L$ implies that $\sqrt{2 + L} = L \Rightarrow L = 2$. Hence, $\lim_{n\to\infty} a_n = 2$.

Section 8.2 Series and Convergence

1. $S_1 = 1$

$S_2 = 1 + \frac{1}{4} = 1.2500$

$S_3 = 1 + \frac{1}{4} + \frac{1}{9} \approx 1.3611$

$S_4 = 1 + \frac{1}{4} + \frac{1}{9} + \frac{1}{16} \approx 1.4236$

$S_5 = 1 + \frac{1}{4} + \frac{1}{9} + \frac{1}{16} + \frac{1}{25} \approx 1.4636$

3. $S_1 = 3$

$S_2 = 3 - \frac{9}{2} = -1.5$

$S_3 = 3 - \frac{9}{2} + \frac{27}{4} = 5.25$

$S_4 = 3 - \frac{9}{2} + \frac{27}{4} - \frac{81}{8} = -4.875$

$S_5 = 3 - \frac{9}{2} + \frac{27}{4} - \frac{81}{8} + \frac{243}{16} = 10.3125$

5. $S_1 = 3$

$S_2 = 3 + \frac{3}{2} = 4.5$

$S_3 = 3 + \frac{3}{2} + \frac{3}{4} = 5.250$

$S_4 = 3 + \frac{3}{2} + \frac{3}{4} + \frac{3}{8} = 5.625$

$S_5 = 3 + \frac{3}{2} + \frac{3}{4} + \frac{3}{8} + \frac{3}{16} = 5.8125$

7. $\sum_{n=0}^{\infty} 3\left(\frac{3}{2}\right)^n$ Geometric series

$r = \frac{3}{2} > 1$

Diverges by Theorem 8.6

9. $\sum_{n=0}^{\infty} 1000(1.055)^n$ Geometric series

$r = 1.055 > 1$

Diverges by Theorem 8.6

11. $\sum_{n=1}^{\infty} \frac{n}{n+1}$

$\lim_{n\to\infty} \frac{n}{n+1} = 1 \neq 0$

Diverges by Theorem 8.9

13. $\sum_{n=1}^{\infty} \frac{n^2}{n^2+1}$

$\lim_{n\to\infty} \frac{n^2}{n^2+1} = 1 \neq 0$

Diverges by Theorem 8.9

15. $\sum_{n=0}^{\infty} \frac{2^n+1}{2^{n+1}}$

$\lim_{n\to\infty} \frac{2^n+1}{2^{n+1}} = \lim_{n\to\infty} \frac{1+2^{-n}}{2} = \frac{1}{2} \neq 0$

Diverges by Theorem 8.9

17. $\sum_{n=0}^{\infty} \frac{9}{4}\left(\frac{1}{4}\right)^n = \frac{9}{4}\left[1 + \frac{1}{4} + \frac{1}{16} + \cdots\right]$

$S_0 = \frac{9}{4}, S_1 = \frac{9}{4} \cdot \frac{5}{4} = \frac{45}{16}, S_2 = \frac{9}{4} \cdot \frac{21}{16} \approx 2.95, \ldots$

Matches graph (c).

Analytically, the series is geometric:

$$\sum_{n=0}^{\infty} \left(\frac{9}{4}\right)\left(\frac{1}{4}\right)^n = \frac{9/4}{1-1/4} = \frac{9/4}{3/4} = 3$$

19. $\sum_{n=0}^{\infty} \frac{15}{4}\left(-\frac{1}{4}\right)^n = \frac{15}{4}\left[1 - \frac{1}{4} + \frac{1}{16} - \cdots\right]$

$S_0 = \frac{15}{4}, S_1 = \frac{45}{16}, S_2 \approx 3.05, \ldots$

Matches graph (a).

Analytically, the series is geometric:

$$\sum_{n=0}^{\infty} \frac{15}{4}\left(-\frac{1}{4}\right)^n = \frac{15/4}{1-(-1/4)} = \frac{15/4}{5/4} = 3$$

21. $\sum_{n=1}^{\infty} \frac{1}{n(n+1)} = \sum_{n=1}^{\infty} \left(\frac{1}{n} - \frac{1}{n+1}\right) = \left(1 - \frac{1}{2}\right) + \left(\frac{1}{2} - \frac{1}{3}\right) + \left(\frac{1}{3} - \frac{1}{4}\right) + \left(\frac{1}{4} - \frac{1}{5}\right) + \cdots,\quad S_n = 1 - \frac{1}{n+1}$

$$\sum_{n=1}^{\infty} \frac{1}{n(n+1)} = \lim_{n\to\infty} S_n = \lim_{n\to\infty} \left(1 - \frac{1}{n+1}\right) = 1$$

23. $\sum_{n=0}^{\infty} 2\left(\frac{3}{4}\right)^n$

Geometric series with $r = \frac{3}{4} < 1$.

Converges by Theorem 8.6

25. $\sum_{n=0}^{\infty} (0.9)^n$

Geometric series with $r = 0.9 < 1$.

Converges by Theorem 8.6

27. (a) $\displaystyle\sum_{n=1}^{\infty} \frac{6}{n(n+3)} = 2\sum_{n=1}^{\infty}\left(\frac{1}{n} - \frac{1}{n+3}\right)$

$$= 2\left[\left(1 - \frac{1}{4}\right) + \left(\frac{1}{2} - \frac{1}{5}\right) + \left(\frac{1}{3} - \frac{1}{6}\right) + \left(\frac{1}{4} - \frac{1}{7}\right) + \cdots\right]$$

$$= 2\left[1 + \frac{1}{2} + \frac{1}{3}\right] = \frac{11}{3} \approx 3.667$$

(b)

n	5	10	20	50	100
S_n	2.7976	3.1643	3.3936	3.5513	3.6078

(c)

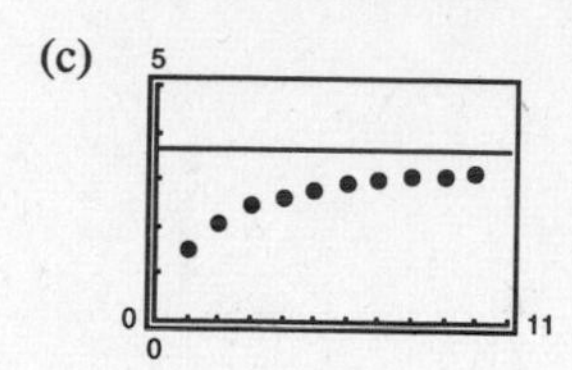

(d) The terms of the series decrease in magnitude slowly. Thus, the sequence of partial sums approaches the sum slowly.

29. (a) $\displaystyle\sum_{n=1}^{\infty} 2(0.9)^{n-1} = \sum_{n=0}^{\infty} 2(0.9)^n = \frac{2}{1-0.9} = 20$

(b)

n	5	10	20	50	100
S_n	8.1902	13.0264	17.5685	19.8969	19.9995

(c)

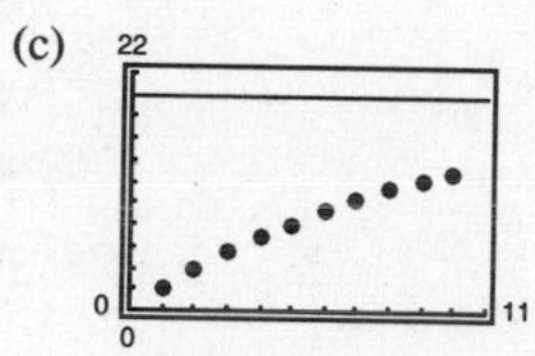

(d) The terms of the series decrease in magnitude slowly. Thus, the sequence of partial sums approaches the sum slowly.

31. (a) $\displaystyle\sum_{n=1}^{\infty} 10(0.25)^{n-1} = \frac{10}{1-0.25} = \frac{40}{3} \approx 13.3333$

(b)

n	5	10	20	50	100
S_n	13.3203	13.3333	13.3333	13.3333	13.3333

(c)

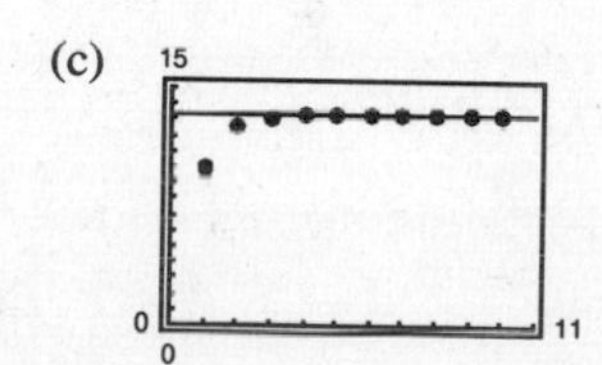

(d) The terms of the series decrease in magnitude rapidly. Thus, the sequence of partial sums approaches the sum rapidly.

33. $\displaystyle\sum_{n=2}^{\infty} \frac{1}{n^2-1} = \sum_{n=2}^{\infty}\left(\frac{1/2}{n-1} - \frac{1/2}{n+1}\right) = \frac{1}{2}\sum_{n=2}^{\infty}\left(\frac{1}{n-1} - \frac{1}{n+1}\right)$

$$= \frac{1}{2}\left[\left(1 - \frac{1}{3}\right) + \left(\frac{1}{2} - \frac{1}{4}\right) + \left(\frac{1}{3} - \frac{1}{5}\right) + \left(\frac{1}{4} - \frac{1}{6}\right) + \cdots\right]$$

$$= \frac{1}{2}\left(1 + \frac{1}{2}\right) = \frac{3}{4}$$

35. $\displaystyle\sum_{n=1}^{\infty} \frac{8}{(n+1)(n+2)} = 8\sum_{n=1}^{\infty}\left(\frac{1}{n+1} - \frac{1}{n+2}\right) = 8\left[\left(\frac{1}{2} - \frac{1}{3}\right) + \left(\frac{1}{3} - \frac{1}{4}\right) + \left(\frac{1}{4} - \frac{1}{5}\right) + \cdots\right] = 8\left(\frac{1}{2}\right) = 4$

37. $\displaystyle\sum_{n=0}^{\infty}\left(\frac{1}{2}\right)^n = \frac{1}{1-(1/2)} = 2$

39. $\displaystyle\sum_{n=0}^{\infty}\left(-\frac{1}{2}\right)^n = \frac{1}{1-(-1/2)} = \frac{2}{3}$

41. $\displaystyle\sum_{n=0}^{\infty}\left(\frac{1}{10}\right)^n = \frac{1}{1-(1/10)} = \frac{10}{9}$

43. $\displaystyle\sum_{n=0}^{\infty} 3\left(-\frac{1}{3}\right)^n = \frac{3}{1-(-1/3)} = \frac{9}{4}$

45. $\displaystyle\sum_{n=0}^{\infty}\left(\frac{1}{2^n}-\frac{1}{3^n}\right)=\sum_{n=0}^{\infty}\left(\frac{1}{2}\right)^n-\sum_{n=0}^{\infty}\left(\frac{1}{3}\right)^n$

$$=\frac{1}{1-(1/2)}-\frac{1}{1-(1/3)}$$

$$=2-\frac{3}{2}=\frac{1}{2}$$

47. $0.\overline{4}=\displaystyle\sum_{n=0}^{\infty}\frac{4}{10}\left(\frac{1}{10}\right)^n$

Geometric series with $a=\frac{4}{10}$ and $r=\frac{1}{10}$

$$S=\frac{a}{1-r}=\frac{4/10}{1-(1/10)}=\frac{4}{9}$$

49. $0.07\overline{575}=\displaystyle\sum_{n=0}^{\infty}\frac{3}{40}\left(\frac{1}{100}\right)^n$

Geometric series with $a=\frac{3}{40}$ and $r=\frac{1}{100}$

$$S=\frac{a}{1-r}=\frac{3/40}{99/100}=\frac{5}{66}$$

51. $\displaystyle\sum_{n=1}^{\infty}\frac{n+10}{10n+1}$

$$\lim_{n\to\infty}\frac{n+10}{10n+1}=\frac{1}{10}\neq 0$$

Diverges by Theorem 8.9

53. $\displaystyle\sum_{n=1}^{\infty}\left(\frac{1}{n}-\frac{1}{n+2}\right)=\left(1-\frac{1}{3}\right)+\left(\frac{1}{2}-\frac{1}{4}\right)+\left(\frac{1}{3}-\frac{1}{5}\right)+\left(\frac{1}{4}-\frac{1}{6}\right)+\cdots=1+\frac{1}{2}=\frac{3}{2}$, converges

55. $\displaystyle\sum_{n=1}^{\infty}\frac{3n-1}{2n+1}$

$$\lim_{n\to\infty}\frac{3n-1}{2n+1}=\frac{3}{2}\neq 0$$

Diverges by Theorem 8.9

57. $\displaystyle\sum_{n=0}^{\infty}\frac{4}{2^n}=4\sum_{n=0}^{\infty}\left(\frac{1}{2}\right)^n$

Geometric series with $r=\frac{1}{2}$

Converges by Theorem 8.6

59. $\displaystyle\sum_{n=0}^{\infty}(1.075)^n$

Geometric series with $r=1.075$

Diverges by Theorem 8.6

61. $\displaystyle\sum_{n=2}^{\infty}\frac{n}{\ln n}$

$$\lim_{n\to\infty}\frac{n}{\ln n}=\lim_{n\to\infty}\frac{1}{1/n}=\infty$$

(by L'Hôpital's Rule) Diverges by Theorem 8.9

63. See definition, page 567.

65. The series given by

$$\sum_{n=0}^{\infty}ar^n=a+ar+ar^2+\cdots+ar^n+\cdots,\ a\neq 0$$

is a geometric series with ratio r. When $0<|r|<1$, the series converges to $\frac{a}{1-r}$. The series diverges if $|r|\geq 1$.

67. (a) x is the common ratio.

(b) $1+x+x^2+\cdots=\displaystyle\sum_{n=0}^{\infty}x^n=\frac{1}{1-x},\ |x|<1$

Geometric series: $a=1$, $r=x$, $|x|<1$

(c) $y_1=\dfrac{1}{1-x},\ |x|<1$

$y_2=S_2=1+x$

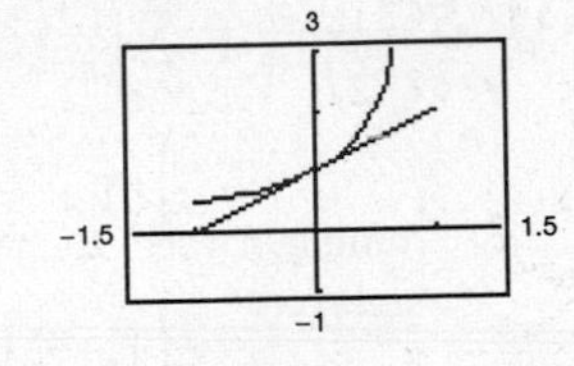

69. $f(x)=3\left[\dfrac{1-0.5^x}{1-0.5}\right]$

Horizontal asymptote: $y=6$

$$\sum_{n=0}^{\infty}3\left(\frac{1}{2}\right)^n$$

$$S=\frac{3}{1-(1/2)}=6$$

The horizontal asymptote is the sum of the series. $f(n)$ is the nth partial sum.

71. $\dfrac{1}{n(n+1)} < 0.001$

$$10{,}000 < n^2 + n$$

$$0 < n^2 + n - 10{,}000$$

$$n = \frac{-1 \pm \sqrt{1^2 - 4(1)(-10{,}000)}}{2}$$

Choosing the positive value for n we have $n \approx 99.5012$. The first *term* that is less than 0.001 is $n = 100$.

$$\left(\frac{1}{8}\right)^n < 0.001$$

$$10{,}000 < 8^n$$

This inequality is true when $n = 5$. This series converges at a faster rate.

73. $\displaystyle\sum_{i=0}^{n-1} 8000(0.9)^i = \frac{8000[1 - (0.9)^{(n-1)+1}]}{1 - 0.9}$

$$= 80{,}000(1 - 0.9^n),\ n > 0$$

75. $\displaystyle\sum_{i=0}^{n-1} 100(0.75)^i = \frac{100[1 - 0.75^{(n-1)+1}]}{1 - 0.75}$

$$= 400(1 - 0.75^n) \text{ million dollars.}$$

Sum = 400 million dollars

77. $D_1 = 16$

$$D_2 = \underbrace{0.81(16)}_{\text{up}} + \underbrace{0.81(16)}_{\text{down}} = 32(0.81)$$

$$D_3 = 16(0.81)^2 + 16(0.81)^2 = 32(0.81)^2$$

$\vdots$

$$D = 16 + 32(0.81) + 32(0.81)^2 + \cdots = -16 + \sum_{n=0}^{\infty} 32(0.81)^n = -16 + \frac{32}{1 - 0.81} \approx 152.42 \text{ ft}$$

79. $P(n) = \dfrac{1}{2}\left(\dfrac{1}{2}\right)^n$

$$P(2) = \frac{1}{2}\left(\frac{1}{2}\right)^2 = \frac{1}{8}$$

$$\sum_{n=0}^{\infty} \frac{1}{2}\left(\frac{1}{2}\right)^n = \frac{1/2}{1 - (1/2)} = 1$$

81. (a) $\displaystyle\sum_{n=1}^{\infty}\left(\frac{1}{2}\right)^n = \sum_{n=0}^{\infty}\frac{1}{2}\left(\frac{1}{2}\right)^n = \frac{1}{2}\frac{1}{(1 - (1/2))} = 1$

(b) No, the series is not geometric.

(c) $\displaystyle\sum_{n=1}^{\infty} n\left(\frac{1}{2}\right)^n = 2$

83. Assuming that the payments are made at the beginning of each year,

$$\text{Present Value} = \sum_{n=0}^{19} 50{,}000\left(\frac{1}{1.06}\right)^n$$

$$= 50{,}000\left(\frac{1 - 1.06^{-20}}{1 - 1.06^{-1}}\right)$$

$$\approx \$607{,}905.82$$

The present value is less than \$1,000,000. After accruing interest over 20 years, it attains its full value.

85. $w = \displaystyle\sum_{i=0}^{n-1} 0.01(2)^i = \frac{0.01(1 - 2^n)}{1 - 2} = 0.01(2^n - 1)$

(a) When $n = 29$: $w = \$5{,}368{,}709.11$

(b) When $n = 30$: $w = \$10{,}737{,}418.23$

(c) When $n = 31$: $w = \$21{,}474{,}836.47$

87. $P = 50, r = 0.03, t = 20$

(a) $A = 50\left(\frac{12}{0.03}\right)\left[\left(1 + \frac{0.03}{12}\right)^{12(20)} - 1\right] \approx \$16,415.10$

(b) $A = \frac{50 - (e^{0.03(20)} - 1)}{e^{0.03/12} - 1} \approx \$16,421.83$

89. $P = 100, r = 0.04, t = 40$

(a) $A = 100\left(\frac{12}{0.04}\right)\left[\left(1 + \frac{0.04}{12}\right)^{12(40)} - 1\right] \approx \$118,196.13$

(b) $A = \frac{100(e^{0.04(40)} - 1)}{e^{0.04/12} - 1} \approx \$118,393.43$

91. (a) $a_n = 6110.1832(1.0544)^x = 6110.1832e^{0.05297n}$

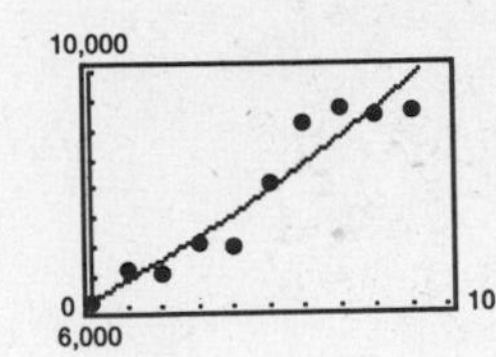

(b) Adding the 10 values $a_0, a_1, \ldots, a_9$, you obtain 78,530 or \$78,530,000,000

(c) Total $= \sum_{n=0}^{9} 6110.1832e^{-0.05297n} \approx 78,449$ or \$78,449,000,000

93. $x = 0.749999\ldots = 0.74 + \sum_{n=0}^{\infty} 0.009(0.1)^n$

$= 0.74 + \frac{0.009}{1 - 0.1}$

$= 0.74 + 0.01 = 0.75$

95. By letting $S_0 = 0$, we have $a_n = \sum_{k=1}^{n} a_k - \sum_{k=1}^{n-1} a_k = S_n - S_{n-1}$. Thus,

$$\sum_{n=1}^{\infty} a_n = \sum_{n=1}^{\infty} (S_n - S_{n-1}) = \sum_{n=1}^{\infty} (S_n - S_{n-1} + c - c) = \sum_{n=1}^{\infty} [(c - S_{n-1}) - (c - S_n)].$$

97. Let $\sum a_n = \sum_{n=0}^{\infty} 1$ and $\sum b_n = \sum_{n=0}^{\infty} (-1)$.

Both are divergent series.

$$\sum(a_n + b_n) = \sum_{n=0}^{\infty} [1 + (-1)] = \sum_{n=0}^{\infty} [1 - 1] = 0$$

99. False. $\lim_{n\to\infty} \frac{1}{n} = 0$, but $\sum_{n=1}^{\infty} \frac{1}{n}$ diverges.

101. False

$$\sum_{n=1}^{\infty} ar^n = \left(\frac{a}{1 - r}\right) - a$$

The formula requires that the geometric series begins with $n = 0$.

103. Let H represent the half-life of the drug. If a patient receives n equal doses of P units each of this drug, administered at equal time interval of length t, the total amount of the drug in the patient's system at the time the last dose is administered is given by

$$T_n = P + Pe^{kt} + Pe^{2kt} + \cdots + Pe^{(n-1)kt}$$

where $k = -(\ln 2)/H$. One time interval *after* the last dose is administered is given by

$$T_{n+1} = Pe^{kt} + Pe^{2kt} + Pe^{3kt} + \cdots + Pe^{nkt}.$$

Two time intervals *after* the last dose is administered is given by

$$T_{n+2} = Pe^{2kt} + Pe^{3kt} + Pe^{4kt} + \cdots + Pe^{(n+1)kt}$$

and so on. Since $k < 0$, $T_{n+s} \to 0$ as $s \to \infty$, where s is an integer.

Section 8.3 The Integral Test and p-Series

1. $\sum_{n=1}^{\infty} \frac{1}{n+1}$

Let $f(x) = \frac{1}{x+1}$.

f is positive, continuous and decreasing for $x \geq 1$.

$$\int_1^{\infty} \frac{1}{x+1}\,dx = \Big[\ln(x+1)\Big]_1^{\infty} = \infty$$

Diverges by Theorem 8.10

3. $\sum_{n=1}^{\infty} e^{-n}$

Let $f(x) = e^{-x}$.

f is positive, continuous, and decreasing for $x \geq 1$.

$$\int_1^{\infty} e^{-x}\,dx = \Big[-e^{-x}\Big]_1^{\infty} = \frac{1}{e}$$

Converges by Theorem 8.10

5. $\sum_{n=1}^{\infty} \frac{1}{n^2+1}$

Let $f(x) = \frac{1}{x^2+1}$.

f is positive, continuous, and decreasing for $x \geq 1$.

$$\int_1^{\infty} \frac{1}{x^2+1}\,dx = \Big[\arctan x\Big]_1^{\infty} = \frac{\pi}{4}$$

Converges by Theorem 8.10

7. $\sum_{n=1}^{\infty} \frac{\ln(n+1)}{n+1}$

Let $f(x) = \frac{\ln(x+1)}{x+1}$

f is positive, continuous, and decreasing for $x \geq 2$ since

$$f'(x) = \frac{1-\ln(x+1)}{(x+1)^2} < 0 \text{ for } x \geq 2.$$

$$\int_1^{\infty} \frac{\ln(x+1)}{x+1}\,dx = \left[\frac{\ln^2(x+1)}{2}\right]_1^{\infty} = \infty$$

Diverges by Theorem 8.10

9. $\sum_{n=1}^{\infty} \frac{n^{k-1}}{n^k+c}$

Let $f(x) = \frac{x^{k-1}}{x^k+c}$.

f is positive, continuous, and decreasing for $x > \sqrt[k]{c(k-1)}$ since

$$f'(x) = \frac{x^{k-2}[c(k-1)-x^k]}{(x^k+c)^2} < 0$$

for $x > \sqrt[k]{c(k-1)}$.

$$\int_1^{\infty} \frac{x^{k-1}}{x^k+c}\,dx = \left[\frac{1}{k}\ln(x^k+c)\right]_1^{\infty} = \infty$$

Diverges by Theorem 8.10

11. $\sum_{n=1}^{\infty} \frac{1}{n^3}$

Let $f(x) = \frac{1}{x^3}$.

f is positive, continuous, and decreasing for $x \geq 1$.

$$\int_1^{\infty} \frac{1}{x^3}\,dx = \left[-\frac{1}{2x^2}\right]_1^{\infty} = \frac{1}{2}$$

Converges by Theorem 8.10

13. $\sum_{n=1}^{\infty} \frac{1}{\sqrt[5]{n}} = \sum_{n=1}^{\infty} \frac{1}{n^{1/5}}$

Divergent p-series with $p = \frac{1}{5} < 1$

15. $\sum_{n=1}^{\infty} \frac{1}{n^{1/2}}$

Divergent p-series with $p = \frac{1}{2} < 1$

17. $\sum_{n=1}^{\infty} \frac{1}{n^{3/2}}$

Convergent p-series with $p = \frac{3}{2} > 1$

19. $\sum_{n=1}^{\infty} \frac{1}{n^{1.04}}$

Convergent p-series with $p = 1.04 > 1$

21. $\sum_{n=1}^{\infty} \frac{2}{\sqrt[4]{n^3}} = \frac{2}{1} + \frac{2}{2^{3/4}} + \frac{2}{3^{3/4}} + \cdots$

$S_1 = 2$

$S_2 \approx 3.189$

$S_3 \approx 4.067$

Matches (a)

Diverges—p-series with $p = \frac{3}{4} < 1$

23. $\sum_{n=1}^{\infty} \frac{2}{n\sqrt{n}} = 2 + 2/2^{3/2} + 2/3^{3/2} + \cdots$

$S_1 = 2$

$S_2 \approx 2.707$

$S_3 \approx 3.092$

Matches (b)

Converges—p-series with $p = 3/2 > 1$

25. No. Theorem 8.9 says that if the series converges, then the terms a_n tend to zero. Some of the series in Exercises 21-24 converge because the terms tend to 0 very rapidly.

27. $\sum_{n=1}^{N} \frac{1}{n} = 1 + \frac{1}{2} + \frac{1}{3} + \frac{1}{4} + \cdots + \frac{1}{N} > M$

(a)

M	2	4	6	8
N	4	31	227	1674

(b) No. Since the terms are decreasing (approaching zero), more and more terms are required to increase the partial sum by 2.

29. $\sum_{n=2}^{\infty} \frac{1}{n(\ln n)^p}$

If $p = 1$, then the series diverges by the Integral Test. If $p \neq 1$,

$$\int_2^{\infty} \frac{1}{x(\ln x)^p}\,dx = \int_2^{\infty} (\ln x)^{-p}\frac{1}{x}\,dx = \left[\frac{(\ln x)^{-p+1}}{-p+1}\right]_2^{\infty}.$$

Converges for $-p + 1 < 0$ or $p > 1$.

31. Let f be positive, continuous, and decreasing for $x \geq 1$ and $a_n = f(n)$. Then,

$$\sum_{n=1}^{\infty} a_n \text{ and } \int_1^{\infty} f(x)\,dx$$

either both converge or both diverge (Theorem 8.10). See Example 1, page 578.

33. Your friend is not correct. The series

$$\sum_{n=10,000}^{\infty} \frac{1}{n} = \frac{1}{10,000} + \frac{1}{10,001} + \cdots$$

is the harmonic series, starting with the 10,000th term, and hence diverges.

35. Since f is positive, continuous, and decreasing for $x \geq 1$ and $a_n = f(n)$, we have,

$$R_N = S - S_N = \sum_{n=1}^{\infty} a_n - \sum_{n=1}^{N} a_n = \sum_{n=N+1}^{\infty} a_n > 0.$$

Also, $R_N = S - S_N = \sum_{n=N+1}^{\infty} a_n \leq a_{N+1} + \int_{N+1}^{\infty} f(x)\,dx \leq \int_N^{\infty} f(x)\,dx$. Thus,

$$0 \leq R_N \leq \int_N^{\infty} f(x)\,dx.$$

37. $S_6 = 1 + \frac{1}{2^4} + \frac{1}{3^4} + \frac{1}{4^4} + \frac{1}{5^4} + \frac{1}{6^4} \approx 1.0811$

$$R_6 \leq \int_6^{\infty} \frac{1}{x^4}\,dx = \left[-\frac{1}{3x^3}\right]_6^{\infty} \approx 0.0015$$

$$1.0811 \leq \sum_{n=1}^{\infty} \frac{1}{n^4} \leq 1.0811 + 0.0015 = 1.0826$$

39. $S_{10} = \frac{1}{2} + \frac{1}{5} + \frac{1}{10} + \frac{1}{17} + \frac{1}{26} + \frac{1}{37} + \frac{1}{50} + \frac{1}{65} + \frac{1}{82} + \frac{1}{101} \approx 0.9818$

$$R_{10} \leq \int_{10}^{\infty} \frac{1}{x^2+1}\,dx = \Big[\arctan x\Big]_{10}^{\infty} = \frac{\pi}{2} - \arctan 10 \approx 0.0997$$

$$0.9818 \leq \sum_{n=1}^{\infty} \frac{1}{n^2+1} \leq 0.9818 + 0.0997 = 1.0815$$

41. $S_4 = \frac{1}{e} + \frac{2}{e^4} + \frac{3}{e^9} + \frac{4}{e^{16}} \approx 0.4049$

$$R_4 \leq \int_4^{\infty} xe^{-x^2}\,dx = \left[-\frac{1}{2}e^{-x^2}\right]_4^{\infty} = \frac{e^{-16}}{2} \approx 5.6 \times 10^{-8}$$

$$0.4049 \leq \sum_{n=1}^{\infty} ne^{-n^2} \leq 0.4049 + 5.6 \times 10^{-8}$$

43. $0 \leq R_N \leq \int_N^{\infty} \frac{1}{x^4}\,dx = \left[-\frac{1}{3x^3}\right]_N^{\infty} = \frac{1}{3N^3} < 0.001$

$\frac{1}{N^3} < 0.003$

$N^3 > 333.33$

$N > 6.93$

$N \geq 7$

45. $R_N \leq \int_N^{\infty} e^{-5x}\,dx = \left[-\frac{1}{5}e^{-5x}\right]_N^{\infty} = \frac{e^{-5N}}{5} < 0.001$

$\frac{1}{e^{5N}} < 0.005$

$e^{5N} > 200$

$5N > \ln 200$

$N > \frac{\ln 200}{5}$

$N > 1.0597$

$N \geq 2$

47. $R_N \leq \int_N^{\infty} \frac{1}{x^2+1}\,dx = \Big[\arctan x\Big]_N^{\infty}$

$= \frac{\pi}{2} - \arctan N < 0.001$

$-\arctan N < -1.5698$

$\arctan N > 1.5698$

$N > \tan 1.5698$

$N \geq 1004$

49. (a) $\sum_{n=2}^{\infty} \frac{1}{n^{1.1}}$. This is a convergent p-series with $p = 1.1 > 1$.

$\sum_{n=2}^{\infty} \frac{1}{n \ln n}$ is a divergent series. Use the Integral Test.

$$\int_2^{\infty} \frac{1}{x \ln x}\,dx = \Big[\ln|\ln x|\Big]_2^{\infty} = \infty$$

(b) $\sum_{n=2}^{6} \frac{1}{n^{1.1}} = \frac{1}{2^{1.1}} + \frac{1}{3^{1.1}} + \frac{1}{4^{1.1}} + \frac{1}{5^{1.1}} + \frac{1}{6^{1.1}} \approx 0.4665 + 0.2987 + 0.2176 + 0.1703 + 0.1393$

$\sum_{n=2}^{6} \frac{1}{n \ln n} = \frac{1}{2 \ln 2} + \frac{1}{3 \ln 3} + \frac{1}{4 \ln 4} + \frac{1}{5 \ln 5} + \frac{1}{6 \ln 6} \approx 0.7213 + 0.3034 + 0.1803 + 0.1243 + 0.0930$

For $n \geq 4$ the terms of the convergent series **seem** to be larger than those of the divergent series!

(c) $\frac{1}{n^{1.1}} < \frac{1}{n \ln n}$

$n \ln n < n^{1.1}$

$\ln n < n^{0.1}$

This inequality holds when $n \geq 3.5 \times 10^{15}$. Or, $n > e^{40}$. Then $\ln e^{40} = 40 < (e^{40})^{0.1} = e^4 \approx 55$.

51. (a) Let $f(x) = 1/x$. f is positive, continuous, and decreasing on $[1, \infty)$.

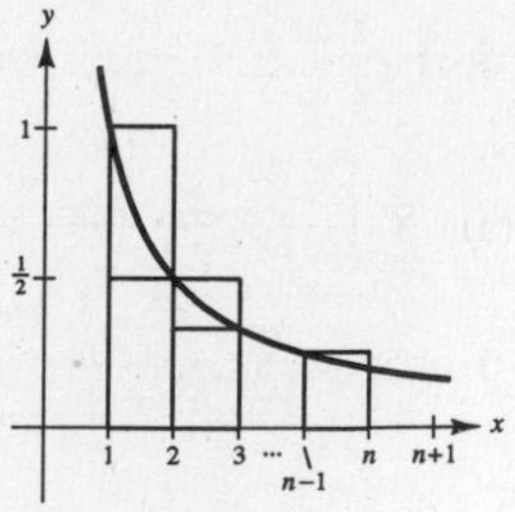

$$S_n - 1 \le \int_1^n \frac{1}{x}\,dx$$

$$S_n - 1 \le \ln n$$

Hence, $S_n \le 1 + \ln n$. Similarly,

$$S_n \ge \int_1^{n+1} \frac{1}{x}\,dx = \ln(n+1).$$

Thus, $\ln(n+1) \le S_n \le 1 + \ln n$.

(b) Since $\ln(n+1) \le S_n \le 1 + \ln n$, we have $\ln(n+1) - \ln n \le S_n - \ln n \le 1$. Also, since $\ln x$ is an increasing function, $\ln(n+1) - \ln n > 0$ for $n \ge 1$. Thus, $0 \le S_n - \ln n \le 1$ and the sequence $\{a_n\}$ is bounded.

(c) $a_n - a_{n+1} = [S_n - \ln n] - [S_{n+1} - \ln(n+1)] = \displaystyle\int_n^{n+1} \frac{1}{x}\,dx - \frac{1}{n+1} \ge 0$

Thus, $a_n \ge a_{n+1}$ and the sequence is decreasing.

(d) Since the sequence is bounded and monotonic, it converges to a limit, γ.

(e) $a_{100} = S_{100} - \ln 100 \approx 0.5822$ (Actually $\gamma \approx 0.577216$.)

53. $\displaystyle\sum_{n=1}^{\infty} \frac{1}{2n-1}$

Let $f(x) = \dfrac{1}{2x-1}$.

f is positive, continuous, and decreasing for $x \ge 1$.

$$\int_1^{\infty} \frac{1}{2x-1}\,dx = \Big[\ln\sqrt{2x-1}\Big]_1^{\infty} = \infty$$

Diverges by Theorem 8.10

55. $\displaystyle\sum_{n=1}^{\infty} \frac{1}{n\sqrt[4]{n}} = \sum_{n=1}^{\infty} \frac{1}{n^{5/4}}$

p-series with $p = \frac{5}{4}$

Converges by Theorem 8.11

57. $\displaystyle\sum_{n=0}^{\infty} \left(\frac{2}{3}\right)^n$

Geometric series with $r = \frac{2}{3}$

Converges by Theorem 8.6

59. $\displaystyle\sum_{n=1}^{\infty} \frac{n}{\sqrt{n^2+1}}$

$$\lim_{n\to\infty} \frac{n}{\sqrt{n^2+1}} = \lim_{n\to\infty} \frac{1}{\sqrt{1+(1/n^2)}} = 1 \ne 0$$

Diverges by Theorem 8.9

61. $\displaystyle\sum_{n=1}^{\infty} \left(1 + \frac{1}{n}\right)^n$

$$\lim_{n\to\infty} \left(1 + \frac{1}{n}\right)^n = e \ne 0$$

Fails nth Term Test

Diverges by Theorem 8.9

63. $\displaystyle\sum_{n=2}^{\infty} \frac{1}{n(\ln n)^3}$

Let $f(x) = \dfrac{1}{x(\ln x)^3}$.

f is positive, continuous and decreasing for $x \ge 2$.

$$\int_2^{\infty} \frac{1}{x(\ln x)^3}\,dx = \int_2^{\infty} (\ln x)^{-3}\frac{1}{x}\,dx = \left[\frac{(\ln x)^{-2}}{-2}\right]_2^{\infty} = \left[-\frac{1}{2(\ln x)^2}\right]_2^{\infty} = \frac{1}{2(\ln 2)^2}$$

Converges by Theorem 8.10. See Exercise 29.

Section 8.4 Comparisons of Series

1. (a) $\sum_{n=1}^{\infty} \frac{6}{n^{3/2}} = \frac{6}{1} + \frac{6}{2^{3/2}} + \cdots \quad S_1 = 6$

$\sum_{n=1}^{\infty} \frac{6}{n^{3/2}+3} = \frac{6}{4} + \frac{6}{2^{3/2}+3} + \cdots \quad S_1 = \frac{3}{2}$

$\sum_{n=1}^{\infty} \frac{6}{n\sqrt{n^2+0.5}} = \frac{6}{1\sqrt{1.5}} + \frac{6}{2\sqrt{4.5}} + \cdots \quad S_1 = \frac{6}{\sqrt{1.5}} \approx 4.9$

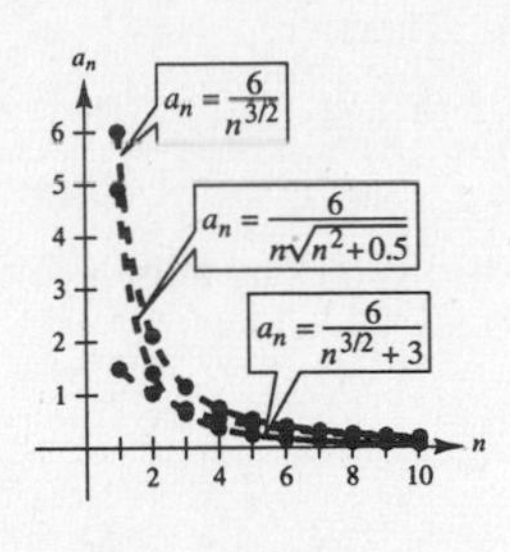

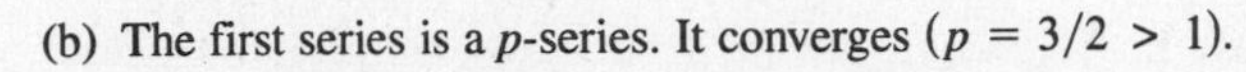
(b) The first series is a p-series. It converges $(p = 3/2 > 1)$.

(c) The magnitude of the terms of the other two series are less than the corresponding terms at the convergent p-series. Hence, the other two series converge.

(d) The smaller the magnitude of the terms, the smaller the magnitude of the terms of the sequence of partial sums.

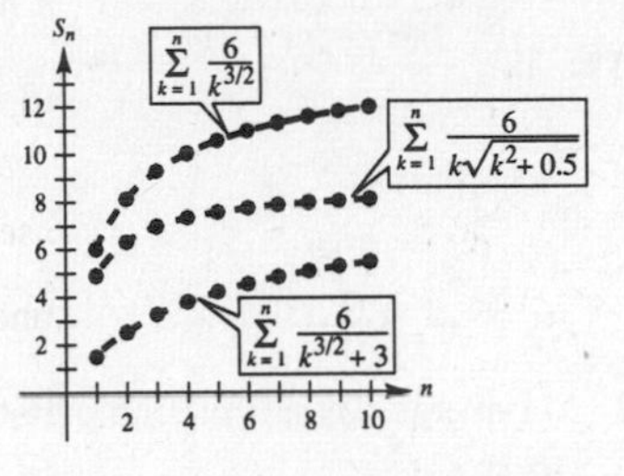

3. $\frac{1}{n^2+1} < \frac{1}{n^2}$

Therefore,

$$\sum_{n=1}^{\infty} \frac{1}{n^2+1}$$

converges by comparison with the convergent p-series

$$\sum_{n=1}^{\infty} \frac{1}{n^2}.$$

5. $\frac{1}{n-1} > \frac{1}{n}$ for $n \geq 2$

Therefore,

$$\sum_{n=2}^{\infty} \frac{1}{n-1}$$

diverges by comparison with the divergent p-series

$$\sum_{n=2}^{\infty} \frac{1}{n}.$$

7. $\frac{1}{3^n+1} < \frac{1}{3^n}$

Therefore,

$$\sum_{n=0}^{\infty} \frac{1}{3^n+1}$$

converges by comparison with the convergent geometric series

$$\sum_{n=0}^{\infty} \left(\frac{1}{3}\right)^n.$$

9. For $n \geq 3$, $\frac{\ln n}{n+1} > \frac{1}{n+1}$.

Therefore,

$$\sum_{n=1}^{\infty} \frac{\ln n}{n+1}$$

diverges by comparison with the divergent series

$$\sum_{n=1}^{\infty} \frac{1}{n+1}.$$

Note: $\sum_{n=1}^{\infty} \frac{1}{n+1}$ **diverges by the integral test.**

11. For $n > 3$, $\frac{1}{n^2} > \frac{1}{n!}$.

Therefore,

$$\sum_{n=0}^{\infty} \frac{1}{n!}$$

converges by comparison with the convergent p-series

$$\sum_{n=1}^{\infty} \frac{1}{n^2}.$$

13. $\frac{1}{e^{n^2}} \leq \frac{1}{e^n}$

Therefore,

$$\sum_{n=0}^{\infty} \frac{1}{e^{n^2}}$$

converges by comparison with the convergent geometric series

$$\sum_{n=0}^{\infty} \left(\frac{1}{e}\right)^n.$$

15. $\lim_{n\to\infty} \frac{n/(n^2+1)}{1/n} = \lim_{n\to\infty} \frac{n^2}{n^2+1} = 1$

Therefore,

$$\sum_{n=1}^{\infty} \frac{n}{n^2+1}$$

diverges by a limit comparison with the divergent p-series

$$\sum_{n=1}^{\infty} \frac{1}{n}.$$

17. $\lim_{n\to\infty} \frac{1/\sqrt{n^2+1}}{1/n} = \lim_{n\to\infty} \frac{n}{\sqrt{n^2+1}} = 1$

Therefore,

$$\sum_{n=0}^{\infty} \frac{1}{\sqrt{n^2+1}}$$

diverges by a limit comparison with the divergent p-series

$$\sum_{n=1}^{\infty} \frac{1}{n}.$$

19. $\lim_{n\to\infty} \frac{\dfrac{2n^2-1}{3n^5+2n+1}}{1/n^3} = \lim_{n\to\infty} \frac{2n^5-n^3}{3n^5+2n+1} = \frac{2}{3}$

Therefore,

$$\sum_{n=1}^{\infty} \frac{2n^2-1}{3n^5+2n+1}$$

converges by a limit comparison with the convergent p-series

$$\sum_{n=1}^{\infty} \frac{1}{n^3}.$$

21. $\lim_{n\to\infty} \frac{\dfrac{n+3}{n(n+2)}}{1/n} = \lim_{n\to\infty} \frac{n^2+3n}{n^2+2n} = 1$

Therefore,

$$\sum_{n=1}^{\infty} \frac{n+3}{n(n+2)}$$

diverges by a limit comparison with the divergent p-series

$$\sum_{n=1}^{\infty} \frac{1}{n}.$$

23. $\lim_{n\to\infty} \frac{1/\left(n\sqrt{n^2+1}\right)}{1/n^2} = \lim_{n\to\infty} \frac{n^2}{n\sqrt{n^2+1}} = 1$

Therefore,

$$\sum_{n=1}^{\infty} \frac{1}{n\sqrt{n^2+1}}$$

converges by a limit comparison with the convergent p-series

$$\sum_{n=1}^{\infty} \frac{1}{n^2}.$$

25. $\lim_{n\to\infty} \frac{(n^{k-1})/(n^k+1)}{1/n} = \lim_{n\to\infty} \frac{n^k}{n^k+1} = 1$

Therefore,

$$\sum_{n=1}^{\infty} \frac{n^{k-1}}{n^k+1}$$

diverges by a limit comparison with the divergent p-series

$$\sum_{n=1}^{\infty} \frac{1}{n}.$$

27. $\lim_{n\to\infty} \frac{\sin(1/n)}{1/n} = \lim_{n\to\infty} \frac{(-1/n^2)\cos(1/n)}{-1/n^2}$

$= \lim_{n\to\infty} \cos\left(\frac{1}{n}\right) = 1$

Therefore,

$$\sum_{n=1}^{\infty} \sin\left(\frac{1}{n}\right)$$

diverges by a limit comparison with the divergent p-series

$$\sum_{n=1}^{\infty} \frac{1}{n}.$$

29. $\sum_{n=1}^{\infty} \frac{\sqrt{n}}{n} = \sum_{n=1}^{\infty} \frac{1}{\sqrt{n}}$

Diverges

p-series with $p = \frac{1}{2}$

31. $\sum_{n=1}^{\infty} \frac{1}{3^n+2}$

Converges

Direct comparison with $\sum_{n=1}^{\infty} \left(\frac{1}{3}\right)^n$

33. $\sum_{n=1}^{\infty} \frac{n}{2n+3}$

Diverges; nth Term Test

$\lim_{n\to\infty} \frac{n}{2n+3} = \frac{1}{2} \neq 0$

35. $\sum_{n=1}^{\infty} \frac{n}{(n^2+1)^2}$

Converges; integral test

37. $\lim_{n\to\infty} \frac{a_n}{1/n} = \lim_{n\to\infty} na_n$ by given conditions $\lim_{n\to\infty} na_n$ is finite and nonzero.

Therefore,

$$\sum_{n=1}^{\infty} a_n$$

diverges by a limit comparison with the p-series

$$\sum_{n=1}^{\infty} \frac{1}{n}.$$

39. $\frac{1}{2} + \frac{2}{5} + \frac{3}{10} + \frac{4}{17} + \frac{5}{26} + \cdots = \sum_{n=1}^{\infty} \frac{n}{n^2+1},$

which diverges since the degree of the numerator is only one less than the degree of the denominator.

41. $\sum_{n=1}^{\infty} \frac{1}{n^3+1}$

converges since the degree of the numerator is three less than the degree of the denominator.

43. $\lim_{n\to\infty} n\left(\frac{n^3}{5n^4+3}\right) = \lim_{n\to\infty} \frac{n^4}{5n^4+3} = \frac{1}{5} \neq 0$

Therefore,

$$\sum_{n=1}^{\infty} \frac{n^3}{5n^4+3} \text{ diverges.}$$

45. See Theorem 8.12, page 583. One example is $\sum_{n=1}^{\infty} \frac{1}{n^2+1}$ converges because

$$\frac{1}{n^2+1} < \frac{1}{n^2} \text{ and } \sum_{n=1}^{\infty} \frac{1}{n^2}$$

converges (p-series).

47.

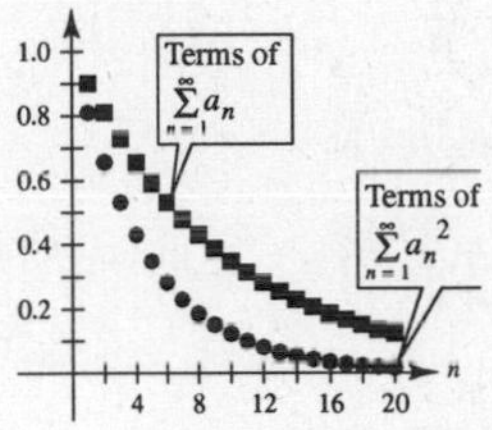

Terms of $\sum_{n=1}^{\infty} a_n$

Terms of $\sum_{n=1}^{\infty} a_n^2$

For $0 < a_n < 1$, $0 < a_n^2 < a_n < 1$. Hence, the lower terms are those of $\Sigma\, a_n^2$.

49. $\frac{1}{200} + \frac{1}{400} + \frac{1}{600} + \cdots = \sum_{n=1}^{\infty} \frac{1}{200n}$, diverges

51. $\frac{1}{201} + \frac{1}{204} + \frac{1}{209} + \frac{1}{216} = \sum_{n=1}^{\infty} \frac{1}{200+n^2}$, converges

53. Some series diverge or converge very slowly. You cannot decide convergence or divergence of a series by comparing the first few terms.

55. False. Let $a_n = 1/n^3$ and $b_n = 1/n^2$. $0 < a_n \le b_n$ and both

$$\sum_{n=1}^{\infty} \frac{1}{n^3} \text{ and } \sum_{n=1}^{\infty} \frac{1}{n^2}$$

converge.

57. True

59. Since $\sum_{n=1}^{\infty} b_n$ converges, $\lim_{n\to\infty} b_n = 0$. There exists N such that $b_n < 1$ for $n > N$. Thus,

$$a_n b_n < a_n \text{ for } n > N \text{ and } \sum_{n=1}^{\infty} a_n b_n$$

converges by comparison to the convergent series $\sum_{i=1}^{\infty} a_n$.

61. $\Sigma \frac{1}{n^2}$ and $\Sigma \frac{1}{n^3}$ both converge, and hence so does $\Sigma\left(\frac{1}{n^2}\right)\left(\frac{1}{n^3}\right) = \Sigma \frac{1}{n^5}$.

63. (a) Suppose Σb_n converges and Σa_n diverges. Then there exists N such that $0 < b_n < a_n$ for $n \geq N$. This means that $1 < a_n/b_n$ for $n \geq N$. Therefore, $\lim_{n\to\infty} a_n/b_n \neq 0$. Thus, Σa_n must also converge.

(b) Suppose Σb_n diverges and Σa_n converges. Then there exists N such that $0 < a_n < b_n$ for $n \geq N$. This means that $0 < a_n/b_n < 1$ for $n \geq N$. Therefore, $\lim_{n\to\infty} a_n/b_n \neq \infty$. Thus, Σa_n must also diverge.

65. Start with one triangle whose sides have length 9. At the nth step, each side is replaced by four smaller line segments each having $\frac{1}{3}$ the length of the original side.

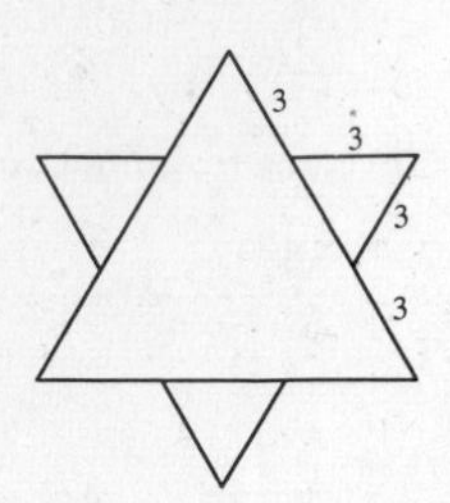

#Sides	Length of sides
3	9
$3 \cdot 4$	$9\left(\frac{1}{3}\right)$
$3 \cdot 4^2$	$9\left(\frac{1}{3}\right)^2$
$\vdots$	
$3 \cdot 4^n$	$9\left(\frac{1}{3}\right)^n$

At the nth step there are $3 \cdot 4^n$ sides, each of length $9\left(\frac{1}{3}\right)^n$. At the next step, there are $3 \cdot 4^n$ new triangles of side $9\left(\frac{1}{3}\right)^{n+1}$. The area of an equilateral triangle of side x is $\frac{1}{4}\sqrt{3}\,x^2$. Thus, the new triangles each have area

$$\frac{\sqrt{3}}{4}\left[9\left(\frac{1}{3^{n+1}}\right)\right]^2 = \frac{9\sqrt{3}}{4} \cdot \frac{1}{3^{2n}} = \frac{9}{4}\sqrt{3}\left(\frac{1}{9}\right)^n$$

The area of the $3 \cdot 4^n$ new triangles is

$$(3 \cdot 4^n)\left[\frac{9\sqrt{3}}{4}\left(\frac{1}{9}\right)^n\right] = \frac{27\sqrt{3}}{4}\left(\frac{4}{9}\right)^n$$

The total area is the infinite sum

$$\frac{81\sqrt{3}}{4} + \sum_{n=0}^{\infty} \frac{27\sqrt{3}}{4}\left(\frac{4}{9}\right)^n = \frac{81\sqrt{3}}{4} + \frac{27\sqrt{3}}{4}\left(\frac{1}{1 - 4/9}\right) = \frac{81\sqrt{3}}{4} + \frac{27\sqrt{3}}{4}\left(\frac{9}{5}\right) = \frac{162\sqrt{3}}{5}.$$

The perimeter is infinite, since at step n there are $3 \cdot 4^n$ sides of length $9\left(\frac{1}{3}\right)^n$. Thus, the perimeter at step n is $27\left(\frac{4}{3}\right)^n \to \infty$.

Section 8.5 Alternating Series

1. $\displaystyle\sum_{n=1}^{\infty} \frac{6}{n^2} = \frac{6}{1} + \frac{6}{4} + \frac{6}{9} + \cdots$

$S_1 = 6, S_2 = 7.5$

Matches (b)

3. $\displaystyle\sum_{n=1}^{\infty} \frac{10}{n2^n} = \frac{10}{2} + \frac{10}{8} + \cdots$

$S_1 = 5, S_2 = 6.25$

Matches (c)

5. $\displaystyle\sum_{n=1}^{\infty} \frac{(-1)^{n-1}}{2n-1} = \frac{\pi}{4} \approx 0.7854$

(a)

n	1	2	3	4	5	6	7	8	9	10
S_n	1	0.6667	0.8667	0.7238	0.8349	0.7440	0.8209	0.7543	0.8131	0.7605

(b)

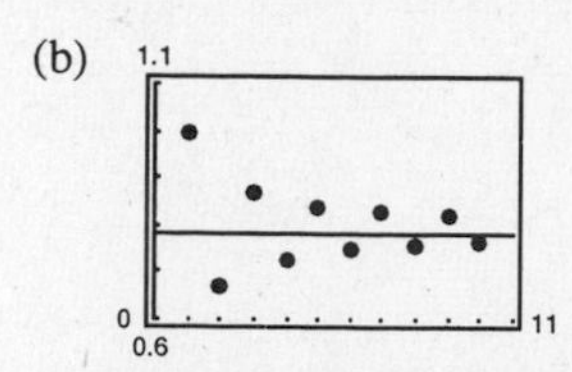

(c) The points alternate sides of the horizontal line that represents the sum of the series. The distance between successive points and the line decreases.

(d) The distance in part (c) is always less than the magnitude of the next term of the series.

7. $\sum_{n=1}^{\infty} \frac{(-1)^{n-1}}{n^2} = \frac{\pi^2}{12} \approx 0.8225$

(a)

n	1	2	3	4	5	6	7	8	9	10
S_n	1	0.75	0.8611	0.7986	0.8386	0.8108	0.8312	0.8156	0.8280	0.8180

(b)

1.1

0 11

0.6

(c) The points alternate sides of the horizontal line that represents the sum of the series. The distance between successive points and the line decreases.

(d) The distance in part (c) is always less than the magnitude of the next term in the series.

9. $\sum_{n=1}^{\infty} \frac{(-1)^{n+1}}{n}$

$a_{n+1} = \frac{1}{n+1} < \frac{1}{n} = a_n$

$\lim_{n\to\infty} \frac{1}{n} = 0$

Converges by Theorem 8.14.

11. $\sum_{n=1}^{\infty} \frac{(-1)^{n+1}}{2n-1}$

$a_{n+1} = \frac{1}{2(n+1)-1} < \frac{1}{2n-1} = a_n$

$\lim_{n\to\infty} \frac{1}{2n-1} = 0$

Converges by Theorem 8.14

13. $\sum_{n=1}^{\infty} \frac{(-1)^n n^2}{n^2+1}$

$\lim_{n\to\infty} \frac{n^2}{n^2+1} = 1$

Diverges by the nth Term Test

15. $\sum_{n=1}^{\infty} \frac{(-1)^n}{\sqrt{n}}$

$a_{n+1} = \frac{1}{\sqrt{n+1}} < \frac{1}{\sqrt{n}} = a_n$

$\lim_{n\to\infty} \frac{1}{\sqrt{n}} = 0$

Converges by Theorem 8.14

17. $\sum_{n=1}^{\infty} \frac{(-1)^{n+1}(n+1)}{\ln(n+1)}$

$\lim_{n\to\infty} \frac{n+1}{\ln(n+1)} = \lim_{n\to\infty} \frac{1}{1/(n+1)} = \lim_{n\to\infty} (n+1) = \infty$

Diverges by the nth Term Test

19. $\sum_{n=1}^{\infty} \sin\left[\frac{(2n-1)\pi}{2}\right] = \sum_{n=1}^{\infty} (-1)^{n+1}$

Diverges by the nth Term Test

21. $\sum_{n=1}^{\infty} \cos n\pi = \sum_{n=1}^{\infty} (-1)^n$

Diverges by the nth Term Test

23. $\sum_{n=0}^{\infty} \frac{(-1)^n}{n!}$

$a_{n+1} = \frac{1}{(n+1)!} < \frac{1}{n!} = a_n$

$\lim_{n\to\infty} \frac{1}{n!} = 0$

Converges by Theorem 8.14

25. $\displaystyle\sum_{n=1}^{\infty} \frac{(-1)^{n+1}\sqrt{n}}{n+2}$

$$a_{n+1} = \frac{\sqrt{n+1}}{(n+1)+2} < \frac{\sqrt{n}}{n+2} \text{ for } n \geq 2$$

$$\lim_{n\to\infty} \frac{\sqrt{n}}{n+2} = 0$$

Converges by Theorem 8.14

27. $\displaystyle\sum_{n=1}^{\infty} \frac{(-1)^{n+1}(2)}{e^n - e^{-n}} = \sum_{n=1}^{\infty} \frac{(-1)^{n+1}(2e^n)}{e^{2n} - 1}$

Let $f(x) = \dfrac{2e^x}{e^{2x} - 1}$. Then

$$f'(x) = \frac{-2e^x(e^{2x} + 1)}{(e^{2x} - 1)^2} < 0.$$

Thus, $f(x)$ is decreasing. Therefore, $a_{n+1} < a_n$, and

$$\lim_{n\to\infty} \frac{2e^n}{e^{2n} - 1} = \lim_{n\to\infty} \frac{2e^n}{2e^{2n}} = \lim_{n\to\infty} \frac{1}{e^n} = 0.$$

The series converges by Theorem 8.14.

29. $\displaystyle S_6 = \sum_{n=1}^{6} \frac{3(-1)^{n+1}}{n^2} = 2.4325$

$$|R_6| = |S - S_6| \leq a_7 = \frac{3}{49} \approx 0.0612;\ 2.3713 \leq S \leq 2.4937$$

31. $\displaystyle S_6 = \sum_{n=0}^{5} \frac{2(-1)^n}{n!} \approx 0.7333$

$$|R_6| = |S - S_6| \leq a_7 = \frac{2}{6!} = 0.002778;\ 0.7305 \leq S \leq 0.7361$$

33. $\displaystyle\sum_{n=0}^{\infty} \frac{(-1)^n}{n!}$

(a) By Theorem 8.15,

$$|R_N| \leq a_{N+1} = \frac{1}{(N+1)!} < 0.001.$$

This inequality is valid when $N = 6$.

(b) We may approximate the series by

$$\sum_{n=0}^{6} \frac{(-1)^n}{n!} = 1 - 1 + \frac{1}{2} - \frac{1}{6} + \frac{1}{24} - \frac{1}{120} + \frac{1}{720}$$
$$\approx 0.368.$$

(7 terms. Note that the sum begins with $n = 0$.)

35. $\displaystyle\sum_{n=0}^{\infty} \frac{(-1)^n}{(2n+1)!}$

(a) By Theorem 8.15,

$$|R_N| \leq a_{N+1} = \frac{1}{[2(N+1)+1]!} < 0.001.$$

This inequality is valid when $N = 2$.

(b) We may approximate the series by

$$\sum_{n=0}^{2} \frac{(-1)^n}{(2n+1)!} = 1 - \frac{1}{6} + \frac{1}{120} \approx 0.842.$$

(3 terms. Note that the sum begins with $n = 0$.)

37. $\displaystyle\sum_{n=1}^{\infty} \frac{(-1)^{n+1}}{n}$

(a) By Theorem 8.15,

$$|R_N| \leq a_{N+1} = \frac{1}{N+1} < 0.001.$$

This inequality is valid when $N = 1000$.

(b) We may approximate the series by

$$\sum_{n=1}^{1000} \frac{(-1)^{n+1}}{n} = 1 - \frac{1}{2} + \frac{1}{3} - \frac{1}{4} + \cdots - \frac{1}{1000}$$
$$\approx 0.693.$$

(1000 terms)

39. $\displaystyle\sum_{n=1}^{\infty} \frac{(-1)^{n+1}}{2n^3 - 1}$

By Theorem 8.15,

$$|R_N| \leq a_{N+1} = \frac{1}{2(N+1)^3 - 1} < 0.001.$$

This inequality is valid when $N = 7$.

41. $\sum_{n=1}^{\infty} \frac{(-1)^{n+1}}{(n+1)^2}$

$\sum_{n=1}^{\infty} \frac{1}{(n+1)^2}$ converges by comparison to the p-series

$$\sum_{n=1}^{\infty} \frac{1}{n^2}.$$

Therefore, the given series converge absolutely.

43. $\sum_{n=1}^{\infty} \frac{(-1)^{n+1}}{\sqrt{n}}$

The given series converges by the Alternating Series Test, but does not converge absolutely since

$$\sum_{n=1}^{\infty} \frac{1}{\sqrt{n}}$$

is a divergent p-series. Therefore, the series converges conditionally.

45. $\sum_{n=1}^{\infty} \frac{(-1)^{n+1} n^2}{(n+1)^2}$

$$\lim_{n\to\infty} \frac{n^2}{(n+1)^2} = 1$$

Therefore, the series diverges by the nth Term Test.

47. $\sum_{n=2}^{\infty} \frac{(-1)^n}{\ln(n)}$

The given series converges by the Alternating Series Test, but does not converge absolutely since the series

$$\sum_{n=2}^{\infty} \frac{1}{\ln n}$$

diverges by comparison to the harmonic series

$$\sum_{n=1}^{\infty} \frac{1}{n}.$$

Therefore, the series converges conditionally.

49. $\sum_{n=2}^{\infty} \frac{(-1)^n n}{n^3 - 1}$

$$\sum_{n=2}^{\infty} \frac{n}{n^3 - 1}$$

converges by a limit comparison to the convergent p-series

$$\sum_{n=2}^{\infty} \frac{1}{n^2}.$$

Therefore, the given series converges absolutely.

51. $\sum_{n=0}^{\infty} \frac{(-1)^n}{(2n+1)!}$

$$\sum_{n=0}^{\infty} \frac{1}{(2n+1)!}$$

is convergent by comparison to the convergent geometric series

$$\sum_{n=0}^{\infty} \left(\frac{1}{2}\right)^n$$

since

$$\frac{1}{(2n+1)!} < \frac{1}{2^n} \text{ for } n > 0.$$

Therefore, the given series converges absolutely.

53. $\sum_{n=0}^{\infty} \frac{\cos n\pi}{n+1} = \sum_{n=0}^{\infty} \frac{(-1)^n}{n+1}$

The given series converges by the Alternating Series Test, but

$$\sum_{n=0}^{\infty} \frac{|\cos n\pi|}{n+1} = \sum_{n=0}^{\infty} \frac{1}{n+1}$$

diverges by a limit comparison to the divergent harmonic series,

$$\sum_{n=1}^{\infty} \frac{1}{n}.$$

$\lim_{n\to\infty} \frac{|\cos n\pi|/(n+1)}{1/n} = 1$, therefore the series converges conditionally.

55. $\sum_{n=1}^{\infty} \frac{\cos n\pi}{n^2} = \sum_{n=1}^{\infty} \frac{(-1)^n}{n^2}$

$\sum_{n=1}^{\infty} \frac{1}{n^2}$ is a convergent p-series. Therefore, the given series converges absolutely.

57. An alternating series is a series whose terms alternate in sign. See Theorem 8.14.

59. $\sum a_n$ is absolutely convergent if $\sum |a_n|$ converges.

$\sum a_n$ is conditionally convergent if $\sum |a_n|$ diverges, but $\sum a_n$ converges.

61. (b). The partial sums alternate above and below the horizontal line representing the sum.

63. Since $\sum_{n=1}^{\infty} |a_n|$ converges we have

$$\lim_{n\to\infty} |a_n| = 0.$$

Thus, there must exist an $N > 0$ such that $|a_N| < 1$ for all $n > N$ and it follows that $a_n^2 \le |a_n|$ for all $n > N$. Hence, by the Comparison Test,

$$\sum_{n=1}^{\infty} a_n^2$$

converges. Let $a_n = 1/n$ to see that the converse is false.

65. $\sum_{n=1}^{\infty} \frac{1}{n^2}$ converges, hence so does $\sum_{n=1}^{\infty} \frac{1}{n^4}$.

67. False

Let $a_n = \frac{(-1)^n}{n}$.

69. $\sum_{n=1}^{\infty} \frac{10}{n^{3/2}} = 10 \sum_{n=1}^{\infty} \frac{1}{n^{3/2}}$ convergent p-series

71. Diverges by nth Term Test. $\lim_{n\to\infty} a_n = \infty$

73. Convergent Geometric Series $\left(r = \frac{7}{8} < 1\right)$

75. Convergent Geometric Series $\left(r = \frac{1}{\sqrt{e}}\right)$ or Integral Test

77. Converges (absolutely) by Alternating Series Test

79. The first term of the series is zero, not one. You cannot regroup series terms arbitrarily.

Section 8.6 The Ratio and Root Tests

1. $$\frac{(n+1)!}{(n-2)!} = \frac{(n+1)(n)(n-1)(n-2)!}{(n-2)!}$$
$$= (n+1)(n)(n-1)$$

3. Use the Principle of Mathematical Induction. When $k = 1$, the formula is valid since $1 = \frac{(2(1))!}{2^1 \cdot 1!}$. Assume that

$$1 \cdot 3 \cdot 5 \cdot \cdot \cdot (2n-1) = \frac{(2n)!}{2^n n!}$$

and show that

$$1 \cdot 3 \cdot 5 \cdot \cdot \cdot (2n-1)(2n+1) = \frac{(2n+2)!}{2^{n+1}(n+1)!}.$$

—CONTINUED—

3. —CONTINUED—

To do this, note that:

$$1 \cdot 3 \cdot 5 \cdot \cdot \cdot (2n-1)(2n+1) = [1 \cdot 3 \cdot 5 \cdot \cdot \cdot (2n-1)](2n+1)$$
$$= \frac{(2n)!}{2^n n!} \cdot (2n+1) \quad \text{(Induction hypothesis)}$$
$$= \frac{(2n)!(2n+1)}{2^n n!} \cdot \frac{(2n+2)}{2(n+1)}$$
$$= \frac{(2n)!(2n+1)(2n+2)}{2^{n+1}n!(n+1)}$$
$$= \frac{(2n+2)!}{2^{n+1}(n+1)}$$

The formula is valid for all $n \geq 1$.

5. $\displaystyle\sum_{n=1}^{\infty} n\left(\frac{3}{4}\right)^n = 1\left(\frac{3}{4}\right) + 2\left(\frac{9}{16}\right) + \cdot \cdot \cdot$

$S_1 = \frac{3}{4}, S_2 \approx 1.875$

Matches (d)

7. $\displaystyle\sum_{n=1}^{\infty} \frac{(-3)^{n+1}}{n!} = 9 - \frac{3^3}{2} + \cdot \cdot \cdot$

$S_1 = 9$

Matches (f)

9. $\displaystyle\sum_{n=1}^{\infty} \left(\frac{4n}{5n-3}\right)^n = \frac{4}{2} + \left(\frac{8}{7}\right)^2 + \cdot \cdot \cdot$

$S_1 = 2$

Matches (a)

11. (a) Ratio Test: $\displaystyle\lim_{n\to\infty} \left|\frac{a_{n+1}}{a_n}\right| = \lim_{n\to\infty} \frac{(n+1)^2(5/8)^{n+1}}{n^2(5/8)^n} = \lim_{n\to\infty} \left(\frac{n+1}{n}\right)^2 \frac{5}{8} = \frac{5}{8} < 1.$ Converges

(b)

n	5	10	15	20	25
S_n	9.2104	16.7598	18.8016	19.1878	19.2491

(c)

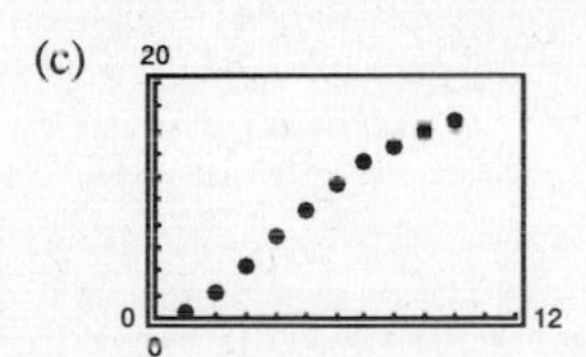

(d) The sum is approximately 19.26.

(e) The more rapidly the terms of the series approach 0, the more rapidly the sequence of the partial sums approaches the sum of the series.

13. $\displaystyle\sum_{n=0}^{\infty} \frac{n!}{3^n}$

$$\lim_{n\to\infty} \left|\frac{a_{n+1}}{a_n}\right| = \lim_{n\to\infty} \left|\frac{(n+1)!}{3^{n+1}} \cdot \frac{3^n}{n!}\right|$$
$$= \lim_{n\to\infty} \frac{n+1}{3} = \infty$$

Therefore, by the Ratio Test, the series diverges.

15. $\displaystyle\sum_{n=1}^{\infty} n\left(\frac{3}{4}\right)^n$

$$\lim_{n\to\infty} \left|\frac{a_{n+1}}{a_n}\right| = \lim_{n\to\infty} \left|\frac{(n+1)(3/4)^{n+1}}{n(3/4)^n}\right|$$
$$= \lim_{n\to\infty} \left|\frac{3(n+1)}{4n}\right| = \frac{3}{4}$$

Therefore, by the Ratio Test, the series converges.

17. $\displaystyle\sum_{n=1}^{\infty} \frac{n}{2^n}$

$$\lim_{n\to\infty} \left|\frac{a_{n+1}}{a_n}\right| = \lim_{n\to\infty} \left|\frac{n+1}{2^{n+1}} \cdot \frac{2^n}{n}\right|$$
$$= \lim_{n\to\infty} \frac{n+1}{2n} = \frac{1}{2}$$

Therefore, by the Ratio Test, the series converges.

19. $\displaystyle\sum_{n=1}^{\infty} \frac{2^n}{n^2}$

$$\lim_{n\to\infty} \left|\frac{a_{n+1}}{a_n}\right| = \lim_{n\to\infty} \left|\frac{2^{n+1}}{(n+1)^2} \cdot \frac{n^2}{2^n}\right|$$
$$= \lim_{n\to\infty} \frac{2n^2}{(n+1)^2} = 2$$

Therefore, by the Ratio Test, the series diverges.

21. $\sum_{n=0}^{\infty} \frac{(-1)^n 2^n}{n!}$

$$\lim_{n\to\infty} \left|\frac{a_{n+1}}{a_n}\right| = \lim_{n\to\infty} \left|\frac{2^{n+1}}{(n+1)!} \cdot \frac{n!}{2^n}\right|$$

$$= \lim_{n\to\infty} \frac{2}{n+1} = 0$$

Therefore, by the Ratio Test, the series converges.

23. $\sum_{n=1}^{\infty} \frac{n!}{n3^n}$

$$\lim_{n\to\infty} \left|\frac{a_{n+1}}{a_n}\right| = \lim_{n\to\infty} \left|\frac{(n+1)!}{(n+1)3^{n+1}} \cdot \frac{n3^n}{n!}\right|$$

$$= \lim_{n\to\infty} \frac{n}{3} = \infty$$

Therefore, by the Ratio Test, the series diverges.

25. $\sum_{n=0}^{\infty} \frac{4^n}{n!}$

$$\lim_{n\to\infty} \left|\frac{a_{n+1}}{a_n}\right| = \lim_{n\to\infty} \left|\frac{4^{n+1}}{(n+1)!} \cdot \frac{n!}{4^n}\right|$$

$$= \lim_{n\to\infty} \frac{4}{n+1} = 0$$

Therefore, by the Ratio Test, the series converges.

27. $\sum_{n=0}^{\infty} \frac{3^n}{(n+1)^n}$

$$\lim_{n\to\infty} \left|\frac{a_{n+1}}{a_n}\right| = \lim_{n\to\infty} \left|\frac{3^{n+1}}{(n+2)^{n+1}} \cdot \frac{(n+1)^n}{3^n}\right| = \lim_{n\to\infty} \frac{3(n+1)^n}{(n+2)^{n+1}} = \lim_{n\to\infty} \frac{3}{n+2}\left(\frac{n+1}{n+2}\right)^n = (0)\left(\frac{1}{e}\right) = 0$$

To find $\lim_{n\to\infty} \left(\frac{n+1}{n+2}\right)^n$, let $y = \lim_{n\to\infty} \left(\frac{n+1}{n+2}\right)^n$. Then,

$$\ln y = \lim_{n\to\infty} n \ln\left(\frac{n+1}{n+2}\right) = \lim_{n\to\infty} \frac{\ln[(n+1)/(n+2)]}{1/n} = \frac{0}{0}$$

$$\ln y = \lim_{n\to\infty} \frac{[(1)/(n+1)] - [(1)/(n+2)]}{-(1/n^2)} = -1 \text{ by L'Hôpital's Rule}$$

$$y = e^{-1} = \frac{1}{e}.$$

Therefore, by the Ratio Test, the series converges.

29. $\sum_{n=0}^{\infty} \frac{4^n}{3^n + 1}$

$$\lim_{n\to\infty} \left|\frac{a_{n+1}}{a_n}\right| = \lim_{n\to\infty} \left|\frac{4^{n+1}}{3^{n+1}+1} \cdot \frac{3^n+1}{4^n}\right| = \lim_{n\to\infty} \frac{4(3^n+1)}{3^{n+1}+1} = \lim_{n\to\infty} \frac{4(1+1/3^n)}{3+1/3^n} = \frac{4}{3}$$

Therefore, by the Ratio Test, the series diverges.

31. $\sum_{n=0}^{\infty} \frac{(-1)^{n+1} n!}{1 \cdot 3 \cdot 5 \cdots (2n+1)}$

$$\lim_{n\to\infty} \left|\frac{a_{n+1}}{a_n}\right| = \lim_{n\to\infty} \left|\frac{(n+1)!}{1 \cdot 3 \cdot 5 \cdots (2n+1)(2n+3)} \cdot \frac{1 \cdot 3 \cdot 5 \cdots (2n+1)}{n!}\right| = \lim_{n\to\infty} \frac{n+1}{2n+3} = \frac{1}{2}$$

Therefore, by the Ratio Test, the series converges.

Note: The first few terms of this series are $-1 + \frac{1}{1 \cdot 3} - \frac{2!}{1 \cdot 3 \cdot 5} + \frac{3!}{1 \cdot 3 \cdot 5 \cdot 7} - \cdots$

33. (a) $\sum_{n=1}^{\infty} \frac{1}{n^{3/2}}$

$$\lim_{n\to\infty} \left|\frac{a_{n+1}}{a_n}\right| = \lim_{n\to\infty} \left|\frac{1}{(n+1)^{3/2}} \cdot \frac{n^{3/2}}{1}\right| = \lim_{n\to\infty} \left(\frac{n}{n+1}\right)^{3/2} = 1$$ Ratio Test is inconclusive.

(b) $\sum_{n=1}^{\infty} \frac{1}{n^{1/2}}$

$$\lim_{n\to\infty} \left|\frac{a_{n+1}}{a_n}\right| = \lim_{n\to\infty} \left|\frac{1}{(n+1)^{1/2}} \cdot \frac{n^{1/2}}{1}\right| = \lim_{n\to\infty} \left(\frac{n}{n+1}\right)^{1/2} = 1$$ Ratio Test is inconclusive.

35. $\sum_{n=1}^{\infty} \left(\frac{n}{2n+1}\right)^n$

$$\lim_{n\to\infty} \sqrt[n]{|a_n|} = \lim_{n\to\infty} \sqrt[n]{\left(\frac{n}{2n+1}\right)^n}$$

$$= \lim_{n\to\infty} \frac{n}{2n+1} = \frac{1}{2}$$

Therefore, by the Root Test, the series converges.

37. $\sum_{n=2}^{\infty} \frac{(-1)^n}{(\ln n)^n}$

$$\lim_{n\to\infty} \sqrt[n]{|a_n|} = \lim_{n\to\infty} \sqrt[n]{\left|\frac{(-1)^n}{(\ln n)^n}\right|}$$

$$= \lim_{n\to\infty} \frac{1}{|\ln n|} = 0$$

Therefore, by the Root Test, the series converges.

39. $\sum_{n=1}^{\infty} (2\sqrt[n]{n} + 1)^n$

$$\lim_{n\to\infty} \sqrt[n]{|a_n|} = \lim_{n\to\infty} \sqrt[n]{(2\sqrt[n]{n} + 1)^n} = \lim_{n\to\infty} (2\sqrt[n]{n} + 1)$$

To find $\lim_{n\to\infty} \sqrt[n]{n}$, let $y = \lim_{n\to\infty} \sqrt[n]{n}$. Then

$$\ln y = \lim_{n\to\infty} (\ln \sqrt[n]{n}) = \lim_{n\to\infty} \frac{1}{n} \ln n = \lim_{n\to\infty} \frac{\ln n}{n} = \lim_{n\to\infty} \frac{1/n}{1} = 0.$$

Thus, $\ln y = 0$, so $y = e^0 = 1$ and $\lim_{n\to\infty} (2\sqrt[n]{n} + 1) = 2(1) + 1 = 3$. Therefore, by the Root Test, the series diverges.

41. $\sum_{n=3}^{\infty} \frac{1}{(\ln n)^n}$

$$\lim_{n\to\infty} \sqrt[n]{|a_n|} = \lim_{n\to\infty} \sqrt[n]{\frac{1}{(\ln n)^n}} = \lim_{n\to\infty} \frac{1}{\ln n} = 0$$

Therefore, by the Root Test, the series converges.

43. $\sum_{n=1}^{\infty} \frac{(-1)^{n+1} 5}{n}$

$$a_{n+1} = \frac{5}{n+1} < \frac{5}{n} = a_n$$

$$\lim_{n\to\infty} \frac{5}{n} = 0$$

Therefore, by the Alternating Series Test, the series converges (conditional convergence).

45. $\sum_{n=1}^{\infty} \frac{3}{n\sqrt{n}} = 3\sum_{n=1}^{\infty} \frac{1}{n^{3/2}}$

This is convergent p-series.

47. $\sum_{n=1}^{\infty} \frac{2n}{n+1}$

$$\lim_{n\to\infty} \frac{2n}{n+1} = 2 \neq 0$$

This diverges by the nth Term Test for Divergence.

49. $\sum_{n=1}^{\infty} \frac{(-1)^n 3^{n-2}}{2^n} = \sum_{n=1}^{\infty} \frac{(-1)^n 3^n 3^{-2}}{2^n} = \sum_{n=1}^{\infty} \frac{1}{9}\left(-\frac{3}{2}\right)^n$

Since $|r| = \frac{3}{2} > 1$, this is a divergent geometric series.

51. $\sum_{n=1}^{\infty} \frac{10n+3}{n2^n}$

$$\lim_{n\to\infty} \frac{(10n+3)/n2^n}{1/2^n} = \lim_{n\to\infty} \frac{10n+3}{n} = 10$$

Therefore, the series converges by a limit comparison test with the geometric series

$$\sum_{n=0}^{\infty} \left(\frac{1}{2}\right)^n.$$

53. $\sum_{n=1}^{\infty} \frac{\cos(n)}{2^n}$

$$\left|\frac{\cos(n)}{2^n}\right| \le \frac{1}{2^n}$$

Therefore, the series

$$\sum_{n=1}^{\infty} \left|\frac{\cos(n)}{2^n}\right|$$

converges by comparison with the geometric series

$$\sum_{n=0}^{\infty} \left(\frac{1}{2}\right)^n.$$

55. $\sum_{n=1}^{\infty} \frac{n7^n}{n!}$

$$\lim_{n\to\infty} \left|\frac{a_{n+1}}{a_n}\right| = \lim_{n\to\infty} \left|\frac{(n+1)7^{n+1}}{(n+1)!} \cdot \frac{n!}{n7^n}\right| = \lim_{n\to\infty} \frac{7}{n} = 0$$

Therefore, by the Ratio Test, the series converges.

57. $\sum_{n=1}^{\infty} \frac{(-1)^n 3^{n-1}}{n!}$

$$\lim_{n\to\infty} \left|\frac{a_{n+1}}{a_n}\right| = \lim_{n\to\infty} \left|\frac{3^n}{(n+1)!} \cdot \frac{n!}{3^{n-1}}\right| = \lim_{n\to\infty} \frac{3}{n+1} = 0$$

Therefore, by the Ratio Test, the series converges.

59. $\sum_{n=1}^{\infty} \frac{(-3)^n}{3 \cdot 5 \cdot 7 \cdots (2n+1)}$

$$\lim_{n\to\infty} \left|\frac{a_{n+1}}{a_n}\right| = \lim_{n\to\infty} \left|\frac{(-3)^{n+1}}{3 \cdot 5 \cdot 7 \cdots (2n+1)(2n+3)} \cdot \frac{3 \cdot 5 \cdot 7 \cdots (2n+1)}{(-3)^n}\right| = \lim_{n\to\infty} \frac{3}{2n+3} = 0$$

Therefore, by the Ratio Test, the series converges.

61. (a) and (c)

$$\sum_{n=1}^{\infty} \frac{n5^n}{n!} = \sum_{n=0}^{\infty} \frac{(n+1)5^{n+1}}{(n+1)!}$$

$$= 5 + \frac{(2)(5)^2}{2!} + \frac{(3)(5)^3}{3!} + \frac{(4)(5)^4}{4!} + \cdots$$

63. (a) and (b) are the same.

65. Replace n with $n + 1$.

$$\sum_{n=1}^{\infty} \frac{n}{4^n} = \sum_{n=0}^{\infty} \frac{n+1}{4^{n+1}}$$

67. Since

$$\frac{3^{10}}{2^{10}\, 10!} \approx 1.59 \times 10^{-5},$$

use 9 terms.

$$\sum_{k=1}^{9} \frac{(-3)^k}{2^k\, k!} \approx -0.7769$$

69. See Theorem 8.17, page 597.

71. No. Let $a_n = \frac{1}{n + 10{,}000}$.

The series $\sum_{n=1}^{\infty} \frac{1}{n + 10{,}000}$ diverges.

73. The series converges absolutely. See Theorem 8.17.

75. First, let

$$\lim_{n\to\infty} \sqrt[n]{|a_n|} = r < 1$$

and choose R such that $0 \le r < R < 1$. There must exist some $N > 0$ such that $\sqrt[n]{|a_n|} < R$ for all $n > N$. Thus, for $n > N$, we $|a_n| < R^n$ and since the geometric series

$$\sum_{n=0}^{\infty} R^n$$

converges, we can apply the Comparison Test to conclude that

$$\sum_{n=1}^{\infty} |a_n|$$

converges which in turn implies that $\sum_{n=1}^{\infty} a_n$ converges.

Second, let

$$\lim_{n\to\infty} \sqrt[n]{|a_n|} = r > R > 1.$$

Then there must exist some $M > 0$ such that $\sqrt[n]{|a_n|} > R$ for infinitely many $n > M$. Thus, for infinitely many $n > M$, we have $|a_n| > R^n > 1$ which implies that $\lim_{n\to\infty} a_n \ne 0$ which in turn implies that

$$\sum_{n=1}^{\infty} a_n \text{ diverges.}$$

Section 8.7 Taylor Polynomials and Approximations

1. $y = -\frac{1}{2}x^2 + 1$

Parabola

Matches (d)

3. $y = e^{-1/2}[(x + 1) + 1]$

Linear

Matches (a)

5. $f(x) = \dfrac{4}{\sqrt{x}} = 4x^{-1/2} \qquad f(1) = 4$

$f'(x) = -2x^{-3/2} \qquad f'(1) = -2$

$P_1(x) = f(1) + f'(1)(x - 1)$

$= 4 + (-2)(x - 1)$

$P_1(x) = -2x + 6$

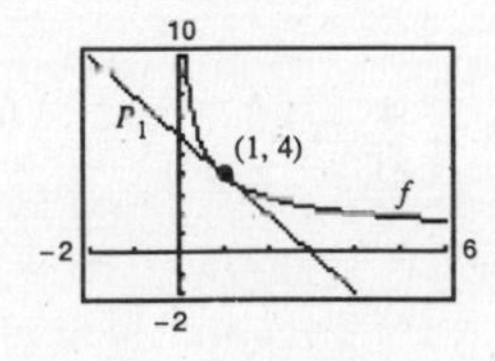

P_1 is called the first degree Taylor polynomial for f at c.

7. $f(x) = \sec x \qquad f\left(\dfrac{\pi}{4}\right) = \sqrt{2}$

$f'(x) = \sec x \tan x \qquad f'\left(\dfrac{\pi}{4}\right) = \sqrt{2}$

$P_1(x) = f\left(\dfrac{\pi}{4}\right) + f'\left(\dfrac{\pi}{4}\right)\left(x - \dfrac{\pi}{4}\right)$

$P_1(x) = \sqrt{2} + \sqrt{2}\left(x - \dfrac{\pi}{4}\right)$

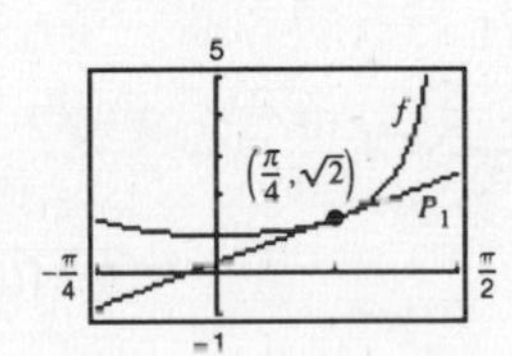

P_1 is called the first degree Taylor polynomial for f at c.

9. $f(x) = \dfrac{4}{\sqrt{x}} = 4x^{-1/2} \qquad f(1) = 4$

$f'(x) = -2x^{-3/2} \qquad f'(1) = -2$

$f''(x) = 3x^{-5/2} \qquad f''(1) = 3$

$P_2 = f(1) + f'(1)(x - 1) + \dfrac{f''(1)}{2}(x - 1)^2$

$= 4 - 2(x - 1) + \dfrac{3}{2}(x - 1)^2$

x	0	0.8	0.9	1.0	1.1	1.2	2
$f(x)$	Error	4.4721	4.2164	4.0	3.8139	3.6515	2.8284
$P_2(x)$	7.5	4.46	4.215	4.0	3.815	3.66	3.5

11. $f(x) = \cos x$

$P_2(x) = 1 - \frac{1}{2}x^2$

$P_4(x) = 1 - \frac{1}{2}x^2 + \frac{1}{24}x^4$

$P_6(x) = 1 - \frac{1}{2}x^2 + \frac{1}{24}x^4 - \frac{1}{720}x^6$

(a)

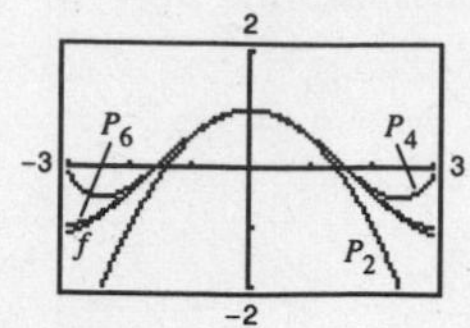

(b) $f'(x) = -\sin x \qquad P_2'(x) = -x$

$f''(x) = -\cos x \qquad P_2''(x) = -1$

$f''(0) = P_2''(0) = -1$

$f'''(x) = \sin x \qquad P_4'''(x) = x$

$f^{(4)}(x) = \cos x \qquad P_4^{(4)}(x) = 1$

$f^{(4)}(0) = 1 = P_4^{(4)}(0)$

$f^{(5)}(x) = -\sin x \qquad P_6^{(5)}(x) = -x$

$f^{(6)}(x) = -\cos x \qquad P^{(6)}(x) = -1$

$f^{(6)}(0) = -1 = P_6^{(6)}(0)$

(c) In general, $f^{(n)}(0) = P_n^{(n)}(0)$ for all n.

13. $f(x) = e^{-x} \qquad f(0) = 1$

$f'(x) = -e^{-x} \qquad f'(0) = -1$

$f''(x) = e^{-x} \qquad f''(0) = 1$

$f'''(x) = -e^{-x} \qquad f'''(0) = -1$

$$P_3(x) = f(0) + f'(0)x + \frac{f''(0)}{2!}x^2 + \frac{f'''(0)}{3!}x^3$$

$$= 1 - x + \frac{x^2}{2} - \frac{x^3}{6}$$

15. $f(x) = e^{2x} \qquad f(0) = 1$

$f'(x) = 2e^{2x} \qquad f'(0) = 2$

$f''(x) = 4e^{2x} \qquad f''(0) = 4$

$f'''(x) = 8e^{2x} \qquad f'''(0) = 8$

$f^{(4)}(x) = 16^{2x} \qquad f^{(4)}(0) = 16$

$$P_4(x) = 1 + 2x + \frac{4}{2!}x^2 + \frac{8}{3!}x^3 + \frac{16}{4!}x^4$$

$$= 1 + 2x + 2x^2 + \frac{4}{3}x^3 + \frac{2}{3}x^4$$

17. $f(x) = \sin x \qquad f(0) = 0$

$f'(x) = \cos x \qquad f'(0) = 1$

$f''(x) = -\sin x \qquad f''(0) = 0$

$f'''(x) = -\cos x \qquad f'''(0) = -1$

$f^{(4)}(x) = \sin x \qquad f^{(4)}(0) = 0$

$f^{(5)}(x) = \cos x \qquad f^{(5)}(0) = 1$

$$P_5(x) = 0 + (1)x + \frac{0}{2!}x^2 + \frac{-1}{3!}x^3 + \frac{0}{4!}x^4 + \frac{1}{5!}x^5$$

$$= x - \frac{1}{6}x^3 + \frac{1}{120}x^5$$

19. $f(x) = xe^x \qquad f(0) = 0$

$f'(x) = xe^x + e^x \qquad f'(0) = 1$

$f''(x) = xe^x + 2e^x \qquad f''(0) = 2$

$f'''(x) = xe^x + 3e^x \qquad f'''(0) = 3$

$f^{(4)}(x) = xe^x + 4e^x \qquad f^{(4)}(0) = 4$

$$P_4(x) = 0 + x + \frac{2}{2!}x^2 + \frac{3}{3!}x^3 + \frac{4}{4!}x^4$$

$$= x + x^2 + \frac{1}{2}x^3 + \frac{1}{6}x^4$$

21. $f(x) = \dfrac{1}{x+1} \qquad f(0) = 1$

$f'(x) = -\dfrac{1}{(x+1)^2} \qquad f'(0) = -1$

$f''(x) = \dfrac{2}{(x+1)^2} \qquad f''(0) = 2$

$f'''(x) = \dfrac{-6}{(x+1)^4} \qquad f'''(0) = -6$

$f^{(4)}(x) = \dfrac{24}{(x+1)^5} \qquad f^{(4)}(0) = 24$

$$P_4(x) = 1 - x + \frac{2}{2!}x^2 + \frac{-6}{3!}x^3 + \frac{24}{4!}x^4$$

$$= 1 - x + x^2 - x^3 + x^4$$

23. $f(x) = \sec x \qquad f(0) = 1$

$f'(x) = \sec x \tan x \qquad f'(0) = 0$

$f''(x) = \sec^3 x + \sec x \tan^2 x \qquad f''(0) = 1$

$$P_2(x) = 1 + 0x + \frac{1}{2!}x^2 = 1 + \frac{1}{2}x^2$$

25. $f(x) = \dfrac{1}{x}$ $\qquad f(1) = 1$

$f'(x) = -\dfrac{1}{x^2}$ $\qquad f'(1) = -1$

$f''(x) = \dfrac{2}{x^3}$ $\qquad f''(1) = 2$

$f'''(x) = -\dfrac{6}{x^4}$ $\qquad f'''(1) = -6$

$f^{(4)}(x) = \dfrac{24}{x^5}$ $\qquad f^{(4)}(1) = 24$

$$P_4(x) = 1 - (x-1) + \frac{2}{2!}(x-1)^2 + \frac{-6}{3!}(x-1)^3 + \frac{24}{4!}(x-1)^4$$

$$= 1 - (x-1) + (x-1)^2 - (x-1)^3 + (x-1)^4$$

27. $f(x) = \sqrt{x}$ $\qquad f(1) = 1$

$f'(x) = \dfrac{1}{2\sqrt{x}}$ $\qquad f'(1) = \dfrac{1}{2}$

$f''(x) = -\dfrac{1}{4x\sqrt{x}}$ $\qquad f''(1) = -\dfrac{1}{4}$

$f'''(x) = \dfrac{3}{8x^2\sqrt{x}}$ $\qquad f'''(1) = \dfrac{3}{8}$

$f^{(4)}(x) = -\dfrac{15}{16x^3\sqrt{x}}$ $\qquad f^{(4)}(1) = -\dfrac{15}{16}$

$$P_4(x) = 1 + \frac{1}{2}(x-1) - \frac{1}{8}(x-1)^2 + \frac{1}{16}(x-1)^3 - \frac{5}{128}(x-1)^4$$

29. $f(x) = \ln x$ $\qquad f(1) = 0$

$f'(x) = \dfrac{1}{x}$ $\qquad f'(1) = 1$

$f''(x) = -\dfrac{1}{x^2}$ $\qquad f''(1) = -1$

$f'''(x) = \dfrac{2}{x^3}$ $\qquad f'''(1) = 2$

$f^{(4)}(x) = -\dfrac{6}{x^4}$ $\qquad f^{(4)}(1) = -6$

$$P_4(x) = 0 + (x-1) - \frac{1}{2}(x-1)^2 + \frac{1}{3}(x-1)^3 - \frac{1}{4}(x-1)^4$$

31. $f(x) = \tan x$

$f'(x) = \sec^2 x$

$f''(x) = 2\sec^2 x \tan x$

$f'''(x) = 4\sec^2 x \tan^2 x + 2\sec^4 x$

$f^{(4)}(x) = 8\sec^2 x \tan^3 x + 16\sec^4 x \tan x$

$f^{(5)}(x) = 16\sec^2 x \tan^4 x + 88\sec^4 x \tan^2 x + 16\sec^6 x$

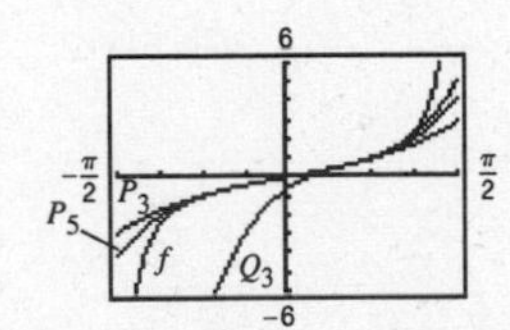

(a) $n = 3,\ c = 0$

$$P_3(x) = 0 + x + \frac{0}{2!}x^2 + \frac{2}{3!}x^3 = x + \frac{1}{3}x^3$$

(b) $n = 5,\ c = 0$

$$P_5(x) = 0 + x + \frac{0}{2!}x^2 + \frac{2}{3!}x^3 + \frac{0}{4!}x^4 + \frac{16}{5!}x^5$$

$$= x + \frac{1}{3}x^3 + \frac{2}{15}x^5$$

(c) $n = 3,\ c = \dfrac{\pi}{4}$

$$Q_3(x) = 1 + 2\left(x - \frac{\pi}{4}\right) + \frac{4}{2!}\left(x - \frac{\pi}{4}\right)^2 + \frac{16}{3!}\left(x - \frac{\pi}{4}\right)^3$$

$$= 1 + 2\left(x - \frac{\pi}{4}\right) + 2\left(x - \frac{\pi}{4}\right)^2 + \frac{8}{3}\left(x - \frac{\pi}{4}\right)^3$$

33. $f(x) = \sin x$

$P_1(x) = x$

$P_3(x) = x - \frac{1}{6}x^3$

$P_5(x) = x - \frac{1}{6}x^3 + \frac{1}{120}x^5$

$P_7(x) = x - \frac{1}{6}x^3 + \frac{1}{120}x^5 - \frac{1}{5040}x^7$

(a)

x	0.00	0.25	0.50	0.75	1.00
$\sin x$	0.0000	0.2474	0.4794	0.6816	0.8415
$P_1(x)$	0.0000	0.2500	0.5000	0.7500	1.0000
$P_3(x)$	0.0000	0.2474	0.4792	0.6797	0.8333
$P_5(x)$	0.0000	0.2474	0.4794	0.6817	0.8417
$P_7(x)$	0.0000	0.2474	0.4794	0.6816	0.8415

(b)

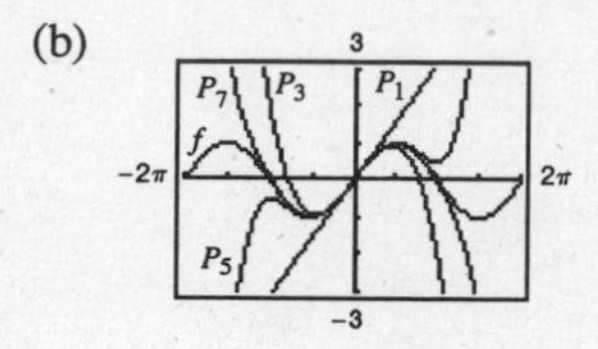

(c) As the distance increases, the accuracy decreases

35. $f(x) = \arcsin x$

(a) $P_3(x) = x + \dfrac{x^3}{6}$

(b)

x	−0.75	−0.50	−0.25	0	0.25	0.50	0.75
$f(x)$	−0.848	−0.524	−0.253	0	0.253	0.524	0.848
$P_3(x)$	−0.820	−0.521	−0.253	0	0.253	0.521	0.820

(c)

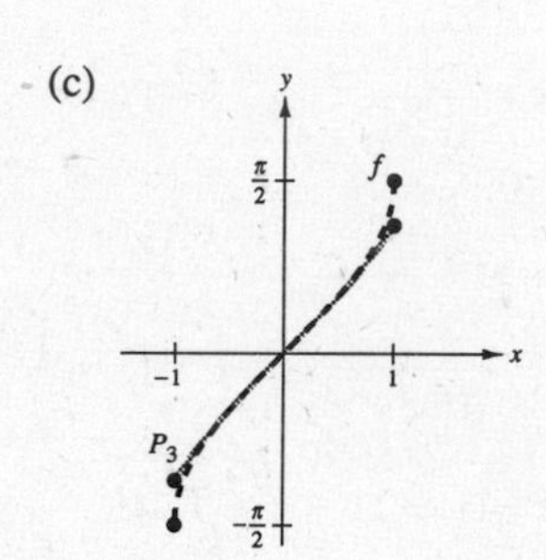

37. $f(x) = \cos x$

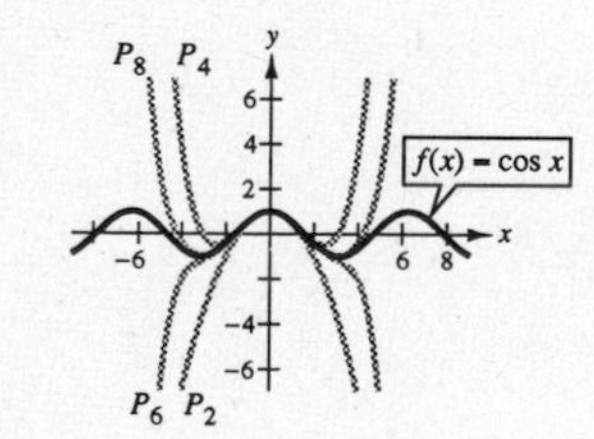

39. $f(x) = \ln(x^2 + 1)$

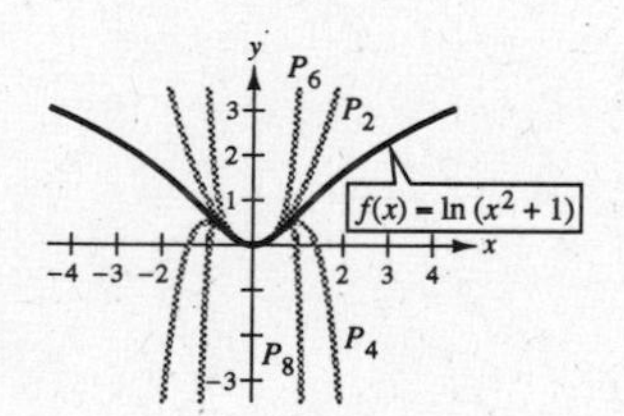

41. $f(x) = e^{-x} \approx 1 - x + \dfrac{x^2}{2} - \dfrac{x^3}{6}$

$f\left(\dfrac{1}{2}\right) \approx 0.6042$

43. $f(x) = \ln x \approx (x - 1) - \frac{1}{2}(x - 1)^2 + \frac{1}{3}(x - 1)^3 - \frac{1}{4}(x - 1)^4$

$f(1.2) \approx 0.1823$

45. $f(x) = \cos x; f^{(5)}(x) = -\sin x \Rightarrow$ Max on $[0, 0.3]$ is 1.

$R_4(x) \le \dfrac{1}{5!}(0.3)^5 = 2.025 \times 10^{-5}$

Note: You could use $R_5(x)$: $f^{(6)}(x) = -\cos x$, max on $[0, 0.3]$ is 1. $R_5(x) \le \dfrac{1}{6!}(0.3)^6 = 1.0125 \times 10^{-6}$

Exact Error: $0.000001 = 1.0 \times 10^{-6}$

47. $f(x) = \arcsin x; f^{(4)}(x) = \dfrac{x(6x^2 + 9)}{(1 - x^2)^{7/2}} \Rightarrow$ Max on $[0, 0.4]$ is $f^{(4)}(0.4) \approx 7.3340$.

$R_3(x) \le \dfrac{7.3340}{4!}(0.4)^4 \approx 0.00782 = 7.82 \times 10^{-3}$

49. $g(x) = \sin x$

$g^{(n+1)}(x) \le 1$ for all x

$R_n(x) \le \dfrac{1}{(n+1)!}(0.3)^{n+1} < 0.001$

By trial and error, $n = 3$.

51. $f(x) = \ln(x + 1)$

$f^{(n+1)}(x) = \dfrac{(-1)^n n!}{(x+1)^{n+1}} \Rightarrow$ Max on $[0, 0.5]$ is $n!$.

$R_n \le \dfrac{n!}{(n+1)!}(0.5)^{n+1} = \dfrac{(0.5)^{n+1}}{n+1} < 0.0001$

By trial and error, $n = 9$. (See Example 9.) Using 9 terms, $\ln(1.5) \approx 0.4055$.

53. $f(x) = e^x \approx 1 + x + \dfrac{x^2}{2} + \dfrac{x^3}{6}, \; x < 0$

$R_3(x) = \dfrac{e^z}{4!}x^4 < 0.001$

$e^z x^4 < 0.024$

$xe^{z/4} < 0.3936$

$x < \dfrac{0.3936}{e^{z/4}} < 0.3936, \; z < 0$

$-0.3936 < x < 0$

55. The graph of the approximating polynomial P and the elementary function f both pass through the point $(c, f(c))$ and the slopes of P and f agree at $(c, f(c))$. Depending on the degree of P, the nth derivatives of P and f agree at $(c, f(c))$.

57. See definition on page 607.

59. The accuracy increases as the degree increases (for values within the interval of convergence).

61. (a) $f(x) = e^x$

$P_4(x) = 1 + x + \dfrac{1}{2}x^2 + \dfrac{1}{6}x^3 + \dfrac{1}{24}x^4$

$g(x) = xe^x$

$Q_5(x) = x + x^2 + \dfrac{1}{2}x^3 + \dfrac{1}{6}x^4 + \dfrac{1}{24}x^5$

$Q_5(x) = x\,P_4(x)$

(b) $f(x) = \sin x$

$P_5(x) = x - \dfrac{x^3}{3!} + \dfrac{x^5}{5!}$

$g(x) = x \sin x$

$Q_6(x) = x\,P_5(x) = x^2 - \dfrac{x^4}{3!} + \dfrac{x^6}{5!}$

(c) $g(x) = \dfrac{\sin x}{x} = \dfrac{1}{x}P_5(x) = 1 - \dfrac{x^2}{3!} + \dfrac{x^4}{5!}$

63. (a) $Q_2(x) = -1 + \dfrac{\pi^2(x+2)^2}{32}$

(b) $R_2(x) = -1 + \dfrac{\pi^2(x-6)^2}{32}$

(c) No. The polynomial will be linear. Translations are possible at $x = -2 + 8n$.

65. Let f be an even function and P_n be the nth Maclaurin polynomial for f. Since f is even, f' is odd, f'' is even, f''' is odd, etc. All of the odd derivatives of f are odd and thus, all of the odd powers of x will have coefficients of zero. P_n will only have terms with even powers of x.

67. As you move away from $x = c$, the Taylor Polynomial becomes less and less accurate.

Section 8.8 Power Series

1. Centered at 0

3. Centered at 2

5. $\sum_{n=0}^{\infty} (-1)^n \frac{x^n}{n+1}$

$$L = \lim_{n\to\infty} \left|\frac{u_{n+1}}{u_n}\right| = \lim_{n\to\infty} \left|\frac{(-1)^{n+1}x^{n+1}}{n+2} \cdot \frac{n+1}{(-1)^n x^n}\right|$$

$$= \lim_{n\to\infty} \left|\frac{n+1}{n+2}\right| |x| = |x|$$

$|x| < 1 \Rightarrow R = 1$

7. $\sum_{n=1}^{\infty} \frac{(2x)^n}{n^2}$

$$L = \lim_{n\to\infty} \left|\frac{u_{n+1}}{u_n}\right| = \lim_{n\to\infty} \left|\frac{(2x)^{n+1}}{(n+1)^2} \cdot \frac{n^2}{(2x)^n}\right|$$

$$= \lim_{n\to\infty} \left|\frac{2n^2 x}{(n+1)^2}\right| = 2|x|$$

$2|x| < 1 \Rightarrow R = \frac{1}{2}$

9. $\sum_{n=0}^{\infty} \frac{(2x)^{2n}}{(2n)!}$

$$L = \lim_{n\to\infty} \left|\frac{u_{n+1}}{u_n}\right| = \lim_{n\to\infty} \left|\frac{(2x)^{2n+2}/(2n+2)!}{(2x)^{2n}/(2n)!}\right|$$

$$= \lim_{n\to\infty} \left|\frac{(2x)^2}{(2n+2)(2n+1)}\right| = 0$$

Thus, the series converges for all x. $R = \infty$.

11. $\sum_{n=0}^{\infty} \left(\frac{x}{2}\right)^n$

Since the series is geometric, it converges only if $|x/2| < 1$ or $-2 < x < 2$.

13. $\sum_{n=1}^{\infty} \frac{(-1)^n x^n}{n}$

$$\lim_{n\to\infty} \left|\frac{u_{n+1}}{u_n}\right| = \lim_{n\to\infty} \left|\frac{(-1)^{n+1}x^{n+1}}{n+1} \cdot \frac{n}{(-1)^n x^n}\right|$$

$$= \lim_{n\to\infty} \left|\frac{nx}{n+1}\right| = |x|$$

Interval: $-1 < x < 1$

When $x = 1$, the alternating series $\sum_{n=1}^{\infty} \frac{(-1)^n}{n}$ converges.

When $x = -1$, the p-series $\sum_{n=1}^{\infty} \frac{1}{n}$ diverges.

Therefore, the interval of convergence is $-1 < x \le 1$.

15. $\sum_{n=0}^{\infty} \frac{x^n}{n!}$

$$\lim_{n\to\infty} \left|\frac{u_{n+1}}{u_n}\right| = \lim_{n\to\infty} \left|\frac{x^{n+1}}{(n+1)!} \cdot \frac{n!}{x^n}\right|$$

$$= \lim_{n\to\infty} \left|\frac{x}{n+1}\right| = 0$$

The series converges for all x. Therefore, the interval of convergence is $-\infty < x < \infty$.

17. $\sum_{n=0}^{\infty} (2n)!\left(\frac{x}{2}\right)^n$

$$\lim_{n\to\infty} \left|\frac{u_{n+1}}{u_n}\right| = \lim_{n\to\infty} \left|\frac{(2n+2)!x^{n+1}}{2^{n+1}} \cdot \frac{2^n}{(2n)!x^n}\right| = \lim_{n\to\infty} \left|\frac{(2n+2)(2n+1)x}{2}\right| = \infty$$

Therefore, the series converges only for $x = 0$.

19. $\sum_{n=1}^{\infty} \frac{(-1)^{n+1}x^n}{4^n}$

Since the series is geometric, it converges only if $|x/4| < 1$ or $-4 < x < 4$.

21. $\sum_{n=1}^{\infty} \frac{(-1)^{n+1}(x-5)^n}{n5^n}$

$$\lim_{n\to\infty}\left|\frac{u_{n+1}}{u_n}\right| = \lim_{n\to\infty}\left|\frac{(-1)^{n+2}(x-5)^{n+1}}{(n+1)5^{n+1}} \cdot \frac{n5^n}{(-1)^{n+1}(x-5)^n}\right| = \lim_{n\to\infty}\left|\frac{n(x-5)}{5(n+1)}\right| = \frac{1}{5}|x-5|$$

$R = 5$

Center: $x = 5$

Interval: $-5 < x - 5 < 5$ or $0 < x < 10$

When $x = 0$, the p-series $\sum_{n=1}^{\infty} \frac{-1}{n}$ diverges.

When $x = 10$, the alternating series $\sum_{n=1}^{\infty} \frac{(-1)^{n+1}}{n}$ converges.

Therefore, the interval of convergence is $0 < x \le 10$.

23. $\sum_{n=0}^{\infty} \frac{(-1)^{n+1}(x-1)^{n+1}}{n+1}$

$$\lim_{n\to\infty}\left|\frac{u_{n+1}}{u_n}\right| = \lim_{n\to\infty}\left|\frac{(-1)^{n+2}(x-1)^{n+2}}{n+2} \cdot \frac{n+1}{(-1)^{n+1}(x-1)^{n+1}}\right| = \lim_{n\to\infty}\left|\frac{(n+1)(x-1)}{n+2}\right| = |x-1|$$

$R = 1$

Center: $x = 1$

Interval: $-1 < x - 1 < 1$ or $0 < x < 2$

When $x = 0$, the series $\sum_{n=0}^{\infty} \frac{1}{n+1}$ diverges by the integral test.

When $x = 2$, the alternating series $\sum_{n=0}^{\infty} \frac{(-1)^{n+1}}{n+1}$ converges.

Therefore, the interval of convergence is $0 < x \le 2$.

25. $\sum_{n=1}^{\infty} \frac{(x-c)^{n-1}}{c^{n-1}}$

$$\lim_{n\to\infty}\left|\frac{u_{n+1}}{u_n}\right| = \lim_{n\to\infty}\left|\frac{(x-c)^n}{c^n} \cdot \frac{c^{n-1}}{(x-c)^{n-1}}\right| = \frac{1}{c}|x-c|$$

$R = c$

Center: $x = c$

Interval: $-c < x - c < c$ or $0 < x < 2c$

When $x = 0$, the series $\sum_{n=1}^{\infty} (-1)^{n-1}$ diverges.

When $x = 2c$, the series $\sum_{n=1}^{\infty} 1$ diverges.

Therefore, the interval of convergence is $0 < x < 2c$.

27. $\sum_{n=1}^{\infty} \frac{n}{n+1}(-2x)^{n-1}$

$$\lim_{n\to\infty}\left|\frac{u_{n+1}}{u_n}\right| = \lim_{n\to\infty}\left|\frac{(n+1)(-2x)^n}{n+2} \cdot \frac{n+1}{n(-2x)^{n-1}}\right|$$

$$= \lim_{n\to\infty}\left|\frac{(-2x)(n+1)^2}{n(n+2)}\right| = 2|x|$$

$R = \frac{1}{2}$

Interval: $-\frac{1}{2} < x < \frac{1}{2}$

When $x = -\frac{1}{2}$, the series $\sum_{n=1}^{\infty} \frac{n}{n+1}$ diverges by the nth Term Test.

When $x = \frac{1}{2}$, the alternating series $\sum_{n=1}^{\infty} \frac{(-1)^{n-1}n}{n+1}$ diverges.

Therefore, the interval of convergence is $-\frac{1}{2} < x < \frac{1}{2}$.

29. $\displaystyle\sum_{n=0}^{\infty} \frac{x^{2n+1}}{(2n+1)!}$

$$\lim_{n\to\infty}\left|\frac{u_{n+1}}{u_n}\right| = \lim_{n\to\infty}\left|\frac{x^{2n+3}}{(2n+3)!}\cdot\frac{(2n+1)!}{x^{2n+1}}\right|$$

$$= \lim_{n\to\infty}\left|\frac{x^2}{(2n+2)(2n+3)}\right| = 0$$

Therefore, the interval of convergence is $-\infty < x < \infty$.

31. $\displaystyle\sum_{n=1}^{\infty} \frac{k(k+1)\cdots(k+n-1)x^n}{n!}$

$$\lim_{n\to\infty}\left|\frac{u_{n+1}}{u_n}\right| = \lim_{n\to\infty}\left|\frac{k(k+1)\cdots(k+n-1)(k+n)x^{n+1}}{(n+1)!}\cdot\frac{n!}{k(k+1)\cdots(k+n-1)x^n}\right| = \lim_{n\to\infty}\left|\frac{(k+n)x}{n+1}\right| = |x|$$

$R = 1$

When $x = \pm 1$, the series diverges and the interval of convergence is $-1 < x < 1$.

$$\left[\frac{k(k+1)\cdots(k+n-1)}{1\cdot 2\cdots n} \geq 1\right]$$

33. $\displaystyle\sum_{n=1}^{\infty} \frac{(-1)^{n+1}3\cdot 7\cdot 11\cdots(4n-1)(x-3)^n}{4^n}$

$$\lim_{n\to\infty}\left|\frac{u_{n+1}}{u_n}\right| = \lim_{n\to\infty}\left|\frac{(-1)^{n+2}\cdot 3\cdot 7\cdot 11\cdots(4n-1)(4n+3)(x-3)^{n+1}}{4^{n+1}}\cdot\frac{4^n}{(-1)^{n+1}\cdot 3\cdot 7\cdot 11\cdots(4n-1)(x-3)^n}\right|$$

$$= \lim_{n\to\infty}\left|\frac{(4n+3)(x-3)}{4}\right| = \infty$$

$R = 0$

Center: $x = 3$

Therefore, the series converges only for $x = 3$.

35. (a) $f(x) = \displaystyle\sum_{n=0}^{\infty}\left(\frac{x}{2}\right)^n$, $-2 < x < 2$ (Geometric)

(b) $f'(x) = \displaystyle\sum_{n=1}^{\infty}\left(\frac{n}{2}\right)\left(\frac{x}{2}\right)^{n-1}$, $-2 < x < 2$

(c) $f''(x) = \displaystyle\sum_{n=2}^{\infty}\left(\frac{n}{2}\right)\left(\frac{n-1}{2}\right)\left(\frac{x}{2}\right)^{n-2}$, $-2 < x < 2$

(d) $\displaystyle\int f(x)\,dx = \sum_{n=0}^{\infty}\frac{2}{n+1}\left(\frac{x}{2}\right)^{n+1}$, $-2 \leq x < 2$

37. (a) $f(x) = \displaystyle\sum_{n=0}^{\infty}\frac{(-1)^{n+1}(x-1)^{n+1}}{n+1}$, $0 < x \leq 2$

(b) $f'(x) = \displaystyle\sum_{n=0}^{\infty}(-1)^{n+1}(x-1)^n$, $0 < x < 2$

(c) $f''(x) = \displaystyle\sum_{n=1}^{\infty}(-1)^{n+1}n(x-1)^{n-1}$, $0 < x < 2$

(d) $\displaystyle\int f(x)\,dx = \sum_{n=1}^{\infty}\frac{(-1)^{n+1}(x-1)^{n+2}}{(n+1)(n+2)}$, $0 \leq x \leq 2$

39. $g(1) = \displaystyle\sum_{n=0}^{\infty}\left(\frac{1}{3}\right)^n = 1 + \frac{1}{3} + \frac{1}{9} + \cdots$

$S_1 = 1$, $S_2 = 1.33$. Matches (c)

41. $g(3.1) = \displaystyle\sum_{n=0}^{\infty}\left(\frac{3.1}{3}\right)^n$ diverges. Matches (b)

43. A series of the form

$$\sum_{n=0}^{\infty} a_n(x-c)^n$$

is called a power series centered at c.

45. A single point, an interval, or the entire real line.

47. (a) $f(x) = \sum_{n=0}^{\infty} \frac{(-1)^n x^{2n+1}}{(2n+1)!}, \; -\infty < x < \infty$ (See Exercise 29.)

$g(x) = \sum_{n=0}^{\infty} \frac{(-1)^n x^{2n}}{(2n)!}, \; -\infty < x < \infty$

(b) $f'(x) = \sum_{n=0}^{\infty} \frac{(-1)^n x^{2n}}{(2n)!} = g(x)$

(c) $g'(x) = \sum_{n=1}^{\infty} \frac{(-1)^n x^{2n-1}}{(2n-1)!} = \sum_{n=0}^{\infty} \frac{(-1)^{n+1} x^{2n+1}}{(2n+1)!} = -\sum_{n=0}^{\infty} \frac{(-1)^n x^{2n+1}}{(2n+1)!} = -f(x)$

(d) $f(x) = \sin x$ and $g(x) = \cos x$

49.

$$y = \sum_{n=0}^{\infty} \frac{x^{2n}}{2^n n!}$$

$$y' = \sum_{n=1}^{\infty} \frac{2nx^{2n-1}}{2^n n!}$$

$$y'' = \sum_{n=1}^{\infty} \frac{2n(2n-1)x^{2n-2}}{2^n n!}$$

$$y'' - xy' - y = \sum_{n=1}^{\infty} \frac{2n(2n-1)x^{2n-2}}{2^n n!} - \sum_{n=1}^{\infty} \frac{2nx^{2n}}{2^n n!} - \sum_{n=0}^{\infty} \frac{x^{2n}}{2^n n!}$$

$$= \sum_{n=1}^{\infty} \frac{2n(2n-1)x^{2n-2}}{2^n n!} - \sum_{n=0}^{\infty} \frac{(2n+1)x^{2n}}{2^n n!}$$

$$= \sum_{n=0}^{\infty} \left[\frac{(2n+2)(2n+1)x^{2n}}{2^{n+1}(n+1)!} - \frac{(2n+1)x^{2n}}{2^n n!} \cdot \frac{2(n+1)}{2(n+1)} \right]$$

$$= \sum_{n=0}^{\infty} \frac{2(n+1)x^{2n}[(2n+1) - (2n+1)]}{2^{n+1}(n+1)!} = 0$$

51. $J_0(x) = \sum_{k=0}^{\infty} \frac{(-1)^k x^{2k}}{2^{2k}(k!)^2}$

(a) $\lim_{k\to\infty} \left| \frac{u_{k+1}}{u_k} \right| = \lim_{k\to\infty} \left| \frac{(-1)^{k+1} x^{2k+2}}{2^{2k+2}[(k+1)!]^2} \cdot \frac{2^{2k}(k!)^2}{(-1)^k x^{2k}} \right| = \lim_{k\to\infty} \left| \frac{(-1)x^2}{2^2(k+1)^2} \right| = 0$

Therefore, the interval of convergence is $-\infty < x < \infty$.

(b)

$$J_0 = \sum_{k=0}^{\infty} (-1)^k \frac{x^{2k}}{4^k (k!)^2}$$

$$J_0' = \sum_{k=1}^{\infty} (-1)^k \frac{2kx^{2k-1}}{4^k (k!)^2} = \sum_{k=0}^{\infty} (-1)^{k+1} \frac{(2k+2)x^{2k+1}}{4^{k+1}[(k+1)!]^2}$$

$$J_0'' = \sum_{k=1}^{\infty} (-1)^k \frac{2k(2k-1)x^{2k-2}}{4^k (k!)^2} = \sum_{k=0}^{\infty} (-1)^{k+1} \frac{(2k+2)(2k+1)x^{2k}}{4^{k+1}[(k+1)!]^2}$$

$$x^2 J_0'' + xJ_0' + x^2 J_0 = \sum_{k=0}^{\infty} (-1)^{k+1} \frac{2(2k+1)x^{2k+2}}{4^{k+1}(k+1)!k!} + \sum_{k=0}^{\infty} (-1)^{k+1} \frac{2x^{2k+2}}{4^{k+1}(k+1)!k!} + \sum_{k=0}^{\infty} (-1)^k \frac{x^{2k+2}}{4^k (k!)^2}$$

$$= \sum_{k=0}^{\infty} \frac{(-1)^k x^{2k+2}}{4^k (k!)^2} \left[(-1)\frac{2(2k+1)}{4(k+1)} + (-1)\frac{2}{4(k+1)} + 1 \right]$$

$$= \sum_{k=0}^{\infty} \frac{(-1)^k x^{2k+2}}{4^k (k!)^2} \left[\frac{-4k-2}{4k+4} - \frac{2}{4k+4} + \frac{4k+4}{4k+4} \right] = 0$$

—CONTINUED—

51. —CONTINUED—

(c) $P_6(x) = 1 - \frac{x^2}{4} + \frac{x^4}{64} - \frac{x^6}{2304}$

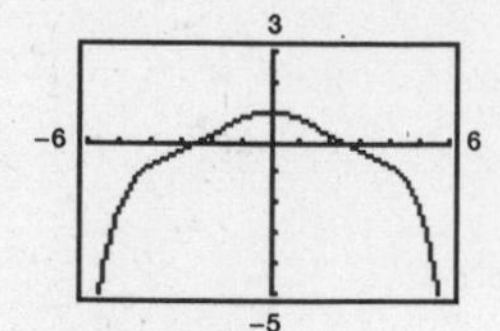

(d) $\int_0^1 J_0 dx = \int_0^1 \sum_{k=0}^{\infty} \frac{(-1)^k x^{2k}}{4^k (k!)^2} dx$

$= \left[\sum_{k=0}^{\infty} \frac{(-1)^k x^{2k+1}}{4^k(k!)^2(2k+1)}\right]_0^1$

$= \sum_{k=0}^{\infty} \frac{(-1)^k}{4^k(k!)^2(2k+1)}$

$= 1 - \frac{1}{12} + \frac{1}{320} \approx 0.92$

(integral is approximately 0.9197304101)

53. $f(x) = \sum_{n=0}^{\infty} (-1)^n \frac{x^{2n}}{(2n)!} = \cos x$

(See Exercise 47.)

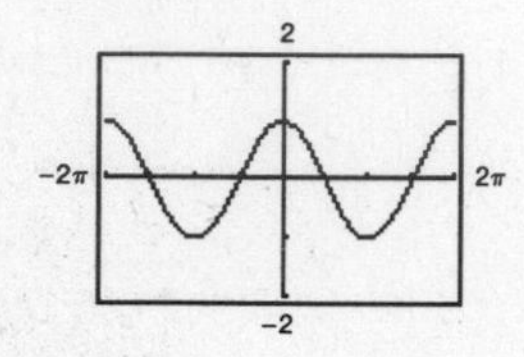

55. $f(x) = \sum_{n=0}^{\infty} (-1)^n x^n = \sum_{n=0}^{\infty} (-x)^n$

$= \frac{1}{1-(-x)} = \frac{1}{1+x}$ for $-1 < x < 1$

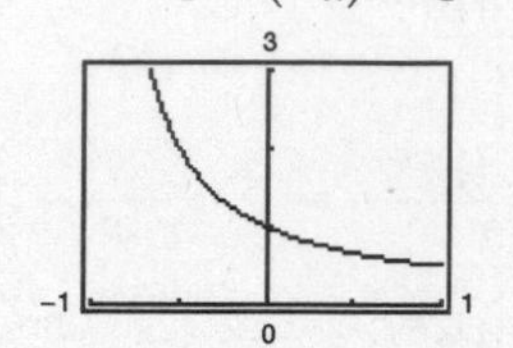

57. $\sum_{n=0}^{\infty} \left(\frac{x}{2}\right)^n$

(a) $\sum_{n=0}^{\infty} \left(\frac{3/4}{2}\right)^n = \sum_{n=0}^{\infty} \left(\frac{3}{8}\right)^n$

$= \frac{1}{1-(3/8)} = \frac{8}{5} = 1.6$

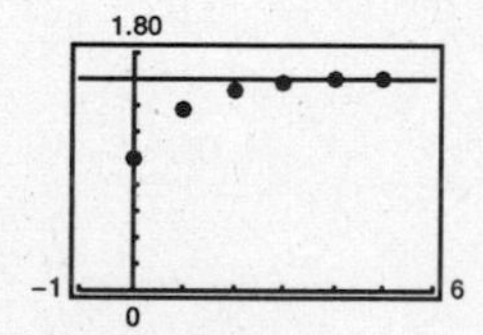

(b) $\sum_{n=0}^{\infty} \left(\frac{-3/4}{2}\right)^n = \sum_{n=0}^{\infty} \left(-\frac{3}{8}\right)^n$

$= \frac{1}{1-(-3/8)} = \frac{8}{11} \approx 0.7272$

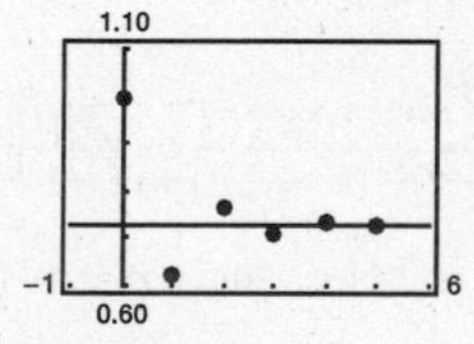

(c) The alternating series converges more rapidly. The partial sums of the series of positive terms approach the sum from below. The partial sums of the alternating series alternate sides of the horizontal line representing the sum.

(d) $\sum_{n=0}^{N} \left(\frac{3}{2}\right)^n > M$

M	10	100	1000	10,000
N	4	9	15	21

59. False;

$$\sum_{n=0}^{\infty} \frac{(-1)^n x^n}{n2^n}$$

converges for $x = 2$ but diverges for $x = -2$.

61. True; the radius of convergence is $R = 1$ for both series.

Section 8.9 Representation of Functions by Power Series

1. (a) $\dfrac{1}{2-x} = \dfrac{1/2}{1-(x/2)} = \dfrac{a}{1-r}$

$$= \sum_{n=0}^{\infty} \frac{1}{2}\left(\frac{x}{2}\right)^n = \sum_{n=0}^{\infty} \frac{x^n}{2^{n+1}}$$

This series converges on $(-2, 2)$.

(b)
$$\begin{array}{r l}
 & \frac{1}{2} + \frac{x}{4} + \frac{x^2}{8} + \frac{x^3}{16} + \cdots \\
2 - x\,\big) & 1 \\
 & 1 - \frac{x}{2} \\ \hline
 & \frac{x}{2} \\
 & \frac{x}{2} - \frac{x^2}{4} \\ \hline
 & \frac{x^2}{4} \\
 & \frac{x^2}{4} - \frac{x^3}{8} \\ \hline
 & \frac{x^3}{8} \\
 & \frac{x^3}{8} - \frac{x^4}{16} \\ \hline
 & \vdots
\end{array}$$

3. (a) $\dfrac{1}{2+x} = \dfrac{1/2}{1-(-x/2)} = \dfrac{a}{1-r}$

$$= \sum_{n=0}^{\infty} \frac{1}{2}\left(-\frac{x}{2}\right)^n = \sum_{n=0}^{\infty} \frac{(-1)^n x^n}{2^{n+1}}$$

This series converges on $(-2, 2)$.

(b)
$$\begin{array}{r l}
 & \frac{1}{2} - \frac{x}{4} + \frac{x^2}{8} - \frac{x^3}{16} + \cdots \\
2 + x\,\big) & 1 \\
 & 1 + \frac{x}{2} \\ \hline
 & -\frac{x}{2} \\
 & -\frac{x}{2} - \frac{x^2}{4} \\ \hline
 & \frac{x^2}{4} \\
 & \frac{x^2}{4} + \frac{x^3}{8} \\ \hline
 & -\frac{x^3}{8} \\
 & -\frac{x^3}{8} - \frac{x^4}{16} \\ \hline
 & \vdots
\end{array}$$

5. Writing $f(x)$ in the form $a/(1-r)$, we have

$$\frac{1}{2-x} = \frac{1}{-3-(x-5)} = \frac{-1/3}{1+(1/3)(x-5)}$$

which implies that $a = -1/3$ and $r = (-1/3)(x-5)$.

Therefore, the power series for $f(x)$ is given by

$$\frac{1}{2-x} = \sum_{n=0}^{\infty} ar^n = \sum_{n=0}^{\infty} -\frac{1}{3}\left[-\frac{1}{3}(x-5)\right]^n$$

$$= \sum_{n=0}^{\infty} \frac{(x-5)^n}{(-3)^{n+1}},\ |x-5| < 3 \text{ or } 2 < x < 8.$$

7. Writing $f(x)$ in the form $a/(1-r)$, we have

$$\frac{3}{2x-1} = \frac{-3}{1-2x} = \frac{a}{1-r}$$

which implies that $a = -3$ and $r = 2x$.

Therefore, the power series for $f(x)$ is given by

$$\frac{3}{2x-1} = \sum_{n=0}^{\infty} ar^n = \sum_{n=0}^{\infty} (-3)(2x)^n$$

$$= -3\sum_{n=0}^{\infty} (2x)^n,\ |2x| < 1 \text{ or } -\frac{1}{2} < x < \frac{1}{2}.$$

9. Writing $f(x)$ in the form $a/(1-r)$, we have

$$\frac{1}{2x-5} = \frac{-1}{11-2(x+3)}$$

$$= \frac{-1/11}{1-(2/11)(x+3)} = \frac{a}{1-r}$$

which implies that $a = -1/11$ and $r = (2/11)(x+3)$. Therefore, the power series for $f(x)$ is given by

$$\frac{1}{2x-5} = \sum_{n=0}^{\infty} ar^n = \sum_{n=0}^{\infty}\left(-\frac{1}{11}\right)\left[\frac{2}{11}(x+3)\right]^n$$

$$= -\sum_{n=0}^{\infty} \frac{2^n(x+3)^n}{11^{n+1}},$$

$$|x+3| < \frac{11}{2} \text{ or } -\frac{17}{2} < x < \frac{5}{2}.$$

11. Writing $f(x)$ in the form $a/(1-r)$, we have

$$\frac{3}{x+2} = \frac{3}{2+x} = \frac{3/2}{1+(1/2)x} = \frac{a}{1-r}$$

which implies that $a = 3/2$ and $r = (-1/2)x$. Therefore, the power series for $f(x)$ is given by

$$\frac{3}{x+2} = \sum_{n=0}^{\infty} ar^n = \sum_{n=0}^{\infty} \frac{3}{2}\left(-\frac{1}{2}x\right)^n$$

$$= 3\sum_{n=0}^{\infty} \frac{(-1)^n x^n}{2^{n+1}} = \frac{3}{2}\sum_{n=0}^{\infty}\left(-\frac{x}{2}\right)^n,$$

$|x| < 2$ or $-2 < x < 2$.

13. $\dfrac{3x}{x^2+x-2} = \dfrac{2}{x+2} + \dfrac{1}{x-1} = \dfrac{2}{2+x} + \dfrac{1}{-1+x} = \dfrac{1}{1+(1/2)x} + \dfrac{-1}{1-x}$

Writing $f(x)$ as a sum of two geometric series, we have

$$\frac{3x}{x^2+x-2} = \sum_{n=0}^{\infty}\left(-\frac{1}{2}x\right)^n + \sum_{n=0}^{\infty}(-1)(x)^n = \sum_{n=0}^{\infty}\left[\frac{1}{(-2)^n} - 1\right]x^n.$$

The interval of convergence is $-1 < x < 1$ since

$$\lim_{n\to\infty}\left|\frac{u_{n+1}}{u_n}\right| = \lim_{n\to\infty}\left|\frac{(1-(-2)^{n+1})x^{n+1}}{(-2)^{n+1}} \cdot \frac{(-2)^n}{(1-(-2)^n)x^n}\right| = \lim_{n\to\infty}\left|\frac{(1-(-2)^{n+1})x}{-2-(-2)^{n+1}}\right| = |x|.$$

15. $\dfrac{2}{1-x^2} = \dfrac{1}{1-x} + \dfrac{1}{1+x}$

Writing $f(x)$ as a sum of two geometric series, we have

$$\frac{2}{1-x^2} = \sum_{n=0}^{\infty} x^n + \sum_{n=0}^{\infty}(-x)^n = \sum_{n=0}^{\infty}(1+(-1)^n)x^n = \sum_{n=0}^{\infty} 2x^{2n}.$$

The interval of convergence is $|x^2| < 1$ or $-1 < x < 1$ since $\displaystyle\lim_{n\to\infty}\left|\frac{u_{n+1}}{u_n}\right| = \lim_{n\to\infty}\left|\frac{2x^{2n+2}}{2x^{2n}}\right| = |x^2|.$

17. $\dfrac{1}{1+x} = \displaystyle\sum_{n=0}^{\infty}(-1)^n x^n$

$$\frac{1}{1-x} = \sum_{n=0}^{\infty}(-1)^n(-x)^n = \sum_{n=0}^{\infty}(-1)^{2n}x^n = \sum_{n=0}^{\infty} x^n$$

$$h(x) = \frac{-2}{x^2-1} = \frac{1}{1+x} + \frac{1}{1-x} = \sum_{n=0}^{\infty}(-1)^n x^n + \sum_{n=0}^{\infty} x^n = \sum_{n=0}^{\infty}[(-1)^n + 1]x^n$$

$$= 2 + 0x + 2x^2 + 0x^3 + 2x^4 + 0x^5 + 2x^6 + \cdots = \sum_{n=0}^{\infty} 2x^{2n},\ -1 < x < 1 \text{ (See Exercise 15.)}$$

19. By taking the first derivative, we have $\dfrac{d}{dx}\left[\dfrac{1}{x+1}\right] = \dfrac{-1}{(x+1)^2}$. Therefore,

$$\frac{-1}{(x+1)^2} = \frac{d}{dx}\left[\sum_{n=0}^{\infty}(-1)^n x^n\right] = \sum_{n=1}^{\infty}(-1)^n n x^{n-1}$$

$$= \sum_{n=0}^{\infty}(-1)^{n+1}(n+1)x^n,\ -1 < x < 1.$$

21. By integrating, we have $\displaystyle\int \frac{1}{x+1}\,dx = \ln(x+1)$. Therefore,

$$\ln(x+1) = \int\left[\sum_{n=0}^{\infty}(-1)^n x^n\right]dx = C + \sum_{n=0}^{\infty}\frac{(-1)^n x^{n+1}}{n+1},\ -1 < x \le 1.$$

To solve for C, let $x = 0$ and conclude that $C = 0$. Therefore,

$$\ln(x+1) = \sum_{n=0}^{\infty}\frac{(-1)^n x^{n+1}}{n+1},\ -1 < x \le 1.$$

23. $\dfrac{1}{x^2+1} = \displaystyle\sum_{n=0}^{\infty}(-1)^n(x^2)^n = \sum_{n=0}^{\infty}(-1)^n x^{2n},\ -1 < x < 1$

25. Since, $\dfrac{1}{x+1} = \displaystyle\sum_{n=0}^{\infty}(-1)^n x^n$, we have $\dfrac{1}{4x^2+1} = \displaystyle\sum_{n=0}^{\infty}(-1)^n(4x^2)^n = \sum_{n=0}^{\infty}(-1)^n 4^n x^{2n} = \sum_{n=0}^{\infty}(-1)^n(2x)^{2n},\ -\frac{1}{2} < x < \frac{1}{2}.$

27. $x - \frac{x^2}{2} \le \ln(x+1) \le x - \frac{x^2}{2} + \frac{x^3}{3}$

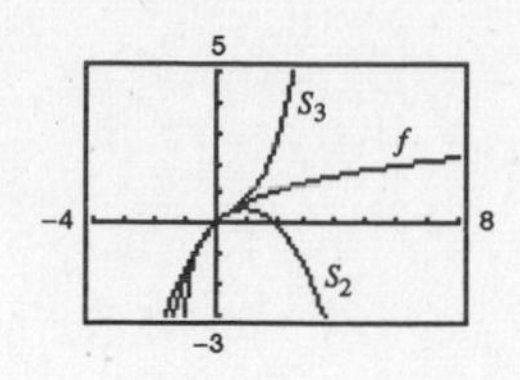

x	0.0	0.2	0.4	0.6	0.8	1.0
$x - \frac{x^2}{2}$	0.000	0.180	0.320	0.420	0.480	0.500
$\ln(x+1)$	0.000	0.182	0.336	0.470	0.588	0.693
$x - \frac{x^2}{2} + \frac{x^3}{3}$	0.000	0.183	0.341	0.492	0.651	0.833

29. $g(x) = x$, line, Matches (c)

31. $g(x) = x - \frac{x^3}{3} + \frac{x^5}{5}$, Matches (a)

33. $f(x) = \arctan x$ is an odd function (symmetric to the origin)

In Exercises 35 and 37, $\arctan x = \sum_{n=0}^{\infty} (-1)^n \frac{x^{2n+1}}{2n+1}$.

35. $\arctan \frac{1}{4} = \sum_{n=0}^{\infty} (-1)^n \frac{(1/4)^{2n+1}}{2n+1} = \sum_{n=0}^{\infty} \frac{(-1)^n}{(2n+1)4^{2n+1}} = \frac{1}{4} - \frac{1}{192} + \frac{1}{5120} + \cdots$

Since $\frac{1}{5120} < 0.001$, we can approximate the series by its first two terms: $\arctan \frac{1}{4} \approx \frac{1}{4} - \frac{1}{192} \approx 0.245$.

37. $$\frac{\arctan x^2}{x} = \frac{1}{x}\sum_{n=0}^{\infty} (-1)^n \frac{(x^2)^{2n+1}}{2n+1} = \sum_{n=0}^{\infty} (-1)^n \frac{x^{4n+1}}{2n+1}$$

$$\int \frac{\arctan x^2}{x}\,dx = \sum_{n=0}^{\infty} (-1)^n \frac{x^{4n+2}}{(4n+2)(2n+1)} + C \quad \text{(Note: } C = 0\text{)}$$

$$\int_0^{1/2} \frac{\arctan x^2}{x}\,dx = \sum_{n=0}^{\infty} (-1)^n \frac{1}{(4n+2)(2n+1)2^{4n+2}} = \frac{1}{8} - \frac{1}{1152} + \cdots$$

Since $\frac{1}{1152} < 0.001$, we can approximate the series by its first term: $\int_0^{1/2} \frac{\arctan x^2}{x}\,dx \approx 0.125$

In Exercises 39 and 41, use $\frac{1}{1-x} = \sum_{n=0}^{\infty} x^n$, $|x| < 1$.

39. (a) $\frac{1}{(1-x)^2} = \frac{d}{dx}\left[\frac{1}{1-x}\right] = \frac{d}{dx}\left[\sum_{n=0}^{\infty} x^n\right] = \sum_{n=1}^{\infty} nx^{n-1}$, $|x| < 1$

(b) $\frac{x}{(1-x)^2} = x\sum_{n=1}^{\infty} nx^{n-1} = \sum_{n=1}^{\infty} nx^n$, $|x| < 1$

(c) $\frac{1+x}{(1-x)^2} = \frac{1}{(1-x)^2} + \frac{x}{(1-x)^2} = \sum_{n=1}^{\infty} n(x^{n-1} + x^n)$, $|x| < 1$

$= \sum_{n=0}^{\infty} (2n+1)x^n$, $|x| < 1$

(d) $\frac{x(1+x)}{(1-x)^2} = x\sum_{n=0}^{\infty} (2n+1)x^n = \sum_{n=0}^{\infty} (2n+1)x^{n+1}$, $|x| < 1$

41. $P(n) = \left(\frac{1}{2}\right)^n$

$$E(n) = \sum_{n=1}^{\infty} nP(n) = \sum_{n=1}^{\infty} n\left(\frac{1}{2}\right)^n = \frac{1}{2}\sum_{n=1}^{\infty} n\left(\frac{1}{2}\right)^{n-1}$$

$$= \frac{1}{2}\frac{1}{[1 - (1/2)]^2} = 2$$

Since the probability of obtaining a head on a single toss is $\frac{1}{2}$, it is expected that, on average, a head will be obtained in two tosses.

43. Replace x with $(-x)$.

45. Replace x with $(-x)$ and multiply the series by 5.

47. Let $\arctan x + \arctan y = \theta$. Then,

$$\tan(\arctan x + \arctan y) = \tan\theta$$

$$\frac{\tan(\arctan x) + \tan(\arctan y)}{1 - \tan(\arctan x)\tan(\arctan y)} = \tan\theta$$

$$\frac{x+y}{1-xy} = \tan\theta$$

$$\arctan\left(\frac{x+y}{1-xy}\right) = \theta. \text{ Therefore, } \arctan x + \arctan y = \arctan\left(\frac{x+y}{1-xy}\right) \text{ for } xy \neq 1.$$

49. (a) $2\arctan\frac{1}{2} = \arctan\frac{1}{2} + \arctan\frac{1}{2} = \arctan\left[\frac{(1/2)+(1/2)}{1-(1/2)^2}\right] = \arctan\frac{4}{3}$

$$2\arctan\frac{1}{2} - \arctan\frac{1}{7} = \arctan\frac{4}{3} + \arctan\left(-\frac{1}{7}\right) = \arctan\left[\frac{(4/3)-(1/7)}{1+(4/3)(1/7)}\right] = \arctan\frac{25}{25} = \arctan 1 = \frac{\pi}{4}$$

(b) $\pi = 8\arctan\frac{1}{2} - 4\arctan\frac{1}{7} \approx 8\left[\frac{1}{2} - \frac{(0.5)^3}{3} + \frac{(0.5)^5}{5} - \frac{(0.5)^7}{7}\right] - 4\left[\frac{1}{7} - \frac{(1/7)^3}{3} + \frac{(1/7)^5}{5} - \frac{(1/7)^7}{7}\right] \approx 3.14$

51. From Exercise 21, we have

$$\ln(x+1) = \sum_{n=0}^{\infty}\frac{(-1)^n x^{n+1}}{n+1} = \sum_{n=1}^{\infty}\frac{(-1)^{n-1}x^n}{n}$$

$$= \sum_{n=1}^{\infty}\frac{(-1)^{n+1}x^n}{n}.$$

Thus, $\sum_{n=1}^{\infty}(-1)^{n+1}\frac{1}{2^n n} = \sum_{n=1}^{\infty}\frac{(-1)^{n+1}(1/2)^n}{n}$

$$= \ln\left(\frac{1}{2}+1\right) = \ln\frac{3}{2} \approx 0.4055$$

53. From Exercise 51, we have

$$\sum_{n=1}^{\infty}(-1)^{n+1}\frac{2^n}{5^n n} = \sum_{n=1}^{\infty}\frac{(-1)^{n+1}(2/5)^n}{n}$$

$$= \ln\left(\frac{2}{5}+1\right) = \ln\frac{7}{5} \approx 0.3365.$$

55. From Exercise 54, we have

$$\sum_{n=0}^{\infty}(-1)^n\frac{1}{2^{2n+1}(2n+1)} = \sum_{n=0}^{\infty}(-1)^n\frac{(1/2)^{2n+1}}{2n+1} = \arctan\frac{1}{2} \approx 0.4636.$$

57. The series in Exercise 54 converges to its sum at a slower rate because its terms approach 0 at a much slower rate.

59. $f(x) = \sum_{n=1}^{\infty}(-1)^{n+1}\frac{(x-1)^n}{n}, \quad 0 < x \le 2$

$$f(0.5) = \sum_{n=1}^{\infty}(-1)^{n+1}\frac{(-0.5)^n}{n} = \sum_{n=1}^{\infty} -\frac{(1/2)^n}{n}$$

$$\sum_{n=1}^{50} -\frac{(1/2)^n}{n} \approx -0.693147$$

$$\ln(0.5) \approx -0.693147$$

Section 8.10 Taylor and Maclaurin Series

1. For $c = 0$, we have:

$$f(x) = e^{2x}$$

$$f^{(n)}(x) = 2^n e^{2x} \Rightarrow f^{(n)}(0) = 2^n$$

$$e^{2x} = 1 + 2x + \frac{4x^2}{2!} + \frac{8x^3}{3!} + \frac{16x^4}{4!} + \cdots = \sum_{n=0}^{\infty}\frac{(2x)^n}{n!}$$

3. For $c = \pi/4$, we have:

$$f(x) = \cos(x) \qquad f\left(\frac{\pi}{4}\right) = \frac{\sqrt{2}}{2}$$

$$f'(x) = -\sin(x) \qquad f'\left(\frac{\pi}{4}\right) = -\frac{\sqrt{2}}{2}$$

$$f''(x) = -\cos(x) \qquad f''\left(\frac{\pi}{4}\right) = -\frac{\sqrt{2}}{2}$$

$$f'''(x) = \sin(x) \qquad f'''\left(\frac{\pi}{4}\right) = \frac{\sqrt{2}}{2}$$

$$f^{(4)}(x) = \cos(x) \qquad f^{(4)}\left(\frac{\pi}{4}\right) = \frac{\sqrt{2}}{2}$$

and so on. Therefore, we have:

$$\begin{aligned}\cos x &= \sum_{n=0}^{\infty} \frac{f^{(n)}(\pi/4)[x - (\pi/4)]^n}{n!} \\ &= \frac{\sqrt{2}}{2}\left[1 - \left(x - \frac{\pi}{4}\right) - \frac{[x - (\pi/4)]^2}{2!} + \frac{[x - (\pi/4)]^3}{3!} + \frac{[x - (\pi/4)]^4}{4!} - \cdots\right] \\ &= \frac{\sqrt{2}}{2} \sum_{n=0}^{\infty} \frac{(-1)^{n(n+1)/2}[x - (\pi/4)]^n}{n!}.\end{aligned}$$

[**Note:** $(-1)^{n(n+1)/2} = 1, -1, -1, 1, 1, -1, -1, 1, \ldots$]

5. For $c = 1$, we have,

$$f(x) = \ln x \qquad f(1) = 0$$

$$f'(x) = \frac{1}{x} \qquad f'(1) = 1$$

$$f''(x) = -\frac{1}{x^2} \qquad f''(1) = -1$$

$$f'''(x) = \frac{2}{x^3} \qquad f'''(1) = 2$$

$$f^{(4)}(x) = -\frac{6}{x^4} \qquad f^{(4)}(1) = -6$$

$$f^{(5)}(x) = \frac{24}{x^5} \qquad f^{(5)}(1) = 24$$

and so on. Therefore, we have:

$$\begin{aligned}\ln x &= \sum_{n=0}^{\infty} \frac{f^{(n)}(1)(x-1)^n}{n!} \\ &= 0 + (x-1) - \frac{(x-1)^2}{2!} + \frac{2(x-1)^3}{3!} - \frac{6(x-1)^4}{4!} + \frac{24(x-1)^5}{5!} - \cdots \\ &= (x-1) - \frac{(x-1)^2}{2} + \frac{(x-1)^3}{3} - \frac{(x-1)^4}{4} + \frac{(x-1)^5}{5} - \cdots \\ &= \sum_{n=0}^{\infty} (-1)^n \frac{(x-1)^{n+1}}{n+1}\end{aligned}$$

7. For $c = 0$, we have:

$$\begin{aligned} f(x) &= \sin 2x & f(0) &= 0 \\ f'(x) &= 2\cos 2x & f'(0) &= 2 \\ f''(x) &= -4\sin 2x & f''(0) &= 0 \\ f'''(x) &= -8\cos 2x & f'''(0) &= -8 \\ f^{(4)}(x) &= 16\sin 2x & f^{(4)}(0) &= 0 \\ f^{(5)}(x) &= 32\cos 2x & f^{(5)}(0) &= 32 \\ f^{(6)}(x) &= -64\sin 2x & f^{(6)}(0) &= 0 \\ f^{(7)}(x) &= -128\cos 2x & f^{(7)}(0) &= -128 \end{aligned}$$

and so on. Therefore, we have:

$$\sin 2x = \sum_{n=0}^{\infty} \frac{f^{(n)}(0)x^n}{n!} = 0 + 2x + \frac{0x^2}{2!} - \frac{8x^3}{3!} + \frac{0x^4}{4!} + \frac{32x^5}{5!} + \frac{0x^6}{6!} - \frac{128x^7}{7!} + \cdots$$

$$= 2x - \frac{8x^3}{3!} + \frac{32x^5}{5!} - \frac{128x^7}{7!} + \cdots = \sum_{n=0}^{\infty} \frac{(-1)^n(2x)^{2n+1}}{(2n+1)!}$$

9. For $c = 0$, we have:

$$\begin{aligned} f(x) &= \sec(x) & f(0) &= 1 \\ f'(x) &= \sec(x)\tan(x) & f'(0) &= 0 \\ f''(x) &= \sec^3(x) + \sec(x)\tan^2(x) & f''(0) &= 1 \\ f'''(x) &= 5\sec^3(x)\tan(x) + \sec(x)\tan^3(x) & f'''(0) &= 0 \\ f^{(4)}(x) &= 5\sec^5(x) + 18\sec^3(x)\tan^2(x) + \sec(x)\tan^4(x) & f^{(4)}(0) &= 5 \end{aligned}$$

$$\sec(x) = \sum_{n=0}^{\infty} \frac{f^{(n)}(0)x^n}{n!} = 1 + \frac{x^2}{2!} + \frac{5x^4}{4!} + \cdots$$

11. The Maclaurin series for $f(x) = \cos x$ is $\displaystyle\sum_{n=0}^{\infty} \frac{(-1)x^{2n}}{(2n)!}$.

Because $f^{(n+1)}(x) = \pm\sin x$ or $\pm\cos x$, we have $|f^{(n+1)}(z)| \le 1$ for all z. Hence by Taylor's Theorem,

$$0 \le |R_n(x)| = \left|\frac{f^{(n+1)}(z)}{(n+1)!}x^{n+1}\right| \le \frac{|x|^{n+1}}{(n+1)!}.$$

Since $\displaystyle\lim_{n\to\infty} \frac{|x|^{n+1}}{(n+1)!} = 0$, it follows that $R_n(x) \to 0$ as $n \to \infty$. Hence, the Maclaurin series for $\cos x$ converges to $\cos x$ for all x.

13. Since $(1+x)^{-k} = 1 - kx + \dfrac{k(k+1)x^2}{2!} - \dfrac{k(k+1)(k+2)x^3}{3!} + \cdots$, we have

$$(1+x)^{-2} = 1 - 2x + \frac{2(3)x^2}{2!} - \frac{2(3)(4)x^3}{3!} + \frac{2(3)(4)(5)x^4}{5!} - \cdots = 1 - 2x + 3x^2 - 4x^3 + 5x^4 - \cdots$$

$$= \sum_{n=0}^{\infty} (-1)^n(n+1)x^n.$$

15. $\dfrac{1}{\sqrt{4+x^2}} = \left(\dfrac{1}{2}\right)\left[1+\left(\dfrac{x}{2}\right)^2\right]^{-1/2}$ and since $(1+x)^{-1/2} = 1 + \displaystyle\sum_{n=1}^{\infty} \frac{(-1)^n 1 \cdot 3 \cdot 5 \cdots (2n-1)x^n}{2^n n!}$, we have

$$\frac{1}{\sqrt{4+x^2}} = \frac{1}{2}\left[1 + \sum_{n=1}^{\infty} \frac{(-1)^n 1 \cdot 3 \cdot 5 \cdots (2n-1)(x/2)^{2n}}{2^n n!}\right] = \frac{1}{2} + \sum_{n=1}^{\infty} \frac{(-1)^n 1 \cdot 3 \cdot 5 \cdots (2n-1)x^{2n}}{2^{3n+1} n!}.$$

17. Since $(1+x)^{1/2} = 1 + \dfrac{x}{2} + \displaystyle\sum_{n=2}^{\infty} \frac{(-1)^{n+1} 1 \cdot 3 \cdot 5 \cdots (2n-3)x^n}{2^n n!}$ (Exercise 14)

we have $(1+x^2)^{1/2} = 1 + \dfrac{x^2}{2} + \displaystyle\sum_{n=2}^{\infty} \frac{(-1)^{n+1} 1 \cdot 3 \cdot 5 \cdots (2n-3)x^{2n}}{2^n n!}$.

19.

$$e^x = \sum_{n=0}^{\infty} \frac{x^n}{n!} = 1 + x + \frac{x^2}{2!} + \frac{x^3}{3!} + \frac{x^4}{4!} + \frac{x^5}{5!} + \cdots$$

$$e^{x^2/2} = \sum_{n=0}^{\infty} \frac{(x^2/2)^n}{n!} = \sum_{n=0}^{\infty} \frac{x^{2n}}{2^n n!} = 1 + \frac{x^2}{2} + \frac{x^4}{2^2 2!} + \frac{x^6}{2^3 3!} + \frac{x^8}{2^4 4!} + \cdots$$

21.

$$\sin x = \sum_{n=0}^{\infty} \frac{(-1)^n x^{2n+1}}{(2n+1)!} = x - \frac{x^3}{3!} + \frac{x^5}{5!} - \frac{x^7}{7!} + \cdots$$

$$\sin 2x = \sum_{n=0}^{\infty} \frac{(-1)^n (2x)^{2n+1}}{(2n+1)!} = \sum_{n=0}^{\infty} \frac{(-1)^n 2^{2n+1} x^{2n+1}}{(2n+1)!} = 2x - \frac{8x^3}{3!} + \frac{32x^5}{5!} - \frac{128x^7}{7!} + \cdots$$

23.

$$\cos x = \sum_{n=0}^{\infty} \frac{(-1)^n x^{2n}}{(2n)!} = 1 - \frac{x^2}{2!} + \frac{x^4}{4!} - \cdots$$

$$\cos x^{3/2} = \sum_{n=0}^{\infty} \frac{(-1)^n (x^{3/2})^{2n}}{(2n)!} = \sum_{n=0}^{\infty} \frac{(-1)^n x^{3n}}{(2n)!} = 1 - \frac{x^3}{2!} + \frac{x^6}{4!} - \cdots$$

25.

$$e^x = 1 + x + \frac{x^2}{2!} + \frac{x^3}{3!} + \frac{x^4}{4!} + \frac{x^5}{5!} + \cdots$$

$$e^{-x} = 1 - x + \frac{x^2}{2!} - \frac{x^3}{3!} + \frac{x^4}{4!} - \frac{x^5}{5!} + \cdots$$

$$e^x - e^{-x} = 2x + \frac{2x^3}{3!} + \frac{2x^5}{5!} + \frac{2x^7}{7!} + \cdots$$

$$\sinh(x) = \frac{1}{2}(e^x - e^{-x}) = x + \frac{x^3}{3!} + \frac{x^5}{5!} + \frac{x^7}{7!} + \cdots = \sum_{n=0}^{\infty} \frac{x^{2n+1}}{(2n+1)!}$$

27.

$$\cos^2(x) = \frac{1}{2}[1 + \cos(2x)]$$

$$= \frac{1}{2}\left[1 + 1 - \frac{(2x)^2}{2!} + \frac{(2x)^4}{4!} - \frac{(2x)^6}{6!} - \cdots\right]$$

$$= \frac{1}{2}\left[1 + \sum_{n=0}^{\infty} \frac{(-1)^n (2x)^{2n}}{(2n)!}\right]$$

29.

$$x \sin x = x\left(x - \frac{x^3}{3!} + \frac{x^5}{5!} - \cdots\right)$$

$$= x^2 - \frac{x^4}{3!} + \frac{x^6}{5!} - \cdots$$

$$= \sum_{n=0}^{\infty} \frac{(-1)^n x^{2n+2}}{(2n+1)!}$$

31.

$$\frac{\sin x}{x} = \frac{x - (x^3/3!) + (x^5/5!) - \cdots}{x}$$

$$= 1 - \frac{x^2}{2!} + \frac{x^4}{4!} - \cdots$$

$$= \sum_{n=0}^{\infty} \frac{(-1)^n x^{2n}}{(2n+1)!},\ x \neq 0$$

33. $$e^{ix} = 1 + ix + \frac{(ix)^2}{2!} + \frac{(ix)^3}{3!} + \frac{(ix)^4}{4!} + \cdots = 1 + ix - \frac{x^2}{2!} - \frac{ix^3}{3!} + \frac{x^4}{4!} + \frac{ix^5}{5!} - \frac{x^6}{6!} - \cdots$$

$$e^{-ix} = 1 - ix + \frac{(-ix)^2}{2!} + \frac{(-ix)^3}{3!} + \frac{(-ix)^4}{4!} + \cdots = 1 - ix - \frac{x^2}{2!} + \frac{ix^3}{3!} + \frac{x^4}{4!} - \frac{ix^5}{5!} - \frac{x^6}{6!} + \cdots$$

$$e^{ix} - e^{-ix} = 2ix - \frac{2ix^3}{3!} + \frac{2ix^5}{5!} - \frac{2ix^7}{7!} + \cdots$$

$$\frac{e^{ix} - e^{-ix}}{2i} = x - \frac{x^3}{3!} + \frac{x^5}{5!} - \frac{x^7}{7!} + \cdots = \sum_{n=0}^{\infty} \frac{(-1)^n x^{2n+1}}{(2n+1)!} = \sin(x)$$

35. $f(x) = e^x \sin x$

$$= \left(1 + x + \frac{x^2}{2} + \frac{x^3}{6} + \frac{x^4}{24} + \cdots\right)\left(x - \frac{x^3}{6} + \frac{x^5}{120} - \cdots\right)$$

$$= x + x^2 + \left(\frac{x^3}{2} - \frac{x^3}{6}\right) + \left(\frac{x^4}{6} - \frac{x^4}{6}\right) + \left(\frac{x^5}{120} - \frac{x^5}{12} + \frac{x^5}{24}\right) + \cdots$$

$$= x + x^2 + \frac{x^3}{3} - \frac{x^5}{30} + \cdots$$

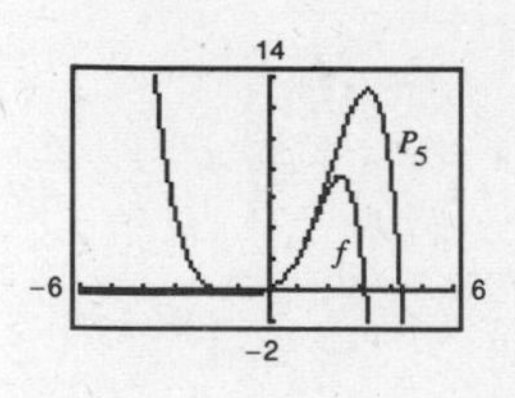

37. $h(x) = \cos x \ln(1 + x)$

$$= \left(1 - \frac{x^2}{2} + \frac{x^4}{24} - \cdots\right)\left(x - \frac{x^2}{2} + \frac{x^3}{3} - \frac{x^4}{4} + \frac{x^5}{5} - \cdots\right)$$

$$= x - \frac{x^2}{2} + \left(\frac{x^3}{3} - \frac{x^3}{2}\right) + \left(\frac{x^4}{4} - \frac{x^4}{4}\right) + \left(\frac{x^5}{5} - \frac{x^5}{6} + \frac{x^5}{24}\right) + \cdots$$

$$= x - \frac{x^2}{2} - \frac{x^3}{6} + \frac{3x^5}{40} + \cdots$$

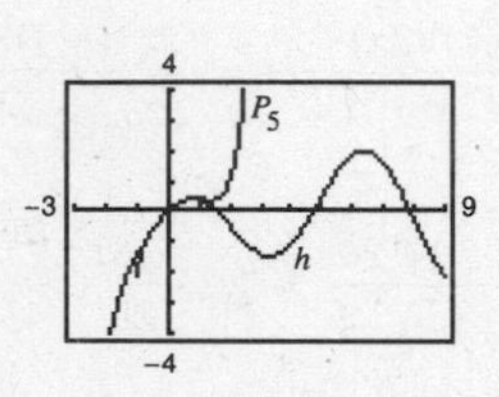

39. $g(x) = \dfrac{\sin x}{1 + x}$. Divide the series for $\sin x$ by $(1 + x)$.

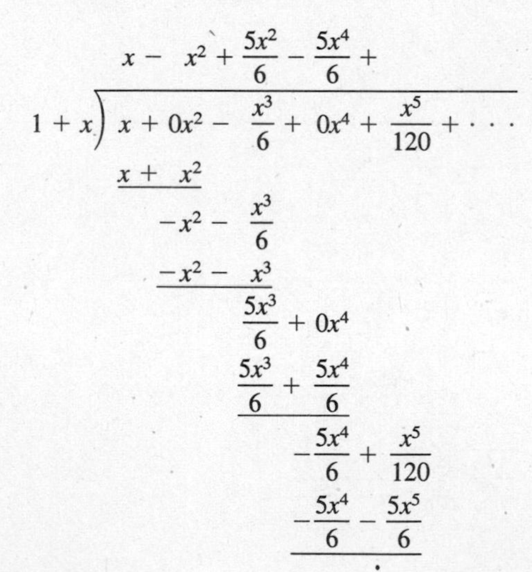

$$g(x) = x - x^2 + \frac{5x^3}{6} - \frac{5x^4}{6} + \cdots$$

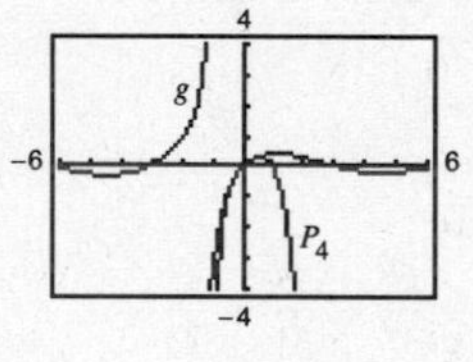

41. $y = x^2 - \dfrac{x^4}{3!} = x\left(x - \dfrac{x^3}{3!}\right) \approx x \sin x.$

Matches (a)

43. $y = x + x^2 + \dfrac{x^3}{2!} = x\left(1 + x + \dfrac{x^2}{2!}\right) \approx xe^x.$

Matches (c)

45. $$\int_0^x (e^{-t^2} - 1)\,dt = \int_0^x \left[\left(\sum_{n=0}^{\infty} \frac{(-1)^n t^{2n}}{n!}\right) - 1\right] dt$$

$$= \int_0^x \left[\sum_{n=0}^{\infty} \frac{(-1)^{n+1} t^{2n+2}}{(n+1)!}\right] dt = \left[\sum_{n=0}^{\infty} \frac{(-1)^{n+1}\, t^{2n+3}}{(2n+3)(n+1)!}\right]_0^x = \sum_{n=0}^{\infty} \frac{(-1)^{n+1} x^{2n+3}}{(2n+3)(n+1)!}$$

47. Since $\ln x = \sum_{n=0}^{\infty} \frac{(-1)^n (x-1)^{n+1}}{n+1} = (x-1) - \frac{(x-1)^2}{2} + \frac{(x-1)^3}{3} - \frac{(x-1)^4}{4} + \cdots \quad (0 < x \le 2)$

we have $\ln 2 = 1 - \frac{1}{2} + \frac{1}{3} - \frac{1}{4} + \cdots = \sum_{n=1}^{\infty} (-1)^{n+1} \frac{1}{n} \approx 0.6931.$ (10,001 terms)

49. Since $e^x = \sum_{n=0}^{\infty} \frac{x^n}{n!} = 1 + x + \frac{x^2}{2!} + \frac{x^3}{3!} + \cdots,$

we have $e^2 = 1 + 2 + \frac{2^2}{2!} + \frac{2^3}{3!} + \cdots = \sum_{n=0}^{\infty} \frac{2^n}{n!} \approx 7.3891.$ (12 terms)

51. Since

$$\cos x = \sum_{n=0}^{\infty} \frac{(-1)^n x^{2n}}{(2n)!} = 1 - \frac{x^2}{2!} + \frac{x^4}{4!} - \frac{x^6}{6!} + \frac{x^8}{8!} - \cdots$$

$$1 - \cos x = \frac{x^2}{2!} - \frac{x^4}{4!} + \frac{x^6}{6!} - \frac{x^8}{8!} + \cdots = \sum_{n=0}^{\infty} \frac{(-1)^n x^{2n+2}}{(2n+2)!}$$

$$\frac{1 - \cos}{x} = \frac{x}{2!} - \frac{x^3}{4!} + \frac{x^5}{6!} - \frac{x^7}{8!} + \cdots = \sum_{n=0}^{\infty} \frac{(-1)^n x^{2n+1}}{(2n+2)!}$$

we have $\lim_{x\to 0} \frac{1 - \cos x}{x} = \lim_{x\to 0} \sum_{n=0}^{\infty} \frac{(-1)x^{2n+1}}{(2n+2)!} = 0.$

53. $$\int_0^1 \frac{\sin x}{x}\,dx = \int_0^1 \left[\sum_{n=0}^{\infty} \frac{(-1)^n x^{2n}}{(2n+1)!}\right] dx = \left[\sum_{n=0}^{\infty} \frac{(-1)^n x^{2n+1}}{(2n+1)(2n+1)!}\right]_0^1 = \sum_{n=0}^{\infty} \frac{(-1)^n}{(2n+1)(2n+1)!}$$

Since $1/(7 \cdot 7!) < 0.0001$, we need three terms:

$$\int_0^1 \frac{\sin x}{x}\,dx = 1 - \frac{1}{3 \cdot 3!} + \frac{1}{5 \cdot 5!} - \cdots \approx 0.9461. \quad \text{(Using three non-zero terms)}$$

Note: We are using $\lim_{x\to 0^+} \frac{\sin x}{x} = 1.$

55. $$\int_0^{\pi/2} \sqrt{x} \cos x\,dx = \int_0^{\pi/2} \left[\sum_{n=0}^{\infty} \frac{(-1)^n x^{(4n+1)/2}}{(2n)!}\right] dx = \left[\sum_{n=0}^{\infty} \frac{(-1)^n x^{(4n+3)/2}}{\left(\frac{4n+3}{2}\right)(2n)!}\right]_0^{\pi/2} = \left[\sum_{n=0}^{\infty} \frac{(-1)^n 2x^{(4n+3)/2}}{(4n+3)(2n)!}\right]_0^{\pi/2}$$

Since $2(\pi/2)^{23/2}/23.10! < 0.0001$, we need five terms

$$\int_0^1 \sqrt{x} \cos x\,dx = 2\left[\frac{(\pi/2)^{3/2}}{3} - \frac{(\pi/2)^{7/2}}{14} + \frac{(\pi/2)^{11/2}}{264} - \frac{(\pi/2)^{15/2}}{10{,}800} + \frac{(\pi/2)^{19/2}}{766{,}080}\right] \approx 0.7040.$$

57. $$\int_{0.1}^{0.3} \sqrt{1 + x^3}\,dx = \int_{0.1}^{0.3} \left(1 + \frac{x^3}{2} - \frac{x^6}{8} + \frac{x^9}{16} - \frac{5x^{12}}{128} + \cdots\right) dx = \left[x + \frac{x^4}{8} - \frac{x^7}{56} + \frac{x^{10}}{160} - \frac{5x^{13}}{1664} + \cdots\right]_{0.1}^{0.3}$$

Since $\frac{1}{56}(0.3^7 - 0.1^7) < 0.0001$, we need two terms

$$\int_{0.1}^{0.3} \sqrt{1 + x^3}\,dx = \left[(0.3 - 0.1) + \frac{1}{8}(0.3^4 - 0.1^4)\right] \approx 0.201.$$

59. From Exercise 19, we have

$$\frac{1}{\sqrt{2\pi}}\int_0^1 e^{-x^2/2}\,dx = \frac{1}{\sqrt{2\pi}}\int_0^1 \sum_{n=0}^{\infty}\frac{(-1)^n x^{2n}}{2^n n!}\,dx = \frac{1}{\sqrt{2\pi}}\left[\sum_{n=0}^{\infty}\frac{(-1)^n x^{2n+1}}{2^n n!(2n+1)}\right]_0^1 = \frac{1}{\sqrt{2\pi}}\sum_{n=0}^{\infty}\frac{(-1)^n}{2^n n!(2n+1)}$$

$$\approx \frac{1}{\sqrt{2\pi}}\left[1 - \frac{1}{2\cdot 1\cdot 3} + \frac{1}{2^2\cdot 2!\cdot 5} - \frac{1}{2^3\cdot 3!\cdot 7}\right] \approx 0.3414.$$

61. $f(x) = x\cos 2x = \displaystyle\sum_{n=0}^{\infty}\frac{(-1)^n 4^n x^{2n+1}}{(2n)!}$

$P_5(x) = x - 2x^3 + \dfrac{2x^5}{3}$

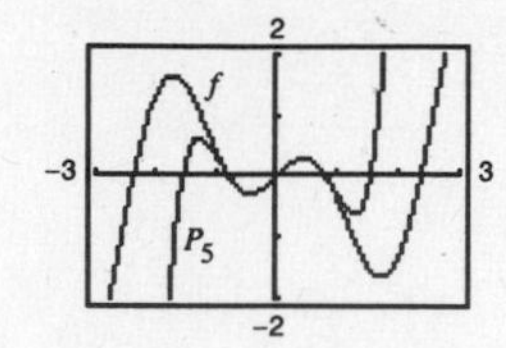

The polynomial is a reasonable approximation on the interval $\left[-\frac{3}{4}, \frac{3}{4}\right]$.

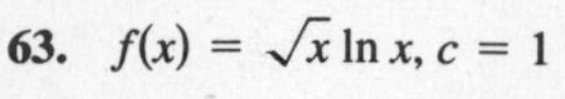

63. $f(x) = \sqrt{x}\ln x,\ c = 1$

$P_5(x) = (x-1) - \dfrac{(x-1)^3}{24} + \dfrac{(x-1)^4}{24} - \dfrac{71(x-1)^5}{1920}$

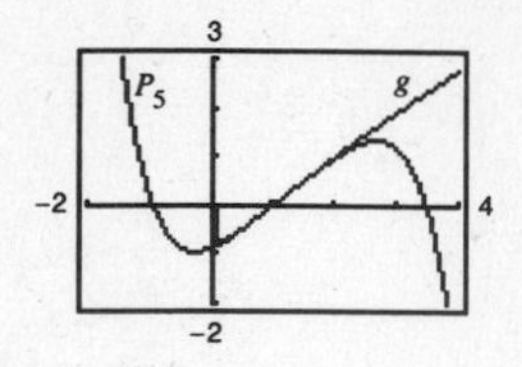

The polynomial is a reasonable approximation on the interval $\left[\frac{1}{4}, 2\right]$.

65. See Guidelines, page 636.

67. (a) Replace x with $(-x)$.

(b) Replace x with $3x$.

(c) Multiply series by x.

(d) Replace x with $2x$, then replace x with $-2x$, and add the two together.

69. $y = \left(\tan\theta - \dfrac{g}{kv_0\cos\theta}\right)x - \dfrac{g}{k^2}\ln\left(1 - \dfrac{kx}{v_0\cos\theta}\right)$

$$= (\tan\theta)x - \frac{gx}{kv_0\cos\theta} - \frac{g}{k^2}\left[-\frac{kx}{v_0\cos\theta} - \frac{1}{2}\left(\frac{kx}{v_0\cos\theta}\right)^2 - \frac{1}{3}\left(\frac{kx}{v_0\cos\theta}\right)^3 - \frac{1}{4}\left(\frac{kx}{v_o\cos\theta}\right)^4 - \cdots\right]$$

$$= (\tan\theta)x - \frac{gx}{kv_0\cos\theta} + \frac{gx}{kv_0\cos\theta} + \frac{gx^2}{2v_0^2\cos^2\theta} + \frac{gkx^3}{3v_0^3\cos^3\theta} + \frac{gk^2x^4}{4v_0^4\cos^4\theta} + \cdots\Big]$$

$$= (\tan\theta)x + \frac{gx^2}{2v_0^2\cos^2\theta} + \frac{kgx^3}{3v_0^3\cos^3\theta} + \frac{k^2gx^4}{4v_0^4\cos^4\theta} + \cdots$$

71. $f(x) = \begin{cases} e^{-1/x^2}, & x \neq 0 \\ 0, & x = 0 \end{cases}$

(a)

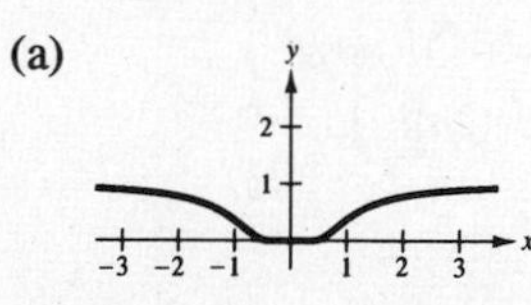

(b) $f'(0) = \displaystyle\lim_{x\to 0}\frac{f(x) - f(0)}{x - 0} = \lim_{x\to 0}\frac{e^{-1/x^2} - 0}{x}$

Let $y = \displaystyle\lim_{x\to 0}\frac{e^{-1/x^2}}{x}$. Then

$$\ln y = \lim_{x\to 0}\ln\left(\frac{e^{-1/x^2}}{x}\right) = \lim_{x\to 0^+}\left[-\frac{1}{x^2} - \ln x\right] = \lim_{x\to 0^+}\left[\frac{-1 - x^2\ln x}{x^2}\right] = -\infty.$$

Thus, $y = e^{-\infty} = 0$ and we have $f'(0) = 0$.

(c) $\displaystyle\sum_{n=0}^{\infty}\frac{f^{(n)}(0)}{n!}x^n = f(0) + \frac{f'(0)x}{1!} + \frac{f''(0)x^2}{2!} + \cdots = 0 \neq f(x)$

This series converges to f at $x = 0$ only.

73. By the Ratio Test: $\displaystyle\lim_{n\to\infty}\left|\frac{x^{n+1}}{(n+1)!}\cdot\frac{n!}{x^n}\right| = \lim_{n\to\infty}\frac{|x|}{n+1} = 0$ which shows that $\displaystyle\sum_{n=0}^{\infty}\frac{x^n}{n!}$ converges for all x.

Review Exercises for Chapter 8

1. $a_n = \dfrac{1}{n!}$

3. $a_n = 4 + \dfrac{2}{n}$: 6, 5, 4.67, . . .

Matches (a)

5. $a_n = 10(0.3)^{n-1}$: 10, 3, . . .

Matches (d)

7. $a_n = \dfrac{5n + 2}{n}$

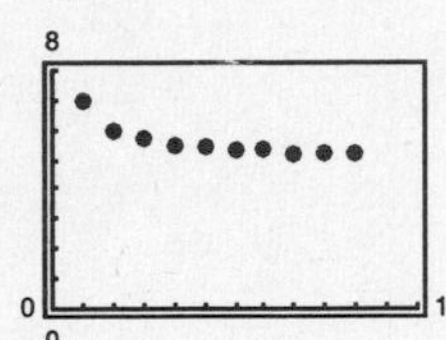

The sequence seems to converge to 5.

$$\lim_{n\to\infty} a_n = \lim_{n\to\infty} \frac{5n + 2}{n}$$

$$= \lim_{n\to\infty} \left(5 + \frac{2}{n}\right) = 5$$

9. $\lim_{n\to\infty} \dfrac{n + 1}{n^2} = 0$

Converges

11. $\lim_{n\to\infty} \dfrac{n^3}{n^2 + 1} = \infty$

13. $\lim_{n\to\infty} \left(\sqrt{n + 1} - \sqrt{n}\right) = \lim_{n\to\infty} \left(\sqrt{n + 1} - \sqrt{n}\right)\dfrac{\sqrt{n + 1} + \sqrt{n}}{\sqrt{n + 1} + \sqrt{n}} = \lim_{n\to\infty} \dfrac{1}{\sqrt{n + 1} + \sqrt{n}} = 0$ Converges

15. $\lim_{n\to\infty} \dfrac{\sin\sqrt{n}}{\sqrt{n}} = 0$

Converges

17. $A_n = 5000\left(1 + \dfrac{0.05}{4}\right)^n = 5000(1.0125)^n$

$n = 1, 2, 3$

(a) $A_1 = 5062.50$ $\quad A_5 \approx 5320.41$

$A_2 \approx 5125.78$ $\quad A_6 \approx 5386.92$

$A_3 \approx 5189.85$ $\quad A_7 \approx 5454.25$

$A_4 \approx 5254.73$ $\quad A_8 \approx 5522.43$

(b) $A_{40} \approx 8218.10$

19. (a)

k	5	10	15	20	25
S_k	13.2	113.3	873.8	6448.5	50,500.3

The series diverges $\left(\text{geometric } r = \frac{3}{2} > 1\right)$

(b)

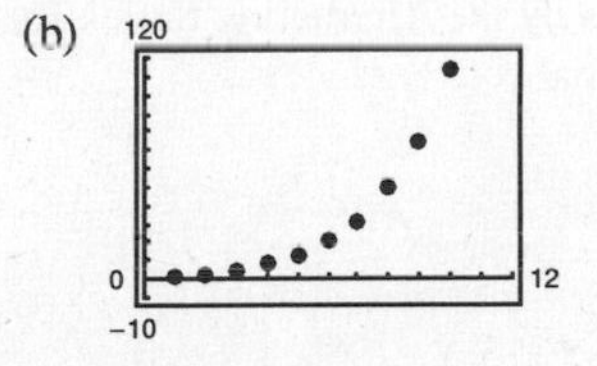

21. (a)

k	5	10	15	20	25
S_k	0.4597	0.4597	0.4597	0.4597	0.4597

The series converges by the Alternating Series Test.

(b)

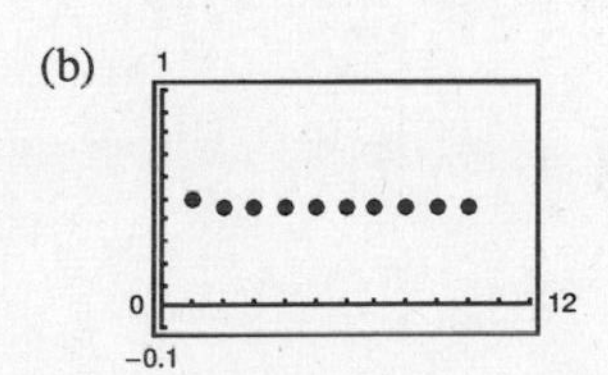

23. Converges. Geometric series, $r = 0.82$, $|r| < 1$.

25. Diverges. nth Term Test. $\lim_{n\to\infty} a_n \neq 0$.

27. $\sum_{n=0}^{\infty}\left(\frac{2}{3}\right)^n$

Geometric series with $a = 1$ and $r = \frac{2}{3}$.

$$S = \frac{a}{1-r} = \frac{1}{1-(2/3)} = \frac{1}{1/3} = 3$$

29. $\sum_{n=0}^{\infty}\left(\frac{1}{2^n} - \frac{1}{3^n}\right) = \sum_{n=0}^{\infty}\left(\frac{1}{2}\right)^n - \sum_{n=0}^{\infty}\left(\frac{1}{3}\right)^n$

$$= \frac{1}{1-(1/2)} - \frac{1}{1-(1/3)} = 2 - \frac{3}{2} = \frac{1}{2}$$

31. $0.\overline{09} = 0.09 + 0.0009 + 0.000009 + \cdots = 0.09(1 + 0.01 + 0.0001 + \cdots) = \sum_{n=0}^{\infty}(0.09)(0.01)^n = \frac{0.09}{1-0.01} = \frac{1}{11}$

33. $D_1 = 8$

$D_2 = 0.7(8) + 0.7(8) = 16(0.7)$

$\vdots$

$D = 8 + 16(0.7) + 16(0.7)^2 + \cdots + 16(0.7)^n + \cdots$

$= -8 + \sum_{n=0}^{\infty} 16(0.7)^n = -8 + \frac{16}{1-0.7} = 45\frac{1}{3}$ meters

35. See Exercise 86 in Section 8.2.

$$A = \frac{P(e^{rt}-1)}{e^{r/12}-1}$$

$$= \frac{200(e^{(0.06)(2)}-1)}{e^{0.06/12}-1}$$

$$\approx \$5087.14$$

37. $\int_1^{\infty} x^{-4}\ln(x)\,dx = \lim_{b\to\infty}\left[-\frac{\ln x}{3x^3} - \frac{1}{9x^3}\right]_1^b$

$$= 0 + \frac{1}{9} = \frac{1}{9}$$

By the Integral Test, the series converges.

39. $\sum_{n=1}^{\infty}\left(\frac{1}{n^2} - \frac{1}{n}\right) = \sum_{n=1}^{\infty}\frac{1}{n^2} - \sum_{n=1}^{\infty}\frac{1}{n}$

Since the second series is a divergent p-series while the first series is a convergent p-series, the difference diverges.

41. $\sum_{n=1}^{\infty}\frac{1}{\sqrt{n^3+2n}}$

$$\lim_{n\to\infty}\frac{1/\sqrt{n^3+2n}}{1/(n^{3/2})} = \lim_{n\to\infty}\frac{n^{3/2}}{\sqrt{n^3+2n}} = 1$$

By a limit comparison test with the convergent p-series $\sum_{n=1}^{\infty}\frac{1}{n^{3/2}}$, the series converges.

43. $\sum_{n=1}^{\infty}\frac{1\cdot 3\cdot 5\cdots(2n-1)}{2\cdot 4\cdot 6\cdots(2n)}$

$$a_n = \frac{1\cdot 3\cdot 5\cdots(2n-1)}{2\cdot 4\cdot 6\cdots(2n)}$$

$$= \left(\frac{3}{2}\cdot\frac{5}{4}\cdots\frac{2n-1}{2n-2}\right)\frac{1}{2n} > \frac{1}{2n}$$

Since $\sum_{n=1}^{\infty}\frac{1}{2n} = \frac{1}{2}\sum_{n=1}^{\infty}\frac{1}{n}$ diverges (harmonic series), so does the original series.

45. Converges by the Alternating Series Test (Conditional convergence)

47. Diverges by the nth Term Test

49. $\sum_{n=1}^{\infty}\frac{n}{e^{n^2}}$

$$\lim_{n\to\infty}\left|\frac{a_{n+1}}{a_n}\right| = \lim_{n\to\infty}\left|\frac{n+1}{e^{(n+1)^2}}\cdot\frac{e^{n^2}}{n}\right|$$

$$= \lim_{n\to\infty}\left|\frac{e^{n^2}(n+1)}{e^{n^2+2n+1}n}\right|$$

$$= \lim_{n\to\infty}\left(\frac{1}{e^{2n+1}}\right)\left(\frac{n+1}{n}\right)$$

$$= (0)(1) = 0 < 1$$

By the Ratio Test, the series converges.

51. $\sum_{n=1}^{\infty}\frac{2^n}{n^3}$

$$\lim_{n\to\infty}\left|\frac{a_{n+1}}{a_n}\right| = \lim_{n\to\infty}\left|\frac{2^{n+1}}{(n+1)^3}\cdot\frac{n^3}{2^n}\right|$$

$$= \lim_{n\to\infty}\frac{2n^3}{(n+1)^3} = 2$$

Therefore, by the Ratio Test, the series diverges.

53. (a) Ratio Test: $\lim_{n\to\infty}\left|\frac{a_{n+1}}{a_n}\right| = \lim_{n\to\infty}\frac{(n+1)(3/5)^{n+1}}{n(3/5)^n}$

$$= \lim_{n\to\infty}\left(\frac{n+1}{n}\right)\left(\frac{3}{5}\right) = \frac{3}{5} < 1$$

Converges

(b)

x	5	10	15	20	25
S_n	2.8752	3.6366	3.7377	3.7488	3.7499

(c)

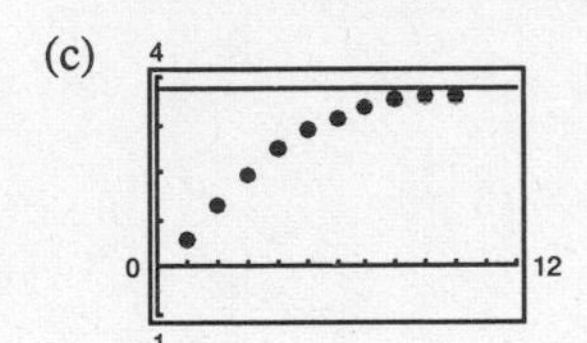

(d) The sum is approximately 3.75.

55. (a) $\int_N^\infty \frac{1}{x^2}\,dx = \left[-\frac{1}{x}\right]_N^\infty = \frac{1}{N}$

N	5	10	20	30	40
$\sum_{n=1}^{N}\frac{1}{n^2}$	1.4636	1.5498	1.5962	1.6122	1.6202
$\int_N^\infty \frac{1}{x^2}\,dx$	0.2000	0.1000	0.0500	0.0333	0.0250

(b) $\int_N^\infty \frac{1}{x^5}\,dx = \left[-\frac{1}{4x^4}\right]_N^\infty = \frac{1}{4N^4}$

N	5	10	20	30	40
$\sum_{n=1}^{N}\frac{1}{n^5}$	1.0367	1.0369	1.0369	1.0369	1.0369
$\int_N^\infty \frac{1}{x^5}\,dx$	0.0004	0.0000	0.0000	0.0000	0.0000

The series in part (b) converges more rapidly. The integral values represent the remainders of the partial sums.

57. $f(x) = e^{-x/2}$ $\qquad f(0) = 1$

$f'(x) = -\frac{1}{2}e^{-x/2}$ $\qquad f'(0) = -\frac{1}{2}$

$f''(x) = \frac{1}{4}e^{-x/2}$ $\qquad f''(0) = \frac{1}{4}$

$f'''(x) = -\frac{1}{8}e^{-x/2}$ $\qquad f'''(0) = -\frac{1}{8}$

$$P_3(x) = f(0) + f'(0)x + f''(0)\frac{x^2}{2!} + f'''(0)\frac{x^3}{3!}$$

$$= 1 - \frac{1}{2}x + \frac{1}{4}\frac{x^2}{2!} - \frac{1}{8}\frac{x^3}{3!}$$

$$= 1 - \frac{1}{2}x + \frac{1}{8}x^2 - \frac{1}{48}x^3$$

59. Since $\frac{(95\pi)^9}{180^9 \cdot 9!} < 0.001$, use four terms $\sin(95°) = \sin\left(\frac{95\pi}{180}\right) \approx \frac{95\pi}{180} - \frac{(95\pi)^3}{180^3 3!} + \frac{(95\pi)^5}{180^5 5!} - \frac{(95\pi)^7}{180^7 7!} \approx 0.99594$

61. $\ln(1.75) \approx (0.75) - \frac{(0.75)^2}{2} + \frac{(0.75)^3}{3} - \frac{(0.75)^4}{4} + \frac{(0.75)^5}{5} - \frac{(0.75)^6}{6} + \cdots - \frac{(0.75)^{14}}{14} \approx 0.559062$

63. $f(x) = \cos x,\ c = 0$

$$R_n(x) = \frac{f^{(n+1)}(z)}{(n+1)!}x^{n+1}$$

$$|f^{(n+1)}(z)| \le 1 \Rightarrow R_n(x) \le \frac{x^{n+1}}{(n+1)!}$$

(a) $R_n(x) \le \dfrac{(0.5)^{n+1}}{(n+1)!} < 0.001$

This inequality is true for $n = 4$.

(b) $R_n(x) \le \dfrac{(1)^{n+1}}{(n+1)!} < 0.001$

This inequality is true for $n = 6$.

(c) $R_n(x) \le \dfrac{(0.5)^{n+1}}{(n+1)!} < 0.0001$

This inequality is true for $n = 5$.

(d) $R_n(x) \le \dfrac{2^{n+1}}{(n+1)!} < 0.0001$

This inequality is true for $n = 10$.

65. $\displaystyle\sum_{n=0}^{\infty}\left(\frac{x}{10}\right)^n$

Geometric series which converges only if $|x/10| < 1$ or $-10 < x < 10$.

67. $\displaystyle\sum_{n=0}^{\infty}\frac{(-1)^n(x-2)^n}{(n+1)^2}$

$$\lim_{n\to\infty}\left|\frac{u_{n+1}}{u_n}\right| = \lim_{n\to\infty}\left|\frac{(-1)^{n+1}(x-2)^{n+1}}{(n+2)^2}\cdot\frac{(n+1)^2}{(-1)^n(x-2)^n}\right|$$
$$= |x-2|$$

$R = 1$
Center: 2

Since the series converges when $x = 1$ and when $x = 3$, the interval of convergence is $1 \le x \le 3$.

69. $\displaystyle\sum_{n=0}^{\infty} n!(x-2)^n$

$$\lim_{n\to\infty}\left|\frac{u_{n+1}}{u_n}\right| = \lim_{n\to\infty}\left|\frac{(n+1)!(x-2)^{n+1}}{n!(x-2)^n}\right| = \infty$$

which implies that the series converges only at the center $x = 2$.

71.

$$y = \sum_{n=0}^{\infty}(-1)^n\frac{x^{2n}}{4^n(n!)^2}$$

$$y' = \sum_{n=1}^{\infty}\frac{(-1)^n(2n)x^{2n-1}}{4^n(n!)^2} = \sum_{n=0}^{\infty}\frac{(-1)^{n+1}(2n+2)x^{2n+1}}{4^{n+1}[(n+1)!]^2}$$

$$y'' = \sum_{n=0}^{\infty}\frac{(-1)^{n+1}(2n+2)(2n+1)x^{2n}}{4^{n+1}[(n+1)!]^2}$$

$$x^2y'' + xy' + x^2y = \sum_{n=0}^{\infty}\frac{(-1)^{n+1}(2n+2)(2n+1)x^{2n+2}}{4^{n+1}[(n+1)!]^2} + \sum_{n=0}^{\infty}\frac{(-1)^{n+1}(2n+2)x^{2n+2}}{4^{n+1}[(n+1)!]^2} + \sum_{n=0}^{\infty}(-1)^n\frac{x^{2n+2}}{4^n(n!)^2}$$

$$= \sum_{n=0}^{\infty}\left[(-1)^{n+1}\frac{(2n+2)(2n+1)}{4^{n+1}[(n+1)!]^2} + \frac{(-1)^{n+1}(2n+2)}{4^{n+1}[(n+1)!]^2} + \frac{(-1)^n}{4^n(n!)^2}\right]x^{2n+2}$$

$$= \sum_{n=0}^{\infty}\left[\frac{(-1)^{n+1}(2n+2)(2n+1+1)}{4^{n+1}[(n+1)!]^2} + (-1)^n\frac{1}{4^n(n!)^2}\right]x^{2n+2}$$

$$= \sum_{n=0}^{\infty}\left[\frac{(-1)^{n+1}4(n+1)^2}{4^{n+1}[(n+1)!]^2} + (-1)^n\frac{1}{4^n(n!)^2}\right]x^{2n+2}$$

$$= \sum_{n=0}^{\infty}\left[\frac{(-1)^{n+1}1}{4^n(n!)^2} + (-1)^n\frac{1}{4^n(n!)^2}\right]x^{2n+2} = 0$$

73. $\dfrac{2}{3-x} = \dfrac{2/3}{1-(x/3)} = \dfrac{a}{1-r}$

$$\sum_{n=0}^{\infty}\frac{2}{3}\left(\frac{x}{3}\right)^n = \sum_{n=0}^{\infty}\frac{2x^n}{3^{n+1}}$$

75. Derivative: $\displaystyle\sum_{n=1}^{\infty}\frac{2nx^{n-1}}{3^{n+1}}$

77. $1 + \frac{2}{3}x + \frac{4}{9}x^2 + \frac{8}{27}x^3 + \cdots = \sum_{n=0}^{\infty}\left(\frac{2x}{3}\right)^n = \frac{1}{1-(2x/3)} = \frac{3}{3-2x}, \quad -\frac{3}{2} < x < \frac{3}{2}$

79. $f(x) = \sin(x)$

$f'(x) = \cos(x)$

$f''(x) = -\sin(x)$

$f'''(x) = -\cos(x), \cdots$

$$\sin(x) = \sum_{n=0}^{\infty} \frac{f^{(n)}(x)[x-(3\pi/4)]^n}{n!}$$

$$= \frac{\sqrt{2}}{2} - \frac{\sqrt{2}}{2}\left(x - \frac{3\pi}{4}\right) - \frac{\sqrt{2}}{2\cdot 2!}\left(x - \frac{3\pi}{4}\right)^2 + \cdots = \frac{\sqrt{2}}{2}\sum_{n=0}^{\infty} \frac{(-1)^{n(n+1)/2}[x-(3\pi/4)]^n}{n!}$$

81. $3^x = (e^{\ln(3)})^x = e^{x\ln(3)}$ and since $e^x = \sum_{n=0}^{\infty} \frac{x^n}{n!}$, we have

$$3^x = \sum_{n=0}^{\infty} \frac{(x \ln 3)^n}{n!}$$

$$= 1 + x\ln 3 + \frac{x^2 \ln^2 3}{2!} + \frac{x^3 \ln^3 3}{3!} + \frac{x^4 \ln^4 3}{4!} + \cdots.$$

83. $f(x) = \frac{1}{x}$

$f'(x) = -\frac{1}{x^2}$

$f''(x) = \frac{2}{x^3}$

$f'''(x) = -\frac{6}{x^4}, \cdots$

$$\frac{1}{x} = \sum_{n=0}^{\infty} \frac{f^{(n)}(-1)(x+1)^n}{n!}$$

$$= \sum_{n=0}^{\infty} \frac{-n!(x+1)^n}{n!} = -\sum_{n=0}^{\infty}(x+1)^n, \quad -2 < x < 0$$

85. $(1+x)^k = 1 + kx + \frac{k(k-1)x^2}{2!} + \frac{k(k-1)(k-2)x^3}{3!} + \cdots$

$$(1+x)^{1/5} = 1 + \frac{x}{5} + \frac{(1/5)(-4/5)x^2}{2!} + \frac{1/5(-4/5)(-9/5)x^3}{3!} + \cdots$$

$$= 1 + \frac{1}{5}x - \frac{1\cdot 4x^2}{5^2 2!} + \frac{1\cdot 4\cdot 9x^3}{5^3 3!} - \cdots$$

$$= 1 + \frac{x}{5} + \sum_{n=2}^{\infty} \frac{(-1)^{n+1} 4\cdot 9\cdot 14\cdots(5n-6)x^n}{5^n n!}$$

$$= 1 + \frac{x}{5} - \frac{2}{25}x^2 + \frac{6}{125}x^3 - \cdots$$

87. $\ln x = \sum_{n=1}^{\infty}(-1)^{n+1}\frac{(x-1)^n}{n}, \quad 0 < x \le 2$

$$\ln\left(\frac{5}{4}\right) = \sum_{n=1}^{\infty}(-1)^{n+1}\left(\frac{(5/4)-1}{n}\right)^n$$

$$= \sum_{n=1}^{\infty}(-1)^{n+1}\frac{1}{4^n n} \approx 0.2231$$

89. $e^x = \sum_{n=0}^{\infty} \frac{x^n}{n!}, \quad -\infty < x < \infty$

$$e^{1/2} = \sum_{n=0}^{\infty} \frac{(1/2)^n}{n!} = \sum_{n=0}^{\infty} \frac{1}{2^n n!} \approx 1.6487$$

91. $\cos x = \sum_{n=0}^{\infty} (-1)^n \frac{x^{2n}}{(2n)!}, \quad -\infty < x < \infty$

$$\cos\left(\frac{2}{3}\right) = \sum_{n=0}^{\infty} (-1)^n \frac{2^{2n}}{3^{2n}(2n)!} \approx 0.7859$$

93. The series for Exercise 41 converges very slowly because the terms approach 0 at a slow rate.

95. (a) $f(x) = e^{2x} \qquad f(0) = 1$

$f'(x) = 2e^{2x} \qquad f'(0) = 2$

$f''(x) = 4e^{2x} \qquad f''(0) = 4$

$f'''(x) = 8e^{2x} \qquad f'''(0) = 8$

$$e^{2x} = 1 + 2x + \frac{4x^2}{2!} + \frac{8x^3}{3!} + \cdots$$

$$= 1 + 2x + 2x^2 + \frac{4}{3}x^3 + \cdots$$

(b) $e^x = \sum_{n=0}^{\infty} \frac{x^n}{n!}$

$$e^{2x} = \sum_{n=0}^{\infty} \frac{(2x)^n}{n!} = 1 + 2x + \frac{4x^2}{2!} + \frac{8x^3}{3!} + \cdots$$

$$= 1 + 2x + 2x^2 + \frac{4}{3}x^3 + \cdots$$

(c) $e^{2x} = e^x \cdot e^x = \left(1 + x + \frac{x^2}{2} + \frac{x^3}{6} + \cdots\right)\left(1 + x + \frac{x^2}{2} + \frac{x^3}{6} + \cdots\right)$

$$= 1 + (x + x) + \left(x^2 + \frac{x^2}{2} + \frac{x^2}{2}\right) + \left(\frac{x^3}{6} + \frac{x^3}{6} + \frac{x^3}{2} + \frac{x^3}{2}\right) + \cdots = 1 + 2x + 2x^2 + \frac{4}{3}x^3 + \cdots$$

97. $\sin t = \sum_{n=0}^{\infty} \frac{(-1)^n t^{2n+1}}{(2n+1)!}$

$$\frac{\sin t}{t} = \sum_{n=0}^{\infty} \frac{(-1)^n t^{2n}}{(2n+1)!}$$

$$\int_0^x \frac{\sin t}{t}\,dt = \left[\sum_{n=0}^{\infty} \frac{(-1)^n t^{2n+1}}{(2n+1)(2n+1)!}\right]_0^x$$

$$= \sum_{n=0}^{\infty} \frac{(-1)^n x^{2n+1}}{(2n+1)(2n+1)!}$$

99. $\frac{1}{1+t} = \sum_{n=0}^{\infty} (-1)^n t^n$

$$\ln(1+t) = \int \frac{1}{1+t}\,dt = \sum_{n=0}^{\infty} \frac{(-1)^n t^{n+1}}{n+1}$$

$$\frac{\ln(t+1)}{t} = \sum_{n=0}^{\infty} \frac{(-1)^n t^n}{n+1}$$

$$\int_0^x \frac{\ln(t+1)}{t}\,dt = \left[\sum_{n=0}^{\infty} \frac{(-1)^n t^{n+1}}{(n+1)^2}\right]_0^x = \sum_{n=0}^{\infty} \frac{(-1)^n x^{n+1}}{(n+1)^2}$$

101. $\arctan x = x - \frac{x^3}{3} + \frac{x^5}{5} - \frac{x^7}{7} + \frac{x^9}{9} - \cdots$

$$\frac{\arctan x}{\sqrt{x}} = \sqrt{x} - \frac{x^{5/2}}{3} + \frac{x^{9/2}}{5} - \frac{x^{13/2}}{7} + \frac{x^{17/2}}{9} - \cdots$$

$$\lim_{x\to 0^+} \frac{\arctan x}{\sqrt{x}} = 0$$

By L'Hôpital's Rule, $\lim_{x\to 0^+} \frac{\arctan x}{\sqrt{x}} = \lim_{x\to 0^+} \frac{\left(\frac{1}{1+x^2}\right)}{\left(\frac{1}{2\sqrt{x}}\right)} = \lim_{x\to 0^+} \frac{2\sqrt{x}}{1+x^2} = 0.$

Problem Solving for Chapter 8

1. (a) $1\left(\frac{1}{3}\right) + 2\left(\frac{1}{9}\right) + 4\left(\frac{1}{27}\right) + \cdots = \sum_{n=0}^{\infty} \frac{1}{3}\left(\frac{2}{3}\right)^n = \frac{1/3}{1 - (2/3)} = 1$

(b) $0, \frac{1}{3}, \frac{2}{3}, 1,$ etc.

(c) $\lim_{n\to\infty} C_n = 1 - \sum_{n=0}^{\infty} \frac{1}{3}\left(\frac{2}{3}\right)^n = 1 - 1 = 0$

3. If there are n rows, then $a_n = \dfrac{n(n+1)}{2}$.

For one circle,

$$a_1 = 1 \text{ and } r_1 = \frac{1}{3}\left(\frac{\sqrt{3}}{2}\right) = \frac{\sqrt{3}}{6} = \frac{1}{2\sqrt{3}}$$

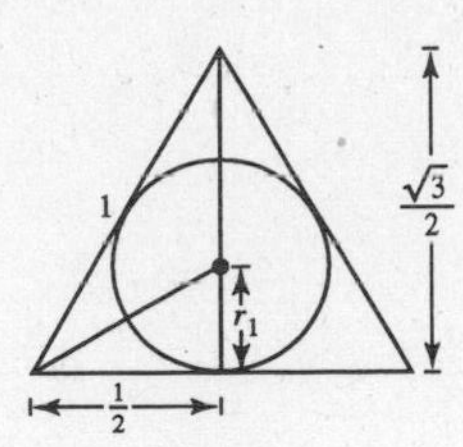

For three circles,

$$a_2 = 3 \text{ and } 1 = 2\sqrt{3}r_2 + 2r_2$$

$$r_2 = \frac{1}{2 + 2\sqrt{3}}$$

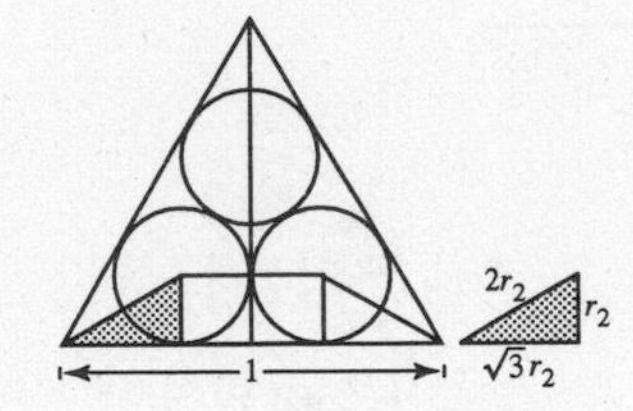

For six circles,

$$a_3 = 6 \text{ and } 1 = 2\sqrt{3}r_3 + 4r_3$$

$$r_3 = \frac{1}{2\sqrt{3} + 4}$$

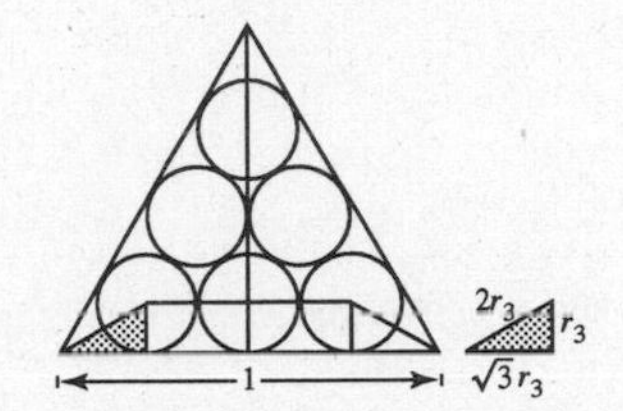

Continuing this pattern, $r_n = \dfrac{1}{2\sqrt{3} + 2(n-1)}$.

$$\text{Total Area} = (\pi r_n^{\,2})a_n = \pi\left(\frac{1}{2\sqrt{3} + 2(n-1)}\right)^2 \frac{n(n+1)}{2}$$

$$A_n = \frac{\pi}{2}\,\frac{n(n+1)}{\left[2\sqrt{3} + 2(n+1)\right]^2}$$

$$\lim_{n\to\infty} A_n = \frac{\pi}{2}\cdot\frac{1}{4} = \frac{\pi}{8}$$

5. (a)
$$\begin{aligned}\sum a_n x^n &= 1 + 2x + 3x^2 + x^3 + 2x^4 + 3x^5 + \cdots \\ &= (1 + x^3 + x^6 + \cdots) + 2(x + x^4 + x^7 + \cdots) + 3(x^2 + x^5 + x^8 + \cdots) \\ &= (1 + x^3 + x^6 + \cdots)[1 + 2x + 3x^2] \\ &= (1 + 2x + 3x^2)\frac{1}{1 - x^3}\end{aligned}$$

$R = 1$ because each series in the second line has $R = 1$.

(b)
$$\begin{aligned}\sum a_n x^n &= (a_0 + a_1 x + \cdots + a_{p-1}x^{p-1}) + (a_0 x^p + a_1 x^{p+1} + \cdots) + \cdots \\ &= a_0(1 + x^p + \cdots) + a_1 x(1 + x^p + \cdots) + \cdots + a_{p-1}x^{p-1}(1 + x^p + \cdots) \\ &= (a_0 + a_1 x + \cdots + a_{p-1}x^{p-1})(1 + x^p + \cdots) \\ &= (a_0 + a_1 x + \cdots + a_{p-1}x^{p-1})\frac{1}{1 - x^p}.\end{aligned}$$

$R = 1$

(Assume all $a_n > 0$.)

7. $$e^x = 1 + x + \frac{x^2}{2!} + \cdots$$

$$xe^x = x + x^2 + \frac{x^3}{2!} + \cdots = \sum_{n=0}^{\infty} \frac{x^{n+1}}{n!}$$

$$\int xe^x\,dx = xe^x - e^x + C = \sum_{n=0}^{\infty} \frac{x^{n+2}}{(n+2)n!}$$

Letting $x = 0$, $C = 1$. Letting $x = 1$,

$$1 = \sum_{n=0}^{\infty} \frac{1}{(n+2)n!} = \frac{1}{2} + \sum_{n=1}^{\infty} \frac{1}{(n+2)n!}.$$

Thus, $\sum_{n=1}^{\infty} \frac{1}{(n+2)n!} = \frac{1}{2}$.

9. Let $a_1 = \int_0^{\pi} \frac{\sin x}{x}\,dx$, $a_2 = -\int_{\pi}^{2\pi} \frac{\sin x}{x}\,dx$, $a_3 = \int_{2\pi}^{3\pi} \frac{\sin x}{x}\,dx$, etc.

Then,

$$\int_0^{\infty} \frac{\sin x}{x}\,dx = a_1 - a_2 + a_3 - a_4 + \cdots.$$

Since $\lim_{n\to\infty} a_n = 0$ and $a_{n+1} < a_n$, this series converges.

11. (a) $a_1 = 3.0$

$a_2 \approx 1.73205$

$a_3 \approx 2.17533$

$a_4 \approx 2.27493$

$a_5 \approx 2.29672$

$a_6 \approx 2.30146$

$\lim_{n\to\infty} a_n = \frac{1 + \sqrt{13}}{2}$ [See part (b) for proof.]

(b) Use mathematical induction to show the sequence is increasing. Clearly, $a_2 = \sqrt{a + a_1} = \sqrt{a\sqrt{a}} > \sqrt{a} = a_1$.

Now assume $a_n > a_{n-1}$. Then

$$a_n + a > a_{n-1} + a$$
$$\sqrt{a_n + a} > \sqrt{a_{n-1} + a}$$
$$a_{n+1} > a_n.$$

Use mathematical induction to show that the sequence is bounded above by a. Clearly, $a_1 = \sqrt{a} < a$.

Now assume $a_n < a$. Then $a > a_n$ and $a - 1 > 1$ implies

$$a(a-1) > a_n(1)$$
$$a^2 - a > a_n$$
$$a^2 > a_n + a$$
$$a > \sqrt{a_n + a} = a_{n+1}.$$

Hence, the sequence converges to some number L. To find L, assume $a_{n+1} \approx a_n \approx L$:

$$L = \sqrt{a + L} \Rightarrow L^2 = a + L \Rightarrow L^2 - L - a = 0$$

$$L = \frac{1 \pm \sqrt{1 + 4a}}{2}.$$

Hence, $L = \frac{1 + \sqrt{1 + 4a}}{2}$.

13. (a) $\displaystyle\sum_{n=1}^{\infty} \frac{1}{2^{n+(-1)^n}} = \frac{1}{2^{1-1}} + \frac{1}{2^{2+1}} + \frac{1}{2^{3-1}} + \frac{1}{2^{4+1}} + \frac{1}{2^{5-1}} + \cdots$

$$S_1 = \frac{1}{2^0} = 1$$

$$S_1 = 1 + \frac{1}{8} = \frac{9}{8}$$

$$S_3 = \frac{9}{8} + \frac{1}{4} = \frac{11}{8}$$

$$S_4 = \frac{11}{8} + \frac{1}{32} = \frac{45}{32}$$

$$S_5 = \frac{45}{32} + \frac{1}{16} = \frac{47}{32}$$

(b) $\displaystyle\frac{a_{n+1}}{a_n} = \frac{2^{n+(-1)^n}}{2^{(n+1)+(-1)^{n+1}}} = \frac{2^{(-1)^n}}{2^{1+(-1)^{n+1}}}$

This sequence is $\frac{1}{8}, 2, \frac{1}{8}, 2, \ldots$ which diverges.

(c) $\displaystyle\sqrt[n]{\frac{1}{2^{n+(-1)^n}}} = \left(\frac{1}{2^n \cdot 2^{(-1)^n}}\right)^{1/n}$

$\displaystyle = \frac{1}{2 \cdot \sqrt[n]{2^{(-1)^n}}} \rightarrow \frac{1}{2} < 1$ converges because $\{2^{(-1)^n}\} = \frac{1}{2}, 2, \frac{1}{2}, 2, \ldots$ and $\sqrt[n]{1/2} \rightarrow 1$ and $\sqrt[n]{2} \rightarrow 1$.

15. $S_6 = 130 + 70 + 40 = 240$

$S_7 = 240 + 130 + 70 = 440$

$S_8 = 440 + 240 + 130 = 810$

$S_9 = 810 + 440 + 240 = 1490$

$S_{10} = 1490 + 810 + 440 = 2740$

CHAPTER 9
Conics, Parametric Equations, and Polar Coordinates

CHAPTER 9
Conics, Parametric Equations, and Polar Coordinates

Section 9.1 Conics and Calculus

Solutions to Odd-Numbered Exercises

1. $y^2 = 4x$

Vertex: $(0, 0)$

$p = 1 > 0$

Opens to the right
Matches graph (h).

3. $(x + 3)^2 = -2(y - 2)$

Vertex: $(-3, 2)$

$p = -\frac{1}{2} < 0$

Opens downward
Matches graph (e).

5. $\dfrac{x^2}{9} + \dfrac{y^2}{4} = 1$

Center: $(0, 0)$
Ellipse
Matches (f)

7. $\dfrac{y^2}{16} - \dfrac{x^2}{1} = 1$

Hyperbola

Center: $(0, 0)$

Vertical transverse axis.
Matches (c)

9. $y^2 = -6x = 4\left(-\frac{3}{2}\right)x$

Vertex: $(0, 0)$

Focus: $\left(-\frac{3}{2}, 0\right)$

Directrix: $x = \frac{3}{2}$

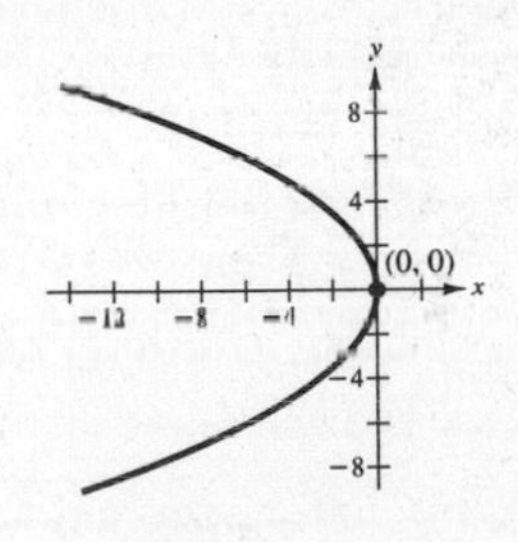

11. $(x + 3) + (y - 2)^2 = 0$

$$(y - 2)^2 = 4\left(-\tfrac{1}{4}\right)(x + 3)$$

Vertex: $(-3, 2)$

Focus: $(-3.25, 2)$

Directrix: $x = -2.75$

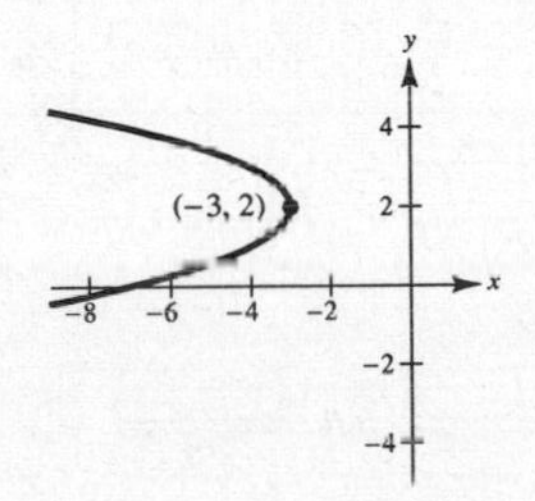

13. $y^2 - 4y - 4x = 0$

$$y^2 - 4y + 4 = 4x + 4$$

$$(y - 2)^2 = 4(1)(x + 1)$$

Vertex: $(-1, 2)$

Focus: $(0, 2)$

Directrix: $x = -2$

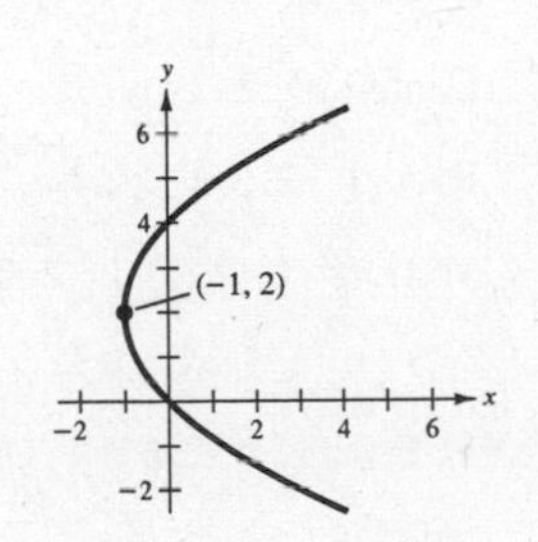

15. $x^2 + 4x + 4y - 4 = 0$

$$x^2 + 4x + 4 = -4y + 4 + 4$$

$$(x + 2)^2 = 4(-1)(y - 2)$$

Vertex: $(-2, 2)$

Focus: $(-2, 1)$

Directrix: $y = 3$

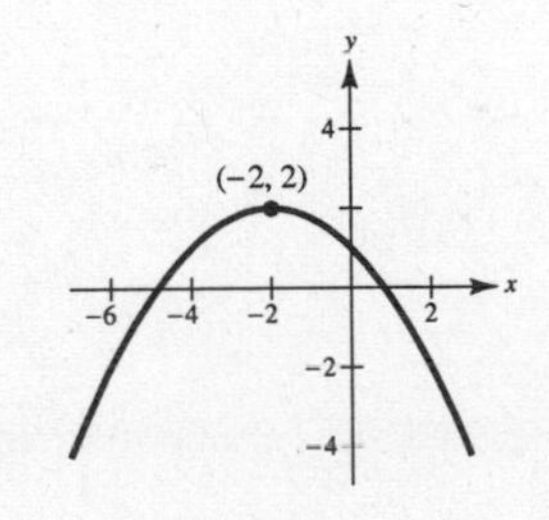

17. $y^2 + x + y = 0$

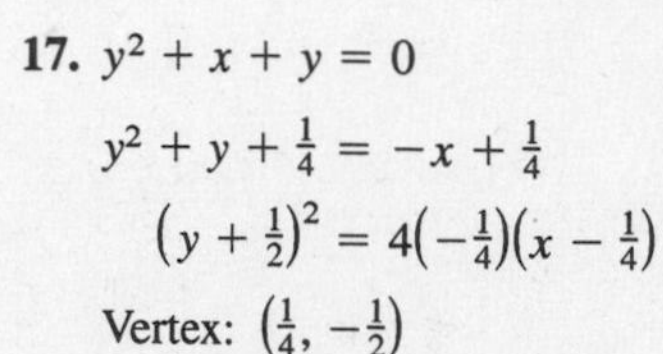

$$y^2 + y + \tfrac{1}{4} = -x + \tfrac{1}{4}$$

$$\left(y + \tfrac{1}{2}\right)^2 = 4\left(-\tfrac{1}{4}\right)\left(x - \tfrac{1}{4}\right)$$

Vertex: $\left(\tfrac{1}{4}, -\tfrac{1}{2}\right)$

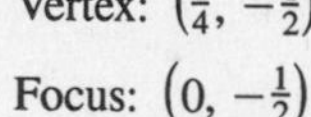

Focus: $\left(0, -\tfrac{1}{2}\right)$

Directrix: $x = \tfrac{1}{2}$

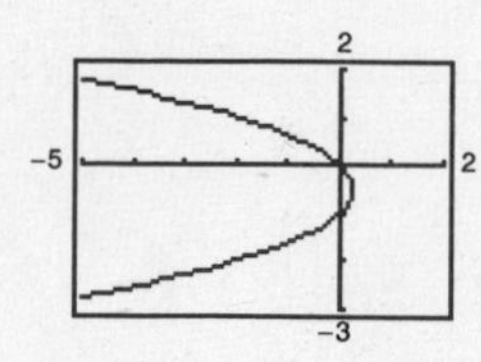

19. $y^2 - 4x - 4 = 0$

$$y^2 = 4x + 4$$

$$= 4(1)(x + 1)$$

Vertex: $(-1, 0)$

Focus: $(0, 0)$

Directrix: $x = -2$

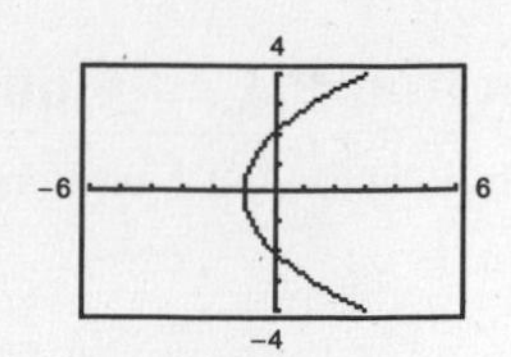

21. $(y - 2)^2 = 4(-2)(x - 3)$

$y^2 - 4y + 8x - 20 = 0$

23. $(x - h)^2 = 4p(y - k)$

$x^2 = 4(6)(y - 4)$

$x^2 - 24y + 96 = 0$

25. $y = 4 - x^2$

$x^2 + y - 4 = 0$

27. Since the axis of the parabola is vertical, the form of the equation is $y = ax^2 + bx + c$. Now, substituting the values of the given coordinates into this equation, we obtain

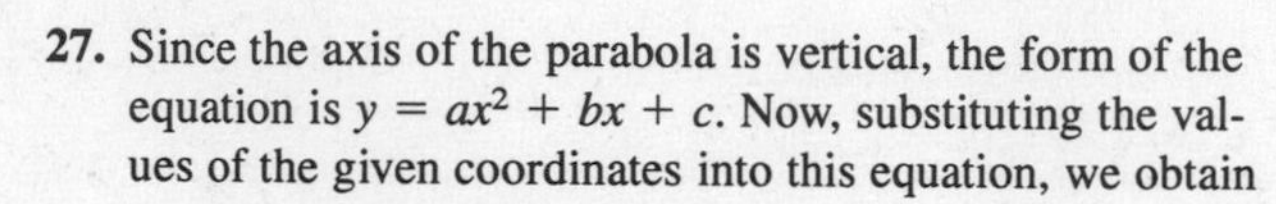

$$3 = c,\ 4 = 9a + 3b + c,\ 11 = 16a + 4b + c.$$

Solving this system, we have $a = \tfrac{5}{3}, b = -\tfrac{14}{3}, c = 3.$ Therefore,

$$y = \tfrac{5}{3}x^2 - \tfrac{14}{3}x + 3 \text{ or } 5x^2 - 14x - 3y + 9 = 0.$$

29. $x^2 + 4y^2 = 4$

$$\frac{x^2}{4} + \frac{y^2}{1} = 1$$

$a^2 = 4, b^2 = 1, c^2 = 3$

Center: $(0, 0)$

Foci: $(\pm\sqrt{3}, 0)$

Vertices: $(\pm 2, 0)$

$$e = \frac{\sqrt{3}}{2}$$

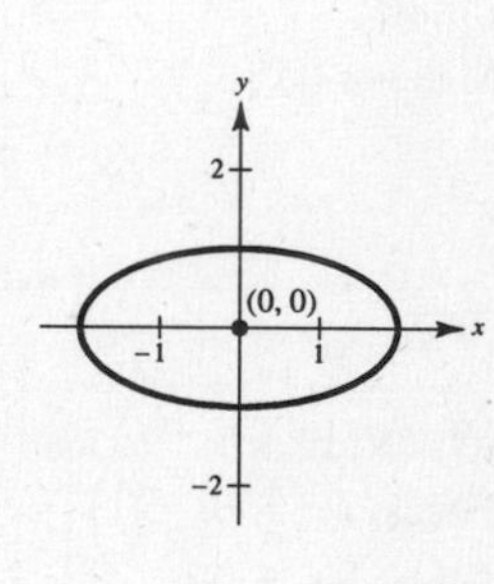

31. $\dfrac{(x - 1)^2}{9} + \dfrac{(y - 5)^2}{25} = 1$

$a^2 = 25, b^2 = 9, c^2 = 16$

Center: $(1, 5)$

Foci: $(1, 9), (1, 1)$

Vertices: $(1, 10), (1, 0)$

$$e = \frac{4}{5}$$

33. $9x^2 + 4y^2 + 36x - 24y + 36 = 0$

$$9(x^2 + 4x + 4) + 4(y^2 - 6y + 9) = -36 + 36 + 36$$

$$= 36$$

$$\frac{(x + 2)^2}{4} + \frac{(y - 3)^2}{9} = 1$$

$a^2 = 9, b^2 = 4, c^2 = 5$

Center: $(-2, 3)$

Foci: $\left(-2, 3 \pm \sqrt{5}\right)$

Vertices: $(-2, 6), (-2, 0)$

$$e = \frac{\sqrt{5}}{3}$$

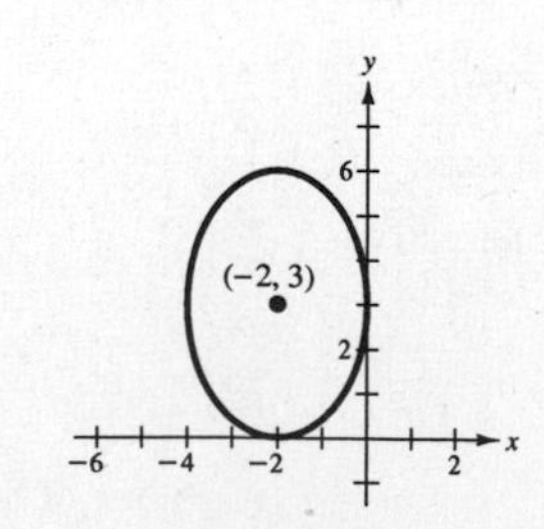

35. $12x^2 + 20y^2 - 12x + 40y - 37 = 0$

$$12\left(x^2 - x + \frac{1}{4}\right) + 20(y^2 + 2y + 1) = 37 + 3 + 20$$

$$= 60$$

$$\frac{[x - (1/2)]^2}{5} + \frac{(y + 1)^2}{3} = 1$$

$a^2 = 5, b^2 = 3, c^2 = 2$

Center: $\left(\frac{1}{2}, -1\right)$

Foci: $\left(\frac{1}{2} \pm \sqrt{2}, -1\right)$

Vertices: $\left(\frac{1}{2} \pm \sqrt{5}, -1\right)$

Solve for y:

$$20(y^2 + 2y + 1) = -12x^2 + 12x + 37 + 20$$

$$(y + 1)^2 = \frac{57 + 12x - 12x^2}{20}$$

$$y = -1 \pm \sqrt{\frac{57 + 12x - 12x^2}{20}}$$

(Graph each of these separately.)

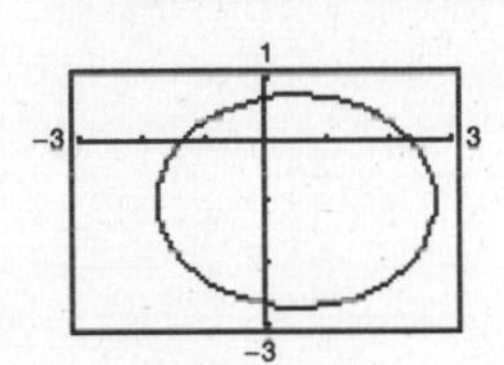

37. $x^2 + 2y^2 - 3x + 4y + 0.25 = 0$

$$\left(x^2 - 3x + \frac{9}{4}\right) + 2(y^2 + 2y + 1) = -\frac{1}{4} + \frac{9}{4} + 2 = 4$$

$$\frac{[x - (3/2)]^2}{4} + \frac{(y + 1)^2}{2} = 1$$

$a^2 = 4, b^2 = 2, c^2 = 2$

Center: $\left(\frac{3}{2}, -1\right)$

Foci: $\left(\frac{3}{2} \pm \sqrt{2}, -1\right)$

Vertices: $\left(-\frac{1}{2}, -1\right), \left(\frac{7}{2}, -1\right)$

Solve for y: $2(y^2 + 2y + 1) = -x^2 + 3x - \frac{1}{4} + 2$

$$(y + 1)^2 = \frac{1}{2}\left(\frac{7}{4} + 3x - x^2\right)$$

$$y = -1 \pm \sqrt{\frac{7 + 12x - 4x^2}{8}}$$

(Graph each of these separately.)

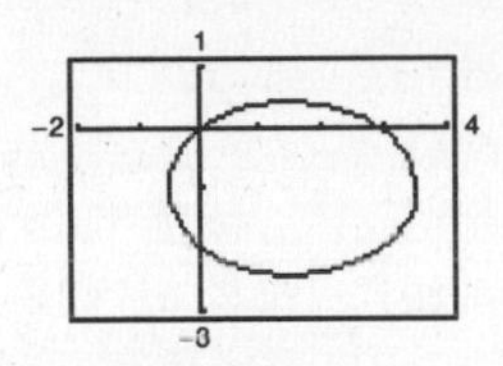

39. Center: $(0, 0)$
Focus: $(2, 0)$
Vertex: $(3, 0)$
Horizontal major axis

$a = 3, c = 2 \Rightarrow b = \sqrt{5}$

$$\frac{x^2}{9} + \frac{y^2}{5} = 1$$

41. Vertices: $(3, 1), (3, 9)$
Minor axis length: 6
Vertical major axis
Center: $(3, 5)$

$a = 4, b = 3$

$$\frac{(x - 3)^2}{9} + \frac{(y - 5)^2}{16} = 1$$

43. Center: $(0, 0)$
Horizontal major axis
Points on ellipse: $(3, 1), (4, 0)$

Since the major axis is horizontal,

$$\left(\frac{x^2}{a^2}\right) + \left(\frac{y^2}{b^2}\right) = 1.$$

Substituting the values of the coordinates of the given points into this equation, we have

$$\left(\frac{9}{a^2}\right) + \left(\frac{1}{b^2}\right) = 1, \text{ and } \frac{16}{a^2} = 1.$$

The solution to this system is $a^2 = 16$, $b^2 = 16/7$.

Therefore,

$$\frac{x^2}{16} + \frac{y^2}{16/7} = 1, \frac{x^2}{16} + \frac{7y^2}{16} = 1.$$

45. $\dfrac{y^2}{1} - \dfrac{x^2}{4} = 1$

$a = 1, b = 2, c = \sqrt{5}$

Center: $(0, 0)$

Vertices: $(0, \pm 1)$

Foci: $\left(0, \pm\sqrt{5}\right)$

Asymptotes: $y = \pm\dfrac{1}{2}x$

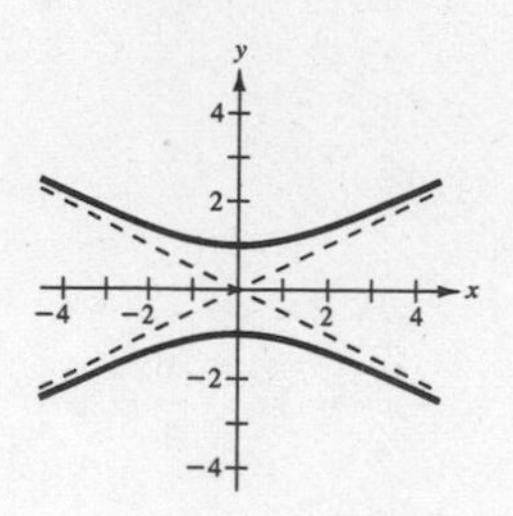

47. $\dfrac{(x-1)^2}{4} - \dfrac{(y+2)^2}{1} = 1$

$a = 2, b = 1, c = \sqrt{5}$

Center: $(1, -2)$

Vertices: $(-1, -2), (3, -2)$

Foci: $\left(1 \pm \sqrt{5}, -2\right)$

Asymptotes: $y = -2 \pm \dfrac{1}{2}(x - 1)$

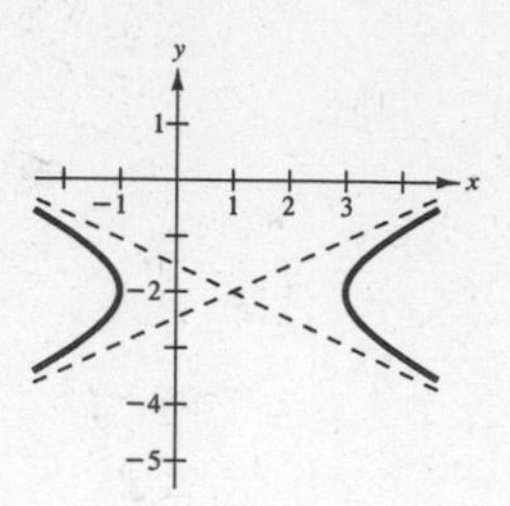

49. $9x^2 - y^2 - 36x - 6y + 18 = 0$

$$9(x^2 - 4x + 4) - (y^2 + 6y + 9) = -18 + 36 - 9$$

$$\frac{(x-2)^2}{1} - \frac{(y+3)^2}{9} = 1$$

$a = 1, b = 3, c = \sqrt{10}$

Center: $(2, -3)$

Vertices: $(1, -3), (3, -3)$

Foci: $\left(2 \pm \sqrt{10}, -3\right)$

Asymptotes: $y = -3 \pm 3(x - 2)$

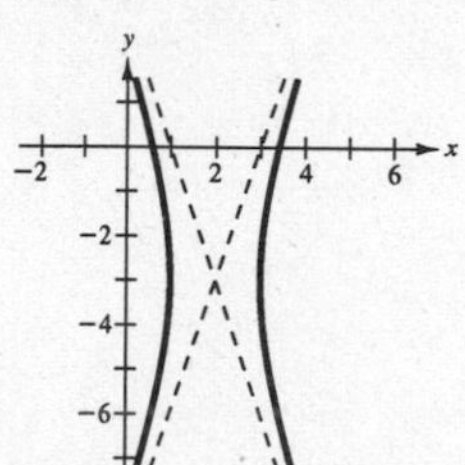

51. $x^2 - 9y^2 + 2x - 54y - 80 = 0$

$$(x^2 + 2x + 1) - 9(y^2 + 6y + 9) = 80 + 1 - 81 = 0$$

$$(x+1)^2 - 9(y+3)^2 = 0$$

$$y + 3 = \pm\frac{1}{3}(x + 1)$$

Degenerate hyperbola is two lines intersecting at $(-1, -3)$.

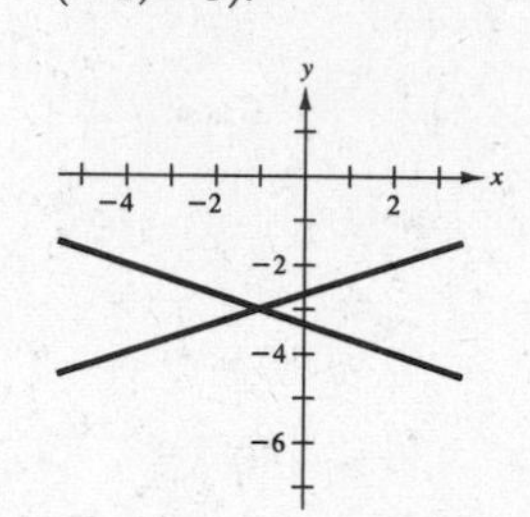

53. $9y^2 - x^2 + 2x + 54y + 62 = 0$

$$9(y^2 + 6y + 9) - (x^2 - 2x + 1) = -62 - 1 + 81 = 18$$

$$\frac{(y+3)^2}{2} - \frac{(x-1)^2}{18} = 1$$

$a = \sqrt{2}, b = 3\sqrt{2}, c = 2\sqrt{5}$

Center: $(1, -3)$

Vertices: $\left(1, -3 \pm \sqrt{2}\right)$

Foci: $\left(1, -3 \pm 2\sqrt{5}\right)$

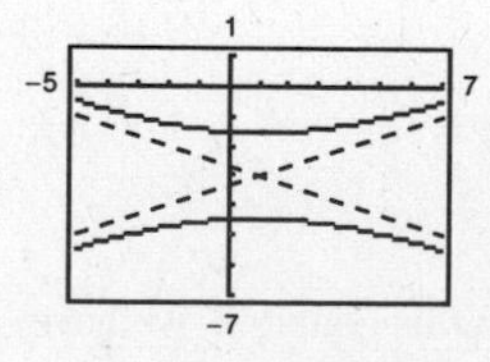

Solve for y:

$$9(y^2 + 6y + 9) = x^2 - 2x - 62 + 81$$

$$(y+3)^2 = \frac{x^2 - 2x + 19}{9}$$

$$y = -3 \pm \frac{1}{3}\sqrt{x^2 - 2x + 19}$$

(Graph each curve separately.)

55. $3x^2 - 2y^2 - 6x - 12y - 27 = 0$

$$3(x^2 - 2x + 1) - 2(y^2 + 6y + 9) = 27 + 3 - 18 = 12$$

$$\frac{(x-1)^2}{4} - \frac{(y+3)^2}{6} = 1$$

$a = 2, b = \sqrt{6}, c = \sqrt{10}$

Center: $(1, -3)$

Vertices: $(-1, -3), (3, -3)$

Foci: $\left(1 \pm \sqrt{10}, -3\right)$

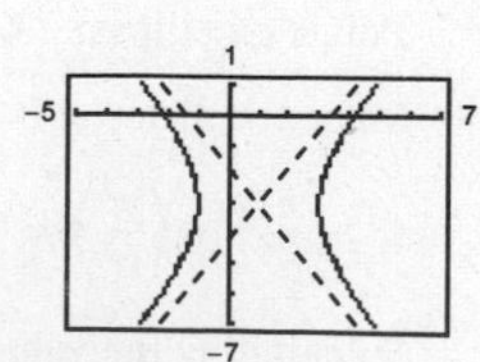

Solve for y:

$$2(y^2 + 6y + 9) = 3x^2 - 6x - 27 + 18$$

$$(y+3)^2 = \frac{3x^2 - 6x - 9}{2}$$

$$y = -3 \pm \sqrt{\frac{3(x^2 - 2x - 3)}{2}}$$

(Graph each curve separately.)

57. Vertices: $(\pm 1, 0)$
Asymptotes: $y = \pm 3x$
Horizontal transverse axis
Center: $(0, 0)$

$$a = 1, \pm\frac{b}{a} = \pm\frac{b}{1} = \pm 3 \Rightarrow b = 3$$

Therefore, $\dfrac{x^2}{1} - \dfrac{y^2}{9} = 1$.

59. Vertices: $(2, \pm 3)$
Point on graph: $(0, 5)$
Vertical transverse axis
Center: $(2, 0)$

$a = 3$

Therefore, the equation is of the form

$$\frac{y^2}{9} - \frac{(x-2)^2}{b^2} = 1.$$

Substituting the coordinates of the point $(0, 5)$, we have

$$\frac{25}{9} - \frac{4}{b^2} = 1 \quad \text{or} \quad b^2 = \frac{9}{4}.$$

Therefore, the equation is $\dfrac{y^2}{9} - \dfrac{(x-2)^2}{9/4} = 1$.

61. Center: $(0, 0)$
Vertex: $(0, 2)$
Focus: $(0, 4)$
Vertical transverse axis

$$a = 2, c = 4, b^2 = c^2 - a^2 = 12$$

Therefore, $\dfrac{y^2}{4} - \dfrac{x^2}{12} = 1$.

63. Vertices: $(0, 2), (6, 2)$

Asymptotes: $y = \dfrac{2}{3}x, y = 4 - \dfrac{2}{3}x$

Horizontal transverse axis

Center: $(3, 2)$

$a = 3$

Slopes of asymptotes: $\pm\dfrac{b}{a} = \pm\dfrac{2}{3}$

Thus, $b = 2$. Therefore,

$$\frac{(x-3)^2}{9} - \frac{(y-2)^2}{4} = 1.$$

65. (a) $\dfrac{x^2}{9} - y^2 = 1, \dfrac{2x}{9} - 2yy' = 0, \dfrac{x}{9y} = y'$

At $x = 6$: $y = \pm\sqrt{3}, y' = \dfrac{\pm 6}{9\sqrt{3}} = \dfrac{\pm 2\sqrt{3}}{9}$

At $(6, \sqrt{3})$: $y - \sqrt{3} = \dfrac{2\sqrt{3}}{9}(x - 6)$

or $2x - 3\sqrt{3}y - 3 = 0$

At $(6, -\sqrt{3})$: $y + \sqrt{3} = \dfrac{-2\sqrt{3}}{9}(x - 6)$

or $2x + 3\sqrt{3}y - 3 = 0$

(b) From part (a) we know that the slopes of the normal lines must be $\pm 9/(2\sqrt{3})$.

At $(6, \sqrt{3})$: $y - \sqrt{3} = -\dfrac{9}{2\sqrt{3}}(x - 6)$

or $9x + 2\sqrt{3}y - 60 = 0$

At $(6, -\sqrt{3})$: $y + \sqrt{3} = \dfrac{9}{2\sqrt{3}}(x - 6)$

or $9x - 2\sqrt{3}y - 60 = 0$

67.
$$\begin{aligned} x^2 + 4y^2 - 6x + 16y + 21 &= 0 \\ x^2 - 6x + 4(y^2 + 4y) &= -21 \\ x^2 - 6x + 9 + 4(y^2 + 4y + 4) &= -21 + 9 + 16 \\ (x-3)^2 + 4(y+2)^2 &= 4 \\ \frac{(x-3)^2}{4} + \frac{(y+2)^2}{1} &= 1 \end{aligned}$$

Ellipse

69.
$$\begin{aligned} y^2 - 4y - 4x &= 0 \\ y^2 - 4y + 4 - 4x &= 4 \\ (y-2)^2 - 4x &= 4 \\ (y-2)^2 &= 4x + 4 \\ (y-2)^2 &= 4(x+1) \end{aligned}$$

Parabola

71.
$$\begin{aligned} 4x^2 + 4y^2 - 16y + 15 &= 0 \\ 4x^2 + 4(y^2 - 4y) &= -15 \\ x^2 + (y^2 - 4y) &= -\frac{15}{4} \\ x^2 + y^2 - 4y + 4 &= -\frac{15}{4} + 4 \\ x^2 + (y-2)^2 &= \frac{1}{4} \end{aligned}$$

Circle

73.
$$9x^2 + 9y^2 - 36x + 6y + 34 = 0$$
$$9x^2 - 36x + 9y^2 + 6y = -34$$
$$9(x^2 - 4x + 4) + 9\left(y^2 + \frac{2}{3}y + \frac{1}{9}\right) = -34 + 36 + 1$$
$$9(x-2)^2 = 9\left(y = \frac{1}{3}\right)^2 = 3$$
$$(x-2)^2 + \left(y + \frac{1}{3}\right)^2 = \frac{1}{3}$$

Circle

75.
$$3(x-1)^2 = 6 + 2(y+1)^2$$
$$3(x-1)^2 - (y+1)^2 = 6$$
$$\frac{(x-1)^2}{2} - \frac{(y+1)^2}{3} = 1$$

Hyperbola

77. (a) A parabola is the set of all points (x, y) that are equidistant from a fixed line (directrix) and a fixed point (focus) not on the line.

(b) $(x-h)^2 = 4p(y-k)$ or $(y-k)^2 = 4p(x-h)$

(c) See Theorem 9.2.

79. (a) A hyperbola is the set of all points (x, y) for which the absolute value of the difference between the distances from two distance fixed points (foci) is constant.

(b) $\dfrac{(x-h)^2}{a^2} - \dfrac{(y-k)^2}{b^2} = 1$ or $\dfrac{(y-k)^2}{a^2} - \dfrac{(x-h)^2}{b^2} = 1$

(c) $y = k \pm \dfrac{b}{a}(x-h)$ or $y = k \pm \dfrac{a}{b}(x-h)$

81. Assume that the vertex is at the origin.

$$x^2 = 4py$$
$$(3)^2 = 4p(1)$$
$$\frac{9}{4} = p$$

The pipe is located $\frac{9}{4}$ meters from the vertex.

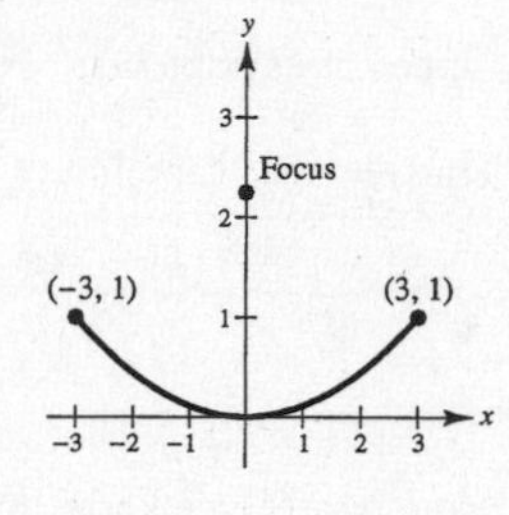

83. $y = ax^2$

$y' = 2ax$

The equation of the tangent line is

$$y - ax_0^2 = 2ax_0(x - x_0)$$

or $y = 2ax_0x - ax_0^2$.

Let $y = 0$. Then:

$$-ax_0^2 = 2ax_0x - 2ax_0^2$$
$$ax_0^2 = 2ax_0x$$
$$x = \frac{x_0}{2}$$

Therefore, $\left(\dfrac{x_0}{2}, 0\right)$ is the x-intercept.

85. (a) Consider the parabola $x^2 = 4py$. Let m_0 be the slope of the one tangent line at (x_1, y_1) and therefore, $-1/m_0$ is the slope of the second at (x_2, y_2). Differentiating, we have $2x = 4py'$ or $y' = \dfrac{x}{2p}$, and

$$m_0 = \frac{1}{2p}x_1 \text{ or } x_1 = 2pm_0$$
$$\frac{-1}{m_0} = \frac{1}{2p}x_2 \text{ or } x_2 = \frac{-2p}{m_0}$$

Substituting these values of x into the equation $x^2 = 4py$, we have the coordinates of the points of tangency $(2pm_0, pm_0^2)$ and $(-2p/m_0, p/m_0^2)$ and the equations of the tangent lines are

$$(y - pm_0^2) = m_0(x - 2pm_0) \quad \text{and} \quad \left(y - \frac{p}{m_0^2}\right) = \frac{-1}{m_0}\left(x + \frac{2p}{m_0}\right).$$

The point of intersection of these lines is

$$\left(\frac{p(m_0^2 - 1)}{m_0}, -p\right)$$ and is on the directrix, $y = -p$.

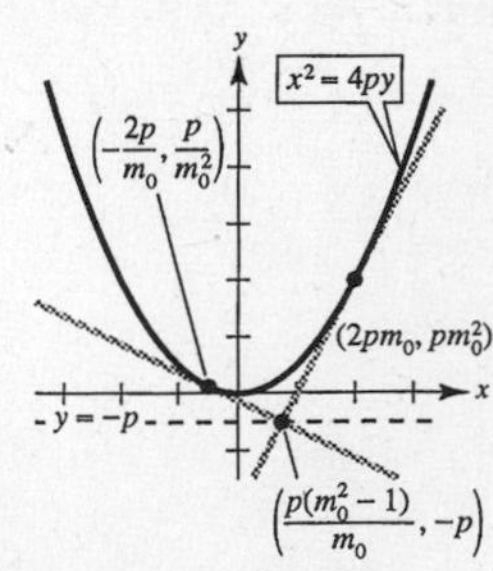

—CONTINUED—

85. —CONTINUED—

(b) $x^2 - 4x - 4y + 8 = 0$

$$(x - 2)^2 = 4(y - 1). \text{ Vertex } (2, 1)$$

$$2x - 4 - 4\frac{dy}{dx} = 0$$

$$\frac{dy}{dx} = \frac{1}{2}x - 1$$

At $(-2, 5)$, $dy/dx = -2$. At $\left(3, \frac{5}{4}\right)$, $dy/dx = \frac{1}{2}$.

Tangent line at $(-2, 5)$: $y - 5 = -2(x + 2) \Rightarrow 2x + y - 1 = 0$.

Tangent line at $\left(3, \frac{5}{4}\right)$: $y - \frac{5}{4} = \frac{1}{2}(x - 3) \Rightarrow 2x - 4y - 1 = 0$.

Since $m_1 m_2 = (-2)\left(\frac{1}{2}\right) = -1$, the lines are perpendicular.

Point of intersection: $-2x + 1 = \frac{1}{2}x - \frac{1}{4}$

$$-\frac{5}{2}x = -\frac{5}{4}$$

$$x = \frac{1}{2}$$

$$y = 0$$

Directrix: $y = 0$ and the point of intersection $\left(\frac{1}{2}, 0\right)$ lies on this line.

87. $y = x - x^2$

$$\frac{dy}{dx} = 1 - 2x$$

At (x_1, y_1) on the mountain, $m = 1 - 2x_1$. Also, $m = \dfrac{y_1 - 1}{x_1 + 1}$.

$$\frac{y_1 - 1}{x_1 + 1} = 1 - 2x_1$$

$$(x_1 - x_1^2) - 1 = (1 - 2x_1)(x_1 + 1)$$

$$-x_1^2 + x_1 - 1 = -2x_1^2 - x_1 + 1$$

$$x_1^2 + 2x_1 - 2 = 0$$

$$x_1 = \frac{-2 \pm \sqrt{2^2 - 4(1)(-2)}}{2(1)} = \frac{-2 \pm 2\sqrt{3}}{2} = -1 \pm \sqrt{3}$$

Choosing the positive value for x_1, we have $x_1 = -1 + \sqrt{3}$.

$$m = 1 - 2\left(-1 + \sqrt{3}\right) = 3 - 2\sqrt{3}$$

$$m = \frac{0 - 1}{x_0 + 1} = -\frac{1}{x_0 + 1}$$

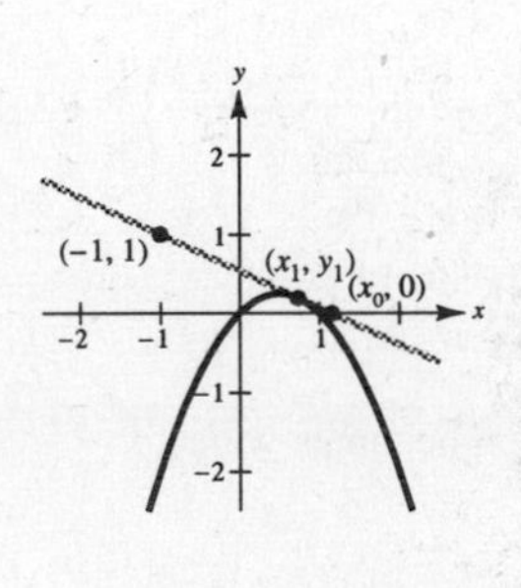

Thus, $-\dfrac{1}{x_0 + 1} = 3 - 2\sqrt{3}$

$$\frac{-1}{3 - 2\sqrt{3}} = x_0 + 1$$

$$\frac{3 + 2\sqrt{3}}{3} - 1 = x_0$$

$$\frac{2\sqrt{3}}{3} = x_0.$$

The closest the receiver can be to the hill is $\left(2\sqrt{3}/3\right) - 1 \approx 0.155$.

89. Parabola

Vertex: $(0, 4)$

$$x^2 = 4p(y - 4)$$
$$4^2 = 4p(0 - 4)$$
$$p = -1$$
$$x^2 = -4(y - 4)$$
$$y = 4 - \frac{x^2}{4}$$

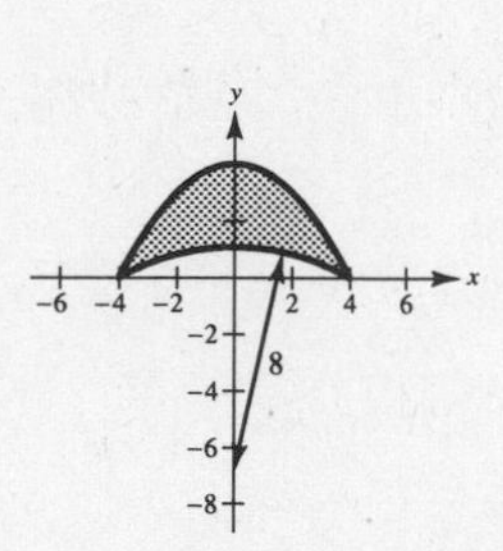

Circle

Center: $(0, k)$

Radius: 8

$$x^2 + (y - k)^2 = 64$$
$$4^2 + (0 - k)^2 = 64$$
$$k^2 = 48$$
$$k = -4\sqrt{3} \quad \text{(Center is on the negative } y\text{-axis.)}$$
$$x^2 + \left(y + 4\sqrt{3}\right)^2 = 64$$
$$y = -4\sqrt{3} \pm \sqrt{64 - x^2}$$

Since the y-value is positive when $x = 0$, we have $y = -4\sqrt{3} + \sqrt{64 - x^2}$.

$$A = 2\int_0^4 \left[\left(4 - \frac{x^2}{4}\right) - \left(-4\sqrt{3} + \sqrt{64 - x^2}\right)\right] dx$$
$$= 2\left[4x - \frac{x^3}{12} + 4\sqrt{3}x - \frac{1}{2}\left(x\sqrt{64 - x^2} + 64 \arcsin \frac{x}{8}\right)\right]_0^4$$
$$= 2\left[16 - \frac{64}{12} + 16\sqrt{3} - 2\sqrt{48} - 32 \arcsin \frac{1}{2}\right]$$
$$= \frac{16\left(4 + 3\sqrt{3} - 2\pi\right)}{3} \approx 15.536 \text{ square feet}$$

91. (a) Assume that $y = ax^2$.

$$20 = a(60)^2 \Rightarrow a = \frac{2}{360} = \frac{1}{180} \Rightarrow y = \frac{1}{180}x^2$$

(b) $f(x) = \frac{1}{180}x^2, f'(x) = \frac{1}{90}x$

$$S = 2\int_0^{60} \sqrt{1 + \left(\frac{1}{90}x\right)^2}\, dx = \frac{2}{90}\int_0^{60} \sqrt{90^2 + x^2}\, dx$$
$$= \frac{2}{90}\frac{1}{2}\left[x\sqrt{90^2 + x^2} + 90^2 \ln\left|x + \sqrt{90^2 + x^2}\right|\right]_0^{60} \quad \text{(formula 26)}$$
$$= \frac{1}{90}\left[60\sqrt{11{,}700} + 90^2 \ln\left(60 + \sqrt{11{,}700}\right) - 90^2 \ln 90\right]$$
$$= \frac{1}{90}\left[1800\sqrt{13} + 90^2 \ln\left(60 + 30\sqrt{13}\right) - 90^2 \ln 90\right]$$
$$= 20\sqrt{13} + 90 \ln\left(\frac{60 + 30\sqrt{13}}{90}\right)$$
$$= 10\left[2\sqrt{13} + 9 \ln\left(\frac{2 + \sqrt{13}}{3}\right)\right] \approx 128.4 \text{ m}$$

93. $x^2 = 4py, p = \frac{1}{4}, \frac{1}{2}, 1, \frac{3}{2}, 2$

As p increases, the graph becomes wider.

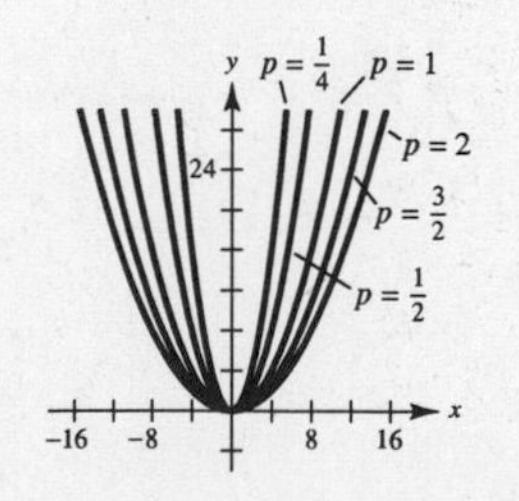

95. 

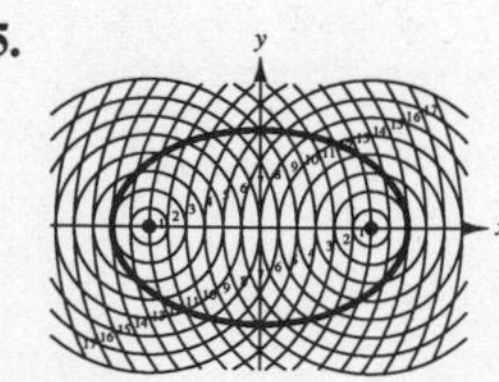

97. $a = \dfrac{5}{2}, b = 2, c = \sqrt{\left(\dfrac{5}{2}\right)^2 - (2)^2} = \dfrac{3}{2}$

The tacks should be placed 1.5 feet from the center. The string should be $2a = 5$ feet long.

99. $$e = \frac{c}{a}$$

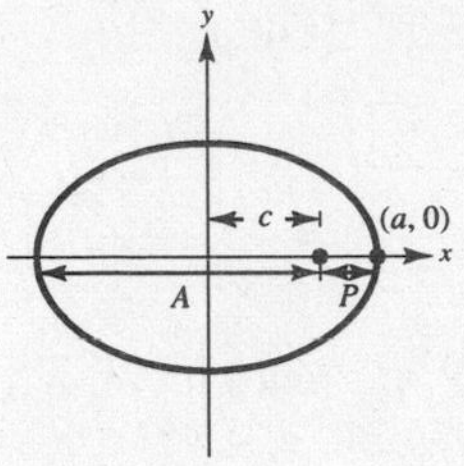

$$A + P = 2a$$

$$a = \frac{A + P}{2}$$

$$c = a - P = \frac{A + P}{2} - P = \frac{A - P}{2}$$

$$e = \frac{c}{a} = \frac{(A - P)/2}{(A + P)/2} = \frac{A - P}{A + P}$$

101. $e = \dfrac{A - P}{A + P} = \dfrac{35.34au - 0.59au}{35.34au + 0.59au} \approx 0.9672$

103. $$\frac{x^2}{10^2} + \frac{y^2}{5^2} = 1$$

$$\frac{2x}{10^2} + \frac{2yy'}{5^2} = 0$$

$$y' = \frac{-5^2x}{10^2y} = \frac{-x}{4y}$$

At $(-8, 3)$: $y' = \dfrac{8}{12} = \dfrac{2}{3}$

The equation of the tangent line is $y - 3 = \frac{2}{3}(x + 8)$. It will cross the y-axis when $x = 0$ and $y = \frac{2}{3}(8) + 3 = \frac{25}{3}$.

105. $$16x^2 + 9y^2 + 96x + 36y + 36 = 0$$

$$32x + 18yy' + 96 + 36y' = 0$$

$$y'(18y + 36) = -(32x + 96)$$

$$y' = \frac{-(32x + 96)}{18y + 36}$$

$y' = 0$ when $x = -3$. y' is undefined when $y = -2$.

At $x = -3$, $y = 2$ or -6.

Endpoints of major axis: $(-3, 2), (-3, -6)$

At $y = -2$, $x = 0$ or -6.

Endpoints of minor axis: $(0, -2), (-6, -2)$

Note: Equation of ellipse is $\dfrac{(x + 3)^2}{9} + \dfrac{(y + 2)^2}{16} = 1$

107. (a) $A = 4\displaystyle\int_0^2 \frac{1}{2}\sqrt{4 - x^2}\,dx = \left[x\sqrt{4 - x^2} + 4\arcsin\left(\frac{x}{2}\right)\right]_0^2 = 2\pi$ [or, $A = \pi ab = \pi(2)(1) = 2\pi$]

(b) **Disk:** $$V = 2\pi\int_0^2 \frac{1}{4}(4 - x^2)\,dx = \frac{1}{2}\pi\left[4x - \frac{1}{3}x^3\right]_0^2 = \frac{8\pi}{3}$$

$$y = \frac{1}{2}\sqrt{4 - x^2}$$

$$y' = \frac{-x}{2\sqrt{4 - x^2}}$$

$$\sqrt{1 + (y')^2} = \sqrt{1 + \frac{x^2}{16 - 4x^2}} = \sqrt{\frac{16 - 3x^2}{4y}}$$

$$S = 2(2\pi)\int_0^2 y\left(\frac{\sqrt{16 - 3x^2}}{4y}\right)dx = \frac{\pi}{2\sqrt{3}}\left[\sqrt{3}x\sqrt{16 - 3x^2} + 16\arcsin\left(\frac{\sqrt{3}x}{4}\right)\right]_0^2 = \frac{2\pi}{9}(9 + 4\sqrt{3}\pi) \approx 21.48$$

—CONTINUED—

107. —CONTINUED—

(c) **Shell:** $V = 2\pi\int_0^2 x\sqrt{4-x^2}\,dx = -\pi\int_0^2 -2x(4-x^2)^{1/2}\,dx = -\frac{2\pi}{3}\Big[(4-x^2)^{3/2}\Big]_0^2 = \frac{16\pi}{3}$

$$x = 2\sqrt{1-y^2}$$

$$x' = \frac{-2y}{\sqrt{1-y^2}}$$

$$\sqrt{1+(x')^2} = \sqrt{1+\frac{4y^2}{1-y^2}} = \frac{\sqrt{1+3y^2}}{\sqrt{1-y^2}}$$

$$S = 2(2\pi)\int_0^1 2\sqrt{1-y^2}\,\frac{\sqrt{1+3y^2}}{\sqrt{1-y^2}}\,dy = 8\pi\int_0^1 \sqrt{1+3y^2}\,dy$$

$$= \frac{8\pi}{2\sqrt{3}}\Big[\sqrt{3}y\sqrt{1+3y^2} + \ln\left|\sqrt{3}y + \sqrt{1+3y^2}\right|\Big]_0^1 = \frac{4\pi}{3}\left|6 + \sqrt{3}\ln\left(2+\sqrt{3}\right)\right| \approx 34.69$$

109. From Example 5,

$$C = 4a\int_0^{\pi/2}\sqrt{1-e^2\sin^2\theta}\,d\theta$$

For $\frac{x^2}{25}+\frac{y^2}{49} = 1$, we have

$a = 7, b = 5, c = \sqrt{49-25} = 2\sqrt{6}, e = \frac{c}{a} = \frac{2\sqrt{6}}{7}$.

$$C = 4(7)\int_0^{\pi/2}\sqrt{1-\frac{24}{49}\sin^2\theta}\,d\theta$$

$$\approx 28(1.3558) \approx 37.9614$$

111. Area circle $= \pi r^2 = 100\pi$

Area ellipse $= \pi ab = \pi a(10)$

$$2(100\pi) = 10\pi a \Rightarrow a = 20$$

Hence, the length of the major axis is $2a = 40$.

113. The transverse axis is horizontal since (2, 2) and (10, 2) are the foci (see definition of hyperbola).

Center: (6, 2)

$c = 4, 2a = 6, b^2 = c^2 - a^2 = 7$

Therefore, the equation is

$$\frac{(x-6)^2}{9} - \frac{(y-2)^2}{7} = 1.$$

115. $2a = 10 \Rightarrow a = 5$

$c = 6 \Rightarrow b = \sqrt{11}$

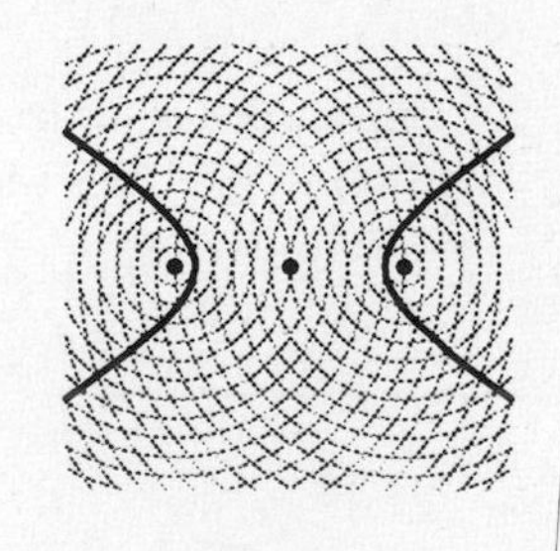

117. Time for sound of bullet hitting target to reach (x, y): $\frac{2c}{v_m} + \frac{\sqrt{(x-c)^2+y^2}}{v_s}$

Time for sound of rifle to reach (x, y): $\frac{\sqrt{(x+c)^2+y^2}}{v_s}$

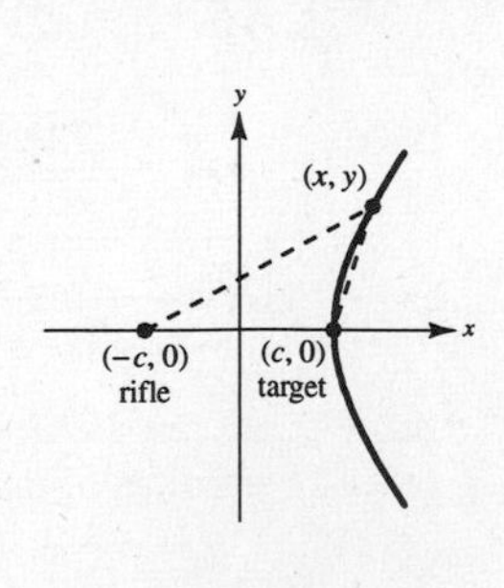

Since the times are the same, we have: $\frac{2c}{v_m} + \frac{\sqrt{(x-c)^2+y^2}}{v_s} = \frac{\sqrt{(x+c)^2+y^2}}{v_s}$

$$\frac{4c^2}{v_m^2} + \frac{4c}{v_m v_s}\sqrt{(x-c)^2+y^2} + \frac{(x-c)^2+y^2}{v_s^2} = \frac{(x+c)^2+y^2}{v_s^2}$$

$$\sqrt{(x-c)^2+y^2} = \frac{v_m^2 x - v_s^2 c}{v_s v_m}$$

$$\left(1-\frac{v_m^2}{v_s^2}\right)x^2 + y^2 = \left(\frac{v_s^2}{v_m^2}-1\right)c^2$$

$$\frac{x^2}{c^2v_s^2/v_m^2} - \frac{y^2}{c^2(v_m^2-v_s^2)/v_m^2} = 1$$

119. The point (x, y) lies on the line between $(0, 10)$ and $(10, 0)$. Thus, $y = 10 - x$. The point also lies on the hyperbola $(x^2/36) - (y^2/64) = 1$. Using substitution, we have:

$$\frac{x^2}{36} - \frac{(10 - x)^2}{64} = 1$$

$$16x^2 - 9(10 - x)^2 = 576$$

$$7x^2 + 180x - 1476 = 0$$

$$x = \frac{-180 \pm \sqrt{180^2 - 4(7)(-1476)}}{2(7)} = \frac{-180 \pm 192\sqrt{2}}{14} = \frac{-90 \pm 96\sqrt{2}}{7}$$

Choosing the positive value for x we have:

$$x = \frac{-90 + 96\sqrt{2}}{7} \approx 6.538 \text{ and } y = \frac{160 - 96\sqrt{2}}{7} \approx 3.462$$

121. $$\frac{x^2}{a^2} + \frac{2y^2}{b^2} = 1 \Rightarrow \frac{2y^2}{b^2} = 1 - \frac{x^2}{a^2},\ c^2 = a^2 - b^2$$

$$\frac{x^2}{a^2 - b^2} - \frac{2y^2}{b^2} = 1 \Rightarrow \frac{2y^2}{b^2} = \frac{x^2}{a^2 - b^2} - 1$$

$$1 - \frac{x^2}{a^2} = \frac{x^2}{a^2 - b^2} - 1 \Rightarrow 2 = x^2\left(\frac{1}{a^2} + \frac{1}{a^2 - b^2}\right)$$

$$x^2 = \frac{2a^2(a^2 - b^2)}{2a^2 - b^2} \Rightarrow x = \pm\frac{\sqrt{2}a\sqrt{a^2 - b^2}}{\sqrt{2a^2 - b^2}} = \pm\frac{\sqrt{2}ac}{\sqrt{2a^2 - b^2}}$$

$$\frac{2y^2}{b^2} = 1 - \frac{1}{a^2}\left(\frac{2a^2c^2}{2a^2 - b^2}\right) \Rightarrow \frac{2y^2}{b^2} = \frac{b^2}{2a^2 - b^2}$$

$$y^2 = \frac{b^4}{2(2a^2 - b^2)} \Rightarrow y = \pm\frac{b^2}{\sqrt{2}\sqrt{2a^2 - b^2}}$$

There are four points of intersection: $\left(\frac{\sqrt{2}ac}{\sqrt{2a^2 - b^2}}, \pm\frac{b^2}{\sqrt{2}\sqrt{2a^2 - b^2}}\right), \left(-\frac{\sqrt{2}ac}{\sqrt{2a^2 - b^2}}, \pm\frac{b^2}{\sqrt{2}\sqrt{2a^2 - b^2}}\right)$

$$\frac{x^2}{a^2} + \frac{2y^2}{b^2} = 1 \Rightarrow \frac{2x}{a^2} + \frac{4yy'}{b^2} = 0 \Rightarrow y'_e = -\frac{b^2x}{2a^2y}$$

$$\frac{x^2}{a^2 - b^2} - \frac{2y^2}{b^2} = 1 \Rightarrow \frac{2x}{c^2} - \frac{4yy'}{b^2} = 0 \Rightarrow y'_h = \frac{b^2x}{2c^2y}$$

At $\left(\frac{\sqrt{2}ac}{\sqrt{2a^2 - b^2}}, \frac{b^2}{\sqrt{2}\sqrt{2a^2 - b^2}}\right)$, the slopes of the tangent lines are:

$$y'_e = \frac{-b^2\left(\frac{\sqrt{2}ac}{\sqrt{2a^2 - b^2}}\right)}{2a^2\left(\frac{b^2}{\sqrt{2}\sqrt{2a^2 - b^2}}\right)} = -\frac{c}{a} \quad \text{and} \quad y'_h = \frac{b^2\left(\frac{\sqrt{2}ac}{\sqrt{2a^2 - b^2}}\right)}{2c^2\left(\frac{b^2}{\sqrt{2}\sqrt{2a^2 - b^2}}\right)} = \frac{a}{c}$$

Since the slopes are negative reciprocals, the tangent lines are perpendicular. Similarly, the curves are perpendicular at the other three points of intersection.

123. False. See the definition of a parabola.

125. True

127. False. $y^2 - x^2 + 2x + 2y = 0$ yields two intersecting lines.

129. True

Section 9.2 Plane Curves and Parametric Equations

1. $x = \sqrt{t}$, $y = 1 - t$

(a)

t	0	1	2	3	4
x	0	1	$\sqrt{2}$	$\sqrt{3}$	2
y	1	0	-1	-2	-3

(b)

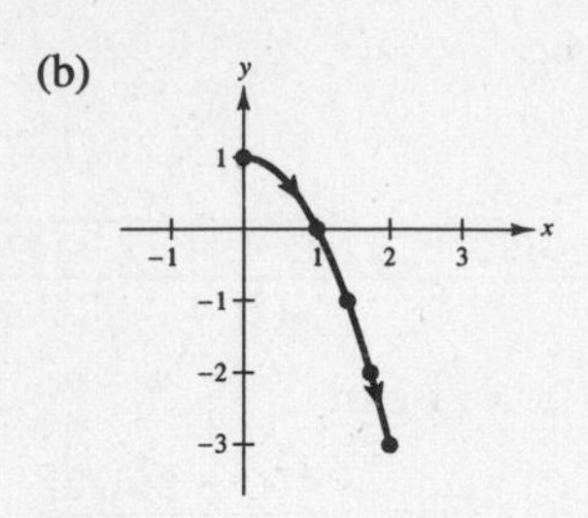

(c)

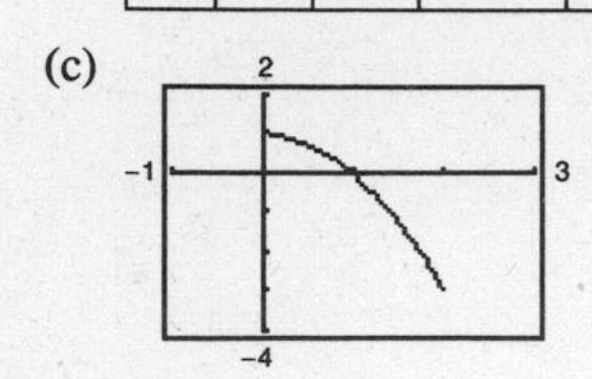

(d) $x^2 = t$

$y = 1 - x^2, x \geq 0$

3. $x = 3t - 1$

$y = 2t + 1$

$y = 2\left(\dfrac{x + 1}{3}\right) + 1$

$2x - 3y + 5 = 0$

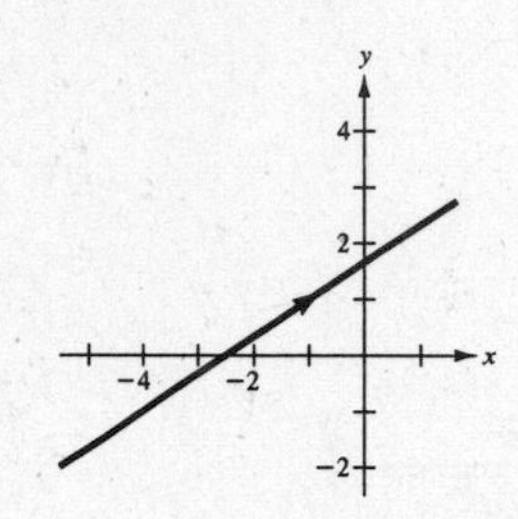

5. $x = t + 1$

$y = t^2$

$y = (x - 1)^2$

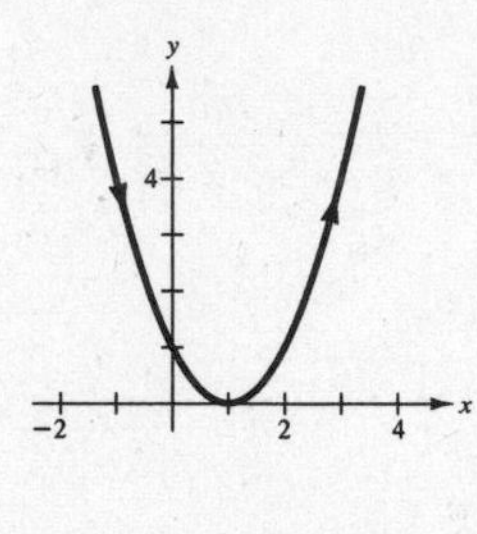

7. $x = t^3$

$y = \frac{1}{2}t^2$

$x = t^3$ implies $t = x^{1/3}$

$y = \frac{1}{2}x^{2/3}$

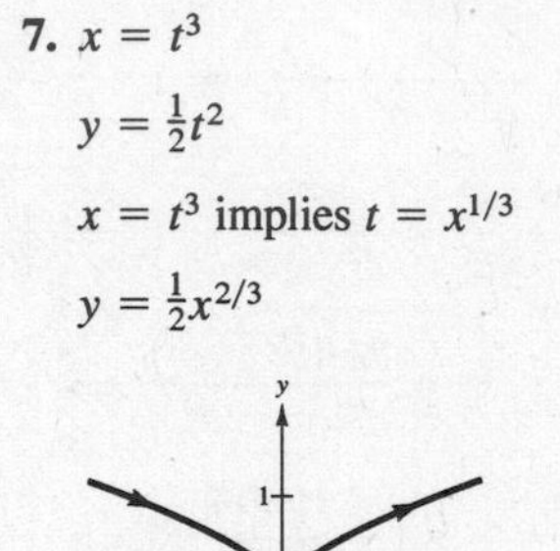

9. $x = \sqrt{t}, t \geq 0$

$y = t - 2$

$y = x^2 - 2, x \geq 0$

11. $x = t - 1$

$y = \dfrac{t}{t - 1}$

$y = \dfrac{x + 1}{x}$

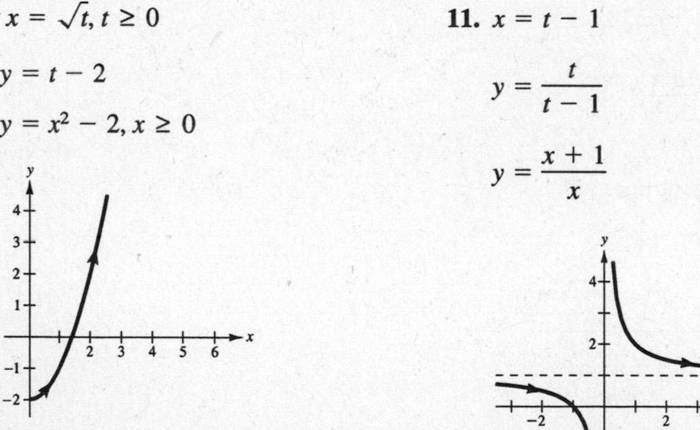

13. $x = 2t$

$y = |t - 2|$

$y = \left|\dfrac{x}{2} - 2\right| = \dfrac{|x - 4|}{2}$

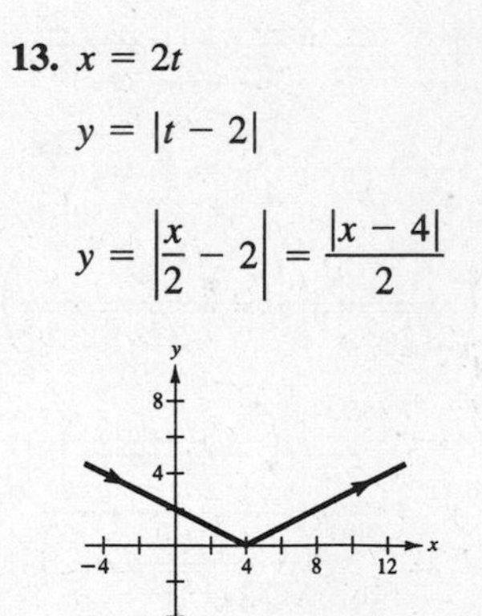

15. $x = e^t, x > 0$

$y = e^{3t} + 1$

$y = x^3 + 1, x > 0$

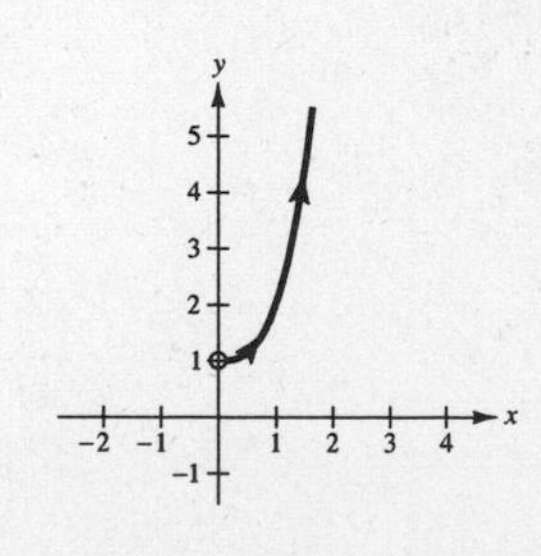

17. $x = \sec\theta$

$y = \cos\theta$

$0 \le \theta < \frac{\pi}{2}, \frac{\pi}{2} < \theta \le \pi$

$xy = 1$

$y = \frac{1}{x}$

$|x| \ge 1, \ |y| \le 1$

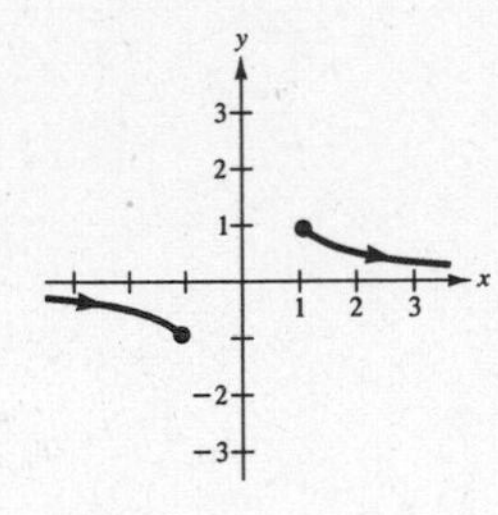

19. $x = 3\cos\theta, \ y = 3\sin\theta$

Squaring both equations and adding, we have

$x^2 + y^2 = 9.$

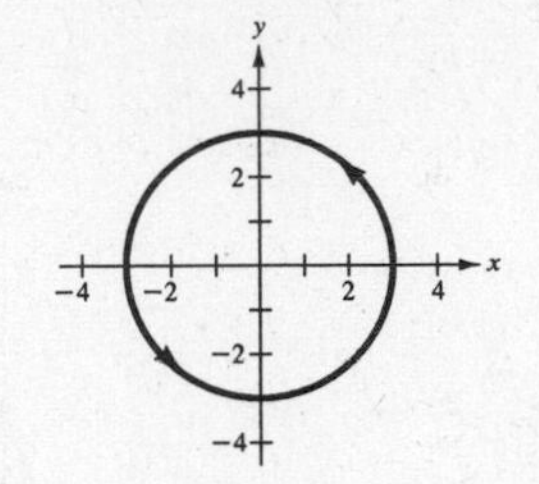

21. $x = 4\sin 2\theta$

$y = 2\cos 2\theta$

$\frac{x^2}{16} = \sin^2 2\theta$

$\frac{y^2}{4} = \cos^2 2\theta$

$\frac{x^2}{16} + \frac{y^2}{4} = 1$

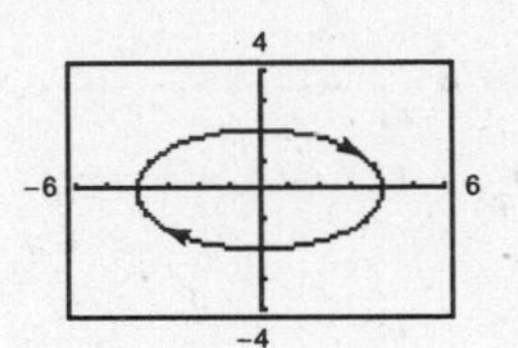

23. $x = 4 + 2\cos\theta$

$y = -1 + \sin\theta$

$\frac{(x-4)^2}{4} = \cos^2\theta$

$\frac{(y+1)^2}{1} = \sin^2\theta$

$\frac{(x-4)^2}{4} + \frac{(y+1)^2}{1} = 1$

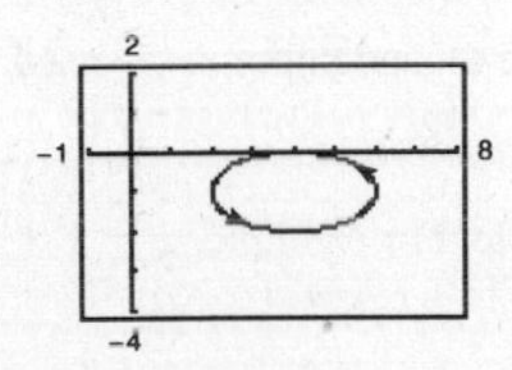

25. $x = 4 + 2\cos\theta$

$y = -1 + 4\sin\theta$

$\frac{(x-4)^2}{4} = \cos^2\theta$

$\frac{(y+1)^2}{16} = \sin^2\theta$

$\frac{(x-4)^2}{4} + \frac{(y+1)^2}{16} = 1$

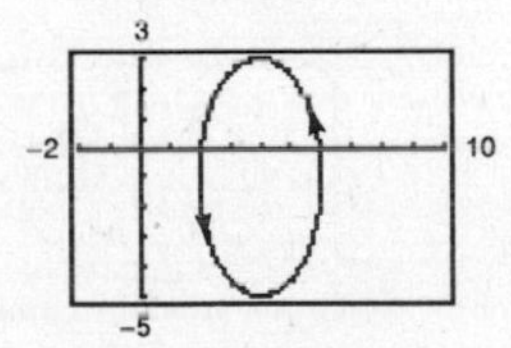

27. $x = 4\sec\theta$

$y = 3\tan\theta$

$\frac{x^2}{16} = \sec^2\theta$

$\frac{y^2}{9} = \tan^2\theta$

$\frac{x^2}{16} - \frac{y^2}{9} = 1$

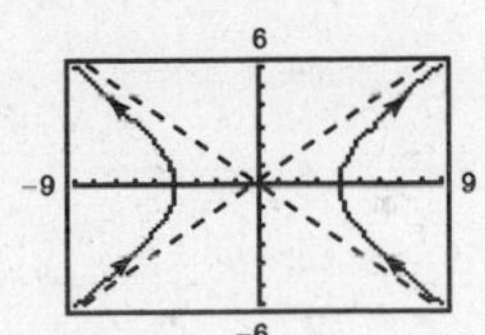

29. $x = t^3$

$y = 3\ln t$

$y = 3\ln\sqrt[3]{x} = \ln x$

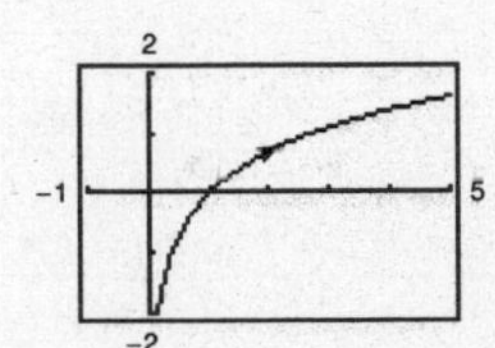

31. $x = e^{-t}$

$y = e^{3t}$

$e^t = \frac{1}{x}$

$e^t = \sqrt[3]{y}$

$\sqrt[3]{y} = \frac{1}{x}$

$y = \frac{1}{x^3}$

$x > 0$

$y > 0$

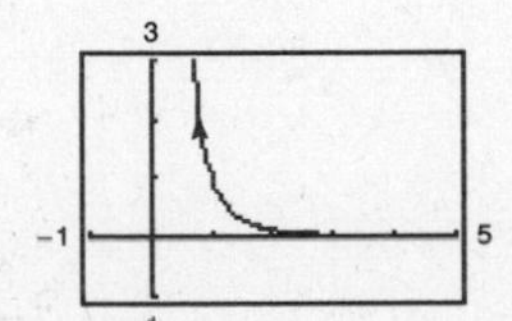

33. By eliminating the parameters in (a) – (d), we get $y = 2x + 1$. They differ from each other in orientation and in restricted domains. These curves are all smooth except for (b).

(a) $x = t,\ y = 2t + 1$

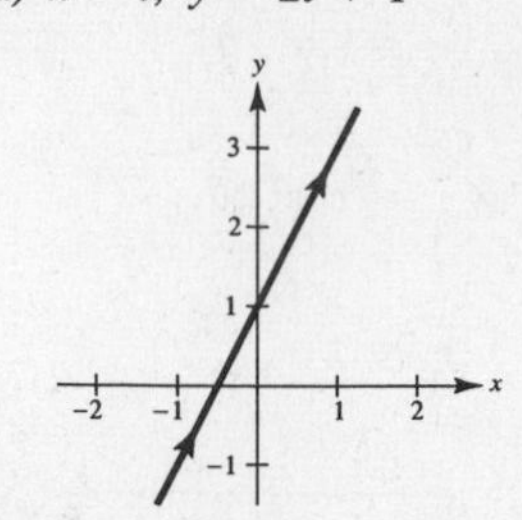

(b) $x = \cos\theta \qquad y = 2\cos\theta + 1$

$-1 \le x \le 1 \qquad -1 \le y \le 3$

$$\frac{dx}{d\theta} = \frac{dy}{d\theta} = 0 \text{ when } \theta = 0, \pm\pi, \pm 2\pi, \ldots$$

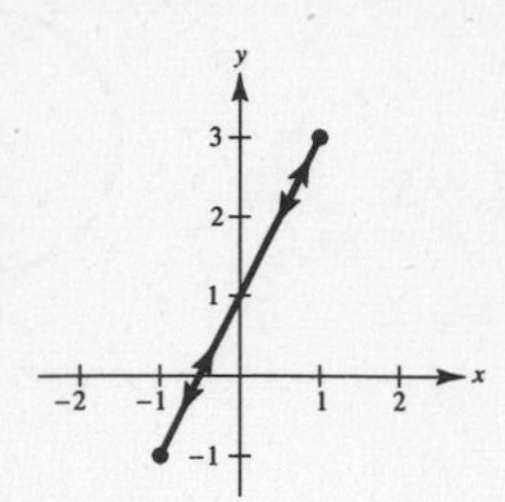

(c) $x = e^{-t} \qquad y = 2e^{-t} + 1$

$x > 0 \qquad y > 1$

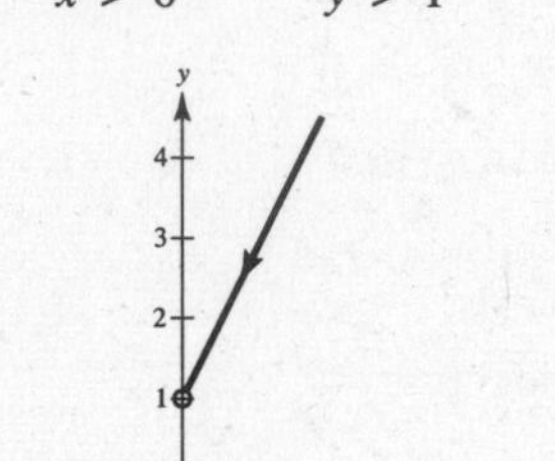

(d) $x = e^{t} \qquad y = 2e^{t} + 1$

$x > 0 \qquad y > 1$

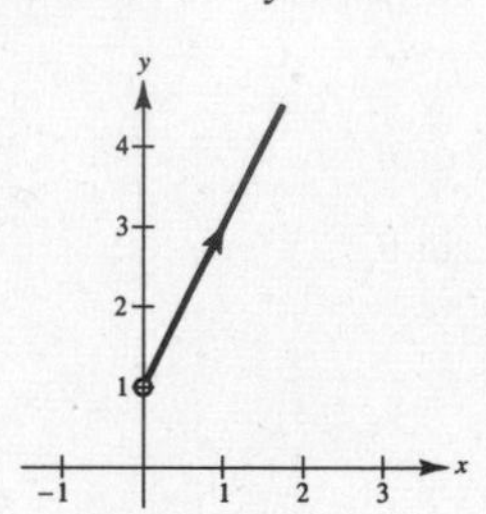

35. The curves are identical on $0 < \theta < \pi$. They are both smooth. Represent $y = 2(1 - x^2)$

37. (a)

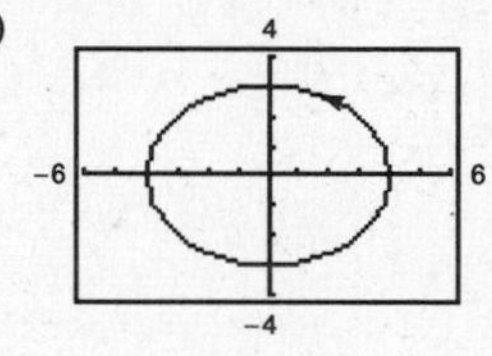

(b) The orientation of the second curve is reversed.

(c) The orientation will be reversed.

(d) Many answers possible. For example, $x = 1 + t$, $y = 1 + 2t$, and $x = 1 - t$, $x = 1 - 2t$.

39.

$$x = x_1 + t(x_2 - x_1)$$
$$y = y_1 + t(y_2 - y_1)$$
$$\frac{x - x_1}{x_2 - x_1} = t$$
$$y = y_1 + \left(\frac{x - x_1}{x_2 - x_1}\right)(y_2 - y_1)$$
$$y - y_1 = \frac{y_2 - y_1}{x_2 - x_1}(x - x_1)$$
$$y - y_1 = m(x - x_1)$$

41.

$$x = h + a\cos\theta$$
$$y = k + b\sin\theta$$
$$\frac{x - h}{a} = \cos\theta$$
$$\frac{y - k}{b} = \sin\theta$$
$$\frac{(x - h)^2}{a^2} + \frac{(y - k)^2}{b^2} = 1$$

43. From Exercise 39 we have

$x = 5t$

$y = -2t.$

Solution not unique

45. From Exercise 40 we have

$x = 2 + 4\cos\theta$

$y = 1 + 4\sin\theta.$

Solution not unique

47. From Exercise 41 we have

$a = 5, c = 4 \Rightarrow b = 3$

$x = 5\cos\theta$

$y = 3\sin\theta.$

Center: $(0, 0)$
Solution not unique

49. From Exercise 42 we have

$a = 4, c = 5 \longrightarrow b = 3$

$x = 4 \sec \theta$

$y = 3 \tan \theta.$

Center: $(0, 0)$
Solution not unique

51. $y = 3x - 2$

Example

$x = t, \quad y = 3t - 2$

$x = t - 3, \quad y = 3t - 11$

53. $y = x^3$

Example

$x = t, \quad y = t^3$

$x = \sqrt[3]{t}, \quad y = t$

$x = \tan t, \quad y = \tan^3 t$

55. $x = 2(\theta - \sin \theta)$

$y = 2(1 - \cos \theta)$

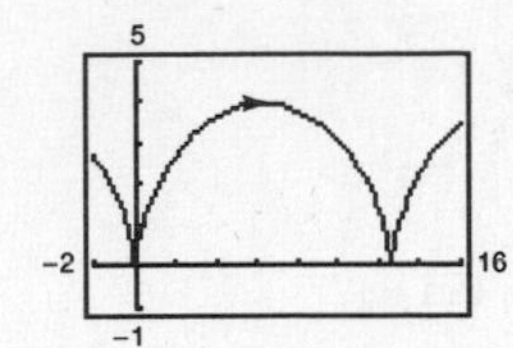

Not smooth at $\theta = 2n\pi$

57. $x = \theta - \frac{3}{2} \sin \theta$

$y = 1 - \frac{3}{2} \cos \theta$

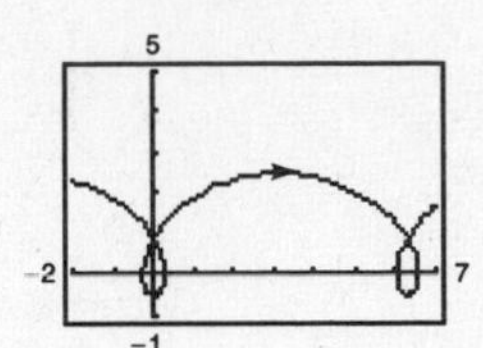

59. $x = 3 \cos^3 \theta$

$y = 3 \sin^3 \theta$

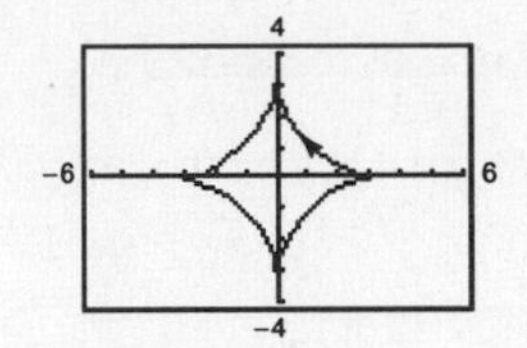

Not smooth at $(x, y) = (\pm 3, 0)$ and $(0, \pm 3)$, or $\theta = \frac{1}{2}n\pi$.

61. $x = 2 \cot \theta$

$y = 2 \sin^2 \theta$

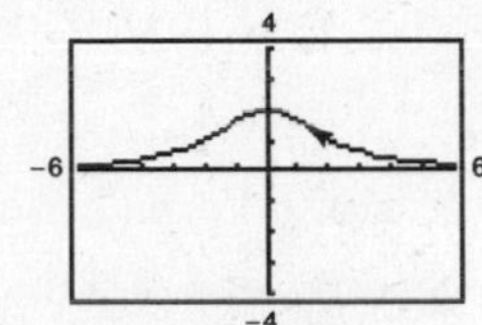

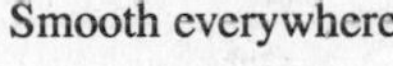
Smooth everywhere

63. See definition on page 665.

65. A plane curve C, represented by $x = f(t)$, $y = g(t)$, is smooth if f' and g' are continuous and not simultaneously 0. See page 670.

67. $x = 4 \cos \theta$

$y = 2 \sin 2\theta$

Matches (d)

69. $x = \cos \theta + \theta \sin \theta$

$y = \sin \theta - \theta \cos \theta$

Matches (b)

71. When the circle has rolled θ radians, we know that the center is at $(a\theta, a)$.

$$\sin \theta = \sin(180^\circ - \theta) = \frac{|AC|}{b} = \frac{|BD|}{b} \quad \text{or} \quad |BD| = b \sin \theta$$

$$\cos \theta = -\cos(180^\circ - \theta) = \frac{|AP|}{-b} \quad \text{or} \quad |AP| = -b \cos \theta$$

Therefore, $x = a\theta - b \sin \theta$ and $y = a - b \cos \theta$.

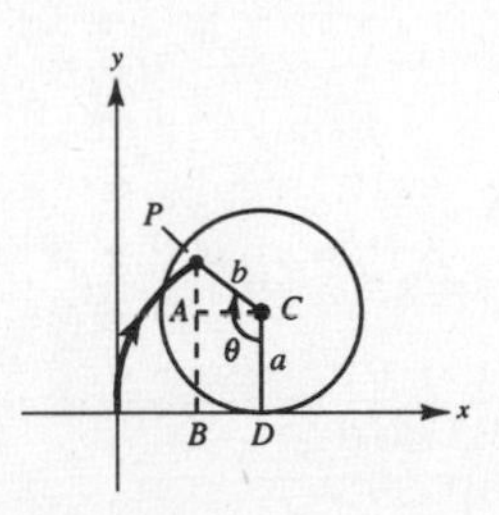

73. False

$x = t^2 \Rightarrow x \geq 0$

$x = t^2 \Rightarrow y \geq 0$

The graph of the parametric equations is only a portion of the line $y = x$.

75. (a) $100 \text{ mi/hr} = \dfrac{(100)(5280)}{3600} = \dfrac{440}{3} \text{ ft/sec}$

$$x = (v_0 \cos \theta)t = \left(\frac{440}{3} \cos \theta\right)t$$

$$y = h + (v_0 \sin \theta)t - 16t^2$$

$$= 3 + \left(\frac{440}{3} \sin \theta\right)t - 16t^2$$

(b)

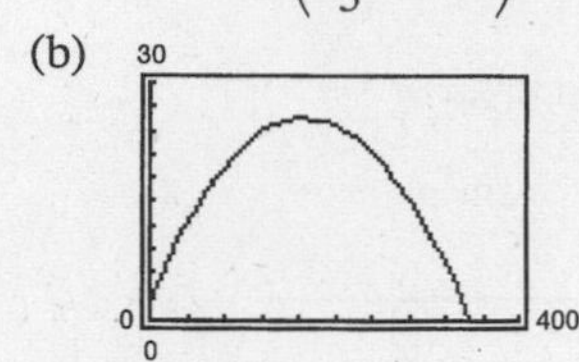

It is not a home run—when $x = 400$, $y < 10$.

(c)

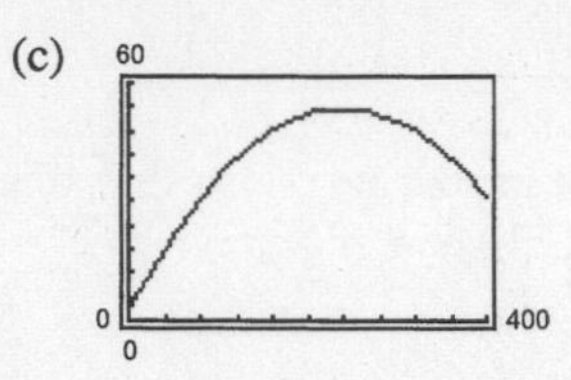

Yes, it's a home run when $x = 400$, $y > 10$.

(d) We need to find the angle θ (and time t) such that

$$x = \left(\frac{440}{3} \cos \theta\right)t = 400$$

$$y = 3 + \left(\frac{440}{3} \sin \theta\right)t - 16t^2 = 10.$$

From the first equation $t = 1200/440 \cos \theta$. Substituting into the second equation,

$$10 = 3 + \left(\frac{440}{3} \sin \theta\right)\left(\frac{1200}{440 \cos \theta}\right) - 16\left(\frac{1200}{440 \cos \theta}\right)^2$$

$$7 = 400 \tan \theta - 16\left(\frac{120}{44}\right)^2 \sec^2 \theta$$

$$= 400 \tan \theta - 16\left(\frac{120}{44}\right)^2 (\tan^2 \theta + 1).$$

We now solve the quadratic for $\tan \theta$:

$$16\left(\frac{120}{44}\right)^2 \tan^2 \theta - 400 \tan \theta + 7 + 16\left(\frac{120}{44}\right)^2 = 0$$

$\tan \theta \approx 0.35185 \implies \theta \approx 19.4°$

Section 9.3 Parametric Equations and Calculus

1. $\dfrac{dy}{dx} = \dfrac{dy/dt}{dx/dt} = \dfrac{-4}{2t} = \dfrac{-2}{t}$

3. $\dfrac{dy}{dx} = \dfrac{dy/d\theta}{dx/d\theta} = \dfrac{-2 \cos \theta \sin \theta}{2 \sin \theta \cos \theta} = -1$

$\left[\text{Note: } x + y = 1 \implies y = 1 - x \text{ and } \dfrac{dy}{d\theta} = -1\right]$

5. $x = 2t,\ y = 3t - 1$

$$\frac{dy}{dx} = \frac{dy/dt}{dx/dt} = \frac{3}{2}$$

$$\frac{d^2y}{dx^2} = 0 \text{ Line}$$

7. $x = t + 1,\ y = t^2 + 3t$

$$\frac{dy}{dx} = \frac{2t + 3}{1} = 1 \text{ when } t = -1.$$

$$\frac{d^2y}{dx^2} = 2 \text{ concave upwards}$$

9. $x = 2 \cos \theta,\ y = 2 \sin \theta$

$$\frac{dy}{dx} = \frac{2 \cos \theta}{-2 \sin \theta} = -\cot \theta = -1 \text{ when } \theta = \frac{\pi}{4}.$$

$$\frac{d^2y}{dx^2} = \frac{\csc^2 \theta}{-2 \sin \theta} = \frac{-\csc^3 \theta}{2} = -\sqrt{2} \text{ when } \theta = \frac{\pi}{4}.$$

concave downward

11. $x = 2 + \sec \theta,\ y = 1 + 2 \tan \theta$

$$\frac{dy}{dx} = \frac{2 \sec^2 \theta}{\sec \theta \tan \theta}$$

$$= \frac{2 \sec \theta}{\tan \theta} = 2 \csc \theta = 4 \text{ when } \theta = \frac{\pi}{6}.$$

$$\frac{d^2y}{dx^2} = \frac{-2 \csc \theta \cot \theta}{\sec \theta \tan \theta}$$

$$= -2 \cot^3 \theta = -6\sqrt{3} \text{ when } \theta = \frac{\pi}{6}.$$

concave downward

13. $x = \cos^3\theta,\ y = \sin^3\theta$

$$\frac{dy}{dx} = \frac{3\sin^2\theta\cos\theta}{-3\cos^2\theta\sin\theta}$$

$$= -\tan\theta = -1 \text{ when } \theta = \frac{\pi}{4}.$$

$$\frac{d^2y}{dx^2} = \frac{-\sec^2\theta}{-3\cos^2\theta\sin\theta} = \frac{1}{3\cos^4\theta\sin\theta}$$

$$= \frac{\sec^4\theta\csc\theta}{3} = \frac{4\sqrt{2}}{3} \text{ when } \theta = \frac{\pi}{4}.$$

concave upward

15. $x = 2\cot\theta,\ y = 2\sin^2\theta$

$$\frac{dy}{dx} = \frac{4\sin\theta\cos\theta}{-2\csc^2\theta} = -2\sin^3\theta\cos\theta$$

At $\left(-\frac{2}{\sqrt{3}}, \frac{3}{2}\right)$, $\theta = \frac{2\pi}{3}$, and $\frac{dy}{dx} = \frac{3\sqrt{3}}{8}$.

Tangent line: $y - \frac{3}{2} = \frac{3\sqrt{3}}{8}\left(x + \frac{2}{\sqrt{3}}\right)$

$$3\sqrt{3}x - 8y + 18 = 0$$

At $(0, 2)$, $\theta = \frac{\pi}{2}$, and $\frac{dy}{dx} = 0$.

Tangent line: $y - 2 = 0$

At $\left(2\sqrt{3}, \frac{1}{2}\right)$, $\theta = \frac{\pi}{6}$, and $\frac{dy}{dx} = -\frac{\sqrt{3}}{8}$.

Tangent line: $y - \frac{1}{2} = -\frac{\sqrt{3}}{8}(x - 2\sqrt{3})$

$$\sqrt{3}x + 8y - 10 = 0$$

17. $x = 2t,\ y = t^2 - 1,\ t = 2$

(a)

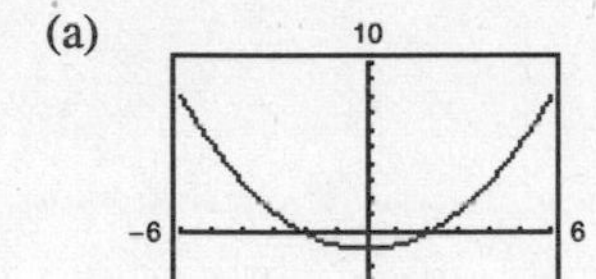

(b) At $t = 2$, $(x, y) = (4, 3)$, and

$$\frac{dx}{dt} = 2,\ \frac{dy}{dt} = 4,\ \frac{dy}{dx} = 2$$

(c) $\frac{dy}{dx} = 2$. At $(4, 3)$, $y - 3 = 2(x - 4)$

$$y = 2x - 5$$

(d)

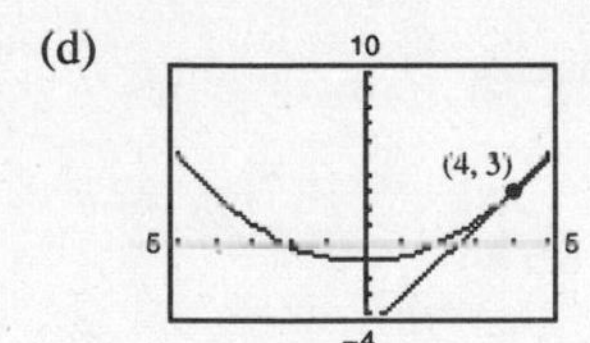

19. $x = t^2 - t + 2,\ y = t^3 - 3t,\ t = -1$

(a)

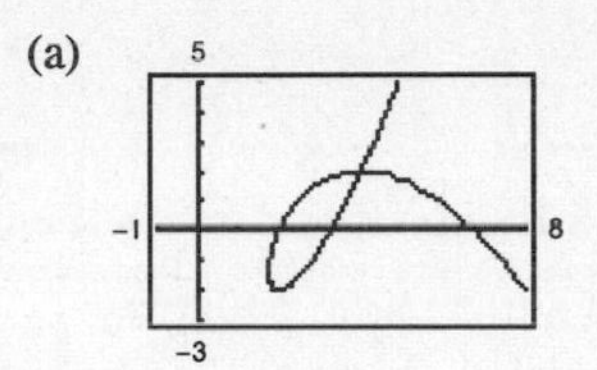

(b) At $t = -1$, $(x, y) = (4, 2)$, and

$$\frac{dx}{dt} = 3,\ \frac{dy}{dt} = 0,\ \frac{dy}{dx} = 0$$

(c) $\frac{dy}{dx} = 0$. At $(4, 2)$, $y - 2 = 0(x - 4)$

$$y = 2$$

(d)

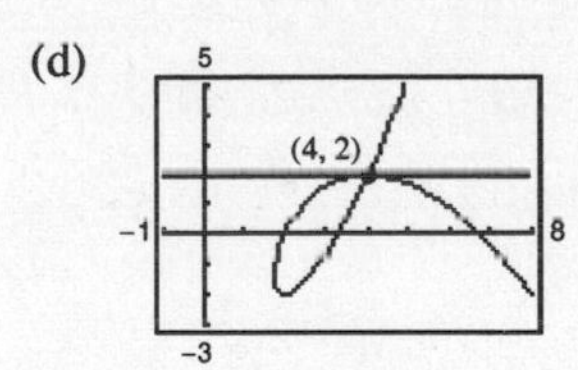

21. $x = 2\sin 2t$, $y = 3\sin t$ crosses itself at the origin, $(x, y) = (0, 0)$.

At this point, $t = 0$ or $t = \pi$.

$$\frac{dy}{dx} = \frac{3\cos t}{4\cos 2t}$$

At $t = 0$: $\frac{dy}{dx} = \frac{3}{4}$ and $y = \frac{3}{4}x$. Tangent Line

At $t = \pi$, $\frac{dy}{dx} = -\frac{3}{4}$ and $y = \frac{-3}{4}x$ Tangent Line

23. $x = \cos\theta + \theta\sin\theta,\ y = \sin\theta - \theta\cos\theta$

Horizontal tangents: $\dfrac{dy}{d\theta} = \theta\sin\theta = 0$ when $\theta = \pm\pi, \pm2\pi, \pm3\pi, \ldots$

Points: $(-1, [2n-1]\pi),\ (1, 2n\pi)$ where n is an integer.

Points shown: $(1, 0),\ (-1, \pi),\ (1, -2\pi)$

Vertical tangents: $\dfrac{dx}{d\theta} = \theta\cos\theta = 0$ when $\theta = \pm\dfrac{\pi}{2}, \pm\dfrac{3\pi}{2}, \pm\dfrac{5\pi}{2}, \ldots$

Note: $\theta = 0$ corresponds to the cusp at $(x, y) = (1, 0)$.

$$\frac{dy}{dx} = \frac{\theta\sin\theta}{\theta\cos\theta} = \tan\theta = 0 \text{ at } \theta = 0.$$

Points: $\left(\dfrac{(-1)^{n+1}(2n-1)\pi}{2}, (-1)^{n+1}\right)$ Points shown: $\left(\dfrac{\pi}{2}, 1\right), \left(-\dfrac{3\pi}{2}, -1\right), \left(\dfrac{5\pi}{2}, 1\right)$

25. $x = 1 - t,\ y = t^2$

Horizontal tangents: $\dfrac{dy}{dt} = 2t = 0$ when $t = 0$.

Point: $(1, 0)$

Vertical tangents: $\dfrac{dx}{dt} = -1 \neq 0$; none

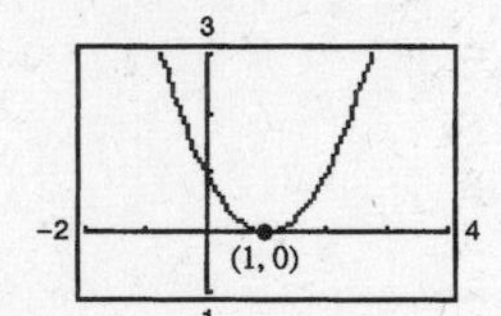

27. $x = 1 - t,\ y = t^3 - 3t$

Horizontal tangents: $\dfrac{dy}{dt} = 3t^2 - 3 = 0$ when $t = \pm1$.

Points: $(0, -2), (2, 2)$

Vertical tangents: $\dfrac{dx}{dt} = -1 \neq 0$; none

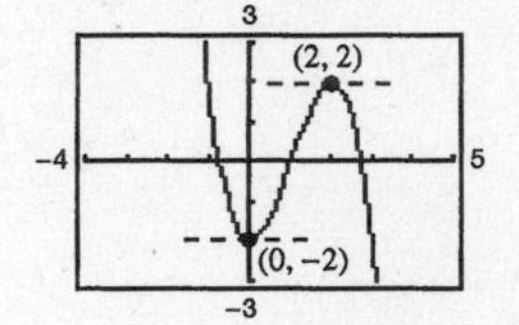

29. $x = 3\cos\theta, y = 3\sin\theta$

Horizontal tangents: $\dfrac{dy}{d\theta} = 3\cos\theta = 0$ when $\theta = \dfrac{\pi}{2}, \dfrac{3\pi}{2}$.

Points: $(0, 3), (0, -3)$

Vertical tangents: $\dfrac{dx}{d\theta} = -3\sin\theta = 0$ when $\theta = 0, \pi$.

Points: $(3, 0), (-3, 0)$

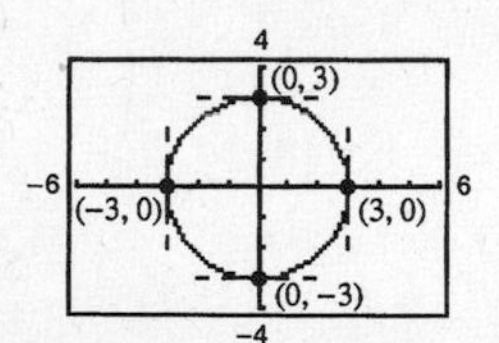

31. $x = 4 + 2\cos\theta,\ y = -1 + \sin\theta$

Horizontal tangents: $\dfrac{dy}{d\theta} = \cos\theta = 0$ when $\theta = \dfrac{\pi}{2}, \dfrac{3\pi}{2}$.

Points: $(4, 0), (4, -2)$

Vertical tangents: $\dfrac{dx}{d\theta} = -2\sin\theta = 0$ when $\theta = 0, \pi$.

Points: $(6, -1), (2, -1)$

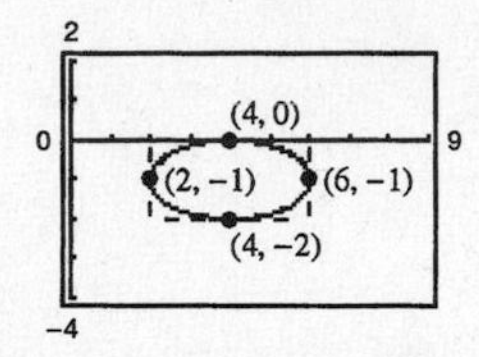

33. $x = \sec\theta,\ y = \tan\theta$

Horizontal tangents: $\dfrac{dy}{d\theta} = \sec^2\theta \neq 0$; none

Vertical tangents: $\dfrac{dx}{d\theta} = \sec\theta\tan\theta = 0$ when $x = 0, \pi$.

Points: $(1, 0), (-1, 0)$

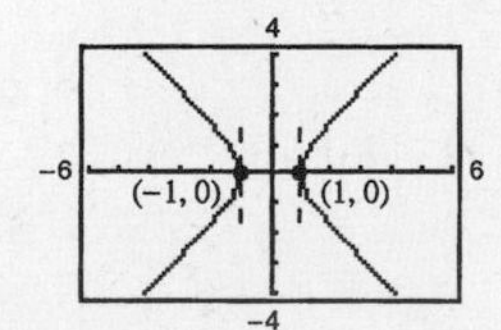

35. $x = t^2, y = 2t, 0 \le t \le 2$

$$\frac{dx}{dt} = 2t,\ \frac{dy}{dt} = 2,\ \left(\frac{dx}{dt}\right)^2 + \left(\frac{dy}{dt}\right)^2 = 4t^2 + 4 = 4(t^2 + 1)$$

$$s = 2\int_0^2 \sqrt{t^2 + 1}\, dt$$

$$= \left[t\sqrt{t^2 + 1} + \ln\left|t + \sqrt{t^2 + 1}\right|\right]_0^2$$

$$= 2\sqrt{5} + \ln\left(2 + \sqrt{5}\right) \approx 5.916$$

37. $x = e^{-t}\cos t,\ y = e^{-t}\sin t,\ 0 \le t \le \frac{\pi}{2}$

$$\frac{dx}{dt} = -e^{-t}(\sin t + \cos t),\ \frac{dy}{dt} = e^{-t}(\cos t - \sin t)$$

$$s = \int_0^{\pi/2} \sqrt{\left(\frac{dx}{dt}\right)^2 + \left(\frac{dy}{dt}\right)^2}\, dt$$

$$= \int_0^{\pi/2} \sqrt{2e^{-2t}}\, dt = -\sqrt{2}\int_0^{\pi/2} e^{-t}(-1)\, dt$$

$$= \left[-\sqrt{2}e^{-t}\right]_0^{\pi/2} = \sqrt{2}(1 - e^{-\pi/2}) \approx 1.12$$

39. $x = \sqrt{t},\ y = 3t - 1,\ \frac{dx}{dt} = \frac{1}{2\sqrt{t}},\ \frac{dy}{dt} = 3$

$$S = \int_0^1 \sqrt{\frac{1}{4t} + 9}\, dt = \frac{1}{2}\int_0^1 \frac{\sqrt{1 + 36t}}{\sqrt{t}}\, dt$$

$$= \frac{1}{6}\int_0^6 \sqrt{1 + u^2}\, du$$

$$= \frac{1}{12}\left[\ln\left(\sqrt{1 + u^2} + u\right) + u\sqrt{1 + u^2}\right]_0^6$$

$$= \frac{1}{12}\left[\ln\left(\sqrt{37} + 6\right) + 6\sqrt{37}\right] \approx 3.249$$

$$u = 6\sqrt{t},\ du = \frac{3}{\sqrt{t}}\, dt$$

41. $x = a\cos^3\theta,\ y = a\sin^3\theta,\ \frac{dx}{d\theta} = -3a\cos^2\theta\sin\theta,$

$$\frac{dy}{d\theta} = 3a\sin^2\theta\cos\theta$$

$$S = 4\int_0^{\pi/2} \sqrt{9a^2\cos^4\theta\sin^2 + 9a^2\sin^4\theta\cos^2\theta}\, d\theta$$

$$= 12a\int_0^{\pi/2} \sin\theta\cos\theta\sqrt{\cos^2\theta + \sin^2\theta}\, d\theta$$

$$= 6a\int_0^{\pi/2} \sin 2\theta\, d\theta = \left[-3a\cos 2\theta\right]_0^{\pi/2} = 6a$$

43. $x = a(\theta - \sin\theta),\ y = a(1 - \cos\theta),$

$$\frac{dx}{d\theta} = a(1 - \cos\theta),\ \frac{dy}{d\theta} = a\sin\theta$$

$$S = 2\int_0^{\pi} \sqrt{a^2(1 - \cos\theta)^2 + a^2\sin^2\theta}\, d\theta$$

$$= 2\sqrt{2}a\int_0^{\pi} \sqrt{1 - \cos\theta}\, d\theta$$

$$= 2\sqrt{2}a\int_0^{\pi} \frac{\sin\theta}{\sqrt{1 + \cos\theta}}\, d\theta$$

$$= \left[-4\sqrt{2}a\sqrt{1 + \cos\theta}\right]_0^{\pi} = 8a$$

45. $x = (90\cos 30°)t,\ y = (90\sin 30°)t - 16t^2$

(a)

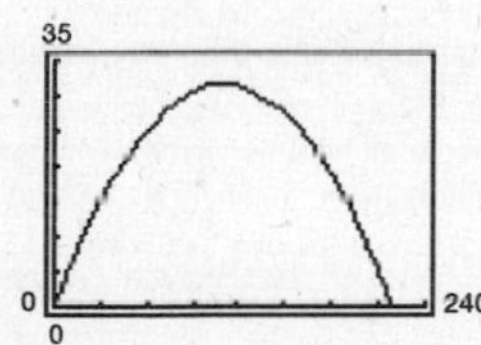

(b) Range: 219.2 ft

(c) $\frac{dx}{dt} = 90\cos 30°,\ \frac{dy}{dt} = 90\sin 30° - 32t.$

$y = 0$ for $t = \frac{45}{16}.$

$$s = \int_0^{45/16} \sqrt{(90\cos 30°)^2 + (90\sin 30° - 32t)^2}\, dt$$

$$= 230.8 \text{ ft}$$

(d) $y = 0 \implies (90\sin\theta)t = 16t^2 \implies t = \frac{90}{16}\sin\theta$

$$x = (90\cos\theta)t = \frac{90^2}{16}\cos\theta\sin\theta = \frac{90^2}{32}\sin 2\theta$$

$$x'(\theta) = \frac{90^2}{32}2\cos 2\theta = 0 \implies \theta = 45°$$

By the First Derivative Test, $\theta = 45°\left(\frac{\pi}{4}\right)$ maximizes the range.

$$\frac{dx}{dt} = 90\cos\theta,$$

$$\frac{dy}{dt} = 90\sin\theta - 32t = 90\sin\theta - 32\left(\frac{90}{16}\sin\theta\right) = -90\sin\theta$$

$$s = \int_0^{(90/16)\sin\theta} \sqrt{(90\cos\theta)^2 + (-90\sin\theta)^2}\, dt$$

$$= \int_0^{(90/16)\sin\theta} 90\, dt = 90t\Big]_0^{(90/16)\sin\theta}$$

$$= \frac{90^2}{16}\sin\theta$$

$$\frac{ds}{d\theta} = \frac{90^2}{16}\cos\theta = 0 \implies \theta = \frac{\pi}{2}$$

By the First Derivative Test, $\theta = 90°$ maximizes the arc length.

47. (a) $x = t - \sin t$

$y = 1 - \cos t$

$0 \le t \le 2\pi$

$x = 2t - \sin(2t)$

$y = 1 - \cos(2t)$

$0 \le t \le \pi$

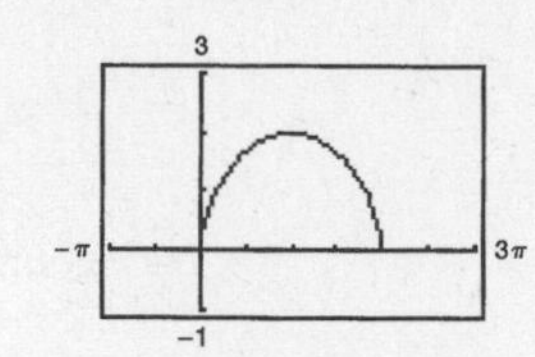

(b) The average speed of the particle on the second path is twice the average speed of a particle on the first path.

(c) $x = \frac{1}{2}t - \sin\left(\frac{1}{2}t\right)$

$y = 1 - \cos\left(\frac{1}{2}t\right)$

The time required for the particle to traverse the same path is $t = 4\pi$.

49. $x = t,\ y = 2t,\ \dfrac{dx}{dt} = 1,\ \dfrac{dy}{dt} = 2$

(a) $S = 2\pi\displaystyle\int_0^4 2t\sqrt{1+4}\,dt = 4\sqrt{5}\pi\int_0^4 t\,dt$

$= \Big[2\sqrt{5}\pi t^2\Big]_0^4 = 32\pi\sqrt{5}$

(b) $S = 2\pi\displaystyle\int_0^4 t\sqrt{1+4}\,dt = 2\sqrt{5}\pi\int_0^4 t\,dt$

$= \Big[\sqrt{5}\pi t^2\Big]_0^4 = 16\pi\sqrt{5}$

51. $x = 4\cos\theta,\ y = 4\sin\theta,\ \dfrac{dx}{d\theta} = -4\sin\theta,\ \dfrac{dy}{d\theta} = 4\cos\theta$

$S = 2\pi\displaystyle\int_0^{\pi/2} 4\cos\theta\sqrt{(-4\sin\theta)^2 + (4\cos\theta)^2}\,d\theta$

$= 32\pi\displaystyle\int_0^{\pi/2}\cos\theta\,d\theta = \Big[32\pi\sin\theta\Big]_0^{\pi/2} = 32\pi$

53. $x = a\cos^3\theta,\ y = a\sin^3\theta,\ \dfrac{dx}{d\theta} = -3a\cos^2\theta\sin\theta,\ \dfrac{dy}{d\theta} = 3a\sin^2\theta\cos\theta$

$S = 4\pi\displaystyle\int_0^{\pi/2} a\sin^3\theta\sqrt{9a^2\cos^4\theta\sin^2\theta + 9a^2\sin^4\theta\cos^2\theta}\,d\theta = 12a^2\pi\int_0^{\pi/2}\sin^4\theta\cos\theta\,d\theta = \frac{12\pi a^2}{5}\Big[\sin^5\theta\Big]_0^{\pi/2} = \frac{12}{5}\pi a^2$

55. $\dfrac{dy}{dx} = \dfrac{dy/dt}{dx/dt}$

See Theorem 9.7, page 675.

57. One possible answer is the graph given by

$x = t,\ y = -t.$

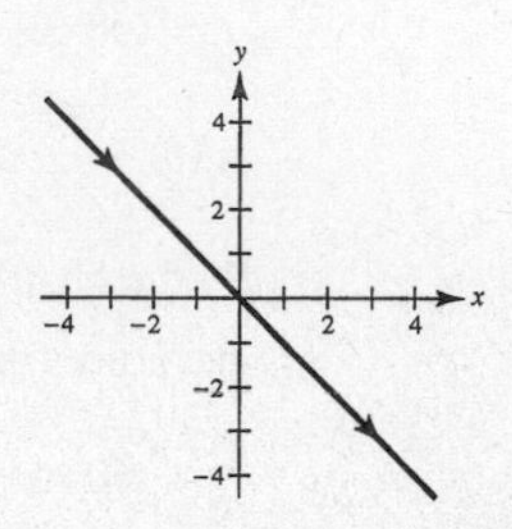

59. $s = \displaystyle\int_a^b \sqrt{\left(\frac{dx}{dt}\right)^2 + \left(\frac{dy}{dt}\right)^2}\,dt$

See Theorem 9.8, page 678.

61. $x = r\cos\phi,\ y = r\sin\phi$

$S = 2\pi\displaystyle\int_0^\theta r\sin\phi\sqrt{r^2\sin^2\phi + r^2\cos^2\phi}\,d\phi$

$= 2\pi r^2\displaystyle\int_0^\theta \sin\phi\,d\phi$

$= \Big[-2\pi r^2\cos\phi\Big]_0^\theta$

$= 2\pi r^2(1 - \cos\theta)$

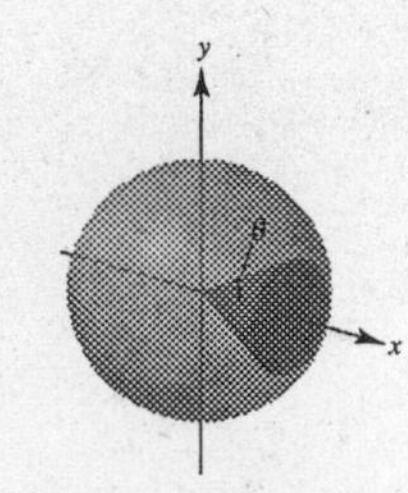

63. $x = \sqrt{t},\ y = 4 - t,\ 0 \le t \le 4$

$$A = \int_0^4 (4-t)\frac{1}{2\sqrt{t}}\,dt = \frac{1}{2}\int_0^4 (4t^{-1/2} - t^{1/2})\,dt = \left[\frac{1}{2}\left(8\sqrt{t} - \frac{2}{3}t\sqrt{t}\right)\right]_0^4 = \frac{16}{3}$$

$$\bar{x} = \frac{3}{16}\int_0^4 (4-t)\sqrt{t}\left(\frac{1}{2\sqrt{t}}\right)dt = \frac{3}{32}\int_0^4 (4-t)\,dt = \left[\frac{3}{32}\left(4t - \frac{t^2}{2}\right)\right]_0^4 = \frac{3}{4}$$

$$\bar{y} = \frac{3}{32}\int_0^4 (4-t)^2\frac{1}{2\sqrt{t}}\,dt = \frac{3}{64}\int_0^4 [16t^{-1/2} - 8t^{1/2} + t^{3/2}]\,dt = \frac{3}{64}\left[32\sqrt{t} - \frac{16}{3}t\sqrt{t} + \frac{2}{5}t^2\sqrt{t}\right]_0^4 = \frac{8}{5}$$

$$(\bar{x}, \bar{y}) = \left(\frac{3}{4}, \frac{8}{5}\right)$$

65. $x = 3\cos\theta,\ y = 3\sin\theta,\ \dfrac{dx}{d\theta} = -3\sin\theta$

$$V = 2\pi\int_{\pi/2}^0 (3\sin\theta)^2(-3\sin\theta)\,d\theta$$

$$= -54\pi\int_{\pi/2}^0 \sin^3\theta\,d\theta$$

$$= -54\pi\int_{\pi/2}^0 (1-\cos^2\theta)\sin\theta\,d\theta$$

$$= -54\pi\left[-\cos\theta + \frac{\cos^3\theta}{3}\right]_{\pi/2}^0 = 36\pi$$

67. $x = 2\sin^2\theta$

$y = 2\sin^2\theta\tan\theta$

$$\frac{dx}{d\theta} = 4\sin\theta\cos\theta$$

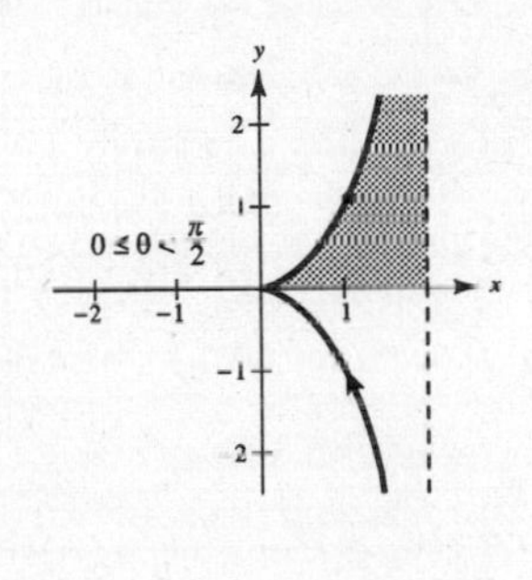

$$A = \int_0^{\pi/2} 2\sin^2\theta\tan\theta(4\sin\theta\cos\theta)\,d\theta = 8\int_0^{\pi/2}\sin^4\theta\,d\theta$$

$$= 8\left[\frac{-\sin^3\theta\cos\theta}{4} - \frac{3}{8}\sin\theta\cos\theta + \frac{3}{8}\theta\right]_0^{\pi/2} = \frac{3\pi}{2}$$

69. πab is area of ellipse (d).

71. $6\pi a^2$ is area of cardioid (f).

73. $\frac{8}{3}ab$ is area of hourglass (a).

75. (a) $x = \dfrac{1-t^2}{1+t^2},\ y = \dfrac{2t}{1+t^2},\ -20 \le t \le 20$

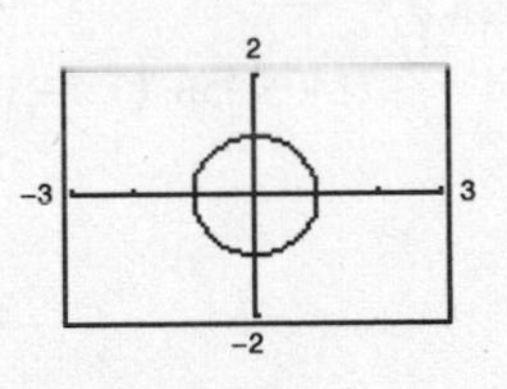

The graph is the circle $x^2 + y^2 = 1$, except the point $(-1, 0)$.

Verify: $x^2 + y^2 = \left(\dfrac{1-t^2}{1+t^2}\right)^2 + \left(\dfrac{2t}{1+t^2}\right)^2 = \dfrac{1 - 2t^2 + t^4 + 4t^2}{(1+t^2)^2} = \dfrac{(1+t^2)^2}{(1+t^2)^2} = 1$

(b) As t increases from -20 to 0, the speed increases, and as t increases from 0 to 20, the speed decreases.

77. False

$$\frac{d^2y}{dx^2} = \frac{\dfrac{d}{dt}\left[\dfrac{g'(t)}{f'(t)}\right]}{f'(t)} = \frac{f'(t)g''(t) - g'(t)f''(t)}{[f'(t)]^3}$$

Section 9.4 Polar Coordinates and Polar Graphs

1. $\left(4, \frac{\pi}{2}\right)$

$x = 4\cos\left(\frac{\pi}{2}\right) = 0$

$y = 4\sin\left(\frac{\pi}{2}\right) = 4$

$(x, y) = (0, 4)$

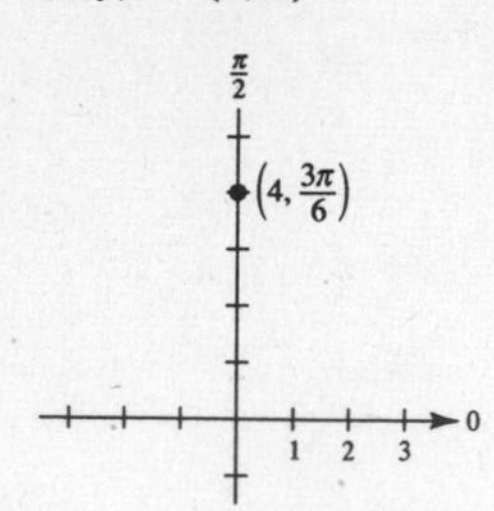

3. $\left(-4, -\frac{\pi}{3}\right)$

$x = -4\cos\left(-\frac{\pi}{3}\right) = -2$

$y = -4\sin\left(-\frac{\pi}{3}\right) = 2\sqrt{3}$

$(x, y) = \left(-2, 2\sqrt{3}\right)$

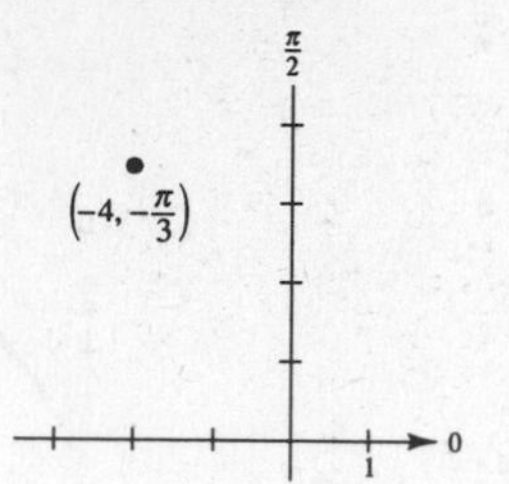

5. $\left(\sqrt{2}, 2.36\right)$

$x = \sqrt{2}\cos(2.36) \approx -1.004$

$y = \sqrt{2}\sin(2.36) \approx 0.996$

$(x, y) = (-1.004, 0.996)$

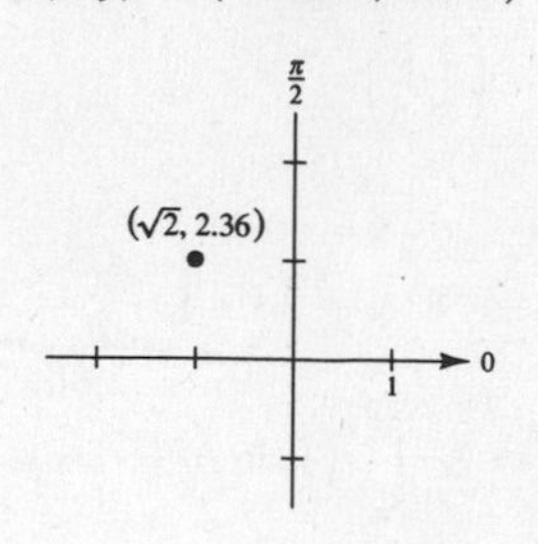

7. $(r, \theta) = \left(5, \frac{3\pi}{4}\right)$

$(x, y) = (-3.5355, 3.5355)$

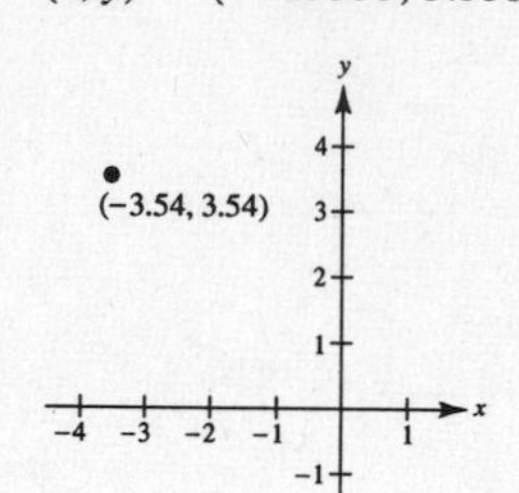

9. $(r, \theta) = (-3.5, 2.5)$

$(x, y) = (2.804, -2.095)$

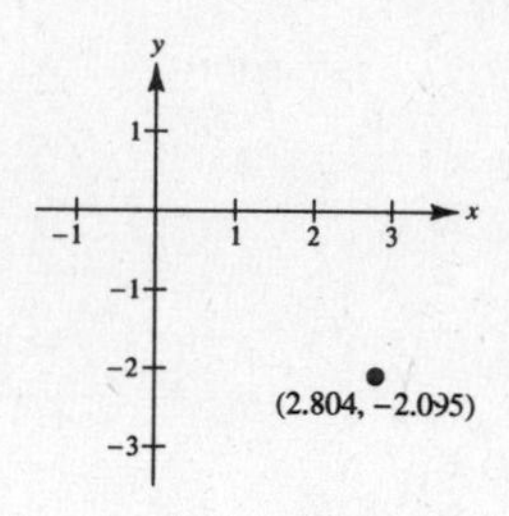

11. $(x, y) = (1, 1)$

$r = \pm\sqrt{2}$

$\tan\theta = 1$

$\theta = \frac{\pi}{4}, \frac{5\pi}{4}, \left(\sqrt{2}, \frac{\pi}{4}\right), \left(-\sqrt{2}, \frac{5\pi}{4}\right)$

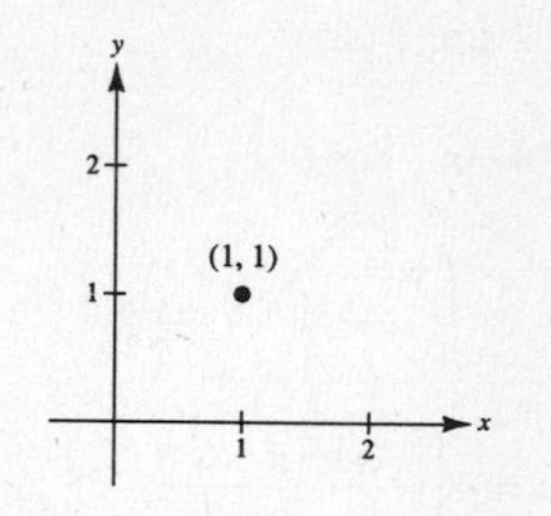

13. $(x, y) = (-3, 4)$

$r = \pm\sqrt{9 + 16} = \pm 5$

$\tan\theta = -\frac{4}{3}$

$\theta \approx 2.214, 5.356, (5, 2.214), (-5, 5.356)$

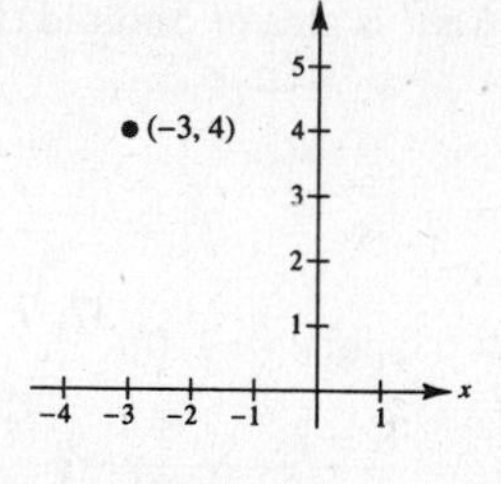

15. $(x, y) = (3, -2)$

$(r, \theta) = (3.606, -0.588)$

17. $(x, y) = \left(\frac{5}{2}, \frac{4}{3}\right)$

$(r, \theta) = (2.833, 0.490)$

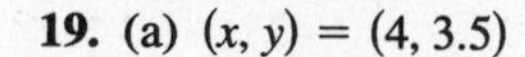

19. (a) $(x, y) = (4, 3.5)$

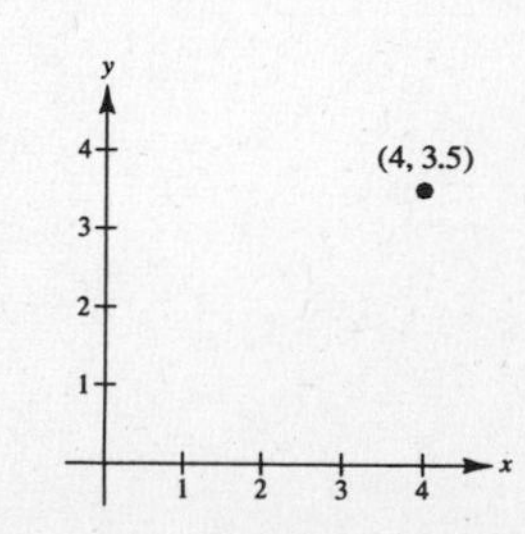

(b) $(r, \theta) = (4, 3.5)$

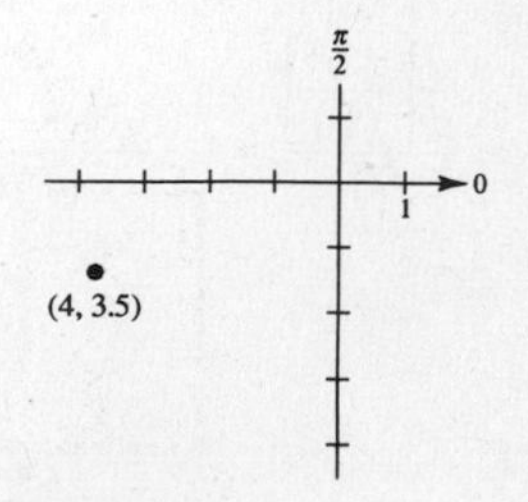

21. $x^2 + y^2 = a^2$

$r = a$

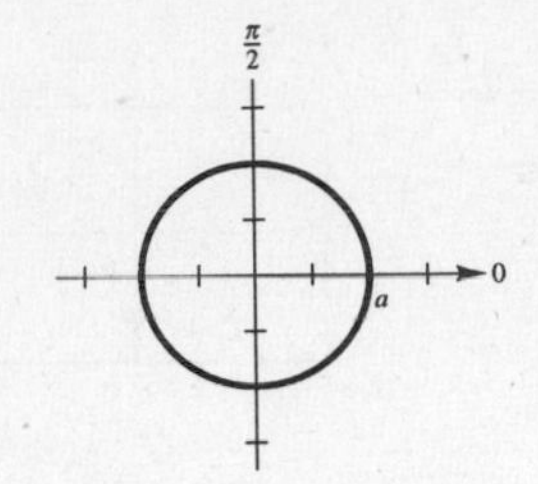

23.

$$y = 4$$
$$r \sin \theta = 4$$
$$r = 4 \csc \theta$$

25.

$$3x - y + 2 = 0$$
$$3r \cos \theta - r \sin \theta + 2 = 0$$
$$r(3 \cos \theta - \sin \theta) = -2$$
$$r = \frac{-2}{3 \cos \theta - \sin \theta}$$

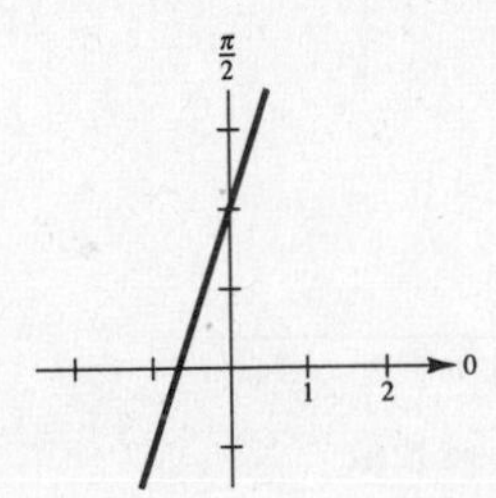

27.

$$y^2 = 9x$$
$$r^2 \sin^2 \theta = 9r \cos \theta$$
$$r = \frac{9 \cos \theta}{\sin^2 \theta}$$
$$r = 9 \csc^2 \theta \cos \theta$$

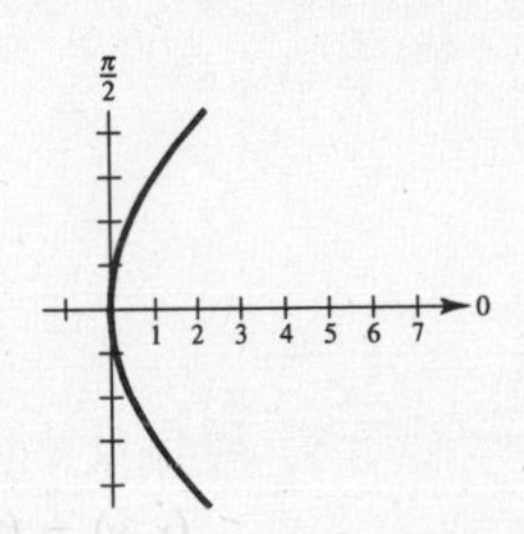

29.

$$r = 3$$
$$r^2 = 9$$
$$x^2 + y^2 = 9$$

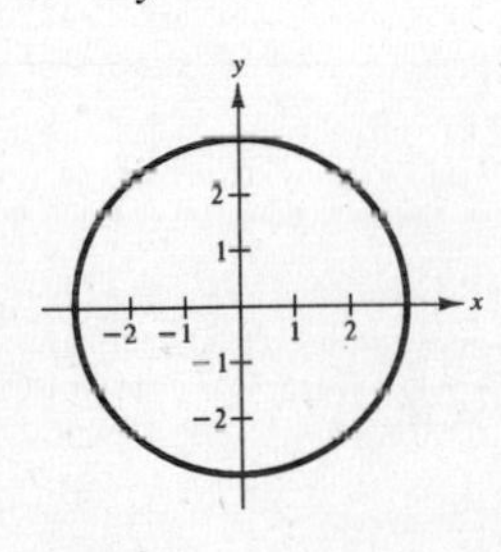

31.

$$r = \sin \theta$$
$$r^2 = r \sin \theta$$
$$x^2 + y^2 = y$$
$$x^2 + \left(y - \frac{1}{2}\right)^2 = \frac{1}{4}$$
$$x^2 + y^2 - y = 0$$

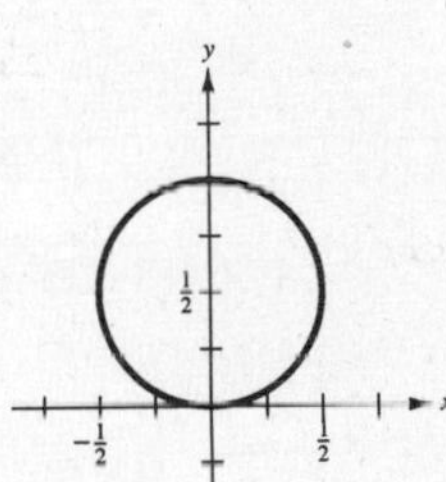

33.

$$r = \theta$$
$$\tan r = \tan \theta$$
$$\tan \sqrt{x^2 + y^2} = \frac{y}{x}$$
$$\sqrt{x^2 + y^2} = \arctan \frac{y}{x}$$

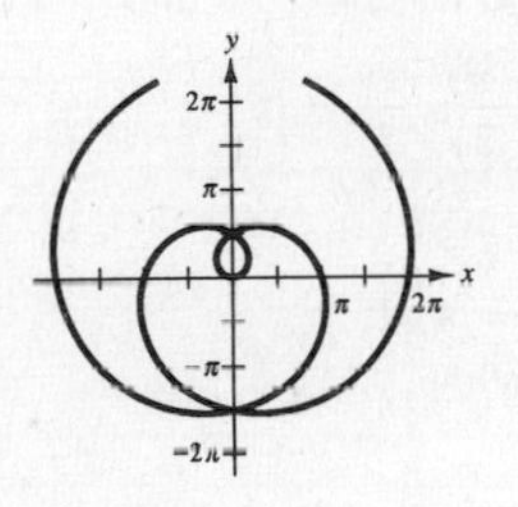

35.

$$r = 3 \sec \theta$$
$$r \cos \theta = 3$$
$$x = 3$$
$$x - 3 = 0$$

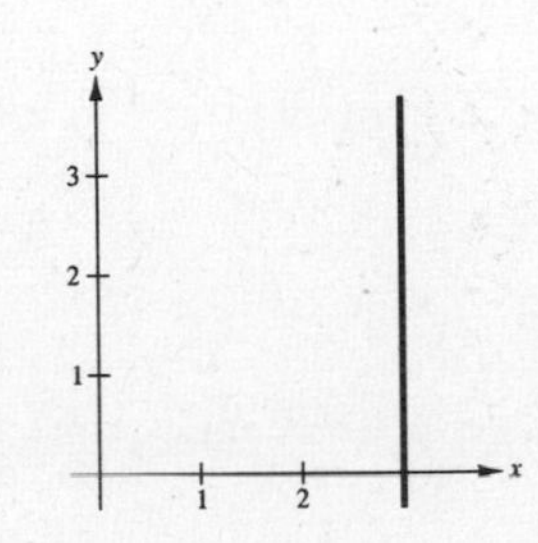

37. $r = 3 - 4 \cos \theta$

$0 \leq \theta < 2\pi$

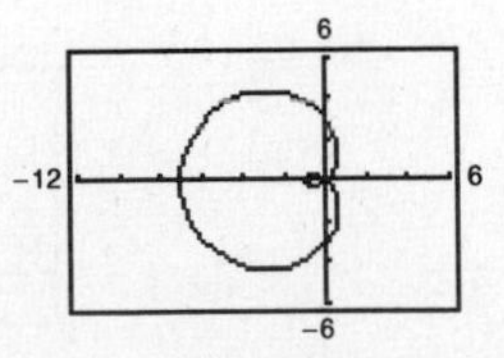

39. $r = 2 + \sin \theta$

$0 \leq \theta < 2\pi$

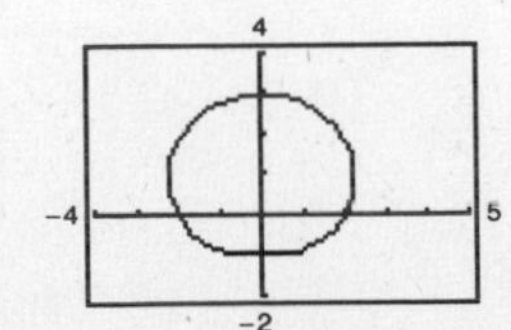

41. $r = \dfrac{2}{1 + \cos \theta}$

Traced out once on $-\pi < \theta < \pi$

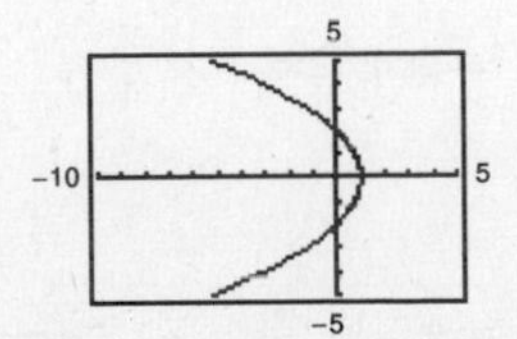

43. $r = 2\cos\left(\frac{3\theta}{2}\right)$

$0 \le \theta < 4\pi$

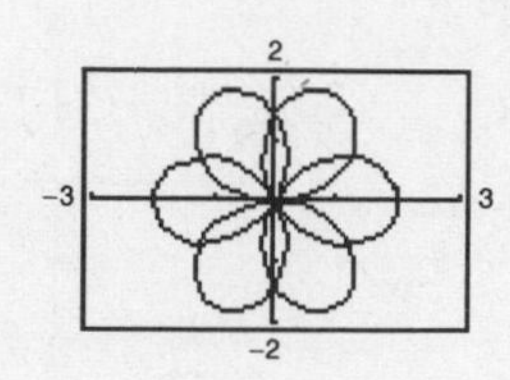

45. $r^2 = 4\sin 2\theta$

$0 \le \theta < \frac{\pi}{2}$

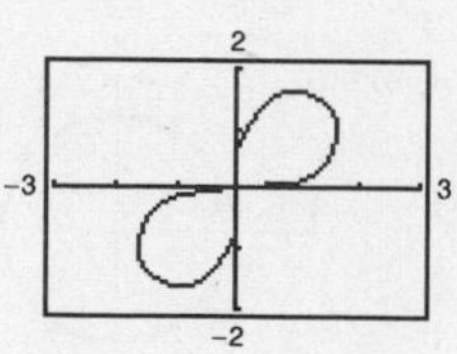

47.

$$\begin{aligned} r &= 2(h\cos\theta + k\sin\theta) \\ r^2 &= 2r(h\cos\theta + k\sin\theta) \\ r^2 &= 2[h(r\cos\theta) + k(r\sin\theta)] \\ x^2 + y^2 &= 2(hx + ky) \\ x^2 + y^2 - 2hx - 2ky &= 0 \\ (x^2 - 2hx + h^2) + (y^2 - 2ky + k^2) &= 0 + h^2 + k^2 \\ (x - h)^2 + (y - k)^2 &= h^2 + k^2 \end{aligned}$$

Radius: $\sqrt{h^2 + k^2}$

Center: (h, k)

49. $\left(4, \frac{2\pi}{3}\right), \left(2, \frac{\pi}{6}\right)$

$$\begin{aligned} d &= \sqrt{4^2 + 2^2 - 2(4)(2)\cos\left(\frac{2\pi}{3} - \frac{\pi}{6}\right)} \\ &= \sqrt{20 - 16\cos\frac{\pi}{2}} = 2\sqrt{5} \approx 4.5 \end{aligned}$$

51. $(2, 0.5), (7, 1.2)$

$$\begin{aligned} d &= \sqrt{2^2 + 7^2 - 2(2)(7)\cos(0.5 - 1.2)} \\ &= \sqrt{53 - 28\cos(-0.7)} \approx 5.6 \end{aligned}$$

53. $r = 2 + 3\sin\theta$

$$\begin{aligned} \frac{dy}{dx} &= \frac{3\cos\theta\sin\theta + \cos\theta(2 + 3\sin\theta)}{3\cos\theta\cos\theta - \sin\theta(2 + 3\sin\theta)} \\ &= \frac{2\cos\theta(3\sin\theta + 1)}{3\cos 2\theta - 2\sin\theta} = \frac{2\cos\theta(3\sin\theta + 1)}{6\cos^2\theta - 2\sin\theta - 3} \end{aligned}$$

At $\left(5, \frac{\pi}{2}\right)$, $\frac{dy}{dx} = 0$.

At $(2, \pi)$, $\frac{dy}{dx} = -\frac{2}{3}$.

At $\left(-1, \frac{3\pi}{2}\right)$, $\frac{dy}{dx} = 0$.

55. (a), (b) $r = 3(1 - \cos\theta)$

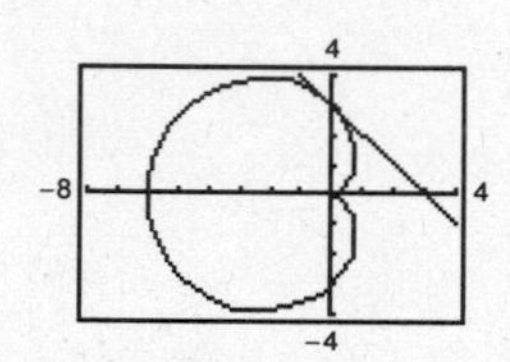

$(r, \theta) = \left(3, \frac{\pi}{2}\right) \Rightarrow (x, y) = (0, 3)$

Tangent line: $y - 3 = -1(x - 0)$

$y = -x + 3$

(c) At $\theta = \frac{\pi}{2}, \frac{dy}{dx} = -1.0$.

57. (a), (b) $r = 3\sin\theta$

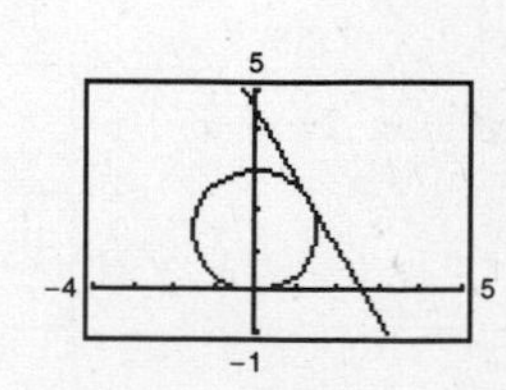

$(r, \theta) = \left(\frac{3\sqrt{3}}{2}, \frac{\pi}{3}\right) \Rightarrow (x, y) = \left(\frac{3\sqrt{3}}{4}, \frac{9}{4}\right)$

Tangent line: $y - \frac{9}{4} = -\sqrt{3}\left(x - \frac{3\sqrt{3}}{4}\right)$

$y = -\sqrt{3}x + \frac{9}{2}$

(c) At $\theta = \frac{\pi}{3}, \frac{dy}{dx} = -\sqrt{3} \approx -1.732$.

59. $r = 1 - \sin\theta$

$$\frac{dy}{d\theta} = (1 - \sin\theta)\cos\theta - \cos\theta\sin\theta$$

$$= \cos\theta(1 - 2\sin\theta) = 0$$

$$\cos\theta = 0, \sin\theta = \frac{1}{2} \Rightarrow \theta = \frac{\pi}{2}, \frac{3\pi}{2}, \frac{\pi}{6}, \frac{5\pi}{6}$$

Horizontal tangents: $\left(2, \frac{3\pi}{2}\right), \left(\frac{1}{2}, \frac{\pi}{6}\right), \left(\frac{1}{2}, \frac{5\pi}{6}\right)$

$$\frac{dx}{d\theta} = (-1 + \sin\theta)\sin\theta - \cos\theta\cos\theta$$

$$= -\sin\theta + \sin^2\theta + \sin^2\theta - 1$$

$$= 2\sin^2\theta - \sin\theta - 1$$

$$= (2\sin\theta + 1)(\sin\theta - 1) = 0$$

$$\sin\theta = 1, \sin\theta = -\frac{1}{2} \Rightarrow \theta = \frac{\pi}{2}, \frac{7\pi}{6}, \frac{11\pi}{6}$$

Vertical tangents: $\left(\frac{3}{2}, \frac{7\pi}{6}\right), \left(\frac{3}{2}, \frac{11\pi}{6}\right)$

61. $r = 2\csc\theta + 3$

$$\frac{dy}{d\theta} = (2\csc\theta + 3)\cos\theta + (-2\csc\theta\cot\theta)\sin\theta$$

$$= 3\cos\theta = 0$$

$$\theta = \frac{\pi}{2}, \frac{3\pi}{2}$$

Horizontal: $\left(5, \frac{\pi}{2}\right), \left(1, \frac{3\pi}{2}\right)$

63. $r = 4\sin\theta\cos^2\theta$

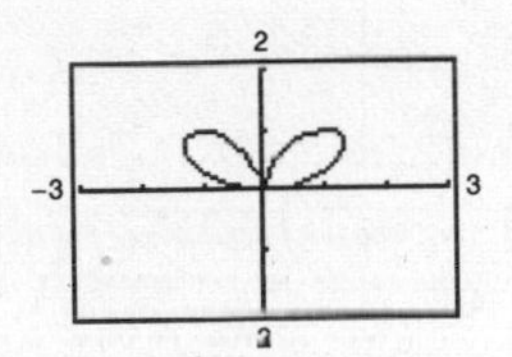

Horizontal tangents:

(0, 0), (1.4142, 0.7854), (1.4142, 2.3562)

65. $r = 2\csc\theta + 5$

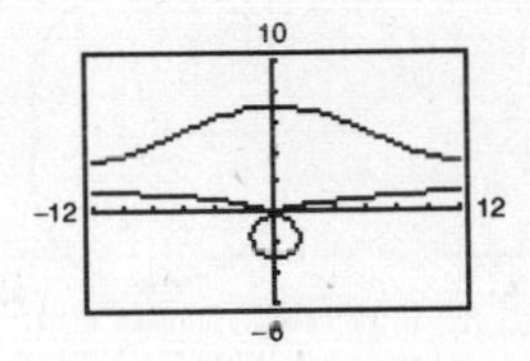

Horizontal tangents: $\left(7, \frac{\pi}{2}\right), \left(3, \frac{3\pi}{2}\right)$

67.

$$r = 3\sin\theta$$

$$r^2 = 3r\sin\theta$$

$$x^2 + y^2 = 3y$$

$$x^2 + \left(y - \frac{3}{2}\right)^2 = \frac{9}{4}$$

Circle $r = \frac{3}{2}$

Center: $\left(0, \frac{3}{2}\right)$

Tangent at the pole: $\theta = 0$

69. $r = 2(1 - \sin\theta)$

Cardioid

Symmetric to y-axis, $\theta = \frac{\pi}{2}$

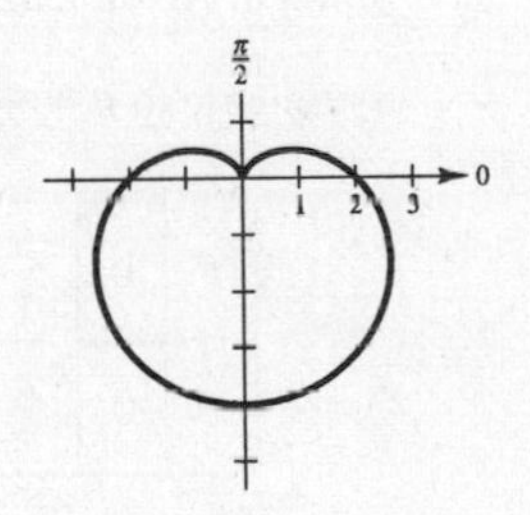

71. $r = 2\cos(3\theta)$

Rose curve with three petals

Symmetric to the polar axis

Relative extrema: $(2, 0), \left(-2, \frac{\pi}{3}\right), \left(2, \frac{2\pi}{3}\right)$

θ	0	$\frac{\pi}{6}$	$\frac{\pi}{4}$	$\frac{\pi}{3}$	$\frac{\pi}{2}$	$\frac{2\pi}{3}$	$\frac{5\pi}{6}$	π
r	2	0	$-\sqrt{2}$	-2	0	2	0	-2

Tangents at the pole: $\theta = \frac{\pi}{6}, \frac{\pi}{2}, \frac{5\pi}{6}$

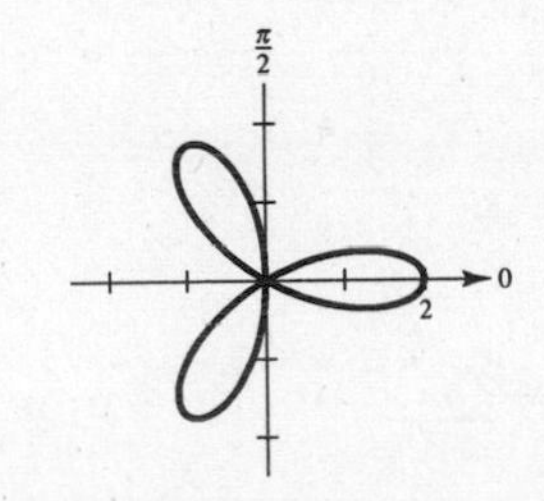

73. $r = 3 \sin 2\theta$

Rose curve with four petals

Symmetric to the polar axis, $\theta = \dfrac{\pi}{2}$, and pole

Relative extrema: $\left(\pm 3, \dfrac{\pi}{4}\right), \left(\pm 3, \dfrac{5\pi}{4}\right)$

Tangents at the pole: $\theta = 0, \dfrac{\pi}{2}$

($\theta = \pi,\ 3\pi/2$ give the same tangents.)

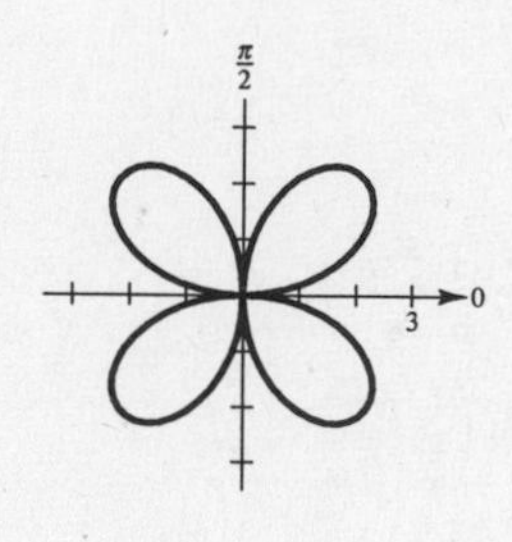

75. $r = 5$

Circle radius: 5

$x^2 + y^2 = 25$

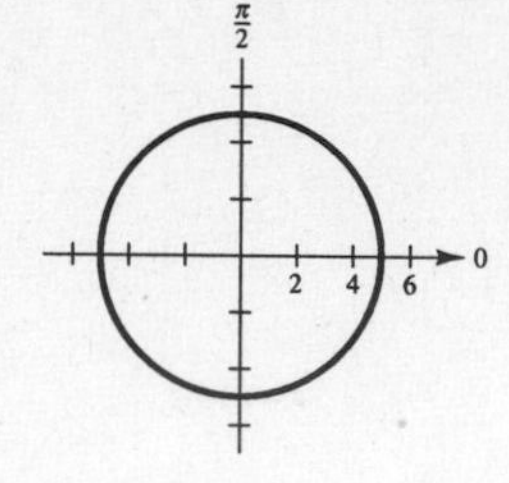

77. $r = 4(1 + \cos \theta)$

Cardioid

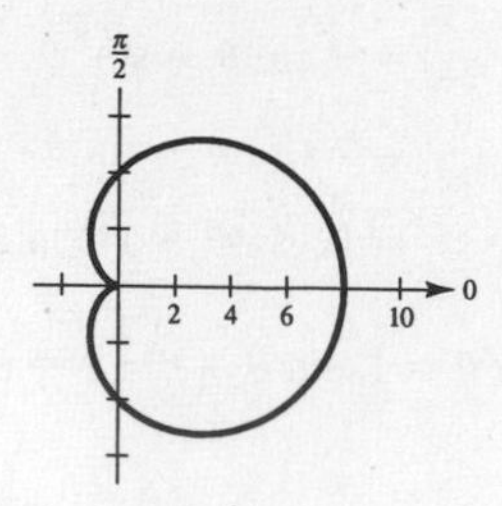

79. $r = 3 - 2 \cos \theta$

Limaçon

Symmetric to polar axis

θ	0	$\frac{\pi}{3}$	$\frac{\pi}{2}$	$\frac{2\pi}{3}$	π
r	1	2	3	4	5

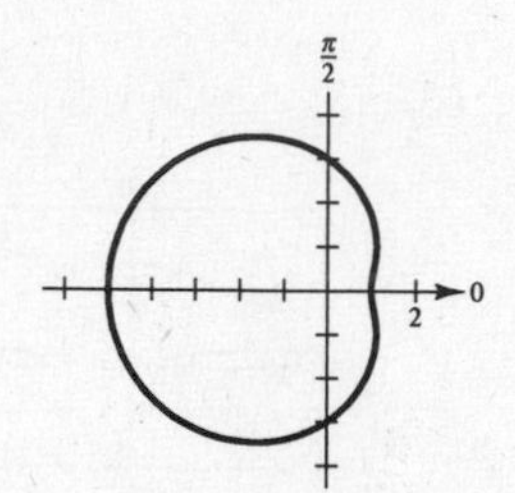

81. $r = 3 \csc \theta$

$r \sin \theta = 3$

$y = 3$

Horizontal line

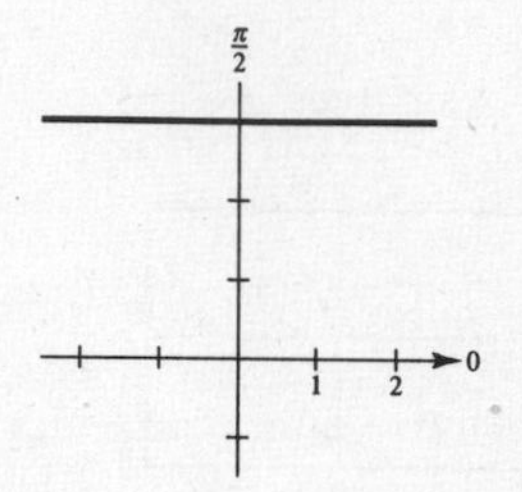

83. $r = 2\theta$

Spiral of Archimedes

Symmetric to $\theta = \dfrac{\pi}{2}$

θ	0	$\frac{\pi}{4}$	$\frac{\pi}{2}$	$\frac{3\pi}{4}$	π	$\frac{5\pi}{4}$	$\frac{3\pi}{2}$
r	0	$\frac{\pi}{2}$	π	$\frac{3\pi}{2}$	2π	$\frac{5\pi}{2}$	3π

Tangent at the pole: $\theta = 0$

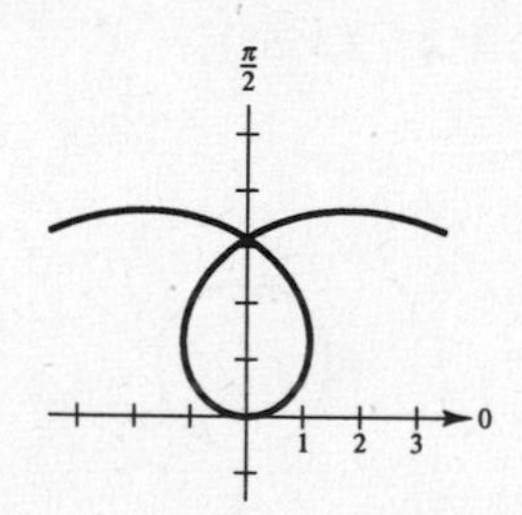

85. $r^2 = 4 \cos(2\theta)$

Lemniscate

Symmetric to the polar axis, $\theta = \dfrac{\pi}{2}$, and pole

Relative extrema: $(\pm 2, 0)$

θ	0	$\frac{\pi}{6}$	$\frac{\pi}{4}$
r	± 2	$\pm\sqrt{2}$	0

Tangents at the pole: $\theta = \dfrac{\pi}{4}, \dfrac{3\pi}{4}$

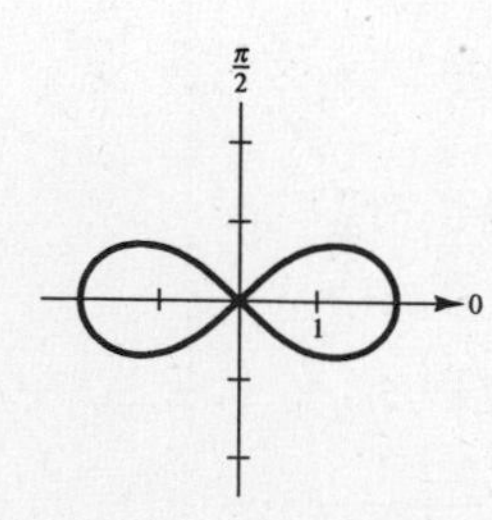

87. Since

$$r = 2 - \sec\theta = 2 - \frac{1}{\cos\theta},$$

the graph has polar axis symmetry and the tangents at the pole are

$$\theta = \frac{\pi}{3}, \frac{-\pi}{3}.$$

Furthermore,

$$r \Rightarrow -\infty \text{ as } \theta \Rightarrow \frac{\pi}{2^-}$$

$$r \Rightarrow \infty \text{ as } \theta \Rightarrow -\frac{\pi}{2^+}.$$

Also, $r = 2 - \dfrac{1}{\cos\theta} = 2 - \dfrac{r}{r\cos\theta} = 2 - \dfrac{r}{x}$

$$rx = 2x - r$$

$$r = \frac{2x}{1 + x}.$$

Thus, $r \Rightarrow \pm\infty$ as $x \Rightarrow -1$.

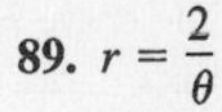

89. $r = \dfrac{2}{\theta}$

Hyperbolic spiral

$r \Rightarrow \infty$ as $\theta \Rightarrow 0$

$$r = \frac{2}{\theta} \Rightarrow \theta = \frac{2}{r} = \frac{2\sin\theta}{r\sin\theta} = \frac{2\sin\theta}{y}$$

$$y = \frac{2\sin\theta}{\theta}$$

$$\lim_{\theta \to 0} \frac{2\sin\theta}{\theta} = \lim_{\theta \to 0} \frac{2\cos\theta}{1} = 2$$

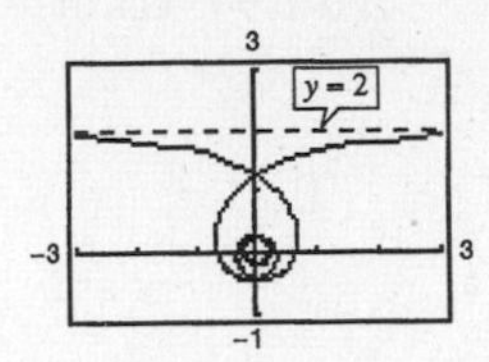

91. The rectangular coordinate system consists of all points of the form (x, y) where x is the directed distance from the y-axis to the point, and y is the directed distance from the x-axis to the point. Every point has a unique representation.

The polar coordinate system uses (r, θ) to designate the location of a point.

r is the directed distance to the origin and θ is the angle the point makes with the positive x-axis, measured clockwise.

Point do not have a unique polar representation.

93. $r = a$ circle

$\theta = b$ line

95. $r = 2\sin\theta$ circle

Matches (c)

97. $r = 3(1 + \cos\theta)$

Cardioid

Matches (a)

99. $r = 4\sin\theta$

(a) $0 \le \theta \le \dfrac{\pi}{2}$

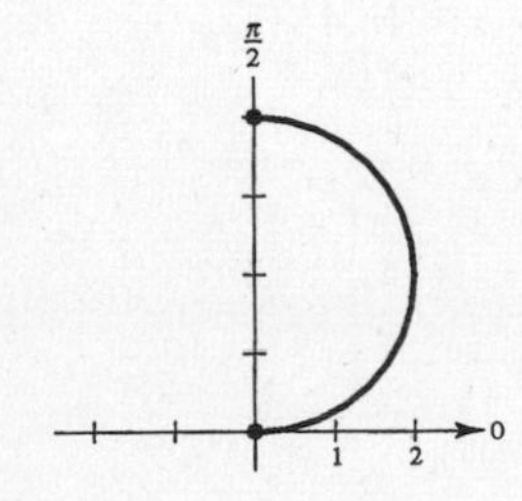

(b) $\dfrac{\pi}{2} \le \theta \le \pi$

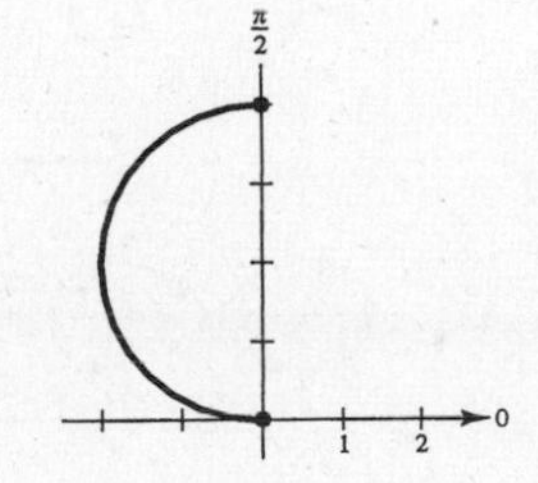

(c) $-\dfrac{\pi}{2} \le \theta \le \dfrac{\pi}{2}$

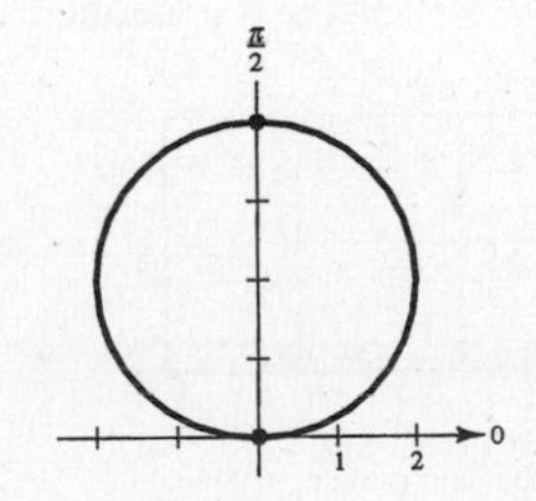

101. Let the curve $r = f(\theta)$ be rotated by ϕ to form the curve $r = g(\theta)$. If (r_1, θ_1) is a point on $r = f(\theta)$, then $(r_1, \theta_1 + \phi)$ is on $r = g(\theta)$. That is,

$$g(\theta_1 + \phi) = r_1 = f(\theta_1).$$

Letting $\theta = \theta_1 + \phi$, or $\theta_1 = \theta - \phi$, we see that

$$g(\theta) = g(\theta_1 + \phi) = f(\theta_1) = f(\theta - \phi).$$

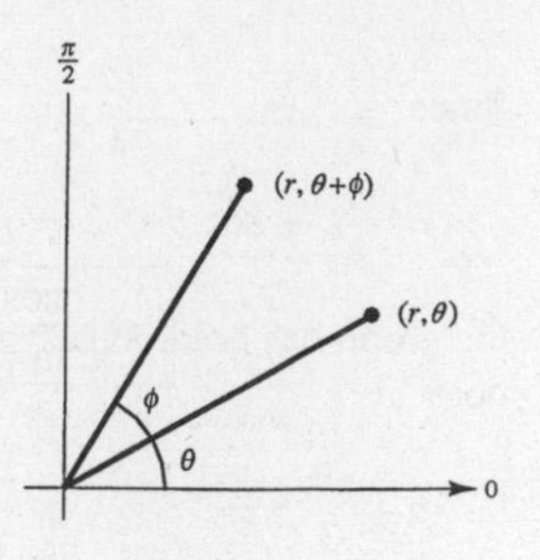

103. $r = 2 - \sin\theta$

(a) $r = 2 - \sin\left(\theta - \dfrac{\pi}{4}\right) = 2 - \dfrac{\sqrt{2}}{2}(\sin\theta - \cos\theta)$

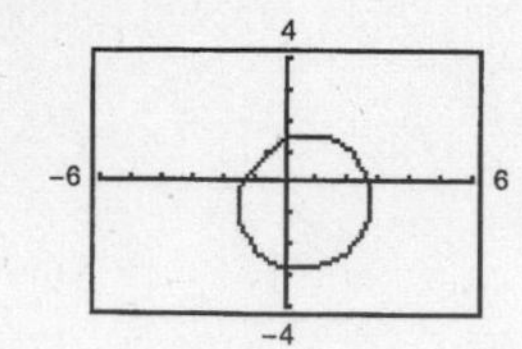

(b) $r = 2 - \sin\left(\theta - \dfrac{\pi}{2}\right) = 2 - (-\cos\theta) = 2 + \cos\theta$

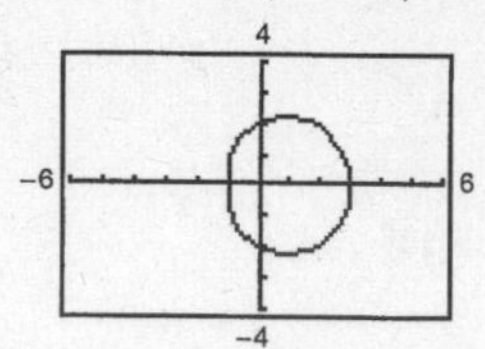

(c) $r = 2 - \sin(\theta - \pi) = 2 - (-\sin\theta) = 2 + \sin\theta$

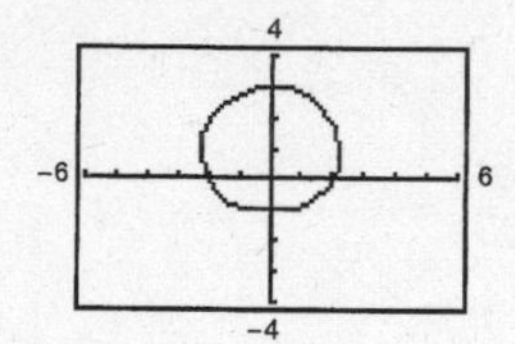

(d) $r = 2 - \sin\left(\theta - \dfrac{3\pi}{2}\right) = 2 - \cos\theta$

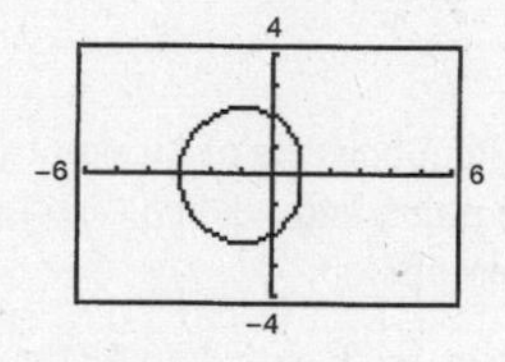

105. (a) $r = 1 - \sin\theta$

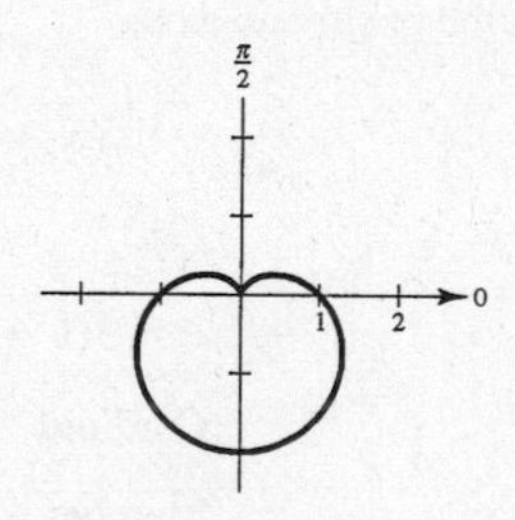

(b) $r = 1 - \sin\left(\theta - \dfrac{\pi}{4}\right)$

Rotate the graph of

$$r = 1 - \sin\theta$$

through the angle $\pi/4$.

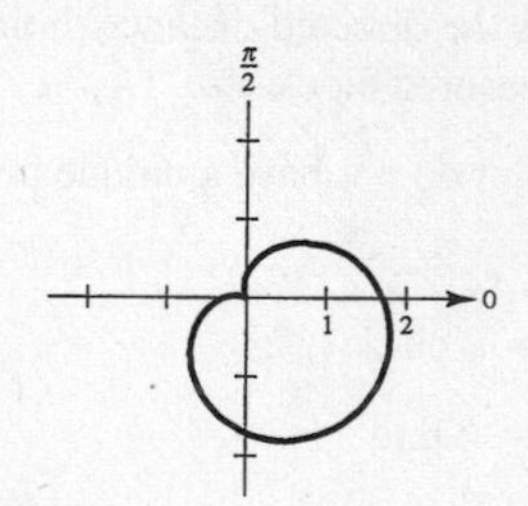

107. $\tan\psi = \dfrac{r}{dr/d\theta} = \dfrac{2(1 - \cos\theta)}{2\sin\theta}$

At $\theta = \pi$, $\tan\psi$ is undefined $\Rightarrow \psi = \dfrac{\pi}{2}$.

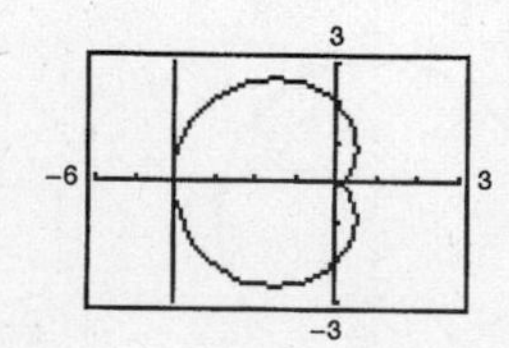

109. $\tan\psi = \dfrac{r}{dr/d\theta} = \dfrac{2\cos 3\theta}{-6\sin 3\theta}$

At $\theta = \dfrac{\pi}{4}$, $\tan\psi = \dfrac{1}{3} \Rightarrow \psi \approx 18.4$

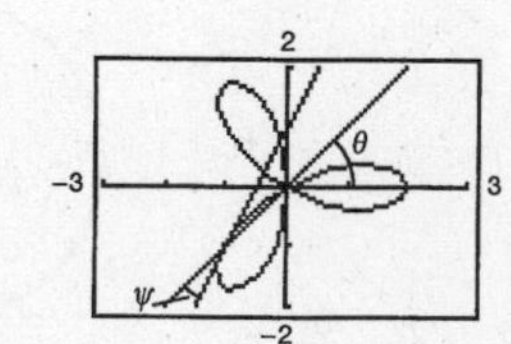

111. $r = \dfrac{6}{1 - \cos\theta} = 6(1 - \cos\theta)^{-1} \Rightarrow \dfrac{dr}{d\theta} = \dfrac{6\sin\theta}{(1 - \cos\theta)^2}$

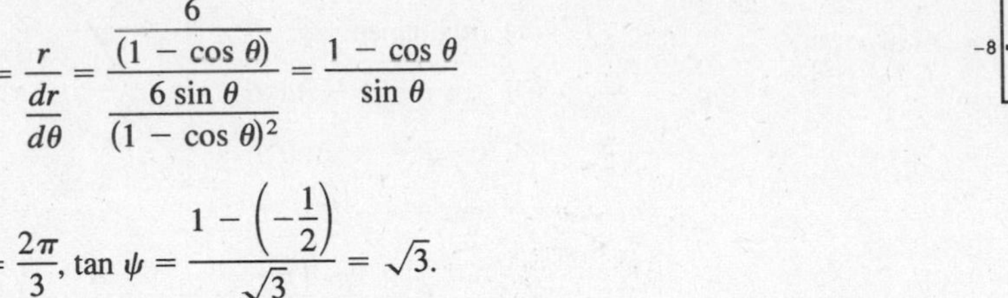

$$\tan\psi = \frac{r}{\dfrac{dr}{d\theta}} = \frac{\dfrac{6}{(1-\cos\theta)}}{\dfrac{6\sin\theta}{(1-\cos\theta)^2}} = \frac{1-\cos\theta}{\sin\theta}$$

At $\theta = \dfrac{2\pi}{3}$, $\tan\psi = \dfrac{1 - \left(-\frac{1}{2}\right)}{\frac{\sqrt{3}}{2}} = \sqrt{3}.$

$\psi = \dfrac{\pi}{3}, (60°)$

113. True

115. True

Section 9.5 Area and Arc Length in Polar Coordinates

1. (a) $r = 8\sin\theta$

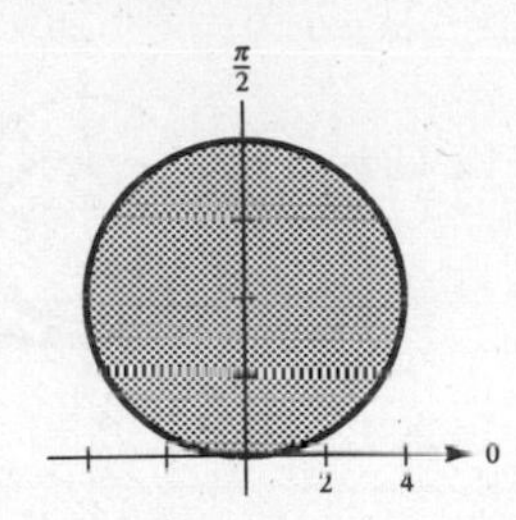

$A = \pi(4)^2 = 16\pi$

(b) $A = 2\left(\dfrac{1}{2}\right)\displaystyle\int_0^{\pi/2}\left[8\sin\theta\right]^2 d\theta$

$= 64\displaystyle\int_0^{\pi/2}\sin^2\theta\,d\theta$

$= 32\displaystyle\int_0^{\pi/2}(1 - \cos 2\theta)\,d\theta$

$= 32\left[\theta - \dfrac{\sin 2\theta}{2}\right]_0^{\pi/2} = 16\pi$

3. $A = 2\left[\dfrac{1}{2}\displaystyle\int_0^{\pi/6}(2\cos 3\theta)^2\,d\theta\right] = 2\left[\theta + \dfrac{1}{6}\sin 6\theta\right]_0^{\pi/6} = \dfrac{\pi}{3}$

5. $A = 2\left[\dfrac{1}{2}\displaystyle\int_0^{\pi/4}(\cos 2\theta)^2\,d\theta\right]$

$= \dfrac{1}{2}\left[\theta + \dfrac{1}{4}\sin 4\theta\right]_0^{\pi/4} = \dfrac{\pi}{8}$

7. $A = 2\left[\dfrac{1}{2}\displaystyle\int_{\pi/2}^{\pi/2}(1 - \sin\theta)^2\,d\theta\right]$

$= \left[\dfrac{3}{2}\theta + 2\cos\theta - \dfrac{1}{4}\sin 2\theta\right]_{-\pi/2}^{\pi/2} = \dfrac{3\pi}{2}$

9. $A = 2\left[\dfrac{1}{2}\displaystyle\int_{2\pi/3}^{\pi}(1 + 2\cos\theta)^2\,d\theta\right]$

$= \left[3\theta + 4\sin\theta + \sin 2\theta\right]_{2\pi/3}^{\pi} = \dfrac{2\pi - 3\sqrt{3}}{2}$

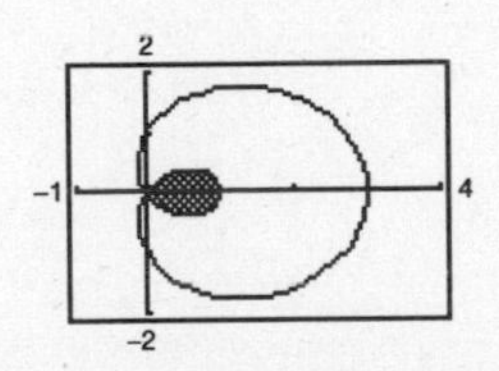

11. The area inside the outer loop is

$2\left[\dfrac{1}{2}\displaystyle\int_0^{2\pi/3}(1 + 2\cos\theta)^2\,d\theta\right] = \left[3\theta + 4\sin\theta + \sin 2\theta\right]_0^{2\pi/3} = \dfrac{4\pi + 3\sqrt{3}}{2}.$

From the result of Exercise 9, the area between the loops is

$A = \left(\dfrac{4\pi + 3\sqrt{3}}{2}\right) - \left(\dfrac{2\pi - 3\sqrt{3}}{2}\right) = \pi + 3\sqrt{3}.$

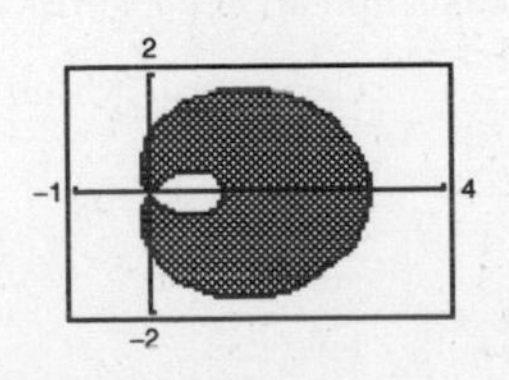

13. $r = 1 + \cos\theta$

$r = 1 - \cos\theta$

Solving simultaneously,

$$1 + \cos\theta = 1 - \cos\theta$$

$$2\cos\theta = 0$$

$$\theta = \frac{\pi}{2}, \frac{3\pi}{2}.$$

Replacing r by $-r$ and θ by $\theta + \pi$ in the first equation and solving, $-1 + \cos\theta = 1 - \cos\theta$, $\cos\theta = 1$, $\theta = 0$. Both curves pass through the pole, $(0, \pi)$, and $(0, 0)$, respectively.

Points of intersection: $\left(1, \frac{\pi}{2}\right), \left(1, \frac{3\pi}{2}\right), (0, 0)$

15. $r = 1 + \cos\theta$

$r = 1 - \sin\theta$

Solving simultaneously,

$$1 + \cos\theta = 1 - \sin\theta$$

$$\cos\theta = -\sin\theta$$

$$\tan\theta = -1$$

$$\theta = \frac{3\pi}{4}, \frac{7\pi}{4}.$$

Replacing r by $-r$ and θ by $\theta + \pi$ in the first equation and solving, $-1 + \cos\theta = 1 - \sin\theta$, $\sin\theta + \cos\theta = 2$, which has no solution. Both curves pass through the pole, $(0, \pi)$, and $(0, \pi/2)$, respectively.

Points of intersection: $\left(\frac{2 - \sqrt{2}}{2}, \frac{3\pi}{4}\right), \left(\frac{2 + \sqrt{2}}{2}, \frac{7\pi}{4}\right), (0, 0)$

17. $r = 4 - 5\sin\theta$

$r = 3\sin\theta$

Solving simultaneously,

$$4 - 5\sin\theta = 3\sin\theta$$

$$\sin\theta = \frac{1}{2}$$

$$\theta = \frac{\pi}{6}, \frac{5\pi}{6}.$$

Both curves pass through the pole, $(0, \arcsin 4/5)$, and $(0, 0)$, respectively.

Points of intersection: $\left(\frac{3}{2}, \frac{\pi}{6}\right), \left(\frac{3}{2}, \frac{5\pi}{6}\right), (0, 0)$

19. $r = \frac{\theta}{2}$

$r = 2$

Solving simultaneously, we have

$\theta/2 = 2$, $\theta = 4$.

Points of intersection:

$(2, 4), (-2, -4)$

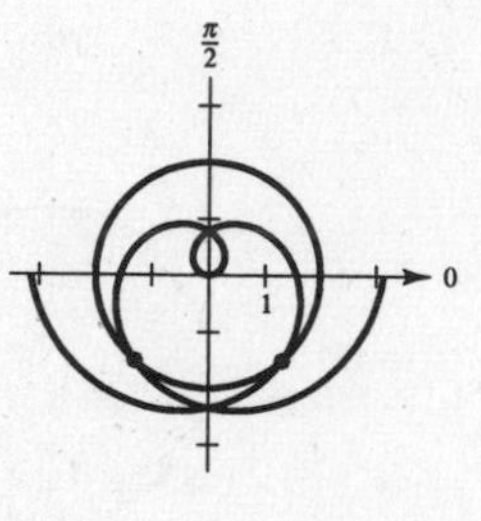

21. $r = 4\sin 2\theta$

$r = 2$

$r = 4\sin 2\theta$ is the equation of a rose curve with four petals and is symmetric to the polar axis, $\theta = \pi/2$, and the pole. Also, $r = 2$ is the equation of a circle of radius 2 centered at the pole. Solving simultaneously,

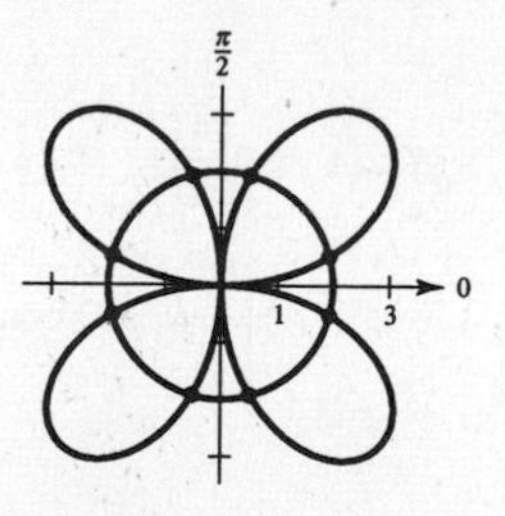

$$4\sin 2\theta = 2$$

$$2\theta = \frac{\pi}{6}, \frac{5\pi}{6}$$

$$\theta = \frac{\pi}{12}, \frac{5\pi}{12}.$$

Therefore, the points of intersection for one petal are $(2, \pi/12)$ and $(2, 5\pi/12)$. By symmetry, the other points of intersection are $(2, 7\pi/12)$, $(2, 11\pi/12)$, $(2, 13\pi/12)$, $(2, 17\pi/12)$, $(2, 19\pi/12)$, and $(2, 23\pi/12)$.

23. $r = 2 + 3\cos\theta$

$r = \dfrac{\sec\theta}{2}$

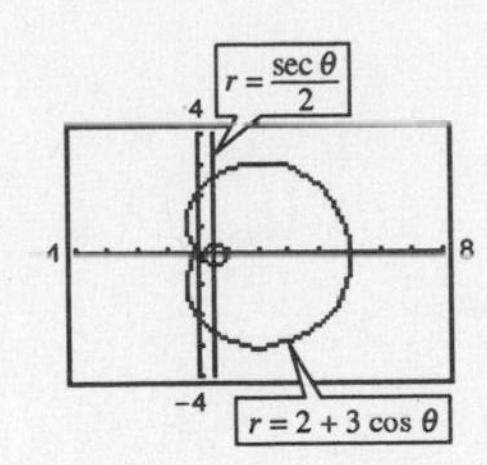

The graph of $r = 2 + 3\cos\theta$ is a limaçon with an inner loop $(b > a)$ and is symmetric to the polar axis. The graph of $r = (\sec\theta)/2$ is the vertical line $x = 1/2$. Therefore, there are four points of intersection. Solving simultaneously,

$$2 + 3\cos\theta = \frac{\sec\theta}{2}$$

$$6\cos^2\theta + 4\cos\theta - 1 = 0$$

$$\cos\theta = \frac{-2 \pm \sqrt{10}}{6}$$

$$\theta = \arccos\left(\frac{-2+\sqrt{10}}{6}\right) \approx 1.376$$

$$\theta = \arccos\left(\frac{-2-\sqrt{10}}{6}\right) \approx 2.6068.$$

Points of intersection: $(-0.581, \pm 2.607)$, $(2.581, \pm 1.376)$

25. $r = \cos\theta$

$r = 2 - 3\sin\theta$

Points of intersection:

$(0, 0)$, $(0.935, 0.363)$, $(0.535, -1.006)$

The graphs reach the pole at different times (θ values).

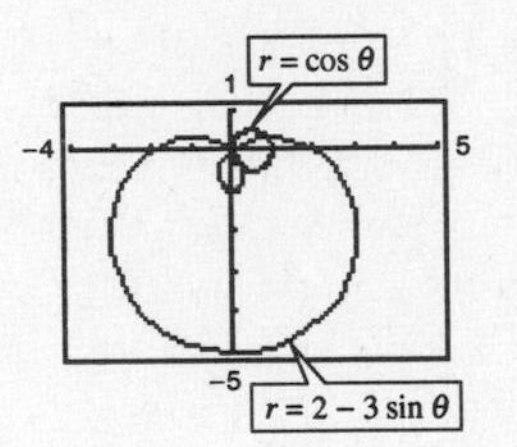

27. From Exercise 21, the points of intersection for one petal are $(2, \pi/12)$ and $(2, 5\pi/12)$. The area within one petal is

$$A = \frac{1}{2}\int_0^{\pi/12} (4\sin 2\theta)^2\,d\theta + \frac{1}{2}\int_{\pi/12}^{5\pi/12} (2)^2\,d\theta + \frac{1}{2}\int_{5\pi/12}^{\pi/2} (4\sin 2\theta)^2\,d\theta$$

$$= 16\int_0^{\pi/12} \sin^2(2\theta)\,d\theta + 2\int_{\pi/12}^{5\pi/12} d\theta \text{ (by symmetry of the petal)}$$

$$= 8\left[\theta - \frac{1}{4}\sin 4\theta\right]_0^{\pi/12} + \left[2\theta\right]_{\pi/12}^{5\pi/12} = \frac{4\pi}{3} - \sqrt{3}.$$

$$\text{Total area} = 4\left(\frac{4\pi}{3} - \sqrt{3}\right) = \frac{16\pi}{3} - 4\sqrt{3} = \frac{4}{3}\left(4\pi - 3\sqrt{3}\right)$$

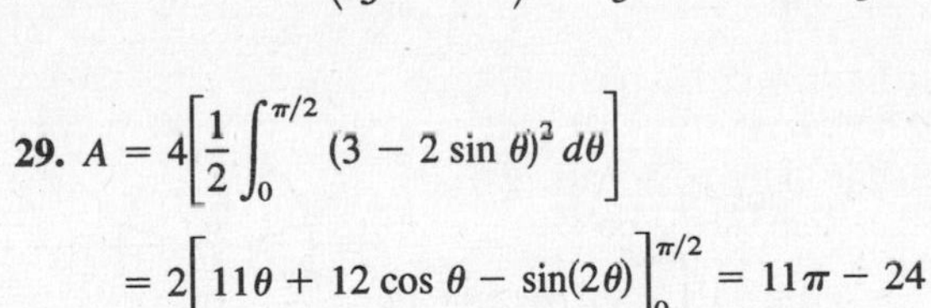

29. $$A = 4\left[\frac{1}{2}\int_0^{\pi/2} (3 - 2\sin\theta)^2\,d\theta\right]$$

$$= 2\left[11\theta + 12\cos\theta - \sin(2\theta)\right]_0^{\pi/2} = 11\pi - 24$$

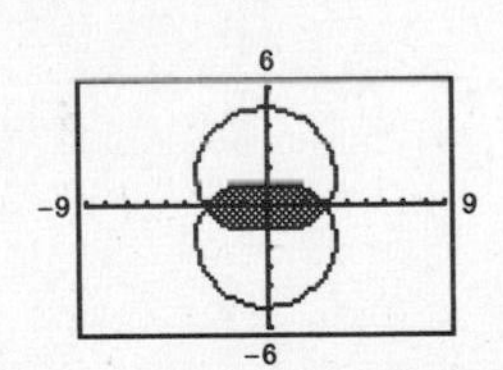

31. $$A = 2\left[\frac{1}{2}\int_0^{\pi/6} (4\sin\theta)^2\,d\theta + \frac{1}{2}\int_{\pi/6}^{\pi/2} (2)^2\,d\theta\right]$$

$$= 16\left[\frac{1}{2}\theta - \frac{1}{4}\sin(2\theta)\right]_0^{\pi/6} + \left[4\theta\right]_{\pi/6}^{\pi/2}$$

$$= \frac{8\pi}{3} - 2\sqrt{3} = \frac{2}{3}\left(4\pi - 3\sqrt{3}\right)$$

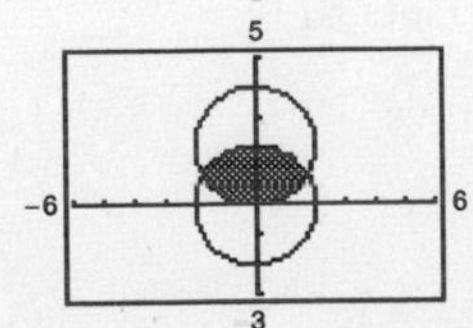

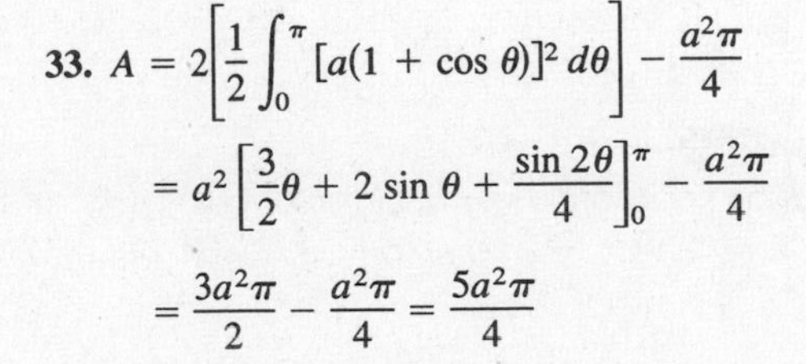

33. $$A = 2\left[\frac{1}{2}\int_0^{\pi} [a(1 + \cos\theta)]^2\,d\theta\right] - \frac{a^2\pi}{4}$$

$$= a^2\left[\frac{3}{2}\theta + 2\sin\theta + \frac{\sin 2\theta}{4}\right]_0^{\pi} - \frac{a^2\pi}{4}$$

$$= \frac{3a^2\pi}{2} - \frac{a^2\pi}{4} = \frac{5a^2\pi}{4}$$

35. $A = \frac{\pi a^2}{8} + \frac{1}{2}\int_{\pi/2}^{\pi} [a(1 + \cos\theta)]^2 \, d\theta$

$= \frac{\pi a^2}{8} + \frac{a^2}{2}\int_{\pi/2}^{\pi} \left(\frac{3}{2} + 2\cos\theta + \frac{\cos 2\theta}{2}\right) d\theta$

$= \frac{\pi a^2}{8} + \frac{a^2}{2}\left[\frac{3}{2}\theta + 2\sin\theta + \frac{\sin 2\theta}{4}\right]_{\pi/2}^{\pi}$

$= \frac{\pi a^2}{8} + \frac{a^2}{2}\left[\frac{3\pi}{2} - \frac{3\pi}{4} - 2\right] = \frac{a^2}{2}[\pi - 2]$

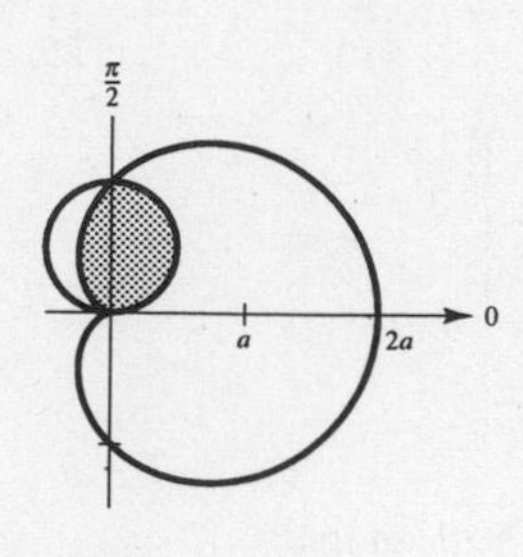

37. (a) $r = a\cos^2\theta$

$r^3 = ar^2\cos^2\theta$

$(x^2 + y^2)^{3/2} = ax^2$

(b)

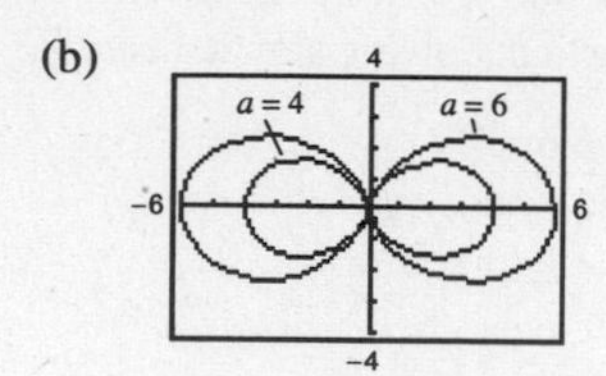

(c) $A = 4\left(\frac{1}{2}\right)\int_0^{\pi/2} [(6\cos^2\theta)^2 - (4\cos^2\theta)^2] \, d\theta = 40\int_0^{\pi/2} \cos^4\theta \, d\theta = 10\int_0^{\pi/2} (1 + \cos 2\theta)^2 \, d\theta$

$= 10\int_0^{\pi/2} \left(1 + 2\cos 2\theta + \frac{1 - \cos 4\theta}{2}\right) d\theta = 10\left[\frac{3}{2}\theta + \sin 2\theta + \frac{1}{8}\sin 4\theta\right]_0^{\pi/2} = \frac{15\pi}{2}$

39. $r = a\cos(n\theta)$

For $n = 1$:

$r = a\cos\theta$

$A = \pi\left(\frac{a}{2}\right)^2 = \frac{\pi a^2}{4}$

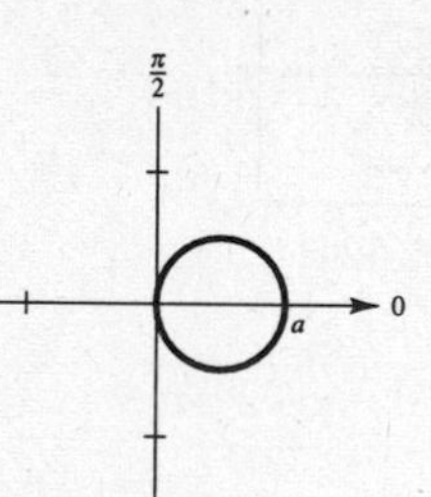

For $n = 2$:

$r = a\cos 2\theta$

$A = 8\left(\frac{1}{2}\right)\int_0^{\pi/4} (a\cos 2\theta)^2 \, d\theta = \frac{\pi a^2}{2}$

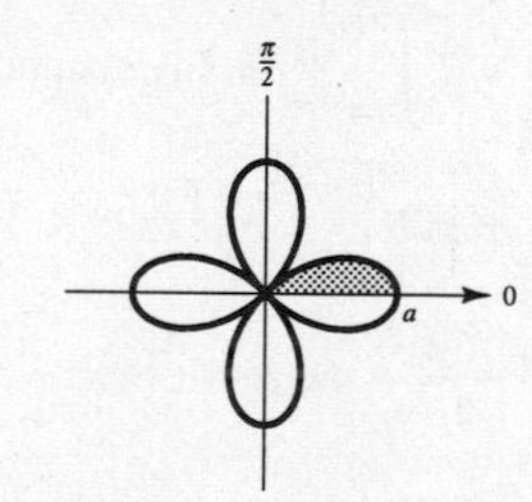

For $n = 3$:

$r = a\cos 3\theta$

$A = 6\left(\frac{1}{2}\right)\int_0^{\pi/6} (a\cos 3\theta)^2 \, d\theta = \frac{\pi a^2}{4}$

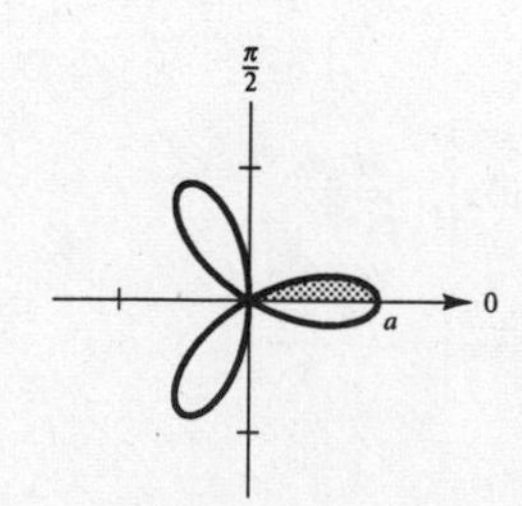

For $n = 4$:

$r = a\cos 4\theta$

$A = 16\left(\frac{1}{2}\right)\int_0^{\pi/8} (a\cos 4\theta)^2 \, d\theta = \frac{\pi a^2}{2}$

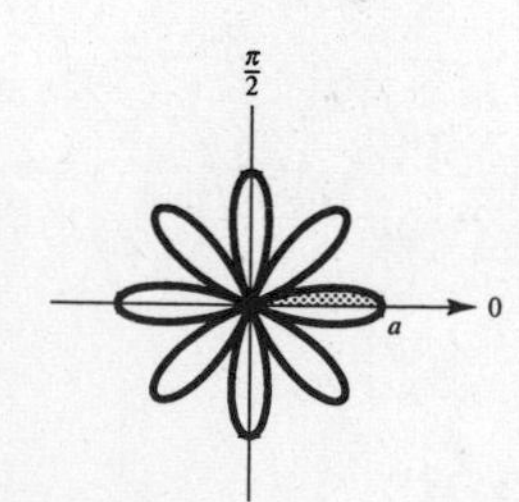

In general, the area of the region enclosed by $r = a\cos(n\theta)$ for $n = 1, 2, 3, \ldots$ is $(\pi a^2)/4$ if n is odd and is $(\pi a^2)/2$ if n is even.

41. $r = a$

$r' = 0$

$$s = \int_0^{2\pi} \sqrt{a^2 + 0^2}\, d\theta = \Big[a\theta\Big]_0^{2\pi} = 2\pi a$$

(circumference of circle of radius a)

43. $r = 1 + \sin\theta$

$r' = \cos\theta$

$$s = 2\int_{\pi/2}^{3\pi/2} \sqrt{(1+\sin\theta)^2 + (\cos\theta)^2}\, d\theta$$

$$= 2\sqrt{2}\int_{\pi/2}^{3\pi/2} \sqrt{1+\sin\theta}\, d\theta$$

$$= 2\sqrt{2}\int_{\pi/2}^{3\pi/2} \frac{-\cos\theta}{\sqrt{1-\sin\theta}}\, d\theta$$

$$= \Big[4\sqrt{2}\sqrt{1-\sin\theta}\Big]_{\pi/2}^{3\pi/2}$$

$$= 4\sqrt{2}\left(\sqrt{2} - 0\right) = 8$$

45. $r = 2\theta,\ 0 \le \theta \le \dfrac{\pi}{2}$

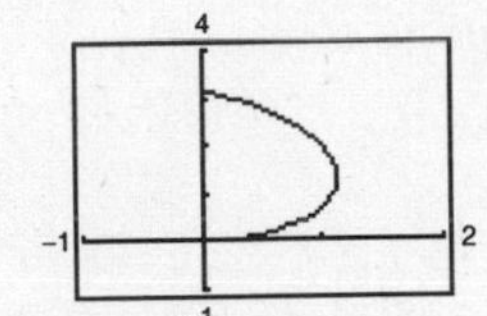

Length ≈ 4.16

47. $r = \dfrac{1}{\theta},\ \pi \le \theta \le 2\pi$

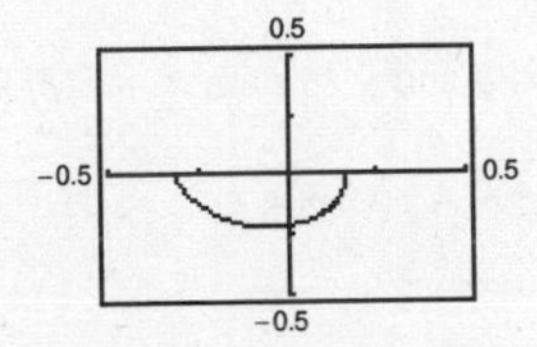

Length ≈ 0.71

49. $r = \sin(3\cos\theta),\ 0 \le \theta \le \pi$

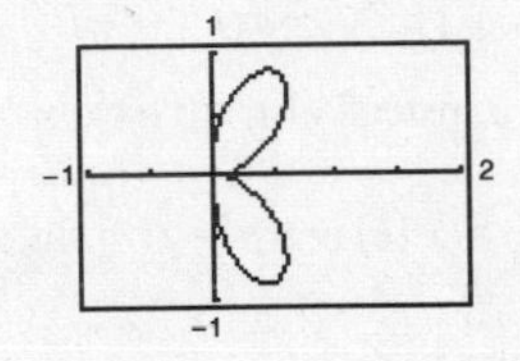

Length ≈ 4.39

51. $r = 6\cos\theta$

$r' = -6\sin\theta$

$$S = 2\pi\int_0^{\pi/2} 6\cos\theta\sin\theta\sqrt{36\cos^2\theta + 36\sin^2\theta}\, d\theta$$

$$= 72\pi\int_0^{\pi/2} \sin\theta\cos\theta\, d\theta$$

$$= \Big[36\pi\sin^2\theta\Big]_0^{\pi/2}$$

$$= 36\pi$$

53. $r = e^{a\theta}$

$r' = ae^{a\theta}$

$$S = 2\pi\int_0^{\pi/2} e^{a\theta}\cos\theta\sqrt{(e^{a\theta})^2 + (ae^{a\theta})^2}\, d\theta$$

$$= 2\pi\sqrt{1+a^2}\int_0^{\pi/2} e^{2a\theta}\cos\theta\, d\theta$$

$$= 2\pi\sqrt{1+a^2}\left[\frac{e^{2a\theta}}{4a^2+1}(2a\cos\theta + \sin\theta)\right]_0^{\pi/2}$$

$$= \frac{2\pi\sqrt{1+a^2}}{4a^2+1}(e^{\pi a} - 2a)$$

55. $r = 4\cos 2\theta$

$r' = -8\sin 2\theta$

$$S = 2\pi\int_0^{\pi/4} 4\cos 2\theta\sin\theta\sqrt{16\cos^2 2\theta + 64\sin^2 2\theta}\, d\theta$$

$$= 32\pi\int_0^{\pi/4} \cos 2\theta\sin\theta\sqrt{\cos^2 2\theta + 4\sin^2 2\theta}\, d\theta \approx 21.87$$

57. $$\text{Area} = \frac{1}{2}\int_\alpha^\beta [f(\theta)]^2 d\theta = \frac{1}{2}\int_\alpha^\beta r^2\, d\theta$$

$$\text{Arc length} = \int_\alpha^\beta \sqrt{f(\theta)^2 + f'(\theta)^2}\, d\theta = \int_\alpha^\beta \sqrt{r^2 + \left(\frac{dr}{d\theta}\right)^2}\, d\theta$$

59. (a) is correct: $s \approx 33.124$.

61. Revolve $r = a$ about the line $r = b \sec \theta$ where $b > a > 0$.

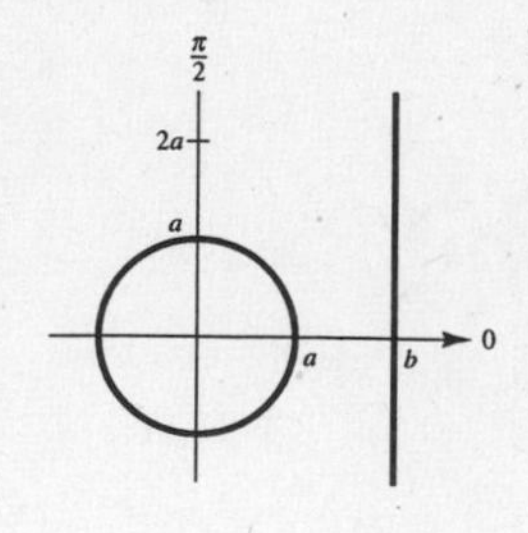

$f(\theta) = a$

$f'(\theta) = 0$

$$S = 2\pi \int_0^{2\pi} [b - a \cos \theta]\sqrt{a^2 + 0^2}\, d\theta$$

$$= 2\pi a\Big[b\theta - a \sin \theta\Big]_0^{2\pi}$$

$$= 2\pi a(2\pi b) = 4\pi^2 ab$$

63. False. $f(\theta) = 1$ and $g(\theta) = -1$ have the same graphs.

65. In parametric form,

$$s = \int_a^b \sqrt{\left(\frac{dx}{dt}\right)^2 + \left(\frac{dy}{dt}\right)^2}\, dt.$$

Using θ instead of t, we have $x = r \cos \theta = f(\theta) \cos \theta$ and $y = r \sin \theta = f(\theta) \sin \theta$. Thus,

$$\frac{dx}{d\theta} = f'(\theta) \cos \theta - f(\theta) \sin \theta \text{ and } \frac{dy}{d\theta} = f'(\theta) \sin \theta + f(\theta) \cos \theta.$$

It follows that

$$\left(\frac{dx}{d\theta}\right)^2 + \left(\frac{dy}{d\theta}\right)^2 = [f(\theta)]^2 + [f'(\theta)]^2.$$

Therefore, $s = \displaystyle\int_\alpha^\beta \sqrt{[f(\theta)]^2 + [f'(\theta)]^2}\, d\theta$

Section 9.6 Polar Equations of Conics and Kepler's Laws

1. $r = \dfrac{2e}{1 + e \cos \theta}$

(a) $e = 1,\ r = \dfrac{2}{1 + \cos \theta}$, parabola

(b) $e = 0.5,\ r = \dfrac{1}{1 + 0.5 \cos \theta} = \dfrac{2}{2 + \cos \theta}$, ellipse

(c) $e = 1.5,\ r = \dfrac{3}{1 + 1.5 \cos \theta} = \dfrac{6}{2 + 3 \cos \theta}$, hyperbola

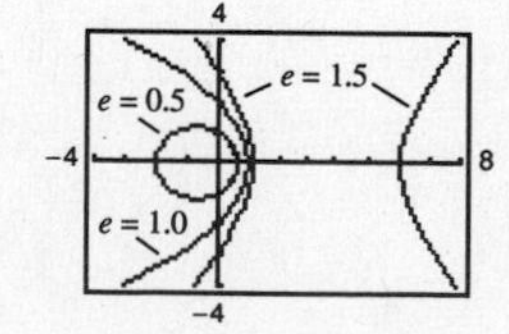

3. $r = \dfrac{2e}{1 - e \sin \theta}$

(a) $e = 1,\ r = \dfrac{2}{1 - \sin \theta}$, parabola

(b) $e = 0.5,\ r = \dfrac{1}{1 - 0.5 \sin \theta} = \dfrac{2}{2 - \sin \theta}$, ellipse

(c) $e = 1.5,\ r = \dfrac{3}{1 - 1.5 \sin \theta} = \dfrac{6}{2 - 3 \sin \theta}$, hyperbola

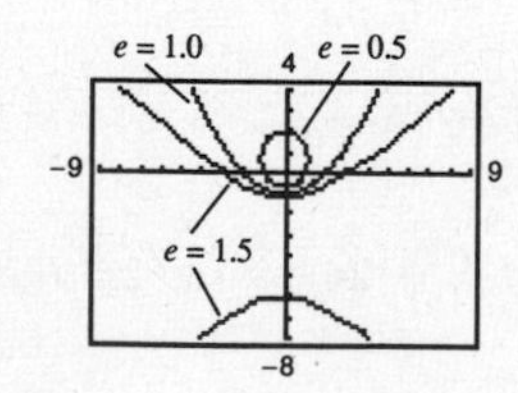

5. $r = \dfrac{4}{1 + e \sin \theta}$

(a)

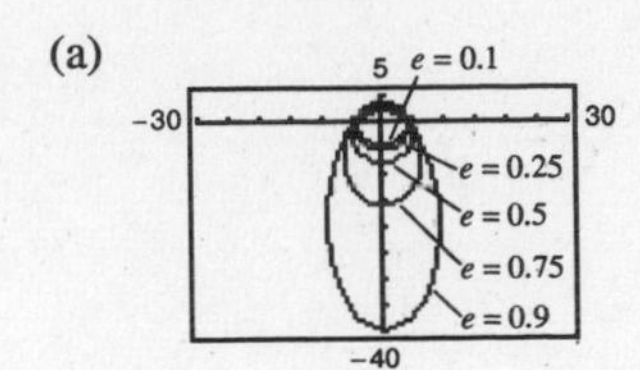

The conic is an ellipse. As $e \to 1^-$, the ellipse becomes more elliptical, and as $e \to 0^+$, it becomes more circular.

(b)

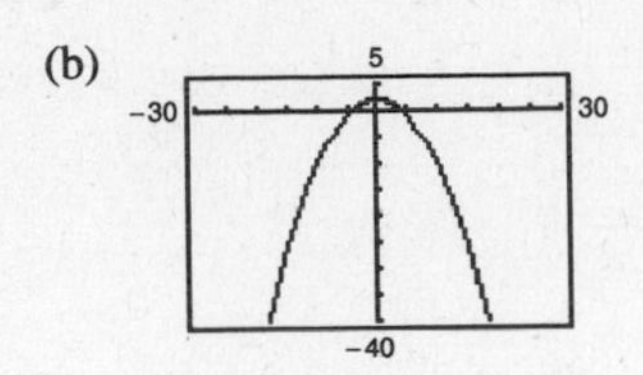

The conic is a parabola.

(c)

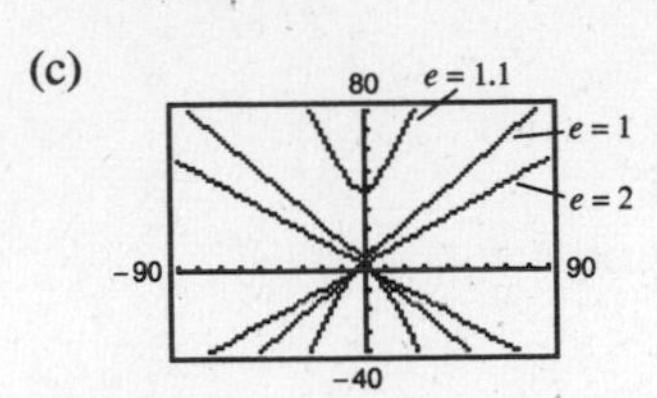

The conic is a hyperbola. As $e \to 1^+$, the hyperbolas opens more slowly, and as $e \to \infty$, they open more rapidly.

7. Parabola; Matches (c)

9. Hyperbola; Matches (a)

11. Ellipse; Matches (b)

13. $r = \dfrac{-1}{1 - \sin \theta}$

Parabola since $e = 1$

Vertex: $\left(-\dfrac{1}{2}, \dfrac{3\pi}{2}\right)$

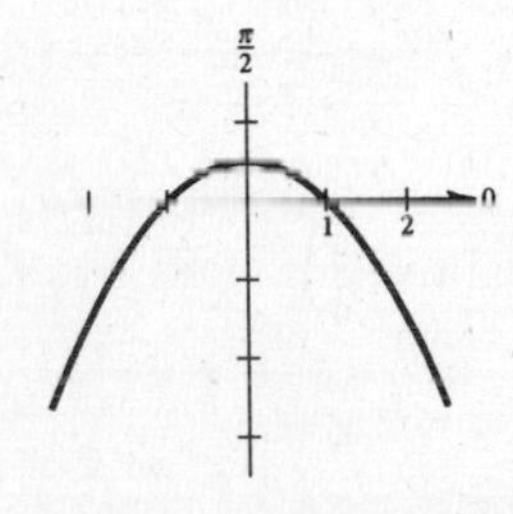

15. $r = \dfrac{6}{2 + \cos \theta}$

$= \dfrac{3}{1 + (1/2) \cos \theta}$

Ellipse since $e = \dfrac{1}{2} < 1$

Vertices: $(2, 0), (6, \pi)$

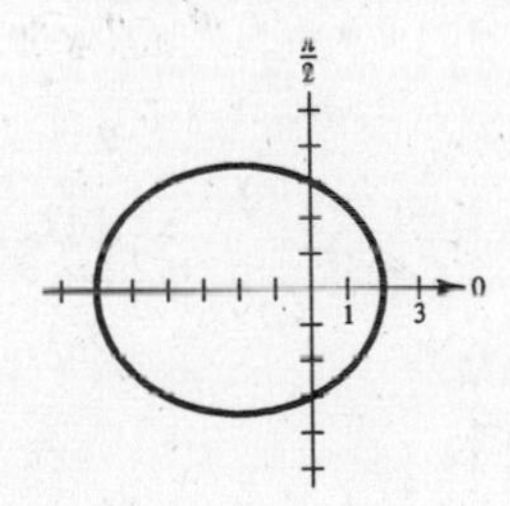

17. $r(2 + \sin \theta) = 4$

$r = \dfrac{4}{2 + \sin \theta}$

$= \dfrac{2}{1 + (1/2) \sin \theta}$

Ellipse since $e = \dfrac{1}{2} < 1$

Vertices: $\left(\dfrac{4}{3}, \dfrac{\pi}{2}\right), \left(4, \dfrac{3\pi}{2}\right)$

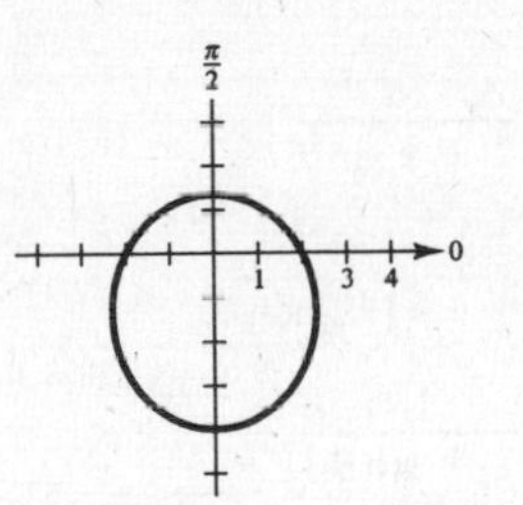

19. $r = \dfrac{5}{-1 + 2 \cos \theta} = \dfrac{-5}{1 - 2 \cos \theta}$

Hyperbola since $e = 2 > 1$

Vertices: $(5, 0), \left(-\dfrac{5}{3}, \pi\right)$

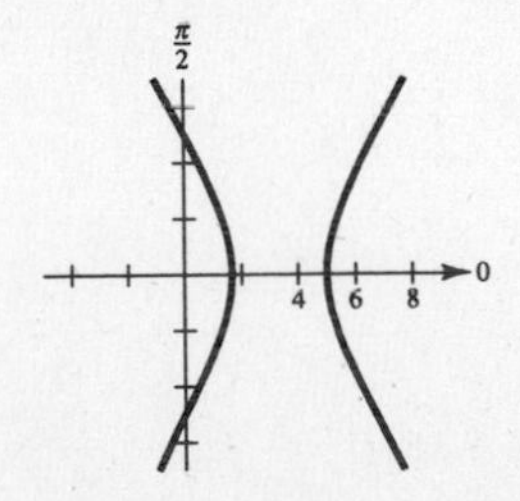

21. $r = \dfrac{3}{2 + 6 \sin \theta} = \dfrac{3/2}{1 + 3 \sin \theta}$

Hyperbola since $e = 3 > 1$

Vertices: $\left(\dfrac{3}{8}, \dfrac{\pi}{2}\right), \left(-\dfrac{3}{4}, \dfrac{3\pi}{2}\right)$

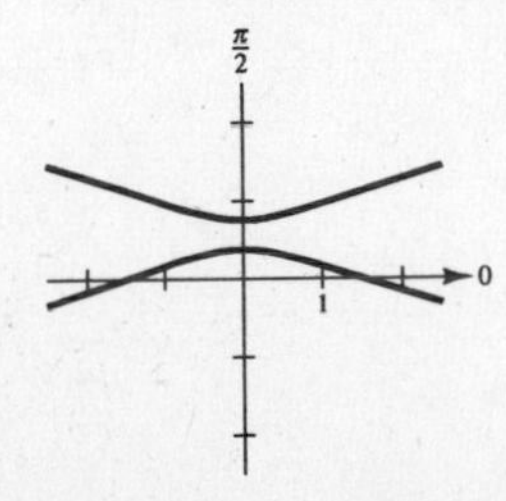

23. Ellipse

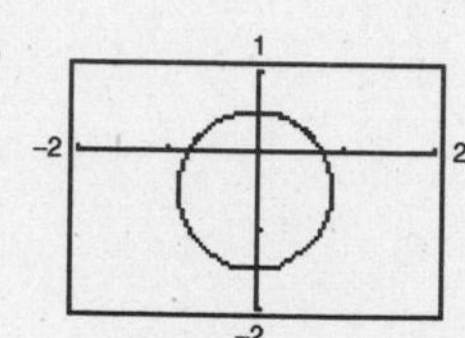

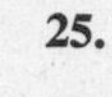
25. Parabola

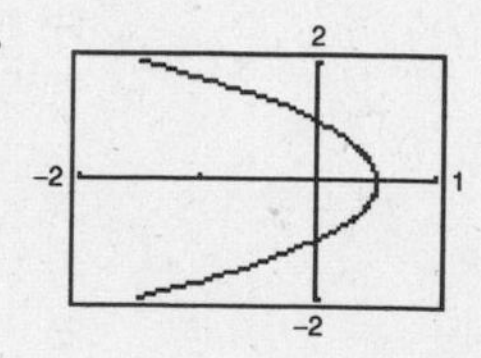

27. $r = \dfrac{-1}{1 - \sin\left(\theta - \frac{\pi}{4}\right)}$

Rotate the graph of

$$r = \frac{-1}{1 - \sin\theta}$$

counterclockwise through the angle $\frac{\pi}{4}$.

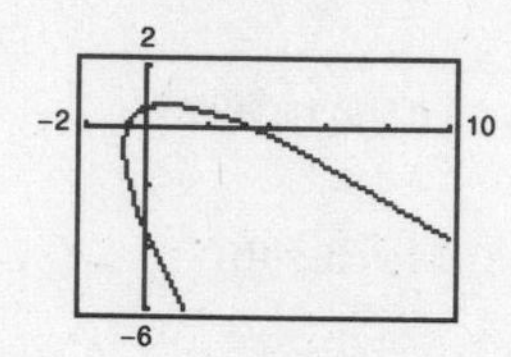

29. $r = \dfrac{6}{2 + \cos\left(\theta + \frac{\pi}{6}\right)}$

Rotate the graph of

$$r = \frac{6}{2 + \cos\theta}$$

clockwise through the angle $\frac{\pi}{6}$.

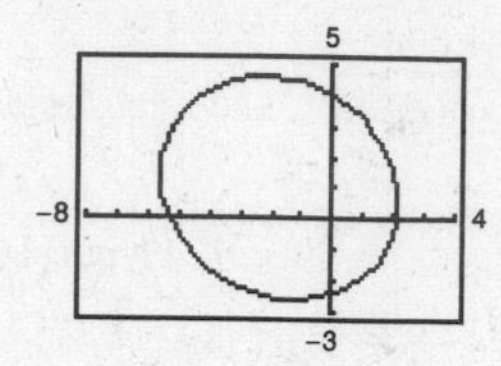

31. Change θ to $\theta + \frac{\pi}{4}$: $r = \dfrac{5}{5 + 3\cos\left(\theta + \frac{\pi}{4}\right)}$.

33. Parabola

$e = 1, x = -1, d = 1$

$$r = \frac{ed}{1 - e\cos\theta} = \frac{1}{1 - \cos\theta}$$

35. Ellipse

$e = \frac{1}{2}, y = 1, d = 1$

$$r = \frac{ed}{1 + e\sin\theta}$$

$$= \frac{1/2}{1 + (1/2)\sin\theta}$$

$$= \frac{1}{2 + \sin\theta}$$

37. Hyperbola

$e = 2, x = 1, d = 1$

$$r = \frac{ed}{1 + e\cos\theta} = \frac{2}{1 + 2\cos\theta}$$

39. Parabola

Vertex: $\left(1, -\frac{\pi}{2}\right)$

$e = 1, d = 2, r = \dfrac{2}{1 - \sin\theta}$

41. Ellipse

Vertices: $(2, 0), (8, \pi)$

$e = \frac{3}{5}, d = \frac{16}{3}$

$$r = \frac{ed}{1 + e\cos\theta}$$

$$= \frac{16/5}{1 + (3/5)\cos\theta}$$

$$= \frac{16}{5 + 3\cos\theta}$$

43. Hyperbola

Vertices: $\left(1, \frac{3\pi}{2}\right), \left(9, \frac{3\pi}{2}\right)$

$e = \frac{5}{4}, d = \frac{9}{5}$

$$r = \frac{ed}{1 - e\sin\theta}$$

$$= \frac{9/4}{1 - (5/4)\sin\theta}$$

$$= \frac{9}{4 - 5\sin\theta}$$

45. Ellipse if $0 < e < 1$, parabola if $e = 1$, hyperbola if $e > 1$.

47. (a) Hyperbola ($e = 2 > 1$)

(b) Ellipse $\left(e = \frac{1}{2} < 1\right)$

(c) Parabola ($e = 1$)

(d) Rotated hyperbola ($e = 3$)

49. $a = 5, c = 4, e = \frac{4}{5}, b = 3$

$$r^2 = \frac{9}{1 - (16/25)\cos^2\theta}$$

51. $a = 3, b = 4, c = 5, e = \frac{5}{3}$

$$r^2 = \frac{-16}{1 - (25/9)\cos^2\theta}$$

53. $$A = 2\left[\frac{1}{2}\int_0^{\pi}\left(\frac{3}{2 - \cos\theta}\right)^2 d\theta\right]$$

$$= 9\int_0^{\pi}\frac{1}{(2 - \cos\theta)^2}\,d\theta \approx 10.88$$

55. Vertices: $(126{,}000, 0), (4119, \pi)$

$$a = \frac{126{,}000 + 4119}{2} = 65{,}059.5,\ c = 65{,}059.5 - 4119 = 60{,}940.5,\ e = \frac{c}{a} = \frac{40{,}627}{43{,}373},\ d = 4119\left(\frac{84{,}000}{40{,}627}\right)$$

$$r = \frac{ed}{1 - e\cos\theta} = \frac{4119(84{,}000/43{,}373)}{1 - (40{,}627/43{,}373)\cos\theta} = \frac{345{,}996{,}000}{43{,}373 - 40{,}627\cos\theta}$$

When $\theta = 60°$, $r = \frac{345{,}996{,}000}{23{,}059.5} \approx 15{,}004.49$.

Distance between the surface of the earth and the satellite is $r - 4000 = 11{,}004.49$ miles.

57. $a = 92.957 \times 10^6$ mi, $e = 0.0167$

$$r = \frac{(1 - e^2)a}{1 - e\cos\theta} = \frac{92{,}931{,}075.2223}{1 - 0.0167\cos\theta}$$

Perihelion distance: $a(1 - e) \approx 91{,}404{,}618$ mi

Aphelion distance: $a(1 + e) \approx 94{,}509{,}382$ mi

59. $a = 5.900 \times 10^9$ km, $e = 0.2481$

$$r = \frac{(1 - e^2)a}{1 - e\cos\theta} \approx \frac{5.537 \times 10^9}{1 - 0.2481\cos\theta}$$

Perihelion distance: $a(1 - e) = 4.436 \times 10^9$ km

Aphelion distance: $a(1 + e) = 7.364 \times 10^9$ km

61. $$r = \frac{5.537 \times 10^9}{1 - 0.2481\cos\theta}$$

(a) $$A = \frac{1}{2}\int_0^{\pi/9}\left[\frac{5.537 \times 10^9}{1 - 0.2481\cos\theta}\right]^2 d\theta \approx 9.341 \times 10^{18}\ \text{km}^2$$

$$248\left[\frac{\frac{1}{2}\int_0^{\pi/9}\left[\frac{5.537 \times 10^9}{1 - 0.2481\cos\theta}\right]^2 d\theta}{\frac{1}{2}\int_0^{2\pi}\left[\frac{5.537 \times 10^9}{1 - 0.2481\cos\theta}\right]^2 d\theta}\right] \approx 21.867\ \text{yr}$$

(b) $$\frac{1}{2}\int_{\pi}^{\alpha - \pi}\left[\frac{5.537 \times 10^9}{1 - 0.2481\cos\theta}\right]^2 d\theta = 9.341 \times 10^{18}$$

$\alpha \approx \pi + 0.8995$ rad

In part (a) the ray swept through a smaller angle to generate the same area since the length of the ray is longer than in part (b).

(c) $$r' = \frac{(-5.537 \times 10^9)(0.2481\sin\theta)}{(1 - 0.2481\cos\theta)^2}$$

$$s = \int_0^{\pi/9}\sqrt{\left(\frac{5.537 \times 10^9}{1 - 0.2481\cos\theta}\right)^2 + \left[\frac{-1.3737297 \times 10^9\sin\theta}{(1 - 0.2481\cos\theta)^2}\right]^2}\,d\theta \approx 2.559 \times 10^9\ \text{km}$$

$$\frac{2.559 \times 10^9\ \text{km}}{21.867\ \text{yr}} \approx 1.17 \times 10^8\ \text{km/yr}$$

$$s = \int_{\pi}^{\pi + 0.899}\sqrt{\left(\frac{5.537 \times 10^9}{1 - 0.2481\cos\theta}\right)^2 + \left[\frac{-1.3737297 \times 10^9\sin\theta}{(1 - 0.2481\cos\theta)^2}\right]^2}\,d\theta \approx 4.119 \times 10^9\ \text{km}$$

$$\frac{4.119 \times 10^9\ \text{km}}{21.867\ \text{yr}} \approx 1.88 \times 10^8\ \text{km/yr}$$

63. $r_1 = \dfrac{ed}{1 + \sin\theta}$ and $r_2 = \dfrac{ed}{1 - \sin\theta}$

Points of intersection: $(ed, 0)$, (ed, π)

$$r_1: \frac{dy}{dx} = \frac{\left(\dfrac{ed}{1 + \sin\theta}\right)(\cos\theta) + \left(\dfrac{-ed\cos\theta}{(1 + \sin\theta)^2}\right)(\sin\theta)}{\left(\dfrac{-ed}{1 + \sin\theta}\right)(\sin\theta) + \left(\dfrac{-ed\cos\theta}{(1 + \sin\theta)^2}\right)(\cos\theta)}$$

At $(ed, 0)$, $\dfrac{dy}{dx} = -1$. At (ed, π), $\dfrac{dy}{dx} = 1$.

$$r_2: \frac{dy}{dx} = \frac{\left(\dfrac{ed}{1 - \sin\theta}\right)(\cos\theta) + \left(\dfrac{ed\cos\theta}{(1 - \sin\theta)^2}\right)(\sin\theta)}{\left(\dfrac{-ed}{1 - \sin\theta}\right)(\sin\theta) + \left(\dfrac{ed\cos\theta}{(1 - \sin\theta)^2}\right)(\cos\theta)}$$

At $(ed, 0)$, $\dfrac{dy}{dx} = 1$. At (ed, π), $\dfrac{dy}{dx} = -1$.

Therefore, at $(ed, 0)$ we have $m_1m_2 = (-1)(1) = -1$, and at (ed, π) we have $m_1m_2 = 1(-1) = -1$. The curves intersect at right angles.

Review Exercises for Chapter 9

1. Matches (d) - ellipse

3. Matches (a) - parabola

5. $16x^2 + 16y^2 - 16x + 24y - 3 = 0$

$$\left(x^2 - x + \frac{1}{4}\right) + \left(y^2 + \frac{3}{2}y + \frac{9}{16}\right) = \frac{3}{16} + \frac{1}{4} + \frac{9}{16}$$

$$\left(x - \frac{1}{2}\right)^2 + \left(y + \frac{3}{4}\right)^2 = 1$$

Circle

Center: $\left(\dfrac{1}{2}, -\dfrac{3}{4}\right)$

Radius: 1

7. $3x^2 - 2y^2 + 24x + 12y + 24 = 0$

$$3(x^2 + 8x + 16) - 2(y^2 - 6y + 9) = -24 + 48 - 18$$

$$\frac{(x + 4)^2}{2} - \frac{(y - 3)^2}{3} = 1$$

Hyperbola

Center: $(-4, 3)$

Vertices: $\left(-4 \pm \sqrt{2}, 3\right)$

Asymptotes: $y = 3 \pm \sqrt{\dfrac{3}{2}}(x + 4)$

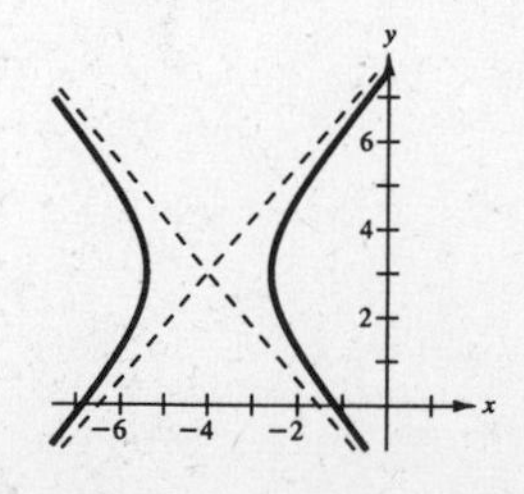

9. $3x^2 + 2y^2 - 12x + 12y + 29 = 0$

$$3(x^2 - 4x + 4) + 2(y^2 + 6y + 9) = -29 + 12 + 18$$

$$\frac{(x - 2)^2}{1/3} + \frac{(y + 3)^2}{1/2} = 1$$

Ellipse

Center: $(2, -3)$

Vertices: $\left(2, -3 \pm \dfrac{\sqrt{2}}{2}\right)$

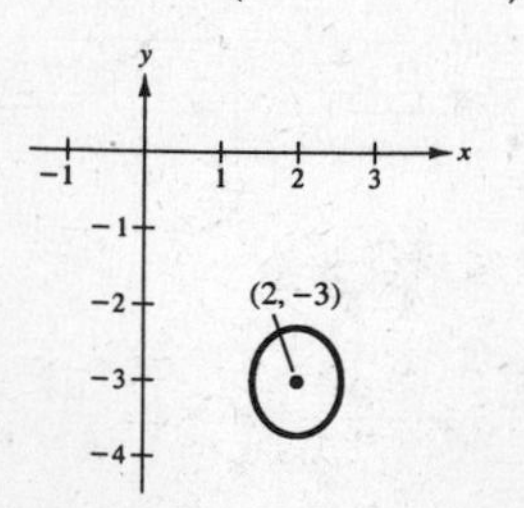

11. Vertex: $(0, 2)$

Directrix: $x = -3$

Parabola opens to the right

$p = 3$

$(y - 2)^2 = 4(3)(x - 0)$

$y^2 - 4y - 12x + 4 = 0$

13. Vertices: $(-3, 0), (7, 0)$

Foci: $(0, 0), (4, 0)$

Horizontal major axis

Center: $(2, 0)$

$a = 5, c = 2, b = \sqrt{21}$

$$\frac{(x-2)^2}{25} + \frac{y^2}{21} = 1$$

15. Vertices: $(\pm 4, 0)$

Foci: $(\pm 6, 0)$

Center: $(0, 0)$

Horizontal transverse axis

$a = 4, c = 6, b = \sqrt{36 - 16} = 2\sqrt{5}$

$$\frac{x^2}{16} - \frac{y^2}{20} = 1$$

17. $\frac{x^2}{9} + \frac{y^2}{4} = 1, a = 3, b = 2, c = \sqrt{5}, e = \frac{\sqrt{5}}{3}$

By Example 5 of Section 9.1,

$$C = 12\int_0^{\pi/2} \sqrt{1 - \left(\frac{5}{9}\right)\sin^2\theta}\, d\theta \approx 15.87.$$

19. $y = x - 2$ has a slope of 1. The perpendicular slope is -1.

$$y = x^2 - 2x + 2$$

$$\frac{dy}{dx} = 2x - 2 = -1 \text{ when } x = \frac{1}{2} \text{ and } y = \frac{5}{4}.$$

Perpendicular line: $y - \frac{5}{4} = -1\left(x - \frac{1}{2}\right)$

$$4x + 4y - 7 = 0$$

21. (a) $V = (\pi ab)(\text{Length}) = 12\pi(16) = 192\pi \text{ ft}^3$

(b) $$F = 2(62.4)\int_{-3}^{3} (3 - y)\frac{4}{3}\sqrt{9 - y^2}\, dy = \frac{8}{3}(62.4)\left[3\int_{-3}^{3} \sqrt{9 - y^2}\, dy - \int_{-3}^{3} y\sqrt{9 - y^2}\, dy\right]$$

$$= \frac{8}{3}(62.4)\left[\frac{3}{2}\left(y\sqrt{9 - y^2} + 9\arcsin\frac{y}{3}\right) + \frac{1}{3}(9 - y^2)^{3/2}\right]_{-3}^{3}$$

$$= \frac{8}{3}(62.4)\left[\frac{3}{2}\left(\frac{9\pi}{2}\right) - \frac{3}{2}\left(-\frac{9\pi}{2}\right)\right] = \frac{8}{3}(62.4)\left(\frac{27\pi}{2}\right) \approx 7057.274$$

(c) You want $\frac{3}{4}$ of the total area of 12π covered. Find h so that

$$2\int_0^h \frac{4}{3}\sqrt{9 - y^2}\, dy = 3\pi$$

$$\int_0^h \sqrt{9 - y^2}\, dy = \frac{9\pi}{8}$$

$$\frac{1}{2}\left[y\sqrt{9 - y^2} + 9\arcsin\left(\frac{y}{3}\right)\right]_0^h = \frac{9\pi}{8}$$

$$h\sqrt{9 - h^2} + 9\arcsin\left(\frac{h}{3}\right) = \frac{9\pi}{4}.$$

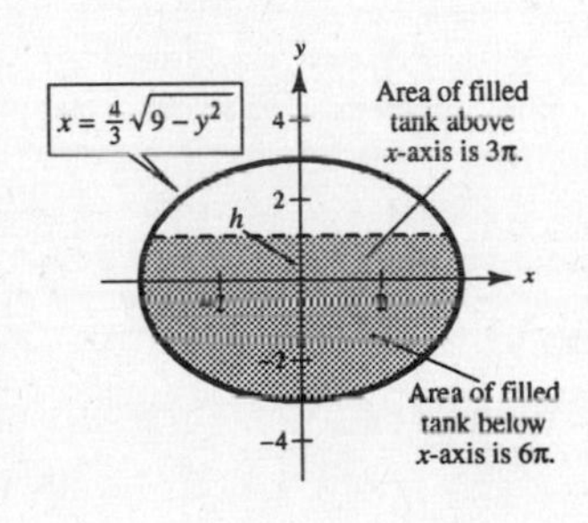

By Newton's Method, $h \approx 1.212$. Therefore, the total height of the water is $1.212 + 3 = 4.212$ ft.

(d) Area of ends $= 2(12\pi) = 24\pi$

Area of sides $=$ (Perimeter)(Length)

$$= 16\int_0^{\pi/2} \left(\sqrt{1 - \left(\frac{7}{16}\right)\sin^2\theta}\right) d\theta(16) \text{ [from Example 5 of Section 9.1]}$$

$$\approx 256\left(\frac{\pi/2}{12}\right)\left[\sqrt{1 - \left(\frac{7}{16}\right)\sin^2(0)} + 4\sqrt{1 - \left(\frac{7}{16}\right)\sin^2\left(\frac{\pi}{8}\right)} + 2\sqrt{1 - \left(\frac{7}{16}\right)\sin^2\left(\frac{\pi}{4}\right)}\right.$$

$$\left. + 4\sqrt{1 - \left(\frac{7}{16}\right)\sin^2\left(\frac{3\pi}{8}\right)} + \sqrt{1 - \left(\frac{7}{16}\right)\sin^2\left(\frac{\pi}{2}\right)}\right] \approx 353.65$$

Total area $= 24\pi + 353.65 \approx 429.05$

23. $x = 1 + 4t, y = 2 - 3t$

$$t = \frac{x-1}{4} \Rightarrow y = 2 - 3\left(\frac{x-1}{4}\right)$$

$$y = -\frac{3}{4}x + \frac{11}{4}$$

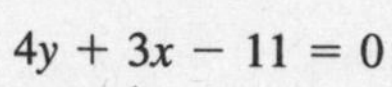

$4y + 3x - 11 = 0$

Line

25. $x = 6\cos\theta, y = 6\sin\theta$

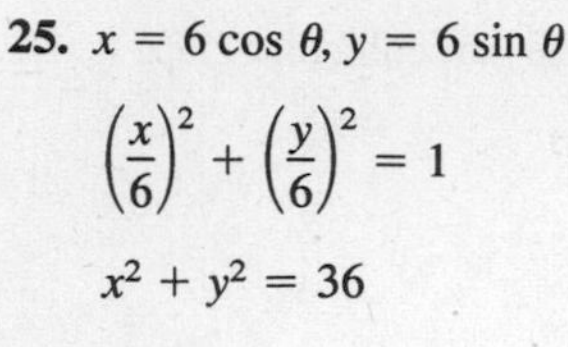

$$\left(\frac{x}{6}\right)^2 + \left(\frac{y}{6}\right)^2 = 1$$

$x^2 + y^2 = 36$

Circle

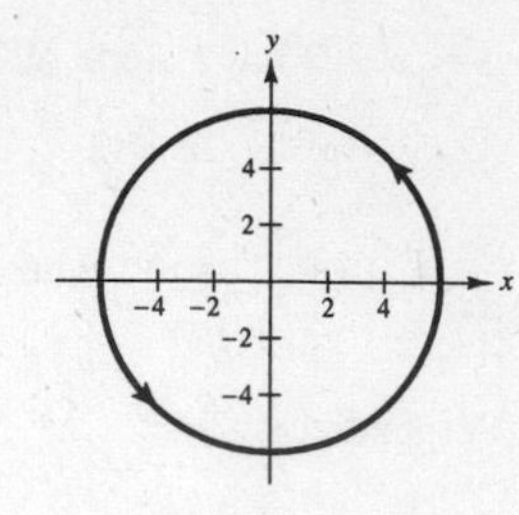

27. $x = 2 + \sec\theta, y = 3 + \tan\theta$

$(x-2)^2 = \sec^2\theta = 1 + \tan^2\theta = 1 + (y-3)^2$

$(x-2)^2 - (y-3)^2 = 1$

Hyperbola

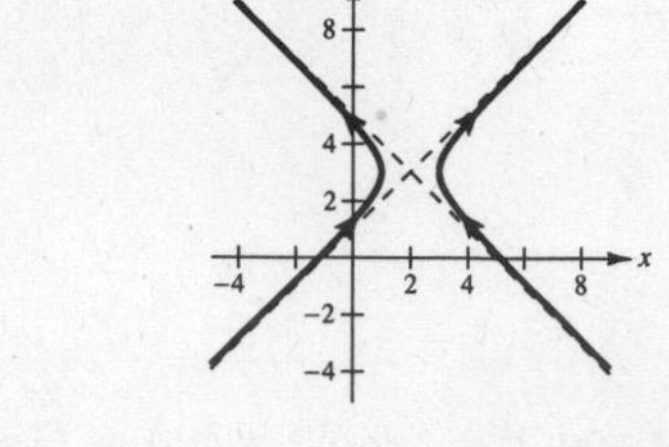

29. $x = 3 + (3 - (-2))t = 3 + 5t$

$y = 2 + (2 - 6)t = 2 - 4t$

(other answers possible)

31. $\frac{(x+3)^2}{16} + \frac{(y-4)^2}{9} = 1$

Let $\frac{(x+3)^2}{16} = \cos^2\theta$ and $\frac{(y-4)^2}{9} = \sin^2\theta$.

Then $x = -3 + 4\cos\theta$ and $y = 4 + 3\sin\theta$.

33. $x = \cos 3\theta + 5\cos\theta$

$y = \sin 3\theta + 5\sin\theta$

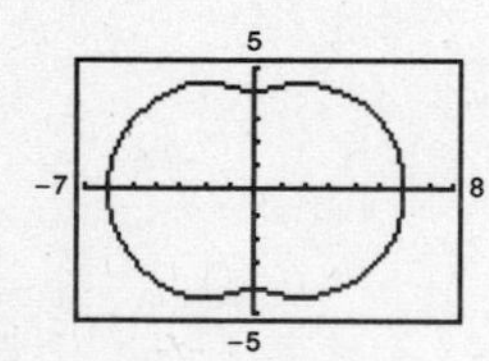

35. (a) $x = 2\cot\theta, y = 4\sin\theta\cos\theta, 0 < \theta < \pi$

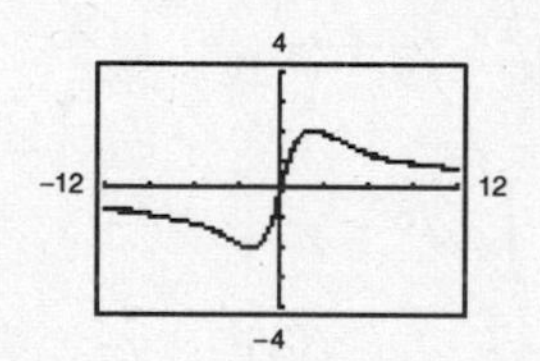

(b) $(4 + x^2)y = (4 + 4\cot^2\theta)4\sin\theta\cos\theta$

$$= 16\csc^2\theta \cdot \sin\theta \cdot \cos\theta$$

$$= 16\frac{\cos\theta}{\sin\theta}$$

$$= 8(2\cot\theta)$$

$$= 8x$$

37. $x = 1 + 4t$

$y = 2 - 3t$

(a) $\frac{dy}{dx} = -\frac{3}{4}$

No horizontal tangents

(b) $t = \frac{x-1}{4}$

$$y = 2 - \frac{3}{4}(x-1) = \frac{-3x+11}{4}$$

(c)

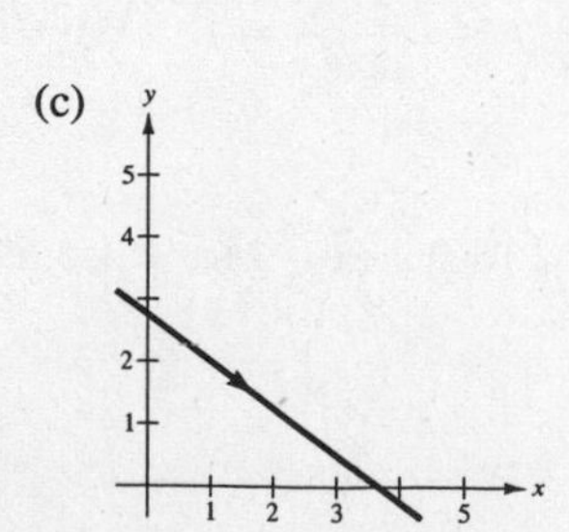

39. $x = \dfrac{1}{t}$

$y = 2t + 3$

(a) $\dfrac{dy}{dx} = \dfrac{2}{-1/t^2} = -2t^2$

No horizontal tangents
$(t \neq 0)$

(b) $t = \dfrac{1}{x}$

$y = \dfrac{2}{x} + 3$

(c)

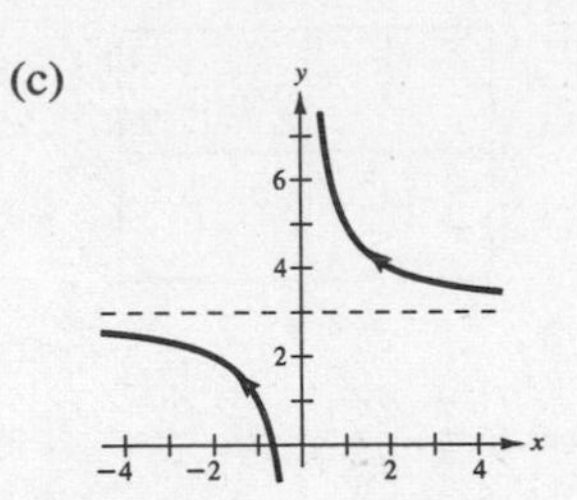

41. $x = \dfrac{1}{2t + 1}$

$y = \dfrac{1}{t^2 - 2t}$

(a) $\dfrac{dy}{dx} = \dfrac{\dfrac{-(2t-2)}{(t^2-2t)^2}}{\dfrac{-2}{(2t+1)^2}} = \dfrac{(t-1)(2t+1)^2}{t^2(t-2)^2} = 0$ when $t = 1$.

Point of horizontal tangency: $\left(\frac{1}{3}, -1\right)$

(b) $2t + 1 = \dfrac{1}{x} \Rightarrow t = \dfrac{1}{2}\left(\dfrac{1}{x} - 1\right)$

$$y = \frac{1}{\frac{1}{2}\left(\frac{1-x}{x}\right)\left[\frac{1}{2}\left(\frac{1-x}{x}\right) - 2\right]}$$

$$= \frac{4x^2}{(1-x)^2 - 4x(1-x)} = \frac{4x^2}{(5x-1)(x-1)}$$

(c)

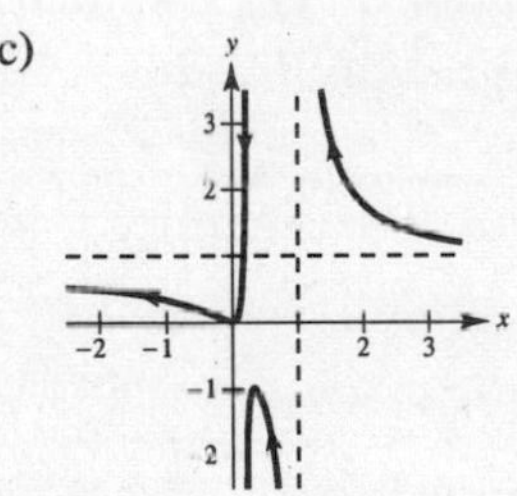

43. $x = 3 + 2\cos\theta$

$y = 2 + 5\sin\theta$

(a) $\dfrac{dy}{dx} = \dfrac{5\cos\theta}{-2\sin\theta} = -2.5\cot\theta = 0$ when $\theta = \dfrac{\pi}{2}, \dfrac{3\pi}{2}$.

Points of horizontal tangency: $(3, 7), (3, -3)$

(b) $\dfrac{(x-3)^2}{4} + \dfrac{(y-2)^2}{25} = 1$

(c)

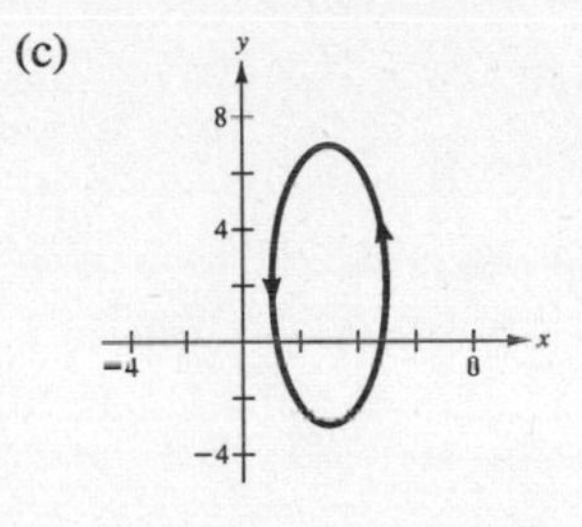

45. $x = \cos^3\theta$

$y = 4\sin^3\theta$

(a) $\dfrac{dy}{dx} = \dfrac{12\sin^2\theta\cos\theta}{3\cos^2\theta(-\sin\theta)} = \dfrac{-4\sin\theta}{\cos\theta} = -4\tan\theta = 0$ when $\theta = 0, \pi$.

But, $\dfrac{dy}{dt} = \dfrac{dx}{dt} = 0$ at $\theta = 0, \pi$. Hence no points of horizontal tangency.

(b) $x^{2/3} + \left(\dfrac{y}{4}\right)^{2/3} = 1$

(c)

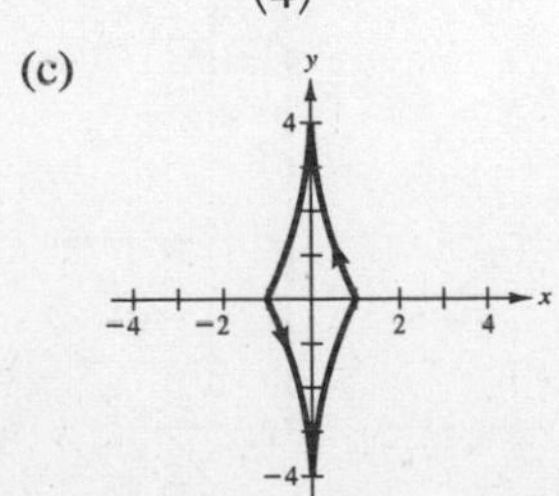

47. $x = \cot\theta$

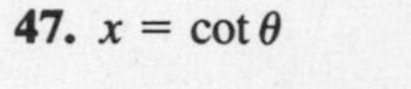

$y = \sin 2\theta = 2\sin\theta\cos\theta$

(a), (c)

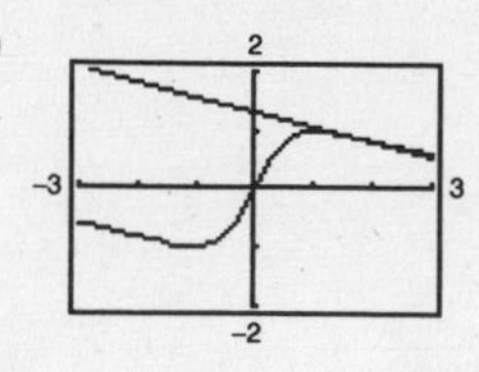

(b) At $\theta = \frac{\pi}{6}$, $\frac{dx}{d\theta} = -4$, $\frac{dy}{d\theta} = 1$, and $\frac{dy}{dx} = -\frac{1}{4}$

49. $x = r(\cos\theta + \theta\sin\theta)$

$y = r(\sin\theta - \theta\cos\theta)$

$\frac{dx}{d\theta} = r\theta\cos\theta$

$\frac{dy}{d\theta} = r\theta\sin\theta$

$$s = r\int_0^{\pi} \sqrt{\theta^2\cos^2\theta + \theta^2\sin^2\theta}\,d\theta$$

$$= r\int_0^{\pi} \theta\,d\theta = \frac{r}{2}\Big[\theta^2\Big]_0^{\pi} = \frac{1}{2}\pi^2 r$$

51. $(x, y) = (4, -4)$

$r = \sqrt{4^2 + (-4)^2} = 4\sqrt{2}$

$\theta = 7\frac{\pi}{4}$

$(r, \theta) = \left(4\sqrt{2}, \frac{7\pi}{4}\right), \left(-4\sqrt{2}, \frac{3\pi}{4}\right)$

53.

$$r = 3\cos\theta$$
$$r^2 = 3r\cos\theta$$
$$x^2 + y^2 = 3x$$
$$x^2 + y^2 - 3x = 0$$

55.

$$r = -2(1 + \cos\theta)$$
$$r^2 = -2r(1 + \cos\theta)$$
$$x^2 + y^2 = -2\left(\pm\sqrt{x^2 + y^2}\right) - 2x$$
$$(x^2 + y^2 + 2x)^2 = 4(x^2 + y^2)$$

57.

$$r^2 = \cos 2\theta = \cos^2\theta - \sin^2\theta$$
$$r^4 = r^2\cos^2\theta - r^2\sin^2\theta$$
$$(x^2 + y^2)^2 = x^2 - y^2$$

59.

$$r = 4\cos 2\theta\sec\theta$$
$$= 4(2\cos^2\theta - 1)\left(\frac{1}{\cos\theta}\right)$$
$$r\cos\theta = 8\cos^2\theta - 4$$
$$x = 8\left(\frac{x^2}{x^2 + y^2}\right) - 4$$
$$x^3 + xy^2 = 4x^2 - 4y^2$$
$$y^2 = x^2\left(\frac{4 - x}{4 + x}\right)$$

61. $(x^2 + y^2)^2 = ax^2y$

$r^4 = a(r^2\cos^2\theta)(r\sin\theta)$

$r = a\cos^2\theta\sin\theta$

63. $x^2 + y^2 = a^2\left(\arctan\frac{y}{x}\right)^2$

$r^2 = a^2\theta^2$

65. $r = 4$

Circle of radius 4

Centered at the pole

Symmetric to polar axis,

$\theta = \pi/2$, and pole

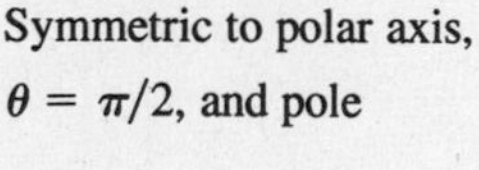

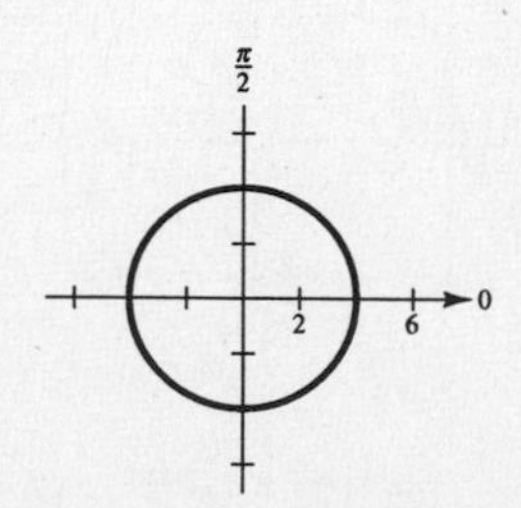

67. $r = -\sec\theta = \frac{-1}{\cos\theta}$

$r\cos\theta = -1,\ x = -1$

Vertical line

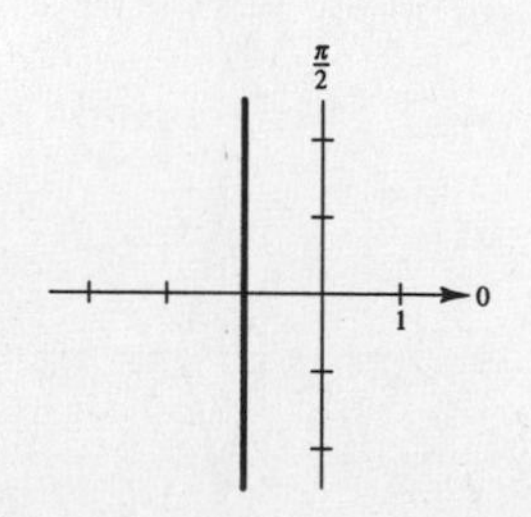

69. $r = -2(1 + \cos\theta)$

Cardioid

Symmetric to polar axis

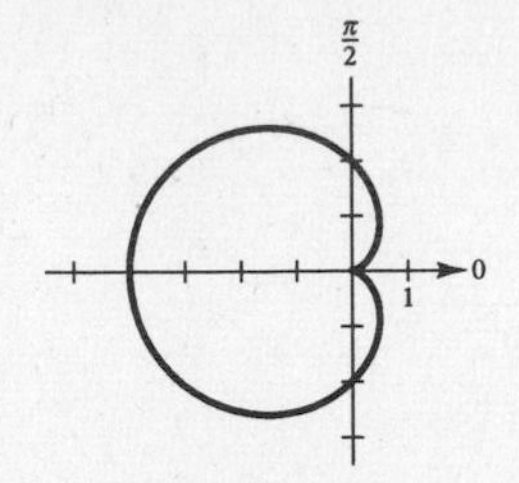

θ	0	$\frac{\pi}{3}$	$\frac{\pi}{2}$	$\frac{2\pi}{3}$	π
r	-4	-3	-2	-1	0

71. $r = 4 - 3\cos\theta$

Limaçon

Symmetric to polar axis

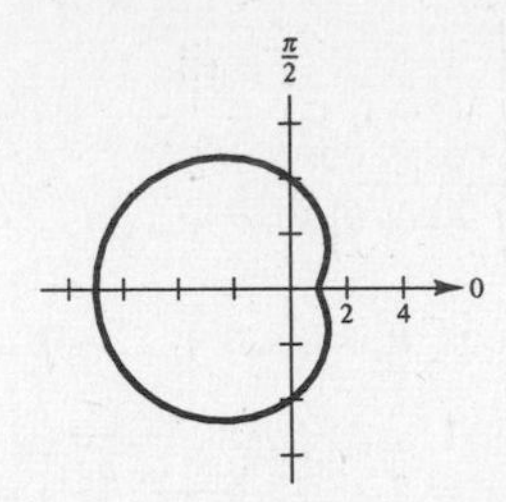

θ	0	$\frac{\pi}{3}$	$\frac{\pi}{2}$	$\frac{2\pi}{3}$	π
r	1	$\frac{5}{2}$	4	$\frac{11}{2}$	7

73. $r = -3\cos(2\theta)$

Rose curve with four petals

Symmetric to polar axis, $\theta = \frac{\pi}{2}$, and pole

Relative extrema: $(-3, 0), \left(3, \frac{\pi}{2}\right), (-3, \pi), \left(3, \frac{3\pi}{2}\right)$

Tangents at the pole: $\theta = \frac{\pi}{4}, \frac{3\pi}{4}$

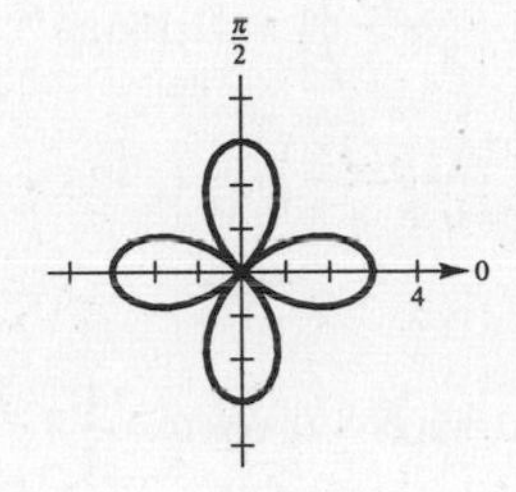

75. $r^2 = 4\sin^2(2\theta)$

$r = \pm 2\sin(2\theta)$

Rose curve with four petals

Symmetric to the polar axis, $\theta = \frac{\pi}{2}$, and pole

Relative extrema: $\left(\pm 2, \frac{\pi}{4}\right), \left(\pm 2, \frac{3\pi}{4}\right)$

Tangents at the pole: $\theta = 0, \frac{\pi}{2}$

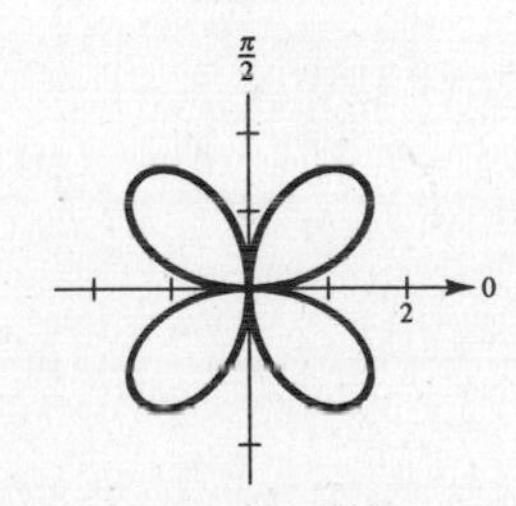

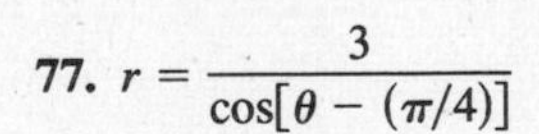

77. $r = \dfrac{3}{\cos[\theta - (\pi/4)]}$

Graph of $r = 3\sec\theta$ rotated through an angle of $\pi/4$

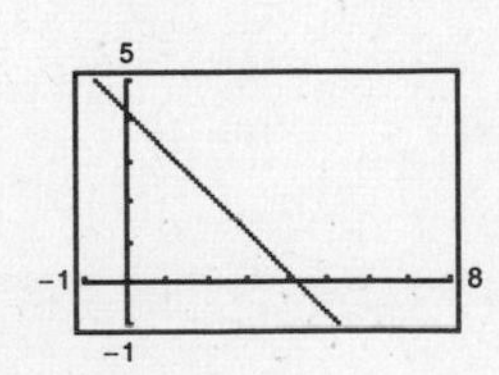

79. $r = 4\cos 2\theta\sec\theta$

Strophoid

Symmetric to the polar axis

$r \Rightarrow -\infty$ as $\theta \Rightarrow \frac{\pi^-}{2}$

$r \Rightarrow -\infty$ as $\theta \Rightarrow \frac{-\pi^+}{2}$

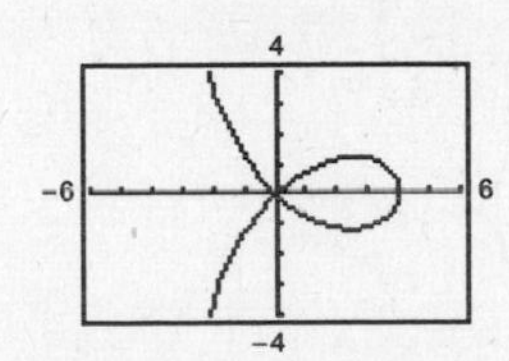

81. $r = 1 - 2\cos\theta$

(a) The graph has polar symmetry and the tangents at the pole are

$$\theta = \frac{\pi}{3}, -\frac{\pi}{3}.$$

(b) $$\frac{dy}{dx} = \frac{2\sin^2\theta + (1 - 2\cos\theta)\cos\theta}{2\sin\theta\cos\theta - (1 - 2\cos\theta)\sin\theta}$$

Horizontal tangents: $-4\cos^2\theta + \cos\theta + 2 = 0$, $\cos\theta = \dfrac{-1 \pm \sqrt{1 + 32}}{-8} = \dfrac{1 \pm \sqrt{33}}{8}$

When $\cos\theta = \dfrac{1 \pm \sqrt{33}}{8}$, $r = 1 - 2\left(\dfrac{1 + \sqrt{33}}{8}\right) = \dfrac{3 \mp \sqrt{33}}{4}$,

$$\left[\frac{3 - \sqrt{33}}{4}, \arccos\left(\frac{1 + \sqrt{33}}{8}\right)\right] \approx (-0.686, 0.568)$$

$$\left[\frac{3 - \sqrt{33}}{4}, -\arccos\left(\frac{1 + \sqrt{33}}{8}\right)\right] \approx (-0.686, -0.568)$$

$$\left[\frac{3 + \sqrt{33}}{4}, \arccos\left(\frac{1 - \sqrt{33}}{8}\right)\right] \approx (2.186, 2.206)$$

$$\left[\frac{3 + \sqrt{33}}{4}, -\arccos\left(\frac{1 - \sqrt{33}}{8}\right)\right] \approx (2.186, -2.206).$$

Vertical tangents:

$\sin\theta(4\cos\theta - 1) = 0$, $\sin\theta = 0$, $\cos\theta = \dfrac{1}{4}$,

$\theta = 0, \pi$, $\theta = \pm\arccos\left(\dfrac{1}{4}\right)$, $(-1, 0)$, $(3, \pi)$

$\left(\dfrac{1}{2}, \pm\arccos\dfrac{1}{4}\right) \approx (0.5, \pm 1.318)$

(c)

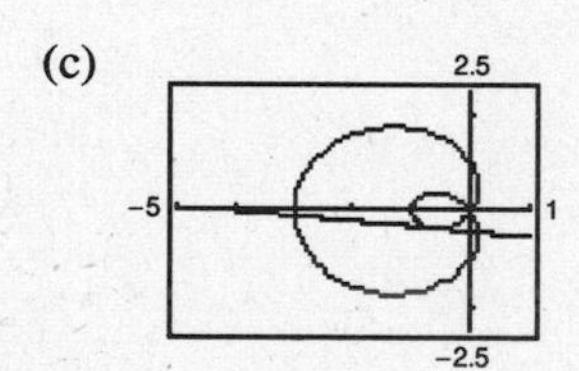

83. Circle: $r = 3\sin\theta$

$$\frac{dy}{dx} = \frac{3\cos\theta\sin\theta + 3\sin\theta\cos\theta}{3\cos\theta\cos\theta - 3\sin\theta\sin\theta} = \frac{\sin 2\theta}{\cos^2\theta - \sin^2\theta} = \tan 2\theta \text{ at } \theta = \frac{\pi}{6}, \frac{dy}{dx} = \sqrt{3}$$

Limaçon: $r = 4 - 5\sin\theta$

$$\frac{dy}{dx} = \frac{-5\cos\theta\sin\theta + (4 - 5\sin\theta)\cos\theta}{-5\cos\theta\cos\theta - (4 - 5\sin\theta)\sin\theta} \text{ at } \theta = \frac{\pi}{6}, \frac{dy}{dx} = \frac{\sqrt{3}}{9}$$

Let α be the angle between the curves:

$$\tan\alpha = \frac{\sqrt{3} - (\sqrt{3}/9)}{1 + (1/3)} = \frac{2\sqrt{3}}{3}.$$

Therefore, $\alpha = \arctan\left(\dfrac{2\sqrt{3}}{3}\right) \approx 49.1°$.

85. $r = 1 + \cos\theta,\ r = 1 - \cos\theta$

The points $(1, \pi/2)$ and $(1, 3\pi/2)$ are the two points of intersection (other than the pole). The slope of the graph of $r = 1 + \cos\theta$ is

$$m_1 = \frac{dy}{dx} = \frac{r'\sin\theta + r\cos\theta}{r'\cos\theta - r\sin\theta} = \frac{-\sin^2\theta + \cos\theta(1 + \cos\theta)}{-\sin\theta\cos\theta - \sin\theta(1 + \cos\theta)}.$$

At $(1, \pi/2)$, $m_1 = -1/-1 = 1$ and at $(1, 3\pi/2)$, $m_1 = -1/1 = -1$. The slope of the graph of $r = 1 - \cos\theta$ is

$$m_2 = \frac{dy}{dx} = \frac{\sin^2\theta + \cos\theta(1 - \cos\theta)}{\sin\theta\cos\theta - \sin\theta(1 - \cos\theta)}.$$

At $(1, \pi/2)$, $m_2 = 1/-1 = -1$ and at $(1, 3\pi/2)$, $m_2 = 1/1 = 1$. In both cases, $m_1 = -1/m_2$ and we conclude that the graphs are orthogonal at $(1, \pi/2)$ and $(1, 3\pi/2)$.

87. $r = 2 + \cos\theta$

$$A = 2\left[\frac{1}{2}\int_0^{\pi} (2 + \cos\theta)^2\, d\theta\right] \approx 14.14 \left(\frac{9\pi}{2}\right)$$

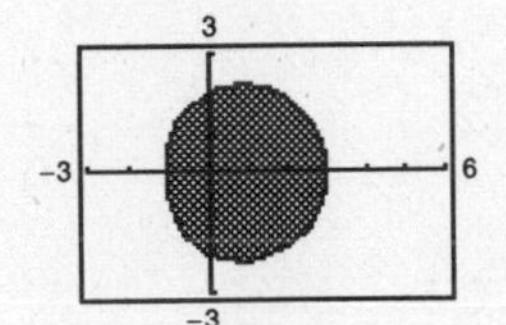

89. $r = \sin\theta \cdot \cos^2\theta$

$$A = 2\left[\frac{1}{2}\int_0^{\pi/2} (\sin\theta\cos^2\theta)^2\, d\theta\right]$$

$$\approx 0.10 \left(\frac{\pi}{32}\right)$$

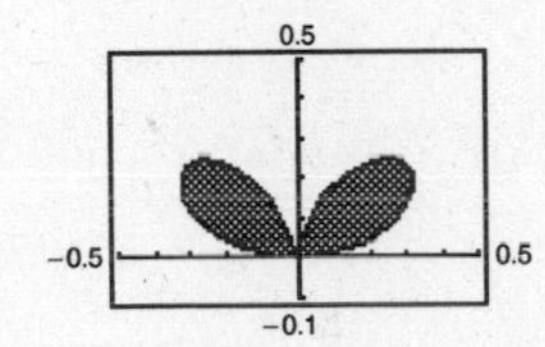

91. $r^2 = 4\sin 2\theta$

$$A = 2\left[\frac{1}{2}\int_0^{\pi/2} 4\sin 2\theta\, d\theta\right] = 4$$

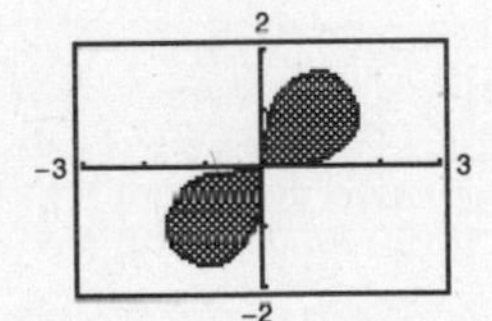

93. $r = 4\cos\theta,\ r = 2$

$$A = 2\left[\frac{1}{2}\int_0^{\pi/3} 4\, d\theta + \frac{1}{2}\int_{\pi/3}^{\pi/2} (4\cos\theta)^2\, d\theta\right] \approx 4.91$$

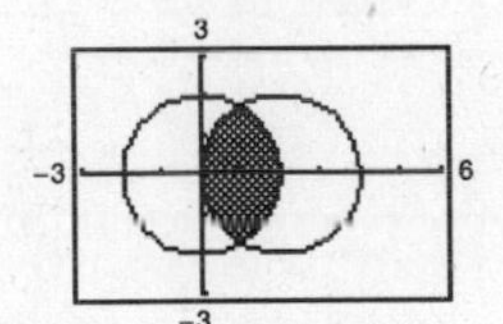

95. $s = 2\int_0^{\pi} \sqrt{a^2(1 - \cos\theta)^2 + a^2\sin^2\theta}\, d\theta$

$$= 2\sqrt{2}\,a\int_0^{\pi} \sqrt{1 - \cos\theta}\, d\theta = 2\sqrt{2}\,a\int_0^{\pi} \frac{\sin\theta}{\sqrt{1 + \cos\theta}}\, d\theta = \left[-4\sqrt{2}\,a(1 + \cos\theta)^{1/2}\right]_0^{\pi} = 8a$$

97. $r = \dfrac{2}{1 - \sin\theta},\ e = 1$

Parabola

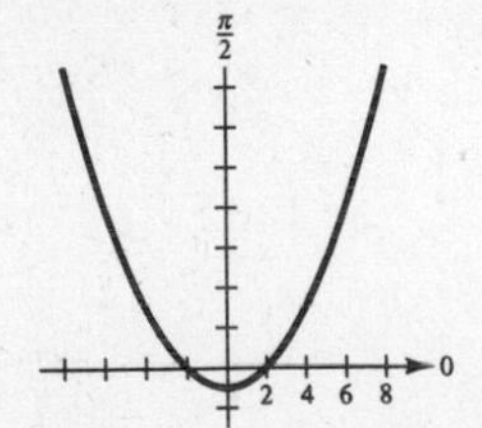

99. $r = \dfrac{6}{3 + 2\cos\theta} = \dfrac{2}{1 + (2/3)\cos\theta},\ e = \dfrac{2}{3}$

Ellipse

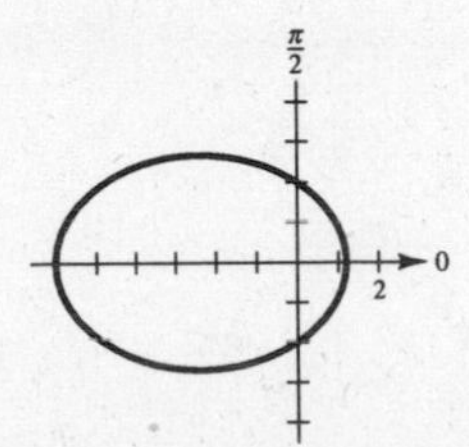

101. $r = \dfrac{4}{2 - 3\sin\theta} = \dfrac{2}{1 - (3/2)\sin\theta}, e = \dfrac{3}{2}$

Hyperbola

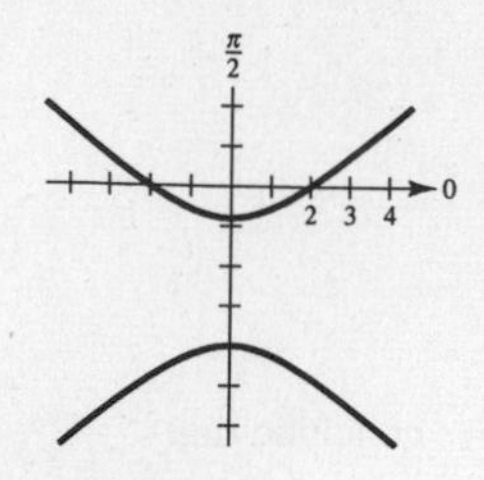

103. Circle

Center: $\left(5, \dfrac{\pi}{2}\right) = (0, 5)$ in rectangular coordinates

Solution point: $(0, 0)$

$$x^2 + (y - 5)^5 = 25$$

$$x^2 + y^2 - 10y = 0$$

$$r^2 - 10r\sin\theta = 0$$

$$r = 10\sin\theta$$

105. Parabola

Vertex: $(2, \pi)$

Focus: $(0, 0)$

$e = 1, d = 4$

$$r = \frac{4}{1 - \cos\theta}$$

107. Ellipse

Vertices: $(5, 0), (1, \pi)$

Focus: $(0, 0)$

$a = 3, c = 2, e = \dfrac{2}{3}, d = \dfrac{5}{2}$

$$r = \frac{\left(\frac{2}{3}\right)\left(\frac{5}{2}\right)}{1 - \left(\frac{2}{3}\right)\cos\theta} = \frac{5}{3 - 2\cos\theta}$$

Problem Solving for Chapter 9

1. (a)

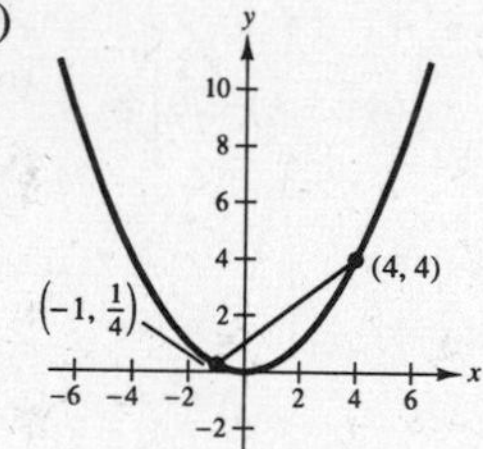

(b) $x^2 = 4y$

$2x = 4y'$

$y' = \dfrac{1}{2}x$

$y - 4 = 2(x - 4) \Rightarrow y = 2x - 4$ Tangent line at $(4, 4)$

$y - \dfrac{1}{4} = -\dfrac{1}{2}(x + 1) \Rightarrow y = -\dfrac{1}{2}x - \dfrac{1}{4}$ Tangent line at $\left(-1, \dfrac{1}{4}\right)$

Tangent lines have slopes of 2 and $-1/2 \Rightarrow$ perpendicular.

(c) Intersection:

$$2x - 4 = -\frac{1}{2}x - \frac{1}{4}$$

$$8x - 16 = -2x - 1$$

$$10x = 15$$

$$x = \frac{3}{2} \Rightarrow \left(\frac{3}{2}, -1\right)$$

Point of intersection, $(3/2, -1)$, is on directrix $y = -1$.

3. Consider $x^2 = 4py$ with focus $F = (0, p)$.

Let $P = (a, b)$ be point on parabola.

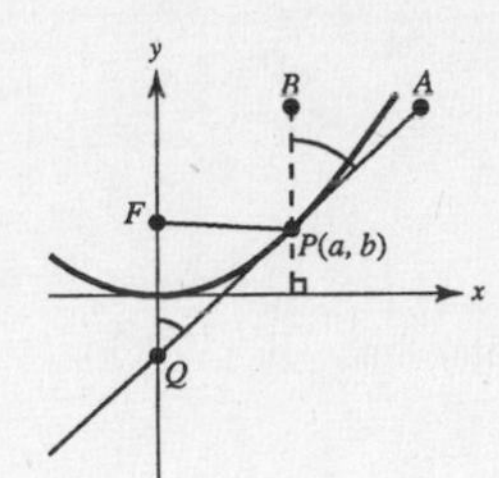

$$zx = 4py' \Rightarrow y' = \frac{x}{2p}$$

$$y - b = \frac{a}{2p}(x - a) \quad \text{Tangent line}$$

For $x = 0$, $y = b + \frac{a}{2p}(-a) = b - \frac{a^2}{2p} = b - \frac{4pb}{2p} = -b.$

Thus, $Q = (0, -b)$.

$\triangle FQP$ is isosceles because

$$|FQ| = p + b$$

$$\begin{aligned} |FP| &= \sqrt{(a-0)^2 + (b-p)^2} = \sqrt{a^2 + b^2 - 2bp + p^2} \\ &= \sqrt{4pb + b^2 - 2bp + p^2} \\ &= \sqrt{(b+p)^2} \\ &= b + p. \end{aligned}$$

Thus, $\measuredangle FQP = \measuredangle BPA = \measuredangle FPQ.$

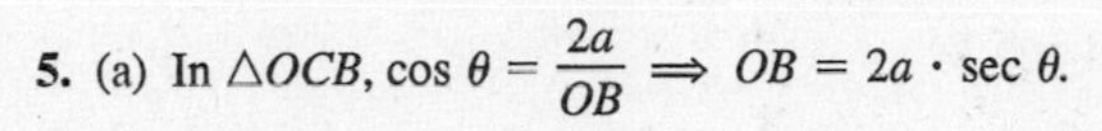

5. (a) In $\triangle OCB$, $\cos\theta = \frac{2a}{OB} \Rightarrow OB = 2a \cdot \sec\theta.$

In $\triangle OAC$, $\cos\theta = \frac{OA}{2a} \Rightarrow OA = 2a \cdot \cos\theta.$

$$\begin{aligned} r = OP = AB = OB - OA &= 2a(\sec\theta - \cos\theta) \\ &= 2a\left(\frac{1}{\cos\theta} - \cos\theta\right) \\ &= 2a \cdot \frac{\sin^2\theta}{\cos\theta} \\ &= 2a \cdot \tan\theta \sin\theta \end{aligned}$$

(b) $x = r\cos\theta = (2a\tan\theta\sin\theta)\cos\theta = 2a\sin^2\theta$

$y = r\sin\theta = (2a\tan\theta\sin\theta)\sin\theta = 2a\tan\theta \cdot \sin^2\theta, \ -\frac{\pi}{2} < \theta < \frac{\pi}{2}$

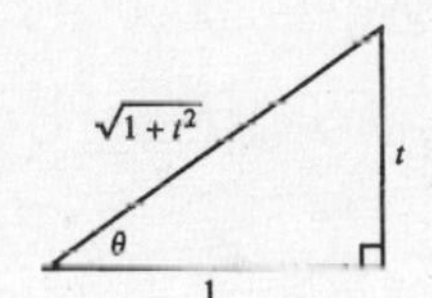

Let $t = \tan\theta$, $-\infty < t < \infty$.

Then $\sin^2\theta = \frac{t^2}{1+t^2}$ and $x = 2a\frac{t^2}{1+t^2}$, $y = 2a\frac{t^3}{1+t^2}$.

(c)

$$\begin{aligned} r &= 2a\tan\theta\sin\theta \\ r\cos\theta &= 2a\sin^2\theta \\ r^3\cos\theta &= 2a\,r^2\sin^2\theta \\ (x^2 + y^2)x &= 2ay^2 \\ y^2 &= \frac{x^3}{(2a - x)} \end{aligned}$$

7. $y = a(1 - \cos\theta) \Rightarrow \cos\theta = \frac{a - y}{a}$

$$\theta = \arccos\left(\frac{a-y}{a}\right)$$

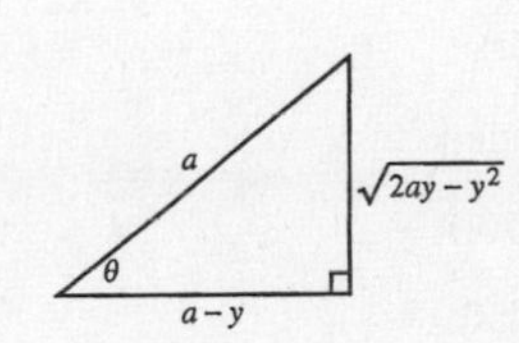

$$\begin{aligned} x &= a(\theta - \sin\theta) \\ &= a\left(\arccos\left(\frac{a-y}{a}\right) - \sin\left(\arccos\left(\frac{a-y}{a}\right)\right)\right) \\ &= a\left(\arccos\left(\frac{a-y}{a}\right) - \frac{\sqrt{2ay - y^2}}{a}\right) \end{aligned}$$

$$x = a \cdot \arccos\left(\frac{a-y}{a}\right) - \sqrt{2ay - y^2}, \ 0 \le y \le 2a$$

9. For $t = \frac{\pi}{2}, \frac{3\pi}{2}, \frac{5\pi}{2}, \frac{7\pi}{2}, \dots$

$$y = \frac{2}{\pi}, \frac{-2}{3\pi}, \frac{2}{5\pi}, \frac{-2}{7\pi}, \dots$$

Hence, the curve has length greater that

$$S = \frac{2}{\pi} + \frac{2}{3\pi} + \frac{2}{5\pi} + \frac{2}{7\pi} + \cdots$$
$$= \frac{2}{\pi}\left(1 + \frac{1}{3} + \frac{1}{5} + \frac{1}{7} + \cdots\right)$$
$$> \frac{2}{\pi}\left(\frac{1}{2} + \frac{1}{4} + \frac{1}{6} + \frac{1}{8} + \cdots\right)$$
$$= \infty.$$

11. (a) Area $= \int_0^{\alpha} \frac{1}{2}r^2\,d\theta$

$$= \frac{1}{2}\int_0^{\alpha} \sec^2\theta\,d\theta$$

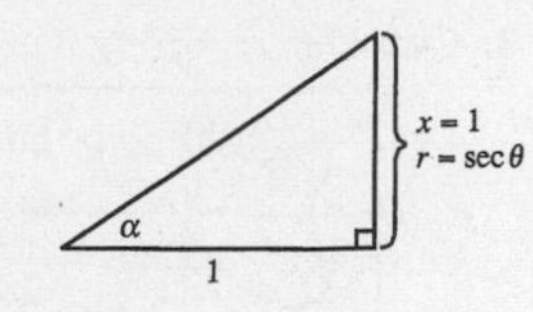

(b) $\tan\alpha = \frac{h}{1} \Rightarrow$ Area $= \frac{1}{2}(1)\tan\alpha$

$$\Rightarrow \tan\alpha = \int_0^{\alpha} \sec^2\theta\,d\theta$$

(c) Differentiating, $\frac{d}{d\alpha}(\tan\alpha) = \sec^2\alpha.$

13. If a dog is located at (r, θ) in the first quadrant, then its neighbor is at $\left(r, \theta + \frac{\pi}{2}\right)$:

$(x, y) = (r\cos\theta, r\sin\theta)$ and $(x, y) = (-r\sin\theta, r\cos\theta)$.

The slope joining these points is

$$\frac{r\cos\theta - r\sin\theta}{-r\sin\theta - r\cos\theta} = \frac{\sin\theta - \cos\theta}{\sin\theta + \cos\theta} = \text{slope of tangent line at } (r, \theta).$$

$$\frac{\frac{dr}{d\theta}\sin\theta + r\cos\theta}{\frac{dr}{d\theta}\cos\theta - r\sin\theta} = \frac{\sin\theta - \cos\theta}{\sin\theta + \cos\theta}$$

$$\Rightarrow \frac{dr}{d\theta} = -r$$

$$\frac{dr}{r} = -d\theta$$

$$\ln r = -\theta + C_1$$

$$r = e^{-\theta + C_1}$$

$$r = Ce^{-\theta}$$

$$r\left(\frac{\pi}{4}\right) = \frac{d}{\sqrt{2}} \Rightarrow r = Ce^{-\pi/4} = \frac{d}{\sqrt{2}} \Rightarrow C = \frac{d}{\sqrt{2}}e^{\pi/4}$$

Finally, $r = \frac{d}{\sqrt{2}}e^{((\pi/4) - \theta)}, \ \theta \ge \frac{\pi}{4}.$

15. (a) The first plane makes an angle of 70° with the positive x-axis, and is 150 miles from P:

$$x_1 = \cos 70°(150 - 375t)$$
$$y_1 = \sin 70°(150 - 375t)$$

Similarly for the second plane,

$$x_2 = \cos 135°(190 - 450t)$$
$$= \cos 45°(-190 + 450t)$$
$$y_2 = \sin 135°(190 - 450t)$$
$$= \sin 45°(190 - 450t)$$

(b) $d = \sqrt{(x_2 - x_1)^2 + (y_2 - y_1)^2}$

$$= [[\cos 45(-190 + 450t) - \cos 70(150 - 375t)]^2 + [\sin 45(190 - 450t) - \sin 70(150 - 375t)]^2]^{1/2}$$

(c)

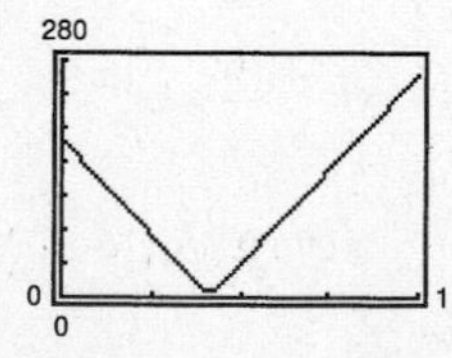

The minimum distance is 7.59 miles when $t = 0.4145$.

17.

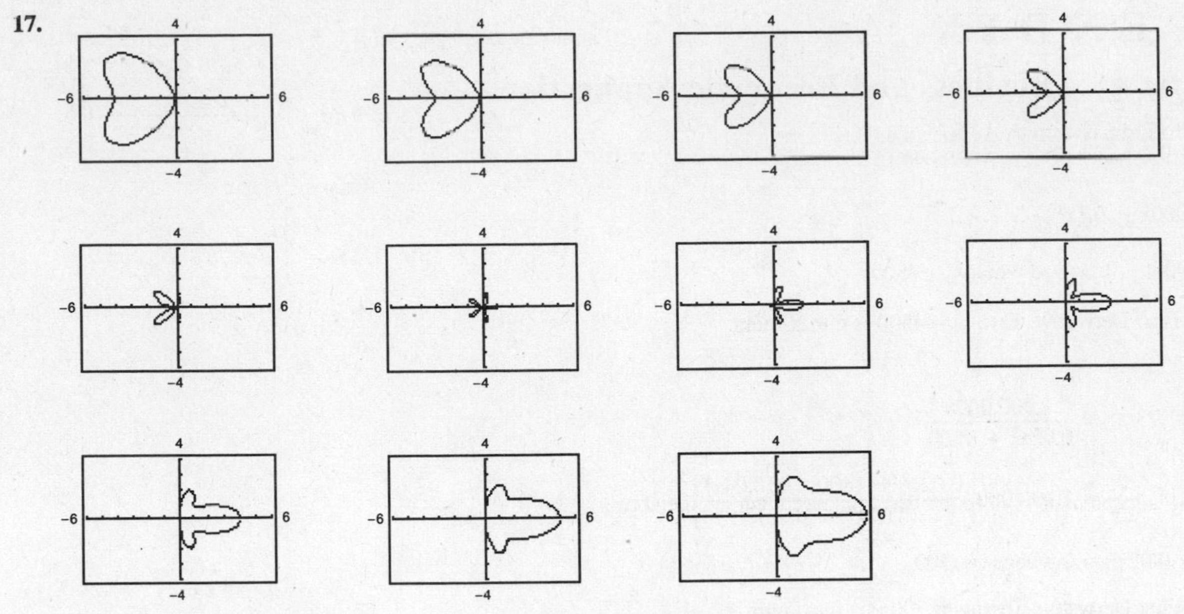

$n = 1, 2, 3, 4, 5$ produce "bells"; $n = -1, -2, -3, -4, -5$ produce "hearts".

APPENDIX A

Appendix A Business and Economic Applications

Solutions to Odd-Numbered Exercises

1. $R = 900x - 0.1x^2$

$\dfrac{dR}{dx} = 900 - 0.2x = 0$ when $x = 4500$.

By the First Derivative Test, $x = 4500$ is a maximum.

3.
$$R = \frac{1,000,000x}{0.02x^2 + 1800}$$
$$\frac{dR}{dx} = 1,000,000\left[\frac{0.02x^2 + 1800 - x(0.04x)}{(0.02x^2 + 1800)^2}\right] = 0$$

$1800 - 0.02x^2 = 0$ when $x = 300$.

By the First Derivative Test, $x = 300$ is a maximum.

5. $\overline{C} = 0.125x + 20 + \dfrac{5000}{x}$

$\dfrac{d\overline{C}}{dx} = 0.125 - \dfrac{5000}{x^2} = 0$ when $x = 200$.

By the First Derivative Test, $x = 200$ yields the minimum average cost.

7. $\overline{C} = 3000 - x(300 - x)^{1/2}$

$\dfrac{d\overline{C}}{dx} = -x\left(\dfrac{1}{2}\right)(300 - x)^{-1/2}(-1) - (300 - x)^{1/2} = -\dfrac{3}{2}(300 - x)^{-1/2}(200 - x) = 0$ when $x = 200$.

By the First Derivative Test, $x = 200$ yields the minimum average cost.

9. $C = 100 + 30x$

$p = 90 - x$

$$\begin{aligned} P &= xp - C \\ &= 90x - x^2 - 30x - 100 \\ &= -x^2 + 60x - 100 \end{aligned}$$

$$\begin{aligned} \frac{dP}{dx} &= -2x + 60 \\ &= 0 \text{ when } x = 30, \text{ so } p = 60. \end{aligned}$$

By the First Derivative Test, $x = 30$ is a maximum.

11. $C = 4000 - 40x + 0.02x^2$

$p = 50 - 0.01x$

$P = xp - C = 50x - 0.01x^2 - 4000 + 40x - 0.02x^2 = -0.03x^2 + 90x - 4000$

$\dfrac{dP}{dx} = -0.06x + 90 = 0$ when $x = 1500$, so $p = 35$.

By the First Derivative Test, $x = 1500$ is a maximum.

13. $P = 60xe^{-x/5}$

$P' = -12e^{-x/5}(x - 5) = 0$ when $x = 5$ units.

$x = 5$ is a maximum by the First Derivative Test.

15. $C = 2x^2 + 5x + 18$

Average cost $= \dfrac{C}{x} = \overline{C} = 2x + 5 + \dfrac{18}{x}$

$\dfrac{d\overline{C}}{dx} = 2 - \dfrac{18}{x^2} = 0$ when $x = 3$.

$\overline{C}(3) = 6 + 5 + 6 = 17$

By the First Derivative Test, $x = 3$ is a minimum.

Marginal cost: $\dfrac{dC}{dx} = 4x + 5$

At $x = 3$: $\dfrac{dC}{dx} = 17 = \overline{C}(3)$

17. Average cost: $\overline{C}(x) = \dfrac{C(x)}{x}$

$$\frac{d\overline{C}}{dx} = \frac{xC'(x) - C(x)}{x^2} = 0 \Longrightarrow xC'(x) - C(x) = 0 \text{ when } C'(x) = \frac{C(x)}{x} = \overline{C}(x).$$

Marginal cost = average cost

This condition will yield a minimum (if it exists).

19. (a)

Order size, x	Price	Profit
102	$90 - 2(0.15)$	$102[90 - 2(0.15)] - 102(60) = 3029.40$
104	$90 - 4(0.15)$	$104[90 - 4(0.15)] - 104(60) = 3057.60$
106	$90 - 6(0.15)$	$106[90 - 6(0.15)] - 106(60) = 3084.60$
108	$90 - 8(0.15)$	$108[90 - 8(0.15)] - 108(60) = 3110.40$
110	$90 - 10(0.15)$	$110[90 - 10(0.15)] - 110(60) = 3135.00$
112	$90 - 12(0.15)$	$112[90 - 12(0.15)] - 112(60) = 3158.40$

(b)

Order size, x	Profit
148	3374.40
149	3374.90
150	3375.00
151	3374.90
152	3374.40

The maximum profit is 3375.00 for $x = 150$.

(c) $P(x) = x[90 - (x - 100)(0.15)] - 60x$

$= x(45 - 0.15x),\ x \geq 100$

(d) $\dfrac{dP}{dx} = 45 - 0.30x = 0$ when $x = 150$.

Since $\dfrac{d^2P}{dx^2} < 0$, an order size of $x = 150$ units yields a maximum profit.

(e)

21. Total cost = (Cost per hour)(Number of hours)

$$T = \left(\frac{v^2}{600} + 5\right)\left(\frac{110}{v}\right) = \frac{11v}{60} + \frac{550}{v}$$

$$\frac{dT}{dv} = \frac{11}{60} - \frac{550}{v^2} = \frac{11v^2 - 33{,}000}{60v^2}$$

$= 0$ when $v = \sqrt{3000} = 10\sqrt{30} \approx 54.8$ mph.

$\dfrac{d^2T}{dv^2} = \dfrac{1100}{v^3} > 0$ when $v = 10\sqrt{30}$ so this value yields a minimum.

23. The total cost $T(x)$ is

$$T(x) = 12(5280)(6 - x) + 16(5280)\sqrt{x^2 + \frac{1}{4}}$$

$$T'(x) = 5280\left[-12 + \frac{16x}{\sqrt{x^2 + (1/4)}}\right] = 0$$

$$12 = \frac{16x}{\sqrt{x^2 + (1/4)}}$$

$$3\sqrt{x^2 + (1/4)} = 4x$$

$$9\left(x^2 + \frac{1}{4}\right) = 16x^2$$

$$\frac{9}{4} = 7x^2 \Longrightarrow x = \frac{3}{2\sqrt{7}} \approx 0.57 \text{ miles.}$$

This is a minimum by the First Derivative Test.

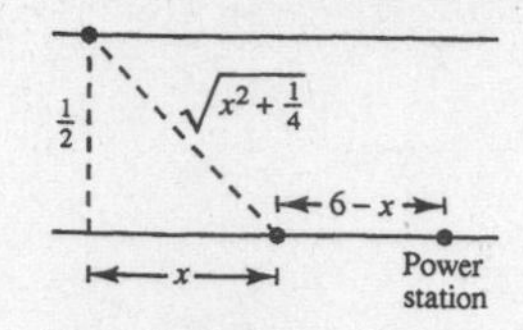

25. For the linear demand function, the rate of change (slope) is $m = -\frac{25}{5}$. Therefore, the demand x is

$$x = 800 - \frac{25}{5}(p - 25) = 925 - 5p$$

$$R = xp = -5p^2 + 925p$$

$$\frac{dR}{dp} = -10p + 925 = 0 \text{ when } p = \$92.50.$$

27. $R = 900x - 0.1x^2$

$x = 3000$

$dx = 100$

$dR = (900 - 0.2x)dx$

$= [900 - 0.2(3000)](100)$

$= \$30,000$

29. 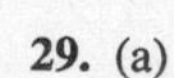
(a)

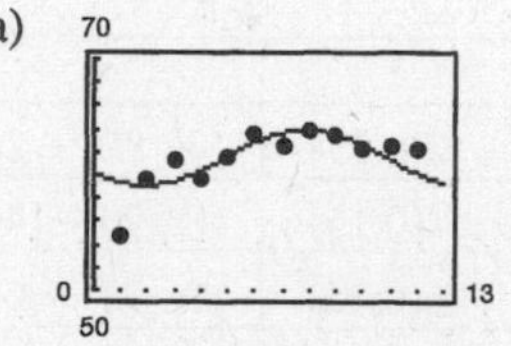

(b) The maximum occurs at $t = 7.9$ (July)

(c) The cosine term reflects the seasonal variation in gasoline sales. The term 61.5 reflects the average sales.

(d) This term might reflect inflation. The maximum in year 2005 $[61 \le t \le 72]$ is $G \approx \$65.27$.

31. (a) Demand function (b) Cost function

(c) Revenue function (d) Profit function

33. $$\eta = \frac{p/x}{dp/dx} = \frac{(400 - 3x)/x}{-3} = 1 - \frac{400}{3x}$$

When $x = 20$, we have

$$\eta = 1 - \frac{400}{3(20)} = -\frac{17}{3}.$$

Since $|\eta| = \frac{17}{3} > 1$, the demand is elastic.

35. $$\eta = \frac{p/x}{dp/dx} = \frac{(400 - 0.5x^2)/x}{-x} = \frac{1}{2} - \frac{400}{x^2}$$

When $x = 20$, we have

$$\eta = \frac{1}{2} - \frac{400}{(20)^2} = -\frac{1}{2}.$$

Since $|\eta| = \frac{1}{2} < 1$, the demand is inelastic.

37. $$P = \frac{200}{1 + 9e^{-s/5}}$$

$$P' = \frac{360e^{-s/5}}{(1 + 9e^{-s/5})^2}$$

$$P'' = \frac{72e^{-s/5}(9e^{-s/5} - 1)}{(1 + 9e^{-s/5})^3}$$

$P'' = 0$ when $x \approx 10.986$ (thousands of dollars)